ORGANIZING A STATISTICAL PROBLEM: A Four-Step Process

STATE: What is the practical question, in the context of the real-world setting?

PLAN: What specific statistical operations does this problem call for?

SOLVE: Make the graphs and carry out the calculations needed for this problem.

CONCLUDE: Give your practical conclusion in the setting of the real-world problem.

CONFIDENCE INTERVALS: The Four-Step Process

STATE: What is the practical question that requires estimating a parameter?

PLAN: Identify the parameter, choose a level of confidence, and select the type of confidence interval that fits your situation.

SOLVE: Carry out the work in two phases:

1. **Check the conditions** for the interval you plan to use.
2. Calculate the **confidence interval.**

CONCLUDE: Return to the practical question to describe your results in this setting.

TESTS OF SIGNIFICANCE: The Four-Step Process

STATE: What is the practical question that requires a statistical test?

PLAN: Identify the parameter, state null and alternative hypotheses, and choose the type of test that fits your situation.

SOLVE: Carry out the test in three phases:

1. **Check the conditions** for the test you plan to use.
2. Calculate the **test statistic.**
3. Find the ***P*-value.**

CONCLUDE: Return to the practical question to describe your results in this setting.

Table entry for C is the critical value t^* required for confidence level C. To approximate one- and two-sided P-values, compare the value of the t statistic with the critical values of t^* that match the P-values given at the bottom of the table.

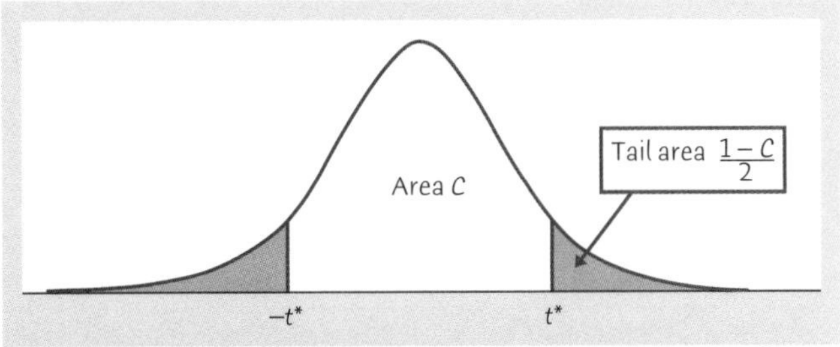

TABLE C *t* DISTRIBUTION CRITICAL VALUES

DEGREES OF FREEDOM	CONFIDENCE LEVEL *C*											
	50%	60%	70%	80%	90%	95%	96%	98%	99%	99.5%	99.8%	99.9%
1	1.000	1.376	1.963	3.078	6.314	12.71	15.89	31.82	63.66	127.3	318.3	636.6
2	0.816	1.061	1.386	1.886	2.920	4.303	4.849	6.965	9.925	14.09	22.33	31.60
3	0.765	0.978	1.250	1.638	2.353	3.182	3.482	4.541	5.841	7.453	10.21	12.92
4	0.741	0.941	1.190	1.533	2.132	2.776	2.999	3.747	4.604	5.598	7.173	8.610
5	0.727	0.920	1.156	1.476	2.015	2.571	2.757	3.365	4.032	4.773	5.893	6.869
6	0.718	0.906	1.134	1.440	1.943	2.447	2.612	3.143	3.707	4.317	5.208	5.959
7	0.711	0.896	1.119	1.415	1.895	2.365	2.517	2.998	3.499	4.029	4.785	5.408
8	0.706	0.889	1.108	1.397	1.860	2.306	2.449	2.896	3.355	3.833	4.501	5.041
9	0.703	0.883	1.100	1.383	1.833	2.262	2.398	2.821	3.250	3.690	4.297	4.781
10	0.700	0.879	1.093	1.372	1.812	2.228	2.359	2.764	3.169	3.581	4.144	4.587
11	0.697	0.876	1.088	1.363	1.796	2.201	2.328	2.718	3.106	3.497	4.025	4.437
12	0.695	0.873	1.083	1.356	1.782	2.179	2.303	2.681	3.055	3.428	3.930	4.318
13	0.694	0.870	1.079	1.350	1.771	2.160	2.282	2.650	3.012	3.372	3.852	4.221
14	0.692	0.868	1.076	1.345	1.761	2.145	2.264	2.624	2.977	3.326	3.787	4.140
15	0.691	0.866	1.074	1.341	1.753	2.131	2.249	2.602	2.947	3.286	3.733	4.073
16	0.690	0.865	1.071	1.337	1.746	2.120	2.235	2.583	2.921	3.252	3.686	4.015
17	0.689	0.863	1.069	1.333	1.740	2.110	2.224	2.567	2.898	3.222	3.646	3.965
18	0.688	0.862	1.067	1.330	1.734	2.101	2.214	2.552	2.878	3.197	3.611	3.922
19	0.688	0.861	1.066	1.328	1.729	2.093	2.205	2.539	2.861	3.174	3.579	3.883
20	0.687	0.860	1.064	1.325	1.725	2.086	2.197	2.528	2.845	3.153	3.552	3.850
21	0.686	0.859	1.063	1.323	1.721	2.080	2.189	2.518	2.831	3.135	3.527	3.819
22	0.686	0.858	1.061	1.321	1.717	2.074	2.183	2.508	2.819	3.119	3.505	3.792
23	0.685	0.858	1.060	1.319	1.714	2.069	2.177	2.500	2.807	3.104	3.485	3.768
24	0.685	0.857	1.059	1.318	1.711	2.064	2.172	2.492	2.797	3.091	3.467	3.745
25	0.684	0.856	1.058	1.316	1.708	2.060	2.167	2.485	2.787	3.078	3.450	3.725
26	0.684	0.856	1.058	1.315	1.706	2.056	2.162	2.479	2.779	3.067	3.435	3.707
27	0.684	0.855	1.057	1.314	1.703	2.052	2.158	2.473	2.771	3.057	3.421	3.690
28	0.683	0.855	1.056	1.313	1.701	2.048	2.154	2.467	2.763	3.047	3.408	3.674
29	0.683	0.854	1.055	1.311	1.699	2.045	2.150	2.462	2.756	3.038	3.396	3.659
30	0.683	0.854	1.055	1.310	1.697	2.042	2.147	2.457	2.750	3.030	3.385	3.646
40	0.681	0.851	1.050	1.303	1.684	2.021	2.123	2.423	2.704	2.971	3.307	3.551
50	0.679	0.849	1.047	1.299	1.676	2.009	2.109	2.403	2.678	2.937	3.261	3.496
60	0.679	0.848	1.045	1.296	1.671	2.000	2.099	2.390	2.660	2.915	3.232	3.460
80	0.678	0.846	1.043	1.292	1.664	1.990	2.088	2.374	2.639	2.887	3.195	3.416
100	0.677	0.845	1.042	1.290	1.660	1.984	2.081	2.364	2.626	2.871	3.174	3.390
1000	0.675	0.842	1.037	1.282	1.646	1.962	2.056	2.330	2.581	2.813	3.098	3.300
z^*	0.674	0.841	1.036	1.282	1.645	1.960	2.054	2.326	2.576	2.807	3.091	3.291
One-sided P	.25	.20	.15	.10	.05	.025	.02	.01	.005	.0025	.001	.0005
Two-sided P	.50	.40	.30	.20	.10	.05	.04	.02	.01	.005	.002	.001

The Basic Practice of Statistics

SEVENTH EDITION

The Basic Practice of Statistics

DAVID S. MOORE • WILLIAM I. NOTZ • MICHAEL A. FLIGNER

Purdue University *The Ohio State University* *University of California at Santa Cruz*

W. H. FREEMAN & COMPANY

A Macmillan Education Imprint

Publisher:	Terri Ward
Senior Acquisitions Editor:	Karen Carson
Marketing Manager:	Cara LeClair
Development Editors:	Leslie Lahr and Jorge Amaral
Associate Editor:	Marie Dripchak
Executive Media Editor:	Laura Judge
Media Editor:	Catriona Kaplan
Associate Media Editor:	Liam Ferguson
Editorial Assistant:	Victoria Garvey
Marketing Assistant:	Bailey James
Photo Editor:	Cecilia Varas
Photo Researcher:	Eileen Liang
Cover and Text Designer:	Vicki Tomaselli
Managing Editor:	Lisa Kinne
Senior Project Manager:	Denise Showers, Aptara®, Inc.
Illustrations and Composition:	Aptara®, Inc.
Production Manager:	Julia DeRosa
Printing and Binding:	RR Donnelley
Cover Credit:	© SoberP/istockphoto

Library of Congress Control Number: 2014950586

Student Edition Hardcover (packaged with EESEE/CrunchIt! access card):
ISBN-13: 978-1-4641-4253-6
ISBN-10: 1-4641-4253-X

Student Edition Loose-leaf (packaged with EESEE/CrunchIt! access card):
ISBN-13: 978-1-4641-7990-7
ISBN-10: 1-4641-7990-5

Instructor Complimentary Copy:
ISBN-13: 978-1-4641-7988-4
ISBN-10: 1-4641-7988-3

Printed in the United States of America

Third printing

W. H. Freeman and Company
41 Madison Avenue
New York, NY 10010
Houndmills, Basingstoke RG21 6XS, England
www.whfreeman.com

BRIEF CONTENTS

CHAPTER 0 **Getting Started** 1

PART I EXPLORING DATA 11

Exploring Data: Variables and Distributions

CHAPTER 1 **Picturing Distributions with Graphs** 13

CHAPTER 2 **Describing Distributions with Numbers** 47

CHAPTER 3 **The Normal Distributions** 75

Exploring Data: Relationships

CHAPTER 4 **Scatterplots and Correlation** 101

CHAPTER 5 **Regression** 127

CHAPTER 6 **Two-Way Tables*** 163

CHAPTER 7 **Exploring Data: Part I Review** 179

PART II PRODUCING DATA 201

CHAPTER 8 **Producing Data: Sampling** 203

CHAPTER 9 **Producing Data: Experiments** 227

CHAPTER 10 **Data Ethics*** 253

CHAPTER 11 **Producing Data: Part II Review** 267

PART III FROM DATA PRODUCTION TO INFERENCE 275

CHAPTER 12 **Introducing Probability** 277

CHAPTER 13 **General Rules of Probability*** 303

CHAPTER 14 **Binomial Distributions*** 327

CHAPTER 15 **Sampling Distributions** 345

CHAPTER 16 **Confidence Intervals: The Basics** 373

*Starred material is not required for later parts of the text.

CHAPTER 17 Tests of Significance: The Basics 391

CHAPTER 18 Inference in Practice 415

CHAPTER 19 From Data Production to Inference: Part III Review 439

PART IV INFERENCE ABOUT VARIABLES 453

Quantitative Response Variable

CHAPTER 20 Inference about a Population Mean 453

CHAPTER 21 Comparing Two Means 485

Categorical Response Variable

CHAPTER 22 Inference about a Population Proportion 517

CHAPTER 23 Comparing Two Proportions 539

CHAPTER 24 Inference about Variables: Part IV Review 559

PART V INFERENCE ABOUT RELATIONSHIPS 575

CHAPTER 25 Two Categorical Variables: The Chi-Square Test 577

CHAPTER 26 Inference for Regression 609

CHAPTER 27 One-Way Analysis of Variance: Comparing Several Means 645

PART VI OPTIONAL COMPANION CHAPTERS (AVAILABLE ONLINE)

CHAPTER 28 Nonparametric Tests 28-1

CHAPTER 29 Multiple Regression 29-1

CHAPTER 30 More about Analysis of Variance 30-1

CHAPTER 31 Statistical Process Control 31-1

CONTENTS

To the Instructor: About This Book xi

Acknowledgments xx

Media and Supplements xxii

About the Authors xxv

CHAPTER 0 **Getting Started 1**
- 0.1 Where the data comes from matters 2
- 0.2 Always look at the data 3
- 0.3 Variation is everywhere 5
- 0.4 What lies ahead in this book 7

PART I **EXPLORING DATA 11**

CHAPTER 1 **Picturing Distributions with Graphs 13**
- 1.1 Individuals and variables 13
- 1.2 Categorical variables: pie charts and bar graphs 16
- 1.3 Quantitative variables: histograms 21
- 1.4 Interpreting histograms 24
- 1.5 Quantitative variables: stemplots 29
- 1.6 Time plots 32

CHAPTER 2 **Describing Distributions with Numbers 47**
- 2.1 Measuring center: the mean 48
- 2.2 Measuring center: the median 49
- 2.3 Comparing the mean and the median 50
- 2.4 Measuring variability: the quartiles 51
- 2.5 The five-number summary and boxplots 53
- 2.6 Spotting suspected outliers and modified boxplots* 55
- 2.7 Measuring variability: the standard deviation 57
- 2.8 Choosing measures of center and variability 59
- 2.9 Using technology 61
- 2.10 Organizing a statistical problem 63

CHAPTER 3 **The Normal Distributions 75**
- 3.1 Density curves 75
- 3.2 Describing density curves 78
- 3.3 Normal distributions 80
- 3.4 The 68–95–99.7 rule 82
- 3.5 The standard Normal distribution 85
- 3.6 Finding Normal proportions 86
- 3.7 Using the standard Normal table 88
- 3.8 Finding a value given a proportion 91

CHAPTER 4 **Scatterplots and Correlation 101**
- 4.1 Explanatory and response variables 101
- 4.2 Displaying relationships: scatterplots 103
- 4.3 Interpreting scatterplots 105
- 4.4 Adding categorical variables to scatterplots 109
- 4.5 Measuring linear association: correlation 111
- 4.6 Facts about correlation 113

CHAPTER 5 **Regression 127**
- 5.1 Regression lines 127
- 5.2 The least-squares regression line 131
- 5.3 Using technology 132
- 5.4 Facts about least-squares regression 135
- 5.5 Residuals 138
- 5.6 Influential observations 143
- 5.7 Cautions about correlation and regression 146
- 5.8 Association does not imply causation 148

CHAPTER 6 **Two-Way Tables* 163**
- 6.1 Marginal distributions 164
- 6.2 Conditional distributions 166
- 6.3 Simpson's paradox 171

CHAPTER 7 **Exploring Data: Part I Review 179**

Part I Summary 181

Test Yourself 183

Supplementary Exercises 195

PART II **PRODUCING DATA 201**

CHAPTER 8 **Producing Data: Sampling 203**
- 8.1 Population versus sample 204
- 8.2 How to sample badly 206

*Starred material is not required for later parts of the text.

8.3 Simple random samples 207
8.4 Inference about the population 211
8.5 Other sampling designs 212
8.6 Cautions about sample surveys 214
8.7 The impact of technology 216

CHAPTER 9 **Producing Data: Experiments 227**
9.1 Observation versus experiment 227
9.2 Subjects, factors, and treatments 230
9.3 How to experiment badly 233
9.4 Randomized comparative experiments 234
9.5 The logic of randomized comparative experiments 237
9.6 Cautions about experimentation 239
9.7 Matched pairs and other block designs 241

CHAPTER 10 **Data Ethics* 253**
10.1 Institutional review boards 254
10.2 Informed consent 256
10.3 Confidentiality 258
10.4 Clinical trials 260
10.5 Behavioral and social science experiments 261

CHAPTER 11 **Producing Data: Part II Review 267**
Part II Summary 268
Test Yourself 269
Supplementary Exercises 272

PART III FROM DATA PRODUCTION TO INFERENCE 275

CHAPTER 12 **Introducing Probability 275**
12.1 The idea of probability 278
12.2 The search for randomness* 280
12.3 Probability models 281
12.4 Probability rules 283
12.5 Finite and discrete probability models 286
12.6 Continuous probability models 289
12.7 Random variables 293
12.8 Personal probability* 294

CHAPTER 13 **General Rules of Probability* 303**
13.1 Independence and the multiplication rule 304
13.2 The general addition rule 307
13.3 Conditional probability 309
13.4 The general multiplication rule 311
13.5 Independence again 313
13.6 Tree diagrams 314
13.7 Bayes' rule* (available online)

CHAPTER 14 **Binomial Distributions* 327**
14.1 The binomial setting and binomial distributions 327
14.2 Binomial distributions in statistical sampling 328
14.3 Binomial probabilities 330
14.4 Using technology 332
14.5 Binomial mean and standard deviation 334
14.6 The Normal approximation to binomial distributions 335

CHAPTER 15 **Sampling Distributions 345**
15.1 Parameters and statistics 346
15.2 Statistical estimation and the law of large numbers 347
15.3 Sampling distributions 350
15.4 The sampling distribution of $\bar{x}$ 352
15.5 The central limit theorem 355
15.6 Sampling distributions and statistical significance 361

CHAPTER 16 **Confidence Intervals: The Basics 373**
16.1 The reasoning of statistical estimation 374
16.2 Margin of error and confidence level 376
16.3 Confidence intervals for a population mean 379
16.4 How confidence intervals behave 383

CHAPTER 17 **Tests of Significance: The Basics 391**
17.1 The reasoning of tests of significance 392
17.2 Stating hypotheses 394
17.3 *P*-value and statistical significance 396
17.4 Tests for a population mean 400
17.5 Significance from a table* 404
17.6 Resampling: significance from a simulation* 406

CHAPTER 18 **Inference in Practice 415**
18.1 Conditions for inference in practice 416
18.2 Cautions about confidence intervals 419
18.3 Cautions about significance tests 421

18.4 Planning studies: sample size for confidence intervals 424
18.5 Planning studies: the power of a statistical test* 426

CHAPTER 19 **From Data Production to Inference: Part III Review 439**
Part III Summary 441
Test Yourself 443
Supplementary Exercises 450

PART IV INFERENCE ABOUT VARIABLES 453

CHAPTER 20 **Inference about a Population Mean 455**
20.1 Conditions for inference about a mean 455
20.2 The *t* distributions 456
20.3 The one-sample *t* confidence interval 458
20.4 The one-sample *t* test 461
20.5 Using technology 464
20.6 Matched pairs *t* procedures 467
20.7 Robustness of *t* procedures 469
20.8 Resampling and standard errors* 472

CHAPTER 21 **Comparing Two Means 485**
21.1 Two-sample problems 485
21.2 Comparing two population means 487
21.3 Two-sample *t* procedures 489
21.4 Using technology 494
21.5 Robustness again 497
21.6 Details of the t approximation* 499
21.7 Avoid the pooled two-sample *t* procedures* 501
21.8 Avoid inference about standard deviations* 501
21.9 Permutation tests* 502

CHAPTER 22 **Inference about a Population Proportion 517**
22.1 The sample proportion $\hat{p}$ 518
22.2 Large-sample confidence intervals for a proportion 520
22.3 Choosing the sample size 523
22.4 Significance tests for a proportion 525
22.5 Plus four confidence intervals for a proportion* 528

CHAPTER 23 **Comparing Two Proportions 539**
23.1 Two-sample problems: proportions 539
23.2 The sampling distribution of a difference between proportions 541
23.3 Large-sample confidence intervals for comparing proportions 542
23.4 Using technology 543
23.5 Significance tests for comparing proportions 545
23.6 Plus four confidence intervals for comparing proportions* 549

CHAPTER 24 **Inference about Variables: Part IV Review 559**
Part IV Summary 562
Test Yourself 564
Supplementary Exercises 571

PART V INFERENCE ABOUT RELATIONSHIPS 575

CHAPTER 25 **Two Categorical Variables: The Chi-Square Test 577**
25.1 Two-way tables 577
25.2 The problem of multiple comparisons 580
25.3 Expected counts in two-way tables 581
25.4 The chi-square test statistic 583
25.5 Cell counts required for the chi-square test 584
25.6 Using technology 585
25.7 Uses of the chi-square test: independence and homogeneity 589
25.8 The chi-square distributions 593
25.9 The chi-square test for goodness of fit* 595

CHAPTER 26 **Inference for Regression 609**
26.1 Conditions for regression inference 611
26.2 Estimating the parameters 612
26.3 Using technology 615
26.4 Testing the hypothesis of no linear relationship 619
26.5 Testing lack of correlation 620
26.6 Confidence intervals for the regression slope 622
26.7 Inference about prediction 624
26.8 Checking the conditions for inference 628

CHAPTER 27 **One-Way Analysis of Variance: Comparing Several Means 645**
27.1 Comparing several means 647
27.2 The analysis of variance *F* test 648
27.3 Using technology 650
27.4 The idea of analysis of variance 653
27.5 Conditions for ANOVA 656
27.6 *F* distributions and degrees of freedom 659
27.7 Some details of ANOVA* 661

Notes and Data Sources 677
Tables 697
TABLE A Standard normal cumulative proportions 698
TABLE B Random digits 700
TABLE C t distribution critical values 701
TABLE D Chi-square distribution critical values 702
TABLE E Critical values of the correlation r 703
Answers to Odd-numbered Exercises 705
Index 759

PART VI OPTIONAL COMPANION CHAPTERS
(AVAILABLE ONLINE)

CHAPTER 28 **Nonparametric Tests 28-1**
28.1 Comparing two samples: the Wilcoxon rank sum test 28-2
28.2 The Normal approximation for *W* 28-6
28.3 Using technology 28-8
28.4 What hypotheses does Wilcoxon test? 28-10
28.5 Dealing with ties in rank tests 28-11
28.6 Matched pairs: the Wilcoxon signed rank test 28-16
28.7 The Normal approximation for W^+ 28-18
28.8 Dealing with ties in the signed rank test 28-20
28.9 Comparing several samples: the Kruskal–Wallis test 28-23
28.10 Hypotheses and conditions for the Kruskal–Wallis test 28-24
28.11 The Kruskal–Wallis test statistic 28-24

CHAPTER 29 **Multiple Regression 29-1**
29.1 Parallel regression lines 29-2
29.2 Estimating parameters 29-5
29.3 Using technology 29-10
29.4 Inference for multiple regression 29-13
29.5 Interaction 29-22
29.6 The general multiple linear regression model 29-28
29.7 The woes of regression coefficients 29-34
29.8 A case study for multiple regression 29-36
29.9 Inference for regression parameters 29-48
29.10 Checking the conditions for inference 29-53

CHAPTER 30 **More about Analysis of Variance 30-1**
30.1 Beyond one-way ANOVA 30-1
30.2 Follow-up analysis: Tukey pairwise multiple comparisons 30-6
30.3 Follow-up analysis: contrasts* 30-10
30.4 Two-way ANOVA: conditions, main effects, and interaction 30-13
30.5 Inference for two-way ANOVA 30-20
30.6 Some details of two-way ANOVA* 30-28

CHAPTER 31 **Statistical Process Control 31-1**
31.1 Processes 31-2
31.2 Describing processes 31-2
31.3 The idea of statistical process control 31-6
31.4 $\bar{x}$ charts for process monitoring 31-7
31.5 *s* charts for process monitoring 31-13
31.6 Using control charts 31-19
31.7 Setting up control charts 31-22
31.8 Comments on statistical control 31-28
31.9 Don't confuse control with capability 31-30
31.10 Control charts for sample proportions 31-32
31.11 Control limits for *p* charts 31-33

TO THE INSTRUCTOR: About This Book

Welcome to the seventh edition of *The Basic Practice of Statistics*. As the name suggests, this text provides an introduction to the practice of statistics that aims to equip students to carry out common statistical procedures and to follow statistical reasoning in their fields of study and in their future employment.

The Basic Practice of Statistics is designed to be accessible to college and university students with limited quantitative background—just "algebra" in the sense of being able to read and use simple equations. It is usable with almost any level of technology for calculating and graphing—from a $15 "two-variable statistics" calculator through a graphing calculator or spreadsheet program through full statistical software. Of course, graphs and calculations are less tedious with good technology, so we recommend making available to your students the most effective technology that circumstances permit.

Despite the lower mathematical level, *The Basic Practice of Statistics* is designed to reflect the actual practice of statistics, where data analysis and design of data production join with probability-based inference to form a coherent science of data. There are good pedagogical reasons for beginning with data analysis (Chapters 1 to 7), then moving to data production (Chapters 8 to 11), and then to probability and inference (Chapters 12 to 27). In studying data analysis, students learn useful skills immediately and get over some of their fear of statistics. Data analysis is a necessary preliminary to inference in practice, because inference requires clean data. Designed data production is the surest foundation for inference, and the deliberate use of chance in random sampling and randomized comparative experiments motivates the study of probability in a course that emphasizes data-oriented statistics. *The Basic Practice of Statistics* gives a full presentation of basic probability and inference (16 of the 27 chapters) but places it in the context of statistics as a whole.

Guiding Principles and the GAISE Guidelines

The Basic Practice of Statistics is based on three principles: balanced content, experience with data, and the importance of ideas. These principles are widely accepted by statisticians concerned about teaching and are directly connected to and reflected by the themes of the College Report of the Guidelines in Assessment and Instruction for Statistics Education (GAISE) Project.

The GAISE Guidelines include six recommendations for the introductory statistics course. The content, coverage, and features of *The Basic Practice of Statistics* are closely aligned to these recommendations:

1. Emphasize statistical literacy and develop statistical thinking. The intent of *The Basic Practice of Statistics* is to be modern *and* accessible. The exposition is straightforward and concentrates on major ideas and skills. One principle of writing for beginners is not to try to tell your students everything you know. Another principle is to offer frequent stopping points, marking off digestible bites of material. Statistical literacy is promoted throughout *The Basic Practice of Statistics* in the many examples and exercises drawn from the popular press and from many fields of study. Statistical thinking is promoted in examples and exercises that give enough background to allow students to consider the meaning of their calculations. Exercises often ask for conclusions that are more than a number (or "reject H_0"). Some exercises require judgment in addition to right-or-wrong calculations and conclusions. Statistics, more than mathematics, depends on judgment for effective use. *The Basic Practice of Statistics* begins to develop students' judgment about statistical studies.

2. Use real data. The study of statistics is supposed to help students work with data in their varied academic disciplines and in their unpredictable later employment. Students learn to work with data by working with data. *The Basic Practice of Statistics* is full of data from many fields of study and from everyday life. Data are more than mere numbers—they are numbers with a context that should play a role in making sense of the numbers and in stating conclusions. Examples and exercises in *The Basic Practice of Statistics,* though intended for beginners, use real data and give enough background to allow students to consider the meaning of their calculations.

3. Stress conceptual understanding rather than mere knowledge of procedures. A first course in statistics introduces many skills, from making a stemplot and calculating a correlation to choosing and carrying out a significance test. In practice (even if not always in the course), calculations and graphs are automated. Moreover, anyone who makes serious use of statistics will need some specific procedures not taught in their college statistics course. *The Basic Practice of Statistics therefore* tries to make clear the larger patterns and big ideas of statistics, not in the abstract, but in the context of learning specific skills and working with specific data. Many of the big ideas are summarized in graphical outlines. Three of the most useful appear inside the front cover. Formulas without guiding principles do students little good once the final exam is past, so it is worth the time to slow down a bit and explain the ideas.

4. Foster active learning in the classroom. Fostering active learning is the business of the teacher, though an emphasis on working with data helps. To this end, we have created interactive applets to our specifications and made them available online. These are designed primarily to help in learning statistics rather than in doing statistics. We suggest using selected applets for classroom demonstrations even if you do not ask students to work with them. *The Correlation and Regression, Confidence Intervals, and P-value of a Test of Significance* applets, for example, convey core ideas more clearly than any amount of chalk and talk.

We also provide web exercises at the end of each chapter. Our intent is to take advantage of the fact that most undergraduates are "web savvy." These exercises require students to search the web for either data or statistical examples and then evaluate what they find. Teachers can use these as classroom activities or assign them as homework projects.

5. Use technology for developing conceptual understanding and analyzing data. Automating calculations increases students' ability to complete problems, reduces their frustration, and helps them concentrate on ideas and problem recognition rather than mechanics. At a minimum, students should have a "two-variable statistics" calculator with functions for correlation and the least-squares regression line as well as for the mean and standard deviation.

Many instructors will take advantage of more elaborate technology, as ASA/MAA and GAISE recommend. And many students who don't use technology in their college statistics course will find themselves using (for example) Excel on the job. *The Basic Practice of Statistics* does not assume or require use of software except in Part V, where the work is otherwise too tedious. It does accommodate software use and tries to convince students that they are gaining knowledge that will enable them to read and use output from almost any source. There are regular "Using Technology" sections throughout the text. Each of these sections displays and comments on output from the same three technologies, representing graphing calculators (the Texas Instruments TI-83 or TI-84), spreadsheets (Microsoft Excel), and statistical software (JMP, Minitab, and CrunchIt!). The output always concerns one of the main teaching examples, so that students can compare text and output.

6. Use assessments to improve and evaluate student learning. Within chapters, a few "Apply Your Knowledge" exercises follow each new idea or skill for a quick check of basic mastery—and also to mark off digestible bites of material. Each of the first four parts of the book ends with a review chapter that includes a point-by-point outline of skills learned, problems students can use to test themselves, and several supplementary exercises. (Instructors can choose to cover any or none of the chapters in Part V, so each of these chapters includes a skills outline.) The review chapters present supplemental exercises without the "I just studied that" context, thus asking for another level of learning. We think it is helpful to assign some supplemental exercises. Many instructors will find that the review chapters appear at the right points for pre-examination review. The "Test Yourself" questions can be used by students to review, self-assess, and prepare for such an examination.

In addition, assessment materials in the form of a test bank and quizzes are available online.

What's New?

The new edition of *The Basic Practice of Statistics* brings many **new examples and exercises**. There are new data sets from a variety of sources, including finance (the relationship between positive articles in the media and the Dow Jones Industrial Average the following week), health (the relationship between salt intake and percent body fat of children), psychology (the relationship between one's attitude about a presidential candidate and how trustworthy the candidate's face appears to be), medicine (the relationship between playing video games and surgical skills), and the environment (global temperatures). Popular examples and exercises such as the Florida manatee regression example return, many with updated data. These are just a few of a large number of new data settings in this edition.

A new edition is also an opportunity to introduce new features and polish the exposition in ways intended to help students learn. Here are some of the changes:

- Each chapter now contains references to online resources to enhance student learning. These include video clips, whiteboard lectures, and technology supplements.
- We have added an introductory chapter, "Getting Started," that instructors may wish to assign to students the first day of classes. This chapter provides an overview of statistical thinking and real examples where the use of statistics can provide valuable insight. It expands on material that was previously included in the Preface, adding motivating examples and exercises.
- Chapter 7 includes descriptions of additional data sets available online that instructors can use for student projects and more extensive data analysis. Along with the description of the data sets, we provide a few suggestions for how they might be used.
- We have added some basic material on resampling and permutation tests in optional sections at the end of Chapters 15, 17, 20, and 21. We hope that instructors who want to introduce students to resampling methods will find this new material useful.
- The essay on data ethics is now Chapter 10, and follows the format of other chapters in the book.
- We have added output from JMP to the "Using Technology" sections.
- The content in Parts I and II has been rewritten to accommodate instructors who prefer to teach data production (Part II) before data exploration (Part I). Instructors can teach these parts in either order while maintaining the continuity of the material.
- Sections are now numbered for easier reference.

FEATURES OF *THE BASIC PRACTICE OF STATISTICS*, Seventh Edition

In this chapter we cover...

Each chapter opener offers a brief overview of where the chapter is heading, often with reference to previous chapters, and includes a section outline of the major topics that will be covered.

In this chapter we cover...

2.1 Measuring center: the mean
2.2 Measuring center: the median
2.3 Comparing the mean and the median
2.4 Measuring variability: the quartiles
2.5 The five-number summary and boxplots
2.6 Spotting suspected outliers and modified boxplots*
2.7 Measuring variability: the standard deviation
2.8 Choosing measures of center and variability
2.9 Using technology
2.10 Organizing a statistical problem

EXAMPLE 2.9 Comparing Graduation Rates

GRADRATE

STATE: Federal law requires all states in the United States to use a common computation of on-time high school graduation rates beginning with the 2010–11 school year. Previously, states chose one of several computation methods that gave answers that could differ by more than 10%. This common computation allows for meaningful comparison of graduation rates between the states.

We know from Table 1.1 (page 22) that the on-time high school graduation rates varied from 59% in the District of Columbia to 88% in Iowa. The U.S. Census Bureau divides the 50 states and the District of Columbia into four geographical regions: the Northeast (NE), Midwest (MW), South (S), and West (W). The region for each state is included in Table 1.1. Do the states in the four regions of the country display distinct distributions of graduation rates? How do the mean graduation rates of the states in each of these regions compare?

PLAN: Use graphs and numerical descriptions to describe and compare the distributions of on-time high school graduation rates of the states in the four regions of the United States.

SOLVE: We might use boxplots to compare the distributions, but stemplots preserve more detail and work well for data sets of these sizes. Figure 2.5 displays the stemplots with the stems lined up for easy comparison. The stems have been split to better display the distributions. The stemplots overlap, and some care is needed when comparing the four stemplots as the sample sizes differ, with some stemplots having more leaves than others. None of the plots shows strong skewness, although the South has one low observation that stands apart from the others with this choice of stems. The states in the Northeast and Midwest have distributions that are similar to each other, as do those in the South and West. The graduation rates tend to be higher for the states in the Northeast and Midwest and more variable for the states in the South and West. With little skewness and no serious outliers, we report $\bar{x}$ and s as our summary measures of center and variability of the distribution of the on-time graduation rates of the states in each region:

Region	Mean	Standard Deviation
Midwest	82.92	4.25
Northeast	82.56	3.47
South	75.93	7.36
West	73.58	6.73

```
Midwest       Northeast     South         West
8 | 66678     8 | 67        8 | 66        8 |
8 | 01334     8 | 33334     8 | 123       8 | 002
7 | 7         7 | 77        7 | 5688      7 | 6668
7 | 4         7 |           7 | 1124      7 | 4
6 |           6 |           6 | 7         6 | 88
6 |           6 |           6 |           6 | 23
5 |           5 |           5 | 9         5 |
```

FIGURE 2.5 Stemplots comparing the distributions of graduation rates for the four census regions from Table 1.1, for Example 2.9.

CONCLUDE: The table of summary statistics confirms what we see in the stemplots. The states in the Midwest and Northeast are quite similar to each other, as are those in the South and West. The states in the Midwest and Northeast have a higher mean graduation rate as well as a smaller standard deviation than those in the South and West. ■

4-Step Examples

In Chapter 2, students learn how to use the four-step process for working through statistical problems: State, Plan, Solve, Conclude. By observing this framework in use in selected examples throughout the text and practicing it in selected exercises, students develop the ability to solve and write reports on real statistical problems encountered outside the classroom.

Apply Your Knowledge

Major concepts are immediately reinforced with problems that are interspersed throughout the chapter (often following examples). These problems allow students to practice their skills concurrently as they work through the text.

Apply Your Knowledge

Photo Researchers/Getty Images

2.10 $\bar{x}$ and s by Hand. Radon is a naturally occurring gas and is the second leading cause of lung cancer in the United States.[9] It comes from the natural breakdown of uranium in the soil and enters buildings through cracks and other holes in the foundations. Found throughout the United States, levels vary considerably from state to state. Several methods can reduce the levels of radon in your home, and the Environmental Protection Agency recommends using one of these if the measured level in your home is above 4 picocuries per liter. Four readings from Franklin County, Ohio, where the county average is 8.4 picocuries per liter, were 6.2, 12.8, 7.6, and 15.4.

(a) Find the mean step-by-step. That is, find the sum of the four observations and divide by 4.

(b) Find the standard deviation step-by-step. That is, find the deviation of each observation from the mean, square the deviations, then obtain the variance and the standard deviation. Example 2.7 shows the method.

(c) Now enter the data into your calculator and use the mean and standard deviation buttons to obtain $\bar{x}$ and s. Do the results agree with your hand calculations?

LaunchPad Online Resources

Many sections end with references to the most relevant and helpful online resources (chosen by the authors and available in LaunchPad) for students to use for further explanation or practice.

> **LaunchPad Online Resources**
>
> - The **Snapshots video**, *Summarizing Quantitative Data,* provides an overview of the need for measures of center and variability as well as some details of the computations.
> - The **StatClips Examples video**, *Summaries of Quantitative Data Example C,* gives the details for the computation of the mean, median, and standard deviation in a small example. You can verify the computations along with the video, either by hand or using your technology.
> - The **StatClips Examples videos**, *Basic Principles of Exploring Data Example B* and *Basic Principles of Exploring Data Example C,* emphasize the need to examine outliers and understand them, rather than simply discarding observations that don't seem to fit.

Using Technology

Located where most appropriate, these special sections display and comment on the output from graphing calculators, spreadsheets, and statistical software in the context of examples from the text.

2.9 Using technology

Although a calculator with "two-variable statistics" functions will do the basic calculations we need, more elaborate tools are helpful. Graphing calculators and computer software will do calculations and make graphs as you command, freeing you to concentrate on choosing the right methods and interpreting your results. Figure 2.4 displays output describing the travel times to work of 20 people in New York State (Example 2.3). Can you find $\bar{x}$, s, and the five-number summary in each output? The big message of this section is: *Once you know what to look for, you can read output from any technological tool.*

The displays in Figure 2.4 come from a Texas Instruments graphing calculator, the Minitab, CrunchIt!, and JMP statistical programs, and the Microsoft Excel spreadsheet program. Minitab and JMP allow you to choose what descriptive measures you want, whereas the descriptive measures in the CrunchIt! output are provided by default. Excel and the calculator give some things we don't need. Just ignore the extras. Excel's "Descriptive Statistics" menu item doesn't give the quartiles. We used the spreadsheet's separate quartile function to get Q_1 and Q_3.

Texas Instruments Graphing Calculator

```
1-Var Stats
 x̄=31.25
 Σx=625
 Σx²=28625
 Sx=21.8773495
 σx=21.32340264
↓n=20
```

```
1-Var Stats
↑n=20
 minX=5
 Q1=15
 Med=22.5
 Q3=42.5
 maxX=85
```

Minitab

Session

Descriptive Statistics: NYtime

	Total								
variable	Count	Mean	StDev	Variance	Minimum	Q1	Median	Q3	Maximum
NYtime	20	31.25	21.88	478.62	5.00	15.00	22.50	43.75	85.00

CrunchIt!

Results - Descriptive Statistics

Export

	n	Sample Mean	Standard Deviation	Min	Q1	Median	Q3	Max
Minutes	20	31.25	21.88	5	15	22.50	42.50	85

Microsoft Excel

Book1

	A	B	C	D
1	minutes			
2				
3	Mean	31.25		
4	Standard Error	4.891924064		
5	Median	22.5	QUARTILE(A2:A21,1)	15
6	Mode	15	QUARTILE(A2:A21,3)	42.5
7	Standard Deviation	21.8773495		
8	Sample Variance	478.6184211		
9	Kurtosis	0.329884126		
10	Skewness	1.040110836		
11	Range	80		
12	Minimum	5		
13	Maximum	85		
14	Sum	625		
15	Count	20		

Sheet4 Sheet1 Sheet2 Sheet

JMP Output

Distributions

NYtime

Quantiles

100%	maximum	85
75%	quartile	43.75
50%	median	22.5
25%	quartile	15
0%	minimum	5

Summary Statistics

Mean	31.25
Std Dev	21.877349
N	20

FIGURE 2.4
Output from a graphing calculator, three statistical software packages, and a spreadsheet program describing the data on travel times to work in New York State.

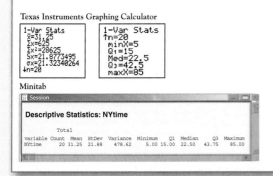

Statistics in Your World

These brief asides in each chapter illustrate major concepts or present cautionary tales through entertaining and relevant stories, allowing students to take a break from the exposition while staying engaged.

HE SAID, SHE SAID.

Height, weight, and body mass distributions in this book come from actual measurements by a government survey. That is a good thing. When *asked* their weight, almost all women say they weigh less than they really do. Heavier men also underreport their weight—but lighter men claim to weigh more than the scale shows. We leave you to ponder the psychology of the two sexes. Just remember that "say-so" is no substitute for measuring.

Chapter Summary and Link It

Each chapter concludes with a summary of the chapter specifics, including major terms and processes, followed by a brief discussion of how the chapter links to material from both previous and upcoming chapters.

CHAPTER 4 SUMMARY

Chapter Specifics

- To study relationships between variables, we must measure the variables on the same group of individuals.
- If we think that a variable x may explain or even cause changes in another variable y, we call x an **explanatory variable** and y a **response variable**.
- A **scatterplot** displays the relationship between two quantitative variables measured on the same individuals. Mark values of one variable on the horizontal axis (x axis) and values of the other variable on the vertical axis (y axis). Plot each individual's data as a point on the graph. Always plot the explanatory variable, if there is one, on the x axis of a scatterplot.
- Plot points with different colors or symbols to see the effect of a categorical variable in a scatterplot.
- In examining a scatterplot, look for an overall pattern showing the **direction**, **form**, and **strength** of the relationship and then for **outliers** or other deviations from this pattern.
- **Direction:** If the relationship has a clear direction, we speak of either **positive association** (high values of the two variables tend to occur together) or **negative association** (high values of one variable tend to occur with low values of the other variable).
- **Form: Linear relationships**, where the points show a straight-line pattern, are an important form of relationship between two variables. Curved relationships and **clusters** are other forms to watch for.
- **Strength:** The **strength** of a relationship is determined by how close the points in the scatterplot lie to a simple form such as a line.
- The **correlation** r measures the direction and strength of the linear association between two quantitative variables x and y. Although you can calculate a correlation for any scatterplot, r measures only straight-line relationships.

Link It

In Chapters 1 to 3, we focused on exploring features of a single variable. In this chapter, we continued our study of exploratory data analysis but for the purpose of examining relationships *between* variables. A useful tool for exploring the relationship between two variables is the scatterplot. When the relationship is linear, correlation is a numerical measure of the strength of the linear relationship.

It is tempting to assume that the patterns we observe in our data hold for values of our variables that we have not observed—in other words, that additional data would continue to conform to these patterns. The process of identifying underlying patterns would seem to assume that this is the case. But is this assumption justified? Parts II to V of the book answer this question.

Check Your Skills and Chapter Exercises

Each chapter ends with a series of multiple-choice problems that test students' understanding of basic concepts and their ability to apply the concepts to real-world statistical situations. The multiple-choice problems are followed by a set of more in-depth exercises that allow students to make judgments and draw conclusions based on real data and real scenarios.

CHECK YOUR SKILLS

4.14 Researchers collect data on 5,134 American adults younger than 60. They measure the reaction times (in seconds) of each subject to a stimulus on a computer screen and how many years later the subject died.[10]

The researchers are interested in whether reaction time can predict time to death (in years). When you make a scatterplot, the explanatory variable on the x axis

(a) is the reaction time.
(b) is the time to death.
(c) can be either reaction time or time to death.

4.15 The researchers in Exercise 4.14 found that people with slower reaction times tended to die sooner. In a scatterplot of the reaction time and the number of years to death, you expect to see

(a) a positive association.
(b) very little association.
(c) a negative association.

4.16 Figure 4.7 is a scatterplot of school GPA against IQ test scores for 15 seventh-grade students. There is one low outlier in the plot. The IQ and GPA scores for this student are

(a) IQ = 0.5, GPA = 103.
(b) IQ = 103, GPA = 0.5.
(c) IQ = 103, GPA = 7.6.

4.17 If we leave out the low outlier, the correlation for the remaining 14 points in Figure 4.7 is closest to

(a) 0.9. (b) −0.9. (c) 0.1.

4.18 What are all the values that a correlation r can possibly take?

(a) $r \geq 0$ (b) $0 \leq r \leq 1$ (c) $-1 \leq r \leq 1$

4.19 If the correlation between two variables is close to 0, you can conclude that a scatterplot would show

(a) a strong straight-line pattern.
(b) a cloud of points with no visible pattern.
(c) no straight-line pattern, but there might be a strong pattern of another form.

4.20 The points on a scatterplot lie very close to a straight line. The correlation between x and y is close to

(a) −1. (b) 1. (c) either −1 or 1, we can't say which.

4.21 A statistics professor warns her class that her second midterm is always harder than the first. She tells her class that students always score 10 points worse on the second midterm compared to their score on the first midterm. This means that the correlation between students' scores on the first and second exam is

(a) 1. (b) −1. (c) Can't tell without seeing the data.

4.22 Researchers asked mothers how much soda (in ounces) their kids drank in a typical day. They also asked these mothers to rate how aggressive their kids were on a scale of 1 to 10, with larger values corresponding to a greater degree of aggression.[11] The correlation between amount of soda consumed and aggression rating was found to be $r = 0.3$. If the researchers had measured amount of soda consumed in liters instead

CHAPTER 4 EXERCISES

4.24 Scores at the Masters. The Masters is one of the four major golf tournaments. Figure 4.8 is a scatterplot of the scores for the first two rounds of the 2013 Masters for all the golfers entered. Only the 60 golfers with the lowest two-round total advance to the final two rounds (unless several people are tied for 60th place, in which case all those tied for 60th place advance). The plot has a grid pattern because golf scores must be whole numbers.[13] MASTR13

(a) Read the graph: What was the lowest score in the first round of play? How many golfers had this low score? What were their scores in the second round?
(b) Read the graph: Alan Dunbar had the highest score in the second round. What was this score? What was Dunbar's score in the first round?

4.25 Happy states. Human happiness or well-being can be assessed either subjectively or objectively. Subjective assessment can be accomplished by listening to what people say. Objective assessment can be made from data related to well-being such as income, climate, availability of entertainment, housing prices, lack of traffic congestion, and so on. Do subjective and objective assessments agree? To study this, investigators made both subjective and objective assessments of happiness for each of the 50 states. The subjective measurement was the mean score on a life-satisfaction question found on the Behavioral Risk Factor Surveillance System (BRFSS), which is a state-based system of health surveys. Lower scores indicate a greater degree of happiness. To objectively assess happiness, the investigators computed a mean well-being score (called

Web Exercises

A final set of exercises asks students to investigate data and statistical issues by researching topics online. These exercises tend to be more involved and provide an opportunity for students to dig deep into contemporary issues and special applications of statistics.

Exploring the Web

6.34 Promoting women. In academics, faculty typically start as assistant professors, are promoted to associate professor (and gain tenure), and finally reach the rank of full professor. Some have argued that women have a harder time gaining promotion to associate and full professor than do men. Do data support this argument? Search the web to find the number of faculty by rank and gender at some university. Do you see a pattern that suggests that the proportion of women decreases as rank increases? We found several sources of data by doing a Google search on "faculty head count by rank and gender." In addition to discussing the pattern you find, provide the data, the name of the school, and the source of the data.

6.35 Accidental deaths and age. Accidental deaths are shocking and tragic. Do the ways in which people die by accident change with age? Look at the most recent *Statistical Abstract of the United States* (`www.census.gov/compendia/statab/`) and make a two-way table that provides the counts of deaths due to accidents from various causes for three different age groups. What do you conclude?

6.36 Simpson's paradox. Find an example of Simpson's paradox and discuss how your example illustrates the paradox. Two examples that we found (thanks to Patricia Humphrey at Georgia Southern University) are `www.nytimes.com/2006/07/15/education/15report.html` and `online.wsj.com/article/SB125970744553071829.html`.

Online Data for Additional Analyses

References to larger data sets are suggested in Chapter 7 to provide an opportunity for students to apply the methods of Chapters 1–6 to explore data on their own. This is intended to reinforce the idea of exploratory data analysis as a tool for exploring data.

Online Data for Additional Analyses

1. SAT, ACT, and teacher salaries for 2013 for each of the 50 states and the District of Columbia are available in the data set SATACT. One could use these data to carry out analyses for ACT scores similar to those for the SAT scores in Chapters 5 and 6. For example, repeat the analyses in Exercises 5.50 (page 157) and 5.51 (page 158) using ACT scores instead of SAT scores. SATACT
2. The data set MLB contains hitting, pitching, fielding, salary, and win–loss performance data from the 2013 season for all major league baseball teams. These data can be used to determine the correlation between payroll and winning percentage. One can also explore what variables are most highly correlated with winning percentage, and whether variables that measure pitching performance are more highly correlated with winning percentage than variables that measure hitting performance. For example, calculate the correlation between winning percentage and number of home runs, between winning percentage and batting average, between winning percentage and ERA, between winning percentage and strikeouts by pitchers, and between winning percentage and payroll. Which has the highest correlation? These data are from `http://www.baseball-reference.com/`. Visit this website for definitions of several of the variables in the data set. MLB
3. Historical temperature data and whether Punxsutawney Phil saw his shadow are available in the data set PHIL. Repeat the analysis in Exercise 6.32 (page 177), but define what constitutes "six more weeks of winter-like weather" differently. For example, you might decide there were six more weeks of winter-like weather if average temperatures for March were at least one degree below historical averages. PHIL
4. Data from the Ohio Department of Health website are available in the pdf "2013OHH Detail Tables." This is a source of many tables that can be used for further analyses using methods discussed in Chapter 6. For example, conduct an analysis like that in Exercise 7.52 to investigate the relationship between sex and strategies about weight (Question 67 in the Tables). HEALTH
5. The data set WHAT contains three variables and 3848 observations on each. At one time, this was considered a large data set and difficult to explore with software. Use various exploratory methods available in software packages such as JMP and Minitab to find the "hidden pattern" in these data. WHAT

Why Did You Do That?

There is no single best way to organize our presentation of statistics to beginners. That said, our choices reflect thinking about both content and pedagogy. Here are comments on several "frequently asked questions" about the order and selection of material in *The Basic Practice of Statistics.*

• **Why does the distinction between population and sample not appear in Part I?** There is more to statistics than inference. In fact, statistical inference is appropriate only in rather special circumstances. The chapters in Part I present tools and tactics for describing data—any data. These tools and tactics do not depend on the idea of inference from sample to population. Many data sets in these chapters (for example, the several sets of data about the 50 states) do not lend themselves to inference because they represent an entire population. John Tukey of Bell Labs and Princeton, the philosopher of modern data analysis, insisted that the population–sample distinction be avoided when it is not relevant. He used the word "batch" for data sets in general. We see no need for a special word, but we think Tukey was right.

• **Why not begin with data production?** We prefer to begin with data exploration (Part I), as most students will use statistics mainly in settings other than planned research studies in their future employment. We place the design of data production (Part II) after data analysis to emphasize that data-analytic techniques apply to any data. However, it is equally reasonable to begin with data production—the natural flow of a planned study is from design to data analysis to inference. Because instructors have strong and differing opinions on this question, these two topics are now the first two parts of the book, with the text written so that it may be started with either Part I or Part II while maintaining the continuity of the material.

• **Why do Normal distributions appear in Part I?** Density curves such as the Normal curves are just another tool to describe the distribution of a quantitative variable, along with stemplots, histograms, and boxplots. Professional statistical software offers to make density curves from data just as it offers histograms. We prefer not to suggest that this material is essentially tied to probability, as the traditional order does. And we find it helpful to break up the indigestible lump of probability that troubles students so much. Meeting Normal distributions early does this and strengthens the "probability distributions are like data distributions" way of approaching probability.

• **Why not delay correlation and regression until late in the course, as was traditional?** *The Basic Practice of Statistics* begins by offering experience working with data and gives a conceptual structure for this nonmathematical but essential part of statistics. Students profit from more experience with data and from seeing the conceptual structure worked out in relations among variables as well as in describing single-variable data. Correlation and least-squares regression are very important descriptive tools and are often used in settings where there is no population–sample distinction, such as studies of all of a firm's employees. Perhaps most important, the approach taken by *The Basic Practice of Statistics* asks students to think about what kind of relationship lies behind the data (confounding, lurking variables, association doesn't imply causation, and so on), without overwhelming them with the demands of formal inference methods. Inference in the correlation and regression setting is a bit complex, demands software, and often comes right at the end of the course. We find that delaying all mention of correlation and regression to that point means that students often don't master the basic uses and properties of these methods. We consider Chapters 4 and 5 (correlation and regression) essential and Chapter 26 (regression inference) optional.

- **Why use the z procedures for a population mean to introduce the reasoning of inference?** This is a pedagogical issue, not a question of statistics in practice. The two most popular choices for introducing inference are z for a mean and z for a proportion. (Another option is resampling and permutation tests. We have included material on these topics, but have not used them to introduce inference.)

We find z for means quite accessible to students. Positively, we can say up front that we are going to explore the reasoning of inference in the overly simple setting described in the box on page 374 titled Simple Conditions for Inference about a Mean. As this box suggests, exactly Normal population and true simple random sample are as unrealistic as known σ. All the issues of practice—robustness against lack of Normality and application when the data aren't an SRS as well as the need to estimate σ—are put off until, with the reasoning in hand, we discuss the practically useful t procedures. This separation of initial reasoning from messier practice works well.

Negatively, starting with inference for p introduces many side issues: no exact Normal sampling distribution, but a Normal approximation to a discrete distribution; use of $\hat{p}$ in both the numerator and denominator of the test statistic to estimate both the parameter p and $\hat{p}$'s own standard deviation; loss of the direct link between test and confidence interval; and the need to avoid small and moderate sample sizes because the Normal approximation for the test is quite unreliable.

There are advantages to starting with inference for p. Starting with z for means takes a fair amount of time and the ideas need to be rehashed with the introduction of the t procedures. Many instructors face pressure from client departments to cover a large amount of material in a single semester. Eliminating coverage of the "unrealistic" z for means with known variance enables instructors to cover additional, more realistic applications of inference. Also, many instructors believe that proportions are simpler and more familiar to students than means. For instructors who would prefer to introduce inference with z for a proportion, we recommend our book, *Statistics in Practice*.

- **Why didn't you cover Topic X?** Introductory texts ought not to be encyclopedic. We chose topics on two grounds: they are the most commonly used in practice, and they are suitable vehicles for learning broader statistical ideas. Students who have completed the core of the book, Chapters 1 to 12 and 15 to 24, will have little difficulty moving on to more elaborate methods. Chapters 25 to 27 offer a choice of slightly more advanced topics, as do the four companion chapters available online.

ACKNOWLEDGMENTS

We have enjoyed the opportunity to once again rethink how to help beginning students achieve a practical grasp of basic statistics. What students actually learn is not identical to what we teachers think we have "covered," so the virtues of concentrating on the essentials are considerable. We hope that the new edition of *The Basic Practice of Statistics* offers a mix of concrete skills and clearly explained concepts that will help many teachers guide their students toward useful knowledge.

We are grateful to colleagues from two-year and four-year colleges and universities who commented on *The Basic Practice of Statistics:*

Faran Ali, *Simon Fraser University*
Michael Allen, *Glendale Community College*
Paul Lawrence Baker, *Catawba College*
Brigitte Baldi, *University of California—Irvine*
Barbara A. Barnet, *University of Wisconsin—Platteville*
Paul R. Bedard, *Saint Clair Community College*
Marjorie E. Bond, *Monmouth College*
Ryan Botts, *Point Loma Nazarene University*
Mine Cetinkaya-Rundel, *Duke University*
Gary Cochell, *Culver-Stockton College*
Patti Collings, *Brigham Young University*
Phyllis Curtis, *Grand Valley State University*
Carolyn Pillers Dobler, *Gustavus Adolphus College*
John Daniel Draper, *The Ohio State University*
Michelle Everson, *The Ohio State University*
Diane G. Fisher, *University of Louisiana at Lafayette*
James Gray, *University of Washington*
Ellen Gundlach, *Purdue University*
James A. Harding, *Green Mountain College*
James Hartman, *The College of Wooster*
Pat Humphrey, *Georgia Southern University*
Dick Jardine, *Keene State College*
Robert W. Jernigan, *American University*
Jennifer Kaplan, *University of Georgia*
Daniel L. King, *Sarah Lawrence College*
William "Sonny" Kirby, *Gadsden State Community College*
Brian Knaeble, *University of Wisconsin—Stout*
Allyn Leon, *Imperial Valley College*
Karen P. Lundberg, *Colorado State University—Pueblo*
Dana E. Madison, *Clarion University of Pennsylvania*
Kimberly Massaro, *University of Texas at San Antonio*
Jackie Miller, *University of Michigan*
Juliann Moore, *Oregon State University*
Penny Ann Morris, *Polk State College*
Kathleen Mowers, *Owensboro Community and Technical College*
Julia Ann Norton, *California State University—East Bay*
Mary R. Parker, *Austin Community College*
Michael Price, *University of Oregon*
David Rangel, *Bellingham Technical College*
Shane Patrick Redmond, *Eastern Kentucky University*
Scott J. Richter, *University of North Carolina at Greensboro*
Laurence David Robinson, *Ohio Northern University*
Caroline Schruth, *Tacoma Community College*
Mack Shelley, *Iowa State University*
Therese N. Shelton, *Southwestern University*
Haskell Sie, *Pennsylvania State University*
Murray H. Siegel, *Arizona State University—Polytechnic Campus*
Sean Simpson, *Westchester Community College*
Robb Sinn, *University of North Georgia*
Karen H. Smith, *University of West Georgia*
Stephen R. Soltys, *Elizabethtown College*
James Stamey, *Baylor University*
Jeanette M. Szwec, *Cape Fear Community College*
Ramin Vakilian, *California State University—Northridge*
Asokan Mulayath Variyath, *Memorial University of Newfoundland*
Lianwen Wang, *University of Central Missouri*
Barbara B. Ward, *Belmont University*
Yajni Warnapala, *Roger Williams University*
Robert E. White, *Allan Hancock College*
Ronald L. White, *Norfolk State University*
Rachelle Curtis Wilkinson, *Austin Community College*

We extend our appreciation to Ruth Baruth, Terri Ward, Karen Carson, Leslie Lahr, Jorge Amaral, Marie Dripchak, Laura Judge, Catriona Kaplan, Liam Ferguson, Victoria Garvey, Cara LeClair, Bailey James, Cecilia Varas, Eileen Liang, Lisa Kinne, Julia DeRosa, Laurel Sparrow, and other publishing professionals who have contributed to the development, production, and cohesiveness of this book and its online resources.

Special thanks are due to Vicki Tomaselli, whose talents were poured into the aesthetic appeal of this book. We extend our appreciation to Denise Showers of Aptara, Inc., who has offered her knowledge, expertise, and patience tirelessly throughout the production process.

We are deeply indebted to our colleagues, Jackie B. Miller and Patricia B. Humphrey, for their many contributions, insights, time, and humor. Their wisdom and experience in the classroom have added to a level of quality that students and instructors alike have come to expect. Each of them brought to the project their individual strengths and talents, but they did so in the spirit of true teamwork and collaboration.

We would also like to specially thank the authors and reviewers of the supplementary materials available with *The Basic Practice of Statistics,* 7e; their work and dedication to quality have resulted in a robust package of resources that complement the ideas and concepts presented in the text:

Solutions manuals written by Pat Humphrey, *Georgia Southern University*
Solutions accuracy reviewed by Jackie Miller, *University of Michigan*
Test bank written by Christiana Drake, *University of California–Davis*
Test bank accuracy reviewed by Catherine Matos, *Clayton State University*
iClicker slides created by Dilshod Achilov, *Tennessee State University*
iClicker slides accuracy reviewed by Jun Ye, *The University of Akron*
Practice Quizzes written by Leslie Hendrix, *University of South Carolina*
Practice Quizzes accuracy reviewed by Jun Ye, *The University of Akron*
Lecture PowerPoints created by Mark Gebert, *University of Kentucky, Lexington*

The team of statistics educators who created the new StatBoards videos deserve our praise and thanks; their creative works offer intuitive approaches to the key concepts in the course:

Doug Tyson, *Central York High School*
Michelle Everson, *The Ohio State University*
Marian Frazier, *Gustavus Adolphus College*
Aimee Schwab, *University of Nebraska–Lincoln*

Finally, we are indebted to the many statistics teachers with whom we have discussed the teaching of our subject over many years; to people from diverse fields with whom we have worked to understand data; and especially to students whose compliments and complaints have changed and improved our teaching. Working with teachers, colleagues in other disciplines, and students constantly reminds us of the importance of hands-on experience with data and of statistical thinking in an era when computer routines quickly handle statistical details.

David S. Moore, William I. Notz, and Michael A. Fligner

MEDIA AND SUPPLEMENTS

LaunchPad

W. H. Freeman's new online homework system, **LaunchPad,** offers our quality content curated and organized for easy assignability in a simple but powerful interface. We've taken what we've learned from thousands of instructors and hundreds of thousands of students to create a new generation of W. H. Freeman/Macmillan technology.

Curated Units. Combining a curated collection of videos, homework sets, tutorials, applets, and e-Book content, LaunchPad's interactive units give instructors a building block to use as is or as a starting point for customized learning units. A majority of exercises from the text can be assigned as online homework, including an abundance of algorithmic exercises. An entire unit's worth of work can be assigned in seconds, drastically reducing the amount of time it takes for instructors to have their course up and running.

Easily customizable. Instructors can customize the LaunchPad units by adding quizzes and other activities from our vast wealth of resources. They can also add a discussion board, a dropbox, and RSS feed, with a few clicks. LaunchPad allows instructors to customize students' experience as much or as little as desired.

Useful analytics. The gradebook quickly and easily allows instructors to look up performance metrics for classes, individual students, and individual assignments.

Intuitive interface and design. The student experience is simplified. Students' navigation options and expectations are clearly laid out at all times, ensuring they can never get lost in the system.

Assets integrated into LaunchPad include the following:

Interactive e-Book. Every LaunchPad e-Book comes with powerful study tools for students, video and multimedia content, and easy customization for instructors. Students can search, highlight, and bookmark, making it easier to study and access key content. And teachers can ensure that their classes get just the book they want to deliver: customize and rearrange chapters, add and share notes and discussions, and link to quizzes, activities, and other resources.

***LEARNING*Curve** **LearningCurve** provides students and instructors with powerful adaptive quizzing, a game-like format, direct links to the e-Book, and instant feedback. The quizzing system features questions tailored specifically to the text and adapts to students' responses, providing material at different difficulty levels and topics based on student performance.

FREEMAN SolutionMaster **SolutionMaster** offers an easy-to-use web-based version of the instructor's solutions, allowing instructors to generate a solution file for any set of homework exercises.

New **StatBoards videos** are brief whiteboard videos that illustrate difficult topics through additional examples, written and explained by a select group of statistics educators.

New Stepped Tutorials are centered on algorithmically generated quizzing with step-by-step feedback to help students work their way toward the correct solution. These new exercise tutorials (two to three per chapter) are easily assignable and assessable.

Statistical Video Series consists of StatClips, StatClips Examples, and Statistically Speaking "Snapshots." View animated lecture videos, whiteboard lessons, and documentary-style footage that illustrate key statistical concepts and help students visualize statistics in real-world scenarios.

New Video Technology Manuals available for TI-83/84 calculators, Minitab, Excel, JMP, SPSS, R, Rcmdr, and CrunchIt!® provide brief instructions for using specific statistical software.

Updated StatTutor Tutorials offer multimedia tutorials that explore important concepts and procedures in a presentation that combines video, audio, and interactive features. The newly revised format includes built-in, assignable assessments and a bright new interface.

Updated Statistical Applets give students hands-on opportunities to familiarize themselves with important statistical concepts and procedures, in an interactive setting that allows them to manipulate variables and see the results graphically. Icons in the textbook indicate when an applet is available for the material being covered.

CrunchIt!® is W. H. Freeman's web-based statistical software that allows users to perform all the statistical operations and graphing needed for an introductory statistics course and more. It saves users time by automatically loading data from BPS, and it provides the flexibility to edit and import additional data.

JMP Student Edition (developed by SAS) is easy to learn and contains all the capabilities required for introductory statistics, including pre-loaded data sets from BPS. JMP is the leading commercial data analysis software of choice for scientists, engineers, and analysts at companies throughout the globe (for Windows and Mac).

Stats@Work Simulations put students in the role of the statistical consultant, helping them better understand statistics interactively within the context of real-life scenarios.

***EESEE* Case Studies** (*Electronic Encyclopedia of Statistical Examples and Exercises*), developed by The Ohio State University Statistics Department, teach students to apply their statistical skills by exploring actual case studies using real data.

Data files are available in CrunchIt!, JMP, ASCII, Excel, TI, Minitab, and SPSS (an IBM Company)* formats.

Student Solutions Manual provides solutions to the odd-numbered exercises in the text. It is available electronically within LaunchPad, as well as in print form.

Interactive Table Reader allows students to use statistical tables interactively to seek the information they need.

Instructor's Guide with Full Solutions includes teaching suggestions, chapter comments, and detailed solutions to all exercises. It is available electronically within LaunchPad.

*SPSS was acquired by IBM in October 2009.

Test Bank offers hundreds of multiple-choice questions. It is also available on CD-ROM (for Windows and Mac), where questions can be downloaded, edited, and resequenced to suit each instructor's needs.

Lecture PowerPoint Slides offer a customizable, detailed lecture presentation of statistical concepts covered in each chapter of BPS.

Additional Resources Available with *BPS*

Companion Website www.whfreeman.com/bps7e This open-access website includes statistical applets, data files, supplementary exercises, and self-quizzes. The website also offers companion chapters covering nonparametric tests, multiple regression, further topics in ANOVA, and statistics for quality control and capability. Instructor access to the Companion Website requires user registration as an instructor and features all the open-access student web materials, plus:

- **Instructor's Guide with Full Solutions**
- **Test Bank**
- **Lecture PowerPoint Slides containing all textbook figures and tables**
- **Instructor version of EESEE with solutions to the exercises in the student version**

Special Software Packages Student versions of JMP and Minitab are available for packaging with the text. JMP is available inside LaunchPad at no additional cost. Contact your W. H. Freeman representative for information or visit www.whfreeman.com.

Course Management Systems W. H. Freeman and Company provides courses for Blackboard, Angel, Desire2Learn, Canvas, Moodle, and Sakai course management systems. These are completely integrated solutions that instructors can customize and adapt to meet teaching goals and course objectives. Visit macmillanhighered.com/Catalog/other/Coursepack for more information.

i›clicker **i-clicker** is a two-way radio-frequency classroom response solution developed by educators for educators. Each step of i-clicker's development has been informed by teaching and learning. To learn more about packaging i-clicker with this textbook, please contact your local sales rep or visit www1.iclicker.com.

ABOUT THE AUTHORS

David S. Moore is Shanti S. Gupta Distinguished Professor of Statistics, Emeritus, at Purdue University and was the 1998 president of the American Statistical Association. He received his A.B. from Princeton and his Ph.D. from Cornell, both in mathematics. He has written many research papers in statistical theory and served on the editorial boards of several major journals. Professor Moore is an elected fellow of the American Statistical Association and of the Institute of Mathematical Statistics and an elected member of the International Statistical Institute. He has served as program director for statistics and probability at the National Science Foundation. Professor Moore has made many contributions to the teaching of statistics. He was the content developer for the Annenberg/Corporation for Public Broadcasting college-level telecourse Against All Odds: Inside Statistics and for the series of video modules Statistics: Decisions through Data, intended to aid the teaching of statistics in schools. He is the author of influential articles on statistics education and of several leading texts. Professor Moore has served as president of the International Association for Statistical Education and has received the Mathematical Association of Americas national award for distinguished college or university teaching of mathematics.

William I. Notz is Professor of Statistics at the Ohio State University. He received his B.S. in physics from the Johns Hopkins University and his Ph.D. in mathematics from Cornell University. His first academic job was as an assistant professor in the Department of Statistics at Purdue University. While there, he taught the introductory concepts course with Professor Moore and as a result of this experience he developed an interest in statistical education. Professor Notz is a co-author of *EESEE* (the *Electronic Encyclopedia of Statistical Examples and Exercises*) and co-author of *Statistics: Concepts and Controversies.*

Professor Notz's research interests have focused on experimental design and computer experiments. He is the author of several research papers and of a book on the design and analysis of computer experiments. He is an elected fellow of the American Statistical Association. He has served as the editor of the journal *Technometrics* and as editor of the *Journal of Statistics Education*. He has served as the Director of the Statistical Consulting Service, as acting chair of the Department of Statistics for a year, and as an Associate Dean in the College of Mathematical and Physical Sciences at the Ohio State University. He is a winner of the Ohio State University's Alumni Distinguished Teaching Award.

Michael A. Fligner is an Adjunct Professor at the University of California at Santa Cruz and a nonresident Professor Emeritus at the Ohio State University. He received his B.S. in mathematics from the State University of New York at Stony Brook and his Ph.D. from the University of Connecticut. He spent most of his professional career at the Ohio State University where he was vice-chair of the department for over 10 years and also served as Director of the Statistical Consulting Service. He has done consulting work with several large corporations in Central Ohio.

Professor Fligner's research interests are in nonparametric statistical methods and he received the Statistics in Chemistry award from the American Statistical Association for work on detecting biologically active compounds. He is co-author of the book *Statistical Methods for Behavioral Ecology* and received a Fulbright scholarship under the American Republics Research program to work at the Charles Darwin Research Station in the Galapagos Islands. He has been an Associate Editor of the *Journal of Statistical Education*. Professor Fligner is currently associated with the Center for Statistical Analysis in the Social Sciences at the University of California at Santa Cruz.

Ryan Etter/Getty Images

CHAPTER 0

Getting Started

In this chapter we cover...

0.1 Where the data comes from matters
0.2 Always look at the data
0.3 Variation is everywhere
0.4 What lies ahead in this book

What's hot in popular music this week? SoundScan knows. SoundScan collects data electronically from the cash registers in more than 14,000 retail outlets and also collects data on download sales from websites. When you buy a CD or download a digital track, the checkout scanner or website is probably telling SoundScan what you bought. SoundScan provides this information to *Billboard* magazine, MTV, and VH1, as well as to record companies and artists' agents.

Should women take hormones such as estrogen after menopause, when natural production of these hormones ends? In 1992, several major medical organizations said "Yes." In particular, women who took hormones seemed to reduce their risk of a heart attack by 35% to 50%. The risks of taking hormones appeared small compared with the benefits. But in 2002, the National Institutes of Health declared these findings wrong. Use of hormones after menopause immediately plummeted. Both recommendations were based on extensive studies. What happened?

Is the climate warming? Is it becoming more extreme? An overwhelming majority of scientists now agree that the earth is undergoing major changes in climate. Enormous quantities of data are continuously being collected from weather stations, satellites, and other sources to monitor factors such as the surface temperature on land and sea, precipitation, solar activity, and the chemical composition of air and water. Climate models incorporate this information to make projections of future climate change and can help us understand the effectiveness of proposed solutions.

SoundScan, medical studies, and climate research all produce data (numerical facts), and lots of them. Using data effectively is a large and growing part of most professions, and reacting to data is part of everyday life. In fact, we define statistics as **the science of learning from data.**

Although data are numbers, they are not "just numbers." *Data are numbers with a context.* The number 8.5, for example, carries no information by itself. But if we hear that a friend's new baby weighed 8.5 pounds at birth, we congratulate her on the healthy size of the child. The context engages our background knowledge and allows us to make judgments. We know that a baby weighing 8.5 pounds is a little above average, and that a human baby is unlikely to weigh 8.5 ounces or 8.5 kilograms (over 18 pounds). The context makes the number informative.

To gain insight from data, we make graphs and do calculations. But graphs and calculations are guided by ways of thinking that amount to educated common sense. Let's begin our study of statistics with an informal look at some aspects of statistical thinking.[1]

0.1 Where the data comes from matters

Although, data can be collected in a variety of ways, the type of conclusion that can be reached from the data depends on how the data were obtained. *Observational studies* and *experiments* are two common methods for collecting data. Let's take a closer look at the hormone replacement data to understand the differences.

EXAMPLE 0.1 Hormone Replacement Therapy

What's behind the flip-flop in the advice offered to women about hormone replacement? The evidence in favor of hormone replacement came from a number of observational studies that compared women who were taking hormones with others who were not. But women who choose to take hormones are very different from women who do not: they are richer and better educated and see doctors more often. These women do many things to maintain their health. It isn't surprising that they have fewer heart attacks.

Large and careful observational studies are expensive, but they are easier to arrange than careful experiments. Experiments don't let women decide what to do. They assign women either to hormone replacement or to dummy pills that look and taste the same as the hormone pills. The assignment is done by a coin toss, so that all kinds of women are equally likely to get either treatment. Part of the difficulty of a good experiment is persuading women to accept the result—invisible to them—of the coin toss. By 2002, several experiments agreed that hormone replacement does *not* reduce the risk of heart attacks, at least for older women. Faced with this better evidence, medical authorities changed their recommendations.[2] ■

Women who chose hormone replacement after menopause were on the average richer and better educated than those who didn't. No wonder they had fewer heart attacks. We can't conclude that hormone replacement reduces heart attacks just because we see this relationship in data. In this example, education and affluence are background factors that help explain the relationship between hormone replacement and good health.

Children who play soccer do better in school (on the average) than children who don't play soccer. Does this mean that playing soccer increases school grades?

Children who play soccer tend to have prosperous and well-educated parents. Once again, education and affluence are background factors that help explain the relationship between soccer and good grades.

Almost all relationships between two observed characteristics or "variables" are influenced by other variables lurking in the background. To understand the relationship between two variables, you must often look at other variables. Careful statistical studies try to think of and measure possible *lurking variables* in order to correct for their influence. As the hormone saga illustrates, this doesn't always work well. News reports often just ignore possible lurking variables that might ruin a good headline like "Playing soccer can improve your grades." The habit of asking, "What might lie behind this relationship?" is part of thinking statistically.

Of course, observational studies are still quite useful. We can learn from observational studies how chimpanzees behave in the wild or which popular songs sold best last week or what percent of workers were unemployed last month. SoundScan's data on popular music and the government's data on employment and unemployment come from *sample surveys,* an important kind of observational study that chooses a part (the sample) to represent a larger whole. Opinion polls interview perhaps 1000 of the 235 million adults in the United States to report the public's views on current issues. Can we trust the results? We'll see that this isn't a simple yes-or-no question. Let's just say that the government's unemployment rate is much more trustworthy than opinion poll results, and not just because the Bureau of Labor Statistics interviews 60,000 people rather than 1000. We can, however, say right away that some samples *can't* be trusted. Consider the following write-in poll.

EXAMPLE 0.2 Would You Have Children Again?

The advice columnist Ann Landers once asked her readers, "If you had it to do over again, would you have children?" A few weeks later, her column was headlined "70% OF PARENTS SAY KIDS NOT WORTH IT." Indeed, 70% of the nearly 10,000 parents who wrote in said they would not have children if they could make the choice again. Those 10,000 parents were upset enough with their children to write Ann Landers. Most parents are happy with their kids and don't bother to write. ■

Statistically designed samples, even opinion polls, don't let people choose themselves for the sample. They interview people selected by impersonal chance so that everyone has an equal opportunity to be in the sample. Such a poll showed that 91% of parents *would* have children again. *Where data come from matters a lot.* If you are careless about how you get your data, you may announce 70% "no" when the truth is close to 90% "yes." Understanding the importance of where data come and its relationship to the conclusions that can be reached is an important part of learning to think statistically.

0.2 Always look at the data

Yogi Berra, the Hall of Fame New York Yankee, said it: "You can observe a lot by just watching." That's a motto for learning from data. *A few carefully chosen graphs are often more instructive than great piles of numbers.* Consider the outcome of the 2000 presidential election in Florida.

EXAMPLE 0.3 Palm Beach County

Elections don't come much closer: after much recounting, state officials declared that George Bush had carried Florida by 537 votes out of almost 6 million votes cast. Florida's vote decided the 2000 presidential election and made George Bush, rather than Al Gore, president. Let's look at some data. Figure 0.1 displays a graph that plots votes for the third-party candidate Pat Buchanan against votes for the Democratic candidate Al Gore in Florida's 67 counties.

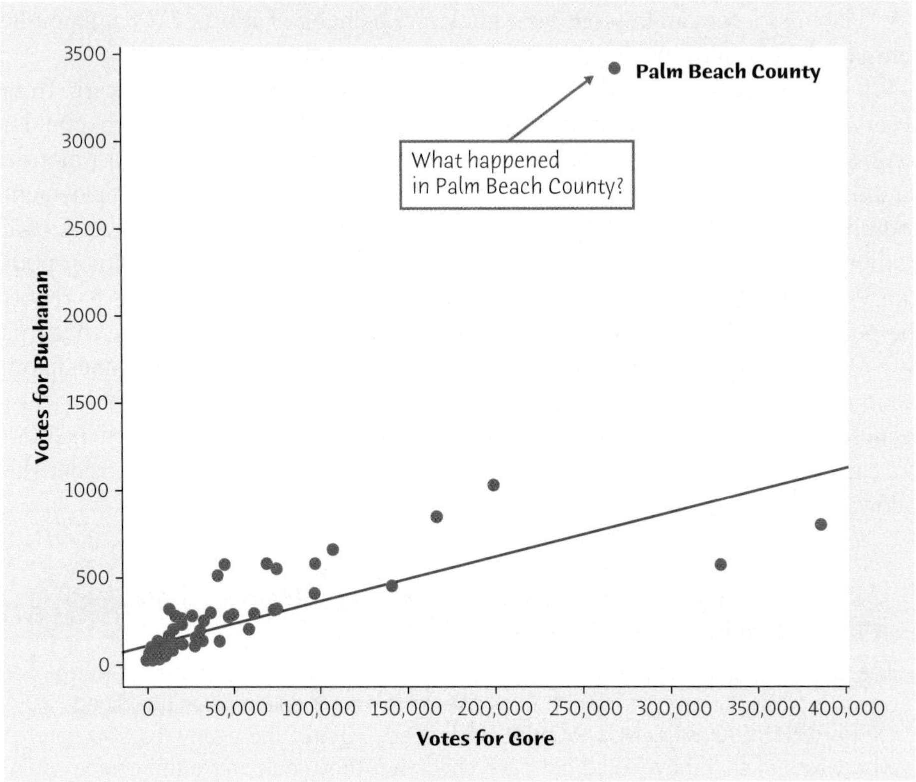

FIGURE 0.1 Votes in the 2000 presidential election for Al Gore and Patrick Buchanan in Florida's 67 counties. What happened in Palm Beach County?

What happened in Palm Beach County? The question leaps out from the graph. In this large and heavily Democratic county, a conservative third-party candidate did far better relative to the Democratic candidate than in any other county. The points for the other 66 counties show votes for both candidates increasing together in a roughly straight-line pattern. Both counts go up as county population goes up. Based on this pattern, we would expect Buchanan to receive around 800 votes in Palm Beach County. He actually received more than 3400 votes. That difference determined the election result in Florida and in the nation. ■

The graph demands an explanation. It turns out that Palm Beach County used a confusing "butterfly" ballot (see photo on page 5), in which candidate names on both left and right pages led to a voting column in the center. It would be easy for a voter who intended to vote for Gore to in fact cast a vote for Buchanan. The graph is convincing evidence that this in fact happened.

Most statistical software will draw a variety of graphs with a few simple commands. Examining your data with appropriate graphs and numerical summaries is the correct place to begin most data analyses. These can often reveal important patterns or trends that will help you understand what your data has to say.

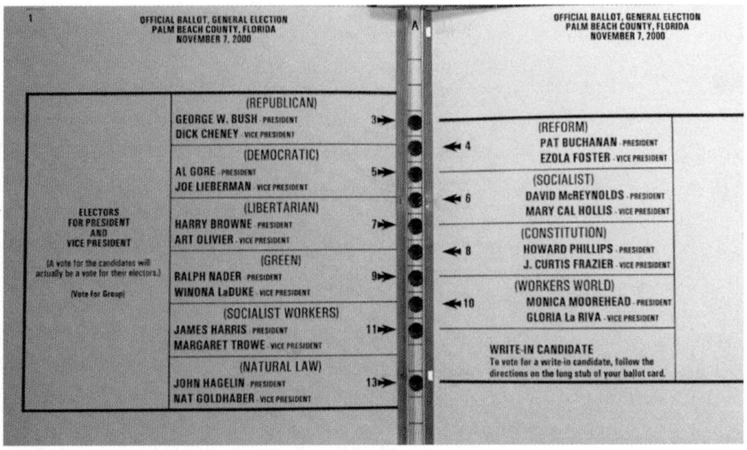

Bruce Weaver/AFP/Getty Images

0.3 Variation is everywhere

The company's sales reps file into their monthly meeting. The sales manager rises. "Congratulations! Our sales were up 2% last month, so we're all drinking champagne this morning. You remember that when sales were down 1% last month I fired half of our reps." This picture is only slightly exaggerated. Many managers overreact to small short-term variations in key figures. Here is Arthur Nielsen, former head of the country's largest market research firm, describing his experience:

> *Too many business people assign equal validity to all numbers printed on paper. They accept numbers as representing Truth and find it difficult to work with the concept of probability. They do not see a number as a kind of shorthand for a range that describes our actual knowledge of the underlying condition.*[3]

Business data such as sales and prices vary from month to month for reasons ranging from the weather to a customer's financial difficulties to the inevitable errors in gathering the data. The manager's challenge is to say when there is a real pattern behind the variation. We'll see that statistics provides tools for understanding variation and for seeking patterns behind the screen of variation. Let's look at some more data.

EXAMPLE 0.4 The Price of Gas

Figure 0.2 plots the average price of a gallon of regular unleaded gasoline each week from September 1990 to June 2013.[4] There certainly is variation! But a close look shows a yearly pattern: gas prices go up during the summer driving season, then down as demand drops in the fall. On top of this regular pattern, we see the effects of international events. For example, prices rose when the 1990 Gulf War threatened oil supplies and dropped when the world economy turned down after the September 11, 2001, terrorist attacks in the United States. The years 2007 and 2008 brought the perfect storm: the ability to produce oil and refine gasoline was overwhelmed by high demand from China and the United States and continued turmoil in the oil-producing areas of the Middle East and Nigeria. Add in a rapid fall in the value of the dollar, and prices at the pump skyrocketed to more than $4 per gallon. In 2010 the Gulf oil spill also affected supply and hence prices. The data carry an important message: because the United States imports much of its oil, we can't control the price we pay for gasoline. ■

Tony Avelar/Bloomberg via Getty Images

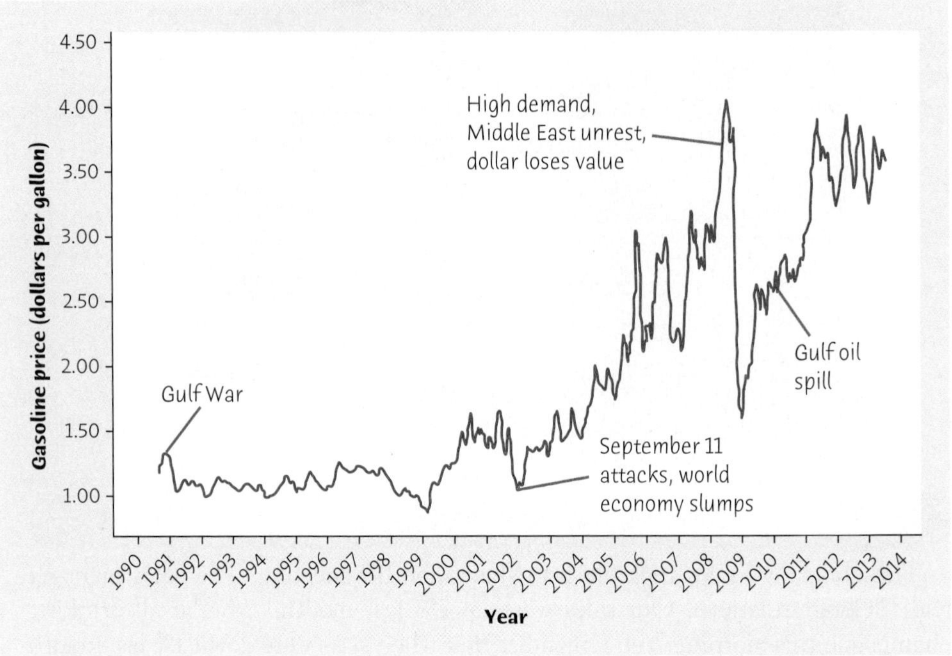

FIGURE 0.2
Variation is everywhere: the average retail price of regular unleaded gasoline, 1990 to mid 2013.

Variation is everywhere. Individuals vary; repeated measurements on the same individual vary; almost everything varies over time. One reason we need to know some statistics is that it helps us deal with variation and to describe the uncertainty in our conclusions. Let's look at another example to see how variation is incorporated into our conclusions.

EXAMPLE 0.5 The HPV Vaccine

Cervical cancer, once the leading cause of cancer deaths among women, is the easiest female cancer to prevent with regular screening tests and follow-up. Almost all cervical cancers are caused by human papillomavirus (HPV). The first vaccine to protect against the most common varieties of HPV became available in 2006. The Centers for Disease Control and Prevention recommend that all girls be vaccinated at age 11 or 12. In 2011, the CDC made the same recommendation for boys, to protect against anal and throat cancers caused by the HPV virus.

How well does the vaccine work? Doctors rely on experiments (called "clinical trials" in medicine) that give some women the new vaccine and others a dummy vaccine. (This is ethical when it is not yet known whether or not the vaccine is safe and effective.) The conclusion of the most important trial was that an estimated 98% of women up to age 26 who are vaccinated before they are infected with HPV will avoid cervical cancers over a three-year period.

Women who get the vaccine are much less likely to get cervical cancer. But because variation is everywhere, the results are different for different women. Some vaccinated women will get cancer, and many who are not vaccinated will escape. Statistical conclusions are "on the average" statements only, and even these "on the average" statements have an element of uncertainty. Although we can't be 100% certain that the vaccine reduces risk on the average, statistics allows us to state how confident we are that this is the case. ■

CHAPTER 0 EXERCISES

0.1 **Observational studies and experiments.** Observational studies have suggested that vitamin E reduces the risk of heart disease. Careful experiments, however, showed that vitamin E has no effect. According to a commentary in the *Journal of the American Medical Association:*

Thus, vitamin E enters the category of therapies that were promising in epidemiologic and observational studies but failed to deliver in adequately powered randomized controlled trials. As in other studies, the "healthy user" bias must be considered; i.e., the healthy lifestyle behaviors that characterize individuals who care enough about their health to take various supplements are actually responsible for the better health, but this is minimized with the rigorous trial design.[6]

(a) Reread Example 0.1 and the comments following it. Explain why observational studies suggest that Vitamin E therapy reduces the risk of heart disease by describing some lurking variables.

(b) A randomized controlled trial is a type of experiment. How does "healthy user bias" explain how people who take vitamin E supplements have better health in observational studies but not in experiments?

0.2 **The price of gas.** In Example 0.4 we examined the variation in the price of gasoline from 1990 to 2013. We saw both a regular pattern and the effects of international events. Figure 0.3 plots the average annual retail price of gasoline from 1929 to 1990.[7] Prices are adjusted for inflation. What overall patterns do you observe? What departures from the overall patterns do you observe? To what international events do these departures correspond?

0.3 **Online polls.** Ed Schultz is a liberal political commentator and host of *The Ed Show*. After an often lengthy and impassioned monologue from Mr. Schultz, viewers are asked to text their replies to a poll question. On June 23, 2013, after a monologue that included various issues on which the Republicans could not be trusted, Ed asked viewers to text in their responses to the question, "Do Republicans care about the personal struggles of undocumented immigrants?" Approximately 96.4% of those responding either by text or online said "no."

(a) This poll has some of the same problems as Ann Landers' poll of Example 0.2. Do you think that the proportion of Americans who feel this way is higher, lower, or close to 96.4%? Explain.

(b) For this poll, 868 people responded. Among those responding, 837 or 96.4% said "no." Do you think the results would have been more trustworthy if 2500 people had responded instead of 868? Explain.

0.4 **Traffic fatalities and 9/11.** Figure 0.4 provides information on the number of fatal traffic accidents by month for the years 1996–2001.[8] The vertical blue line above each month gives the lowest to highest number of fatal crashes for that month for the years 1996–2000. For example, in January the number of fatal crashes for the five years from 1996 through 2000 was between about 2600 and 2900. The blue dots give the number of fatal

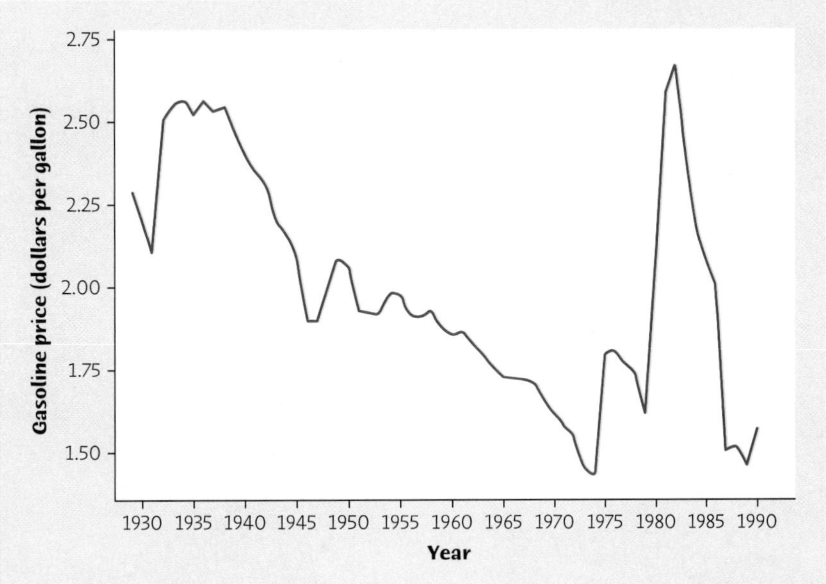

FIGURE 0.3
The average annual retail price of gasoline, 1929 to 1990. Prices are adjusted for inflation.

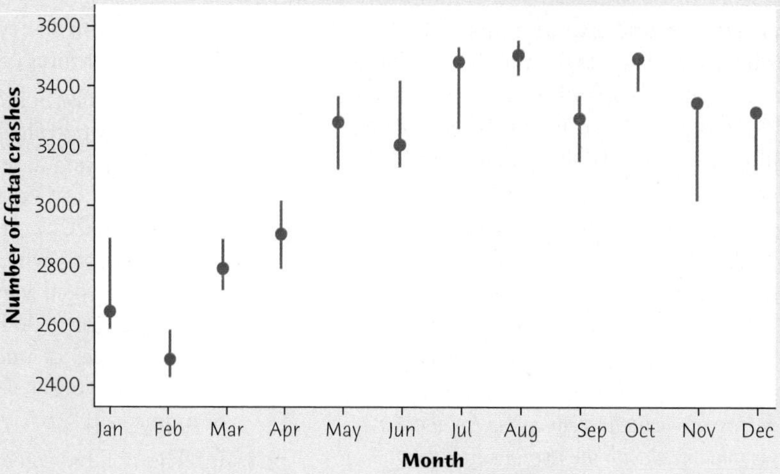

FIGURE 0.4

Number of fatal traffic accidents in the United States in 1996 through 2000 versus 2001. For each month, the blue lines represent the range of the number of fatal accidents from 1996 through 2000, and the blue dot gives the number of fatal accidents in 2001.

crashes for each month in 2001. The numbers of fatal crashes from January through August of 2001 follow the general pattern for the five preceding years as we see the blue dots are well within the blue lines for each month.

(a) What happened in the last three months of 2001? The numbers of fatal crashes in October through December of 2001 are consistently at or above the values for the previous five years. How can you tell this from the graph?

(b) On September 11, 2001, terrorists hijacked four U.S. airplanes and used them to strike various targets on the East Coast. In part (a), we saw from the graph that fatal traffic accidents seemed to be unusually high in the three months following the attacks. Did the terrorists cause fatal crashes to increase? Can you give a simple explanation for the apparent increase in fatal crashes during these months?

Moments by Mullineux/Shutterstock

PART I

Exploring Data

"What do the data say?" is the first question we ask in any statistical study. *Data analysis* answers this question by open-ended exploration of the data. The tools of data analysis are graphs such as histograms and scatterplots, and numerical measures such as means and correlations. At least as important as the tools are principles that organize our thinking as we examine data. The seven chapters in Part I present the principles and tools of statistical data analysis. They equip you with skills that are immediately useful whenever you deal with numbers.

These chapters reflect the strong emphasis on exploring data that characterizes modern statistics. Sometimes we hope to draw conclusions that apply to a setting that goes beyond the data in hand. This is *statistical inference*, the topic of much of the rest of the book. Data analysis is essential if we are to trust the results of inference, but data analysis isn't just preparation for inference. Roughly speaking, you can always do data analysis but inference requires rather special conditions.

One of the organizing principles of data analysis is to first look at one thing at a time and then at relationships. Our presentation follows this principle. In Chapters 1, 2, and 3 you will study *variables and their distributions*. Chapters 4, 5, and 6 concern *relationships among variables*. Chapter 7 reviews this part of the text.

EXPLORING DATA: Variables and Distributions

CHAPTER 1 Picturing Distributions with Graphs

CHAPTER 2 Describing Distributions with Numbers

CHAPTER 3 The Normal Distributions

EXPLORING DATA: Relationships

CHAPTER 4 Scatterplots and Correlation

CHAPTER 5 Regression

CHAPTER 6 Two-Way Tables*

CHAPTER 7 Exploring Data: Part I Review

Photograph by the U.S. Census Bureau, Public Information Office (PIO)

CHAPTER 1

Picturing Distributions with Graphs

In this chapter we cover...

1.1 Individuals and variables

1.2 Categorical variables: pie charts and bar graphs

1.3 Quantitative variables: histograms

1.4 Interpreting histograms

1.5 Quantitative variables: stemplots

1.6 Time plots

Statistics is the science of data. The volume of data available to us is overwhelming. For example, the U.S. Census Bureau's American Community Survey collects data from about 3,000,000 housing units each year. Astronomers work with data on tens of millions of galaxies. The checkout scanners at Walmart's 8000 stores in 15 countries record hundreds of millions of transactions every week, all saved to inform both Walmart and its suppliers. The first step in dealing with such a flood of data is to organize our thinking about data. Fortunately, we can do this without looking at millions of data points.

1.1 Individuals and variables

Any set of data contains information about some group of *individuals*. The information is organized in *variables*.

Individuals and Variables

Individuals are the objects described by a set of data. Individuals may be people, but they may also be animals or things.

A **variable** is any characteristic of an individual. A variable can take different values for different individuals.

A college's student database, for example, includes data about every currently enrolled student. The students are the individuals described by the data set. For each individual, the data contain the values of variables such as date of birth, choice of major, and grade point average (GPA). In practice, any set of data is accompanied by background information that helps us understand the data. When you plan a statistical study or explore data from someone else's work, ask yourself the following questions:

DATA! The documentary *Particle Fever* recreates the excitement of the Large Hadron Collider (LHC) experiment. The LHC is a 17 mile tunnel, designed to accelerate a proton to close to the speed of light, and then have protons collide to help physicists understand how the universe works. When the first collisions are recorded live in the film, American physicist Monica Dunford exclaims, "We have data. It's unbelievable how fantastic this data is."

1. **Who?** What **individuals** do the data describe? **How many** individuals appear in the data?
2. **What?** How many **variables** do the data contain? What are the **exact definitions** of these variables? In what **unit of measurement** is each variable recorded? Weights, for example, might be recorded in pounds, in thousands of pounds, or in kilograms.
3. **Where?** Student GPAs and SAT scores (or lack of them) will vary from college to college depending on many variables, including admissions "selectivity" for the college.
4. **When?** Students change from year to year, as do prices, salaries, and so forth.
5. **Why? What purpose** do the data have? Do we hope to answer some specific questions? Do we want answers for just these individuals or for some larger group that these individuals are supposed to represent? Are the individuals and variables suitable for the intended purpose?

Some variables, such as a person's sex or college major, simply place individuals into categories. Others, like height and GPA, take numerical values for which we can do arithmetic. It makes sense to give an average income for a company's employees, but it does not make sense to give an "average" sex. We can, however, count the numbers of female and male employees and do arithmetic with these counts.

Categorical and Quantitative Variables

A **categorical variable** places an individual into one of several groups or categories.

A **quantitative variable** takes numerical values for which arithmetic operations such as adding and averaging make sense. The values of a quantitative variable are usually recorded with a **unit of measurement** such as seconds or kilograms.

EXAMPLE 1.1 The American Community Survey

At the U.S. Census Bureau website, you can view the detailed data collected by the American Community Survey, though of course the identities of people and housing units are protected. If you choose the file of data on people, the *individuals* are the people living in the housing units contacted by the survey. Over 100 variables are recorded for each individual. Figure 1.1 displays a very small part of the data.

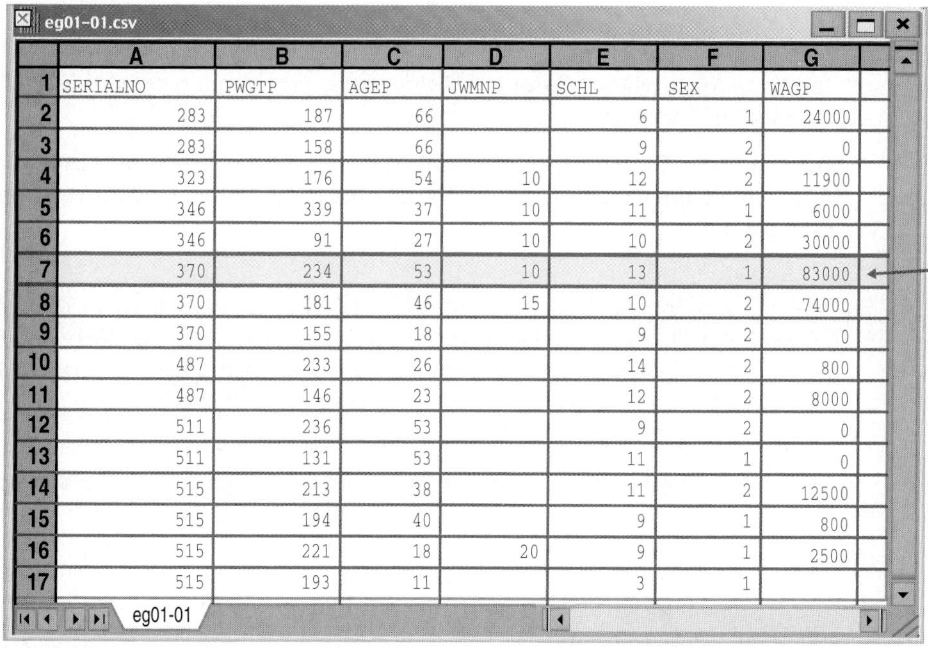

	A	B	C	D	E	F	G
1	SERIALNO	PWGTP	AGEP	JWMNP	SCHL	SEX	WAGP
2	283	187	66		6	1	24000
3	283	158	66		9	2	0
4	323	176	54	10	12	2	11900
5	346	339	37	10	11	1	6000
6	346	91	27	10	10	2	30000
7	370	234	53	10	13	1	83000
8	370	181	46	15	10	2	74000
9	370	155	18		9	2	0
10	487	233	26		14	2	800
11	487	146	23		12	2	8000
12	511	236	53		9	2	0
13	511	131	53		11	1	0
14	515	213	38		11	2	12500
15	515	194	40		9	1	800
16	515	221	18	20	9	1	2500
17	515	193	11		3	1	

eg01-01.csv
eg01-01
Each row in the spreadsheet contains data on one individual.

FIGURE 1.1 A spreadsheet displaying data from the American Community Survey, for Example 1.1.

Each row records data on one individual. Each column contains the values of one *variable* for all the individuals. Translated from the U.S. Census Bureau's abbreviations, the variables are

SERIALNO	An identifying number for the household.
PWGTP	Weight in pounds.
AGEP	Age in years.
JWMNP	Travel time to work in minutes.
SCHL	Highest level of education. The numbers designate categories, *not* specific grades. For example, 9 = high school graduate, 10 = some college but no degree, and 13 = bachelor's degree.
SEX	Sex, designated by 1 = male and 2 = female.
WAGP	Wage and salary income last year, in dollars.

Look at the highlighted row in Figure 1.1. This individual is a 53-year-old man who weighs 234 pounds, travels 10 minutes to work, has a bachelor's degree, and earned $83,000 last year.

In addition to the household serial number, there are six variables. Education and sex are categorical variables. The values for education and sex are stored as numbers, but these numbers are just labels for the categories and have no units of measurement. The other four variables are quantitative. Their values do have units. These variables are weight in pounds, age in years, travel time in minutes, and income in dollars.

The *purpose* of the American Community Survey is to collect data that represent the entire nation to guide government policy and business decisions. To do this, the households contacted are chosen at random from all households in the country. We will see in Chapter 8 why choosing at random is a good idea. ■

Most data tables follow this format—each row is an individual, and each column is a variable. The data set in Figure 1.1 appears in a **spreadsheet** program that has rows and columns ready for your use. Spreadsheets are commonly used to enter and transmit data and to do simple calculations.

spreadsheet

LaunchPad Online Resources

- The **Snapshots video**, *Data and Distributions*, provides a nice introduction to the ideas of this section.
- The **StatClips Examples video**, *Basic Principles of Exploring Data Example A*, describes the variables collected in the American Community Survey from Example 1.1.

Apply Your Knowledge

1.1 **Fuel Economy.** Here is a small part of a data set that describes the fuel economy (in miles per gallon (mpg)) of model year 2014 motor vehicles:

Make and Model	Vehicle Class	Transmission Type	Number of Cylinders	City mpg	Highway mpg	Annual Fuel Cost
⋮						
Chevrolet Corvette	Two-seater	Manual	8	17	29	$2,650
Nissan Cube	Small station wagon	Manual	4	25	30	$1,850
Ford Fusion	Midsize	Automatic	4	23	36	$1,800
Chevrolet Impala	Large	Automatic	6	18	28	$2,400
⋮						

The annual fuel cost is an estimate assuming 15,000 miles of travel a year (55% city and 45% highway) and an average fuel price.

(a) What are the individuals in this data set?

(b) For each individual, what variables are given? Which of these variables are categorical, and which are quantitative? In what units are the quantitative variables measured?

1.2 **Students and Exercise.** You are preparing to study the exercise habits of college students. Describe two categorical variables and two quantitative variables that you might measure for each student. Give the units of measurement for the quantitative variables.

1.2 Categorical variables: pie charts and bar graphs

exploratory data analysis

Statistical tools and ideas help us examine data in order to describe their main features. This examination is called **exploratory data analysis.** Like an explorer crossing unknown lands, we want first to simply describe what we see. Here are two principles that help us organize our exploration of a set of data.

Exploring Data

1. Begin by examining each variable by itself. Then move on to study the relationships among the variables.
2. Begin with a graph or graphs. Then add numerical summaries of specific aspects of the data.

We will follow these principles in organizing our learning. Chapters 1 to 3 present methods for describing a single variable. We study relationships among several variables in Chapters 4 to 6. In each case, we begin with graphical displays, then add numerical summaries for more complete description.

The proper choice of graph depends on the nature of the variable. To examine a single variable, we usually want to display its *distribution*.

Distribution of a Variable

The **distribution** of a variable tells us what values it takes and how often it takes these values.

The values of a categorical variable are labels for the categories. The **distribution of a categorical variable** lists the categories and gives either the count or the percent of individuals who fall into each category.

EXAMPLE 1.2 Which Major?

DATA MAJORS

Approximately 1.5 million full-time, first-year students enrolled in colleges and universities in 2013. What do they plan to study? Here are data on the percents of first-year students who plan to major in several discipline areas:[1]

Field of Study	Percent of Students
Arts and humanities	10.6
Biological sciences	14.7
Business	14.5
Education	5.2
Engineering	11.2
Health professions	12.8
Math and computer science	3.7
Physical sciences	2.4
Social sciences	10.1
Other majors and undeclared	14.9
Total	100.1

It's a good idea to check data for consistency. The percents should add to 100%. In fact, they add to 100.1%. What happened? Each percent is rounded to the nearest tenth. The exact percents would add to 100, but the rounded percents only come close. This is **roundoff error**. Roundoff errors don't point to mistakes in our work, just to the effect of rounding off results. ■

roundoff error

Columns of numbers take time to read. You can use a pie chart or a bar graph to display the distribution of a categorical variable more vividly. Figures 1.2 and 1.3 illustrate these displays for the distribution of intended college majors.

pie charts

Pie charts show the distribution of a categorical variable as a "pie" whose slices are sized by the counts or percents for the categories. Pie charts are awkward to make by hand, but software will do the job for you. *A pie chart must include all the categories that make up a whole. Use a pie chart only when you want to emphasize each category's relation to the whole.* We need the "Other majors and undeclared" category in Example 1.2 to complete the whole (all intended majors) and allow us to make the pie chart in Figure 1.2.

CAUTION

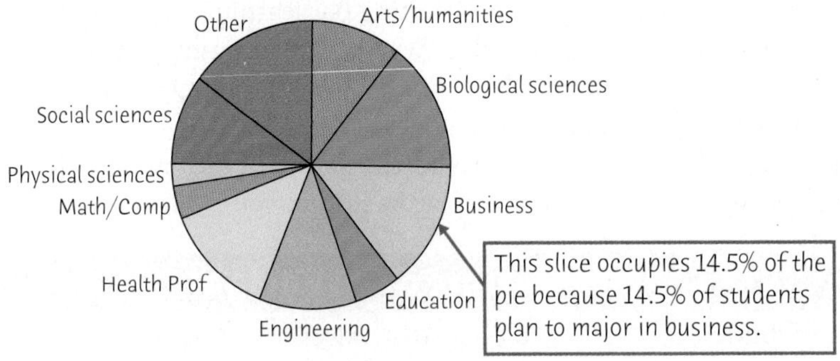

FIGURE 1.2
You can use a pie chart to display the distribution of a categorical variable. This pie chart shows the distribution of intended majors of students entering college.

bar graphs

Bar graphs represent each category as a bar. The bar heights show the category counts or percents. Bar graphs are easier to make than pie charts and also easier to read. Figure 1.3 displays two bar graphs of the data on intended majors. The first orders the bars alphabetically by field of study (with "Other" at the end). It is often better to arrange the bars in order of height, as in Figure 1.3(b). This helps us immediately see which majors appear most often.

Bar graphs are more flexible than pie charts. Both graphs can display the distribution of a categorical variable, but a bar graph can also compare any set of quantities that are measured in the same units.

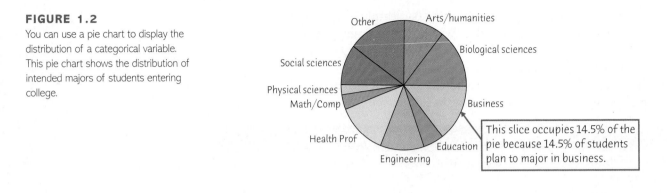

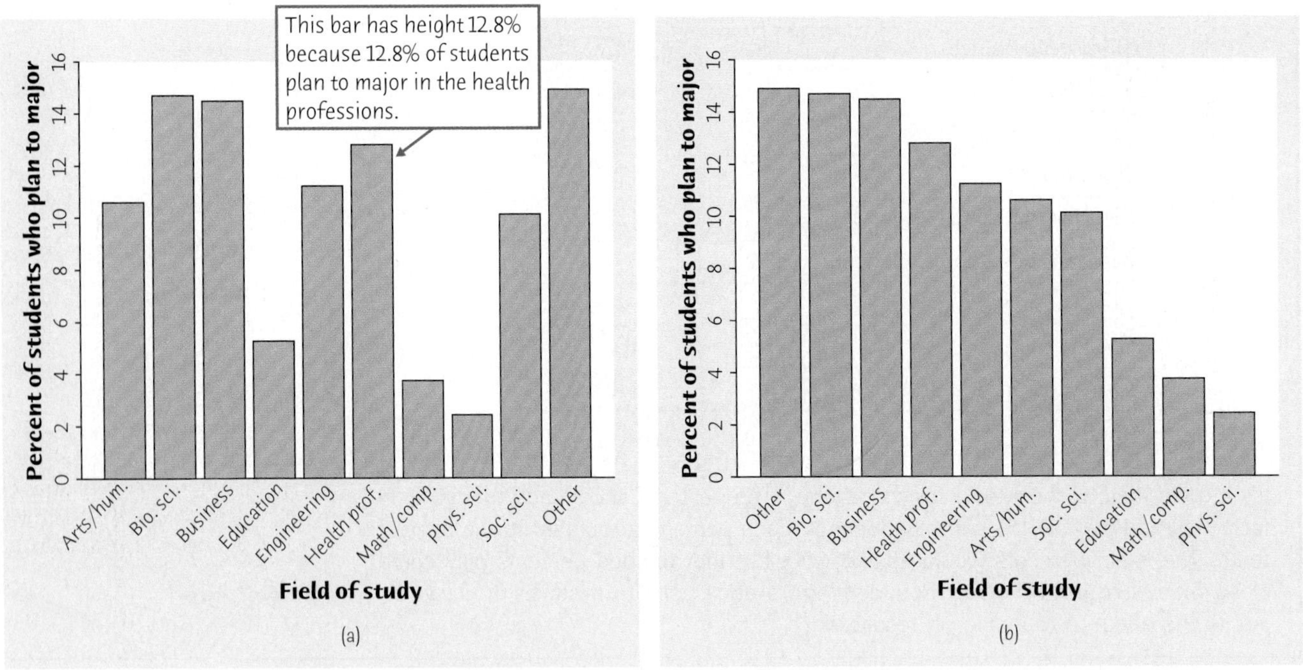

FIGURE 1.3
Bar graphs of the distribution of intended majors of students entering college. In part (a), the bars follow the alphabetical order of fields of study. In part (b), the same bars appear in order of height.

EXAMPLE 1.3 How Do 12–24s Learn about New Music?

DATA MUSIC

What sources do Americans aged 12–24 years use to keep up-to-date and learn about music? Among those saying it was important to keep up with music, Arbitron asked which of several sources they had ever used. Here are the percents who have used each source.[2]

Source	Percent of 12–24s Who Have Used Each Source
AM/FM radio	72
Friends/family	79
YouTube	77
Music television channels	53
Facebook	57
Pandora	58
Apple iTunes	47
Information at local stores	29
SiriusXM Satellite Radio	12
Music blogs	20
iHeartRadio	22
Spotify	13

We can't make a pie chart to display these data. Each percent in the table refers to a different source, not to parts of a single whole. Figure 1.4 is a bar graph comparing the 12 sources. We have again arranged the bars in order of height. ■

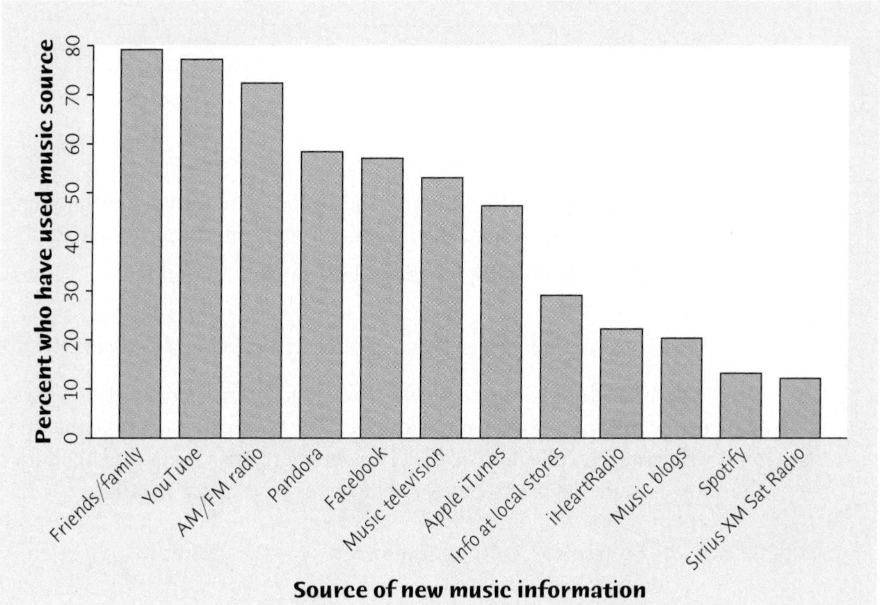

FIGURE 1.4 You can use a bar graph to compare quantities that are not part of a whole. This bar graph compares the percents of Americans aged 12–24 who have used each source to learn about new music, for Example 1.3.

Bar graphs and pie charts are mainly tools for presenting data: they help your audience grasp data quickly. Because it is easy to understand data on a single categorical variable without a graph, bar graphs and pie charts are of limited use for data analysis. We will move on to quantitative variables, where graphs are essential tools.

LaunchPad Online Resources

- The **Snapshots video**, *Visualizing and Summarizing Categorical Data*, illustrates the ideas of both pie charts and bar graphs.
- The **StatClips Examples video**, *Summaries and Pictures for Categorical Data Example B*, provides the details of constructing pie charts and bar graphs through an example.

Apply Your Knowledge

1.3 Do You Listen to Country Radio? The rating service Arbitron places U.S. radio stations into more than 50 categories that describe the kinds of programs they broadcast. Which formats attract the largest audiences? Here are Arbitron's measurements of the share of the listening audience (aged 12 and over) at a given time for the most popular formats:[3] DATA RADIO

Format	Audience Share
Country	14.2%
News/talk/information	11.4%
Pop contemporary hit	8.2%
Adult contemporary	8.1%
Classic hits	5.2%
Classic rock	5.2%
Hot adult contemporary	4.7%
Urban adult contemporary	4.1%
Rhythmic contemporary hit	3.4%
All sports	3.1%
Urban contemporary	3.0%
Contemporary Christian	2.9%
Mexican regional	2.9%

(a) What is the sum of the audience shares for these formats? What percent of the radio audience listens to stations with other formats?

(b) Make a bar graph to display these data. Be sure to include an "Other format" category.

(c) Would it be correct to display these data in a pie chart? Why or why not?

1.4 How Do Students Pay for College? The Higher Education Research Institute's Freshman Survey includes over 200,000 first-time, full-time freshmen who entered college in 2013.[4] The survey reports the following data on the sources students use to pay for college expenses. DATA EXPENSE

Source for College Expenses	Students
Family resources	77.8%
Student resources	62.3%
Aid—not to be repaid	72.9%
Aid—to be repaid	50.8%

(a) Explain why it is *not* correct to use a pie chart to display these data.

(b) Make a bar graph of the data. Notice that because the data contrast groups such as family and student resources, it is better to keep these bars next to each other rather than to arrange the bars in order of height.

1.5 Never on Sunday? Births are not, as you might think, evenly distributed across the days of the week. Here are the average numbers of babies born on each day of the week in 2012:[5] DATA BIRTHS

Day	Births
Sunday	7,134
Monday	11,430
Tuesday	12,387
Wednesday	12,230
Thursday	12,308
Friday	12,047
Saturday	8,124

Present these data in a well-labeled bar graph. Would it also be correct to make a pie chart? Suggest some possible reasons why there are fewer births on weekends.

1.3 Quantitative variables: histograms

Quantitative variables often take many values. The distribution tells us what values the variable takes and how often it takes these values. A graph of the distribution is clearer if nearby values are grouped together. The most common graph of the distribution of one quantitative variable is a **histogram.**

histogram

EXAMPLE 1.4 Making a Histogram

DATA GRADRATE

What percent of your home state's high school students graduate within four years? The No Child Left Behind Act of 2001 uses on-time high school graduation rates as one of its monitoring requirements. However, in 2001 most states were not collecting the necessary data to compute these rates accurately. The Freshman Graduation Rate (FGR) counts the number of high school graduates in a given year for a state and divides this by the number of ninth graders enrolled four years previously. Although the FGR can be computed from readily available data, it neglects high school students moving into and out of a state and may include students who have repeated a grade. Several alternative measures are available that partially correct for these deficiencies, but states have been free to choose their own measure, and the resulting rates can differ by more than 10%. Federal law now requires all states to use a common, more rigorous computation, the *Adjusted Cohort Graduation Rate,* that tracks individual students. Table 1.1 presents the data for 2010–11, the first year in which states used a common formula and for which graduation rates could be compared between states.[6] Idaho, Kentucky, and Oklahoma received "timeline extensions" and were not required to file in 2010–11.

The *individuals* in this data set are the states. The *variable* is the percent of a state's high school students who graduate within four years. The states vary quite a bit on this variable, from 59% in the District of Columbia to 88% in Iowa. It's much easier to see how your state compares with other states from a graph like a histogram than from the table. To make a histogram of the distribution of this variable, proceed as follows:

Step 1. Choose the classes. Divide the range of the data into classes of equal width. The data in Table 1.1 range from 59 to 88, so we decide to use these classes:

percent on-time graduates between 55 and 60 (55 to <60)

percent on-time graduates between 60 and 65 (60 to <65)

⋮

percent on-time graduates between 85 and 90 (85 to <90)

Statistics In Your World

WHAT'S THAT NUMBER?

You might think that numbers, unlike words, are universal. Think again. A "billion" in the United States means 1,000,000,000 (nine zeros). In Europe, a "billion" is 1,000,000,000,000 (12 zeros). OK, those are words that describe numbers. But those commas in big numbers are periods in many other languages. This is so confusing that international standards call for spaces instead, so that an American billion is written 1 000 000 000. And the decimal point of the English-speaking world is the decimal comma in many other languages, so that 3.1416 in the United States becomes 3,1416 in Europe. So what is the number 10,642.389? It depends on where you are.

TABLE 1.1 PERCENT OF STATE HIGH SCHOOL STUDENTS GRADUATING ON TIME

STATE	PERCENT	REGION	STATE	PERCENT	REGION	STATE	PERCENT	REGION
Alabama	72	S	Louisiana	71	S	Ohio	80	MW
Alaska	68	W	Maine	84	NE	Oklahoma	—	S
Arizona	78	W	Maryland	83	S	Oregon	68	W
Arkansas	81	S	Massachusetts	83	NE	Pennsylvania	83	NE
California	76	W	Michigan	74	MW	Rhode Island	77	NE
Colorado	74	W	Minnesota	77	MW	South Carolina	74	S
Connecticut	83	NE	Mississippi	75	S	South Dakota	83	MW
Delaware	78	S	Missouri	81	MW	Tennessee	86	S
Florida	71	S	Montana	82	W	Texas	86	S
Georgia	67	S	Nebraska	86	MW	Utah	76	W
Hawaii	80	W	Nevada	62	W	Vermont	87	NE
Idaho	—	W	New Hampshire	86	NE	Virginia	82	S
Illinois	84	MW	New Jersey	83	NE	Washington	76	W
Indiana	86	MW	New Mexico	63	W	West Virginia	76	S
Iowa	88	MW	New York	77	NE	Wisconsin	87	MW
Kansas	83	MW	North Carolina	78	S	Wyoming	80	W
Kentucky	—	S	North Dakota	86	MW	Dist. of Columbia	59	S

It is important to specify the classes carefully so that each individual falls into exactly one class. Our notation 55 to <60 indicates that the first class includes states with graduation rates starting at 55% and up to, but not including, graduation rates of 60%. Thus, a state with an on-time graduation rate of 60% falls into the second class, whereas a state with an on-time graduation rate of 59% falls into the first class. It is equally correct to use classes 56 to <62, 62 to <68, and so forth. Just be sure to specify the classes precisely so that each individual falls into exactly one class.

Step 2. Count the individuals in each class. Here are the counts:

Class	Count	Class	Count
55 to <60	1	75 to <80	11
60 to <65	2	80 to <85	16
65 to <70	3	85 to <90	9
70 to <75	6		

Check that the counts add to 48, the number of individuals in the data set (the 47 states reporting and the District of Columbia).

Step 3. Draw the histogram. Mark the scale for the variable whose distribution you are displaying on the horizontal axis. That's the percent of a state's high school students who graduate within four years. The scale runs from 55 to 90 because that is the span of the classes we chose. The vertical axis contains the scale of counts. Each bar represents a class. The base of the bar covers the class, and the bar height is the class count. Draw the bars with no horizontal space between them unless a class is empty, so that its bar has height zero. Figure 1.5 is our histogram. Remember, an observation on the boundary of the bars—say, 65—is counted in the bar to its right. ■

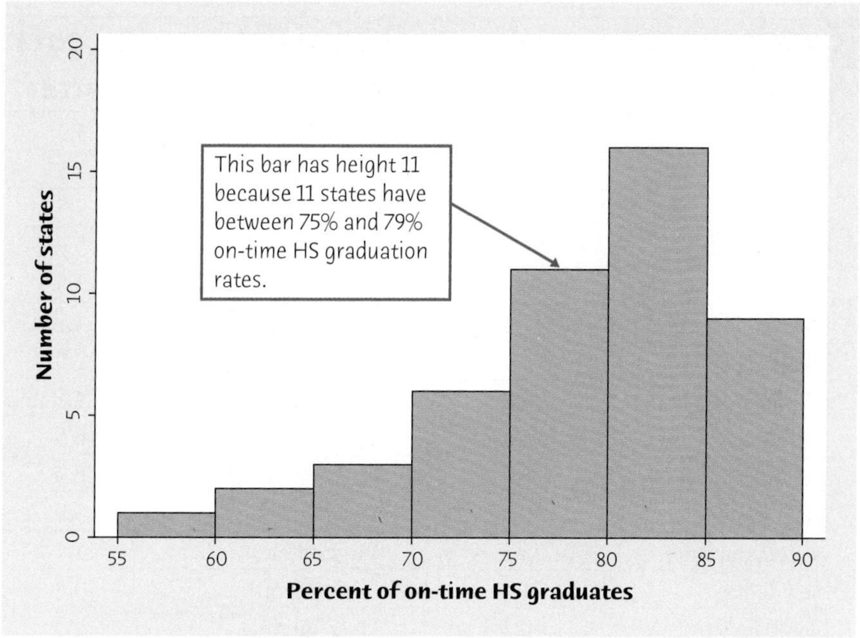

FIGURE 1.5
Histogram of the distribution of the percent of on-time high school graduates in 47 states and the District of Columbia, for Example 1.4.

CAUTION

Although histograms resemble bar graphs, their details and uses are different. A histogram displays the distribution of a quantitative variable. The horizontal axis of a histogram is marked in the units of measurement for the variable. A bar graph compares the sizes of different quantities. The horizontal axis of a bar graph simply identifies the quantities being compared and need not have any measurement scale. These quantities may be the values of a categorical variable, but they may also be unrelated, like the sources used to learn about music in Example 1.3. Draw bar graphs with blank space between the bars to separate the quantities being compared. Draw histograms with no space, to indicate that all values of the variable are covered. A gap between bars in a histogram indicates that there are no values for that class.

Our eyes respond to the *area* of the bars in a histogram.[7] Because the classes are all the same width, area is determined by height, and all classes are fairly represented. There is no one right choice of the classes in a histogram. Too few classes will give a "skyscraper" graph, with all values in a few classes with tall bars. Too many will produce a "pancake" graph, with most classes having one or no observations. Neither choice will give a good picture of the shape of the distribution. You must use your judgment in choosing classes to display the shape. Statistics software will choose the classes for you. The software's choice is usually a good one, but you can change it if you want. The histogram function in the *One-Variable Statistical Calculator* applet on the text website allows you to change the number of classes by dragging with the mouse, so that it is easy to see how the choice of classes affects the histogram.

APPLET

Apply Your Knowledge

1.6 Foreign Born. How are foreign-born residents distributed in the United States? The country as a whole has 13.0% foreign-born residents, but the states vary from 1.3% in West Virginia to 27.1% in California. Table 1.2 presents the data for all 50 states and the District of Columbia.[8] Make a histogram of

TABLE 1.2 PERCENT OF STATE POPULATION BORN OUTSIDE THE UNITED STATES, 2011

STATE	PERCENT	STATE	PERCENT	STATE	PERCENT
Alabama	3.4	Louisiana	3.8	Ohio	3.9
Alaska	6.2	Maine	3.3	Oklahoma	5.5
Arizona	13.4	Maryland	13.7	Oregon	9.5
Arkansas	4.3	Massachusetts	14.9	Pennsylvania	5.9
California	27.1	Michigan	6.1	Rhode Island	13.5
Colorado	9.7	Minnesota	7.4	South Carolina	4.7
Connecticut	13.3	Mississippi	2.3	South Dakota	2.9
Delaware	8.6	Missouri	4.1	Tennessee	4.7
Florida	19.4	Montana	2.0	Texas	16.5
Georgia	9.6	Nebraska	6.2	Utah	8.4
Hawaii	18.2	Nevada	19.2	Vermont	3.9
Idaho	5.9	New Hampshire	5.3	Virginia	11.1
Illinois	13.9	New Jersey	21.3	Washington	13.4
Indiana	4.6	New Mexico	10.2	West Virginia	1.3
Iowa	4.3	New York	22.2	Wisconsin	4.8
Kansas	6.7	North Carolina	7.3	Wyoming	2.9
Kentucky	3.3	North Dakota	2.4	Dist. of Columbia	13.6

the percents using classes of width 5% starting at 0.0%. That is, the first bar covers 0.0% to <5.0%, the second covers 5.0% to <10.0%, and so on. (Make this histogram by hand, even if you have software, to be sure you understand the process. You may then want to compare your histogram with your software's choice.) FOREIGN

1.7 Choosing Classes in a Histogram. The data set menu that accompanies the *One-Variable Statistical Calculator* applet includes the data on foreign-born residents in the states from Table 1.2. Choose these data, then click on the "Histogram" tab to see a histogram.

(a) How many classes does the applet choose to use? (You can click on the graph outside the bars to get a count of classes.)

(b) Click on the graph and drag to the left. What is the smallest number of classes you can get? What are the lower and upper bounds of each class? (Click on the bar to find out.) Make a rough sketch of this histogram.

(c) Click and drag to the right. What is the greatest number of classes you can get? How many observations does the largest class have?

(d) You see that the choice of classes changes the appearance of a histogram. Drag back and forth until you get the histogram that you think best displays the distribution. How many classes did you use? Why do you think this is best?

1.4 Interpreting histograms

Making a statistical graph is not an end in itself. *The purpose of graphs is to help us understand the data.* After you make a graph, always ask, "What do I see?" Once you have displayed a distribution, you can see its important features as follows.

Examining a Histogram

In any graph of data, look for the **overall pattern** and for striking deviations from that pattern.

You can describe the overall pattern of a histogram by its **shape**, **center**, and **variability**. You will sometimes see variability referred to as **spread**.

An important kind of deviation is an **outlier**, an individual value that falls outside the overall pattern.

One way to describe the center of a distribution is by its *midpoint,* the value with roughly half the observations taking smaller values and half taking larger values. To find the midpoint, order the observations from smallest to largest, making sure to include repeated observations as many times as they appear in the data. First cross off the largest and smallest observations, then the largest and smallest of those remaining, and continue this process. If there were an odd number of observations initially, you will be left with a single observation, which is the midpoint. If there were an even number of observations initially, you will be left with two observations, and their average is the midpoint.

For now, we will describe the variability of a distribution by giving the *smallest and largest values.* We will learn better ways to describe center and variability in Chapter 2. The overall shape of a distribution can often be described in terms of symmetry or skewness, defined as follows.

Symmetric and Skewed Distributions

A distribution is **symmetric** if the right and left sides of the histogram are approximately mirror images of each other.

A distribution is **skewed to the right** if the right side of the histogram (containing the half of the observations with larger values) extends much farther out than the left side. It is **skewed to the left** if the left side of the histogram extends much farther out than the right side.

EXAMPLE 1.5 Describing a Distribution

Look again at the histogram in Figure 1.5. To describe the distribution, we want to look at its overall pattern and any deviations.

DATA GRADRATE

Shape: The distribution has a *single peak,* which represents states in which between 80% and 85% of students graduate high school on time. The distribution is *skewed to the left.* A majority of states have more than 75% of students graduating high school on time, but several states have much lower percents, so the graph extends quite far to the left of its peak.

Center: Arranging the observations from Table 1.1 in order of size shows that 80% is the midpoint of the distribution. There are a total of 48 observations, and if we cross off the 23 highest graduation rates and the 23 lowest graduation rates, we are left with two graduation rates, both of which are 80%. The center is their average, which is 80%.

Variability: The graduation rates range from 59% to 88%, which shows considerable variability in graduation rates among the states.

Outliers: Figure 1.5 shows no observations outside the overall single-peaked, left-skewed pattern of the distribution. Figure 1.6 is another histogram of the same distribution, with classes of width 3% rather than 5%. Now there are three states that

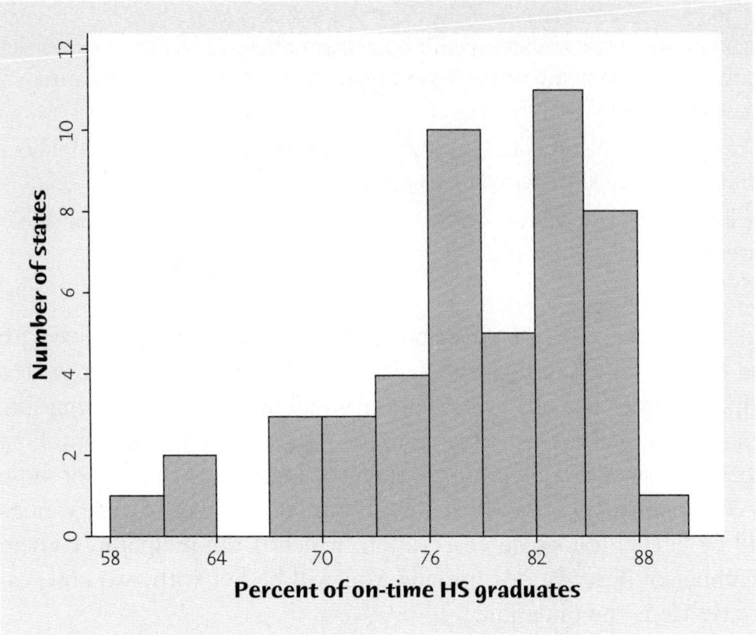

FIGURE 1.6
Another histogram of the distribution of the percent of on-time high school graduates, with narrower class widths than in Figure 1.5. Histograms with more classes show more detail but may have a less clear pattern.

stand a bit apart to the left of the rest of the distribution: the District of Columbia at 58%, Nevada at 62%, and New Mexico at 63%. Are these states outliers or just the smallest observations in a strongly skewed distribution? Unfortunately, there is no rule. Let's agree to call attention to only strong outliers that suggest something special about an observation—or an error such as typing 10.1 as 101. These states do not appear to be strong outliers. ■

CAUTION

Figures 1.5 and 1.6 remind us that interpreting graphs calls for judgment. We also see that *the choice of classes in a histogram can influence the appearance of a distribution.* Because of this, and to avoid worrying about minor details, concentrate on the main features of a distribution that persist with several choices of class intervals. Look for major peaks, not for minor ups and downs, in the bars of the histogram. For example, don't immediately conclude that Figure 1.6 shows two important peaks, one between 76% and 79% and the second between 82% and 85%. When you choose a larger number of class intervals, the histogram can become more jagged, leading to the appearance of multiple peaks that are close together. If Arizona, Delaware, and North Carolina had graduation rates of 79% instead of 78%, these states would have been in the class interval 79% to <82% rather than the class interval from 76% to <79%. This small change would have eliminated the second peak between 76% and 79% in Figure 1.6, leaving only one peak as in Figure 1.5. Be sure to check for clear outliers, not just for the smallest and largest observations, and look for rough *symmetry* or clear *skewness.*

Here are more examples of describing the overall pattern of a histogram.

EXAMPLE 1.6 Iowa Tests Scores

DATA
IOWATEST

Figure 1.7 displays the scores of all 947 seventh-grade students in the public schools of Gary, Indiana, on the vocabulary part of the Iowa Tests of Basic Skills.[9] The distribution is *single-peaked* and *symmetric.* In mathematics, the two sides of symmetric patterns are exact mirror images. Real data are almost never exactly symmetric. We are content to describe Figure 1.7 as symmetric. The center (half above, half below) is close to 7. This is seventh-grade reading level. The scores range from 2.0 (second-grade level) to 12.1 (twelfth-grade level).

Notice that the vertical scale in Figure 1.7 is not the *count* of students but the *percent* of students in each histogram class. A histogram of percents rather than counts is convenient when we want to compare several distributions. To compare Gary with Los Angeles, a much bigger city, we would use percents so that both histograms have the same vertical scale. ■

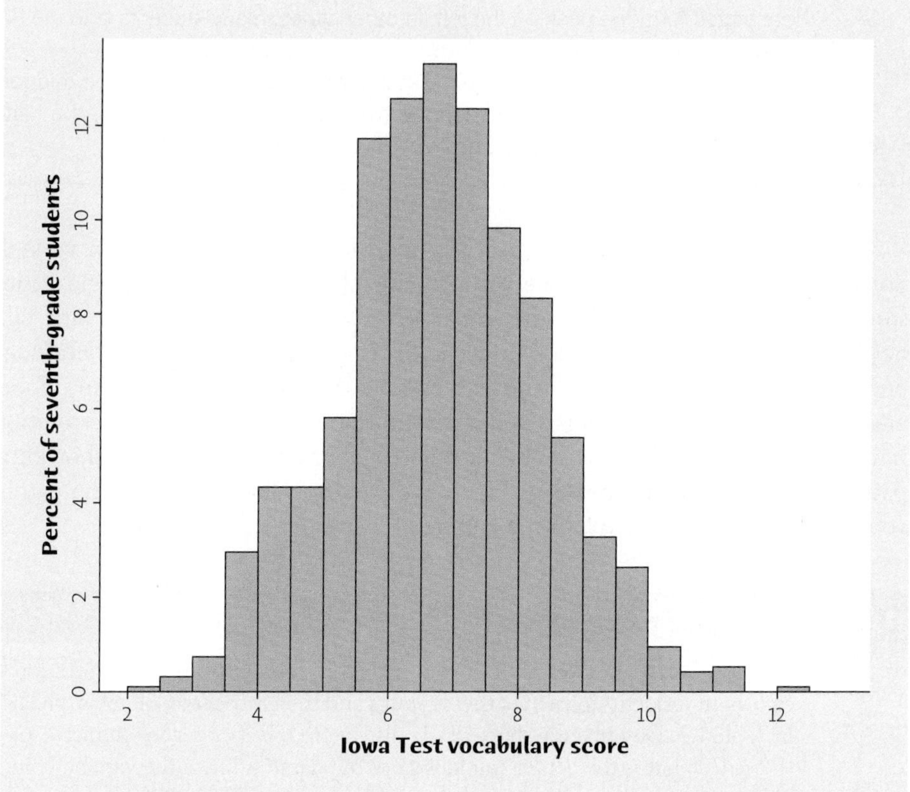

FIGURE 1.7 Histogram of the Iowa Tests vocabulary scores of all seventh-grade students in Gary, Indiana, for Example 1.6. This distribution is single peaked and symmetric.

EXAMPLE 1.7 Who Takes the SAT?

Depending on where you went to high school, the answer to this question may be "almost everybody" or "almost nobody." Figure 1.8 is a histogram of the percent of high school graduates in each state who took the SAT test in 2013.[10]

DATA SATPCT

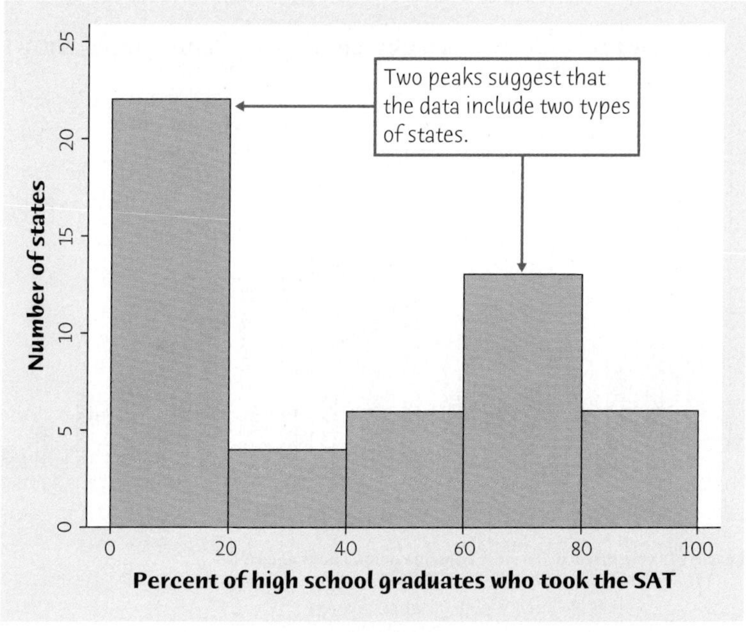

FIGURE 1.8 Histogram of the percent of high school graduates in each state who took the SAT Reasoning test, for Example 1.7. The graph shows two groups of states: ACT states (where few students take the SAT) at the left and SAT states at the right.

The histogram shows two peaks: a high peak at the left and a lower but broader peak centered in the 60% to <80% class. The presence of more than one peak suggests that a distribution mixes several kinds of individuals. That is the case here. There are two major tests of readiness for college, the ACT and the SAT. Most states have a strong preference for one or the other. In some states, many students take the ACT exam and few take the SAT—these states form the peak on the left. In other states, many students take the SAT and few choose the ACT—these states form the broader peak at the right.

Giving the center and variability of this distribution is not very useful. The midpoint falls in the 40% to <60% class, between the two peaks. The story told by the histogram is in the two peaks corresponding to ACT states and SAT states. ■

The overall shape of a distribution is important information about a variable. Some variables have distributions with predictable shapes. Many biological measurements on specimens from the same species and sex—lengths of bird bills, heights of young women—have symmetric distributions. On the other hand, data on people's incomes are usually strongly skewed to the right. There are many moderate incomes, some large incomes, and a few enormous incomes. Many distributions have irregular shapes that are neither symmetric nor skewed. Some data show other patterns, such as the two peaks in Figure 1.8. Use your eyes, describe the pattern you see, and then try to explain the pattern.

Apply Your Knowledge

1.8 Foreign Born. In Exercise 1.6, you made a histogram of the percent of foreign-born residents in each of the 50 states and the District of Columbia, given in Table 1.2. Describe the shape of the distribution. Is it closer to symmetric or skewed? What is the center (midpoint) of the data? What is the variability in terms of the smallest and largest values? Are there any states with an unusually large or small percent of foreign-born residents? DATA FOREIGN

Kallista Images/SuperStock

1.9 Lyme Disease. Lyme disease is caused by a bacteria called *Borrelia burgdorferi* and is spread through the bite of an infected black legged tick, generally found in woods and grassy areas. There were 213,515 confirmed cases reported to the Centers for Disease Control (CDC) between 2001 and 2010, and these are broken down by age and sex in Figure 1.9.[11]

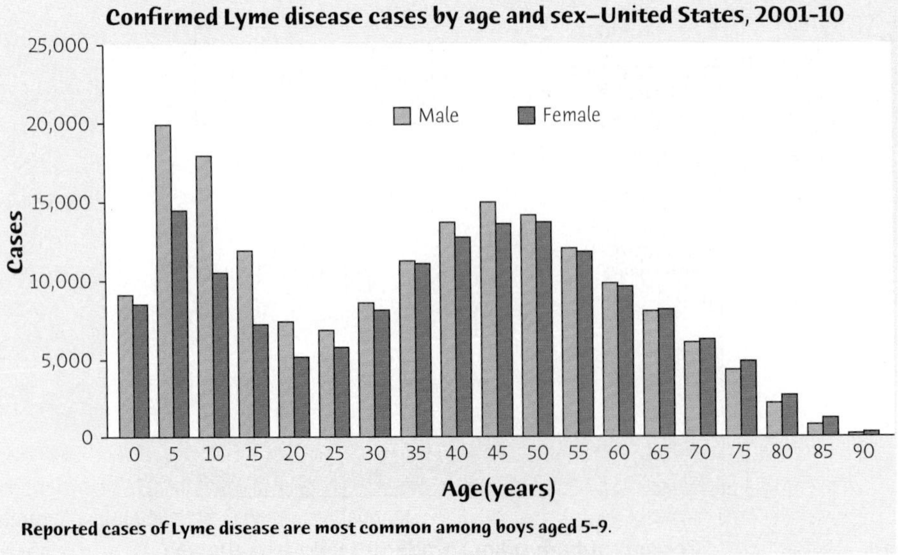

FIGURE 1.9 Histogram of the ages of infected individuals with Lyme disease for cases reported between 2001 and 2010 in the United States, for males and females, for Exercise 1.9.

Here is how Figure 1.9 relates to what we have been studying. The individuals are the 213,515 people with confirmed cases, and two of the variables measured on each individual are sex and age. Considering males and females separately, we could draw a histogram of the variable age using class intervals 0 to <5 years old, 5 to <10 years old, and so forth. Look at the leftmost two bars in Figure 1.9. The light blue bar shows that approximately 9000 of the males were between 0 and 5 years old, and the dark blue bar shows slightly fewer females were in this age range. If we were to take all the light blue bars and put them side by side, we would have the histogram of age for males using the class intervals stated. Similarly the dark blue bars show females. Because we are trying to display both histograms in the same graph, the bars for males and females within each class interval have been placed alongside each other for ease of comparison, with the bars for different class intervals separated by small spaces.

(a) Describe the main features of the distribution of age for males. Why would describing this distribution in terms of only the center and variability be misleading?

(b) Suppose that different age groups of males spend differing amounts of time outdoors. How could this fact be used to explain the pattern that you found in part (a)? Remember to use your eyes to describe the pattern you see, and then try to explain the pattern.

(c) A 45-year-old male friend of yours looks at the histogram and tells you that he is planning on giving up hiking because this graph suggests he is in a high-risk group for getting Lyme disease. He will resume hiking when he is 65, as he will be less likely to get Lyme disease at that age. Is this a correct interpretation of the histogram?

(d) Comparing the histograms for males and females, how are they similar? What is the main difference, and why do you think it occurs?

1.5 Quantitative variables: stemplots

Histograms are not the only graphical display of distributions. For small data sets, a *stemplot* is quicker to make and presents more detailed information.

Stemplot

To make a **stemplot**:

1. Separate each observation into a **stem**, consisting of all but the final (rightmost) digit, and a **leaf**, the final digit. Stems may have as many digits as needed, but each leaf contains only a single digit.
2. Write the stems in a vertical column with the smallest at the top, and draw a vertical line at the right of this column. Be sure to include all the stems needed to span the data, even when some stems will have no leaves.
3. Write each leaf in the row to the right of its stem, in increasing order out from the stem.

EXAMPLE 1.8 Making a Stemplot

Table 1.2 presents the percents of state residents who were born outside the United States. To make a stemplot of these data, take the whole-number part of the percent as the stem and the final digit (tenths) as the leaf. Write stems from 1 for West Virginia to 27 for California. Now add leaves. Nevada, 19.2%, has leaf 2 on the 19 stem.

DATA FOREIGN

Florida, at 19.4%, places leaf 4 on the same stem. These are the only observations on this stem. Arrange the leaves in order, so that 19|24 is one row in the stemplot. Figure 1.10 is the complete stemplot for the data in Table 1.2. ■

1	3
2	03499
3	334899
4	1336778
5	3599
6	1227
7	34
8	46
9	567
10	2
11	1
12	
13	3445679
14	9
15	
16	5
17	
18	2
19	24
20	
21	3
22	2
23	
24	
25	
26	
27	1

This 19 stem contains the values 19.2 for Nevada and 19.4 for Florida.

FIGURE 1.10
Stemplot of the percents of foreign-born residents in the states, for Example 1.8. Each stem is a percent, and leaves are tenths of a percent.

THE VITAL FEW

Skewed distributions can show us where to concentrate our efforts. Ten percent of the cars on the road account for half of all carbon dioxide emissions. A histogram of CO_2 emissions would show many cars with small or moderate values and a few with very high values. Cleaning up or replacing these cars would reduce pollution at a cost much lower than that of programs aimed at all cars. Statisticians who work at improving quality in industry make a principle of this: Distinguish "the vital few" from "the trivial many."

A stemplot looks like a histogram turned on end, with the stems corresponding to the class intervals. The first stem in Figure 1.10 contains all states with percents between 1.0% and 1.9%. In this example, the stemplot is like a histogram with many classes. Compare the stemplot to Figure 1.11 (page 31), which is a histogram of the same data using class intervals 1.0% to <2.0%, 2.0% to <3.0%, and so forth. Although Figures 1.10 and 1.11 display exactly the same pattern, the stemplot, unlike the histogram, preserves the actual value of each observation.

CAUTION In a stemplot, the classes (the stems of a stemplot) are given to you. Histograms are more flexible than stemplots because you can choose the classes more easily. In both Figures 1.10 and 1.11, California (27.1%) stands slightly apart from the long right tail of the skewed distribution. *Stemplots do not work well for large data sets, where each stem must hold a large number of leaves.* Don't try to make a stemplot of a large data set, such as the 947 Iowa Tests scores in Figure 1.7.

rounding

When there are too many stems, there are often no leaves or just one or two leaves on many of the stems as in Figure 1.10. The number of stems can be reduced by first **rounding** the data. In this example, we can round the data for each state to the nearest percent before drawing the stemplot. Here is the result:

0	12222233333344444445555566667788999
1	0133333333468899
2	127

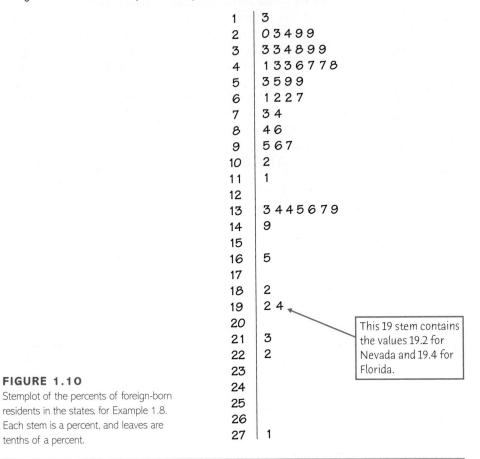

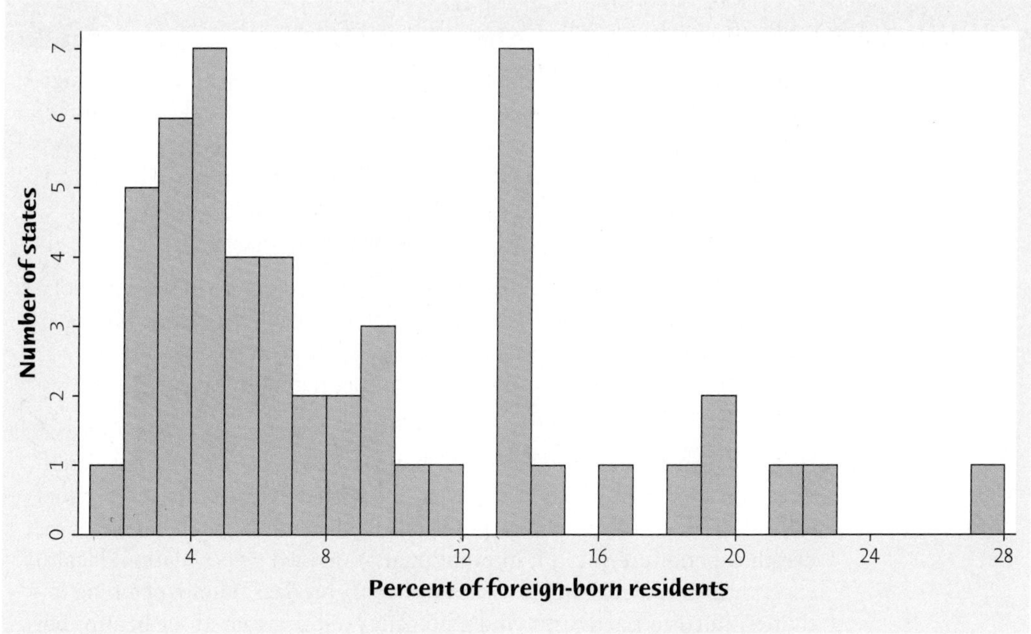

FIGURE 1.11
Histogram of the percents of foreign-born residents in the states, for Example 1.8. The class widths have been chosen to agree with the widths of the stems in the stemplot in Figure 1.10.

splitting stems

Now it seems that there are too few stems. **Splitting stems** in a stemplot will double the number of stems when all the leaves would otherwise fall on just a few stems, as occurred when we rounded to the nearest percent. Each stem then appears twice. Leaves 0 to 4 go on the upper stem, and leaves 5 to 9 go on the lower stem. If you split the stems with the data rounded to the nearest percent, the stemplot becomes

```
0 | 1 2 2 2 2 2 3 3 3 3 3 3 4 4 4 4 4 4 4
0 | 5 5 5 5 6 6 6 6 7 7 8 8 9 9 9
1 | 0 1 3 3 3 3 3 3 3 4
1 | 6 8 9 9
2 | 1 2
2 | 7
```

which makes the right skew pattern clearer. Rounding and splitting stems are matters for judgment, like choosing the classes in a histogram. Some data require rounding but don't require splitting stems, some require just splitting stems, and other data require both. The *One-Variable Statistical Calculator* applet on the text website allows you to decide whether to split stems so that it is easy to see the effect.

CAUTION Comparing Figures 1.11 (right-skewed) and 1.5 (left-skewed) reminds us that *the direction of skewness is the direction of the long tail, not the direction where most observations are clustered.*

APPLET

LaunchPad Online Resources

- The **Snapshots video**, *Visualizing Quantitative Data*, illustrates the ideas of both stemplots and histograms.
- The **StatBoards videos**, *Creating and Interpreting a Histogram* and *Creating and Interpreting a Stemplot*, provide the details of constructing both stemplots and histograms through an example.

Apply Your Knowledge

1.10 Graduation Rates. Table 1.1 presents the data for the on-time high school graduation rates in each of the 50 states and the District of Columbia. GRADRATE

(a) Make a stemplot of the data as it appears in Table 1.1. Because the stemplot preserves the actual value of the observations, it is easy to find the midpoint and the variability. What are they?

(b) For this data, splitting the stems gives a clearer picture of the shape of the distribution. Make a second stemplot splitting the stems, placing leaves 0 to 4 on the first stem and leaves 5 to 9 on the second stem of the same value. How does the resulting stemplot compare to the histogram in Figure 1.5?

1.11 Health Care Spending. Table 1.3 shows the 2011 per capita total expenditure on health in 35 countries with the highest gross domestic product in 2011.[12] Health expenditure per capita is the sum of public and private health expenditure (in PPP, international $) divided by population. Health expenditures include the provision of health services, family-planning activities, nutrition activities, and emergency aid designated for health, but exclude the provision of water and sanitation. Make a stemplot of the data after rounding to the nearest $100 (so that stems are thousands of dollars and leaves are hundreds of dollars). Split the stems, placing leaves 0 to 4 on the first stem and leaves 5 to 9 on the second stem of the same value. Describe the shape, center, and variability of the distribution. Which country is the high outlier? HEALTH

TABLE 1.3 PER CAPITA TOTAL EXPENDITURE ON HEALTH (INTERNATIONAL DOLLARS)

COUNTRY	DOLLARS	COUNTRY	DOLLARS	COUNTRY	DOLLARS
Argentina	1434	India	141	Saudi Arabia	901
Australia	3692	Indonesia	127	South Africa	943
Austria	4482	Iran	929	Spain	3041
Belgium	4119	Italy	3130	Sweden	3870
Brazil	1043	Japan	3174	Switzerland	5564
Canada	4520	Korea, South	2181	Thailand	353
China	432	Malaysia	559	Turkey	1161
Colombia	618	Mexico	940	United Arab Emirates	1732
Denmark	4564	Netherlands	5123	United Kingdom	3322
France	4086	Norway	5674	United States	8608
Germany	4371	Poland	1423	Venezuela	659
Greece	2918	Russia	1316		

1.6 Time plots

Many variables are measured at intervals over time. We might, for example, measure the height of a growing child or the price of a stock at the end of each month. In these examples, our main interest is change over time. To display change over time, make a *time plot*.

Time Plot

A **time plot** of a variable plots each observation against the time at which it was measured. Always put time on the horizontal scale of your plot and the variable you are measuring on the vertical scale. Connecting the data points by lines helps emphasize any change over time.

EXAMPLE 1.9 Water Levels in the Everglades

DATA WATERLEV

Water levels in Everglades National Park are critical to the survival of this unique region. The photo shows a water-monitoring station in Shark River Slough, the main path for surface water moving through the "river of grass" that is the Everglades. Each day the mean gauge height, the height in feet of the water surface above the gauge datum, is measured at the Shark River Slough monitoring station. (The gauge datum is a vertical control measure established in 1929 and is used as a reference for establishing varying elevations. It establishes a zero point from which to measure the gauge height.) Figure 1.12 is a time plot of mean daily gauge height at this station from January 1, 2000, to December 31, 2012.[13] ■

When you examine a time plot, look once again for an overall pattern and for strong deviations from the pattern. Figure 1.12 shows strong **cycles**, regular up-and-down movements in water level. The cycles show the effects of Florida's wet season (roughly June to November) and dry season (roughly December to May). Water levels are highest in late fall. If you look closely, you can see the year-to-year variation. The dry season in 2003 ended early, with the first-ever April tropical storm. In consequence, the dry-season water level in 2003 did not dip as low as in other years. The drought in the southeastern portion of the country in 2008 and 2009 shows up in the steep drop in the mean gauge height in 2009, whereas the lower peaks in 2006 and 2007 reflect lower water levels during the wet seasons in these years. Finally, in 2011, an extra-long dry season and a slow start to the 2011 rainy season compounded into the worst drought in the southwest Florida area in 80 years, which shows up as the steep drop in the mean gauge height in 2011.

cycles

Courtesy U.S. Geological Survey

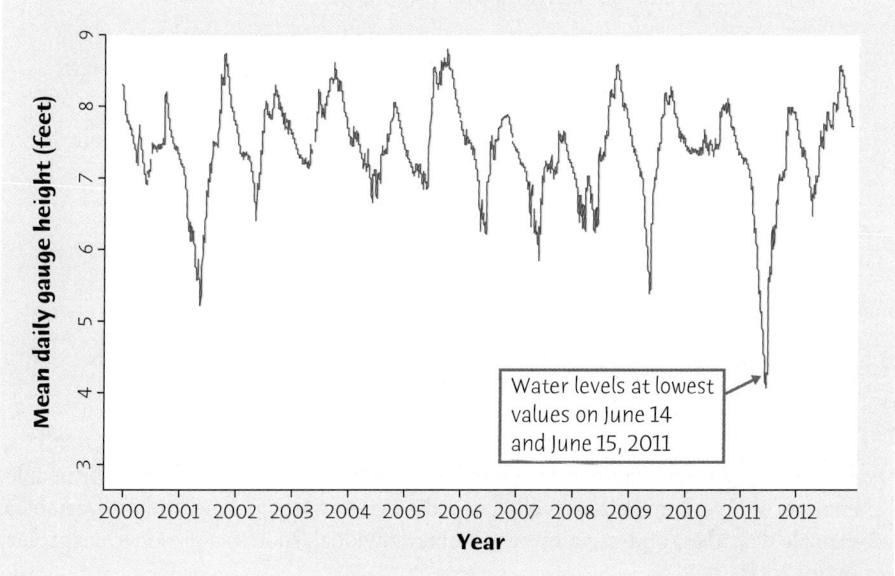

FIGURE 1.12 Time plot of average gauge height at a monitoring station in Everglades National Park over a twelve-year period, for Example 1.9. The yearly cycles reflect Florida's wet and dry seasons.

trend Another common overall pattern in a time plot is a **trend**, a long-term upward or downward movement over time. Many economic variables show an upward trend. Incomes, house prices, and (alas) college tuitions tend to move generally upward over time.

time series data ***cross-sectional data*** Histograms and time plots give different kinds of information about a variable. The time plot in Figure 1.12 presents **time series data** that show the change in water level at one location over time. A histogram displays **cross-sectional data**, such as water levels at many locations in the Everglades at the same time.

Apply Your Knowledge

1.12 The Cost of College. Here are data on the average tuition and fees charged to in-state students by public four-year colleges and universities for the 1980 to 2013 academic years. Because almost any variable measured in dollars increases over time due to inflation (the falling buying power of a dollar), the values are given in "constant dollars," adjusted to have the same buying power that a dollar had in 2013.[14] DATA COLLEGEX

Year	Tuition	Year	Tuition	Year	Tuition	Year	Tuition
1980	$2274	1989	$3191	1998	$4656	2007	$6943
1981	$2321	1990	$3424	1999	$4719	2008	$7008
1982	$2475	1991	$3621	2000	$4751	2009	$7672
1983	$2689	1992	$3887	2001	$4966	2010	$8174
1984	$2761	1993	$4108	2002	$5324	2011	$8557
1985	$2861	1994	$4266	2003	$5900	2012	$8821
1986	$3022	1995	$4314	2004	$6322	2013	$8893
1987	$3054	1996	$4434	2005	$6566		
1988	$3116	1997	$4536	2006	$6662		

(a) Make a time plot of average tuition and fees.

(b) What overall pattern does your plot show?

(c) Some possible deviations from the overall pattern are outliers, periods when charges went down (in 2013 dollars) and periods of particularly rapid increase. Which are present in your plot, and during which years?

(d) In looking for patterns, do you think that it would be better to study a time series of the tuition for each year or the percent increase for each year? Why?

CHAPTER 1 SUMMARY

Chapter Specifics

- A data set contains information on a number of **individuals**. Individuals may be people, animals, or things. For each individual, the data give values for one or more **variables**. A variable describes some characteristic of an individual, such as a person's height, sex, or salary.

- Some variables are **categorical**, and others are **quantitative**. A categorical variable places each individual into a category, such as male or female. A quantitative variable has numerical values that describe some characteristic of each individual using a **unit of measurement**, such as height in centimeters or salary in dollars.
- **Exploratory data analysis** uses graphs and numerical summaries to describe the variables in a data set and the relations among them.
- After you understand the background of your data (individuals, variables, units of measurement), almost always the first thing to do is **plot your data**.
- The **distribution** of a variable describes what values the variable takes and how often it takes these values. **Pie charts** and **bar graphs** display the distribution of a categorical variable. Bar graphs can also compare any set of quantities measured in the same units. **Histograms** and **stemplots** graph the distribution of a quantitative variable.
- When examining any graph, look for an **overall pattern** and for notable **deviations** from the pattern.
- **Shape, center, and variability** (or **spread**) describe the overall pattern of the distribution of a quantitative variable. Some distributions have simple shapes, such as **symmetric** or **skewed**. Not all distributions have a simple overall shape, especially when there are few observations.
- **Outliers** are observations that lie outside the overall pattern of a distribution. Always look for outliers, and try to explain them.
- When observations on a variable are taken over time, make a **time plot** that graphs time horizontally and the values of the variable vertically. A time plot can reveal **trends**, **cycles**, or other changes over time.

Link It

Practical statistics uses data to draw conclusions about some broader universe. You should reread Example 1.1, as it will help you understand this basic idea. For the American Community Survey described in the example, the data are the responses from those households responding to the survey, although the broader universe of interest is the entire nation.

In our study of practical statistics, we will divide the subject into three main areas. In exploratory data analysis, graphs and numerical summaries are used for exploring, organizing, and describing data so that the patterns become apparent. Data production concerns where the data come from and helps us to understand whether what we learn from our data can be generalized to a wider universe. And statistical inference provides tools for generalizing what we learn to a wider universe.

In this chapter we have begun to learn about data analysis. A data set can consist of hundreds of observations on many variables. Even if we consider only one variable at a time, it is difficult to see what the data have to say by scanning a list containing many data values. Graphs provide a visual tool for organizing and identifying patterns in data and are a good starting point in the exploration of the distribution of a variable.

Pie charts and bar graphs can summarize the information in a categorical variable by giving us the percent of the distribution in the various categories. Although a table containing the categories and percent gives the same information as a bar graph, a substantial advantage of the bar graph over a tabular presentation is that the bar graph allows us to visually compare percents among all categories simultaneously by means of the heights of the bars.

Histograms and stemplots are graphical tools for summarizing the information provided by a quantitative variable. The overall pattern in a histogram or stemplot illustrates some of the important features of the distribution of a variable that will be of interest as we continue our study of practical statistics. The center of the histogram tells us about the value of a "typical" observation on this variable, whereas the variability gives us a sense of how close most of the observations are to this value. Other interesting features are the presence of outliers and the general shape of the plot. For data collected over time, time plots can show patterns such as seasonal variation and trends in the variable. In the next chapter we will see how the information about the distribution of a variable can also be described using numerical summaries.

LaunchPad Online Resources

If you are having difficulty with any of the sections of this chapter, these online resources should help prepare you to solve the exercises at the end of this chapter.

- **StatTutor** starts with a video review of each section and asks a series of questions to check your understanding.
- **LearningCurve** provides you with a series of questions about the chapter geared to your level of understanding.

CHECK YOUR SKILLS

The multiple-choice exercises in Check Your Skills ask straightforward questions about basic facts from the chapter. Answers to all these exercises appear in the back of the book. You should expect almost all your answers to be correct.

1.13 Here are the first lines of a professor's data set at the end of a statistics course:

Name	Major	Total Points	Grade
ADVANI, SURA	COMM	397	B
BARTON, DAVID	HIST	323	C
BROWN, ANNETTE	BIOL	446	A
CHIU, SUN	PSYC	405	B
CORTEZ, MARIA	PSYC	461	A

The individuals in these data are

(a) the students.
(b) the total points.
(c) the course grades.

1.14 To display the distribution of grades (A, B, C, D, F) for all students in the course, it would be correct to use

(a) a pie chart but not a bar graph.
(b) a bar graph but not a pie chart.
(c) either a pie chart or a bar graph.

1.15 A description of different houses on the market includes the variables square footage of the house and average monthly gas bill.

(a) Square footage and average monthly gas bill are both categorical variables.
(b) Square footage and average monthly gas bill are both quantitative variables.
(c) Square footage is a categorical variable, and average monthly gas bill is a quantitative variable.

1.16 A political party's data bank includes the zip codes of past donors, such as

47906 34236 53075 10010 90210 75204
30304 99709

Zip code is a

(a) quantitative variable.
(b) categorical variable.
(c) unit of measurement.

1.17 Figure 1.6 is a histogram of the percent of on-time high school graduates in each state. The leftmost bar in the histogram covers percents of on-time high school graduates ranging from about

(a) 58% to 64%.
(b) 58% to 61%.
(c) 0% to 58%.

1.18 Here are the exam scores of 10 students in a statistics class:

50 35 41 97 76 69 94 91 23 65

To make a stemplot of these data, you would use stems

(a) 2, 3, 4, 5, 6, 7, 9.
(b) 2, 3, 4, 5, 6, 7, 8, 9.
(c) 20, 30, 40, 50, 60, 70, 80, 90.

1.19 Where do students go to school? Although 81.4% of first time first-year students attended college in the state in which they lived, this percent varied considerably over the states. Here is a stemplot of the percent of first-year students in each of the 50 states who were from the state where they enrolled. The stems are 10s and the leaves are 1s. The stems have been split in the plot.[15] DATA INSTATE

```
3 | 6
4 | 4
4 |
5 |
5 | 559
6 | 0233
6 | 889
7 | 0334
7 | 5666799
8 | 0000111112233333444
8 | 556
9 | 11222
```

The midpoint of this distribution is

(a) 65%
(b) 80%
(c) 83%

1.20 The shape of the distribution in Exercise 1.19 is

(a) clearly skewed to the left with two low outliers.
(b) roughly symmetric.
(c) clearly skewed to the right with two low outliers.

1.21 The state with the largest percent of first-year students enrolled in the state has

(a) 91.1% enrolled.
(b) 92% enrolled.
(c) 92.22% enrolled.

1.22 You look at real estate ads for houses in Naples, Florida. There are many houses ranging from $200,000 to $500,000 in price. The few houses on the water, however, have prices up to $15 million. The distribution of house prices will be

(a) skewed to the left.
(b) roughly symmetric.
(c) skewed to the right.

CHAPTER 1 EXERCISES

1.23 Medical students. Students who have finished medical school are assigned to residencies in hospitals to receive further training in a medical specialty. Here is part of a hypothetical database of students seeking residency positions. USMLE is the student's score on Step 1 of the national medical licensing examination.

NAME	MEDICAL SCHOOL	SEX	AGE	USMLE	SPECIALTY SOUGHT
Abrams, Laurie	Florida	F	28	238	Family medicine
Brown, Gordon	Meharry	M	25	205	Radiology
Cabrera, Maria	Tufts	F	26	191	Pediatrics
Ismael, Miranda	Indiana	F	32	245	Internal medicine

(a) What individuals does this data set describe?
(b) In addition to the student's name, how many variables does the data set contain? Which of these variables are categorical, and which are quantitative? If a variable is quantitative, what units is it measured in?

1.24 Buying a refrigerator. *Consumer Reports* will have an article comparing refrigerators in the next issue. Some of the characteristics to be included in the report are the brand name and model; whether it has a top, bottom, or side-by-side freezer; the estimated energy consumption per year (kilowatts); whether or not it is Energy Star compliant; the width, depth, and height in inches; and both the freezer and refrigerator net capacity in cubic feet. Which of these variables are categorical, and which are quantitative? Give the units for the quantitative variables and the categories for the categorical variables. What are the individuals in the report?

1.25 What color is your car? The most popular colors for cars and light trucks vary with region and over time. In North America white remains the top color choice, with silver the top choice in South America and white the top choice worldwide for the third consecutive year. Here is the distribution of the top colors for vehicles sold globally in 2013:[16] CARCOLOR

COLOR	POPULARITY
White	25%
Black	18%
Silver	18%
Gray	12%
Red	9%
Beige, brown	8%
Blue	7%
Other colors	

Fill in the percent of vehicles that are in other colors. Make a graph to display the distribution of color popularity.

1.26 Facebook, Twitter, and LinkedIn users. After years of explosive growth in number of users of social networking sites in all age ranges and demographics, it is hard to argue that social media haven't changed forever how we interact and connect online. Although Facebook is still the dominant player in social networking, both Twitter and LinkedIn have continued to increase their usage. Here is the age distribution of the users for the three sites in 2013:[17] SOCIALNT

AGE GROUP	FACEBOOK USERS	TWITTER USERS	LINKEDIN USERS
13 to 17 years	10%	10%	4%
18 to 24 years	14%	18%	10%
25 to 34 years	19%	22%	20%
35 to 44 years	17%	17%	18%
45 to 54 years	17%	15%	20%
55 to 64 years	13%	11%	17%
Over 65 years	10%	7%	11%

(a) Draw a bar graph for the age distribution of Facebook users. The leftmost bar should correspond to "13 to 17," the next bar to "18 to 24," and so on. Do the same for Twitter and LinkedIn, using the same scale for the percent axis.
(b) Describe the most important difference in the age distribution of the audience for these three social networking sites. How does this difference show up in the bar graphs? Do you think it was important to order the bars by age to make the comparison easier? Why or why not?
(c) Explain why it *is* appropriate to use a pie chart to display any of these distributions. Draw a pie chart for each distribution. Do you think it is easier to compare the three distributions with bar graphs or pie charts? Explain your reasoning.

1.27 Deaths among young people. Among persons aged 15 to 24 years in the United States, the leading causes of death and number of deaths in 2011 were: accidents, 12,032; suicide, 4688; homicide, 4508; cancer, 1609; heart disease, 948; congenital defects, 429.[18]

(a) Make a bar graph to display these data.
(b) To make a pie chart, you need one additional piece of information. What is it?

1.28 Hispanic origins. According to the 2010 U.S. Census, 308.7 million people resided in the United States on April 1, 2010, of which 50.5 million (or 16%) were of Hispanic origin. What countries do they come from? Figure 1.13 is a pie chart to show the country of origin of Hispanics in the United States in 2010.[19] About what percent of Hispanics are Mexican? Puerto Rican? You see that it is hard to determine numbers from a pie chart. Bar graphs are much easier to use. (The U.S. Census Bureau includes the percents in many of its pie charts.)

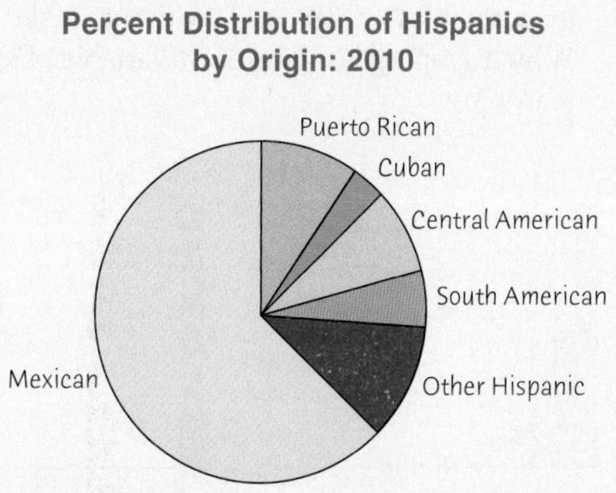

FIGURE 1.13
Pie chart of the national origins of Hispanic residents of the United States, for Exercise 1.28.

1.29 Canadian students rate their universities. The National Survey of Student Engagement asked students at many universities, "How would you evaluate your entire educational experience at this university? " Here are the percents of senior-year students at Canada's 10 largest primarily English-speaking universities who responded "Excellent":[20] CANADA

UNIVERSITY	EXCELLENT RATING
Toronto	21%
York	18%
Alberta	23%
Ottawa	11%
Western Ontario	38%
British Columbia	18%
Calgary	14%
McGill	26%
Waterloo	36%
Concordia	21%

(a) The list is arranged in order of undergraduate enrollment. Make a bar graph with the bars in order of student rating.
(b) Explain carefully why it is not correct to make a pie chart of these data.

1.30 Do adolescent girls eat fruit? We all know that fruit is good for us. Many of us don't eat enough. Figure 1.14 is a histogram of the number of servings of fruit per day claimed by 74 seventeen-year-old girls in a study in Pennsylvania.[21] Describe the shape, center, and variability of this distribution. Are there any outliers? What percent of these girls ate six or more servings per day? How many of these girls ate fewer than two servings per day?

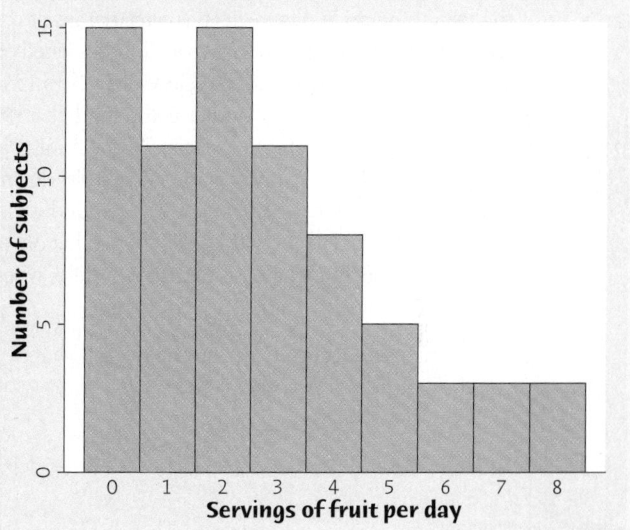

FIGURE 1.14
The distribution of fruit consumption in a sample of 74 seventeen-year-old girls, for Exercise 1.30.

1.31 IQ test scores. Figure 1.15 (see page 40) is a stemplot of the IQ test scores of 78 seventh-grade students in a rural midwestern school.[22] IQ

(a) Four students had low scores that might be considered outliers. Ignoring these, describe the shape, center, and variability of the remainder of the distribution.
(b) We often read that IQ scores for large populations are centered at 100. What percent of these 78 students have scores above 100?

```
 7 | 24
 7 | 79
 8 |
 8 | 69
 9 | 0133
 9 | 6778
10 | 0022333344
10 | 555666777789
11 | 0000111122223334444
11 | 55688999
12 | 003344
12 | 677888
13 | 02
13 | 6
```

FIGURE 1.15
The distribution of IQ scores for 78 seventh-grade students, for Exercise 1.31.

1.32 **Returns on common stocks.** The return on a stock is the change in its market price plus any dividend payments made. Total return is usually expressed as a percent of the beginning price. Figure 1.16 is a histogram of the distribution of the monthly returns for all stocks listed on U.S. markets from January 1985 to December 2013 (348 months).[23] The extreme low outlier is the market crash of October 1987, when stocks lost 23% of their value in one month. The other two low outliers are 16% during August 1998, a month when the Dow Jones Industrial Average experienced its second-largest drop in history to that time, and the financial crisis in October 2008, when stocks lost 17% of their value. STOCKRET

(a) Ignoring the outliers, describe the overall shape of the distribution of monthly returns.

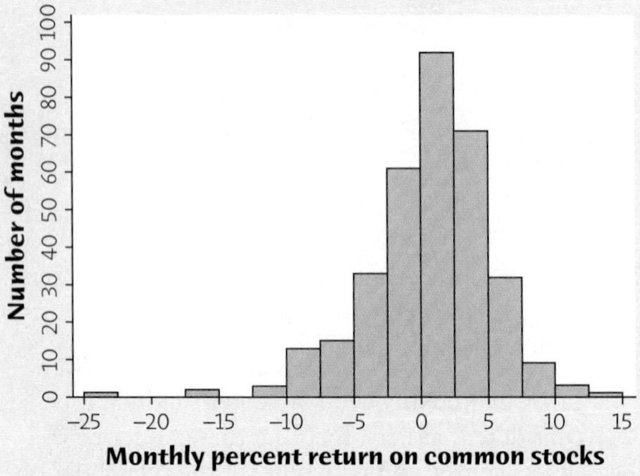

FIGURE 1.16
The distribution of monthly percent returns on U.S. common stocks from January 1985 to December 2013, for Exercise 1.32.

(b) What is the approximate center of this distribution? (For now, take the center to be the value with roughly half the months having lower returns and half having higher returns.)

(c) Approximately what were the smallest and largest monthly returns, leaving out the outliers? (This is one way to describe the variability of the distribution.)

(d) A return less than zero means that stocks lost value in that month. About what percentage of all months had returns less than zero?

1.33 **Name that variable.** A survey of a large college class asked the following questions:

1. Are you female or male? (In the data, male = 0, female = 1.)
2. Are you right-handed or left-handed? (In the data, right = 0, left = 1.)
3. What is your height in inches?
4. How many minutes do you study on a typical weeknight?

Figure 1.17 shows histograms of the student responses, in scrambled order and without scale markings. Which graph goes with each variable? Explain your reasoning.

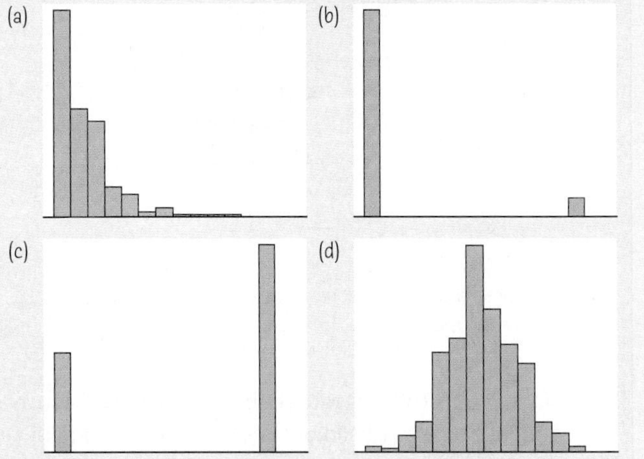

FIGURE 1.17
Histograms of four distributions, for Exercise 1.33.

1.34 **Food oils and health.** Fatty acids, despite their unpleasant name, are necessary for human health. Two types of essential fatty acids, called omega-3 and omega-6, are not produced by our bodies and so must be obtained from our food. Food oils, widely used in food processing and cooking, are major sources of these compounds. There is some evidence that a healthy diet should have more omega-3 than omega-6. Table 1.4 gives the ratio of omega-3 to omega-6 in some common food oils.[24] Values greater than 1 show that an oil has more omega-3 than omega-6. FOODOILS

TABLE 1.4 OMEGA-3 FATTY ACIDS AS A FRACTION OF OMEGA-6 FATTY ACIDS IN FOOD OILS

OIL	RATIO	OIL	RATIO
Perilla	5.33	Flaxseed	3.56
Walnut	0.20	Canola	0.46
Wheat germ	0.13	Soybean	0.13
Mustard	0.38	Grape seed	0.00
Sardine	2.16	Menhaden	1.96
Salmon	2.50	Herring	2.67
Mayonnaise	0.06	Soybean, hydrogenated	0.07
Cod liver	2.00	Rice bran	0.05
Shortening (household)	0.11	Butter	0.64
Shortening (industrial)	0.06	Sunflower	0.03
Margarine	0.05	Corn	0.01
Olive	0.08	Sesame	0.01
Shea nut	0.06	Cottonseed	0.00
Sunflower (oleic)	0.05	Palm	0.02
Sunflower (linoleic)	0.00	Cocoa butter	0.04

(a) Make a histogram of these data, using classes bounded by the whole numbers from 0 to 6.
(b) What is the shape of the distribution? How many of the 30 food oils have more omega-3 than omega-6? What does this distribution suggest about the possible health effects of modern food oils?
(c) Table 1.4 contains entries for several fish oils (cod, herring, menhaden, salmon, sardine). How do these values support the idea that eating fish is healthy?

1.35 **Where are the nurses?** Table 1.5 gives the number of active nurses per 100,000 people in each state.[25] NURSES

(a) Why is the number of nurses per 100,000 people a better measure of the availability of nurses than a simple count of the number of nurses in a state?
(b) Make a histogram that displays the distribution of nurses per 100,000 people. Write a brief description of the distribution. Are there any outliers? If so, can you explain them?

TABLE 1.5 NURSES PER 100,000 PEOPLE, BY STATE

STATE	NURSES	STATE	NURSES	STATE	NURSES
Alabama	912	Louisiana	894	Ohio	1001
Alaska	756	Maine	1053	Oklahoma	712
Arizona	544	Maryland	869	Oregon	795
Arkansas	774	Massachusetts	1210	Pennsylvania	1017
California	641	Michigan	841	Rhode Island	1007
Colorado	761	Minnesota	1017	South Carolina	795
Connecticut	994	Mississippi	868	South Dakota	1215
Delaware	977	Missouri	958	Tennessee	894
Florida	814	Montana	748	Texas	662
Georgia	653	Nebraska	1010	Utah	625
Hawaii	753	Nevada	574	Vermont	912
Idaho	642	New Hampshire	970	Virginia	750
Illinois	812	New Jersey	907	Washington	774
Indiana	864	New Mexico	580	West Virginia	938
Iowa	990	New York	859	Wisconsin	905
Kansas	867	North Carolina	886	Wyoming	812
Kentucky	923	North Dakota	1097	Dist. of Columbia	1380

TABLE 1.6 ANNUAL CARBON DIOXIDE EMISSIONS IN 2010 (METRIC TONS PER PERSON)

COUNTRY	CO_2	COUNTRY	CO_2	COUNTRY	CO_2
Algeria	3.3315	Indonesia	1.8032	Poland	8.3086
Argentina	4.4710	Iran	7.6765	Russia	12.2255
Bangladesh	0.3716	Iraq	3.7034	South Africa	9.2041
Brazil	2.1503	Italy	6.7177	Spain	5.8535
Canada	14.6261	Japan	9.1857	Sudan	0.3109
China	6.1949	Kenya	0.3038	Tanzania	0.1522
Colombia	1.6295	Korea, South	11.4869	Thailand	4.4469
Congo	0.4932	Mexico	3.7636	Turkey	4.1310
Egypt	2.6228	Morocco	1.5994	Uganda	0.1113
Ethiopia	0.0746	Myanmar	0.1732	Ukraine	6.6449
France	5.5554	Nigeria	0.4941	United Kingdom	7.9251
Germany	9.1148	Pakistan	0.9321	United States	17.5642
India	1.6662	Philippines	0.8731	Vietnam	1.7281

1.36 Carbon dioxide emissions. Burning fuels in power plants and motor vehicles emits carbon dioxide (CO_2), which contributes to global warming. Table 1.6 displays the 2010 CO_2 emissions per person from countries with populations of at least 30 million in that year.[26] DATA CO2EMISS

(a) Why do you think we choose to measure emissions per person rather than total CO_2 emissions for each country?

(b) Make a stemplot to display the data of Table 1.6. The data will first need to be rounded (see page 30). What units are you going to use for the stems? The leaves? You should round the data to the units you are planning to use for the leaves before drawing the stemplot. Describe the shape, center, and variability of the distribution. Which countries are outliers?

1.37 Fur seals on St. Paul Island. Every year hundreds of thousands of northern fur seals return to their haul-outs in the Pribilof Islands in Alaska to breed, give birth, and teach their pups to swim, hunt, and survive in the Bering Sea. U.S. commercial fur sealing operations continued on St. Paul until 1984, but despite a reduction in harvest, the population of fur seals has continued to decline. Possible reasons include climate shifts in the North Pacific, changes in the availability of prey, and new or increased interaction with commercial fisheries that increase mortality. Here are data on the estimated number of fur seal pups born on St. Paul Island (in thousands) from 1979 to 2012, where a dash indicates a year in which no data were collected:[27]

©Arco Images GmbH/Alamy Images

YEAR	PUPS BORN (THOUSANDS)	YEAR	PUPS BORN (THOUSANDS)
1979	245.93	1996	170.12
1980	203.82	1997	–
1981	179.44	1998	179.15
1982	203.58	1999	–
1983	165.94	2000	158.74
1984	173.27	2001	–
1985	182.26	2002	145.72
1986	167.66	2003	–
1987	171.61	2004	122.82
1988	202.23	2005	–
1989	171.53	2006	109.96
1990	201.30	2007	–
1991	—	2008	102.67
1992	182.44	2009	–
1993	—	2010	94.50
1994	192.10	2011	–
1995	—	2012	96.83

Make a stemplot to display the distribution of pups born per year. Describe the shape, center, and variability of the distribution. Are there any outliers? DATA FURSEALS

1.38 Nintendo and laparoscopic skills. In laparoscopic surgery, a video camera and several thin instruments are inserted into the patient's abdominal cavity. The surgeon uses the image from the video camera positioned inside the patient's body to perform the procedure by manipulating the instruments that have been inserted. It has been found that the Nintendo Wii™ reproduces the movements required in laparoscopic surgery more closely than other video games with its motion-sensing interface. If training with a Nintendo Wii™ can improve laparoscopic skills, it can complement the more expensive training on a laparoscopic simulator. Forty-two medical residents were chosen, and all were tested on a set of basic laparoscopic skills. Twenty-one were selected at random to undergo systematic Nintendo Wii™ training for one hour a day, five days a week, for four weeks. The remaining 21 residents were given no Nintendo Wii™ training and asked to refrain from video games during this period. At the end of four weeks, all 42 residents were tested again on the same set of laparoscopic skills. One of the skills involved a virtual gall bladder removal, with several performance measures including time to complete the task recorded. Here are the improvement (before − after) times in seconds after four weeks for the two groups:[28] DATA NINTENDO

TREATMENT						CONTROL					
281	134	186	128	84	243	21	66	54	85	229	92
212	121	134	221	59	244	43	27	77	−29	−14	88
79	333	−13	−16	71	−16	145	110	32	90	46	−81
71	77	144				68	61	44			

(a) In the context of this study, what do the negative values in the data set mean?

back-to-back stemplots

(b) **Back-to-back stemplots** can be used to compare the two samples. That is, use one set of stems with two sets of leaves, one to the right and one to the left of the stems. (Draw a line on either side of the stems to separate stems and leaves.) Order both sets of leaves from smallest at the stem to largest away from the stem. Complete the back-to-back stemplot started below. The data have been rounded to the nearest 10, with stems being 100s and leaves being 10s. The stems have been split. The first control observation corresponds to −80 and the next two to −30 and −10.

Treatment		Control
	−0	8
122	−0	31
	0	
	0	
	1	
	1	
	2	
	2	
	3	

(c) Report the approximate midpoints of both groups. Does it appear that the treatment has resulted in a greater improvement in times than seen in the control group? (To better understand the magnitude of the improvements, note that the median time to complete the task on the first occasion was 11 minutes and 40 seconds, using the times of all 42 residents.)

1.39 Fur seals on St. Paul Island. Make a time plot of the number of fur seals born per year from Exercise 1.37. What does the time plot show that your stemplot in Exercise 1.37 did not show? When you have data collected over time, a time plot is often needed to understand what is happening. DATA FURSEALS

1.40 Marijuana and traffic accidents. Researchers in New Zealand interviewed 907 drivers at age 21. They had data on traffic accidents, and they asked the drivers about marijuana use. Here are data on the numbers of accidents caused by these drivers at age 19, broken down by marijuana use at the same age:[29]

	MARIJUANA USE PER YEAR			
	NEVER	1–10 TIMES	11–50 TIMES	51+ TIMES
Accidents caused	59	36	15	50
Drivers	452	229	70	156

(a) Explain carefully why a useful graph must compare *rates* (accidents per driver) rather than *counts* of accidents in the four marijuana use classes.
(b) Compute the accident rates in the four marijuana use classes. After you have done this, make a graph that displays the accident rate for each class. What do you conclude? (You can't conclude that marijuana use *causes* accidents, because risk takers are more likely both to drive aggressively and to use marijuana.)

1.41 Dates on coins. Sketch a histogram for a distribution that is skewed to the left. Suppose that you and your friends emptied your pockets of coins and recorded the year marked on each coin. The distribution of dates would be skewed to the left. Explain why.

1.42 She sounds tall! Presented with recordings of a pair of people of the same sex speaking the same phrase, can a listener determine which speaker is taller simply from the sound of their voice? Twenty four young adults at Washington University listened to 100 pairs of speakers, and within each pair were asked to indicate which of the two speakers was the taller. Here are the number correct (out of 100) for each of the 24 participants:[30] HEARING

65	61	67	59	58	62	56	67	61	67	63	53
68	49	66	58	69	70	65	56	68	56	58	70

Researchers believe that the key to correct discrimination is contained in a particular type of sound produced in the lower airways or the lungs, known as subglottal resonances, whose frequency is lower for taller people. Despite the masking of these resonances by other voice sounds, researchers wondered whether the information they contained could still be heard by listeners and used to identify the taller person.

(a) Make two stemplots, with and without splitting the stems. Which plot do you prefer and why?
(b) Describe the shape, center, and variability of the distribution. Are there any outliers?
(c) If the experimental subjects are just guessing which speaker is taller, they should correctly identify the taller person about 50% of the time. Does this data support the researchers, conjecture that there is information in a person's voice to help identify the taller person? Why or why not?

1.43 Watch those scales! Figures 1.18(a) and 1.18(b) both show time plots of tuition charged to in-state students from 1980 through 2013.[31]

(a) Which graph appears to show the biggest increase in tuition between 2000 and 2013?
(b) Read the graphs and compute the actual increase in tuition between 2000 and 2013 in each graph. Do you think these graphs are for the same or different data sets? Why?

The impression that a time plot gives depends on the scales you use on the two axes. Changing the scales can make tuition appear to increase very rapidly or to have only a gentle increase. The moral of this exercise is: Always pay close attention to the scales when you look at a time plot.

1.44 Housing starts. Figure 1.19 (see page 45) is a time plot of the number of single-family homes started by builders each month from January 1990 through December 2013.[32] The counts are in thousands of homes. HOUSING

(a) The most notable pattern in this time plot is yearly up-and-down cycles. At what season of the year are housing starts highest? Lowest? The cycles are explained by the weather in the northern part of the country.
(b) Is there a longer-term trend visible in addition to the cycles? If so, describe it.
(c) The big economic news of 2007 was a severe downturn in housing that began in mid-2006. This was followed by the financial crisis in 2008. How are these economic events reflected in the time plot?
(d) How would you describe the behavior of the time plot since January 2011?

1.45 Ozone hole. The ozone hole is a region in the stratosphere over the Antarctic with exceptionally depleted ozone. The size of the hole is not constant

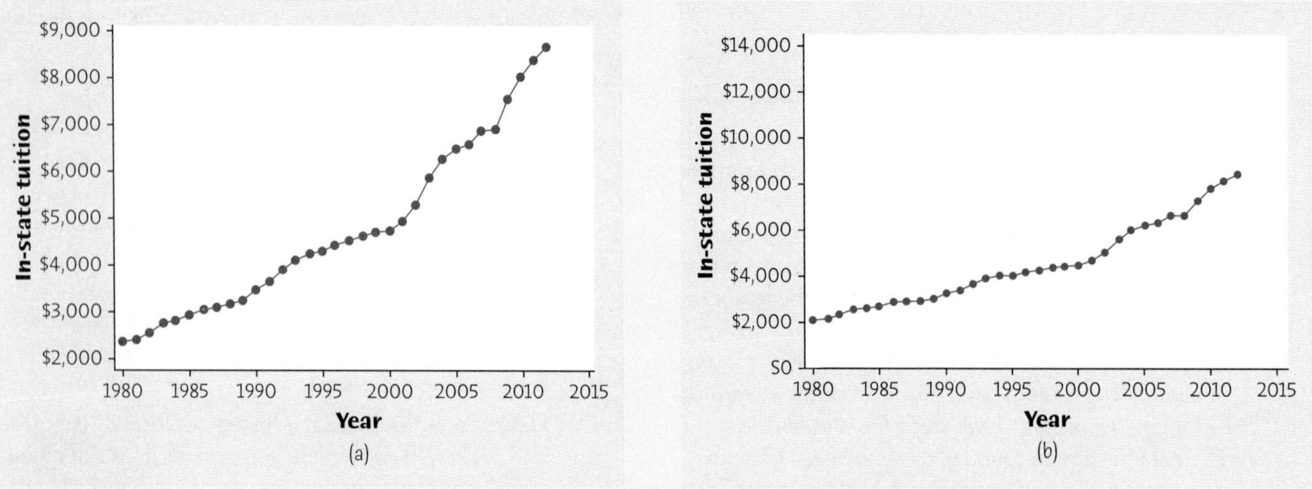

FIGURE 1.18
Time plots of in-state tuition between 1980 and 2013, for Exercise 1.43.

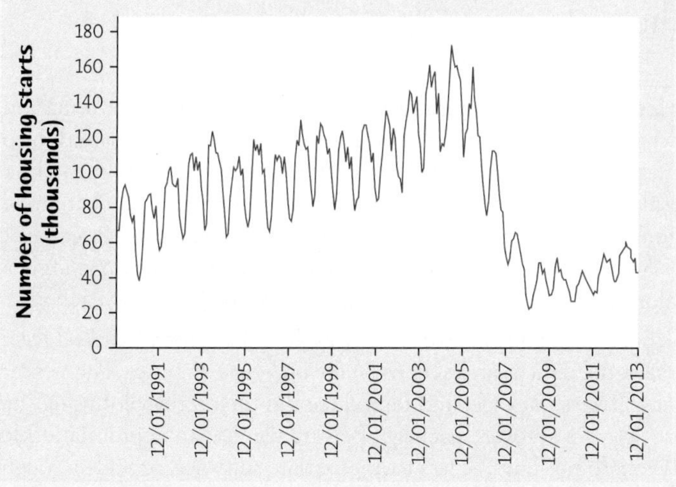

FIGURE 1.19
Time plot of the monthly count of new single-family houses started (in thousands) between January 1990 and December 2013, for Exercise 1.44.

over the year but is largest at the beginning of the Southern Hemisphere spring (August–October). The increase in the size of the ozone hole led to the Montreal Protocol in 1987, an international treaty designed to protect the ozone layer by phasing out the production of substances, such as chlorofluorocarbons (CFCs), believed to be responsible for ozone depletion. The following table gives the average ozone hole size for the period September 7 to October 13 for each of the years from 1979 through 2013 (note that no data were acquired in 1995).[33] To get a better feel for the magnitude of the numbers, the area of North America is approximately 24.5 million square kilometers (km^2). OZONE

The two parts of this exercise will have you draw two graphs of these data.

(a) First make a time plot of the data. The severity of the ozone hole will vary from year to year depending on the meteorology of the atmosphere above Antarctica. Does the time plot illustrate only year-to-year variation or are other patterns apparent? Specifically, is there a trend over any period of years? What about cyclical fluctuation? Explain in words the change in the average size of the ozone hole over this 35-year period.

(b) Now make a stemplot of the data. What is the midpoint of the distribution of ozone hole size? Do you think that the stemplot and the midpoint are a good description of this data set? Is there important information in the time plot that is not contained in the stemplot? When data are collected over time, you should always make a time plot.

1.46 **Choosing class intervals.** Student engineers learn that, although handbooks give the strength of a material as a single number, in fact the strength varies from piece to piece. A vital lesson in all fields of study is that "variation is everywhere." Here are data from a typical student laboratory exercise: the load in pounds needed to pull apart pieces of Douglas fir 4 inches long and 1.5 inches square.

33,190	31,860	32,590	26,520	33,280
32,320	33,020	32,030	30,460	32,700
23,040	30,930	32,720	33,650	32,340
24,050	30,170	31,300	28,730	31,920

The data sets in the *One-Variable Statistical Calculator* applet on the text website include the "pulling wood apart" data given in this exercise. How many class intervals does the applet choose when drawing the histogram? Use the applet to make several histograms with a larger number of class intervals. Are there any important features of the data that are revealed using a larger number of class intervals? Which histogram do you prefer? Explain your choice.

YEAR	AREA (MILLIONS OF km^2)	YEAR	AREA (MILLIONS OF km^2)	YEAR	AREA (MILLIONS OF km^2)
1979	0.1	1991	18.8	2003	25.8
1980	1.4	1992	22.3	2004	19.5
1981	0.6	1993	24.2	2005	24.4
1982	4.8	1994	23.6	2006	26.6
1983	7.9	1995	—	2007	22.0
1984	10.1	1996	22.7	2008	25.2
1985	14.2	1997	22.1	2009	22.0
1986	11.3	1998	25.9	2010	19.4
1987	19.3	1999	23.2	2011	24.7
1988	10.0	2000	24.8	2012	17.8
1989	18.7	2001	25.0	2013	21.0
1990	19.2	2002	12.0		

Exploring the Web

1.47 Natural gas prices. The Department of Energy website contains information about monthly wholesale and retail prices for natural gas in each state. Go to `www.eia.doe.gov/naturalgas/data.cfm` and then click on the link Monthly Wholesale and Retail Prices. Under Area, choose a state of interest to you, make sure the Period is monthly, and then under Residential Price click on View History. A window will open with a time plot covering approximately a 25-year period, along with a table of the monthly residential prices for each year.

(a) If you have access to statistical software, you should use the *Download Data (XLS file)* link to save the data as an Excel (.xls) file on your computer. Then enter the data into your software package, and reproduce the time series plot using the graphic capabilities of your software package. Be sure you use an appropriate title and axis labels. If you do not have access to appropriate software, provide a rough sketch of the time plot that is given on the website.

(b) Is there a regular pattern of seasonal variation that repeats each year? Describe it. Are the prices increasing over time?

1.48 Hank Aaron's home run record. The all-time home run leader prior to 2007 was Hank Aaron. You can find his career statistics by going to the website `www.baseball-reference.com` and then clicking on the Players tab at the top of the page and going to Hank Aaron.

(a) Make a stemplot or a histogram of the number of home runs that Hank Aaron hit in each year during his career. Is the distribution roughly symmetric, clearly skewed, or neither? About how many home runs did Aaron hit in a typical year? Are there any outliers?

(b) Would a time plot be appropriate for these data? If so, what information would be included in the time plot that is not in the stemplot?

(c) How has the average number of home runs per game changed over time? You can find information about average home runs per game on the website `www.baseball-reference.com/leagues/MLB/bat.shtml`. Give an appropriate plot and describe any patterns that you see. Do you see any difficulties comparing the home run records of players whose careers spanned different eras? Explain.

Logan Mock-Bunting/Getty Images

CHAPTER 2

Describing Distributions with Numbers

In this chapter we cover...

2.1 Measuring center: the mean
2.2 Measuring center: the median
2.3 Comparing the mean and the median
2.4 Measuring variability: the quartiles
2.5 The five-number summary and boxplots
2.6 Spotting suspected outliers and modified boxplots*
2.7 Measuring variability: the standard deviation
2.8 Choosing measures of center and variability
2.9 Using technology
2.10 Organizing a statistical problem

We saw in Chapter 1 (page 14) that the American Community Survey asks, among much else, workers' travel times to work. Here are the travel times in minutes for 15 workers in North Carolina, chosen at random by the U.S. Census Bureau:[1]

30 20 10 40 25 20 10 60 15 40 5 30 12 10 10

We aren't surprised that most people estimate their travel time in multiples of 5 minutes. Here is a stemplot of these data:

0	5
1	000025
2	005
3	00
4	00
5	
6	0

The distribution is single peaked and right-skewed. The longest travel time (60 minutes) may be an outlier. Our goal in this chapter is to describe with numbers the center and variability of this and other distributions.

2.1 Measuring center: the mean

The most common measure of center is the ordinary arithmetic average, or *mean*.

The Mean $\bar{x}$

To find the **mean** of a set of observations, add their values and divide by the number of observations. If the n observations are $x_1, x_2, \ldots, x_n$, their mean is

$$\bar{x} = \frac{x_1 + x_2 + \cdots + x_n}{n}$$

or, in more compact notation,

$$\bar{x} = \frac{1}{n}\sum x_i$$

DON'T HIDE THE OUTLIERS

Data from an airliner's control surfaces, such as the vertical tail rudder, go to cockpit instruments and then to the "black box" flight data recorder. To avoid confusing the pilots, short erratic movements in the data are "smoothed" so that the instruments show overall patterns. When a crash killed 260 people, investigators suspected a catastrophic movement of the tail rudder. But the black box contained only the smoothed data. Sometimes outliers are more important than the overall pattern.

The Σ (capital Greek sigma) in the formula for the mean is short for "add them all up." The subscripts on the observations x_i are just a way of keeping the n observations distinct. They do not necessarily indicate order or any other special facts about the data. The bar over the x indicates the mean of all the x-values. Pronounce the mean $\bar{x}$ as "x-bar." This notation is very common. When writers who are discussing data use $\bar{x}$ or $\bar{y}$, they are talking about a mean.

EXAMPLE 2.1 Travel Times to Work

NCTRAVEL

The mean travel time of our 15 North Carolina workers is

$$\bar{x} = \frac{x_1 + x_2 + \cdots + x_n}{n} = \frac{30 + 20 + \cdots + 10}{15} = \frac{337}{15} = 22.5 \text{ minutes}$$

In practice, you can enter the data into your calculator and ask for the mean. You don't have to actually add and divide. But you should know that this is what the calculator is doing.

Notice that only 6 of the 15 travel times are larger than the mean. If we leave out the longest single travel time, 60 minutes, the mean for the remaining 14 people is 19.8 minutes. That one observation raises the mean by 2.7 minutes. ■

Example 2.1 illustrates an important fact about the mean as a measure of center: it is sensitive to the influence of a few extreme observations. These may be outliers, but a skewed distribution that has no outliers will also pull the mean toward its long tail. Because the mean cannot resist the influence of extreme observations, we say that it is not a **resistant measure** of center.

resistant measure

Apply Your Knowledge

2.1 **E. coli in Swimming Areas.** To investigate water quality, the *Columbus Dispatch* took water specimens at 16 Ohio State Park swimming areas in central Ohio. Those specimens were taken to laboratories and tested for *E. coli*, which are bacteria that can cause serious gastrointestinal problems. For reference, if a 100-milliliter specimen (about 3.3 ounces) of water contains more than 130 *E. coli* bacteria, it is considered unsafe. Here are the *E. coli* levels per 100 milliliters found by the laboratories:[2]

291.0	10.9	47.0	86.0	44.0	18.9	1.0	50.0
190.4	45.7	28.5	18.9	16.0	34.0	8.6	9.6

Find the mean *E. coli* level. How many of the lakes have *E. coli* levels greater than the mean? What feature of the data explains the fact that the mean is greater than most of the observations? DATA ECOLI

2.2 **Health Care Spending.** Table 1.3 (page 32) gives the 2011 health care expenditure per capita in 35 countries with the highest gross domestic product in 2011. The United States, at 8608 international dollars per person, is a high outlier. Find the mean health care spending in these nations with and without the United States. How much does the one outlier increase the mean? DATA HEALTH

2.2 Measuring center: the median

In Chapter 1, we used the midpoint of a distribution as an informal measure of center and gave a method for its computation. The *median* is the formal version of the midpoint, and we now provide a more detailed rule for its calculation.

The Median *M*

The **median *M*** is the midpoint of a distribution, the number such that half the observations are smaller and the other half are larger. To find the median of a distribution:

1. Arrange all observations in order of size, from smallest to largest.
2. If the number of observations n is odd, the median M is the center observation in the ordered list. If the number of observations n is even, the median M is midway between the two center observations in the ordered list.
3. You can always locate the median in the ordered list of observations by counting up $(n + 1)/2$ observations from the start of the list.

CAUTION *Note that the formula $(n + 1)/2$ does not give the median, just the location of the median in the ordered list.* Medians require little arithmetic, so they are easy to find by hand for small sets of data. Arranging even a moderate number of observations in order is very tedious, however, so finding the median by hand for larger sets of data is unpleasant. Even simple calculators have an $\bar{x}$ button, but you will need to use software or a graphing calculator to automate finding the median.

EXAMPLE 2.2 Finding the Median: Odd *n*

What is the median travel time for our 15 North Carolina workers? Here are the data arranged in order:

DATA NCTRAVEL

5 10 10 10 10 12 15 **20** 20 25 30 30 40 40 60

The count of observations, $n = 15$, is odd. The bold **20** is the center observation in the ordered list, with seven observations to its left and seven to its right. This is the median, $M = 20$ minutes.

Because $n = 15$, our rule for the location of the median gives

$$\text{location of } M = \frac{n+1}{2} = \frac{16}{2} = 8$$

That is, the median is the eighth observation in the ordered list. It is faster to use this rule than to locate the center by eye. ■

EXAMPLE 2.3 Finding the Median: Even *n*

NYTRAVEL

Travel times to work in New York State are (on the average) longer than in North Carolina. Here are the travel times in minutes of 20 randomly chosen New York workers:

10 30 5 25 40 20 10 15 30 20 15 20 85 15 65 15 60 60 40 45

A stemplot not only displays the distribution but also makes finding the median easy because it arranges the observations in order:

Mitchell Funk/Getty Images

0	5
1	0 0 5 5 5 5
2	0 0 **0 5**
3	0 0
4	0 0 5
5	
6	0 0 5
7	
8	5

The distribution is single peaked and right-skewed, with several travel times of an hour or more. There is no center observation, but there is a center pair. These are the bold **20** and **25** in the stemplot, which have nine observations before them in the ordered list and nine after them. The median is midway between these two observations:

$$M = \frac{20 + 25}{2} = 22.5 \text{ minutes}$$

With $n = 20$, the rule for locating the median in the list gives

$$\text{location of } M = \frac{n+1}{2} = \frac{21}{2} = 10.5$$

The location 10.5 means "halfway between the 10th and 11th observations in the ordered list." That agrees with what we found by eye. ■

2.3 Comparing the mean and the median

Examples 2.1 and 2.2 illustrate an important difference between the mean and the median. The median travel time (the midpoint of the distribution) is 20 minutes. The mean travel time is higher, 22.5 minutes. The mean is pulled toward the right tail of this right-skewed distribution. The median, unlike the mean, is *resistant*. If the longest travel time were 600 minutes rather than 60 minutes, the mean would increase to more than 58 minutes but the median would not change at all. The outlier just counts as one observation above the center, no matter how far above the center it lies. The mean uses the actual value of each observation and so will chase a single large observation upward. The *Mean and Median* applet is an excellent way to compare the resistance of M and $\bar{x}$.

APPLET

Comparing the Mean and the Median

The mean and median of a roughly symmetric distribution are close together. If the distribution is exactly symmetric, the mean and median are exactly the same. In a skewed distribution, the mean is usually farther out in the long tail than the median.[3]

Many economic variables have distributions that are skewed to the right. For example, the median endowment of colleges and universities in the United States and Canada in 2013 was about $101 million—but the mean endowment was almost $537 million. Most institutions have modest endowments, but a few are very wealthy. Harvard's endowment was over $32 billion.[4] The few wealthy institutions pull the mean up but do not affect the median. Reports about incomes and other strongly skewed distributions usually give the median ("midpoint") rather than the mean ("arithmetic average"). However, a county that is about to impose a tax of 1% on the incomes of its residents cares about the mean income, not the median. The tax revenue will be 1% of total income, and the total is the mean times the number of residents. The mean and median measure center in different ways, and both are useful. *Don't confuse the "average" value of a variable (the mean) with its "typical" value, which we might describe by the median.*

CAUTION

Apply Your Knowledge

2.3 New York Travel Times. Find the mean of the travel times to work for the 20 New York workers in Example 2.3. Compare the mean and median for these data. What general fact does your comparison illustrate?

NYTRAVEL

2.4 New House Prices. The mean and median sales prices of new homes sold in the United States in November 2013 were $270,900 and $340,300.[5] Which of these numbers is the mean and which is the median? Explain how you know.

2.5 Carbon Dioxide Emissions. Table 1.6 (page 42) gives the 2010 carbon dioxide (CO_2) emissions per person for countries with populations of at least 30 million. Find the mean and the median for these data. Make a histogram of the data. What features of the distribution explain why the mean is larger than the median? CO2EMISS

Albert Valles Photography/Getty Images

2.4 Measuring variability: the quartiles

The mean and median provide two different measures of the center of a distribution. But a measure of center alone can be misleading. The U.S. Census Bureau reports that in 2012 the median income of American households was $51,017. Half of all households had incomes below $51,017, and half had higher incomes. The mean was much higher, $71,274, because the distribution of incomes is skewed to the right. But the median and mean don't tell the whole story. The bottom 20% of households had incomes less than $20,593, and households in the top 5% took in more than $191,156.[6] We are interested in the *variability* of incomes as well as their center. *The simplest useful numerical description of a distribution requires both a measure of center and a measure of variability.*

CAUTION

One way to measure variability is to give the smallest and largest observations. For example, the travel times of our 15 North Carolina workers range from 5 minutes to 60 minutes. These single observations show the full variability of the data, but they may be outliers. We can improve our description of variability by also looking at the variability of the middle half of the data. The *quartiles* mark out the middle half. Count up the ordered list of observations, starting from the smallest. The *first quartile* lies one-quarter of the way up the list. The *third quartile* lies three-quarters of the way up the list. In other words, the first quartile is larger than 25% of the observations, and the third quartile is larger than 75% of the observations. The second quartile is the median, which is larger than 50% of the observations. That is the idea of quartiles. We need a rule to make the idea exact. The rule for calculating the quartiles uses the rule for the median.

The Quartiles Q_1 and Q_3

To calculate the **quartiles:**

1. Arrange the observations in increasing order and locate the median, M, in the ordered list of observations.
2. The **first quartile, Q_1**, is the median of the observations whose position in the ordered list is to the left of the location of the overall median.
3. The **third quartile, Q_3**, is the median of the observations whose position in the ordered list is to the right of the location of the overall median.

Here are examples that show how the rules for the quartiles work for both odd and even numbers of observations.

EXAMPLE 2.4 Finding the Quartiles: Odd *n*

NCTRAVEL

Our North Carolina sample of 15 workers' travel times, arranged in increasing order, is

5 10 10 10 10 12 15 **20** 20 25 30 30 40 40 60

There is an odd number of observations, so the median is the middle one, the bold **20** in the list. The first quartile is the median of the seven observations to the left of the median. This is the fourth of these seven observations, so $Q_1 = 10$ minutes. If you want, you can use the rule for the location of the median with $n = 7$:

$$\text{location of } Q_1 = \frac{n+1}{2} = \frac{7+1}{2} = 4$$

The third quartile is the median of the seven observations to the right of the median, $Q_3 = 30$ minutes. *When there is an odd number of observations, leave out the overall median when you locate the quartiles in the ordered list.*

The quartiles are *resistant* because they are not affected by a few extreme observations. For example, Q_3 would still be 30 if the outlier were 600 rather than 60. ■

EXAMPLE 2.5 Finding the Quartiles: Even *n*

NYTRAVEL

Here are the travel times to work of the 20 New York workers from Example 2.3, arranged in increasing order:

5 10 10 15 15 15 15 20 20 20 | 25 30 30 40 40 45 60 60 65 85

There is an even number of observations, so the median lies midway between the middle pair, the 10th and 11th in the list. Its value is $M = 22.5$ minutes. We have marked the location of the median by |. The first quartile is the median of the first 10 observations because these are the observations to the left of the location of the median. Check that $Q_1 = 15$ minutes and $Q_3 = 42.5$ minutes. *When the number of observations is even, include all the observations when you locate the quartiles.* ■

Be careful when, as in these examples, several observations take the same numerical value. Write down all the observations, arrange them in order, and apply the rules just as if they all had distinct values.

2.5 The five-number summary and boxplots

The smallest and largest observations tell us little about the distribution as a whole, but they give information about the tails of the distribution that is missing if we know only the median and the quartiles. To get a quick summary of both center and variability, combine all five numbers.

The Five-Number Summary

The **five-number summary** of a distribution consists of the smallest observation, the first quartile, the median, the third quartile, and the largest observation, written in order from smallest to largest. In symbols, the five-number summary is

$$\text{Minimum} \quad Q_1 \quad M \quad Q_3 \quad \text{Maximum}$$

These five numbers offer a reasonably complete description of center and variability. The five-number summaries of travel times to work from Examples 2.4 and 2.5 are

North Carolina	5	10	20	30	60
New York	5	15	22.5	42.5	85

The five-number summary of a distribution leads to a new graph, the *boxplot*. Figure 2.1 shows boxplots comparing travel times to work in North Carolina and New York.

Boxplot

A **boxplot** is a graph of the five-number summary.

- A central box spans the quartiles Q_1 and Q_3.
- A line in the box marks the median M.
- Lines extend from the box out to the smallest and largest observations.

Because boxplots show less detail than histograms or stemplots, they are best used for side-by-side comparison of more than one distribution, as in Figure 2.1. Be sure to include a numerical scale in the graph. When you look at a boxplot, first locate the median, which marks the center of the distribution. Then look at the variability. The span of the central box shows the variability of the middle half of the data, and the extremes (the smallest and largest observations) show the variability of the entire data set. We see from Figure 2.1 that travel times to work are in general a bit longer in

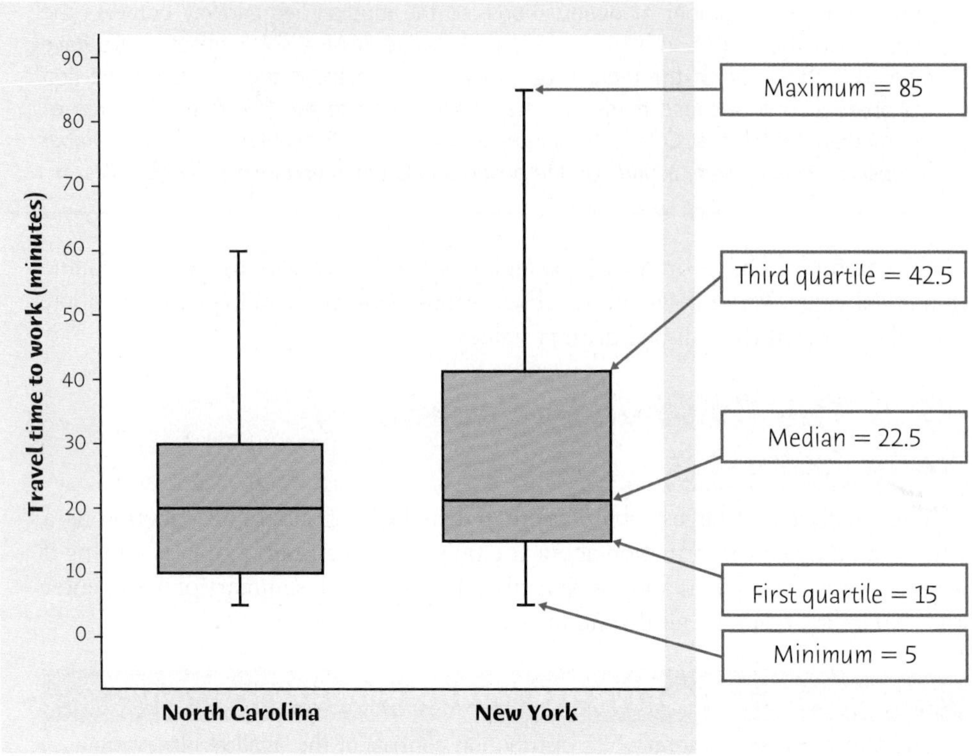

FIGURE 2.1 Boxplots comparing the travel times to work of samples of workers in North Carolina and New York.

New York than in North Carolina. The median, both quartiles, and the maximum are all larger in New York. New York travel times are also more variable, as shown by the span of the box and the difference between the extremes. Note that the boxes with arrows in Figure 2.1 that indicate the location of the five-number summary are *not* part of the boxplot, but are included purely for illustration.

Finally, the New York data are more strongly right-skewed. In a symmetric distribution, the first and third quartiles are equally distant from the median. In most distributions that are skewed to the right, on the other hand, the third quartile will be farther above the median than the first quartile is below it. The extremes behave the same way, but remember that they are just single observations and may say little about the distribution as a whole.

LaunchPad Online Resources

- The **StatClips Examples video**, *Exploratory Pictures for Quantitative Data Example C*, provides a worked out example of computing the five-number summary for both odd and even sample sizes, and then draws a comparative boxplot.

Apply Your Knowledge

Chip Somodevilla/Getty Images

2.6 The Seattle Seahawks. The 2013–14 roster of the Seattle Seahawks, winners of the 2014 NFL Super Bowl, included 10 defensive linemen and 9 offensive linemen. The weights in pounds of the defensive linemen were DATA SEAHAWKS

254 311 297 323 260 242 300 252 303 274

and the weights of the offensive linemen were

310 315 305 318 298 301 321 332 320

(a) Make a stemplot of the weights of the defensive linemen and find the five-number summary.

(b) Make a stemplot of the weights of the offensive linemen and find the five-number summary.

(c) Does either group contain one or more clear outliers? Which group of players tends to be heavier?

2.7 Fuel Economy for Midsize Cars. The Department of Energy provides fuel economy ratings for all cars and light trucks sold in the United States. Here are the estimated miles per gallon for city driving for the 205 cars classified as midsize in 2014, arranged in increasing order:[7] DATA MIDCARS

11	11	11	12	12	12	13	14	14	14	14	14	15	15	15	15	15	15
15	15	16	16	16	16	16	16	16	16	16	16	16	16	16	16	17	17
17	17	17	17	17	17	17	18	18	18	18	18	18	18	18	18	18	18
18	18	18	18	18	19	19	19	19	19	19	19	19	19	19	19	19	19
20	20	20	20	20	20	20	20	20	20	20	20	20	20	20	20	20	21
21	21	21	21	21	21	21	21	21	21	21	21	21	21	22	22	22	22
22	22	22	22	22	22	22	22	23	23	23	23	23	24	24	24	24	24
24	24	24	24	24	24	24	24	24	24	25	25	25	25	25	25	25	25
25	25	25	25	25	25	26	26	26	26	26	26	26	26	26	27	27	27
27	27	27	27	27	27	27	27	28	28	28	28	28	28	28	28	28	29
29	29	29	29	29	30	30	30	30	30	30	30	31	35	36	36	36	40
40	40	43	45	47	50	51											

(a) Give the five-number summary of this distribution.

(b) Draw a boxplot of these data. What is the shape of the distribution shown by the boxplot? Which features of the boxplot led you to this conclusion? Are any observations unusually small or large?

2.6 Spotting suspected outliers and modified boxplots*

Look again at the stemplot of travel times to work in New York in Example 2.3. The five-number summary for this distribution is

$$5 \quad 15 \quad 22.5 \quad 42.5 \quad 85$$

How shall we describe the variability of this distribution? The smallest and largest observations are extremes that don't describe the variability of the majority of the data. The distance between the quartiles (the range of the center half of the data) is a more resistant measure of variability. This distance is called the *interquartile range.*

The Interquartile Range (*IQR*)

The **interquartile range (*IQR*)** is the distance between the first and third quartiles,

$$IQR = Q_3 - Q_1$$

CAUTION

For our data on New York travel times, $IQR = 42.5 - 15 = 27.5$ minutes. However, *no single numerical measure of variability, such as IQR, is very useful for describing skewed distributions.* The two sides of a skewed distribution

*This short section is optional.

have different variability, so one number can't summarize them. That's why we give the full five-number summary. The interquartile range is mainly used as the basis for a rule of thumb for identifying suspected outliers.

The 1.5 × *IQR* Rule For Outliers

Call an observation a suspected outlier if it falls more than 1.5 × *IQR* above the third quartile or below the first quartile.

EXAMPLE 2.6 Using the 1.5 × *IQR* Rule

NYTRAVEL

For the New York travel time data, *IQR* = 27.5 and

$$1.5 \times IQR = 1.5 \times 27.5 = 41.25$$

Any values not falling between

$$Q_1 - (1.5 \times IQR) = 15.0 - 41.25 = -26.25 \quad \text{and}$$
$$Q_3 + (1.5 \times IQR) = 42.5 + 41.25 = 83.75$$

are flagged as suspected outliers. Look again at the stemplot in Example 2.3: The only suspected outlier is the longest travel time, 85 minutes. The 1.5 × *IQR* rule suggests that the three next-longest travel times (60 and 65 minutes) are just part of the long right tail of this skewed distribution. ■

HOW MUCH IS THAT HOUSE WORTH?

The town of Manhattan, Kansas, is sometimes called "the Little Apple" to distinguish it from that other Manhattan, "the Big Apple." A few years ago, a house there appeared in the county appraiser's records valued at $200,059,000. That would be quite a house even on Manhattan Island. As you might guess, the entry was wrong: the true value was $59,500. But before the error was discovered, the county, the city, and the school board had based their budgets on the total appraised value of real estate, which the one outlier jacked up by 6.5%. It can pay to spot outliers before you trust your data.

In a modified boxplot, which is provided by many software packages, the suspected outliers are identified in the boxplot with a special plotting symbol such as an asterisk (*). Comparing Figure 2.2 with Figure 2.1, we see that the largest observation from New York is flagged as an outlier. The line beginning at the third quartile no longer extends to the maximum but now ends at 65, which is the largest observation from New York that is not identified as an outlier. Figure 2.2 also displays the modified boxplots horizontally rather than vertically, an option available in some software packages which does not change the interpretation of the plot. Finally, the 1.5 × *IQR* rule is not a replacement for looking at the data. It is most useful when large volumes of data are processed automatically.

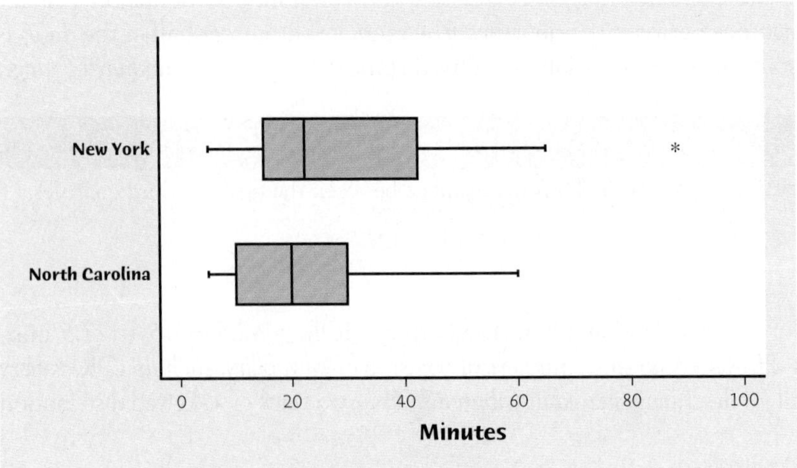

FIGURE 2.2
Horizontal modified boxplots comparing the travel times to work of samples of workers in North Carolina and New York.

Apply Your Knowledge

2.8 Travel Time to Work. In Example 2.1, we noted the influence of one long travel time of 60 minutes in our sample of 15 North Carolina workers. Does the $1.5 \times IQR$ rule identify this travel time as a suspected outlier? DATA NCTRAVEL

2.9 Fuel Economy for Midsize Cars. Exercise 2.7 gives the estimated miles per gallon (mpg) for city driving for the 205 cars classified as midsize in 2014. In that exercise we noted that several of the mpg values were unusually large. Which of these are suspected outliers by the $1.5 \times IQR$ rule? Although outliers can be produced by errors or incorrectly recorded observations, they are often observations that differ from the others in some particular way. In this case, the cars producing the high outliers share a common feature. What do you think that is? DATA MIDCARS

2.7 Measuring variability: the standard deviation

The five-number summary is not the most common numerical description of a distribution. That distinction belongs to the combination of the mean to measure center and the *standard deviation* to measure variability. The standard deviation and its close relative, the *variance*, measure variability by looking at how far the observations are from their mean.

The Standard Deviation s

The **variance s^2** of a set of observations is an average of the squares of the deviations of the observations from their mean. In symbols, the variance of n observations $x_1, x_2, \ldots, x_n$ is

$$s^2 = \frac{(x_1 - \bar{x})^2 + (x_2 - \bar{x})^2 + \cdots + (x_n - \bar{x})^2}{n - 1}$$

or, more compactly,

$$s^2 = \frac{1}{n-1}\sum (x_i - \bar{x})^2$$

The **standard deviation s** is the square root of the variance s^2:

$$s = \sqrt{\frac{1}{n-1}\sum (x_i - \bar{x})^2}$$

In practice, use software or your calculator to obtain the standard deviation from keyed-in data. Doing an example step-by-step will help you understand how the variance and standard deviation work, however.

EXAMPLE 2.7 Calculating the Standard Deviation

DATA SATMATH

Georgia Southern University had 3219 students with regular admission in its freshman class of 2013. For each student, data are available on their SAT and ACT scores (if taken), high school GPA, and the college within the university to which they were admitted.[8] In Exercise 3.49 (page 98), the full data set for the SAT

Mathematics scores will be examined. Here are the first five observations from that data set:

490 580 450 570 650

We will compute $\bar{x}$ and s for these students. First find the mean:

$$\bar{x} = \frac{490 + 580 + 450 + 570 + 650}{5}$$

$$= \frac{2740}{5} = 548$$

Figure 2.3 displays the data as points above the number line, with their mean marked by an asterisk (*). The arrows mark two of the deviations from the mean. The deviations show how variable the data are about their mean. They are the starting point for calculating the variance and the standard deviation.

Observations x_i	Deviations $x_i - \bar{x}$	Squared deviations $(x_i - \bar{x})^2$
490	490 − 548 = −58	$(-58)^2$ = 3,364
580	580 − 548 = 32	32^2 = 1,024
450	450 − 548 = −98	$(-98)^2$ = 9,604
570	570 − 548 = 22	22^2 = 484
650	650 − 548 = 102	102^2 = 10,404
	sum = 0	sum = 24,880

The variance is the sum of the squared deviations divided by one less than the number of observations:

$$s^2 = \frac{1}{n-1} \sum (x_i - \bar{x})^2 = \frac{24{,}880}{4} = 6220$$

The standard deviation is the square root of the variance:

$$s = \sqrt{6220} = 78.87$$ ■

FIGURE 2.3
SAT Critical Reading scores for five students, with their mean (*) and the deviations of two observations from the mean shown, for Example 2.7.

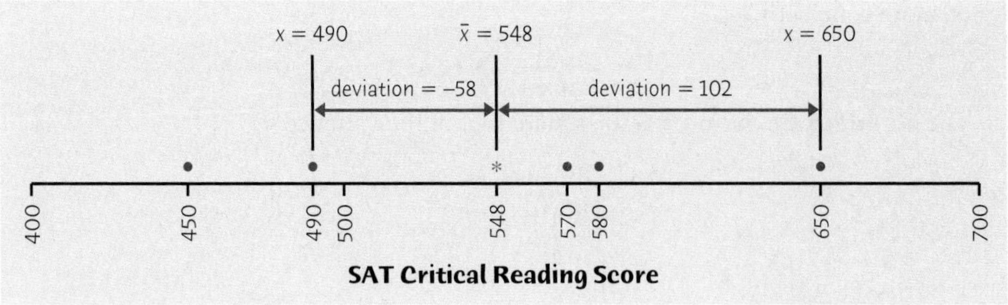

Notice that the "average" in the variance s^2 divides the sum by one fewer than the number of observations, that is, $n - 1$ rather than n. The reason is that the deviations $x_i - \bar{x}$ always sum to exactly 0, so that knowing $n - 1$ of them determines the last one. Only $n - 1$ of the squared deviations can vary freely, and we average by dividing the total by $n - 1$. The number $n - 1$ is called the **degrees of freedom** of the variance or standard deviation. Some calculators offer a choice between dividing by n and dividing by $n - 1$, so be sure to use $n - 1$.

degrees of freedom

More important than the details of hand calculation are the properties that determine the usefulness of the standard deviation:

- s measures *variability about the mean* and should be used only when the mean is chosen as the measure of center.
- s is *always zero or greater than zero.* $s = 0$ only when there is no variability. This happens only when all observations have the same value. Otherwise, $s > 0$. As the observations become more variable about their mean, s gets larger.
- s has the *same units of measurement as the original observations.* For example, if you measure weight in kilograms, both the mean $\bar{x}$ and the standard deviation s are also in kilograms. This is one reason to prefer s to the variance s^2, which would be in squared kilograms.
- Like the mean $\bar{x}$, s is *not resistant.* A few outliers can make s very large.

CAUTION

The use of squared deviations renders s even more sensitive than $\bar{x}$ to a few extreme observations. For example, the standard deviation of the travel times for the 15 North Carolina workers in Example 2.1 is 15.23 minutes. (Use your calculator or software to verify this.) If we omit the high outlier, the standard deviation drops to 11.56 minutes.

If you feel that the importance of the standard deviation is not yet clear, you are right. We will see in Chapter 3 that the standard deviation is the natural measure of variability for a very important class of symmetric distributions, the Normal distributions. The usefulness of many statistical procedures is tied to distributions of particular shapes. This is certainly true of the standard deviation.

2.8 Choosing measures of center and variability

We now have a choice between two descriptions of the center and variability of a distribution: the five-number summary, or $\bar{x}$ and s. Because $\bar{x}$ and s are sensitive to extreme observations, they can be misleading when a distribution is strongly skewed or has outliers. In fact, because the two sides of a skewed distribution have different variability, no single number describes the variability well. The five-number summary, with its two quartiles and two extremes, does a better job.

Choosing a Summary

The five-number summary is usually better than the mean and standard deviation for describing a skewed distribution or a distribution with strong outliers. Use $\bar{x}$ and s only for reasonably symmetric distributions that are free of outliers.

CAUTION

Outliers can greatly affect the values of the mean $\bar{x}$ and the standard deviation s, the most common measures of center and variability. Many more elaborate statistical procedures also can't be trusted when outliers are present. *Whenever you find outliers in your data, try to find an explanation for them.* Sometimes the explanation is as simple as a typing error, such as typing 10.1 as 101; if this is the case, correct the typing error. Sometimes a measuring device broke down or a subject gave a frivolous response, like the student in a class survey who claimed to study 30,000 minutes per night. (Yes, that really happened.) In all these cases, you can simply remove the outlier from your data. When outliers are "real data," like the long travel times of some New York workers, you should choose statistical methods that are not greatly disturbed by the outliers. For example, use the five-number summary rather than $\bar{x}$ and s to describe a distribution with extreme outliers. We will encounter other examples later in the book.

CAUTION *Remember that a graph gives the best overall picture of a distribution.* If data have been entered into a calculator or statistical program, it is very simple and quick to create several graphs to see all the different features of a distribution. Numerical measures of center and variability report specific facts about a distribution, but they do not describe its entire shape. Numerical summaries do not disclose the presence of multiple peaks or clusters, for example. Exercise 2.11 shows how misleading numerical summaries can be. **Always plot your data.**

LaunchPad Online Resources

- The **Snapshots video**, *Summarizing Quantitative Data,* provides an overview of the need for measures of center and variability as well as some details of the computations.
- The **StatClips Examples video**, *Summaries of Quantitative Data Example C,* gives the details for the computation of the mean, median, and standard deviation in a small example. You can verify the computations along with the video, either by hand or using your technology.
- The **StatClips Examples videos**, *Basic Principles of Exploring Data Example B* and *Basic Principles of Exploring Data Example C,* emphasize the need to examine outliers and understand them, rather than simply discarding observations that don't seem to fit.

Apply Your Knowledge

Photo Researchers/Getty Images

2.10 $\bar{x}$ and s by Hand. Radon is a naturally occurring gas and is the second leading cause of lung cancer in the United States.[9] It comes from the natural breakdown of uranium in the soil and enters buildings through cracks and other holes in the foundations. Found throughout the United States, levels vary considerably from state to state. Several methods can reduce the levels of radon in your home, and the Environmental Protection Agency recommends using one of these if the measured level in your home is above 4 picocuries per liter. Four readings from Franklin County, Ohio, where the county average is 8.4 picocuries per liter, were 6.2, 12.8, 7.6, and 15.4.

(a) Find the mean step-by-step. That is, find the sum of the four observations and divide by 4.

(b) Find the standard deviation step-by-step. That is, find the deviation of each observation from the mean, square the deviations, then obtain the variance and the standard deviation. Example 2.7 shows the method.

(c) Now enter the data into your calculator and use the mean and standard deviation buttons to obtain $\bar{x}$ and s. Do the results agree with your hand calculations?

2.11 $\bar{x}$ and s Are Not Enough. The mean $\bar{x}$ and standard deviation s measure center and variability but are not a complete description of a distribution. Data sets with different shapes can have the same mean and standard deviation. To demonstrate this fact, use your calculator to find $\bar{x}$ and s for these two small data sets. Then make a stemplot of each and comment on the shape of each distribution. DATASET2

Data A:	9.14	8.14	8.74	8.77	9.26	8.10	6.13	3.10	9.13	7.26	4.74
Data B:	6.58	5.76	7.71	8.84	8.47	7.04	5.25	5.56	7.91	6.89	12.50

2.12 Choose a Summary. The shape of a distribution is a rough guide to whether the mean and standard deviation are a helpful summary of center and variability. For which of the following distributions would $\bar{x}$ and s be useful? In each case, give a reason for your decision.

(a) Percents of high school graduates in the states taking the SAT, Figure 1.8 (page 27)

(b) Iowa Tests scores, Figure 1.7 (page 27)

(c) New York travel times, Figure 2.1 (page 54)

2.9 Using technology

Although a calculator with "two-variable statistics" functions will do the basic calculations we need, more elaborate tools are helpful. Graphing calculators and computer software will do calculations and make graphs as you command, freeing you to concentrate on choosing the right methods and interpreting your results. Figure 2.4 displays output describing the travel times to work of 20 people in New York State (Example 2.3). Can you find $\bar{x}$, s, and the five-number summary in each output? The big message of this section is: *Once you know what to look for, you can read output from any technological tool.*

The displays in Figure 2.4 come from a Texas Instruments graphing calculator, the Minitab, CrunchIt!, and JMP statistical programs, and the Microsoft Excel spreadsheet program. Minitab and JMP allow you to choose what descriptive measures you want, whereas the descriptive measures in the CrunchIt! output are provided by default. Excel and the calculator give some things we don't need. Just ignore the extras. Excel's "Descriptive Statistics" menu item doesn't give the quartiles. We used the spreadsheet's separate quartile function to get Q_1 and Q_3.

EXAMPLE 2.8 What Is the Third Quartile?

CAUTION

In Example 2.5, we saw that the quartiles of the New York travel times are $Q_1 = 15$ and $Q_3 = 42.5$. Look at the output displays in Figure 2.4. The calculator, CrunchIt!, and Excel agree with our work. Minitab and JMP say that $Q_3 = 43.75$. What happened? *There are several rules for finding the quartiles. Some calculators and software use rules that give results different from ours for some sets of data.* This is true of Minitab, JMP, and also Excel, though Excel agrees with our work in this example. Results from the various rules are always close to each other, so the differences are never important in practice. Our rule is the simplest for hand calculation. ■

Texas Instruments Graphing Calculator

```
1-Var Stats
 x̄=31.25
 Σx=625
 Σx²=28625
 Sx=21.8773495
 σx=21.32340264
↓n=20
```

```
1-Var Stats
↑n=20
 minX=5
 Q1=15
 Med=22.5
 Q3=42.5
 maxX=85
```

Minitab

Session

Descriptive Statistics: NYtime

variable	Total Count	Mean	StDev	Variance	Minimum	Q1	Median	Q3	Maximum
NYtime	20	31.25	21.88	478.62	5.00	15.00	22.50	43.75	85.00

CrunchIt!

Results - Descriptive Statistics

Export ▾

	n	Sample Mean	Standard Deviation	Min	Q1	Median	Q3	Max
Minutes	20	31.25	21.88	5	15	22.50	42.50	85

Microsoft Excel

Book1

	A	B	C	D
1	minutes			
2				
3	Mean	31.25		
4	Standard Error	4.891924064		
5	Median	22.5	QUARTILE(A2:A21,1)	15
6	Mode	15	QUARTILE(A2:A21,3)	42.5
7	Standard Deviation	21.8773495		
8	Sample Variance	478.6184211		
9	Kurtosis	0.329884126		
10	Skewness	1.040110836		
11	Range	80		
12	Minimum	5		
13	Maximum	85		
14	Sum	625		
15	Count	20		

Sheet4 / Sheet1 / Sheet2 / Sheet

JMP Output

Distributions

NYtime

Quantiles

100%	maximum	85
75%	quartile	43.75
50%	median	22.5
25%	quartile	15
0%	minimum	5

Summary Statistics

Mean	31.25
Std Dev	21.877349
N	20

FIGURE 2.4
Output from a graphing calculator, three statistical software packages, and a spreadsheet program describing the data on travel times to work in New York State.

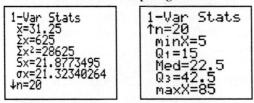

LaunchPad Online Resources

There are online resources to help you use technology.

- The software CrunchIt! is available online, as are CrunchIt! help videos.
- The **Video Technology Manuals** for Minitab, the TI graphing calculator, Excel, JMP, and SPSS provide explicit instructions for producing output similar to that provided in Figure 2.4 for the particular technology you are using.

2.10 Organizing a statistical problem

Most of our examples and exercises have aimed to help you learn basic tools (graphs and calculations) for describing and comparing distributions. You have also learned principles that guide use of these tools, such as "start with a graph" and "look for the overall pattern and striking deviations from the pattern." The data you work with are not just numbers—they describe specific settings such as water depth in the Everglades or travel time to work. Because data come from a specific setting, the final step in examining data is a *conclusion for that setting*. Water depth in the Everglades has a yearly cycle that reflects Florida's wet and dry seasons. Travel times to work are generally longer in New York than in North Carolina.

Let's return to the on-time high school graduation rates discussed in Example 1.4, page 21. We know from the example that the on-time graduation rates vary from 59% in the District of Columbia to 88% in Iowa, with a median of 80%. State graduation rates are related to many factors, and in a statistical problem we often try to explain the differences or variation in a variable such as graduation rate by some of these factors. For example, do states with lower household incomes tend to have lower high school graduation rates? Or, do the states in some regions of the country tend to have lower high school graduation rates than in other regions?

As you learn more statistical tools and principles, you will face more complex statistical problems. Although no framework accommodates all the varied issues that arise in applying statistics to real settings, we find the following four-step thought process gives useful guidance. In particular, the first and last steps emphasize that statistical problems are tied to specific real-world settings and therefore involve more than doing calculations and making graphs.

Organizing a Statistical Problem: A Four-Step Process

STATE: What is the practical question, in the context of the real-world setting?

PLAN: What specific statistical operations does this problem call for?

SOLVE: Make the graphs and carry out the calculations needed for this problem.

CONCLUDE: Give your practical conclusion in the setting of the real-world problem.

To help you master the basics, many exercises will continue to tell you what to do—make a histogram, find the five-number summary, and so on. Real statistical problems don't come with detailed instructions. From now on, especially in the later chapters of the book, you will meet some exercises that are more realistic. Use the four-step process as a guide to solving and reporting these problems. They are marked with the four-step icon, as the following example illustrates.

EXAMPLE 2.9 Comparing Graduation Rates

GRADRATE

STATE: Federal law requires all states in the United States to use a common computation of on-time high school graduation rates beginning with the 2010–11 school year. Previously, states chose one of several computation methods that gave answers that could differ by more than 10%. This common computation allows for meaningful comparison of graduation rates between the states.

We know from Table 1.1 (page 22) that the on-time high school graduation rates varied from 59% in the District of Columbia to 88% in Iowa. The U.S. Census Bureau divides the 50 states and the District of Columbia into four geographical regions: the Northeast (NE), Midwest (MW), South (S), and West (W). The region for each state is included in Table 1.1. Do the states in the four regions of the country display distinct distributions of graduation rates? How do the mean graduation rates of the states in each of these regions compare?

PLAN: Use graphs and numerical descriptions to describe and compare the distributions of on-time high school graduation rates of the states in the four regions of the United States.

SOLVE: We might use boxplots to compare the distributions, but stemplots preserve more detail and work well for data sets of these sizes. Figure 2.5 displays the stemplots with the stems lined up for easy comparison. The stems have been split to better display the distributions. The stemplots overlap, and some care is needed when comparing the four stemplots as the sample sizes differ, with some stemplots having more leaves than others. None of the plots shows strong skewness, although the South has one low observation that stands apart from the others with this choice of stems. The states in the Northeast and Midwest have distributions that are similar to each other, as do those in the South and West. The graduation rates tend to be higher for the states in the Northeast and Midwest and more variable for the states in the South and West. With little skewness and no serious outliers, we report $\bar{x}$ and s as our summary measures of center and variability of the distribution of the on-time graduation rates of the states in each region:

Region	Mean	Standard Deviation
Midwest	82.92	4.25
Northeast	82.56	3.47
South	75.93	7.36
West	73.58	6.73

Midwest		Northeast		South		West	
8	66678	8	67	8	66	8	
8	01334	8	33334	8	123	8	002
7	7	7	77	7	5688	7	6668
7	4	7		7	1124	7	4
6		6		6	7	6	88
6		6		6		6	23
5		5		5	9	5	

FIGURE 2.5 Stemplots comparing the distributions of graduation rates for the four census regions from Table 1.1, for Example 2.9.

CONCLUDE: The table of summary statistics confirms what we see in the stemplots. The states in the Midwest and Northeast are quite similar to each other, as are those in the South and West. The states in the Midwest and Northeast have a higher mean graduation rate as well as a smaller standard deviation than those in the South and West. ■

It is important to remember that the individuals in Example 2.9 are the states. For example, the mean of 82.56 is the mean of the on-time graduation rates for the nine Northeastern states, and the standard deviation tells us how much these state rates vary about this mean. However, the mean of these nine states *is not* the same as the graduation rate for all high school students in the Northeast, unless the states have the same number of high school graduates. The graduation rate for all high school students in the Northeast would be a *weighted* average of the state rates, with the larger states receiving more weight. For example, since New York is the most populous state in the Northeast and also has the lowest graduation rate, we would expect the graduation rate of all high school students in the Northeast to be lower than 82.56, as New York would pull down the overall graduation rate. See Exercise 2.37 for a similar example.

LaunchPad Online Resources

The StatBoards video, *The 4-Step Process*, provides an additional example of the four-step process. Many exercises throughout the book are marked with the four-step icon and this process will be necessary in solving these problems. If you are unclear about any of the steps, you should watch the video to become more comfortable with the process.

Apply Your Knowledge

2.13 Logging in the Rain Forest. "Conservationists have despaired over destruction of tropical rain forest by logging, clearing, and burning." These words begin a report on a statistical study of the effects of logging in Borneo.[10] Charles Cannon of Duke University and his co-workers compared forest plots that had never been logged (Group 1) with similar plots nearby that had been logged one year earlier (Group 2) and eight years earlier (Group 3). All plots were 0.1 hectare in area. Here are the counts of trees for plots in each group: DATA LOGGING

Group 1:	27	22	29	21	19	33	16	20	24	27	28	19
Group 2:	12	12	15	9	20	18	17	14	14	2	17	19
Group 3:	18	4	22	15	18	19	22	12	12			

To what extent has logging affected the count of trees? Follow the four-step process in reporting your work.

2.14 Diplomatic Scofflaws. Until Congress allowed some enforcement in 2002, the thousands of foreign diplomats in New York City could freely violate parking laws. Two economists looked at the number of unpaid parking tickets per diplomat over a five-year period ending when enforcement reduced the problem.[11] They concluded that large numbers of unpaid tickets indicated a "culture of corruption" in a country and lined up well with more elaborate measures of corruption. The data set for 145 countries is too large to print here, but look at the data file on the text website. The first 32 countries in the list (Australia to Trinidad and Tobago) are classified by the World Bank as "developed." The remaining countries (Albania to Zimbabwe) are "developing." The World Bank classification is based only on national income and does not take into account measures of social development. DATA SCOFFLAW

Give a full description of the distribution of unpaid tickets for both groups of countries and identify any high outliers. Compare the two groups. Does national income alone do a good job of distinguishing countries whose diplomats do and do not obey parking laws?

© James Leynse/Corbis

CHAPTER 2 SUMMARY

Chapter Specifics

- A numerical summary of a distribution should report at least its **center** and its **variability**.
- The **mean** $\bar{x}$ and the **median M** describe the center of a distribution in different ways. The mean is the arithmetic average of the observations, and the median is the midpoint of the values.
- When you use the median to indicate the center of the distribution, describe its variability by giving the **quartiles**. The **first quartile**, Q_1, has one-fourth of the observations below it, and the **third quartile**, Q_3, has three-fourths of the observations below it.
- The **five-number summary** consisting of the median, the quartiles, and the smallest and largest individual observations provides a quick overall description of a distribution. The median describes the center, and the quartiles and extremes show the variability.
- **Boxplots** based on the five-number summary are useful for comparing several distributions. The box spans the quartiles and shows the variability of the central half of the distribution. The median is marked within the box. Lines extend from the box to the extremes and show the full variability of the data.
- The **variance** s^2 and especially its square root, the **standard deviation** s, are common measures of variability about the mean as center. The standard deviation s is zero when there is no variability and gets larger as the variability increases.
- A **resistant measure** of any aspect of a distribution is relatively unaffected by changes in the numerical value of a small proportion of the total number of observations, no matter how large these changes are. The median and quartiles are resistant, but the mean and the standard deviation are not.
- The mean and standard deviation are good descriptions for symmetric distributions without outliers. They are most useful for the Normal distributions introduced in the next chapter. The five-number summary is a better description for skewed distributions.
- Numerical summaries do not fully describe the shape of a distribution. Always plot your data.
- A statistical problem has a real-world setting. You can organize many problems using the following four steps: **State, Plan, Solve,** and **Conclude**.

Link It

In this chapter we have continued our study of exploratory data analysis. Graphs are an important visual tool for organizing and identifying patterns in data. They give a fairly complete description of a distribution, although for many problems the important information in your data can be described by a few numbers. These numerical summaries can be useful for describing a single distribution as well as for comparing the distributions from several groups of observations.

Two important features of a distribution are the center and the variability. For distributions that are approximately symmetric without outliers, the mean and standard deviation are important numeric summaries for describing and comparing distributions. But if the distribution is not symmetric and/or has outliers, the five-number summary often provides a better description.

The boxplot gives a picture of the five-number summary that is useful for a simple comparison of several distributions. Remember that the boxplot is based only on the five-number summary and does not have any information beyond these five numbers. Certain features of a distribution that are revealed in histograms and stemplots will not be evident from a boxplot alone. These include gaps in the data and the presence of several peaks. You must be careful when reducing a distribution to a few numbers to make sure that important information has not been lost in the process.

LaunchPad Online Resources

If you are having difficulty with any of the sections of this chapter, these online resources should help prepare you to solve the exercises at the end of this chapter.

- **StatTutor** starts with a video review of each section and asks a series of questions to check your understanding.
- **LearningCurve** provides you with a series of questions about the chapter geared to your level of understanding.

CHECK YOUR SKILLS

2.15 The respiratory system can be a limiting factor in maximal exercise performance. Researchers from the United Kingdom studied the effect of two breathing frequencies on both performance times and several physiological parameters in swimming.[12] Subjects were 10 male collegiate swimmers. Here are their times in seconds to swim 200 meters at 90% of race pace when breathing every second stroke in front crawl swimming: DATA SWIMTIME

151.6 165.1 159.2 163.5 174.8 173.2 177.6
174.3 164.1 171.4

The mean of these data is

(a) 165.10. (b) 167.48. (c) 168.25.

2.16 The median of the data in Exercise 2.15 is

(a) 167.48. (b) 168.25. (c) 174.00.

2.17 The five-number summary of the data in Exercise 2.15 is

(a) 151.6, 163.5, 167.48, 174.3, 177.6.
(b) 151.6, 163.5, 168.25, 174.3, 177.6.
(c) 151.6, 159.2, 168.25, 174.3, 177.6.

2.18 If a distribution is skewed to the left,

(a) the mean is less than the median.
(b) the mean and median are equal.
(c) the mean is greater than the median.

2.19 What percent of the observations in a distribution are greater than the third quartile?

(a) 25% (b) 50% (c) 75%

2.20 To make a boxplot of a distribution, you must know

(a) all the individual observations.
(b) the mean and the standard deviation.
(c) the five-number summary.

2.21 The standard deviation of the 10 swim times in Exercise 2.15 (use your calculator) is about

(a) 7.4. (b) 7.8. (c) 8.2.

2.22 What are all the values that a standard deviation s can possibly take?

(a) $0 \leq s$ (b) $0 \leq s \leq 1$ (c) $-1 \leq s \leq 1$

2.23 The correct units for the standard deviation in Exercise 2.21 are

(a) no units—it's just a number. (b) seconds.
(c) seconds squared.

2.24 Which of the following is least affected if an extreme high outlier is added to your data?

(a) The median (b) The mean
(c) The standard deviation

CHAPTER 2 EXERCISES

2.25 **Incomes of college grads.** According to the U.S. Census Bureau's Current Population Survey, the mean and median 2012 income of people at least 25 years old who had a bachelor's degree but no higher degree were \$50,281 and \$62,597.[13] Which of these numbers is the mean and which is the median? Explain your reasoning.

2.26 **Saving for retirement.** Retirement seems a long way off, and we need money now, so saving for retirement is hard. Once every three years, the Board of Governors of the Federal Reserve System collects data on household assets and liabilities through the Survey of Consumer Finances (SCF). The most recent such survey was conducted in 2010, and the survey results were released to the public in April 2013. The survey presents data on retirement assets, which include defined contribution and Individual Retirement Account (IRA) balances. For married households the mean value per household is \$123,968, but the median value is just \$10,000. For single households, the mean is \$33,585, and the median is \$0.[14] What explains the differences between the two measures of center, both for married and single households? What does a median of \$0 say about the percentage of single households with retirement assets?

2.27 **University endowments.** The National Association of College and University Business Officers collects data on college endowments. In 2013, its report included the endowment values of 849 colleges and universities in the United States and Canada. When the endowment values are arranged in order, what are the locations of the median and the quartiles in this ordered list?

2.28 **Pulling wood apart.** Exercise 1.46 (page 45) gives the breaking strengths of 20 pieces of Douglas fir. DATA WOOD

(a) Give the five-number summary of the distribution of breaking strengths.

(b) Here is a stemplot of the data rounded to the nearest hundred pounds. The stems are thousands of pounds, and the leaves are hundreds of pounds.

23	0
24	1
25	
26	5
27	
28	7
29	
30	259
31	399
32	033677
33	0237

The stemplot shows that the distribution is skewed to the left. Does the five-number summary show the skew? Remember that only a graph gives a clear picture of the shape of a distribution.

2.29 Comparing graduation rates. An alternative presentation to compare the graduation rates in Table 1.1 (page 22) by region of the country reports five-number summaries and uses boxplots to display the distributions. Do the boxplots fail to reveal any important information visible in the stemplots of Figure 2.5? Which plots make it simpler to compare the regions? Why? DATA GRADRATE

2.30 How much fruit do adolescent girls eat? Figure 1.14 (page 39) is a histogram of the number of servings of fruit per day claimed by 74 seventeen-year-old girls.

(a) With a little care, you can find the median and the quartiles from the histogram. What are these numbers? How did you find them?

(b) You can also find the mean number of servings of fruit claimed per day from the histogram. First use the information in the histogram to compute the sum of the 74 observations, and then use this to compute the mean. What is the relationship between the mean and median? Is this what you expected?

(c) In general, you cannot find the exact values of the median, quartiles, or mean from the histogram. What is special about the histogram of the number of servings of fruit that allows you to do this?

2.31 Guinea pig survival times. Here are the survival times in days of 72 guinea pigs after they were injected with infectious bacteria in a medical experiment.[15] Survival times, whether of machines under stress or cancer patients after treatment, usually have distributions that are skewed to the right. DATA GUINPIGS

43	45	53	56	56	57	58	66	67	73	74	79
80	80	81	81	81	82	83	83	84	88	89	91
91	92	92	97	99	99	100	100	101	102	102	102
103	104	107	108	109	113	114	118	121	123	126	128
137	138	139	144	145	147	156	162	174	178	179	184
191	198	211	214	243	249	329	380	403	511	522	598

(a) Graph the distribution and describe its main features. Does it show the expected right-skew?

(b) Which numerical summary would you choose for these data? Calculate your chosen summary. How does it reflect the skewness of the distribution?

2.32 Maternal age at childbirth. How old are women when they have their first child? Here is the distribution of the age of the mother for all firstborn children in the United States in 2012:[16]

AGE	COUNT	AGE	COUNT
10 to 14 years	3,578	30 to 34 years	299,857
15 to 19 years	251,022	35 to 39 years	106,892
20 to 24 years	461,553	40 to 44 years	24,251
25 to 29 years	421,704	45 to 49 years	1,952

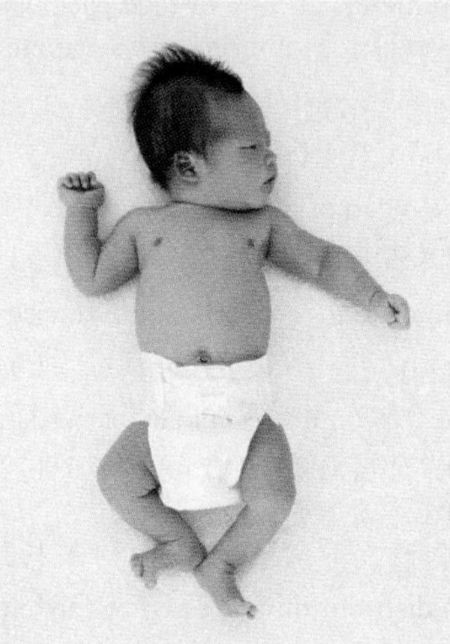

Photodisc/Getty Images

The number of firstborn children to mothers under 10 or over 50 years of age represent a negligible percentage of all first births, and are not included in the table.

(a) For comparison with other years and with other countries, we prefer a histogram of the *percents* in each age class rather than the counts. Explain why.

(b) How many babies were there?

(c) Make a histogram of the distribution, using percents on the vertical scale. Using this histogram, describe the distribution of the age at which women have their first child.

(d) What are the locations of the median and quartiles in the ordered list of all maternal ages? In which age classes do the median and quartiles fall?

2.33 More on Nintendo and laparoscopic surgery. In Exercise 1.38 (page 43) you examined the improvement in times to complete a virtual gall bladder removal for those with and without four weeks of Nintendo Wii™ training. The most common methods for formal comparison of two groups use $\bar{x}$ and s to summarize the data. DATA NINTENDO

(a) What kinds of distributions are best summarized by $\bar{x}$ and s? Do you think these summary measures are appropriate in this case?

(b) In the control group, one subject improved their time by 229 seconds. How much does removing this observation change $\bar{x}$ and s for the control group? You will need to compute $\bar{x}$ and s for the control group, both with and without the high outlier.

(c) Compute the median for the control group with and without the high outlier. What does this show about the resistance of the median and $\bar{x}$?

2.34 Making resistance visible. In the *Mean and Median* applet, place three observations on the line by clicking below it: two close together near the center of the line, and one somewhat to the right of these two.

APPLET

(a) Pull the single rightmost observation out to the right. (Place the cursor on the point, hold down a mouse button, and drag the point.) How does the mean behave? How does the median behave? Explain briefly why each measure acts as it does.

(b) Now drag the single rightmost point to the left as far as you can. What happens to the mean? What happens to the median as you drag this point past the other two (watch carefully)?

2.35 Behavior of the median. Place five observations on the line in the *Mean and Median* applet by clicking below it.

APPLET

(a) Add one additional observation *without changing the median*. Where is your new point?

(b) Use the applet to convince yourself that when you add yet another observation (there are now seven in all), the median does not change, no matter where you put the seventh point. Explain why this must be true.

2.36 Never on Sunday: also in Canada? Exercise 1.5 (page 20) gives the number of births in the United States on each day of the week during an entire year. The boxplots in Figure 2.6 are based on more detailed data from Toronto, Canada: the number of births on each of the 365 days in a year, grouped by day of the week.[17] Based on these plots, compare the day-of-the-week distributions using shape, center, and variability. Summarize your findings.

2.37 Thinking about means. Table 1.2 (page 24) gives the percent of foreign-born residents in each of the states. For the nation as a whole, 13.0% of residents are foreign-born. Find the mean of the 51 entries in Table 1.2. It is *not* 13.0%. Explain carefully why this happens. (*Hint:* The states with the largest populations are California, Texas, New York, and Florida. Look at their entries in Table 1.2.) DATA FOREIGN

2.38 Thinking about medians. A report says that "the median credit card debt of American households is zero." We know that many households have large amounts of credit card debt. In fact, the mean household credit card debt is close to $8000. Explain how the median debt can nonetheless be zero.

2.39 Thinking about means and medians. In 2012, approximately 5% of employed wage and salary workers were being paid hourly rates at the federal minimum wage level. Would federal legislation to increase the minimum wage have a greater effect on the mean or the median income of all workers. Explain your answer.

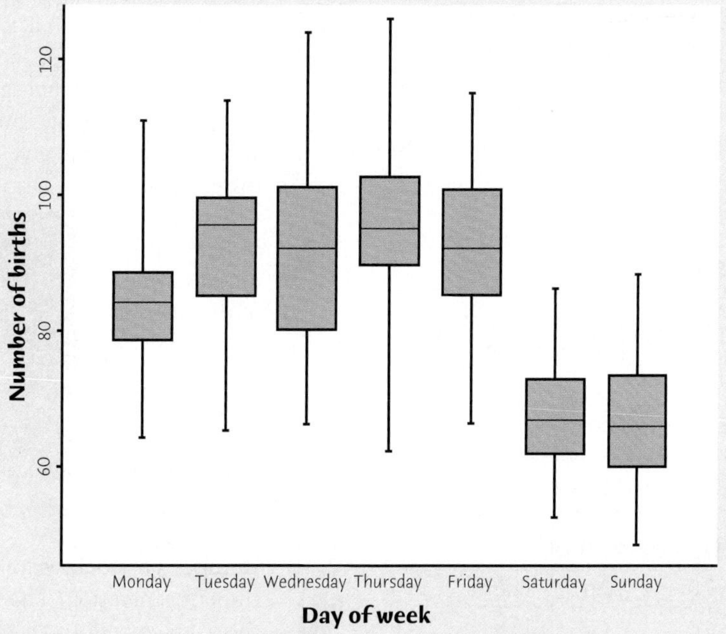

FIGURE 2.6
Boxplots of the distributions of numbers of births in Toronto, Canada, on each day of the week during a year, for Exercise 2.36.

2.40 **A standard deviation contest.** This is a standard deviation contest. You must choose four numbers from the whole numbers 0 to 10, with repeats allowed.

(a) Choose four numbers that have the smallest possible standard deviation.
(b) Choose four numbers that have the largest possible standard deviation.
(c) Is more than one choice possible in either (a) or (b)? Explain.

2.41 **You create the data.** Create a set of seven numbers (repeats allowed) that have the five-number summary

$$\text{Minimum} = 4 \quad Q_1 = 7 \quad M = 12 \quad Q_3 = 14 \quad \text{Maximum} = 19$$

There is more than one set of seven numbers with this five-number summary. What must be true about the seven numbers to have this five-number summary?

2.42 **You create the data.** Give an example of a small set of data for which the mean is smaller than the first quartile.

2.43 **Adolescent obesity.** Adolescent obesity is a serious health risk affecting more than 5 million young people in the United States alone. Laparoscopic adjustable gastric banding has the potential to provide a safe and effective treatment. Fifty adolescents between 14 and 18 years old with a body mass index (BMI) higher than 35 were recruited from the Melbourne, Australia, community for the study.[18] Twenty-five were randomly selected to undergo gastric banding, and the remaining twenty-five were assigned to a supervised lifestyle intervention program involving diet, exercise, and behavior modification. All subjects were followed for two years. Here are the weight losses in kilograms for the subjects who completed the study: GASTRIC

GASTRIC BANDING

35.6	81.4	57.6	32.8	31.0	37.6
36.5	−5.4	27.9	49.0	64.8	39.0
43.0	33.9	29.7	20.2	15.2	41.7
53.4	13.4	24.8	19.4	32.3	22.0

LIFESTYLE INTERVENTION

6.0	2.0	−3.0	20.6	11.6	15.5
−17.0	1.4	4.0	−4.6	15.8	34.6
6.0	−3.1	−4.3	−16.7	−1.8	−12.8

(a) In the context of this study, what do the negative values in the data set mean?
(b) Give a graphical comparison of the weight loss distribution for both groups using side-by-side boxplots. Provide appropriate numerical summaries for the two distributions and identify any high outliers in either group. What can you say about the effects of gastric banding versus lifestyle intervention on weight loss for the subjects in this study?
(c) The measured variable was weight loss in kilograms. Would two subjects with the same weight loss always have similar benefits from a weight-reduction program? Does it depend on their initial weights? Other variables considered in this study were the percent of excess weight lost and the reduction in BMI. Do you see any advantages to either of these variables when comparing weight loss for two groups?
(d) One subject from the gastric-banding group dropped out of the study, and seven subjects from the lifestyle group dropped out. Of the seven dropouts in the lifestyle group, six had gained weight at the time they dropped out. If all subjects had completed the study, how do you think it would have affected the comparison between the two groups?

Exercises 2.44 to 2.49 ask you to analyze data without having the details outlined for you. The exercise statements give you the ***State*** *step of the four-step process. In your work, follow the* ***Plan, Solve,*** *and* ***Conclude*** *steps as illustrated in Example 2.9 (page 64).*

2.44 **Athletes' salaries.** The Montreal Canadiens were founded in 1909 and are the longest continuously operating professional ice hockey team. They have won 24 Stanley Cups, making them one of the most successful professional sports teams of the traditional four major sports of Canada and the United States. Table 2.1 (page 71) gives the salaries of the 2013–14 roster.[19] Provide the team owner with a full description of the distribution of salaries and a brief summary of its most important features. HOCKEY

Oliver Samson Arcand/ NHLI via Getty Images

2.45 **Returns on stocks.** How well have stocks done over the past generation? The Wilshire 5000 index describes the average performance of all U.S. stocks. The average is weighted by the total market value of each company's stock, so think of the index as measuring the

TABLE 2.1 SALARIES FOR THE 2013–14 MONTREAL CANADIENS

PLAYER	SALARY	PLAYER	SALARY	PLAYER	SALARY
Andrei Markov	\$5,750,000	Brandon Prust	\$2,500,000	George Parros	\$950,000
Carey Price	\$5,750,000	Alexei Emelin	\$2,000,000	Alex Galchenyuk	\$925,000
Brian Gionta	\$5,000,000	Travis Moen	\$1,850,000	Yannick Weber	\$900,000
Tomas Plekanec	\$5,000,000	Francis Bouillon	\$1,500,000	Brendan Gallagher	\$715,000
Rene Bourque	\$4,000,000	Douglas Murray	\$1,500,000	Ryan White	\$700,000
Daniel Briere	\$4,000,000	Lars Eller	\$1,500,000	Michael Bournival	\$690,000
Max Pacioretty	\$4,000,000	Peter Budaj	\$1,400,000	Michael Blunden	\$575,000
P. K. Suban	\$3,750,000	Raphael Diaz	\$1,250,000	Petteri Nokelainen	\$575,000
David Desharnais	\$3,500,000	Colby Armstrong	\$1,000,000		

performance of the average investor. Here are the percent returns on the Wilshire 5000 index for the years from 1971 to 2013: WILSHIRE

YEAR	RETURN	YEAR	RETURN	YEAR	RETURN
1971	16.19	1986	15.61	2001	−10.97
1972	17.34	1987	1.75	2002	−20.86
1973	−18.78	1988	17.59	2003	31.64
1974	−27.87	1989	28.53	2004	12.48
1975	37.38	1990	−6.03	2005	6.38
1976	26.77	1991	33.58	2006	15.77
1977	−2.97	1992	9.02	2007	5.62
1978	8.54	1993	10.67	2008	−37.23
1979	24.40	1994	0.06	2009	28.30
1980	33.21	1995	36.41	2010	17.16
1981	−3.98	1996	21.56	2011	0.98
1982	20.43	1997	31.48	2012	16.06
1983	22.71	1998	24.31	2012	16.06
1984	3.27	1999	24.23	2013	33.07
1985	31.46	2000	−10.89		

What can you say about the distribution of yearly returns on stocks?

2.46 **Do good smells bring good business?** Businesses know that customers often respond to background music. Do they also respond to odors? Nicolas Guéguen and his colleagues studied this question in a small pizza restaurant in France on Saturday evenings in May. On one evening, a relaxing lavender odor was spread through the restaurant; on another evening, a stimulating lemon odor; a third evening served as a control, with no odor. Table 2.2 (page 72) shows the amounts (in euros) that customers spent on each of these evenings.[20] Compare the three distributions. Were both odors associated with increased customer spending? ODORS

2.47 **Good weather and tipping.** Favorable weather has been shown to be associated with increased tipping. Will just the belief that future weather will be favorable lead to higher tips? The researchers gave 60 index cards to a waitress at an Italian restaurant in New Jersey. Before delivering the bill to each customer, the waitress randomly selected a card and wrote on the bill the same message that was printed on the index card. Twenty of the cards had the message "The weather is supposed to be really good tomorrow. I hope you enjoy the day!" Another 20 cards contained the message "The weather is supposed to be not so good tomorrow. I hope you enjoy the day anyway!" The remaining 20 cards were blank, indicating that the waitress was not supposed to write any message. Choosing a card at random ensured that there was a random assignment of the customers to the three experimental conditions. Here are the tip percents for the three messages:[21] TIPPING

Good weather report:	20.8	18.7	19.9	20.6	22.0	23.4	22.8	24.9	22.2	20.3
	24.9	22.3	27.0	20.4	22.2	24.0	21.2	22.1	22.0	22.7
Bad weather report:	18.0	19.0	19.2	18.8	18.4	19.0	18.5	16.1	16.8	14.0
	17.0	13.6	17.5	19.9	20.2	18.8	18.0	23.2	18.2	19.4
No weather report:	19.9	16.0	15.0	20.1	19.3	19.2	18.0	19.2	21.2	18.8
	18.5	19.3	19.3	19.4	10.8	19.1	19.7	19.8	21.3	20.6

TABLE 2.2 AMOUNT SPENT (EUROS) BY CUSTOMERS IN A RESTAURANT WHEN EXPOSED TO ODORS

NO ODOR									
15.9	18.5	15.9	18.5	18.5	21.9	15.9	15.9	15.9	15.9
15.9	18.5	18.5	18.5	20.5	18.5	18.5	15.9	15.9	15.9
18.5	18.5	15.9	18.5	15.9	18.5	15.9	25.5	12.9	15.9
LEMON ODOR									
18.5	15.9	18.5	18.5	18.5	15.9	18.5	15.9	18.5	18.5
15.9	18.5	21.5	15.9	21.9	15.9	18.5	18.5	18.5	18.5
25.9	15.9	15.9	15.9	18.5	18.5	18.5	18.5		
LAVENDER ODOR									
21.9	18.5	22.3	21.9	18.5	24.9	18.5	22.5	21.5	21.9
21.5	18.5	25.5	18.5	18.5	21.9	18.5	18.5	24.9	21.9
25.9	21.9	18.5	18.5	22.8	18.5	21.9	20.7	21.9	22.5

Compare the three distributions. How did the tip percents vary with the weather report information?

2.48 Does playing video games improve surgical skill? In laparoscopic surgery a video camera and several thin instruments are inserted into the patient's abdominal cavity. The surgeon uses the image from the video camera positioned inside the patient's body to perform the procedure by manipulating the instruments that have been inserted. The Top Gun Laparoscopic Skills and Suturing Program was developed to help surgeons develop the skill set necessary for laparoscopic surgery. Because of the similarity in many of the skills involved in video games and laparoscopic surgery, it was hypothesized that surgeons with greater prior video game experience might acquire the skills required in laparoscopic surgery more easily. Thirty three surgeons participated in the study and were classified into the three categories—never used, under three hours, and more than three hours—depending on the number of hours they played video games at the height of their video game use. They also performed Top Gun drills, and received a score based on the time to complete the drill and the number of errors made, with lower scores indicating better performance. Here are the Top Gun scores and video game categories for the 33 participants.[22] TOPGUN

Never played:	9379	8302	5489	5334	4605	4789	9185	7216	9930
	4828	5655	4623	7778	8837	5947			
Under three hours:	5540	6259	5163	6149	4398	3968	7367	4217	5716
Three or more hours:	7288	4010	4859	4432	4845	5394	2703	5797	3758

Compare the distributions for the three groups. How is prior video game experience related to Top Gun scores?

2.49 Cholesterol levels and age. The National Health and Nutrition Examination Survey (NHANES) is a unique survey that combines interviews and physical examinations.[23] It includes basic demographic information, questions about topics such as diet, physical activity, and prescription medications, as well as the results of a physical examination measuring a variety of variables, including blood pressure and cholesterol levels. The program began in the early 1960s, and the survey currently examines a nationally representative sample of about 5000 persons each year. You will work with the total cholesterol measurements (mg/dL) obtained from participants in the survey in 2009–10. CHOLEST

To examine changes in cholesterol with age, we consider only the 3044 participants between 20 and 50 years of age and have classified them into the three age categories 20s, 30s, and 40s. The full data set is too large to print here, but here are the first 10 individuals:

Age category:	30s	20s	20s	40s	30s	40s	20s	30s	30s	20s
Total cholesterol:	135	160	299	197	196	202	175	216	181	149

The first individual is in the 30s with a total cholesterol of 135, the second in the 20s with total cholesterol of 160, and so forth.

(a) Use graphical and numerical summaries to compare the three distributions. How does cholesterol change with age?

(b) The ideal range of total cholesterol is below 200 mg/dL. For individuals with elevated cholesterol levels, prescription drugs are often recommended to lower levels. Among the 3044 participants between 20 and 50 years of age, 4 individuals in their 20s, 24 individuals in their 30s, and 117 individuals in their 40s were taking prescription medications to reduce their cholesterol levels. How do you think your comparison of the distribution would be changed if none of the individuals was taking medication? Explain.

Exercises 2.50 to 2.53 make use of the optional material on the 1.5 × IQR rule for suspected outliers.

2.50 Graduation rates. In Exercise 1.10 (page 32) you were asked to use a stemplot to display the distribution of the percents of on-time high school graduates in the states. Stemplots help you find the five-number summary because they arrange the observations in increasing order.

(a) Give the five-number summary of this distribution.

(b) Use the five-number summary to draw a boxplot of all the data. What is the shape of the distribution?

(c) Which observations does the 1.5 × *IQR* rule flag as suspected outliers? Is there a simple explanation for the outlier(s)? DATA GRADRATE

2.51 Foreign born. Figure 1.10 (page 30) gives a stemplot of the percent of state residents who were born outside the United States for the 50 states and the District of Columbia. DATA FOREIGN

(a) Give the five-number summary of this distribution.

(b) Is California an outlier or just the largest observation in a strongly skewed distribution? What does the 1.5 × *IQR* rule say?

2.52 The *Fortune* Global 500. The *Fortune* Global 500, also known as the Global 500, is an annual ranking of the top 500 corporations worldwide as measured by revenue. Table 2.3 provides a list of the 31 companies with the highest revenues (in billions of dollars) in 2013.[24] A stemplot or histogram shows that the distribution is strongly skewed to the right. DATA GLOBE500

(a) Give the five-number summary. Explain why this summary suggests that the distribution is right-skewed.

(b) Which companies are outliers according to the 1.5 × *IQR* rule? Make a stemplot of the data. Do you agree with the rule's suggestions about which companies are and are not outliers?

(c) If you consider *all* 500 companies, the 31 companies in Table 2.3 represent the high outliers among all Global 500 companies. Is there a common feature shared by many of the 31 companies in the table?

2.53 Cholesterol for people in their 20s. Exercise 2.49 contains the cholesterol levels of individuals in their 20s from the NHANES survey in 2009–10. The cholesterol levels are right-skewed, with a few large cholesterol levels. Which cholesterol levels are suspected outliers by the1.5 × *IQR* rule? DATA CHOLES20

TABLE 2.3 REVENUES FOR THE TOP GLOBAL 500 COMPANIES IN 2013

COMPANY NAME	REVENUES ($B)	COMPANY NAME	REVENUES ($B)
Royal Dutch Shell	481.7	ENI	167.9
Wal-Mart Stores	469.2	Berkshire Hathaway	162.5
Exxon Mobil	449.9	Apple	156.5
Sinopec Group	428.2	AXA	154.6
China National Petroleum	408.6	Gazprom	153.5
BP	388.3	General Motors	152.3
State Grid	298.4	Daimler	146.9
Toyota Motor	265.7	General Electric	146.9
Volkswagen	247.6	Petrobras	144.1
Total	234.3	Exor Group	142.2
Chevron	233.9	Valero Energy	138.3
Glencore Xstrata	214.4	Ford Motor	134.3
Japan Post Holdings	190.9	Industrial & Commercial Bank of China	133.6
Samsung Electronics	178.6	Hon Hai Precision Industry	132.1
E.ON	169.8	Allianz	130.8
Phillips 66	169.6		

Exploring the Web

2.54 Home run leaders. The three top players on the career home run list are Barry Bonds, Hank Aaron, and Babe Ruth. You can find their home run statistics by going to the website `www.baseball-reference.com` and then clicking on the Players tab at the top of the page. Construct three side-by-side boxplots comparing the yearly home run production of Barry Bonds, Hank Aaron, and Babe Ruth. Describe any differences that you observe. It is worth noting that in his first four seasons, Babe Ruth was primarily a pitcher. If these four seasons are ignored, how does Babe Ruth compare with Barry Bonds and Hank Aaron?

2.55 Crime rates and outliers. The *Statistical Abstract of the United States* is a comprehensive summary of statistics on the social, political, and economic organization of the United States. It can be found at the website `www.census.gov/compendia/statab/`. Go to the section Law Enforcement, Courts and Prisons, and then to the subsection Crimes and Crime Rates. Several tables of data will be available.

(a) Open the Table on Crime Rates by State and Type for the latest year given. Why do you think they use rates per 100,000 population rather than the number of crimes committed? The District of Columbia is a high outlier in several crime categories.

(b) Open the Table on Crime Rates by Type for Selected Large Cities. This table includes the District of Columbia, which is listed as Washington, DC. Without doing any formal calculations, does the District of Columbia look like a high outlier in the table for large cities? Whether or not the District of Columbia is an outlier depends on more than its crime rate. It also depends on the other observations included in the data set. Which data set do you feel is more appropriate for the District of Columbia?

(c) Using the Table on Crime Rates by State and Type for the latest year given, choose a crime category, and give a full description of its distribution over the 50 states, omitting the District of Columbia. Your description of the distribution should include appropriate graphical and numerical summaries and a brief report describing the main features of the distribution. You can open the data as an Excel file and import it into your statistical software.

ImageSource/Photolibrary/Getty Images

CHAPTER 3

The Normal Distributions

In this chapter we cover...

3.1 Density curves
3.2 Describing density curves
3.3 Normal distributions
3.4 The 68–95–99.7 rule
3.5 The standard Normal distribution
3.6 Finding Normal proportions
3.7 Using the standard Normal table
3.8 Finding a value given a proportion

We now have a tool box of graphical and numerical methods for describing distributions. What is more, we have a clear strategy for exploring data on a single quantitative variable.

Exploring a Distribution

1. Always plot your data: make a graph, usually a histogram or a stemplot.
2. Look for the overall pattern (shape, center, variability) and for striking deviations such as outliers.
3. Calculate a numerical summary to briefly describe center and variability.

In this chapter, we add one more step to this strategy:

4. Sometimes the overall pattern of a large number of observations is so regular that we can describe it by a smooth curve.

3.1 Density curves

Figure 3.1 is a histogram of the scores of all 947 seventh-grade students in Gary, Indiana, on the vocabulary part of the *Iowa Tests of Basic Skills*.[1] Scores of many students on this national test have a quite regular distribution. The histogram is

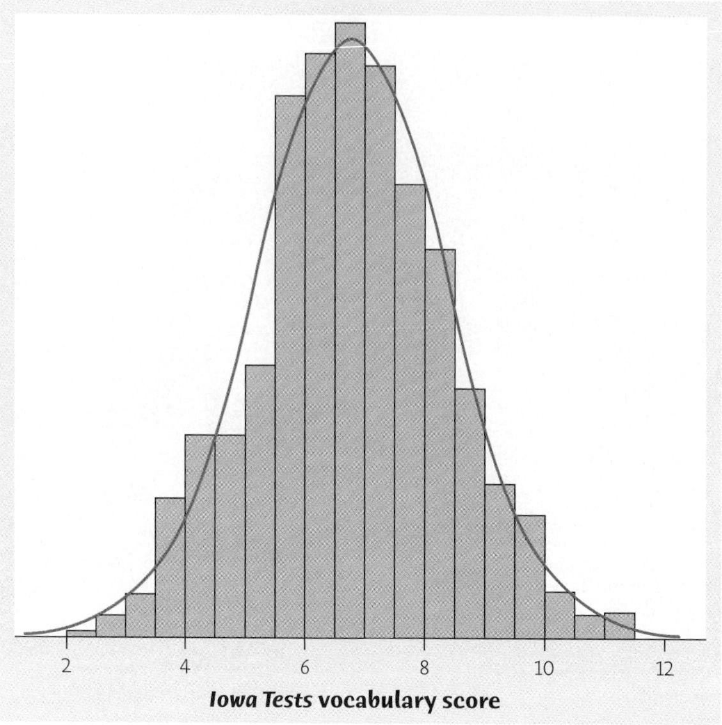

FIGURE 3.1
Histogram of the *Iowa Tests* vocabulary scores of all seventh-grade students in Gary, Indiana. The smooth curve shows the overall shape of the distribution.

symmetric, and both tails fall off smoothly from a single center peak. There are no large gaps or obvious outliers. The smooth curve drawn through the tops of the histogram bars in Figure 3.1 is a good description of the overall pattern of the data.

EXAMPLE 3.1 From Histogram to Density Curve

DATA IOWATEST

Our eyes respond to the *areas* of the bars in a histogram. The bar areas represent proportions of the observations. Figure 3.2(a) is a copy of Figure 3.1 with the left-most bars shaded. The area of the shaded bars in Figure 3.2(a) represents the students with vocabulary scores of 6.0 or lower. There are 287 such students, who make up the proportion $287/947 = 0.303$ of all Gary seventh-graders.

Now look at the curve drawn through the bars. In Figure 3.2(b), the area under the curve to the left of 6.0 is shaded. We can draw histogram bars taller or shorter by adjusting the vertical scale. In moving from histogram bars to a smooth curve, we make a specific choice: adjust the scale of the graph so that *the total area under the curve is exactly 1.* The total area represents the proportion 1, that is, all the observations. We can then interpret areas under the curve as proportions of the observations. The curve is now a *density curve.* The shaded area under the density curve in Figure 3.2(b) represents the proportion of students with score 6.0 or lower. This area is 0.293, only 0.010 away from the actual proportion 0.303. The method for finding this area will be presented shortly. For now, note that the areas under the density curve give quite good approximations to the actual distribution of the 947 test scores. ■

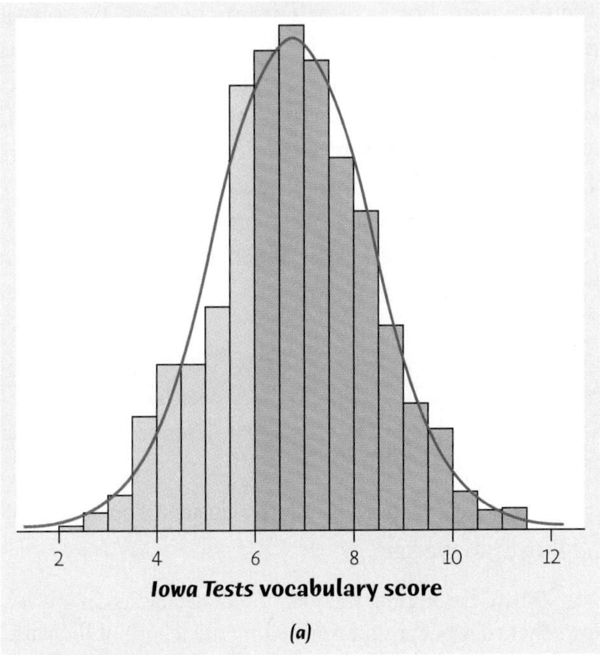

FIGURE 3.2(a)
The proportion of scores less than or equal to 6.0 in the actual data is 0.303.

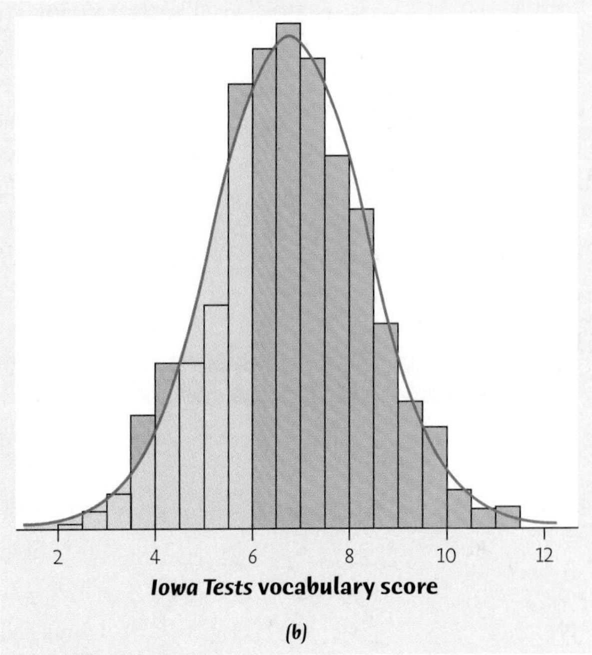

FIGURE 3.2(b)
The proportion of scores less than or equal to 6.0 from the density curve is 0.293. The density curve is a good approximation to the distribution of the data.

Density Curve

A **density curve** is a curve that

- Is always on or above the horizontal axis.
- Has area exactly 1 underneath it.

A density curve describes the overall pattern of a distribution. The area under the curve and above any range of values is the proportion of all observations that fall in that range.

Density curves, like distributions, come in many shapes. Figure 3.3 shows a strongly skewed distribution, the survival times of guinea pigs from Exercise 2.31 (page 68).

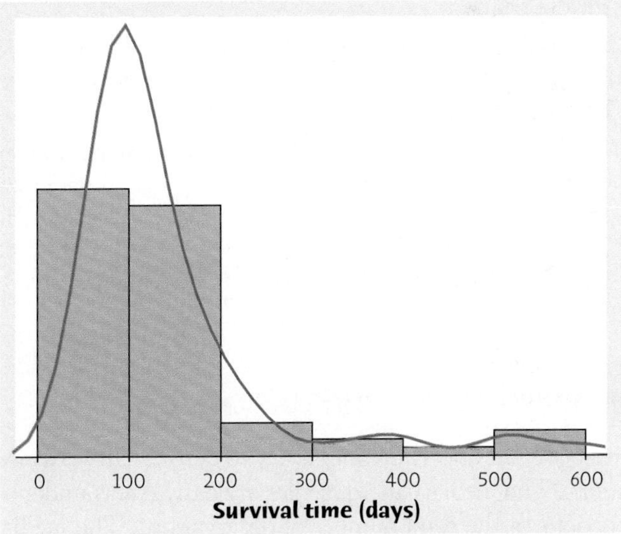

FIGURE 3.3
A right-skewed distribution pictured by both a histogram and a density curve.

The histogram and density curve were both created from the data by software. Both show the overall shape and the "bumps" in the long right tail. The density curve shows a single high peak as a main feature of the distribution. The histogram divides the observations near the peak between two bars, thus reducing the height of the peak. A density curve is often a good description of the overall pattern of a distribution. Outliers, which are deviations from the overall pattern, are not described by the curve. *Of course, no set of real data is exactly described by a density curve. The curve is an idealized description that is easy to use and accurate enough for practical use.*

CAUTION

Apply Your Knowledge

3.1 Sketch Density Curves. Sketch density curves that describe distributions with the following shapes:

(a) Symmetric, but with two peaks (that is, two strong clusters of observations)

(b) Single peak and skewed to the left

3.2 Accidents on a Bike Path. Examining the location of accidents on a level, 5-mile bike path shows that they occur uniformly along the length of the path. Figure 3.4 displays the density curve that describes the distribution of accidents.

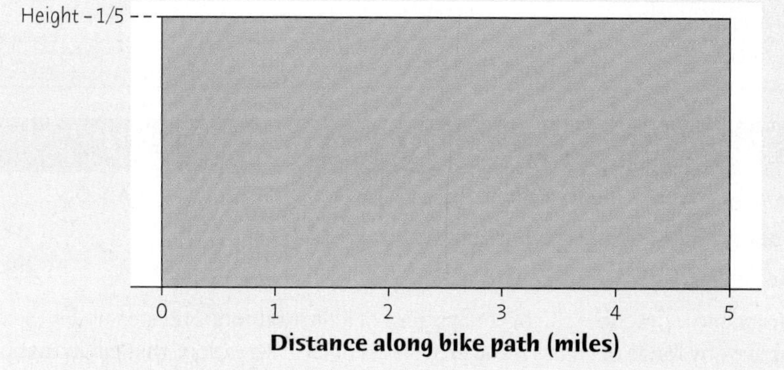

FIGURE 3.4
The density curve for the location of accidents along a 5-mile bike path, for Exercise 3.2.

(a) Explain why this curve satisfies the two requirements for a density curve.

(b) The proportion of accidents that occur in the first 2 miles of the path is the area under the density curve between 0 miles and 2 miles. What is this area?

(c) There is a stream alongside the bike path between the 0.7 mile mark and the 1.5 mile mark. What proportion of accidents happen on the bike path alongside the stream?

(d) There are roads at the two ends of the bike path, but the remainder of the paved bike path goes through the woods. What proportion of accidents happen more than 1 mile from either road? (*Hint:* First determine where on the bike path the accident needs to occur to be more than 1 mile from either road, and then find the area.)

3.2 Describing density curves

Our measures of center and variability apply to density curves as well as to actual sets of observations. The median and quartiles are easy. Areas under a density curve represent proportions of the total number of observations. The median is the point with half the observations on either side. So *the median of a density curve is the equal-areas point*, the point with half the area under the curve to its left and the remaining half of the area to its right. The quartiles divide the area under the curve into quarters. One-fourth of the area under the curve is to the left of the first quartile, and three-fourths of the area is to the left of the third quartile. You can roughly locate the median and quartiles of any density curve by eye by dividing the area under the curve into four equal parts.

Because density curves are idealized patterns, a symmetric density curve is exactly symmetric. The median of a symmetric density curve is therefore at its center. Figure 3.5(a) shows a symmetric density curve with the median marked. It isn't so easy to spot the equal-areas point on a skewed curve. There are mathematical ways of finding the median for any density curve. That's how we marked the median on the skewed curve in Figure 3.5(b).

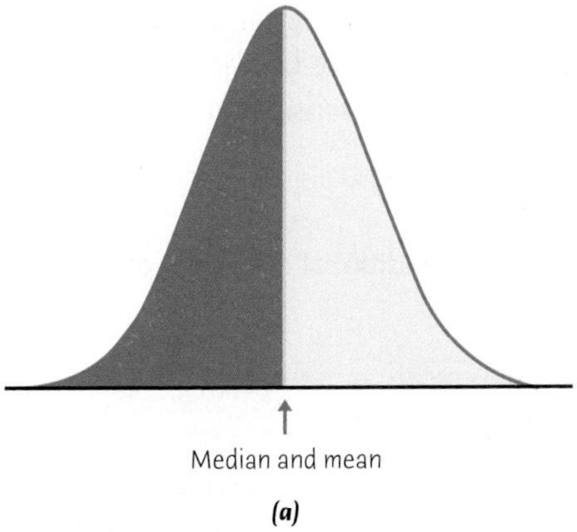
Median and mean
(a)

FIGURE 3.5(a)
The median and mean of a symmetric density curve both lie at the center of symmetry.

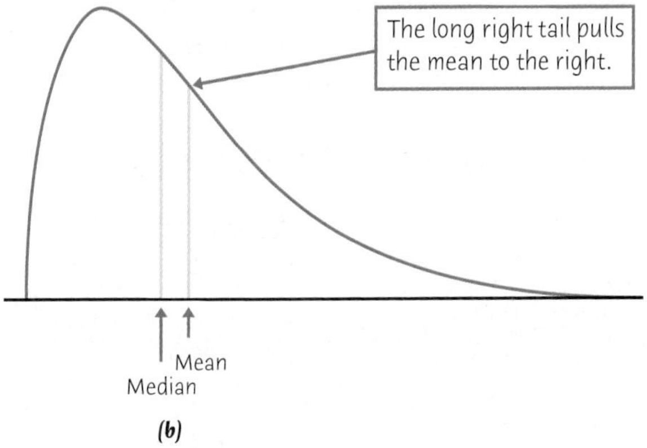
The long right tail pulls the mean to the right.
Mean
Median
(b)

FIGURE 3.5(b)
The median and mean of a right-skewed density curve. The mean is pulled away from the median toward the long tail.

What about the mean? The mean of a set of observations is their arithmetic average. If we think of the observations as weights strung out along a thin rod, the mean is the point at which the rod would balance. This fact is also true of density curves. *The mean is the point at which the curve would balance if made of solid material.* Figure 3.6 illustrates this fact about the mean. A symmetric curve balances at its center because the two sides are identical. *The mean and median of a symmetric density curve are equal,* as in Figure 3.5(a). We know that the mean of a skewed distribution is pulled toward the long tail. Figure 3.5(b) shows how the mean of a skewed density curve is pulled toward the long tail more than is the median. It's hard to locate the balance point by eye on a skewed curve. There are mathematical ways of calculating the mean for any density curve, so we are able to mark the mean as well as the median in Figure 3.5(b).

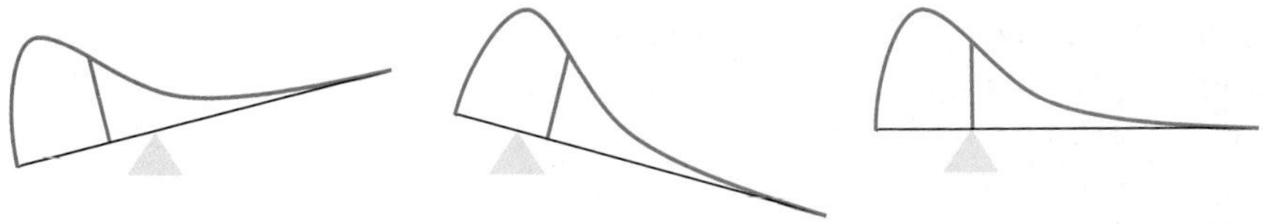

FIGURE 3.6
The mean is the balance point of a density curve.

Median and Mean of a Density Curve

The **median** of a density curve is the equal-areas point, the point that divides the area under the curve in half.

The **mean** of a density curve is the balance point at which the curve would balance if made of solid material.

The median and mean are the same for a symmetric density curve. They both lie at the center of the curve. The mean of a skewed curve is pulled away from the median in the direction of the long tail.

mean of a density curve μ

standard deviation of a density curve σ

Because a density curve is an idealized description of a distribution of data, we need to distinguish between the mean and standard deviation of the density curve and the mean $\bar{x}$ and standard deviation s computed from the actual observations. The usual notation for the **mean of a density curve** is μ (the Greek letter mu). We write the **standard deviation of a density curve** as σ (the Greek letter sigma). We can roughly locate the mean μ of any density curve by eye, as the balance point. There is no easy way to locate the standard deviation σ by eye for density curves in general.

LaunchPad Online Resources

- The **StatBoards video**, *Density Curves*, provides additional examples of density curves.

Apply Your Knowledge

3.3 Mean and Median. What is the mean μ of the density curve pictured in Figure 3.4 on page 78? (That is, where would the curve balance?) What is the median? (That is, where is the point with area 0.5 on either side?)

3.4 Mean and Median. Figure 3.7 displays three density curves, each with three points marked on it. At which of these points on each curve do the mean and the median fall?

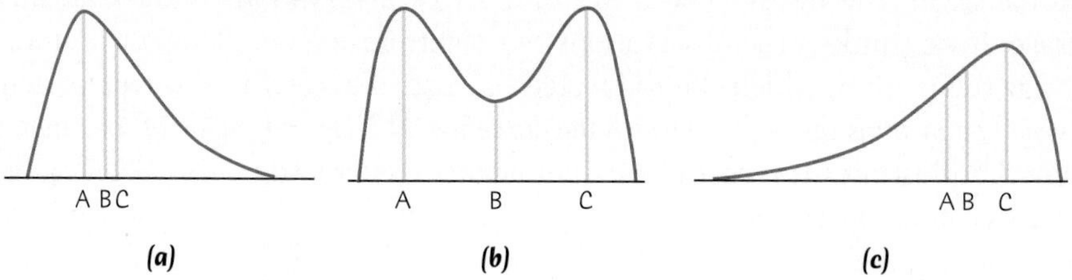

FIGURE 3.7
Three density curves, for Exercise 3.4.

3.3 Normal distributions

Normal curves

Normal distributions

One particularly important class of density curves has already appeared in Figures 3.1 and 3.2. They are called **Normal curves.** The distributions they describe are called **Normal distributions.** Normal distributions play a large role in statistics, but they are rather special and not at all "normal" in the sense of being usual or average. We capitalize Normal to remind you that these curves are special. Look at the two Normal curves in Figure 3.8. They illustrate several important facts:

- All Normal curves have the same overall shape: symmetric, single-peaked, bell-shaped.
- Any specific Normal curve is completely described by giving its mean μ and its standard deviation σ.

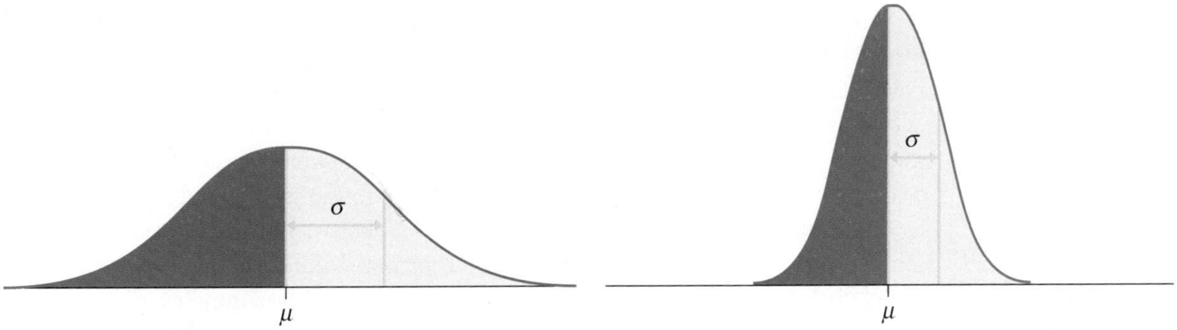

FIGURE 3.8
Two Normal curves, showing the mean μ and standard deviation σ.

- The mean is located at the center of the symmetric curve and is the same as the median. Changing μ without changing σ moves the Normal curve along the horizontal axis without changing its variability.
- The standard deviation σ controls the variability of a Normal curve. When the standard deviation is larger, the area under the normal curve is less concentrated about the mean.

The standard deviation σ is the natural measure of variability for Normal distributions. Not only do μ and σ completely determine the shape of a Normal curve, but we can also locate σ by eye on a Normal curve. Here's how. Imagine that you are skiing down a mountain that has the shape of a Normal curve. At first, you descend at an ever-steeper angle as you go out from the peak:

Fortunately, before you find yourself going straight down, the slope begins to grow flatter rather than steeper as you go out and down:

The points at which this change of curvature takes place are located at distance σ on either side of the mean μ. You can feel the change as you run a pencil along a Normal curve, and so find the standard deviation. Remember that *μ and σ alone do not specify the shape of most distributions,* and that the shape of density curves in general does not reveal σ. These are special properties of Normal distributions.

Normal Distributions

A **Normal distribution** is described by a Normal density curve. Any particular Normal distribution is completely specified by two numbers: its mean μ and standard deviation σ.

The mean of a Normal distribution is at the center of the symmetric Normal curve. The standard deviation is the distance from the center to the change-of-curvature points on either side.

Why are the Normal distributions important in statistics? Here are three reasons. First, Normal distributions are good descriptions for some distributions of *real data*. Distributions that are often close to Normal include scores on tests taken by many people (such as *Iowa Tests* and SAT exams), repeated careful measurements of the same quantity, and characteristics of biological populations (such as lengths of crickets and yields of corn). Second, Normal distributions are good approximations to the results of many kinds of *chance outcomes*, such as the proportion of heads in many tosses of a coin. Third, we will see that many *statistical inference* procedures based on Normal distributions work well for other roughly symmetric distributions. However, many sets of data do not follow a Normal distribution. Most income distributions, for example, are skewed to the right and so are not Normal. Non-Normal data, like non-normal people, not only are common but also are sometimes more interesting than their Normal counterparts.

3.4 The 68–95–99.7 rule

Although there are many Normal curves, they all have common properties. In particular, all Normal distributions obey the following rule.

The 68–95–99.7 Rule

In the Normal distribution with mean μ and standard deviation σ:

- Approximately **68%** of the observations fall within σ of the mean μ.
- Approximately **95%** of the observations fall within 2σ of μ.
- Approximately **99.7%** of the observations fall within 3σ of μ.

Figure 3.9 illustrates the 68–95–99.7 rule. By remembering these three numbers, you can think about Normal distributions without constantly making detailed calculations.

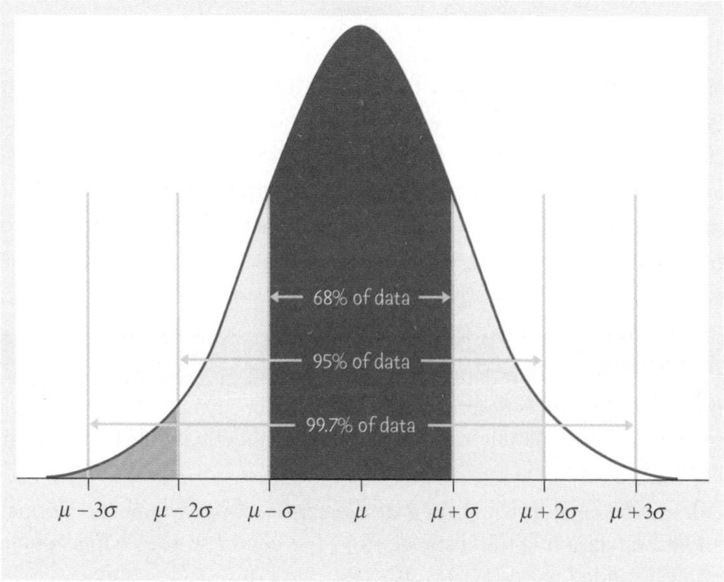

FIGURE 3.9
The 68–95–99.7 rule for Normal distributions.

EXAMPLE 3.2 Iowa Tests Scores

IOWATEST

Figures 3.1 and 3.2 (see pages 76 and 77) show that the distribution of *Iowa Tests* vocabulary scores for seventh-grade students in Gary, Indiana, is close to Normal. Suppose that the distribution is exactly Normal with mean $\mu = 6.84$ and standard deviation $\sigma = 1.55$. (These are the mean and standard deviation of the 947 actual scores.)

Figure 3.10 applies the 68–95–99.7 rule to the *Iowa Tests* scores. The 95 part of the rule says that approximately 95% of all scores are between

$$\mu - 2\sigma = 6.84 - (2)(1.55) = 6.84 - 3.10 = 3.74$$

and

$$\mu + 2\sigma = 6.84 + (2)(1.55) = 6.84 + 3.10 = 9.94$$

The other 5% of scores are outside this range. Because Normal distributions are symmetric, half of these scores are lower than 3.74 and half are higher than 9.94. That is, about 2.5% of the scores are below 3.74 and 2.5% are above 9.94. ■

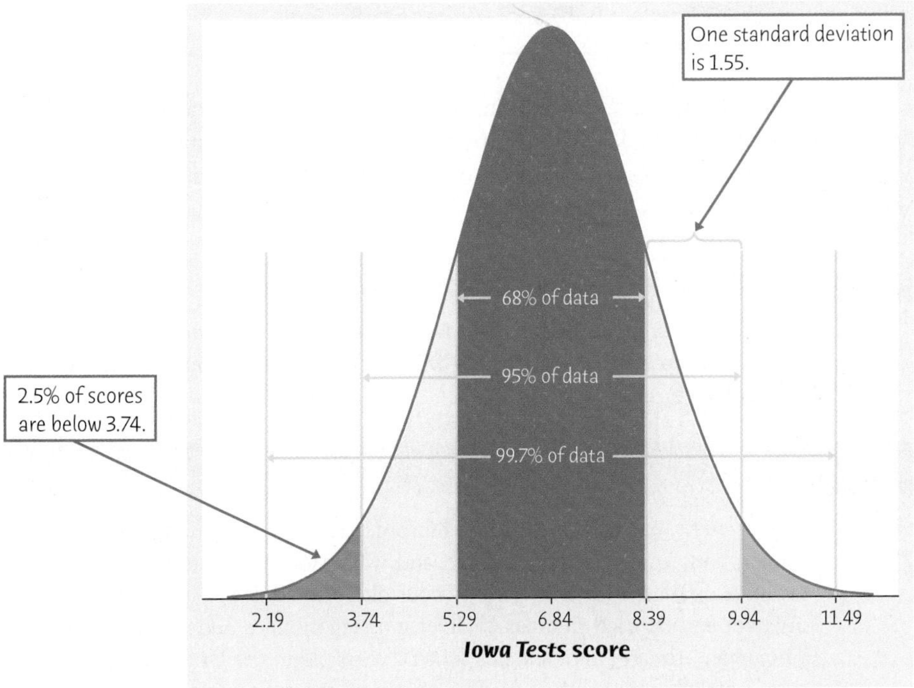

FIGURE 3.10 The 68–95–99.7 rule applied to the distribution of Iowa Tests scores for seventh-grade students in Gary, Indiana, for Example 3.2. The mean and standard deviation are $\mu = 6.84$ and $\sigma = 1.55$.

CAUTION

The 68–95–99.7 rule describes distributions that are exactly Normal. Real data such as the actual Gary scores are never exactly Normal. For one thing, *Iowa Tests* scores are reported only to the nearest tenth. A score can be 9.9 or 10.0, but not 9.94. We use a Normal distribution because it's a good approximation and because we think the knowledge that the test measures is continuous rather than stopping at tenths.

How well does our work in Example 3.2 describe the actual *Iowa Tests* scores? Well, 900 of the 947 scores are between 3.74 and 9.94. That's 95.04%, very accurate indeed. Of the remaining 47 scores, 20 are below 3.74 and 27 are above 9.94. The tails of the actual data are not quite equal, as they would be in an exactly Normal distribution. Normal distributions often describe real data better in the center of the distribution than in the extreme high and low tails.

EXAMPLE 3.3 *Iowa Tests* Scores

Look again at Figure 3.10. A score of 5.29 is one standard deviation below the mean. What percent of scores are higher than 5.29? Find the answer by adding areas in the figure. Here is the calculation in pictures:

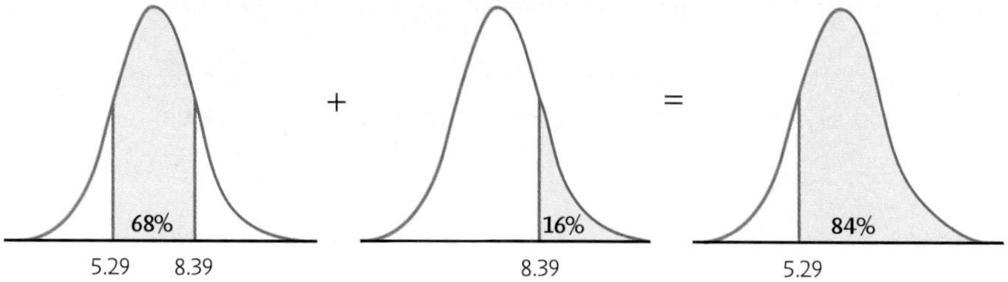

percent between 5.29 and 8.39 + percent above 8.39 = percent above 5.29

68% + 16% = 84%

So approximately 84% of scores are higher than 5.29. Be sure you understand where the 16% came from. We know that 68% of scores are between 5.29 and 8.39, so 32% of scores are outside that range. These are equally split between the two tails, with 16% below 5.29 and 16% above 8.39. ■

Because we will mention Normal distributions often, a short notation is helpful. We abbreviate the Normal distribution with mean μ and standard deviation σ as $N(\mu, \sigma)$. For example, the distribution of Gary *Iowa Tests* scores is approximately $N(6.84, 1.55)$.

LaunchPad Online Resources

- The **StatClips Examples video**, *The Normal Distribution Example B,* works through an example of using the 68–95–99.7 rule to compute areas under the Normal curve.

Apply Your Knowledge

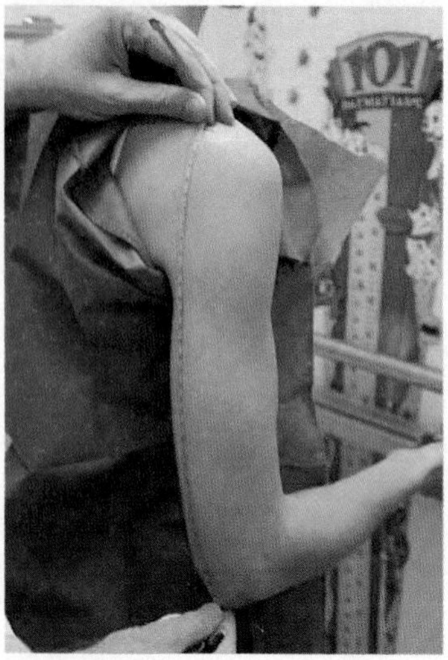

cdc.gov

FIGURE 3.11
Correct tape placement when measuring upper arm length, for Exercise 3.5.

3.5 Upper Arm Lengths. Anthropomorphic data are measurements on the human body that can track growth and weight of infants and children and evaluate changes in the body that occur over the adult life span. The resulting data can be used in areas as diverse as ergonomics and clothing design. The upper arm length of females over 20 years old in the United States is approximately Normal with mean 35.8 centimeters (cm) and standard deviation 2.1 cm. Draw a Normal curve on which this mean and standard deviation are correctly located. (*Hint:* Draw an unlabeled Normal curve, locate the points where the curvature changes, then add number labels on the horizontal axis.)

As seen in Figure 3.11, the upper arm length is measured from the acromion process, the highest point of the shoulder, down the posterior surface of the arm to the tip of the olecranon process, the bony part of the mid-elbow.[2]

3.6 Upper Arm Lengths. The upper arm length of females over 20 years old in the United States is approximately Normal with mean 35.8 centimeters (cm) and standard deviation 2.1 cm. Use the 68–95–99.7 rule to answer the following questions. (Start by making a sketch like Figure 3.10.)

(a) What range of lengths covers almost all (99.7%) of this distribution?

(b) What percent of women over 20 have upper arm lengths less than 33.7 cm?

3.7 **Monsoon Rains.** The summer monsoon rains bring 80% of India's rainfall and are essential for the country's agriculture. Records going back more than a century show that the amount of monsoon rainfall varies from year to year according to a distribution that is approximately Normal with mean 852 millimeters (mm) and standard deviation 82 mm.[3] Use the 68–95–99.7 rule to answer the following questions.

(a) Between what values do the monsoon rains fall in 95% of all years?

(b) How small are the monsoon rains in the driest 2.5% of all years?

3.5 The standard Normal distribution

As the 68–95–99.7 rule suggests, all Normal distributions share many properties. In fact, all Normal distributions are the same if we measure in units of size σ about the mean μ as center. Changing to these units is called *standardizing.* To standardize a value, subtract the mean of the distribution and then divide by the standard deviation.

Standardizing and z-Scores

If x is an observation from a distribution that has mean μ and standard deviation σ, the **standardized value** of x is

$$z = \frac{x - \mu}{\sigma}$$

A standardized value is often called a **z-score.**

Statistics in Your World: HE SAID, SHE SAID.

Height, weight, and body mass distributions in this book come from actual measurements by a government survey. That is a good thing. When *asked* their weight, almost all women say they weigh less than they really do. Heavier men also underreport their weight—but lighter men claim to weigh more than the scale shows. We leave you to ponder the psychology of the two sexes. Just remember that "say-so" is no substitute for measuring.

A z-score tells us how many standard deviations the original observation falls away from the mean, and in which direction. Observations larger than the mean are positive when standardized, and observations smaller than the mean are negative.

EXAMPLE 3.4 Standardizing Women's Heights

The heights of women aged 20 to 29 in the United States are approximately Normal with $\mu = 64.2$ inches and $\sigma = 2.8$ inches.[4] The standardized height is

$$z = \frac{\text{height} - 64.2}{2.8}$$

A woman's standardized height is the number of standard deviations by which her height differs from the mean height of all women aged 20 to 29. A woman 70 inches tall, for example, has standardized height

$$z = \frac{70 - 64.2}{2.8} = 2.07$$

or 2.07 standard deviations above the mean. Similarly, a woman 5 feet (60 inches) tall has standardized height

$$z = \frac{60 - 64.2}{2.8} = -1.50$$

or 1.50 standard deviations less than the mean height. ■

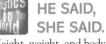

We often standardize observations from symmetric distributions to express them in a common scale. We might, for example, compare the heights of two children of different ages by calculating their z-scores. The standardized heights tell us where each child stands in the distribution for his or her age group.

If the variable we standardize has a Normal distribution, standardizing does more than give a common scale. It makes all Normal distributions into a single distribution, and this distribution is still Normal. Standardizing a variable that has any Normal distribution produces a new variable that has the *standard Normal distribution.*

Standard Normal Distribution

The **standard Normal distribution** is the Normal distribution $N(0, 1)$ with mean 0 and standard deviation 1.

If a variable x has any Normal distribution $N(\mu, \sigma)$ with mean μ and standard deviation σ, then the standardized variable

$$z = \frac{x - \mu}{\sigma}$$

has the standard Normal distribution.

LaunchPad Online Resources

- The **Snapshots video**, *Normal Distributions*, gives a nice example of comparing two Normal curves using both the 68–95–99.7 rule and z-scores.

Apply Your Knowledge

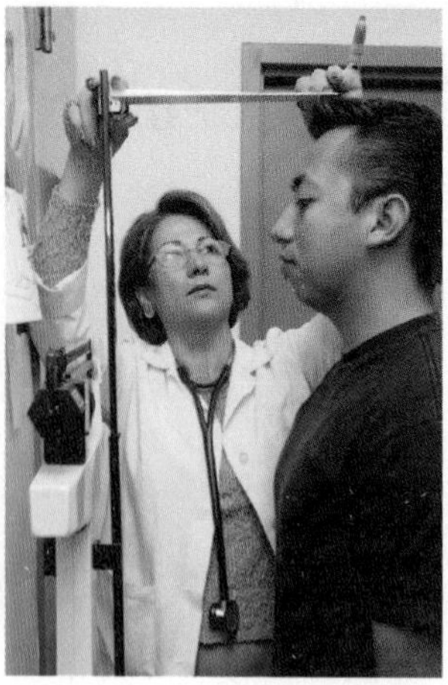

3.8 SAT versus ACT. In 2013, when she was a high school senior, Idonna scored 670 on the mathematics part of the SAT.[5] The distribution of SAT math scores in 2013 was Normal with mean 514 and standard deviation 118. Jonathan took the ACT and scored 26 on the mathematics portion. ACT math scores for 2013 were Normally distributed with mean 20.9 and standard deviation 5.3. Find the standardized scores for both students. Assuming that both tests measure the same kind of ability, who had the higher score?

3.9 Men's and Women's Heights. The heights of women aged 20 to 29 in the United States are approximately Normal with mean 64.2 inches and standard deviation 2.8 inches. Men the same age have mean height 69.4 inches with standard deviation 3.0 inches.[6] What are the z-scores for a woman 6 feet tall and a man 6 feet tall? Say in simple language what information the z-scores give that the original nonstandardized heights do not.

3.6 Finding Normal proportions

Areas under a Normal curve represent proportions of observations from that Normal distribution. There is no formula for areas under a Normal curve. Calculations use either software that calculates areas or a table of areas. Most tables and software calculate one kind of area, *cumulative proportions.* The idea of "cumulative" is "everything that came before." Here is the exact statement.

Cumulative Proportions

The **cumulative proportion** for a value x in a distribution is the proportion of observations in the distribution that are less than or equal to x.

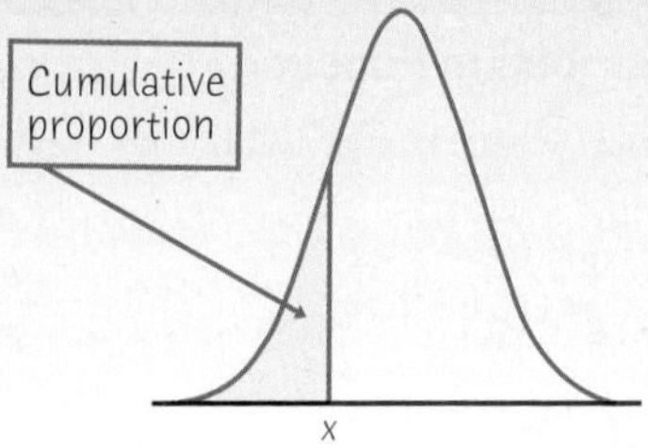

The key to calculating Normal proportions is to match the area you want with areas that represent cumulative proportions. If you make a sketch of the area you want, you will almost never go wrong. Find areas for cumulative proportions either from software or (with an extra step) from a table. The following example shows the method in a picture.

EXAMPLE 3.5 Who Qualifies for College Sports?

The National Collegiate Athletic Association (NCAA) uses a sliding scale for eligibility for Division I athletes.[7] Those students with a 2.5 high school GPA must score at least 820 on the combined mathematics and critical reading parts of the SAT to compete in their first college year. The combined scores of the almost 1.7 million high school seniors taking the SAT in 2013 were approximately Normal with mean 1011 and standard deviation 216. What percent of high school seniors meet this SAT requirement of a combined score of 820 or better?

Here is the calculation in a picture: the proportion of scores above 820 is the area under the curve to the right of 820. That's the total area under the curve (which is always 1) minus the cumulative proportion up to 820.

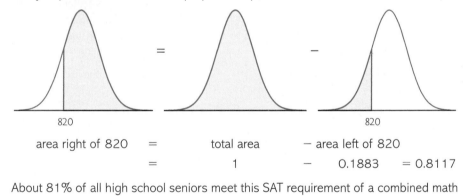

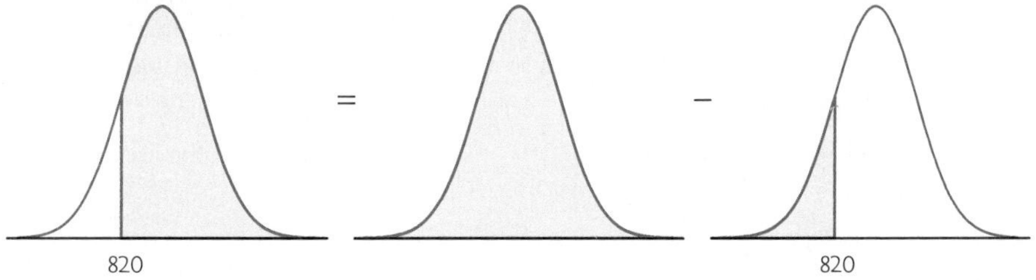

area right of 820	=	total area	− area left of 820	
	=	1	− 0.1883	= 0.8117

About 81% of all high school seniors meet this SAT requirement of a combined math and reading score of 820 or higher. ■

There is *no* area under a smooth curve and exactly over the point 820. Consequently, the area to the right of 820 (the proportion of scores >820) is the same as the area at or to the right of this point (the proportion of scores ≥820). The actual data may contain a student who scored exactly 820 on the SAT. That the proportion of scores exactly equal to 820 is 0 for a Normal distribution is a consequence of the idealized smoothing of Normal distributions for data.

To find the numerical value 0.1883 of the cumulative proportion in Example 3.5 using software, enter the mean 1011 and standard deviation 216 and ask for the cumulative proportion for 820. Software often uses terms such as "cumulative distribution" or "cumulative probability." We will learn in Chapter 12 why the language of probability fits. Here, for example, is Minitab's output:

```
Session
Cumulative Distribution Function

Normal with mean = 1011 and standard deviation = 216

  x   P ( X ≤ x )
820     0.188278
```

APPLET

The P in the output stands for "probability," but we can read it as "proportion of the observations." The *Normal Density Curve* applet is even handier because it draws pictures as well as finding areas. If you are not using software, you can find cumulative proportions for Normal curves from a table. This requires an extra step.

3.7 Using the standard Normal table

The extra step in finding cumulative proportions from a table is that we must first standardize to express the problem in the standard scale of z-scores. This allows us to get by with just one table, a table of *standard Normal cumulative proportions.* Table A in the back of the book gives cumulative proportions for the standard Normal distribution. The pictures at the top of the table remind us that the entries are cumulative proportions, areas under the curve to the left of a value z.

EXAMPLE 3.6 The Standard Normal Table

What proportion of observations on a standard Normal variable z take values less than 1.47?

Solution: To find the area to the left of 1.47, locate 1.4 in the left-hand column of Table A, then locate the remaining digit 7 as 0.07 in the top row. The entry opposite 1.4 and under 0.07 is 0.9292. This is the cumulative proportion we seek. Figure 3.12 illustrates this area. ■

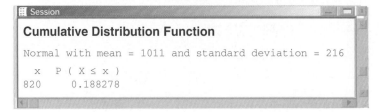

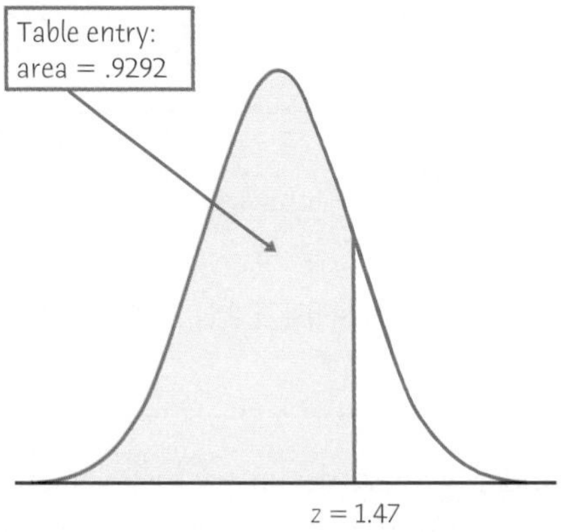

FIGURE 3.12
The area under a standard Normal curve to the left of the point $z = 1.47$ is 0.9292. Table A gives areas under the standard Normal curve.

Now that you see how Table A works, let's redo Example 3.5 using the table. We can break Normal calculations using the table into three steps.

EXAMPLE 3.7 Who Qualifies for College Sports?

Scores of high school seniors on the SAT follow the Normal distribution with mean $\mu = 1011$ and standard deviation $\sigma = 216$. What percent of seniors score at least 820?

Step 1. Draw a picture. The picture is exactly as in Example 3.5. It shows that

area to the right of 820 = 1 − area to the left of 820

Step 2. Standardize. Call the SAT score x. Subtract the mean, then divide by the standard deviation, to transform the problem about x into a problem about a standard Normal z:

$$x \geq 820$$

$$\frac{x - 1011}{216} \geq \frac{820 - 1011}{216}$$

$$z \geq -0.88$$

Step 3. Use the table. The picture shows that we need the cumulative proportion for $x = 820$. Step 2 says this is the same as the cumulative proportion for $z = -0.88$. The Table A entry for $z = -0.88$ says that this cumulative proportion is 0.1894. The area to the right of -0.88 is therefore $1 - 0.1894 = 0.8106$. ■

The area from the table in Example 3.7 (0.8106) is slightly less accurate than the area from software in Example 3.5 (0.8117) because we must round z to two decimal places when we use Table A. The difference is rarely important in practice. Here's the method in outline form.

Using Table a to Find Normal Proportions

Step 1. State the problem in terms of the observed variable x. **Draw a picture** that shows the proportion you want in terms of cumulative proportions.

Step 2. Standardize x to restate the problem in terms of a standard Normal variable z.

Step 3. Use Table A and the fact that the total area under the curve is 1 to find the required area under the standard Normal curve.

EXAMPLE 3.8 Who Qualifies for College Sports?

Recall that the NCAA uses a sliding scale for eligibility for Division I athletics. Students with a 2.5 GPA must have a combined SAT score of 820 or higher to be eligible. Students with lower GPAs will require higher SAT scores for eligibility, whereas students with higher GPAs can have a lower SAT score and still be eligible. For example, students with a 2.75 GPA are only required to have a combined SAT score that is at least 720. What proportion of all students who take the SAT would meet an SAT requirement of at least 720, but not 820?

Step 1. State the problem and draw a picture. Call the SAT score x. The variable x has the $N(1011, 216)$ distribution. What proportion of SAT scores fall between 720 and 820? Here is the picture:

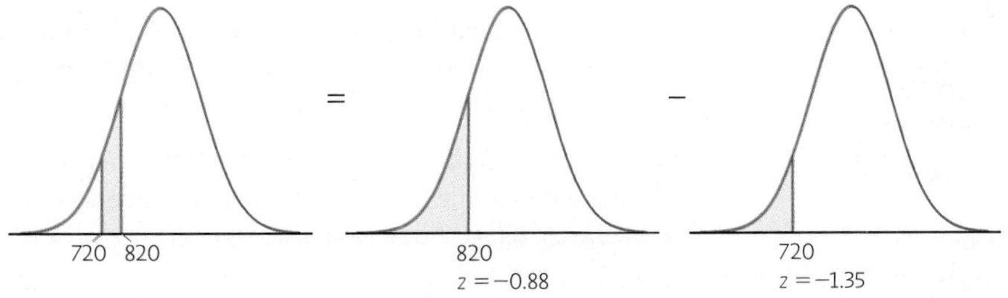

Step 2. Standardize. Subtract the mean, then divide by the standard deviation, to turn *x* into a standard Normal *z*:

$$720 \le x < 820$$

$$\frac{720 - 1011}{216} \le \frac{x - 1011}{216} < \frac{820 - 1011}{216}$$

$$-1.35 \le z < -0.88$$

Step 3. Use the table. Follow the picture (we added the *z*-scores to the picture label to help you):

$$\begin{aligned}\text{area between } -1.35 \text{ and } -0.88 &= (\text{area left of } -0.88) - (\text{area left of } -1.35)\\ &= 0.1894 - 0.0885 = 0.1009\end{aligned}$$

About 10% of high school seniors have SAT scores between 720 and 820. ■

Sometimes we encounter a value of *z* more extreme than those appearing in Table A. For example, the area to the left of $z = -4$ is not given directly in the table. The *z*-values in Table A leave only area 0.0002 in each tail unaccounted for. For practical purposes, we can act as if there is approximately zero area outside the range of Table A. Specifically, we act as if there is approximately zero area below $z = -3.5$ and approximately zero area above $z = 3.5$. While saying the area above $z = 3.5$ is 0.0002 versus saying the area is approximately 0 makes little difference in statistical practice, conceptually it is important to remember that these areas are not actually zero.

LaunchPad Online Resources

- The **StatBoards video**, *Finding a Proportion Given a Value*, provides the details of finding a proportion for a Normal distribution using the standard Normal curve.

Apply Your Knowledge

3.10 Use the Normal Table. Use Table A to find the proportion of observations from a standard Normal distribution that satisfies each of the following statements. In each case, sketch a standard Normal curve and shade the area under the curve that is the answer to the question.

(a) $z < -0.76$ (b) $z > -0.76$

(c) $z < 1.45$ (d) $-0.76 < z < 1.45$

© Robert Harding Picture Library Ltd/Alamy

3.11 Monsoon Rains. The summer monsoon rains in India follow approximately a Normal distribution with mean 852 millimeters (mm) of rainfall and standard deviation 82 mm.

(a) In the drought year 1987, 697 mm of rain fell. In what percent of all years will India have 697 mm or less of monsoon rain?

(b) "Normal rainfall" means within 20% of the long-term average, or between 682 mm and 1022 mm. In what percent of all years is the rainfall normal?

3.12 The Medical College Admissions Test. Almost all medical schools in the United States require students to take the Medical College Admission Test (MCAT).[8] The exam is composed of three multiple-choice sections (Physical Sciences, Verbal Reasoning, and Biological Sciences). The score on each section is converted to a 15-point scale so that your total score has a maximum value of 45. The total scores follow a Normal distribution, and in 2013 the mean was 25.3 with a standard deviation of 6.5. There is little change in the distribution of scores from year to year.

(a) What proportion of students taking the MCAT had a score over 30?

(b) What proportion had scores between 20 and 25?

3.8 Finding a value given a proportion

Examples 3.5 to 3.8 illustrated the use of software or Table A to find what proportion of the observations satisfies some condition, such as "SAT score above 820." We may instead want to find the observed value with a given proportion of the observations above or below it. Statistical software will do this directly.

EXAMPLE 3.9 Find the Top 10% Using Software

Scores on the SAT critical reading test in 2013 follow approximately the $N(496, 115)$ distribution. How high must a student score to place in the top 10% of all students taking the SAT?

We want to find the SAT score x with area 0.1 to its *right* under the Normal curve with mean $\mu = 496$ and standard deviation $\sigma = 115$. That's the same as finding the SAT score x with area 0.9 to its *left*. Figure 3.13 poses the question in graphical form. Most software will tell you x when you enter mean 496, standard deviation 115, and cumulative proportion 0.9. Here is Minitab's output:

```
Session
Inverse Cumulative Distribution Function

Normal with mean = 496 and standard deviation = 115

P( X ≤ x )        x
       0.9  643.378
```

Minitab gives $x = 643.378$. So scores above 644 are in the top 10%. (Round up because SAT scores can only be whole numbers.) ■

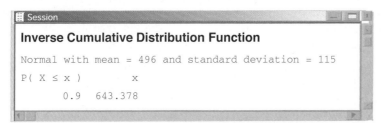

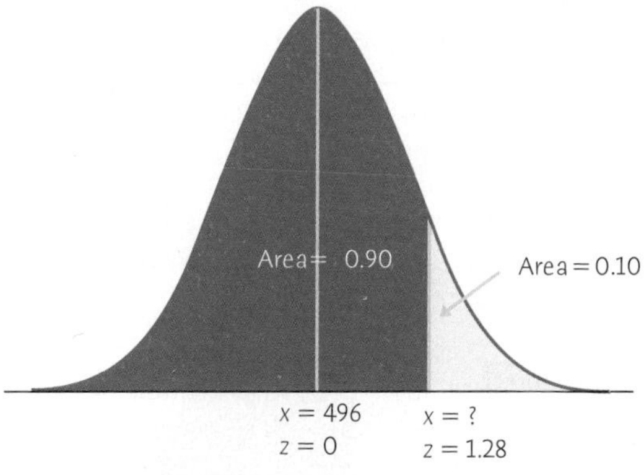

FIGURE 3.13 Locating the point on a Normal curve with area 0.10 to its right, for Examples 3.9 and 3.10.

Without software, use Table A backward. Find the given proportion in the body of the table and then read the corresponding z from the left column and top row. There are again three steps.

EXAMPLE 3.10 Find the Top 10% Using Table A

Scores on the SAT critical reading test in 2013 follow approximately the $N(496, 115)$ distribution. How high must a student score in order to place in the top 10% of all students taking the SAT?

Step 1. State the problem and draw a picture. This step is exactly as in Example 3.9. The picture is Figure 3.13. The x-value that puts a student in the top 10% is the same as the x-value for which 90% of the area is to the left of x.

Step 2. Use the table. Look in the body of Table A for the entry closest to 0.9. It is 0.8997. This is the entry corresponding to $z = 1.28$. So $z = 1.28$ is the standardized value with area 0.9 to its left.

Step 3. Unstandardize to transform z back to the original x scale. We know that the standardized value of the unknown x is $z = 1.28$. This means that x itself lies 1.28 standard deviations above the mean on this particular Normal curve. That is,

$$\begin{aligned} x &= \text{mean} + (1.28)(\text{standard deviation}) \\ &= 496 + (1.28)(115) = 643.2 \end{aligned}$$

A student must score at least 644 to place in the highest 10%. ■

EXAMPLE 3.11 Find the First Quartile

© Vladimirs Koskins | Dreamstime.com

High levels of cholesterol in the blood increase the risk of heart disease. For 14-year-old boys, the distribution of blood cholesterol is approximately Normal with mean $\mu = 170$ milligrams of cholesterol per deciliter of blood (mg/dL) and standard deviation $\sigma = 30$ mg/dL.[9] What is the first quartile of the distribution of blood cholesterol?

Step 1. State the problem and draw a picture. Call the cholesterol level x. The variable x has the $N(170, 30)$ distribution. The first quartile is the value with 25% of the distribution to its left. Figure 3.14 is the picture.

Step 2. Use the table. Look in the body of Table A for the entry closest to 0.25. It is 0.2514. This is the entry corresponding to $z = -0.67$. So $z = -0.67$ is the standardized value with area 0.25 to its left.

Step 3. Unstandardize. The cholesterol level corresponding to $z = -0.67$ lies 0.67 standard deviations below the mean, so

$$\begin{aligned} x &= \text{mean} - (0.67)(\text{standard deviation}) \\ &= 170 - (0.67)(30) = 149.9 \end{aligned}$$

The first quartile of blood cholesterol levels in 14-year-old boys is about 150 mg/dL. ■

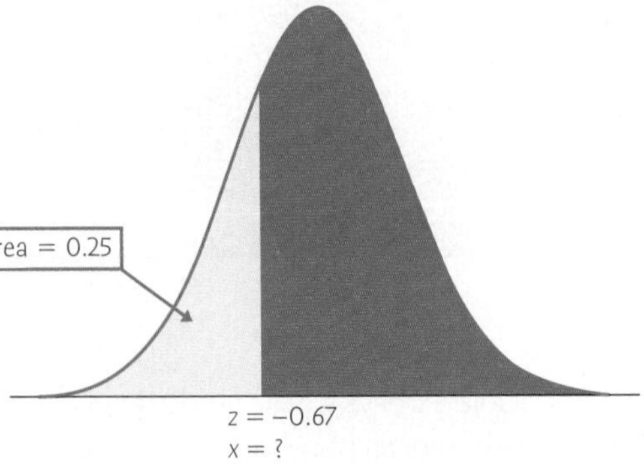

FIGURE 3.14 Locating the first quartile of a Normal curve, for Example 3.11.

LaunchPad Online Resources

- The **StatBoards video**, *Finding the Value Given a Proportion*, provides the details of finding the percentiles (values given in a proportion) for a Normal distribution using the standard Normal curve.

Apply Your Knowledge

3.13 Table A. Use Table A to find the value z of a standard Normal variable that satisfies each of the following conditions. (Use the value of z from Table A that comes closest to satisfying the condition.) In each case, sketch a standard Normal curve with your value of z marked on the axis.

(a) The point z with 65% of the observations falling below it

(b) The point z with 20% of the observations falling above it

3.14 The Medical College Admissions Test. The total scores on the Medical College Admission Test (MCAT) in 2013 follow a Normal distribution with mean 25.3 and standard deviation 6.5.

(a) What are the median and the first and third quartiles of the MCAT scores? What is the interquartile range?

(b) Give the interval that contains the central 80% of the MCAT scores.

CHAPTER 3 SUMMARY

Chapter Specifics

- We can sometimes describe the overall pattern of a distribution by a **density curve.** A density curve has total area 1 underneath it. An area under a density curve gives the proportion of observations that fall in a range of values.
- A density curve is an idealized description of the overall pattern of a distribution that smooths out the irregularities in the actual data. We write the **mean of a density curve** as μ and the **standard deviation of a density curve** as σ to distinguish them from the mean $\bar{x}$ and standard deviation s of the actual data.
- The mean, the median, and the quartiles of a density curve can be located by eye. The **mean** μ is the balance point of the curve. The **median** divides the area under the curve in half. The **quartiles** and the median divide the area under the curve into quarters. The **standard deviation** σ cannot be located by eye on most density curves.
- The mean and median are equal for symmetric density curves. The mean of a skewed curve is located farther toward the long tail than is the median.
- The **Normal distributions** are described by a special family of bell-shaped, symmetric density curves, called **Normal curves.** The mean μ and standard deviation σ completely specify a Normal distribution $N(\mu, \sigma)$. The mean is the center of the curve, and σ is the distance from μ to the change-of-curvature points on either side.
- To **standardize** any observation x, subtract the mean of the distribution and then divide by the standard deviation. The resulting **z-score**

$$z = \frac{x - \mu}{\sigma}$$

says how many standard deviations x lies from the distribution mean.

- All Normal distributions are the same when measurements are transformed to the standardized scale. In particular, all Normal distributions satisfy the **68–95–99.7 rule,**

Statistics in Your World — THE BELL CURVE? Does the distribution of human intelligence follow the "bell curve" of a Normal distribution? Scores on IQ tests do roughly follow a Normal distribution. That is because a test score is calculated from a person's answers in a way that is designed to produce a Normal distribution. To conclude that intelligence follows a bell curve, we must agree that the test scores directly measure intelligence. Many psychologists don't think there is one human characteristic that we can call "intelligence" and can measure by a single test score.

which describes what percent of observations lie within one, two, and three standard deviations of the mean, respectively.

- If x has the $N(\mu, \sigma)$ distribution, then the **standardized variable** $z = (x - \mu)/\sigma$ has the **standard Normal distribution** $N(0, 1)$ with mean 0 and standard deviation 1. Table A gives the **cumulative proportions** of standard Normal observations that are less than z for many values of z. By standardizing, we can use Table A for any Normal distribution.

Link It

When exploring data, some data sets can be shown to closely follow the Normal distribution. When this is true, the description of the data can be greatly simplified without much loss of information. We can calculate the percentage of the distribution in an interval for *any* Normal distribution if we know its mean and standard deviation. This also shows why the mean and standard deviation can be important numerical summaries. For distributions that are approximately Normal, these two numerical summaries give a complete description of the distribution of our data. It is important to remember that not all distributions can be well approximated by a Normal curve. In these cases, calculations based on the Normal distribution can be misleading.

Normal distributions are also good approximations to many kinds of chance outcomes such as the proportion of heads in many tosses of a coin—this setting will be described in more detail in Chapter 22. And when we discuss statistical inference in Parts III and IV of the text, we will find that many procedures based on Normal distributions work well for other roughly symmetric distributions.

LaunchPad Online Resources

If you are having difficulty with any of the sections of this chapter, these online resources should help prepare you to solve the exercises at the end of this chapter.

- **StatTutor** starts with a video review of each section and asks a series of questions to check your understanding.
- **LearningCurve** provides you with a series of questions about the chapter geared to your level of understanding.

CHECK YOUR SKILLS

3.15 Which of these variables is most likely to have a Normal distribution?

(a) Income per person for 150 different countries
(b) Sale prices of 200 homes in Santa Barbara, CA
(c) Lengths of 100 newborns in Connecticut

3.16 To completely specify the shape of a Normal distribution, you must give

(a) the mean and the standard deviation.
(b) the five-number summary.
(c) the median and the quartiles.

3.17 Figure 3.15 shows a Normal curve. The mean of this distribution is

(a) 0. (b) 2. (c) 3.

3.18 The standard deviation of the Normal distribution in Figure 3.15 is

(a) 2. (b) 3. (c) 5.

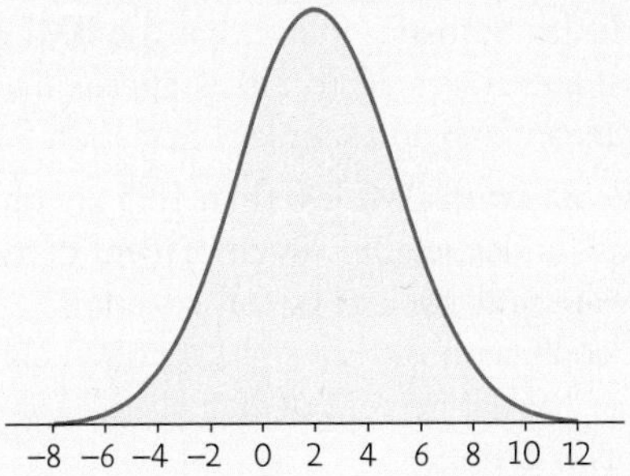

FIGURE 3.15
A Normal curve, for Exercises 3.17 and 3.18.

3.19 The length of human pregnancies from conception to birth varies according to a distribution that is approximately Normal with mean 266 days and standard deviation 16 days. About 95% of all pregnancies last between

(a) 250 and 282 days. (b) 234 and 298 days.
(c) 218 and 314 days.

3.20 The scores of adults on an IQ test are approximately Normal with mean 100 and standard deviation 15. The organization MENSA, which calls itself "the high IQ society," requires an IQ score of 130 or higher for membership. What percent of adults would qualify for membership?

(a) 95% (b) 5% (c) 2.5%

3.21 The scores of adults on an IQ test are approximately Normal with mean 100 and standard deviation 15. Clara scores 132 on such a test. Her z-score is about

(a) 2.13. (b) 2.80. (c) 8.47.

3.22 The proportion of observations from a standard Normal distribution that take values greater than 1.45 is about

(a) 0.9265. (b) 0.0735. (c) 0.0808.

3.23 The proportion of observations from a standard Normal distribution that take values less than −1.25 is about

(a) 0.1151. (b) 0.1056. (c) 0.8849.

3.24 The scores of adults on an IQ test are approximately Normal with mean 100 and standard deviation 15. Clara scores 132 on such a test. She scores higher than what percent of all adults?

(a) About 10% (b) About 90%
(c) About 98%

CHAPTER 3 EXERCISES

3.25 **Understanding density curves.** Remember that it is areas under a density curve, not the height of the curve, that give proportions in a distribution. To illustrate this, sketch a density curve that has a tall, thin peak at 0 on the horizontal axis but has most of its area close to 1 on the horizontal axis without a high peak at 1.

3.26 **Daily activity.** It appears that people who are mildly obese are less active than leaner people. One study looked at the average number of minutes per day that people spend standing or walking.[10] Among mildly obese people, minutes of activity varied according to the $N(373, 67)$ distribution. Minutes of activity for lean people had the $N(526, 107)$ distribution. Within what limits do the active minutes for about 95% of the people in each group fall? Use the 68–95–99.7 rule.

3.27 **Low IQ test scores.** Scores on the Wechsler Adult Intelligence Scale (WAIS) are approximately Normal with mean 100 and standard deviation 15. People with WAIS scores below 70 are considered intellectually disabled when, for example, applying for Social Security disability benefits. According to the 68–95–99.7 rule, about what percent of adults are intellectually disabled by this criterion?

3.28 **Standard Normal drill.** Use Table A to find the proportion of observations from a standard Normal distribution that falls in each of the following regions. In each case, sketch a standard Normal curve and shade the area representing the region.

(a) $z \leq -2.15$ (b) $z \geq -2.15$ (c) $z > 1.57$
(d) $-2.15 < z < 1.57$

3.29 Standard Normal drill.

(a) Find the number z such that the proportion of observations that are less than z in a standard Normal distribution is 0.3.

(b) Find the number z such that 35% of all observations from a standard Normal distribution are greater than z.

3.30 Fruit flies. The common fruit fly *Drosophila melanogaster* is the most studied organism in genetic research because it is small, is easy to grow, and reproduces rapidly. The length of the thorax (where the wings and legs attach) in a population of male fruit flies is approximately Normal with mean 0.800 millimeters (mm) and standard deviation 0.078 mm.

© Plastique1/Dreamstime.com

(a) What proportion of flies have thorax length less than 0.6 mm?

(b) What proportion have thorax length greater than 0.9 mm?

(c) What proportion have thorax length between 0.6 mm and 0.9 mm?

3.31 Acid rain? Emissions of sulfur dioxide by industry set off chemical changes in the atmosphere that result in "acid rain." The acidity of liquids is measured by pH on a scale of 0 to 14. Distilled water has pH 7.0, and lower pH values indicate acidity. Normal rain is somewhat acidic, so acid rain is sometimes defined as rainfall with a pH below 5.0. The pH of rain at one location varies among rainy days according to a Normal distribution with mean 5.43 and standard deviation 0.54. What proportion of rainy days have rainfall with pH below 5.0?

3.32 Runners. In a study of exercise, a large group of male runners walk on a treadmill for 6 minutes. Their heart rates in beats per minute at the end vary from runner to runner according to the $N(104, 12.5)$ distribution. The heart rates for male nonrunners after the same exercise have the $N(130, 17)$ distribution.

(a) What percent of the runners have heart rates above 135?

(b) What percent of the nonrunners have heart rates above 135?

3.33 A milling machine. Automated manufacturing operations are quite precise but still vary, often with distributions that are close to Normal. The width in inches of slots cut by a milling machine follows approximately the $N(0.8750, 0.0012)$ distribution. The specifications allow slot widths between 0.8725 and 0.8775 inch. What proportion of slots meet these specifications?

3.34 Body mass index. Your body mass index (BMI) is your weight in kilograms divided by the square of your height in meters. Many online BMI calculators allow you to enter weight in pounds and height in inches. High BMI is a common but controversial indicator of overweight or obesity. A study by the National Center for Health Statistics found that the BMI of American young men (ages 20 to 29) is approximately Normal with mean 26.8 and standard deviation 5.2.[11]

(a) People with BMI less than 18.5 are often classified as "underweight." What percent of men aged 20 to 29 are underweight by this criterion?

(b) People with BMI over 30 are often classified as "obese." What percent of men aged 20 to 29 are obese by this criterion?

Miles per gallon. *In its* Fuel Economy Guide *for 2014 model vehicles, the Environmental Protection Agency gives data on 1160 vehicles. There are a number of high outliers, mainly hybrid gas–electric vehicles. If we ignore the vehicles identified as outliers, however, the combined city and highway gas mileage of the other 1134 vehicles is approximately Normal with mean 22.2 miles per gallon (mpg) and standard deviation 5.2 mpg. Exercises 3.35 to 3.38 concern this distribution.*

3.35 I love my bug! The 2014 Volkswagen Beetle with a four-cylinder 1.8 L engine and automatic transmission has combined gas mileage of 28 mpg. What percent of all vehicles have better gas mileage than the Beetle?

3.36 The top 5%. How high must a 2014 vehicle's gas mileage be to fall in the top 5% of all vehicles?

3.37 The middle half. The quartiles of any distribution are the values with cumulative proportions 0.25 and 0.75. They span the middle half of the distribution. What are the quartiles of the distribution of gas mileage?

3.38 Quintiles. The quintiles of any distribution are the values with cumulative proportions 0.20, 0.40, 0.60, and 0.80. What are the quintiles of the distribution of gas mileage?

3.39 What's your percentile? Reports on a student's test score such as the SAT or a child's height or weight usually give the percentile as well as the actual value of the variable. The percentile is just the cumulative proportion stated as a percent: the percent of all values of the variable that were lower than this one. The upper arm lengths of females in the United States are approximately Normal with mean 35.8 cm and standard deviation 2.1 cm, and those for males are approximately Normal with mean 39.1 cm and standard deviation 2.3 cm.

(a) Larry, a 60-year-old male in the United States, has an upper arm length of 37.2 cm. What is his percentile?
(b) Measure your upper arm length to the nearest tenth of a centimeter, referring to Exercise 3.5 (page 84) for the measurement instructions. What is your arm length in centimeters? What is your percentile?

3.40 Perfect SAT scores. It is possible to score higher than 1600 on the combined mathematics and reading portions of the SAT, but scores 1600 and above are reported as 1600. The distribution of SAT scores (combining mathematics and reading) in 2013 was close to Normal with mean 1011 and standard deviation 216. What proportion of SAT scores for these two parts were reported as 1600? (that is, what proportion of SAT scores were actually higher than 1600?)

3.41 Heights of women. The heights of women aged 20 to 29 follow approximately the $N(64.2, 2.8)$ distribution. Men the same age have heights distributed as $N(69.4, 3.0)$. What percent of women aged 20 to 29 are taller than the mean height of men aged 20 to 29?

3.42 Weights aren't Normal. The heights of people of the same sex and similar ages follow a Normal distribution reasonably closely. Weights, on the other hand, are not Normally distributed. The weights of women aged 20 to 29 in the United States have mean 161.9 pounds and median 149.4 pounds. The first and third quartiles are 126.3 pounds and 181.2 pounds. What can you say about the shape of the weight distribution? Why?

3.43 A surprising calculation. Changing the mean and standard deviation of a Normal distribution by a moderate amount can greatly change the percent of observations in the tails. Suppose a college is looking for applicants with SAT math scores 750 and above.

(a) In 2013, the scores of men on the math SAT followed the $N(531, 121)$ distribution. What percent of men scored 750 or better?
(b) Women's SAT math scores that year had the $N(499, 112)$ distribution. What percent of women scored 750 or better? You see that the percent of men above 750 is almost three times the percent of women with such high scores. Why this is true is controversial. (On the other hand, women score higher than men on the new SAT writing test, though by a smaller amount.)

3.44 Grading managers. Some companies "grade on a bell curve" to compare the performance of their managers and professional workers. This forces the use of some low performance ratings so that not all workers are listed as "above average." Ford Motor Company's "performance management process" for this year assigned 10% A grades, 80% B grades, and 10% C grades to the company's managers. Suppose Ford's performance scores really are Normally distributed. This year, managers with scores less than 25 received C grades and those with scores above 475 received A grades. What are the mean and standard deviation of the scores?

3.45 Osteoporosis. Osteoporosis is a condition in which the bones become brittle due to loss of minerals. To diagnose osteoporosis, an elaborate apparatus measures bone mineral density (BMD). BMD is usually reported in standardized form. The standardization is based on a population of healthy young adults. The World Health Organization (WHO) criterion for osteoporosis is a BMD 2.5 standard deviations below the mean for healthy young adults. BMD measurements in a population of people similar in age and sex roughly follow a Normal distribution.

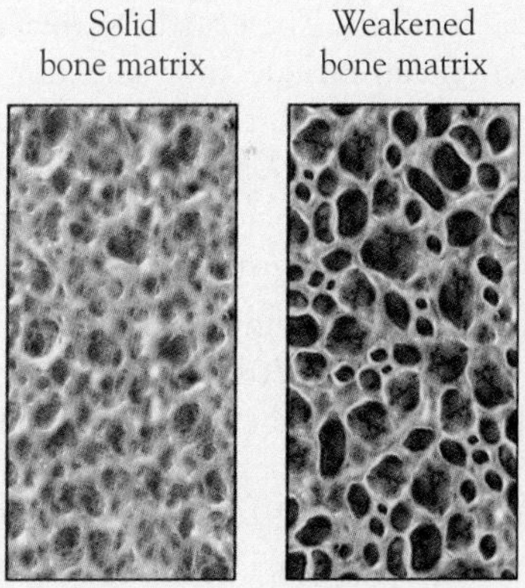

© Nucleus Medical Art, Inc/Phototake

(a) What percent of healthy young adults have osteoporosis by the WHO criterion?
(b) Women aged 70 to 79 are of course not young adults. The mean BMD in this age is about −2 on the standard scale for young adults. Suppose the standard deviation is the same as for young adults. What percent of this older population have osteoporosis?

In later chapters we will meet many statistical procedures that work well when the data are "close enough to Normal." Exercises 3.46 to 3.50 concern data that are mostly close enough to Normal for statistical work, whereas Exercise 3.51 concerns data for which the data are not close to

Normal. These exercises ask you to do data analysis and Normal calculations to investigate how close to Normal real data are.

3.46 Normal is only approximate: IQ test scores. Here are the IQ test scores of 31 seventh-grade girls in a Midwest school district:[12] MIDWSTIQ

114	100	104	89	102	91	114	114
103	105	108	130	120	132	111	128
118	119	86	72	111	103	74	112
107	103	98	96	112	112	93	

(a) We expect IQ scores to be approximately Normal. Make a stemplot to check that there are no major departures from Normality.
(b) Nonetheless, proportions calculated from a Normal distribution are not always very accurate for small numbers of observations. Find the mean $\bar{x}$ and standard deviation s for these IQ scores. What proportions of the scores are within one standard deviation and within two standard deviations of the mean? What would these proportions be in an exactly Normal distribution?

3.47 Normal is only approximate: ACT scores. Composite scores on the ACT test for the 2013 high school graduating class had mean 20.9 and standard deviation 5.4. In all, 1,799,243 students in this class took the test. Of these, 170,777 had scores higher than 28, and another 56,351 had scores exactly 28. ACT scores are always whole numbers. The exactly Normal $N(20.9, 5.4)$ distribution can include any value, not just whole numbers. What is more, there is *no* area exactly above 28 under the smooth Normal curve. So ACT scores can be only approximately Normal. To illustrate this fact, find
(a) the percent of 2013 ACT scores greater than 28 using the actual counts reported.
(b) the percent of 2013 ACT scores greater than or equal to 28, using the actual counts reported.
(c) the percent of observations that are greater than 28 using the $N(20.9, 5.4)$ distribution. (The percent greater than or equal to 28 is the same, because there is no area exactly over 28.)

3.48 Are the data Normal? Acidity of rainfall. Exercise 3.31 concerns the acidity (measured by pH) of rainfall. A sample of 105 rainwater specimens had mean pH 5.43, standard deviation 0.54, and five-number summary 4.33, 5.05, 5.44, 5.79, 6.81.[13]
(a) Compare the mean and median and also the distances of the two quartiles from the median. Does it appear that the distribution is quite symmetric? Why?
(b) If the distribution is really $N(5.43, 0.54)$, what proportion of observations would be less than 5.05? Less than 5.79? Do these proportions suggest that the distribution is close to Normal? Why?

3.49 Are the data Normal? SAT mathematics scores. Georgia Southern University (GSU) had 3219 students with regular admission in its freshman class of 2013. For each student, data is available on their SAT and ACT scores, if taken, high school GPA, and the college within the university to which they were admitted.[14] Here are the first 20 SAT mathematics scores from that data set: SATMATH

650	490	580	450	570	540	510	530
510	560	560	590	470	690	530	570
460	590	530	490				

The complete data is in the file SATMATH, which contains both the original and ordered mathematics scores.
(a) Make a histogram of the distribution (if your software allows it, superimpose a normal curve over the histogram as in Figure 3.1). Although the resulting histogram depends a bit on your choice of classes, the distribution appears roughly symmetric with no outliers.
(b) Find the mean, median, standard deviation, and quartiles for these data. Comparing the mean and the median and comparing the distances of the two quartiles from the median suggest that the distribution is quite symmetric. Why?
(c) In 2013, the mean score on the mathematics portion of the SAT for all college-bound seniors was 514. If the distribution were exactly Normal with the mean and standard deviation you found in part (b), what proportion of GSU freshmen scored above the mean for all college-bound seniors?
(d) Compute the exact proportion of GSU freshmen who scored above the mean for all college-bound seniors. It will be simplest to use the ordered scores in the SATMATH file to calculate this. How does this percentage compare with the percentage calculated in part (c)? Despite the discrepancy, this distribution is "close enough to Normal" for statistical work in later chapters.

3.50 Are the data Normal? Monsoon rains. Here are the amounts of summer monsoon rainfall (millimeters) for India in the 100 years 1901 to 2000:[15] MONSOON

722.4	792.2	861.3	750.6	716.8	885.5	777.9	897.5	889.6	935.4
736.8	806.4	784.8	898.5	781.0	951.1	1004.7	651.2	885.0	719.4
866.2	869.4	823.5	863.0	804.0	903.1	853.5	768.2	821.5	804.9
877.6	803.8	976.2	913.8	843.9	908.7	842.4	908.6	789.9	853.6
728.7	958.1	868.6	920.8	911.3	904.0	945.9	874.3	904.2	877.3
739.2	793.3	923.4	885.8	930.5	983.6	789.0	889.6	944.3	839.9
1020.5	810.0	858.1	922.8	709.6	740.2	860.3	754.8	831.3	940.0
887.0	653.1	913.6	748.3	963.0	857.0	883.4	909.5	708.0	882.9
852.4	735.6	955.9	836.9	760.0	743.2	697.4	961.7	866.9	908.8
784.7	785.0	896.6	938.4	826.4	857.3	870.5	873.8	827.0	770.2

(a) Make a histogram of these rainfall amounts. Find the mean and the median.
(b) Although the distribution is reasonably Normal, your work shows some departure from Normality. In what way are the data not Normal?

3.51 **Are the data Normal? Weight of females in their 20s.** Many body measurements of people of the same sex and similar ages such as height and upper arm length follow a Normal distribution reasonably closely. Weights, on the other hand, are not Normally distributed. The NHANES survey of 2009–10[16] includes the weights of a representative sample of 548 females in the United States aged 20 to 29. The mean of the weights was 161.58 pounds, and the standard deviation was 48.96 pounds. Figure 3.16 gives a histogram of the data along with a smooth curve representing an $N(161.58, 48.96)$ distribution. From the figure, the Normal curve does not appear to follow the pattern in the histogram that closely. Because of this, the use of areas under the Normal curve may not provide a good approximation to weights in various intervals. FEMWEIGH

(a) Using the data file on the text website, what proportion of females aged 20 to 29 weighed under 100 pounds? What percent of the $N(161.58, 48.96)$ distribution is below 100?
(b) What proportion of females aged 20 to 29 weighed over 250 pounds? What percent of the $N(161.58, 48.96)$ distribution is above 250?
(c) Based on your answers in parts (a) and (b), do you think it is a good idea to summarize the distribution of weights by an $N(161.58, 48.96)$ distribution?

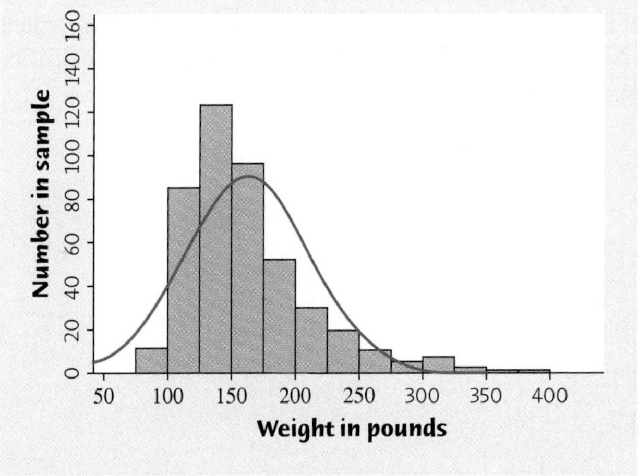

FIGURE 3.16
Histogram of the weights of 548 females aged 20 to 29 in the 2009–10 NHANES survey, with a Normal curve superimposed, for Exercise 3.51.

The Normal Density Curve *applet allows you to do Normal calculations quickly. It is somewhat limited by the number of pixels available for use, so that it can't hit every value exactly. In the following exercises, use the closest available values. In each case, make a sketch of the curve from the applet marked with the values you used to answer the questions.*

3.52 **How accurate is 68–95–99.7?** The 68–95–99.7 rule for Normal distributions is a useful approximation. To see how accurate the rule is, drag one flag across the other so that the applet shows the area under the curve between the two flags.
APPLET
(a) Place the flags one standard deviation on either side of the mean. What is the area between these two values? What does the 68–95–99.7 rule say this area is?
(b) Repeat for locations two and three standard deviations on either side of the mean. Again compare the 68–95–99.7 rule with the area given by the applet.

3.53 **Where are the quartiles?** How many standard deviations above and below the mean do the quartiles of any Normal distribution lie? (Use the standard Normal distribution to answer this question.)
APPLET

3.54 **Grading managers.** In Exercise 3.44, we saw that Ford Motor Company once graded its managers in such a way that the top 10% received an A grade, the bottom 10% a C, and the middle 80% a B. Let's suppose that performance scores follow a Normal distribution. How many standard deviations above and below the mean do the A/B and B/C cutoffs lie? (Use the standard Normal distribution to answer this question.)
APPLET

Exploring the Web

3.55 **Are the data Normal? Comparing quartiles.** The website `http://research.collegeboard.org/programs/sat/data` presents data for high school seniors who participated in the SAT Program from the current year as well as previous years. Click on the link for College Bound Seniors for the most recent year given. In the window that opens, click on the link for Total Group Report for this year. The Total Group Profile will open and contains several tables, each giving different summary information. Go to the table for Overall Mean Scores. How many students took the Critical Reading portion of the SAT? What were the mean and standard deviation of the scores? Assuming the distribution of scores is Normal with the mean and standard deviation given in the Overall Mean Scores table, what would be the first and third quartiles of the distribution? Now go to the table for Percentiles for the Total Group, and compare the actual first and third quartiles from the data to the values obtained from the Normal curve. Does this give any evidence that the distribution of Critical Reading scores is not Normal?

3.56 Are the data Normal? Comparing proportions. The website `http://research.collegeboard.org/programs/sat/data` presents data for high school seniors who participated in the SAT Program from the current year and previous years. Click on the link for College Bound Seniors for the most recent year given. In the window that opens, click on the link for Total Group Report for this year. The Total Group Profile will open and contains several tables, each giving different summary information. Go to the table for Overall Mean Scores. How many students took the Critical Reading portion of the SAT? What were the mean and standard deviation of the scores? Now go to the table on the Score Distribution. The scores are broken into intervals 200–290, 300–390, and so on. Using the Total column, find the actual percentage of students who scored in each of the six intervals reported for the Critical Reading portion of the SAT. Now, assuming the distribution of scores is Normal with the mean and standard deviation given in the Overall Mean Scores table, find the area under the Normal curve for each interval. Because of the discreteness of the actual scores, take the interval 200–290 as the interval 200–300, the interval 300–390 as the interval 300–400, and so on, when finding the areas under the Normal curve. How do the actual percentages compare to the areas under the Normal curve? Does this give any evidence that the distribution of Critical Reading scores is not Normal?

© Reinhard Dirscherl/age fotostock

CHAPTER 4

Scatterplots and Correlation

In this chapter we cover...

4.1 Explanatory and response variables

4.2 Displaying relationships: scatterplots

4.3 Interpreting scatterplots

4.4 Adding categorical variables to scatterplots

4.5 Measuring linear association: correlation

4.6 Facts about correlation

A medical study finds that short women are more likely to have heart attacks than women of average height, whereas tall women have the fewest heart attacks. An insurance group reports that heavier cars have fewer deaths per 10,000 vehicles registered than do lighter cars. These and many other statistical studies look at the *relationship between two variables.* Statistical relationships are overall tendencies, not ironclad rules. They allow individual exceptions. Although smokers on the average die younger than nonsmokers, some people live to 90 while smoking three packs a day.

To understand a statistical relationship between two variables, we measure both variables on the same individuals. Often, we must examine other variables as well. To conclude that shorter women have higher risk from heart attacks, for example, the researchers had to eliminate the effect of other variables such as weight and exercise habits. In this and the following chapter, we study relationships between variables. One of our main themes is that the relationship between two variables can be strongly influenced by other variables that are lurking in the background.

4.1 Explanatory and response variables

We think that car weight helps explain accident deaths and that smoking influences life expectancy. In each of these relationships, the two variables play different roles: one explains or influences the other.

Response Variable, Explanatory Variable

A **response variable** measures an outcome of a study. An **explanatory variable** may explain or influence changes in a response variable.

You will often find explanatory variables called *independent variables* or *predictor variables* and response variables called *dependent variables.* The idea behind this language is that the response variable depends on the explanatory variable. Because "independent" and "dependent" have other meanings in statistics that are unrelated to the explanatory–response distinction, we prefer to avoid those words.

It is easiest to identify explanatory and response variables when we actually set values of one variable to see how it affects another variable.

EXAMPLE 4.1 Beer and Blood Alcohol

How does drinking beer affect the level of alcohol in our blood? The legal limit for driving in all states is 0.08%. Student volunteers at The Ohio State University drank different numbers of cans of beer. Thirty minutes later, a police officer measured their blood alcohol content. Number of beers consumed is the explanatory variable, and percent of alcohol in the blood is the response variable. ■

When we don't set the values of either variable but just observe both variables, there may or may not be explanatory and response variables. Whether there are depends on how we plan to use the data.

EXAMPLE 4.2 College Debts

A college student aid officer looks at the findings of the National Student Loan Survey. She notes data on the amount of debt of recent graduates, their current income, and how stressed they feel about college debt. She isn't interested in predictions but is simply trying to understand the situation of recent college graduates. The distinction between explanatory and response variables does not apply.

A sociologist looks at the same data with an eye to using amount of debt and income, along with other variables, to explain the stress caused by college debt. Now amount of debt and income are explanatory variables, and stress level is the response variable. ■

AFTER YOU PLOT YOUR DATA, THINK!

The statistician Abraham Wald (1902–1950) worked on war problems during World War II. Wald invented some statistical methods that were military secrets until the war ended. Here is one of his simpler ideas. Asked where extra armor should be added to airplanes, Wald studied the location of enemy bullet holes in planes returning from combat. He plotted the locations on an outline of the plane. As data accumulated, most of the outline filled up. Put the armor in the few spots with no bullet holes, said Wald. That's where bullets hit the planes that didn't make it back.

In many studies, the goal is to show that changes in one or more explanatory variables actually *cause* changes in a response variable. Other explanatory–response relationships do not involve direct causation. Nations with more television sets per person have greater life expectancies, but shipping many television sets to Botswana won't *cause* life expectancy to increase.

Most statistical studies examine data on more than one variable. Fortunately, statistical analysis of several-variable data builds on the tools we used to examine individual variables. The principles that guide our work also remain the same:

- Plot your data. Look for overall patterns and deviations from those patterns.
- Based on what your plot shows, choose numerical summaries for some aspects of the data.

Apply Your Knowledge

4.1 Explanatory and Response Variables? You have data on a large group of college students. Here are four pairs of variables measured on these students. For each pair, is it more reasonable to simply explore the relationship between the two variables, or to view one of the variables as an explanatory variable and the other as a response variable? In the latter case, which is the explanatory variable, and which is the response variable?

(a) Number of lectures attended in your statistics course and grade on the final exam for the course

(b) Number of hours per week spent exercising and calories burned per week

(c) Hours per week spent online using Facebook and grade point average

(d) Hours per week spent online using Facebook and IQ

4.2 Coral Reefs. How sensitive to changes in water temperature are coral reefs? To find out, scientists examined data on sea surface temperatures and coral growth per year at locations in the Red Sea.[1] What are the explanatory and response variables? Are they categorical or quantitative?

© Georgie Holland/age fotostock

4.3 Beer and Blood Alcohol. Example 4.1 describes a study in which college students drank different amounts of beer. The response variable was their blood alcohol content (BAC). BAC for the same amount of beer might depend on other facts about the students. Name two other variables that could influence BAC.

4.2 Displaying relationships: scatterplots

The most useful graph for displaying the relationship between two quantitative variables is a *scatterplot.*

EXAMPLE 4.3 State SAT Mathematics Scores

Figure 1.8 (page 27) reminded us that in some states most high school graduates take the SAT to test their readiness for college, and in other states most take the ACT. Who takes a test may infuence the average score. Let's follow our four-step process (page 63) to examine this influence.[2]

MATHSAT

STATE: The percent of high school students who take the SAT varies from state to state. Does this fact help explain differences among the states in average SAT Mathematics score?

PLAN: Examine the relationship between percent taking the SAT and state mean score on the Mathematics part of the SAT. Choose the explanatory and response variables. Make a *scatterplot* to display the relationship between the variables. Interpret the plot to understand the relationship.

SOLVE (make the plot): We suspect that "percent taking" will help explain "mean score." So "percent taking" is the explanatory variable, and "mean score" is the response variable. We want to see how mean score changes when percent taking changes, so we put percent taking (the explanatory variable) on the horizontal axis.

Figure 4.1 is the scatterplot. Each point represents a single state. In Colorado, for example, 14% took the SAT, and their mean SAT Math score was 581. Find 14 on the x (horizontal) axis and 581 on the y (vertical) axis. Colorado appears as the point (14,581) above 14 and to the right of 581.

CONCLUDE: We will explore conclusions in Example 4.4. ■

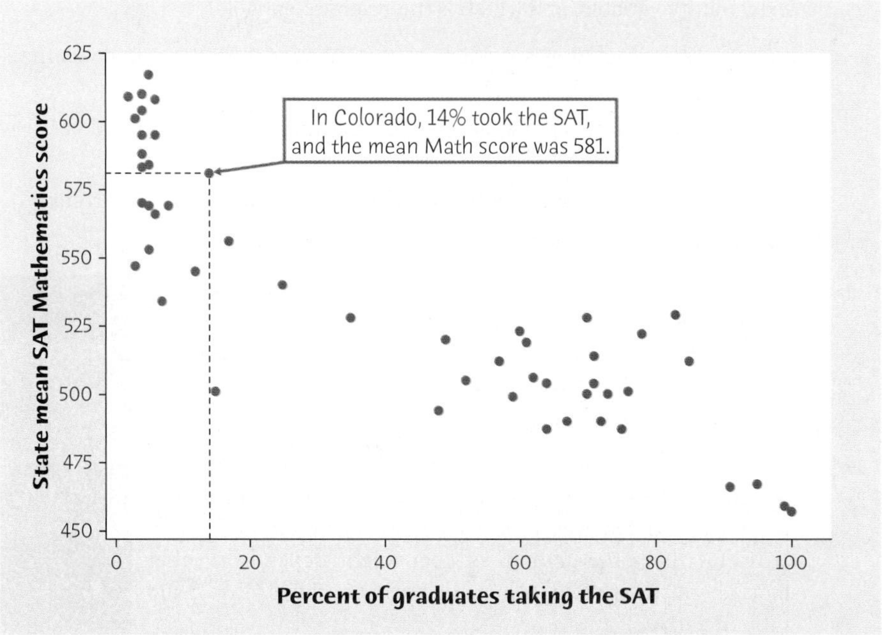

FIGURE 4.1 Scatterplot of the mean SAT Mathematics score in each state against the percent of that state's high school graduates who take the SAT, for Example 4.3. The dotted lines intersect at the point (14, 581), the data point for Colorado.

Scatterplot

A **scatterplot** shows the relationship between two quantitative variables measured on the same individuals. The values of one variable appear on the horizontal axis, and the values of the other variable appear on the vertical axis. Each individual in the data appears as the point in the plot fixed by the values of both variables for that individual.

Always plot the explanatory variable, if there is one, on the horizontal axis (the x axis) of a scatterplot. As a reminder, we usually call the explanatory variable x and the response variable y. If there is no explanatory-response distinction, either variable can go on the horizontal axis.

LaunchPad Online Resources

- The **StatBoards video**, *Creating and Interpreting Scatterplots*, provides details of making a scatterplot through an example.

Apply Your Knowledge

4.4 Death by Intent. Homicide and suicide are both intentional means of ending a life. However, the reason for committing a homicide is different from that for suicide, and we might expect homicide and suicide rates to be uncorrelated. On the other hand, both can involve some degree of violence, so perhaps we might expect some level of correlation in the rates. Here are data from 2008–10 for 26 counties in Ohio.[3] Rates are per 100,000 people. DATA DEATH

County	Homicide Rate	Suicide Rate	County	Homicide Rate	Suicide Rate
Allen	4.2	9.5	Lorain	3.1	11.0
Ashtabula	1.8	15.5	Lucas	7.4	13.3
Butler	2.6	12.7	Mahoning	10.9	12.4
Clermont	1.0	16.0	Medina	0.5	10.0
Clark	5.6	14.5	Miami	2.6	9.2
Columbiana	3.5	16.6	Montgomery	9.5	15.2
Cuyahoga	9.2	9.5	Portage	1.6	9.6
Delaware	0.8	7.6	Stark	4.7	13.5
Franklin	8.7	11.4	Summit	4.9	11.5
Greene	2.7	12.8	Trumbull	5.8	16.6
Hamilton	8.9	10.8	Warren	0.7	11.3
Lake	1.8	11.3	Wayne	1.8	8.9
Licking	4.5	12.9	Wood	1.0	7.4

Make a scatterplot to examine whether homicide and suicide rates are correlated. For these data we are simply interested in exploring the relationship between the two variables, so neither variable is an obvious choice for the explanatory variable. For convenience, use homicide rate as the explanatory variable and suicide rate as the response. (The *Two-Variable Statistical Calculator* Applet provides an easy way to make scatterplots. Click "Data" to enter your data, then "Scatterplot" to see the plot.)

APPLET

4.5 **Outsourcing by Airlines.** Airlines have increasingly outsourced the maintenance of their planes to other companies. A concern voiced by critics is that the maintenance may be less carefully done, so that outsourcing creates a safety hazard. In addition, flight delays are often due to maintenance problems, so one might look at government data on percent of major maintenance outsourced and percent of flight delays blamed on the airline to determine if these concerns are justified. This was done, and data from 2005 and 2006 appeared to justify the concerns of the critics. Do more recent data still support the concerns of the critics? Here are data from 2012:[4] DATA AIRLINE

Airline	Outsource Percent	Delay Percent	Airline	Outsource Percent	Delay Percent
Alaska	52.1	10.51	JetBlue	66.3	19.60
American	25.7	20.41	Southwest	60.8	19.58
Delta	41.3	12.46	United	47.3	23.17
Frontier	0.1	19.80	US Airways	56.5	10.80
Hawaiian	79.1	4.85			

Make a scatterplot that shows the relation between delays and outsourcing.

4.3 Interpreting scatterplots

To interpret a scatterplot, adapt the strategies of data analysis learned in Chapters 1 and 2 to the new two-variable setting. Describe the overall pattern of the plot by its **direction**, **form**, and **strength**. *Direction* indicates whether the overall pattern moves from lower left to upper right, from upper right to lower left, or neither. *Form* refers to the approximate functional form. For example, is it roughly a straight line, is it curved, or does it oscillate in some way? *Strength* refers to how closely the points in the plot follow the form. If they fall almost perfectly on a straight line, we say there is a strong straight line relationship. If they are widely scattered around a straight line, we say the relationship is weak.

Examining a Scatterplot

In any graph of data, look for the **overall pattern** and for striking **deviations** from that pattern.

You can describe the overall pattern of a scatterplot by the **direction, form,** and **strength** of the relationship.

An important kind of deviation is an **outlier,** an individual value that falls outside the overall pattern of the relationship.

Be careful not to confuse ways we describe patterns for distributions of a single variable, such as symmetric or skewed, with ways we describe patterns in scatterplots.

EXAMPLE 4.4 Understanding State SAT Scores

4 step

DATA MATHSAT

We continue to explore the state SAT Mathematics scores by interpreting what the scatterplot tells us about the variation in scores from state to state.

SOLVE (interpret the plot): Figure 4.1 shows a clear *direction:* the overall pattern moves from upper left to lower right. That is, states in which higher percents of high school graduates take the SAT tend to have lower mean SAT Mathematics scores. We call this a *negative association* between the two variables.

clusters

The *form* of the relationship is roughly a straight line with a slight curve to the right as it moves down. What is more, most states fall into two distinct **clusters.** As in the histogram in Figure 1.8, the ACT states cluster at the left and the SAT states at the right. In 23 states, fewer than 20% of seniors took the SAT; in another 26 states, more than 45% took the SAT.

The *strength* of a relationship in a scatterplot is determined by how closely the points follow a clear form. The overall relationship in Figure 4.1 is moderately strong: states with similar percents taking the SAT tend to have roughly similar mean SAT Math scores.

CONCLUDE: Percent taking explains much of the variation among states in average SAT Mathematics score. States in which a higher percent of students take the SAT tend to have lower mean scores because the mean includes a broader group of students. SAT states as a group have lower mean SAT scores than ACT states. So average SAT score says almost nothing about the quality of education in a state. It is foolish to "rank" states by their average SAT scores. ■

When discussing the direction of the relationship between two variables, we will use the word **association**. *Association* and *relationship* are often treated as synonyms by statisticians.

Positive Association, Negative Association

Two variables are **positively associated** when above-average values of one tend to accompany above-average values of the other, and below-average values also tend to occur together.

Two variables are **negatively associated** when above-average values of one tend to accompany below-average values of the other, and vice versa.

Douglas Faulkner/Science Source

Of course, not all relationships have a clear direction that we can describe as positive association or negative association. Exercise 4.8 gives an example that does not have a single direction. Here is an example of a strong positive association with a simple and important form.

EXAMPLE 4.5 The Endangered Manatee

STATE: Manatees are large, gentle, slow-moving creatures found along the coast of Florida. Many manatees are injured or killed by boats. Table 4.1 contains data on the number of boats registered in Florida (in thousands) and the number of manatees killed by boats for the years between 1977 and 2013.[5] Examine the relationship. Is it plausible that restricting the number of boats would help protect manatees?

PLAN: Make a scatterplot with "boats registered" as the explanatory variable and "manatees killed" as the response variable. Describe the form, direction, and strength of the relationship.

TABLE 4.1 FLORIDA BOAT REGISTRATIONS (THOUSANDS) AND MANATEES KILLED BY BOATS

YEAR	BOATS	MANATEES	YEAR	BOATS	MANATEES	YEAR	BOATS	MANATEES
1977	447	13	1990	719	47	2003	978	73
1978	460	21	1991	681	53	2004	983	69
1979	481	24	1992	679	38	2005	1010	79
1980	498	16	1993	678	35	2006	1024	92
1981	513	24	1994	696	49	2007	1027	73
1982	512	20	1995	713	42	2008	1010	90
1983	526	15	1996	732	60	2009	982	97
1984	559	34	1997	755	54	2010	942	83
1985	585	33	1998	809	66	2011	922	88
1986	614	33	1999	830	82	2012	902	81
1987	645	39	2000	880	78	2013	897	72
1988	675	43	2001	944	81			
1989	711	50	2002	962	95			

SOLVE: Figure 4.2 is the scatterplot. There is a positive association—more boats goes with more manatees killed. This form is a **linear relationship.** That is, the overall pattern follows a straight line from lower left to upper right. The relationship is strong because the points don't deviate greatly from a line.

linear relationship

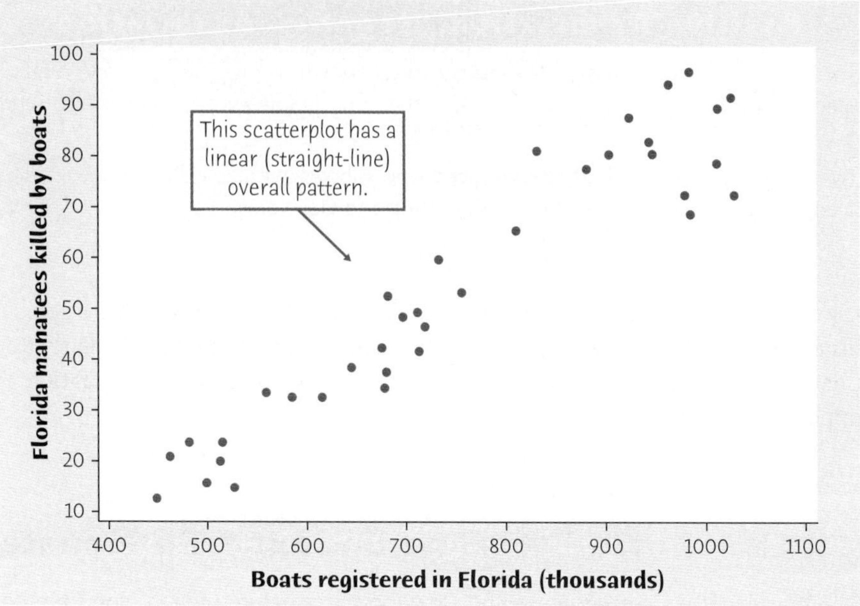

FIGURE 4.2
Scatterplot of the number of Florida manatees killed by boats in the years 1977 to 2013 against the number of boats registered in Florida that year, for Example 4.5. There is a strong linear (straight-line) pattern.

CONCLUDE: As more boats are registered, the number of manatees killed by boats goes up linearly. Data from the Florida Wildlife Commission indicate that in 2013 boats accounted for 9% of manatee deaths and 13% of deaths whose causes could be determined. Although many manatees die from other causes, it appears that fewer boats would mean fewer manatee deaths. ■

CAUTION

As the following chapter will emphasize, *it is wise to always ask what other variables lurking in the background might contribute to the relationship displayed in a scatterplot.* Because both boats registered and manatees killed are recorded year by year, any change in conditions over time might affect the relationship. For example, if boats in Florida have tended to go faster over the years, that might result in more manatees killed by the same number of boats.

LaunchPad Online Resources

- The **Snapshots video**, *Correlation and Causation,* discusses scatterplots, correlation, and their interpretation.
- The **StatBoards video**, *Creating and Interpreting Scatterplots,* provides details of interpreting a scatterplot through an example.

Apply Your Knowledge

4.6 **Death by Intent.** Describe the direction, form, and strength of the relationship between homicide rate and suicide rate, as displayed in your plot for Exercise 4.4. Are there any deviations from the overall pattern? DATA DEATH

4.7 **Outsourcing by Airlines.** Does your plot for Exercise 4.5 show a positive, negative, or no association between maintenance outsourcing and delays caused by the airline? If it shows association, is the relationship very strong? Are there any outliers? DATA AIRLINE

4.8 Does Fast Driving Waste Fuel? How does the fuel consumption of a car change as its speed increases? Here are data for a British Ford Escort. Speed is measured in kilometers per hour, and fuel consumption is measured in liters of gasoline used per 100 kilometers traveled.[6] DATA FASTDR

Speed	10	20	30	40	50	60	70	80
Fuel	21.00	13.00	10.00	8.00	7.00	5.90	6.30	6.95
Speed	90	100	110	120	130	140	150	
Fuel	7.57	8.27	9.03	9.87	10.79	11.77	12.83	

(a) Make a scatterplot. (Which is the explanatory variable?)

(b) Describe the form of the relationship. It is not linear. Explain why the form of the relationship makes sense.

(c) It does not make sense to describe the variables as either positively associated or negatively associated. Why?

(d) Is the relationship reasonably strong or quite weak? Explain your answer.

4.4 Adding categorical variables to scatterplots

The U.S. Census Bureau groups the states into four broad regions, named Midwest, Northeast, South, and West. We might ask about regional patterns in SAT exam scores. Figure 4.3 repeats part of Figure 4.1, with an important difference: we have plotted only the Midwest and Northeast groups of states, using circles for the Midwest states and plus signs for the Northeast states.

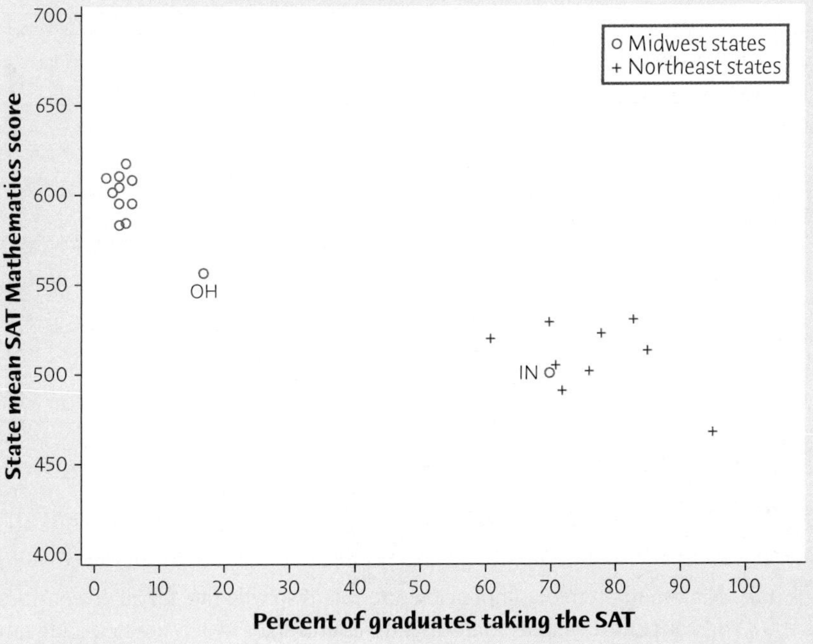

FIGURE 4.3 Mean SAT Mathematics score and percent of high school graduates who take the test for only the Midwest (o) and Northeast (+) states.

The regional comparison is striking. The 9 Northeast states are all SAT states—at least 61% of high school graduates in each of these states take the SAT. The

13 Midwest states are mostly ACT states. In 11 of these states, fewer than 7% of high school graduates take the SAT. One Midwest state is clearly an outlier within the region: Indiana is an SAT state (70% take the SAT) that falls close to the Northeast cluster. Ohio, where 17% take the SAT, also lies outside the Midwest cluster.

Dividing the states into regions introduces a third variable into the scatterplot. "Region" is a categorical variable that has four values, although we plotted data from only two of the four regions. The two regions are identified by the two different plotting symbols.

Categorical Variables in Scatterplots

To add a categorical variable to a scatterplot, use a different plot color or symbol for each category.

LaunchPad Online Resources

- The **Video Technology Manuals** discuss how to add categorical variables to a scatterplot using software for several statistical software packages.

Apply Your Knowledge

4.9 Death by Intent The data described in Exercise 4.4 also indicated that the homicide rates for some counties should be treated with caution because of low counts: DATA DEATH2

County	Homicide Rate	Suicide Rate	Caution	County	Homicide Rate	Suicide Rate	Caution
Allen	4.2	9.2	Y	Lorain	3.1	11.0	Y
Ashtabula	1.8	15.5	Y	Lucas	7.4	13.3	N
Butler	2.6	12.7	Y	Mahoning	10.9	12.4	N
Clermont	1.0	16.0	Y	Medina	0.5	10.0	Y
Clark	5.6	14.5	N	Miami	2.6	9.2	Y
Columbiana	3.5	16.6	N	Montgomery	9.5	15.2	N
Cuyahoga	9.2	9.5	N	Portage	1.6	9.6	Y
Delaware	0.8	7.6	Y	Stark	4.7	13.5	N
Franklin	8.7	11.4	N	Summit	4.9	11.5	N
Greene	2.7	12.8	Y	Trumbull	5.8	16.6	N
Hamilton	8.9	10.8	N	Warren	0.7	11.3	Y
Lake	1.8	11.3	Y	Wayne	1.8	8.9	Y
Licking	4.5	12.9	N	Wood	1.0	7.4	Y

(a) Make a scatterplot of homicide rate versus suicide rate for all 26 counties. Use separate symbols to distinguish counties for which the homicide rate is to be treated with caution from those for which there is no caution.

(b) Does the same overall pattern hold for both types of counties? What is the most important difference between the two types of counties?

4.5 Measuring linear association: correlation

A scatterplot displays the direction, form, and strength of the relationship between two quantitative variables. Linear (straight-line) relations are particularly important because a straight line is a simple pattern that is quite common. A linear relation is strong if the points lie close to a straight line, and weak if they are widely scattered about a line. Our eyes are not good judges of how strong a linear relationship is. The two scatterplots in Figure 4.4 depict exactly the same data, but the lower plot is drawn smaller in a large field. The lower plot seems to show a stronger linear relationship. Our eyes can be fooled by changing the plotting scales or the amount of space around the cloud of points in a scatterplot.[7] We need to follow our strategy for data analysis by using a numerical measure to supplement the graph. *Correlation* is the measure we use.

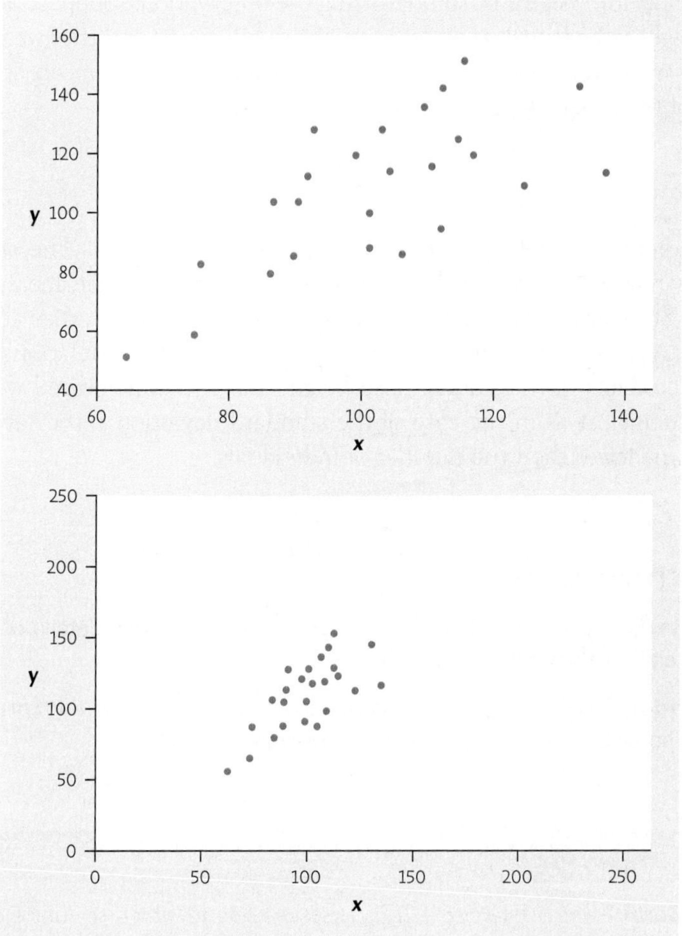

FIGURE 4.4 Two scatterplots of the same data. The straight-line pattern in the lower plot appears stronger because of the surrounding space.

Correlation

The **correlation** measures the direction and strength of the linear relationship between two quantitative variables. Correlation is usually written as r.

Suppose that we have data on variables x and y for n individuals. The values for the first individual are x_1 and y_1, the values for the second individual are x_2 and y_2, and so on. The means and standard deviations of the two variables are $\bar{x}$ and s_x for the x-values, and $\bar{y}$ and s_y for the y-values. The correlation r between x and y is

$$r = \frac{1}{n-1}\left[\left(\frac{x_1 - \bar{x}}{s_x}\right)\left(\frac{y_1 - \bar{y}}{s_y}\right) + \left(\frac{x_2 - \bar{x}}{s_x}\right)\left(\frac{y_2 - \bar{y}}{s_y}\right) + \cdots + \left(\frac{x_n - \bar{x}}{s_x}\right)\left(\frac{y_n - \bar{y}}{s_y}\right)\right]$$

or, more compactly,

$$r = \frac{1}{n-1}\sum\left(\frac{x_i - \bar{x}}{s_x}\right)\left(\frac{y_i - \bar{y}}{s_y}\right)$$

Statistics in Your World

DEATH FROM SUPERSTITION?

Is there a relationship between superstitious beliefs and bad things happening? Apparently there is. Chinese and Japanese people think that the number 4 is unlucky because when pronounced it sounds like the word for "death." Sociologists looked at 15 years' worth of death certificates for Chinese and Japanese Americans and for white Americans. Deaths from heart disease were notably higher on the fourth day of the month among Chinese and Japanese but not among whites. The sociologists think the explanation is increased stress on "unlucky days."

The formula for the correlation r is a bit complex. It helps us see what correlation is, but in practice you should use software or a calculator that finds r from keyed-in values of two variables, x and y. Exercise 4.10 asks you to calculate a correlation step-by-step from the definition to solidify its meaning.

The formula for r begins by standardizing the observations. Suppose, for example, that x is height in centimeters and y is weight in kilograms and that we have height and weight measurements for n people. Then $\bar{x}$ and s_x are the mean and standard deviation of the n heights, both in centimeters. The value

$$\frac{x_i - \bar{x}}{s_x}$$

is the standardized height of the ith person, from Chapter 3. The standardized height says how many standard deviations above or below the mean a person's height lies. Standardized values have no units—in this example, they are no longer measured in centimeters. Standardize the weights also. The correlation r is an average of the products of the standardized height and the standardized weight for all the individuals. Just as in the case of the standard deviation s, the "average" here divides by one fewer than the number of individuals.

LaunchPad Resources

- The **StatBoards video**, *Computing a Correlation*, provides details of calculating a correlation through an example.
- The **Video Technology Manuals** discuss how to compute the correlation using software for several statistical software packages.

Apply Your Knowledge

4.10 Coral Reefs. Exercise 4.2 discusses a study in which scientists examined data on mean sea surface temperatures (in degrees Celsius) and mean coral growth (in millimeters per year) over a several-year period at locations in the Red Sea. Here are the data:[8] DATA CORAL

Sea Surface Temperature	29.68	29.87	30.16	30.22	30.48	30.65	30.90
Growth	2.63	2.58	2.68	2.60	2.48	2.38	2.26

(a) Make a scatterplot. Which is the explanatory variable? The plot shows a negative linear pattern.

(b) Find the correlation r step by step. You may wish to round off to two decimal places in each step. First, find the mean and standard deviation of each variable. Then, find the seven standardized values for each variable. Finally, use the formula for r. Explain how your value for r matches your graph in part (a).

(c) Enter these data into your calculator or software, and use the correlation function to find r. Check that you get the same result as in part (b), up to roundoff error.

4.6 Facts about correlation

The formula for correlation helps us see that r is positive when there is a positive association between the variables. Height and weight, for example, have a positive association. People who are above average in height tend also to be above average in weight. Both the standardized height and the standardized weight are positive. People who are below average in height tend also to have below-average weight. Then both standardized height and standardized weight are negative. In both cases, the products in the formula for r are mostly positive, and so r is positive. In the same way, we can see that r is negative when the association between x and y is negative. More detailed study of the formula gives more detailed properties of r. Here is what you need to know to interpret correlation.

1. *Correlation makes no distinction between explanatory and response variables.* It makes no difference which variable you call x and which you call y in calculating the correlation.
2. Because r uses the standardized values of the observations, *r does not change when we change the units of measurement of x, y, or both.* Measuring height in inches rather than centimeters and weight in pounds rather than kilograms does not change the correlation between height and weight. The correlation r itself has no unit of measurement; it is just a number.
3. *Positive r indicates positive association between the variables, and negative r indicates negative association.*
4. *The correlation r is always a number between -1 and 1.* Values of r near 0 indicate a very weak linear relationship. The strength of the linear relationship increases as r moves away from 0 toward either -1 or 1. Values of r close to -1 or 1 indicate that the points in a scatterplot lie close to a straight line. The extreme values $r = -1$ and $r = 1$ occur only in the case of a perfect linear relationship, when the points lie exactly along a straight line.

EXAMPLE 4.6 From Scatterplot to Correlation

The scatterplots in Figure 4.5 illustrate how values of r closer to 1 or -1 correspond to stronger linear relationships. To make the meaning of r clearer, the standard deviations of both variables in these plots are equal, and the horizontal and vertical scales are the same. In general, it is not so easy to guess the value of r from the appearance of a scatterplot. Remember that changing the plotting scales in a scatterplot may mislead our eyes, but it does not change the correlation.

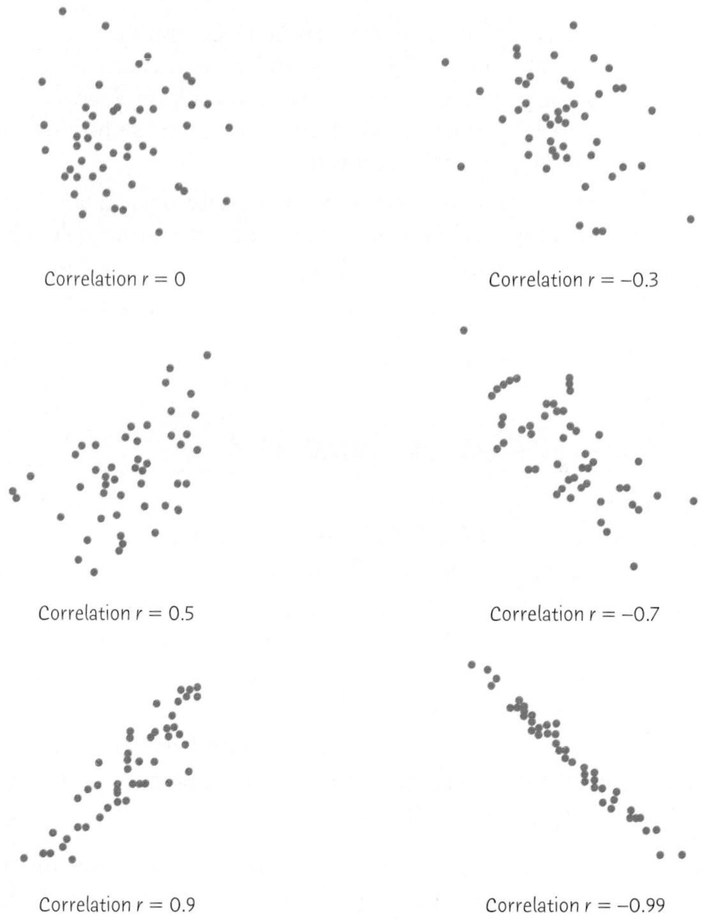

FIGURE 4.5
How correlation measures the strength of a linear relationship, for Example 4.6. Patterns closer to a straight line have correlations closer to 1 or −1.

The scatterplots in Figure 4.6 show four sets of real data. The patterns are less regular than those in Figure 4.5, but they also illustrate how correlation measures the strength of linear relationships.[9]

(a) This repeats the manatee plot in Figure 4.2. There is a strong positive linear relationship, $r = 0.954$.

DATA STORMS

(b) Here are the number of named tropical storms each year between 1984 and 2013 plotted against the number predicted before the start of hurricane season by William Gray of Colorado State University. There is a moderate linear relationship, $r = 0.609$.

(c) These data come from an experiment that studied how quickly cuts in the limbs of newts heal. Each point represents the healing rate in micrometers (millionths of a meter) per hour for the two front limbs of the same newt. This relationship is weaker than those in parts (a) and (b), with $r = 0.358$.

(d) Does last year's stock market performance help predict how stocks will do this year? No. The correlation between last year's percent return and this year's percent return over 56 years is only $r = -0.081$. The scatterplot shows a cloud of points with no visible linear pattern. ■

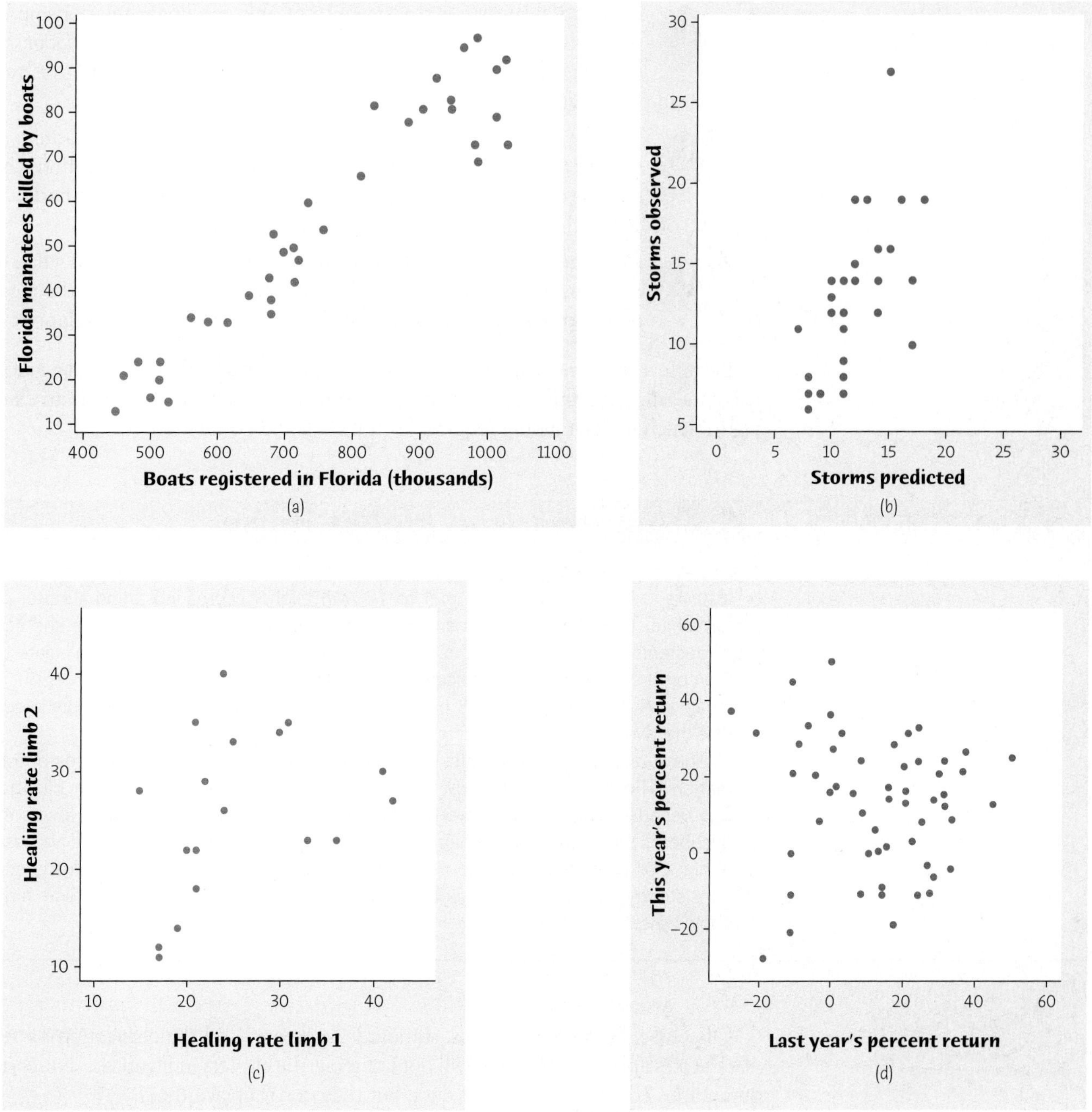

FIGURE 4.6
How correlation measures the strength of a linear relationship, for Example 4.6. Four sets of real data with (a) $r = 0.954$, (b) $r = 0.609$, (c) $r = 0.358$, and (d) $r = -0.081$.

Describing the relationship between two variables is a more complex task than describing the distribution of one variable. Here are some more facts about correlation, cautions to keep in mind when you use r:

1. CAUTION *Correlation requires that both variables be quantitative, so that it makes sense to do the arithmetic indicated by the formula for r.* We cannot calculate a correlation between the incomes of a group of people and what city they live in, because city is a categorical variable.

2. CAUTION Correlation measures the strength of only the linear relationship between two variables. *Correlation does not describe curved relationships between variables, no matter how strong they are.* Exercise 4.13 illustrates this important fact.

3. CAUTION *Like the mean and standard deviation, the correlation is not resistant: r is strongly affected by a few outlying observations.* Use r with caution when outliers appear in the scatterplot. Reporting the correlation both with the outliers included and with the outliers removed is informative.

4. CAUTION *Correlation is not a complete summary of two-variable data,* even when the relationship between the variables is linear. You should give the means and standard deviations of both x and y along with the correlation.

Because the formula for correlation uses the means and standard deviations, these measures are the proper choice to accompany a correlation. Here is an example in which understanding requires both means and correlation.

EXAMPLE 4.7 Scoring *American Idol* at Home

One website recommends that fans of the television show *American Idol* score contestants at home on a scale from 1 to 10, with higher scores indicating a better performance. Two friends, Angela and Elizabeth, decide to follow this advice and score contestants over the course of a season. How well do they agree? We calculate that the correlation between their scores is $r = 0.9$, suggesting that they agree. But the mean of Angela's scores is 0.8 point lower than Elizabeth's mean. Does this suggest that the two friends disagree?

These facts do not contradict each other. They are simply different kinds of information. The mean scores show that Angela awards lower scores than Elizabeth. But because Angela gives *every* contestant a score about 0.8 point lower than Elizabeth, the correlation remains high. Adding the same number to all values of either x or y does not change the correlation. Angela and Elizabeth actually score consistently because they agree on which performances are better. The high r shows their agreement. ■

Of course, even giving means, standard deviations, and the correlation for state SAT scores and percent taking will not point out the clusters in Figure 4.1. Numerical summaries complement plots of data, but they don't replace them.

Apply Your Knowledge

4.11 Changing the Units. The sea surface temperatures in Exercise 4.10 are measured in degrees Celsius and growth is in millimeters per year. The correlation between sea surface temperature and coral growth is $r = -0.8636$. If the measurements were made in degrees Fahrenheit and inches per year, would the correlation change? Explain your answer.

APPLET **4.12 Changing the Correlation.** Use your calculator, software, or the *Two-Variable Statistical Calculator* applet to demonstrate how outliers can affect correlation.

(a) What is the correlation between homicide rate and suicide rate for the 26 counties in Exercise 4.4? DATA DEATH

(b) Make a scatterplot of the data with one new point added. Point A: homicide rate 30, suicide rate 30. Find the new correlation for the original data plus Point A.

(c) Make a scatterplot of the data with a different new point added. Point B: homicide rate 14, suicide rate 6.5. Find the new correlation for the original data plus Point B. DATA DEATH3

(d) By looking at your plot, explain why adding Point A makes the correlation stronger (closer to 1) and adding Point B makes the correlation weaker (closer to 0).

4.13 Strong Association But No Correlation. The gas mileage of an automobile first increases and then decreases as the speed increases. Suppose this relationship is very regular, as shown by the following data on speed (miles per hour) and mileage (miles per gallon): DATA MPG

Speed	30	40	50	60	70
Mileage	24	28	30	28	24

Make a scatterplot of mileage versus speed. Show that the correlation between speed and mileage is $r = 0$. Explain why the correlation is 0 even though there is a strong relationship between speed and mileage.

CHAPTER 4 SUMMARY

Chapter Specifics

- To study relationships between variables, we must measure the variables on the same group of individuals.
- If we think that a variable x may explain or even cause changes in another variable y, we call x an **explanatory variable** and y a **response variable**.
- A **scatterplot** displays the relationship between two quantitative variables measured on the same individuals. Mark values of one variable on the horizontal axis (x axis) and values of the other variable on the vertical axis (y axis). Plot each individual's data as a point on the graph. Always plot the explanatory variable, if there is one, on the x axis of a scatterplot.
- Plot points with different colors or symbols to see the effect of a categorical variable in a scatterplot.
- In examining a scatterplot, look for an overall pattern showing the **direction**, **form**, and **strength** of the relationship and then for **outliers** or other deviations from this pattern.
- **Direction:** If the relationship has a clear direction, we speak of either **positive association** (high values of the two variables tend to occur together) or **negative association** (high values of one variable tend to occur with low values of the other variable).
- **Form: Linear relationships**, where the points show a straight-line pattern, are an important form of relationship between two variables. Curved relationships and **clusters** are other forms to watch for.
- **Strength:** The **strength** of a relationship is determined by how close the points in the scatterplot lie to a simple form such as a line.
- The **correlation** r measures the direction and strength of the linear association between two quantitative variables x and y. Although you can calculate a correlation for any scatterplot, r measures only straight-line relationships.

- Correlation indicates the direction of a linear relationship by its sign: $r > 0$ for a positive association and $r < 0$ for a negative association. Correlation always satisfies $-1 \leq r \leq 1$ and indicates the strength of a relationship by how close it is to -1 or 1. Perfect correlation, $r = \pm 1$, occurs only when the points on a scatterplot lie exactly on a straight line.
- Correlation ignores the distinction between explanatory and response variables. The value of r is not affected by changes in the unit of measurement of either variable. Correlation is not resistant, so outliers can greatly change the value of r.

Link It

In Chapters 1 to 3, we focused on exploring features of a single variable. In this chapter, we continued our study of exploratory data analysis but for the purpose of examining relationships *between* variables. A useful tool for exploring the relationship between two variables is the scatterplot. When the relationship is linear, correlation is a numerical measure of the strength of the linear relationship.

It is tempting to assume that the patterns we observe in our data hold for values of our variables that we have not observed—in other words, that additional data would continue to conform to these patterns. The process of identifying underlying patterns would seem to assume that this is the case. But is this assumption justified? Parts II to V of the book answer this question.

LaunchPad Online Resources

If you are having difficulty with any of the sections of this chapter, these online resources should help prepare you to solve the exercises at the end of this chapter.

- **StatTutor** starts with a video review of each section and asks a series of questions to check your understanding.
- **LearningCurve** provides you with a series of questions about the chapter geared to your level of understanding.

CHECK YOUR SKILLS

4.14 Researchers collect data on 5,134 American adults younger than 60. They measure the reaction times (in seconds) of each subject to a stimulus on a computer screen and how many years later the subject died.[10]

The researchers are interested in whether reaction time can predict time to death (in years). When you make a scatterplot, the explanatory variable on the x axis

(a) is the reaction time.
(b) is the time to death.
(c) can be either reaction time or time to death.

4.15 The researchers in Exercise 4.14 found that people with slower reaction times tended to die sooner. In a scatterplot of the reaction time and the number of years to death, you expect to see

(a) a positive association.
(b) very little association.
(c) a negative association.

4.16 Figure 4.7 is a scatterplot of school GPA against IQ test scores for 15 seventh-grade students. There is one low outlier in the plot. The IQ and GPA scores for this student are

(a) IQ = 0.5, GPA = 103.
(b) IQ = 103, GPA = 0.5.
(c) IQ = 103, GPA = 7.6.

4.17 If we leave out the low outlier, the correlation for the remaining 14 points in Figure 4.7 is closest to

(a) 0.9. (b) −0.9. (c) 0.1.

4.18 What are all the values that a correlation r can possibly take?

(a) $r \geq 0$ (b) $0 \leq r \leq 1$ (c) $-1 \leq r \leq 1$

4.19 If the correlation between two variables is close to 0, you can conclude that a scatterplot would show

(a) a strong straight-line pattern.
(b) a cloud of points with no visible pattern.
(c) no straight-line pattern, but there might be a strong pattern of another form.

4.20 The points on a scatterplot lie very close to a straight line. The correlation between x and y is close to

(a) −1. (b) 1. (c) either −1 or 1, we can't say which.

4.21 A statistics professor warns her class that her second midterm is always harder than the first. She tells her class that students always score 10 points worse on the second midterm compared to their score on the first midterm. This means that the correlation between students' scores on the first and second exam is

(a) 1. (b) −1. (c) Can't tell without seeing the data.

4.22 Researchers asked mothers how much soda (in ounces) their kids drank in a typical day. They also asked these mothers to rate how aggressive their kids were on a scale of 1 to 10, with larger values corresponding to a greater degree of aggression.[11] The correlation between amount of soda consumed and aggression rating was found to be $r = 0.3$. If the researchers had measured amount of soda consumed in liters instead

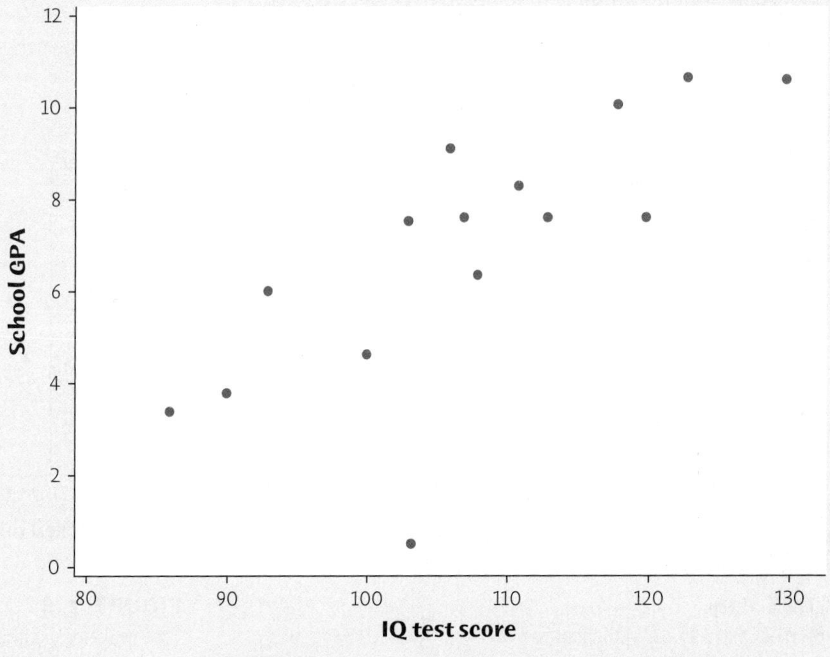

FIGURE 4.7
Scatterplot of school GPA against IQ test scores for seventh-grade students, for Exercises 4.16 and 4.17.

of ounces, what would be the correlation? (There are 35 ounces in a liter.)

(a) 0.3/35 = 0.009 (b) 0.3
(c) (0.3)(35) = 10.5

4.23 Researchers measured the percent body fat and the preferred amount of salt (percent weight/volume) for several children. Here are data for seven children:[12] DATA SALT

Preferred amount of salt x	0.2	0.3	0.4	0.5	0.6	0.8	1.1
Percent body fat y	20	30	22	30	38	23	30

Use your calculator or software: The correlation between percent body fat and preferred amount of salt is about
(a) $r = 0.08$. (b) $r = 0.3$. (c) $r = 0.8$.

CHAPTER 4 EXERCISES

4.24 Scores at the Masters. The Masters is one of the four major golf tournaments. Figure 4.8 is a scatterplot of the scores for the first two rounds of the 2013 Masters for all the golfers entered. Only the 60 golfers with the lowest two-round total advance to the final two rounds (unless several people are tied for 60th place, in which case all those tied for 60th place advance). The plot has a grid pattern because golf scores must be whole numbers.[13] DATA MASTR13

(a) Read the graph: What was the lowest score in the first round of play? How many golfers had this low score? What were their scores in the second round?
(b) Read the graph: Alan Dunbar had the highest score in the second round. What was this score? What was Dunbar's score in the first round?
(c) Is the correlation between first-round scores and second-round scores closest to $r = 0.01$, $r = 0.25$, $r = 0.75$, or $r = 0.99$? Explain your choice. Does the graph suggest that knowing a professional golfer's score for one round is much help in predicting his score for another round on the same course?

4.25 Happy states. Human happiness or well-being can be assessed either subjectively or objectively. Subjective assessment can be accomplished by listening to what people say. Objective assessment can be made from data related to well-being such as income, climate, availability of entertainment, housing prices, lack of traffic congestion, and so on. Do subjective and objective assessments agree? To study this, investigators made both subjective and objective assessments of happiness for each of the 50 states. The subjective measurement was the mean score on a life-satisfaction question found on the Behavioral Risk Factor Surveillance System (BRFSS), which is a state-based system of health surveys. Lower scores indicate a greater degree of happiness. To objectively assess happiness, the investigators computed a mean well-being score (called the compensating-differentials score) for each state, based on objective measures that have been found to be related to happiness or well-being. The states were then ranked according to this score (Rank 1 being the happiest). Figure 4.9 is a scatterplot of mean BRFSS scores (response) against the rank based on the compensating-differentials (explanatory).[14] DATA HAPPY

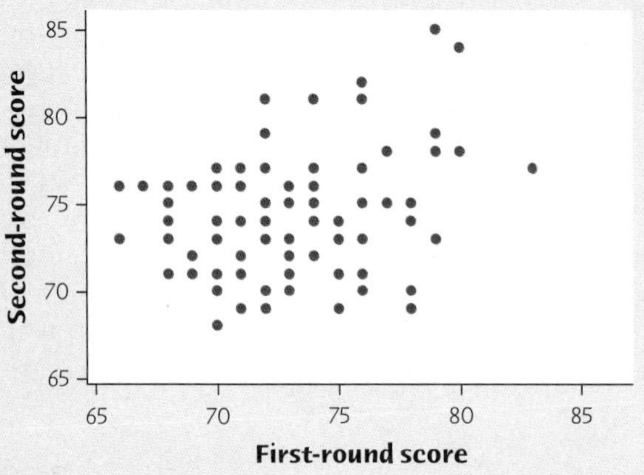

FIGURE 4.8
Scatterplot of the scores in the first two rounds of the 2013 Masters Tournament, for Exercise 4.24.

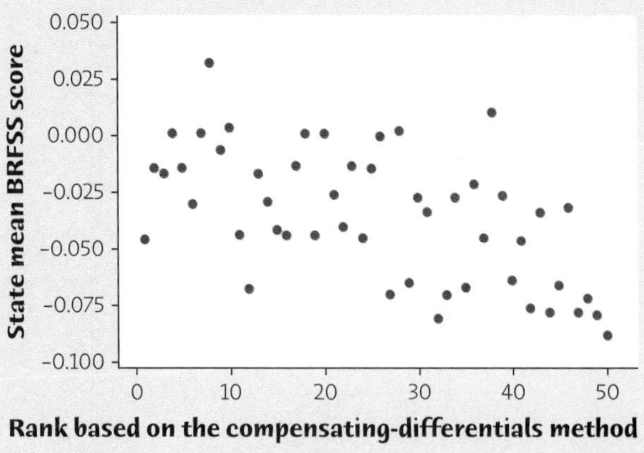

FIGURE 4.9
Scatterplot of mean BRFSS score in each state against each state's well-being rank, for Exercise 4.25.

(a) Is there an overall positive association or an overall negative association between mean BRFSS score and rank based on the compensating-differentials method?
(b) Does the overall association indicate agreement or disagreement between the mean subjective BRFSS score and the ranking based on objective data used in the compensating-differentials method?
(c) Are there any outliers? If so, what are the BRFSS scores corresponding to these outliers?

4.26 Wine and cancer in women. Some studies have suggested that a nightly glass of wine may not only take the edge off a day but also improve health. Is wine good for your health? A study of nearly 1.3 million middle-aged British women examined wine consumption and the risk of breast cancer. The researchers were interested in how risk changed as wine consumption increased. Risk is based on breast cancer rates in drinkers relative to breast cancer rates in nondrinkers in the study, with higher values indicating greater risk. In particular, a value greater than 1 indicates a greater breast cancer rate than that of nondrinkers. Wine intake is the mean wine intake, in grams of alcohol per day (1 glass of wine is approximately 10 grams of alcohol), of all women in the study who drank approximately the same amount of wine per week. Here are the data (for drinkers only):[15] DATA CANCER

Wine intake x (grams of alcohol per day)	2.5	8.5	15.5	26.5
Relative risk y	1.00	1.08	1.15	1.22

(a) Make a scatterplot of these data. Based on the scatterplot, do you expect the correlation to be positive or negative? Near ± 1 or not?
(b) Find the correlation r between wine intake and relative risk. Do the data show that women who consume more wine tend to have higher relative risks of breast cancer?

4.27 Ebola and gorillas. The deadly Ebola virus is a threat to both people and gorillas in Central Africa. An outbreak in 2002 and 2003 killed 91 of the 95 gorillas in seven home ranges in the Congo. To study the spread of the virus, measure "distance" by the number of home ranges separating a group of gorillas from the first group infected. Here are data on distance and time in number of days until deaths began in each later group:[16] DATA EBOLA

Distance	1	3	4	4	4	5
Time	4	21	33	41	43	46

(a) Make a scatterplot. Which is the explanatory variable? What kind of pattern does your plot show?
(b) Find the correlation r between distance and time.
(c) If time in days were replaced by time in number of weeks until death began in each later group (fractions allowed so that 4 days becomes 4/7 weeks), would the correlation between distance and time change? Explain your answer.

4.28 Sparrowhawk colonies. One of nature's patterns connects the percent of adult birds in a colony that return from the previous year and the number of new adults that join the colony. Here are data for 13 colonies of sparrowhawks:[17] DATA SPARROW

age fotostock/Superstock

Percent return	74	66	81	52	73	62	52	45	62	46	60	46	38
New adults	5	6	8	11	12	15	16	17	18	18	19	20	20

(a) Plot the count of new adults (response) against the percent of returning birds (explanatory). Describe the direction and form of the relationship. Is the correlation r an appropriate measure of the strength of this relationship? If so, find r.
(b) For short-lived birds, the association between these variables is positive: changes in weather and food supply drive the populations of new and returning birds up or down together. For long-lived territorial birds, on the other hand, the association is negative because returning birds claim their territories in the colony and don't leave room for new recruits. Which type of species is the sparrowhawk?

4.29 Our brains don't like losses. Most people dislike losses more than they like gains. In money terms, people are about as sensitive to a loss of \$10 as to a gain of \$20. To discover what parts of the brain are active in decisions about gain and loss, psychologists presented subjects with a series of gambles with different odds and different amounts of winnings and losses. From a subject's choices, they constructed a measure of "behavioral loss aversion." Higher scores show greater sensitivity to losses. Observing brain activity while subjects made their decisions pointed to specific brain regions. Here are data for 16 subjects on behavioral loss aversion and "neural loss aversion," a measure of activity in one region of the brain:[18] DATA LOSSES

Neural	−50.0	−39.1	−25.9	−26.7	−28.6	−19.8	−17.6	5.5
Behavioral	0.08	0.81	0.01	0.12	0.68	0.11	0.36	0.34
Neural	2.6	20.7	12.1	15.5	28.8	41.7	55.3	155.2
Behavioral	0.53	0.68	0.99	1.04	0.66	0.86	1.29	1.94

(a) Make a scatterplot that shows how behavior responds to brain activity.
(b) Describe the overall pattern of the data. There is one clear outlier. What is the behavioral score associated with this outlier?
(c) Find the correlation r between neural and behavioral loss aversion both with and without the outlier. Does the outlier have a strong influence on the value of r? By looking at your plot, explain why adding the outlier to the other data points causes r to increase.

4.30 **Sulfur, the ocean, and the sun.** Sulfur in the atmosphere affects climate by influencing formation of clouds. The main natural source of sulfur is dimethyl sulfide (DMS) produced by small organisms in the upper layers of the oceans. DMS production is in turn influenced by the amount of energy the upper ocean receives from sunlight. Here are monthly data on solar radiation dose (SRD, in watts per square meter) and surface DMS concentration (in nanomolars) for a region in the Mediterranean:[19] DATA SULFUR

SRD	12.55	12.91	14.34	19.72	21.52	22.41	37.65	48.41
DMS	0.796	0.692	1.744	1.062	0.682	1.517	0.736	0.720
SRD	74.41	94.14	109.38	157.79	262.67	268.96	289.23	
DMS	1.820	1.099	2.692	5.134	8.038	7.280	8.872	

(a) Make a scatterplot that shows how DMS responds to SRD.
(b) Describe the overall pattern of the data. Find the correlation r between DMS and SRD. Because SRD changes with the seasons of the year, the close relationship between SRD and DMS helps explain other seasonal patterns.

4.31 **Alcohol and cancer in women.** Exercise 4.26 discusses a study of the relationship between wine consumption and the risk of breast cancer in women. The researchers were also interested in how risk changed as consumption of alcoholic beverages other than wine increased. Intake of alcoholic beverages other than wine is the mean intake, in grams of alcohol per day, of all women in the study who drank approximately the same amount of alcohol other than wine per week. Here are the data for both women who drank wine and women who drank alcoholic beverages other than wine: DATA CANCER

Wine intake (grams of alcohol per day)	2.5	8.5	15.5	26.5
Relative risk	1.00	1.08	1.15	1.22
Alcohol intake (grams of alcohol per day)	2.0	7.0	13.0	24.0
Relative risk	0.96	1.06	1.11	1.20

(a) Make a scatterplot of the relative risk versus intake, using separate symbols for the two types of drinks.
(b) What does your plot show about the pattern of risk? What does it show about the effect of type of drink on risk?

4.32 **Feed the birds.** Canaries provide more food to their babies when the babies beg more intensely. Researchers wondered if begging was the main factor determining how much food baby canaries receive, or if parents also take into account whether the babies are theirs or not. To investigate, researchers conducted an experiment allowing canary parents to raise two broods: one of their own and one fostered from a different pair of parents. If begging determines how much food babies receive, then differences in the "begging intensities" of the broods should be strongly associated with differences in the amount of food the broods receive. The researchers decided to use the relative growth rates (the growth rate of the foster babies relative to that of the natural babies, with values greater than 1 indicating that the foster babies grew more rapidly than the natural babies) as a measure of the difference in the amount of food received. They recorded the difference in begging intensities (the begging intensity of the foster babies minus that of the natural babies) and relative growth rates. Here are data from the experiment:[20] DATA CANARY

Difference in begging intensity	−14.0	−12.5	−12.0	−8.0	−8.0	−6.5	−5.5
Relative growth rate	0.85	1.00	1.33	0.85	0.90	1.15	1.00
Difference in begging intensity	−3.5	−3.0	−2.0	−1.5	−1.5	0.0	0.0
Relative growth rate	1.30	1.33	1.03	0.95	1.15	1.13	1.00
Difference in begging intensity	2.00	2.00	3.00	4.50	7.00	8.00	8.50
Relative growth rate	1.07	1.14	1.00	0.83	1.15	0.93	0.70

(a) Make a scatterplot that shows how relative growth rate responds to the difference in begging intensity.
(b) Describe the overall pattern of the relationship. Is it linear? Is there a positive or negative association, or neither? Find the correlation r. Is r a helpful description of this relationship?
(c) If begging intensity is the main factor determining food received, with higher intensity leading to more food, one would expect the relative growth rate to increase as the difference in begging intensity increases. However, if both begging intensity and a preference for their own babies determine the amount of food received (and hence the relative growth rate), we might expect growth rate to increase initially as begging intensity increases but then to level off (or even decrease) as the parents begin to ignore increases in begging by the foster babies. Which of these theories do the data appear to support? Explain your answer.

4.33 Good weather and tipping. Favorable weather has been shown to be associated with increased tipping. Will just the belief that future weather will be favorable lead to higher tips? Researchers gave 60 index cards to a waitress at an Italian restaurant in New Jersey. Before delivering the bill to each customer, the waitress randomly selected a card and wrote on the bill the same message that was printed on the index card. Twenty of the cards had the message "The weather is supposed to be really good tomorrow. I hope you enjoy the day! " Another 20 cards contained the message "The weather is supposed to be not so good tomorrow. I hope you enjoy the day anyway!" The remaining 20 cards were blank, indicating that the waitress was not supposed to write any message. Choosing a card at random ensured that there was a random assignment of the diners to the three experimental conditions. Here are the percentage tips for the three messages:[21] DATA TIPPING

Jupiterimages/Getty Images/Thinkstock

WEATHER REPORT	PERCENTAGE TIP									
Good	20.8	18.7	19.9	20.6	22.0	23.4	22.8	24.9	22.2	20.3
	24.9	22.3	27.0	20.4	22.2	24.0	21.2	22.1	22.0	22.7
Bad	18.0	19.0	19.2	18.8	18.4	19.0	18.5	16.1	16.8	14.0
	17.0	13.6	17.5	19.9	20.2	18.8	18.0	23.2	18.2	19.4
None	19.9	16.0	15.0	20.1	19.3	19.2	18.0	19.2	21.2	18.8
	18.5	19.3	19.3	19.4	10.8	19.1	19.7	19.8	21.3	20.6

(a) Make a plot of percentage tip against the weather report on the bill (space the three weather reports equally on the horizontal axis). Which weather report appears to lead to the best tip?

(b) Does it make sense to speak of a positive or negative association between weather report and percentage tip? Why? Is correlation r a helpful description of the relationship? Why?

4.34 Thinking about correlation. Exercise 4.26 presents data on wine intake and the relative risk of breast cancer in women.

(a) If wine intake is measured in ounces of alcohol per day rather than grams per day, how would the correlation change? (There are 0.035 ounces in a gram.)

(b) How would r change if all the relative risks were 0.25 less than the values given in the table? Does the correlation tell us that among women who drink, those who drink more wine tend to have a greater relative risk of cancer than women who don't drink at all?

(c) If drinking an additional gram of alcohol each day raised the relative risk of breast cancer by exactly 0.01, what would be the correlation between alcohol in wine intake and relative risk of breast cancer? (*Hint*: Draw a scatterplot for several values of alcohol in wine intake.)

4.35 The effect of changing units. Changing the units of measurement can dramatically alter the appearance of a scatterplot. Return to the data on percent body fat and preferred amount of salt in Exercise 4.23: DATA SALT2

Preferred amount of salt x	0.2	0.3	0.4	0.5	0.6	0.8	1.1
Percent body fat y	20	30	22	30	38	23	30

In calculating the preferred amount of salt, the weight of the salt was in milligrams. A mad scientist decides to measure weight in tenths of milligrams. The same data in these units are

Preferred amount of salt x	2	3	4	5	6	8	11
Percent body fat y	20	30	22	30	38	23	30

(a) Make a plot with the x axis extending from 0 to 12 and the y axis from 15 to 35. Plot the original data on these axes. Then plot the new data using a different color or symbol. The two plots look very different.

(b) Nonetheless, the correlation is exactly the same for the two sets of measurements. Why do you know that this is true without doing any calculations? Find the two correlations to verify that they are the same.

4.36 Statistics for investing. Investment reports now often include correlations. Following a table of correlations among mutual funds, a report adds: "Two funds can have perfect correlation, yet different levels of risk. For example, Fund A and Fund B may be perfectly correlated, yet Fund A moves 20% whenever Fund B moves 10%." Write a brief explanation, for someone who knows no statistics, of how this can happen. Include a sketch to illustrate your explanation.

4.37 Statistics for investing. A mutual funds company's newsletter says, "A well diversified portfolio includes assets with low correlations." The newsletter includes a table of correlations between the returns on various classes of investments. For example, the correlation between municipal bonds and large-cap stocks is 0.50, and the correlation between municipal bonds and small-cap stocks is 0.21.

(a) Rachel invests heavily in municipal bonds. She wants to diversify by adding an investment whose

returns do not closely follow the returns on her bonds. Should she choose large-cap stocks or small-cap stocks for this purpose? Explain your answer.
(b) If Rachel wants an investment that tends to increase when the return on her bonds drops, what kind of correlation should she look for?

4.38 Teaching and research. A college newspaper interviews a psychologist about student ratings of the teaching of faculty members. The psychologist says, "The evidence indicates that the correlation between the research productivity and teaching rating of faculty members is close to zero." The paper reports this as "Professor McDaniel said that good researchers tend to be poor teachers, and vice versa." Explain why the paper's report is wrong. Write a statement in plain language (don't use the word *correlation*) to explain the psychologist's meaning.

4.39 Sloppy writing about correlation. Each of the following statements contains a blunder. Explain in each case what is wrong.

(a) "There is a high correlation between the sex of American workers and their income."
(b) "We found a high correlation ($r = 1.09$) between students' ratings of faculty teaching and ratings made by other faculty members."
(c) "The correlation between height and weight of the subjects was $r = 0.63$ centimeter per kilogram."

4.40 More about scatterplots and correlation. Here are two sets of data.

Data set A

x	1	2	3	4
y	1	1.5	0.5	4

Data set B

x	1	1	1	2	3	4	4	4
y	1	1	1	1.5	0.5	4	4	4

(a) Make a scatterplot of both sets of data. Comment on any differences you see in the two plots.
(b) Compute the correlation for both sets of data. Comment on any differences in the two values. Are these differences what you would expect from the plots in part (a)?

4.41 Correlation is not resistant. Go to the *Correlation and Regression* applet. Click on the scatterplot to create a group of 10 points in the lower-left corner of the scatterplot with a strong straight-line pattern (correlation about 0.9).

(a) Add one point at the upper right that is in line with the first 10. How does the correlation change?
(b) Drag this last point down until it is opposite the group of 10 points. How small can you make the correlation? Can you make the correlation negative? You see that a single outlier can greatly strengthen or weaken a correlation. Always plot your data to check for outlying points.

4.42 Match the correlation. You are going to use the *Correlation and Regression* applet to make scatterplots with 10 points that have correlation close to 0.7. The lesson is that many patterns can have the same correlation. Always plot your data before you trust a correlation.

(a) Click on the scatterplot to add the first two points. What is the value of the correlation? Why does it have this value?
(b) Make a lower-left to upper-right pattern of 10 points with correlation about $r = 0.7$. (You can drag points up or down to adjust r after you have 10 points.) Make a rough sketch of your scatterplot.
(c) Make another scatterplot with nine points in a vertical stack at the left of the plot. Add one point far to the right, and move it until the correlation is close to 0.7. Make a rough sketch of your scatterplot.
(d) Make yet another scatterplot with 10 points in a curved pattern that starts at the lower left, rises to the right, then falls again at the far right. Adjust the points up or down until you have a fairly smooth curve with correlation close to 0.7. Make a rough sketch of this scatterplot also.

The following exercises ask you to answer questions from data without having the details outlined for you. The exercise statements give you the **State** *step of the four-step process. In your work, follow the* **Plan, Solve,** *and* **Conclude** *steps of the process, described on page 63.*

4.43 Global warming. Have average global temperatures been increasing in recent years? Here are annual average global temperatures for the last 20 years in degrees Celsius.[22] GTEMPS

Year	1994	1995	1996	1997	1998	1999	2000	2001	2002	2003
Temperature	14.23	14.35	14.22	14.42	14.54	14.36	14.33	14.45	14.51	14.52
Year	2004	2005	2006	2007	2008	2009	2010	2011	2012	2013
Temperature	14.48	14.55	14.50	14.49	14.41	14.50	14.56	14.43	14.48	14.52

Discuss what the data show about change in average global temperatures over time.

4.44 Will women outrun men? Does the physiology of women make them better suited than men to long-distance running? Will women eventually outperform men in long-distance races? Researchers examined data on world record times (in seconds) for men and women in the marathon. Here are data for women:[23] RUNNING

Year	1926	1964	1967	1970	1971	1974	1975
Time	13,222.0	11,973.0	11,246.0	10,973.0	9990.0	9834.5	9499.0
Year	1977	1980	1981	1982	1983	1985	
Time	9287.5	9027.0	8806.0	8771.0	8563.0	8466.0	

Here are data for men:

Year	1908	1909	1913	1920	1925	1935	1947
Time	10,518.4	9751.0	9366.6	9155.8	8941.8	8802.0	8739.0
Year	1952	1953	1954	1958	1960	1963	1964
Time	8442.2	8314.8	8259.4	8117.0	8116.2	8068.0	7931.2
Year	1965	1967	1969	1981	1984	1985	1988
Time	7920.0	7776.4	7713.6	7698.0	7685.0	7632.0	7610.0

(a) What do the data show about women's and men's times in the marathon? (Start by plotting both sets of data on the same plot, using two different plotting symbols.)
(b) Based on these data, researchers (in 1992) predicted that women would outrun men in the marathon in 1998. How do you think they arrived at this date? Was their prediction accurate? (You may want to look on the web; try doing a Google search on "women's world record marathon times.")

4.45 **Toucan's beak.** The toco toucan, the largest member of the toucan family, possesses the largest beak relative to body size of all birds. This exaggerated feature has received various interpretations, such as being a refined adaptation for feeding. However, the large surface area may also be an important mechanism for radiating heat (and hence cooling the bird) as outdoor temperature increases. Here are data for beak heat loss, as a percent of total body heat loss, at various temperatures in degrees Celsius:[24] TOUCAN

© Kevin Schafer/Alamy

Temperature (°C)	15	16	17	18	19	20	21	22
Percent heat loss from beak	32	34	35	33	37	46	55	51
Temperature (°C)	23	24	25	26	27	28	29	30
Percent heat loss from beak	43	52	45	53	58	60	62	62

Investigate the relationship between outdoor temperature and beak heat loss, as a percentage of total body heat loss.

4.46 **Does social rejection hurt?** We often describe our emotional reaction to social rejection as "pain." Does social rejection cause activity in areas of the brain that are known to be activated by physical pain? If it does, we really do experience social and physical pain in similar ways. Psychologists first included and then deliberately excluded individuals from a social activity while they measured changes in brain activity. After each activity, the subjects filled out questionnaires that assessed how excluded they felt. Here are data for 13 subjects:[25] REJECT

SUBJECT	SOCIAL DISTRESS	BRAIN ACTIVITY
1	1.26	−0.055
2	1.85	−0.040
3	1.10	−0.026
4	2.50	−0.017
5	2.17	−0.017
6	2.67	0.017
7	2.01	0.021
8	2.18	0.025
9	2.58	0.027
10	2.75	0.033
11	2.75	0.064
12	3.33	0.077
13	3.65	0.124

The explanatory variable is "social distress" measured by each subject's questionnaire score after exclusion relative to the score after inclusion. (So values greater than 1 show the degree of distress caused by exclusion.) The response variable is change in activity in a region of the brain that is activated by physical pain. Negative values show a decrease in activity, suggesting less distress. Discuss what the data show about the relationship between social distress and brain activity.

4.47 **Yukon squirrels.** The population density of North American red squirrels in the Yukon, Canada, fluctuates annually. Researchers believe one reason for the fluctuation may be the availability of white spruce cones in the spring, a significant source of food for the squirrels. To explore this, researchers measured red squirrel population density in the spring and spruce cone production the previous autumn over a 23-year period. The data for one study area appear in Table 4.2.[26] Squirrel population density is measured in squirrels per hectare. Spruce cone production is an index on a logarithmic scale, with larger values indicating larger spruce cone production. Discuss whether the data support the idea that higher spruce cone production in the autumn leads to a higher squirrel population density the following spring. SQRLCO

TABLE 4.2 RED SQUIRREL POPULATION DENSITY AND SPRUCE CONE PRODUCTION IN YUKON, CANADA

SQUIRREL DENSITY (SQUIRRELS PER ha)	CONE PRODUCTION (INDEX)	SQUIRREL DENSITY (SQUIRRELS PER ha)	CONE PRODUCTION (INDEX)
1.0	0.1	1.3	2.3
1.5	0.3	2.3	2.5
1.4	0.4	2.0	2.9
1.5	0.5	1.0	3.1
0.7	0.5	1.4	3.3
1.2	0.8	0.9	3.6
0.9	1.2	1.3	3.8
1.0	1.4	1.9	4.2
1.2	1.8	1.8	5.1
1.4	2.0	1.9	5.2
1.5	2.1	3.4	5.3
0.8	2.2		

4.48 **Teacher salaries.** For each of the 50 states and the District of Columbia, average Mathematics SAT scores and average high school teacher salaries for 2013 are available.[27] Discuss whether the data support the idea that higher teacher salaries lead to higher Mathematics SAT scores. TCHSAL

Exploring the Web

4.49 **Drive for show, putt for dough.** A popular saying in golf is "You drive for show, but you putt for dough." The point is that hitting the golf ball a long way with a driver looks impressive, but putting well is more important for the final score and hence the amount of money you win. You can find this season's Professional Golfers, Association (PGA) Tour statistics at the PGA Tour website: `www.pgatour.com/stats.html` (click on "View All" under any category displayed to see the statistics for all golfers). You can also find these statistics at the ESPN website: `espn.go.com/golf/statistics/_/year`. Look at the most recent putting, driving, and money earnings data for the current season on the PGA Tour.

(a) Make a scatterplot of earnings and putting average. Use earnings as the response variable. Describe the direction, form, and strength of the relationship in the plot. Are there any outliers?

(b) Make a scatterplot of earnings and driving distance. Use earnings as the response variable. Describe the direction, form, and strength of the relationship in the plot. Are there any outliers?

(c) Do your plots support the maxim "You drive for show but you putt for dough"?

4.50 **Olympic medals.** Go to the *Chance News* website at `www.causeweb.org/wiki/chance/index.php/Chance_News_61#Predicting_medal_counts` and read the article "Predicting Medal Counts." Next, search the web and locate the Winter Olympics medal counts for 2010 and 2014 (I found Winter Olympics medal counts on Wikipedia). Make a scatterplot that is similar to the one in the *Chance News* article but that uses the 2010 medal counts to predict the 2014 medal counts. How does your plot compare with the plot in the *Chance News* article?

CHAPTER 5

Masakazu Watanabe/Aflo/Getty Images

Regression

In this chapter we cover...

5.1 Regression lines
5.2 The least-squares regression line
5.3 Using technology
5.4 Facts about least-squares regression
5.5 Residuals
5.6 Influential observations
5.7 Cautions about correlation and regression
5.8 Association does not imply causation

Linear (straight-line) relationships between two quantitative variables are easy to understand and quite common. In Chapter 4, we found linear relationships in settings as varied as Florida manatee deaths, the risk of cancer, and predicting tropical storms. Correlation measures the direction and strength of these linear relationships. When a scatterplot shows a linear relationship, we would like to summarize the overall pattern by drawing a line on the scatterplot.

5.1 Regression lines

A *regression line* summarizes the relationship between two variables, but only in a specific setting: one of the variables helps explain or predict the other. That is, regression describes a relationship between an explanatory variable and a response variable.

Regression Line

A **regression line** is a straight line that describes how a response variable y changes as an explanatory variable x changes. We often use a regression line to predict the value of y for a given value of x when we believe the relationship between y and x is linear.

EXAMPLE 5.1 Does Fidgeting Keep You Slim?

4 step

DATA FATGAIN

Why is it that some people find it easy to stay slim? Here, following our four-step process (page 63), is an account of a study that sheds some light on gaining weight.

STATE: Some people don't gain weight even when they overeat. Perhaps fidgeting and other "nonexercise activity" (NEA) explains why. Some people may spontaneously increase nonexercise activity when fed more. Researchers deliberately overfed 16 healthy young adults for eight weeks. They measured fat gain (in kilograms) and, as an explanatory variable, change in energy use (in calories) from activity other than deliberate exercise—fidgeting, daily living, and the like. Change in energy use was energy use measured the last day of the eight-week period minus energy use measured the day before the overfeeding began. Here are the data:[1]

NEA Change (cal)	−94	−57	−29	135	143	151	245	355
Fat Gain (kg)	4.2	3.0	3.7	2.7	3.2	3.6	2.4	1.3
NEA Change (cal)	392	473	486	535	571	580	620	690
Fat Gain (kg)	3.8	1.7	1.6	2.2	1.0	0.4	2.3	1.1

Do people with larger increases in NEA tend to gain less fat?

PLAN: Make a scatterplot of the data, and examine the pattern. If it is linear, use correlation to measure its strength and draw a regression line on the scatterplot to predict fat gain from change in NEA.

SOLVE: Figure 5.1 is a scatterplot of these data. The plot shows a moderately strong negative linear association with no outliers. The correlation is

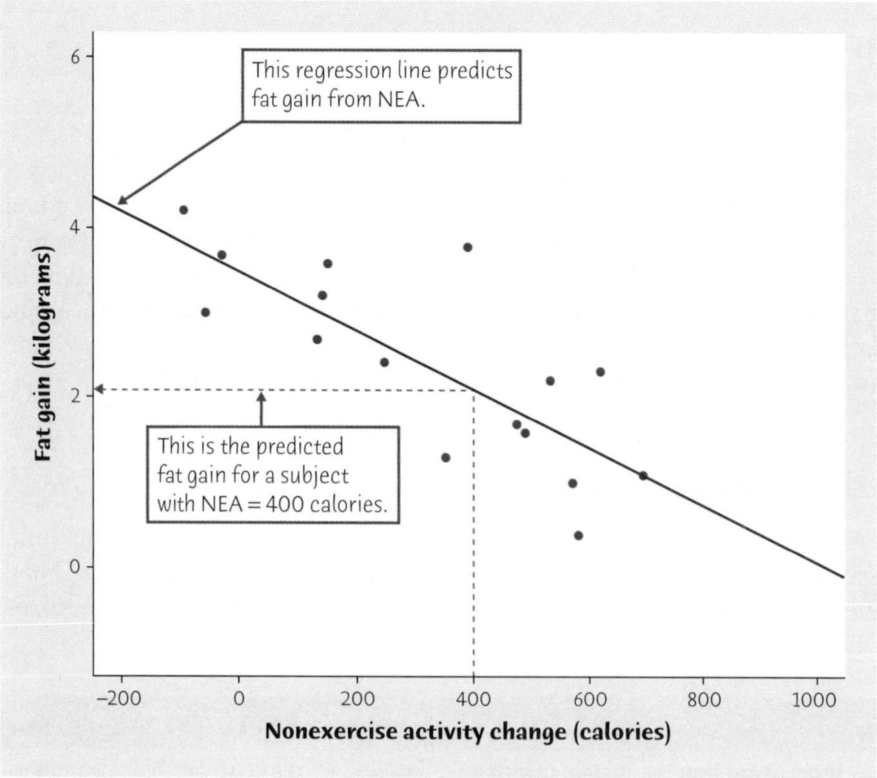

FIGURE 5.1
Fat gain after eight weeks of overeating, plotted against increase in nonexercise activity over the same period, for Example 5.1.

$r = -0.7786$. The line on the plot is a regression line for predicting fat gain from change in NEA.

CONCLUDE: People with larger increases in NEA do indeed gain less fat. To add to this conclusion, we must study regression lines in more detail.

We can, however, already use the regression line to predict fat gain from NEA. Suppose that an individual's NEA increases by 400 calories when she overeats. Go "up and over" on the graph in Figure 5.1. From 400 calories on the x axis, go up to the regression line and then over to the y axis. The graph shows that the predicted gain in fat is a bit more than 2 kilograms. ■

Statistics in Your World: REGRESSION TOWARD THE MEAN

To "regress" means to go backward. Why are statistical methods for predicting a response from an explanatory variable called "regression"? Sir Francis Galton (1822–1911), who was the first to apply regression to biological and psychological data, looked at examples such as the heights of children versus the heights of their parents. He found that taller-than-average parents tended to have children who were also taller than average but not as tall as their parents. Galton called this fact "regression toward the mean," and the name came to be applied to the statistical method.

Many calculators and software programs will give you the equation of a regression line from keyed-in data. Understanding and using the line are more important than the details of where the equation comes from.

Review of Straight Lines

Suppose that y is a response variable (plotted on the vertical axis) and x is an explanatory variable (plotted on the horizontal axis). A straight line relating y to x has an equation of the form

$$y = a + bx$$

In this equation, b is the **slope**, the amount by which y changes when x increases by one unit. The number a is the **intercept**, the value of y when $x = 0$.

EXAMPLE 5.2 Using a Regression Line

Any straight line describing the NEA data has the form

$$\text{fat gain} = a + (b \times \text{NEA change})$$

The line in Figure 5.1 is the regression line with the equation

$$\text{fat gain} = 3.505 - 0.00344 \times \text{NEA change}$$

Be sure you understand the role of the two numbers in this equation:

- The slope $b = -0.00344$ tells us that, on average, fat gained goes down by 0.00344 kilogram for each added calorie of NEA change. The slope of a regression line is the *rate of change* in the response, on average, as the explanatory variable changes.
- The intercept, $a = 3.505$ kilograms, is the estimated fat gain if NEA does not change when a person overeats.

The equation of the regression line makes it easy to predict fat gain. If a person's NEA change increases by 400 calories when she overeats, substitute $x = 400$ in the equation. The predicted fat gain is

$$\text{fat gain} = 3.505 - (0.00344 \times 400) = 2.13 \text{ kilograms}$$

This is a bit more than 2 kilograms, as we estimated directly from the plot in Example 5.1.

plotting a line

When **plotting a line** on the scatterplot, use the equation to find the predicted y for two values of x, one near each end of the range of x in the data. Plot each y above its x value, and draw the line through the two points. ■

The slope of a regression line is an important numerical description of the relationship between the two variables. Although we need the value of the intercept to draw the line, this value is statistically meaningful only when, as in Example 5.2, the explanatory variable can actually take values close to zero. The slope $b = -0.00344$ in Example 5.2 is small. This does *not* mean that change in NEA has little effect on fat gain. The size of the slope depends on the units in which we measure the two variables. In this example, the slope is the change in fat gain in kilograms when NEA increases by one calorie. There are 1000 grams in a kilogram. If we measured fat gain in grams, the slope would be 1000 times larger, $b = -3.44$. *You can't say how important a relationship is by looking at the size of the slope of the regression line.*

CAUTION

LaunchPad Online Resources

- The **StatClips Examples videos**: *Regression—Introduction and Motivation Examples A, B, and C,* discuss examples of regression lines.

Apply Your Knowledge

5.1 **City Mileage, Highway Mileage.** We expect a car's highway gas mileage to be related to its city gas mileage (in miles per gallon, mpg). Data for all 1137 vehicles in the government's *2013 Fuel Economy Guide* give the regression line

$$\text{highway mpg} = 6.785 + (1.033 \times \text{city mpg})$$

for predicting highway mileage from city mileage.

(a) What is the slope of this line? Say in words what the numerical value of the slope tells you.

(b) What is the intercept? Explain why the value of the intercept is not statistically meaningful.

(c) Find the predicted highway mileage for a car that gets 16 miles per gallon in the city. Do the same for a car with city mileage of 28 mpg.

(d) Draw a graph of the regression line for city mileages between 10 and 50 mpg. (Be sure to show the scales for the x and y axes.)

5.2 **What's the Line?** An online article suggested that for each additional person who took up regular running for exercise, the number of cigarettes smoked daily would decrease by 0.178.[2] If we assume that 48 million cigarettes would be smoked per day if nobody ran, what is the equation of the regression line for predicting number of cigarettes smoked per day from the number of people who regularly run for exercise?

5.3 **Shrinking Forests.** Scientists measured the annual forest loss (in square kilometers) in Indonesia from 2000 to 2012.[3] They found the regression line

$$\text{forest loss} = 7500 + (1021 \times \text{year since 2000})$$

for predicting forest loss in square kilometers from years since 2000.

(a) What is the slope of this line? Say in words what the numerical value of the slope tells you.

(b) If we measured forest loss in square meters per year, what would the slope be? Note that there are 10^6 square meters in a square kilometer.

(c) If we measured forest loss in thousands of square kilometers per year, what would the slope be?

5.2 The least-squares regression line

In most cases, no line will pass exactly through all the points in a scatterplot. Different people will draw different lines by eye. We need a way to draw a regression line that doesn't depend on our guess of where the line should go. Because we use the line to predict y from x, the prediction errors we make are errors in y, the vertical direction in the scatterplot. *A good regression line makes the vertical distances of the points from the line as small as possible.*

Figure 5.2 illustrates the idea. This plot shows three of the points from Figure 5.1, along with the line, on an expanded scale. The line passes above one of the points and below two of them. The three prediction errors appear as vertical line segments.

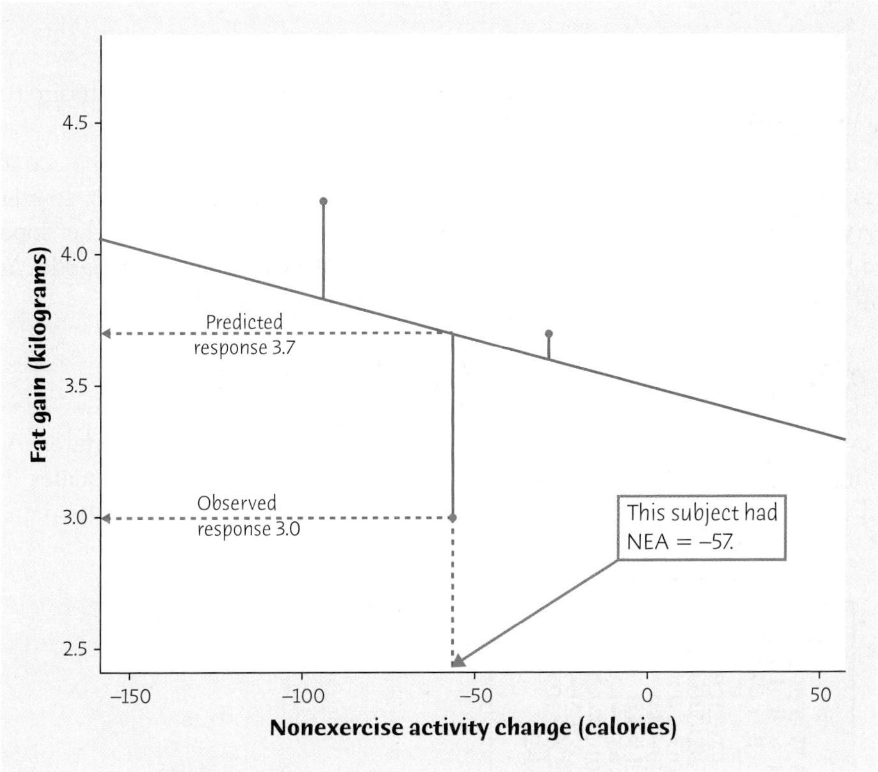

FIGURE 5.2 The least-squares idea. For each observation, find the vertical distance of each point on the scatterplot from a regression line. The least-squares regression line makes the sum of the squares of these distances as small as possible.

For example, one subject had $x = -57$, a decrease of 57 calories in NEA. The line predicts a fat gain of 3.7 kilograms, but the actual fat gain for this subject was 3.0 kilograms. The prediction error is

$$\begin{aligned} \text{error} &= \text{observed response} - \text{predicted response} \\ &= 3.0 - 3.7 = -0.7 \text{ kilogram} \end{aligned}$$

There are many ways to make the collection of vertical distances "as small as possible." The most common is the *least-squares* method.

Least-Squares Regression Line

The **least-squares regression line** of y on x is the line that makes the sum of the squares of the vertical distances of the data points from the line as small as possible.

One reason for the popularity of the least-squares regression line is that the problem of finding the line has a simple answer. We can give the equation for the least-squares line in terms of the means and standard deviations of the two variables and the correlation between them.

Equation of the Least-Squares Regression Line

We have data on an explanatory variable x and a response variable y for n individuals. From the data, calculate the means $\bar{x}$ and $\bar{y}$ and the standard deviations s_x and s_y of the two variables and their correlation r. The least-squares regression line is the line

$$\hat{y} = a + bx$$

with **slope**

$$b = r\frac{s_y}{s_x}$$

and **intercept**

$$a = \bar{y} - b\bar{x}$$

We write $\hat{y}$ (read "y hat") in the equation of the regression line to emphasize that the line gives a *predicted* response $\hat{y}$ for any x. Because of the scatter of points about the line, the predicted response will usually not be exactly the same as the actually *observed* response y. In practice, you don't need to calculate the means, standard deviations, and correlation first. Software or your calculator will give the slope b and intercept a of the least-squares line from the values of the variables x and y. You can then concentrate on understanding and using the regression line.

5.3 Using technology

Least-squares regression is one of the most common statistical procedures. Any technology you use for statistical calculations will give you the least-squares line and related information. Figure 5.3 displays the regression output for the data of

Texas Instruments Graphing Calculator

```
LinReg
 y=a+bx
 a=3.505122916
 b=-.003441487
 r²=.6061492049
 r=-.7785558457
```

Minitab

```
Session

Regression Analysis: fat versus nea

The regression equation is
fat = 3.51 - 0.00344 nea


Predictor         Coef     SE Coef       T      P
Constant        3.5051      0.3036   11.54  0.000
nea         -0.0034415   0.0007414   -4.64  0.000


S = 0.739853   R-Sq = 60.6%   R-Sq (adj) = 57.8%
```

FIGURE 5.3 Least-squares regression for the nonexercise activity data: output from a graphing calculator, three statistical programs, and a spreadsheet program.

(Continued)

FIGURE 5.3
(*Continued*)

CrunchIt!

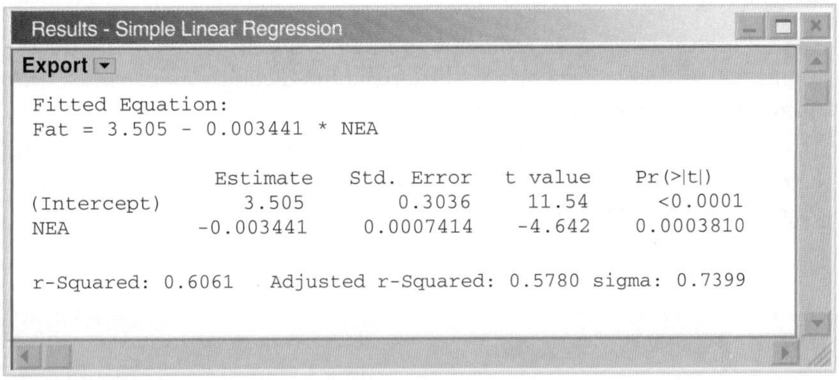

Results - Simple Linear Regression

Export

Fitted Equation:
Fat = 3.505 - 0.003441 * NEA

	Estimate	Std. Error	t value	Pr(>|t|)
(Intercept)	3.505	0.3036	11.54	<0.0001
NEA	-0.003441	0.0007414	-4.642	0.0003810

r-Squared: 0.6061 Adjusted r-Squared: 0.5780 sigma: 0.7399

Microsoft Excel

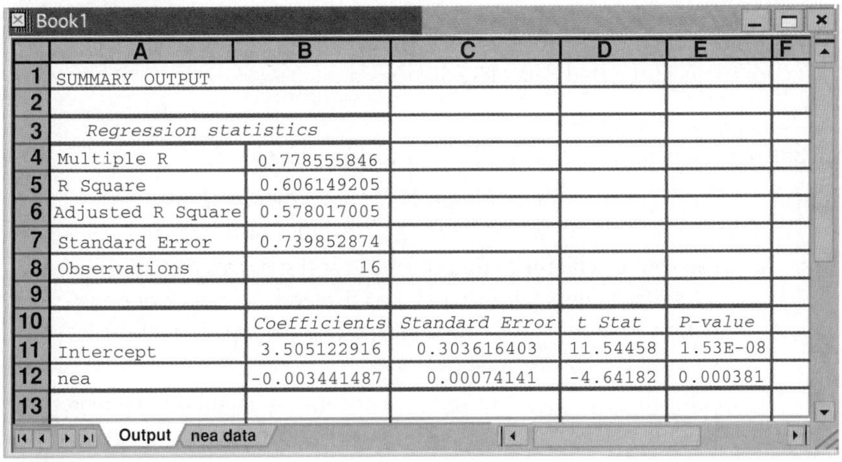

Book1

	A	B	C	D	E	F
1	SUMMARY OUTPUT					
2						
3	*Regression statistics*					
4	Multiple R	0.778555846				
5	R Square	0.606149205				
6	Adjusted R Square	0.578017005				
7	Standard Error	0.739852874				
8	Observations	16				
9						
10		*Coefficients*	*Standard Error*	*t Stat*	*P-value*	
11	Intercept	3.505122916	0.303616403	11.54458	1.53E-08	
12	nea	-0.003441487	0.00074141	-4.64182	0.000381	
13						

Output / nea data

JMP

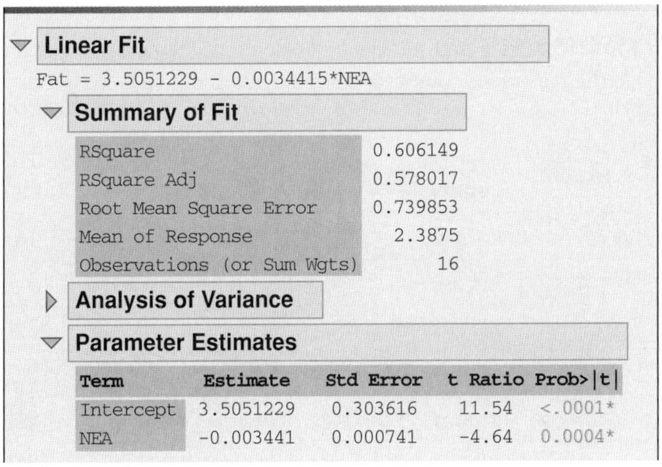

Linear Fit

Fat = 3.5051229 - 0.0034415*NEA

Summary of Fit

RSquare	0.606149
RSquare Adj	0.578017
Root Mean Square Error	0.739853
Mean of Response	2.3875
Observations (or Sum Wgts)	16

Analysis of Variance

Parameter Estimates

Term	Estimate	Std Error	t Ratio	Prob>|t|
Intercept	3.5051229	0.303616	11.54	<.0001*
NEA	-0.003441	0.000741	-4.64	0.0004*

Examples 5.1 and 5.2 from a graphing calculator, three statistical programs, and a spreadsheet program. Each output records the slope and intercept of the least-squares line. The software also provides information that we do not yet need, although we will use much of it later. (In fact, we left out part of the Minitab and Excel outputs.) Be sure that you can locate the slope and intercept on all four outputs. *Once you understand the statistical ideas, you can read and work with almost any software output.*

LaunchPad Online Resources

- The **Snapshots video**, *Introduction to Regression,* discusses least-squares regression in the context of a real example (predicting surface water supply from the snowpack in Colorado).
- The software CrunchIt! is available in the student resources, as are CrunchIt! help videos. Also among the online student resources are **Video Technology Manuals** for Minitab, the TI graphing calculator, Excel, JMP, and SPSS.

Apply Your Knowledge

5.4 Coral Reefs. Exercises 4.2 and 4.10 discuss a study in which scientists examined data on mean sea surface temperatures (in degrees Celsius) and mean coral growth (in millimeters per year) over a several-year period at locations in the Red Sea. Here are the data:[4] DATA CORAL

Sea Surface Temperature	29.68	29.87	30.16	30.22	30.48	30.65	30.90
Growth	2.63	2.58	2.68	2.60	2.48	2.38	2.26

(a) Use your calculator to find the mean and standard deviation of both sea surface temperature x and growth y and the correlation r between x and y. Use these basic measures to find the equation of the least-squares line for predicting y from x.

(b) Enter the data into your software or calculator, and use the regression function to find the least-squares line. The result should agree with your work in part (a) up to roundoff error.

(c) Say in words what the numerical value of the slope tells you.

5.5 Death by Intent. Homicide and suicide are both intentional means of ending a life. However, the reason for committing a homicide is different from that for

County	Homicide Rate	Suicide Rate	County	Homicide Rate	Suicide Rate
Allen	4.2	9.2	Lorain	3.1	11.0
Ashtabula	1.8	15.5	Lucas	7.4	13.3
Butler	2.6	12.7	Mahoning	10.9	12.4
Clermont	1.0	16.0	Medina	0.5	10.0
Clark	5.6	14.5	Miami	2.6	9.2
Columbiana	3.5	16.6	Montgomery	9.5	15.2
Cuyahoga	9.2	9.5	Portage	1.6	9.6
Delaware	0.8	7.6	Stark	4.7	13.5
Franklin	8.7	11.4	Summit	4.9	11.5
Greene	2.7	12.8	Trumbull	5.8	16.6
Hamilton	8.9	10.8	Warren	0.7	11.3
Lake	1.8	11.3	Wayne	1.8	8.9
Licking	4.5	12.9	Wood	1.0	7.4

suicide and we might expect homicide and suicide rates to be uncorrelated. On the other hand, both can involve some degree of violence, so perhaps we might expect some level of correlation in the rates. The data from 2008–10 for 26 counties in Ohio are shown in the table.[5] Rates are per 100,000 people. DEATH

(a) Make a scatterplot that shows how suicide rate can be predicted from homicide rate. There is a weak linear relationship, with correlation $r = 0.23$.

(b) Find the least-squares regression line for predicting suicide rate from homicide rate. Add this line to your scatterplot.

(c) Explain in words what the slope of the regression line tells us.

(d) Another Ohio county has a homicide rate of 8.0 per 100,000 people. What is the county's predicted suicide rate?

5.4 Facts about least-squares regression

One reason for the popularity of least-squares regression lines is that they have many convenient properties. Here are some facts about least-squares regression lines.

Fact 1. The distinction between explanatory and response variables is essential in regression. Least-squares regression makes the distances of the data points from the line small only in the y direction. If we reverse the roles of the two variables, we get a different least-squares regression line.

EXAMPLE 5.3 Predicting Fat Gain, Predicting Change in NEA

Figure 5.4 repeats the scatterplot of the NEA data in Figure 5.1, but with *two* least-squares regression lines. The solid line is the regression line for predicting fat gain from change in NEA. This is the line that appeared in Figure 5.1.

We might also use the data on these 16 subjects to predict the change in NEA for another subject from that subject's fat gain when overfed for eight weeks. Now the roles of the variables are reversed: fat gain is the explanatory variable, and change in NEA is the response variable. The dashed line in Figure 5.4 is the least-squares line for predicting NEA change from fat gain. The two regression lines are not the same. *In the regression setting, you must know clearly which variable is explanatory.* ■

CAUTION

Fact 2. There is a close connection between correlation and the slope of the least-squares line. The slope is

$$b = r\frac{s_y}{s_x}$$

You see that **the slope and the correlation always have the same sign.** For example, if a scatterplot shows a positive association, then both b and r are positive. The formula for the slope b says more: along the regression line, **a change of one standard deviation in x corresponds to a change of r standard deviations in y.** When the variables are perfectly correlated ($r = 1$ or $r = -1$), the change in the predicted response $\hat{y}$ is the same (in standard deviation units) as the change in x.

FIGURE 5.4
Two least-squares regression lines for the nonexercise activity data, for Example 5.3. The solid line predicts fat gain from change in nonexercise activity. The equation for this line is fat = 3.505 − 0.00344 NEA. The dashed line predicts change in nonexercise activity from fat gain. The equation for this line is NEA = 745.3 − 176 fat (or after rearranging terms, fat = 4.23 − 0.00568 NEA).

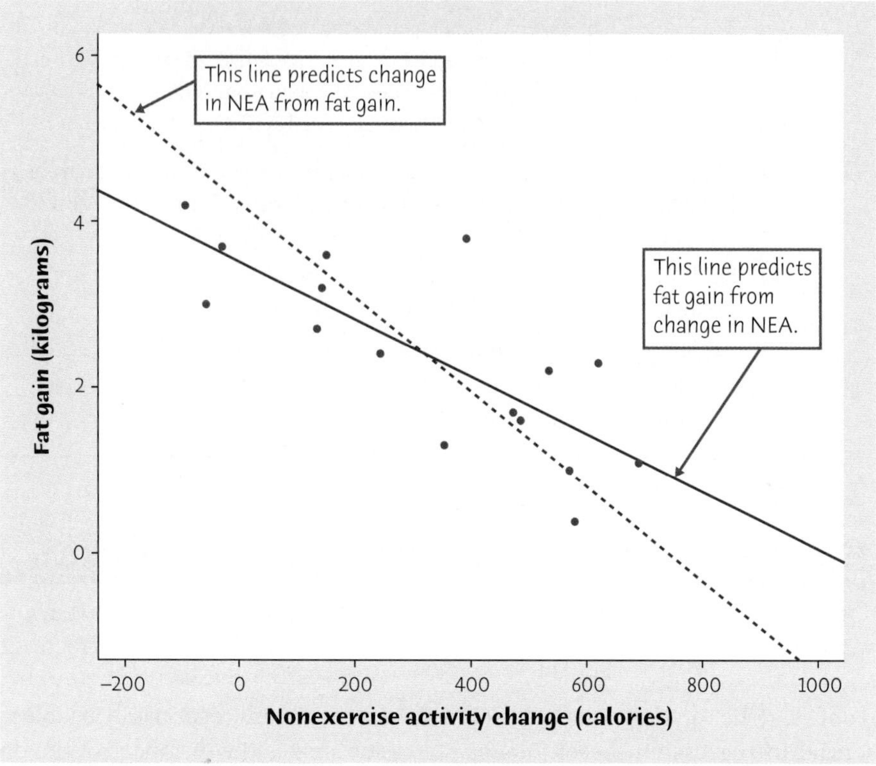

Otherwise, because $-1 \le r \le 1$, the change in $\hat{y}$ (in standard deviation units) is less than the change in x (in standard deviation units). As the correlation grows less strong, the prediction $\hat{y}$ moves less in response to changes in x.

Fact 3. The least-squares regression line always passes through the point $(\bar{x}, \bar{y})$ on the graph of y against x. This is a consequence of the equation of the least-squares regression line (box on page 132). In Exercise 5.54 we ask you to confirm this.

Fact 4. The correlation r describes the strength of a straight-line relationship. In the regression setting, this description takes a specific form: **the square of the correlation, r^2, is the fraction of the variation in the values of y that is explained by the least-squares regression of y on x.**

The idea is that when there is a linear relationship, some of the variation in y is accounted for by the fact that as x changes, y changes along with it. Look again at Figure 5.1, the scatterplot of the NEA data. The variation in y appears as the spread of fat gains from 0.4 to 4.2 kg. Some of this variation is explained by the fact that x (change in NEA) varies from a loss of 94 calories to a gain of 690 calories. As x changes from −94 to 690, y changes along the line. You would predict a smaller fat gain for a subject whose NEA increased by 600 calories than for someone with 0 change in NEA. But the straight-line tie of y to x doesn't explain *all* of the variation in y. The remaining variation appears as the scatter of points above and below the line.

Although we won't do the algebra, it is possible to break the variation in the observed values of y into two parts. One part measures the variation in $\hat{y}$ along the least-squares regression line as x varies. The other measures the vertical scatter of the data points above and below the line. The squared correlation r^2 is the first of these as a fraction of the whole:

$$r^2 = \frac{\text{variation in } \hat{y} \text{ along the regression line as } x \text{ varies}}{\text{total variation in observed values of } y}$$

EXAMPLE 5.4 Using r^2

For the NEA data, $r = -0.7786$ and $r^2 = (-0.7786)^2 = 0.6062$. About 61% of the variation in fat gained is accounted for by the linear relationship with change in NEA. The other 39% is individual variation among subjects that is not explained by the linear relationship.

Figure 4.2 (page 108) shows a stronger linear relationship between boat registrations in Florida and manatees killed by boats. The correlation is $r = 0.954$ and $r^2 = (0.954)^2 = 0.910$. Ninety-one percent of the year-to-year variation in number of manatees killed by boats is explained by regression on number of boats registered. Only about 9% is variation among years with similar numbers of boats registered. ■

CAUTION *You can find a regression line for any relationship between two quantitative variables, but the usefulness of the line for prediction depends on the strength of the linear relationship.* So r^2 is almost as important as the equation of the line in reporting a regression. All the outputs in Figure 5.3 include r^2, either in decimal form or as a percent. When you see a correlation, square it to get a better feel for the strength of the association. Perfect correlation ($r = -1$ or $r = 1$) means the points lie exactly on a line. Then $r^2 = 1$, and all the variation in one variable is accounted for by the linear relationship with the other variable. If $r = -0.7$ or $r = 0.7$, $r^2 = 0.49$ and about half the variation is accounted for by the linear relationship. In the r^2 scale, correlation ± 0.7 is about halfway between 0 and ± 1.

Facts 2, 3, and 4 are special properties of least-squares regression. They are not true for other methods of fitting a line to data that are discussed in more advanced courses.

Apply Your Knowledge

5.6 How Useful Is Regression? Figure 4.8 (page 120) displays the relationship between golfers' scores on the first and second rounds of the 2013 Masters Tournament. The correlation is $r = 0.246$. Exercise 4.30 (page 122) gives data on solar radiation (SRD) and concentration of dimethyl sulfide (DMS) over a region of the Mediterranean. The correlation is $r = 0.969$. Explain in simple language why knowing only these correlations enables you to say that prediction of DMS from SRD by a regression line will be much more accurate than prediction of a golfer's second-round score from his first-round score.

© Arco Images GmbH/Alamy

5.7 Feed the Birds. Exercise 4.32 (page 122) gives data from a study in which canary parents cared for both their own babies and those of other parents. Investigators looked at how the growth rate of the foster babies relative to the growth rate of the natural babies changed as the begging intensity for food by the foster babies increased over the begging intensity of the natural babies. If begging intensity is the main factor determining food received, with higher intensity leading to more food, one would expect the relative growth rate to increase as the difference in begging intensity increases. However, if both begging intensity and a preference for their own babies determine the amount of food received (and hence the relative growth rate), we might expect growth rate to increase initially as begging intensity increases but then to level off (or even decrease) as the parents begin to ignore further increases in begging by the foster babies. DATA CANARY

(a) Make a scatterplot of the data. Find the least-squares regression line for predicting relative growth rate of the foster brood from the difference in begging intensity between the foster brood and the actual babies of the parents, and add this line to your plot. Should we *not* use the regression line for prediction in this setting?

(b) What is r^2? What does this value say about the success of the regression line in predicting relative growth rate?

5.5 Residuals

One of the first principles of data analysis is to look for an overall pattern and also for striking deviations from the pattern. A regression line describes the overall pattern of a linear relationship between an explanatory variable and a response variable. We see deviations from this pattern by looking at the scatter of the data points about the regression line. The vertical distances from the points to the least-squares regression line are as small as possible, in the sense that they have the smallest possible sum of squares. Because they represent "leftover" variation in the response after fitting the regression line, these distances are called *residuals*.

Residuals

A **residual** is the difference between an observed value of the response variable and the value predicted by the regression line. That is, a residual is the prediction error that remains after we have chosen the regression line:

$$\text{residual} = \text{observed } y - \text{predicted } y$$
$$= y - \hat{y}$$

EXAMPLE 5.5 I Feel Your Pain

Jacobs Stock Photography/ Getty Images

"Empathy" means being able to understand what others feel. To see how the brain expresses empathy, researchers recruited 16 couples in their midtwenties who were married or had been dating for at least two years. They zapped the man's hand with an electrode while the woman watched, and measured the activity in several parts of the woman's brain that would respond to her own pain. Brain activity was recorded as a fraction of the activity observed when the woman herself was zapped with the electrode. The women also completed a psychological test that measures empathy. Will women who score higher in empathy respond more strongly when their partner has a painful experience? Here are data for one brain region:[6]

DATA EMPATHY

Subject	1	2	3	4	5	6	7	8
Empathy Score	38	53	41	55	56	61	62	48
Brain Activity	−0.120	0.392	0.005	0.369	0.016	0.415	0.107	0.506
Subject	9	10	11	12	13	14	15	16
Empathy Score	43	47	56	65	19	61	32	105
Brain Activity	0.153	0.745	0.255	0.574	0.210	0.722	0.358	0.779

Figure 5.5 is a scatterplot, with empathy score as the explanatory variable x and brain activity as the response variable y. The plot shows a positive association. That is, women who score higher in empathy do indeed react more strongly to their partner's pain. The overall pattern is moderately linear, with correlation $r = 0.515$.

The line on the plot is the least-squares regression line of brain activity on empathy score. Its equation is

$$\text{brain activity} = -0.0578 + 0.00761(\text{empathy score})$$

For Subject 1, with empathy score 38, we predict

$$\text{brain activity} = -0.0578 + (0.00761)(38) = 0.231$$

This subject's actual brain activity level was −0.120. The residual is

$$\begin{aligned}\text{residual} &= \text{observed brain activity} - \text{predicted brain activity}\\ &= -0.120 - 0.231 = -0.351\end{aligned}$$

The residual is negative because the data point lies below the regression line. The dashed line segment in Figure 5.5 shows the size of the residual. ■

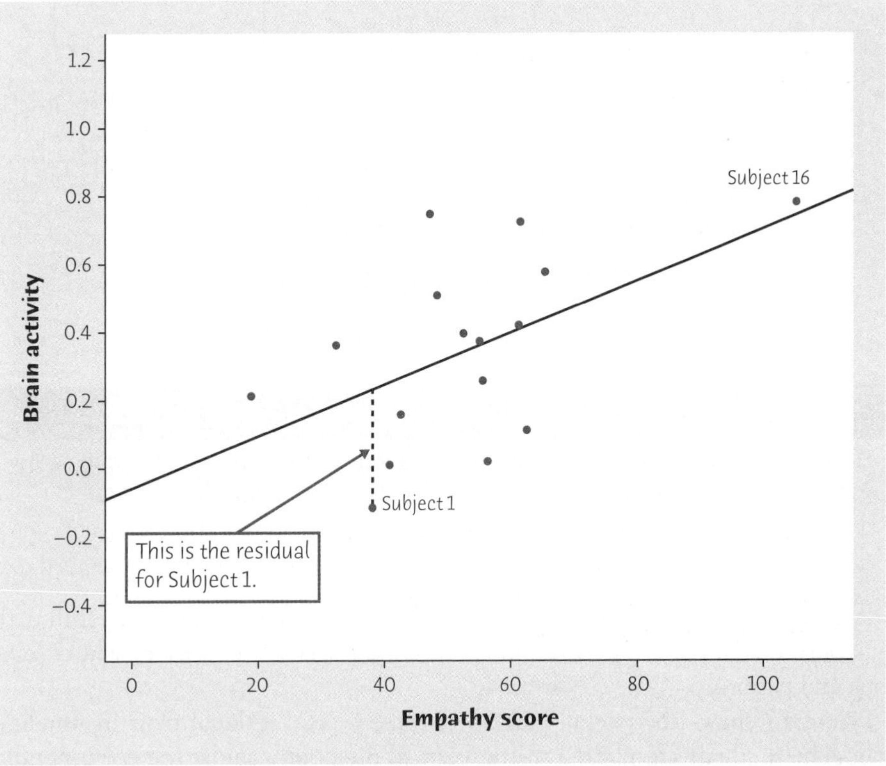

FIGURE 5.5 Scatterplot of activity in a region of the brain that responds to pain versus score on a test of empathy, for Example 5.5. Brain activity is measured as the subject watches her partner experience pain. The line is the least-squares regression line.

There is a residual for each data point. Finding the residuals is a bit unpleasant because you must first find the predicted response for every x. Software or a graphing calculator gives you the residuals all at once. Here are the 16 residuals for the empathy study data, from software:

```
residuals:
-0.3515 -0.2494 -0.3526 -0.3072 -0.1166 -0.1136  0.1231  0.1721
 0.0463  0.0080  0.0084  0.1983  0.4449  0.1369  0.3154  0.0374
```

Because the residuals show how far the data fall from our regression line, examining the residuals helps us assess how well the line describes the data. Although residuals can be calculated from any curve or line fitted to the data, the residuals from the least-squares line have a special property: **the mean of the least-squares residuals is always zero.**

Compare the scatterplot in Figure 5.5 with the *residual plot* for the same data in Figure 5.6. The horizontal line at zero in Figure 5.6 helps orient us. This "residual = 0" line corresponds to the regression line in Figure 5.5.

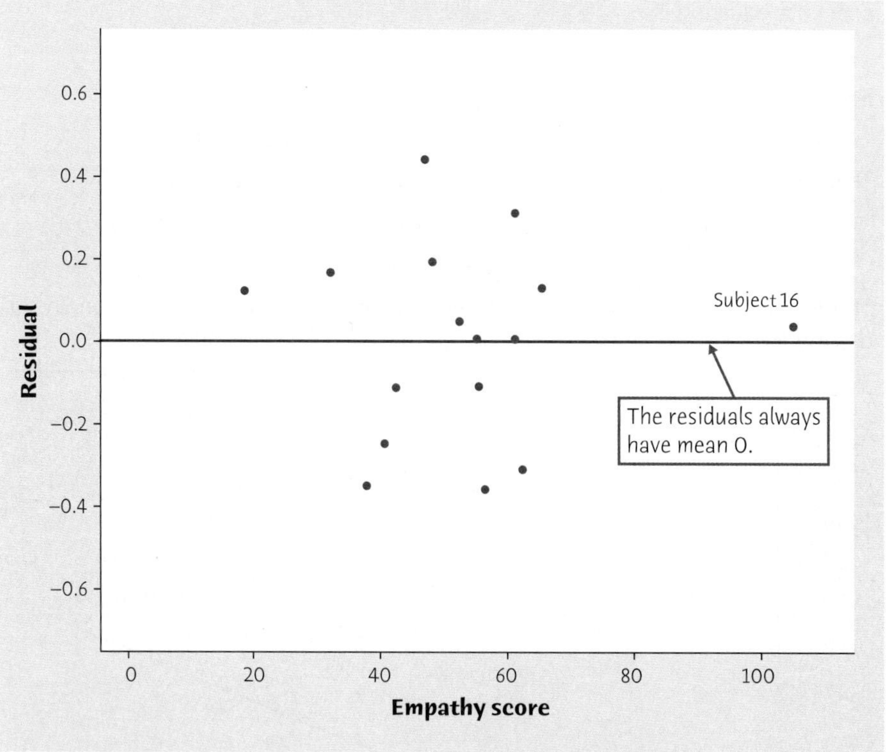

FIGURE 5.6
Residual plot for the data shown in Figure 5.5. The horizontal line at zero residual corresponds to the regression line in Figure 5.5.

Residual Plots

A **residual plot** is a scatterplot of the regression residuals against the explanatory variable. Residual plots help us assess how well a regression line fits the data.

A residual plot in effect turns the regression line horizontal. It magnifies the deviations of the points from the line and makes it easier to see unusual observations and patterns.

Figure 5.7 shows the overall pattern of some typical residual plots in simplified form. The residuals are plotted in the vertical direction against the corresponding values of the explanatory variable in the horizontal direction. If our assumptions hold, the pattern of this plot will be an unstructured horizontal band centered at 0 (the mean of the residuals) and symmetric about 0, as in Figure 5.7(a). A curved pattern, like the one in Figure 5.7(b), indicates that the relationship between the response and explanatory variable is curved rather than linear. A straight line is not a good description of such a relationship. A fan-shaped pattern like the one in Figure 5.7(c) shows that the variation of the response about the least-squares line increases as the explanatory variable increases. Predictions of the response will be more precise for smaller values of the explanatory variable, where the response shows less variability about the line.

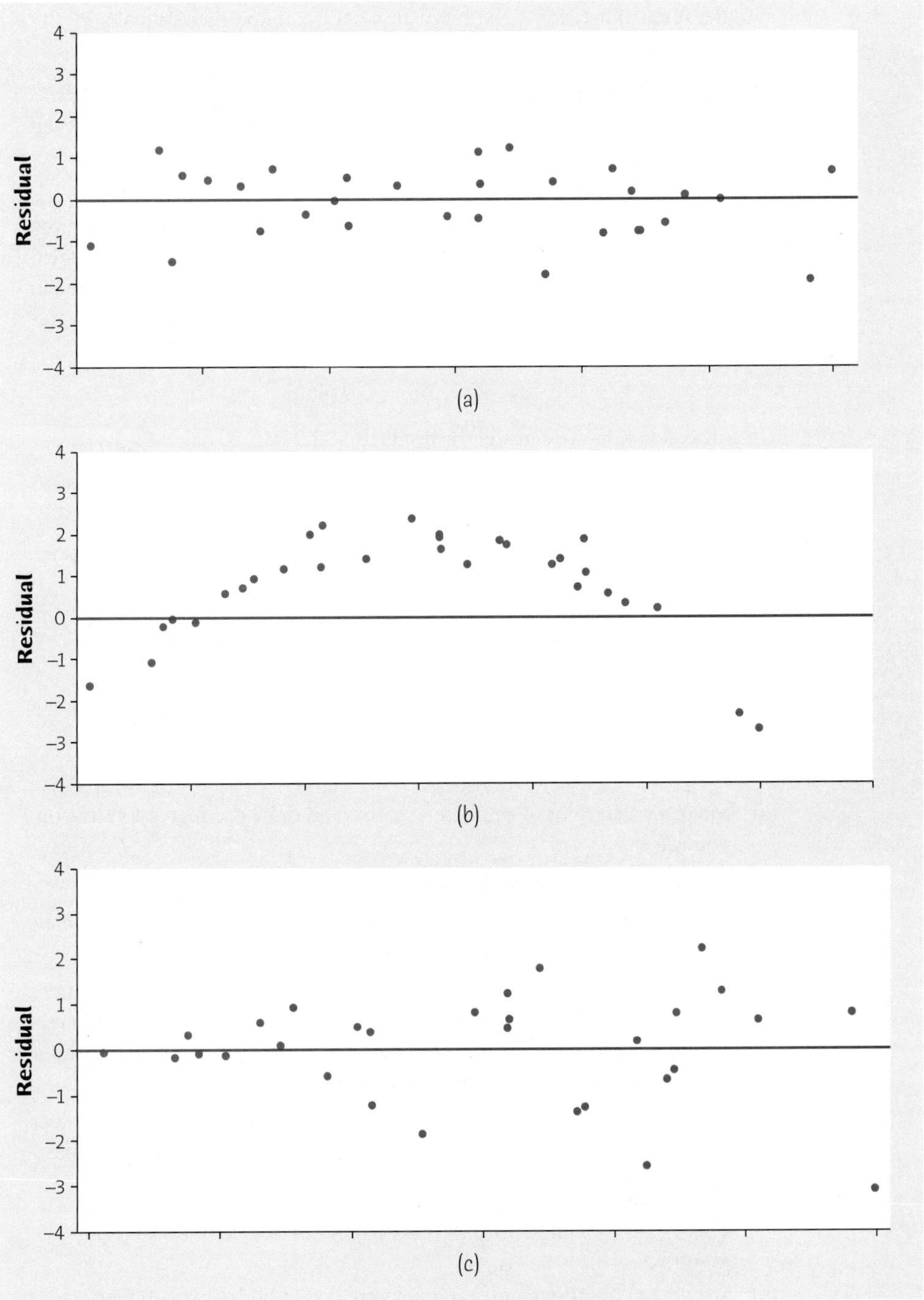

FIGURE 5.7
Some typical residual plots in simplified form.

LaunchPad Online Resources

- The **StatBoards video**, *Calculating and Plotting Residuals,* provides the details of calculating residuals and making a residual plot.

Apply Your Knowledge

5.8 Residuals by Hand. In Exercise 5.4 you found the equation of the least-squares line for predicting coral growth y from mean sea surface temperature x.

(a) Use the equation to obtain the seven residuals step by step. That is, find the prediction $\hat{y}$ for each observation and then find the residual $y - \hat{y}$.

(b) Check that (up to roundoff error) the residuals add to 0.

(c) The residuals are the part of the response y left over after the straight-line tie between y and x is removed. Show that the correlation between the residuals and x is 0 (up to roundoff error). That this correlation is always 0 is another special property of least-squares regression.

5.9 Does Fast Driving Waste Fuel? Exercise 4.8 (page 109) gives data on the fuel consumption y of a car at various speeds x. Fuel consumption is measured in liters of gasoline per 100 kilometers driven, and speed is measured in kilometers per hour. Software tells us that the equation of the least-squares regression line is

$$\hat{y} = 11.058 - 0.01466x$$

Using this equation, we can add the residuals to the original data: DATA FASTDR2

Speed	10	20	30	40	50	60	70	80
Fuel	21.00	13.00	10.00	8.00	7.00	5.90	6.30	6.95
Residual	10.09	2.24	−0.62	−2.47	−3.33	−4.28	−3.73	−2.94
Speed	90	100	110	120	130	140	150	
Fuel	7.57	8.27	9.03	9.87	10.79	11.77	12.83	
Residual	−2.17	−1.32	−0.42	0.57	1.64	2.76	3.97	

(a) Make a scatterplot of the observations, and draw the regression line on your plot.

(b) Would you use the regression line to predict y from x? Explain your answer.

(c) Verify the value of the first residual, for $x = 10$. Verify that the residuals have sum zero (up to roundoff error).

(d) Make a plot of the residuals against the values of x. Draw a horizontal line at height zero on your plot. How does the pattern of the residuals about this line compare with the pattern of the data points about the regression line in your scatterplot from part (a)?

5.10 Not Obvious to the Naked Eye. The data set "resids" contains the values of a response y, an explanatory variable x, and the residuals from the least-squares regression line for predicting y from x. DATA RESIDS

(a) Make a scatterplot of the observations, and draw the regression line on your plot.

(b) Would you use the regression line to predict y from x? Explain your answer.

(c) Make a plot of the residuals against the values of x. Draw a horizontal line at height zero on your plot. How does the pattern of the residuals about this line compare with the pattern of the data points about the regression line in your scatterplot from part (a)?

5.6 Influential observations

Figures 5.5 and 5.6 show one unusual observation: Subject 16 is an outlier in the x direction, with empathy score 40 points higher than any other subject. Because of its extreme position on the empathy scale, this point has a strong influence on the correlation. Dropping Subject 16 reduces the correlation from $r = 0.515$ to $r = 0.331$. You can see that this point extends the linear pattern in Figure 5.5 and so increases the correlation. We say that Subject 16 is *influential* for calculating the correlation.

Influential Observations

An observation is **influential** for a statistical calculation if removing it would markedly change the result of the calculation.

The result of a statistical calculation may be of little practical use if it depends strongly on a few influential observations.

Points that are outliers in either the x or the y direction of a scatterplot are often influential for the correlation. Points that are outliers in the x direction are often influential for the least-squares regression line.

What constitutes a "marked" change? This is somewhat subjective. Changes in a calculation that are the same size as round-off error are often not influential. Changes in a calculation that differ by a factor of 1.5 or more are often influential. If the least-squares regression line computed after removing an observation still fits the original data in the scatterplot, the observation is probably not influential. However, one can find exceptions to these guidelines and statisticians may disagree as to whether an observation should be considered influential.

If an observation is influential, or if there is some doubt as to whether an observation is influential, it can be informative to report statistical calculations with both the observation included and the observation removed. This provides readers with the ability to assess the effect of the observation.

EXAMPLE 5.6 An Influential Observation?

Subject 16 in Example 5.5 is influential for the correlation between empathy score and brain activity, because removing it reduces r from 0.515 to 0.331. Calculating that $r = 0.515$ is not a very useful description of the data, because the value depends so strongly on just one of the 16 subjects.

Is this observation also influential for the least-squares line? Figure 5.8 shows that it is not. The regression line calculated without Subject 16 (dashed) differs little from the line that uses all the observations (solid). The reason that the outlier has little influence on the regression line is that it lies close to the dashed regression line calculated from the other observations. ■

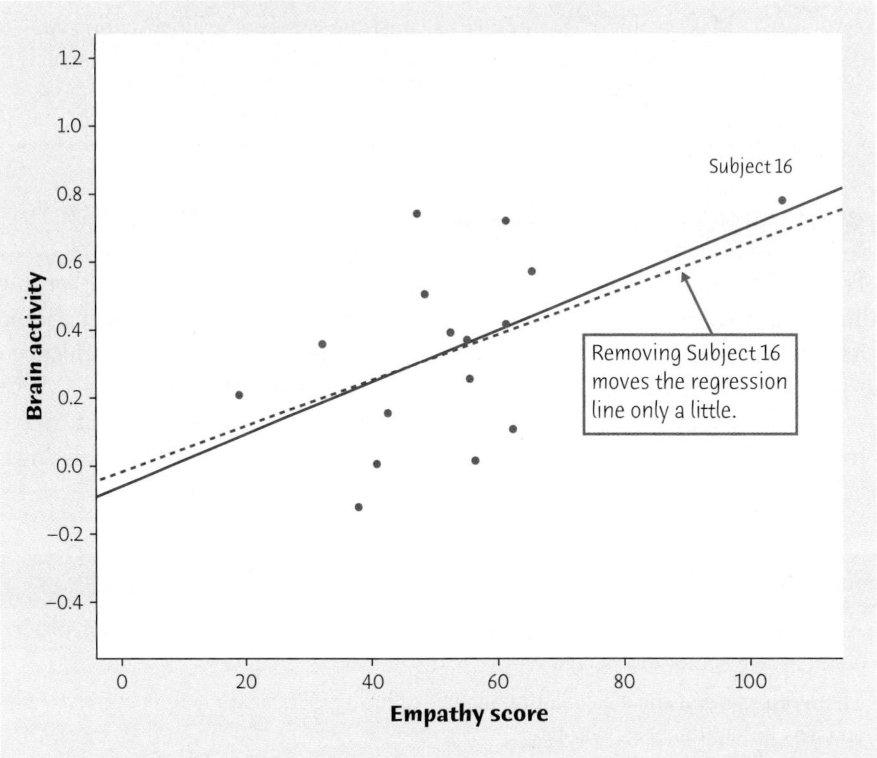

FIGURE 5.8
Subject 16 is an outlier in the x direction. The outlier is not influential for least-squares regression because removing it moves the regression line only a little.

To see why points that are outliers in the x direction are often influential for regression, let's try an experiment. Suppose that Subject 16's point in the scatterplot moves straight down. What happens to the regression line? Figure 5.9 gives the answer. The dashed line is the regression line with the outlier in its new, lower position. Because there are no other points with similar x-values, the line chases the outlier.

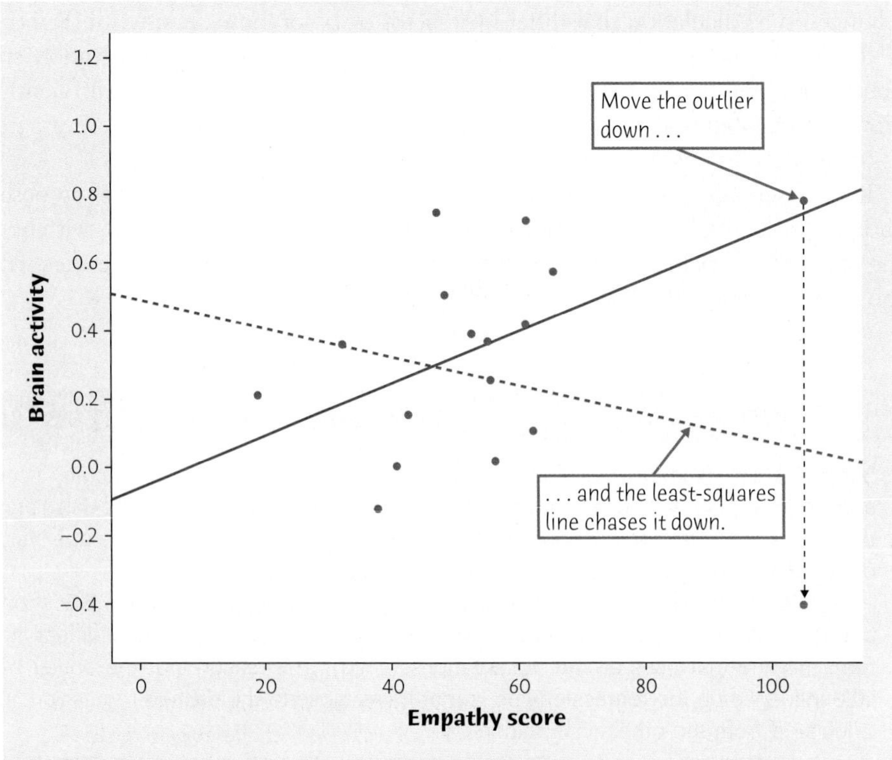

FIGURE 5.9
An outlier in the x direction pulls the least-squares line to itself because there are no other observations with similar values of x to hold the line in place. When the outlier moves down, the regression line chases it down. The original regression line is solid, and the final position of the regression line is dashed.

APPLET

The *Correlation and Regression* applet allows you to try this experiment yourself—see Exercise 5.11. *An outlier in x pulls the least-squares line toward itself. If the outlier does not lie close to the line calculated from the other observations, it will be influential.*

We did not need the distinction between outliers and influential observations in Chapter 2. A single high salary that pulls up the mean salary $\bar{x}$ for a group of workers is an outlier because it lies far above the other salaries. It is also influential because the mean changes when it is removed. In the regression setting, however, not all outliers are influential.

Apply Your Knowledge

5.11 Influence in Regression. The *Correlation and Regression* applet allows you to animate Figure 5.9. Click to create a group of 10 points in the lower-left corner of the scatterplot with a strong straight-line pattern (correlation about 0.9). Click the "Show least-squares line" box to display the regression line.

APPLET

(a) Add one point at the upper right that is far from the other 10 points but exactly on the regression line. Why does this outlier have no effect on the line even though it changes the correlation?

(b) Now use the mouse to drag this last point straight down. You see that one end of the least-squares line chases this single point, whereas the other end remains near the middle of the original group of 10. What makes the last point so influential?

5.12 Death by Intent. Return to the data of Exercise 5.5 on homicide rate and suicide rate. We will use these data to illustrate influence. DATA DEATH3

(a) Make a scatterplot of the data that is suitable for predicting suicide rate from homicide rate, with two new points added. Point A: homicide rate 10.8, suicide rate 27.0. Point B: homicide rate 18.0, suicide rate 14.6. In which direction is each of these points an outlier?

(b) Add three least-squares regression lines to your plot: for the original 26 counties, for the original 26 counties plus Point A, and for the original 26 counties plus Point B. Which new point is more influential for the regression line? Explain in simple language why each new point moves the line in the way your graph shows.

5.13 Outsourcing by Airlines. Exercise 4.5 (page 105) gives data for nine airlines on the percent of major maintenance outsourced and the percent of flight delays blamed on the airline. DATA AIRLINE

(a) Make a scatterplot with outsourcing percent as x and delay percent as y. Would you consider Hawaiian Airlines to be influential? Would you consider Frontier Airlines to be influential?

(b) Find the correlation r with and without Hawaiian Airlines. How influential is Hawaiian Airlines for correlation?

(c) Find the correlation r without Frontier Airlines. How influential is Frontier Airlines for correlation?

(d) Find the correlation r without both Hawaiian and Frontier Airlines. How influential is the pair of airlines for correlation?

(e) Find the least-squares line for predicting y from x with and without both Hawaiian and Frontier Airlines. Draw both lines on your scatterplot. Use both lines to predict the percent of delays blamed on an airline that has outsourced 79.1% of its major maintenance, and on an airline that has outsourced 0.1% of its major maintenance. How influential are Hawaiian and Frontier Airlines for the least-squares line?

5.7 Cautions about correlation and regression

Correlation and regression are powerful tools for describing the relationship between two variables. When you use these tools, you must be aware of their limitations. You already know that

- *Correlation and regression lines describe only linear relationships.* You can do the calculations for any relationship between two quantitative variables, but the results are useful only if the scatterplot shows a linear pattern.
- *Correlation and least-squares regression lines are not resistant.* Always plot your data and look for observations that may be influential.

Here are three more things to keep in mind when you use correlation and regression.

Beware ecological correlation There is a large positive correlation between *average* income and number of years of education. The correlation is smaller if we compare the incomes of *individuals* with number of years of education. The correlation based on average income ignores the large variation in the incomes of individuals having the same amount of education. The variation from individual to individual increases the scatter in a scatterplot, reducing the correlation. The correlation between average income and education overstates the strength of the relation between the incomes of individuals and number of years of education. *Correlations based on averages can be misleading if they are interpreted to be about individuals.*

Ecological Correlation

A correlation based on averages rather than on individuals is called an **ecological correlation**.

DO LEFT-HANDERS DIE EARLY?

Yes, said a study of 1000 deaths in California. Left-handed people died at an average age of 66 years; right-handers died at 75 years of age. Should left-handed people fear an early death? No—the lurking variable has struck again. Older people grew up in an era when many natural left-handers were forced to use their right hands. So right-handers are more common among older people, and left-handers are more common among the young. When we look at deaths, the left-handers who die are younger on the average because left-handers in general are younger. Mystery solved.

Beware extrapolation Suppose that you have data on a child's growth between three and eight years of age. You find a strong linear relationship between age x and height y. If you fit a regression line to these data and use it to predict height at age 25 years, you will predict that the child will be eight feet tall. Growth slows down and then stops at maturity, so extending the straight line to adult ages is foolish. *Few relationships are linear for all values of x. Don't make predictions far outside the range of x that actually appears in your data.*

Extrapolation

Extrapolation is the use of a regression line for prediction far outside the range of values of the explanatory variable x that you used to obtain the line. Such predictions are often not accurate.

Beware the lurking variable Another caution is even more important: *the relationship between two variables can often be understood only by taking other variables into account. Lurking variables* can make a correlation or regression misleading.

Lurking Variable

A **lurking variable** is a variable that is not among the explanatory or response variables in a study and yet may influence the interpretation of relationships among those variables.

You should always think about possible lurking variables before you draw conclusions based on correlation or regression.

EXAMPLE 5.7 Magic Mozart?

The Kalamazoo (Michigan) Symphony once advertised a "Mozart for Minors" program with this statement: "Question: Which students scored 51 points higher in verbal skills and 39 points higher in math? Answer: Students who had experience in music."[7]

We could as well answer "Students who played soccer." Why? Children with prosperous and well-educated parents are more likely than poorer children to have experience with music and also to play soccer. They are also likely to attend good schools, get good health care, and be encouraged to study hard. These advantages lead to high test scores. Family background is a lurking variable that explains why test scores are related to experience with music. ■

LaunchPad Online Resources

- The **StatBoards video**, *Beware Extrapolation!*, provides a discussion of the dangers of extrapolation.

Apply Your Knowledge

5.14 SAT Scores. The correlation between mean 2013 Math SAT scores and mean 2013 Writing SAT scores for all 50 states and the District of Columbia is 0.972. Would you expect the correlation between the mean state SAT scores for these two tests to be lower, about the same, or higher than the correlation between the scores of individuals on these two tests? Explain your answer.

5.15 The Endangered Manatee. Table 4.1 gives 37 years of data on boats registered in Florida and manatees killed by boats. Figure 4.2 (page 108) shows a strong positive linear relationship. The correlation is $r = 0.954$.

DATA MANATEE

PhotoDisc

(a) Find the equation of the least-squares line for predicting manatees killed from thousands of boats registered. Because the linear pattern is so strong, we expect predictions from this line to be quite accurate—but only if conditions in Florida remain similar to those of the past 37 years.

(b) Suppose we expect that the number of boats registered in Florida will be 890,000 in 2014. What would you predict the number of manatees killed by boats to be if there are 890,000 boats registered? Explain why we can trust this prediction.

(c) Predict manatee deaths if there were *no* boats registered in Florida. Explain why the predicted count of deaths is impossible. (We use $x = 0$ to find the intercept of the regression line, but unless the explanatory variable x actually takes values near 0, prediction for $x = 0$ is an example of extrapolation.)

5.16 **Is Math the Key to Success in College?** A College Board study of 15,941 high school graduates found a strong correlation between how much math minority students took in high school and their later success in college. News articles quoted the head of the College Board as saying that "math is the gatekeeper for success in college."[8] Maybe so, but we should also think about lurking variables. What might lead minority students to take more or fewer high school math courses? Would these same factors influence success in college?

5.8 Association does not imply causation

Thinking about lurking variables leads to the most important caution about correlation and regression. When we study the relationship between two variables, we often hope to show that changes in the explanatory variable *cause* changes in the response variable. *A strong association between two variables is not enough to draw conclusions about cause and effect.* A least-squares regression line that fits the data well and gives accurate predictions is also not enough to draw conclusions about cause and effect. Sometimes an observed association really does reflect cause and effect. A household that heats with natural gas uses more gas in colder months because cold weather requires burning more gas to stay warm. In other cases, an association is explained by lurking variables, and the conclusion that x causes y is either wrong or not proven.

EXAMPLE 5.8 Does Having More Cars Make You Live Longer?

© America/Alamy

A serious study once found that people with two cars live longer than people who own only one car.[9] Owning three cars is even better, and so on. There is a substantial positive correlation between number of cars x and length of life y.

The basic meaning of causation is that by changing x we can bring about a change in y. Could we lengthen our lives by buying more cars? No. The study used number of cars as a quick indicator of affluence. Well-off people tend to have more cars. They also tend to live longer, probably because they are better educated, take better care of themselves, and get better medical care. The cars have nothing to do with it. There is no cause-and-effect tie between number of cars and length of life. ■

THE SUPER BOWL EFFECT
The Super Bowl is the most-watched TV broadcast in the United States. Data show that, on Super Bowl Sunday, we consume three times as many potato chips as on an average day, and 17 times as much beer. What's more, the number of fatal traffic accidents goes up in the hours after the game ends. Could that be celebration? Or catching up with tasks left undone? Or maybe it's the beer.

Correlations such as that in Example 5.8 are sometimes called "nonsense correlations." The correlation is real. What is nonsense is the conclusion that changing one of the variables causes changes in the other. A lurking variable—such as personal affluence in Example 5.8—that influences both x and y can create a high correlation even though there is no direct connection between x and y.

Association Does Not Imply Causation

An association between an explanatory variable x and a response variable y, even if it is very strong, is not by itself good evidence that changes in x actually cause changes in y.

EXAMPLE 5.9 SAT Scores and Teacher Salaries

Exercise 4.48 asks you to explore the relationship between Mathematics SAT scores and average teacher salaries using 2013 data for each of the 50 states and the District of Columbia. A scatterplot of these data shows that SAT scores and salaries are negatively correlated and $r = -0.278$.

Average scores on the Mathematics SAT are partly determined by the percent of test takers. States that don't require the SAT have fewer test takers. Typically, only better students take the SAT exam in these states, hence average scores are high. States that do require the SAT tend to be states with higher costs of living (for example, states in the Northeastern U.S. such as New York, Connecticut, and Massachusetts), and hence higher average teacher salaries. These lurking variables explain the negative correlation, and you will explore this in Exercises 5.50 and 5.51. In fact, you will discover that, after accounting for the percent of test takers, the association actually reverses! ■

EXAMPLE 5.10 Overweight Mothers, Overweight Daughters

Overweight parents tend to have overweight children. The results of a study of Mexican American girls aged nine to 12 years are typical. The investigators measured body mass index (BMI), a measure of weight relative to height, for both the girls and their mothers. People with high BMI are overweight. The correlation between the BMI of daughters and the BMI of their mothers was $r = 0.506$.[10]

Body type is in part determined by heredity. Daughters inherit half their genes from their mothers. There is therefore a direct cause-and-effect link between the BMI of mothers and daughters. But perhaps mothers who are overweight also set an example of little exercise, poor eating habits, and lots of television viewing. Their daughters may pick up these habits, so the influence of heredity is mixed up with influences from the girls' environment. Both contribute to the mother–daughter correlation. ■

The lesson of Example 5.9 is that association does not imply causation. A lurking variable may not only help explain the observed association, but once accounted for may reverse the observed association. In Chapter 6, we refer to this as "Simpson's paradox."

CAUTION The lesson of Example 5.10 is more subtle than just "association does not imply causation." *Even when direct causation is present, it may not be the whole explanation for a correlation.* You must still worry about lurking variables. Careful statistical studies try to anticipate lurking variables and measure them. The mother–daughter study did measure TV viewing, exercise, and diet. Elaborate statistical analysis can remove the effects of these variables to come closer to the direct effect of mother's BMI on daughter's BMI. This remains a second-best approach to causation. The best way to get good evidence that x causes y is to do an **experiment** in which we change x and keep lurking variables under control. We will discuss experiments in Chapter 9.

experiment

When experiments cannot be done, explaining an observed association can be difficult and controversial. Many of the sharpest disputes in which statistics plays a role involve questions of causation that cannot be settled by experiment. Do gun control laws reduce violent crime? Does using cell phones cause brain tumors? Has increased free trade widened the gap between the incomes of more educated and less educated American workers? All of these questions have become public issues.

All concern associations among variables. And all have this in common: they try to pinpoint cause and effect in a setting involving complex relations among many interacting variables.

EXAMPLE 5.11 Does Smoking Cause Lung Cancer?

© James Leynse/Corbis

Despite the difficulties, it is sometimes possible to build a strong case for causation in the absence of experiments. The evidence that smoking causes lung cancer is about as strong as nonexperimental evidence can be.

Doctors had long observed that most lung cancer patients were smokers. Comparison of smokers and "similar" nonsmokers showed a very strong association between smoking and death from lung cancer. Could the association be explained by lurking variables? Might there be, for example, a genetic factor that predisposes people both to nicotine addiction and to lung cancer? Smoking and lung cancer would then be positively associated even if smoking had no direct effect on the lungs. How were these objections overcome? ■

Let's answer this question in general terms: what are the criteria for establishing causation when we cannot do an experiment?

- *The association is strong.* The association between smoking and lung cancer is very strong.
- *The association is consistent.* Many studies of different kinds of people in many countries link smoking to lung cancer. That reduces the chance that a lurking variable specific to one group or one study explains the association.
- *Higher doses are associated with stronger responses.* People who smoke more cigarettes per day or who smoke over a longer period get lung cancer more often. People who stop smoking reduce their risk.
- *The alleged cause precedes the effect in time.* Lung cancer develops after years of smoking. The number of men dying of lung cancer rose as smoking became more common, with a lag of about 30 years. Lung cancer kills more men than any other form of cancer. Lung cancer was rare among women until women began to smoke. Lung cancer in women rose along with smoking, again with a lag of about 30 years, and has now passed breast cancer as the leading cause of cancer death among women.
- *The alleged cause is plausible.* Experiments with animals show that tars from cigarette smoke do cause cancer.

Medical authorities do not hesitate to say that smoking causes lung cancer. The U.S. Surgeon General has long stated that cigarette smoking is "the largest avoidable cause of death and disability in the United States."[11] The evidence for causation is overwhelming—but it is not as strong as the evidence provided by well-designed experiments.

Apply Your Knowledge

5.17 Another Reason Not to Smoke? A stop-smoking booklet says, "Children of mothers who smoked during pregnancy scored nine points lower on intelligence tests at ages three and four than children of nonsmokers." Suggest some lurking variables that may help explain the association between smoking during pregnancy and children's later test scores. The association by itself is not good evidence that mothers' smoking *causes* lower scores.

5.18 Education and Income. There is a strong positive association between workers' education and their income. For example, the U.S. Census Bureau reported in 2012 that the mean income of young adults (ages 25 to 34) who worked full-time increased from \$19,174 for those with less than a ninth-grade education, to \$29,058 for high school graduates, to \$49,796 for holders of a bachelor's degree, and on up for yet more education. In part, this association reflects causation—education helps people qualify for better jobs. Suggest several lurking variables that also contribute. (Ask yourself what kinds of people tend to get more education.)

5.19 To Earn More, Get Married? Data show that men who are married, and also divorced or widowed men, earn quite a bit more than men the same age who have never been married. This does not mean that a man can raise his income by getting married, because men who have never been married are different from married men in many ways other than marital status. Suggest several lurking variables that might help explain the association between marital status and income.

CHAPTER 5 SUMMARY

Chapter Specifics

- A **regression line** is a straight line that describes how a response variable y changes as an explanatory variable x changes. You can use a regression line to **predict** the value of y for any value of x by substituting this x into the equation of the line.
- The **slope** b of a regression line $\hat{y} = a + bx$ is the rate at which the predicted response $\hat{y}$ changes along the line as the explanatory variable x changes. Specifically, b is the change in $\hat{y}$ when x increases by 1.
- The **intercept** a of a regression line $\hat{y} = a + bx$ is the predicted response $\hat{y}$ when the explanatory variable $x=0$. This prediction is of no statistical interest unless x can actually take values near 0.
- The most common method of fitting a line to a scatterplot is least squares. The **least-squares regression line** is the straight line $\hat{y} = a + bx$ that minimizes the sum of the squares of the vertical distances of the observed points from the line.
- The least-squares regression line of y on x is the line with slope $b = rs_y/s_x$ and intercept $a = \bar{y} - b\bar{x}$. This line always passes through the point $(\bar{x}, \bar{y})$.
- **Correlation and regression** are closely connected. The correlation r is the slope of the least-squares regression line when we measure both x and y in standardized units. The **square of the correlation** r^2 is the fraction of the variation in one variable that is explained by least-squares regression on the other variable.
- Correlation and regression must be **interpreted with caution. Plot the data** to be sure the relationship is roughly linear and to detect outliers and influential observations. A plot of the **residuals** makes these effects easier to see.
- Look for **influential observations**, individual points that substantially change the correlation or the regression line. Outliers in the x direction are often influential for the regression line.
- Be aware of **ecological correlation**, the tendency for correlations based on averages to be stronger than correlations based on individuals. Be careful not to misinterpret correlations based on averages as applying to individuals.
- Avoid **extrapolation,** the use of a regression line for prediction for values of the explanatory variable far outside the range of the data from which the line was calculated.

- **Lurking variables** may explain the relationship between the explanatory and response variables. Correlation and regression can be misleading if you ignore important lurking variables.
- Most of all, be careful not to conclude that there is a cause-and-effect relationship between two variables just because they are strongly associated. **High correlation does not imply causation**. The best evidence that an association is due to causation comes from an **experiment** in which the explanatory variable is directly changed and other influences on the response are controlled.

Link It

In this chapter we continued our study of exploring relationships between two variables, begun in Chapter 4. We use the least-squares regression line to describe the straight-line relationship between two variables when such a pattern is seen in a scatterplot. The equation of the least-squares regression line is a numerical summary that makes precise the notion of "straight-line relationship."

Even if the least-squares regression line is a good description of the relationship between the observed values of two variables, we must exercise caution in how we interpret this relationship. Such interpretations rest on the assumption that the relationship is valid in some broader sense. We will explore this more carefully later in this book, but in this chapter we content ourselves with providing some cautions. Association, as indicated by a large correlation, does not imply that there is an underlying cause-and-effect relation between the response and explanatory variables. There may, in fact, be a lurking variable that influences the interpretation of any relation between the response and explanatory variables. Correlations based on averages of measurements tend to be higher than correlations based on the individual observations used to compute the averages. Be careful not to misinterpret correlations based on averages as applying to individuals. Finally, be careful not to use the least-squares regression line to make predictions outside the range of values of the explanatory variable that you used to obtain the line.

In this chapter, we studied relationships between two quantitative variables. In the next chapter, we explore the relationship between two qualitative variables.

LaunchPad Online Resources

If you are having difficulty with any of the sections of this chapter, these online resources should help prepare you to solve the exercises at the end of this chapter.

- **StatTutor** starts with a video review of each section and asks a series of questions to check your understanding.
- **LearningCurve** provides you with a series of questions about the chapter geared to your level of understanding.

CHECK YOUR SKILLS

5.20 Figure 5.10 is a scatterplot of school GPA against IQ test scores for 15 seventh-grade students. The line is the least-squares regression line for predicting school GPA from IQ score. If another child in this class has IQ score 110, you predict the school GPA to be close to

(a) 2. (b) 7.5. (c) 11.

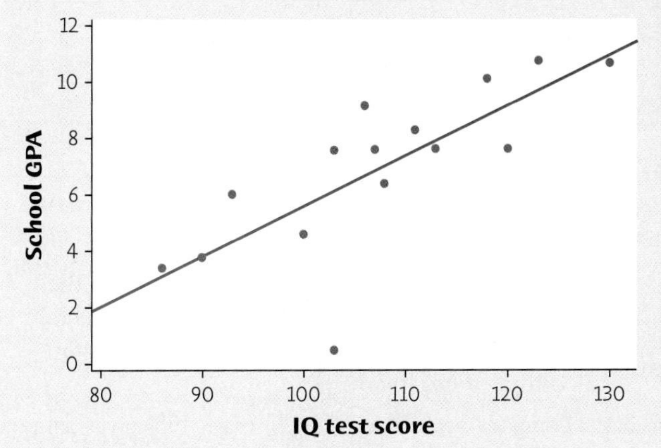

FIGURE 5.10
Scatterplot of IQ test scores and school GPA for 15 seventh-grade students, for Exercises 5.20 and 5.21.

5.21 The slope of the line in Figure 5.10 is closest to

(a) −11. (b) 0.2. (c) 2.0.

5.22 The points on a scatterplot lie close to the line whose equation is $y = 4 - 3x$. The slope of this line is

(a) 4. (b) 3. (c) −3.

5.23 Fred keeps his savings in his mattress. He began with \$1000 from his mother and adds \$100 each year. His total savings y after x years are given by the equation

(a) $y = 1000 + 100x$.
(b) $y = 100 + 1000x$.
(c) $y = 1000 + x$.

5.24 Smokers don't live as long (on the average) as nonsmokers, and heavy smokers don't live as long as light smokers. You regress the age at death of a group of male smokers on the number of packs per day they smoked. The slope of your regression line

(a) will be greater than 0.
(b) will be less than 0.
(c) can't be determined without seeing the data.

5.25 An owner of a home in the Midwest installed solar panels to reduce heating costs. After installing the solar panels, he measured the amount of natural gas used y (in cubic feet) to heat the home and outside temperature x (in degree-days, where a day's degree-days are the number of degrees its average temperature falls below 65°F) over a 23-month period. He then computed the least-squares regression line for predicting y from x and found it to be[12]

$$\hat{y} = 85 + 16x$$

How much, on average, does gas used increase for each additional degree-day?

(a) 23 cubic feet (b) 85 cubic feet
(c) 16 cubic feet

5.26 According to the regression line in Exercise 5.25, the predicted amount of gas used when the outside temperature is 20 degree-days is about

(a) 405 cubic feet.
(b) 320 cubic feet.
(c) 105 cubic feet.

5.27 By looking at the equation of the least-squares regression line in Exercise 5.25, you can see that the correlation between amount of gas used and degree-days is

(a) greater than zero.
(b) less than zero.
(c) unable to be determined without seeing the data.

5.28 The software used to compute the least-squares regression line in Exercise 5.25 says that $r^2 = 0.98$. This suggests that

(a) although degree-days and gas used are correlated, degree-days do not predict gas used very accurately.
(b) gas used increases by $\sqrt{0.98} = 0.99$ cubic feet for each additional degree-day.
(c) prediction of gas used from degree-days will be quite accurate.

5.29 Researchers measured the percent body fat and the preferred amount of salt (percent weight/volume) for several children. Here are data for seven children:[13] DATA SALT

Preferred amount of salt x	0.2	0.3	0.4	0.5	0.6	0.8	1.1
Percent body fat y	20	30	22	30	38	23	30

Use your calculator or software: what is the equation of the least-squares regression line for predicting percent body fat from preferred amount of salt?

(a) $\hat{y} = 24.2 + 6.0x$
(b) $\hat{y} = 0.15 + 0.01x$
(c) $\hat{y} = 6.0 + 24.2x$

CHAPTER 5 EXERCISES

5.30 Penguins diving. A study of king penguins looked for a relationship between how deep the penguins dive to seek food and how long they stay underwater.[14]

© Paul A. Souders/Corbis

For all but the shallowest dives, there is a linear relationship that is different for different penguins. The study report gives a scatterplot for one penguin titled "The relation of dive duration (DD) to depth (D)." Duration DD is measured in minutes, and depth D is in meters. The report then says, "The regression equation for this bird is: DD = 2.69 + 0.0138D."

(a) What is the slope of the regression line? Explain in specific language what this slope says about this penguin's dives.
(b) According to the regression line, how long does a typical dive to a depth of 200 meters last?
(c) The dives varied from 40 meters to 300 meters in depth. Use the regression equation to determine DD for $D = 40$ and $D = 300$ and then plot the regression line from $D = 40$ to $D = 300$.

5.31 The price of diamond rings. A newspaper advertisement in the *Straits Times* of Singapore contained pictures of diamond rings and listed their prices, diamond weight (in carats), and gold purity. Based on data for only the 20-carat gold ladies' rings in the advertisement, the least-squares regression line for predicting price (in Singapore dollars) from the weight of the diamond (in carats) is[15]

$$\text{Price} = 259.63 + 3721.02 \text{ Carats}$$

(a) What does the slope of this line say about the relationship between price and number of carats?
(b) What is the predicted price when number of carats = 0? How would you interpret this price?

5.32 Does social rejection hurt? Exercise 4.46 (page 125) gives data from a study that shows that social exclusion causes "real pain." That is, activity in an area of the brain that responds to physical pain goes up as distress from social exclusion goes up. A scatterplot shows a moderately strong linear relationship. Figure 5.11 shows Minitab regression output for these data. DATA REJECT

(a) What is the equation of the least-squares regression line for predicting brain activity from social distress score? Interpret the slope in the context of the problem. Use the equation to predict brain activity for a social distress score of 2.0.
(b) What percent of the variation in brain activity among these subjects is explained by the straight-line relationship with social distress score?

```
Session

Regression Analysis: Brain versus Distress

The regression equation is
Brain = -0.126 + 0.0608 distress

Predictor        Coef    SE Coef      T      P
Constant     -0.12608    0.02465  -5.12  0.000
distress     0.060782   0.009979   6.09  0.000

S = 0.0250896   R-Sq = 77.1%  R-Sq (adj) = 75.1%
```

FIGURE 5.11
Minitab regression output for a study of the effects of social rejection on brain activity, for Exercise 5.32.

(c) Use the information in Figure 5.11 to find the correlation r between social distress score and brain activity. How do you know whether the sign of r is positive or negative?

5.33 Toucan's beak. Exercise 4.45 (page 125) gives data on beak heat loss, as a percent of total body heat loss from all sources, at various temperatures. The data show that beak heat loss is higher at higher temperatures and that the relationship is roughly linear. Figure 5.12 shows Minitab regression output for these data. DATA TOUCAN

(a) What is the equation of the least-squares regression line for predicting beak heat loss, as a percent of total body heat loss from all sources, from temperature? Explain in specific language what this slope says about the relationship between beak heat loss and temperature.
(b) Use the equation of the least-squares regression line to predict beak heat loss, as a percent of total body heat loss from all sources, at a temperature of 25 degrees Celsius.

```
Session

Regression Analysis: Percent heat loss versus Temperature

The regression equation is
Percent heat loss = 0.92 + 2.06 Temperature

Predictor     Coef   SE Coef      T      P
Constant     0.919     5.613   0.16  0.872
Pairs       2.0647    0.2444   8.45  0.111

S = 4.50655   R-Sq = 83.6%  R-Sq(adj) = 82.4%
```

FIGURE 5.12
Minitab regression output for a study of how temperature affects beak heat loss in toucans, for Exercise 5.33.

(c) What percent of the variation in beak heat loss is explained by the straight-line relationship with temperature?
(d) Use the information in Figure 5.12 to find the correlation r between beak heat loss and temperature. How do you know whether the sign of r is positive or negative?

5.34 Husbands and wives. The mean height of American women in their twenties is about 64.3 inches, and the standard deviation is about 2.7 inches. The mean height of men the same age is about 69.9 inches, with standard deviation about 3.1 inches. Suppose that the correlation between the heights of husbands and wives is about $r = 0.5$.

(a) What are the slope and intercept of the regression line of the husband's height on the wife's height in young couples? Interpret the slope in the context of the problem.
(b) Draw a graph of this regression line for heights of wives between 56 and 72 inches. Predict the height of the husband of a woman who is 67 inches tall, and plot the wife's height and predicted husband's height on your graph.
(c) You don't expect this prediction for a single couple to be very accurate. Why not?

5.35 What's my grade? In Professor Krugman's economics course, the correlation between the students' total scores prior to the final examination and their final-examination scores is $r = 0.5$. The pre-exam totals for all students in the course have mean 280 and standard deviation 40. The final-exam scores have mean 75 and standard deviation 8. Professor Krugman has lost Julie's final exam but knows that her total before the exam was 300. He decides to predict her final-exam score from her pre-exam total.

(a) What is the slope of the least-squares regression line of final-exam scores on pre-exam total scores in this course? What is the intercept? Interpret the slope in the context of the problem.
(b) Use the regression line to predict Julie's final-exam score.
(c) Julie doesn't think this method accurately predicts how well she did on the final exam. Use r^2 to argue that her actual score could have been much higher (or much lower) than the predicted value.

5.36 Going to class. A study of class attendance and grades among first-year students at a state university showed that, in general, students who attended a higher percent of their classes earned higher grades. Class attendance explained 16% of the variation in grade index among the students. What is the numerical value of the correlation between percent of classes attended and grade index?

5.37 Sisters and brothers. How strongly do physical characteristics of sisters and brothers correlate? Here are data on the heights (in inches) of 12 adult pairs:[16] DATA BROSIS

Brother	71	68	66	67	70	71	70	73	72	65	66	70
Sister	69	64	65	63	65	62	65	64	66	59	62	64

(a) Use your calculator or software to find the correlation and the equation of the least-squares line for predicting sister's height from brother's height. Make a scatterplot of the data, and add the regression line to your plot.
(b) Damien is 70 inches tall. Predict the height of his sister Tonya. Based on the scatterplot and the correlation r, do you expect your prediction to be very accurate? Why?

5.38 Keeping water clean. Keeping water supplies clean requires regular measurement of levels of pollutants. The measurements are indirect—a typical analysis involves forming a dye by a chemical reaction with the dissolved pollutant, then passing light through the solution and measuring its "absorbence." To calibrate such measurements, the laboratory measures known standard solutions and uses regression to relate absorbence and pollutant concentration. This is usually done every day. Here is one series of data on the absorbence for different levels of nitrates. Nitrates are measured in milligrams per liter of water.[17] DATA NITRATES

Nitrates	50	50	100	200	400	800	1200	1600	2000	2000
Absorbence	7.0	7.5	12.8	24.0	47.0	93.0	138.0	183.0	230.0	226.0

(a) Chemical theory says that these data should lie on a straight line. If the correlation is not at least 0.997, something went wrong, and the calibration procedure is repeated. Plot the data and find the correlation. Must the calibration be done again?
(b) The calibration process sets nitrate level and measures absorbence. The linear relationship that results is used to estimate the nitrate level in water from a measurement of absorbence. What is the equation of the line used to estimate nitrate level? What does the slope of this line say about the relationship between nitrate level and absorbence? What is the estimated nitrate level in a water specimen with absorbence 40?
(c) Do you expect estimates of nitrate level from absorbence to be quite accurate? Why?

5.39 Sparrowhawk colonies. One of nature's patterns connects the percent of adult birds in a colony that return from the previous year and the number of new adults that join the colony. Here are data for 13 colonies of sparrowhawks:[18] DATA SPARROW

Percent return x	74	66	81	52	73	62	52	45	62	46	60	46	38
New adults y	5	6	8	11	12	15	16	17	18	18	19	20	20

You saw in Exercise 4.28 that there is a moderately strong linear relationship, with correlation $r = -0.748$.

(a) Find the least-squares regression line for predicting y from x. Make a scatterplot and draw your line on the plot.
(b) Explain in words what the slope of the regression line tells us.
(c) An ecologist uses the line, based on 13 colonies, to predict how many new birds will join another colony, to which 60% of the adults from the previous year return. What is the prediction?

5.40 Global warming. Exercise 4.43 (page 124) gives data on annual average global temperatures for the last 20 years in degrees Celsius. GTEMPS

(a) Find the least-squares regression line for predicting average global temperature from year. Make a scatterplot and draw your line on the plot.
(b) Explain in words what the slope of the regression line tells us.
(c) An environmentalist uses the line, based on the 20 years, to predict average global temperatures in 2050. What is the prediction? How reliable do you think your prediction is?

5.41 Our brains don't like losses. Exercise 4.29 (page 121) describes an experiment that showed a linear relationship between how sensitive people are to monetary losses ("behavioral loss aversion") and activity in one part of their brains ("neural loss aversion"). LOSSES

(a) Make a scatterplot with neural loss aversion as x and behavioral loss aversion as y. One point is a high outlier in both the x and y directions.
(b) Find the least-squares line for predicting y from x, *leaving out the outlier,* and add the line to your plot.
(c) The outlier lies very close to your regression line. Looking at the plot, you now expect that adding the outlier will increase the correlation but will have little effect on the least-squares line. Explain why.
(d) Find the correlation with and without the outlier. Your results verify the expectations from part (c).

5.42 Always plot your data! Table 5.1 presents four sets of data prepared by the statistician Frank Anscombe to illustrate the dangers of calculating without first plotting the data.[19]

(a) Without making scatterplots, find the correlation and the least-squares regression line for all four data sets. What do you notice? Use the regression line to predict y for $x = 10$. ANSCOMB A, B, C, and D

(b) Make a scatterplot for each of the data sets, and add the regression line to each plot.

(c) In which of the four cases would you be willing to use the regression line to describe the dependence of y on x? Explain your answer in each case.

5.43 Managing diabetes. People with diabetes must manage their blood sugar levels carefully. They measure their fasting plasma glucose (FPG) several times a day with a glucose meter. Another measurement, made at regular medical checkups, is called HbA. This is roughly the percent of red blood cells that have a glucose molecule attached. It measures average exposure to glucose over a period of several months. Table 5.2 gives data on both HbA and FPG for 18 diabetics five months after they had completed a diabetes education class.[20] DIABETES

© Glow Wellness/Alamy

(a) Make a scatterplot with HbA as the explanatory variable. There is a positive linear relationship, but it is surprisingly weak.
(b) Subject 15 is an outlier in the y direction. Subject 18 is an outlier in the x direction. Find the correlation

TABLE 5.1 FOUR DATA SETS FOR EXPLORING CORRELATION AND REGRESSION

DATA SET A

x	10	8	13	9	11	14	6	4	12	7	5
y	8.04	6.95	7.58	8.81	8.33	9.96	7.24	4.26	10.84	4.82	5.68

DATA SET B

x	10	8	13	9	11	14	6	4	12	7	5
y	9.14	8.14	8.74	8.77	9.26	8.10	6.13	3.10	9.13	7.26	4.74

DATA SET C

x	10	8	13	9	11	14	6	4	12	7	5
y	7.46	6.77	12.74	7.11	7.81	8.84	6.08	5.39	8.15	6.42	5.73

DATA SET D

x	8	8	8	8	8	8	8	8	8	8	19
y	6.58	5.76	7.71	8.84	8.47	7.04	5.25	5.56	7.91	6.89	12.50

TABLE 5.2 TWO MEASURES OF GLUCOSE LEVEL IN DIABETICS

SUBJECT	HbA (%)	FPG (mg/mL)	SUBJECT	HbA (%)	FPG (mg/mL)	SUBJECT	HbA (%)	FPG (mg/mL)
1	6.1	141	7	7.5	96	13	10.6	103
2	6.3	158	8	7.7	78	14	10.7	172
3	6.4	112	9	7.9	148	15	10.7	359
4	6.8	153	10	8.7	172	16	11.2	145
5	7.0	134	11	9.4	200	17	13.7	147
6	7.1	95	12	10.4	271	18	19.3	255

for all 18 subjects, for all except Subject 15, and for all except Subject 18. Are either or both of these subjects influential for the correlation? Explain in simple language why r changes in opposite directions when we remove each of these points.

5.44 The effect of changing units. The equation of a regression line, unlike the correlation, depends on the units we use to measure the explanatory and response variables. Return to the data on percent body fat and preferred amount of salt in Exercise 4.23: DATA SALT3

Preferred amount of salt x	0.2	0.3	0.4	0.5	0.6	0.8	1.1
Percent body fat y	20	30	22	30	38	23	30

In calculating the preferred amount of salt, the weight of the salt was in milligrams.

(a) Find the equation of the regression line for predicting percent body fat from preferred amount of salt when weight is in milligrams.

(b) A mad scientist decides to measure weight in tenths of milligrams. The same data in these units are

Preferred amount of salt x	2	3	4	5	6	8	11
Percent body fat y	20	30	22	30	38	23	30

Find the equation of the regression line for predicting percent body fat from preferred amount of salt when weight is in tenths of milligrams.

(c) Use both lines to predict the percent body fat from preferred amount of salt for a child with preferred amount of salt 0.7 when weight is measured in milligrams, which is the same as 7 when weight is in tenths of milligrams. Are the two predictions the same (up to any roundoff error)?

5.45 Managing diabetes, continued. Add three regression lines for predicting FPG from HbA to your scatterplot from Exercise 5.43: for all 18 subjects, for all except Subject 15, and for all except Subject 18. Is either Subject 15 or Subject 18 strongly influential for the least-squares line? Explain in simple language what features of the scatterplot explain the degree of influence. DATA DIABETES

5.46 Are you happy? Exercise 4.25 (page 120) discusses a study in which the mean BRFSS life-satisfaction score of individuals in each state was compared with the mean of an objective measure of well-being (based on the "compensating-differentials method") for each state. Suppose that instead of the means for the states, the BRFSS life-satisfaction scores for individuals were compared with the corresponding measure of well-being (based on the compensating-differentials method) for these individuals. Would you expect the correlation between the mean state scores on these two measures to be lower, about the same, or higher than the correlation between the scores of individuals on these two measures? Explain your answer.

5.47 One more inch, three more pounds. Data on the *average* weight of men who are between 5 feet 2 inches and 6 feet 4 inches tall (rounded to the nearest inch) show a very high positive correlation with height. Would the correlation be greater, smaller, or about the same if you calculated the correlation between the weights of individual men and their heights (rounded to the nearest inch)? Explain your answer.

5.48 Do artificial sweeteners cause weight gain? People who use artificial sweeteners in place of sugar tend to be heavier than people who use sugar. Does this mean that artificial sweeteners cause weight gain? Give a more plausible explanation for this association.

5.49 Learning online. Many colleges offer online versions of courses that are also taught in the classroom. It often happens that the students who enroll in the online version do better than the classroom students on the course exams. This does not show that online instruction is more effective than classroom teaching, because the people who sign up for online courses are often quite different from the classroom students. Suggest some differences between online and classroom students that might explain why online students do better.

5.50 SAT scores and teacher salaries. The data set "tchsal" gives the mean Mathematics SAT score and mean salary of teachers in each of the 50 states and the District of Columbia in 2013.[21] The correlation between mean Mathematics SAT score and mean teacher salary is −0.2782. DATA TCHSAL

(a) Find the least-squares line for predicting mean Mathematics SAT score from mean teacher salary. Interpret the slope in the context of the problem.

(b) Is it reasonable to conclude that states can increase the mean Mathematics SAT score in their state by reducing teacher salaries? (*Hint:* Which states have the highest and lowest mean Mathematics SAT scores? Which states have the highest and lowest cost of living?)

5.51 SAT scores and teacher salaries, continued. The data set "tchsal2" gives the mean Mathematics SAT score and mean salary of teachers in each of the 50 states and the District of Columbia in 2013. It also includes a categorical variable, pct. taking, that indicates whether the percentage taking is above 40% (Y) or below 40% (N). DATA TCHSAL2

(a) Find the least-squares line for predicting mean Mathematics SAT score from mean teacher salary for only the cases where the percent taking is above 40%. Interpret the slope in the context of the problem.
(b) Find the least-squares line for predicting mean Mathematics SAT score from mean teacher salary for only the cases where the percent taking is below 40%. Interpret the slope in the context of the problem.
(c) If you did Exercise 4.48, compare your results with those in part (a) of Exercise 4.48. What do you conclude?

5.52 Grade inflation and the SAT. The effect of a lurking variable can be surprising when individuals are divided into groups. In recent years, the mean SAT score of all high school seniors has increased. But the mean SAT score has decreased for students at each level of high school grades (A, B, C, and so on). Explain how grade inflation in high school (the lurking variable) can account for this pattern.

5.53 Workers' incomes. Here is another example of the group effect cautioned about in the previous exercise. Explain how, as a nation's population grows older, median income can go down for workers in each age group, yet still go up for all workers.

5.54 Some regression math. Use the equation of the least-squares regression line (box on page 132) to show that the regression line for predicting y from x always passes through the point $(\bar{x}, \bar{y})$. That is, when $x = \bar{x}$, the equation gives $\hat{y} = \bar{y}$.

5.55 Regression to the mean. Figure 4.8 (page 120) displays the relationship between golfers' scores on the first and second rounds of the 2013 Masters Tournament. The least-squares line for predicting second-round scores from first-round scores has equation $\hat{y} = 56.47 + 0.243$. Find the predicted second-round scores for a player who shot 80 in the first round and for a player who shot 70. The mean second-round score for all players was 74.20. So, a player who does well in the first round is predicted to do less well, but still better than average, in the second round. In addition, a player who does poorly in the first is predicted to do better, but still worse than average, in the second.

regression to the mean

(*Comment:* This is **regression to the mean.** If you select individuals with extreme scores on some measure, they tend to have less extreme scores when measured again. That's because their extreme position is partly merit and partly luck, and the luck will be different next time. Regression to the mean contributes to lots of "effects." The rookie of the year often doesn't do as well the next year; the best player in an orchestral audition may play less well once hired than the runners-up; a student who feels she needs coaching after taking the SAT often does better on the next try without coaching.)

5.56 Regression to the mean. We expect that students who do well on the midterm exam in a course will usually also do well on the final exam. Gary Smith of Pomona College looked at the exam scores of all 346 students who took his statistics class over a 10-year period.[22] The least-squares line for predicting final exam score from midterm-exam score was $\hat{y} = 46.6 + 0.41x$. (Both exams have a 100-point scale.)

Octavio scores 10 points above the class mean on the midterm. How many points above the class mean do you predict that he will score on the final? (*Hint:* Use the fact that the least-squares line passes through the point $(\bar{x}, \bar{y})$ and the fact that Octavio's midterm score is $\bar{x} + 10$.) This is another example of regression to the mean: students who do well on the midterm will, in general, do less well, but still above average, on the final.

5.57 Is regression useful? APPLET In Exercise 4.42 (page 124) you used the *Correlation and Regression* applet to create three scatterplots having correlation about $r = 0.7$ between the horizontal variable x and the vertical variable y. Create three similar scatterplots again, and click the "Show least-squares line" box to display the regression lines. Correlation $r = 0.7$ is considered reasonably strong in many areas of work. Because there is a reasonably strong correlation, we might use a regression line to predict y from x. In which of your three scatterplots does it make sense to use a straight line for prediction?

5.58 Guessing a regression line. APPLET In the *Correlation and Regression* applet, click on the scatterplot to create a group of 15 to 20 points from lower left to upper right with a clear positive straight-line pattern (correlation around 0.7). Click the "Draw line" button and use the mouse (right-click and drag) to draw a line through the middle of the cloud of points from lower left to upper right. Note the "thermometer" above the plot. The red portion is the sum of the squared vertical distances from the points in the plot to the least-squares line. The green portion is the "extra" sum of squares for your line—it shows by how much your line misses the smallest possible sum of squares.

(a) You drew a line by eye through the middle of the pattern. Yet the right-hand part of the bar is probably almost entirely green. What does that tell you?
(b) Now click the "Show least-squares line" box. Is the slope of the least-squares line smaller (the new

TABLE 5.3 REACTION TIMES (IN MILLISECONDS) IN A COMPUTER GAME

TIME	DISTANCE	HAND	TIME	DISTANCE	HAND
115	190.70	right	240	190.70	left
96	138.52	right	190	138.52	left
110	165.08	right	170	165.08	left
100	126.19	right	125	126.19	left
111	163.19	right	315	163.19	left
101	305.66	right	240	305.66	left
111	176.15	right	141	176.15	left
106	162.78	right	210	162.78	left
96	147.87	right	200	147.87	left
96	271.46	right	401	271.46	left
95	40.25	right	320	40.25	left
96	24.76	right	113	24.76	left
96	104.80	right	176	104.80	left
106	136.80	right	211	136.80	left
100	308.60	right	238	308.60	left
113	279.80	right	316	279.80	left
123	125.51	right	176	125.51	left
111	329.80	right	173	329.80	left
95	51.66	right	210	51.66	left
108	201.95	right	170	201.95	left

line is less steep) or larger (line is steeper) than that of your line? If you repeat this exercise several times, you will consistently get the same result. The least-squares line minimizes the *vertical* distances of the points from the line. It is *not* the line through the "middle" of the cloud of points. This is one reason why it is hard to draw a good regression line by eye.

*The following exercises ask you to answer questions from data without having the details outlined for you. The exercise statements give you the **State** step of the four-step process. In your work, follow the **Plan, Solve,** and **Conclude** steps of the process, described on page 63.*

5.59 **Beavers and beetles.** Do beavers benefit beetles? Researchers laid out 23 circular plots, each four meters in diameter, in an area where beavers were cutting down cottonwood trees. In each plot, they counted the number of stumps from trees cut by beavers and the number of clusters of beetle larvae. Ecologists think that the new sprouts from stumps are more tender than other cottonwood growth, so that beetles prefer them. If so, more stumps should produce more beetle larvae. Here are the data:[23] BEAVERS

Stumps	2	2	1	3	3	4	3	1	2	5	1	3
Beetle larvae	10	30	12	24	36	40	43	11	27	56	18	40
Stumps	2	1	2	2	1	1	4	1	2	1	4	
Beetle larvae	25	8	21	14	16	6	54	9	13	14	50	

Analyze these data to see if they support the "beavers benefit beetles" idea.

5.60 **A computer game.** A multimedia statistics learning system includes a test of skill in using the computer's mouse. The software displays a circle at a random location on the computer screen. The subject clicks in the circle with the mouse as quickly as possible. A new circle appears as soon as the subject clicks in the old one. Table 5.3 gives data for one subject's trials, 20 with each hand. Distance is the distance from the cursor location to the center of the new circle, in units whose actual size depends on the size of the screen. Time is the time required to click in the new circle, in milliseconds.[24] We suspect that time depends on distance. We also suspect that performance will not be the same with the right and left hands. Analyze the data with a view to predicting performance separately for the two hands. COMGAME

5.61 Predicting tropical storms. William Gray heads the Tropical Meteorology Project at Colorado State University (well away from the hurricane belt). His forecasts before each year's hurricane season attract lots of attention. Here are data on the number of named Atlantic tropical storms predicted by Dr. Gray and the actual number of storms for the years 1984 to 2013:[25] STORMS2

NASA/Goddard Space Flight Center/ Scientific Visualization Studio

YEAR	FORECAST	ACTUAL	YEAR	FORECAST	ACTUAL
1984	10	12	1999	14	12
1985	11	11	2000	12	14
1986	8	6	2001	12	15
1987	8	7	2002	11	12
1988	11	12	2003	14	16
1989	7	11	2004	14	14
1990	11	14	2005	15	27
1991	8	8	2006	17	10
1992	8	6	2007	17	14
1993	11	8	2008	15	16
1994	9	7	2009	11	9
1995	12	19	2010	18	19
1996	10	13	2011	16	19
1997	11	7	2012	14	19
1998	10	14	2013	18	13

Analyze these data. How accurate are Dr. Gray's forecasts? How many tropical storms would you expect in a year when his preseason forecast calls for 16 storms? What is the effect of the disastrous 2005 season on your answers?

5.62 Great Arctic rivers. One effect of global warming is to increase the flow of water into the Arctic Ocean from rivers. Such an increase may have major effects on the world's climate. Six rivers (Yenisey, Lena, Ob, Pechora, Kolyma, and Severnaya Dvina) drain two-thirds of the Arctic in Europe and Asia. Several of these are among the largest rivers on earth. Table 5.4 presents the total discharge from these rivers each year from 1936 to 2010.[26] Discharge is measured in cubic kilometers of water. Analyze these data to uncover the nature and strength of the trend in total discharge over time. ARCTIC

TABLE 5.4 ARCTIC RIVER DISCHARGE (CUBIC KILOMETERS), 1936 TO 2010

YEAR	DISCHARGE	YEAR	DISCHARGE	YEAR	DISCHARGE	YEAR	DISCHARGE
1936	1721	1955	1656	1974	2000	1993	1845
1937	1713	1956	1721	1975	1928	1994	1902
1938	1860	1957	1762	1976	1653	1995	1842
1939	1739	1958	1936	1977	1698	1996	1849
1940	1615	1959	1906	1978	2008	1997	2007
1941	1838	1960	1736	1979	1970	1998	1903
1942	1762	1961	1970	1980	1758	1999	1970
1943	1709	1962	1849	1981	1774	2000	1905
1944	1921	1963	1774	1982	1728	2001	1890
1945	1581	1964	1606	1983	1920	2002	2085
1946	1834	1965	1735	1984	1823	2003	1780
1947	1890	1966	1883	1985	1822	2004	1900
1948	1898	1967	1642	1986	1860	2005	1930
1949	1958	1968	1713	1987	1732	2006	1910
1950	1830	1969	1742	1988	1906	2007	2270
1951	1864	1970	1751	1989	1932	2008	2078
1952	1829	1971	1879	1990	1861	2009	1900
1953	1652	1972	1736	1991	1801	2010	1813
1954	1589	1973	1861	1992	1793		

5.63 Will women outrun men? Does the physiology of women make them better suited than men to long-distance running? Will women eventually outperform men in long-distance races? Researchers examined data on world record times (in seconds) for men and women in the marathon. Based on these data, researchers (in 1992) attempted to predict when women would outrun men in the marathon. Here are data for women:[27] MARATHON

Year	1926	1964	1967	1970	1971	1974	1975
Time	13,222.0	11,973.0	11,246.0	10,973.0	9990.0	9834.5	9499.0
Year	1977	1980	1981	1982	1983	1985	
Time	9287.5	9027.0	8806.0	8771.0	8563.0	8466.0	

Here are data for men:

Year	1908	1909	1913	1920	1925	1935	1947
Time	10,518.4	9751.0	9366.6	9155.8	8941.8	8802.0	8739.0
Year	1952	1953	1954	1958	1960	1963	1964
Time	8442.2	8314.8	8259.4	8117.0	8116.2	8068.0	7931.2
Year	1965	1967	1969	1981	1984	1985	1988
Time	7920.0	7776.4	7713.6	7698.0	7685.0	7632.0	7610.0

Analyze these data using least-squares regression to estimate when men and women's record times will be equal. How reliable is your estimate? (You may wish to look online to find the current men and women's record times.)

5.64 Will women outrun men?, continued. The data set "run2" contains all the world record times (in seconds) for men and women in the marathon up to 2013. Use these data to repeat the analysis in Exercise 5.63; that is, use least-squares regression to estimate when men and women's record times will be equal. How reliable s this estimate? Explain your answer. RUN2

Exploring the Web

5.65 Association and causation. Find an example of a study in which the issue of association and causation is present (either association is confused with causation, or association is not confused with causation). Summarize the study and its conclusions in your own words. Be sure to include a copy of the actual article or at least the web source, title, and location where the article was published. The *Chance News* website at `www.causeweb.org/wiki/chance/index.php/Main_Page` is a good place to look for examples.

5.66 Predicting batting averages. Go to `www.mlb.com/` and find the batting averages for a diverse set of 30 players for both the 2012 and 2013 seasons. You can click on the "Stats" tab to find the results for the current season as well as historical data. You should select only players who played in at least 50 games both seasons. Make a scatterplot of the batting averages using the 2012 season average as the explanatory variable and the 2013 season average as the response. Is it reasonable to fit a straight line to these data? If so, find the least-squares regression line for predicting batting average in 2013 from that in 2012 based on your sample of

30 players. In 2012, the major league leader in batting was Buster Posey, who had a batting average of 0.336. What does your least-squares regression line predict for the 2013 batting average of someone who hit 0.336 in 2012? Is the 2013 predicted batting average higher or lower than 0.336?

5.67 Predicting the federal budget. Go to the Congressional Budget Office website, `www.cbo.gov/topics/`. Click on the "topics and subtopics" button and select "Budget and Economic Outlook." Then click on "apply." What is the current prediction for the federal budget in five years' time? Is a surplus or a deficit predicted? Do you think the prediction is accurate? Why or why not?

© Purestock/age fotostock

CHAPTER 6

Two-Way Tables*

In this chapter we cover...

6.1 Marginal distributions
6.2 Conditional distributions
6.3 Simpson's paradox

We have concentrated on relationships in which at least the response variable is quantitative. Now we will describe relationships between two categorical variables. Some variables—such as sex, race, and occupation—are categorical by nature. Other categorical variables are created by grouping values of a quantitative variable into classes. Published data often appear in grouped form to save space. To analyze categorical data, we use the *counts* or *percents* of individuals that fall into various categories.

EXAMPLE 6.1 A Job Outside the Home

A sample survey of adults (aged 18 and over) asked, "If you were free to do either, would you prefer to have a job outside the home, or would you prefer to stay home and take care of the house and family?" Table 6.1 shows the responses.[1] This is a **two-way table** because it describes two categorical variables. One is the sex and education level of the respondent. The other is the preferred lifestyle (a job outside the home, stay home, or no preference). Sex and education is the **row variable** because each row in the table describes a combination of sex and education level. Because the education level has a natural order from "No college" to "College," the rows are also in this order for women and men. Preferred lifestyle

two-way table

row variable

*This material is important in statistics, but it is needed later in this book only for Chapter 25. You may omit it if you do not plan to read Chapter 25, or delay reading it until you reach Chapter 25.

TABLE 6.1 ADULTS BY PREFERRED LIFESTYLE AND SEX AND EDUCATION

	PREFERRED LIFESTYLE			
SEX AND EDUCATION	JOB OUTSIDE HOME	STAY HOME AND CARE FOR HOUSE, FAMILY	NO PREFERENCE	TOTAL
Women no college	81	104	10	195
Women college	173	115	15	303
Men no college	92	32	2	126
Men college	299	81	8	388
Total	645	332	35	1012

column variable

is the **column variable** because each column describes one choice. The entries in the table are the counts of individuals in each sex-and-education-level-by-preferred-lifestyle class. The entries in the right margin are the total of the row entries, the entries in the bottom margin are the total of the column entries, and the entry at the bottom right is the total of all adults in the study. ■

6.1 Marginal distributions

How can we best grasp the information contained in Table 6.1? First, *look at the distribution of each variable separately.* The distribution of a categorical variable says how often each outcome occurred. The "Total" column at the right of the table contains the totals for each of the rows. These row totals give the distribution of sex and education level in the entire group of 1012 adults: 195 were women with no college education, 303 were women with a college education, and so on.

marginal distributions

If the row and column totals are missing, the first thing to do in studying a two-way table is to calculate them. The distributions of sex and education alone and preferred lifestyle alone are called **marginal distributions** because they appear at the right and bottom margins of the two-way table.

Percents are often more informative than counts. We can display the marginal distribution of sex and education in percents by dividing each row total by the table total and converting to a percent.

EXAMPLE 6.2 Calculating a Marginal Distribution

The percent of these adults who were women with no college is

$$\frac{\text{women with no college total}}{\text{table total}} = \frac{195}{1012} = 0.193 = 19.3\%$$

Do three more such calculations to obtain the marginal distribution of sex and education in percents for each group. Here is the complete distribution:

Response	Percent
Women no college	$\frac{195}{1012} = 19.3\%$
Women college	$\frac{303}{1012} = 29.9\%$
Men no college	$\frac{126}{1012} = 12.5\%$
Men college	$\frac{388}{1012} = 38.3\%$

It seems that more women and more men have attended college than not attended college. The total is 100% because everyone belongs to one of the four sex and education classes. ■

Each marginal distribution from a two-way table is a distribution for a single categorical variable. As we saw in Chapter 1, we can use a bar graph or a pie chart to display such a distribution. Figure 6.1 is a bar graph of the distribution of sex and education among adults in the sample.

In working with two-way tables, you must calculate lots of percents. Here's a tip to help you decide what fraction gives the percent you want. Ask, "What group represents the total of which I want a percent?" The count for that group is the denominator of the fraction that leads to the percent. In Example 6.2, we want a percent "of adults," so the count of adults (the table total) is the denominator.

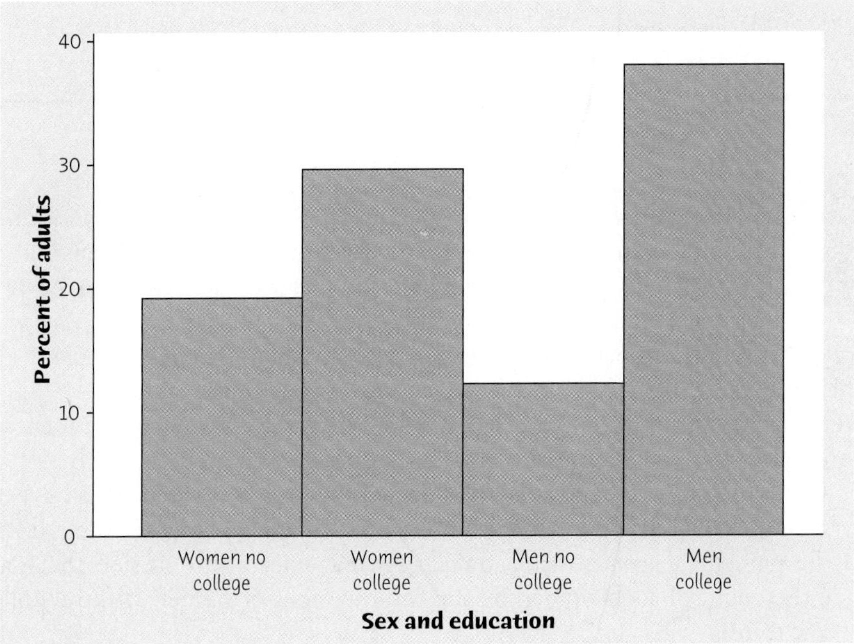

FIGURE 6.1
Bar graph of the distribution of sex and education of adults who participated in the survey. This is one of the marginal distributions for Table 6.1.

LaunchPad Online Resources

- The **StatBoards video**, *Marginal Distributions*, provides the details of calculating marginal distributions through an example.

Apply Your Knowledge

6.1 Video Gaming and Grades. The popularity of computer, video, online, and virtual reality games has raised concerns about their ability to negatively affect youth. The data in this exercise are based on a recent survey of 14- to 18-year-olds in Connecticut high schools. Here are the grade distributions of boys who have and have not played video games.[2] DATA GAMING

	Grade Average		
	As and Bs	Cs	Ds and Fs
Played games	736	450	193
Never played games	205	144	80

(a) How many people does this table describe? How many of these have played video games?

(b) Give the marginal distribution of the grades. What percent of the boys represented in the table received a grade of C or lower?

Dirk Kruell/Laif/Redux

6.2 Ages of College Students. Here is a two-way table of U.S. Census Bureau data describing the age and sex of all American students enrolled in college. The table entries are counts in thousands of students.[3] AGES

Age Group	Female	Male
18 to 24 years	6432	5640
25 to 34 years	2450	1843
35 years or older	2124	1069

(a) How many college undergraduates are there?

(b) Find the marginal distribution of age group. What percent of undergraduates are in the 18- to 24-year-old college age group?

6.2 Conditional distributions

CAUTION

Table 6.1 contains much more information than the two marginal distributions of sex and education alone and preferred lifestyle alone. *Marginal distributions tell us nothing about the relationship between two variables.* To describe a relationship between two categorical variables, we must calculate some well-chosen percents from the counts given in the body of the table.

Let's say we want to compare the preferred lifestyles of women and men with different levels of education. To do this, compare percents for each sex and education category. To study the preferred lifestyles of women who have not been to college, we look only at the "Women no college" row in Table 6.1. To find the percent *of women with no college* who prefer a job outside the home, divide the count of such women by the total number of women with no college (the row total):

$$\frac{\text{women with no college who prefer a job outside the home}}{\text{row total}} = \frac{81}{195} = 0.415 = 41.5\%$$

Doing this for all three entries in the "Women no college" row gives the *conditional distribution* of preferred lifestyles among women with no college. We use the term *conditional* because this distribution describes only adults who satisfy the condition that they are women with no college.

SMILING FACES Women smile more than men. The same data that produce this fact allow us to link smiling to other variables in two-way tables. For example, as the second variable, add whether or not the person thinks they are being observed. If yes, that's when women smile more. If no, there's no difference between women and men. Next, take the second variable to be the person's social role (for example, is he or she the boss in an office?). Within each role, there is very little difference in smiling between women and men.

Marginal and Conditional Distributions

The **marginal distribution** of one of the categorical variables in a two-way table of counts is the distribution of values of that variable among all individuals described by the table.

A **conditional distribution** of a variable is the distribution of values of that variable among only individuals who have a given value of the other variable. There is a separate conditional distribution for each value of the other variable.

EXAMPLE 6.3 Comparing Different Sex and Education Groups

STATE: How do women and men with different levels of education differ in their responses to the question, "If you were free to do either, would you prefer to have a job outside the home, or would you prefer to stay home and take care of the house and family?"

PLAN: Make a two-way table of response by sex and education category. Find the four conditional distributions of response for each sex and education category. Compare these four distributions.

SOLVE: Table 6.1 is the two-way table we need. Look first at just the "Women no college" row to find the conditional distribution for women with no college, then at just the "Women college" row to find the conditional distribution for women who have been to college, and so on. Here are the calculations and the four conditional distributions:

Response	Job Outside Home	Stay Home	No Preference
Women no college	$\frac{81}{195} = 41.5\%$	$\frac{104}{195} = 53.3\%$	$\frac{10}{195} = 5.1\%$
Women college	$\frac{173}{303} = 57.1\%$	$\frac{115}{303} = 38.0\%$	$\frac{15}{303} = 5.0\%$
Men no college	$\frac{92}{126} = 73.0\%$	$\frac{32}{126} = 25.4\%$	$\frac{2}{126} = 1.6\%$
Men college	$\frac{299}{388} = 77.1\%$	$\frac{81}{388} = 20.9\%$	$\frac{8}{388} = 2.1\%$

The percents in each row should be 100% because for each sex and education category, everyone holds one of the three lifestyle preferences. In fact, the percents add to 99.9%, 100.0%, or 100.1% because we round each one to the nearest tenth. This is **roundoff error.**

roundoff error

Each set of percents adds to 100% because everyone holds one of the three opinions.

CONCLUDE: For those in the sample, men are more likely to prefer a job outside the home than are women, whether they went to college or not. Adults who attended college are more likely to prefer a job outside the home, whether they are women or men. ■

Software will do these calculations for you. Most programs allow you to choose which conditional distributions you want to compare. The output in Figure 6.2 presents the four conditional distributions of preferred lifestyle, for each sex and education category, and also the marginal distribution of opinion for all the adults. The distributions agree (up to roundoff) with the results in Examples 6.2 and 6.3.

Remember that there are two sets of conditional distributions for any two-way table. Example 6.3 looked at the conditional distributions of preferred lifestyles for the four sex and education categories. We could also examine the three conditional distributions of sex and education, one for each of the three preferred lifestyles, by looking separately at the three columns in Table 6.1. Figure 6.3 makes this comparison in a bar graph. Each bar is divided (segmented) into four parts, represented by four colors. The upper portion of each bar represents the percent of women with no college among adults who prefer each lifestyle. The other portions represent the percents of each of the other sex and education categories. Each bar has a height of 100% because each bar represents all the adults in each different group of people. Bar graphs like that in Figure 6.3 in which each bar is divided into parts, each part representing a different category, are sometimes called **segmented bar graphs.**

segmented bar graphs

FIGURE 6.2
Minitab and CrunchIt! output for the two-way table of adults by preferred lifestyle and sex and education. Each entry in the Minitab output includes the percent of its row total. The "Men college," "Men no college," "Women college," and "Women no college" rows give the conditional distributions of responses for each sex and education category, and the "All" row shows the marginal distribution of responses for all these adults. Notice that Minitab orders variables in the table alphabetically. Each entry in the CrunchIt! output includes the percent of its row total, the percent of its column total, and the percent of the entire table total. The second entry in each cell gives the conditional distribution of responses for the different sex and education categories. The third entry in each cell for the preferred lifestyle columns gives the conditional distributions of responses for each category of preferred living. The "All" row and column show the corresponding marginal distribution of responses for all these adults.

Session

Rows: Sex and education Columns: Preferred lifestyle

	Job outside home	No preference	Stay home	All
Men college	299 77.06	8 2.06	81 20.88	388 100.00
Men no college	92 73.02	2 1.59	32 25.40	126 100.00
Women college	173 57.10	15 4.95	115 37.95	303 100.00
Women no college	81 41.54	10 5.13	104 53.33	195 100.00
All	645 63.74	35 3.46	332 32.81	1012 100.00

Cell Contents: Count
% of Row

	Stay home	Job outside home	No preference	All
Women no college	104 53.33 31.33 10.28	81 41.54 12.56 8.004	10 5.128 28.57 0.9881	195 100 19.27 19.27
Men no college	32 25.40 9.639 3.162	92 73.02 14.26 9.091	2 1.587 5.714 0.1976	126 100 12.45 12.45
Men college	81 20.88 24.40 8.004	299 77.06 46.36 29.55	8 2.062 22.86 0.7905	388 100 38.34 38.34
Women college	115 37.95 34.64 11.36	173 57.10 26.82 17.09	15 4.950 42.86 1.482	303 100 29.94 29.94
All	332 3281 100 3281	645 63.74 100 63.74	35 3.458 100 3.458	1012 100 100 100

Count % of Row % of Col % of Total

Chi-squared statistic:	83.10
df:	6
P-value:	<0.0001

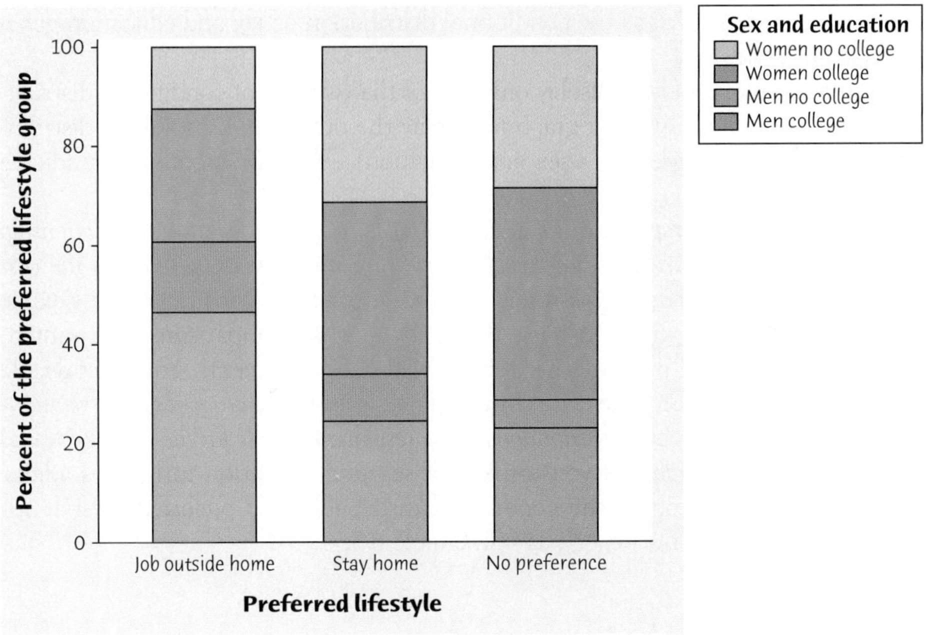

FIGURE 6.3
Segmented bar graph comparing the percents of women and no college (orange), women and college (blue), men and no college (green), and men and college (red) among those who prefer each lifestyle.

mosaic plot

Figure 6.4 is called a **mosaic plot** and is a variation of a segmented bar graph. The bars now have different widths, and these widths correspond to the proportion of people in each of the three preferred lifestyle categories. Thus, the widths display the marginal distribution of preferred lifestyle. Each bar is again divided (segmented) into four parts, represented by four colors. The upper portion of each bar represents the proportion of women with no college among adults who prefer each lifestyle. The other portions represents the proportions of each of the other sex and education categories. Each bar has a height of 1 because each bar represents all the adults in each different group of people. The mosaic plot is more informative than the segmented bar plot because it displays the marginal distribution of

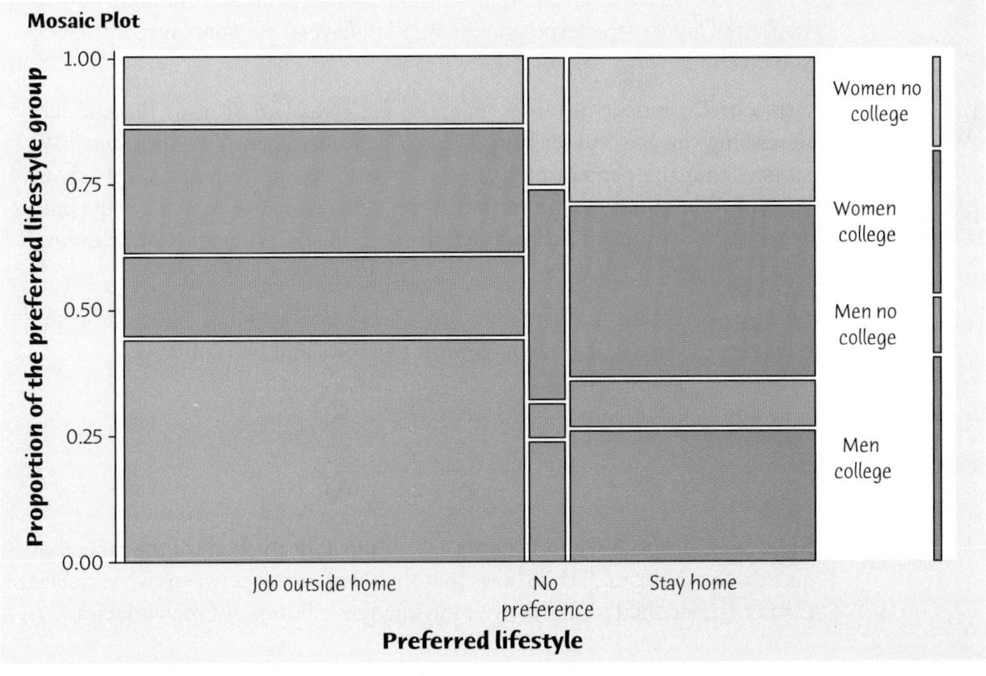

FIGURE 6.4
Mosaic plot comparing the proportions of women and no college (orange), women and college (blue), men and no college (green), and men and college (red) among those who prefer each lifestyle.

preferred lifestyle as well as the conditional distribution of sex and education given preferred lifestyle.

Both Figures 6.3 and 6.4 display only one of the two sets of conditional distributions. We would need another graph to display the other (the conditional distribution of preferred lifestyle given sex and education). Also, the graphs only indicate percents or proportions, not total counts.

CAUTION *No single graph (such as a scatterplot) portrays the form of the relationship between categorical variables. No single numerical measure (such as the correlation) summarizes the strength of the association.* Bar graphs are flexible enough to be helpful, but you must think about what comparisons you want to display. For numerical measures, we rely on well-chosen percents. You must decide which percents you need. Here is a hint: *If there is an explanatory-response relationship, compare the conditional distributions of the response variable for the separate values of the explanatory variable.* If you think that sex and education influences adults' preferred lifestyle, compare the conditional distributions of preferred lifestyle for each sex and education category, as in Example 6.3.

LaunchPad Online Resources

- **The StatBoards video, *Conditional Distributions*, provides the details of calculating conditional distributions through an example.**
- **The StatBoards video, *Graphing a Two-Way Table*, provides the details of creating a mosaic plot through an example.**
- **The Video Technology Manuals discuss how to make a segmented bar graph using software for several statistical software packages.**

Apply Your Knowledge

Keith Bedford/The New York Times/Redux

6.3 Video Gaming and Grades. Exercise 6.1 gives data on the grade distribution of boys who have and have not played video games. To see the relationship between grades and game-playing experience, find the conditional distributions of grades (the response variable) for players and nonplayers. What do you conclude? GAMING

6.4 Ages of College Students. Exercise 6.2 gives U.S. Census Bureau data describing the age and sex of all American students enrolled in college. We suspect that the percent of women is higher among students in the 25- to 34-year-old age group than in the 18- to 24-year-old age group. Do the data support this suspicion? Follow the four-step process as illustrated in Example 6.3. AGES

6.5 Marginal Distributions Aren't the Whole Story. Here are the row and column totals for a two-way table with two rows and two columns:

a	b	50
c	d	50
60	40	100

Make up *two different* sets of counts a, b, c, and d for the body of the table that give these same totals. This shows that the relationship between two variables cannot be obtained from the two individual distributions of the variables.

6.3 Simpson's paradox

As is the case with quantitative variables, the effects of lurking variables can change or even reverse relationships between two categorical variables. Here is an example that demonstrates the surprises that can await the unsuspecting user of data.

EXAMPLE 6.4 Do Medical Helicopters Save Lives?

Accident victims are sometimes taken by helicopter from the accident scene to a hospital. Helicopters save time. Do they also save lives? Let's compare the percents of accident victims who die with helicopter evacuation and with the usual transport to a hospital by road. Here are hypothetical data that illustrate a practical difficulty:[4]

© Ashley Cooper/Corbis

	Helicopter	Road
Victim died	64	260
Victim survived	136	840
Total	200	1100

We see that 32% (64 out of 200) of helicopter patients died, but only 24% (260 out of 1100) of the others did. That seems discouraging.

The explanation is that the helicopter is sent mostly to serious accidents, so that the victims transported by helicopter are more often seriously injured. They are more likely to die with or without helicopter evacuation. Here are the same data broken down by the seriousness of the accident:

Serious Accidents		
	Helicopter	Road
Died	48	60
Survived	52	40
Total	100	100

Less Serious Accidents		
	Helicopter	Road
Died	16	200
Survived	84	800
Total	100	1000

Inspect these tables to convince yourself that they describe the same 1300 accident victims as the original two-way table. For example, 200 (= 100 + 100) were moved by helicopter, and 64 (= 48 + 16) of these died.

Among victims of serious accidents, the helicopter saves 52% (52 out of 100) compared with 40% for road transport. If we look only at less serious accidents, 84% of those transported by helicopter survive, versus 80% of those transported by road. Both groups of victims have a higher survival rate when evacuated by helicopter. ■

How can it happen that the helicopter does better for both groups of victims but worse when all victims are lumped together? Examining the data makes the explanation clear. Half the helicopter transport patients are from serious accidents, compared with only 100 of the 1100 road transport patients. So the helicopter carries patients who are more likely to die. The seriousness of the accident was a lurking variable that, until we uncovered it, hid the true relationship between survival and mode of transport to a hospital. Example 6.4 illustrates *Simpson's paradox*.

Simpson's Paradox

An association or comparison that holds for all of several groups can reverse direction when the data are combined to form a single group. This reversal is called **Simpson's paradox.**

The lurking variable in Simpson's paradox is categorical. That is, it breaks the individuals into groups, as when accident victims are classified as injured in a "serious accident" or a "less serious accident." Simpson's paradox is just an extreme form of the fact that observed associations can be misleading when there are lurking variables.

Apply Your Knowledge

6.6 Field Goal Shooting. Here are data on field goal shooting for two members of the Benedict College 2012–13 men's basketball team:[5] DATA BSKTBALL

	Alex Brailsford		Rickie Jackson	
	Made	**Missed**	**Made**	**Missed**
Two-pointers	15	15	30	29
Three-pointers	5	15	65	130

(a) What percent of all field goal attempts did Alex Brailsford make? What percent of all field goal attempts did Rickie Jackson make?

(b) Now find the percent of all two-point field goals and all three-point field goals that Alex made. Do the same for Rickie.

(c) Alex had a lower percent for *both* types of field goals, but had a better overall percent. That sounds impossible. Explain carefully, referring to the data, how this can happen.

Mike Powell/Getty Images

6.7 Bias in the Jury Pool? The New Zealand Department of Justice did a study of the composition of juries in court cases. Of interest was whether Maori, the indigenous people of New Zealand, were adequately represented in jury pools. Here are the results for two districts, Rotorua and Nelson, in New Zealand (similar results were found in all districts):[6] DATA JURY

Rotorua

	Maori	Non-Maori
In jury pool	79	258
Not in jury pool	8810	23,751
Total	8889	24,009

Nelson

	Maori	Non-Maori
In jury pool	1	56
Not in jury pool	1328	32,602
Total	1329	32,658

(a) Compare percents to show that the percent of all Maori in the jury pool in each district is less than the percent of non-Maori in the jury pool.

(b) Combine the data into a single two-way table of outcome ("in jury pool" or "not in jury pool") by ethnicity (Maori or non-Maori). The original study only reported such an overall rate. Which ethnic group has a higher percent of its people in the jury pool?

(c) Explain from the data, in language that a reporter can understand, how Maori can have a higher percent overall even though non-Maori have higher percents for both districts.

CHAPTER 6 SUMMARY

Chapter Specifics

- A **two-way table** of counts organizes data about two categorical variables. Values of the **row variable** label the rows that run across the table, and values of the **column variable** label the columns that run down the table. Two-way tables are often used to summarize large amounts of information by grouping outcomes into categories.
- The **row totals** and **column totals** in a two-way table give the **marginal distributions** of the two individual variables. It is clearer to present these distributions as percents of the table total. Marginal distributions tell us nothing about the relationship between the variables.
- There are two sets of **conditional distributions** for a two-way table: the distributions of the row variable for each fixed value of the column variable and the distributions of the column variable for each fixed value of the row variable. Comparing one set of conditional distributions is one way to describe the association between the row and the column variables.
- To find the **conditional distribution** of the row variable for one specific value of the column variable, look only at that one column in the table. Find each entry in the column as a percent of the column total.
- **Bar graphs** are a flexible means of presenting categorical data. There is no single best way to describe an association between two categorical variables.
- A comparison between two variables that holds for each individual value of a third variable can be changed or even reversed when the data for all values of the third variable are combined. This is **Simpson's paradox.** Simpson's paradox is an example of the effect of lurking variables on an observed association.

Link It

In Chapters 4 and 5 we considered relationships between two quantitative variables. In this chapter we use two-way tables to describe relationships between two *categorical* variables. To explore relationships between two categorical variables, we examine their conditional distributions. Changes in the pattern of the conditional distribution of one variable as the value of the other varies provides information about the relationship between the variables. No change in this pattern suggests that there is no relationship.

As in Chapters 4 and 5, we must be careful not to assume that the patterns we observe would continue to hold for additional data or in a broader setting. Simpson's paradox is an example of how such an assumption could mislead us.

An important question is whether a pattern we observe while exploring a given set of data holds for values of our variables that we have not observed—in other words, that additional data would continue to conform to these patterns. We will begin to answer this question in Part III. And in Chapter 25 we will learn the answer to this question for two-way tables.

LaunchPad Online Resources

If you are having difficulty with any of the sections of this chapter, these online resources should help prepare you to solve the exercises at the end of this chapter.

- **StatTutor** starts with a video review of each section and asks a series of questions to check your understanding. You may find this helpful if you need additional help understanding how to calculate marginal distributions or conditional distributions, or if you want to see more examples of Simpson's paradox.
- **LearningCurve** provides you with a series of questions about the chapter geared to your level of understanding.

CHECK YOUR SKILLS

The Pew Internet and American Life Project interviewed several hundred teens (aged 12 to 17). One question asked was, "How often do you take your cell phone to school?" Here is a two-way table of the responses by how permissive the school is with regard to cell phone use:[7]

FREQUENCY	FORBID	ALLOW IN SCHOOL BUT NOT IN CLASS	ALLOW IN CLASS
Never	25	19	4
Less often	14	31	6
At least several times per week	14	23	8
Every day	97	314	57

Exercises 6.8 to 6.16 are based on this table. DATA PHONE

6.8 How many individuals are described by this table?

(a) 468 (b) 612 (c) Need more information

6.9 How many teens from schools that forbid cell phones were among the respondents?

(a) 48 (b) 150 (c) Need more information

6.10 The percent of teens from schools that forbid cell phones among the respondents was

(a) about 8%.
(b) about 25%.
(c) about 48%.

6.11 Your percent from Exercise 6.10 is part of

(a) the marginal distribution of school permissiveness.
(b) the marginal distribution of the frequency at which a teen brought a cell phone to school.
(c) the conditional distribution of the frequency at which a teen brought a cell phone to school among schools with a given level of permissiveness.

6.12 What percent of teens from schools that forbid cell phones brought their cell phone to school every day?

(a) about 16% (b) about 21% (c) about 65%

6.13 Your percent from Exercise 6.12 is part of

(a) the marginal distribution of the frequency at which a teen brought a cell phone to school.
(b) the conditional distribution of school permissiveness among those who brought a cell phone to school every day.
(c) the conditional distribution of the frequency at which a teen brought a cell phone to school among schools that forbid cell phones.

6.14 What percent of those who brought their cell phone to school every day were from schools that forbid cell phones?

(a) about 16% (b) about 21% (c) about 65%

6.15 Your percent from Exercise 6.14 is part of

(a) the marginal distribution of the frequency at which a teen brought a cell phone to school.
(b) the conditional distribution of school permissiveness among those who brought a cell phone to school every day.
(c) the conditional distribution of the frequency at which a teen brought a cell phone to school among schools with a given level of permissiveness.

6.16 A bar graph showing the conditional distribution of the frequency at which a teen brought a cell phone to school among schools with a given level of permissiveness would have

(a) 3 bars. (b) 4 bars. (c) 12 bars.

6.17 A college looks at the grade point average (GPA) of its full-time and part-time students. Grades in science courses are generally lower than grades in other courses. There are few science majors among part-time students, but many science majors among full-time students. The college finds that full-time students who are science majors have higher GPAs than part-time students who are science majors. Full-time students who are not science majors also have higher GPAs than part-time students who are not science majors. Yet part-time students as a group have higher GPAs than full-time students. This finding is

(a) not possible: if both science and other majors who are full-time have higher GPAs than those who are part-time, then all full-time students together must have higher GPAs than all part-time students together.
(b) an example of Simpson's paradox: full-time students do better in both kinds of courses but worse overall because they take more science courses.
(c) due to comparing two conditional distributions that should not be compared.

CHAPTER 6 EXERCISES

6.18 Is astrology scientific? The University of Chicago's General Social Survey (GSS) is the nation's most important social science sample survey. The GSS asked a random sample of adults their opinion about whether astrology is very scientific, sort of scientific, or not at all scientific. Here is a two-way table of counts for people in the sample who had three levels of higher education degrees:[8] DATA ASTRLGY

	DEGREE HELD		
	JUNIOR COLLEGE	BACHELOR	GRADUATE
Not at all scientific	44	122	71
Very or sort of scientific	31	62	27

Find the two conditional distributions of degree held: one for those who hold the opinion that astrology is not at all scientific, and one for those who say astrology is very or sort of scientific. Based on your calculations, describe with a graph and in words the differences between those who say astrology is not at all scientific and those who say it is very or sort of scientific.

6.19 Weight-lifting injuries. Resistance training is a popular form of conditioning aimed at enhancing sports performance, and is widely used among high school, college, and professional athletes, although its use for younger athletes is controversial. A random sample of 4111 patients between the ages of 8 and 30 admitted to U.S. emergency rooms with the injury code "weightlifting" was obtained. These injuries were classified as "accidental" if caused by dropped weight or improper equipment use. The patients were also classified into the four age categories 8 to 13 years, 14 to 18, 19 to 22, and 23 to 30. Here is a two-way table of the results:[9] DATA LIFTING

AGE	ACCIDENTAL	NOT ACCIDENTAL
8–13	295	102
14–18	655	916
19–22	239	533
23–30	363	1008

Compare the distributions of ages for accidental and nonaccidental injuries. Use percents and draw a bar graph. What do you conclude?

Marital status and income. *We sometimes hear that getting married is good for your career. Table 6.2 presents data from the U.S. Census Bureau that classifies men, ages 45–64, according to marital status and annual income in 2011–12. We include only data on men, ages 45–64, to avoid sex bias and reduce the effect of age.*[10] *Exercises 6.20 to 6.24 are based on these data.*

6.20 Marginal distributions. Give (in percents) the two marginal distributions, for marital status and for income. Do each of your two sets of percents add to exactly 100%? If not, why not? DATA STATUS

6.21 Percents. What percent of single men have no income? What percent of men with no income are single men? DATA STATUS

6.22 Conditional distribution. Give (in percents) the conditional distribution of income level among single men. Should your percents add to 100% (up to roundoff error)? Explain your reasoning. DATA STATUS

6.23 Marital status and income. One way to see the relationship is to look at who has no income. DATA STATUS

(a) There are 995,000 married men with no income, and 513,000 single men with no income. Explain why these counts by themselves don't describe the relationship between marital status and job grade.
(b) Find the percent of men in each marital status group who have no income. Then find the percent in each marital group who have an income of $100,000 and over. What do these percents say about the relationship?

6.24 Association is not causation. The data in Table 6.2 show that single men are more likely to hold lower-income jobs than are married men. We should not conclude that single men can increase their income by getting married. What lurking variables might help

TABLE 6.2 MARITAL STATUS AND SALARY LEVEL (THOUSANDS OF MEN)

	MARITAL STATUS				
INCOME	SINGLE (NEVER MARRIED)	MARRIED	DIVORCED	WIDOWED	TOTAL
No income	513	995	385	25	1,918
$1–$49,999	3,323	13,478	3,678	453	20,932
$50,000–$99,999	814	8,492	1,316	128	10,750
$100,000 and over	288	5,167	544	49	6,048
Total	4,938	28,132	5,923	655	39,648

explain the association between marital status and income?

6.25 **Race and the death penalty.** Whether a convicted murderer gets the death penalty seems to be influenced by the race of the victim. Several researchers studied this issue in the 1970s and 1980s, resulting in several landmark, oft-cited, and controversial papers. Here are data on 326 cases in which the defendant was convicted of murder from one of these studies:[11] DATA DISCRIM

WHITE DEFENDANT		
	WHITE VICTIM	BLACK VICTIM
Death	19	0
Not	132	9

BLACK DEFENDANT		
	WHITE VICTIM	BLACK VICTIM
Death	11	6
Not	52	97

(a) Use these data to make a two-way table of defendant's race (white or black) versus death penalty (yes or no).
(b) Show that Simpson's paradox holds: a higher percent of white defendants are sentenced to death overall, but for both black and white victims a higher percent of black defendants are sentenced to death.
(c) Use the data to explain why the paradox holds, in language that a judge could understand.

6.26 **Obesity and health.** To estimate the health risks of obesity, we might compare how long obese and nonobese people live. Smoking is a lurking variable that may reduce the gap between the two groups because smoking tends to both reduce weight and lead to earlier death. So if we ignore smoking, we may underestimate the health risks of obesity. Illustrate Simpson's paradox by a simplified version of this situation: make up two-way tables of obese (yes or no) by early death (yes or no) separately for smokers and nonsmokers such that

- Obese smokers and obese nonsmokers are both more likely to die earlier than those who are not obese.
- But when smokers and nonsmokers are combined into a two-way table of obese by early death, persons who are not obese are more likely to die earlier because more of them are smokers.

The following exercises ask you to answer questions from data without having the details outlined for you. The exercise statements give you the State step of the four-step process. In your work, follow the ***Plan, Solve,*** *and* ***Conclude*** *steps of the process as illustrated in Example 6.3.*

6.27 **Smoking cessation.** A large randomized trial was conducted to assess the efficacy of Chantix for smoking cessation compared with bupropion (more commonly known as Wellbutrin or Zyban) and a placebo. Chantix is different from most other quit-smoking products in that it targets nicotine receptors in the brain, attaches to them, and blocks nicotine from reaching them, whereas bupropion is an antidepressant often used to help people stop smoking. Generally healthy smokers who smoked at least 10 cigarettes per day were assigned at random to take Chantix ($n = 352$), bupropion ($n = 329$), or a placebo ($n = 344$). The response measure is continuous cessation from smoking for Weeks 9 through 12 of the study. Here is a two-way table of the results:[12] DATA SMOKE

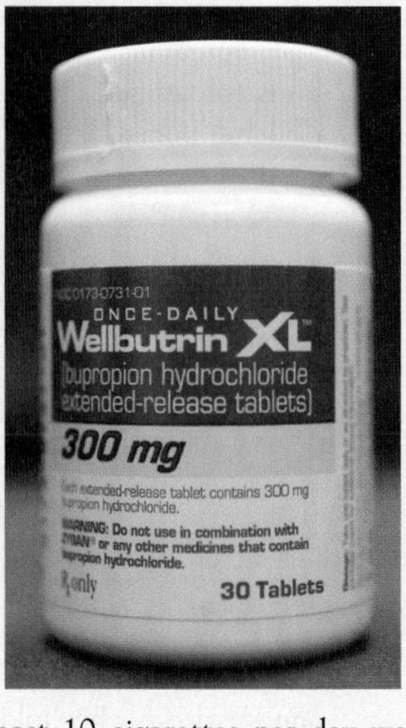

Joe Raedle/Getty Images

	TREATMENT		
	CHANTIX	BUPROPION	PLACEBO
No smoking in Weeks 9–12	155	97	61
Smoked in Weeks 9–12	197	232	283

6.28 **Animal testing.** "It is right to use animals for medical testing if it might save human lives." The General Social Survey asked 1152 adults to react to this statement. Here is the two-way table of their responses: DATA ANTEST

RESPONSE	MALE	FEMALE
Strongly agree	76	59
Agree	270	247
Neither agree nor disagree	87	139
Disagree	61	123
Strongly disagree	22	68

How do the distributions of opinion differ between men and women?

6.29 **College degrees.** "Colleges and universities across the country are grappling with the case of the mysteriously vanishing male." So said an article in the *Washington Post.* Here are projections of the numbers of degrees that will be earned in 2021–22, as projected by the National Center for Education Statistics. The table entries are counts of degrees in thousands.[13] DATA DEGREES

DEGREE	FEMALE	MALE
Associate's	646	383
Bachelor's	1160	844
Master's	576	354
Professional or Doctor's	106	92

Briefly contrast the counts and distributions of men and women in earning degrees. Are men projected to be "vanishing" from colleges and universities across the country?

6.30 **Complications of bariatric surgery.** Bariatric surgery, or weight-loss surgery, includes a variety of procedures performed on people who are obese. Weight loss is achieved by reducing the size of the stomach with an implanted medical device (gastric banding), by removing a portion of the stomach (sleeve gastrectomy), or by resecting and rerouting the small intestines to a small stomach pouch (gastric bypass surgery). Because there can be complications using any of these methods, the National Institute of Health recommends bariatric surgery for obese people with a body mass index (BMI) of at least 40, and for people with BMI 35 and serious coexisting medical conditions such as diabetes. Serious complications include potentially life-threatening, permanently disabling, and fatal outcomes. Here is a two-way table for data collected in Michigan over several years, giving counts of non-life-threatening complications, serious complications, and no complications for these three types of surgeries:[14] DATA BARI

	TYPE OF COMPLICATION			
	NON-LIFE-THREATENING	SERIOUS	NONE	TOTAL
Gastric banding	81	46	5253	5380
Sleeve gastrectomy	31	19	804	854
Gastric bypass	606	325	8110	9041

What do the data say about differences in complications for the three types of surgeries?

6.31 **Smokers rate their health.** The University of Michigan Health and Retirement Study (HRS) surveys more than 22,000 Americans over the age of 50 every two years. A subsample of the HRS participated in a 2009 Internet-based survey that collected information on a number of topical areas, including health (physical and mental health behaviors), psychosocial items, economics (income, assets, expectations, and consumption), and retirement.[15] Two of the questions asked were, "Would you say your health is excellent, very good, good, fair, or poor?" and "Do you smoke cigarettes now?" The two-way table summarizes the answers on these two questions. DATA SMRATE

	CURRENT SMOKER	
HEALTH	YES	NO
Excellent	25	484
Very good	115	1557
Good	145	1309
Fair	90	545
Poor	29	11

What do the data say about differences in self-evaluation of health for current smokers and nonsmokers?

6.32 **Punxsutawney Phil.** In the United States, Groundhog Day is celebrated on February 2. On February 2, Punxsutawney Phil, a mythical groundhog in Punxsutawney, PA, emerges from his home and if he sees his shadow and returns to his hole, he has predicted six more weeks of winter-like weather. How accurate is Phil? We have data for 113 years (up to 2013) indicating whether Phil saw his shadow and average temperature in March. It is not clear what constitutes "six more weeks of winter-like weather," so here we define it to occur if the average temperature for March is not above the historical average. The results are: 52 years where Phil saw his shadow and the average March temperature was not above the historical average, 46 years where Phil saw his shadow and the average March temperature was above the historical average, 5 years where Phil did not see his shadow and the average March temperature was not above the historical average, and 10 years where Phil did not see his shadow and the average March temperature was above the historical average.[16]

(a) Make a two-way table of "Phil saw his shadow or not" against "above historical average temperatures in March or not."

(b) What do the data tell us about Phil as a weather forecaster?

6.33 **Sleep quality.** A random sample of 871 students between the ages of 20 and 24 at a large midwestern university completed a survey including questions about their sleep quality, moods, academic performance, physical health, and psychoactive drug use. Sleep quality was measured using the Pittsburgh Sleep Quality Index (PSQI), with students scoring less than or equal to 5 on the index classified as optimal sleepers, those scoring a 6 or 7 classified as borderline, and those scoring over 7 classified as poor sleepers. The following table looks at the relationship between sleep quality classification and the use of over-the-counter (OTC) or prescription (Rx) stimulant medication more than once a month to help keep awake.[17] DATA SLEEPQ

USE OF OTC/RX MEDS TO WAKE > 1X/MONTH	SLEEP QUALITY ON PSQI INDEX		
	OPTIMAL	BORDERLINE	POOR
Yes	37	53	84
No	266	186	245

What do the data say about differences in sleep quality for those who use over-the-counter or prescription stimulant medication medication more than once a month to keep awake, and those who don't?

Exploring the Web

6.34 Promoting women. In academics, faculty typically start as assistant professors, are promoted to associate professor (and gain tenure), and finally reach the rank of full professor. Some have argued that women have a harder time gaining promotion to associate and full professor than do men. Do data support this argument? Search the web to find the number of faculty by rank and gender at some university. Do you see a pattern that suggests that the proportion of women decreases as rank increases? We found several sources of data by doing a Google search on "faculty head count by rank and gender." In addition to discussing the pattern you find, provide the data, the name of the school, and the source of the data.

6.35 Accidental deaths and age. Accidental deaths are shocking and tragic. Do the ways in which people die by accident change with age? Look at the most recent *Statistical Abstract of the United States* (`www.census.gov/compendia/statab/`) and make a two-way table that provides the counts of deaths due to accidents from various causes for three different age groups. What do you conclude?

6.36 Simpson's paradox. Find an example of Simpson's paradox and discuss how your example illustrates the paradox. Two examples that we found (thanks to Patricia Humphrey at Georgia Southern University) are `www.nytimes.com/2006/07/15/education/15report.html` and `online.wsj.com/article/SB125970744553071829.html`.

Moments by Mullineux/Shutterstock

CHAPTER 7

Exploring Data: Part I Review

In this chapter we cover...

- Part I Summary
- Test Yourself
- Supplementary Exercises
- Online Data for Additional Analyses

Data analysis is the art of describing data using graphs and numerical summaries. The purpose of exploratory data analysis is to help us see and understand the most important features of a set of data. Chapter 1 commented on graphs to display distributions: pie charts and bar graphs for categorical variables, histograms and stemplots for quantitative variables. In addition, time plots show how a quantitative variable changes over time. Chapter 2 presented numerical tools for describing the center and variability of the distribution of one variable. Chapter 3 discussed density curves for describing the overall pattern of a distribution, with emphasis on the Normal distributions.

The first STATISTICS IN SUMMARY figure on the next page organizes the big ideas for exploring a quantitative variable. Plot your data, then describe their center and variability using either the mean and standard deviation or the five-number summary. The last step, which makes sense only for some data, is to summarize the data in compact form by using a Normal curve as a description of the overall pattern. The question marks at the last two stages remind us that the usefulness of numerical summaries and Normal distributions depends on what we find when we examine graphs of our data. No short summary does justice to irregular shapes or to data with several distinct clusters.

STATISTICS IN SUMMARY

Analyzing Data for One Variable

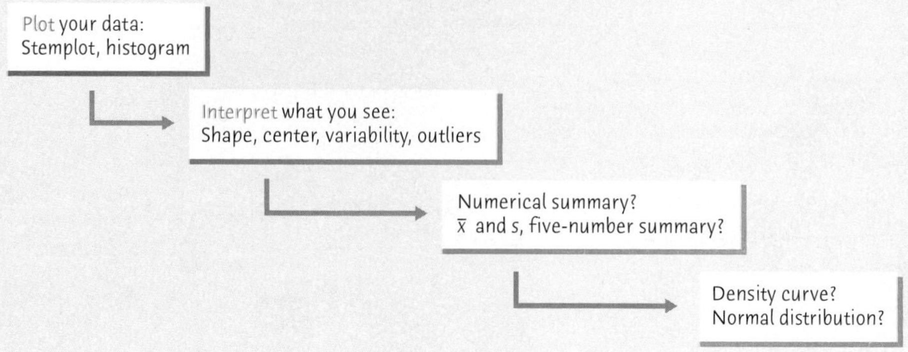

Chapters 4 and 5 applied the same ideas to relationships between two quantitative variables. The second STATISTICS IN SUMMARY figure retraces the big ideas, with details that fit the new setting. Always begin by making graphs of your data. In the case of a scatterplot, we have learned a numerical summary only for data that show a roughly linear pattern on the scatterplot. The summary is then the means and standard deviations of the two variables and their correlation. A regression line drawn on the plot gives a compact description of the overall pattern that we can use for prediction. Once again there are question marks at the last two stages to remind us that correlation and regression describe only straight-line relationships. Chapter 6 shows how to understand relationships between two categorical variables; comparing well-chosen percents is the key.

You can organize your work in any open-ended data analysis setting by following the four-step **State**, **Plan**, **Solve**, and **Conclude** process first introduced in Chapter 2. After you have mastered the extra background needed for statistical inference, this process will also guide practical work on inference later in the book.

STATISTICS IN SUMMARY

Analyzing Data for Two Variables

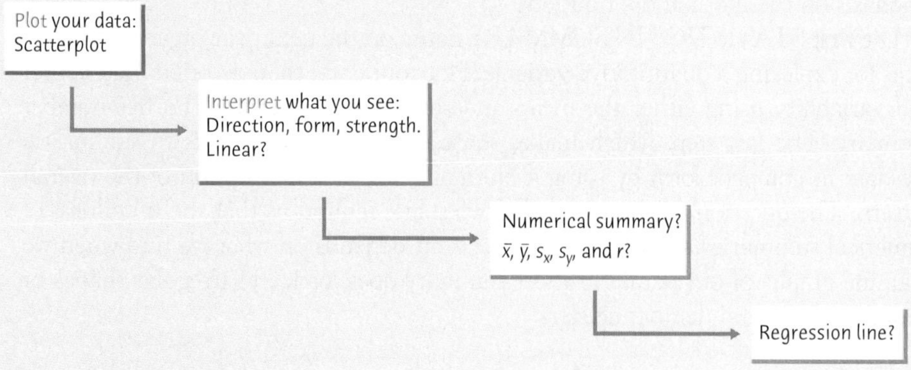

Part I Summary

Here are the most important skills you should have acquired from reading Chapters 1 to 6.

A. Data

1. Identify the individuals and variables in a set of data.
2. Identify each variable as categorical or quantitative. Identify the units in which each quantitative variable is measured.
3. Identify the explanatory and response variables in situations where one variable explains or influences another.

B. Displaying Distributions

1. Recognize when a pie chart can and cannot be used.
2. Make a bar graph of the distribution of a categorical variable or, in general, to compare related quantities.
3. Interpret pie charts and bar graphs.
4. Make a histogram of the distribution of a quantitative variable.
5. Make a stemplot of the distribution of a small set of observations. Round leaves or split stems as needed to make an effective stemplot.
6. Make a time plot of a quantitative variable over time. Recognize patterns such as trends and cycles in time plots.

C. Describing Distributions (Quantitative Variable)

1. Look for the overall pattern and for major deviations from the pattern.
2. Assess from a histogram or stemplot whether the shape of a distribution is roughly symmetric, distinctly skewed, or neither. Assess whether the distribution has one or more major peaks.
3. Describe the overall pattern by giving numerical measures of center and variability in addition to a verbal description of shape.
4. Decide which measures of center and variability are more appropriate: the mean and standard deviation (especially for symmetric distributions) or the five-number summary (especially for skewed distributions).
5. Recognize outliers and give plausible explanations for them.

D. Numerical Summaries of Distributions

1. Find the median M and the quartiles Q_1 and Q_3 for a set of observations.
2. Find the five-number summary and draw a boxplot; assess center, variability, symmetry, and skewness from a boxplot.
3. Find the mean $\bar{x}$ and the standard deviation s for a set of observations.
4. Understand that the median is more resistant than the mean. Recognize that skewness in a distribution moves the mean away from the median toward the long tail.
5. Know the basic properties of the standard deviation: $s \geq 0$ always; $s = 0$ only when all observations are identical and increases as the variability increases; s has the same units as the original measurements; s is pulled strongly up by outliers or skewness.

E. Density Curves and Normal Distributions

1. Know that areas under a density curve represent proportions of all observations and that the total area under a density curve is 1.

2. Approximately locate the median (equal-areas point) and the mean (balance point) on a density curve.
3. Know that the mean and median both lie at the center of a symmetric density curve and that the mean moves farther toward the long tail of a skewed curve.
4. Recognize the shape of Normal curves and estimate by eye both the mean and standard deviation from such a curve.
5. Use the 68–95–99.7 rule and symmetry to state what percent of the observations from a Normal distribution fall between two points when both points lie at the mean or one, two, or three standard deviations on either side of the mean.
6. Find the standardized value (z-score) of an observation. Interpret z-scores and understand that any Normal distribution becomes the standard Normal $N(0, 1)$ distribution when standardized.
7. Given that a variable has a Normal distribution with a stated mean μ and standard deviation σ, calculate the proportion of values above a stated number, below a stated number, or between two stated numbers.
8. Given that a variable has a Normal distribution with a stated mean μ and standard deviation σ, calculate the point having a stated proportion of all values above it or below it.

F. Scatterplots and Correlation

1. Make a scatterplot to display the relationship between two quantitative variables measured on the same subjects. Place the explanatory variable (if any) on the horizontal scale of the plot.
2. Add a categorical variable to a scatterplot by using a different plotting symbol or color.
3. Describe the direction, form, and strength of the overall pattern of a scatterplot. In particular, recognize positive or negative association and linear (straight-line) patterns. Recognize outliers in a scatterplot.
4. Judge whether it is appropriate to use correlation to describe the relationship between two quantitative variables. Find the correlation r.
5. Know the basic properties of correlation: r measures the direction and strength of only straight-line relationships; r is always a number between -1 and 1; $r = \pm 1$ only for perfect straight-line relationships; r moves away from 0 toward ± 1 as the straight-line relationship gets stronger.

G. Regression Lines

1. Understand that regression requires an explanatory variable and a response variable. Correctly identifying which variable is the explanatory variable and which is the response variable is important. Switching these will result in different regression lines. Use a calculator or software to find the least-squares regression line of a response variable y on an explanatory variable x from data.
2. Explain what the slope b and the intercept a mean in the equation $\hat{y} = a + bx$ of a regression line.
3. Draw a graph of a regression line when you are given its equation.
4. Use a regression line to predict y for a given x. Recognize extrapolation and be aware of its dangers.

5. Find the slope and intercept of the least-squares regression line from the means and standard deviations of x and y and their correlation.
6. Use r^2, the square of the correlation, to describe how much of the variation in one variable can be accounted for by a straight-line relationship with another variable.
7. Recognize outliers and potentially influential observations from a scatterplot with the regression line drawn on it.
8. Calculate the residuals and plot them against the explanatory variable x. Recognize that a residual plot magnifies the pattern of the scatterplot of y versus x.

H. Cautions about Correlation and Regression

1. Understand that both r and the least-squares regression line can be strongly influenced by a few extreme observations.
2. Recognize possible lurking variables that may explain the observed association between two variables x and y.
3. Understand that even a strong correlation does not mean there is a cause-and-effect relationship between x and y.
4. Give plausible explanations for an observed association between two variables: direct cause and effect, the influence of lurking variables, or both.

I. Categorical Data

1. From a two-way table of counts, find the marginal distributions of both variables by obtaining the row sums and column sums.
2. Express any distribution in percents by dividing the category counts by their total.
3. Describe the relationship between two categorical variables by computing and comparing percents. Often this involves comparing the conditional distributions of one variable for the different categories of the other variable.
4. Recognize Simpson's paradox and be able to explain it.

DRIVING IN CANADA

Canada is a civilized and restrained nation, at least in the eyes of Americans. A survey sponsored by the Canada Safety Council suggests that driving in Canada may be more adventurous than expected. Of the Canadian drivers surveyed, 88% admitted to aggressive driving in the past year, and 76% said that sleep-deprived drivers were common on Canadian roads. What really alarms us is the name of the survey: the Nerves of Steel Aggressive Driving Study.

Test Yourself

The questions below include multiple-choice, calculations, and short-answer questions. They will help you review the basic ideas and skills presented in Chapters 1 to 6.

7.1 As part of a database on new births at a hospital, some variables recorded are the age of the mother, marital status of the mother (single, married, divorced, other), weight of the baby, and sex of the baby. Of these variables

(a) age, marital status, and weight are quantitative variables.
(b) age and weight are categorical variables.
(c) sex and marital status are categorical variables.
(d) sex, marital status, and age are categorical variables.

7.2 You are interested in obtaining information about the performance of students in your statistics class and seeing how this performance is affected by several factors such as sex. To do this, you are going to give a questionnaire to all students in the class. Give two questions for which the response is categorical and two questions for which the response is quantitative. For the categorical variables, give the possible values; for the quantitative variables, give the unit of measurement.

FIGURE 7.1
Histogram of the number of seeds produced by velvetleaf plants when no herbicide was used, for Questions 7.3 to 7.5.

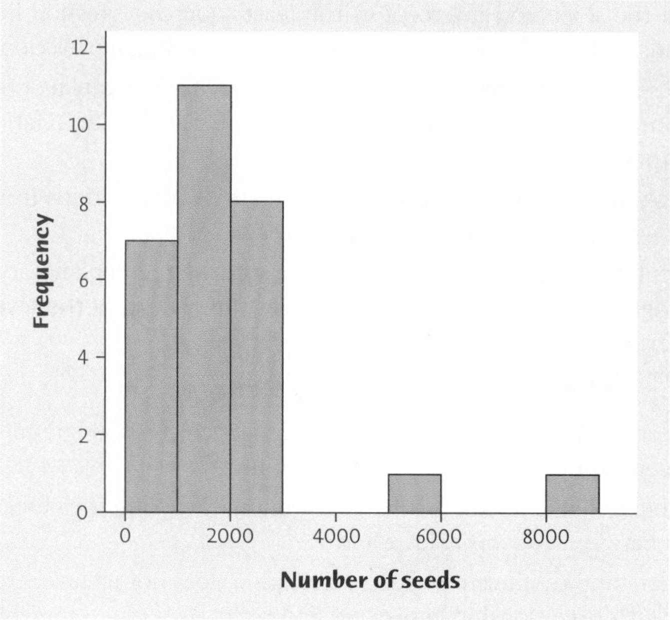

Nigel Cattlin/Science Source

Weeds among the corn. Velvetleaf is a particularly annoying weed in corn fields. It produces lots of seeds, and the seeds wait in the soil for years until conditions are right. How many seeds do velvetleaf plants produce? Figure 7.1 is a histogram of the number of seeds produced from 28 velvetleaf plants that sprouted in a corn field where no herbicide was used.[1] Use this histogram to answer Questions 7.3 to 7.5.

7.3 The histogram

(a) is skewed right. (b) has outliers.
(c) is asymmetric. (d) is all of the above.

7.4 The median number of seeds produced is

(a) under 1000. (b) between 1000 and 2000.
(c) between 2000 and 3000. (d) over 3000.

7.5 The *percent* of plants that produced over 2000 seeds is

(a) about 15%. (b) about 25%.
(c) about 35%. (d) about 45%.

7.6 A reporter wishes to portray baseball players as overpaid. Which measure of center should he report as the average salary of major league players?

(a) The mean
(b) The median
(c) Either the mean or the median. It doesn't matter since they will be equal.
(d) Neither the mean nor the median. Both will be much lower than the actual average salary.

El Niño and the monsoon. The earth is interconnected. For example, it appears that El Niño, the periodic warming of the Pacific Ocean west of South America, affects the monsoon rains that are essential for agriculture in India. Here are the monsoon rains (in millimeters) for the 23 strong El Niño years between 1871 and 2004, arranged in increasing order:[2]

604	628	651	653	669	698	710	717	736	740	781	784
790	790	792	806	811	830	858	858	872	896	957	

Use these data to answer Questions 7.7 to 7.9.

7.7 What is the median amount of rainfall for the strong El Niño years?

(a) 669 millimeters (b) 698 millimeters
(c) 784 millimeters (d) 830 millimeters

7.8 What is the first quartile for these data?

(a) 669 millimeters (b) 698 millimeters
(c) 784 millimeters (d) 830 millimeters

7.9 The average monsoon rainfall for all years from 1871 to 2004 is about 850 millimeters. What effect does El Niño appear to have on monsoon rains?

(a) Strong El Niño years tend to have higher monsoon rainfalls than in other years.
(b) Strong El Niño years tend to have the same monsoon rainfalls as in other years.
(c) Strong El Niño years tend to have lower monsoon rainfalls than in other years.
(d) None of the above.

7.10 Which of the following is likely to have a mean that is smaller than the median?

(a) The salaries of all National Football League players
(b) The scores of students (out of 100 points) on a very easy exam in which most students score perfectly, but a few do very poorly
(c) The prices of homes in a large city
(d) The scores of students (out of 100 points) on a very difficult exam in which most students score poorly, but a few do very well

7.11 For a biology project, you measure the tail length in centimeters and weight in grams of 12 mice of the same variety. What units of measurement do each of the following have?

(a) The mean tail length
(b) The first quartile of the tail lengths
(c) The standard deviation of the tail lengths
(d) The variance of the weights

Travel times. *How long must you travel each day to get to work? Here is a stemplot of the average travel times to work for workers in the 50 states and the District of Columbia who are at least 16 years of age and don't work at home.*[3] *The stems are whole minutes, and the leaves are tenths of a minute. Use the stemplot to answer Questions 7.12 and 7.13.*

15	5 9
16	
17	6 7 7 9
18	2 5
19	
20	0 1 7 8 8 9
21	2 8
22	0 1 3 3 3 4 9 9
23	4 4 5 6 6 9
24	0 1 2 6 6
25	0 0 1 2 5 6 9
26	6 8 9
27	3 9
28	
29	1 2
30	6 9

7.12 The shape of the distribution is

(a) clearly skewed to the right.
(b) roughly symmetric.
(c) clearly skewed to the left.

7.13 The *percent* of states for which the average travel time is 20 minutes or more is closest to

(a) 16%. (b) 27%. (c) 73%. (d) 84%.

Search engine use. *In 2012, approximately 91% of Internet users used a search engine to find information on the web. Search engine users were asked, "How often do you use a search engine to find information online?" The bar graph in Figure 7.2 gives the percentages for the most frequent use categories.*[4] *Use the bar graph to help answer Questions 7.14 and 7.15.*

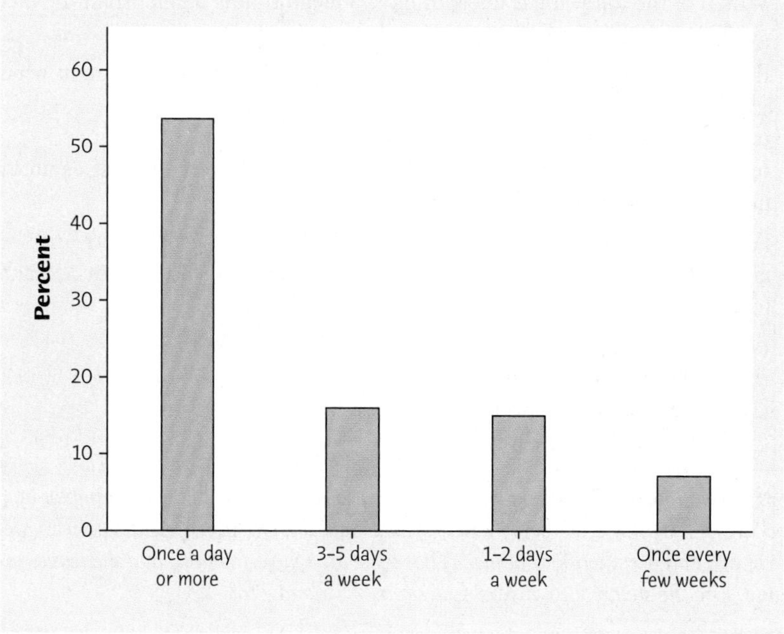

FIGURE 7.2 Bar graph of the distribution of search engine use for the most frequent use categories in 2012, for Questions 7.14 and 7.15.

7.14 Approximately what percent of search engine users used a search engine one to two days per week?

(a) 5% (b) 10% (c) 15% (d) 20%

7.15 The total of the percents of the bars in the graph is 92%. Which of the following statements is true?

(a) The percent of search engine users who use a search engine less than once every few weeks is greater than 10%.
(b) A pie chart could be drawn for the categories given in the bar graph.
(c) The percent of search engine users who use a search engine *at least* three to five days a week is about 70%.
(d) None of the above.

7.16 Reports on a student's ACT, SAT, or MCAT usually give the percentile as well as the actual score. The percentile is just the cumulative proportion stated as a percent: the percent of all scores that were lower than this one. In 2013, the total

MCAT scores were close to Normal with mean 25.3 and standard deviation 6.5.[5] William scored 32. What was his percentile?

7.17 The length of the thorax in a population of male fruit flies is approximately Normal with mean 0.800 millimeters (mm) and standard deviation 0.078 mm. Use the 68–95–99.7 rule to answer the following questions.

(a) What range of thorax lengths covers almost all (99.7%) of this distribution?
(b) What percent of male fruit flies have a thorax length exceeding 0.878 mm?

7.18 A professor knows from past experience that the time for students to complete a quiz is normally distributed with mean 21 minutes and standard deviation 3 minutes.

(a) If he allows 25 minutes for the quiz, what percent of the students will not complete the quiz?
(b) Suppose that he wants to allow sufficient time so that 95% of the students will complete the quiz in the allotted time. How much time should he allow for the quiz?

7.19 The Aleppo pine and the Torrey pine are widely planted as ornamental trees in Southern California. Here are the lengths (in centimeters) of 15 Aleppo pine needles:[6]

10.2 7.2 7.6 9.3 12.1 10.9 9.4 11.3 8.5 8.5 12.8 8.7 9.0 9.0 9.4

(a) Find the five-number summary for the distribution of Aleppo pine needles.

Figure 7.3 gives a boxplot for the distribution of the lengths (in centimeters) of 18 Torrey pine needles. Use this information to help answer the remainder of this question.

(b) The median of the distribution of Torrey pine needles is closest to which of the following values?

24 25 27 30

(c) Twenty-five percent of the Torrey pine needles exceed what value?
(d) Given only the length of a needle, do you think you could say which pine species it comes from? Explain briefly.

Statistics in Your World: BEER IN SOUTH DAKOTA

Take a break from doing exercises to apply your math to beer cans in South Dakota. A newspaper there reported that, every year, an average of 650 beer cans per mile are tossed onto the state's highways. South Dakota has about 83,000 miles of roads. How many beer cans is that in all? The U.S. Census Bureau says that there are about 810,000 people in South Dakota. How many beer cans does each man, woman, and child in the state toss on the road each year? That's pretty impressive. Maybe the paper got its numbers wrong.

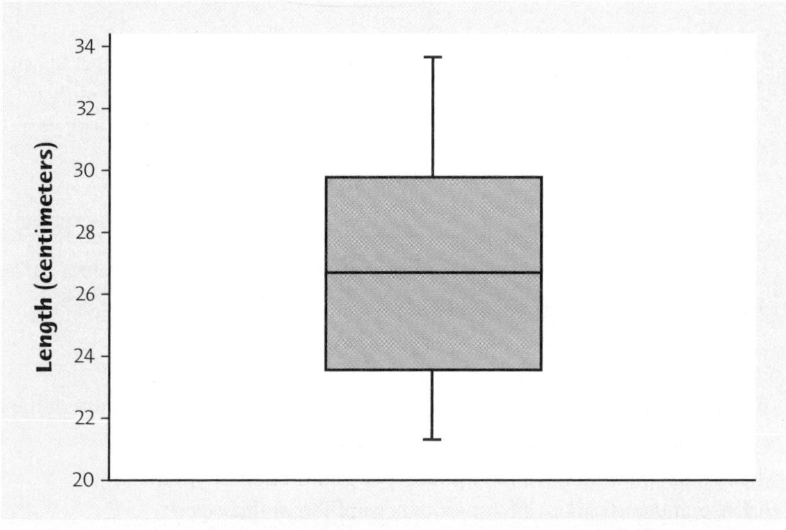

FIGURE 7.3
Boxplot for the distribution of the lengths (in centimeters) of 18 Torrey pine needles, for Question 7.19.

TEAM	SPENDING	POINTS	TEAM	SPENDING	POINTS	TEAM	SPENDING	POINTS
Flyers	66.9	103	Rangers	61.9	109	Panthers	55.5	94
Capitals	66.4	92	Devils	61.7	102	Coyotes	55.1	97
Sabres	65.3	89	Sharks	61.4	96	Blues	54.9	109
Canucks	64.9	111	Blue Jackets	61.3	65	Predators	52.2	104
Flames	64.2	90	Lightning	60.9	84	Jets	51.8	84
Penguins	64.2	108	Oilers	60.5	74	Senators	51.7	92
Kings	63.7	95	Blackhawks	60.2	101	Hurricanes	50.5	82
Maple Leafs	63.5	80	Red Wings	59.3	102	Stars	49.9	89
Bruins	62.2	102	Ducks	59.3	80	Avalanche	49.4	88
Canadiens	62.2	78	Wild	55.8	81	Islanders	49.1	79

NHL salaries. *One can find online[7] the amount each NHL team spent on players (in millions of dollars) for the 2011–12 NHL season and the total points each team earned by the end of the season. Questions 7.20 to 7.23 are based on the NHL data set.*

7.20 Figure 7.4 is a scatterplot of points earned against spending. How would you describe the overall pattern?

(a) Sharply curved
(b) Two distinct clusters that are widely separated
(c) A very weak association
(d) A strong positive association

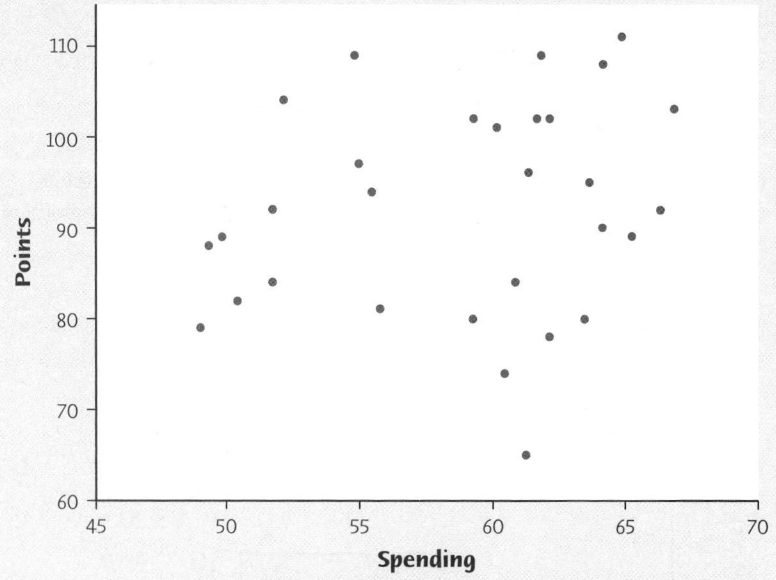

FIGURE 7.4
Scatterplot of the points earned in the 2011–12 NHL regular season against spending on player salaries for Question 7.20.

7.21 The equation of the least-squares regression line for predicting points earned from spending is

$$\text{points} = 66.5 + 0.43 \times \text{spending}$$

What does this tell us about points gained for each additional million dollars spent?

(a) A team gains about 0.43 points per million dollars spent.
(b) A team gains about 0.665 points per million dollars spent.
(c) A team gains about 66.5 points per million dollars spent.
(d) A team gains about 66.93 points per million dollars spent.

7.22 The equation of the least-squares regression line for predicting points earned from spending is

$$\text{points} = 66.5 + 0.43 \times \text{spending}$$

Use the regression equation to predict the points earned if a team spent $50 million.

(a) 21.5 (b) 66.5 (c) 88.0 (d) 3325.4

7.23 The equation of the least-squares regression line for predicting points earned from spending is

$$\text{points} = 66.5 + 0.43 \times \text{spending}$$

Use the regression equation to predict the points a team would earn if it spent no money. Which conclusion do you agree with?

(a) The team will earn 66.5 points.
(b) The prediction is not sensible, because the prediction is far outside the range of values of the response variable.
(c) The prediction is not sensible, because no money is far outside the range of values of the explanatory variable.
(d) The prediction is not sensible, because of the outlier present in the data.

Measuring heights of the elderly. *Because elderly people may have difficulty standing to have their heights measured, a study looked at predicting overall height from height to the knee. Here are data (in centimeters) for six elderly men:*

Knee height x	57.7	47.4	43.5	44.8	55.2	54.6
Height y	192.1	153.3	146.4	162.7	169.1	177.8

Questions 7.24 to 7.26 are based on these data.

7.24 Figure 7.5 is a scatterplot of height of a subject against knee height. Which of the following is a plausible value of the correlation, r, between height and knee height?

(a) 0 (b) −0.25 (c) 0.90 (d) −0.90

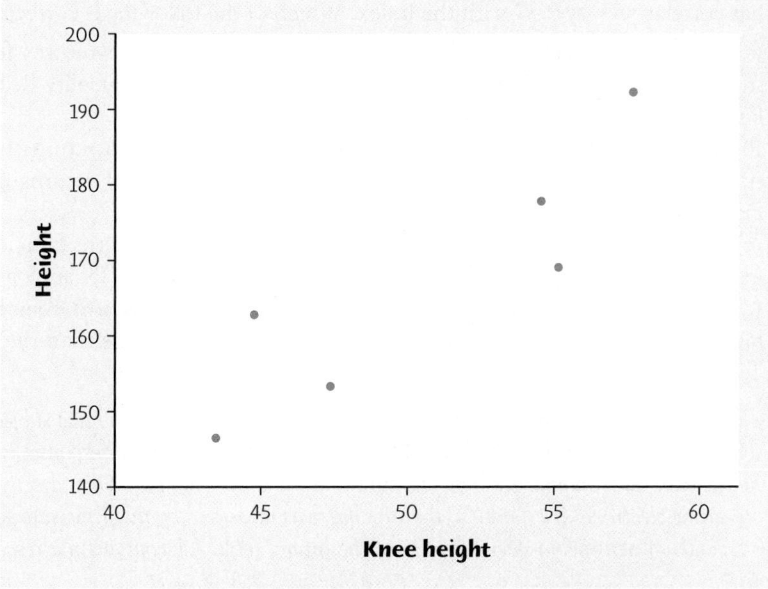

FIGURE 7.5
Scatterplot of the heights of six elderly men against knee height, for Question 7.24.

7.25 The equation of the least-squares regression line for predicting the height of a subject from his knee height is

$$\text{height} = 42.93 + 2.45 \times \text{knee height}$$

What does the slope of 2.45 tell us?

(a) The average height of the elderly subjects in the study is 2.45 centimeters per centimeter of knee height.
(b) The predicted height of the elderly subjects in the study is 2.45 centimeters per centimeter of knee height.
(c) The predicted height of a subject with a knee height of 0 is 42.92 centimeters.
(d) For each additional centimeter of knee height, the predicted height of a subject increases by 2.45 centimeters.

7.26 The equation of the least-squares regression line for predicting the height of a subject from their knee height is

$$\text{height} = 42.93 + 2.45 \times \text{knee height}$$

Use this to predict the height of a person with a knee height of 50 cm.

(a) 165.43 cm (b) 122.5 cm (c) 95.38 cm (d) 92.93 cm

7.27 How well do people remember their past diet? Data are available for 91 people who were asked about their diet when they were 18 years old. Researchers asked them at about age 55 to describe their eating habits at age 18. For each subject, the researchers calculated the correlation between actual intakes of many foods at age 18 and the intakes the subjects now remember. The median of the 91 correlations was $r = 0.217$.[8] Which of the following conclusions is consistent with this correlation?

(a) Subjects remember approximately 21.7% of their food intakes at age 18.
(b) Subjects remember approximately $r^2 = 0.217^2 = 0.047$ of their food intakes at age 18.
(c) Food intake at age 55 is about 21.7% of food intake at age 18.
(d) Memory of food intake in the distant past is fair to poor.

7.28 Joe's retirement plan invests in stocks through an "index fund" that follows the behavior of the stock market as a whole, as measured by the Standard & Poor's (S&P) 500 stock index. Joe wants to buy a mutual fund that does not track the index closely. He reads that monthly returns from Fidelity Technology Fund have correlation $r = 0.77$ with the S&P 500 index, and that Fidelity Real Estate Fund has correlation $r = 0.37$ with the index. Which of the following is correct?

(a) The Fidelity Technology Fund has a closer relationship to returns from the stock market as a whole and also has higher returns than the Fidelity Real Estate Fund.
(b) The Fidelity Technology Fund has a closer relationship to returns from the stock market as a whole, but we cannot say that it has higher returns than the Fidelity Real Estate Fund.
(c) The Fidelity Real Estate Fund has a closer relationship to returns from the stock market as a whole and also has higher returns than the Fidelity Technology Fund.
(d) The Fidelity Real Estate Fund has a closer relationship to returns from the stock market as a whole, but we cannot say that it has higher returns than the Fidelity Technology Fund.

Monkey calls. *The usual way to study the brain's response to sounds is to have subjects listen to "pure tones." The response to recognizable sounds may differ. To compare responses, researchers anesthetized macaque monkeys. They fed pure tones and also monkey calls directly to their brains by inserting electrodes. Response to the stimulus was measured by the firing rate (electrical spikes per second) of neurons in various areas of the brain. Table 7.1 contains the responses for 37 neurons.*[9]

TABLE 7.1 NEURON RESPONSE (ELECTRICAL FIRING RATE PER SECOND) TO PURE TONES AND MONKEY CALLS

NEURON	TONE	CALL	NEURON	TONE	CALL	NEURON	TONE	CALL
1	474	500	14	145	42	26	71	134
2	256	138	15	141	241	27	68	65
3	241	485	16	129	194	28	59	182
4	226	338	17	113	123	29	59	97
5	185	194	18	112	182	30	57	318
6	174	159	19	102	141	31	56	201
7	176	341	20	100	118	32	47	279
8	168	85	21	74	62	33	46	62
9	161	303	22	72	112	34	41	84
10	150	208	23	20	193	35	26	203
11	19	66	24	21	129	36	28	192
12	20	54	25	26	135	37	31	70
13	35	103						

Figure 7.6 is a scatterplot of monkey call response against pure-tone response (explanatory variable). Questions 7.29 and 7.30 refer to these data and the scatterplot.

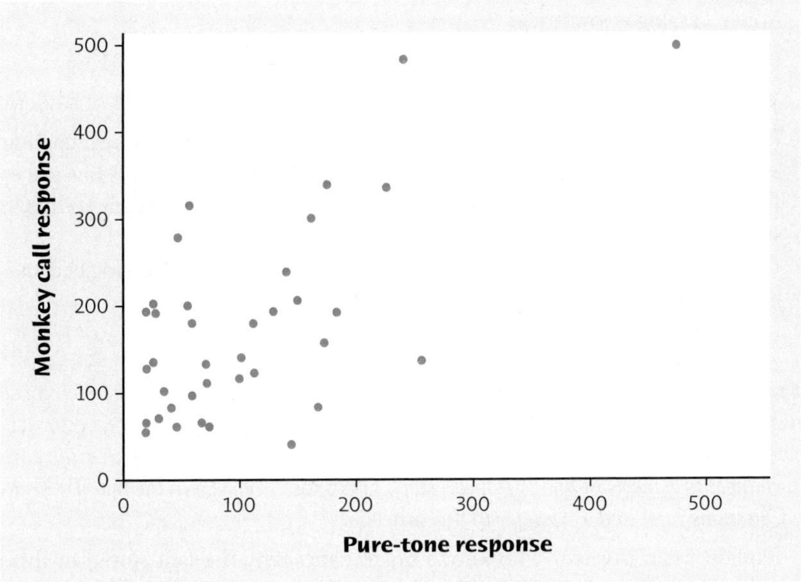

FIGURE 7.6
Scatterplot of monkey tone response against pure-tone response, for Questions 7.29 and 7.30.

7.29 We might expect some neurons to have strong responses to any stimulus and others to have consistently weak responses. There would then be a strong relationship between tone response and call response. From the scatterplot of monkey call response against pure-tone response in Figure 7.6, what would you estimate the correlation r to be?

DATA MONKEY

(a) −0.6 (b) −0.1 (c) 0.1 (d) 0.6

7.30 Which of the following statements about the scatterplot in Figure 7.6 is correct?

(a) There is moderate evidence that pure-tone response causes monkey call response.
(b) There is moderate evidence that monkey call response causes pure-tone response.
(c) There are one or two outliers and at least one of these may also be influential.
(d) None of the above.

Catalog shopping. *What is the most important reason that students buy from catalogs? The answer may differ for different groups of students. Here are counts for samples of American and East Asian students at a large midwestern university.*[10] *Use these counts to answer Questions 7.31 to 7.33, which are based on optional material.*

REASON	AMERICAN	ASIAN
Save time	29	10
Easy	28	11
Low price	17	34
Live far from stores	11	4
No pressure to buy	10	3
Other	20	7
Total	115	69

SHOP

7.31 What percent of all students say that the most important reason to buy from a catalog is to save time?

(a) 74% (b) 25% (c) 21% (d) 14%

SHOP

7.32 What percent of East Asian students say that the most important reason to buy from a catalog is low price?

(a) 67% (b) 49% (c) 28% (d) 18%

7.33 What are the most important differences between the two groups of students?

(a) The most important reasons for American students to buy from a catalog are to save time and because it is easy, while for East Asian students it is low price.
(b) American students appear to be almost three times more likely to live far from stores than East Asian students.
(c) East Asian students are twice as likely to purchase from a catalog because of low price than American students.
(d) All of the above.

Investment strategies. *One reason to invest abroad is that markets in different countries don't move in step. When American stocks go down, foreign stocks may go up. So an investor who holds both bears less risk. That's the theory. But then we read in a magazine article that the correlation between changes in American and European stock prices rose from 0.4 in the mid-1990s to 0.8 in 2000.*[11] *Questions 7.34 and 7.35 refer to this article.*

7.34 Explain to an investor who knows no statistics why the fact stated in this article reduces the protection provided by buying European stocks.

7.35 The same article claiming that the correlation between changes in stock prices in Europe and the United States is 0.8 goes on to say: "Crudely, that means that movements on Wall Street can explain 80% of price movements in Europe."

(a) Is this true?
(b) What is the correct percent explained if $r = 0.8$?

INTROSP

7.36 Researchers wished to determine whether individual differences in introspective ability are reflected in the anatomy of brain regions responsible for this function.

They measured introspective ability (using a score on a test of introspective ability, with larger values indicating greater introspective ability) and gray-matter volume in milliliters (the Brodmann area) in the anterior prefrontal cortex of the brain of 29 subjects. Here are the data:

Introspective ability	0.55	0.58	0.59	0.59	0.59	0.61	0.62	0.63	0.63	0.63
Volume	59	62	43	63	83	61	55	57	57	67
Introspective ability	0.63	0.64	0.65	0.65	0.65	0.65	0.65	0.66	0.66	0.67
Volume	72	62	58	62	65	70	75	60	63	71
Introspective ability	0.67	0.67	0.68	0.69	0.70	0.70	0.71	0.72	0.75	
Volume	71	80	68	72	66	73	61	80	75	

The researchers wished to determine the equation of the least-squares regression line for predicting gray-matter volume ability (y) from introspective ability (x). To do this, they calculated the following summary statistics:

$$\bar{x} = 0.649,\ s_x = 0.045$$
$$\bar{y} = 65.90,\ s_y = 8.69$$
$$r = 0.448$$

(a) Use these summary statistics to calculate the equation of the least-squares regression line.
(b) Based on the least-squares regression line, what would you predict gray-matter volume to be for someone with introspective ability 0.60?
(c) Based on the least-squares regression line, what would you predict gray-matter volume to be for someone with introspective ability 0.99? How reliable do you think this prediction is? Explain your answer.

7.37 Animals and people that take in more energy than they expend will get fatter. Here are data on 12 rhesus monkeys: 6 lean monkeys (4% to 9% body fat) and 6 obese monkeys (13% to 44% body fat). The data report the energy expended in 24 hours (kilojoules per minute) and the lean body mass (kilograms, leaving out fat) for each monkey.[12]

THINFAT

LEAN		OBESE	
MASS	**ENERGY**	**MASS**	**ENERGY**
6.6	1.17	7.9	0.93
7.8	1.02	9.4	1.39
8.9	1.46	10.7	1.19
9.8	1.68	12.2	1.49
9.7	1.06	12.1	1.29
9.3	1.16	10.8	1.31

(a) Compute the mean lean body mass of the lean monkeys.
(b) Compute the mean lean body mass of the obese monkeys.
(c) The goal of the study is to compare the energy expended in 24 hours by the lean monkeys with that of the obese monkeys. However, animals with higher lean mass usually expend more energy. Based on your calculations in parts (a) and (b), would it make sense to simply compute the mean energy expended by lean and obese monkeys and compare the means? Explain.
(d) To investigate how energy expended is related to body mass, make a scatterplot of energy versus mass, using different plot symbols for lean and obese monkeys.
(e) What do the trends in your scatterplot suggest about the monkeys?

DATA SMOKERS

7.38 The number of adult Americans who smoke continues to drop. Here are estimates of the percents of adults (aged 18 and over) who were smokers in the years between 1965 and 2011:[13]

Year x	1965	1974	1979	1983	1987	1990	1993
Smokers y	41.9	37.0	33.3	31.9	28.6	25.3	24.8
Year x	1997	2000	2002	2006	2009	2011	
Smokers y	24.6	23.1	22.5	20.8	20.6	19.0	

(a) Make a scatterplot of these data.
(b) Describe the direction, form, and strength of the relationship between percent of smokers and year. Are there any outliers?
(c) Here are the means and standard deviations for both variables and the correlation between percent of smokers and year:

$$\bar{x} = 1992.0,\ s_x = 14.0$$
$$\bar{y} = 27.2,\ s_y = 7.0$$
$$r = -0.98$$

Use this information to find the least-squares regression line for predicting percent of smokers from year and add the line to your plot.
(d) According to your regression line, how much did smoking decline per year during this period, on the average?
(e) What percent of the observed variation in percent of adults who smoke can be explained by linear change over time?
(f) One of the government's national health objectives is to reduce smoking to no more than 12% of adults by 2020. Use your regression line to predict the percent of adults who will smoke in 2020.
(g) Use your regression line to predict the percent of adults who will smoke in 2075. Why is your result impossible? Why was it foolish to use the regression line for this prediction?

DATA ANGER

7.39 **(Optional topic)** People who get angry easily tend to have more heart disease. That's the conclusion of a study that followed a random sample of 12,986 people from three locations for about four years. All subjects were free of heart disease at the beginning of the study. The subjects took the Spielberger Trait Anger Scale test, which measures how prone a person is to sudden anger. Here are data for the 8474 people in the sample who had normal blood pressure. CHD stands for "coronary heart disease." This includes people who had heart attacks and those who needed medical treatment for heart disease.[14]

	LOW ANGER	MODERATE ANGER	HIGH ANGER	TOTAL
CHD	53	110	27	190
No CHD	3057	4621	606	8284
Total	3110	4731	633	8474

(a) What percent of all 8474 people with normal blood pressure had CHD?
(b) What percent of all 8474 people were classified as having high anger?
(c) What percent of those classified as having high anger had CHD?
(d) What percent of those with no CHD were classified as having moderate anger?
(e) Do these data provide any evidence that as anger score increases, the percent who suffer CHD increases? Explain.

Supplementary Exercises

*Supplementary exercises apply the skills you have learned in ways that require more thought or more elaborate use of technology. Some of these exercises ask you to follow the **Plan**, **Solve**, and **Conclude** steps of the four-step process introduced on page 63.*

7.40 The Mississippi River. Table 7.2 gives the volume of water discharged by the Mississippi River into the Gulf of Mexico for each year from 1954 to 2001.[15] The units are cubic kilometers of water—the Mississippi is a big river. DATA MISSIP

AP Photo/The Gleaner, Mike Lawrence

(a) Make a graph of the distribution of water volume. Describe the overall shape of the distribution and any outliers.

(b) Based on the shape of the distribution, do you expect the mean to be close to the median, clearly less than the median, or clearly greater than the median? Why? Find the mean and the median to check your answer.

TABLE 7.2 YEARLY DISCHARGE (CUBIC KILOMETERS OF WATER) OF THE MISSISSIPPI RIVER

YEAR	DISCHARGE	YEAR	DISCHARGE	YEAR	DISCHARGE
1954	290	1970	540	1986	600
1955	420	1971	480	1987	450
1956	390	1972	600	1988	420
1957	610	1973	880	1989	630
1958	550	1974	710	1990	680
1959	440	1975	670	1991	700
1960	470	1976	420	1992	510
1961	600	1977	430	1993	900
1962	550	1978	560	1994	640
1963	360	1979	800	1995	590
1964	390	1980	500	1996	670
1965	500	1981	420	1997	680
1966	410	1982	640	1998	690
1967	460	1983	770	1999	580
1968	510	1984	710	2000	390
1969	560	1985	680	2001	580

(c) Based on the shape of the distribution, does it seem reasonable to use $\bar{x}$ and s to describe the center and variability of this distribution? Why? Find $\bar{x}$ and s if you think they are a good choice. Otherwise, find the five-number summary.

7.41 More on the Mississippi River. The data in Table 7.2 are a time series. Make a time plot that shows how the volume of water in the Mississippi changed between 1954 and 2001. What does the time plot reveal that the histogram from Exercise 7.40 does not? It is a good idea to always make a time plot of time series data because a histogram cannot show changes over time.

Falling through the ice. *The Nenana Ice Classic is an annual contest to guess the exact time in the spring thaw when a tripod erected on the frozen Tanana River near Nenana, Alaska, will fall through the ice. The 2013 jackpot prize was $318,500. The contest has been run since 1917. Table 7.3 gives simplified data that record*

TABLE 7.3 DAYS FROM APRIL 20 FOR THE TANANA RIVER TRIPOD TO FALL

YEAR	DAY	YEAR	DAY	YEAR	DAY	YEAR	DAY
1917	11	1942	11	1967	15	1992	25
1918	22	1943	9	1968	19	1993	4
1919	14	1944	15	1969	9	1994	10
1920	22	1945	27	1970	15	1995	7
1921	22	1946	16	1971	19	1996	16
1922	23	1947	14	1972	21	1997	11
1923	20	1948	24	1973	15	1998	1
1924	22	1949	25	1974	17	1999	10
1925	16	1950	17	1975	21	2000	12
1926	7	1951	11	1976	13	2001	19
1927	23	1952	23	1977	17	2002	18
1928	17	1953	10	1978	11	2003	10
1929	16	1954	17	1979	11	2004	5
1930	19	1955	20	1980	10	2005	9
1931	21	1956	12	1981	11	2006	13
1932	12	1957	16	1982	21	2007	8
1933	19	1958	10	1983	10	2008	16
1934	11	1959	19	1984	20	2009	12
1935	26	1960	13	1985	23	2010	10
1936	11	1961	16	1986	19	2011	14
1937	23	1962	23	1987	16	2012	3
1938	17	1963	16	1988	8	2013	31
1939	10	1964	31	1989	12		
1940	1	1965	18	1990	5		
1941	14	1966	19	1991	12		

only the date on which the tripod fell each year. The earliest date so far is April 20. To make the data easier to use, the table gives the date each year in days starting with April 20. That is, April 20 is 1, April 21 is 2, and so on. Exercises 7.42 to 7.44 concern these data.[16]

7.42 When does the ice break up? We have 97 years of data on the date of ice breakup on the Tanana River. Describe the distribution of the breakup date with both a graph or graphs and appropriate numerical summaries. What is the median date (month and day) for ice breakup? TANANA

7.43 Global warming? Because of the high stakes, the falling of the tripod has been carefully observed for many years. If the date the tripod falls has been getting earlier, that may be evidence for the effects of global warming.

(a) Make a time plot of the date the tripod falls against year.

(b) There is a great deal of year-to-year variation. Fitting a regression line to the data may help us see the trend. Fit the least-squares line and add it to your time plot. What do you conclude?

(c) There is much variation about the line. Give a numerical description of how much of the year-to-year variation in ice breakup time is accounted for by the time trend represented by the regression line. (This simple example is typical of more complex evidence for the effects of global warming: large year-to-year variation requires many years of data to see a trend.)

7.44 More on global warming. Side-by-side boxplots offer a different look at the data. Group the data into periods of roughly equal length: 1917 to 1940, 1941 to 1964, 1965 to 1988, and 1989 to 2013. Make boxplots to compare ice breakup dates in these four time periods. Write a brief description of what the plots show.

7.45 Teachers' salaries. The Organisation for Economic Co-operation and Development (OECD) began in 1961 to stimulate economic progress and world trade. It originally consisted of European countries, the U.S., and Canada, but has now grown to include 34 countries spanning the globe. The data file *TEACHSL* on the website gives the average starting salaries in 2012 of primary public school teachers (PPP, US$) for the member nations.[17] TEACHSL

(a) Make a stemplot or a histogram to display the distribution of teachers' salaries in the OECD countries.

(b) There is one high outlier. What country is this? What is the overall shape of the distribution if you ignore the outlier?

(c) Based on your work in part (b), give a numerical summary of the center and variability of the distribution, omitting the outlier.

(d) Some Americans complain about teachers being overpaid. Where does the United States ($36,858) stand in this international comparison?

7.46 Cicadas as fertilizer? Every 17 years, swarms of cicadas emerge from the ground in the eastern United States, live for about six weeks, then die. (There are several "broods," so we experience cicada eruptions more often than every 17 years.) There are so many cicadas that their dead bodies can serve as fertilizer and increase plant growth. In an experiment, a researcher added 10 cicadas under some plants in a natural plot of American bellflowers in a forest, leaving other plants undisturbed. One of the response variables was the size of seeds produced by the plants. Here are data (seed mass in milligrams) for 39 cicada plants and 33 undisturbed (control) plants:[18] CICADA

© Alastair Shay; Papilio/Corbis

CICADA PLANTS				CONTROL PLANTS			
0.237	0.277	0.241	0.142	0.212	0.188	0.263	0.253
0.109	0.209	0.238	0.277	0.261	0.265	0.135	0.170
0.261	0.227	0.171	0.235	0.203	0.241	0.257	0.155
0.276	0.234	0.255	0.296	0.215	0.285	0.198	0.266
0.239	0.266	0.296	0.217	0.178	0.244	0.190	0.212
0.238	0.210	0.295	0.193	0.290	0.253	0.249	0.253
0.218	0.263	0.305	0.257	0.268	0.190	0.196	0.220
0.351	0.245	0.226	0.276	0.246	0.145	0.247	0.140
0.317	0.310	0.223	0.229	0.241			
0.192	0.201	0.211					

Describe and compare the two distributions. Do the data support the idea that dead cicadas can serve as fertilizer?

7.47 A big-toe problem. Hallux abducto valgus (call it HAV) is a deformation of the big toe that is not common in youth and often requires surgery. Doctors used X-rays to measure the angle (in degrees) of deformity in 38 consecutive patients under the age of 21 who came to a medical center for surgery to correct HAV.[19] The angle is a measure of the seriousness of the deformity. The data appear in Table 7.4 as "HAV angle." Describe the distribution of the angle of deformity among young patients needing surgery for this condition. BIGTOE

TABLE 7.4 ANGLE OF DEFORMITY (DEGREES) FOR TWO TYPES OF FOOT DEFORMITY

HAV ANGLE	MA ANGLE	HAV ANGLE	MA ANGLE	HAV ANGLE	MA ANGLE
28	18	21	15	16	10
32	16	17	16	30	12
25	22	16	10	30	10
34	17	21	7	20	10
38	33	23	11	50	12
26	10	14	15	25	25
25	18	32	12	26	30
18	13	25	16	28	22
30	19	21	16	31	24
26	10	22	18	38	20
28	17	20	10	32	37
13	14	18	15	21	23
20	20	26	16		

7.48 Prey attract predators. Here is one way in which nature regulates the size of animal populations: High population density attracts predators, who remove a higher proportion of the population than when the density of the prey is low. One study looked at kelp perch and their common predator, the kelp bass. The researcher set up four large circular pens on sandy ocean bottom in Southern California. He chose young perch at random from a large group and placed 10, 20, 40, and 60 perch in the four pens. Then he dropped the nets protecting the pens, allowing bass to swarm in, and counted the perch left after 2 hours. Here are data on the proportions of perch eaten in four repetitions of this setup:[20] PREY

PERCH	PROPORTION KILLED			
10	0.0	0.1	0.3	0.3
20	0.2	0.3	0.3	0.6
40	0.075	0.3	0.6	0.725
60	0.517	0.55	0.7	0.817

Do the data support the principle that "more prey attract more predators, who drive down the number of prey"?

7.49 Predicting foot problems. Metatarsus adductus (call it MA) is a turning in of the front part of the foot that is common in adolescents and usually corrects itself. Table 7.4 gives the severity of MA ("MA angle"). Doctors speculate that the severity of MA can help predict the severity of HAV. Describe the relationship between MA and HAV. Do you think the data confirm the doctors' speculation? Why or why not? BIGTOE

7.50 Change in the Serengeti. Long-term records from the Serengeti National Park in Tanzania show interesting ecological relationships. When wildebeest are more abundant, they graze the grass more heavily, so there are fewer fires and more trees grow. Lions feed more successfully when there are more trees, so the lion population increases. Here are data on one part of this cycle, wildebeest abundance (in thousands of animals) and the percent of the grass area that burned in the same year:[21] SERENG

Gallo Images-Anthony Bannister/Getty Images

WILDEBEEST (1000s)	PERCENT BURNED	WILDEBEEST (1000s)	PERCENT BURNED
396	56	622	60
476	50	600	56
698	25	902	45
1049	16	1440	21
1178	7	1147	32
1200	5	1173	31
1302	7	1178	24
360	88	1253	24
444	88	1249	53
524	75		

To what extent do these data support the claim that more wildebeest reduce the percent of grasslands that burn? How rapidly does burned area decrease as the number of wildebeest increases? Include a graph and suitable calculations.

7.51 Casting aluminum. In casting metal parts, molten metal flows through a "gate" into a die that shapes the part. The gate velocity (the speed at which metal is forced through the gate) plays a critical role in die casting. A firm that casts cylindrical aluminum pistons examined 12 types formed from the same alloy. How does the piston wall thickness (inches) influence the gate velocity (feet per second) chosen by the skilled workers who do the casting? If there is a clear pattern, it can be used to direct new workers or to automate the process. Analyze these data and report your findings.[22] ALUM

THICKNESS	VELOCITY	THICKNESS	VELOCITY
0.248	123.8	0.628	326.2
0.359	223.9	0.697	302.4
0.366	180.9	0.697	145.2
0.400	104.8	0.752	263.1
0.524	228.6	0.806	302.4
0.552	223.8	0.821	302.4

7.52 Texting and driving. The Ohio Youth Risk Behavior Survey is conducted every two years in a sample of high schools across the state of Ohio. One of the questions on the 2013 survey was, "During the past 30 days, on how many did you text or email while driving a car or other vehicle?" Here are the survey results:[23] TEXTING

DAYS	FEMALE	MALE
0	263	229
1–9	96	100
10–19	19	27
20–30	53	53
Total	431	409

Write a brief analysis of these results that focuses on the relationship between sex and driving behavior.

7.53 Influence: hot sector funds? Investment advertisements always warn that "past performance does not guarantee future results." Here is an example that shows why you should pay attention to this warning. Stocks fell sharply in 2002, then rose sharply in 2003. The table below gives the percent returns from 23 Fidelity Investments "sector funds" in these two years. Sector funds invest in narrow segments of the stock market. They often rise and fall faster than the market as a whole. MFUNDS

2002 RETURN	2003 RETURN	2002 RETURN	2003 RETURN	2002 RETURN	2003 RETURN
−17.1	23.9	−0.7	36.9	−37.8	59.4
−6.7	14.1	−5.6	27.5	−11.5	22.9
−21.1	41.8	−26.9	26.1	−0.7	36.9
−12.8	43.9	−42.0	62.7	64.3	32.1
−18.9	31.1	−47.8	68.1	−9.6	28.7
−7.7	32.3	−50.5	71.9	−11.7	29.5
−17.2	36.5	−49.5	57.0	−2.3	19.1
−11.4	30.6	−23.4	35.0		

(a) Make a scatterplot of 2003 return (response) against 2002 return (explanatory). The funds with the best performance in 2002 tend to have the worst performance in 2003. Fidelity Gold Fund, the only fund with a positive return in both years, is an extreme outlier.
(b) To demonstrate that correlation is not resistant, find r for all 23 funds and then find r for the 22 funds other than Gold. Explain from Gold's position in your plot why omitting this point makes r more negative.
(c) Find the equations of two least-squares lines for predicting 2003 return from 2002 return, one for all 23 funds and one omitting Fidelity Gold Fund. Add both lines to your scatterplot. Starting with the least-squares idea, explain why adding Fidelity Gold Fund to the other 22 funds moves the line in the direction that your graph shows.

7.54 Influence: monkey calls. Table 7.1 contains data on the response of 37 monkey neurons to pure tones and to monkey calls. Figure 7.6 is a scatterplot of these data. MONKEY

(a) Find the least-squares line for predicting a neuron's call response from its pure tone response. Add the line to your scatterplot. Mark on your plot the point (call it A) with the largest residual (either positive or negative) and also the point (call it B) that is an outlier in the x direction.
(b) How influential are each of these points for the correlation r?
(c) How influential are each of these points for the regression line?

7.55 Influence: bushmeat. Table 7.5 gives data on fish catches in a region of West Africa and the percent change in the biomass (total weight) of 41 animals in nature reserves. It appears that years with smaller fish catches see greater declines in animals, probably because local people turn to "bushmeat" when other sources of protein are not available. The next year (1999) had a fish catch of 23.0 kilograms per person and animal biomass change of −22.9%. BUSHMT

(a) Make a scatterplot that shows how change in animal biomass depends on fish catch. Be sure to include the additional data point. Describe the overall pattern. The added point is a low outlier in the y direction.
(b) Find the correlation between fish catch and change in animal biomass both with and without the outlier. The outlier is influential for correlation. Explain from your plot why adding the outlier makes the correlation smaller.
(c) Find the least-squares line for predicting change in animal biomass from fish catch both with and without the additional data point for 1999. Add both lines to your scatterplot from part (a). The outlier is not influential for the least-squares line. Explain from your plot why this is true.

TABLE 7.5 FISH SUPPLY AND WILDLIFE DECLINE IN WEST AFRICA

YEAR	FISH SUPPLY (KILOGRAMS PER PERSON)	BIOMASS CHANGE (PERCENT)	YEAR	FISH SUPPLY (KILOGRAMS PER PERSON)	BIOMASS CHANGE (PERCENT)
1971	34.7	2.9	1985	21.3	−5.5
1972	39.3	3.1	1986	24.3	−0.7
1973	32.4	−1.2	1987	27.4	−5.1
1974	31.8	−1.1	1988	24.5	−7.1
1975	32.8	−3.3	1989	25.2	−4.2
1976	38.4	3.7	1990	25.9	0.9
1977	33.2	1.9	1991	23.0	−6.1
1978	29.7	−0.3	1992	27.1	−4.1
1979	25.0	−5.9	1993	23.4	−4.8
1980	21.8	−7.9	1994	18.9	−11.3
1981	20.8	−5.5	1995	19.6	−9.3
1982	19.7	−7.2	1996	25.3	−10.7
1983	20.8	−4.1	1997	22.0	−1.8
1984	21.1	−8.6	1998	21.0	−7.4

7.56 Python eggs. How is the hatching of water python eggs influenced by the temperature of the snake's nest? Researchers placed 104 newly laid eggs in a hot environment, 56 in a neutral environment, and 27 in a cold environment. Hot duplicates the warmth provided by the mother python. Neutral and cold are cooler, as when the mother is absent. The results: 75 of the hot eggs hatched, along with 38 of the neutral eggs and 16 of the cold eggs.[24]

(a) Make a two-way table of "environment temperature" against "hatched or not."

(b) The researchers anticipated that eggs would hatch less well at cooler temperatures. Do the data support that anticipation?

Online Data for Additional Analyses

1. SAT, ACT, and teacher salaries for 2013 for each of the 50 states and the District of Columbia are available in the data set SATACT. One could use these data to carry out analyses for ACT scores similar to those for the SAT scores in Chapters 5 and 6. For example, repeat the analyses in Exercises 5.50 (page 157) and 5.51 (page 158) using ACT scores instead of SAT scores. SATACT

2. The data set MLB contains hitting, pitching, fielding, salary, and win–loss performance data from the 2013 season for all major league baseball teams. These data can be used to determine the correlation between payroll and winning percentage. One can also explore what variables are most highly correlated with winning percentage, and whether variables that measure pitching performance are more highly correlated with winning percentage than variables that measure hitting performance. For example, calculate the correlation between winning percentage and number of home runs, between winning percentage and batting average, between winning percentage and ERA, between winning percentage and strikeouts by pitchers, and between winning percentage and payroll. Which has the highest correlation? These data are from `http://www.baseball-reference.com/`. Visit this website for definitions of several of the variables in the data set. MLB

3. Historical temperature data and whether Punxsutawney Phil saw his shadow are available in the data set PHIL. Repeat the analysis in Exercise 6.32 (page 177), but define what constitutes "six more weeks of winter-like weather" differently. For example, you might decide there were six more weeks of winter-like weather if average temperatures for March were at least one degree below historical averages. PHIL

4. Data from the Ohio Department of Health website are available in the pdf "2013OHH Detail Tables." This is a source of many tables that can be used for further analyses using methods discussed in Chapter 6. For example, conduct an analysis like that in Exercise 7.52 to investigate the relationship between sex and strategies about weight (Question 67 in the Tables). HEALTH

5. The data set WHAT contains three variables and 3848 observations on each. At one time, this was considered a large data set and difficult to explore with software. Use various exploratory methods available in software packages such as JMP and Minitab to find the "hidden pattern" in these data. WHAT

© Cultura RM/Alamy

PART II

Producing Data

PRODUCING DATA

CHAPTER 8 Producing Data: Sampling

CHAPTER 9 Producing Data: Experiments

CHAPTER 10 Data Ethics*

CHAPTER 11 Producing Data: Part II Review

The purpose of statistics is to gain understanding from data. We can seek understanding in different ways, depending on the circumstances. We have studied one approach to data, *exploratory data analysis,* in some detail. Now we begin the move from data analysis toward *statistical inference*. Both types of reasoning are essential to effective work with data. Here is a brief sketch of the differences between them:

EXPLORATORY DATA ANALYSIS	STATISTICAL INFERENCE
The purpose is unrestricted exploration of the data, searching for interesting patterns.	The purpose is to answer specific questions, posed before the data were produced.
Conclusions apply only to the individuals and circumstances for which we have data in hand.	Conclusions apply to a larger group of individuals or a broader class of circumstances.
Conclusions are informal, based on what we see in the data.	Conclusions are formal, backed by a statement of our confidence in them.

Our journey toward inference begins in Chapters 8–10, which describe statistical designs for *producing data* by samples and experiments, and the ethical issues involved. The important lesson of Part II is that the quality of the inferences we make from data depends heavily on how the data are produced.

Frank Fell/Getty Images

CHAPTER 8

Producing Data: Sampling

In this chapter we cover...

8.1 Population versus sample

8.2 How to sample badly

8.3 Simple random samples

8.4 Inference about the population

8.5 Other sampling designs

8.6 Cautions about sample surveys

8.7 The impact of technology

Statistics, the science of data, provides ideas and tools that we can use in many settings. Sometimes we have data that describe a group of individuals, and want to learn what the data say. That's the job of exploratory data analysis. Sometimes we have specific questions, but no data to answer them. To get sound answers, we must *produce data* in a way that is designed to answer our questions.

Suppose our question is, "What percent of college students think that people should not obey laws that violate their personal values?" To answer the question, we interview undergraduate college students. We can't afford to ask all students, so we put the question to a *sample* chosen to represent the entire student *population*. How shall we choose a sample that truly represents the opinions of the entire population? Statistical designs for choosing samples are the topic of this chapter. We will see that

- A sound statistical design is necessary if we are to trust data from a sample for drawing sound conclusions about the population.
- In sampling from large human populations, however, "practical problems" can overwhelm even sound designs.
- The impact of technology (particularly cell phones and the Internet) is making it harder to produce trustworthy national data by sampling.

8.1 Population versus sample

A political scientist wants to know what percent of college-age adults consider themselves conservatives. An automaker hires a market research firm to learn what percent of adults aged 18 to 35 recall seeing television advertisements for a new gas–electric hybrid car. Government economists inquire about average household income. In all these cases, we want to gather information about a large group of individuals. Time, cost, and inconvenience preclude contacting every individual. So we gather information about only part of the group to draw conclusions about the whole.

Population, Sample, Sampling Design

The **population** in a statistical study is the entire group of individuals about which we want information.

A **sample** is a part of the population from which we actually collect information. We use a sample to draw conclusions about the entire population.

A **sampling design** describes exactly how to choose a sample from the population.

Pay careful attention to the details of the definitions of "population" and "sample." Read and answer Exercise 8.1 right now to check your understanding.

We often draw conclusions about a whole on the basis of a sample. Everyone has tasted a sample of ice cream and ordered a cone on the basis of that taste. But ice cream is uniform, so that the single taste represents the whole. Choosing a representative sample from a large and varied population is not so easy. The first step in planning a **sample survey** is to say exactly *what population* we want to describe. The second step is to say exactly *what we want to measure,* that is, to give exact definitions of our variables. These preliminary steps can be complicated, as the following example illustrates.

sample survey

EXAMPLE 8.1 The Current Population Survey

The most important government sample survey in the United States is the monthly Current Population Survey (CPS) conducted by the Bureau of the Census for the Bureau of Labor Statistics. The CPS contacts about 60,000 households each month. It produces the monthly unemployment rate and much other economic and social information. (See Figure 8.1.) To measure unemployment, we must first specify the population we want to describe. Which age groups will we include? Will we include illegal immigrants or people in prisons? The CPS defines its population as all U.S. residents (legal or not) 16 years of age and over who are civilians and are not in an institution such as a prison. The unemployment rate announced in the news refers to this specific population.

The second question is harder: what does it mean to be "unemployed"? Someone who is not looking for work—for example, a full-time student—should not be called unemployed just because she is not working for pay. If you are chosen for the CPS sample, the interviewer first asks whether you are available to work and whether you actually looked for work in the past four weeks. If not, you are neither employed nor unemployed—you are not in the labor force. So discouraged workers who haven't looked for a job in four weeks are excluded from the count.

If you are in the labor force, the interviewer goes on to ask about employment. If you did any work for pay or in your own business during the week of the survey,

U.S. Bureau of Labor Statistics

FIGURE 8.1 The home page of the Current Population Survey at the Bureau of Labor Statistics.

you are employed. If you worked at least 15 hours in a family business without pay, you are employed. You are also employed if you have a job but didn't work because of vacation, being on strike, or other good reason. An unemployment rate of 6.7% means that 6.7% of the sample was unemployed, using the exact CPS definitions of both "labor force" and "unemployed." ■

The final step in planning a sample survey is the sampling design. We will now introduce basic statistical designs for sampling.

LaunchPad Online Resources

- The **Snapshots video**, *Sampling*, provides a good overview of why we need to sample and the distinction between populations and samples.

Apply Your Knowledge

8.1 Sampling Students. A political scientist wants to know how college students feel about the Social Security system. She obtains a list of the 3456 undergraduates at her college and mails a questionnaire to 250 students selected at random. Only 104 questionnaires are returned.

(a) What is the population in this study? Be careful: about what group does she *want information?*

(b) What is the sample? Be careful: from what group does she *actually obtain information?*

The important message in this problem is that the sample can redefine the population about which information is obtained.

© John Elk III/Alamy

8.2 **Student Archaeologists.** An archaeological dig turns up large numbers of pottery shards, broken stone implements, and other artifacts. Students working on the project classify each artifact and assign it a number. The counts in different categories are important for understanding the site, so the project director chooses 2% of the artifacts at random and checks the students' work. What are the population and the sample here?

8.3 **Software Survey.** A statistical software company is planning on updating Version 8.1 of its software and wants to know what features are most important to users. The company's managers have the email addresses of 1100 individuals, mostly faculty at universities, for whom they have supplied free courtesy copies of Version 8.1. They email these 1100 individuals and ask them to complete a survey online. A total of 186 of these individuals complete the survey.

(a) What is the population of interest to the software company? Do you think the 1100 individuals contacted are representative of the population? Explain your reasons.

(b) What is the sample? From what group is information actually obtained?

8.2 How to sample badly

How can we choose a sample that we can trust to represent the population? A sampling design is a specific method for choosing a sample from the population. The easiest—but not the best—design just chooses individuals close at hand. If we are interested in finding out how many people have jobs, for example, we might go to a shopping mall and ask people passing by if they are employed. A sample selected by taking the members of the population that are easiest to reach is called a **convenience sample.** Convenience samples often produce unrepresentative data.

convenience sample

EXAMPLE 8.2 Sampling at the Mall

A sample of mall shoppers is fast and cheap. But people at shopping malls tend to be more prosperous than typical Americans. They are also more likely to be teenagers or retired. Moreover, unless interviewers are carefully trained, they tend to question well-dressed, respectable-looking people and avoid poorly dressed or tough-looking individuals. The types of people at the mall will also vary by time of day and day of week. In short, mall interviews will not contact a sample that is representative of the entire population. ■

Interviews at shopping malls will almost surely overrepresent middle-class and retired people and underrepresent the poor. This will happen almost every time we take such a sample. That is, it is a systematic error caused by a bad sampling design, not just bad luck on one sample. This is *bias:* the outcomes of mall surveys will repeatedly miss the truth about the population in the same ways.

Bias

The design of a statistical study is **biased** if it systematically favors certain outcomes.

EXAMPLE 8.3 Online Polls

Former CNN evening commentator Lou Dobbs doesn't like illegal immigration. One of his broadcasts in 2007 was largely devoted to attacking a proposal by the governor of New York State to offer driver's licenses to illegal immigrants as a public safety measure. During the show, Mr. Dobbs invited his viewers to go to `loudobbs.com` to vote on the question, "Would you be more or less likely to vote for a presidential candidate who supports giving drivers' licenses to illegal aliens?" We aren't surprised that 97% of the 7350 people who voted by the end of the broadcast said, "Less likely." ■

The `loudobbs.com` poll was biased because people chose whether or not to participate (see Exercise 8.33). Most who voted were viewers of Lou Dobbs's program who had just heard him denounce the governor's idea. *People who take the trouble to respond to an open invitation are usually not representative of any clearly defined population.* That's true of the people who bother to respond to write-in, call-in, or online polls in general. Polls like these are examples of *voluntary response sampling.*

CAUTION

Voluntary Response Sample

A **voluntary response sample** consists of people who choose themselves by responding to a broad appeal. Voluntary response samples are biased because people with strong opinions are most likely to respond.

Apply Your Knowledge

8.4 **Sampling on Campus.** You see a student standing in front of the student center, now and then stopping other students to ask them questions. She says that she is collecting student opinions for a class assignment. Explain why this sampling method is almost certainly biased.

8.5 **More Sampling on Campus.** You would like to start a club for psychology majors on campus, and you are interested in finding out what proportion of psychology majors would join. The dues would be $35 and used to pay for speakers to come to campus. You ask five psychology majors from your senior psychology honors seminar whether they would be interested in joining this club and find that four of the five students questioned are interested. Is this sampling method biased, and if so, what is the likely direction of bias?

8.3 Simple random samples

In a voluntary response sample, people choose whether to respond. In a convenience sample, the interviewer makes the choice. In both cases, personal choice produces bias. The statistician's remedy is to allow impersonal chance to choose the sample. A sample chosen by chance rules out both favoritism by the sampler and self-selection by respondents. Choosing a sample by chance attacks bias by giving all individuals an equal chance to be chosen. Rich and poor, young and old, liberal and conservative, all have the same chance to be in the sample.

The simplest way to use chance to select a sample is to place names in a hat (the population) and draw out a handful (the sample). This is the idea of *simple random sampling.* Although the idea of drawing names from a hat is a good way to conceptualize a simple random sample, it is generally *not* a good method for obtaining a simple random sample. Writing names on slips of paper can lead to bias if the slips of paper are not well mixed or there is a tendency to select them from, say, the top or bottom. Drawing names from a hat would be particularly difficult to implement if the population size is large, possibly requiring thousands of slips of paper.

Simple Random Sample

A **simple random sample (SRS)** of size *n* consists of *n* individuals from the population chosen in such a way that every set of *n* individuals has an equal chance to be the sample actually selected.

An SRS not only gives each individual an equal chance to be chosen, but also gives every possible sample an equal chance to be chosen. There are other random sampling designs that give each individual, but not each sample, an equal chance. Exercise 8.43 describes one such design.

When you think of an SRS, you can still picture the conceptual situation of drawing names from a hat to remind yourself that an SRS doesn't favor any part of the population. That's why an SRS is a better method of choosing samples than convenience or voluntary response sampling. But writing names on slips of paper, mixing them well, and drawing them from a hat is slow and inconvenient. That's especially true if, like the Current Population Survey, we must draw a sample of size 60,000. In practice, samplers use software. The *Simple Random Sample* applet makes choosing an SRS very fast. If you don't use the applet or other software, you can randomize by using a *table of random digits.* In fact, software for choosing samples starts by generating random digits, so using a table just does by hand what the software does more quickly.

APPLET

Random Digits

A **table of random digits** is a long string of the digits 0, 1, 2, 3, 4, 5, 6, 7, 8, and 9 with these two properties:

1. Each entry in the table is equally likely to be any of the 10 digits 0 through 9.
2. The entries are independent of each other. That is, knowledge of one part of the table gives no information about any other part.

ARE THESE RANDOM DIGITS REALLY RANDOM?

Not a chance. The random digits in Table B were produced by a computer program. Computer programs do exactly what you tell them to do. Give the program the same input, and it will produce exactly the same "random" digits. Of course, clever people have devised computer programs that produce output that *looks* like random digits. These are called "pseudo-random numbers," and that's what Table B contains. Pseudo-random numbers work fine for statistical randomizing, but they have hidden nonrandom patterns that can mess up more refined uses.

Table B at the back of the book is a table of random digits. Table B begins with the digits 19223950340575628713. To make the table easier to read, the digits appear in groups of five and in numbered rows. The groups and rows have no meaning—the table is just a long list of randomly chosen digits. There are two steps in using the table to choose a simple random sample.

Using Table B to Choose an SRS

Label: Give each member of the population a numerical label of the *same length.*

Table: To choose an SRS, read from Table B successive groups of digits of the length you used as labels. Your sample contains the individuals whose labels you find in the table.

You can label up to 100 items with two digits: 01, 02, . . . , 99, 00. Up to 1000 items can be labeled with three digits, and so on. Always use the shortest labels that will cover your population. As standard practice, we recommend that you begin with label 1 (or 01 or 001, as needed). Reading groups of digits from the table gives all individuals the same chance to be chosen because all labels of the same length have the same chance to be found in the table. For example, any pair of digits in the table is equally likely to be any of the 100 possible labels 01, 02, . . . , 99, 00. Ignore any group of digits that was not used as a label or that duplicates a label already in the sample. You can read digits from Table B in any order—across a row, down a column, and so on—because the table has no order. As standard practice, we recommend reading across rows.

EXAMPLE 8.4 Sampling Spring Break Resorts

A campus newspaper plans a major article on spring break destinations. The reporters intend to call four randomly chosen resorts at each destination to ask about their attitudes toward groups of students as guests. Here are the resorts listed in one city:

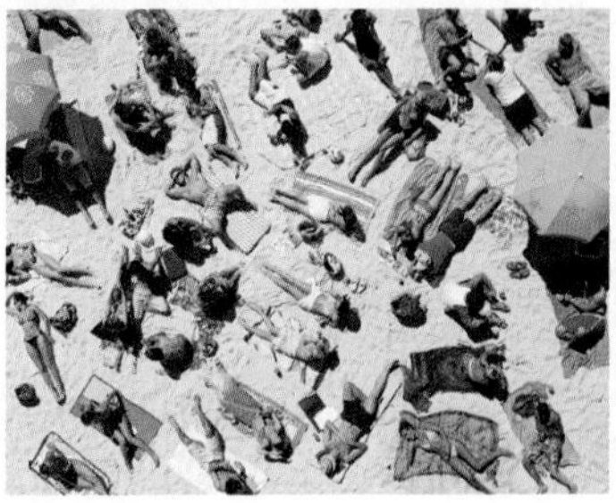

Robert Daly/Getty Images

01	Aloha Kai	08	Captiva	15	Palm Tree	22	Sea Shell
02	Anchor Down	09	Casa del Mar	16	Radisson	23	Silver Beach
03	Banana Bay	10	Coconuts	17	Ramada	24	Sunset Beach
04	Banyan Tree	11	Diplomat	18	Sandpiper	25	Tradewinds
05	Beach Castle	12	Holiday Inn	19	Sea Castle	26	Tropical Breeze
06	Best Western	13	Lime Tree	20	Sea Club	27	Tropical Shores
07	Cabana	14	Outrigger	21	Sea Grape	28	Veranda

Label: Because two digits are needed to label the 28 resorts, all labels will have two digits. We have added labels 01 to 28 in the list of resorts. Always say how you labeled the members of the population. To sample from the 1240 resorts in a major vacation area, you would label the resorts 0001, 0002, . . . , 1239, 1240.

Select sample: To use the *Simple Random Sample* applet, just enter 28 in the "Population =" box and 4 in the "Select a sample of size:" box, and click "Sample." Figure 8.2 shows the result of one sample which contains the resorts labeled 20, 16, 21, and 10. These are Sea Club, Radisson, Sea Grape, and Coconuts.

APPLET

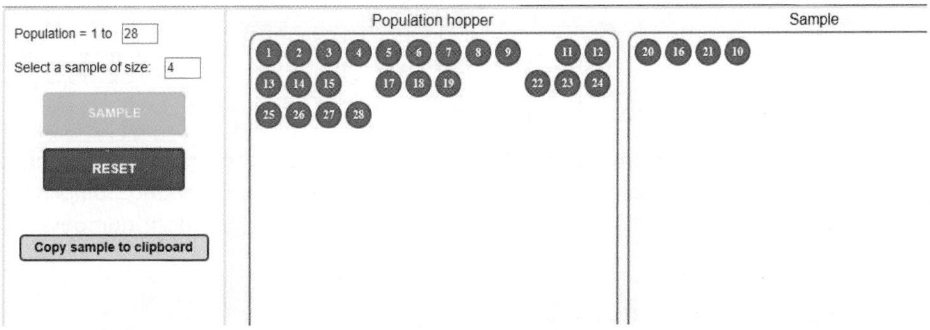

FIGURE 8.2
The *Simple Random Sample* applet used to choose an SRS of size $n = 4$ from a population of size 28.

To use Table B, read two-digit groups until you have chosen four resorts. Starting at line 130 (any line will do), we find

69051 64817 87174 09517 84534 06489 87201 97245

Because the labels are two digits long, read successive two-digit groups from the table: the first three two-digit groups here are 69, 05, and 16. Ignore groups not used as labels, like the initial 69. Also ignore any repeated labels, like the second and third 17s in this row, because you can't choose the same resort twice. Your sample contains the resorts labeled 05, 16, 17, and 20. These are Beach Castle, Radisson, Ramada, and Sea Club. ■

We can trust results from an SRS, as well as from other types of random samples that we will meet later, because the use of impersonal chance avoids bias. Online polls and mall interviews also produce samples. We can't trust results from these samples, because they are chosen in ways that invite bias. *The first question to ask about any sample is whether it was chosen at random.*

CAUTION

EXAMPLE 8.5 Gun Control Laws

"In general, do you think gun control laws should be made more strict, less strict, or kept as they are now?" When the *New York Times* and CBS News asked this question of 1644 adults in February 2014, 54% said "More," 9% said "Less," and 36% said "Kept as they are now." Can we trust the opinions of this sample to fairly represent the opinions of all adults? Here's part of the statement by the *Times* on how the poll was conducted:

> *The latest* New York Times/*CBS News poll is based on telephone interviews conducted February 19 through February 23 with 1644 adults throughout the United States.*
>
> *The sample of land line telephone exchanges called was randomly selected by a computer from a complete list of more than 81,000 active residential exchanges across the country. The exchanges were chosen so as to ensure that each region of the country was represented in its proper proportion.*
>
> *Within each exchange, random digits were added to form a complete telephone number, thus permitting access to listed and unlisted numbers alike. Within each household, one adult was designated by a random procedure to be the respondent for the survey.*
>
> *Cellphone numbers were generated by a similar random process. The two samples were then combined and adjusted to assure the proper ration of landine-only, cellphone-only, and dual phone users.*[1]

random digit dialing

The selection of landline phone numbers is a good description of a common method for choosing national samples, called **random digit dialing.** Although the information about the conduct of the poll provides the reader with some basic details, data collection and analysis of a national survey is a great deal more complex than this short description suggests. We'll come back to random digit dialing and its problems later (see Exercise 8.16), but this statement on the conduct of the poll does contain important information. We know the size of the sample, when the poll was taken, and that the comforting word "random" appears four times. ■

LaunchPad Online Resources

- The StatClips video, *Data and Sampling*, discusses the importance of simple random samples, and the StatClips Examples video, *Data and Sampling: Example C*, provides an illustration of using a table of random numbers to select an SRS.

Apply Your Knowledge

8.6 **Apartment Living.** You are planning a report on apartment living in a college town. You decide to select three apartment complexes at random for in-depth interviews with residents. Use the *Simple Random Sample* applet, other software, or Table B to select a simple random sample of four of the following apartment complexes. If you use Table B, start at line 128.

Ashley Oaks	Country View	Mayfair Village
Bay Pointe	Country Villa	Nobb Hill
Beau Jardin	Crestview	Pemberly Courts
Bluffs	Del-Lynn	Peppermill
Brandon Place	Fairington	Pheasant Run
Briarwood	Fairway Knolls	River Walk
Brownstone	Fowler	Sagamore Ridge
Burberry Place	Franklin Park	Salem Courthouse
Cambridge	Georgetown	Village Square
Chauncey Village	Greenacres	Waterford Court

8.7 Minority Managers. A firm wants to understand the attitudes of its minority managers toward its system for assessing management performance. Following is a list of all the firm's managers who are members of minority groups. Use the *Simple Random Sample* applet, other software, or Table B at line 141 to choose five managers to be interviewed in detail about the performance appraisal system.

Adelaja	Draguljic	Huo	Modur
Ahmadiani	Fernandez	Ippolito	Rettiganti
Barnes	Fox	Jiang	Rodriguez
Bonds	Gao	Jung	Sanchez
Burke	Gemayel	Mani	Sgambellone
Deis	Gupta	Mazzeo	Yajima
Ding	Hernandez		

8.8 Sampling Gravestones. The local genealogical society in Coles County, Illinois, has compiled records on all 55,914 gravestones in cemeteries in the county for the years 1825 to 1985. Historians plan to use these records to learn about African Americans in Coles County's history. They first choose an SRS of 395 records to check their accuracy by visiting the actual gravestones.[2]

(a) How would you label the 55,914 records?

(b) Use software or Table B, starting at line 137, to choose the first five records for the SRS.

The Photo Works

8.4 Inference about the population

The purpose of a sample is to give us information about a larger population. The process of drawing conclusions about a population on the basis of sample data is called **inference** because we *infer* information about the population from what we *know* about the sample.

inference

Inference from convenience samples or voluntary response samples would be misleading because these methods of choosing a sample are biased. We are almost certain that the sample does *not* fairly represent the population. *The first reason to rely on random sampling is to eliminate bias in selecting samples from the list of available individuals.*

Nonetheless, it is unlikely that results from a random sample are exactly the same as for the entire population. Sample results, like the unemployment rate obtained from the monthly Current Population Survey, are only estimates of the truth about the population. If we select two samples at random from the same population, we will almost certainly draw different individuals. So the sample results will differ somewhat, just by chance. Properly designed samples avoid systematic bias, but their results are rarely exactly correct, and they vary from sample to sample.

Why can we trust random samples? The big idea is that the results of random sampling don't change haphazardly from sample to sample. Because we deliberately use chance, the results obey the laws of probability that govern chance behavior. These laws allow us to say how likely it is that sample results are close to the truth about the population. *The second reason to use random sampling is that the laws of probability allow trustworthy inference about the population.* Results from random samples come with a margin of error that sets bounds on the size of the likely error. How to do this is part of the technique of statistical inference. We will describe the reasoning in Chapter 16 and present details throughout the rest of the book.

One point is worth making now: *larger random samples give more accurate results than smaller random samples.* By taking a very large sample, you can be confident that the sample result is very close to the truth about the population. The Current Population Survey contacts about 60,000 households, so it estimates the national unemployment rate very accurately. Opinion polls that contact 1000 or 1500 people give less accurate results. Of course, only samples chosen by chance carry this guarantee. Lou Dobbs's online sample tells us little about overall American public opinion even though 7350 people clicked a response.

Apply Your Knowledge

8.9 **Ask More People.** In the 2012 presidential pre-election surveys, Pew Research sampled 1,112 likely voters during October 4–7, 2012, and asked if they were planning to vote for Obama, and then asked the same question of a sample of 1,495 likely voters taken from October 24–28, 2012. However, in their last survey taken October 31–November 3, 2012, just before the election held on November 6, 2012, they asked this question of a sample of 2,709 likely voters. Why do you think Pew did this?

8.10 **How Accurate Is the Poll?** The *New York Times*/CBS News poll conducted during February 19–23, 2014, included 1644 adults, of which 519 were Republican, 515 were Democrats, 550 were Independent, and 60 didn't know or didn't respond.[3] Each person sampled was asked their opinion on a variety of issues facing the nation, such as, "Do you feel that the distribution of money and wealth in this country is fair, or do you feel that the money and wealth in this country should be more evenly distributed among more people?" The margin of error (we will give more detail in later chapters) was reported as ±3% for the entire sample. When considering the opinions of only the Republicans in the sample, the margin of error was reported as ±6%. What do you think explains the fact that estimates for Republicans were less precise than for the entire sample?

8.5 Other sampling designs

Random sampling, the use of chance to select the sample, is the essential principle of statistical sampling. Designs for random sampling from large populations spread out over a wide area are usually more complex than an SRS. For example, it is common to sample important groups within the population separately, then combine these samples. This is the idea of a *stratified random sample.*

GOLFING AT RANDOM

Random drawings give everyone the same chance to be chosen, so they offer a fair way to decide who gets a scarce good—like a round of golf. Lots of golfers want to play the famous Old Course at St. Andrews, Scotland. Some can reserve in advance, at considerable expense. Most must hope that chance favors them in the daily random drawing for tee times. At the height of the summer season, only one in six wins the right to pay $250 for a round.

Stratified Random Sample

To select a **stratified random sample**, first classify the population into groups of similar individuals, called **strata**. Then, choose a separate SRS in each stratum and combine these SRSs to form the full sample.

Choose the strata based on facts known before the sample is taken. For example, a population of election districts might be divided into urban, suburban, and rural strata. A stratified design can produce more precise information than an SRS of the same size by taking advantage of the fact that individuals in the same stratum are similar to one another.

EXAMPLE 8.6 Seat Belt Use in Hawaii

Each state conducts an annual survey of seat belt use by drivers, following guidelines set by the federal government. The guidelines require random sampling. Seat belt use is observed at randomly chosen road locations at random times during daylight hours. The locations are not an SRS of all locations in the state, but rather a stratified sample using the state's counties as strata.

In Hawaii, the counties are the islands that make up the state's territory. The seat belt survey sample consists of 135 road locations in the four most populated islands: 66 in Oahu, 24 in Maui, 23 in Hawaii, and 22 in Kauai. The sample sizes on the islands are proportional to the amount of road traffic.[4] ■

Ryan McVay/Getty Images

multistage sample

Most large-scale sample surveys use a **multistage sample**. For example, the opinion poll described in Example 8.5 has three stages: choose a random sample of telephone exchanges (stratified by region of the country), then an SRS of household telephone numbers within each exchange, and then a random adult in each household.

Analysis of data from sampling designs more complex than an SRS takes us beyond basic statistics. But the SRS is the building block of more elaborate designs, and analysis of other designs differs more in complexity of detail than in fundamental concepts.

Apply Your Knowledge

8.11 Sampling Metro Chicago. Cook County, Illinois, has the second-largest population of any county in the United States (after Los Angeles County, California). Cook County has 30 suburban townships and an additional eight townships that make up the city of Chicago. The suburban townships are

Barrington	Elk Grove	Maine	Orland	Riverside
Berwyn	Evanston	New Trier	Palatine	Schaumburg
Bloom	Hanover	Niles	Palos	Stickney
Bremen	Lemont	Northfield	Proviso	Thornton
Calumet	Leyden	Norwood Park	Rich	Wheeling
Cicero	Lyons	Oak Park	River Forest	Worth

The Chicago townships are

Hyde Park	Lake	North Chicago	South Chicago
Jefferson	Lake View	Rogers Park	West Chicago

Because city and suburban areas may differ, the first stage of a multistage sample chooses a stratified sample of five suburban townships and three of the more heavily populated Chicago townships. Use software, the *Simple Random Sample* applet, or Table B to choose this sample. (If you use Table B, assign labels in alphabetical order and start at line 116 for the suburbs and at line 126 for Chicago.)

8.12 Academic Dishonesty. A study of academic dishonesty among college students used a two-stage sampling design. The first stage chose a sample of 30 colleges and universities. Then, the study authors mailed questionnaires to a stratified sample of 200 seniors, 100 juniors, and 100 sophomores at each school.[5] One of the schools chosen has 1127 freshmen, 989 sophomores, 943 juniors, and 895 seniors. You have alphabetical lists of the students in each class. Explain how you would assign labels for stratified sampling. Then use software or Table B, starting at line 140, to select the first five students in the sample from each stratum. After selecting five students for a stratum, continue to select the students for the next stratum.

8.6 Cautions about sample surveys

Random selection eliminates bias in the choice of a sample from a list of the population. When the population consists of human beings, however, accurate information from a sample requires more than a good sampling design.

To begin, we need an accurate and complete list of the population. Because such a list is rarely available, most samples suffer from some degree of *undercoverage.* A sample survey of households, for example, will miss not only homeless people but also prison inmates and students in dormitories. An opinion poll conducted by calling landline telephone numbers will miss households that have only cell phones as well as households without a phone. The results of national sample surveys therefore have some bias if the people not covered differ from the rest of the population.

A more serious source of bias in most sample surveys is *nonresponse,* which occurs when a selected individual cannot be contacted or refuses to cooperate. Nonresponse to sample surveys often exceeds 50%, even with careful planning and several callbacks. Because nonresponse is higher in urban areas, most sample surveys substitute other people in the same area to avoid favoring rural areas in the final sample. If the people contacted differ from those who are rarely at home or who refuse to answer questions, some bias remains.

Undercoverage and Nonresponse

Undercoverage occurs when some groups in the population are left out of the process of choosing the sample.

Nonresponse occurs when an individual chosen for the sample can't be contacted or refuses to participate.

EXAMPLE 8.7 How Bad Is Nonresponse?

The U.S. Census Bureau's American Community Survey (ACS) has the lowest nonresponse rate of any poll we know: in 2012, only about 1.2% of the households in the sample refused to respond; the overall nonresponse rate, including "never at home" and other causes, was just 2.7%.[6] This monthly survey of almost 300,000 housing units replaces the "long form" that in the past was sent to some households in the every-10-years national census. Participation in the ACS is mandatory, and the U.S. Census Bureau follows up by telephone and then in person if a household fails to return the mail questionnaire.

The University of Chicago's General Social Survey (GSS) is the nation's most important social science survey. (See Figure 8.3.) The GSS contacts its sample in person, and it is run by a university. Its response rates are about 70%, among the highest in the world.

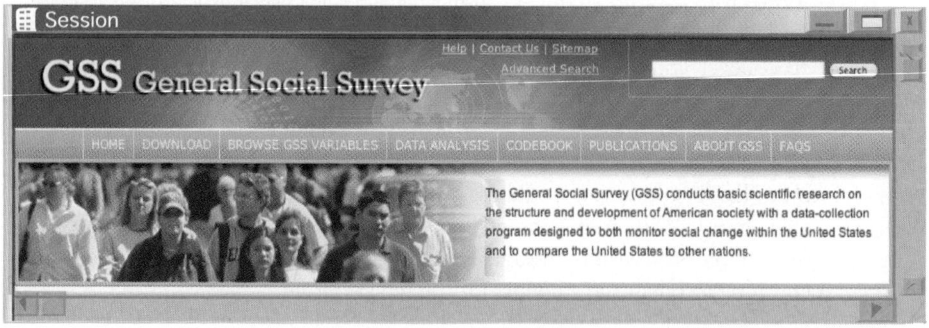

GSS General Social Survey

FIGURE 8.3
The home page of the General Social Survey at the University of Chicago's National Opinion Research Center. The GSS has tracked opinions about a wide variety of issues since 1972.

What about opinion polls by news media and opinion-polling firms? We don't know the rates of nonresponse for many of these surveys because they don't say, which in itself is a bad sign. In a December 2013 survey on social media use, the Pew Research Center provided a full disposition of the sampled phone numbers. Here are the details. Initially, 40,985 landline and 27,000 cell phones were dialed, with 11,260 of the landlines and 15,758 of the cell numbers being working numbers. Of these 27,018 working numbers, they were able to contact 17,335, or about 65%, as a large portion of the calls went to voicemail. Among those contacted, about 15% cooperated. Of those cooperating, some numbers were ineligible due to language barriers or contacting a child's cell phone, and some calls were eventually broken off without being completed. To sum it up, the 27,018 working numbers dialed resulted in a final sample of 1,801, giving a response rate of about 9%, an estimate of the fraction of all *eligible* respondents in the sample who were ultimately interviewed.[7] ■

In addition, the behavior of the respondent or of the interviewer can cause **response bias** in sample results. People know that they should take the trouble to vote, for example, so many who didn't vote in the last election will tell an interviewer that they did. The race or sex of the interviewer can influence responses to questions about race relations or attitudes toward feminism. Answers to questions that ask respondents to recall past events are often inaccurate because of faulty memory. For example, many people "telescope" events in the past, bringing them forward in memory to more recent time periods. "Have you visited a dentist in the last six months?" will often draw a "yes" from someone who last visited a dentist eight months ago.[8] Careful training of interviewers and careful supervision to avoid variation among the interviewers can reduce response bias. Good interviewing technique is another aspect of a well-done sample survey.

response bias

Wording effects are the most important influence on the answers given to a sample survey. Confusing or leading questions can introduce strong bias, and variations in wording can greatly change a survey's outcome. Even the order in which questions are asked matters. Here are some examples.[9]

wording effects

EXAMPLE 8.8 What Was That Question?

How do Americans feel about illegal immigrants? "Should illegal immigrants be prosecuted and deported for being in the United States illegally, or shouldn't they?" Asked this question in an opinion poll, 69% favored deportation. But when the very same sample was asked whether illegal immigrants who have worked in the United States for two years "should be given a chance to keep their jobs and eventually apply for legal status," 62% said that they should. Different questions give quite different impressions of attitudes toward illegal immigrants.

What about government help for the poor? Only 13% think we are spending too much on "assistance to the poor," but 44% think we are spending too much on "welfare." ■

EXAMPLE 8.9 Are You Happy?

Ask a sample of college students these two questions:

"How happy are you with your life in general?" (Answers on a scale of 1 to 5)

"How many dates did you have last month?"

The correlation between answers is $r = -0.012$ when asked in this order. It appears that dating has little to do with happiness. Reverse the order of the questions, however, and $r = 0.66$. Asking a question that brings dating to mind makes dating success a big factor in happiness. ■

Don't trust the results of a sample survey until you have read the exact questions asked. The amount of nonresponse and the date of the survey are also important. Good statistical design is a part, but only a part, of a trustworthy survey.

Apply Your Knowledge

8.13 A Survey of 100,000 Physicians. In 2010, the Physicians Foundation conducted a survey of physicians' attitude about health care reform, calling the report "a survey of 100,000 physicians." The survey was sent to 100,000 randomly selected physicians practicing in the United States: 40,000 via post-office mail and 60,000 via email. A total of 2,379 completed surveys were received.[10]

(a) State carefully what population is sampled in this survey and what is the sample size. Could you draw conclusions from this study about all physicians practicing in the United States?

(b) What is the rate of nonresponse for this survey? How might this affect the credibility of the survey results?

(c) Why is it misleading to call the report "a survey of 100,000 physicians"?

8.14 Gays in the Military. In 2010, a Quinnipiac University Poll and a CNN Poll each asked a nationwide sample about their views on openly gay men and women serving in the military.[11] Here are the two questions:

Question A: *Federal law currently prohibits openly gay men and women from serving in the military. Do you think this law should be repealed or not?*

Question B: *Do you think people who are openly gay or homosexual should or should not be allowed to serve in the U.S. military?*

One of these questions had 78% responding "should," and the other question had only 57% responding "should." Which wording is slanted toward a more negative response on gays in the military? Why?

8.7 The impact of technology

A few national sample surveys, including the General Social Survey and the U.S. government's American Community Survey and Current Population Survey, interview some or all of their subjects in person. This is expensive and time-consuming, so most national surveys contact subjects by telephone using the random digit dialing (RDD) method described in Example 8.5. Technology, especially the spread of cell phones, is making traditional RDD methods outdated.

First, *call screening* is now common. A large majority of American households have answering machines, voicemail, or caller ID, and many use these methods to screen their calls. Calls from polling organizations are rarely returned.

More seriously, the percent of *cell-phone-only households* is increasing rapidly. By mid-2007, 14% of American households had a cell phone but no landline phone; by the end of 2009, that percent had increased to almost 25%; and in 2012, the percent was almost 36%. It's clear from these numbers that RDD reaching only landline numbers is in trouble. Can surveys just add cell phone numbers? Not easily. Federal regulations prohibit automated dialing to cell phones, which rules out computerized RDD sampling and requires hand dialing of cell phone numbers, which is expensive. A cell phone can be anywhere, and many people keep their cell number despite moving, so stratifying by location becomes difficult. And a cell phone user may be driving or otherwise unable to talk safely.

Statistics In Your World

DO NOT CALL! People who do sample surveys hate telemarketing. We all get so many unwanted sales pitches by phone that many people hang up before learning that the caller is conducting a survey rather than selling vinyl siding. You can eliminate calls from commercial telemarketers by placing your phone number on the National Do Not Call Registry. Just go to `www.donotcall.gov` to sign up.

People who screen calls and people who have only a cell phone tend to be younger than the general population. In 2012, just over 60% of adults aged 25 to 29 years lived in households with no landline phone. So RDD surveys using only landlines may be biased (see Exercise 8.16). Careful surveys weight their responses to reduce bias. For example, if a sample contains too few young adults, the responses of the young adults who do respond are given extra weight. But with response rates steadily dropping and cell-phone-only use steadily growing, the future of RDD landline telephone surveys is not promising. Many polling organizations now include a minimum quota of cell phone users in their samples to help adjust for bias[12] (see Example 8.5 and Exercise 8.45).

One alternative is to use *web surveys*, an increasingly popular survey method, rather than telephone surveys. Web surveys have several advantages over more traditional survey methods. It is possible to collect large amounts of survey data at lower costs than traditional methods allow. Anyone can put survey questions on dedicated sites offering free services; thus large-scale data collection is available to almost every person with access to the Internet. Furthermore, web surveys allow delivery of multimedia survey content to respondents, opening up new realms of survey possibilities that would be extremely difficult to implement using traditional methods. Some argue that eventually web surveys will replace traditional survey methods.

Although web surveys are easy to do, they are not easy to do well. Three major problems are voluntary response, undercoverage, and nonresponse. Voluntary response appears in several forms in online surveys. Example 8.3 is a survey that invited individuals to a particular website to participate in a poll. Other web surveys solicit participation through announcements in news groups, email invitations, and banner ads on high-traffic sites.[13]

EXAMPLE 8.10 More to Makeup than Meets the Eye

Dougal Waters/Getty Images

The Renfrew Center Foundation, a nonprofit charitable organization, reports that 44% of women have negative feelings when they don't wear makeup.[14] How did they come up with this number? It is based on an online survey of 1292 women, 18 years of age and older, conducted by Harris Interactive during December 20–22, 2011. This is a more sophisticated example of voluntary response. It occurs when the polling organization, in this case Harris Interactive, maintains a panel of volunteers. The panel is recruited to fill out questionnaires on the Internet, often in exchange for points redeemable for cash and gifts. Panelists join by going to the polling company website and filling out some personal information later used to select them for specific surveys. Harris Interactive uses an online research panel of over 6 million volunteers worldwide, from which it selected the sample for the Renfrew Center Foundation poll.

In its report on online panels,[15] the American Association of Public Opinion Research states, "Researchers should avoid nonprobability online panels when one of the research objectives is to accurately estimate population values. From a total survey error perspective, the principal source of error in estimates from these types of sample sources is a combination of the lack of Internet access in roughly one in three U.S. households and the self-selection bias inherent in the panel recruitment processes." It also states that, "The reporting of a margin of sampling error associated with an opt-in or self-identified sample is misleading." In the methodology section of the poll, the Renfew Center Foundation does state, "This online survey is not based on a probability sample, and therefore, no estimate of theoretical sampling error can be calculated," although the issue of potential bias in the estimate is not addressed. ■

Undercoverage is a serious problem for even careful web surveys, because about 25% of Americans lack Internet access and only about 70% have broadband access. People without Internet access are more likely to be poor, elderly, minority, or rural than the overall population, so the potential for bias in a web survey is clear. There is no easy way to choose a random sample even from people with web access because there is no technology that generates personal email addresses at random in the way that RDD generates residential telephone numbers, and individuals may have several email addresses. Even if such technology existed, etiquette and regulations aimed at spammers would prevent mass emailing. For the present, web surveys work well only for restricted populations, for example, surveying students at your university using the school's list of student email addresses. Here is an example of a successful web survey.

EXAMPLE 8.11 Doctors and Placebos

A placebo is a dummy treatment such as a pill that has no direct effect on a patient but may bring about a response because patients expect it to. Do academic physicians who maintain private practices sometimes give their patients placebos? A web survey of doctors in internal medicine departments at Chicago-area medical schools was possible because almost all the doctors had listed email addresses.

An email was sent to each doctor, explaining the purpose of the study, promising anonymity, and giving an individual web link for response. In all, 231 of 443 doctors responded. The response rate was helped by the fact that the email came from a team at a medical school. Result: 45% said they sometimes used placebos in their clinical practice.[16] ■

Apply Your Knowledge

8.15 NPR Facebook Survey. In 2010, National Public Radio (NPR) conducted a survey of preferences and habits of its Facebook fans by recruiting respondents through messages posted on its Facebook page. The survey was conducted online and deployed July 12–19. A total of 40,043 respondents began the survey, with 33,304 completing all questions. It was found that people accessed NPR on the radio, at NPR.org, through iPhone apps, and several other platforms. Asked about time spent with NPR, about 20% of respondents indicated that they spent more than three hours per day, including radio listening.

(a) Here is what NPR says about the survey methodology: "Respondents were self-selected and the resulting sample is non-random—therefore a margin of error cannot be calculated, and the survey results cannot be projected to any population other than the sample itself."[17] Why can't inference about any population be made?

(b) Suppose that people who spent more time with NPR were more likely to respond to the survey. Do you think the true percentage of NPR's Facebook fans who spend more than three hours with NPR is higher or lower than the 20% found from the survey? Explain why.

8.16 More on Random Digit Dialing. In the first half of 2013, about 38% of adults lived in households with a cell phone and no landline phone. Among adults aged 25 to 29, this percent was about 65%, while among adults over 65, the percent was only 13%.[18]

(a) Write a survey question for which the opinions of adults with landline phones only are likely to differ from the opinions of adults with cell phones only. Give the direction of the difference of opinion.

(b) For the survey question in part (a), suppose a survey was conducted using random digit dialing of landline phones only. Would the results be biased? What would be the direction of bias?

(c) Most surveys now supplement the landline sample contacted by RDD with a second sample of respondents reached through random dialing of cell phone numbers. The landline respondents are weighted to take account of household size and number of telephone lines into the residence, whereas the cell phone respondents are weighted according to whether they were reachable only by cell phone or also by landline. Explain why it is important to include both a landline sample and a cell phone sample. Why is the number of telephone lines into the residence important? (*Hint:* How does the number of telephone lines into the residence affect the chance of the household being included in the RDD sample?)

CHAPTER 8 SUMMARY

Chapter Specifics

- A **sample survey** selects a **sample** from the **population** of all individuals about which we desire information. We base conclusions about the population on data from the sample. It is important to specify exactly what population you are interested in and what variables you will measure.
- The **design** of a sample describes the method used to select the sample from the population. **Random sampling** designs use chance to select a sample.
- The basic random sampling design is a **simple random sample (SRS)**. An SRS gives every possible sample of a given size the same chance to be chosen.
- Choose an SRS by labeling the members of the population and using **random digits** to select the sample. Software can automate this process.
- To choose a **stratified random sample**, classify the population into **strata**, groups of individuals that are similar in some way that is important to the response. Then choose a separate SRS from each stratum.
- Failure to use random sampling often results in **bias**, or systematic errors in the way the sample represents the population. **Voluntary response samples**, in which the respondents choose themselves, are particularly prone to large bias.
- In human populations, even random samples can suffer from bias due to **undercoverage** or **nonresponse**, from **response bias**, or from misleading results due to **poorly worded questions**. Sample surveys must deal expertly with these potential problems in addition to using a random sampling design.
- Most national sample surveys are carried out by telephone, using **random digit dialing** to choose residential telephone numbers at random. Because call screening is increasing nonresponse to such surveys, and the rise of cell-phone-only households is increasing undercoverage, many surveys include a minimum quota of cell phone users in their samples to help adjust for bias. **Web surveys** are becoming more frequent, but many suffer from volunteer response, undercoverage, and nonresponse.

Link It

The methods of Chapters 1 to 6 can be used to describe data regardless of how the data were obtained. However, if we want to reason from data to give answers to specific questions or to draw conclusions about the larger population, then the method that was used to collect the data is important. Sampling is one way to collect data, but it does not guarantee that we can draw meaningful conclusions. Biased sampling methods, such as convenience sampling and voluntary response samples, produce data that can be misleading, resulting in incorrect conclusions. Simple random sampling avoids bias and produces data that can lead to valid conclusions regarding the population. Even with perfect sampling methods, there is still sample-to-sample variation; we will begin our study of the connection between sampling variation and drawing conclusions in Chapter 15.

Even when we take a simple random sample, our conclusions can be weakened by undercoverage, nonresponse, and poor wording of questions. Careful attention must be given to all aspects of the sampling process to ensure that the conclusions we make are valid. In some cases, more complex designs are required, such as stratified sampling or multistage sampling. But the use of impersonal chance to select the sample remains a key ingredient in the sampling process. And issues such as undercoverage and nonresponse still remain for these more complex designs.

The next chapter is about statistical designs for experiments, a quite different means of producing data than sample surveys.

LaunchPad Online Resources

If you are having difficulty with any of the sections of this chapter, these online resources should help prepare you to solve the exercises at the end of this chapter.

- **StatTutor** starts with a video review of each section and asks a series of questions to check your understanding.
- **LearningCurve** provides you with a series of questions about the chapter geared to your level of understanding.

CHECK YOUR SKILLS

8.17 An online store contacts 1000 customers from its list of customers who have purchased in the last year. In all, 696 of the 1000 say that they are very satisfied with the store's website. The population in this setting is

(a) all customers who have purchased something in the last year.
(b) the 1000 customers contacted.
(c) the 696 customers who were very satisfied with the store's website.

8.18 A state representative wants to know how voters in his district feel about enacting a statewide smoking ban in all enclosed public places, including bars and restaurants, as well as several other current statewide issues. He mails a questionnaire addressing these issues to an SRS of 800 voters in his district. Of the 800 questionnaires mailed, 152 were returned. The sample is

(a) the 800 voters receiving the questionnaire.
(b) the 152 voters returning the questionnaire.
(c) all voters in his district.

8.19 In the survey for Exercise 8.18, there are 8741 registered voters in his district. You label the voters 0001 to 8741 in alphabetical order. Using line 123 of Table B to select the sample, the first five voters in your sample would be

(a) 5458, 0815, 7271, 2560, 2755.
(b) 5458, 0815, 0727, 1025, 6027.
(c) 5458, 8150, 7271, 2560, 2755.

8.20 The website `www.twiigs.com` allows you to vote on polls that interest you or to post one of your own. Once you have found a poll of interest, you just click on "Vote," and your response becomes part of the sample. One of the questions in July 2010 was, "How many times have you been pulled over by the police?" Of the 780 people responding, 70% said "1–5 times." You can conclude that

(a) about 70% of Americans have been pulled over by the police "1–5 times."
(b) the poll uses voluntary response, so the results tell us little about the population of all adults.
(c) more people still need to vote on the question, as a larger sample is required to reduce bias.

8.21 Archaeologists plan to examine a sample of two-meter-square plots near an ancient Greek city for artifacts visible in the ground. They choose separate samples of plots from floodplain, coast, foothills, and high hills. What kind of sample is this?

(a) A simple random sample
(b) A stratified random sample
(c) A voluntary response sample

8.22 You must choose an SRS of 10 of the 440 retail outlets in New York that sell your company's products. How would you label this population to select a simple random sample?

(a) 001, 002, 003, . . . , 439, 440
(b) 000, 001, 002, . . . , 439, 440
(c) 1, 2, . . . , 439, 440

8.23 You are using the table of random digits to choose a simple random sample of six students from a class of 30 students. You label the students 01 to 30 in alphabetical order, and then select a simple random sample. Which of the following is a possible sample that could be obtained?

(a) 45, 74, 04, 18, 07, 65
(b) 04, 18, 07, 13, 02, 07
(c) 04, 18, 07, 13, 02, 05

8.24 A sample of households in a community is selected at random from the telephone directory. In this community, 4% of households have no telephone, 10% have only cell phones, and another 25% have unlisted telephone numbers. The sample will certainly suffer from

(a) nonresponse.
(b) undercoverage.
(c) false responses.

8.25 The Pew Research Centers Report entitled "How Americans value public libraries in their communities," released December 11, 2013, asked a random sample of 6224 Americans aged 16 and over, "Have you used a Public Library website in the last 12 months?" In the entire sample, 30% said "yes." But only 17% of those in the sample over 65 years of age said "yes." Which of these two sample percents will be more accurate as an estimate of the truth about the population?

(a) The result for those over 65 is more accurate because it is easier to estimate a proportion for a small group of people.
(b) The result for the entire sample is more accurate because it comes from a larger sample.
(c) Both are equally accurate because both come from the same sample.

CHAPTER 8 EXERCISES

In all exercises asking for an SRS, you may use software, the Simple Random Sample applet, or Table B.

8.26 **Environmental problems.** A Gallup Poll found that Americans, when asked about a list of environmental problems, were increasingly worried about problems such as pollution of drinking water, soil, air, and waterways, but were least worried about climate change or global warming. Gallup's report said, "Results are based on telephone interviews conducted March 6–9, 2014, with a random sample of 1,048 adults, aged 18 and older, living in all 50 U.S. states and the District of Columbia."[19] What is the population for this sample survey? What is the sample?

8.27 Sampling stuffed envelopes. A large retailer prepares its customers' monthly credit card bills using an automatic machine that folds the bills, stuffs them into envelopes, and seals the envelopes for mailing. Are the envelopes completely sealed? Inspectors choose 40 envelopes from the 1000 stuffed each hour for visual inspection. What is the population for this sample survey? What is the sample?

8.28 Stormwater pollution. The administrators of the city's Stormwater Program are interested in evaluating the public's knowledge of the causes and effects of stormwater pollution. They set up an information booth at the city's Earth Day celebration, and people who visit the booth are offered the opportunity to complete a survey on their knowledge of stormwater pollution.

(a) If the population of interest is adult residents of the city, discuss any sources of bias that might limit the usefulness in using this survey to draw conclusions about this population.
(b) The survey at the Earth Day celebration was intended to be used as a pilot study to make sure the questions were not ambiguous. The full survey is to be an online survey. Residents are to be made aware of the survey through information mailed with their utility bills, and asked to go online to complete the survey. Discuss any sources of bias that might be contained in the full survey.
(c) Suppose we select a random sample of utility customers. *Only* the residents in the sample are to be made aware of the survey through information mailed with their utility bills, and asked to go online to complete the survey. Does this have any advantages or disadvantages over the full survey in part (b) in terms of possible bias? Explain.

8.29 Sampling telephone area codes. The United States currently has approximately 287 Numbering Plan Areas (NPAs) in service, corresponding to geographic regions. Each NPA is identified by a three-digit code, commonly called an area code. (More are created regularly.)[20] You want to choose an SRS of 20 of these area codes for a study of available telephone numbers. Label the codes 001 to 287, and use the *Simple Random Sample* applet or other software to choose your sample. (If you use Table B, start at line 122 and choose only the first five area codes in the sample.)

8.30 Paying taxes. In April 2014, a Gallup Poll asked two questions about the amount one pays in federal income taxes.[21] Here are the two questions:

Question A: *Do you regard the income tax which you will have to pay this year as fair?*

Question B: *Do you consider the amount of federal income tax you have to pay as too high, about right, or too low?*

One of these questions drew 52% saying the amount was fair or about right; the other, only 41%. Which wording produced the higher percentage? Why?

8.31 Wording of questions. Do the opinions of Canadians and Americans differ on the right of citizens to bear arms? Although the population of the U.S. is about 10 times that of Canada, the estimated number of privately owned firearms is 25 times higher in the U.S. When Gallup asked a random sample of 1002 American adults the question, "Do you think there should or should not be a law that would ban the possession of handguns, except by the police and other authorized persons?", 35% said there should be a law banning possession of handguns. At the same time, when Gallup asked a random sample of 1011 Canadians the question, "Do you think the general public should be allowed by law to own a gun?", 63% said "no."[22] What problems do you see in using the percentages in these two surveys to make a comparison on the opinion on the right to bear arms in the two countries? Explain.

8.32 Movie viewing. An opinion poll calls 2000 randomly chosen residential telephone numbers and asks to speak with an adult member of the household. The interviewer asks, "How many movies have you watched in a movie theater in the past 12 months?"

(a) What population do you think the poll has in mind?
(b) In all, 831 people respond. What is the rate (percent) of nonresponse?
(c) What source of response error is likely for the question asked?

8.33 Online polls. Example 8.3 reports an online poll in which 97% of the respondents opposed issuing driver's licenses to illegal immigrants. National random samples taken at the same time showed about 70% of the respondents opposed to such licenses. Explain briefly to someone who knows no statistics why the random samples report public opinion more reliably than the online poll.

8.34 Nonresponse. Academic sample surveys, unlike commercial polls, often discuss nonresponse. A survey of drivers began by randomly sampling all listed residential telephone numbers in the United States. Of 45,956 calls to these numbers, 5029 were completed.[23] What was the rate of nonresponse for this sample? (Only one call was made to each number. Nonresponse would be lower if more calls were made.)

8.35 Running red lights. The sample described in Exercise 8.34 produced a list of 5024 licensed drivers. The investigators then chose an SRS of 880 of these drivers to answer questions about their driving habits.

(a) How would you assign labels to the 5024 drivers? Choose the first five drivers in the sample. If you use Table B, start at line 114.

(b) One question asked was, "Recalling the last ten traffic lights you drove through, how many of them were red when you entered the intersections?" Of the 880 respondents, 171 admitted that at least one light had been red. A practical problem with this survey is that people may not give truthful answers. What is the likely direction of the bias: do you think more or fewer than 171 of the 880 respondents really ran a red light? Why?

8.36 Seat belt use. A study in El Paso, Texas, looked at seat belt use by drivers. Drivers were observed at randomly chosen convenience stores. After they left their cars, they were invited to answer questions that included questions about seat belt use. In all, 75% said they always used seat belts, yet only 61.5% were wearing seat belts when they pulled into the store parking lots.[24] Explain the reason for the bias observed in responses to the survey. Do you expect bias in the same direction in most surveys about seat belt use?

8.37 Sampling at a party. At a large block party, there are 40 men and 30 women. You want to ask opinions about how to improve the next party. You choose at random four of the men and separately choose at random three of the women to interview.

(a) What is the probability that any of the 40 men is in your random sample of four men to be interviewed? What is the probability that any of the 30 women is in your random sample of three women to be interviewed?

(b) If you have done the calculations correctly in part (a), the probability of any person at the party being interviewed is the same. Why is your sample of seven men and women not an SRS of people from the party?

8.38 Ring-no-answer. A common form of nonresponse in telephone surveys is "ring-no-answer." That is, a call is made to an active number but no one answers. The Italian National Statistical Institute looked at nonresponse to a government survey of households in Italy during the periods January 1 to Easter and July 1 to August 31. All calls were made between 7 P.M. and 10 P.M., but 21.4% gave "ring-no-answer" in one period versus 41.5% "ring-no-answer" in the other period.[25] Which period do you think had the higher rate of no answers? Why? Explain why a high rate of nonresponse makes sample results less reliable.

8.39 Retweeters. Twitter and Compete, a marketing services company, conducted a survey to investigate some of the characteristics of those who retweet (reposting someone else's tweet). Among other findings, it was found that Twitter users who retweet are demographically similar to those who don't, use Twitter more often during the day, and are more likely to use Twitter on a mobile phone. Here is the methodology section contained with the survey results:

> The findings are based on data from surveys fielded in the United States during 2012. Twitter and Compete worked together to build a questionnaire that asked respondents about their propensity to use Twitter and other services as well as the when, where, how and why of their usage patterns. Compete interviewed 655 Internet users in the U.S. for this study.[26]

(a) Explain in simple language why it is important to know how the sample was selected when drawing conclusions about a survey.

(b) Do you feel that the methodology section adequately explains how this sample was selected? If not, what information is lacking and why is it important?

8.40 Sampling pharmacists. All pharmacists in the Canadian province of Ontario are required to be members of the Ontario College of Pharmacists. In 2013, there were 13,317 members of the college divided geographically into six voting districts. The number of members in each district follow:[27]

District	H	K	L	M	N	P
Membership	2013	1602	4316	2771	1928	687

Suppose the college is interested in obtaining members' views and understanding of the 2012 Expanded Scope of Practice Regulations, which authorizes pharmacists to provide additional services including prescribing drug products for smoking cessation and administering the publicly funded influenza vaccine. To be sure that the opinions of all districts are represented, you choose a stratified random sample of five pharmacists from each district. Explain how you will assign labels within each district, and then give the label numbers for the five pharmacists in each of Districts H and K who will be part of your sample. Use software or Table B. If you use Table B, start at line 131 for District H and at line 124 for District K. Why should you not start at the same line in Table B to obtain your samples for Districts H and K?

8.41 Sampling Amazon forests. Stratified samples are widely used to study large areas of forest. Based on satellite images, a forest area in the Amazon basin is divided into 14 types. Foresters studied the four most

age fotostock/SuperStock

commercially valuable types: alluvial climax forests of quality levels 1, 2, and 3, and mature secondary forest. They divided the area of each type into large parcels, chose parcels of each type at random, and counted tree species in a 20- by 25-meter rectangle randomly placed within each parcel selected. Here is some detail:

FOREST TYPE	TOTAL PARCELS	SAMPLE SIZE
Climax 1	36	4
Climax 2	72	7
Climax 3	31	3
Secondary	42	4

Choose the stratified sample of 18 parcels. Be sure to explain how you assigned labels to parcels. If you use Table B, start at line 112.

8.42 Canadian health care survey. The Tenth Annual Health Care in Canada Survey is a survey of the opinions of the Canadian public and health care providers on a variety of health care issues, including quality of health care, access to health care, health and the environment, and so forth. A description of the survey follows:

> The 10th edition of the Health Care in Canada Survey was conducted by POLLARA Research between October 3rd and November 8th, 2007. Results for the survey are based on telephone interviews with nationally representative samples of 1,223 members of the Canadian public, 202 doctors, 201 nurses, 202 pharmacists and 201 health managers. Public results are considered to be accurate within ±2.8%, while the margin of error for results for doctors, nurses, pharmacists and managers is ±6.9%.[28]

(a) Why is the accuracy greater for the public than for health care providers and managers?

(b) Why do you think they sampled the public as well as health care providers and managers?

8.43 Systematic random samples. *Systematic random samples* go through a list of the population at fixed intervals from a randomly chosen starting point. For example, a study of dating among college students chose a systematic sample of 200 single male students at a university as follows.[29] Start with a list of all 9000 single male students. Because 9000/200 = 45, choose one of the first 45 names on the list at random and then every 45th name after that. For example, if the first name chosen is at position 23, the systematic sample consists of the names at positions 23, 68, 113, 158, and so on up to 8978.

(a) Choose a systematic random sample of five names from a list of 200. If you use Table B, enter the table at line 120.

(b) Like an SRS, a systematic sample gives all individuals the same chance to be chosen. Explain why this is true, then explain carefully why a systematic sample is nonetheless *not* an SRS.

8.44 More on systematic sampling. Foresters were interested in studying a remote sensing measure of standing timber as an alternative to taking measurements on the ground. The study area was a 1200 acre pine forest in Louisiana on which the U.S. Forest Service first created a grid of 1410 equally spaced circular plots of 0.05 acre in size over a map of the forest. The ground survey then visited every 10th plot and took measurements of tree volume.[30]

(a) Assuming the plots are numbered from 1 to 1410, using the information in Exercise 8.43, explain how you will select a systematic sample of every 10th plot.

(b) Now choose a systematic random sample of 141 plots for the ground survey. If you use Table B, start at line 125.

8.45 Why random digit dialing is common. The list of individuals from which a sample is actually selected is called the *sampling frame.* Ideally, the frame should list every individual in the population, but in practice this is often difficult. A frame that leaves out part of the population is a common source of undercoverage.

(a) Suppose that a sample of households in a community is selected at random from the telephone directory. What households are omitted from this frame? What types of people do you think are likely to live in these households? These people will probably be underrepresented in the sample.

(b) It is usual in telephone surveys to use random digit dialing equipment that selects the last four digits of a telephone number at random after being given the exchange (the first three digits), as described in Example 8.5. Which of the households that you mentioned in your answer to part (a) will be included in the sampling frame by random digit dialing?

8.46 Alternative energy favored. A Gallup poll finds that, among all Americans, 64% prefer an emphasis on the development of alternative energy production, such as wind and solar power, to an emphasis on production of traditional fossil fuels. This preference is strongest among young Americans, with older age groups each successively favoring it less. According to the survey methods section, "Results for this Gallup poll are based on telephone interviews conducted March 6–9, 2014, with a random sample of 1,048 adults, aged 18 and older, living in all 50 U.S. states and the District of Columbia. Interviews are conducted with respondents on landline telephones and cellular phones. Landline and cellphone numbers are selected using random-digit-dial methods. Landline respondents are chosen at random within each household on the basis of which member had the most recent birthday."[31]

(a) The survey wants the opinion of an individual adult, but a landline phone reaches a household in which several adults may live. In that case, the survey interviewed the adult with the most recent birthday. Why is this preferable to simply interviewing the person who answers the phone?

(b) What is the population that this survey wants to describe? Why do you think it is important to include both landline and cellular phones in your sample?

8.47 Wording survey questions. Comment on each of the following as a potential sample survey question. Is the question sufficiently clear? Is it slanted toward a desired response?

(a) "In light of skyrocketing gasoline prices, we should consider opening up a very small amount of Alaskan wilderness for oil exploration as a way of reducing our dependence on foreign oil. Do you agree or disagree?"

(b) "Do you agree that a national system of health insurance should be favored because it would provide health insurance for everyone and would reduce administrative costs?"

(c) "In view of the negative externalities in parent labor force participation and pediatric evidence associating increased group size with morbidity of children in day care, do you support government subsidies for day care programs?"

8.48 Your own bad questions. Write your own examples of bad sample survey questions.

(a) Write a biased question designed to get one answer rather than another.

(b) Write a question to which many people may not give truthful answers.

8.49 The Canadian census. The Canadian government's decision to eliminate the mandatory long-form version of the census and to move these questions to an optional survey has many concerned. Many members of the business community and economists stressed the importance of the census data for crafting public policy. The minister of industry was given the task of defending the government's decision. In response to an argument that making the long form of the census voluntary would skew the data by eliminating the statistical randomness of the survey, the minister replied: "Wrong. Statisticians can ensure validity with a larger sample size."[32] Is the minister correct? If not, explain in simple terms the error in his statement.

Exploring the Web

8.50 Poor survey designs. The website for the American Association for Public Opinion Research discusses several issues about polls. This information can be found at www.aapor.org. Under Resources & Education, click on the link Poll & Survey FAQ. Click on the link What Information Should Survey Researchers Disclose, and read Section A for recommendations about the information that should be available about how the poll was conducted. After reading Section A, go back to the previous web page and click on the link What Is a Random Sample? and then the link Bad Samples for some examples of flawed samples.

(a) You are going to design a survey at your university. Give a question of interest and two examples of bad ways to collect your sample, along with the likely direction of bias that would result. Explain your answers.

(b) How would you modify your examples in part (a) to produce a better sample? What are some difficulties you might encounter when collecting your sample?

8.51 Find a survey. The website for the Pew Research Center for the People and the Press is www.people-press.org. Go to the website and read one of the featured surveys. What information under Section A of What Information Should Survey Researchers Disclose from Exercise 8.50 is included in the featured survey you have read? You may find the concluding section entitled About the Survey and some of the links at the end of the featured survey helpful for answering this question.

Ambrophoto/Shutterstock

CHAPTER 9

Producing Data: Experiments

In this chapter we cover...

9.1 Observation versus experiment
9.2 Subjects, factors, and treatments
9.3 How to experiment badly
9.4 Randomized comparative experiments
9.5 The logic of randomized comparative experiments
9.6 Cautions about experimentation
9.7 Matched pairs and other block designs

A sample survey aims to gather information about a population without disturbing the population in the process. Sample surveys are one kind of *observational study*. Other observational studies observe the behavior of animals in the wild or the interactions between teacher and students in the classroom. This chapter is about statistical designs for *experiments*, a quite different means of producing data.

9.1 Observation versus experiment

In contrast to observational studies, experiments don't just observe individuals or ask them questions. They actively impose some treatment to observe the response. Experiments can answer questions such as, "Does aspirin reduce the chance of a heart attack?" and "Do a majority of college students prefer Pepsi to Coke when they taste both without knowing which they are drinking?"

YOU JUST DON'T UNDERSTAND

A sample survey of journalists and scientists found quite a communication gap. Journalists think that scientists are arrogant, and scientists think that journalists are ignorant. We won't take sides, but here is one interesting result from the survey: 82% of the scientists agree that the "media do not understand statistics well enough to explain new findings" in medicine and other fields.

Observation Versus Experiment

An **observational study** observes individuals and measures variables of interest, but does not attempt to influence the responses. The purpose of an observational study is to describe some group or situation.

An **experiment**, on the other hand, deliberately imposes some treatment on individuals to observe their responses. The purpose of an experiment is to study whether the treatment causes a change in the response.

An observational study, even one based on a statistical sample, is a poor way to gauge the effect of a treatment. To see the response to a change, we must actually impose the change. *When our goal is to understand cause and effect, experiments are the only source for fully convincing data.* For this reason, the distinction between observation and experiment is one of the most important in statistics.

EXAMPLE 9.1 Does Playing Video Games Improve Surgical Skills?

In laparoscopic surgery, a video camera and several thin instruments are inserted into the patient's abdominal cavity. The surgeon uses the image from the video camera positioned inside the patient's body to perform the procedure by manipulating the instruments that have been inserted. Because of the similarity in many of the skills involved in video games and laparoscopic surgery, it was hypothesized that surgeons with greater prior video game experience might acquire the skills required in laparoscopic surgery more easily. Thirty-three surgeons participated in the study and were classified into the three categories—never used, under three hours per day, and more than three hours per day—depending on the number of hours they played video games at the height of their video game use. It was found that those with more video game playing did better in a simulator program measuring laparoscopic skills. However, in their conclusions the authors correctly point out, "This is a correlational (observational) study and, therefore, causality cannot be definitely determined."[1] Although the data showed a clear association between prior video game experience and improved scores in the simulator program, we cannot conclude that more video game playing caused the improvement. People who play more video games may be different from those who don't, in terms of both their interest and the natural skills that are required in video games. Those who played more video games may have scored better simply because they have more ability in areas such as fine motor skills, eye–hand coordination, and depth perception that are required in both video games and laparoscopic surgery.

It is easy to imagine an experiment that would settle the issue of whether playing video games really causes improvement in laparoscopic skills. Choose half of a group of surgeons at random to be the "treatment" group. The remaining half becomes the "control" group. Require the treatment group to play video games on a regular basis for several weeks and require the control group to abstain from video games. This experiment isolates the effect of playing video games. See Exercise 9.8 for a description of such an experiment. ■

The point of Example 9.1 is the contrast between observing people who chose for themselves how many hours of video games to play, and an experiment that requires some people to play video games and others to abstain. When we simply observe people's video game choices, the effect of choosing to play more video games is confounded with (mixed up with) the characteristics of people who choose to play more. These characteristics are lurking variables (see page 147) that make it hard to see the true relationship between the explanatory and response variables. Figure 9.1 shows the confounding in picture form.

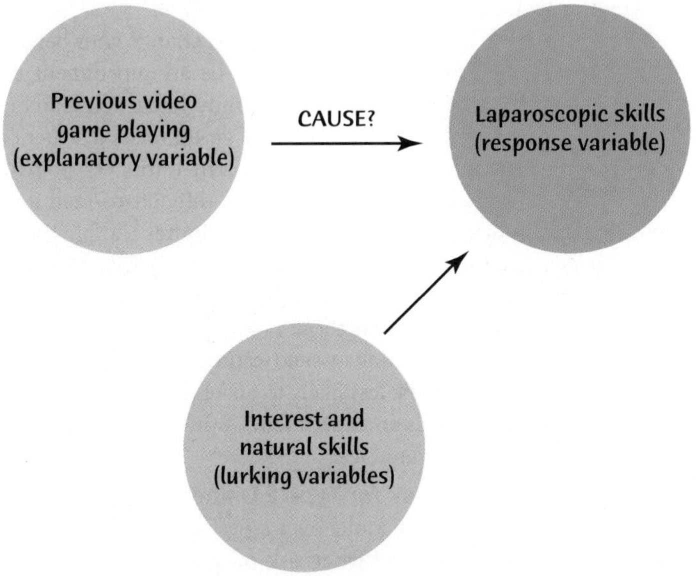

FIGURE 9.1
Confounding: We can't distinguish the effects of previous video game experience from the effects of interest and natural skills.

Confounding

Two variables (explanatory variables or lurking variables) are **confounded** when their effects on a response variable cannot be distinguished from each other.

CAUTION *Observational studies of the effect of one variable on another often fail because the explanatory variable is confounded with lurking variables.* Well-designed experiments take steps to prevent confounding.

LaunchPad Online Resources

- **The Snapshots video, *Types of Studies*, and the StatClips video, *Types of Studies*, both review the important distinctions between observational studies and experiments.**

Apply Your Knowledge

9.1 **Drink a Little, but Not a Lot.** Many observational studies show that people who drink a moderate amount of alcohol have less heart disease than people who drink no alcohol or who drink heavily.[2] ("Moderate" means one or two drinks a day for men and one drink a day for women.)

(a) What are the explanatory and response variables?

(b) Moderate drinkers as a group tend to be more likely to maintain a healthy weight, get enough sleep, and exercise regularly than those who drink no alcohol or drink heavily. Are the variables maintaining a healthy weight, getting enough sleep, and exercising regularly explanatory variables, response variables, or lurking variables? Explain your reason.

(c) Is the *association* between drinking and heart disease good reason to think that moderate drinking actually *causes* less heart disease? Explain why or why not?

9.2 **The Font Matters!** In general, when trying to change your behavior, if the effort required is perceived as high, this will be an impediment to change, whether it is modifying your diet or your study habits. Researchers divide 40 students into two groups of 20. The first group reads instructions for an exercise program printed in an easy-to-read font (Arial, 12 point), and the second group reads identical instructions in a difficult-to-read font (Brush, 12 point). Each subject estimates how many minutes the program will take and also uses a seven-point rating scale to report whether they are likely to include the exercise program as part of their daily routine (7 = very likely). The researchers hypothesize that those reading about the exercise program in the more difficult-to-read font would estimate that the program would take longer and they would be less likely to make the exercise program part of their regular routine.[3] Is this an experiment? Why or why not? What are the explanatory and response variables?

© Paula Solloway/Alamy

9.3 **Quitting Smoking and Risk for Type 2 Diabetes.** Researchers studied a group of 10,892 middle-aged adults over a period of nine years. They found that smokers who quit had a higher risk of diabetes within three years of quitting than either nonsmokers or continuing smokers.[4] Does this show that stopping smoking causes the short-term risk for type 2 diabetes to increase? (Weight gain has been shown to be a major risk factor for developing type 2 diabetes and is often a side effect of quitting smoking.) Based on this research, should you tell a middle-aged adult who smokes that stopping smoking can *cause* diabetes and advise him or her to continue smoking? Carefully explain your answers to both questions.

9.2 Subjects, factors, and treatments

A study is an experiment when we actually do something to people, animals, or objects to observe the response. Because the purpose of an experiment is to reveal the response of one variable to changes in other variables, the distinction between explanatory and response variables is essential. Here is the basic vocabulary of experiments.

Subjects, Factors, and Treatments

The **individuals** studied in an experiment are often called **subjects**, particularly when they are people.

The explanatory variables in an experiment are often called **factors**.

A **treatment** is any specific experimental condition applied to the subjects. If an experiment has more than one factor, a treatment is a combination of specific values of each factor.

EXAMPLE 9.2 Foster Care Versus Orphanages

Do abandoned children placed in foster homes do better than similar children placed in an institution? The Bucharest Early Intervention Project found that the answer is a clear "yes." The *subjects* were 136 young children abandoned at birth and living in orphanages in Bucharest, Romania. Half of the children, chosen at random, were placed in foster homes. The other half remained in the orphanages. The experiment compared these two *treatments*. There is a single *factor*, "type of care," with two values, foster and institutional care. When there is only one factor, the levels or values of the factor correspond to the treatments. The *response variables* included measures of mental and physical development.[5] (Foster care was not easily available in Romania at the time and so was paid for by the study. Exercise 10.22 on page 265 in the Data Ethics chapter for ethical questions concerning this experiment.) ■

EXAMPLE 9.3 Effects of TV Advertising

What are the effects of repeated exposure to an advertising message? The answer may depend on both the length of the ad and how often it is repeated. An experiment investigated this question using undergraduate students as *subjects*. All subjects viewed a 40-minute television program that included ads for a digital camera. Some subjects saw a 30-second commercial; others, a 90-second version. The same commercial was shown either one, three, or five times during the program.

This experiment has two *factors:* length of the commercial, with two values, and repetitions, with three values. The six combinations of one value of each factor form six *treatments.* Figure 9.2 shows the layout of the treatments. After viewing, all the subjects answered questions about their recall of the ad, their attitude toward the camera, and their intention to purchase it. These are the *response variables.* ■

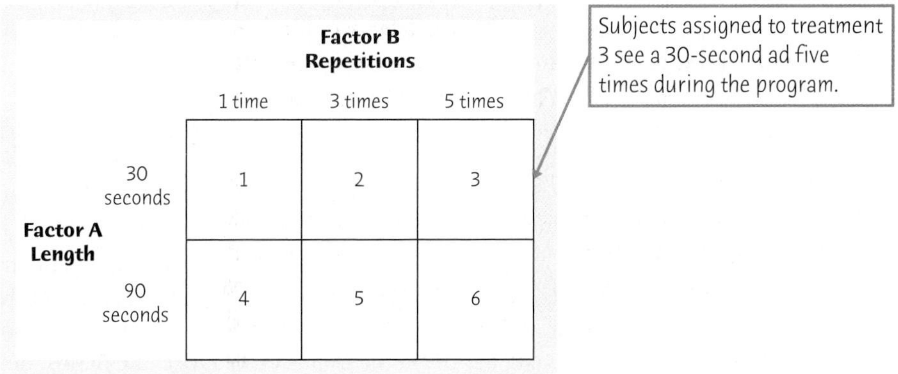

FIGURE 9.2 The treatments in the experimental design of Example 9.3. Combinations of values of the two factors form six treatments.

Examples 9.2 and 9.3 illustrate the advantages of experiments over observational studies. In an experiment, we can study the effects of the specific treatments we are interested in. By assigning subjects to treatments, we can avoid confounding. For example, observational studies of the effects of foster homes versus institutions on the development of children have often been biased because healthier or more alert children tend to be placed in homes. The random assignment in Example 9.2 eliminates bias in placing the children. Moreover, we can control the environment of the subjects to hold constant factors that are of no interest to us, such as the specific product advertised in Example 9.3.

Another advantage of experiments is that we can study the combined effects of several factors simultaneously. The interaction of several factors can produce effects that could not be predicted from looking at the effect of each factor alone. Perhaps longer commercials increase interest in a product, and more commercials also increase interest, but if we both make a commercial longer and show it more often, viewers get annoyed and their interest in the product drops. The two-factor experiment in Example 9.3 will help us find out.

LaunchPad Online Resources

- The **StatBoards video**, *Factors and Treatments,* identifies subjects, factors, treatments, and response variables in additional experiments.

Apply Your Knowledge

For the experiments described in Exercises 9.4 and 9.5, identify the subjects, the factors, the treatments, and the response variables.

9.4 Adolescent Obesity. Adolescent obesity is a serious health risk affecting more than 5 million young people in the United States alone. Laparoscopic adjustable gastric banding has the potential to provide a safe and effective treatment. Fifty adolescents between 14 and 18 years old with a body mass index higher than 35 were recruited from the Melbourne, Australia, community for the study. Twenty-five were randomly selected to undergo gastric banding, and the remaining 25 were assigned to a supervised lifestyle intervention program involving diet, exercise, and behavior modification. All subjects were followed for two years, and their weight loss was recorded.[6]

9.5 Reactions to Simulated News Reports. A sample of University of Colorado students each viewed one of two simulated news reports about a terrorist bombing against the United States by a fictitious country. One report showed the bombing attack on a military target and the other on a cultural/educational site. In addition, before viewing the news report, each student read one of two "primes." The first was a prime for *forgiveness* based on the biblical saying "Love thy enemy," and the second was a *retaliatory* prime based on the biblical saying "An eye for an eye, and a tooth for a tooth." After viewing the news report, the students were asked to rate on a scale of 1 to 12 what the U.S. reaction should be, with the lowest score (1) corresponding to the United States sending a special ambassador to the country and the highest score (12) corresponding to an all-out nuclear attack against the country.[7]

9.6 Ripening Mangoes. The mango is considered the "king of fruits" in many parts of the world. Mangoes are generally harvested at the mature green stage and ripen during the marketing process of transport, storage, and so on. During this process, about 30% of the fruit is wasted. Because of this, the impact of harvest stage and storage conditions on the postharvest quality are of interest. In this experiment, the fruit was harvested at 80, 95, or 110 days after the fruit setting (the transition from flower to fruit) and then stored at temperatures of 20, 30, or 40 degrees centigrade. For each harvest time and storage temperature, a random sample of mangoes was selected, and the time to ripening was measured.[8]

(a) What are the factors, the treatments, and the response variables? Use a diagram like Figure 9.2 to display the factors and treatments.

(b) For simplicity, the experimenter thought about selecting nine mango trees and randomly assigning one tree to each of the nine treatments. All the mangoes on a tree would then receive the same treatment. Do you think this would be a good way to assign the mangoes to the treatments? Explain your reasoning in simple English.

9.3 How to experiment badly

Experiments are the preferred method for examining the effect of one variable on another. By imposing the specific treatment of interest and controlling other influences, we can pin down cause and effect. Statistical designs are often essential for effective experiments. To see why, let's look at an example in which an experiment suffers from confounding just as observational studies do.

EXAMPLE 9.4 An Uncontrolled Experiment

A college regularly offers a review course to prepare candidates for the Graduate Management Admission Test (GMAT), which is required by most graduate business schools. This year, it offers only an online version of the course. The average GMAT score of students in the online course is 10% higher than the longtime average for those who took the classroom review course. Is the online course more effective?

This experiment has a very simple design. A group of subjects (the students) were exposed to a treatment (the online course), and the outcome (GMAT scores) was observed. Here is the design:

$$\text{Subjects} \longrightarrow \text{Online course} \longrightarrow \text{GMAT scores}$$

A closer look at the GMAT review course showed that the students in the online review course were quite different from the students who in past years took the classroom course. In particular, they were older and more likely to be employed. An online course appeals to these mature people, but we can't compare their performance with that of the undergraduates who previously dominated the course. The online course might even be less effective than the classroom version. The effect of online versus in-class instruction is confounded with the effect of lurking variables. As a result of confounding, the experiment is biased in favor of the online course.

Would the situation have been different if both the online and the classroom courses had been given this year? If students still chose the course they wanted, with older students tending to sign up for the online course and younger students tending to sign up for the classroom course, then the effect of type of course would still be confounded with the lurking variable age. The solution will be described in the next section. ■

Most laboratory experiments use a design like that in Example 9.4:

$$\text{Subjects} \longrightarrow \text{Treatment} \longrightarrow \text{Measure response}$$

CAUTION

In the controlled environment of the laboratory, simple designs often work well. Field experiments and experiments with living subjects are exposed to more variable conditions and deal with more variable subjects. *Outside the laboratory, uncontrolled experiments often yield worthless results because of confounding with lurking variables.*

Apply Your Knowledge

9.7 **Reducing Unemployment.** Will cash bonuses speed the return to work of unemployed people? A state department of labor notes that, last year, 41% of people who filed claims for unemployment insurance found a new job within 15 weeks. As an experiment, the state offers $500 to people filing unemployment claims if they find a job within 15 weeks. The percent who do so increases to 53%. Suggest some conditions that might make it easier or harder to find a job this year as opposed to last year. Confounding with these lurking variables makes it impossible to say whether the $500 bonus really caused the increase.

9.4 Randomized comparative experiments

The remedy for the confounding in Example 9.4 is to be sure that we do a *comparative experiment* in which some students are taught in the classroom and other, similar students take the course online. In Example 9.4, the classroom group is called a **control group**. A control group receives either a "standard" treatment or in some cases a "sham" treatment, and provides a basis for comparison with the other treatment groups. Although in many experiments one of the treatments is a control treatment, a comparative experiment does not require this. In Example 9.4, a comparative experiment could simply compare two newly developed online courses without including a control or classroom treatment. Most well-designed experiments compare two or more treatments. Part of the design of an experiment is a description of the factors (explanatory variables) and the layout of the treatments, with comparison as the leading principle.

control group

However, as discussed at the end of Example 9.4, comparison alone isn't enough to produce results we can trust. If the treatments are given to groups that differ markedly when the experiment begins, bias will result. If we allow students to select online or classroom instruction, students who are older and employed are likely to sign up for the online course. Personal choice will bias our results in the same way that volunteers bias the results of online opinion polls. The solution to the problem of bias in sampling is random selection, and the same is true in experiments. The subjects assigned to any treatment should be chosen at random from the available subjects.

Randomized Comparative Experiment

An experiment that uses both comparison of two or more treatments and random assignment of subjects to treatments is a **randomized comparative experiment**.

EXAMPLE 9.5 Classroom Versus Online

The college decides to compare the progress of 25 on-campus students taught in the classroom with that of 25 students taught the same material online. Select the students who will be taught online by taking a simple random sample of size 25 from the 50 available subjects. The remaining 25 students form the control group. They will receive classroom instruction. The result is a randomized comparative experiment with two groups. Figure 9.3 outlines the design in graphical form.

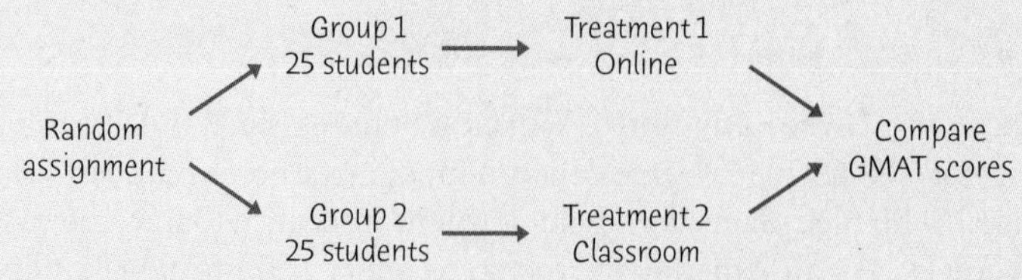

FIGURE 9.3
Outline of a randomized comparative experiment to compare online and classroom instruction, for Example 9.5.

The selection procedure is exactly the same as it is for sampling.

Label: Label the 50 students 01 to 50.

Table: Go to the table of random digits and read successive two-digit groups. The first 25 labels encountered select the online group. As usual, ignore repeated labels and groups of digits not used as labels. For example, if you begin at line 125 in Table B, the first five students chosen are those labeled 21, 49, 37, 18, and 44. Software such as the *Simple Random Sample* applet makes it particularly easy to choose treatment groups at random. ■

APPLET

The design in Example 9.5 is *comparative* because it compares two treatments (the two instructional settings). It is *randomized* because the subjects are assigned to the treatments by chance. This "flowchart" outline in Figure 9.3 presents all the essentials: randomization, the sizes of the groups and which treatment they receive, and the response variable. There are, as we will see later, statistical reasons for generally using treatment groups that are about equal in size. We call designs like that in Figure 9.3 *completely randomized.*

Completely Randomized Design

In a **completely randomized** experimental design, all the subjects are allocated at random among all the treatments.

Completely randomized designs can compare any number of treatments. Here is an example that compares three treatments.

EXAMPLE 9.6 Conserving Energy

Many utility companies have introduced programs to encourage energy conservation among their customers. An electric company considers placing small digital displays in households to show current electricity use and what the cost would be if this use continued for a month. Will the displays reduce electricity use? Would cheaper methods work almost as well? The company decides to conduct an experiment.

One cheaper approach is to give customers a chart and information about monitoring their electricity use from their outside meter. The experiment compares these two approaches (display, chart) and also a control. The control group of customers receives information about energy conservation but no help in monitoring electricity use. The response variable is total electricity used in a year. The company finds 60 single-family residences in the same city willing to participate, so it assigns 20 residences at random to each of the three treatments. Figure 9.4 outlines the design.

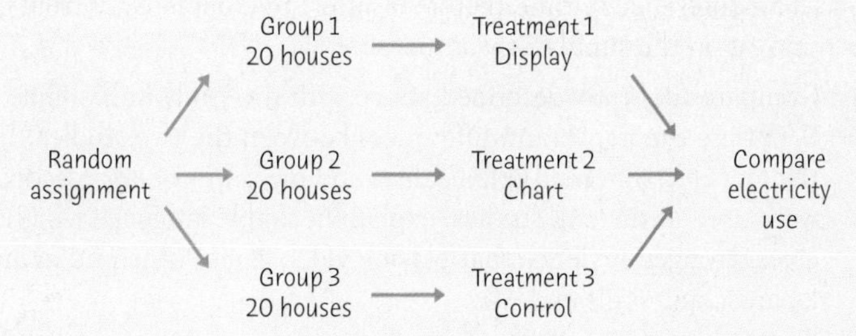

FIGURE 9.4
Outline of a completely randomized design comparing three energy-saving programs, for Example 9.6.

APPLET

To use the *Simple Random Sample* applet, set the population labels as 1 to 60 and the sample size to 20 and click "Sample." The 20 households chosen receive the displays. The "Population hopper" now contains the 40 remaining households, in scrambled order. Click "Sample" again to choose 20 of these to receive charts. The 20 households remaining in the "Population hopper" form the control group.

To use Table B, label the 60 households 01 to 60. Enter the table to select an SRS of 20 to receive the displays. Continue in Table B, selecting 20 more to receive charts. The remaining 20 form the control group. ■

Examples 9.5 and 9.6 describe completely randomized designs that compare values of a single factor. In Example 9.5, the factor is the type of instruction. In Example 9.6, it is the method used to encourage energy conservation. Completely randomized designs can have more than one factor. The advertising experiment of Example 9.3 has two factors: the length and the number of repetitions of a television commercial. Their combinations form the six treatments outlined in Figure 9.2. A completely randomized design assigns subjects at random to these six treatments. Once the layout of treatments is set, the randomization needed for a completely randomized design is tedious but straightforward.

LaunchPad Online Resources

- **The StatBoards video, *Outlining an Experiment,* provides additional examples of outlining the design of an experiment using figures similar to those given in this section.**

Apply Your Knowledge

9.8 **Does Playing Video Games Improve Surgical Skills? Another Look.** In laparoscopic surgery, a video camera and several thin instruments are inserted into the patient's abdominal cavity. The surgeon uses the image from the video camera positioned inside the patient's body to perform the procedure by manipulating the instruments that have been inserted. It has been found that the Nintendo Wii™, with its motion-sensing interface, reproduces the movements required in laparoscopic surgery more closely than other video games. If training with a Nintendo Wii™ can improve laparoscopic skills, it can complement the more expensive training on a laparoscopic simulator. Forty-two medical residents were chosen and all were tested on a set of basic laparoscopic skills. Twenty-one were selected at random to undergo systematic Nintendo Wii™ training for one hour a day, five days a week, for four weeks. The remaining 21 residents were given no Nintendo Wii™ training and asked to refrain from video games during this period. At the end of four weeks, all 42 residents were tested again on the same set of laparoscopic skills. The difference in time (before − after) to complete a virtual gall bladder removal on the simulator was measured.[9]

(a) Compare the study described above with the study in Example 9.1. What are the important differences between the two studies? Ignoring the fact that different video games and measures of laparoscopic skill were used in the two studies, explain in simple language which study gives stronger evidence that playing video games is helpful in improving laparoscopic skills and why.

(b) Outline the design of this experiment, following the model of Figure 9.3. What is the response variable?

(c) Carry out the random assignment of 21 residents to the Nintendo training group, using the *Simple Random Sample* applet, other software, or Table B, starting at line 130.

APPLET

9.9 More Rain for California? The changing climate will probably bring more rain to California, but we don't know whether the additional rain will come during the winter wet season or extend into the long dry season in spring and summer. Kenwyn Suttle of the University of California at Berkeley and his co-workers carried out a randomized controlled experiment to study the effects of more rain in either season. They randomly assigned plots of open grassland to three treatments: added water equal to 20% of annual rainfall either during January to March (winter) or during April to June (spring), and no added water (control). Thirty-six circular plots of area 70 square meters were available (see the photo), of which 18 were used for this study. One response variable was total plant biomass, in grams per square meter, produced in a plot over a year.[10]

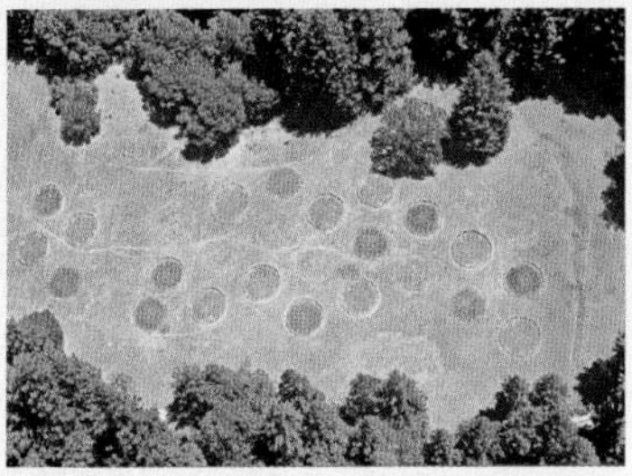

Suttle, K. B., M. A. Thomsen, and M. E. Power. "Species Interactions Reverse Grassland Responses to Changing Climate." Science 315.5812 (2007): 640–42.

(a) Outline the design of the experiment, following the model of Figure 9.4.

(b) Number all 36 plots and choose six at random for each of the three treatments. Be sure to explain how you did the random selection.

9.10 Effects of TV Advertising. Figure 9.2 displays the six treatments for the two-factor experiment on TV advertising described in Example 9.3. The 24 students named here will serve as subjects. Outline the design and randomly assign the subjects to the six treatments, an equal number of subjects to each treatment. If you use Table B, start at line 135.

Abramson	Weingold	Cohen	Cressie	Linder	Stanley
Anthony	Blake	Santner	Kessis	Minor	Tory
Pearl	Brower	Delp	Lahr	Carson	Verducci
Baker	Knab	Disbro	Kruger	Shi	Walsh

9.5 The logic of randomized comparative experiments

APPLET

Randomized comparative experiments are designed to give good evidence that differences in the treatments actually *cause* the differences we see in the response. The logic is as follows:

- Random assignment of subjects forms groups that should be similar in all respects before the treatments are applied. Exercise 9.52 uses the *Simple Random Sample* applet to demonstrate this.
- A comparative experiment with randomization ensures that influences other than the experimental treatments operate equally on all groups.
- Therefore, differences in average response must be due either to the treatments or to the play of chance in the random assignment of subjects to the treatments.

That "either–or" deserves more thought. In Example 9.5, we cannot say that *any* difference between the average GMAT scores of students enrolled online and in the classroom must be caused by a difference in the effectiveness of the two types of instruction. There would be some difference even if both groups received the same instruction, because of variation among students in background and study habits. Chance assigns students to one group or the other, and this creates a chance difference between the groups. We would not trust an experiment with just one student in each group, for example. The results would depend too much on which group got lucky and received the stronger student. If we replicate the experiment by assigning

Statistics in Your World

WHAT'S NEWS?

Randomized comparative experiments provide the best evidence for medical advances. Do newspapers care? Maybe not. University researchers looked at 1192 articles in medical journals, of which 7% were turned into stories by the two newspapers examined. Of the journal articles, 37% concerned observational studies, and 25% described randomized experiments. Among the articles publicized by the newspapers, 58% were observational studies, and only 6% were randomized experiments. Conclusion: the newspapers want exciting stories, especially bad-news stories, whether or not the evidence is good.

many subjects to each group, the effects of chance will average out, and there will be little difference in the average responses in the two groups unless the treatments themselves cause a difference. **Replication**, or the use of enough subjects to reduce chance variation, is the third big idea of statistical design of experiments.

replication

Principles of Experimental Design

The basic principles of statistical design of experiments are

1. **Control**—restrict the effects of lurking variables on the response, most simply by comparing two or more treatments.
2. **Randomization**—use chance to assign subjects to treatments.
3. **Replication**—use enough subjects in each group to reduce chance variation in the results.

We hope to see a difference in the responses between the two groups so large that it is unlikely to happen just because of chance variation. We can use the laws of probability, which describe chance behavior, to learn if the treatment effects are larger than we would expect to see if only chance were operating. If they are, we call them *statistically significant.*

Statistical Significance

An observed effect so large that it would rarely occur by chance is called **statistically significant.**

If we observe statistically significant differences among the groups in a randomized comparative experiment, we have good evidence that the treatments actually caused these differences. You will often see the phrase "statistically significant" in reports of investigations in many fields of study. The great advantage of randomized comparative experiments is that they can produce data that give good evidence for a cause-and-effect relationship between the explanatory and response variables. We know that in general a strong association does not imply causation. A statistically significant association in data from a well-designed experiment *does* imply causation.

LaunchPad Online Resources

- The **Snapshots video**, *Experimental Design*, reviews the important ideas of randomized controlled experiments in the context of a study with five treatments.

Apply Your Knowledge

9.11 Prayer and Meditation. You read in a magazine that "nonphysical treatments such as meditation and prayer have been shown to be effective in controlled scientific studies for such ailments as high blood pressure, insomnia, ulcers, and asthma." Explain in simple language what the article means by "controlled scientific studies." Why can such studies in principle provide good evidence that, for example, meditation is an effective treatment for high blood pressure?

9.12 Conserving Energy. Example 9.6 describes an experiment to learn whether providing households with digital displays or charts will reduce their electricity consumption. An executive of the electric company objects to including a control group. He says: "It would be simpler to just compare electricity use last year (before the display or chart was provided) with consumption in the same period this year. If households use less electricity this year, the display or chart must be working." Explain clearly why this design is inferior to that in Example 9.6.

9.13 Healthy Diet and Cataracts. The relationship between healthy diet and prevalence of cataracts was assessed using a sample of 1808 participants from the Women's Health Initiative Observational Study. Having a high Healthy Eating Index score was the strongest predictor of a reduced risk of cataracts, among modifiable behaviors considered. The Healthy Eating Index score was created by the U.S. Department of Agriculture and measures how well a person's diet conforms to recommended healthy eating patterns. The report concludes: "These data add to the body of evidence suggesting that eating foods rich in a variety of vitamins and minerals may contribute to postponing the occurrence of the most common type of cataract in the United States."[11]

(a) Explain why this is an observational study rather than an experiment.

(b) Although the result was statistically significant, the authors did not use strong language in stating their conclusions, instead using words such as "suggesting" and "may." Do you think that their language is appropriate given the nature of the study? Why?

9.6 Cautions about experimentation

The logic of a randomized comparative experiment depends on our ability to treat all the subjects identically in every way except for the actual treatments being compared. Good experiments therefore require careful attention to details to ensure that all subjects really are treated identically.

If some subjects in a medical experiment take a pill each day and a control group takes no pill, the subjects are not treated identically. Many medical experiments are therefore "placebo controlled." A study of the effects of taking vitamin E on heart disease is a good example. All the subjects receive the same medical attention during the several years of the experiment. All take a pill every day: vitamin E in the treatment group, and a placebo in the control group. A **placebo** is a dummy treatment that is as similar to the treatment as possible but contains no "active ingredient." In this experiment, the placebo would be a pill that looked like the vitamin E pill, yet contained no active ingredient. As a second example, a study compared arthroscopic surgery versus no surgery on several recovery outcomes for a partial meniscus tear.[12] Patients randomly assigned to the "no surgery" group received a sham surgery in which the surgeon asked for all instruments, manipulated the knee as in surgery and kept the sham surgery sufficiently realistic so that afterwards patients were unaware of whether they received the actual surgery or the sham surgery.

placebo

Many patients respond favorably to any treatment, even a placebo, perhaps because they trust the doctor or they believe that the treatment will work. The favorable response to a placebo treatment or a treatment with no therapeutic value is called the *placebo effect*. If the control group did not take any pills, the effect of vitamin E in the treatment group would be confounded with the placebo effect, the effect of simply taking pills. That is, if the vitamin E group improved, we wouldn't know whether it was due to simply taking a pill, or the actual vitamin E content of the pill. Having a placebo group for comparison allows us to see if the effect of

Statistics in Your World — SCRATCH MY FURRY EARS

Rats and rabbits, specially bred to be uniform in their inherited characteristics, are the subjects in many experiments. Animals, like people, are quite sensitive to how they are treated. This can create opportunities for hidden bias. For example, human affection can change the cholesterol level of rabbits. Choose some rabbits at random and regularly remove them from their cages to have their heads scratched by friendly people. Leave other rabbits unloved. All the rabbits eat the same diet, but the rabbits that receive affection have lower cholesterol.

taking a vitamin E pill in the treatment group is larger than the effect of simply taking a pill in the placebo group.

In addition, such studies are usually *double-blind.* The subjects don't know whether they are taking vitamin E or a placebo. Neither do the medical personnel who work with them. The double-blind method avoids unconscious bias by, for example, a doctor who is convinced that a vitamin must be better than a placebo. In many medical studies, only the statistician who does the randomization knows which treatment each patient is receiving.

Double-Blind Experiments

In a **double-blind** experiment, neither the subjects nor the people who interact with them know which treatment each subject is receiving.

When testing a drug against a placebo, we indicated that the placebo contains no "active ingredient," but the situation can be more complex than this would suggest. Since many drugs being tested have known side effects, such as dry mouth, an "active placebo," may be used. An active placebo contains ingredients designed to mimic the side effects of the drug, but does not treat the particular disease.[13] This is important in blinding as it can prevent the subject from being aware of whether they received the drug or the placebo. Drug companies create these placebo drugs and many do not provide information regarding the contents of an active placebo. Unfortunately, this then allows drug companies to make advertising claims such as that the occurrence of dry mouth was similar in the drug and the placebo group.

lack of realism

Placebo controls and the double-blind method are more ways to eliminate possible confounding. But even well-designed experiments often face another problem: **lack of realism.** Practical constraints may mean that the subjects or treatments or setting of an experiment don't realistically duplicate the conditions we really want to study. Here are two examples.

EXAMPLE 9.7 Response to Advertising

The study of television advertising in Example 9.3 showed a 40-minute video to students who knew an experiment was going on. We can't be sure that the results apply to everyday television viewers. Many behavioral science experiments use as subjects students or other volunteers who know they are subjects in an experiment. That's not a realistic setting. ■

EXAMPLE 9.8 Center Brake Lights

© Corbis

Do those high center brake lights, required on all cars sold in the United States since 1986, really reduce rear-end collisions? Randomized comparative experiments with fleets of rental and business cars, done before the lights were required, showed that the third brake light reduced rear-end collisions by as much as 50%. Alas, requiring the third light in all cars led to only a 5% drop.

What happened? Most cars did not have the extra brake light when the experiments were carried out, so it caught the eye of following drivers. Now that almost all cars have the third light, they no longer capture attention. ■

CAUTION Lack of realism can limit our ability to apply the conclusions of an experiment to the settings of greatest interest. Most experimenters want to generalize their conclusions to some setting wider than that of the actual experiment. *Statistical analysis of an experiment cannot tell us how far the results will generalize.* Nonetheless, the randomized comparative experiment, because of its ability to give convincing evidence for causation, is one of the most important ideas in statistics.

Apply Your Knowledge

9.14 Testosterone for Older Men. As men age, their testosterone levels gradually decrease. This may cause a reduction in lean body mass, an increase in fat, and other undesirable changes. Do testosterone supplements reverse some of these effects? A study in the Netherlands assigned 237 men aged 60 to 80 with low or low-normal testosterone levels to either a testosterone supplement or a placebo. The report in the *Journal of the American Medical Association* described the study as a "double-blind, randomized, placebo-controlled trial."[14] Explain each of these terms to someone who knows no statistics.

9.15 Does Meditation Reduce Anxiety? An experiment that claimed to show that meditation reduces anxiety proceeded as follows. The experimenter interviewed the subjects and rated their level of anxiety. Then the subjects were randomly assigned to two groups. The experimenter taught one group how to meditate, and they meditated daily for a month. The other group was simply told to relax more. At the end of the month, the experimenter interviewed all the subjects again and rated their anxiety level. The meditation group now had less anxiety. Psychologists said that the results were suspect because the ratings were not blind. Explain what this means and how lack of blindness could bias the reported results.

9.7 Matched pairs and other block designs

Completely randomized designs are the simplest statistical designs for experiments. They illustrate clearly the principles of control, randomization, and adequate number of subjects. However, completely randomized designs are often inferior to more elaborate statistical designs. In particular, matching the subjects in various ways can produce more precise results than simple randomization.

matched pairs design

One common design that combines matching with randomization is the **matched pairs design**. A matched pairs design compares just two treatments. Choose pairs of subjects that are as closely matched as possible. Use chance to decide which subject in a pair gets the first treatment. The other subject in that pair gets the other treatment. That is, the random assignment of subjects to treatments is done within each matched pair, not for all subjects at once. Sometimes each "pair" in a matched pairs design consists of just one subject, who gets both treatments one after the other. Each subject serves as his or her own control. The *order* of the treatments can influence the subject's response, so we randomize the order for each subject.

EXAMPLE 9.9 Cell Phones and Driving

© Tom Grill/Corbis

Does talking on a hands-free cell phone distract drivers? Undergraduate students "drove" in a high-fidelity driving simulator equipped with a hands-free cell phone. The car ahead brakes: how quickly does the subject react? Let's compare two designs for this experiment. There are 40 student subjects available.

In design 1, the *completely randomized design*, all 40 subjects are assigned at random: 20 simply to drive, and the other 20 to talk on the cell phone while driving. In design 2, the *matched pairs design* that was actually used, all subjects drive both with and without using the cell phone. The two drives are on separate days to reduce carryover effects. The *order* of the two treatments is assigned at random: 20 subjects are chosen to drive first with the phone, and the remaining 20 drive first without the phone.[15]

Some subjects naturally react faster than others. The completely randomized design relies on chance to distribute the faster subjects roughly evenly between the two groups. The matched pairs design compares each subject's reaction time with and without the cell phone. This makes it easier to see the effects of using the phone. ■

Matched pairs designs use the principles of comparison of treatments and randomization. However, the randomization is not complete—we do not randomly assign all the subjects at once to the two treatments. Instead, we randomize only within each matched pair. This allows matching to reduce the effect of variation among the subjects. Matched pairs are one kind of *block design*, with each pair forming a *block*.

Block Design

A **block** is a group of individuals that are known before the experiment to be similar in some way that is expected to affect the response to the treatments.

In a **block design**, the random assignment of individuals to treatments is carried out separately within each block.

A block design combines the idea of creating equivalent treatment groups by matching with the principle of forming treatment groups at random. Blocks are another form of *control*. They control the effects of some outside variables by bringing those variables into the experiment to form the blocks. Here are some typical examples of block designs.

EXAMPLE 9.10 Men, Women, and Advertising

Women and men respond differently to advertising. An experiment to compare the effectiveness of three advertisements for the same product will want to look separately at the reactions of men and women, as well as to assess the overall response to the ads.

A *completely randomized design* considers all subjects, both men and women, as a single pool. The randomization assigns subjects to three treatment groups without regard to their sex. This ignores the differences between men and women. A *block design* considers women and men separately. Randomly assign the women to three groups, with one to view each advertisement. Then separately assign the men at random to three groups. Figure 9.5 outlines this improved design. ■

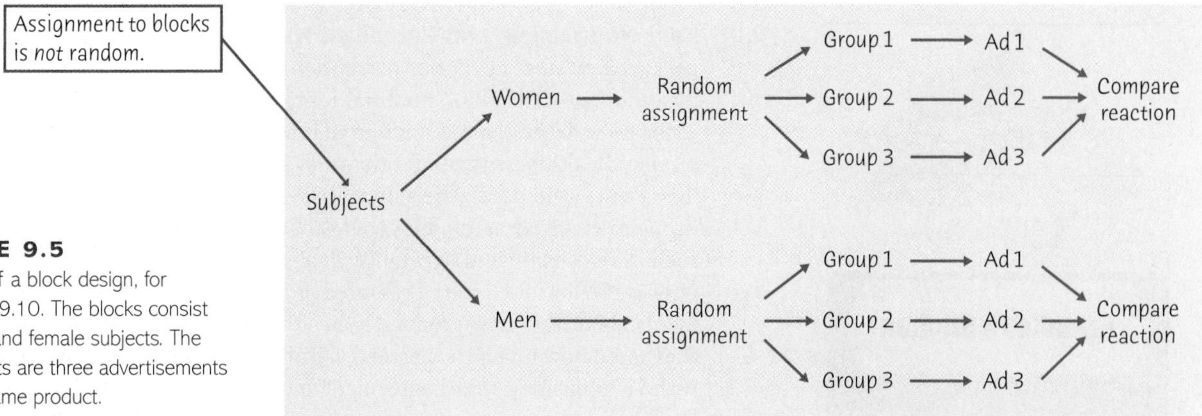

FIGURE 9.5
Outline of a block design, for Example 9.10. The blocks consist of male and female subjects. The treatments are three advertisements for the same product.

EXAMPLE 9.11 Comparing Welfare Policies

A social policy experiment will assess the effect on family income of several proposed new welfare systems and compare them with the present welfare system. Because the future income of a family is strongly related to its present income, the families who agree to participate are divided into blocks of similar income levels. The families in each block are then allocated at random among the welfare systems. ■

A block design allows us to draw separate conclusions about each block, for example, about men and women in Example 9.10. Blocking also allows more precise overall conclusions because the systematic differences between men and women can be removed when we study the overall effects of the three advertisements. The idea of blocking is an important additional principle of statistical design of experiments. A wise experimenter will form blocks based on the most important unavoidable sources of variability among the subjects. Randomization will then average out the effects of the remaining variation and allow an unbiased comparison of the treatments.

Like the design of samples, the design of complex experiments is a job for experts. Now that we have seen a bit of what is involved, we will concentrate for the most part on completely randomized experiments.

Apply Your Knowledge

9.16 Comparing Breathing Frequencies in Swimming. Researchers from the United Kingdom studied the effect of two breathing frequencies on both performance times and several physiological parameters in front crawl swimming.[16] The breathing frequencies were one breath every second stroke (B2) and one breath every fourth stroke (B4). Subjects were 10 male collegiate swimmers. Each subject swam 200 meters, once with breathing frequency B2 and once on a different day with breathing frequency B4.

(a) Describe the design of this matched pairs experiment, including the randomization required by this design.

(b) Could this experiment be conducted using a completely randomized design? How would the design differ from the matched pairs experiment?

(c) Suppose we allow each swimmer to choose their own breathing frequency and then swim 200 meters using their selected frequency. Are there any problems with then comparing the performance of the two breathing frequencies?

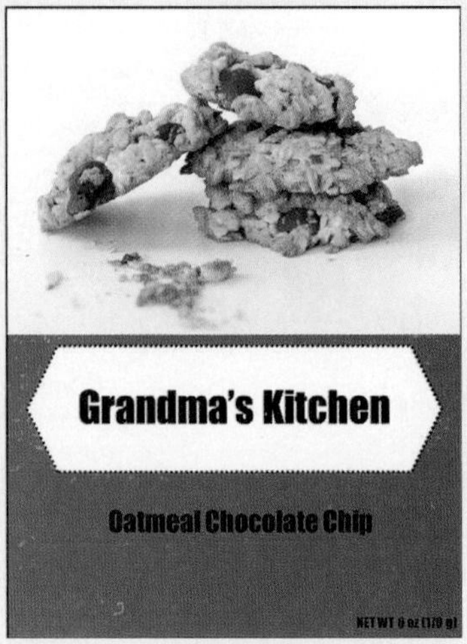

Alysha Fligner

Alysha Fligner

9.17 Font Naturalness and Perceived Healthiness. Can the font used in a packaged product affect our perception of the product's healthiness? It was hypothesized that use of a natural font, which looks more handwritten and tends to be more slanted and curved, would lead to a higher perception of product healthiness than an unnatural font. Two fonts, Impact and Sketchflow Print, were used. These fonts were shown in a previous study to differ in their perceived naturalness, but otherwise were rated similarly on factors such as readability and likeability. Images of two identical packages differing only in the font used were presented to the subjects. Participants read statements such as, "This product is healthy"/natural/wholesome/organic, and they rated how much they agreed with the statement on a seven-point scale with "1" indicating strong agreement and "7" indicating strong disagreement. Each subject's responses were combined to create a perceived healthiness score.[17] The researcher had 100 students available to serve as subjects.

(a) What are the response and the explanatory variables?

(b) Describe the design of a completely randomized experiment to learn the effect of font naturalness on perceived healthiness.

(c) Describe the design of a matched pairs experiment using the same 100 subjects.

9.18 Technology for Teaching Statistics. The Brigham Young University statistics department is performing randomized comparative experiments to compare teaching methods. Response variables include students' final-exam scores and a measure of their attitude toward statistics. One study compares two levels of technology for large lectures: standard (overhead projectors and chalk) and multimedia. The individuals in the study are the eight lectures in a basic statistics course. There are four instructors, each of whom teaches two lectures. Because the lecturers differ, their lectures form four blocks.[18] Suppose the lectures and lecturers are as follows:

Lecture	Lecturer	Lecture	Lecturer
1	Hilton	5	Tolley
2	Christensen	6	Hilton
3	Hadfield	7	Tolley
4	Hadfield	8	Christensen

Outline a block design and do the randomization that your design requires.

CHAPTER 9 SUMMARY

Chapter Specifics

- We can produce data intended to answer specific questions by **observational studies** or **experiments**. Sample surveys that select a part of a population of interest to represent the whole are one type of observational study. **Experiments**, unlike observational studies, actively impose some treatment on the subjects of the experiment.
- Variables are **confounded** when their effects on a response can't be distinguished from each other. Observational studies and uncontrolled experiments often fail to show that changes in an explanatory variable actually cause changes in a response variable because the explanatory variable is confounded with lurking variables.
- In an experiment, we impose one or more **treatments** on individuals, often called **subjects**. Each treatment is a combination of values of the explanatory variables, which we call **factors**.
- The **design** of an experiment describes the choice of treatments and the manner in which the subjects are assigned to the treatments. The basic principles of statistical

design of experiments are **control** and **randomization** to combat bias and **replication** (using enough subjects) to reduce chance variation.

- The simplest form of control is **comparison**. Experiments should compare two or more treatments to avoid confounding of the effect of a treatment with other influences, such as lurking variables.
- **Randomization** uses chance to assign subjects to the treatments. Randomization creates treatment groups that are similar (except for chance variation) before the treatments are applied. Randomization and comparison together prevent **bias**, or systematic favoritism, in experiments.
- You can carry out randomization by using software or by giving numerical labels to the subjects and using a **table of random digits** to choose treatment groups.
- Applying each treatment to many subjects reduces the role of chance variation and makes the experiment more sensitive to differences among the treatments.
- Good experiments also require attention to detail as well as good statistical design. Many behavioral and medical experiments are **double-blind**. Some give a **placebo** to a control group. **Lack of realism** in an experiment can prevent us from generalizing its results.
- In addition to comparison, a second form of control is to restrict randomization by forming **blocks** of individuals that are similar in some way that is important to the response. Randomization is then carried out separately within each block.
- **Matched pairs** are a common form of blocking for comparing just two treatments. In some matched pairs designs, each subject receives both treatments in a random order. In others, the subjects are matched in pairs as closely as possible, and each subject in a pair receives one of the treatments.

Link It

Observational studies and experiments are two methods for producing data. Observational studies are useful when the conclusion involves describing a group or situation without disturbing the scene we observe. Sample surveys, discussed in Chapter 8, are an important type of observational study in which we draw conclusions about a population by observing only a part of the population (the sample). In contrast, experiments are used when the situation calls for a conclusion about whether a treatment *causes* a change in a response. The distinction between observational studies and experiments will be important when stating your conclusions in later chapters.

Only well-designed experiments provide a sound basis for concluding cause-and-effect relationships. In a simple comparative experiment, two treatments are imposed on two groups of individuals. Reaching the conclusion that the difference between the groups is caused by the treatments, rather than lurking variables, requires that the two groups of individuals be similar at the outset. A randomized comparative experiment is used to create groups that are similar. If there is a sufficiently large difference between the groups after imposing the treatments, we can say that the results are statistically significant and conclude that the differences in the response were *caused* by the treatments. In later chapters, the specific statistical procedures for reaching these conclusions will be described.

As with sampling, when conducting an experiment, attention to detail is important because our conclusions can be weakened by several factors. A lack of blinding can result in the expectations of the researcher influencing the results, while the placebo effect can confound the comparison between a treatment and a control group. In many instances, a more complex design is required to overcome difficulties and can produce more precise results.

LaunchPad Online Resources

If you are having difficulty with any of the sections of this chapter, these online resources should help prepare you to solve the exercises at the end of this chapter.

- **StatTutor** starts with a video review of each section and asks a series of questions to check your understanding.
- **LearningCurve** provides you with a series of questions about the chapter geared to your level of understanding.

CHECK YOUR SKILLS

9.19 A large representative random sample of 6906 U.S. adults collected over 20 years showed that "parents reported higher levels of life satisfaction than nonparents," with the observed difference in life satisfaction between the two groups being statistically significant.[19] This is an example of

(a) an observational study.
(b) a randomized comparative experiment.
(c) a matched pairs experiment.

9.20 In the study described in Exercise 9.19, we can conclude that

(a) having children leads to higher levels of life satisfaction. We can reach this conclusion because we have a representative sample.
(b) having children leads to higher levels of life satisfaction. We can reach this conclusion because we have both a large and a representative sample.
(c) parents tend to have higher satisfaction in their lives than nonparents.

9.21 What electrical changes occur in muscles as they get tired? Student subjects hold their arms above their shoulders until they have to drop them. Meanwhile, the electrical activity in their arm muscles is measured. This is

(a) an observational study.
(b) an uncontrolled experiment.
(c) a randomized comparative experiment.

9.22 Do violence and sex in television programs help sell products in advertisements? Subjects were randomly assigned to watch one of four types of TV shows: (1) neither sex nor violence in the content code, (2) violence but no sex in the content code, (3) sex but no violence in the content code, and (4) both sex and violence in the content code. For each TV show, the original advertisements were replaced with the same set of 12 advertisements. Subjects were not told the purpose of the study, but were instead told that the researchers were studying attitudes toward TV shows. After viewing the show, subjects received a surprise memory test to check their recall of the products advertised.[20] This experiment has

(a) four factors, the four TV shows being compared.
(b) 12 factors, the advertisements being shown.
(c) two factors, with/without violent content and with/without sexual content.

9.23 In the experiment of Exercise 9.22, the 336 subjects are labeled 001 to 336. Labels are selected at random by software, with the first 84 selected assigned to view TV show 1, the next 84 to view TV show 2, and the next 84 to view TV show 3. The 84 remaining subjects view TV show 4. This is a

(a) matched pairs design because subjects are matched to the TV shows.
(b) completely randomized design.
(c) block design with TV shows representing the four blocks.

9.24 In the experiment described in Exercise 9.22,

(a) it would have been better to have subjects choose the type of TV show they preferred to view to improve their recall and reduce confounding.
(b) the score on the memory test of their recall of advertisements is the response.
(c) the experimenters should have used different advertisements for each type of TV show to reduce confounding.

9.25 Does exposure to aircraft noise increase the risk of hospitalization for cardiovascular disease in older people ($\geq$65 years) residing near airports? Selecting a random sample of approximately 650,000 Medicare claims, it was found that about 75,000 people had zip codes near airports and the remaining 575,000 did not. The proportions of hospital admissions related to cardiovascular diseases were computed for those with zip codes near airports and those who did not have zip codes near airports. A larger proportion of admissions for cardiovascular disease was found for older people living in zip codes near airports. Which of the following statements is correct?

(a) Since this is an observational study, living in a zip code near an airport may or may not be causing the increase in the proportions of admissions for cardiovascular disease.
(b) Because of the large sample sizes from each group, we can claim that living in a zip code near an airport is causing the increase in the proportion of admissions for cardiovascular disease.
(c) Since this is an experiment, but not a randomized experiment, we can still conclude that living in a zip code near an airport is causing the increase in the proportions of admissions for cardiovascular disease.

9.26 The Community Intervention Trial for Smoking Cessation asked whether a community-wide advertising campaign would reduce smoking. The researchers located 11 pairs of communities, with each pair similar in location, size, economic status, and so on. One community in each pair was chosen at random to participate in the advertising campaign and the other was not. This is

(a) an observational study.
(b) a matched pairs experiment.
(c) a completely randomized experiment.

9.27 To decide which community in each pair in Exercise 9.26 should get the advertising campaign, it is best to

(a) toss a coin.
(b) choose the community that will help pay for the campaign.
(c) choose the community with a mayor who will participate.

9.28 A marketing class designs two videos advertising an expensive Mercedes sports car. They test the videos by asking fellow students to view both (in random order) and say which makes them more likely to buy the car. Mercedes should be reluctant to agree that the video favored in this study will sell more cars because

(a) the study used a matched pairs design instead of a completely randomized design.
(b) results from students may not generalize to the older and richer customers who might buy a Mercedes.
(c) this is an observational study, not an experiment.

CHAPTER 9 EXERCISES

In all exercises that require randomization, you may use software, the Simple Random Sample applet, or Table B. See Example 9.6 (page 235) for directions on using the applet for more than two treatment groups.

9.29 Red meat and mortality. Many studies have found an association between red meat consumption and an increased risk of chronic diseases. What is the relationship between red meat consumption and mortality? A large study followed 120,000 men and women who were free of coronary heart disease and cancer at the beginning of the study. Participants were asked detailed questions about their eating habits every four years, and the study spanned almost 30 years. It was found that the risk of dying at an early age—from heart disease, cancer, or any other cause, rises with the amount of red meat that they consumed.[21]

(a) Is this an observational study or an experiment? What are the explanatory and response variables?
(b) The authors noted that, "Men and women with higher intake of red meat were less likely to be physically active and were more likely to be current smokers, to drink alcohol, and to have a higher body mass index." Explain carefully why differences in these variables make it more difficult to conclude that higher intake of red meat explains the increased death rate. What are the variables physical activity, smoking status, drinking behavior, and body mass index called?
(c) Suggest at least one lurking variable related to diet that may be confounded with higher intake of red meat. Explain why you chose these variables.

9.30 Neighborhood's effect on grades. To study the effect of neighborhood on academic performance, 1000 families were given federal housing vouchers to move out of their low-income neighborhoods. No improvement in the academic performance of the children in the families was found one year after the move. Explain clearly why the lack of improvement in academic performance after one year does not necessarily mean that neighborhood does not affect academic performance. In particular, identify some lurking variables whose effect on academic performance may be confounded with the effect of neighborhood. Use a figure like Figure 9.1 (page 229) to illustrate your explanation.

9.31 Reducing nonresponse. How can we reduce the rate of refusals in telephone surveys? Most people who answer at all listen to the interviewer's introductory remarks and then decide whether to continue. One study made telephone calls to randomly selected households to ask opinions about the next election. In some calls, the interviewer gave her name, in others she identified the university she was representing, and in still others she identified both herself and the university. The study recorded what percent of each group of interviews was completed. Is this an observational study or an experiment? Why? What are the explanatory and response variables?

9.32 Running and sleep. Sufficient sleep is important for adolescents for both their neural and psychological development. Despite this, daytime sleepiness and poor physical and psychological functioning related to chronic sleep disturbances are common. There is a growing body of evidence that exercise is associated with both better sleep and improved psychological functioning. Sixty participants were recruited from a high school in northwestern Switzerland. They were randomly assigned to either a running group or a control group, 30 to each group. The running group ran every morning for a little over 30 minutes on weekdays for a three-week period. All participants used a sleep log for subjective evaluation of sleep, and sleep was also objectively assessed at the beginning and end of the study using a sleep electroencephalographic device that measured quantities such as sleep efficiency and time spent in the four different sleep phases. Running was found to impact positively on both objective and subjective measures of sleep functioning.[22]

(a) What are the explanatory variable(s) and the response variable(s)?

(b) Outline the design of the experiment.
(c) Here are some more details on the treatment and control groups. All participants arrived at school at 7 A.M., and the running group did two laps on the track and then ran cross country in groups of least four people for 30 minutes. The control group remained seated at the track, worked on homework, and interacted with each other. When the runners returned, all participants prepared for school and ate a breakfast that was provided. Why do you think the experimenters had the control group arrive at 7 A.M., interact with classmates, and have breakfast together? Explain. Do you think having the control group do these activities is important for the types of conclusions that can be reached? How?
(d) Time to sleep onset was measured before the beginning of the study and again at the end of the study for participants in both groups. Can this be considered a randomized controlled experiment with time to sleep onset as the response and four treatments (runners before, runners after, controls before, and controls after)? Explain why or why not.

9.33 Observation versus experiment. An *LA Times* article reported that the artery walls of people living within 100 meters of a highway thicken more than twice as fast as the average person's.[23] Researchers used ultrasound to measure the carotid artery wall thickness of 1483 people living near freeways in the Los Angeles area. The artery wall thickness among those living within 100 meters of a highway increased by 5.5 micrometers (roughly 1/20th the thickness of a human hair) each year during the three-year study, which is more than twice the progression observed in participants who did not live within this distance of a highway.
(a) The study compared artery thickening of subjects in the study who were living within 100 meters of a highway with those who were not. Without reading any further details of this study, how do you know that this was an observational study?
(b) Suggest some variables that might differ between the subjects in the study living within 100 meters of a highway and those who were further away. Are any of these possible confounding variables? Explain. (Think about whether living very close to a highway is a desirable neighborhood.)
(c) Could this study be conducted as a randomized comparative experiment? What would be the difficulties?

9.34 Attitudes toward homeless people. Negative attitudes toward poor people are common. Are attitudes more negative when a person is homeless? To find out, a description of a poor person is read to subjects. There are two versions of this description. One begins

Jim is a 30-year-old single man. He is currently living in a small single-room apartment.

The other description begins

Jim is a 30-year-old single man. He is currently homeless and lives in a shelter for homeless people.

Otherwise, the descriptions are the same. After reading the description, you ask subjects what they believe about Jim and what they think should be done to help him. The subjects are 544 adults interviewed by telephone.[24] Outline the design of this experiment.

9.35 Getting teachers to come to school. Elementary schools in rural India are usually small, with a single teacher. The teachers often fail to show up for work. Here is an idea for improving attendance: give the teacher a digital camera with a tamper-proof time and date stamp and ask a student to take a photo of the teacher and class at the beginning and end of the day. Offer the teacher better pay for good attendance—verified by the photos. Will this work? A randomized comparative experiment started with 120 rural schools in Rajasthan and assigned 60 to this treatment and 60 to a control group. Random checks for teacher attendance showed that 21% of teachers in the treatment group were absent, as opposed to 42% in the control group.[25]
(a) Outline the design of this experiment.
(b) Label the schools, and choose the first 10 schools for the treatment group. If you use Table B, start at line 108.

9.36 Ibuprofen and atherosclerosis. The theory of atherosclerosis (hardening and narrowing of the arteries) emphasizes the role of inflammation in the vascular walls. Since ibuprofen is known to possess a wide range of anti-inflammatory actions, it was hypothesized that it might help in the prevention of atherosclerotic lesion development. Both a low cholesterol and a high cholesterol diet were used, as the extent of atherosclerosis is also affected by diet. Thirty-two New Zealand rabbits served as the subjects in the experiment and, after three months, the percentage of the surface covered by atherosclerotic plaques in a region of the aorta was evaluated. Although ibuprofen did suppress the expression of the MCP-1 gene thought to be related to atherosclerosis, it was not shown to have an effect on the extent of fat-induced atherosclerotic lesions.[26]
(a) Use a diagram like Figure 9.2 (page 231) to display the treatments in a design with two factors: "ibuprofen, yes or no" and "high cholesterol diet, yes or no." Then outline the design of a completely randomized experiment to compare these treatments.
(b) Explain how you would number the rabbits and then randomly assign the rabbits to the treatments. If you use the *Simple Random Sample* applet or other software,

assign all the rabbits. If you use Table B, start at line 128 and assign rabbits to only the first treatment group.

9.37 Marijuana and work. How does smoking marijuana affect willingness to work? Canadian researchers persuaded young adult men who used marijuana to live for 98 days in a "planned environment." The men earned money by weaving belts. They used their earnings to pay for meals and other consumption and could keep any money left over. One group smoked two potent marijuana cigarettes every evening. The other group smoked two weak marijuana cigarettes. All subjects could buy more cigarettes but were given strong or weak cigarettes depending on their group. Did the weak and strong groups differ in work output and earnings?[27]

(a) Outline the design of this experiment.
(b) Here are the names of the 30 subjects. Use software or Table B at line 120 to carry out the randomization your design requires.

Abel	DeVore	Kennedy	Reichert	Stout
Aeffner	Fleming	Lamone	Riddle	Williams
Birkel	Fritz	Mani	Sawant	Wilson
Bower	Giriunas	Mattos	Scannell	Worbis
Burke	Glosup	Molnar	Sheldon	Zaccai
Deis	Heaton	Newlen	Simmons	Zelaski

(c) Do you think this can be run as a double-blind experiment? Explain.

9.38 c u L8r 2day: Textese and spelling ability. Textese is a sound-based form of spelling to reduce the time of text messaging and the number of characters, using textisms such as "2day" for "today." Educators and parents have long been concerned about the effect of textese on spelling ability. A study of Australian children aged 10–12 considered the effect of text entry method on the spelling subtest of the Wide Range Achievement Test (WRAT). *Multipress* is the original entry method in which three or four letters are assigned a key and the key must be pressed one to four times to produce each letter. *Predictive* is a single key-press per letter with suggested completion of the word. A total of 84 children took the WRAT and were classified according to text entry method usually used: Multipress (27 students), Predictive (45 students), or non-texter (12 students).[28]

(a) What are the explanatory and response variables?
(b) Is this an observational study or an experiment? Explain why.
(c) The differences among the three texting methods were not statistically significant. What does no significant difference mean in describing the outcome of this study?

9.39 Can low-fat food labels lead to obesity? What are the effects of low-fat food labels on food consumption? Do people eat more of a snack food when the food is labeled as low-fat? The answer may depend both on whether the snack food is labeled low-fat and whether the label includes serving-size information. An experiment investigated this question using university staff, graduate students, and undergraduate students at a large university as subjects. Subjects were asked to evaluate a pilot episode for an upcoming TV show in a theater on campus and were given a cold 24-ounce bottle of water and a bag of granola from a respected campus restaurant called The Spice Box. They were told to enjoy as much or as little of the granola as they wanted. Depending on the condition randomly assigned to the subjects, the granola was labeled as either "Regular Rocky Mountain Granola" or "Low-Fat Rocky Mountain Granola." Below this, the label indicated "Contains 1 Serving," or "Contains 2 Servings," or it provided no serving-size information.[29] Twenty subjects were assigned to each treatment, and their granola bags were weighed at the end of the session to determine how much granola was eaten.

(a) What are the factors and the treatments? How many subjects does the experiment require?
(b) Outline a completely randomized design for this experiment. (You need not actually do the randomization.)

Treating sinus infections. *Sinus infections are common, and doctors commonly treat them with antibiotics. Another treatment is to spray a steroid solution into the nose. A well-designed clinical trial found that these treatments, alone or in combination, do not reduce the severity or the length of sinus infections.[30] Exercises 9.40 to 9.42 concern this trial.*

9.40 Experimental design. The clinical trial was a completely randomized experiment that assigned 240 patients at random among four treatments as follows:

	ANTIBIOTIC PILL	PLACEBO PILL
Steroid spray	53	64
Placebo spray	60	63

(a) Outline the design of the experiment.
(b) How will you label the 240 subjects?
(c) Explain briefly how you would do the random assignment of patients to treatments. Assign the first five patients who will receive the first treatment.

9.41 Describing the design. The report of this study in the *Journal of the American Medical Association* describes it as a "double-blind, randomized, placebo-controlled factorial trial." "Factorial" means that the treatments are formed from more than one factor. What are the factors? What do "double-blind" and "placebo-controlled" mean?

9.42 Checking the randomization. If the random assignment of patients to treatments did a good job of eliminating bias, possible lurking variables such as smoking history, asthma, and hay fever should be similar in all four groups. After recording and comparing many such variables, the investigators said that "all showed no significant difference between groups." Explain to someone who knows no statistics what "no significant difference" means. Does it mean that the presence of all these variables was exactly the same in all four treatment groups?

9.43 Liquid water enhancers? Bottled water, flavored and plain, is expected to become the largest segment of the liquid refreshment market by the end of this decade, surpassing traditional carbonated soft drinks.[31] Kraft's MiO, a liquid water enhancer, comes in a variety of flavors and a few drops added to water gives a zero calorie flavored water drink. You wonder if those who drink flavored water like the taste of MiO as well as they like the taste of a competing flavored water product that comes ready to drink.

(a) Describe a matched pairs design to answer this question. Be sure to include proper blinding of your subjects. What is your response variable going to be?
(b) You have 20 people on hand who prefer to drink flavored water. Use the *Simple Random Sample* applet, software, or Table B at line 138 to do the randomization that your design requires.

9.44 Growing trees faster. The concentration of carbon dioxide (CO_2) in the atmosphere is increasing rapidly due to our use of fossil fuels. Because green plants use CO_2 to fuel photosynthesis, more CO_2 may cause trees to grow faster. An elaborate apparatus allows researchers to pipe extra CO_2 to a 30-meter circle of forest. We want to compare the growth in base area of trees in treated and untreated areas to see if extra CO_2 does in fact increase growth. We can afford to treat three circular areas.[32]

(a) Describe the design of a completely randomized experiment using six well-separated 30-meter circular areas in a pine forest. Sketch the circles, and carry out the randomization your design calls for.
(b) Areas within the forest may differ in soil fertility. Describe a matched pairs design using three pairs of circles that will reduce the extra variation due to different fertility. Sketch the circles and carry out the randomization your design calls for.

9.45 Athletes taking oxygen. We often see players on the sidelines of a football game inhaling oxygen. Their coaches think this will speed their recovery. We might measure recovery from intense exertion as follows: Have a football player run 100 yards three times in quick succession. Then allow three minutes to rest before running 100 yards again. Time the final run.

AP Photo/Wade Payne

Because players vary greatly in speed, you plan a matched pairs experiment using 25 football players as subjects. Discuss the design of such an experiment to investigate the effect of inhaling oxygen during the rest period.

9.46 Protecting ultramarathon runners. An ultramarathon, as you might guess, is a footrace longer than the 26.2 miles of a marathon. Runners commonly develop respiratory infections after an ultramarathon. Will taking 600 milligrams of vitamin C daily reduce these infections? Researchers randomly assigned ultramarathon runners to receive either vitamin C or a placebo. Separately, they also randomly assigned these treatments to a group of nonrunners the same age as the runners. All subjects were watched for 14 days after the big race to see if infections developed.[33]

(a) What is the name for this experimental design?
(b) Use a diagram to outline the design.

9.47 Wine, beer, or spirits? There is good evidence that moderate alcohol use improves health. Some people think that red wine is better for your health than other alcoholic drinks. You have recruited 300 adults aged 45–65 who are willing to follow your orders about alcohol consumption over the next five years. You want to compare the effects on heart disease of moderate drinking of just red wine, just beer, or just spirits. Outline the design of a completely randomized experiment to do this. (No such experiment has been done because subjects aren't willing to have their drinking regulated for years.)

9.48 Wine, beer, or spirits? Women as a group develop heart disease much later than men. We can improve the completely randomized design of Exercise 9.47 by using women and men as blocks. Your 300 subjects include 120 women and 180 men. Outline a block design for comparing wine, beer, and spirits. Be sure to say how many subjects you will put in each group in your design.

9.49 Quick randomizing. Here's a quick and easy way to randomize. You have 100 subjects: 50 women and 50 men. Toss a coin. If it's heads, assign all the men to the treatment group and all the women to the control group. If the coin comes up tails, assign all the women to treatment and all the men to control. This gives every individual subject a 50–50 chance of being assigned to treatment or control. Why isn't this a good way to randomly assign subjects to treatment groups?

9.50 Do antioxidants prevent cancer? People who eat lots of fruits and vegetables have lower rates of colon cancer than those who eat little of these foods. Fruits and vegetables are rich in "antioxidants" such as vitamins A, C, and E. Will taking antioxidants help prevent colon cancer? A medical experiment studied this question with 864 people who were at risk of colon cancer. The subjects were divided into four groups: daily beta-carotene, daily vitamins C and E, all three vitamins every day, or daily placebo. After four years, the researchers were surprised to find no significant difference in colon cancer among the groups.[34]

(a) What are the explanatory and response variables in this experiment?
(b) Outline the design of the experiment. Use your judgment in choosing the group sizes.
(c) The study was double-blind. What does this mean?
(d) What does "no significant difference" mean in describing the outcome of the study?
(e) Suggest some lurking variables that could explain why people who eat lots of fruits and vegetables have lower rates of colon cancer. The experiment suggests that these variables, rather than the antioxidants, may be responsible for the observed benefits of fruits and vegetables.

9.51 SAMe for depression? S-adenosyl methionine, (SAMe), a naturally occurring molecule found throughout the body, has been used as an antidepressant with some success. It has been available commercially in Europe since the late 1970s and is now available over-the-counter in the United States. Participants in the current study were 73 individuals with major depressive disorder who had not responded to a standard treatment using serotonin reuptake inhibitors (SRI) to relieve their symptoms. The effect of augmenting their SRI treatment with SAMe was investigated.[35]

(a) The study was a *randomized, double-blind* trial conducted over six weeks, with 34 participants receiving a placebo (dummy pills) and the remaining 39 receiving pills containing SAMe (a trial is a medical experiment using actual patients as subjects). Explain why it is important to have a placebo group rather than having all participants receive pills containing SAMe. What is the purpose of the two italicized terms in the context of this study?
(b) A 50% reduction in the Hamilton Rating scale for depression over the treatment period was considered a positive response to treatment. It was found that 36.1% of the SAMe group had a positive response versus 17.6% in the placebo group, a statistically significant difference. Explain what statistical significance means in the context of this trial.
(c) From the information given, use a diagram to outline the design of this trial.

9.52 Randomization avoids bias. Suppose that the 25 even-numbered students among the 50 students available for the comparison of classroom and online instruction (Example 9.5) are older, employed students. We hope that randomization will distribute these students roughly equally between the classroom and online groups. Use the *Simple Random Sample* applet to take 20 samples of size 25 from the 50 students. These 25 students will be the classroom instruction group. (Be sure to click "Reset" after each sample.) Record the counts of even-numbered students in each of your 20 samples.

(a) How many older students would you expect to see in the classroom instruction group?
(b) You see that there is considerable chance variation in the number of older (even-numbered) students assigned to the classroom group. Draw a stem-and-leaf plot of the number of older students assigned to the classroom group. Do you see any systematic bias in favor of one or the other group being assigned the older students? Larger samples from a larger population will, on the average, do an even better job of creating two similar groups.

Exploring the Web

9.53 Smoking cessation. Go to the *New England Journal of Medicine* website, www.nejm.org, and find the article "A Randomized, Controlled Trial of Financial Incentives for Smoking Cessation" by Volpp et al. in the February 12, 2009, issue. Under the Issues link, you need to go to the Browse Full Index link and then to the February 12, 2009, issue. You can then download the pdf of the article for free. Was this a comparative study? Was randomization used? How many subjects took part? There were 22 subjects in the control group and 64 in the incentive group who were still not smoking six months after they quit. What were the percents in each group? This difference is statistically significant. Explain in simple language what this means.

9.54 Find an experiment. You can find the latest medical research in the *Journal of the American Medical Association* at www.jama.ama-assn.org and the *New England Journal of Medicine* at www.nejm.org. Many of the articles describe randomized comparative experiments and use the language of statistical significance when giving conclusions. Look through the abstracts and find an experiment of interest to you. If your institution has a subscription to these journals, you should be able to view the entire article. Otherwise, use the information in the abstract to answer as many of these questions as you can. What was the purpose of the experiment? How many factors were in the experiment, and what were the levels of the factors? What response(s) were measured? How many subjects were assigned to each of the treatments, and was randomization used? Was it a double-blind experiment? What were the conclusions, and were the results statistically significant?

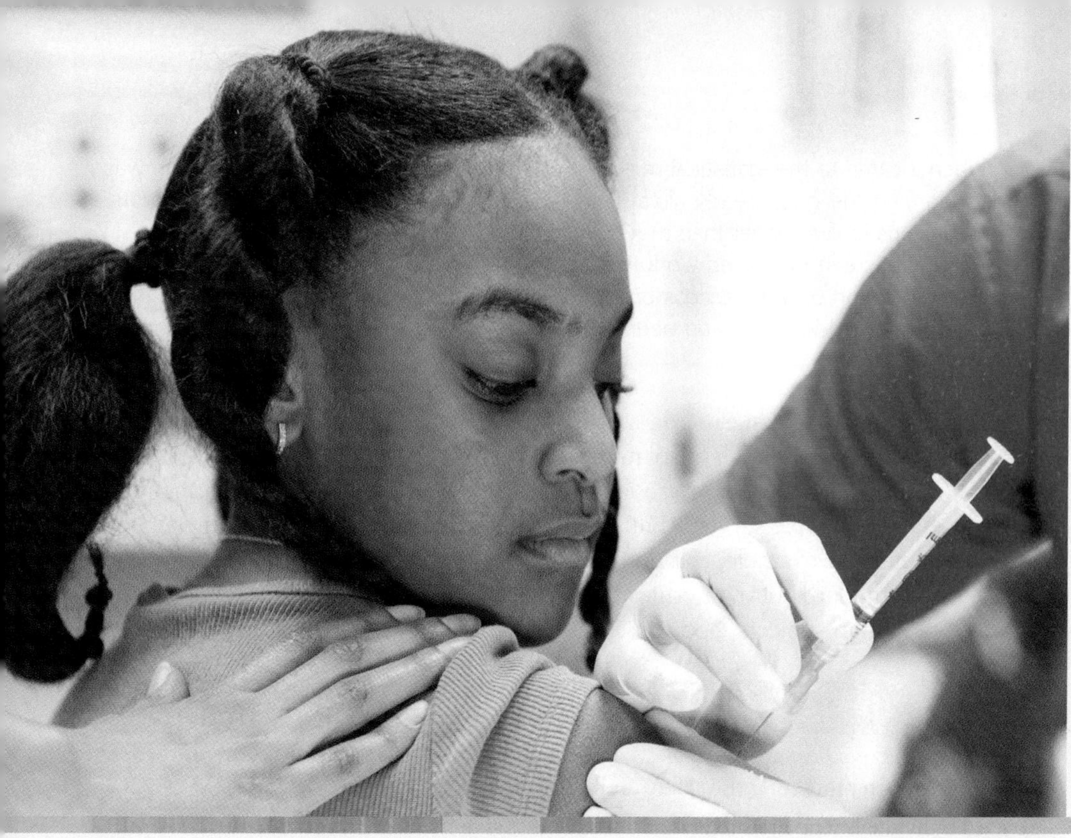

ERproductions Ltd./Getty Images

CHAPTER 10

Data Ethics*

In this chapter we cover...

10.1 Institutional review boards

10.2 Informed consent

10.3 Confidentiality

10.4 Clinical trials

10.5 Behavioral and social science experiments

The production and use of data, like all human endeavors, raise ethical questions. We won't discuss the telemarketer who begins a telephone sales pitch with, "I'm conducting a survey." Such deception is clearly unethical. It enrages legitimate survey organizations, which find the public less willing to talk with them. Neither will we discuss those few researchers who, in the pursuit of professional advancement, publish fake data. There is no ethical question here—faking data to advance your career is just wrong.[1] It will end your career when uncovered. But just how honest must researchers be about real, unfaked data? Here is an example suggesting the answer is, "More honest than they often are."

EXAMPLE 10.1 The Whole Truth?

Papers reporting scientific research are supposed to be short, with no extra baggage. Brevity, however, can allow researchers to avoid complete honesty about their data. Did they choose their subjects in a biased way? Did they report data on only some of their subjects? Did they try several statistical analyses and report only the ones that looked best? The statistician John Bailar screened more than 4000 medical papers in more than a decade as consultant to the *New England Journal of Medicine*. He says,

*This short chapter concerns a very important topic, but the material is not needed to understand the rest of the book.

"When it came to the statistical review, it was often clear that critical information was lacking, and the gaps nearly always had the practical effect of making the authors' conclusions look stronger than they should have."[2] The situation is no doubt worse in fields that screen published work less carefully. This problem continues to grow with the proliferation of open-access online journals that "will print seemingly anything for a fee" and provide little or no peer review.[3] ■

The most complex issues of data ethics arise when we collect data from people (but research with animals also raises ethical issues—see Exercise 10.25). The ethical difficulties are more severe for experiments that impose some treatment on people than for sample surveys that simply gather information. Trials of new medical treatments, for example, can do harm as well as good to their subjects. Here are some basic standards of data ethics that must be obeyed by all studies that gather data from human subjects, both observational studies and experiments.

Basic Data Ethics for Human Subjects

All planned studies must be reviewed in advance by an **institutional review board** charged with protecting the safety and well-being of the subjects.

All individuals who are subjects in a study must give their **informed consent** before data are collected.

All individual data must be kept **confidential**. Only statistical summaries for groups of subjects may be made public.

If subjects are children, then their consent is needed in addition to that of the parents or guardians.

Many journals have a formal requirement of explicitly addressing human subjects issues if the study is classified as human subjects research. For example, here is a statement from the instructions for authors for JAMA (*The Journal of the American Medical Association*):[4]

> *For all manuscripts reporting data from studies involving human participants or animals, formal review and approval, or formal review and waiver, by an appropriate institutional review board or ethics committee is required and should be described in the Methods section. For those investigators who do not have formal ethics review committees, the principles outlined in the Declaration of Helsinki should be followed. For investigations of humans, state in the Methods section the manner in which informed consent was obtained from the study participants (ie, oral or written). Editors may request that authors provide documentation of the formal review and recommendation from the institutional review board or ethics committee responsible for oversight of the study.*

Also, the law requires that studies carried out or funded by the federal government obey these principles.[5] But neither the law nor the consensus of experts is completely clear about the details of their application.

10.1 Institutional review boards

The purpose of an institutional review board is not to decide whether a proposed study will produce valuable information or whether it is statistically sound. The board's purpose is, in the words of one university's board, "to protect the rights and welfare of human subjects (including patients) recruited to participate in research activities." The board reviews the plan of the study and can require changes. It reviews the consent form to ensure that subjects are informed about the nature of the study and about any potential risks. Once research begins, the board monitors the study's progress at least once a year. See Figure 10.1 for the web page of the institutional review board for the Mayo Clinic.

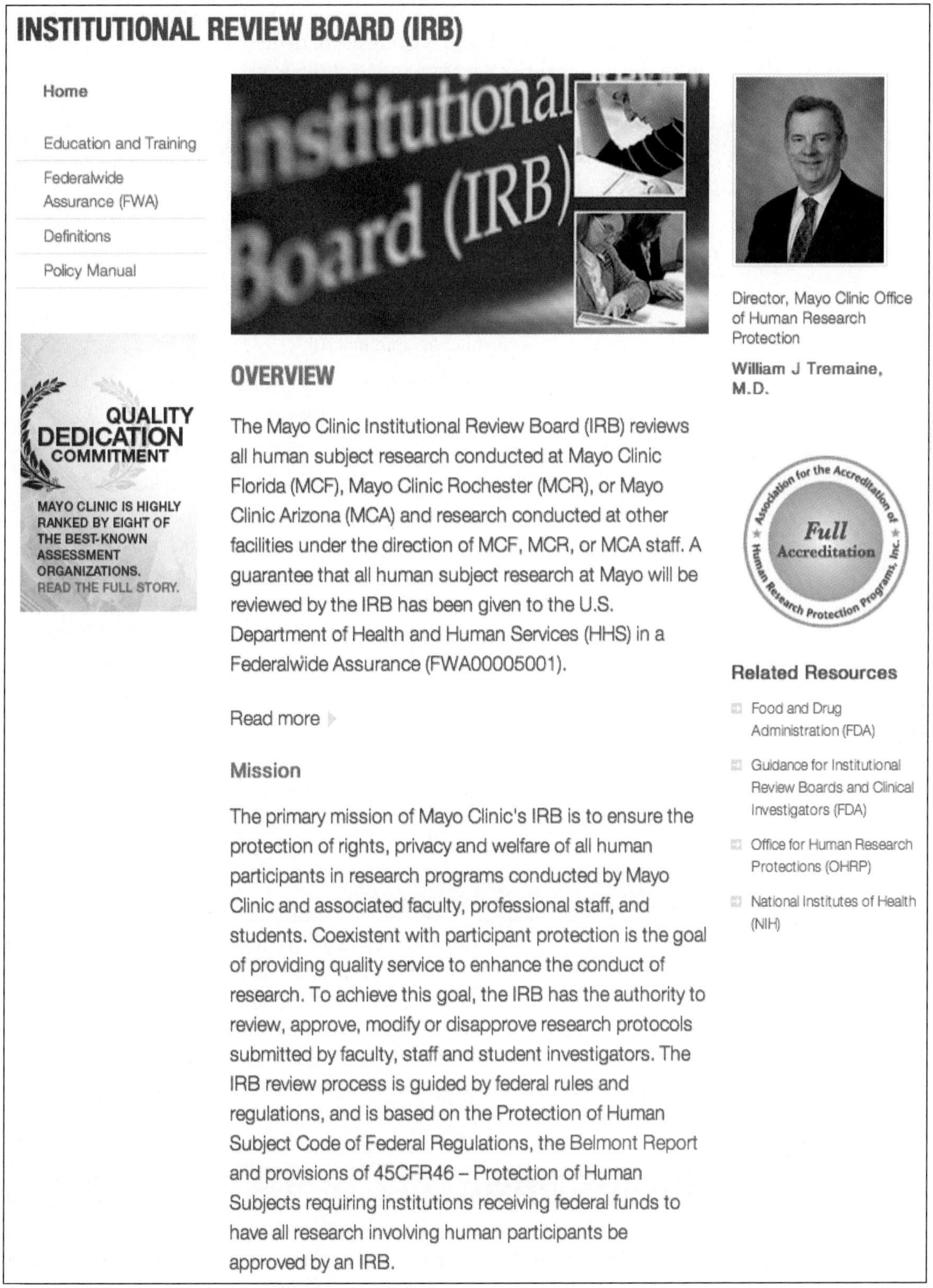
INSTITUTIONAL REVIEW BOARD (IRB)
Home
Education and Training
Federalwide Assurance (FWA)
Definitions
Policy Manual
Institutional Review Board (IRB)
QUALITY DEDICATION COMMITMENT
MAYO CLINIC IS HIGHLY RANKED BY EIGHT OF THE BEST-KNOWN ASSESSMENT ORGANIZATIONS.
READ THE FULL STORY.
OVERVIEW
The Mayo Clinic Institutional Review Board (IRB) reviews all human subject research conducted at Mayo Clinic Florida (MCF), Mayo Clinic Rochester (MCR), or Mayo Clinic Arizona (MCA) and research conducted at other facilities under the direction of MCF, MCR, or MCA staff. A guarantee that all human subject research at Mayo will be reviewed by the IRB has been given to the U.S. Department of Health and Human Services (HHS) in a Federalwide Assurance (FWA00005001).
Read more
Mission
The primary mission of Mayo Clinic's IRB is to ensure the protection of rights, privacy and welfare of all human participants in research programs conducted by Mayo Clinic and associated faculty, professional staff, and students. Coexistent with participant protection is the goal of providing quality service to enhance the conduct of research. To achieve this goal, the IRB has the authority to review, approve, modify or disapprove research protocols submitted by faculty, staff and student investigators. The IRB review process is guided by federal rules and regulations, and is based on the Protection of Human Subject Code of Federal Regulations, the Belmont Report and provisions of 45CFR46 – Protection of Human Subjects requiring institutions receiving federal funds to have all research involving human participants be approved by an IRB.
Director, Mayo Clinic Office of Human Research Protection
William J Tremaine, M.D.
Association for the Accreditation of Human Research Protection Programs, Inc.
Full Accreditation
Related Resources
Food and Drug Administration (FDA)
Guidance for Institutional Review Boards and Clinical Investigators (FDA)
Office for Human Research Protections (OHRP)
National Institutes of Health (NIH)

Courtesy of Mayo Clinic

FIGURE 10.1
The web page of the Mayo Clinic's institutional review board. It begins by describing the job of such boards.

The most pressing issue concerning institutional review boards is whether their workload has become so large that their effectiveness in protecting subjects drops. When the government temporarily stopped human subject research at Duke University Medical Center in 1999 due to inadequate protection of subjects, more than 2000 studies were going on. That's a lot of review work. There are shorter review procedures for projects that involve only minimal risks to subjects, such as most sample surveys. When a board is overloaded, there is a temptation to put more proposals in the minimal-risk category to speed the work.

Apply Your Knowledge

10.1 Minimal Risk? You are a member of your college's institutional review board. You must decide whether several research proposals qualify for less rigorous review because they involve only minimal risk to subjects. Federal regulations say that "minimal risk" means the risks are no greater than "those ordinarily encountered in daily life or during the performance of routine physical or psychological examinations or tests." That's vague. Which of these do you think qualifies as "minimal risk"?

(a) Take hair and nail clippings in a nondisfiguring manner.

(b) Draw a drop of blood by pricking a finger to measure blood sugar.

(c) Collect data on subjects through the use of X rays or microwaves.

(d) Insert a tube that remains in the arm so that blood can be drawn regularly.

(e) Take permanent teeth if routine patient care indicates a need for extraction.

(f) Take extra specimens from a subject who is undergoing an invasive clinical procedure such as a bronchoscopy (a procedure in which a physician views the inside of the airways for diagnostic and therapeutic purposes using an instrument that is inserted into the airways, usually through the nose or mouth).

10.2 Does This Really Need to Be Reviewed? A college professor would like to investigate a new method for teaching statistics. He teaches two lectures. He will use the standard approach to teaching in one lecture, and the new approach in the other. Should he seek institutional review board approval before proceeding? Discuss.

10.2 Informed consent

Both words in the phrase "informed consent" are important, and both can be controversial. Subjects must be *informed* in advance about the nature of a study and any risk of harm it may bring. In the case of a sample survey, physical harm is not possible. The subjects should be told what kinds of questions the survey will ask and about how much of their time it will take. Experimenters must tell subjects the nature and purpose of the study and outline possible risks. Subjects must then *consent* in writing.

EXAMPLE 10.2 Who Can Consent?

Are there some subjects who can't give informed consent? It was once common, for example, to test new vaccines on prison inmates who gave their consent in return for good-behavior credit. Now we worry that prisoners are not really free to refuse, and the law forbids almost all medical research in prisons.

Children can't give fully informed consent, so the usual procedure is to ask their parents. A study of new ways to teach reading is about to start at a local elementary school, so the study team sends consent forms home to parents. Many parents don't return the forms. Can their children take part in the study because the parents did not say "no," or should we allow only children whose parents returned the form and said "yes"?

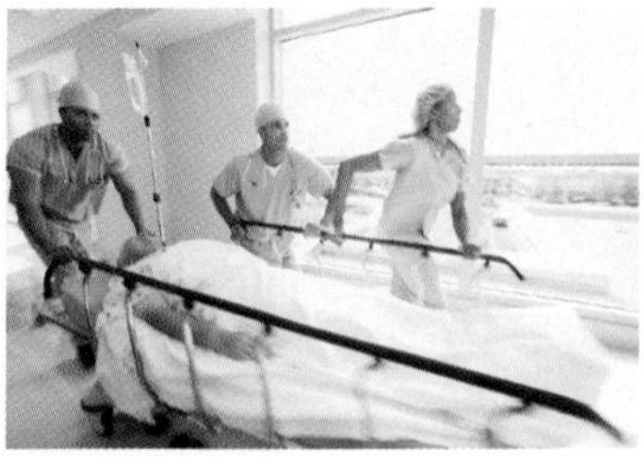

© Tim Pannell/Corbis

What about research into new medical treatments for people with mental disorders? What about studies of new ways to help emergency room patients who may be unconscious? In most cases, there is not time to get the consent of the family. Does the principle of informed consent bar realistic trials of new treatments for unconscious patients?

These are questions without clear answers. Reasonable people differ strongly on all of them. There is nothing simple about informed consent.[6] ■

The difficulties of informed consent do not vanish even for capable subjects. Some researchers, especially in medical trials, regard consent as a barrier to getting patients to participate in research. They may not explain all possible risks; they may not point out that there are other therapies that might be better than those being studied; they may be too optimistic in talking with patients, even when the consent form has all the right details. On the other hand, mentioning every possible risk leads to very long consent forms that really are barriers. "They are like rental car contracts," one lawyer said. Some subjects don't read forms that run five or six printed pages. Others are frightened by the large number of possible (but unlikely) disasters that might happen and so refuse to participate. Of course, unlikely disasters sometimes happen. When they do, lawsuits follow, and the consent forms become yet longer and more detailed.

LaunchPad Online Resources

- The StatBoards video, *Informed Consent and Psychological Experimentation*, discusses a real example involving issues of informed consent.

Apply Your Knowledge

10.3 Coercion? The U.S. Department of Health and Human Services regulations[7] for informed consent state that, "An investigator shall seek such consent only under circumstances that provide the prospective subject or the representative sufficient opportunity to consider whether or not to participate and that minimize the possibility of coercion or undue influence." Coercion occurs when an overt or implicit threat of harm is intentionally presented by one person to another in order to obtain compliance. Which of the following circumstances do you believe constitutes coercion? Discuss.

(a) An investigator tells a prospective subject that she or he will lose access to needed health services if she or he does not participate in the research.

(b) An employer asks employees to participate in a research study. Although the employer has assured employees that participation is voluntary, several employees are concerned that a decision not to participate could affect performance evaluations or job advancement.

10.4 Undue Influence? Undue influence in obtaining informed consent often occurs through an offer of an excessive or inappropriate reward or other overture in order to obtain compliance. Which of the following circumstances do you believe constitutes undue influence? Discuss.

(a) The patients of a physician are asked to participate in a study in which the physician is also the investigator.

(b) A professor asks a student to participate in a research study. He tells the student that everyone else in the class has agreed to participate.

(c) Research subjects are to be paid in exchange for their participation.

10.3 Confidentiality

Ethical problems do not disappear once a study has been cleared by the review board, has obtained consent from its subjects, and has actually collected data about the subjects. It is important to protect the subjects' privacy. Privacy refers to a person's interest in controlling the access of others to himself or herself, including information about himself or herself. One way this is done is by keeping all data about individuals confidential. Confidentiality refers to the agreement between the investigator and participant about how data will be managed and used. The report of an opinion poll may say what percent of the 1200 respondents felt that legal immigration should be reduced. It may not report what *you* said about this or any other issue. However the investigator who collected the data will know what you said about this or other issues in the poll.

anonymity

Confidentiality is not the same as **anonymity.** Anonymity means that subjects are anonymous—their names are not known even to the director of the study. Anonymity provides a high degree of privacy, but anonymity is rare in statistical studies. Even where it is possible (mainly in surveys conducted by mail), anonymity prevents any follow-up to improve nonresponse or inform subjects of results.

Any breach of confidentiality is a serious violation of data ethics. The best practice is to separate the identity of the subjects from the rest of the data at once. Sample surveys, for example, use the identification only to check on who did or did not respond. In an era of advanced technology, however, it is no longer enough to be sure that each individual set of data protects people's privacy. The government, for example, maintains a vast amount of information about citizens in many separate databases—census responses, tax returns, Social Security information, data from surveys such as the Current Population Survey, and so on. Many of these databases can be searched by computers for statistical studies. A clever computer search of several databases might be able, by combining information, to identify you and learn a great deal about you, even if your name and other identification have been removed from the data available for search. A colleague from Germany once remarked that "female full professor of statistics with a PhD from the United States" was enough to identify her among all the 83 million residents of Germany. Privacy and confidentiality of data are hot issues among statisticians in the computer age. Computer hacking and thefts of laptops containing data add to the difficulties. Is it even possible to guarantee confidentiality of data stored in databases that can be hacked or stolen? The U.S. Social Security Administration has devised a comprehensive Internet privacy policy, part of which can be seen in Figure 10.2.

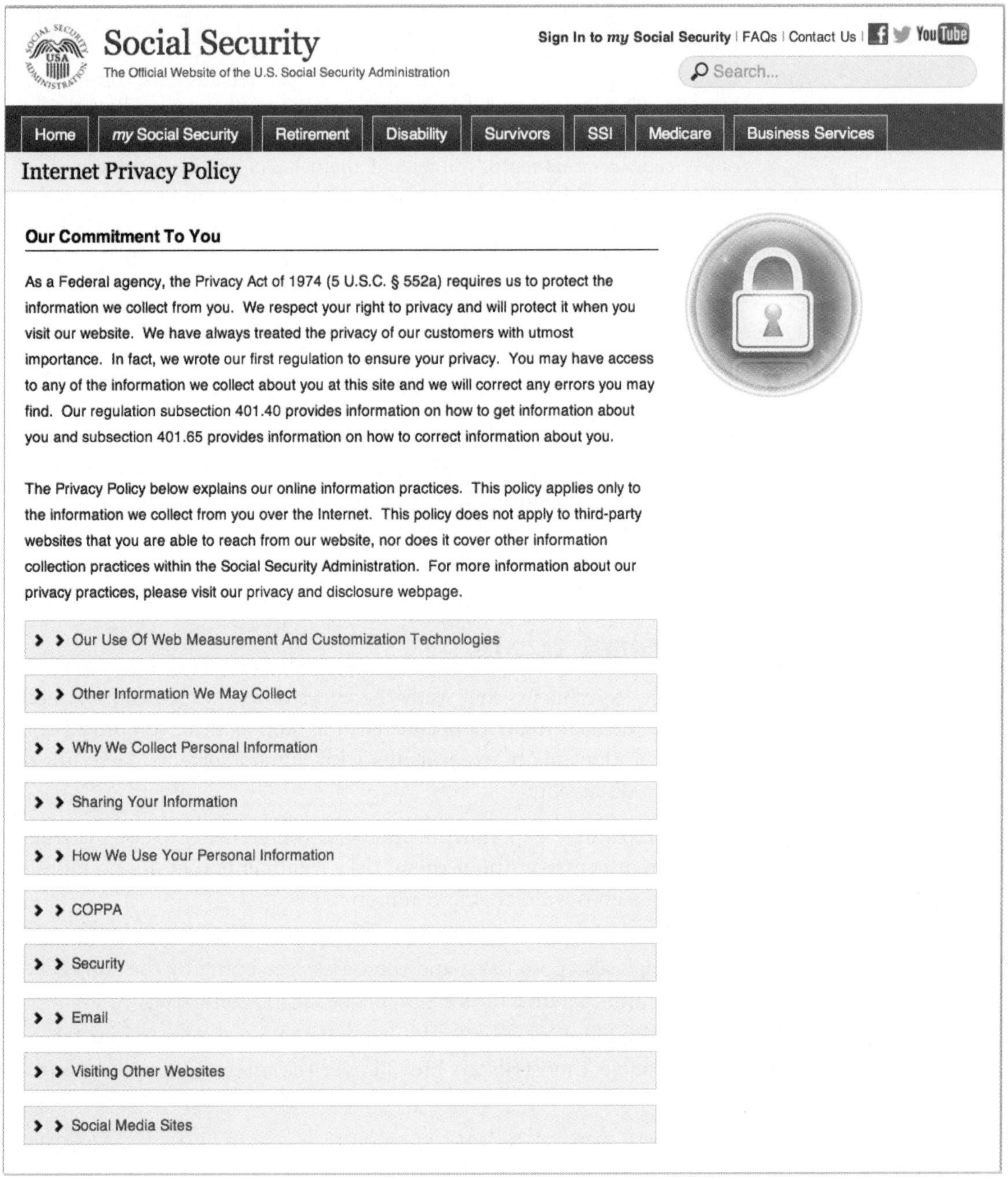
Social Security
The Official Website of the U.S. Social Security Administration
Sign In to my Social Security | FAQs | Contact Us |
Search...
Home
my Social Security
Retirement
Disability
Survivors
SSI
Medicare
Business Services
Internet Privacy Policy
Our Commitment To You
As a Federal agency, the Privacy Act of 1974 (5 U.S.C. § 552a) requires us to protect the information we collect from you. We respect your right to privacy and will protect it when you visit our website. We have always treated the privacy of our customers with utmost importance. In fact, we wrote our first regulation to ensure your privacy. You may have access to any of the information we collect about you at this site and we will correct any errors you may find. Our regulation subsection 401.40 provides information on how to get information about you and subsection 401.65 provides information on how to correct information about you.
The Privacy Policy below explains our online information practices. This policy applies only to the information we collect from you over the Internet. This policy does not apply to third-party websites that you are able to reach from our website, nor does it cover other information collection practices within the Social Security Administration. For more information about our privacy practices, please visit our privacy and disclosure webpage.
Our Use Of Web Measurement And Customization Technologies
Other Information We May Collect
Why We Collect Personal Information
Sharing Your Information
How We Use Your Personal Information
COPPA
Security
Email
Visiting Other Websites
Social Media Sites

Social Security Administration

FIGURE 10.2
The privacy policy of the government's Social Security Administration website.

EXAMPLE 10.3 Uncle Sam Knows

Citizens are required to give information to the government. Think of tax returns and Social Security contributions. The government needs these data for administrative purposes—to see if you paid the right amount of tax and how large a Social Security benefit you are owed when you retire. Some people feel that individuals should be able to forbid any other use of their data, even with all identification removed. This would prevent using government records to study, say, the ages, incomes, and household sizes of Social Security recipients. Such a study could well be vital to debates on reforming Social Security. ■

Apply Your Knowledge

10.5 **Sunshine Laws.** All states in the U.S. have open records laws, sometimes known as "Sunshine Laws," that give citizens access to government meetings and records.[8] This includes, for example, reports of crimes and recordings of 911 calls. A crime report will include the name of anyone accused of the crime. Suppose a 10-year-old juvenile is accused of committing a crime. A reporter from the local newspaper asks for a copy of the crime report. The sheriff refuses to provide the report because the accused is a juvenile and he believes the name of the accused should be confidential. Is this an issue of confidentiality? Discuss.

10.6 **https.** Generally, secure websites use encryption and authentication standards to protect the confidentiality of web transactions. The most commonly used protocol for web security has been TLS, or Transport Layer Security. This technology is still commonly referred to as SSL. Web sites with addresses beginning with https use this protocol. Do you believe that https websites provide true confidentiality?[9] Do you think it is possible to guarantee the confidentiality of data on any website? Discuss.

10.4 Clinical trials

Clinical trials are experiments that study the effectiveness of medical treatments on actual patients. Medical treatments can harm as well as heal, so clinical trials spotlight the ethical problems of experiments with human subjects. Here are the starting points for a discussion:

- Randomized comparative experiments are by far the best way to see the true effects of new treatments. Without them, risky treatments that are no more effective than placebos will become common.[10]
- Clinical trials produce great benefits, but most of these benefits go to future patients. The trials also pose risks, and these risks are borne by the subjects of the trial. So we must balance future benefits against present risks.
- Both medical ethics and international human rights standards say that "the interests of the subject must always prevail over the interests of science and society."

The quoted words are from the 1964 Helsinki Declaration of the World Medical Association, the most respected international standard. The most outrageous examples of unethical experiments are those that ignore the interests of the subjects.

EXAMPLE 10.4 The Tuskegee Study

In the 1930s, syphilis was common among black men in the rural South, a group that had almost no access to medical care. The Public Health Service Tuskegee study recruited 399 poor black sharecroppers with syphilis and 201 others without the disease to observe how syphilis progressed when no treatment was given. Beginning in 1943, penicillin became available to treat syphilis. The study subjects were not treated. In fact, the Public Health Service prevented any treatment until word leaked out and forced an end to the study in the 1970s.

The Tuskegee study is an extreme example of investigators following their own interests and ignoring the well-being of their subjects. A 1996 review said, "It has come to symbolize racism in medicine, ethical misconduct in human research, paternalism by physicians, and government abuse of vulnerable people." In 1997, President Clinton formally apologized to the surviving participants in a White House ceremony.[11] ■

Because "the interests of the subject must always prevail," medical treatments can be tested in clinical trials only when there is reason to hope that they will help the patients who are subjects in the trials. Future benefits aren't enough to justify experiments with human subjects. Of course, if there is already strong evidence that a treatment works and is safe, it is unethical *not* to give it. Here are the words of Dr. Charles Hennekens of the Harvard Medical School, who directed the large clinical trial that showed that aspirin reduces the risk of heart attacks:

> *There's a delicate balance between when to do or not do a randomized trial. On the one hand, there must be sufficient belief in the agent's potential to justify exposing half the subjects to it. On the other hand, there must be sufficient doubt about its efficacy to justify withholding it from the other half of subjects who might be assigned to placebos.*[12]

Why is it ethical to give a control group of patients a placebo? Well, we know that placebos often work. Moreover, placebos have no harmful side effects. So in the state of balanced doubt described by Dr. Hennekens, the placebo group may be getting a better treatment than the drug group. If we *knew* which treatment was better, we would give it to everyone. When we don't know, it is ethical to try both and compare them.[13]

Apply Your Knowledge

10.7 Ethics and Scientific Validity. The authors of a paper on clinical research and ethics[14] stated the following.

> *For a clinical research protocol to be ethical, the methods must be valid and practically feasible: the research must have a clear scientific objective; be designed using accepted principles, methods, and reliable practices; have sufficient power to definitively test the objective; and offer a plausible data analysis plan. In addition, it must be possible to execute the proposed study.*

Do you think this rules out observational studies as "ethical"? Discuss. You might wish to refer to the discussion on correlation and causation in Chapter 5 (page 148).

10.5 Behavioral and social science experiments

When we move from medicine to the behavioral and social sciences, the direct risks to experimental subjects are less acute, but so are the possible benefits to the subjects. Consider, for example, the experiments conducted by psychologists in their study of human behavior.

EXAMPLE 10.5 Psychologists in the Men's Room

© David Pollack/Corbis

Psychologists observe that people have a "personal space" and are uneasy if others come too close to them. We don't like strangers to sit at our table in a coffee shop if other tables are available, and we see people move apart in elevators if there is room to do so. Americans tend to require more personal space than people in most other cultures. Can violations of personal space have physical, as well as emotional, effects?

Investigators set up shop in a men's public restroom. They blocked off urinals to force men walking in to use either a urinal next to an experimenter (treatment group) or a urinal separated from the experimenter (control group). Another experimenter, using a periscope from a toilet stall, measured how long the subject took to start urinating and how long he continued.[15] ■

This personal space experiment illustrates the difficulties facing those who plan and review behavioral studies.

- There is no risk of harm to the subjects, although they would certainly object to being watched through a periscope. What should we protect subjects from when physical harm is unlikely? Possible emotional harm? Undignified situations? Invasion of privacy?
- What about informed consent? The subjects did not even know they were participating in an experiment. Many behavioral experiments rely on hiding the true purpose of the study. The subjects would change their behavior if told in advance what the investigators were looking for. Subjects are asked to consent on the basis of vague information. They receive full information only after the experiment.

The "Ethical Principles" of the American Psychological Association require consent unless a study merely observes behavior in a public place. They allow deception only when it is necessary to the study, does not hide information that might influence a subject's willingness to participate, and is explained to subjects as soon as possible. The personal space study (from the 1970s) does not meet current ethical standards.

We see that the basic requirement for informed consent is understood differently in medicine and psychology. Here is an example of another setting with yet another interpretation of what is ethical. The subjects get no information and give no consent. They don't even know that an experiment may be sending them to jail for the night.

EXAMPLE 10.6 Reducing Domestic Violence

How should police respond to domestic violence calls? In the past, the usual practice was to remove the offender and order him to stay out of the household overnight. Police were reluctant to make arrests because the victims rarely pressed charges. Women's groups argued that arresting offenders would help prevent future violence even if no charges were filed. Is there evidence that arrest will reduce future offenses? That's a question that experiments have tried to answer.

A typical domestic violence experiment compares two treatments: arrest the suspect and hold him overnight, or warn the suspect and release him. When police officers reach the scene of a domestic violence call, they calm the participants and investigate. Weapons or death threats require an arrest. If the facts permit an arrest but do not require it, an officer radios headquarters for instructions. The person on duty opens the next envelope in a file prepared in advance by a statistician. The envelopes contain the treatments in random order. The police either arrest the suspect or warn and release him, depending on the contents of the envelope. The researchers then watch police records and visit the victim to see if the domestic violence recurs.

Such experiments show that arresting domestic violence suspects does reduce their future violent behavior.[16] As a result of this evidence, arrest has become the common police response to domestic violence. ■

The domestic violence experiments shed light on an important issue of public policy. Because there is no informed consent, the ethical rules that govern clinical trials and most social science studies would forbid these experiments. They were cleared by review boards because, in the words of one domestic violence researcher, "These people became subjects by committing acts that allow the police to arrest them. You don't need consent to arrest someone."

Apply Your Knowledge

10.8 Deceiving Subjects. Students sign up to be subjects in a psychology experiment. When they arrive, they are placed in a room and assigned a task. During the task, the subject hears a loud thud from an adjacent room and then a piercing cry for help. Some subjects are placed in a room by themselves. Others are placed in a room with "confederates" who have been instructed by the researcher to look up on hearing the cry, then return to their task. The treatments being compared are whether the subject is alone in the room or in the room with confederates. Will the subject ignore the cry for help?

The students had agreed to take part in an unspecified study, and the true nature of the experiment is explained to them afterward. Do you think this study is ethically okay? Discuss.

CHAPTER 10 SUMMARY

Chapter Specifics

- All planned studies must be reviewed in advance by an **institutional review board** charged with protecting the safety and well-being of the subjects.
- All individuals who are subjects in a study must give their **informed consent** before data are collected.
- All individual data must be kept **confidential.** Only statistical summaries for groups of subjects may be made public. The goal is to protect subjects' privacy.

Link It

Chapters 8 and 9 discuss methods for producing data. Applying these methods in practice raises ethical questions. In this chapter, we present some of these ethical issues. Many ethical questions do not have a clear, correct answer, but there are some basic standards that must be obeyed by all studies.

CHAPTER 10 EXERCISES

Most of these exercises pose issues for discussion. There are no right or wrong answers, but there are more and less thoughtful answers.

10.9 **Who reviews?** Government regulations require that institutional review boards consist of at least five people, including at least one scientist, one nonscientist, and one person from outside the institution. Most boards are larger, but many contain just one outsider.

(a) Why should review boards contain people who are not scientists?
(b) Do you think that one outside member is enough? How would you choose that member? (For example, would you prefer a medical doctor? A member of the clergy? An activist for patients' rights?)

10.10 **Informed consent.** A researcher suspects that people with fundamentalist religious beliefs tend to be more prone to depression. She prepares a questionnaire that measures depression and also asks many religious questions. Write a description of the purpose of this research to be read by subjects to obtain their informed consent. You must balance the conflicting goals of not deceiving the subjects about what the questionnaire will tell about them, and not biasing the sample by scaring off people with fundamentalist religious views.

10.11 **Is consent needed?** In which of the following circumstances would you allow collecting personal information without the subjects' consent?

(a) A government agency takes a random sample of income tax returns to obtain information on the marital status and average income of people who identify themselves as belonging to an ultraconservative political group. Only the marital status and income are recorded from the returns, not the names.
(b) A social psychologist attends public meetings of an ultraconservative political group to study the behavior patterns of members.
(c) A social psychologist pretends to be converted to membership of an ultraconservative political group and attends private meetings to study the behavior patterns of members.

10.12 **Studying your blood.** Long ago, doctors drew a blood specimen from you as part of treating minor anemia. Unknown to you, the sample was stored. Now researchers plan to use stored samples from you and many other people to look for genetic factors that may influence anemia. It is no longer possible to ask your consent. Modern technology can read your entire genetic makeup from the blood sample.

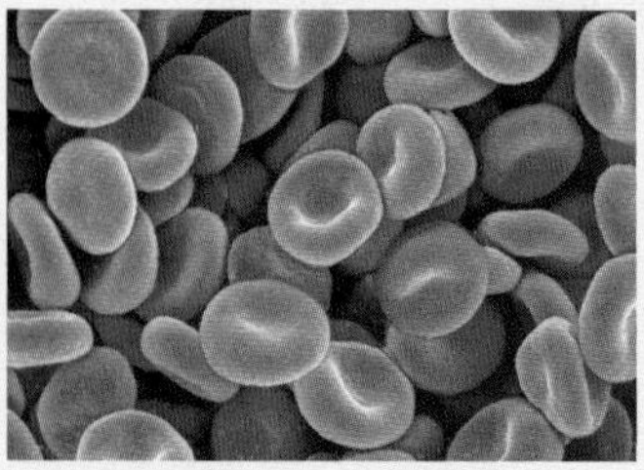

Susumu Nishinaga/Science Source

(a) Do you think it violates the principle of informed consent to use your blood sample if your name is on it but you were not told that it might be saved and studied later?
(b) Suppose that your identity is not attached. The blood sample is known only to come from (say) "a 20-year-old white female being treated for anemia." Is it now okay to use the sample for research?
(c) Perhaps we should use biological materials such as blood samples only from patients who have agreed to allow the material to be stored for later use in research. It isn't possible to say in advance what kind of research, so this falls short of the usual standard for informed consent. Is it nonetheless acceptable, given complete confidentiality and the fact that using the sample can't physically harm the patient?

10.13 **Anonymous? Confidential?** One of the most important nongovernment surveys in the United States is the National Opinion Research Center's General Social Survey. The GSS regularly monitors public opinion on a wide variety of political and social issues. Interviews are conducted in person in the subject's home. Are a subject's responses to GSS questions anonymous, confidential, or both? Explain your answer.

10.14 **Anonymous or confidential?** The University of Wisconsin LaCrosse, like many universities, offers free screening for HIV, the virus that causes AIDS. The announcement at the Student Life Office website says that in the free testing program "you are never asked your name, address or social security number, only your age and sex. You will be assigned a 12-digit code number which only you will know." Does this practice offer anonymity or just confidentiality? Explain your answer.

10.15 **Political polls.** Suppose the presidential election campaign is in full swing, and the candidates have hired polling organizations to take sample surveys to find out what the voters think about the issues. What information should the pollsters be required to give out?

(a) What does the standard of informed consent require the pollsters to tell potential respondents?
(b) The standards accepted by polling organizations also require giving respondents the name and address of the organization that carries out the poll. Why do you think this is required?
(c) The polling organization usually has a professional name, such as "Samples Incorporated," so respondents don't know that the poll is being paid for by a political party or candidate. Would revealing the sponsor to respondents bias the poll? Should the sponsor always be announced whenever poll results are made public?

10.16 Making poll results public. Some people think that the law should require all political poll results to be made public. Otherwise, the possessors of poll results can use the information to their own advantage. They can act on the information, release only selected parts of it, or time the release for best effect. A candidate's organization replies that they are paying for the poll to gain information for their own use, not to amuse the public. Do you favor requiring complete disclosure of political poll results? What about other private surveys, such as market research surveys of consumer tastes?

10.17 Student subjects. Students taking Psychology 001 are required to serve as experimental subjects. Students in Psychology 002 are not required to serve, but they are given extra credit if they do so. Students in Psychology 003 are required either to sign up as subjects or to write a term paper. Serving as an experimental subject may be educational, but current ethical standards frown on using "dependent subjects" such as prisoners or charity medical patients. Students are certainly somewhat dependent on their teachers. Do you object to any of these course policies? If so, which ones, and why?

10.18 The Willowbrook hepatitis studies. In the 1960s, children entering the Willowbrook State School, an institution for the intellectually disabled, were deliberately infected with hepatitis. The researchers argued that almost all children in the institution quickly became infected anyway. The studies showed for the first time that two strains of hepatitis existed. This finding contributed to the development of effective vaccines. Despite these valuable results, the Willowbrook studies are now considered an example of unethical research. Explain why, according to current ethical standards, useful results are not enough to allow a study.

10.19 Unequal benefits. Researchers on aging proposed to investigate the effect of supplemental health services on the quality of life of older people. Eligible patients on the rolls of a large medical clinic were to be randomly assigned to treatment and control groups. The treatment group would be offered hearing aids, dentures, transportation, and other services not available without charge to the control group. The review board felt that providing these services to some but not other persons in the same institution raised ethical questions. Do you agree?

10.20 How many have HIV? Researchers from Yale, working with medical teams in Tanzania, wanted to know how common infection with HIV, the virus that causes AIDS, is among pregnant women in that African country. To do this, they planned to test blood samples drawn from pregnant women.

Yale's institutional review board insisted that the researchers get the informed consent of each woman and tell her the results of the test. This is the usual procedure in developed nations. The Tanzanian government did not want to tell the women why blood was drawn or tell them the test results. The government feared panic if many people turned out to have an incurable disease for which the country's medical system could not provide care. The study was canceled. Do you think that Yale was right to apply its usual standards for protecting subjects?

10.21 AIDS trials in Africa. The drug programs that treat AIDS in rich countries are very expensive, so some African nations cannot afford to give them to large numbers of people. Yet AIDS is more common in parts of Africa than anywhere else. "Short-course" drug programs that are much less expensive might help, for example, in preventing infected pregnant women from passing the infection to their unborn children. Is it ethical to compare a short-course program with a placebo in a clinical trial? Some say "no": this is a double standard, because in rich countries the full drug program would be the control treatment. Others say "yes": the intent is to find treatments that are practical in Africa, and the trial does not withhold any treatment that subjects would otherwise receive. What do you think?

10.22 Abandoned children in Romania. The study described in Example 9.2 (page 231) randomly assigned abandoned children in Romanian orphanages to move to foster homes or to remain in an orphanage. All the children would otherwise have remained in an orphanage. The foster care was paid for by the study. There was no informed consent because the children had been abandoned and had no adult to speak for them. The experiment was considered ethical because "people who cannot consent can be protected by enrolling them only in minimal-risk research, whose risks do not exceed those of everyday life," and because the study "aimed to produce results that would primarily benefit abandoned, institutionalized children."[17] Do you agree?

10.23 Asking teens about sex. The Centers for Disease Control and Prevention, in a survey of teenagers, asked the subjects if they had ever had sexual intercourse. Males who said "yes" were then asked, "That very first time that you had sexual intercourse with a female, how old were you?" and "Please tell me the name or initials of your first sexual partner so that I can refer to her during the interview." Should consent of parents be required to ask minors about sex, drugs, and other such issues, or is consent of the minors themselves enough? Give reasons for your opinion.

10.24 Deceiving subjects. A psychologist conducts the following experiment: he measures the attitude of subjects toward cheating, then has them take a mathematics skills exam in which the subjects are tempted to cheat. Subjects are told that high scores will receive a $100 gift certificate and that the purpose of the experiment is to see if rewards affect performance. The exam is computer-based and multiple choice. Subjects are left alone in a room with a computer on which the exam is available and are told that they are to click on the answer they believe is correct. However, when subjects click on an answer, a small pop-up window appears with the correct answer indicated. When the pop-up window is closed, it is possible to change the answer selected. The computer records—unknown to the subjects—whether or not they change their answers after closing the pop-up window. After the exam is finished, attitude toward cheating is retested.

Subjects who cheat tend to change their attitudes to find cheating more acceptable. Those who resist the temptation to cheat tend to condemn cheating more strongly on the second test of attitude. These results confirm the psychologist's theory.

This experiment tempts subjects to cheat. The subjects are led to believe that they can cheat secretly when in fact they are observed. Is this experiment ethically objectionable? Explain your position.

Exploring the Web

10.25 Research on animals. In Section 10.4, we quoted the 1964 Helsinki Declaration of the World Medical Association as saying "the interests of the subject must always prevail over the interests of science and society." This applies to experiments with human subjects. What about experiments with animals?

(a) Do you think there should be similar or different standards for research involving animals? In a few sentences, state your opinion. (There are no wrong answers.)

(b) Visit the website `www.stanford.edu/group/hopes/cgi-bin/wordpress/2010/07/animal-research/`. What positions regarding research with animals are described at this site? Which of these positions do you favor?

(c) Visit either the website `www.nc3rs.org.uk/page.asp?id=7` or the site `www.ccac.ca/en_/education/niaut/stream/cs-3rs`. What are the three Rs of humane animal experimentation? Give an example of one way each of the three Rs might be satisfied.

© Cultura RM/Alamy

CHAPTER 11

Producing Data: Part II Review

In this chapter we cover...

- Part II Summary
- Test Yourself
- Supplementary Exercises

In Part I of this book, you mastered **data analysis**, the use of graphs and numerical summaries to organize and explore any set of data. Part II has introduced designs for data production. Part III will discuss basic probability and the foundations of inference. Parts IV and V will deal in detail with statistical inference.

Designs for producing data are essential if the data are intended to represent some wider population or process. Figures 11.1 and 11.2 display the big ideas visually. Random sampling and randomized comparative experiments are perhaps the most important statistical inventions of the 20th century. Both were slow to gain acceptance, and you will still see many voluntary response samples and uncontrolled experiments. You should now understand good designs for producing data and also why bad designs often produce data that are worthless for inference. The deliberate use of chance in producing data is a central idea in statistics. It not only reduces bias but also allows us to use **probability**, the mathematics of chance, as the basis for inference.

FIGURE 11.1
Statistics in summary.

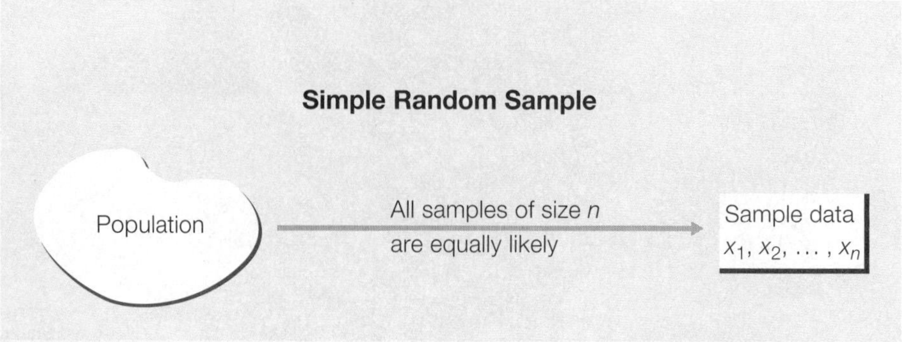

FIGURE 11.2
Statistics in summary.

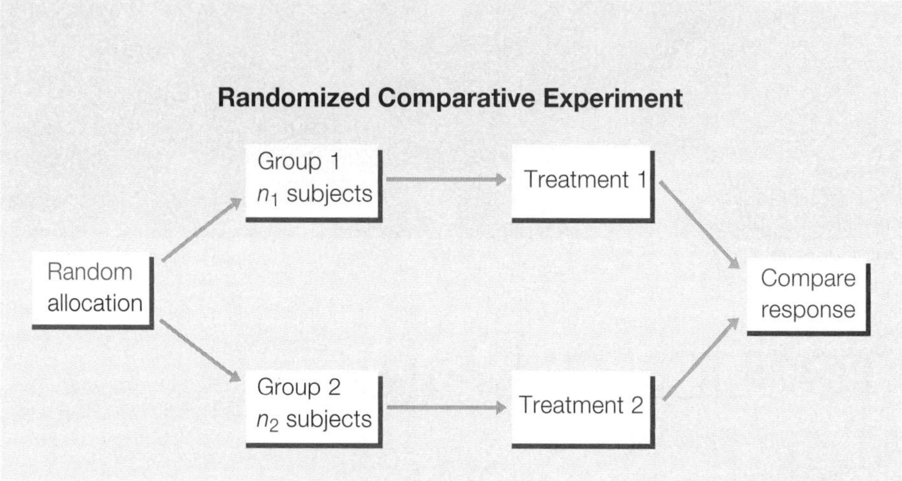

Part II Summary

Here are the most important skills you should have acquired from reading Chapters 8 to 10.

A. Sampling

1. Identify the population in a sampling situation.
2. Recognize bias due to voluntary response samples and other inferior sampling methods.
3. Use software or Table B of random digits to select a simple random sample (SRS) from a population.
4. Recognize the presence of undercoverage and nonresponse as sources of error in a sample survey. Recognize the effect of the wording of questions on the responses.
5. Use software or Table B of random digits to select a stratified random sample from a population when the strata are identified.

B. Experiments

1. Recognize whether a study is an observational study or an experiment.
2. Recognize bias due to confounding of explanatory variables with lurking variables in either an observational study or an experiment.
3. Identify the factors (explanatory variables), treatments, response variables, and individuals or subjects in an experiment.
4. Outline the design of a completely randomized experiment using a diagram like that in Figure 9.3 (page 234). The diagram in a specific case

should show the sizes of the groups, the specific treatments, and the response variable.

5. Use software or Table B of random digits to carry out the random assignment of subjects to groups in a completely randomized experiment.
6. Recognize the placebo effect. Recognize when the double-blind technique should be used.
7. Explain why randomized comparative experiments can give good evidence for cause-and-effect relationships.

C. Data Ethics (Optional topic)

1. Understand the purpose of institutional review boards.
2. Understand what informed consent means.
3. Explain the difference between confidentiality and anonymity in research studies.

ICING THE KICKER

The football team lines up for what they hope will be the winning field goal . . . and the other team calls time-out. "Make the kicker think about it" is their motto. Does "icing the kicker" really work? That is, does the probability of making a field goal go down when the kicker must wait around during the time-out? This isn't a simple question. A detailed statistical study considered the distance, the weather, the kicker's skill, and so on. The conclusion is cheering to coaches: yes, icing the kicker does reduce the probability of success.

Test Yourself

The following questions include both multiple-choice and short-answer questions and calculations. They will help you review the basic ideas and skills presented in Chapters 8–10.

Elephants and bees. *Elephants sometimes damage crops in Africa. It turns out that elephants dislike bees. They recognize beehives in areas where they are common and avoid them. Can this be used to keep elephants away from trees? A group in Kenya placed active beehives in some trees and empty beehives in others, whereas others received no beehives.*[1] *Will elephant damage be less in trees with hives? Will even empty hives keep elephants away? Use this information to answer Questions 11.1 and 11.2.*

© David Paynter/age fotostock

11.1 This experiment has

(a) two factors, beehives present or absent.
(b) matched pairs.
(c) three treatments.
(d) stratification by beehive.

11.2 The response in this experiment is

(a) the type of crop.
(b) the presence or absence of bees.
(c) the presence or absence of hives.
(d) elephant damage.

11.3 **Do you trust the Internet?** You want to ask a sample of college students the question, "How much do you trust information about health that you find on the Internet—a great deal, somewhat, not much, or not at all?" You try out this and other questions on a pilot group of six students chosen from your class. The class members are

Adams	Devore	Guo	Newberg	Shoepf
Aeffner	Ding	Heaton	Paulsen	Spagnola
Barnes	Drake	Huling	Payton	Terry
Bower	Eckstein	Kahler	Prince	Vore
Burke	Fassnacht	Kessis	Pulak	Wallace
Cao	Fullmer	Lu	Rabin	Wanner
Cisse	Gandhi	Mattos	Roberts	Zhang

(a) Describe how you will label the students to select the sample.
(b) Use the *Simple Random Sample* applet, other software, or Table B, beginning at line 115, to select the six students in the sample.
(c) What is the response variable in this study?

American Community Survey. *Each month the U.S. Census Bureau's American Community Survey mails survey forms to 300,000 households asking questions about demographic, social, economic, and housing characteristics such as mortgage and utility costs. Telephone calls are made to households that don't return the form. In one month, responses were obtained from 295,000 of the households contacted. Use this information to answer Questions 11.4 to 11.6.*

11.4 The sample is

(a) the 300,000 households initially contacted.
(b) the 295,000 households that responded.
(c) the 5,000 households that did not respond.
(d) all U.S. households.

11.5 The population of interest is

(a) all households with mortgages.
(b) the 300,000 households contacted.
(c) only U.S. households with phones.
(d) all U.S. households.

11.6 A source of bias in this survey is

(a) voluntary response.
(b) nonresponse.
(c) the fact that the survey was not double-blind.
(d) that only U.S. households were contacted.

11.7 The Nurses' Health Study has interviewed a sample of more than 100,000 female registered nurses every two years since 1976. The study finds that "light-to-moderate drinkers had a significantly lower risk of death" than either nondrinkers or heavy drinkers. The Nurses' Health Study is

(a) an observational study.
(b) an experiment.
(c) uncategorizable without more information.

11.8 At a local health club, a researcher samples 75 people whose primary exercise is cardiovascular and 75 people whose primary exercise is strength training. The researcher's objective is to assess the effect of type of exercise on cholesterol. Each subject reported to a clinic to have his or her cholesterol measured. The subjects were unaware of the purpose of the study, and the technician measuring the cholesterol was not aware of the subjects' type of exercise. This is

(a) an observational study.
(b) an experiment, but not a double-blind experiment.
(c) a double-blind experiment.
(d) a matched pairs experiment.

11.9 A common definition of "binge drinking" is five or more drinks at one setting for men, and four or more for women. An observational study finds that students who binge have lower average GPA than those who don't. Suggest two lurking variables that may be confounded with binge drinking, and be sure to give a reason why you have chosen each of these variables. The possibility of confounding means that we can't conclude that binge drinking *causes* lower GPA.

11.10 In 2000, when the federal budget showed a large surplus, the Pew Research Center asked random samples of adults two questions about using the remaining surplus. Both questions stated that Social Security would be "fixed."

> Question A: *Should the money be used for a tax cut, or should it be used to fund new government programs?*
>
> Question B: *Should the money be used for a tax cut, or should it be spent on programs for education, the environment, health care, crime-fighting, and military defense?*

One of these questions drew 60% favoring a tax cut. The other drew only 22%. Which wording pulls respondents toward a tax cut? Why?

Effects of TV advertising. *What are the effects of repeated exposure to an advertising message? The answer may depend both on the length of the ad and on how often it is repeated. To investigate this question, assign undergraduate students (the subjects) to view a 40-minute television program that includes commercials for a digital camera. There are three versions of the commercial: a 30-second commercial, a 45-second commercial, and a 60-second commercial. For each version of the commercial, some subjects will see the commercial once during the program and others will see the commercial twice during the program. After watching the TV program, all the subjects fill out a questionnaire that produces an "intention to buy" score with values between 0 and 100. Use this information to answer Questions 11.11 and 11.12.*

11.11 The number of treatments in this study is
(a) 2.
(b) 3.
(c) 4.
(d) 6.

11.12 The response in this experiment is
(a) the length of time of the commercial.
(b) the number of times the commercial is shown.
(c) the "intention to buy" score.
(d) the type of television program shown.

11.13 The website of the PBS television program *NOVA Science Now* invites viewers to vote on issues such as re-creating the virus responsible for the deadly flu epidemic of 1918. This online poll is unusual in offering detailed arguments for both sides. Of the 790 viewers who read the arguments and voted, 64% said that re-creating the virus was justified.[2] Explain to someone who knows no statistics why these 790 responses probably don't represent the opinions of all American adults.

11.14 A study attempts to determine whether a football filled with helium travels farther when kicked than one filled with air. Each subject kicks twice: once with a football filled with helium, and once with a football filled with air. The order of the type of football kicked is randomized. This is an example of
(a) a matched pairs experiment.
(b) a randomized controlled experiment.
(c) a stratified experiment.
(d) the placebo effect.

11.15 In many countries, it has been found that people with higher IQs tend to have a greater incidence of myopia (nearsightedness). One of the more recent studies in Singapore looked at a sample of 1453 children aged 10 to 12 years. Participants were given both a nonverbal IQ test and an eye exam.[3] This is as an example of
(a) a two factor study with factors IQ and myopia.
(b) an observational study.
(c) a single blind experiment as only the experimenters were unaware of the children's IQs and vision.
(d) a matched pairs experiment.

11.16 **(Optional topic)** A researcher promises subjects that he will only release statistical summaries of the results of the study to the public, although he will privately retain information about your data. This researcher is promising subjects
(a) anonymity.
(b) confidentiality.
(c) minimal risk.

11.17 (Optional topic) Informed consent means

(a) subjects are informed in advance about the nature of a study and any risk of harm it may bring.
(b) subjects are volunteers and are adults.
(c) subjects have been interviewed by an institutional review board prior to the start of the study.

Supplementary Exercises

Supplementary exercises apply the skills you have learned in ways that require more thought or more elaborate use of technology.

11.18 Sampling students. You want to investigate the attitudes of students at your school toward the school's policy on sexual harassment. You have a grant that will pay the costs of contacting about 500 students.

(a) Specify the exact population for your study. For example, will you include part-time students?
(b) Describe your sample design. Will you use a stratified sample?
(c) Briefly discuss the practical difficulties that you anticipate. For example, how will you contact the students in your sample?

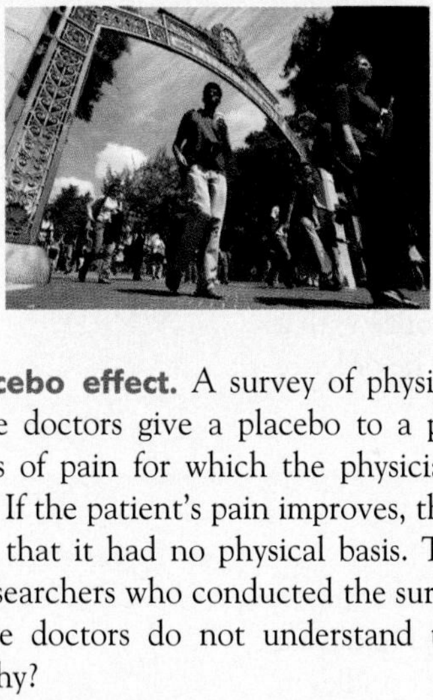
Justin Sullivan/Getty Images

11.19 The placebo effect. A survey of physicians found that some doctors give a placebo to a patient who complains of pain for which the physician can find no cause. If the patient's pain improves, these doctors conclude that it had no physical basis. The medical school researchers who conducted the survey claimed that these doctors do not understand the placebo effect. Why?

11.20 Informed consent. The requirement that human subjects give their informed consent to participate in an experiment can greatly reduce the number of available subjects. For example, a study of new teaching methods asks the consent of parents for their children to be taught by either a new method or the standard method. Many parents do not return the forms, so their children must continue to follow the standard curriculum. Why is it not correct to consider these children as part of the control group along with children who are randomly assigned to the standard method?

11.21 Fixing health care. The cost of health care and health insurance is the biggest health concern among Americans, even ahead of cancer and other diseases. Changing to a national government health insurance system is controversial. An opinion poll will give different results depending on the wording of the question asked. For each of the following claims, say whether including it in the question would *increase* or *decrease* the percent of a poll sample who support a government health insurance system.

(a) A national system would mean that everybody has health insurance.
(b) A national system would probably require an increase in taxes.
(c) Eliminating private insurance companies and their profits would reduce insurance costs.
(d) A national system would limit the medical treatments available to contain costs.

11.22 Market research. Stores advertise price reductions to attract customers. What type of price cut is most attractive? Market researchers prepared ads for athletic shoes announcing different levels of discounts (20%, 40%, or 60%). The student subjects who read the ads were also given "inside information" about the fraction of shoes on sale (50% or 100%). Each subject then rated the attractiveness of the sale on a scale of 1 to 7.[4]

(a) There are two factors. Make a sketch like Figure 9.2 (page 231) that displays the treatments formed by all combinations of levels of the factors.
(b) Outline a completely randomized design using 60 student subjects. Use software or Table B at line 111 to choose the subjects for the first treatment.

11.23 Making french fries. Few people want to eat discolored french fries. Potatoes are kept refrigerated before being cut for french fries to prevent spoiling and preserve flavor. But immediate processing of cold potatoes causes discoloring due to complex chemical reactions. The potatoes must therefore be brought to room temperature before processing. Design an experiment in which tasters will rate the color and flavor of french fries prepared from several groups of potatoes. The potatoes will be freshly picked or stored for a month at room temperature or stored for a month refrigerated. They will then be sliced and cooked either immediately or after an hour at room temperature.

(a) What are the factors and their levels, the treatments, and the response variables?
(b) Describe and outline the design of this experiment.
(c) It is efficient to have each taster rate fries from all treatments. How will you use randomization in presenting fries to the tasters?

11.24 How long did I work? A psychologist wants to know if the difficulty of a task influences our estimate of how long we spend working at it. She designs two sets of mazes that subjects can work through on a computer. One set has easy mazes, and the other has hard mazes. Subjects work until told to stop (after six minutes, but subjects do not know this). They are then asked to estimate how long they worked. The psychologist has 30 students available to serve as subjects.

(a) Describe the design of a completely randomized experiment to learn the effect of difficulty on estimated time.
(b) Describe the design of a matched pairs experiment using the same 30 subjects.

11.25 Alcohol and heart attacks. Many studies have found that people who drink alcohol in moderation have lower risk of heart attack than either nondrinkers or heavy drinkers. Does alcohol consumption also improve survival after a heart attack? One study followed 1913 people who were hospitalized after severe heart attacks. In the year before their heart attacks, 47% of these people did not drink, 36% drank moderately, and 17% drank heavily. After four years, fewer of the moderate drinkers had died.[5]

(a) Is this an observational study or an experiment? Why? What are the explanatory and response variables?
(b) Suggest some lurking variables that may be confounded with the drinking habits of the subjects. The possible confounding makes it difficult to conclude that drinking habits explain death rates.

PART III

Chris Clinton/Getty Images

From Data Production to Inference

PROBABILITY AND SAMPLING DISTRIBUTIONS

CHAPTER 12 Introducing Probability

CHAPTER 13 General Rules of Probability*

CHAPTER 14 Binomial Distributions*

CHAPTER 15 Sampling Distributions

FOUNDATIONS OF INFERENCE

CHAPTER 16 Confidence Intervals: The Basics

CHAPTER 17 Tests of Significance: The Basics

CHAPTER 18 Inference in Practice

CHAPTER 19 From Data Production to Inference: Part III Review

Armed with designs for producing trustworthy data, we continue our journey toward *statistical inference*. Exploratory data analysis allows us to examine the data obtained from sampling or experiments, but simply describing or looking for patterns in the data at hand is often not the primary goal. Usually data are used to answer specific questions, posed before the data are collected. If the sample has been selected using the principles presented in Chapter 8, the sample can tell us about important aspects of the population from which it was obtained. In a comparative experiment, the data can indicate how strong the evidence is that our treatment would be superior to the placebo for a broader class of circumstances.

Generalizing the results of sampling or experiments to a larger group of individuals or a broader class of circumstances is one goal of statistical inference. The conclusions of inference use the language of *probability*, the mathematics of chance. Chapters 12 and 15 present the ideas we need, and the optional Chapters 13 and 14 add more detail. Armed with designs for producing trustworthy data, data analysis to examine the data, and the language of probability, we are prepared to understand the big ideas of inference in Chapters 16, 17, and 18. These chapters are the foundation for the discussion of inference in practice that occupies the rest of the book.

© Eyebyte/Alamy

CHAPTER 12

Introducing Probability

In this chapter we cover...

12.1 The idea of probability
12.2 The search for randomness*
12.3 Probability models
12.4 Probability rules
12.5 Finite and discrete probability models
12.6 Continuous probability models
12.7 Random variables
12.8 Personal probability*

Why is probability, the mathematics of chance behavior, needed to understand statistics, the science of data? Let's look at a typical sample survey.

EXAMPLE 12.1 Do You Own a Gun?

What proportion of all U.S. adults own a gun? We don't know, but we do have results from the Gallup Poll. Gallup took a random sample of 1005 adults. The poll found that 342 of the people in the sample own a gun. The proportion who own a gun is

$$\text{sample proportion} = \frac{342}{1005} = 0.34 \text{ (that is, 34\%)}$$

If the sample was a simple random sample of all adults,[1] then all adults had the same chance to be among the chosen 1005. It would be reasonable to use this 34% as an estimate of the unknown proportion in the population. It's a *fact* that 34% of the sample claimed to own a gun—we know because Gallup asked them. We don't know what percent of all adults own a gun, but we *estimate* that about 34% did. This is a basic move in statistics: use a result from a sample to estimate something about a population. ■

What if Gallup took a second random sample of 1005 adults? The new sample would have different people in it. It is almost certain that there would not be exactly 342 positive responses. That is, Gallup's estimate of the proportion of adults who own a gun will vary from sample to sample. Could it happen that one random sample finds that 34% of adults own a gun, and a second random sample finds that 56% own a gun? *Random samples eliminate bias from the act of choosing a sample, but they can still be wrong because of the variability that results when we choose at random.* If the variation when we take repeated samples from the same population is too great, we can't trust the results of any one sample.

CAUTION

This is where we need facts about probability to make progress in statistics. Because Gallup uses chance to choose its samples, the laws of probability govern the behavior of the samples. Gallup says that the probability is 0.95 that an estimate from one of its samples comes within ±4 percentage points of the truth about the population of all adults. The first step toward understanding this statement is to understand what "probability 0.95" means. Our purpose in this chapter is to understand the language of probability, but without going into the full mathematics of probability theory.

12.1 The idea of probability

To understand why we can trust random samples and randomized comparative experiments, we must look closely at chance behavior. The big fact that emerges is this: **chance behavior is unpredictable in the short run but has a regular and predictable pattern in the long run.**

Toss a coin, or choose a random sample. The result can't be predicted in advance because the result will vary when you toss the coin or choose the sample repeatedly. But there is still a regular pattern in the results, a pattern that emerges clearly only after many repetitions. This remarkable fact is the basis for the idea of probability.

EXAMPLE 12.2 Coin Tossing

When you toss a coin, there are only two possible outcomes: heads or tails. Figure 12.1 shows the results of tossing a coin 5000 times, twice. For each number of tosses from 1 to 5000, we have plotted the proportion of those tosses that gave a head. Trial A (solid red line) begins tail, head, tail, tail. You can see that the proportion of heads for Trial A starts at 0 on the first toss, rises to 0.5 when the second toss gives a head, then falls to 0.33 and 0.25 as we get two more tails. Trial B, on the other hand, starts with five straight heads, so the proportion of heads is 1 until the sixth toss.

The proportion of tosses that produce heads is quite variable at first. Trial A starts low, and Trial B starts high. As we make more and more tosses, however, the proportion of heads for both trials gets close to 0.5 and stays there. If we made yet a third trial at tossing the coin a great many times, the proportion of heads would again settle down to 0.5 in the long run. This is the intuitive idea of probability. Probability 0.5 means "occurs half the time in a very large number of trials." The probability 0.5 appears as a horizontal line on the graph. ■

We might suspect that a coin has probability 0.5 of coming up heads just because the coin has two sides. But we can't be sure. In fact, spinning a penny on a flat surface, rather than tossing the coin, gives heads probability about 0.45 rather than 0.5.[2] The idea of probability is empirical. That is, it is based on observation rather than theorizing. Probability describes what happens in very many trials, and we must actually observe many trials to pin down a probability. In the case of tossing a coin, some diligent people have in fact made thousands of tosses.

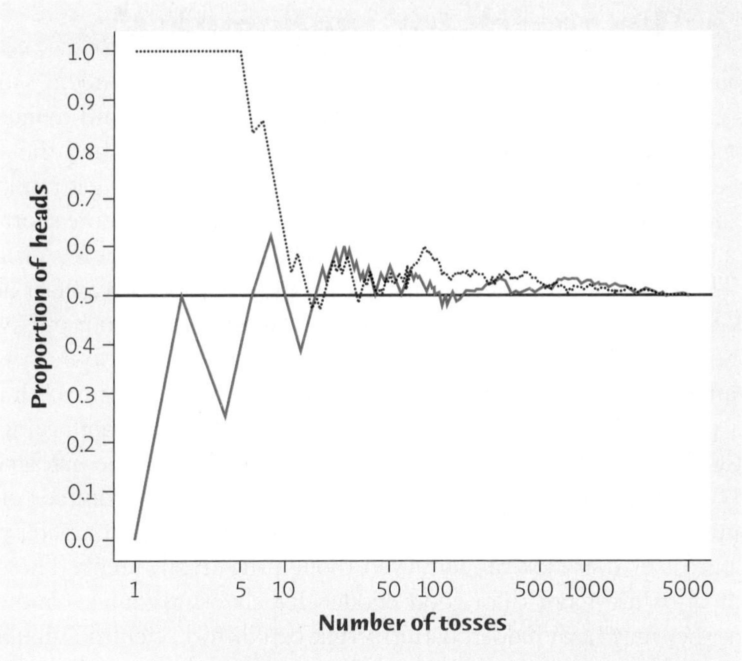

FIGURE 12.1 The proportion of tosses of a coin that give a head changes as we make more tosses. Eventually, however, the proportion approaches 0.5, the probability of a head. This figure shows the results of two trials of 5000 tosses each.

EXAMPLE 12.3 Some Coin Tossers

The French naturalist Count Buffon (1707–1788) tossed a coin 4040 times. Result: 2048 heads, or proportion 2048/4040 = 0.5069 for heads.

Around 1900, the English statistician Karl Pearson heroically tossed a coin 24,000 times. Result: 12,012 heads, a proportion of 0.5005.

While imprisoned by the Germans during World War II, the South African mathematician John Kerrich tossed a coin 10,000 times. Result: 5067 heads, a proportion of 0.5067. ■

Randomness and Probability

We call a phenomenon **random** if individual outcomes are uncertain but there is nonetheless a regular distribution of outcomes in a large number of repetitions.

The **probability** of any outcome of a random phenomenon is the proportion of times the outcome would occur in a very long series of repetitions.

The best way to understand randomness is to observe random behavior, as in Figure 12.1. You can do this with physical devices like coins, but computer simulations (imitations) of random behavior allow faster exploration. The *Probability* applet is a computer simulation that animates Figure 12.1. It allows you to choose the probability of a head and simulate any number of tosses of a coin with that probability. Experience shows that the proportion of heads gradually settles down close to the probability. Equally important, it also shows that *the proportion in a small or moderate number of tosses can be far from the probability. Probability describes only what happens in the long run.* Of course, we can never observe a probability exactly. We could always continue tossing the coin, for example. Mathematical probability is an idealization based on imagining what would happen in an indefinitely long series of trials.

DOES GOD PLAY DICE?

Few things in the world are truly random in the sense that no amount of information will allow us to predict the outcome. But, according to the branch of physics called quantum mechanics, randomness does rule events inside individual atoms. Although Albert Einstein helped quantum theory get started, he always insisted that nature must have some fixed reality, not just probabilities. "I shall never believe that God plays dice with the world," said the great scientist. A century after Einstein's first work on quantum theory, it appears that he was wrong.

12.2 The search for randomness*

Random numbers are valuable. They are used to choose random samples, to shuffle the cards in online poker games, to encrypt our credit card numbers when we buy online, and as part of simulations of the flow of traffic and the spread of epidemics. Where does randomness come from, and how can we get random numbers? We defined randomness by how it behaves: unpredictably in the short run, but showing a regular pattern in the long run. Probability describes the long-run regular pattern. That many things are random in this sense is an observed fact about the world. Not all these things are "really" random. Here's a quick tour of how to find random behavior and get random numbers.

The easiest way to get random numbers is from a *computer program.* Of course, a computer program just does what it is told to do. Run the program again, and you get exactly the same result. The random numbers in Table B, the outcomes of the *Probability* applet, and the random numbers that shuffle cards for online poker come from computer programs, so they aren't "really" random. Clever computer programs produce outcomes that look random even though they really aren't. These *pseudo-random numbers* are more than good enough for choosing samples and shuffling cards. But they may have hidden patterns that can distort scientific simulations.

You might think that *physical devices such as coins and dice* produce really random outcomes. But a tossed coin obeys the laws of physics. If we knew all the inputs of the toss (forces, angles, and so on), then we could say in advance whether the outcome will be heads or tails. The outcome of a toss is predictable rather than random. Why do the results of tossing a coin *look* random? The outcomes are extremely sensitive to the inputs, so that very small changes in the forces you apply when you toss a coin change the outcome from heads to tails and back again. In practice, the outcomes are not predictable. Probability is a lot more useful than physics for describing coin tosses.

We call a phenomenon with "small changes in, big changes out" behavior *chaotic.* If we can feed chaotic behavior into a computer, we can do better than pseudorandom numbers. Coins and dice are awkward, but you can go to the website `www.random.org` to get random numbers from radio noise in the atmosphere, a chaotic phenomenon that is easy to feed to a computer.

Is anything really random? As far as current science can say, behavior inside atoms really is random—that is, there isn't any way to predict behavior in advance no matter how much information we have. It was this "really, truly random" idea that Einstein disliked as he watched the new science of quantum mechanics emerge. Really, truly random numbers generated from the radioactive decay of atoms is available at the HotBits website, `www.fourmilab.ch/hotbits`.

LaunchPad Online Resources

- The **Snapshots video**, *Probability*, discusses some examples of probability and its interpretation.

Apply Your Knowledge

12.1 A Straight Flush. You read online that the probability of being dealt a straight flush in a five-card poker hand is 1/64,974. Explain carefully what this means. In particular, explain why it does *not* mean that if you are dealt 64,974 five-card poker hands, one will be a straight flush.

*This short discussion is optional.

12.2 Probability Says . . . Probability is a measure of how likely an event is to occur. Match one of the probabilities that follow with each statement of likelihood given. (The probability is usually a more exact measure of likelihood than is the verbal statement.)

0 0.01 0.45 0.50 0.55 0.99 1

(a) This event is impossible. It can never occur.

(b) This event is certain. It will occur on every trial.

(c) This event is very likely, but it will not occur once in a while in a long sequence of trials.

(d) This event will occur slightly less often than not.

12.3 Random Digits. The table of random digits (Table B) was produced by a random mechanism that gives each digit probability 0.1 of being a 0.

APPLET

(a) What proportion of the first 200 digits (those in the first five lines) in the table are 0s? This proportion is an estimate, based on 200 repetitions, of the true probability, which we know is 0.1.

(b) The *Probability* applet can imitate random digits. Set the probability of heads in the applet to 0.1. Check "Show true probability" to show this value on the graph. A head stands for a 0 in the random digit table, and a tail stands for any other digit. Simulate 200 digits (set the "Number of Tosses" to 200 and click on "Toss"). What was the result of your 200 tosses?

12.4 The Long Run but Not the Short Run. Our intuition about chance behavior is not very accurate. In particular, we tend to expect that the long-run pattern described by probability will show up in the short run as well. For example, we tend to think that tossing a coin 10 times will give close to five heads.

APPLET

(a) Set the probability of heads in the *Probability* applet to 0.5 and the number of tosses to 10. Click "Toss" to simulate 10 tosses of a balanced coin. What was the proportion of heads?

(b) Click "Reset" and toss again. The simulation is fast, so do it 25 times and keep a record of the proportion of heads in each set of 10 tosses. Make a stemplot of your results. You see that the result of tossing a coin 10 times is quite variable and need not be very close to the probability 0.5 of heads.

12.3 Probability models

Gamblers have known for centuries that the fall of coins, cards, and dice displays clear patterns in the long run. The idea of probability rests on the observed fact that the average result of many thousands of chance outcomes can be known with near certainty. How can we give a mathematical description of long-run regularity?

To see how to proceed, think first about a very simple random phenomenon, tossing a coin once. When we toss a coin, we cannot know the outcome in advance. What *do* we know? We are willing to say that the outcome will be either heads or tails. We believe that each of these outcomes has probability 1/2. This description of coin tossing has two parts:

- A list of possible outcomes
- A probability for each outcome

Such a description is the basis for all *probability models.* Here is the basic vocabulary we use.

Probability Models

The **sample space *S*** of a random phenomenon is the set of all possible outcomes.

An **event** is an outcome or a set of outcomes of a random phenomenon. That is, an event is a subset of the sample space.

A **probability model** is a mathematical description of a random phenomenon consisting of two parts: a sample space S and a way of assigning probabilities to events.

A sample space *S* can be very simple or very complex. When we toss a coin once, there are only two outcomes: heads and tails. The sample space is $S = \{H, T\}$. When Gallup draws a random sample of 1005 adults, the sample space contains all possible choices of 1005 of the 241 million adults in the United States. This *S* is extremely large. Each member of *S* is a possible sample, so *S* is the collection or "space" of all possible samples. This explains the term *sample space.*

EXAMPLE 12.4 Rolling Dice

Rolling two dice is a common way to lose money in casinos. There are 36 possible outcomes when we roll two dice and record the up-faces in order (first die, second die). Figure 12.2 displays these outcomes. They make up the sample space *S*. "Roll a 5" is an event, call it *A*, that contains four of these 36 outcomes:

$A =$ 

How can we assign probabilities to this sample space? We can find the actual probabilities for two specific dice only by actually tossing the dice many times, and even then only approximately. So we will give a probability model that assumes ideal, perfectly balanced dice. This model will be quite accurate for carefully made casino dice and less accurate for the cheap dice that come with a board game.

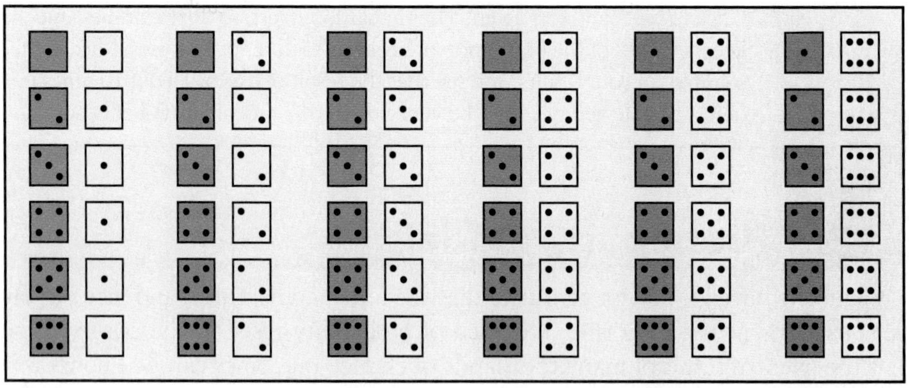

FIGURE 12.2
The 36 possible outcomes in rolling two dice. If the dice are carefully made, all these outcomes have the same probability.

If the dice are perfectly balanced, all 36 outcomes in Figure 12.2 will be *equally likely*. That is, each of the 36 outcomes will come up on one thirty-sixth of all rolls in the long run. So each outcome has probability 1/36. There are four outcomes in the event *A* ("roll a 5"), so this event has probability 4/36. In this way we can assign a probability to any event. So we have a complete probability model. ■

In general, if all outcomes in a sample space are equally likely, we find the probability of any event by

$$\frac{\text{number of ways the event could occur}}{\text{total number of outcomes in the sample space}}$$

EXAMPLE 12.5 Rolling Dice and Counting the Spots

Gamblers care only about the total number of spots on the up-faces of the dice. The sample space for rolling two dice and counting the spots is

$$S = \{2, 3, 4, 5, 6, 7, 8, 9, 10, 11, 12\}$$

CAUTION Comparing this S with Figure 12.2 reminds us that *we can change S by changing the detailed description of the random phenomenon we are describing.*

What are the probabilities for this new sample space? The 11 possible outcomes are *not* equally likely because there are six ways to roll a 7 and only one way to roll a 2 or a 12. That's the key: each outcome in Figure 12.2 has probability 1/36. So "roll a 7" has probability 6/36 because this event contains six of the 36 outcomes. Similarly, "roll a 2" has probability 1/36, and "roll a 5" (four outcomes from Figure 12.2) has probability 4/36. Here is the complete probability model:

Total Spots	2	3	4	5	6	7	8	9	10	11	12
Probability	1/36	2/36	3/36	4/36	5/36	6/36	5/36	4/36	3/36	2/36	1/36

■

Apply Your Knowledge

12.5 Sample Space. Choose a student at random from a large statistics class. Describe a sample space S for each of the following. (In some cases, you may have some freedom in specifying S.)

(a) Does the student live on campus or off campus?

(b) What is the student's age in years?

(c) What are the last four digits of the student's cell phone number?

(d) Record the student's letter grade at the end of the course.

12.6 Role-Playing Games. Computer games in which the players take the roles of characters are very popular. They go back to earlier tabletop games such as Dungeons & Dragons. These games use many different types of dice. A four-sided die has faces with one of the numbers 1, 2, 3, or 4 appearing at the bottom of each visible face.

(a) What is the sample space for rolling a four-sided die twice (numbers on first and second rolls)? Follow the example of Figure 12.2.

(b) What is the assignment of probabilities to outcomes in this sample space? Assume that the die is perfectly balanced, and follow the method of Example 12.4.

Slpix/Dreamstime.com

12.7 Role-Playing Games. The intelligence of a character in a game is determined by rolling the four-sided die twice and adding 1 to the sum of the numbers. Start with your work in Exercise 12.6 to give a probability model (sample space and probabilities of outcomes) for the character's intelligence. Follow the method of Example 12.5.

12.4 Probability rules

In Examples 12.4 and 12.5, we found probabilities for tossing dice. As random phenomena go, dice are pretty simple. Even so, we had to assume idealized, perfectly balanced dice. In most situations, it isn't easy to give a "correct" probability model. We can make progress by listing some facts that must be true for *any* assignment of probabilities. These facts follow from the idea of probability as "the long-run proportion of repetitions on which an event occurs."

1. **Any probability is a number between 0 and 1.** Any proportion is a number between 0 and 1, so any probability is also a number between 0 and 1. An event with probability 0 never occurs, and an event with probability 1 occurs in every trial. An event with probability 0.5 occurs in half the trials in the long run.
2. **All possible outcomes together must have probability 1.** Because some outcome must occur on every trial, the sum of the probabilities for all possible outcomes must be exactly 1.
3. **If two events have no outcomes in common, the probability that one or the other occurs is the sum of their individual probabilities.** If one event occurs in 40% of all trials, a different event occurs in 25% of all trials, and the two can never occur together, then one or the other occurs on 65% of all trials because 40% + 25% = 65%.
4. **The probability that an event does not occur is 1 minus the probability that the event does occur.** If an event occurs in (say) 70% of all trials, then it fails to occur in the other 30%. The probability that an event occurs and the probability that it does not occur always add to 100%, or 1.

EQUALLY LIKELY?

A game of bridge begins by dealing all 52 cards in the deck to the four players, 13 to each. If the deck is well shuffled, all the immense number of possible hands will be equally likely. But don't expect the hands that appear in newspaper bridge columns to reflect the equally likely probability model. Writers on bridge choose "interesting" hands, especially those that lead to high bids that are rare in actual play.

We can use mathematical notation to state Facts 1 to 4 more concisely. Capital letters near the beginning of the alphabet denote events. If A is any event, we write its probability as $P(A)$. Here are our probability facts in formal language. As you apply these rules, remember that they are just another form of intuitively true facts about long-run proportions.

Probability Rules

Rule 1. The probability $P(A)$ of any event A satisfies $0 \leq P(A) \leq 1$.

Rule 2. If S is the sample space in a probability model, then $P(S) = 1$.

Rule 3. Two events A and B are **disjoint** if they have no outcomes in common and so can never occur together. If A and B are disjoint,

$$P(A \text{ or } B) = P(A) + P(B)$$

This is the **addition rule for disjoint events.**

Rule 4. For any event A,

$$P(A \text{ does not occur}) = 1 - P(A)$$

The addition rule extends to more than two events that are disjoint in the sense that no two have any outcomes in common. If events A, B, and C are disjoint, the probability that one of these events occurs is $P(A) + P(B) + P(C)$.

EXAMPLE 12.6 Using the Probability Rules

We already used the addition rule for disjoint events, without calling it by that name, to find the probabilities in Example 12.5. The event "roll a 5" contains the four disjoint outcomes displayed in Example 12.4, so the addition rule (Rule 3) says that its probability is

P(roll a 5) = P(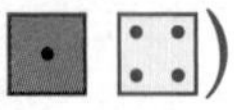) + P(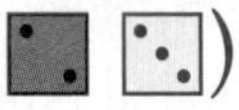) + P(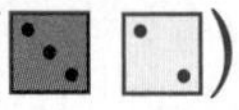) + P(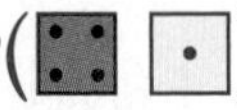)

$$= \frac{1}{36} + \frac{1}{36} + \frac{1}{36} + \frac{1}{36}$$

$$= \frac{4}{36} = 0.111$$

Check that the probabilities in Example 12.5, found using the addition rule, are all between 0 and 1 and add to exactly 1. That is, this probability model obeys Rules 1 and 2.

What is the probability of rolling anything other than a 5? By Rule 4,

$$P(\text{roll does not give a 5}) = 1 - P(\text{roll a 5})$$
$$= 1 - 0.111 = 0.889$$

Our model assigns probabilities to individual outcomes. To find the probability of an event, just add the probabilities of the outcomes that make up the event. For example:

$$P(\text{outcome is odd}) = P(3) + P(5) + P(7) + P(9) + P(11)$$
$$= \frac{2}{36} + \frac{4}{36} + \frac{6}{36} + \frac{4}{36} + \frac{2}{36}$$
$$= \frac{18}{36} = \frac{1}{2} \blacksquare$$

© Image Source Plus/Alamy

LaunchPad Online Resources

The **StatBoards video**, *The Four Basic Probability Rules*, discusses some examples of the use of the four probability rules, including examples of disjoint events.

Apply Your Knowledge

12.8 **Who Takes the GMAT?** In many settings, the "rules of probability" are just basic facts about percents. The Graduate Management Admission Test (GMAT) website provides the following information about the geographic regions of those who took the test in 2012–13: 2.3% were from Africa; 0.4% were from Australia and the Pacific Islands; 2.7% were from Canada; 11.5% were from Central and South Asia; 29.1% were from East and Southeast Asia; 2.1% were from Eastern Europe; 2.8% were from Mexico, the Caribbean, and Latin America; 3.6% were from the Middle East; 38.0% were from the United States; and 7.4% were from Western Europe.[3]

(a) What percent of those who took the test in 2012–13 were from North America (either Canada, the United States, Mexico, the Caribbean, or Latin America)? Which rule of probability did you use to find the answer?

(b) What percent of those who took the test in 2012–13 were from some other region than the United States? Which rule of probability did you use to find the answer?

12.9 **Overweight?** Although the rules of probability are just basic facts about percents or proportions, we need to be able to use the language of events and their probabilities. Choose an American adult aged 20 years and over at random. Define two events:

A = the person chosen is obese

B = the person chosen is overweight, but not obese

According to the National Center for Health Statistics, $P(A) = 0.36$ and $P(B) = 0.33$.

(a) Explain why events A and B are disjoint.

(b) Say in plain language what the event "A or B" is. What is $P(A \text{ or } B)$?

(c) If C is the event that the person chosen has normal weight or less, what is $P(C)$?

12.10 Languages in Canada. Canada has two official languages: English and French. Choose a Canadian at random and ask, "What is your mother tongue?" Here is the distribution of responses, combining many separate languages from the province of Quebec:[4]

Language	English	French	Other
Probability	0.083	0.789	?

(a) What is the probability that a Canadian's mother tongue is either English or French?

(b) What probability should replace "?" in the distribution?

(c) What is the probability that a Canadian's mother tongue is not English?

12.11 Are They Disjoint? Which of the following pairs of events, A and B, are disjoint? Explain your answers.

(a) A person is selected at random. A is the event "sex of the person selected is male" and B is the event "sex of the person selected is female."

(b) A person is selected at random. A is the event "the person selected earns more than \$100,000 per year" and B is the event "the person selected earns more than \$250,000 per year."

(c) A pair of dice are tossed. A is the event "one of the dice is a 3" and B is the event "the sum of the two dice is 3."

12.5 Finite and discrete probability models

Examples 12.4, 12.5, and 12.6 illustrate one way to assign probabilities to events: assign a probability to every individual outcome, then add these probabilities to find the probability of any event. This idea works well when there are only a finite (fixed and limited) number of outcomes.

Finite Probability Model

A probability model with a finite sample space is called **finite**.

To assign probabilities in a finite model, list the probabilities of all the individual outcomes. These probabilities must be numbers between 0 and 1 that add to exactly 1. The probability of any event is the sum of the probabilities of the outcomes making up the event.

Finite probability models are sometimes called **discrete** probability models. However, discrete probability models include finite sample spaces as well as sample spaces that are infinite and equivalent to the set of all positive integers. An example of a discrete but not finite sample space would be the sample space for the number of free-throw attempts until a basketball player makes her first free throw. This could occur on her first attempt, her second attempt, her third attempt, and so on. Assigning probabilities to individual outcomes in an infinite discrete sample space is more complicated than for a finite sample space. In this book we will often refer to finite probability models as discrete, and in practice statisticians often refer to finite probability models as discrete.

EXAMPLE 12.7 Benford's Law

Faked numbers in tax returns, invoices, or expense account claims often display patterns that aren't present in legitimate records. Some patterns, such as too many round numbers, are obvious and easily avoided by a clever crook. Others are more subtle. It is a striking fact that the first digits of numbers in legitimate records often follow a model known as Benford's law.[5] Call the first digit of a randomly chosen record X for short. Benford's law gives this probability model for X (note that a first digit can't be 0):

First Digit X	1	2	3	4	5	6	7	8	9
Probability	0.301	0.176	0.125	0.097	0.079	0.067	0.058	0.051	0.046

Check that the probabilities of the outcomes sum to exactly 1. This is therefore a valid finite (or discrete) probability model. With these probabilities, investigators can detect fraud by comparing the first digits in records such as invoices paid by a business.

The probability that a first digit is equal to or greater than 6 is

$$\begin{aligned} P(X \geq 6) &= P(X = 6) + P(X = 7) + P(X = 8) + P(X = 9) \\ &= 0.067 + 0.058 + 0.051 + 0.046 = 0.222 \end{aligned}$$

This is less than the probability that a record has first digit 1,

$$P(X = 1) = 0.301$$

Fraudulent records tend to have too few 1s and too many higher first digits.

CAUTION *Note that the probability that a first digit is greater than or equal to 6 is not the same as the probability that a first digit is strictly greater than 6.* The latter probability is

$$P(X > 6) = 0.058 + 0.051 + 0.046 = 0.155$$

The outcome $X = 6$ is included in "greater than or equal to" and is not included in "strictly greater than." ■

EXAMPLE 12.8 A Completely Randomized Design

In Chapter 9 (page 235) we discussed completely randomized experimental designs. Suppose you have three men—Ari, Luis, Troy—and three women—Ana, Deb, and Hui—for an experiment, Three of the six subjects are to be assigned completely at random to a new experimental weight loss treatment and three to a placebo. Here all

Treatment Group	Treatment Group
Ari, Luis, Troy	Luis, Troy, Ana
Ari, Luis, Ana	Luis, Troy, Deb
Ari, Luis, Deb	Luis, Troy, Hui
Ari, Luis, Hui	Luis, Ana, Deb
Ari, Troy, Ana	Luis, Ana, Hui
Ari, Troy, Deb	Luis, Deb, Hui
Ari, Troy, Hui	Troy, Ana, Deb
Ari, Ana, Deb	Troy, Ana, Hui
Ari, Ana, Hui	Troy, Deb, Hui
Ari, Deb, Hui	Ana, Deb, Hui

20 possible ways of selecting three of these subjects for the treatment group (the remaining three are in the placebo group).

With a completely randomized design, each of these 20 possible treatment groups is equally likely, thus each has probability 1/20 of being the actual group assigned to the treatment. Notice that the chance that all the men are assigned to the treatment group is 1/20, and the chance that the treatment group consists of either all men or all women is 2/20. ■

LaunchPad Online Resources

- The **StatClips Examples video**, *Probability Distributions Example A*, discusses Benford's law.

Apply Your Knowledge

12.12 Rolling a Die. Figure 12.3 displays several possible finite probability models for rolling a die. We can learn which model is actually *accurate* for a particular die only by rolling the die many times. However, some of the models are not *valid*. That is, they do not obey the rules. Which are valid and which are not? In the case of the invalid models, explain what is wrong.

	Probability			
Outcome	Model 1	Model 2	Model 3	Model 4
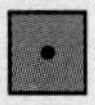	1/7	1/3	1/3	1
	1/7	1/6	1/6	1
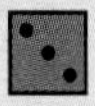	1/7	1/6	1/6	2
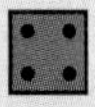	1/7	0	1/6	1
	1/7	1/6	1/6	1
	1/7	1/6	1/6	2

FIGURE 12.3
Four assignments of probabilities to the six faces of a die, for Exercise 12.12.

12.13 Benford's Law. The first digit of a randomly chosen expense account claim follows Benford's law (Example 12.7). Consider the events

$$A = \{\text{first digit is 4 or greater}\}$$
$$B = \{\text{first digit is even}\}$$

(a) What outcomes make up the event A? What is $P(A)$?

(b) What outcomes make up the event B? What is $P(B)$?

(c) What outcomes make up the event "A or B"? What is $P(A \text{ or } B)$? Why is this probability not equal to $P(A) + P(B)$?

12.14 Dinner at Home. Choose a parent with children under age 18 in the United States at random and ask, "How many nights a week out of seven does your family eat dinner together at home?" Call the response X for short. Based on a large sample survey, here is a probability model for the answer you will get:[6]

Nights	0	1	2	3	4	5	6	7
Probability	0.05	0.00	0.07	0.08	0.10	0.17	0.15	0.38

(a) Verify that this is a valid finite probability model.

(b) Describe the event $X < 4$ in words. What is $P(X < 4)$?

(c) Express the event "eat dinner together at home at least once a week" in terms of X. What is the probability of this event?

12.6 Continuous probability models

When we use the table of random digits to select a digit between 0 and 9, the finite probability model assigns probability 1/10 to each of the 10 possible outcomes. Suppose that we want to choose a number at random between 0 and 1, allowing *any* number between 0 and 1 as the outcome. Software random number generators will do this. For example, here is the result of asking software to produce five random numbers (to a fixed number of decimal places) between 0 and 1:

0.2893511   0.3213787   0.5816462   0.9787920   0.4475373

The sample space is now an entire interval of numbers:

$$S = \{\text{all numbers between 0 and 1}\}$$

Call the outcome of the random number generator Y for short. How can we assign probabilities to such events as $\{0.3 \le Y \le 0.7\}$? As in the case of selecting a random digit, we would like all possible outcomes to be equally likely. But we cannot assign probabilities to each individual value of Y and then add them, because there is an infinite continuum of possible values. In fact, we cannot even make a list of the individual values of Y. For example, what is the next largest value of Y after 0.3?

We use a new way of assigning probabilities directly to events—as *areas under a density curve.* Any density curve has area exactly 1 underneath it, corresponding to total probability 1. We met density curves as models for data in Chapter 3 (page 75).

REALLY RANDOM DIGITS

For purists, the RAND Corporation long ago published a book titled *One Million Random Digits*. The book lists 1,000,000 digits that were produced by a very elaborate physical randomization and really are random. An employee of RAND once said that this is not the most boring book that RAND has ever published.

Continuous Probability Model

A **continuous probability model** assigns probabilities as areas under a density curve. The area under the curve and above any range of values is the probability of an outcome in that range.

EXAMPLE 12.9 Random Numbers

The random number generator will spread its output uniformly across the entire interval from 0 to 1 as we allow it to generate a long sequence of numbers. Figure 12.4 is a histogram of 10,000 random numbers. They are quite uniform, but not exactly so. The bar heights would all be exactly equal (1000 numbers for each bar) if the 10,000 numbers were exactly uniform. In fact, the counts vary from a low of 978 to a high of 1060.

As in Chapter 3, we have adjusted the histogram scale so that the total area of the bars is exactly 1. Now we can add the density curve that describes the distribution of

FIGURE 12.4
The probability model for the outcomes of a software random number generator, for Example 12.9. Compare the histogram of 10,000 actual outcomes with the uniform density curve that spreads probability evenly between 0 and 1.

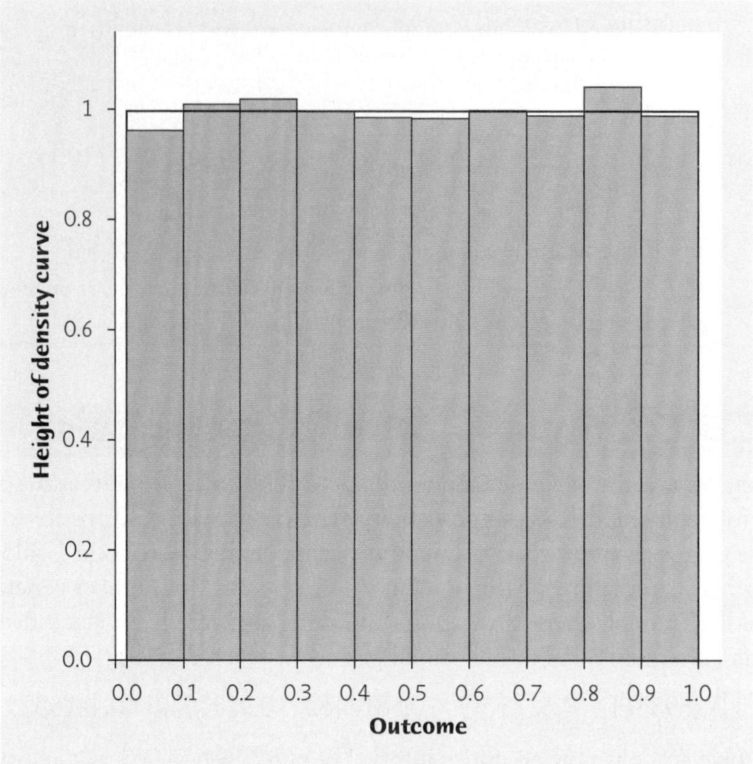

uniform distribution

perfectly random numbers. This density curve also appears in Figure 12.4. It has height 1 over the interval from 0 to 1. This is the density curve of a **uniform distribution**. It is the continuous probability model for the results of generating very many random numbers. Like the probability models for perfectly balanced coins and dice, the density curve is an idealized description of the outcomes of a perfectly uniform random number generator. It is a good approximation for software outcomes, but even 10,000 tries isn't enough for actual outcomes to look exactly like the idealized model. ■

The uniform density curve has height 1 over the interval from 0 to 1. The area under the curve is 1, and the probability of any event is the area under the curve and above the interval that corresponds to the event in question. Figure 12.5 illustrates finding probabilities as areas under the density curve. The probability that the random number generator produces a number between 0.3 and 0.7 is

$$P(0.3 \leq Y \leq 0.7) = 0.4$$

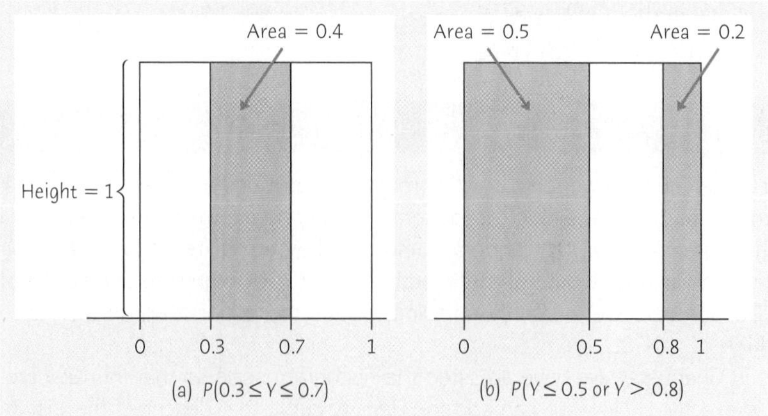

FIGURE 12.5
Probability as area under a density curve. The uniform density curve spreads probability evenly between 0 and 1.

because the area under the density curve and above the interval from 0.3 to 0.7 is 0.4. The height of the curve is 1, and the area of a rectangle is the product of height and length, so the probability of any interval of outcomes is just the length of the interval. Similarly,

$$P(Y \leq 0.5) = 0.5$$
$$P(Y > 0.8) = 0.2$$
$$P(Y \leq 0.5 \text{ or } Y > 0.8) = 0.7$$

The last event consists of two nonoverlapping intervals, so the total area above the event is found by adding two areas, as illustrated by Figure 12.5(b). This assignment of probabilities obeys all our rules for probability.

Continuous probability models assign probabilities to intervals of outcomes rather than to individual outcomes. In fact, *all continuous probability models assign probability 0 to every individual outcome.* Only intervals of values have positive probability. To see that this is true, consider a specific outcome such as $P(Y = 0.8)$. The probability of any interval is the same as its length. The point 0.8 has no length, so its probability is 0. Put another way, $P(Y > 0.8)$ and $P(Y \geq 0.8)$ are both 0.2 because that is the area in Figure 12.5(b) between 0.8 and 1.

We can use any density curve to assign probabilities. The density curves that are most familiar to us are the Normal curves. **Normal distributions are continuous probability models** as well as descriptions of data. There is a close connection between a Normal distribution as an idealized description for data and a Normal probability model. If we look at the heights of all young women, we find that they closely follow the Normal distribution with mean $\mu = 64.3$ inches and standard deviation $\sigma = 2.7$ inches. This is a distribution for a large set of data. Now choose one young woman at random. Call her height X. If we repeat the random choice very many times, the distribution of values of X is the same Normal distribution that describes the heights of all young women.

EXAMPLE 12.10 The Heights of Young Women

APPLET

What is the probability that a randomly chosen young woman has height between 68 and 70 inches? The height X of the woman we choose has the $N(64.3, 2.7)$ distribution. We want $P(68 \leq X \leq 70)$. This is the area under the Normal curve in Figure 12.6. Software or the *Normal Density Curve* applet will give us the answer at once: $P(68 \leq X \leq 70) = 0.0679$.

We can also find the probability by standardizing and using Table A, the table of standard Normal probabilities. We will reserve capital Z for a standard Normal variable.

Blend Images/Getty Images

$$P(68 \leq X \leq 70) = P\left(\frac{68 - 64.3}{2.7} \leq \frac{X - 64.3}{2.7} \leq \frac{70 - 64.3}{2.7}\right)$$
$$= P(1.37 \leq Z \leq 2.11)$$
$$= P(Z \leq 2.11) - P(Z \leq 1.37)$$
$$= 0.9826 - 0.9147 = 0.0679$$

The calculation is the same as those we did in Chapter 3. Only the language of probability is new. ■

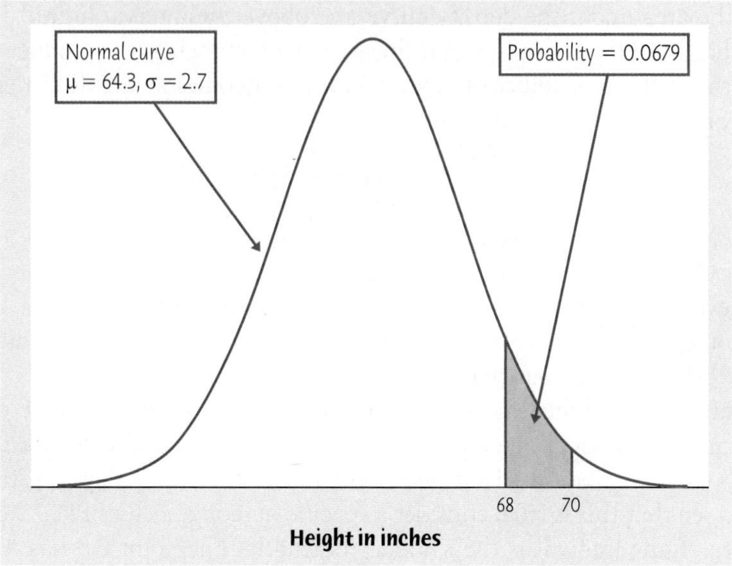

FIGURE 12.6
The probability in Example 12.10 as an area under a Normal curve.

LaunchPad Online Resources

- The **StatClips Examples videos**, *Probability Distributions Example B and C*, discuss examples of calculating probabilities for continuous probability models.

Apply Your Knowledge

12.15 Random Numbers. Let Y be a random number between 0 and 1 produced by the idealized random number generator described in Example 12.9 and Figure 12.4. Find the following probabilities:

(a) $P(Y \leq 0.6)$

(b) $P(Y < 0.6)$

(c) $P(0.4 \leq Y \leq 0.8)$

(d) $P(0.4 < Y \leq 0.8)$

12.16 Adding Random Numbers. Generate two random numbers between 0 and 1 and take X to be their sum. The sum X can take any value between 0 and 2. The density curve of X is the triangle shown in Figure 12.7.

(a) Verify by geometry that the area under this curve is 1.

(b) What is the probability that X is less than 1? (Sketch the density curve, shade the area that represents the probability, then find that area. Do this for part (c) also.)

(c) What is the probability that X is less than 0.5?

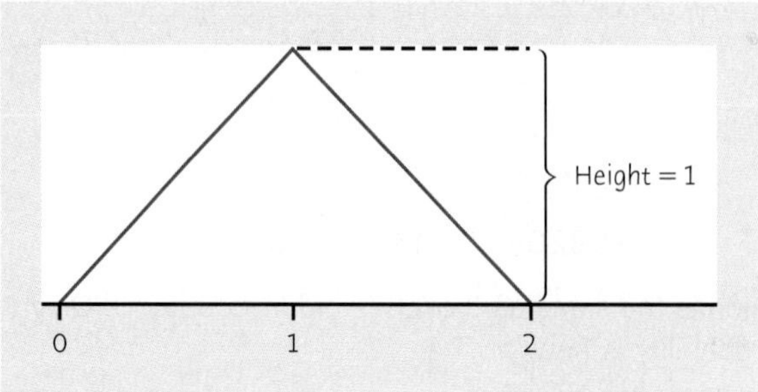

FIGURE 12.7
The density curve for the sum of two random numbers, for Exercise 12.16. This density curve spreads probability between 0 and 2.

12.17 The Medical College Admission Test. The Normal distribution with mean $\mu = 25.3$ and standard deviation $\sigma = 6.5$ is a good description of the total score on the Medical College Admission Test (MCAT).[7] This is a continuous probability model for the score of a randomly chosen student. Call the score of a randomly chosen student X for short.

(a) Write the event "the student chosen has a score of 35 or higher" in terms of X.

(b) Find the probability of this event.

12.7 Random variables

Examples 12.7, 12.9, and 12.10 use a shorthand notation that is often convenient. In Example 12.10, we let X stand for the result of choosing a woman at random and measuring her height. We know that X would take a different value if we made another random choice. Because its value changes from one random choice to another, we call the height X a *random variable.*

Random Variable

A **random variable** is a variable whose value is a numerical outcome of a random phenomenon.

The **probability distribution** of a random variable X tells us what values X can take and how to assign probabilities to those values.

We usually denote random variables by capital letters near the end of the alphabet, such as X or Y. Of course, the random variables of greatest interest to us are outcomes such as the mean $\bar{x}$ of a random sample, for which we will keep the familiar notation. There are two main types of random variables, corresponding to two types of probability models: *discrete* and *continuous.* Notice that neither a finite sample space nor a sample space consisting of all positive integers would be considered continuous, because neither sample space is an interval of all possible numbers (to an arbitrary number of decimal places) between two given values. Thus, we classify random variables as either discrete or continuous, rather than as either finite or continuous.

EXAMPLE 12.11 Discrete and Continuous Random Variables

The first digit X in Example 12.7 is a random variable whose possible values are the whole numbers $\{1, 2, 3, 4, 5, 6, 7, 8, 9\}$. The distribution of X assigns a probability to each of these outcomes. A **discrete random variable** has a finite list of possible outcomes.

discrete random variable

Compare the output Y of the random number generator in Example 12.8. The values of Y fill the entire interval of numbers between 0 and 1. The probability distribution of Y is given by its density curve, shown in Figure 12.4. A **continuous random variable** can take on any value in an interval, with probabilities given as areas under a density curve. ■

continuous random variable

We have defined a random variable to be a variable whose value is a *numerical* outcome of a random phenomenon. However, there are situations in which the outcomes of a random phenomenon are not numerical. In Example 12.2, the possible outcomes of a coin toss are heads and tails. We can represent these outcomes by numbers, letting 1 represent heads and 0 represent tails. Using numbers to represent the nonnumerical outcomes of a random phenomenon is a common practice in statistics. For mathematical purposes it is convenient to restrict random variables to take on numerical values, even if this means representing nonnumerical outcomes by numbers.

Apply Your Knowledge

12.18 Grades in an Economics Course. Indiana University posts the grade distributions for its courses online.[8] Students in Economics 201 in the fall 2013 semester received these grades: 18% A, 8% A−, 7% B+, 16% B, 11% B−, 6% C+, 12% C, 4% C−, 3% D+, 6% D, 4% D−, and 5% F. Choose an Economics 201 student at random. To "choose at random" means to give every student the same chance to be chosen. The student's grade on a four-point scale (with A = 4, A− = 3.7, B+ = 3.3, B = 3.0, B− = 2.7, C+ = 2.3, C = 2.0, C− = 1.7, D+ = 1.3, D = 1.0, D− = 0.7, and F = 0.0) is a discrete random variable X with this probability distribution:

Value of X	0.0	0.7	1.0	1.3	1.7	2.0	2.3	2.7	3.0	3.3	3.7	4.0
Probability	0.05	0.04	0.06	0.03	0.04	0.12	0.06	0.11	0.16	0.07	0.08	0.18

(a) Say in words what the meaning of $P(X \geq 3.0)$ is. What is this probability?

(b) Write the event "the student got a grade poorer than B−" in terms of values of the random variable X. What is the probability of this event?

12.19 Running a Mile. A study of 12,000 able-bodied male students at the University of Illinois found that their times for the mile run were approximately Normal with mean 7.11 minutes and standard deviation 0.74 minute.[9] Choose a student at random from this group and call his time for the mile Y.

(a) Say in words what the meaning of $P(Y \geq 8)$ is. What is this probability?

(b) Write the event "the student could run a mile in less than 6 minutes" in terms of values of the random variable Y. What is the probability of this event?

12.8 Personal probability*

We began our discussion of probability with one idea: the probability of an outcome of a random phenomenon is the proportion of times that outcome would occur in a very long series of repetitions. This idea ties probability to actual outcomes. It allows us, for example, to estimate probabilities by simulating random phenomena. Yet we often meet another, quite different, idea of probability.

EXAMPLE 12.12 Joe and the Chicago Cubs

Joe sits staring into his beer as his favorite baseball team, the Chicago Cubs, loses another game. The Cubbies have some good young players, so let's ask Joe, "What's the chance that the Cubs will go to the World Series next year?" Joe brightens up. "Oh, about 10%," he says.

Does Joe assign probability 0.10 to the Cubs appearing in the World Series? The outcome of next year's pennant race is certainly unpredictable, but we can't reasonably ask what would happen in many repetitions. Next year's baseball season

*This short section is optional.

will happen only once and will differ from all other seasons in players, weather, and many other ways. If probability measures "what would happen if we did this many times," Joe's 0.10 is not a probability. Probability is based on data about many repetitions of the same random phenomenon. Joe is giving us something else, his personal judgment. ■

Although Joe's 0.10 isn't a probability in our usual sense, it gives useful information about Joe's opinion. More seriously, a company asking, "How likely is it that building this plant will pay off within five years?" can't employ an idea of probability based on many repetitions of the same thing. The opinions of company officers and advisers are nonetheless useful information, and these opinions can be expressed in the language of probability. These are *personal probabilities.*

Personal Probability

A **personal probability** of an outcome is a number between 0 and 1 that expresses an individual's judgment of how likely the outcome is.

Rachel's opinion about the Cubs may differ from Joe's, and the opinions of several company officers about the new plant may differ. Personal probabilities are indeed personal: they vary from person to person. Moreover, if two people assign different personal probabilities to an event, it may be difficult or impossible to determine who is more correct. If we say, "In the long run, this coin will come up heads 60% of the time," we can find out if we are right by actually tossing the coin several thousand times. If Joe says, "I think the Cubs have a 10% chance of going to the World Series next year," that's just Joe's opinion. Why think of personal probabilities as probabilities? Because *any set of personal probabilities that makes sense obeys the same basic Rules 1 to 4 that describe any legitimate assignment of probabilities to events.* If Joe thinks there's a 10% chance that the Cubs will go to the World Series, he must also think that there's a 90% chance that they won't go. There is just one set of rules of probability, even though we now have two interpretations of what probability means.

Statistics in Your World: WHAT ARE THE ODDS?

Gamblers often express chance in terms of *odds* rather than probability. Odds of A to B against an outcome means that the probability of that outcome is $B/(A + B)$. So "odds of 5 to 1" is another way of saying "probability 1/6." A probability is always between 0 and 1, but odds range from 0 to infinity. Although odds are mainly used in gambling, they give us a way to make very small probabilities clearer. "Odds of 999 to 1" may be easier to understand than "probability 0.001."

Apply Your Knowledge

12.20 Will You Have an Accident? The probability that a randomly chosen driver will be involved in an accident in the next year is about 0.2. This is based on the proportion of millions of drivers who have accidents. "Accident" includes things like crumpling a fender in your own driveway, not just highway accidents.

(a) What do you think is your own probability of being in an accident in the next year? This is a personal probability.

(b) Give some reasons why your personal probability might be a more accurate prediction of your "true chance" of having an accident than the probability for a random driver.

(c) Almost everyone says their personal probability is lower than the random driver probability. Why do you think this is true?

12.21 Winning the ACC Tournament. The annual Atlantic Coast Conference men's basketball tournament has temporarily taken Joe's mind off of the

Chicago Cubs. He says to himself, "I think that Syracuse has probability 0.1 of winning. North Carolina's probability is twice Syracuse's, and Duke's probability is three times Syracuse's."

(a) What are Joe's personal probabilities for North Carolina and Duke?

(b) What is Joe's personal probability that one of the 12 teams other than Syracuse, North Carolina, and Duke will win the tournament?

CHAPTER 12 SUMMARY

Chapter Specifics

- A **random phenomenon** (chance experiment) has outcomes that we cannot predict but that nonetheless have a regular distribution in very many repetitions.
- The **probability** of an event is the proportion of times the event occurs in many repeated trials of a random phenomenon.
- A **probability model** for a random phenomenon consists of a sample space S and an assignment of probabilities P.
- The **sample space** S is the set of all possible outcomes of the random phenomenon. Sets of outcomes are called **events.** P assigns a number $P(A)$ to an event A as its probability.
- Any assignment of probability must obey the rules that state the basic properties of probability:
  1. $0 \leq P(A) \leq 1$ for any event A.
  2. $P(S) = 1$.
  3. **Addition rule for disjoint events:** Events A and B are **disjoint** if they have no outcomes in common. If A and B are disjoint, then $P(A \text{ or } B) = P(A) + P(B)$.
  4. For any event A, $P(A \text{ does not occur}) = 1 - P(A)$.
- When a sample space S contains finitely many possible values, a **finite probability model** assigns each of these values a probability between 0 and 1 such that the sum of all the probabilities is exactly 1. The probability of any event is the sum of the probabilities of all the values that make up the event. Finite probability models are also referred to as discrete probability models.
- A sample space can contain all values in some interval of numbers. A **continuous probability model** assigns probabilities as areas under a **density curve.** The probability of any event is the area under the curve above the values that make up the event.
- A **random variable** is a variable taking numerical values determined by the outcome of a random phenomenon. The **probability distribution** of a random variable X tells us what the possible values of X are and how probabilities are assigned to those values.
- A random variable X and its distribution can be **discrete** or **continuous.** The distribution of a **discrete random variable** with finitely many possible values gives the probability of each value. A **continuous random variable** takes all values in some interval of numbers. A density curve describes the probability distribution of a continuous random variable.

Link It

This chapter begins our study of probability. The important fact is that random phenomena are unpredictable in the short run, but have a regular and predictable behavior in the long run. Probability rules and probability models provide the tools for describing and predicting the long-run behavior of random phenomena.

Probability helps us understand why we can trust random samples and randomized comparative experiments, the subjects of Chapters 8 and 9. It is the key to generalizing what we learn from data produced by random samples and randomized comparative experiments to some wider universe or population. How we use probability to do this will be the topic of the remainder of this book.

LaunchPad Online Resources

If you are having difficulty with any of the sections of this chapter, these online resources should help prepare you to solve the exercises at the end of this chapter.

- **StatTutor** starts with a video review of each section and asks a series of questions to check your understanding.
- **LearningCurve** provides you with a series of questions about the chapter geared to your level of understanding.

CHECK YOUR SKILLS

12.22 You read in a book on poker that the probability of being dealt two pairs in a five-card poker hand is 1/21. This means that

(a) if you deal thousands of poker hands, the fraction of them that contain two pairs will be very close to 1/21.
(b) if you deal 21 poker hands, exactly one of them will contain two pairs.
(c) if you deal 21,000 poker hands, exactly 1000 of them will contain two pairs.

12.23 A basketball player shoots five free throws during a game. The sample space for counting the number she makes is

(a) S = any number between 0 and 1.
(b) S = whole numbers 0 to 5.
(c) S = all sequences of five hits or misses, like HMMHH.

Here is the probability model for the political affiliation of a randomly chosen adult in the United States.[10] *Exercises 12.24 to 12.27 use this information.*

Political affiliation	Republican	Independent	Democrat	Other
Probability	0.25	0.42	0.30	?

12.24 This probability model is

(a) continuous. (b) finite. (c) equally likely.

12.25 The probability that a randomly chosen American adult's political affiliation is "Other" must be

(a) any number between 0 and 1.
(b) 0.03. (c) 0.3.

12.26 What is the probability that a randomly chosen American adult is a member of one of the two major political parties (Republicans and Democrats)?

(a) 0.42 (b) 0.55 (c) 0.97

12.27 What is the probability that a randomly chosen American adult is not a Republican?

(a) 0.25 (b) 0.75 (c) 0.03

12.28 In a table of random digits such as Table B, each digit is equally likely to be any of 0, 1, 2, 3, 4, 5, 6, 7, 8, or 9. What is the probability that a digit in the table is a 7?

(a) 1/9 (b) 1/10 (c) 9/10

12.29 In a table of random digits such as Table B, each digit is equally likely to be any of 0, 1, 2, 3, 4, 5, 6, 7, 8, or 9. What is the probability that a digit in the table is 7 or greater?

(a) 7/10 (b) 4/10 (c) 3/10

12.30 Choose an American household at random, and let the random variable X be the number of cars (including SUVs and light trucks) the residents own. Here is the probability model if we ignore the few households that own more than six cars:[11]

Number of cars X	0	1	2	3	4	5	6
Probability	0.09	0.33	0.36	0.14	0.05	0.02	0.01

A housing company builds houses with two-car garages. What percent of households have more cars than the garage can hold?

(a) 14% (b) 22% (c) 42%

12.31 Choose a common fruit fly *Drosophila melanogaster* at random. Call the length of the thorax (where the wings and legs attach) Y. The random variable Y has the Normal distribution with mean $\mu = 0.800$ millimeter (mm) and standard deviation $\sigma = 0.078$ mm. The probability $P(Y > 1)$ that the fly you choose has a thorax more than 1 mm long is about

(a) 0.995. (b) 0.5. (c) 0.005.

CHAPTER 12 EXERCISES

12.32 **Sample space.** In each of the following situations, describe a sample space S for the random phenomenon.

(a) A basketball player shoots four free throws. You record the sequence of hits and misses.

(b) A basketball player shoots four free throws. You record the number of baskets she makes.

Darrell Walker/HWMS/Icon SMI/Newscom

12.33 **Probability models?** In each of the following situations, state whether or not the given assignment of probabilities to individual outcomes is legitimate, that is, satisfies the rules of probability. Remember, a legitimate model need not be a practically reasonable model. If the assignment of probabilities is not legitimate, give specific reasons for your answer.

(a) Roll a six-sided die, and record the count of spots on the up-face:

P(1) = 0	P(2) = 1/6	P(3) = 1/3
P(4) = 1/3	P(5) = 1/6	P(6) = 0

(b) Deal a card from a shuffled deck:

P(clubs) = 12/52   P(diamonds) = 12/52
P(hearts) = 12/52   P(spades) = 16/52

(c) Choose a college student at random and record sex and enrollment status:

P(female full-time) = 0.56   P(male full-time) = 0.44
P(female part-time) = 0.24   P(male part-time) = 0.17

12.34 Education among young adults. Choose a young adult (aged 25 to 29) at random. The probability is 0.10 that the person chosen did not complete high school, 0.27 that the person has a high school diploma but no further education, and 0.34 that the person has at least a bachelor's degree.

(a) What must be the probability that a randomly chosen young adult has some education beyond high school but does not have a bachelor's degree?
(b) What is the probability that a randomly chosen young adult has at least a high school education?

12.35 Land in Canada. Canada's national statistics agency, Statistics Canada, says that the land area of Canada is 9,094,000 square kilometers. Of this land, 4,176,000 square kilometers are forested. Choose a square kilometer of land in Canada at random.

(a) What is the probability that the area you chose is forested?
(b) What is the probability that it is not forested?

12.36 Foreign-language study. Choose a student in a U.S. public high school at random and ask if he or she is studying a language other than English. Here is the distribution of results:

Language	Spanish	French	German	All others	None
Probability	0.30	0.08	0.02	0.03	0.57

(a) Explain why this is a legitimate probability model.
(b) What is the probability that a randomly chosen student is studying a language other than English?
(c) What is the probability that a randomly chosen student is studying French, German, or Spanish?

12.37 Car colors. Choose a new car or light truck at random and note its color. Here are the probabilities of the most popular colors for vehicles sold globally in 2013:[12]

Color	White	Black	Silver	Gray	Red	Beige, brown	Blue
Probability	0.25	0.18	0.18	0.12	0.09	0.08	0.07

(a) What is the probability that the vehicle you chose has any color other than those listed?
(b) What is the probability that a randomly chosen vehicle is neither white nor silver?

12.38 Drawing cards. You are about to draw a card at random (that is, all choices have the same probability) from a set of seven cards. Although you can't see the cards, here they are:

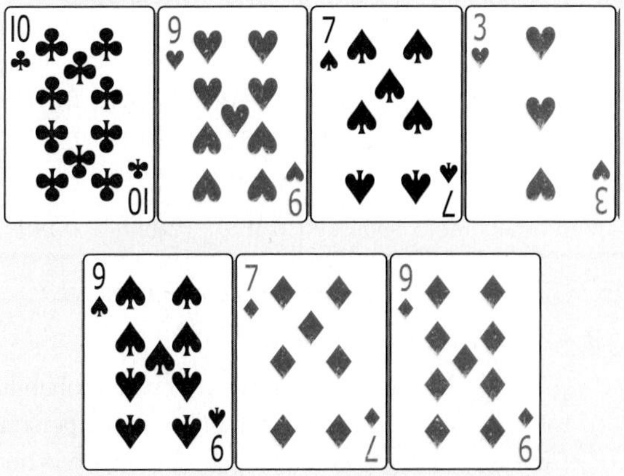

(a) What is the probability that you draw a 9?
(b) What is the probability that you draw a red 9?
(c) What is the probability that you do not draw a 7?

12.39 Loaded dice. There are many ways to produce crooked dice. To *load* a die so that 6 comes up too often and 1 (which is opposite 6) comes up too seldom, add a bit of lead to the filling of the spot on the 1 face. If a die is loaded so that 6 comes up with probability 0.2 and the probabilities of the 2, 3, 4, and 5 faces are not affected, what is the assignment of probabilities to the six faces?

12.40 A door prize. A party host gives a door prize to one guest chosen at random. There are 48 men and 42 women at the party. What is the probability that the prize goes to a woman? Explain how you arrived at your answer.

12.41 Race and ethnicity. The U.S. Census Bureau allows each person to choose from a long list of races. That is, in the eyes of the U.S. Census Bureau, you belong to whatever race you say you belong to. "Hispanic/Latino" is a separate category; Hispanics may be of any race. If we choose a resident of the United States at random, the U.S. Census Bureau gives these probabilities:[13]

	HISPANIC	NOT HISPANIC
Asian	0.002	0.051
Black	0.008	0.126
White	0.153	0.645
Other	0.006	0.009

(a) Verify that this is a legitimate assignment of probabilities.
(b) What is the probability that a randomly chosen American is Hispanic?
(c) Non-Hispanic whites are the historical majority in the United States. What is the probability that a randomly chosen American is not a member of this group?

Choose at random a person aged 15 to 44 years. Ask their age and who they live with (alone, with spouse, with other persons). Here is the probability model for 12 possible answers:[14]

	AGE IN YEARS			
	15–19	**20–24**	**25–34**	**35–44**
Alone	0.001	0.010	0.032	0.028
With spouse	0.001	0.021	0.151	0.208
With others (not a spouse)	0.168	0.139	0.149	0.092

Exercises 12.42 to 12.44 use this probability model.

12.42 Living arrangements.

(a) Why is this a legitimate finite probability model?
(b) What is the probability that the person chosen is a 15- to 19-year-old who lives with others (not a spouse)?
(c) What is the probability that the person is 15 to 19 years old?
(d) What is the probability that the person chosen lives with others (not a spouse)?

12.43 Living arrangements, continued.

(a) List the outcomes that make up the event

A = {The person chosen is *either* 15 to 19 years old *or* lives with others, or both}

(b) What is $P(A)$? Explain carefully why $P(A)$ is not the sum of the probabilities you found in parts (c) and (d) of Exercise 12.42.

12.44 Living arrangements, continued.

(a) What is the probability that the person chosen is 20 years old or older?
(b) What is the probability that the person chosen does not live alone?

12.45 Spelling errors. Spell-checking software catches "nonword errors" that result in a string of letters that is not a word, as when "the" is typed as "teh." When undergraduates are asked to type a 250-word essay (without spell-checking), the number X of nonword errors has the following distribution:

Value of X	0	1	2	3	4
Probability	0.1	0.2	0.3	0.3	0.1

(a) Is the random variable X discrete or continuous? Why?
(b) Write the event "at least one nonword error" in terms of X. What is the probability of this event?
(c) Describe the event $X \le 2$ in words. What is its probability? What is the probability that $X < 2$?

12.46 First digits again. A crook who never heard of Benford's law might choose the first digits of his faked invoices so that all of 1, 2, 3, 4, 5, 6, 7, 8, and 9 are equally likely. Call the first digit of a randomly chosen fake invoice W for short.

(a) Write the probability distribution for the random variable W.
(b) Find $P(W \ge 6)$ and compare your result with the Benford's law probability from Example 12.7.

12.47 Who gets interviewed? Abby, Deborah, Mei-Ling, Sam, and Roberto are students in a small seminar course. Their professor decides to choose two of them to interview about the course. To avoid unfairness, the choice will be made by drawing two names from a hat. (This is an SRS of size 2.)

(a) Write down all possible choices of two of the five names. This is the sample space.
(b) The random drawing makes all choices equally likely. What is the probability of each choice?
(c) What is the probability that Mei-Ling is chosen?
(d) Abby, Deborah, and Mei-Ling liked the course. Sam and Roberto did not like the course. What is the probability that both people selected liked the course?

12.48 Birth order. A couple plans to have three children. There are eight possible arrangements of girls and boys. For example, GGB means the first two children are girls and the third child is a boy. All eight arrangements are (approximately) equally likely.

© Picture Press/Alamy

(a) Write down all eight arrangements of the sexes of three children. What is the probability of any one of these arrangements?
(b) Let X be the number of girls the couple has. What is the probability that $X = 2$?
(c) Starting from your work in part (a), find the distribution of X. That is, what values can X take, and what are the probabilities for each value?

12.49 Unusual dice. Nonstandard dice can produce interesting distributions of outcomes. You have two balanced, six-sided dice. One is a standard die, with faces having 1, 2, 3, 4, 5, and 6 spots. The other die has three faces with 0 spots and three faces with 6 spots. Find the probability distribution for the total number of spots Y on the up-faces when you roll these two dice. (*Hint:* Start with a picture like Figure 12.2 for the possible up-faces. Label the three 0 faces on the second die 0a, 0b, 0c in your picture, and similarly distinguish the three 6 faces.)

12.50 A taste test. A tea-drinking Canadian friend of yours claims to have a very refined palate. She tells you that she can tell if, in preparing a cup of tea, milk is first added to the cup and then hot tea poured into the cup, or the hot tea is first poured into the cup and then the milk is added.[15] To test her claims, you prepare six cups of tea. Three have the milk added first and the other three the tea first. In a blind taste test, your friend tastes all six cups and is asked to identify the three that had the milk added first.

(a) How many different ways are there to select three of the six cups? (*Hint:* See Example 12.8.)
(b) If your friend is just guessing, what is the probability that she correctly identifies the three cups with the milk added first?

12.51 Random numbers. Many random number generators allow users to specify the range of the random numbers to be produced. Suppose you specify that the random number Y can take any value between 0 and 2. Then the density curve of the outcomes has constant height between 0 and 2, and height 0 elsewhere.

(a) Is the random variable Y discrete or continuous? Why?
(b) What is the height of the density curve between 0 and 2? Draw a graph of the density curve.
(c) Use your graph from part (b) and the fact that probability is area under the curve to find $P(Y \leq 1)$.

12.52 More random numbers. Find these probabilities as areas under the density curve you sketched in Exercise 12.51.

(a) $P(0.5 < Y < 1.3)$
(b) $P(Y \geq 0.8)$

12.53 Survey accuracy. A sample survey contacted an SRS of 2854 registered voters shortly before the 2012 presidential election and asked respondents whom they planned to vote for. Election results show that 51% of registered voters voted for Barack Obama. We will see later that in this situation the proportion of the sample who planned to vote for Barack Obama (call this proportion V) has approximately the Normal distribution with mean $\mu = 0.51$ and standard deviation $\sigma = 0.009$.

(a) If the respondents answer truthfully, what is $P(0.49 \leq V \leq 0.53)$? This is the probability that the sample proportion V estimates the population proportion 0.51 within plus or minus 0.02.
(b) In fact, 49% of the respondents said they planned to vote for Barack Obama ($V = 0.49$. If respondents answer truthfully, what is $P(V \leq 0.49)$?

12.54 Friends. How many close friends do you have? Suppose that the number of close friends adults claim to have varies from person to person with mean $\mu = 9$ and standard deviation $\sigma = 2.5$. An opinion poll asks this question of an SRS of 1100 adults. We will see later, in Chapter 19, that in this situation the sample mean response $\bar{x}$ has approximately the Normal distribution with mean 9 and standard deviation 0.075. What is $P(8.9 \leq \bar{x} \leq 9.1)$, the probability that the sample result $\bar{x}$ estimates the population truth $\mu = 9$ to within ± 0.1?

12.55 Playing Pick 4. The Pick 4 games in many state lotteries announce a four-digit winning number each day. Each of the 10,000 possible numbers 0000 to 9999 has the same chance of winning. You win if your choice matches the winning digits. Suppose your chosen number is 5974.

(a) What is the probability that the winning number matches your number exactly?
(b) What is the probability that the winning number has the same digits as your number *in any order?*

12.56 Nickels falling over. You may feel that it is obvious that the probability of a head in tossing a coin is about 1/2 because the coin has two faces. Such opinions are not always correct. Stand a nickel on edge on a hard, flat surface. Pound the surface with your hand so that the nickel falls over. What is the probability that it falls with heads upward? Make at least 50 trials to estimate the probability of a head.

12.57 What probability doesn't say. The idea of probability is that the *proportion* of heads in many tosses of a balanced coin eventually gets close to 0.5. But does the actual *count* of heads get close to one-half the number of tosses? Let's find out. Set the "Probability of heads" in the *Probability* applet to 0.5 and the number of tosses to 50. You can extend the number of tosses by clicking "Toss" again to get 50 more. Don't click "Reset" during this exercise.

(a) After 50 tosses, what is the proportion of heads? What is the count of heads? What is the difference between the count of heads and 25 (one-half the number of tosses)?
(b) Keep going to 150 tosses. Again record the proportion and count of heads and the difference between the count and 75 (half the number of tosses).
(c) Keep going. Stop at 250 tosses and again at 500 tosses to record the same facts. Although it may take a long time, the laws of probability say that the proportion of heads will always get close to 0.5 and also that the difference between the count of heads and half the number of tosses will always grow without limit.

12.58 LeBron's free throws. The basketball player LeBron James makes about three-quarters of his free throws over an entire season. Use the *Probability* applet or statistical software to simulate 100 free throws shot by a player who has probability 0.75 of making each shot. (In most software, the key phrase to look for is "Bernoulli trials." This is the technical term for independent trials with yes/no outcomes. Our outcomes here are "Hit" and "Miss.")

(a) What percent of the 100 shots did he hit?
(b) Examine the sequence of hits and misses. How long was the longest run of shots made? Of shots missed? (Sequences of random outcomes often show runs longer than our intuition thinks likely.)

12.59 **Simulating an opinion poll.** A 2013 opinion poll showed that about 33% of the American public have very little or no confidence in big business. Suppose that this is exactly true. Choosing a person at random then has probability 0.33 of getting one who has very little or no confidence in big business. Use the *Probability* applet or statistical software to simulate choosing many people at random. (In most software, the key phrase to look for is "Bernoulli trials." This is the technical term for independent trials with Yes/No outcomes. Our outcomes here are "Favorable" or "Not Favorable.")

(a) Simulate drawing 50 people, then 100 people, then 400 people. What proportion have very little or no confidence in big business in each case? We expect (but because of chance variation we can't be sure) that the proportion will be closer to 0.38 in longer runs of trials.
(b) Simulate drawing 50 people 10 times and record the percents in each sample who have very little or no confidence in big business. Then simulate drawing 400 people 10 times and again record the 10 percents. Which set of 10 results is less variable? We expect the results of samples of size 400 to be more predictable (less variable) than the results of samples of size 50. That is "long-run regularity" showing itself.

Exploring the Web

12.60 **Super Bowl odds.** Oddsmakers often list the odds for certain sporting events on the web. For example, one can find the current odds of winning the next Super Bowl for each NFL team. We found a list of such odds at `www.vegasinsider.com/nfl/odds/futures/`. When an oddsmaker says the odds are A to B of winning, he or she means that the probability of winning is $B/(A + B)$. For example, when we checked the website listed, the odds that the Seattle Seahawks would win Super Bowl XLIX were 6 to 1. This corresponds to a probability of winning of $1/(6 + 1) = 1/7$.

On the web, find the current odds, according to an oddsmaker, of winning the Super Bowl for each NFL team. Convert these odds to probabilities. Do these probabilities satisfy Rules 1 and 2 given in this chapter? If they don't, can you think of a reason why?

Lonely Planet Images/Getty Images

CHAPTER 13

General Rules of Probability*

In this chapter we cover...

13.1 Independence and the multiplication rule
13.2 The general addition rule
13.3 Conditional probability
13.4 The general multiplication rule
13.5 Independence again
13.6 Tree diagrams

Probability models can describe the flow of traffic through a highway system, a telephone interchange, or a computer processor; the genetic makeup of populations; the energy states of subatomic particles; the spread of epidemics or rumors; and the rate of return on risky investments. Although we are interested in probability mainly because it is the foundation for statistical inference, the mathematics of chance is important in many fields of study. Our introduction to probability in Chapter 12 concentrated on basic ideas and facts. Now we look at some further details. With more probability at our command, we can model more complex random phenomena.

Although we won't emphasize the math, everything in this chapter (and much more) follows from the four rules we met in Chapter 12. Here they are again.

*This chapter introduces some of the mathematics of probability. The material is not needed to understand the rest of the book.

Probability Rules

Rule 1. For any event A, $0 \leq P(A) \leq 1$.

Rule 2. If S is the sample space, $P(S) = 1$.

Rule 3. Addition rule for disjoint events: If A and B are **disjoint** events,

$$P(A \text{ or } B) = P(A) + P(B)$$

Rule 4. For any event A,

$$P(A \text{ does not occur}) = 1 - P(A)$$

13.1 Independence and the multiplication rule

Rule 3, the addition rule for disjoint events, describes the probability that *one or the other* of two events A and B occurs in the special situation when A and B cannot occur together. Now we will describe the probability that *both* events A and B occur, again only in a special situation.

You may find it helpful to draw a picture to display relations among several events. A picture like Figure 13.1 that shows the sample space S as a rectangular area and events as areas within S is called a **Venn diagram**. The events A and B in Figure 13.1 are disjoint because they do not overlap. The Venn diagram in Figure 13.2 illustrates two events that are not disjoint. The event {A and B} appears as the overlapping area that is common to both A and B. Can we find the probability P(A and B) that both events occur if we know the individual probabilities P(A) and P(B)?

Venn diagram

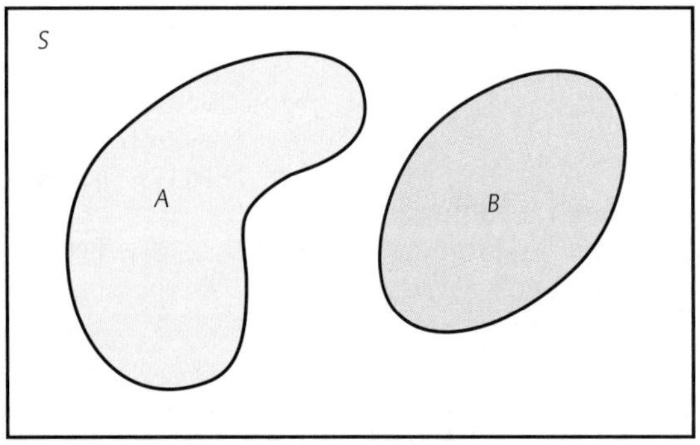

FIGURE 13.1
Venn diagram showing disjoint events *A* and *B*.

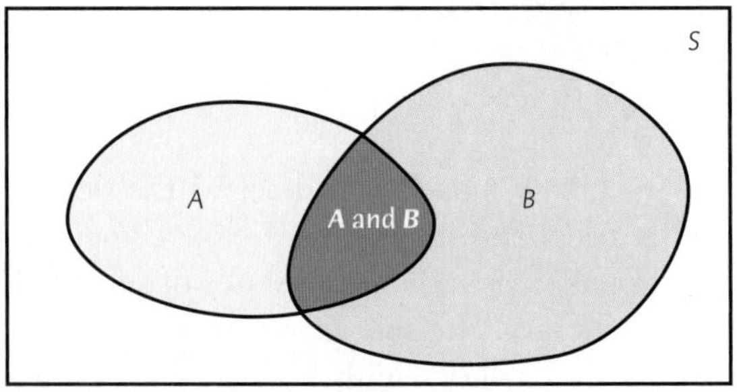

FIGURE 13.2
Venn diagram showing events *A* and *B* that are not disjoint. The event {*A* and *B*} consists of outcomes common to *A* and *B*.

EXAMPLE 13.1 Can You Taste PTC?

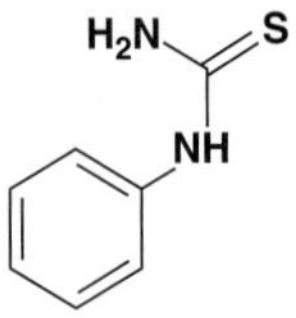

That molecule in the diagram is PTC (phenylthiocarbamide), a substance with an unusual property: 70% of people find that it has a bitter taste, and the other 30% can't taste it at all. The difference is genetic, depending on a single gene. Ask two people chosen at random to taste PTC. We are interested in the events

$$A = \{\text{first person can taste PTC}\}$$
$$B = \{\text{second person can taste PTC}\}$$

We know that $P(A) = 0.7$ and $P(B) = 0.7$. What is the probability $P(A \text{ and } B)$ that both can taste PTC?

We can think our way to the answer. The first person chosen can taste PTC in 70% of all samples and then the second person can taste it in 70% of those samples. We will get two tasters in 70% of 70% of all samples. That's $P(A \text{ and } B) = 0.7 \times 0.7 = 0.49$. ■

The argument in Example 13.1 works because knowing that the first person can taste PTC tells us nothing about the second person. The probability is still 0.7 that the second person can taste PTC whether or not the first person can. We say that the events "first person can taste PTC" and "second person can taste PTC" are **independent events**. Now we have another rule of probability.

independent events

Multiplication Rule for Independent Events

Two events A and B are **independent** if knowing that one occurs does not change the probability that the other occurs. If A and B are independent,

$$P(A \text{ and } B) = P(A)P(B)$$

EXAMPLE 13.2 Independent or Not?

To use this multiplication rule, we must decide whether events are independent. Here are some examples to help you recognize when you can assume events are independent.

In Example 13.1, we think that the ability of one randomly chosen person to taste PTC tells us nothing about whether or not a second person, also randomly chosen, can taste PTC. That's independence. But if the two people are members of the same family, the fact that ability to taste PTC is inherited warns us that they are not independent.

Independence is clearly recognized in artificial settings such as games of chance. Because a coin has no memory and most coin tossers cannot influence the fall of the coin, it is safe to assume that successive coin tosses are independent, so that the probability of three heads in succession is $0.5 \times 0.5 \times 0.5 = 0.125$.

On the other hand, the colors of successive cards dealt from the same deck are not independent. A standard 52-card deck contains 26 red and 26 black cards. For the first card dealt from a shuffled deck, the probability of a red card is $26/52 = 0.50$. Once we see that the first card is red, we know that there are only 25 reds among the remaining 51 cards. The probability that the second card is red is therefore only $25/51 = 0.49$. Knowing the outcome of the first deal changes the probabilities for the second. ■

Statistics in Your World: CONDEMNED BY INDEPENDENCE

Assuming independence when it isn't true can lead to disaster. Several mothers in England were convicted of murder simply because two of their children had died in their cribs with no visible cause. An "expert witness" for the prosecution said that the probability of an unexplained crib death in a nonsmoking middle-class family is 1/8500. He then multiplied 1/8500 by 1/8500 to claim that there is only a 1 in 73 million chance that two children in the same family could have died naturally. This is nonsense: it assumes that crib deaths are independent, and data suggest that they are not. Some common genetic or environmental cause, not murder, probably explains the deaths.

The multiplication rule extends to collections of more than two events, provided that all are independent. Independence of events A, B, and C means that no information about any one or any two can change the probability of the remaining events. Independence is often assumed in setting up a probability model when the events we are describing seem to have no connection.

If two events A and B are independent, the event that A does not occur is also independent of B, and so on. For example, choose two people at random and ask if they can taste PTC. Because 70% can taste PTC and 30% cannot, the probability that the first person is a taster and the second is not is $(0.7)(0.3) = 0.21$.

EXAMPLE 13.3 Surviving?

© CORBIS

During World War II, the British found that the probability of a bomber's being lost through enemy action on a mission over occupied Europe was 0.05. The probability that the bomber returned safely from a mission was therefore 0.95. It is reasonable to assume that missions were independent. Take A_i to be the event that a bomber survived its ith mission. The probability of surviving two missions is

$$\begin{aligned} P(A_1 \text{ and } A_2) &= P(A_1)P(A_2) \\ &= (0.95)(0.95) = 0.9025 \end{aligned}$$

The multiplication rule also applies to more than two independent events, so the probability of surviving three missions is

$$\begin{aligned} P(A_1 \text{ and } A_2 \text{ and } A_3) &= P(A_1)P(A_2)P(A_3) \\ &= (0.95)(0.95)(0.95) = 0.8574 \end{aligned}$$

In 1941, the tour of duty for an airman was established as 30 missions. The probability of surviving 30 missions is only

$$\begin{aligned} P(A_1 \text{ and } A_2 \text{ and} \ldots \text{and } A_{30}) &= P(A_1)P(A_2)\cdots P(A_{30}) \\ &= (0.95)(0.95)\cdots(0.95) \\ &= (0.95)^{30} = 0.2146 \end{aligned}$$

The probability of surviving two tours of duty was much smaller. ■

Here is another example of using the multiplication rule for independent events to compute probabilities.

EXAMPLE 13.4 Rapid HIV Testing

STATE: Many people who come to clinics to be tested for HIV, the virus that causes AIDS, don't come back to learn the test results. Clinics now use "rapid HIV tests" that give a result while the client waits. In a clinic in Malawi, for example, use of rapid tests increased the percent of clients who learned their test results from 69% to 99.7%.

The trade-off for fast results is that rapid tests are less accurate than slower laboratory tests. Applied to people who have no HIV antibodies, one rapid test has probability about 0.004 of producing a false positive (that is, of falsely indicating that antibodies are present).[1] If a clinic tests 200 people who are free of HIV antibodies, what is the chance that at least one false positive will occur?

PLAN: It is reasonable to assume that the test results for different individuals are independent. We have 200 independent events, each with probability 0.004. What is the probability that at least one of these events occurs?

SOLVE: "At least one" combines many outcomes. It is much easier to use Rule 4 (page 304) which says that

$$P(\text{at least one positive}) = 1 - P(\text{no positives})$$

and find $P(\text{no positives})$ first.

The probability of a negative result for any one person is $1 - 0.004 = 0.996$. To find the probability that all 200 people tested have negative results, use the multiplication rule:

$$\begin{aligned} P(\text{no positives}) &= P(\text{all 200 negative}) \\ &= (0.996)(0.996)\cdots(0.996) \\ &= 0.996^{200} = 0.4486 \end{aligned}$$

The probability we want is therefore

$$P(\text{at least one positive}) = 1 - 0.4486 = 0.5514$$

CONCLUDE: The probability is greater than 1/2 that at least one of the 200 people will test positive for HIV, even though no one has the virus. ■

CAUTION

CAUTION

The multiplication rule P(A and B) = P(A)P(B) holds if A and B are independent but not otherwise. The addition rule P(A or B) = P(A) + P(B) holds if A and B are disjoint but not otherwise. Resist the temptation to use these simple rules when the circumstances that justify them are not present. *You must also be careful not to confuse disjointness and independence.* If A and B are disjoint, then the fact that A occurs tells us that B cannot occur—look again at Figure 13.1. So disjoint events are not independent. Unlike disjointness, we cannot depict independence in a Venn diagram because it involves the probabilities of the events rather than just the outcomes that make up the events.

LaunchPad Online Resources

- The **StatBoards video**, *Independence and the Multiplication Rule*, discusses independence and provides an example of using the multiplication rule for independent events.

Apply Your Knowledge

13.1 Older College Students. Government data show that 8% of adults are full-time college students and that 30% of adults are aged 55 or older. Nonetheless, we can't conclude that because $(0.08)(0.30) = 0.024$, about 2.4% of adults are college students 55 or older. Why not?

13.2 Common Names. The U.S. Census Bureau says that the 10 most common names in the United States are (in order) Smith, Johnson, Williams, Brown, Jones, Miller, Davis, Garcia, Rodriguez, and Wilson. These names account for 9.6% of all U.S. residents. Out of curiosity, you look at the authors of the textbooks for your current courses. There are 9 authors in all. Would you be surprised if none of the names of these authors were among the 10 most common? (Assume that authors' names are independent and follow the same probability distribution as the names of all residents.)

13.3 Lost Internet Sites. Internet sites often vanish or move, so that references to them can't be followed. In fact, 13% of Internet sites referenced in major scientific journals are lost within two years after publication.[2] If a paper contains seven Internet references, what is the probability that all seven are still good two years later? What specific assumptions did you make to calculate this probability?

13.2 The general addition rule

We know that if A and B are disjoint events, then $P(A \text{ or } B) = P(A) + P(B)$. If events A and B are *not* disjoint, they can occur together. The probability that one or the other occurs is then *less* than the sum of their probabilities. As Figure 13.3 illustrates, outcomes common to both are counted twice when we add probabilities, so we must subtract this probability once. Here is the addition rule for any two events, disjoint or not.

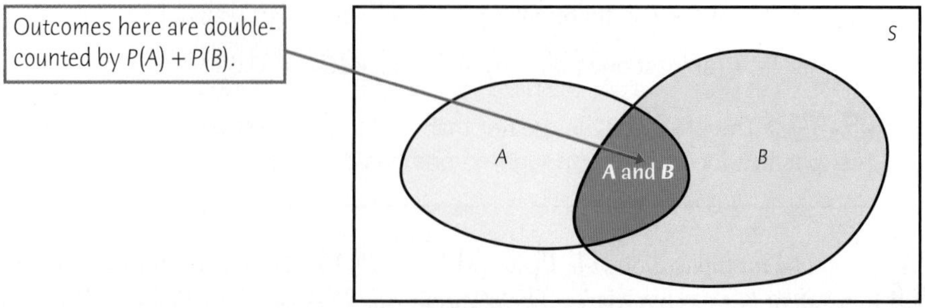

FIGURE 13.3
The general addition rule: for any events A and B, $P(A \text{ or } B) = P(A) + P(B) - P(A \text{ and } B)$.

Addition Rule for Any Two Events

For any two events A and B,

$$P(\text{A or B}) = P(\text{A}) + P(\text{B}) - P(\text{A and B})$$

If A and B are disjoint, the event $\{A \text{ and } B\}$ that both occur contains no outcomes and therefore has probability 0. Because the general addition rule includes Rule 3, the addition rule for disjoint events, we can always use the general addition rule to find $\{A \text{ or } B\}$.

EXAMPLE 13.5 Motor Vehicle Sales

Motor vehicles sold in the United States (ignoring heavy trucks) are classified as either cars or light trucks and as either domestic or imported. "Light trucks" include SUVs and minivans. "Domestic" means made in Canada, Mexico, or the United States, so that a Toyota made in Canada counts as domestic.

In 2012, 78.3% of the new vehicles sold to individuals were domestic, 49.5% were light trucks, and 42.1% were domestic light trucks.[3] Choose a vehicle sale at random. Then

$$\begin{aligned} P(\text{domestic or light truck}) &= P(\text{domestic}) + P(\text{light truck}) - P(\text{domestic light truck}) \\ &= 0.783 + 0.495 - 0.421 = 0.857 \end{aligned}$$

That is, 85.7% of vehicles sold were either domestic or light trucks (or both). A vehicle is an imported car if it is *neither* domestic nor a light truck. So

$$P(\text{imported car}) = 1 - 0.857 = 0.143 \ ■$$

Venn diagrams clarify events and their probabilities because you can just think of adding and subtracting areas. Figure 13.4 shows all the events formed from "domestic"

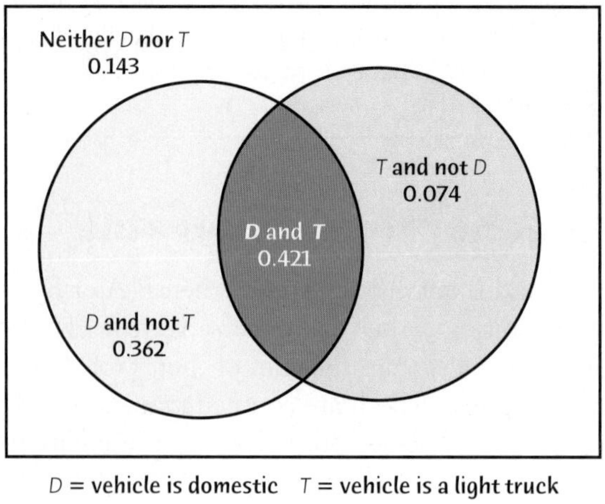

FIGURE 13.4
Venn diagram and probabilities for motor vehicle sales, for Example 13.5.

and "truck" in Example 13.5. The four probabilities that appear in the figure add to 1 because they refer to four disjoint events that make up the entire sample space. All of these probabilities come from the information in Example 13.5. For example, the probability that a randomly chosen vehicle sale is a domestic car ("D and not T" in the figure) is

$$P(\text{domestic car}) = P(\text{domestic}) - P(\text{domestic light truck})$$
$$= 0.783 - 0.421 = 0.362$$

LaunchPad Online Resources

- The **StatClips Examples videos**, *Probability: Basic Rules Example A* and *Probability: Basic Rules Example C*, both illustrate Venn diagrams and the general addition rule.

Apply Your Knowledge

13.4 Photo and Video Sharing. Photos and videos have become an important part of the online social experience, with more than half of Internet users posting photos or videos online that they have taken themselves. Let A be the event an Internet user posts photos that they have taken themselves, and B be the event an Internet user posts videos that they have taken themselves. Pew Research Center finds that $P(A) = 0.52$, $P(B) = 0.26$, and $P(A \text{ or } B) = 0.54$.[4]

(a) Make a Venn diagram similar to Figure 13.4 showing the events $\{A \text{ and } B\}$, $\{A \text{ and not } B\}$, $\{B \text{ and not } A\}$, and $\{\text{neither } A \text{ nor } B\}$.

(b) Describe each of these events in words.

(c) Find the probabilities of all four events, and add the probabilities to your Venn diagram.

13.5 College Degrees. Of all college degrees awarded in the United States, 50% are bachelor's degrees, 59% are earned by women, and 29% are bachelor's degrees earned by women. Make a Venn diagram and use it to answer these questions.

(a) What percent of all degrees are earned by men?

(b) What percent of all degrees are bachelor's degrees earned by men?

(c) Are the events "earning a bachelor's degree" and "being a man" independent? Why?

Barry Austin Photography/Getty Images

13.3 Conditional probability

The probability we assign to an event can change if we know that some other event has occurred. This idea is the key to many applications of probability.

EXAMPLE 13.6 Truck among Imported Motor Vehicles

Figure 13.4, based on the information in Example 13.5, gives the following probabilities for a randomly chosen light motor vehicle sold at retail in the United States:

	Domestic	Imported	Total
Light truck	0.421	0.074	0.495
Car	0.362	0.143	0.505
Total	0.783	0.217	1

The four probabilities in the body of the table add to 1 because they describe all vehicles sold. We obtain the "Total" row and column from these probabilities by the addition rule. For example, the probability that a randomly chosen vehicle is a light truck is

$$P(\text{truck}) = P(\text{truck and domestic}) + P(\text{truck and imported}) = 0.421 + 0.074 = 0.495$$

Suppose we are told that the vehicle chosen is imported. That is, it is one of the 21.7% in the "Imported" column of the table. The probability that a vehicle is a light truck, *given the information that it is imported*, is the proportion of trucks in the "Imported" column,

$$P(\text{truck} \mid \text{imported}) = \frac{0.074}{0.217} = 0.341$$

conditional probability

This is a **conditional probability**. You can read the bar | as "given the information that." ■

Although 49.5% of all vehicles sold are trucks, only 34.1% of imported vehicles are trucks. It's common sense that knowing that one event (the vehicle is imported) occurs often changes the probability of another event (the vehicle is a truck). The example also shows how we should define conditional probability. The idea of a conditional probability $P(B \mid A)$ of one event B given that another event A occurs is the proportion *of all occurrences of* A for which B also occurs.

Conditional Probability

When $P(A) > 0$, the **conditional probability** of B given A is

$$P(B \mid A) = \frac{P(A \text{ and } B)}{P(A)}$$

CAUTION

The conditional probability $P(B \mid A)$ makes no sense if the event A can never occur, so we require that $P(A) > 0$ whenever we talk about $P(B \mid A)$. *Be sure to keep in mind the distinct roles of the events A and B in $P(B \mid A)$.* Event A represents the information we are given, and B is the event whose probability we are calculating. Here is an example that emphasizes this distinction.

EXAMPLE 13.7 Imports among Trucks

What is the conditional probability that a randomly chosen vehicle is imported, *given the information that it is a truck?* Using the definition of conditional probability,

$$P(\text{imported} \mid \text{truck}) = \frac{P(\text{imported and truck})}{P(\text{truck})} = \frac{0.074}{0.495} = 0.149$$

Only 14.9% of trucks sold are imports. ■

Be careful not to confuse the two different conditional probabilities

$$P(\text{truck} \mid \text{imported}) = 0.341$$
$$P(\text{imported} \mid \text{truck}) = 0.149$$

The first answers the question, "What proportion of imports are trucks?" The second answers, "What proportion of trucks are imports?"

LaunchPad Online Resources

- The **StatBoards video**, *Conditional Probabilities*, illustrates the computation of conditional probabilities.

Apply Your Knowledge

13.6 Photo and Video Sharing. In the setting of Exercise 13.4, what is the conditional probability that an Internet user posts photos that they have taken themselves, given that they post videos that they have taken themselves? (*Hint:* First, organize the four probabilities in Exercise 13.4 in a table as in Example 13.6.)

13.7 College Degrees. In the setting of Exercise 13.5, what is the conditional probability that a degree is earned by a woman, given that it is a bachelor's degree?

13.8 Computer Games. Here is the distribution of computer games sold by type of game:[5]

Game Type	Probability
Strategy	0.354
Role playing	0.139
Family entertainment	0.127
Shooters	0.109
Children's	0.057
Other	0.214

What is the conditional probability that a computer game is a role-playing game, given that it is not a strategy game?

13.4 The general multiplication rule

The definition of conditional probability reminds us that in principle all probabilities, including conditional probabilities, can be found from the assignment of probabilities to events that describes a random phenomenon. More often, however, conditional probabilities are part of the information given to us in a probability model. The definition of conditional probability then turns into a rule for finding the probability that both of two events occur.

Multiplication Rule for Any Two Events

The probability that both of two events A and B happen together can be found by

$$P(A \text{ and } B) = P(A)P(B \mid A)$$

Here $P(B \mid A)$ is the conditional probability that B occurs, given the information that A occurs.

In words, this rule says that for both of two events to occur, first one must occur and then, given that the first event has occurred, the second must occur. This is just common sense expressed in the language of probability, as the following example illustrates.

Statistics in Your World — WINNING THE LOTTERY TWICE

In 1986, Evelyn Marie Adams won the New Jersey lottery for the second time, adding $1.5 million to her previous $3.9 million jackpot. The *New York Times* claimed that the odds of one person winning the big prize twice were 1 in 17 trillion. Nonsense, said two statisticians in a letter to the *Times*. The chance that Evelyn Marie Adams would win twice is indeed tiny, but it is almost certain that *someone* among the millions of lottery players would win two jackpots. Sure enough, Robert Humphries won his second Pennsylvania lottery jackpot ($6.8 million total) in 1988, and more recently Ernest Pullen of St. Louis won the Missouri lottery in June of 2010 and then again in September of 2010 ($3 million total). When commenting on the double win, Ernest Pullen said he considers himself to be a "lucky guy."

EXAMPLE 13.8 Teens Use of Social Media

The Pew Internet and American Life Project finds that 95% of teenagers (aged 12 to 17) use the Internet, and that 81% of online teens use some kind of social media.[6] What percent of teens are online *and* use some kind of social media?

Use the multiplication rule:

$$\begin{aligned} P(\text{online}) &= 0.95 \\ P(\text{use social media} \mid \text{online}) &= 0.81 \\ P(\text{online and use social media}) &= P(\text{online}) \times P(\text{use social media} \mid \text{online}) \\ &= (0.95)(0.81) = 0.7695 \end{aligned}$$

That is, about 77% of all teens are online and use some kind of social media.

You should think your way through this: if 95% of teens are online and 81% *of these* use some kind of social media, then 81% of 95% are both online and use some kind of social media.

It is important to remember that the conditional probability of an event generally depends on the event we condition on. Although we have shown

$$P(\text{use social media} \mid \text{online}) = 0.81,$$

since someone who is not online cannot be using social media, we have

$$P(\text{use social media} \mid \text{not online}) = 0. \blacksquare$$

We can extend the multiplication rule to find the probability that all of several events occur. The key is to condition each event on the occurrence of *all* of the preceding events. So for any three events A, B, and C,

$$P(A \text{ and } B \text{ and } C) = P(A)P(B \mid A)P(C \mid \text{both } A \text{ and } B)$$

Here is an example of the extended multiplication rule.

EXAMPLE 13.9 Fundraising by Telephone

STATE: A charity raises funds by calling a list of prospective donors to ask for pledges. It is able to talk with 40% of the names on its list. Of those the charity reaches, 30% make a pledge. But only half of those who pledge actually make a contribution. What percent of the donor list contributes?

PLAN: Express the information we are given in terms of events and their probabilities:

If A = {the charity reaches a prospect}	then	$P(A) = 0.4$
If B = {the prospect makes a pledge}	then	$P(B \mid A) = 0.3$
If C = {the prospect makes a contribution}	then	$P(C \mid \text{both } A \text{ and } B) = 0.5$

We want to find $P(A \text{ and } B \text{ and } C)$.

SOLVE: Use the multiplication rule:

$$\begin{aligned} P(A \text{ and } B \text{ and } C) &= P(A)P(B \mid A)P(C \mid \text{both } A \text{ and } B) \\ &= 0.4 \times 0.3 \times 0.5 = 0.06 \end{aligned}$$

CONCLUDE: Only 6% of the prospective donors make a contribution. ■

As Example 13.9 illustrates, formulating a problem in the language of probability is often the key to success in applying probability ideas.

Apply Your Knowledge

13.9 **At the Gym.** Suppose that 10% of adults belong to health clubs, and 40% of these health club members go to the club at least twice a week. What percent of all adults go to a health club at least twice a week? Write the information given in terms of probabilities and use the general multiplication rule.

13.10 **Teens, Use of Social Media.** We saw in Example 13.8 that 95% of teenagers are online, and that 81% of online teens use some kind of social media. Of online teens who use some kind of social media, 91% have posted a photo of themselves. What percent of all teens are online, use social media, and have posted a photo of themselves? Define events and probabilities, and follow the pattern of Example 13.9.

13.11 **The Probability of a Flush.** A poker player holds a flush when all five cards in the hand belong to the same suit (clubs, diamonds, hearts, or spades). We will find the probability of a flush when five cards are drawn in succession from the top of the deck. Remember that a deck contains 52 cards, 13 of each suit, and that when the deck is well shuffled, each card drawn is equally likely to be any of those that remain in the deck.

The Photo Works

(a) Concentrate on spades. What is the probability that the first card drawn is a spade? What is the conditional probability that the second card drawn is a spade, given that the first is a spade? (*Hint:* How many cards remain? How many of these are spades?)

(b) Continue to count the remaining cards to find the conditional probabilities of a spade for the third, the fourth, and the fifth card drawn, given in each case that all previous cards are spades.

(c) The probability of drawing five spades in succession from the top of the deck is the product of the five probabilities you have found. Why? What is this probability?

(d) The probability of drawing five hearts or five diamonds or five clubs is the same as the probability of drawing five spades. What is the probability that the five cards drawn all belong to the same suit?

13.5 Independence again

The conditional probability $P(B \mid A)$ is generally not equal to the unconditional probability $P(B)$. That's because the occurrence of event A generally gives us some additional information about whether or not event B occurs. If knowing that A occurs gives no additional information about B, then A and B are independent events. The precise definition of independence is expressed in terms of conditional probability.

Independent Events

Two events A and B that both have positive probability are **independent** if

$$P(B \mid A) = P(B)$$

We now see that the multiplication rule for independent events, $P(A \text{ and } B) = P(A)P(B)$, is a special case of the general multiplication rule, $P(A \text{ and } B) = P(A)P(B \mid A)$, just as the addition rule for disjoint events is a special case of the general addition rule. We rarely use the definition of independence because most often independence is part of the information given to us in a probability model.

Apply Your Knowledge

13.12 Independent? The Clemson University Fact Book and Data Center for 2013 shows that 101 of the university's 253 assistant professors were women, along with 97 of the 301 associate professors and 71 of the 350 full professors.

(a) What is the probability that a randomly chosen Clemson professor (of any rank) is a woman?

(b) What is the conditional probability that a randomly chosen professor is a woman, given that the person chosen is a full professor?

(c) Are the rank and sex of Clemson professors independent? How do you know?

13.6 Tree diagrams

Probability models often have several stages, with probabilities at each stage conditional on the outcomes of earlier stages. These models require us to combine several of the basic rules into a more elaborate calculation. Here is an example.

EXAMPLE 13.10 Who Shares Images and Videos Online?

4 step

STATE: The number of adult Internet users who take photos or videos they have found online and post them on sites designed for sharing images with many people continues to increase. Looking only at adult Internet users, aged 18 and over, about 15% are 18 to 29 years old, another 26% are 30 to 49 years old, and the remaining 59% are 50 and over. The Pew Internet and American Life Project finds that 68% of Internet users aged 18 to 29 have posted photos or videos they have found online, along with 54% of those aged 30 to 49 and 26% of those 50 or older.[7] What percent of all adult Internet users post photos or videos that they have found online?

PLAN: To use the tools of probability, restate all these percents as probabilities. If we choose an adult Internet user at random,

$$P(\text{aged 18 to 29}) = 0.15$$
$$P(\text{aged 30 to 49}) = 0.26$$
$$P(\text{aged 50 and older}) = 0.59$$

These three probabilities add to 1 because all adult Internet users are in one of the three age groups. The percents of each group who post photos or videos they have found online are *conditional* probabilities:

$$P(\text{post yes} \mid \text{aged 18 to 29}) = 0.68$$
$$P(\text{post yes} \mid \text{aged 30 to 49}) = 0.54$$
$$P(\text{post yes} \mid \text{aged 50 and older}) = 0.26$$

We want to find the unconditional probability $P(\text{post yes})$.

tree diagram

SOLVE: The **tree diagram** in Figure 13.5 organizes this information. Each segment in the tree is one stage of the problem. Each complete branch shows a path through the two stages. The probability written on each segment is the conditional probability of an Internet user following that segment, given that he or she has reached the node from which it branches.

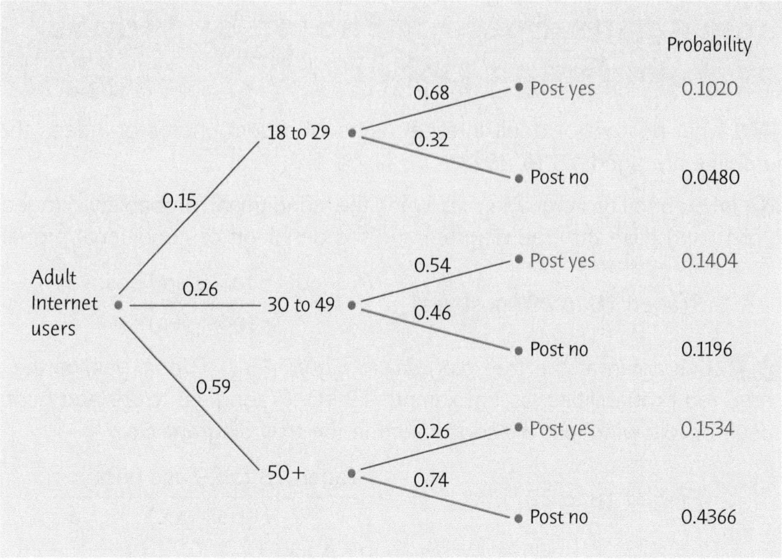

FIGURE 13.5
Tree diagram for who shares images and videos online, for Example 13.10. The three disjoint paths to the outcome that an adult Internet user posts photos or videos that they have found online are colored red.

Starting at the left, an Internet user falls into one of the three age groups. The probabilities of these groups mark the leftmost segments in the tree. Look at age 18 to 29, the top branch. The two segments going out from the "18 to 29" branch point carry the conditional probabilities

$$P(\text{post yes} \mid \text{aged 18 to 29}) = 0.68$$
$$P(\text{post no} \mid \text{aged 18 to 29}) = 0.32$$

The full tree shows the probabilities for all three age groups.

Now use the multiplication rule. The probability that a randomly chosen Internet user is an 18- to 29-year-old who posts photos or images they have found online is

$$\begin{aligned} P(\text{aged 18 to 29 and post yes}) &= P(\text{aged 18 to 29})P(\text{post yes} \mid \text{aged 18 to 29}) \\ &= (0.15)(0.68) = 0.1020 \end{aligned}$$

This probability appears at the end of the topmost branch. The multiplication rule says that the probability of any complete branch in the tree is the product of the probabilities of the segments in that branch.

There are three disjoint paths to "post yes," one for each of the three age groups. These paths are colored red in Figure 13.5. Because the three paths are disjoint, the probability that an adult Internet user posts photos or videos they have found online is the sum of their probabilities:

$$\begin{aligned} P(\text{post yes}) &= (0.15)(0.68) + (0.26)(0.54) + (0.59)(0.26) \\ &= 0.1020 + 0.1404 + 0.1534 = 0.3958 \end{aligned}$$

CONCLUDE: Almost 40% of all adult Internet users have posted photos or videos they have found online. ■

It takes longer to explain a tree diagram than it does to use it. Once you have understood a problem well enough to draw the tree, the rest is easy. Here is another question about posting photos or videos online that the tree diagram helps us answer.

EXAMPLE 13.11 Young Adults Posting Photos or Videos They Have Found Online

STATE: What percent of adult Internet users who post photos or videos they have found online are aged 18 to 29?

PLAN: In probability language, we want the conditional probability $P(\text{aged 18 to 29} \mid \text{post yes})$. Use the tree diagram and the definition of conditional probability:

$$P(\text{aged 18 to 29} \mid \text{post yes}) = \frac{P(\text{aged 18 to 29 and post yes})}{P(\text{post yes})}$$

SOLVE: Look again at the tree diagram in Figure 13.5. $P(\text{post yes})$ is the sum of the three red probabilities, as in Example 13.10. $P(\text{aged 18 to 29 and post yes})$ is the result of following just the top branch in the tree diagram. So

$$P(\text{aged 18 to 29} \mid \text{post yes}) = \frac{P(\text{aged 18 to 29 and post yes})}{P(\text{post yes})} = \frac{0.1020}{0.3958} = 0.2577$$

CONCLUDE: About 26% of adults who post photos or videos that they have found online are between 18 and 29 years old. Compare this conditional probability with the original information (unconditional) that 15% of adult Internet users are between 18 and 29 years old. Knowing that a person posts photos or videos found online increases the probability that he or she is young. ■

Examples 13.10 and 13.11 illustrate a common setting for tree diagrams. Some outcome (such as posting photos or videos found online) has several sources (such as the three age groups). Starting from

- the probability of each source, and
- the conditional probability of the outcome given each source

the tree diagram leads to the overall probability of the outcome. Example 13.10 does this. You can then use the probability of the outcome and the definition of conditional probability to find the conditional probability of one of the sources, given that the outcome occurred. Example 13.11 shows how.

LaunchPad Online Resources

- **The StatBoards video, *General Mutiplication Rule and Tree Diagrams*, provides an example of creating a tree diagram to visually show how to organize a set of conditional probabilities and then use the general multiplication rule to compute probabilities along the paths of the tree.**

Apply Your Knowledge

13.13 PSA Screening for Prostate Cancer. The prostate-specific antigen (PSA) test is a simple blood test to screen for prostate cancer. It has been used in men over 50 as a routine part of a physical exam, with levels above 4 ng/mL indicating possible prostate cancer. The test result is not always correct, sometimes indicating prostate cancer when it is not present and often missing prostate cancer that is present. Here are the approximate conditional probabilities of a positive (above 4 ng/mL) and negative test result given cancer is present or cancer is absent:[8]

	Test Result	
	Positive	**Negative**
Cancer present	0.21	0.79
Cancer absent	0.06	0.94

In a large study of prostate cancer screening, it was found that about 6.3% of the population has prostate cancer.

(a) Draw a tree diagram for selecting a person from this population (outcomes: cancer present or absent) and testing his blood (outcomes: test positive or negative).

(b) What is the probability that the test is positive for a randomly chosen person from this population?

13.14 Eye Color, Hair Color, and Freckles. A large study of children of Caucasian descent in Germany looked at the effect of eye color, hair color, and freckles on the reported extent of burning from sun exposure.[9] The population's distribution of hair color, eye color, and freckles was as shown in the tree diagram of Figure 13.6. Find the following probabilities and describe them in plain English:

(a) P(blue eyes | red hair) and P (blue eyes and red hair).

(b) P(freckles | red hair and blue eyes) and P(freckles and red hair and blue eyes).

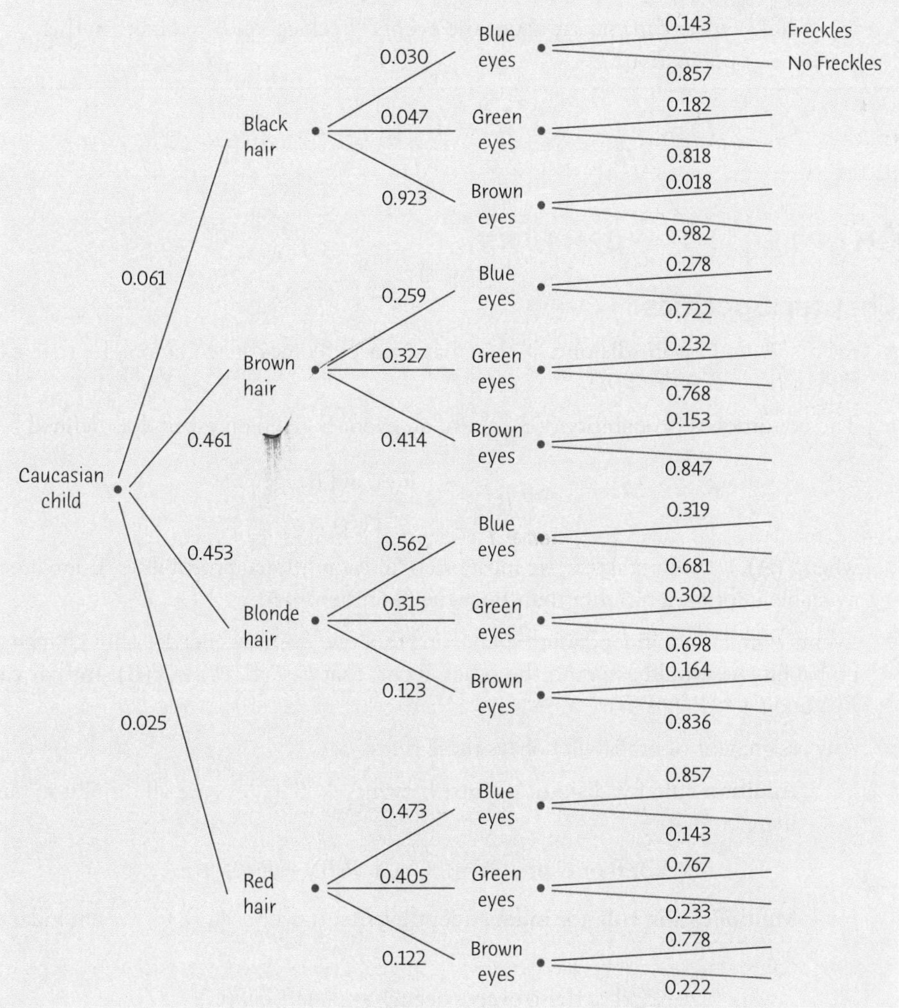

FIGURE 13.6
Tree diagram of the features of children of Caucasian descent in Germany, for Exercise 13.14. The three stages are hair color, eye color, and whether or not the child has freckles.

13.15 False PSA Positives. Continue your work from Exercise 13.13. The probabilities given in the table of Exercise 13.13 are properties of the PSA test. The false-positive rate is a property of the PSA test as applied to a population.

(a) Suppose that 6.3% of the population has prostate cancer. What is the probability that a person does not have cancer, given that the PSA test is positive? This is the false-positive rate.

(b) Treatment for prostate cancer can have serious side effects, including incontinence and impotence, and many men diagnosed with prostate cancer would eventually die of another cause if the cancer were left untreated. In fact, it has been found in some large clinical trials that screening did not reduce overall mortality.[10] In October 2011, the U.S. Preventive Services Task Force (USPSTF) released a draft report in which it recommended against using the PSA test to screen for prostate cancer in the general population. Based on the facts given about prostate cancer and the calculations that you have done, explain simply the reasoning behind the USPSTF recommendation.

13.16 Eye Color, Hair Color, and Freckles, Continued. Continue your work from Exercise 13.14 using the tree diagram of Figure 13.6. Find the following probabilities and describe them in plain English:

(a) P(red hair), P(blue eyes), and P(freckles).

(b) P(freckles and red hair) and P(freckles | red hair).

(c) What can you say about the events "freckles" and "red hair" in this population?

POLITICALLY CORRECT

In 1950, the Soviet mathematician B. V. Gnedenko (1912–1995) wrote *The Theory of Probability*, a text that was popular around the world. The introduction contains a mystifying paragraph that begins, "We note that the entire development of probability theory shows evidence of how its concepts and ideas were crystallized in a severe struggle between materialistic and idealistic conceptions." It turns out that "materialistic" is jargon for "Marxist-Leninist." It was good for the health of Soviet scientists in the Stalin era to add such statements to their books.

CHAPTER 13 SUMMARY

Chapter Specifics

- Events A and B are **disjoint** if they have no outcomes in common. In that case, $P(A \text{ or } B) = P(A) + P(B)$.
- The **conditional probability** $P(B \mid A)$ of an event B given an event A is defined by

$$P(B \mid A) = \frac{P(A \text{ and } B)}{P(A)}$$

  when $P(A) > 0$. In practice, we most often find conditional probabilities from directly available information rather than from the definition.
- Events A and B are **independent** if knowing that one event occurs does not change the probability we would assign to the other event; that is, $P(B \mid A) = P(B)$. In that case, $P(A \text{ and } B) = P(A)P(B)$.
- Any assignment of probability obeys these rules:

  Addition rule for disjoint events: If events A, B, C, . . . are all disjoint in pairs, then

$$P(A \text{ or } B \text{ or } C \text{ or} \ldots) = P(A) + P(B) + P(C) + \cdots$$

  Multiplication rule for independent events: If events A, B, C, . . . are independent, then

$$P(\text{all of these events occur}) = P(A)P(B)P(C)\cdots$$

General addition rule: For any two events A and B,

$$P(A \text{ or } B) = P(A) + P(B) - P(A \text{ and } B)$$

General multiplication rule: For any two events A and B,

$$P(A \text{ and } B) = P(A)P(B \mid A)$$

- **Tree diagrams** organize probability models that have several stages.

Link It

Probability models provide the important connection between the observed data and the process that generated the data. Probability is also the foundation of statistical inference and provides the language by which we answer questions and draw conclusions from our data. For these reasons, it is important that we have at least a basic understanding of what we mean by probability and some of the rules and properties of probabilities.

Chapter 12 introduced basic ideas and facts about probability and in this chapter we have considered some further details. The conditional distributions discussed in Chapter 6 are an example of the computation of conditional probabilities. Conditional probabilities play a central role when studying the relationship between two categorical variables and will be revisited in Chapter 24. In terms of later chapters, the most important idea of this chapter is independence. This is an assumption that many statistical procedures make about the observations in a data set, and it is an important assumption to fully understand.

LaunchPad Online Resources

If you are having difficulty with any of the sections of this chapter, these online resources should help prepare you to solve the exercises at the end of this chapter.

- **StatTutor** starts with a video review of each section and asks a series of questions to check your understanding.
- **LearningCurve** provides you with a series of questions about the chapter geared to your level of understanding.

CHECK YOUR SKILLS

13.17 An instant lottery game gives you probability 0.02 of winning on any one play. Plays are independent of each other. If you play three times, the probability that you win on none of your plays is about

(a) 0.98. (b) 0.94. (c) 0.000008.

13.18 The probability that you win on one or more of your three plays of the game in Exercise 13.17 is about

(a) 0.02. (b) 0.06. (c) 0.999992.

13.19 An athlete suspected of having used steroids is given two tests that operate independently of each other. Test A has probability 0.9 of being positive if steroids have been used. Test B has probability 0.8 of being positive if steroids have been used. What is the probability that *at least one* test is positive if steroids have been used?

(a) 0.98 (b) 0.72 (c) 0.28

What is the distribution of doctorates conferred by field and sex? Here are the counts from the most popular fields in 2012.[11] *The physical sciences include mathematics and computer and information sciences; the life sciences include agricultural sciences/natural resources, biological/biomedical sciences, and health sciences; and the social sciences include psychology.*

	MALE	FEMALE	TOTAL
Engineering	6,527	1,883	8,410
Physical sciences	6,393	2,551	8,944
Life sciences	5,331	6,698	12,029
Social sciences	3,488	4,861	8,349
Education	1,501	3,297	4,798
Humanities	2,654	2,847	5,501
Other	1,496	1,425	2,921
Total	27,390	23,562	50,592

Exercises 13.20 to 13.23 are based on this table.

13.20 Choose a doctoral recipient at random from this group. The probability that the recipient is male is about

(a) 0.46. (b) 0.54. (c) 0.56.

13.21 The conditional probability that the recipient is male, given that the degree is in education, is about

(a) 0.28. (b) 0.31. (c) 0.46.

13.22 The conditional probability that the degree is in education, given that the recipient is female, is about

(a) 0.05. (b) 0.14. (c) 0.18.

13.23 Let A be the event that the degree is in engineering and B the event that the recipient is a female. The proportion of engineering doctorates conferred on females is expressed in probability notation as

(a) $P(A \text{ and } B)$. (b) $P(A \mid B)$. (c) $P(B \mid A)$.

13.24 Choose an American adult at random. The probability that you choose a woman is 0.52. The probability that the person you choose has never married is 0.25. The probability that you choose a woman who has never married is 0.11. The probability that the person you choose is either a woman or never married (or both) is therefore about

(a) 0.77. (b) 0.66. (c) 0.13.

13.25 Of people who died in the United States in recent years, 86% were white, 12% were black, and 2% were Asian. (This ignores a small number of deaths among other races.) Diabetes caused 2.8% of deaths among whites, 4.4% among blacks, and 3.5% among Asians. The probability that a randomly chosen death is a white person who died of diabetes is about

(a) 0.107. (b) 0.030. (c) 0.024.

13.26 Using the information in Exercise 13.25, the probability that a randomly chosen death was due to diabetes is about

(a) 0.107. (b) 0.030. (c) 0.024.

CHAPTER 13 EXERCISES

13.27 Playing the lottery. New York State's "Quick Draw" lottery moves right along. Players choose between one and 10 numbers from the range 1 to 80; 20 winning numbers are displayed on a screen every four minutes. If you choose just one number, your probability of winning is 20/80, or 0.25. Lester plays one number eight times as he sits in a bar. What is the probability that all eight bets lose?

13.28 Universal blood donors. People with type O-negative blood are referred to as universal donors, although if you give type O-negative blood to any patient you run the risk of a transfusion reaction due to certain antibodies present in the blood. However, any patient can receive a transfusion of O-negative *red blood cells*. Only 7.2% of the American population have O-negative blood. If 10 people appear at random to give blood, what is the probability that at least one of them is a universal donor?

13.29 Playing the slots. Slot machines are now video games, with outcomes determined by random number generators. In the old days, slot machines were like this: you pull the lever to spin three wheels; each

Peter Dazeley/Getty Images

wheel has 20 symbols, all equally likely to show when the wheel stops spinning; the three wheels are independent of each other. Suppose that the middle wheel has nine cherries among its 20 symbols, and the left and right wheels have one cherry each.

(a) You win the jackpot if all three wheels show cherries. What is the probability of winning the jackpot?
(b) There are three ways that the three wheels can show two cherries and one symbol other than a cherry. Find the probability of each of these ways.
(c) What is the probability that the wheels stop with exactly two cherries showing among them?

13.30 **A whale of a time.** Hacksaw's Boats of St. Lucia takes tourists on a daily dolphin/whale watch cruise. Its brochure claims an 85% chance of sighting a whale or a dolphin, and you can assume that sightings from day to day are independent.

© Mark Conlin/Alamy

(a) If you take the dolphin/whale watch cruise on two consecutive days, what is the probability that you see a whale or a dolphin on both days?
(b) If you take the dolphin/whale watch cruise on two consecutive days, what is the probability you will see a dolphin or a whale on at least one day? (*Hint*: First compute the probability that this event does not occur.)
(c) If you want to have a 99% probability of seeing a whale or a dolphin at least once, what is the minimum number of days that you will need to take the cruise?

13.31 **Tendon surgery.** You have torn a tendon and are facing surgery to repair it. The surgeon explains the risks to you: infection occurs in 3% of such operations, the repair fails in 14%, and both infection and failure occur together in 1%. What percent of these operations succeed and are free from infection? Follow the four-step process in your answer.

13.32 **A whale of a time, continued.** Hacksaw's Boats of St. Lucia takes tourists on a daily dolphin/whale watch cruise. Its brochure claims an 85% chance of sighting a whale or a dolphin. Suppose there is a 65% chance of seeing dolphins and a 10% chance of seeing both dolphins and whales. Make a Venn diagram. Then answer these questions.

(a) What is the probability of seeing a whale on the cruise?
(b) What is the probability of seeing a whale but not a dolphin?
(c) Are "seeing a whale" and "seeing a dolphin" independent events?

13.33 **Tendon surgery, continued.** You have torn a tendon and are facing surgery to repair it. The surgeon explains the risks to you: infection occurs in 3% of such operations, the repair fails in 14%, and both infection and failure occur together in 1%. What is the probability of infection, given that the repair is successful? Follow the four-step process in your answer.

13.34 **Screening job applicants.** A company retains a psychologist to assess whether job applicants are suited for assembly-line work. The psychologist classifies applicants as one of A (well suited), B (marginal), or C (not suited). The company is concerned about the event D that an employee leaves the company within a year of being hired. Data on all people hired in the past five years give these probabilities:

$P(A) = 0.4$	$P(B) = 0.3$	$P(C) = 0.3$
$P(A \text{ and } D) = 0.1$	$P(B \text{ and } D) = 0.1$	$P(C \text{ and } D) = 0.2$

Sketch a Venn diagram of the events A, B, C, and D and mark on your diagram the probabilities of all combinations of psychological assessment and leaving (or not) within a year. What is $P(D)$, the probability that an employee leaves within a year?

13.35 **Cancer-detecting dogs.** Research has shown that specific biochemical markers are found exclusively in the breath of patients with lung cancer. However, no lab test can currently distinguish the breath of lung cancer patients from that of other subjects. Could dogs be trained to identify these markers in specimens of human breath, as they can be to detect illegal substances or to follow a person's scent? An experiment trained dogs to distinguish breath specimens of lung cancer patients from breath specimens of control individuals by using a food-reward training method. After the training was complete, the dogs were tested on new breath specimens without any reward or clue using a double-blind, completely randomized design. Here are the results for a random sample of 1286 breath specimens:[12]

	BREATH SPECIMEN FROM A		
DOG TEST RESULT	**CONTROL SUBJECT**	**CANCER SUBJECT**	**TOTAL**
Negative	708	10	718
Positive	4	564	568
Total	712	574	1286

(a) The *sensitivity* of a diagnostic test is its ability to correctly give a positive result when a person tested has the disease, or P(positive test | disease). Find the sensitivity of the dog cancer-detection test for lung cancer.
(b) The *specificity* of a diagnostic test is the conditional probability that the subject tested doesn't have the disease, given that the test has come up negative. Find the specificity of the dog cancer-detection test for lung cancer.

13.36 **Income tax returns.** Here is the distribution of the adjusted gross income (in thousands of dollars) reported on individual federal income tax returns in 2011:[13]

Income	<15	15–29	30–99	100–199	≥200
Probability	0.263	0.214	0.388	0.102	0.033

(a) What is the probability that a randomly chosen return shows an adjusted gross income of $30,000 or more?
(b) Given that a return shows an income of at least $30,000, what is the conditional probability that the income is at least $100,000?

13.37 **Thomas's pizza.** You work at Thomas's pizza shop. You have the following information about the seven pizzas in the oven: three of the seven have thick crust, and of these one has only sausage and two have only mushrooms; the remaining four pizzas have regular crust, and of these two have only sausage and two have only mushrooms. Choose a pizza at random from the oven.

(a) Are the events "getting a thick crust pizza" and "getting a pizza with mushrooms" independent? Explain.
(b) You add an eighth pizza to the oven. This pizza has thick crust with only cheese. Now are the events "getting a thick crust pizza" and "getting a pizza with mushrooms" independent? Explain.

13.38 **A probability teaser.** Suppose (as is roughly correct) that each child born is equally likely to be a boy or a girl and that the sexes of successive children are independent. If we let BG mean that the older child is a boy and the younger child is a girl, then each of the combinations BB, BG, GB, GG has probability 0.25. Ashley and Brianna each have two children.

(a) You know that at least one of Ashley's children is a boy. What is the conditional probability that she has two boys?
(b) You know that Brianna's older child is a boy. What is the conditional probability that she has two boys?

13.39 **College degrees.** A striking trend in higher education is that more women than men reach each level of attainment. The National Center for Educational Statistics provides projections for the number of degrees earned, classified by level and by the sex of the degree recipient. Here are the projected number of earned degrees (in thousands) in the United States for the 2015–16 academic year:[14]

	ASSOCIATE'S	BACHELOR'S	MASTER'S	DOCTORATE	TOTAL
Female	601	1066	494	95	2256
Male	357	803	322	87	1569
Total	958	1869	816	182	3825

(a) If you choose a degree recipient at random, what is the probability that the person you choose is a man?
(b) What is the conditional probability that you choose a man, given that the person chosen received a master's?
(c) Are the events "choose a man" and "choose a master's degree recipient" independent? How do you know?

13.40 **College degrees.** Exercise 13.39 gives the projected counts (in thousands) of earned degrees in the United States in the 2015–16 academic year. Use these data to answer the following questions.

(a) What is the probability that a randomly chosen degree recipient is a woman?
(b) What is the conditional probability that the person chosen received an associate's degree, given that she is a woman?
(c) Use the general multiplication rule to find the probability of choosing a female associate's degree recipient. Check your result by finding this probability directly from the table of counts.

13.41 **Deer and pine seedlings.** As suburban gardeners know, deer will eat almost anything green. In a study of pine seedlings at an environmental center in Ohio, researchers noted how deer damage varied with how much of the seedling was covered by thorny undergrowth:[15]

	DEER DAMAGE	
THORNY COVER	YES	NO
None	60	151
<1/3	76	158
1/3 to 2/3	44	177
>2/3	29	176

(a) What is the probability that a randomly selected seedling was damaged by deer?
(b) What are the conditional probabilities that a randomly selected seedling was damaged, given each level of cover?
(c) Does knowing about the amount of thorny cover on a seedling change the probability of deer damage? If so, cover and damage are not independent.

Peter Skinner/Science Source

13.42 **Deer and pine seedlings.** In the setting of Exercise 13.41, what percent of the trees that were not damaged by deer were more than two-thirds covered by thorny plants?

13.43 **Deer and pine seedlings.** In the setting of Exercise 13.41, what percent of the trees that were damaged by deer were less than one-third covered by thorny plants?

Julie is graduating from college. She has studied biology, chemistry, and computing and hopes to use her science background in crime investigation. Late one night she thinks about some jobs for which she has applied. Let A, B, and C be the events that Julie is offered a job by

A = the Connecticut Office of the Chief Medical Examiner

B = the New Jersey Division of Criminal Justice

C = the Federal Disaster Mortuary Operations Response Team

Julie writes down her personal probabilities for being offered these jobs:

$P(A) = 0.5$ $\quad P(B) = 0.4$ $\quad P(C) = 0.2$

$P(A \text{ and } B) = 0.1$ $\quad P(A \text{ and } C) = 0.05$ $\quad P(B \text{ and } C) = 0.05$

$P(A \text{ and } B \text{ and } C) = 0$

Make a Venn diagram of the events A, B, and C. As in Figure 13.4 (see page 308), mark the probabilities of every intersection involving these events. Use this diagram for Exercises 13.44 to 13.46.

13.44 **Will Julie get a job offer?** What is the probability that Julie is not offered any of the three jobs?

13.45 **Will Julie get just these offers?** What is the probability that Julie is offered the Connecticut job, but not the New Jersey or federal job?

13.46 **Julie's conditional probabilities.** If Julie is offered the federal job, what is the conditional probability that she is also offered the New Jersey job? If Julie is offered the New Jersey job, what is the conditional probability that she is also offered the federal job?

13.47 **The geometric distributions.** You are rolling a pair of balanced dice in a board game. Rolls are independent. You land in a danger zone that requires you to roll doubles (both faces show the same number of spots) before you are allowed to play again. How long will you wait to play again?

(a) What is the probability of rolling doubles on a single toss of the dice? (If you need review, the possible outcomes appear in Figure 12.2 (page 282). All 36 outcomes are equally likely.)

(b) What is the probability that you do not roll doubles on the first toss, but you do on the second toss?

(c) What is the probability that the first two tosses are not doubles and the third toss is doubles? This is the probability that the first doubles occurs on the third toss.

(d) Now you see the pattern. What is the probability that the first doubles occurs on the fourth toss? On the fifth toss? Give the general result: what is the probability that the first doubles occurs on the kth toss?

(e) What is the probability that you get to go again within three turns?

(*Comment:* The distribution of the number of trials to the first success is called a *geometric distribution*. In this problem, you have found geometric distribution probabilities when the probability of a success on each trial is 1/6. The same idea works for any probability of success.)

13.48 **Winning at tennis.** A player serving in tennis has two chances to get a serve into play. If the first serve is out, the player serves again. If the second serve is also out, the player loses the point. Here are probabilities based on four years of the Wimbledon Championship:[16]

$$P(\text{1st serve in}) = 0.59$$
$$P(\text{win point} \mid \text{1st serve in}) = 0.73$$
$$P(\text{2nd serve in} \mid \text{1st serve out}) = 0.86$$
$$P(\text{win point} \mid \text{1st serve out and 2nd serve in}) = 0.59$$

Make a tree diagram for the results of the two serves and the outcome (win or lose) of the point. (The branches in your tree have different numbers of stages depending on the outcome of the first serve.) What is the probability that the serving player wins the point?

13.49 **Peanut allergies among children.** About 2% of children in the United States are allergic to peanuts.[17] Choose three children at random and let the random variable X be the number in this sample who are allergic to peanuts. The possible values X can take are 0, 1, 2, and 3. Make a three-stage tree diagram of the outcomes (allergic or not allergic) for the three individuals and use it to find the probability distribution of X.

13.50 **Winning at tennis, continued.** Based on your work in Exercise 13.48, in what percent of points won by the server was the first serve in? (Write this as a conditional probability and use the definition of conditional probability.)

13.51 **Peanut allergies among children, continued.** Continue your work from Exercise 13.49. What is the conditional probability that exactly two of the children will be allergic to peanuts, given that at least one of the three children suffers from this allergy?

13.52 **Lactose intolerance.** Lactose intolerance causes difficulty digesting dairy products that contain lactose (milk sugar). It is particularly common among people of African and Asian ancestry. In the United States (ignoring other groups and people who consider themselves to belong to more than one race), 82% of the population is white, 14% is black, and 4% is Asian. Moreover, 15% of whites, 70% of blacks, and 90% of Asians are lactose intolerant.[18]

(a) What percent of the entire population is lactose intolerant?

(b) What percent of people who are lactose intolerant are Asian?

13.53 **Fundraising by telephone.** Tree diagrams can organize problems having more than two stages. Figure 13.7 shows probabilities for a charity calling potential donors by telephone.[19] Each person called is either a recent donor, a past donor, or a new prospect. At the next stage, the person called either does or does not pledge to contribute, with conditional probabilities that depend on the donor class the person belongs to. Finally, those who make a pledge either do or don't actually make a contribution.

(a) What percent of calls result in a contribution?

(b) What percent of those who contribute are recent donors?

DNA Forensics. *When a suspect's DNA is compared to a sample of DNA collected at a crime scene, the comparison is made between certain sections of the DNA called loci. Each locus has two alleles (gene forms), one inherited from our mother and the other from our father. Suppose there are two alleles for a particular locus called A and B. These alleles can be present at the locus in three combinations. A person could have both alleles at the locus be A, one allele be A and the other B, or both alleles be B, giving the three combinations (A and A), (A and B), and (B and B). Here's how the math works. If the proportion of the population with allele A as one of their alleles at the locus is a, and the proportion of the population with allele B as one of their alleles at the locus is b, then the proportion of the population with the three combinations of these allele types at the locus follows:*

ALLELES AT THE LOCUS	POPULATION PROPORTION WITH ALLELE COMBINATION
A and A	a^2
A and B	$2ab$
B and B	b^2

Use this information in Exercises 13.54 and 13.55. The numbers used in the exercises are from the FBI database.[20]

13.54 Suppose the locus D21S11 has two alleles called 29 and 31. The proportion of the Caucasian population with allele 29 is 0.181 and with allele 31 is 0.071. What proportion of the Caucasian population has the combination (29, 31) at the locus D21S11? What proportion has the combination (29, 29)?

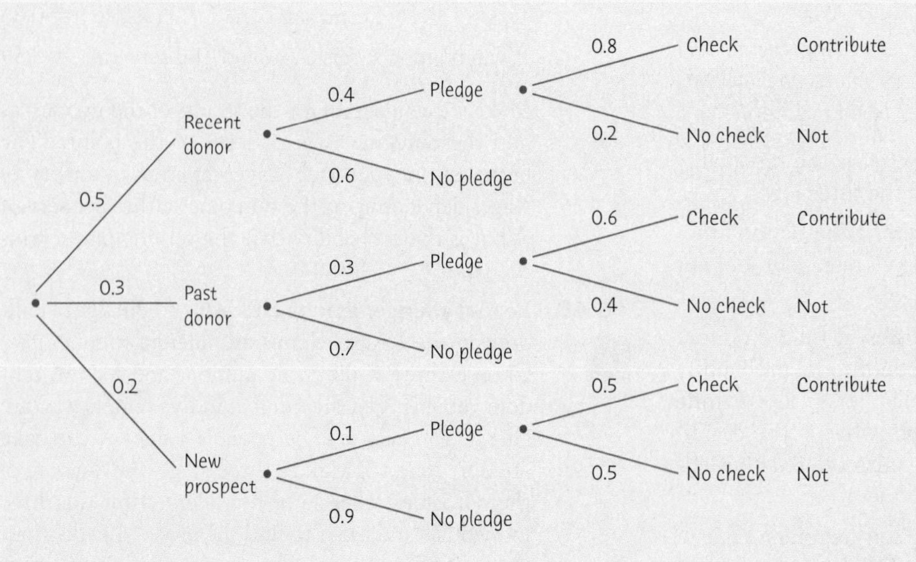

FIGURE 13.7
Tree diagram for fundraising by telephone, for Exercise 13.53. The three stages are the type of prospect called, whether or not the person makes a pledge, and whether or not a person who pledges actually makes a contribution.

13.55 Suppose the locus D3S1358 has two alleles called 16 and 17. The proportion of the Caucasian population with allele 16 is 0.232 and with allele 17 is 0.213. What proportion of the Caucasian population has the combination (16, 17) at the locus D3S1358?

One important fact regarding the loci evaluated in such forensic tests is that the allele combinations at each locus have been shown to be independent. Use this information in Exercise 13.56.

13.56 What proportion of the Caucasian population has the combination (29, 31) at the locus D21S11 and combination (16, 17) at the locus D3S1358? As we specify the alleles present at more loci, what will happen to the proportion of the Caucasian population that matches the allele combinations at all the loci?

A defendant in Ohio was indicted on December 17, 2009, on the charges of Aggravated Burglary and Assault (State of Ohio vs. Myers, Case No. 09 CR 666). In this case, a hair found at the crime scene was tested at six loci and demonstrated a specific combination of alleles found in a proportion of about one in 1.6 million individuals in the population. Comparison of the DNA profile found on the hair to a database of convicted felons revealed a match between the allelic profile found on the hair and an individual in the database (the defendant). Defense attorneys in the case requested that the state perform additional DNA testing because several previously untested loci were available to test. The results of this testing revealed that the defendant did not match at some of these newly tested markers, indicating that the DNA from the hair was not the defendant's and the charges were dropped. Use this information in Exercise 13.57.

13.57 If the DNA profile (or combination of alleles) found on the hair is possessed by one in 1.6 million individuals, and the database of convicted felons contains 4.5 million individuals, approximately how many individuals in the database would demonstrate a match between their DNA and that found on the hair?

Exploring the Web

13.58 **SAT Mathematics scores.** The website `http://research.collegeboard.org/programs/sat/data` presents data for high school seniors who participated in the SAT program from both the current year and previous years. Under Data, click on the link for College Bound Seniors for the most recent year given. In the window that opens, click on the link for Total Group Report for this year. The Total Group Profile Report presents data for high school graduates who participated in the SAT program that year. Students are counted only once, no matter how often they are tested, and only their latest scores are summarized. Suppose a high school graduate is selected at random. Use the information in the tables available to answer the following questions.

(a) What is the probability that the selected student is female?

(b) What is the probability that the selected student scores 600 or over on the Mathematics section of the SAT?

(c) What is the conditional probability that the selected student scores 600 or over on the Mathematics section given that the student is male?

(d) What is the conditional probability that the selected student scores 600 or over on the Mathematics section given that the student is female?

(e) Are scoring over 600 on the Mathematics section and sex independent? If not, explain in a simple sentence the nature of the dependence.

13.59 **Let's make a deal.** The Monty Hall Problem is an example of a simple probability problem with an answer that is counterintuitive. The problem was made popular when Marilyn vos Savant published it in her *Parade Magazine* column.[21] Here is the question:

Suppose you're on a game show, and you're given a choice of three doors: Behind one door is a car; behind the others, goats. You pick a door, say number 1, and the host, who knows what's behind the doors, opens another door, say number 3, which has a goat. He says to you, "Do you want to pick door number 2?" Is it to your advantage to switch your choice of doors?

(a) Go to the website `www.nytimes.com/2008/04/08/science/08monty.html` and click on Let's Play. Choose a door and click on it, and then click on Continue. Now click on Switch. Did you win? Click on Try Again and play the game 30 times, switching doors each time. What percent of the time did you win by switching? What percent of the time would you have won if you didn't switch? Based on your simulation, does it seem better to switch or stay with your initial choice?

(b) The website provides an explanation of why it is better to switch doors. Click on See How It Works. The probability of winning if you switch is 2/3, and if you don't switch it is 1/3. How well do these probabilities agree with the proportions you found in your simulation?

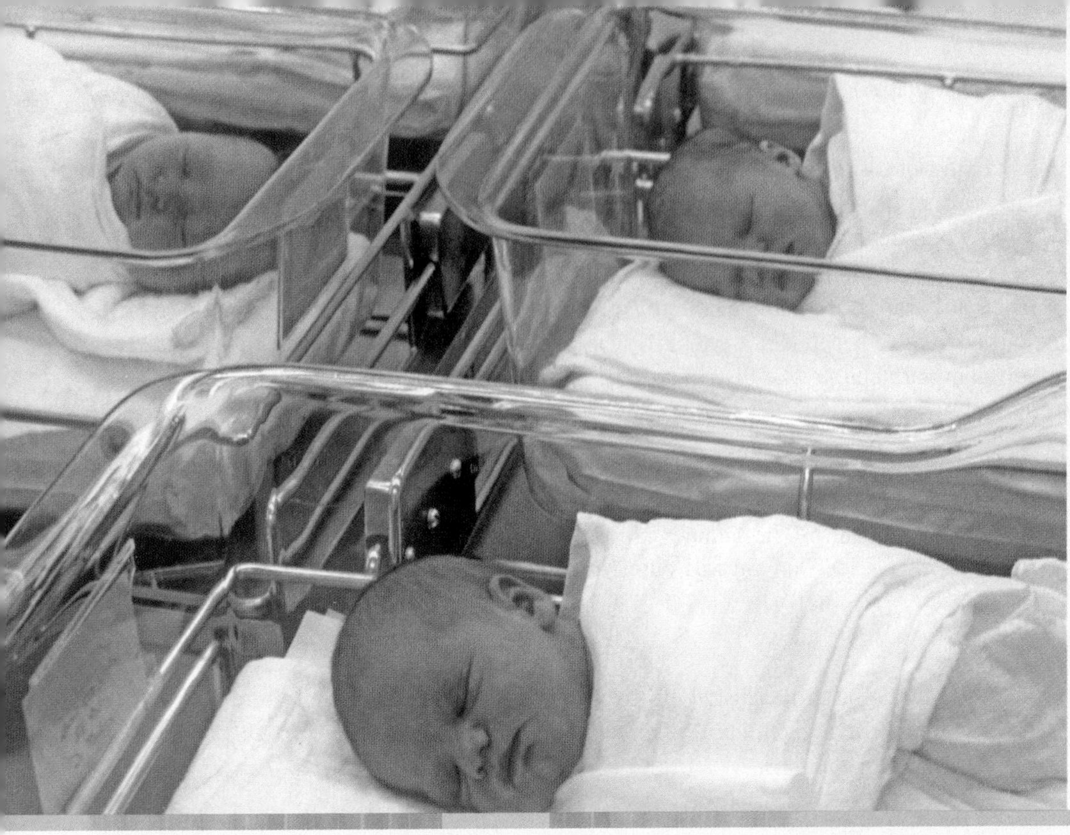

© Randy Duchaine/Alamy

CHAPTER 14

Binomial Distributions*

In this chapter we cover...

14.1 The binomial setting and binomial distributions
14.2 Binomial distributions in statistical sampling
14.3 Binomial probabilities
14.4 Using technology
14.5 Binomial mean and standard deviation
14.6 The Normal approximation to binomial distributions

A basketball player shoots five free throws; how many does she make? A sample survey dials 1200 landline phone numbers at random; how many of the numbers dialed correspond to working residential numbers? You plant 10 dogwood trees; how many live through the winter? In all these situations, we want a probability model for a *count* of successful outcomes out of a fixed (known) number of trials.

14.1 The binomial setting and binomial distributions

The distribution of a count depends on how the data are produced. Here is a common situation.

The Binomial Setting

1. There are a fixed number n of observations.
2. The n observations are all **independent**. That is, knowing the result of one observation does not change the probabilities we assign to other observations.
3. Each observation falls into one of just two categories, which for convenience we call "success" and "failure."
4. The probability of a success, call it p, is the same for each observation.

*This chapter concerns a special topic in probability. The material is not needed to understand the rest of the book.

WAS HE GOOD OR WAS HE LUCKY?

When a baseball player hits .300, everyone applauds. A .300 hitter gets a hit in 30% of times at bat. Could a .300 year just be luck? Typical major leaguers bat about 500 times a season and hit about .260. A hitter's successive tries seem to be independent, so we have a binomial setting. From this model, we can calculate or simulate the probability of hitting .300. It is about 0.025. Out of 100 run-of-the-mill major league hitters, two or three each year will bat .300 because they were lucky.

Think of tossing a coin n times as an example of the binomial setting. Each toss gives either heads or tails. Knowing the outcome of one toss doesn't change the probability of a head on any other toss, so the tosses are independent. If we call heads a success, then p is the probability of a head and remains the same as long as we toss the same coin. For tossing a coin, p is close to 0.5. If we spin the coin on a flat surface rather than toss it, p is not equal to 0.5. The number of heads we count is a discrete random variable X. The distribution of X is called a *binomial distribution*.

Binomial Distribution

The count X of successes in the binomial setting has the **binomial distribution** with parameters n and p. The parameter n is the number of observations, and p is the probability of a success on any one observation. The possible values of X are the whole numbers from 0 to n.

CAUTION The binomial distributions are an important class of discrete probability models. *Pay attention to the binomial setting, because not all counts have binomial distributions.*

EXAMPLE 14.1 Blood Types

Genetics says that children receive genes from their parents independently. Each child of a particular pair of parents has probability 0.25 of having type O blood. If these parents have five children, the number who have type O blood is the count X of successes in five independent observations with probability 0.25 of a success on each observation. So X has the binomial distribution with $n = 5$ and $p = 0.25$. ■

EXAMPLE 14.2 Counting Boys

Here is a set of genetic examples that require more thought.

Choose two births at random from the last year's births at a large hospital and count the number of boys (0, 1, or 2). The sexes of children born to different mothers are surely independent. The probability that a randomly chosen birth in Canada and the United States is a boy is about 0.52. (Why it is not 0.5 is something of a mystery.) So the count of boys has a binomial distribution with $n = 2$ and $p = 0.52$.

Next, observe successive births at a large hospital, and let X be the number of births until the first boy is born. Births are independent, and each has probability 0.52 of being a boy. Yet X is *not* binomial because there is no fixed number of observations. "Count observations until the first success" is a different setting than "count the number of successes in a fixed number of observations."

Finally, choose at random a family with exactly two children and count the number of boys. Careful study of such families shows that the count of boys is *not* binomial: the probability of exactly one boy is too high.[1] Families are less likely to have a third child if the first two are a boy and a girl, so when we look at families that stopped at two children, "one of each" is more common than if we look at randomly chosen births. The sexes of successive children in two-child families are *not independent* because the parents' choices interfere with the genetics. ■

14.2 Binomial distributions in statistical sampling

The binomial distributions are important in statistics when we wish to make inferences about the proportion p of "successes" in a population. Here is a typical example.

EXAMPLE 14.3 Choosing an SRS of Tomatoes

lola1960/Shutterstock

Physical damage to tomatoes, which can occur throughout the distribution system from field to consumer, has a major impact on market loss of fresh tomatoes. At the packing house, a distributor inspects an SRS of 10 tomatoes from a shipment of 10,000 tomatoes. Suppose that (unknown to the distributor) 11% of tomatoes in the shipment can be considered unmarketable due to physical damage, most generally bruising.[2] Count the number X of unmarketable tomatoes in the sample.

This is not quite a binomial setting. Removing one tomato changes the proportion of unmarketable tomatoes remaining in the shipment. So the probability that the second tomato chosen is unmarketable changes when we know whether the first is unmarketable or not. But removing one tomato from a shipment of 10,000 changes the makeup of the remaining 9999 tomatoes very little. In practice, the distribution of X is very close to the binomial distribution with $n = 10$ and $p = 0.11$. ■

Example 14.3 shows how we can use the binomial distributions in the statistical setting of selecting an SRS. When the population is much larger than the sample, a count of successes in an SRS of size n has approximately the binomial distribution with n equal to the sample size and p equal to the proportion of successes in the population.

Sampling Distribution of a Count

Choose an SRS of size n from a population with proportion p of successes. When the population is much larger than the sample, the count X of successes in the sample has approximately the binomial distribution with parameters n and p.

LaunchPad Online Resources

- **The StatBoards video, *The Binomial Scenario*, reviews the assumptions required to use the binomial distribution and how to recognize when these assumptions are met.**

Apply Your Knowledge

In each of Exercises 14.1 to 14.3, X is a count. Does X have a binomial distribution? Give your reasons in each case.

14.1 Working Numbers. When an opinion poll calls landline telephone numbers at random, approximately 30% of the numbers are working residential phone numbers. The remainder are either non-residential, non-working, or computer/fax numbers.[3] You watch the random dialing machine make 20 calls. X is the number that reach a working residential number.

14.2 Working Numbers. When an opinion poll calls landline telephone numbers at random, approximately 30% of the numbers are working residential phone numbers. The remainder are either non-residential, non-working, or computer/fax numbers. You watch the random dialing machine make calls. X is the number of calls until the first working residential number is reached.

14.3 Boxes of Tiles. Boxes of six-inch slate flooring tile contain 40 tiles per box. The count X is the number of cracked tiles in a box. You have noticed that most boxes contain no cracked tiles, but if there are cracked tiles in a box, then there are usually several.

14.4 Smoking in Canada. Statistics Canada reports that 19.9% of Canadians aged 12 and older smoke either daily or occasionally.[4] If you take an SRS of 1500 Canadians aged 12 and over, what is the approximate distribution of the number in your sample who smoke either daily or occasionally?

14.3 Binomial probabilities

We can find a formula for the probability that a binomial random variable takes any value by adding probabilities for the different ways of getting exactly that many successes in n observations. Here is an example that illustrates the idea.

EXAMPLE 14.4 Inheriting Blood Type

The blood types of successive children born to the same parents are independent and have fixed probabilities that depend on the genetic makeup of the parents. Each child born to a particular set of parents has probability 0.25 of having blood type O. If these parents have five children, what is the probability that exactly two of them have type O blood?

The count of children with type O blood is a binomial random variable X with $n = 5$ tries and probability $p = 0.25$ of a success on each try. We want $P(X = 2)$. ■

Because the method doesn't depend on the specific example, let's use "S" for success and "F" for failure for short, with a success representing type O blood. Do the work in two steps.

Step 1. Find the probability that a specific two of the five tries—say, the first and the third—give successes. This is the outcome SFSFF. Because tries are independent, the multiplication rule for independent events applies. The probability we want is

$$\begin{aligned} P(\text{SFSFF}) &= P(S)P(F)P(S)P(F)P(F) \\ &= (0.25)(0.75)(0.25)(0.75)(0.75) \\ &= (0.25)^2(0.75)^3 \end{aligned}$$

Step 2. Observe that *any one arrangement* of two Ss and three Fs has this same probability. This is true because we multiply together 0.25 twice and 0.75 three times whenever we have two Ss and three Fs. The probability that $X = 2$ is the probability of getting two Ss and three Fs in any arrangement whatsoever. Here are all the possible arrangements:

SSFFF	SFSFF	SFFSF	SFFFS	FSSFF
FSFSF	FSFFS	FFSSF	FFSFS	FFFSS

There are 10 of them, all with the same probability. The overall probability of two successes is therefore

$$P(X = 2) = 10(0.25)^2(0.75)^3 = 0.2637$$

The pattern of this calculation works for any binomial probability. To use it, we must count the number of arrangements of k successes in n observations. We use the following fact to do the counting without actually listing all the arrangements.

WHAT LOOKS RANDOM?

Toss a coin six times and record heads (H) or tails (T) on each toss. Which of these outcomes is more probable: HTHTTH or TTTHHH? Almost everyone says that HTHTTH is more probable because TTTHHH does not "look random." In fact, both are equally probable. That heads has probability 0.5 says that about half of a very long sequence of tosses will be heads. It doesn't say that heads and tails must come close to alternating in the short run. The coin doesn't know what past outcomes were, and it can't try to create a balanced sequence.

Binomial Coefficient

The number of ways of arranging k successes among n observations is given by the **binomial coefficient**

$$\binom{n}{k} = \frac{n!}{k!(n-k)!}$$

for $k = 0, 1, 2, \ldots, n$.

The formula for binomial coefficients uses the **factorial** notation. For any positive whole number n, its factorial $n!$ is

factorial

$$n! = n \times (n - 1) \times (n - 2) \times \cdots \times 3 \times 2 \times 1$$

In addition, we define $0! = 1$.

The larger of the two factorials in the denominator of a binomial coefficient will cancel much of the $n!$ in the numerator. For example, the binomial coefficient we need for Example 14.4 is

$$\begin{aligned}\binom{5}{2} &= \frac{5!}{2!3!} \\ &= \frac{(5)(4)(3)(2)(1)}{(2)(1) \times (3)(2)(1)} \\ &= \frac{(5)(4)}{(2)(1)} = \frac{20}{2} = 10\end{aligned}$$

CAUTION The *binomial coefficient* $\binom{5}{2}$ *is not related to the fraction* $\frac{5}{2}$. A helpful way to remember its meaning is to read it as "5 choose 2." Binomial coefficients have many uses, but we are interested in them only as an aid to finding binomial probabilities. The binomial coefficient $\binom{n}{k}$ counts the number of different ways in which k successes can be arranged among n observations. The binomial probability $P(X = k)$ is this count multiplied by the probability of any one specific arrangement of the k successes. Here is the result we seek.

Binomial Probability

If X has the binomial distribution with n observations and probability p of success on each observation, the possible values of X are $0, 1, 2, \ldots, n$. If k is any one of these values,

$$P(X = k) = \binom{n}{k} p^k (1 - p)^{n-k}$$

EXAMPLE 14.5 Inspecting Tomatoes

The number X of unmarketable tomatoes in Example 14.3 has approximately the binomial distribution with $n = 10$ and $p = 0.11$.

The probability that the sample contains no more than one unmarketable tomato is

$$\begin{aligned}P(X \leq 1) &= P(X = 1) + P(X = 0) \\ &= \binom{10}{1}(0.11)^1(0.89)^9 + \binom{10}{0}(0.11)^0(0.89)^{10} \\ &= \frac{10!}{1!9!}(0.11)(0.3504) + \frac{10!}{0!10!}(1)(0.3118) \\ &= (10)(0.11)(0.3504) + (1)(1)(0.3118) \\ &= 0.3854 + 0.3118 = 0.6972\end{aligned}$$

This calculation uses the facts that $0! = 1$ and that $a^0 = 1$ for any number a other than 0. We see that about 70% of all samples will contain no more than one unmarketable tomato. In fact, about 31% of the samples will contain no unmarketable tomatoes. A sample of size 10 cannot be trusted to alert the distributor to the presence of unacceptable tomatoes in the shipment.

The complement rule described in Chapter 12 can make the computation of certain binomial probabilities simpler. For example, the probability that the sample contains at least one unmarketable tomato is

$$\begin{aligned} P(X \geq 1) &= P(X = 1) + P(X = 2) + \cdots + P(X = 10) \\ &= 1 - P(X = 0) \\ &= 1 - 0.3118 = 0.6882 \end{aligned}$$

When computing binomial probabilities by hand, it is useful to keep the complement rule in mind. ■

14.4 Using technology

The binomial probability formula is awkward to use unless the number of observations n is quite small. You can find tables of binomial probabilities $P(X = k)$ and cumulative probabilities $P(X \leq k)$ for selected values of n and p, but the most efficient way to do binomial calculations is to use technology.

Figure 14.1 shows output for the calculation in Example 14.5 from a graphing calculator, two statistical programs, and a spreadsheet program. We asked all four to give cumulative probabilities. The calculator, Minitab, and CrunchIt! have menu entries for binomial cumulative probabilities. Excel has no menu entry, but the worksheet function BINOM.DIST is available. All the outputs agree with the result 0.6972 of Example 14.5. JMP can also be used to obtain binomial probabilities, but you need to access the formula editor.

Texas Instruments Graphing Calculator

```
binomcdf(10,0.11,1)
                .6972
```

Minitab

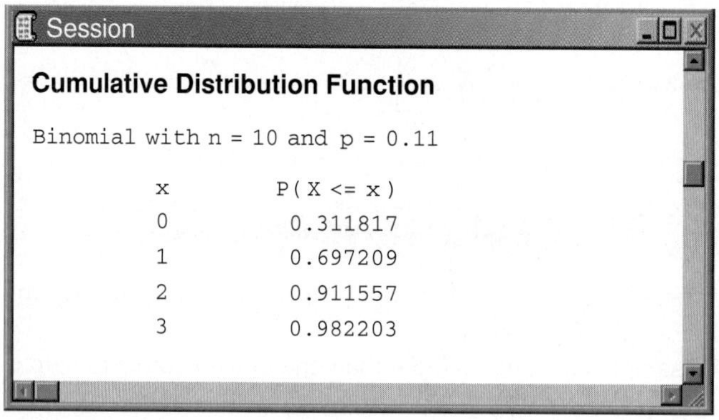

CrunchIt!

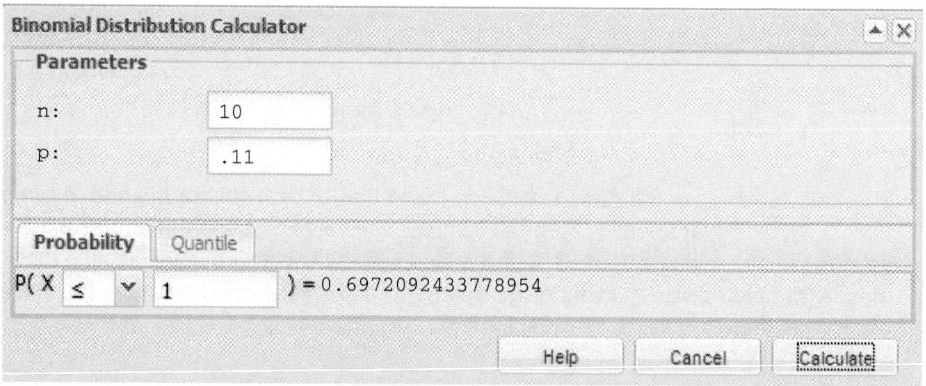

FIGURE 14.1
The binomial probability $P(X \leq 1)$ for Example 14.5: output from a graphing calculator, two statistical programs, and a spreadsheet program.

Microsoft Excel

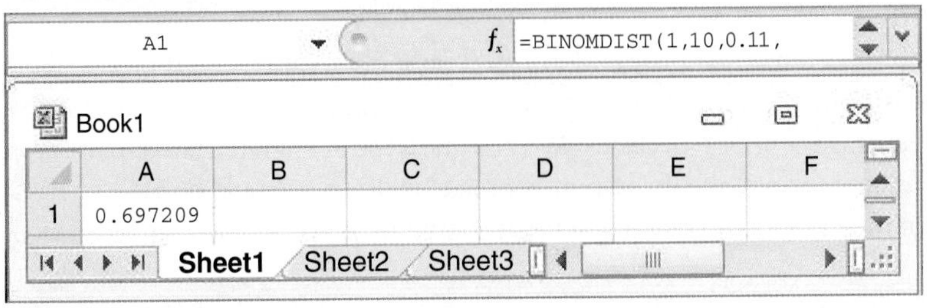

FIGURE 14.1
(Continued)

LaunchPad Online Resources

- The **StatBoards video**, *Calculating Binomial Probabilities*, illustrates the use of the formula to compute binomial probabilities.

Apply Your Knowledge

14.5 Proofreading. Typing errors in a text are either nonword errors (as when "the" is typed as "teh") or word errors that result in a real but incorrect word. Spell-checking software will catch nonword errors but not word errors. Human proofreaders catch 70% of word errors. You ask a fellow student to proofread an essay in which you have deliberately made 10 word errors.

(a) If the student matches the usual 70% rate, what is the distribution of the number of errors caught? What is the distribution of the number of errors missed?

(b) Missing three or more out of 10 errors seems a poor performance. What is the probability that a proofreader who catches 70% of word errors misses exactly three out of 10? If you use software, also find the probability of missing three or more out of 10.

14.6 Working Numbers. When an opinion poll calls landline telephone numbers at random, approximately 30% of the numbers are working residential phone numbers. The remainder are either non-residential, non-working, or computer/fax numbers. You watch the random dialing machine make 20 calls.

(a) What is the probability that exactly three calls reach working residential numbers?

(b) What is the probability that at most three calls reach working residential numbers?

(c) What is the probability that at least three calls reach working residential numbers?

(d) What is the probability that fewer than three calls reach working residential numbers?

(e) What is the probability that more than three calls reach working residential numbers?

14.7 Google Does Binomial. Point your web browser to `www.google.com`. Instead of searching the web or looking for images, you can request a calculation in the Search box.

(a) Enter `5 choose 2` and click Search. What does Google return?

(b) You see that Google calculates the binomial coefficient "5 choose 2." What are the values of the binomial coefficients for "500 choose 2" and "500 choose 100"? We expect that there are more ways to choose 100 than to choose 2, but how many more may be a surprise. That e+107 in Google's answer means a 1 followed by 107 zeros.

(c) Google also does binomial probabilities. Enter `(10 choose 1)*0.11*0.89^9` to find $P(X = 1)$ in Example 14.5. What is Google's answer with all its decimal places?

14.5 Binomial mean and standard deviation

If a count X has the binomial distribution based on n observations with probability p of success, what is its mean μ? That is, in very many repetitions of the binomial setting, what will be the average count of successes? We can guess the answer. If a basketball player makes 80% of her free throws, the mean number made in 10 tries should be 80% of 10, or 8. In general, the mean of a binomial distribution should be $\mu = np$. Here are the facts.

Binomial Mean and Standard Deviation

If a count X has the binomial distribution with number of observations n and probability of success p, the **mean** and **standard deviation** of X are

$$\mu = np$$
$$\sigma = \sqrt{np(1-p)}$$

CAUTION *Remember that these short formulas are good only for binomial distributions.* They can't be used for other distributions.

EXAMPLE 14.6 Inspecting Tomatoes

STATISTICS IN YOUR WORLD — RANDOMNESS TURNS SILVER TO BRONZE

After many charges of favoritism by judges, the rules for scoring international figure skating competitions changed in 2004. The big change is that 12 judges score all performances, then scores from three judges chosen at random are dropped for each part of the program. So there are $\binom{12}{9} = 220$ possible panels of nine judges for (say) the "Free Skate," and these panels will have slightly different scores. Result: at the 2006 World Figure Skating Championships, the Russian pair Maria Petrova and Alexei Tikhonov received the bronze medal when the consensus of all 12 judges would have given them the silver medal. Perhaps the system needs another change.

Continuing Example 14.5, the count X of unmarketable tomatoes is binomial with $n = 10$ and $p = 0.11$. The histogram in Figure 14.2 displays this probability distribution. (Because probabilities are long-run proportions, using probabilities as the heights of the bars shows what the distribution of X would be in very many repetitions.) The distribution is strongly right skewed. Although X can take any whole-number value from zero to 10, the probabilities of values larger than five are so small that they do not appear in the histogram.

The mean and standard deviation of the binomial distribution in Figure 14.2 are

$$\begin{aligned}\mu &= np\\ &= (10)(0.11) = 1.1\\ \sigma &= \sqrt{np(1-p)}\\ &= \sqrt{(10)(0.11)(0.89)} = \sqrt{0.979} = 0.9894\end{aligned}$$

The mean is marked on the probability histogram in Figure 14.2. ■

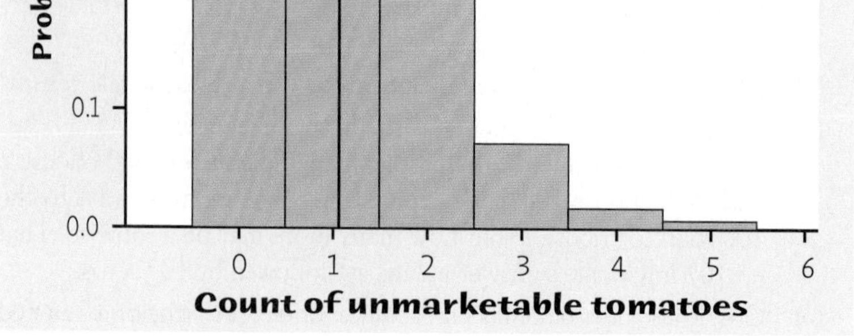

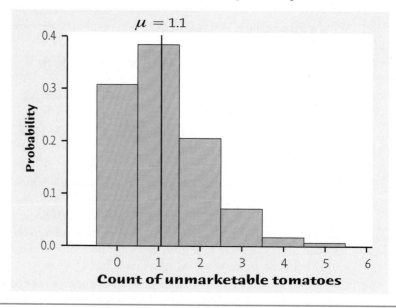

FIGURE 14.2 Probability histogram for the binomial distribution with $n = 10$ and $p = 0.11$, for Example 14.6.

LaunchPad Online Resources

- The **StatBoards video**, *The Mean and Standard Deviation of Binomials*, provides an additional example of the computation of the mean and standard deviation for a binomial distribution.

Apply Your Knowledge

14.8 Working Numbers. When an opinion poll calls landline telephone numbers at random, approximately 30% of the numbers are working residential phone numbers. The remainder are either non-residential, non-working, or computer/fax numbers. You watch the random dialing machine make 20 calls.

(a) What is the mean number of calls that reach working residential numbers?

(b) What is the standard deviation σ of the count of calls that reach working residential numbers?

(c) Suppose that the probability of reaching a working residential number was only $p = 0.20$. How does this new p affect the standard deviation? What would be the standard deviation if $p = 0.10$? What does your work show about the behavior of the standard deviation of a binomial distribution as the probability of a success gets closer to zero?

14.9 Proofreading. Return to the proofreading setting of Exercise 14.5.

(a) If X is the number of word errors missed, what is the distribution of X? If Y is the number of word errors caught, what is the distribution of Y?

(b) What is the mean number of errors caught? What is the mean number of errors missed? The mean counts of successes and of failures always add to n, the number of observations.

(c) What is the standard deviation of the number of errors caught? What is the standard deviation of the number of errors missed? The standard deviations of the count of successes and the count of failures are always the same.

14.6 The Normal approximation to binomial distributions

It isn't practical to use the formula for binomial probabilities when the number of observations n is large. (Look at part (b) of Exercise 14.7 to see why.) Software or a graphing calculator will handle many problems that are beyond the scope of hand calculation. If technology does not rescue you, there is another alternative: *as the number of observations n gets larger, the binomial distribution gets close to a Normal distribution.* When n is large, we can use Normal probability calculations to approximate binomial probabilities. Here are the facts.

Normal Approximation for Binomial Distributions

Suppose that a count X has the binomial distribution with n observations and success probability p. When n is large, the distribution of X is approximately Normal, $N(np, \sqrt{np(1-p)})$.

As a rule of thumb, we will use the Normal approximation when n is so large that $np \geq 10$ and $n(1-p) \geq 10$.

The Normal approximation is easy to remember because it says to act as if X is Normal with exactly the same mean and standard deviation as the binomial. The accuracy of the Normal approximation improves as the sample size n increases. It is most accurate for any fixed n when p is close to 1/2, and least accurate when p is near 0 or 1. This is why the rule of thumb in the box depends on p as well as n.

EXAMPLE 14.7 Tracking Your Health

Sample surveys show that more people are tracking changes in their health on paper, spreadsheet, mobile device, or just "in their heads." A survey asked a nationwide random sample of 3014 adults, "Now thinking about your health overall, do you currently keep track of your own weight, diet, or exercise routine, or is this not something you currently do?"[5] The population that the poll wants to draw conclusions about is all U.S. residents aged 18 and over. Suppose that in fact 62% of all adult U.S. residents would say "Yes" if asked this question. What is the probability that 1900 or more of the sample say "Yes"? ■

Because there are about 235 million adults in the United States, the responses of 3014 randomly chosen adults are very close to independent. So the number in our sample who are tracking their weight, diet, or exercise routine is a random variable X having the binomial distribution with $n = 3014$ and $p = 0.62$. To find the probability $P(X \geq 1900)$ that at least 1900 of the people in the sample are tracking their weight, diet, or exercise routine, we must add the binomial probabilities of all outcomes from $X = 1900$ to $X = 3014$. Figure 14.3 is a probability histogram of this binomial distribution, from Minitab. As the Normal approximation suggests, the shape of the distribution looks Normal. The probability we want is the sum of the heights of the shaded bars. Here are three ways to find this probability.

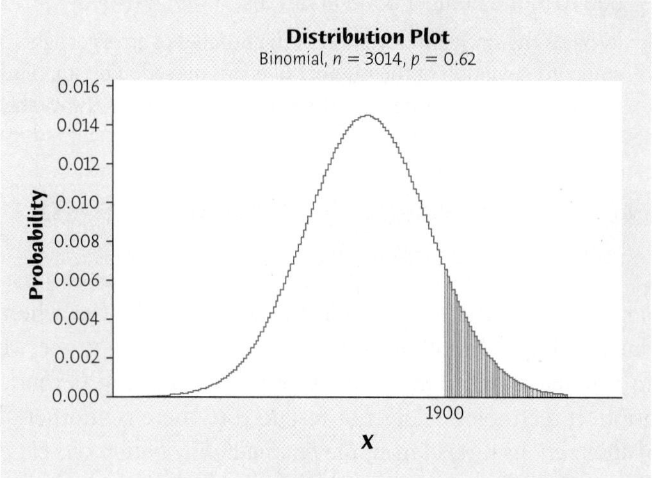

FIGURE 14.3 Probability histogram for the binomial distribution with $n = 3014$, $p = 0.62$. The bars at and above 1900 are shaded to highlight the probability of getting at least 1900 successes. The shape of this binomial probability distribution closely resembles a Normal curve.

1. Use technology. Statistical software can find the exact binomial probability. In most cases, software finds cumulative probabilities $P(X \leq x)$. So start by writing

$$P(X \geq 1900) = 1 - P(X \leq 1899)$$

Here is Minitab's answer for $P(X \leq 1899)$:

```
Binomial with n = 3014 and p = 0.62

   x     P(X<=x)
1899    0.876384
```

The probability we want is $1 - 0.876384 = 0.123616$, correct to six decimal places.

2. Simulate a large number of samples. Figure 14.4 displays a histogram of the counts X from 5000 samples of size 3014 when the truth about the population is

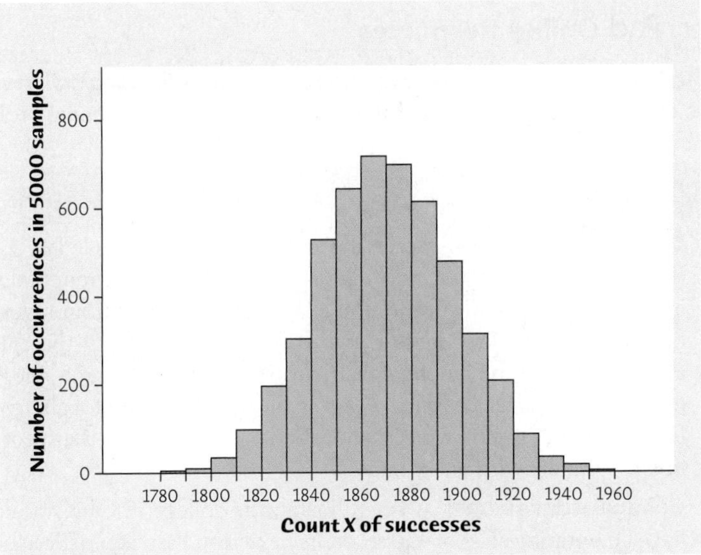

FIGURE 14.4
Histogram of 5000 simulated binomial counts ($n = 3014$, $p = 0.62$).

$p = 0.62$. The simulated distribution, like the exact distribution in Figure 14.3, looks Normal. Because 649 of these 5000 samples have X at least 1900, the probability estimated from the simulation is

$$P(X \geq 1900) = \frac{649}{5000} = 0.1298$$

This estimate misses the true probability by about 0.006. The law of large numbers says that the results of such simulations always get closer to the true probability as we simulate more and more samples.

3. Both of the previous methods require software. We can avoid the need for software by using the Normal approximation.

EXAMPLE 14.8 Normal Approximation of a Binomial Probability

Act as though the count X in Example 14.7 has the Normal distribution with the same mean and standard deviation as the binomial distribution:

$$\mu = np = (3014)(0.62) = 1868.68$$
$$\sigma = \sqrt{np(1-p)} = \sqrt{(3014)(0.62)(0.38)} = 26.648$$

Standardizing X gives a standard Normal variable Z. The probability we want is

$$\begin{aligned} P(X \geq 1900) &= P\left(\frac{X - 1868.68}{26.648} \geq \frac{1900 - 1868.68}{26.648}\right) \\ &= P(Z \geq 1.18) \\ &= 1 - 0.8810 = 0.1190 \end{aligned}$$

The Normal approximation 0.1190 misses the true probability calculated in Example 14.7 by about 0.005. ■

The *Normal Approximation to Binomial* applet shows in visual form how well the Normal approximation fits the binomial distribution for any n and p. You can slide n and watch the approximation get better. Whether or not the Normal approximation is satisfactory depends on how accurate your calculations need to be. For most statistical purposes, great accuracy is not required. Our rule of thumb for use of the Normal approximation reflects this judgment.

APPLET

LaunchPad Online Resources

- The **StatBoards video**, *The Normal Approximation to the Binomial*, reviews the details of using the Normal distribution to approximate a binomial probability.

Apply Your Knowledge

14.10 Using Benford's Law. According to Benford's law (Example 12.7, page 287) the probability that the first digit of the amount of a randomly chosen invoice is a 1 or a 2 is 0.477. You examine 90 invoices from a vendor and find that 29 have first digits 1 or 2. If Benford's law holds, the count of 1s and 2s will have the binomial distribution with $n = 90$ and $p = 0.477$. Too few 1s and 2s suggests fraud. What is the approximate probability of 29 or fewer 1s and 2s if the invoices follow Benford's law? Do you suspect that the invoice amounts are not genuine?

14.11 College Admissions. A small liberal arts college in Ohio would like to have an entering class of 450 students next year. Past experience shows that about 37% of the students admitted will decide to attend. The college is planning to admit 1175 students. Suppose that students make their decisions independently and that the probability is 0.37 that a randomly chosen student will accept the offer of admission.

(a) What are the mean and standard deviation of the number of students who accept the admissions offer from this college?

(b) Use the Normal approximation to approximate the probability that the college gets more students than it wants. Be sure to check that you can safely use the approximation.

(c) Use software or an online binomial calculator to compute the exact probability that the college gets more students than it wants. How good is the approximation in part (b)?

(d) To decrease the probability of getting more students than are wanted, does the college need to increase or decrease the number of students it admits? Using software or an online binomial calculator, what is the largest number of students that the college can admit if administrators want the exact probability of getting more students than they want to be no larger than 5%?

14.12 Checking for Survey Errors. One way of checking the effect of undercoverage, nonresponse, and other sources of error in a sample survey is to compare the sample with known facts about the population. About 24% of the Canadian population over 15 are first generation, that is, they were born outside Canada. The number X of first-generation Canadians in random samples of 1000 persons over 15 should therefore vary with the binomial ($n = 1000, p = 0.24$) distribution.

(a) What are the mean and standard deviation of X?

(b) Use the Normal approximation to find the probability that the sample will contain between 210 and 270 first-generation Canadians. Be sure to check that you can safely use the approximation.

CHAPTER 14 SUMMARY

Chapter Specifics

- A count X of successes has a **binomial distribution** in the **binomial setting:** there are n observations; the observations are independent of each other; each observation results in a success or a failure; and each observation has the same probability p of a success.

- The binomial distribution with n observations and probability p of success gives a good approximation to the sampling distribution of the count of successes in an SRS of size n from a large population containing proportion p of successes.
- If X has the binomial distribution with parameters n and p, the possible values of X are the whole numbers $0, 1, 2, \ldots, n$. The **binomial probability** that X takes any of these values is

$$P(X = k) = \binom{n}{k} p^k (1 - p)^{n-k}$$

Binomial probabilities in practice are best found using software.

- The **binomial coefficient**

$$\binom{n}{k} = \frac{n!}{k!(n - k)!}$$

counts the number of ways k successes can be arranged among n observations. Here the **factorial $n!$** is

$$n! = n \times (n - 1) \times (n - 2) \times \cdots \times 3 \times 2 \times 1$$

for positive whole numbers n, and $0! = 1$.

- The **mean** and **standard deviation** of a binomial count X are

$$\mu = np$$
$$\sigma = \sqrt{np(1 - p)}$$

- The **Normal approximation** to the binomial distribution says that if X is a count having the binomial distribution with parameters n and p, then when n is large, X is approximately $N(np, \sqrt{np(1 - p)})$. Use this approximation only when $np \geq 10$ and $n(1 - p) \geq 10$.

Link It

The binomial distribution is used to compute probabilities for the count of successes among n observations that are produced under the binomial setting. An important situation for which the binomial setting can be used is when we choose an SRS from a population with a proportion p of successes. When the success probability p is known, probabilities associated with the number of successes among n observations can be computed using either the binomial formula, or the Normal approximation when the sample size is large enough.

An important application of the binomial distribution is in making inferences about the *proportion* of some outcome in a population. This is described in Chapter 22 where our data is collected under the binomial setting, but the proportion with the given outcome in the population is not known. For example, we may be interested in learning about the proportion of young adults (aged 19 to 25) that still live at home with their parents based on a sample from the population of young adults. When we are interested in making inferences about an unknown proportion in a population, we generally work with the *proportion of successes in the sample* rather than the count of successes in the sample. When using the proportion of successes in the sample to answer questions and draw conclusions about this unknown proportion in the population, the statistical methods are still related to the binomial distribution and the Normal approximation to the binomial.

LaunchPad Online Resources

If you are having difficulty with any of the sections of this chapter, these online resources should help prepare you to solve the exercises at the end of this chapter.

- **StatTutor** starts with a video review of each section and asks a series of questions to check your understanding.
- **LearningCurve** provides you with a series of questions about the chapter geared to your level of understanding.

CHECK YOUR SKILLS

14.13 Larry reads that one out of four eggs contains salmonella bacteria. So he never uses more than three eggs in cooking. If eggs do or don't contain salmonella independent of each other, the number of contaminated eggs when Larry uses three chosen at random has the distribution

(a) binomial with $n = 4$ and $p = 1/4$.
(b) binomial with $n = 3$ and $p = 1/4$.
(c) binomial with $n = 3$ and $p = 1/3$.

14.14 In Exercise 14.13, the probability that at least one of Larry's three eggs contains salmonella is about

(a) 0.68. (b) 0.58. (c) 0.30.

14.15 In a group of 10 college students, four are business majors. You choose three of the 10 students at random and ask their major. The distribution of the number of business majors you choose is

(a) binomial with $n = 10$ and $p = 0.4$.
(b) binomial with $n = 3$ and $p = 0.4$.
(c) not binomial.

In a test for ESP (extrasensory perception), a subject is told that cards the experimenter can see but he cannot contain either a star, a circle, a wave, or a square. As the experimenter looks at each of five cards in turn, the subject names the shape on the card. A subject who is just guessing has probability 0.25 of guessing correctly on each card. Exercises 14.16 to 14.18 use this information.

14.16 If the subject guesses two shapes correctly and three incorrectly, in how many ways can you arrange the sequence of correct and incorrect guesses?

(a) $\binom{3}{2} = 3$ (b) $\binom{5}{2} = 20$ (c) $\binom{5}{2} = 10$

14.17 Assume the subject's guesses are independent of each other. The probability that the subject guesses the shape correctly on the first and last card, but incorrectly on the other three cards, is about

(a) 0.264. (b) 0.026. (c) 0.090.

14.18 Assume the subject's guesses are independent of each other. The probability that the subject guesses the shape correctly on two out of the five cards is about

(a) 0.264. (b) 0.026. (c) 0.090.

Each entry in a table of random digits like Table B has probability 0.1 of being any of the 10 digits 0 to 9, and digits are independent of each other. Exercises 14.19 to 14.21 use this setting.

14.19 The probability of an entry being either an 8 or a 9 is

(a) 0.1. (b) 0.2. (c) 0.4.

14.20 Each line in Table B has 40 digits. The number of times an 8 or a 9 occurs in *two lines* of the table is

(a) binomial with $n = 80$ and $p = 0.2$.
(b) binomial with $n = 80$ and $p = 0.1$.
(c) binomial with $n = 40$ and $p = 0.2$.

14.21 The mean number of times an 8 or a 9 occurs in *two lines* of the table is

(a) 16. (b) 12.8. (c) 8.

CHAPTER 14 EXERCISES

14.22 **Binomial setting?** In each of the following situations, is it reasonable to use a binomial distribution for the random variable X? Give reasons for your answer in each case.

(a) An auto manufacturer chooses one car from each hour's production for a detailed quality inspection. One variable recorded is the count X of finish defects (dimples, ripples, etc.) in the car's paint.
(b) The pool of potential jurors for a murder case contains 100 persons chosen at random from the adult residents of a large city. Each person in the pool is asked whether he or she opposes the death penalty; X is the number who say "Yes."
(c) Joe buys a ticket in his state's Pick 3 lottery game every week; X is the number of times in a year that he wins a prize.

14.23 **Binomial setting?** A binomial distribution will be approximately correct as a model for one of these two sports settings and not for the other. Explain why by briefly discussing both settings.

© David Bergman/Corbis

(a) A National Football League kicker has made 80% of his field goal attempts in the past. This season he attempts 20 field goals. The attempts differ widely in distance, angle, wind, and so on.
(b) A National Basketball Association player has made 80% of his free-throw attempts in the past. This season he takes 150 free throws. Basketball free throws are always attempted from 15 feet away from the basket with no interference from other players.

14.24 Antibiotic resistance. Antibiotic resistance occurs when disease-causing microbes become resistant to antibiotic drug therapy. Because this resistance is typically genetic and transferred to the next generations of microbes, it is a very serious public health problem. According to the CDC, approximately 30% of gonorrhea cases tested in 2011 were resistant to at least one of the three major antibiotics commonly used to treat gonorrhea.[6] A physician treated 10 cases of gonorrhea during one week of 2011.

(a) What is the distribution of the cases resistant to at least one of the three major antibiotics?
(b) What is the probability that exactly one out of the 10 cases was resistant to at least one of the three major antibiotics? What is the probability for exactly two out of 10?
(c) Also find the probability that one or more out of the 10 were resistant to at least one of the three major antibiotics. (*Hint:* It is easier to first find the probability that exactly zero of the 10 cases were resistant.)

14.25 Random stock prices. A believer in the random walk theory of stock markets thinks that an index of stock prices has probability 0.65 of increasing in any year. Moreover, the change in the index in any given year is not influenced by whether it rose or fell in earlier years. Let X be the number of years among the next five years in which the index rises.

(a) X has a binomial distribution. What are n and p?
(b) What are the possible values that X can take?
(c) Find the probability of each value of X. Draw a probability histogram for the distribution of X. (See Figure 14.2 (page 334) for an example of a probability histogram.)
(d) What are the mean and standard deviation of this distribution? Mark the location of the mean on your histogram.

14.26 Roulette—betting on red. A roulette wheel has 38 slots, numbered 0, 00, and 1 to 36. The slots 0 and 00 are colored green, 18 of the others are red, and 18 are black. The dealer spins the wheel and at the same time rolls a small ball along the wheel in the opposite direction. The wheel is carefully balanced so that the ball is equally likely to land in any slot when the wheel slows. Gamblers can bet on various combinations of numbers and colors.

(a) If you bet on "red," you win if the ball lands in a red slot. What is the probability of winning with a bet on red in a single play of roulette?
(b) You decide to play roulette four times, each time betting on red. What is the distribution of X, the number of times you win?
(c) If you bet the same amount on each play and win on exactly two of the four plays, then you will "break even." What is the probability that you will break even?
(d) If you win on *fewer than* two of the four plays, then you will lose money. What is the probability that you will lose money?

14.27 The birth control shot. The birth control shot is one of the most effective methods of birth control available, and it works best when you get the shot regularly, every 12 weeks. Under ideal conditions, only 1% of women taking the shot become pregnant within one year. In typical use, however, 3% become pregnant.[7] Choose at random 20 women using the shot as their method of birth control. We count the number who become pregnant in the next year.

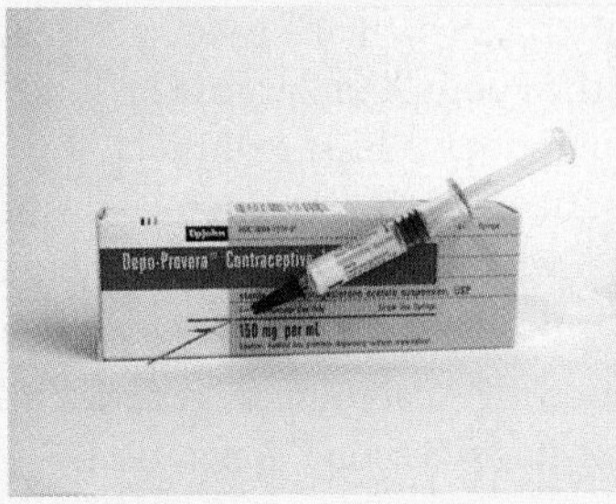

Urbano Delvalle/The LIFE Images Collection/Getty Images

(a) Explain why this is a binomial setting.
(b) What is the probability that at least one of the women becomes pregnant under ideal conditions? What is the probability in typical use?

14.28 Roulette, continued. You decide to play roulette 200 times, each time betting the same amount on red. You will lose money if you win on fewer than 100 of the plays. Based on the information in Exercise 14.26, what is the probability that you will lose money? (Check that the Normal approximation is permissible and use it to find this probability. If your software allows, find the exact binomial probability and compare the two results.) In general, if you bet the same amount on red every time, you will lose money if you win on fewer than half of the plays. What do you think happens to the probability of making money the longer you continue to play?

14.29 The birth control shot, continued. A study of the effectiveness of the birth control shot interviews a random sample of 600 women who are using the shot as their method of birth control.

(a) Based on the information about typical use in Exercise 14.27, what is the probability that at least 20 of these women become pregnant in the next year? (Check that the Normal approximation is permissible, and use it to find this probability. If your software allows, find the exact binomial probability, and compare the two results.)
(b) We can't use the Normal approximation to the binomial distributions to find this probability under ideal conditions as described in Exercise 14.27. Why not?

14.30 Hitting the fairway. One statistic used to assess professional golfers is driving accuracy, the percent of drives that land in the fairway. In 2013, driving accuracy for PGA Tour professionals ranged from about 45% to about 70%. Tiger Woods, the highest money winner on the PGA tour in 2013, only hits the fairway about 63% of the time.[8] Although Woods's average driving distance has decreased almost 10 yards over the last 10 years, his percent of drives that hit the fairway has increased about 7%, which is not surprising as increased distance is generally associated with decreased accuracy.

(a) Woods hits 14 drives in a round. What assumptions must you make to use a binomial distribution for the count X of fairways he hits? Which of these assumptions is least realistic?
(b) Assuming that a binomial distribution can be used, what is the expected number of fairways that Woods hits in a round in which he hits 14 drives?

14.31 Genetics. According to genetic theory, the blossom color in the second generation of a certain cross of sweet peas should be red or white in a 3:1 ratio. That is, each plant has probability 3/4 of having red blossoms, and the blossom colors of separate plants are independent.

© blickwinkel/Alamy

(a) What is the probability that exactly three out of four of these plants have red blossoms?
(b) What is the mean number of red-blossomed plants when 60 plants of this type are grown from seeds?
(c) What is the probability of obtaining at least 45 red-blossomed plants when 60 plants are grown from seeds? Use the Normal approximation. If your software allows, find the exact binomial probability and compare the two results.

14.32 False positives in testing for HIV. A rapid test for the presence in the blood of antibodies to HIV, the virus that causes AIDS, gives a positive result with probability about 0.004 when a person who is free of HIV antibodies is tested. A clinic tests 1000 people who are all free of HIV antibodies.

(a) What is the distribution of the number of positive tests?
(b) What is the mean number of positive tests?
(c) You cannot safely use the Normal approximation for this distribution. Explain why.

14.33 Hyundai sales in 2013. Hyundai Motor America sold 720,783 vehicles in the U.S. in 2013, with the U.S.-built Elantra leading sales with 247,912 cars sold, up 23% from 2012. The other top selling nameplates in 2013 were the Sonata with 203,648 sold, the Santa Fe with 88,844 sold, and the Accent with 60,458 sold.[9] The company wants to undertake a survey of 2013 Hyundai buyers to ask them about satisfaction with their purchase.

(a) What proportion of the Hyundais sold in 2013 were Elantras?
(b) If they plan to survey a simple random sample of 1200 Hyundai buyers, what is the expected number and standard deviation of the number of Elantra buyers in the sample?
(c) What is the probability they will get fewer than 400 Elantra buyers in their sample?

14.34 Retention rates in a weight loss program. Americans spend over $30 billion annually on a variety of weight loss products and services. In a study of retention rates of those using the Rewards Program at Jenny Craig in 2005, it was found that about 18% of those who began the program dropped out in the first four weeks.[10] Assume we have a random sample of 300 people beginning the program.

(a) What is the mean number of people who would drop out of the Rewards Program within four weeks in a sample of this size? What is the standard deviation?
(b) What is the approximate probability that at least 235 people in the sample will still be in the Rewards Program after the first four weeks?

14.35 Multiple-choice tests. Here is a simple probability model for multiple-choice tests. Suppose each student has probability p of correctly answering a question chosen at random from a universe of possible questions. (A strong student has a higher p than a weak student.) Answers to different questions are independent.

(a) Stacey is a good student for whom $p = 0.75$. Use the Normal approximation to find the probability that Stacey scores between 70% and 80% on a 100-question test.
(b) If the test contains 250 questions, what is the probability that Stacey will score between 70% and

80%? You see that Stacey's score on the longer test is more likely to be close to her "true score."

14.36 Is this coin balanced? While he was a prisoner of war during World War II, John Kerrich tossed a coin 10,000 times. He got 5067 heads. If the coin is perfectly balanced, the probability of a head is 0.5. Is there reason to think that Kerrich's coin was not balanced? To answer this question, find the probability that tossing a balanced coin 10,000 times would give a count of heads at least this far from 5000 (that is, at least 5067 heads or no more than 4933 heads.)

14.37 Binomial variation. Never forget that probability describes only what happens in the long run. Example 14.5 concerns the count of unmarketable tomatoes in inspection samples of size 10. The count has the binomial distribution with $n = 10$ and $p = 0.11$. The *Probability* applet simulates inspecting an SRS of 10 tomatoes if you set the probability of heads to 0.11, toss 10 times, and let each head stand for an unmarketable tomato.

APPLET

(a) The mean number of unmarketable tomatoes in a sample is 1.1. Click "Toss" and "Reset" repeatedly to simulate 20 samples. How many unmarketable tomatoes did you find in each sample? How close to the mean 1.1 is the average number of unmarketable tomatoes in these samples?

(b) Example 14.5 shows that the probability of exactly one unmarketable tomato is 0.3854. How close to the probability is the proportion of the 20 samples that have exactly one unmarketable tomato?

Whooping cough. *Whooping cough (pertussis) is a highly contagious bacterial infection that was a major cause of childhood deaths before the development of vaccines. About 80% of unvaccinated children who are exposed to whooping cough will develop the infection, as opposed to only about 5% of vaccinated children. Exercises 14.38 to 14.41 are based on this information.*

14.38 Vaccination at work. A group of 20 children at a nursery school are exposed to whooping cough by playing with an infected child.

(a) If all 20 have been vaccinated, what is the mean number of new infections? What is the probability that no more than two of the 20 children develop infections?

(b) If none of the 20 have been vaccinated, what is the mean number of new infections? What is the probability that 18 or more of the 20 children develop infections?

14.39 A whooping cough outbreak. In 2007, Bob Jones University in Greenville, SC, ended its fall semester a week early because of a whooping cough outbreak; 158 students were isolated and another 1200 given antibiotics as a precaution.[11] Authorities react strongly to whooping cough outbreaks because the disease is so contagious. Because the effect of childhood vaccination often wears off by late adolescence, treat the Bob Jones students as if they were unvaccinated. It appears that about 1400 students were exposed. What is the probability that at least 75% of these students develop infections if not treated? (Fortunately, whooping cough is much less serious after infancy.)

14.40 A mixed group: means. A group of 20 children at a nursery school are exposed to whooping cough by playing with an infected child. Of these children, 17 have been vaccinated and three have not.

(a) What is the distribution of the number of new infections among the 17 vaccinated children? What is the mean number of new infections?

(b) What is the distribution of the number of new infections among the three unvaccinated children? What is the mean number of new infections?

(c) Add your means from parts (a) and (b). This is the mean number of new infections among all 20 exposed children.

14.41 A mixed group: probabilities. We would like to find the probability that exactly two of the 20 exposed children in Exercise 14.40 develop whooping cough.

(a) One way to get two infections is to get one among the 17 vaccinated children and one among the three unvaccinated children. Find the probability of exactly one infection among the 17 vaccinated children. Find the probability of exactly one infection among the three unvaccinated children. These events are independent: what is the probability of exactly one infection in each group?

(b) Write down all the ways in which two infections can be divided between the two groups of children. Follow the pattern of part (a) to find the probability of each of these possibilities. Add all of your results (including the result of part (a)) to obtain the probability of exactly two infections among the 20 children.

14.42 Estimation π from random numbers. Kenyon College student Eric Newman used basic geometry to evaluate software random number generators as part of a summer research project. He generated 2000 independent random points (X, Y) in the unit square. (That is, X and Y are independent random numbers between 0 and 1, each having the density function illustrated in Figure 12.4 (page 290). The probability that (X, Y) falls in any region within the unit square is the area of the region.)[12]

(a) Sketch the unit square, the region of possible values for the point (X, Y).

(b) The set of points (X, Y) where $X^2 + Y^2 < 1$ describes a circle of radius 1. Add this circle to your sketch in part (a), and label the intersection of the two regions A.

(c) Let T be the total number of the 2000 points that fall into the region A. T follows a binomial distribution. Identify n and p. (*Hint:* Recall that the area of a circle is πr^2).
(d) What are the mean and standard deviation of T?
(e) Explain how Eric used a random number generator and the facts given here to estimate π.

14.43 The continuity correction. One reason why the Normal approximation may fail to give accurate estimates of binomial probabilities is that the binomial distributions are discrete and the Normal distributions are continuous. That is, counts take only whole number values, but Normal variables can take any value. We can improve the Normal approximation by treating each whole number count as if it occupied the interval from 0.5 below the number to 0.5 above the number. For example, approximate a binomial probability $P(X \geq 10)$ by finding the Normal probability $P(X \geq 9.5)$. Be careful: binomial $P(X > 10)$ is approximated by Normal $P(X \geq 10.5)$.

We saw in Exercise 14.30 that Tiger Woods hits the fairway in 63% of his drives. We will assume that his drives are independent and that each has probability 0.63 of hitting the fairway. Suppose Woods drives 30 times. The exact binomial probability that he hits 23 or more fairways is 0.084.

(a) Show that this setting satisfies the rule of thumb for use of the Normal approximation (just barely).
(b) What is the Normal approximation to $P(X \geq 23)$?
(c) What is the Normal approximation using the continuity correction? That's a lot closer to the true binomial probability.

Exploring the Web

14.44 MCAT physical science score. Go to the website `www.aamc.org/students/applying/mcat/data/mcat_stats/`, which reports the distribution of the total scaled MCAT composite score as well as the scores on the individual areas of assessment for recent years. The areas of assessment include three multiple-choice portions scored on a 15-point scale, which are combined to give the overall composite score.

(a) Open the PDF with the percentage and scaled score tables for the most recent year provided. Find the percentage of students who obtained a physical science section score of 9 or greater.

(b) A survey organization is planning on contacting an SRS of 300 of the examinees from the most recent year to see how they prepared for the MCAT exam. What is the distribution of the number of examinees in the sample that had a physical science section score of 9 or greater?

(c) Use the Normal distribution to approximate the probability that at least half of the examinees in the sample had a physical science section score of 9 or greater.

(d) Use software or a calculator to compute the exact probability that at least half of the examinees in the sample had a physical science section score of 9 or greater. How do the exact probability and the Normal approximation to this probability compare? Which of the two answers would you report to the survey organization? Why?

14.45 Binomial calculators. A number of websites will do exact binomial probability calculations for you.

(a) Find a website with a binomial calculator and give its URL.

(b) In Example 14.7, the number of people in the sample who track their health is a binomial random variable X with $n = 3014$ and $p = 0.62$. Use the binomial calculator to compute the probability that 1900 or more of the sample agree.

age fotostock/SuperStock

CHAPTER 15

Sampling Distributions

In this chapter we cover...

15.1 Parameters and statistics
15.2 Statistical estimation and the law of large numbers
15.3 Sampling distributions
15.4 The sampling distribution of $\bar{x}$
15.5 The central limit theorem
15.6 Sampling distributions and statistical significance

What is the average income of American households? Each March, the government's Current Population Survey asks detailed questions about income. The 98,095 households contacted in March 2013 had a mean "total money income" of $71,274 in 2012.[1] (The median income was of course lower, $51,017.) That $71,274 describes the sample, but we use it to estimate the mean income of all households. This is an example of statistical inference: we use information from a sample to infer something about a wider population.

Because the results of random samples and randomized comparative experiments include an element of chance, we can't guarantee that our inferences are correct. What we can guarantee is that our methods usually give correct answers. The reasoning of statistical inference rests on asking, "How often would this method give a correct answer if I used it very many times?" If our data come from random sampling or randomized comparative experiments, the laws of probability answer the question "What would happen if we did this many times?" This chapter presents some facts about probability that help answer this question.

15.1 Parameters and statistics

As we begin to use sample data to draw conclusions about a wider population, we must take care to keep straight whether a number describes a sample or a population. Here is the vocabulary we use.

Parameter, Statistic

A **parameter** is a number that describes the population. In practice, the value of a parameter is not known because we can rarely examine the entire population.

A **statistic** is a number that can be computed from the sample data without making use of any unknown parameters. In practice, we often use a statistic to estimate an unknown parameter.

EXAMPLE 15.1 Household Earnings

The mean income of the sample of 98,095 households contacted by the Current Population Survey was $\bar{x} = \$71,274$. The number $71,274 is a *statistic* because it describes this one Current Population Survey sample. The population that the poll wants to draw conclusions about is all 122 million U.S. households. The *parameter* of interest is the mean income of all these households. We don't know the value of this parameter. ■

population mean μ

population standard deviation σ

sample mean $\bar{x}$

sample standard deviation s

Remember *s* and ***p***: **s**tatistics come from **s**amples, and **p**arameters come from **p**opulations. As long as we were just doing data analysis, searching for patterns, or summarizing features of our data, the distinction between population and sample was not important. Now, as we begin to understand what our data (sample) tell us about a population, it is essential. The notation we use must reflect this distinction. We write μ (the Greek letter mu) for the **population mean** and σ (the Greek letter sigma) for the **population standard deviation.** These are fixed parameters that are unknown when we use a sample for inference. The **sample mean** is the familiar $\bar{x}$, the average of the observations in the sample. The **sample standard deviation** is denoted by s, the standard deviation of the observations in the sample. These are statistics that would almost certainly take different values if we chose another sample from the same population. The sample mean $\bar{x}$ and sample standard deviation s from a sample or an experiment are estimates of the mean μ and standard deviation σ of the underlying population.

LaunchPad Online Resources

- The **StatClips video**, *Introduction to Statistics—Populations, Parameters, Samples, and Sample Statistics,* discusses the meaning of these four terms.

Apply Your Knowledge

15.1 **Genetic Engineering.** Here's an idea for treating advanced melanoma, the most serious kind of skin cancer: genetically engineer white blood cells to better recognize and destroy cancer cells, then infuse these cells into patients. The subjects in a small initial study of this approach were 11 patients whose melanoma had not responded to existing treatments. One outcome of this experiment was measured by a test for the presence of cells that trigger an immune response in the body and so may help fight cancer. The mean counts of active cells per 100,000 cells for the 11 subjects were **3.8** before infusion and **160.2** after infusion. Is each of the boldface numbers a parameter or a statistic?

15.2 Florida Voters. Florida played a key role in recent presidential elections. Voter registration records in February 2014 show that **39%** of Florida voters are registered as Democrats and **35%** as Republicans. (Most of the others did not choose a party.) To test a random digit dialing device that you plan to use to poll voters for the 2014 Senate elections, you use it to call 250 randomly chosen residential telephones in Florida. Of the registered voters contacted, **34%** are registered Democrats. Is each of the boldface numbers a parameter or a statistic?

15.3 Steroid Use. Researchers surveyed 500 American anabolic androgenic steroid users, ranging in age from 16 to 62, and found that **98.8%** of them were male. The proportion of all Americans between the ages of 16 and 62 who are male is **50.0%**. The median age at which those surveyed began using steroids was **22**. Is each of the boldface numbers a parameter or a statistic?

15.2 Statistical estimation and the law of large numbers

Statistical inference uses sample data to draw conclusions about the entire population. Because good samples are chosen randomly, statistics such as $\bar{x}$ computed from these samples are random variables. We can describe the behavior of a sample statistic by a probability model that answers the question, "What would happen if we did this many times?" Here is an example that will lead us toward the probability ideas most important for statistical inference.

Statistics In Your World: HIGH-TECH GAMBLING

There are twice as many slot machines as bank ATMs in the United States. Once upon a time, you put in a coin and pulled the lever to spin three wheels, each with 20 symbols. No longer. Now the machines are video games with flashy graphics and outcomes produced by random number generators. Machines can accept many coins at once, can pay off on a bewildering variety of outcomes, and can be networked to allow common jackpots. Gamblers still search for systems, but in the long run the law of large numbers guarantees the house its 5% profit.

EXAMPLE 15.2 Does This Wine Smell Bad?

One of the reasons that winemaking is referred to as an art is because so many things can go wrong during production. Wine is chemically delicate and must be carefully supervised and nurtured. Sulfur compounds such as dimethyl sulfide (DMS) are formed naturally in the process of making wine. It is present in all wines. At low levels, it contributes to roundness, fruitiness, and the complexity of wine. DMS increases with wine age, and can lead to interesting sensory characteristics, such as increased complexity. Unfortunately, at higher levels it can contribute a vegetative, cooked cabbage, onion-like, or sulfur smell. In restaurants, when you order a bottle of wine, you are often presented with a small sample of the newly opened bottle to confirm that there are no unpleasant odors.

Winemakers need to control DMS in wine and understand at what levels it causes a wine to smell bad. Thus, winemakers want to know the "odor threshold," the lowest concentration of DMS that the human nose can detect. People vary in their ability to detect DMS, and it is important to understand this variation. For example, wine aficionados often have sensitive palates and may have a lower-than-average odor threshold.

Because different people have different thresholds, we start by asking about the mean threshold μ in the population of all adults. The number μ is a parameter that describes this population.

To estimate μ, we present tasters with both natural wine and the same wine spiked with DMS at different concentrations to find the lowest concentration at which they identify the spiked wine. Here are the odor thresholds (measured in micrograms of DMS per liter of wine) for 10 randomly chosen subjects:

28 40 28 33 20 31 29 27 17 21

Jupiterimages/Getty Images

The mean threshold for these subjects is $\bar{x} = 27.4$. It seems reasonable to use the sample result $\bar{x} = 27.4$ to estimate the unknown μ. An SRS should fairly represent the population, so the mean $\bar{x}$ of the sample should be somewhere near the mean μ of the population. Of course, we don't expect $\bar{x}$ to be exactly equal to μ. We realize that if we choose another SRS, the luck of the draw will probably produce a different $\bar{x}$. ■

If $\bar{x}$ is rarely exactly right and varies from sample to sample, why is it nonetheless a reasonable estimate of the population mean μ? Here is one answer: *if we keep on taking larger and larger samples, the statistic $\bar{x}$ is guaranteed to get closer and closer to the parameter μ.* We have the comfort of knowing that if we can afford to keep on measuring more subjects, eventually we will estimate the mean odor threshold of all adults very accurately. This remarkable fact is called the *law of large numbers.* It is remarkable because it holds for *any* population, not just for some special class such as Normal distributions.

Law of Large Numbers

Draw observations at random from any population with finite mean μ. As the number of observations drawn increases, the mean $\bar{x}$ of the observed values tends to get closer and closer to the mean μ of the population.

The law of large numbers can be proven mathematically starting from the basic laws of probability. The behavior of $\bar{x}$ is similar to the idea of probability. In the long run, the *proportion* of outcomes taking any value gets close to the probability of that value, and the *average* outcome gets close to the population mean. Figure 12.1 (page 279) shows how proportions approach probability in one example. Here is an example of how sample means approach the population mean.

EXAMPLE 15.3 The Law of Large Numbers in Action

Suppose the distribution of odor thresholds among all adults has mean 25. The mean $\mu = 25$ is the true value of the parameter we seek to estimate. Figure 15.1 shows how the sample mean $\bar{x}$ of an SRS drawn from this population changes as we add more subjects to our sample.

The first subject in Example 15.2 had threshold 28, so the line in Figure 15.1 starts there. The mean for the first two subjects is

$$\bar{x} = \frac{28 + 40}{2} = 34$$

This is the second point on the graph. At first, the graph shows that the mean of the sample changes as we take more observations. Eventually, however, the mean of the observations gets close to the population mean $\mu = 25$ and settles down at that value.

If we started over, again choosing people at random from the population, we would get a different path from left to right in Figure 15.1. The law of large numbers says that whatever path we get will always settle down at 25 as we draw more and more people. ■

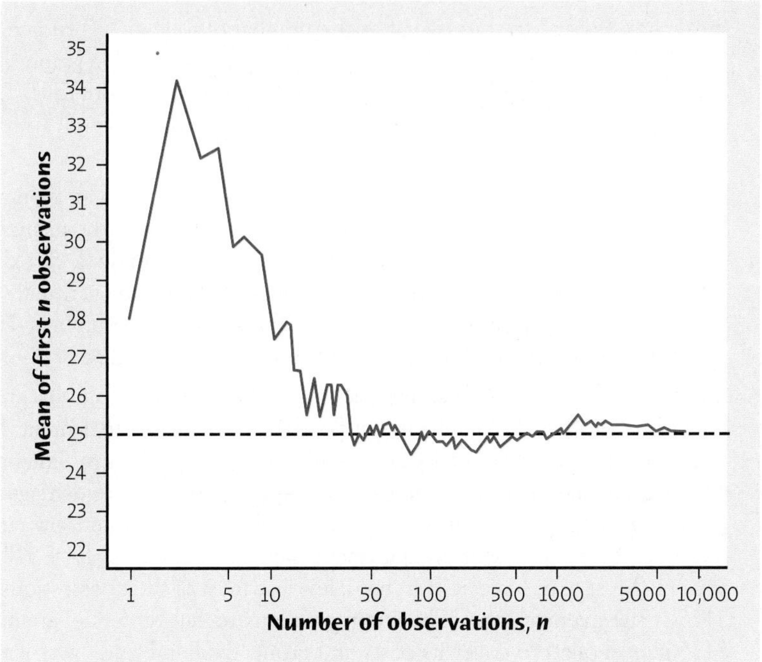

FIGURE 15.1
The law of large numbers in action: as we take more observations, the sample mean $\bar{x}$ always approaches the mean μ of the population.

The *Law of Large Numbers* applet animates Figure 15.1 in a different setting. You can use the applet to watch $\bar{x}$ change as you average more observations until it eventually settles down at the mean μ.

APPLET

The law of large numbers is the foundation of such business enterprises as gambling casinos and insurance companies. The winnings (or losses) of a gambler on a few plays are uncertain—that's why some people find gambling exciting. In Figure 15.1, the mean of even 100 observations is not yet very close to μ. It is only *in the long run* that the mean outcome is predictable. The house plays tens of thousands of times. So the house, unlike individual gamblers, can count on the long-run regularity described by the law of large numbers. The average winnings of the house on tens of thousands of plays will be very close to the mean of the distribution of winnings. Needless to say, this mean guarantees the house a profit. That's why gambling can be a business.

Apply Your Knowledge

15.4 The Law of Large Numbers Made Visible. Roll two balanced dice and count the total spots on the up-faces. The probability model appears in Example 12.5 (page 283). You can see that this distribution is symmetric with 7 as its center, so it's no surprise that the mean is $\mu = 7$. This is the population mean for the idealized population that contains the results of rolling two dice forever. The law of large numbers says that the average $\bar{x}$ from a finite number of rolls tends to get closer and closer to 7 as we do more and more rolls.

APPLET

(a) Click "More dice" once in the *Law of Large Numbers* applet to get two dice. Click "Show mean" to see the mean 7 on the graph. Leaving the number of rolls at 1, click "Roll dice" three times. How many spots did each roll produce? What is the average for the three rolls? You see that the graph displays at each point the average number of spots for all rolls up to the last one. This is exactly like Figure 15.1.

(b) Click "Reset" to start over. Set the number of rolls to 100 and click "Roll dice." The applet rolls the two dice 100 times. The graph shows how the average count of spots changes as we make more rolls. That is, the graph shows $\bar{x}$ as we continue to roll the dice. Sketch (or print out) the final graph.

(c) Repeat your work from part (b). Click "Reset" to start over, then roll two dice 100 times. Make a sketch of the final graph of the mean $\bar{x}$ against the number of rolls. Your two graphs will often look very different. What they have in common is that the average eventually gets close to the population mean $\mu = 7$. The law of large numbers says that this will *always* happen if you keep on rolling the dice.

15.5 **Insurance.** The idea of insurance is that we all face risks that are unlikely but carry high cost. Think of a fire or flood destroying your apartment. Insurance spreads the risk: we all pay a small amount, and the insurance policy pays a large amount to those few of us whose apartments are damaged. An insurance company looks at the records for millions of apartment owners and sees that the mean loss from apartment damage in a year is $\mu = \$125$ per person. (Most of us have no loss, but a few lose most of their possessions. The \$125 is the average loss.) The company plans to sell renters insurance for \$125 plus enough to cover its costs and profit. Explain clearly why it would be unwise to sell only 12 policies. Then explain why selling thousands of such policies is a safe business.

15.3 Sampling distributions

The law of large numbers assures us that if we measure enough subjects, the statistic $\bar{x}$ will eventually get very close to the unknown parameter μ. But the odor threshold study in Example 15.2 had just 10 subjects. What can we say about estimating μ by $\bar{x}$ from a sample of 10 subjects? Put this one sample in the context of all such samples by asking, "What would happen if we took many samples of 10 subjects from this population?" Here's how to answer this question:

- Take a large number of samples of size 10 from the population.
- Calculate the sample mean $\bar{x}$ for each sample.
- Make a histogram of the values of $\bar{x}$.
- Examine the shape, center, and variability of the distribution displayed in the histogram.

In practice it is too expensive to take many samples from a large population such as all adult U.S. residents. But we can imitate many samples by using software. Using software to imitate chance behavior is called **simulation**.

simulation

EXAMPLE 15.4 What Would Happen in Many Samples?

Extensive studies have found that the DMS odor threshold of adults follows roughly a Normal distribution with mean $\mu = 25$ micrograms per liter and standard deviation $\sigma = 7$ micrograms per liter. We call this the *population distribution* of odor threshold.

Figure 15.2 illustrates the process of choosing many samples and finding the sample mean threshold $\bar{x}$ for each one. Follow the flow of the figure from the population at the left, to choosing an SRS and finding the $\bar{x}$ for this sample, to collecting

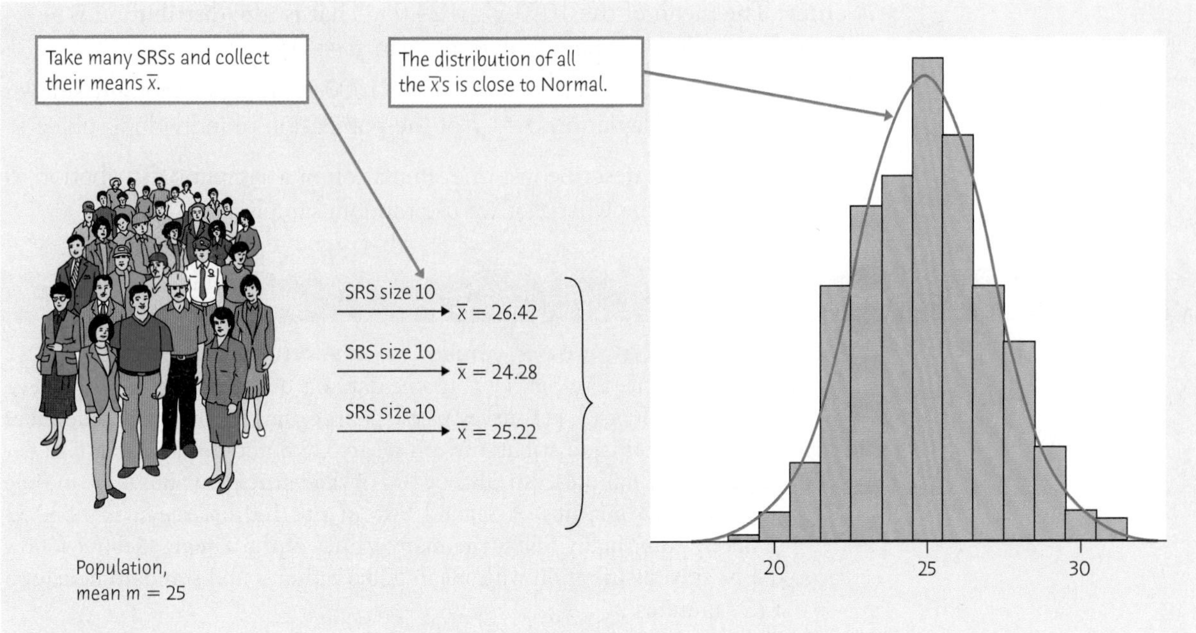

FIGURE 15.2
The idea of a sampling distribution: take many samples from the same population, collect the $\bar{x}$'s from all the samples, and display the distribution of the $\bar{x}$'s. The histogram shows the results of 1000 samples.

together the $\bar{x}$'s from many samples. The first sample has $\bar{x} = 26.42$. The second sample contains a different 10 people, with $\bar{x} = 24.28$, and so on. The histogram at the right of the figure shows the distribution of the values of $\bar{x}$ from 1000 separate SRSs of size 10. This histogram displays the *sampling distribution* of the statistic $\bar{x}$. ■

Population Distribution, Sampling Distribution

The **population distribution** of a variable is the distribution of values of the variable among all the individuals in the population.

The **sampling distribution** of a statistic is the distribution of values taken by the statistic in all possible samples of the same size from the same population.

CAUTION

Be careful: The population distribution describes the *individuals* that make up the population. A sampling distribution describes how a *statistic* varies in many samples from the population.

Strictly speaking, the sampling distribution is the ideal pattern that would emerge if we looked at all possible samples of size 10 from our population. A distribution obtained from a fixed number of trials, like the 1000 trials in Figure 15.2, is only an approximation to the sampling distribution. One of the uses of probability theory in statistics is to obtain sampling distributions without simulation. The interpretation of a sampling distribution is the same, however, whether we obtain it by simulation or by the mathematics of probability.

We can use the tools of data analysis to describe any distribution. Let's apply those tools to Figure 15.2. What can we say about the shape, center, and variability of this distribution?

- **Shape:** It looks Normal! Detailed examination confirms that the distribution of $\bar{x}$ from many samples is very close to Normal.

- **Center:** The mean of the 1000 $\bar{x}$'s is 24.95. That is, the distribution is centered very close to the population mean $\mu = 25$.
- **Variability:** The standard deviation of the 1000 $\bar{x}$'s is 2.217, notably smaller than the standard deviation $\sigma = 7$ of the population of individual subjects.

Although these results describe just one simulation of a sampling distribution, they reflect facts that are true whenever we use random sampling.

Apply Your Knowledge

15.6 Sampling Distribution versus Population Distribution. The 2012 American Time Use Survey contains data on how many minutes of sleep per night each of 12,443 survey participants estimated they get.[2] The times follow the Normal distribution with mean 528.8 minutes and standard deviation 137.2 minutes. An SRS of 100 of the participants has a mean time of $\bar{x} = 509.23$ minutes. A second SRS of size 100 has mean $\bar{x} = 530.32$ minutes. After many SRSs, the many values of the sample mean $\bar{x}$ follow the Normal distribution with mean 528.8 minutes and standard deviation 13.72 minutes.

(a) What is the population? What values does the population distribution describe? What is this distribution?

(b) What values does the sampling distribution of $\bar{x}$ describe? What is the sampling distribution?

APPLET

15.7 Generating a Sampling Distribution. Let's illustrate the idea of a sampling distribution in the case of a very small sample from a very small population. The population is the scores of 10 students on an exam:

Student	1	2	3	4	5	6	7	8	9	10
Score	86	63	81	55	72	72	65	66	75	59

The parameter of interest is the mean score μ in this population. The sample is an SRS of size $n = 4$ drawn from the population. The *Simple Random Sample* applet can be used to select simple random samples of four single digit numbers between 1 and 10, corresponding to the students.

(a) Make a histogram of these 10 scores.

(b) Find the mean of the 10 scores in the population. This is the population mean μ.

(c) Use the *Simple Random Sample* applet to draw an SRS of size 4 from this population. What are the four scores in your sample? What is their mean $\bar{x}$? This statistic is an estimate of μ.

(d) Repeat this process nine more times, using the applet. Make a histogram of the 10 values of $\bar{x}$. You are constructing the sampling distribution of $\bar{x}$. Is the center of your histogram close to μ? How does the shape of this histogram compare with the histogram you made in part (a)?

15.4 The sampling distribution of $\bar{x}$

Figure 15.2 suggests that when we choose many SRSs from a population, the sampling distribution of the sample means is centered at the mean of the original population and is less variable (spread out) than the distribution of individual observations. Here are the facts.

Mean and Standard Deviation of a Sample Mean[3]

Suppose that $\bar{x}$ is the mean of an SRS of size n drawn from a large population with mean μ and standard deviation σ. Then the sampling distribution of $\bar{x}$ has **mean μ** and **standard deviation $\sigma/\sqrt{n}$**.

By "large population" we mean that the size of the population is much larger than the size of the sample—say, at least 20 times as large.

These facts about the mean and the standard deviation of the sampling distribution of $\bar{x}$ are true for *any* population, not just for some special class such as Normal distributions. They have important implications for statistical inference:

- The mean of the statistic $\bar{x}$ is always equal to the mean μ of the population. That is, the sampling distribution of $\bar{x}$ is centered at μ. In repeated sampling, $\bar{x}$ will sometimes fall above the true value of the parameter μ and sometimes below, but there is no systematic tendency to overestimate or underestimate the parameter. This makes the idea of lack of bias in the sense of "no favoritism" more precise. Because the mean of $\bar{x}$ is equal to μ, we say that the statistic $\bar{x}$ is an **unbiased estimator** of the parameter μ.
- An unbiased estimator is "correct on the average" in many samples. How close the estimator falls to the parameter in most samples is determined by the variability of the sampling distribution. If individual observations have standard deviation σ, then sample means $\bar{x}$ from samples of size n have standard deviation $\sigma/\sqrt{n}$. That is, **averages are less variable than individual observations.**
- Not only is the standard deviation of the distribution of $\bar{x}$ smaller than the standard deviation of individual observations, but it gets smaller as we take larger samples. **The results of large samples are less variable than the results of small samples.**

unbiased estimator

CAUTION

The upshot of all this is that we can trust the sample mean from a large random sample to estimate the population mean accurately. If the sample size n is large, the standard deviation of $\bar{x}$ is small, and almost all samples will give values of $\bar{x}$ that lie very close to the true parameter μ. *However, the standard deviation of the sampling distribution gets smaller only at the rate $\sqrt{n}$. To cut the standard deviation of $\bar{x}$ in half, we must take four times as many observations, not just twice as many.* So very precise estimates (estimates with very small standard deviation) may be expensive.

We have described the center and variability of the sampling distribution of a sample mean $\bar{x}$, but not its shape. The shape of the sampling distribution depends on the shape of the population distribution. In one important case there is a simple relationship between the two distributions: if the population distribution is Normal, then so is the sampling distribution of the sample mean.

SAMPLE SIZE MATTERS

The new thing in baseball is using statistics to evaluate players, with new measures of performance to help decide which players are worth the high salaries they demand. This challenges traditional subjective evaluation of young players and the usefulness of traditional measures such as batting average. But success has led many major league teams to hire statisticians. The statisticians say that sample size matters in baseball also: the 162-game regular season is long enough for the better teams to come out on top, but five game and seven game play-off series are so short that luck has a lot to do with who wins.

Sampling Distribution of a Sample Mean

If individual observations have the $N(\mu, \sigma)$ distribution, then the sample mean $\bar{x}$ of an SRS of size n has the $N(\mu, \sigma/\sqrt{n})$ distribution.

Notice that if the population distribution is Normal, then the sampling distribution of the sample mean is Normal regardless of the sample size n.

EXAMPLE 15.5 Population Distribution, Sampling Distribution

If we measure the DMS odor thresholds of individual adults (as described in Example 15.2), the values follow the Normal distribution with mean $\mu = 25$ micrograms per liter and standard deviation $\sigma = 7$ micrograms per liter. This is the population distribution of odor threshold.

Take many SRSs of size 10 from this population and find the sample mean $\bar{x}$ for each sample, as in Figure 15.2. The sampling distribution describes how the values of $\bar{x}$ vary among samples. That sampling distribution is also Normal, with mean $\mu = 25$ and standard deviation

$$\frac{\sigma}{\sqrt{n}} = \frac{7}{\sqrt{10}} = 2.2136$$

Figure 15.3 contrasts these two Normal distributions. Both are centered at the population mean, but sample means are much less variable than individual observations.

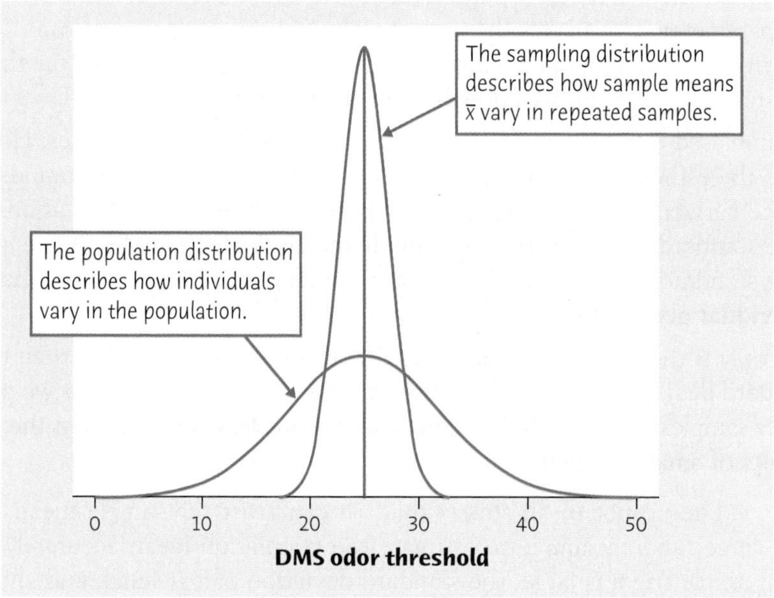
The sampling distribution describes how sample means $\bar{x}$ vary in repeated samples.
The population distribution describes how individuals vary in the population.
0
10
20
30
40
50
DMS odor threshold

FIGURE 15.3 The distribution of single observations (the population distribution) compared with the sampling distribution of the means $\bar{x}$ of 10 observations, for Example 15.5. Both have the same mean, but averages are less variable than individual observations.

The smaller variation of sample means shows up in probability calculations. You can show (using software or standardizing and using Table A) that about 52% of all adults have odor thresholds between 20 and 30. But almost 98% of means of samples of size 10 lie in this range. ■

LaunchPad Online Resources

- The **StatBoards video**, *$\bar{x}$ Is Normal for All Sample Sizes If the Population Is Normal*, discusses examples of computing the sampling distribution of $\bar{x}$ when the population is normal.

Apply Your Knowledge

15.8 A Sample of Young Men. A government sample survey plans to measure the LDL (bad) cholesterol level of an SRS of men aged 20 to 34. The researchers will report the mean $\bar{x}$ from their sample as an estimate of the mean LDL cholesterol level μ in this population.

(a) Explain to someone who knows no statistics what it means to say that $\bar{x}$ is an "unbiased" estimator of μ.

(b) The sample result $\bar{x}$ is an unbiased estimator of the population truth μ no matter what size SRS the study uses. Explain to someone who knows no statistics why a large sample gives more trustworthy results than a small sample.

15.9 Larger Sample, More Accurate Estimate. Suppose that in fact the LDL cholesterol level of all men aged 20 to 34 follows the Normal distribution with mean $\mu = 115$ milligrams per deciliter (mg/dL) and standard deviation $\sigma = 25$ mg/dL.

(a) Choose an SRS of 100 men from this population. What is the sampling distribution of $\bar{x}$? What is the probability that $\bar{x}$ takes a value between 112 and 118 mg/dL? This is the probability that $\bar{x}$ estimates μ within ± 3 mg/dL.

(b) Choose an SRS of 1000 men from this population. Now what is the probability that $\bar{x}$ falls within ± 3 mg/dL of μ? The larger sample is much more likely to give an accurate estimate of μ.

15.10 Measurements in the Lab. Juan makes a measurement in a chemistry laboratory and records the result in his lab report. Suppose that if Juan makes this measurement repeatedly, the standard deviation of his measurements will be $\sigma = 10$ milligrams. Juan repeats the measurement four times and records the mean $\bar{x}$ of his four measurements.

(a) What is the standard deviation of Juan's mean result? (That is, if Juan kept on making four measurements and averaging them, what would be the standard deviation of all his $\bar{x}$ values?)

(b) How many times must Juan repeat the measurement to reduce the standard deviation of $\bar{x}$ to 2? Explain to someone who knows no statistics the advantage of reporting the average of several measurements rather than the result of a single measurement.

15.5 The central limit theorem

The facts about the mean and standard deviation of $\bar{x}$ are true no matter what the shape of the population distribution may be. But what is the shape of the sampling distribution when the population distribution is not Normal? *It is a remarkable fact that as the sample size increases, the distribution of $\bar{x}$ changes shape: it looks less like that of the population and more like a Normal distribution.* When the sample is large enough, the distribution of $\bar{x}$ is very close to Normal. This is true no matter what shape the population distribution has, as long as the population has a finite standard deviation σ. This famous fact of probability theory is called the *central limit theorem*. It is much more useful than the fact that the distribution of $\bar{x}$ is exactly Normal if the population is exactly Normal.

Central Limit Theorem

Draw an SRS of size n from any population with mean μ and finite standard deviation σ. The **central limit theorem** says that when n is large, the sampling distribution of the sample mean $\bar{x}$ is approximately Normal:

$$\bar{x} \text{ is approximately } N\left(\mu, \frac{\sigma}{\sqrt{n}}\right)$$

WHAT WAS THAT PROBABILITY AGAIN?

Wall Street uses fancy mathematics to predict the probabilities that fancy investments will go wrong. The probabilities are always too low—sometimes because something was assumed to be Normal but was not. Probability predictions in other areas also go wrong. In mid-September 2007, the New York Mets had probability 0.998 of making the National League play-offs, or so an elaborate calculation said. Then the Mets lost 12 of their final 17 games, the Phillies won 13 of their final 17, and the Mets were out.

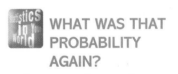

The central limit theorem allows us to use Normal probability calculations to answer questions about sample means from many observations even when the population distribution is not Normal.

More general versions of the central limit theorem say that the distribution of any sum or average of many small random quantities is close to Normal. This is true even if the quantities are correlated with each other (as long as they are not too highly correlated) and even if they have different distributions (as long as no one random quantity is so large that it dominates the others). The central limit theorem suggests why the Normal distributions are common models for observed data. Any variable that is a sum of many small influences will have approximately a Normal distribution.

How large a sample size n is needed for $\bar{x}$ to be close to Normal depends on the population distribution. More observations are required if the shape of the population distribution is far from Normal. Here are two examples in which the population is far from Normal.

EXAMPLE 15.6 The Central Limit Theorem in Action

In March 2013, the Current Population Survey contacted 98,095 households. Figure 15.4(a) is a histogram of the earnings of the 73,769 households that had earned income greater than zero in 2012.[4] As we expect, the distribution of earned incomes is strongly skewed to the right and very spread out. The right tail of the distribution is even longer than the histogram shows because there are too few high incomes for their bars to be visible on this scale. In fact, we cut off the earnings scale at \$400,000 to save space—a few households earned even more than \$400,000. The mean earnings for these 73,769 households was \$74,621.

Regard these 73,769 households as a population with mean $\mu = \$74{,}621$. Take an SRS of 100 households. The mean earnings in this sample is $\bar{x} = \$75{,}847$. That's

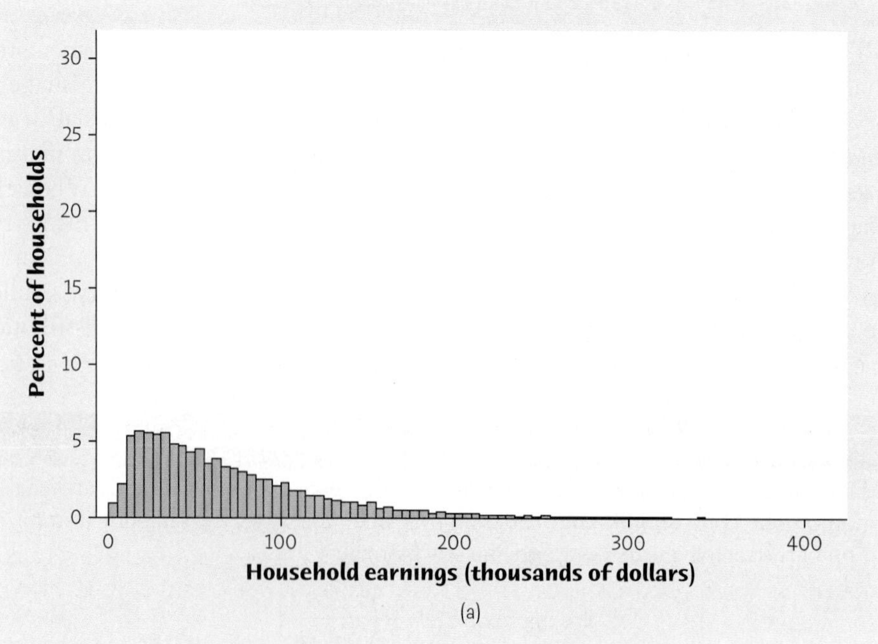

FIGURE 15.4
The central limit theorem in action, for Example 15.6. (a) The distribution of earned income in a population of 73,769 households. (b) The distribution of the mean earnings for 500 SRSs of 100 households each from this population. (c) The distribution of the sample means in more detail: the shape is close to Normal.

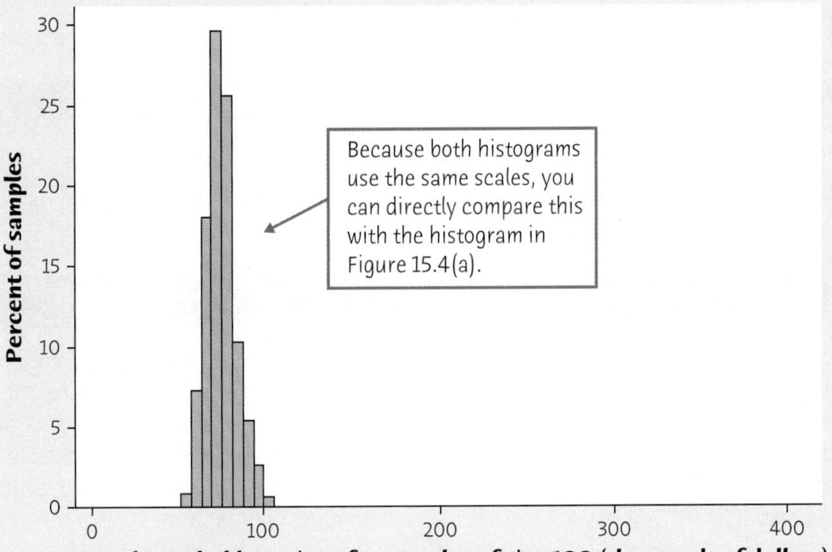

(b)

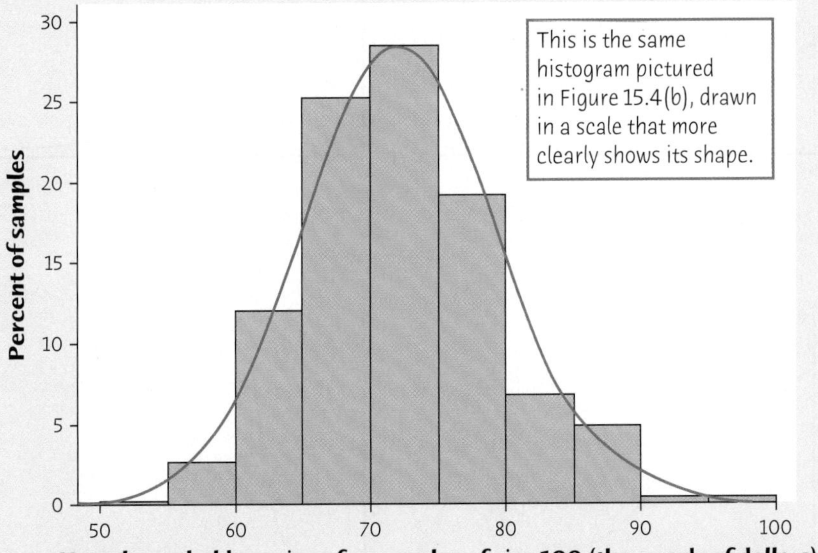

(c)

FIGURE 15.4
(Continued)

higher than the mean of the population. Take another SRS of size 100. The mean for this sample is $\bar{x} = \$66{,}860$. That's less than the mean of the population. *What would happen if we did this many times?* Figure 15.4(b) is a histogram of the mean earnings for 500 samples, each of size 100. The scales in Figures 15.4(a) and 15.4(b) are the same, for easy comparison. Although the distribution of individual earnings is skewed and with large variability, the distribution of sample means is roughly symmetric and shows much less variability.

Figure 15.4(c) zooms in on the center part of the histogram in Figure 15.4(b) to more clearly show its shape. Although $n = 100$ is not a very large sample size and the population distribution is extremely skewed, we can see that the distribution of sample means is close to Normal. ■

Comparing Figure 15.4(a) with Figures 15.4(b) and 15.4(c) illustrates the two most important ideas of this chapter.

Thinking about Sample Means

Means of random samples are **less variable** than individual observations.

Means of random samples are **more Normal** than individual observations.

EXAMPLE 15.7 The Central Limit Theorem in Action

Exponential distributions are used as models for the lifetime of electronic components and for the time required to serve a customer or repair a machine. Figure 15.5(a) shows the exponential population distribution, that is, the density curve of a single observation. This distribution is strongly right-skewed, and the most probable outcomes are near 0. The mean μ of this distribution is 1, and its standard deviation σ is also 1.

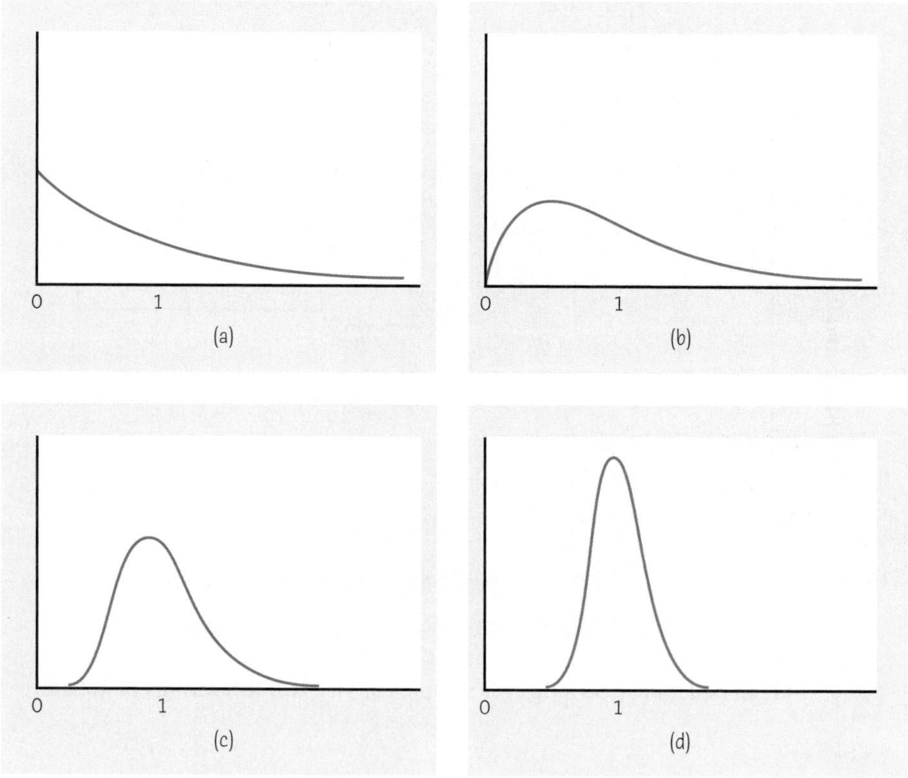

FIGURE 15.5 The central limit theorem in action, for Example 15.7. The distribution of sample means $\bar{x}$ from a strongly non-Normal population becomes more Normal as the sample size increases. (a) The distribution of 1 observation. (b) The distribution of $\bar{x}$ for 2 observations. (c) The distribution of $\bar{x}$ for 10 observations. (d) The distribution of $\bar{x}$ for 25 observations.

Mathematics can be used to derive the theoretical sampling distribution of $\bar{x}$ when sampling from an exponential distribution. Figures 15.5(b), 15.5(c), and 15.5(d) are the theoretical density curves of the sample means of 2, 10, and 25 observations from this population. As n increases, the shape becomes more Normal. The mean remains at $\mu = 1$, and the standard deviation decreases, taking the value $1/\sqrt{n}$. The density curve for 10 observations is still somewhat skewed to the right but already resembles a Normal curve having $\mu = 1$ and $\sigma = 1/\sqrt{10} = 0.32$. The density curve for $n = 25$ is yet more Normal. The contrast between the shapes of the population distribution and of the distribution of the mean of 10 or 25 observations is striking. ■

The *Central Limit Theorem* applet allows you to watch the central limit theorem in action. The applet simulates the sampling distribution of $\bar{x}$ for several population distributions and you can see how the sampling distribution changes shape as the sample size increases.

APPLET

Let's use Normal calculations based on the central limit theorem to answer a question about the very non-Normal distribution in Figure 15.5(a).

EXAMPLE 15.8 Maintaining Air Conditioners

4 step

STATE: The time (in hours) that a technician requires to perform preventive maintenance on an air-conditioning unit is governed by the exponential distribution whose density curve appears in Figure 15.5(a). The exponential distribution arises in many engineering and industrial problems, such as time until failure of a machine or time until a success. The mean time is $\mu = 1$ hour and the standard deviation is $\sigma = 1$ hour. Your company has a contract to maintain 70 of these units in an apartment building. You must schedule technicians' time for a visit to this building. Is it safe to budget an average of 1.1 hours for each unit? Or should you budget an average of 1.25 hours?

PLAN: We believe that the manufacturing and distribution process associated with this type of air-conditioning unit is such that variation from one unit to the next is random. Thus, we treat these 70 air conditioners as an SRS from all units of this type. What is the probability that the average maintenance time for 70 units exceeds 1.1 hours? That the average time exceeds 1.25 hours?

SOLVE: The central limit theorem says that the sample mean time $\bar{x}$ spent working on 70 units has approximately the Normal distribution with mean equal to the population mean $\mu = 1$ hour and standard deviation

$$\frac{\sigma}{\sqrt{70}} = \frac{1}{\sqrt{70}} = 0.12 \text{ hour}$$

The distribution of $\bar{x}$ is therefore approximately $N(1, 0.12)$. This Normal curve is the solid curve in Figure 15.6.

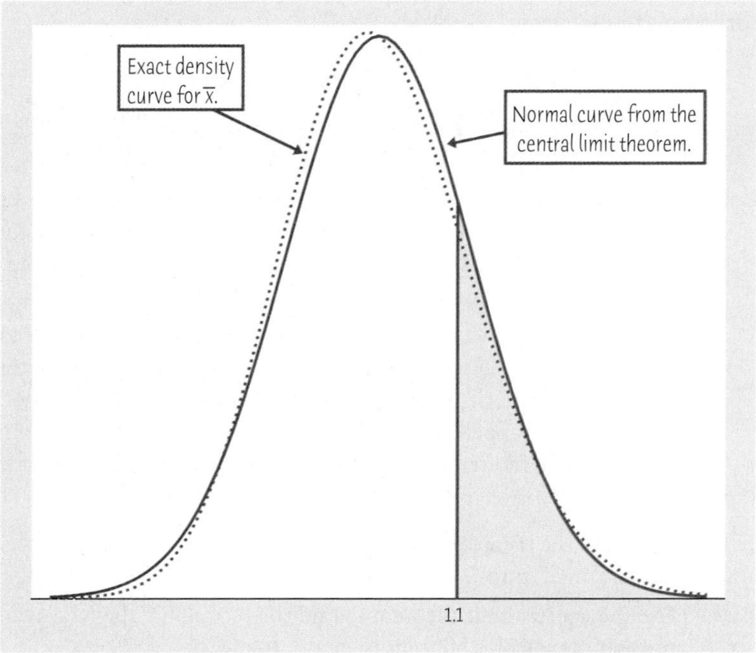

FIGURE 15.6
The exact distribution (dotted) and the Normal approximation from the central limit theorem (solid) for the average time needed to maintain an air conditioner, for Example 15.8. The probability we want is the area to the right of 1.1.

Using this Normal distribution, the probabilities we want are

$$P(\bar{x} > 1.10 \text{ hours}) = 0.2014$$
$$P(\bar{x} > 1.25 \text{ hours}) = 0.0182$$

Software gives these probabilities immediately, or you can standardize and use Table A. For example,

$$P(\bar{x} > 1.10) = P\left(\frac{\bar{x} - 1}{0.12}\right) > P\left(\frac{1.10 - 1}{0.12}\right)$$
$$= P(Z > 0.83) = 1 - 0.7967 = 0.2033$$

with the usual roundoff error. Don't forget to use standard deviation 0.12 in your software or when you standardize $\bar{x}$.

CONCLUDE: If you budget 1.1 hours per unit, there is a 20% chance that the technicians will not complete the work in the building within the budgeted time. This chance drops to 2% if you budget 1.25 hours. You should therefore budget 1.25 hours per unit. ■

Using more mathematics, we can start with the exponential distribution and find the actual density curve of $\bar{x}$ for 70 observations. This is the dotted curve in Figure 15.6. You can see that the solid Normal curve is a good approximation. The exactly correct probability for 1.1 hours is an area to the right of 1.1 under the dotted density curve. It is 0.1977. The central limit theorem Normal approximation 0.2014 is off by only about 0.004.

LaunchPad Online Resources

- **The StatClips Examples video, *Central Limit Theorem Part I Example A*, and the Snapshots video, *Sampling Distributions and CLT*, discuss examples of the central limit theorem in action.**
- **The StatBoards video, *$\bar{x}$ Is Approximately Normal for Large n for Any Distribution*, discusses examples of computing the sampling distribution of $\bar{x}$ for large sample sizes.**

Apply Your Knowledge

15.11 What Does the Central Limit Theorem Say? Asked what the central limit theorem says, a student replies, "As you take larger and larger samples from a population, the histogram of the sample values looks more and more Normal." Is the student right? Explain your answer.

© Bruce Coleman Inc./Alamy

15.12 Detecting Gypsy Moths. The gypsy moth is a serious threat to oak and aspen trees. A state agriculture department places traps throughout the state to detect the moths. When traps are checked periodically, the mean number of moths trapped is only 0.5, but some traps have several moths. The distribution of moth counts is finite and strongly skewed, with standard deviation 0.7.

(a) What are the mean and standard deviation of the average number of moths $\bar{x}$ in 50 traps?

(b) Use the central limit theorem to find the probability that the average number of moths in 50 traps is greater than 0.6.

15.13 **More on Insurance.** An insurance company knows that, in the entire population of millions of apartment owners, the mean annual loss from damage is $\mu = \$125$ and the standard deviation of the loss is $\sigma = \$300$. The distribution of losses is strongly right-skewed: most policies have \$0 loss, but a few have large losses. If the company sells 10,000 policies, can it safely base its rates on the assumption that its average loss will be no greater than \$135? Follow the four-step process as illustrated in Example 15.8.

15.6 Sampling distributions and statistical significance

We have looked carefully at the sampling distribution of a sample mean. However, any statistic we can calculate from a sample will have a sampling distribution.

EXAMPLE 15.9 Median, Variance and Standard Deviation

In Example 15.5, we took 1000 SRSs of size 10 from a Normal population with mean $\mu = 25$ micrograms per liter and standard deviation $\sigma = 7$ micrograms per liter. This Normal distribution is the distribution of the DMS odor thresholds of all adults. Figure 15.3 is a histogram of the distribution of the sample means.

Now take 1000 SRSs of size 5 from a Normal population with mean $\mu = 25$ and standard deviation $\sigma = 7$. For each sample, compute the sample median, variance, and standard deviation. Figure 15.7 displays histograms of the 1000 sample results. These histograms show the sampling distributions of the three statistics. The sampling distribution of the sample median is symmetric, centered at 25, and approximately Normal. The sampling distribution of the sample variance is strongly skewed to the right. The sampling distribution of the sample standard deviation is very slightly skewed to the right. ■

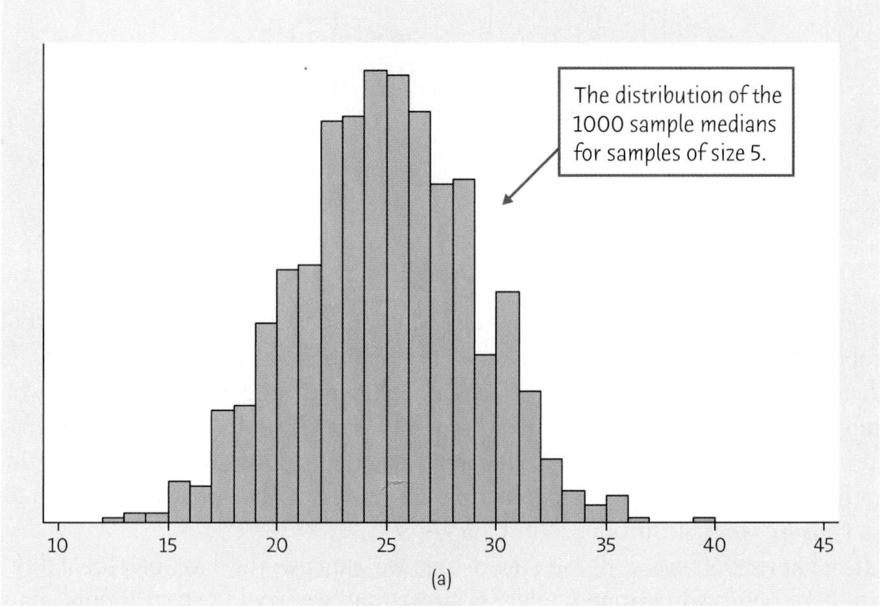

FIGURE 15.7 (a) The distribution of 1000 sample medians for samples of size 5 from a Normal population with $\mu = 25$ and $\sigma = 7$. (b) The distribution of 1000 sample variances for samples of size 5 from a Normal population with $\mu = 25$ and $\sigma = 7$. (c) The distribution of 1000 sample standard deviations for samples of size 5 from a Normal population with $\mu = 25$ and $\sigma = 7$.

FIGURE 15.7
(Continued)

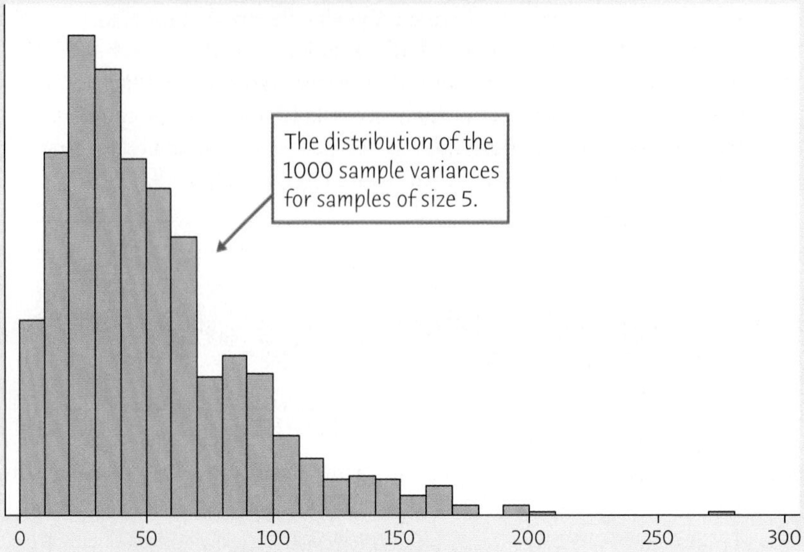

(b)

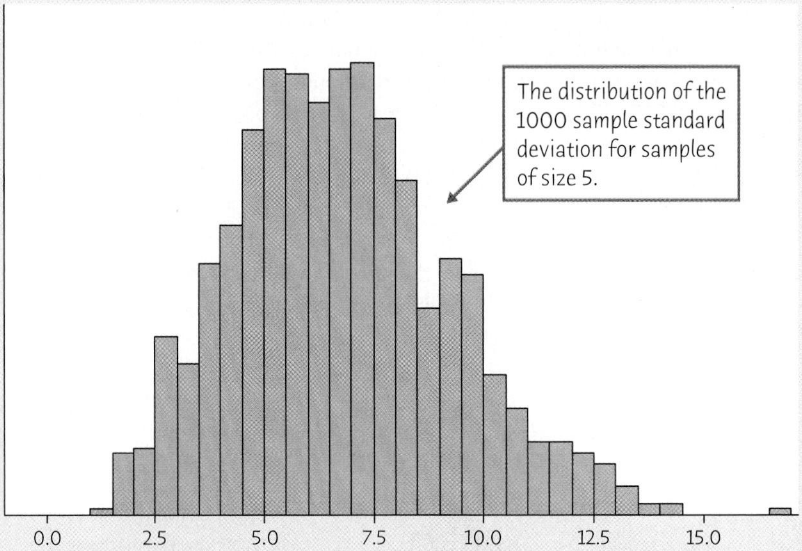

(c)

The sampling distribution of a sample statistic is determined by the particular sample statistic we are interested in, the distribution of the population of individual values from which the sample statistic is computed, and the method by which samples are selected from the population. The sampling distribution allows us to determine the probability of observing any particular value of the sample statistic in another such sample from the population. Let's look at one application of this idea to a concept we first introduced in Chapter 9.

In Chapter 9 (page 238) we stated that we can use the laws of probability to learn if an observed treatment effect is larger than we would expect if only chance were operating. We said that an observed effect that is so large that it would rarely occur by chance is called statistically significant. How does one determine whether an observed effect would rarely occur if only chance were operating? By "only chance operating" we mean "under the assumption that there is no treatment effect." If there is no treatment effect, the response of any subject will be the same regardless of whether the subject received the treatment. If the observed treatment effect is represented by the value of a sample statistic—for example, the mean of the responses of those in a treatment group minus the mean of those in a control group—this can be done by considering the sampling distribution of the sample statistic under the assumption that no effect is actually present. Use this sampling distribution to determine the probability that we would observe values as extreme as we actually observed. The next two examples illustrate this calculation.

EXAMPLE 15.10 A Sampling Distribution and Statistical Significance

Do adults aged 65 and older have palates similar to other adults? Are they just as capable as other adults of appreciating a fine wine? Suppose we take an SRS of five adults from the population of all adults aged 65 and older. We present them with both natural wine and the same wine spiked with DMS at different concentrations to find the lowest concentration (odor threshold) at which they identify the spiked wine. We compute the median of the odor thresholds for the five subjects and find the median to be 35. If the palates of those aged 65 and older are the same as those of other adults (in other words, there is no "treatment effect" where the treatment is being aged 65 and older), we would expect the sample median to be similar to those computed from SRSs of size 5 from the population of all adults. We constructed this sampling distribution in Example 15.9. Looking at Figure 15.7(a), we see that 35 is an unusually large value for the median, if the palates of adults aged 65 and older are the same as other adults. Thus, the observed median of 35 might be regarded as evidence that the palates of adults aged 65 and over are different from those of adults in the general population. ■

We can use the same type of reasoning to determine statistical significance for statistics computed from completely randomized designs.

EXAMPLE 15.11 A Completely Randomized Design

Example 12.8 (page 287) discusses a weight-loss experiment with six subjects, Ari, Luis, Troy, Ana, Deb, and Hui. Three are assigned completely at random to a new experimental weight-loss treatment and three to a placebo. If the new treatment has no effect, then the weight lost by any subject will be the same regardless of whether the subject received the new treatment or a placebo. Here are the observed weight losses (in pounds) for each subject:

Subject	Ari	Luis	Troy	Ana	Deb	Hui
Weight Loss	2	15	8	1	12	9

We emphasize that we would have observed these same weight losses regardless of who was assigned the treatment and who the placebo.

The observed treatment effect is the mean weight lost by those receiving the new treatment minus the mean weight lost by those receiving the placebo. If there is no difference between the treatment and the placebo, the observed treatment effect is the result of the "luck of the draw," namely which three subjects just happened, by chance, to be assigned to the treatment group. Here are all 20 possible ways of selecting three of these subjects for the treatment group (the remaining three are in the placebo group).

Treatment Group	Treatment Group
Ari, Luis, Troy	Luis, Troy, Ana
Ari, Luis, Ana	Luis, Troy, Deb
Ari, Luis, Deb	Luis, Troy, Hui
Ari, Luis, Hui	Luis, Ana, Deb
Ari, Troy, Ana	Luis, Ana, Hui
Ari, Troy, Deb	Luis, Deb, Hui
Ari, Troy, Hui	Troy, Ana, Deb
Ari, Ana, Deb	Troy, Ana, Hui
Ari, Ana, Hui	Troy, Deb, Hui
Ari, Deb, Hui	Ana, Deb, Hui

With a completely randomized design, each of these 20 possible treatment groups is equally likely, thus each has probability 1/20 of being the actual group assigned to the treatment. We can use this information to determine the sampling distribution of the possible observed treatment effects. For example, if Ari, Luis, and Troy are assigned to the treatment group, the mean weight loss for the group is $\frac{2 + 15 + 8}{3} = 8.33$. The mean weight loss for the placebo group (Ana, Deb, and Hui) is then $\frac{1 + 12 + 9}{3} = 7.33$. The difference in mean weight losses (treatment group weight loss minus control group weight loss) is $8.33 - 7.33 = 1.00$.

We can repeat this calculation for each possible assignment of subjects to experimental groups. For the 20 possible ways we can assign subjects to the treatment and placebo groups, the differences in the mean weight losses for the two groups are:

Treatment Group	Control Group	Difference in Mean Weight Loss
Ari, Luis, Troy	Ana, Deb, Hui	1.00
Ari, Luis, Ana	Troy, Deb, Hui	−3.67
Ari, Luis, Deb	Troy, Ana, Hui	3.67
Ari, Luis, Hui	Troy, Ana, Deb	1.67
Ari, Troy, Ana	Luis, Deb, Hui	−8.33
Ari, Troy, Deb	Luis, Ana, Hui	−1.00
Ari, Troy, Hui	Luis, Ana, Deb	−3.00
Ari, Ana, Deb	Luis, Troy, Hui	−6.33
Ari, Ana, Hui	Luis, Troy, Deb	−7.67
Ari, Deb, Hui	Luis, Troy, Ana	−0.33
Luis, Troy, Ana	Ari, Deb, Hui	0.33
Luis, Troy, Deb	Ari, Ana, Hui	7.67
Luis, Troy, Hui	Ari, Ana, Deb	6.33
Luis, Ana, Deb	Ari, Troy, Hui	3.00
Luis, Ana, Hui	Ari, Troy, Deb	1.00
Luis, Deb, Hui	Ari, Troy, Ana	8.33
Troy, Ana, Deb	Ari, Luis, Hui	−1.67
Troy, Ana, Hui	Ari, Luis, Deb	−3.67
Troy, Deb, Hui	Ari, Luis, Ana	3.67
Ana, Deb, Hui	Ari, Luis, Troy	−1.00

Ordering the differences in mean weight losses from low to high, recalling each assignment of subjects to the treatment groups has probability 1/20 = 0.05, and combining duplicates, we obtain:

Weight Loss	−8.33	−7.67	−6.33	−3.67	−3.00	−1.67	−1.00	−0.33
Probability	0.05	0.05	0.05	0.10	0.05	0.05	0.10	0.05
Weight Loss	0.33	1.00	1.67	3.00	3.67	6.33	7.67	8.33
Probability	0.05	0.10	0.05	0.10	0.10	0.05	0.10	0.05

This is the sampling distribution of the differences in mean weight losses for the two groups under the assumption that weight lost by a subject does not depend on the group to which the subject was assigned. From this sampling distribution we can determine whether the actual treatment effect observed is statistically significant. ■

What would we conclude if we conducted this experiment and Troy, Deb, and Hui were assigned to the group receiving the new treatment? In this case, the observed treatment effect (the difference in mean weight lost for the two groups) would be 3.67. The mean weight lost for the treatment group is greater than for the placebo group. So this might be regarded as evidence that the new treatment is effective. However, if there is no difference in the effects of the new treatment and a placebo, Example 15.10 suggests that the chance of observing a difference as large as or larger than 3.67 is 0.3. Would you regard an effect that has probability 0.3 of occurring by chance rare? If not, then you would not consider the results to be statistically significant.

LaunchPad Online Resources

- The **StatBoards video**, *Simulating a Sampling Distribution*, discusses examples of simulating sampling distributions.

Apply Your Knowledge

15.14 Statistical Significance of a Variance. In Example 15.9, we constructed the sampling distribution (see Figure 15.7 (b)) of the sample variance of an SRS of size 5 from the distribution of DMS odor thresholds of individual adults. We assume that this population distribution is Normal with mean $\mu = 25$ micrograms per liter and standard deviation $\sigma = 7$ micrograms per liter. In Example 15.10, we selected an SRS of size 5 from the members of a national wine-tasting organization. Suppose the variance of this SRS is 9.2. Would you consider this value to be statistically significant, if the palates of the members of the wine-tasting organization are no different than those of all adults? Explain your answer.

15.15 Statistical Significance from a Sampling Distribution. In Exercise 15.7, you generated 10 samples of size 4 from the population of 10 students, calculated $\bar{x}$ for each sample, and constructed a histogram of these 10 values of $\bar{x}$.

APPLET

(a) Use the *Simple Random Sample* applet to generate 25 new samples of size 4, calculate $\bar{x}$ for each, and construct a histogram of the 25 values. Once again, you are constructing the sampling distribution of $\bar{x}$.

(b) Based on your histogram in part (a), what would you estimate to be the chance of obtaining a simple random sample of four students with $\bar{x} \geq 76$?

(c) Suppose you learn that students 1, 3, 5, and 7 are honors students. Would you regard their mean score to be "statistically significant"?

APPLET

15.16 **A Completely Randomized Design and Statistical Significance.** Eight students volunteer for an experiment to compare two methods (A and B) for preparing for the Mathematics SAT exam. Four students are assigned completely at random to Method A, and the other four to Method B. All eight take the Mathematics SAT, study using the assigned method for three months, and retake the exam. The change in score (score after three months using the assigned method minus initial score) for all eight students is:

Student	1	2	3	4	5	6	7	8
Change in Score	0	0	0	0	24	24	24	24

Suppose that there is no difference in the effects of the two methods, so that the change in scores for the eight students would be the same regardless of which method they were assigned.

(a) Use the *Simple Random Sample* applet to assign students to the two methods. Compute the difference in mean scores for those you assigned to Method A minus those you assigned to Method B.

(b) Repeat part (a) 24 more times. Make a histogram of the 25 differences in mean scores.

(c) Based on your histogram in part (b), what would you estimate to be the probability of observing a difference in mean scores of 12 or greater?

(d) Suppose you didn't know whether the two methods had the same effect on Mathematics SAT scores. Suppose you conducted the experiment and students 4, 5, 6, and 7 were assigned to Method A. The difference in the mean scores of students assigned to Method A minus the mean scores of those assigned to Method B is 12. Should this result be regarded as evidence that Method A is better than Method B.? Explain.

CHAPTER 15 SUMMARY

Chapter Specifics

- A **parameter** in a statistical problem is a number that describes a population, such as the **population mean** μ. To estimate an unknown parameter, use a **statistic** calculated from a sample, such as the **sample mean** $\bar{x}$.
- The **law of large numbers** states that the actual observed mean outcome $\bar{x}$ must approach the mean μ of the population as the number of observations increases.
- The **population distribution** of a variable describes the values of the variable for all individuals in a population.
- The **sampling distribution** of a statistic describes the values of the statistic in all possible samples of the same size from the same population.
- When the sample is an SRS from the population, the **mean** of the sampling distribution of the sample mean $\bar{x}$ is the same as the population mean μ. That is, $\bar{x}$ is an **unbiased estimator** of μ.
- The **standard deviation** of the sampling distribution of $\bar{x}$ is $\sigma/\sqrt{n}$ for an SRS of size n if the population has standard deviation σ. That is, averages are less variable than individual observations.

- When the sample is an SRS from a population that has a Normal distribution, the sample mean $\bar{x}$ also has a Normal distribution.
- Choose an SRS of size n from any population with mean μ and finite standard deviation σ. The **central limit theorem** states that when n is large, the sampling distribution of $\bar{x}$ is approximately Normal. That is, averages are more Normal than individual observations. We can use the $N(\mu, \sigma/\sqrt{n})$ distribution to calculate approximate probabilities for events involving $\bar{x}$.

Link It

As we mentioned in Chapter 12, probability is the tool we will use to generalize from data produced by random samples and randomized comparative experiments to some wider population. In this chapter we begin to formalize this process. We use a statistic to estimate an unknown parameter. We use the sampling distribution to summarize the behavior of a statistic in all possible random samples of the same size from a population.

More specifically, in this chapter we begin to think about how a sample mean, $\bar{x}$, can provide information about a population mean, μ. When the sample mean is computed from an SRS drawn from a large population, its sampling distribution has properties that help us understand how the sample mean can be used to draw conclusions about a population mean. The law of large numbers tells us that a sample mean computed from a *random* sample from some population gets closer and closer to the population mean as the sample size increases. The central limit theorem describes the sampling distribution of the sample mean for "large" SRSs and allows us to make probability statements about possible values of the sample mean. Beginning with Chapter 16, we will develop specific methods for drawing conclusions about a population mean based on a sample mean computed from an SRS. These methods will use the tools developed in this chapter.

LaunchPad Online Resources

If you are having difficulty with any of the sections of this chapter, these online resources should help prepare you to solve the exercises at the end of this chapter.

- **StatTutor** starts with a video review of each section and asks a series of questions to check your understanding.
- **LearningCurve** provides you with a series of questions about the chapter geared to your level of understanding.

CHECK YOUR SKILLS

15.17 The Bureau of Labor Statistics announces that last month it interviewed all members of the labor force in a sample of 60,000 households; **6.7%** of the people interviewed were unemployed. The boldface number is a

(a) sampling distribution.
(b) statistic.
(c) parameter.

15.18 An April 2011 poll of Canadian adults found that 57% were certain that they would vote in the May 2011 elections. Election records show that **61.1%** of registered voters voted in the election. The boldface number is a

(a) sampling distribution.
(b) statistic.
(c) parameter.

15.19 Annual returns on common small stocks available to investors vary a lot. In a recent year, the mean return was 11.7% and the standard deviation of returns was 34.1%. The law of large numbers says that

(a) you can get an average return higher than the mean 11.7% by investing in a large number of stocks.
(b) as you invest in more and more stocks chosen at random, your average return on these stocks gets ever closer to 11.7%.
(c) if you invest in a large number of stocks chosen at random, your average return will have approximately a Normal distribution.

15.20 Scores on the Critical Reading part of the SAT exam in a recent year were roughly Normal with mean 496 and standard deviation 115. You choose an SRS of 100 students and average their SAT Critical Reading scores. If you do this many times, the mean of the average scores you get will be close to

(a) 496.
(b) $496/100 = 4.96$.
(c) $496/\sqrt{100} = 49.6$.

15.21 Scores on the Critical Reading part of the SAT exam in a recent year were roughly Normal with mean 496 and standard deviation 115. You choose an SRS of 100 students and average their SAT Critical Reading scores. If you do this many times, the standard deviation of the average scores you get will be close to

(a) 115.
(b) $115/100 = 1.15$.
(c) $115/\sqrt{100} = 11.5$.

15.22 A newborn baby has extremely low birth weight (ELBW) if it weighs less than 1000 grams. A study of the health of such children in later years examined a random sample of 219 children. Their mean weight at birth was $\bar{x} = 810$ grams. This sample mean is an *unbiased estimator* of the mean weight μ in the population of all ELBW babies. This means that

(a) in many samples from this population, the mean of the many values of $\bar{x}$ will be equal to μ.
(b) as we take larger and larger samples from this population, $\bar{x}$ will get closer and closer to μ.
(c) in many samples from this population, the many values of $\bar{x}$ will have a distribution that is close to Normal.

15.23 The number of hours a battery lasts before failing varies from battery to battery. The distribution of failure times follows an exponential distribution (see Example 15.8), which is strongly skewed to the right. The central limit theorem says that

(a) as we look at more and more batteries, their average failure time gets ever closer to the mean μ for all batteries of this type.
(b) the average failure time of a large number of batteries has a distribution of the same shape (strongly skewed) as the distribution for individual batteries.
(c) the average failure time of a large number of batteries has a distribution that is close to Normal.

15.24 The length of human pregnancies from conception to birth varies according to a distribution that is approximately Normal with mean 266 days and standard deviation 16 days. The probability that the average pregnancy length for six randomly chosen women exceeds 270 days is about

(a) 0.40. (b) 0.27. (c) 0.07.

CHAPTER 15 EXERCISES

15.25 **Testing glass.** How well materials conduct heat matters when designing houses. As a test of a new measurement process, 10 measurements are made on pieces of glass known to have conductivity **1**. The average of the 10 measurements is **1.07**. For each of the boldface numbers, indicate whether it is a parameter or a statistic. Explain your answer.

15.26 **Statistics anxiety.** What can teachers do to alleviate statistics anxiety in their students? To explore this question, statistics anxiety for students in two classes was compared. In one class, the instructor lectured in a formal manner, including dressing formally. In the other, the instructor was less formal, dressed informally, was more personal, used humor, and called on students by their first names. Anxiety was measured using a questionnaire. Higher scores indicate a greater level of anxiety. The mean anxiety score for students in the formal lecture class was **25.40**; in the informal class, the mean was **20.41**. For each of the boldface numbers, indicate whether it is a parameter or a statistic. Explain your answer.

15.27 **Roulette.** A roulette wheel has 38 slots, of which 18 are black, 18 are red, and 2 are green. When the wheel is spun, the ball is equally likely to come to rest in any of the slots. One of the simplest wagers chooses red or black. A bet of $1 on red returns $2 if the ball lands in a red slot. Otherwise, the player loses his dollar. When gamblers bet on red or black, the two green slots belong to the house. Because the probability of winning $2 is 18/38, the mean payoff from a $1 bet is twice 18/38, or 94.7 cents. Explain what the law of large numbers tells us about what will happen if a gambler makes very many bets on red.

15.28 **The Medical College Admission Test.** Almost all medical schools in the United States require students to take the Medical College Admission Test (MCAT). To estimate the mean score μ of those who took the MCAT on your campus, you will obtain the scores of an SRS of students. The scores follow a Normal distribution, and from published information you know that the standard deviation is 6.5. Suppose that (unknown to you) the mean score of those taking the MCAT on your campus is 25.0.

(a) If you choose one student at random, what is the probability that the student's score is between 20 and 30?
(b) You sample 25 students. What is the sampling distribution of their average score $\bar{x}$?
(c) What is the probability that the mean score of your sample is between 20 and 30?

15.29 **Glucose testing.** Shelia's doctor is concerned that she may suffer from gestational diabetes (high blood glucose levels during pregnancy). There is variation both in the actual glucose level and in the blood test that measures the level. In a test to screen for gestational diabetes, a patient is classified as needing further testing for gestational diabetes if the glucose level is above 130 milligrams per deciliter (mg/dL) one hour after having a sugary drink. Shelia's measured glucose level one hour after the sugary drink varies according to the Normal distribution with $\mu = 122$ mg/dL and $\sigma = 12$ mg/dL.

(a) If a single glucose measurement is made, what is the probability that Shelia is diagnosed as needing further testing for gestational diabetes?
(b) If measurements are made on four separate days and the mean result is compared with the criterion 130 mg/dL, what is the probability that Shelia is diagnosed as needing further testing for gestational diabetes?

15.30 **Daily activity.** It appears that people who are mildly obese are less active than leaner people. One study looked at the average number of minutes per day that people spend standing or walking.[5] Among mildly obese people, the mean number of minutes of daily activity (standing or walking) is approximately Normally distributed with mean 373 minutes and standard deviation 67 minutes. The mean number of minutes of daily activity for lean people is approximately Normally distributed with mean 526 minutes and standard deviation 107 minutes. A researcher records the minutes of activity for an SRS of five mildly obese people and an SRS of five lean people.

(a) What is the probability that the mean number of minutes of daily activity of the five mildly obese people exceeds 420 minutes?
(b) What is the probability that the mean number of minutes of daily activity of the five lean people exceeds 420 minutes?

15.31 **Glucose testing, continued.** Shelia's measured glucose level one hour after having a sugary drink varies according to the Normal distribution with $\mu = 122$ mg/dL and $\sigma = 12$ mg/dL. What is the level L such that there is probability only 0.05 that the mean glucose level of four test results falls above L? (*Hint:* This requires a backward Normal calculation. See page 91 in Chapter 3 if you need to review.)

15.32 **Pollutants in auto exhausts.** Light vehicles sold in the United States must emit an average of no more than 0.05 grams per mile (g/mi) of nitrogen oxides (NOX) after 50,000 or fewer miles of driving. NOX emissions after 50,000 or fewer miles of driving for one car model vary Normally with mean 0.03 g/mi and standard deviation 0.01 g/mi.

(a) What is the probability that a single car of this model emits more than 0.05 g/mi of NOX?
(b) A company has 25 cars of this model in its fleet. What is the probability that the average NOX level $\bar{x}$ of these cars is above 0.05 g/mi?

15.33 **Runners.** In a study of exercise, a large group of male runners walk on a treadmill for six minutes. After this exercise, their heart rates vary with mean 8.8 beats per five seconds and standard deviation 1.0 beats per five seconds. This distribution takes only whole-number values, so it is certainly not Normal.

Bruce Laurance/Getty Images

(a) Let $\bar{x}$ be the mean number of beats per five seconds after measuring heart rate for 24 five-second intervals (two minutes). What is the approximate distribution of $\bar{x}$ according to the central limit theorem?
(b) What is the approximate probability that $\bar{x}$ is less than 8?
(c) What is the approximate probability that the heart rate of a runner is less than 100 beats per minute? (*Hint:* Restate this event in terms of $\bar{x}$.)

15.34 **Pollutants in auto exhausts, continued.** The level of nitrogen oxides (NOX) in the exhaust after 50,000 miles or fewer of driving of cars of a particular model varies Normally with mean 0.03 g/mi and standard deviation 0.01 g/mi. A company has 25 cars of this model in its fleet. What is the level L such that the probability that the average NOX level $\bar{x}$ for the fleet is greater than L is only 0.01? (*Hint:* This requires a backward Normal calculation. See page 91 in Chapter 3 if you need to review.)

15.35 **Returns on stocks.** Andrew plans to retire in 40 years. He plans to invest part of his retirement funds in stocks, so he seeks out information on past returns. He learns that from 1964 to 2013, the annual returns on U.S. common stocks had mean 13.3% and standard deviation 17.0%.[6] The distribution of annual returns on common stocks is roughly symmetric, so the mean return over even a moderate number of years is close to Normal. What is the probability (assuming that the past pattern of variation continues) that the mean annual return on common stocks over the next 40 years will exceed 10%? What is the probability that the mean return will be less than 5%? Follow the four-step process as illustrated in Example 15.8.

15.36 **Airline passengers get heavier.** In response to the increasing weight of airline passengers, the Federal Aviation Administration (FAA) in 2003 told airlines to assume that passengers average 190 pounds in the summer, including clothing and carry-on baggage. But passengers vary, and the FAA did not specify a standard deviation. A reasonable standard deviation is 35 pounds. Weights are not Normally distributed, especially when the population includes both men and women, but they are not very non-Normal. A commuter plane carries 22 passengers. What is the approximate probability that the total weight of the passengers exceeds 4500 pounds? Use the four-step process to guide your work. (*Hint:* To apply the central limit theorem, restate the problem in terms of the mean weight.)

© Jeff Greenberg/The Image Works

15.37 **Sampling students.** To estimate the mean score μ of those who took the Medical College Admission Test on your campus, you will obtain the scores of an SRS of students. From published information, you know that the scores are approximately Normal with standard deviation about 6.5. How large an SRS must you take to reduce the standard deviation of the sample mean score to 1?

15.38 **Sampling students, continued.** To estimate the mean score μ of those who took the Medical College Admission Test on your campus, you will obtain the scores of an SRS of students. From published information, you know that the scores are approximately Normal with standard deviation about 6.5. You want your sample mean $\bar{x}$ to estimate μ with an error of no more than 1 point in either direction.

(a) What standard deviation must $\bar{x}$ have so that 99.7% of all samples give an $\bar{x}$ within 1 point of μ? (Use the 68–95–99.7 rule.)
(b) How large an SRS do you need in order to reduce the standard deviation of $\bar{x}$ to the value you found in part (a)?

15.39 **Playing the numbers.** The numbers racket is a well-entrenched illegal gambling operation in most large cities. One version works as follows: you choose one of the 1000 three-digit numbers 000 to 999 and pay your local numbers runner a dollar to enter your bet. Each day, one three-digit number is chosen at random and pays off \$600. The mean payoff for the population of thousands of bets is μ = 60 cents. Joe makes one bet every day for many years. Explain what the law of large numbers says about Joe's results as he keeps on betting.

15.40 **Playing the numbers: a gambler gets chance outcomes.** The law of large numbers tells us what happens in the long run. Like many games of chance, the numbers racket has outcomes so variable—one three-digit number wins $600 and all others win nothing—that gamblers never reach "the long run." Even after many bets, their average winnings may not be close to the mean. For the numbers racket, the mean payout for single bets is $0.60 (60 cents) and the standard deviation of payouts is about $18.96. If Joe plays 350 days a year for 40 years, he makes 14,000 bets.

(a) What are the mean and standard deviation of the average payout $\bar{x}$ that Joe receives from his 14,000 bets?
(b) The central limit theorem says that his average payout is approximately Normal with the mean and standard deviation you found in part (a). What is the approximate probability that Joe's average payout per bet is between $0.50 and $0.70? You see that Joe's average may not be very close to the mean $0.60 even after 14,000 bets.

15.41 **Playing the numbers: the house has a business.** Unlike Joe (see Exercise 15.40) the operators of the numbers racket can rely on the law of large numbers. It is said that the New York City mobster Casper Holstein took as many as 25,000 bets per day in the Prohibition era. That's 150,000 bets in a week if he takes Sunday off. Casper's mean winnings per bet are $0.40 (he pays out 60 cents of each dollar bet to people like Joe and keeps the other 40 cents.) His standard deviation for single bets is about $18.96, the same as Joe's.

NY Daily News Archive via Getty Images

(a) What are the mean and standard deviation of Casper's average winnings $\bar{x}$ on his 150,000 bets?
(b) According to the central limit theorem, what is the approximate probability that Casper's average winnings per bet are between $0.30 and $0.50? After only a week, Casper can be pretty confident that his winnings will be quite close to $0.40 per bet.

15.42 **Can we trust the central limit theorem?** The central limit theorem says that "when n is large" we can act as if the distribution of a sample mean $\bar{x}$ is close to Normal. How large a sample we need depends on how far the population distribution is from being Normal. Example 15.8 shows that we can trust this Normal approximation for quite moderate sample sizes even when the population has a strongly skewed continuous distribution.

The central limit theorem requires much larger samples for Joe's bets with his local numbers racket. The population of individual bets has a finite distribution with only two possible outcomes: $600 (probability 0.001) and $0 (probability 0.999). This distribution has mean $\mu = 0.6$ and standard deviation about $\sigma = 18.96$. With more math and good software, we can find exact probabilities for Joe's average winnings.

(a) If Joe makes 14,000 bets, the exact probability $P(0.5 \leq \bar{x} \leq 0.7) = 0.4961$. How accurate was your Normal approximation from part (b) of Exercise 15.40?
(b) If Joe makes only 3500 bets, $P(0.5 \leq \bar{x} \leq 0.7) = 0.4048$. How accurate is the Normal approximation for this probability?
(c) If Joe and his buddies make 150,000 bets, $P(0.5 \leq \bar{x} \leq 0.7) = 0.9629$. How accurate is the Normal approximation?

15.43 APPLET **What's the mean?** Suppose that you roll three balanced dice. We wonder what the mean number of spots on the up-faces of the three dice is. The law of large numbers says that we can find out by experience: roll three dice many times, and the average number of spots will eventually approach the true mean. Set up the *Law of Large Numbers* applet to roll three dice. Don't click "Show mean" yet. Roll the dice until you are confident you know the mean quite closely, then click "Show mean" to verify your discovery. What is the mean? Make a rough sketch of the path the averages $\bar{x}$ followed as you kept adding more rolls.

15.44 **Statistical significance?** Look again at Exercise 12.50 (page 301). If your Canadian friend correctly identifies all three cups with the milk added first, would you regard the result as statistically significant?

Exploring the Web

15.45 Online videos. There are several online videos of the law of large numbers and central limit theorem. Locate one such video, watch it, and write a brief summary of the video. We did a Google search of "videos for law of large numbers" and "videos for the central limit theorem" and found several links.

15.46 Work the law of large numbers. Read the online article at `ezinearticles.com/?Work-The-Law-of-Large-Numbers-But-Remember-It-Only-Takes-One-to-Succeed!&id=932026`. Does this article accurately describe the law of large numbers? Explain your answer.

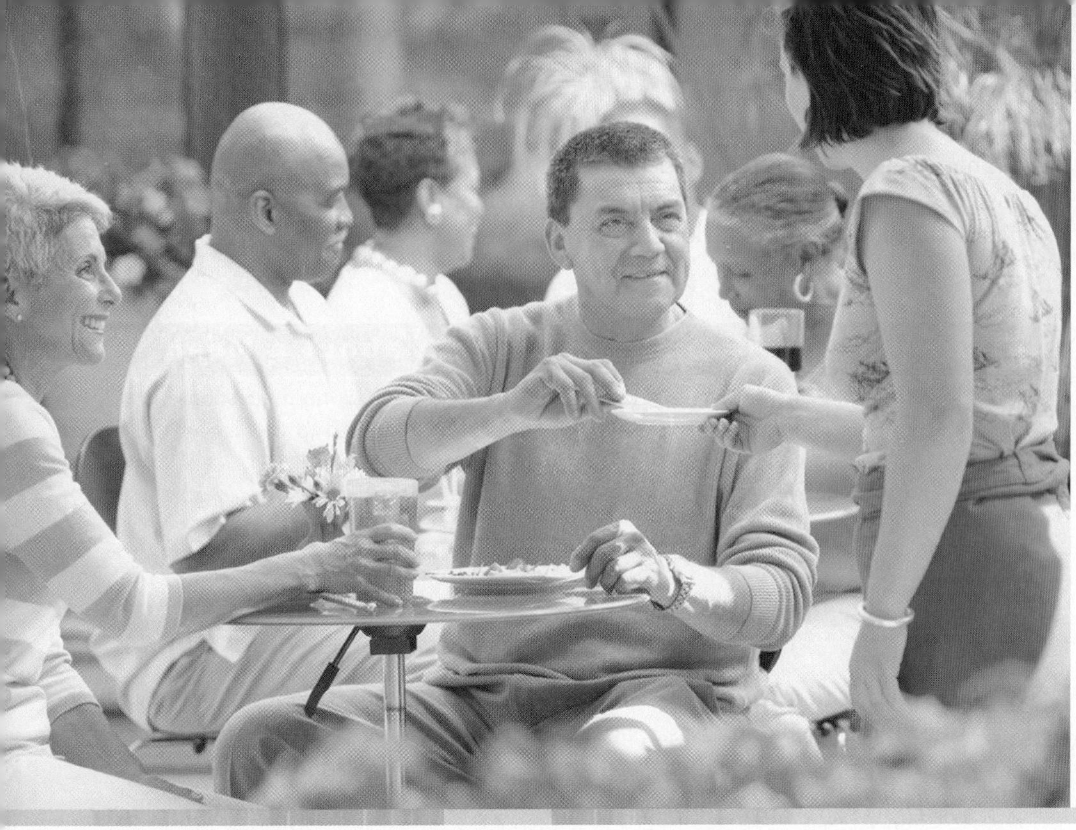

CHAPTER

16

© Ariel Skelley/age fotostock

Confidence Intervals: The Basics

In this chapter we cover...

16.1 The reasoning of statistical estimation

16.2 Margin of error and confidence level

16.3 Confidence intervals for a population mean

16.4 How confidence intervals behave

After we have selected a sample, we know the responses of the individuals in the sample. The usual reason for taking a sample is not to learn about the individuals in the sample but to *infer* from the sample data some conclusion about the wider population that the sample represents.

Statistical Inference

Statistical inference provides methods for drawing conclusions about a population from sample data.

Because a different sample might lead to different conclusions, we can't be certain that our conclusions are correct. Statistical inference uses the language of probability to say how trustworthy our conclusions are. This chapter introduces one of the two most common types of inference, *confidence intervals* for estimating the value of a population parameter. The next chapter discusses the other common type of inference, *tests of significance* for assessing the evidence for a claim about a population parameter. Both types of inference are based on the sampling distributions of statistics. That is, both use probability to say what would happen if we applied the inference method many times.

This chapter presents the basic reasoning of statistical inference. To make the reasoning as clear as possible, we start with a setting that is too simple to be realistic. Here is the setting for our work in this chapter.

Simple Conditions for Inference about a Mean

1. We have an SRS from the population of interest. There is no nonresponse or other practical difficulty. The population is large compared to the size of the sample.
2. The variable we measure has an exactly Normal distribution $N(\mu, \sigma)$ in the population.
3. We don't know the population mean μ. But we do know the population standard deviation σ.

The condition that the population is large relative to the size of the sample will be satisfied if the population is, say, at least 20 times as large. *The conditions that we have a perfect SRS, that the population is exactly Normal, and that we know the population σ are all unrealistic.* Chapter 18 begins to move from the "simple conditions" toward the reality of statistical practice. Later chapters deal with inference in fully realistic settings.

CAUTION

If these "simple conditions" are unrealistic, why study them? One reason is that under these simple conditions we can apply what we have learned in previous chapters about the Normal distribution and the sampling distribution of a sample mean to develop, step-by-step, methods for inference about a mean. The reasoning used under simple conditions applies to more realistic settings where the mathematics is more complicated.

Another reason for studying inference under these simple conditions is that we can carry out calculations using what we have already learned about the Normal distribution in previous chapters. This includes the calculations used to plan statistical studies, which we discuss in Chapter 18. Unfortunately, under more realistic conditions, such calculations are more complicated, and the connection to previous material about the Normal distribution is less clear.

Although we never know whether a population is exactly Normal and we never know the population σ, the methods we discuss in this and the next two chapters are approximately correct for sufficiently large sample sizes, provided we treat the sample standard deviation as though it were the population σ. Thus, there are situations (admittedly rare) where these methods can be used in practice.

Statistics In Your World: RANGES ARE FOR STATISTICS?

Many people like to think that statistical estimates are exact. The Nobel Prize–winning economist Daniel McFadden tells a story of his time on the Council of Economic Advisers. Presented with a range of forecasts for economic growth, President Lyndon Johnson replied: "Ranges are for cattle; give me one number."

16.1 The reasoning of statistical estimation

Body mass index (BMI) is used to screen for possible weight problems. It is calculated as weight divided by the square of height, measuring weight in kilograms and height in meters. Many online BMI calculators allow you to enter weight in pounds and height in inches. Adults with BMI less than 18.5 are considered underweight, and those with BMI greater than 25 may be overweight. For data about BMI, we turn to the National Health and Nutrition Examination Survey (NHANES), a continuing government sample survey that monitors the health of the American population.

EXAMPLE 16.1 Body Mass Index of Young Women

Thomas Northcut/Getty Images

An NHANES report gives data for 654 women aged 20 to 29 years.[1] The mean BMI of these 654 women was $\bar{x} = 26.8$. On the basis of this sample, we want to estimate the mean BMI μ in the population of all 20.6 million women in this age group.

To match the "simple conditions," we will treat the NHANES sample as an SRS from a Normal population with known standard deviation $\sigma = 7.5$.

Here is the reasoning of statistical estimation in a nutshell:

1. To estimate the unknown population mean BMI μ, use the mean $\bar{x} = 26.8$ of the random sample. We don't expect $\bar{x}$ to be exactly equal to μ, so we want to say how accurate this estimate is.
2. We know the sampling distribution of $\bar{x}$. In repeated samples, $\bar{x}$ has the Normal distribution with mean μ and standard deviation $\sigma/\sqrt{n}$. So the average BMI $\bar{x}$ of an SRS of 654 young women has standard deviation
$$\frac{\sigma}{\sqrt{n}} = \frac{7.5}{\sqrt{654}} = 0.3 \text{ (rounded off)}$$
3. The 95 part of the 68–95–99.7 rule for Normal distributions says that $\bar{x}$ is within 0.6 (that's two standard deviations) of the mean μ in 95% of all samples. That is, for 95% of all samples of size 654, the distance between the sample mean $\bar{x}$ and the population mean μ is less than 0.6. So if we estimate that μ lies somewhere in the interval from $\bar{x} - 0.6$ to $\bar{x} + 0.6$, we'll be right for 95% of all possible samples. For this particular sample, this interval is
$$\bar{x} - 0.6 = 26.8 - 0.6 = 26.2$$
to
$$\bar{x} + 0.6 = 26.8 + 0.6 = 27.4$$
4. Because we got the interval 26.2 to 27.4 from a method that captures the population mean for 95% of all possible samples, we say that we are *95% confident* that the mean BMI μ of all young women is some value in that interval, no lower than 26.2 and no higher than 27.4. ■

The big idea is that the sampling distribution of $\bar{x}$ tells us how close to μ the sample mean $\bar{x}$ is likely to be. Statistical estimation just turns that information around to say how close to $\bar{x}$ the unknown population mean μ is likely to be. We call the interval of numbers between the values $\bar{x} \pm 0.6$ a *95% confidence interval* for μ.

LaunchPad Online Resources

- **The Snapshots video, *Statistical Inference,* provides a good general introduction to statistical inference with several interesting examples.**

Apply Your Knowledge

16.1 Number Skills of Eighth Graders. The National Assessment of Educational Progress (NAEP) includes a mathematics test for eighth-grade students.[2] Scores on the test range from 0 to 500. Demonstrating the ability to use the mean to solve a problem is an example of the skills and knowledge associated with performance at the Basic level. An example of the knowledge and skills associated with the Proficient level is being able to read and interpret a stem-and-leaf plot.

In 2013, 170,100 eighth-graders were in the NAEP sample for the mathematics test. The mean mathematics score was $\bar{x} = 285$. We want to estimate the mean score μ in the population of all eighth-graders. Consider the NAEP sample as an SRS from a Normal population with standard deviation $\sigma = 125$.

(a) If we take many samples, the sample mean $\bar{x}$ varies from sample to sample according to a Normal distribution with mean equal to the unknown mean score μ in the population. What is the standard deviation of this sampling distribution?

(b) According to the 95 part of the 68–95–99.7 rule, 95% of all values of $\bar{x}$ fall within ______ on either side of the unknown mean μ. What is the missing number?

(c) What is the 95% confidence interval for the population mean score μ based on this one sample?

16.2 Retaking the SAT. An SRS of 400 high school seniors gained an average of $\bar{x} = 12$ points in their second attempt at the SAT Mathematics exam. Assume that the change in score has a Normal distribution with standard deviation $\sigma = 42$. We want to estimate the mean change in score μ in the population of all high school seniors.

(a) Give a 95% confidence interval for μ based on this sample.

(b) Based on your confidence interval in part (a), how certain are you that the mean change in score μ in the population of all high school seniors is greater than 0? (*Hint:* Does the interval in part (a) include 0?).

16.2 Margin of error and confidence level

The 95% confidence interval for the mean BMI of young women, based on the NHANES sample, is $\bar{x} \pm 0.6$. Once we have the sample results in hand, we know that for this sample $\bar{x} = 26.8$, so that our confidence interval is 26.8 ± 0.6. Most confidence intervals have a form similar to this,

$$\text{estimate} \pm \text{margin of error}$$

margin of error

The estimate ($\bar{x} = 26.8$ in our example) is our guess for the value of the unknown parameter. The **margin of error** ± 0.6 shows how accurate we believe our guess to be, based on the variability of the estimate. We have a 95% confidence interval because the interval $\bar{x} \pm 0.6$ catches the unknown parameter in 95% of all possible samples.

Confidence Interval

A **level *C* confidence interval** for a parameter has two parts:

- An interval calculated from the data, usually of the form

$$\text{estimate} \pm \text{margin of error}$$

- **A confidence level** C, which gives the probability that the interval will capture the true parameter value in repeated samples. That is, the confidence level is the success rate for the method.

Users can choose the confidence level, usually 90% or higher because we generally want to be quite sure of our conclusions. The most common confidence level is 95%.

Interpreting a Confidence Level

The confidence level is the success rate of the method that produces the interval. We don't know whether the 95% confidence interval from a particular sample is one of the 95% that capture μ or one of the unlucky 5% that miss.

To say that we are **95% confident** that the unknown μ lies between 26.2 and 27.4 is shorthand for **"We got these numbers using a method that gives correct results 95% of the time."**

EXAMPLE 16.2 Statistical Estimation in Pictures

Figures 16.1 and 16.2 illustrate the behavior of confidence intervals. Study these figures carefully. If you understand what they say, you have mastered one of the big ideas of statistics.

Figure 16.1 illustrates the behavior of the interval $\bar{x} \pm 0.6$ for the mean BMI of young women. Starting with the population, imagine taking many SRSs of 654 young women. The first sample has $\bar{x} = 26.8$, the second has $\bar{x} = 27.0$, the third has $\bar{x} = 26.2$, and so on. The sample mean varies from sample to sample, but when we use the formula $\bar{x} \pm 0.6$ to get an interval based on each sample, *95% of these intervals capture the unknown population mean* μ. Notice we use the same margin of error, 0.6, for each interval because the sample size and standard deviation σ are the same.

Figure 16.2 illustrates the idea of a 95% confidence interval in a different form. It shows the result of drawing many SRSs from the same population and calculating a 95% confidence interval from each sample. The center of each interval is at $\bar{x}$ and therefore varies from sample to sample. The sampling distribution of $\bar{x}$ appears at the top of the figure to show the long-term pattern of this variation. The population mean μ is at the center of the sampling distribution. The 95% confidence intervals from 25 SRSs appear underneath. The center $\bar{x}$ of each interval is marked by a dot. The arrows on either side of the dot span the confidence interval. All except one of these 25 intervals capture the true value of μ. If we take a very large number of samples, 95% of the confidence intervals will contain μ. ■

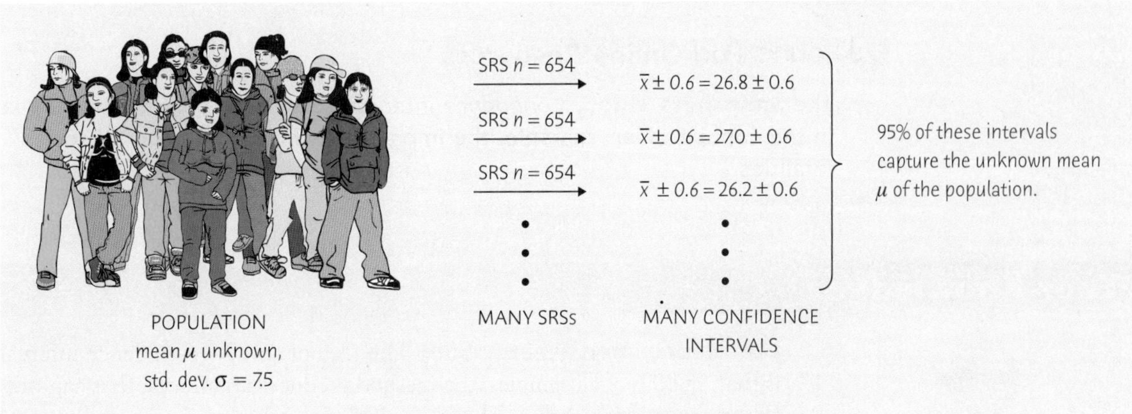

FIGURE 16.1
To say that $\bar{x} \pm 0.6$ is a 95% confidence interval for the population mean μ is to say that, in repeated samples, 95% of these intervals capture.

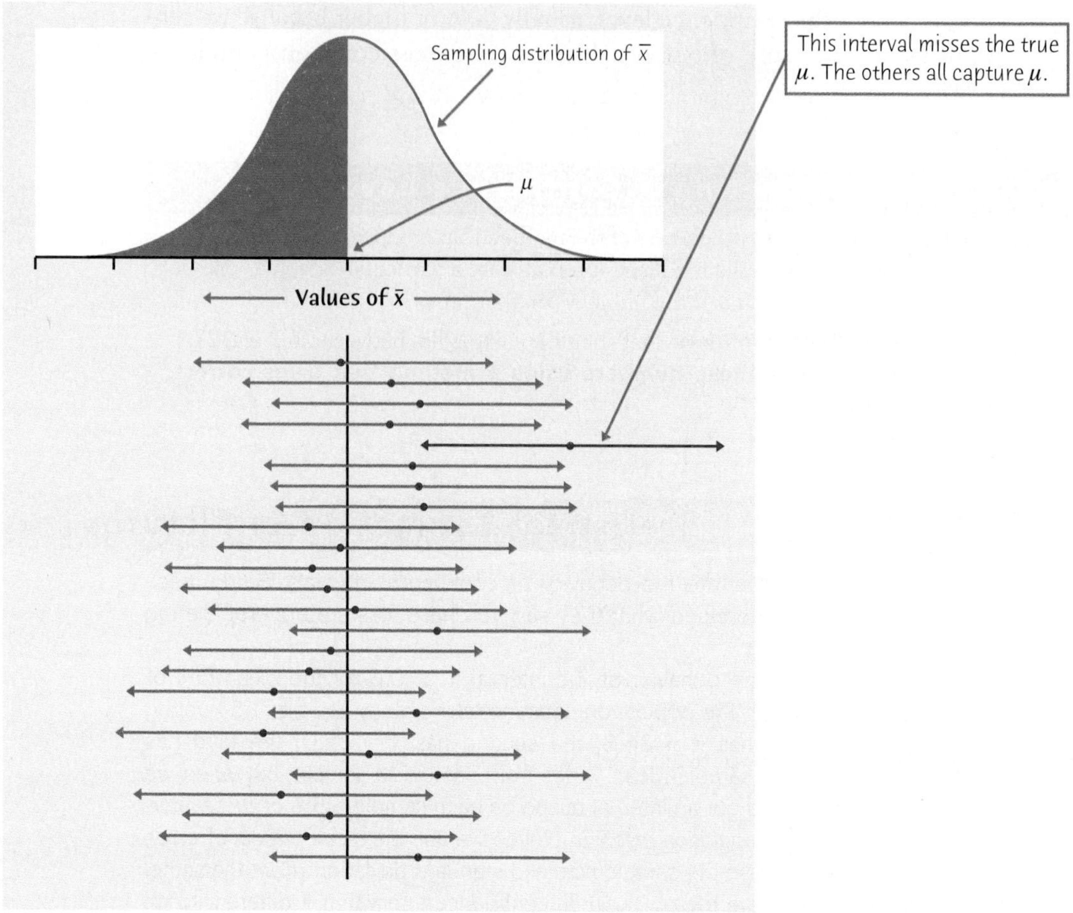

FIGURE 16.2
Twenty-five samples from the same population gave these 95% confidence intervals. In the long run, 95% of all samples give an interval that contains the population mean μ.

APPLET

The *Confidence Intervals* applet creates figures similar to Figure 16.2. You can use the applet to watch confidence intervals from one sample after another capture or fail to capture the true parameter.

LaunchPad Online Resources

- The **Snapshots video**, *Confidence Intervals,* discusses confidence intervals in the context of an example: the impact of wind turbines on raptor fatalities.

Apply Your Knowledge

APPLET

16.3 **Confidence Intervals in Action.** The idea of an 80% confidence interval is that, in 80% of all samples, the method produces an interval that captures the true parameter value. That's not high enough confidence for practical use, but 80% hits and 20% misses make it easy to see how a confidence interval behaves in repeated samples from the same population. Go to the *Confidence Intervals* applet.

(a) Set the confidence level to 80% and the sample size to 50. Click "Sample" to choose an SRS and calculate the confidence interval. Do this 10 times to simulate 10 SRSs with their 10 confidence intervals. How many of the 10 intervals captured the true mean μ? How many missed?

(b) You see that we can't predict whether the next sample will hit or miss. The confidence level, however, tells us what percent will hit in the long run. Reset the applet and click "Sample 25" to get the confidence intervals from 25 SRSs. How many hit?

(c) Click "Sample 25" repeatedly and write down the number of hits each time. What was the percent of hits among 100, 200, 300, 400, 500, 600, 700, 800, and 1000 SRSs? Even 1000 samples is not truly "the long run," but we expect the percent of hits in 1000 samples to be fairly close to the confidence level, 80%.

16.4 Losing Weight. A Gallup Poll in November 2013 found that 51% of the people in the sample said they want to lose weight. Gallup announced, "For results based on the total sample of national adults, one can say with 95% confidence that the maximum margin of sampling error is ± 4 percentage points."

(a) What is the 95% confidence interval for the percent of all adults who want to lose weight?

(b) What does it mean to have 95% confidence in this interval?

16.3 Confidence intervals for a population mean

In the setting of Example 16.1, we outlined the reasoning that leads to a 95% confidence interval for the unknown mean μ of a population. Now we will reduce the reasoning to a formula.

In Example 16.1, what role did 95% play in determining the confidence interval? To find a 95% confidence interval for the mean BMI of young women, we first caught the central 95% of the Normal sampling distribution by going out *two* standard deviations in both directions from the mean. The value 95% determined how many standard deviations we go out in both directions from the mean to capture this central 95%. To find a level C confidence interval, we first catch the central area C under the Normal sampling distribution. How many standard deviations must we go out in both directions from the mean to capture this central area C? Because all Normal distributions are the same in the standard scale, we can obtain everything we need from the standard Normal curve.

Figure 16.3 shows how the central area C under a standard Normal curve is marked off by two points z^* and $-z^*$. Numbers like z^* that mark off specified areas are called **critical values** of the standard Normal distribution. Values of z^* for many choices of C appear at the bottom of Table C in the back of the book, in the row labeled z^*. Here are the entries for the most common confidence levels:

critical values

Confidence Level C	90%	95%	99%
Critical Value z^*	1.645	1.960	2.576

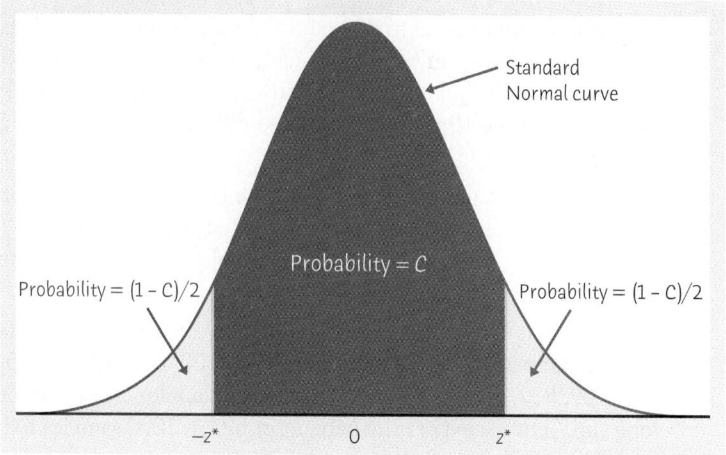

FIGURE 16.3
The critical value z^* is the number that catches central probability C under a standard Normal curve between $-z^*$ and z^*.

You see that, for C = 95%, the table gives $z^* = 1.960$. This is a bit more precise than the approximate value $z^* = 2$ based on the 68–95–99.7 rule. You can, of course, use software to find critical values z^*, as well as the entire confidence interval.

Figure 16.3 shows that there is area C under the standard Normal curve between $-z^*$ and z^*. So *any* Normal curve has area C within z^* standard deviations on either side of its mean. The Normal sampling distribution of $\bar{x}$ has area C within $z^*\sigma/\sqrt{n}$ on either side of the population mean μ because it has mean μ and standard deviation $\sigma/\sqrt{n}$. If we start at $\bar{x}$ and go out $z^*\sigma/\sqrt{n}$ in both directions, we get an interval that contains the population mean μ in a proportion C of all samples. This interval is

$$\text{from } \bar{x} - z^*\frac{\sigma}{\sqrt{n}} \text{ to } \bar{x} + z^*\frac{\sigma}{\sqrt{n}}$$

or

$$\bar{x} \pm z^*\frac{\sigma}{\sqrt{n}}$$

It is a level C confidence interval for μ.

Confidence Interval for the Mean of a Normal Population

Draw an SRS of size n from a Normal population having unknown mean μ and known standard deviation σ. A level C **confidence interval for** μ is

$$\bar{x} \pm z^*\frac{\sigma}{\sqrt{n}}$$

The critical value z^* corresponding to the confidence level C is illustrated in Figure 16.3 and found at the bottom of Table C.

The steps in finding a confidence interval mirror the overall four-step process for organizing statistical problems.

Confidence Intervals: The Four-Step Process

STATE: What is the practical question that requires estimating a parameter?

PLAN: Identify the parameter, choose a level of confidence, and select the type of confidence interval that fits your situation.

SOLVE: Carry out the work in two phases:

1. **Check the conditions** for the interval you plan to use.
2. Calculate the **confidence interval.**

CONCLUDE: Return to the practical question to describe your results in this setting.

EXAMPLE 16.3 Good Weather, Good Tips?

STATE: Does the expectation of good weather lead to more generous behavior? Psychologists studied the size of the tip in a restaurant when a message indicating that the next day's weather would be good was written on the bill. Here are tips from 20 patrons, measured in percent of the total bill:[3] TIP2

20.8	18.7	19.9	20.6	21.9	23.4	22.8	24.9	22.2	20.3
24.9	22.3	27.0	20.4	22.2	24.0	21.1	22.1	22.0	22.7

This is one of three sets of measurements made, the others being tips received when the message on the bill said that the next day's weather would not be good or there was no message on the bill. We want to estimate the mean tip for comparison with tips under the other conditions.

PLAN: We will estimate the mean percentage tip μ for all patrons of this restaurant when they receive a message on their bill indicating that the next day's weather will be good by giving a 95% confidence interval. The confidence interval just introduced fits this situation.

SOLVE: We should start by checking the conditions for inference. For this example, we will first find the interval and then discuss how statistical practice deals with conditions that are never perfectly satisfied.

The mean percentage tip of the sample is $\bar{x} = 22.21$. As part of the "simple conditions," suppose that, from past experience with patrons of this restaurant, we know that the standard deviation of percentage tip is $\sigma = 2$. For 95% confidence, the critical value is $z^* = 1.960$. A 95% confidence interval for μ is therefore

$$\begin{aligned}\bar{x} \pm z^*\frac{\sigma}{\sqrt{n}} &= 22.21 \pm 1.960\frac{2}{\sqrt{20}} \\ &= 22.21 \pm 0.88 \\ &= 21.33 \text{ to } 23.09\end{aligned}$$

CONCLUDE: We are 95% confident that the mean percentage tip for all patrons of this restaurant when their bill contains a message that the next day's weather will be good is between 21.33 and 23.09. ■

In practice, the first part of the "Solve" step is to check the conditions for inference. The "simple conditions" are as follows:

1. **SRS:** We don't have an actual SRS from the population of all patrons of this restaurant. Scientists often act as if subjects are SRSs if there is nothing special

```
27 | 0
26 |
25 |
24 | 099
23 | 4
22 | 0122378
21 | 19
20 | 3468
19 | 9
18 | 7
```

FIGURE 16.4
Stemplot of the percentage tips in Example 16.3.

about how the subjects were obtained. But it is always better to have an actual SRS because otherwise we can never be sure that hidden biases aren't present. This study was actually a randomized comparative experiment in which these 20 patrons were assigned at random from a larger group of patrons to get one of the treatments being compared.

2. **Normal distribution:** The psychologists expect from past experience that measurements like this on patrons of the same restaurant under the same conditions will follow approximately a Normal distribution. We can't look at the population, but we can examine the sample. Figure 16.4 is a stemplot that is roughly bell-shaped, with perhaps a modest outlier but no strong skewness. Shapes like this often occur in small samples from Normal populations, so we have no reason to doubt that the population distribution is Normal.
3. **Known σ:** It really is unrealistic to suppose that we know that $\sigma = 2$. We will see in Chapter 20 that it is easy to do away with the need to know σ.

As this discussion suggests, inference methods are often used when conditions like SRS and Normal population are not exactly satisfied. In this introductory chapter, we act as though the "simple conditions" are satisfied. In reality, wise use of inference requires judgment. Chapter 17 and the later chapters on each inference method will give you a better basis for judgment.

LaunchPad Online Resources

- The StatBoards video, *Creating and Interpreting Confidence Intervals,* provides examples of computing and interpreting confidence intervals.

Apply Your Knowledge

16.5 Find a Critical Value. The critical value z^* for confidence level 75% is not in Table C. Use software or Table A of standard Normal probabilities to find z^*. Include in your answer a sketch like Figure 16.3 with C = 0.75 and your critical value z^* marked on the axis.

16.6 Measuring Conductivity. The National Institute of Standards and Technology (NIST) supplies "standard materials" whose physical properties are supposed to be known. For example, you can buy from NIST an iron rod whose electrical conductivity is supposed to be 10.1 at 293 kelvin. (The units for conductivity are microsiemens per centimeter. Distilled water has conductivity 0.5.) Of course, no measurement is exactly correct. NIST knows the variability of its measurements very well, so it is quite realistic to assume that the population of all measurements of the same rod has the Normal distribution with mean μ equal to the true conductivity and standard deviation $\sigma = 0.1$. Here are six measurements on the same standard iron rod, which is supposed to have conductivity 10.1: COND

10.08 9.89 10.05 10.16 10.21 10.11

NIST wants to give the buyer of this iron rod a 90% confidence interval for its true conductivity. What is this interval? Follow the four-step process as illustrated in Example 16.3.

16.7 IQ Test Scores. Here are the IQ test scores of 31 seventh-grade girls in a Midwest school district:[4] MWSTIQ

114	100	104	89	102	91	114	114	103	105	
108	130	120	132	111	128	118	119	86	72	
111	103	74	112	107	103	98	96	112	112	93

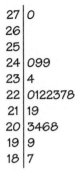

(a) These 31 girls are an SRS of all seventh-grade girls in the school district. Suppose that the standard deviation of IQ scores in this population is known to be $\sigma = 15$. We expect the distribution of IQ scores to be close to Normal. Make a stemplot of the distribution of these 31 scores (split the stems) to verify that there are no major departures from Normality. You have now checked the "simple conditions" to the extent possible.

(b) Estimate the mean IQ score for all seventh-grade girls in the school district, using a 99% confidence interval. Follow the four-step process as illustrated in Example 16.3.

16.4 How confidence intervals behave

The z confidence interval $\bar{x} \pm z^*\sigma/\sqrt{n}$ for the mean of a Normal population illustrates several important properties that are shared by all confidence intervals in common use. The user chooses the confidence level, and the margin of error follows from this choice. We would like high confidence and also a small margin of error. High confidence says that our method almost always gives correct answers. A small margin of error says that we have pinned down the parameter quite precisely. The factors that influence the margin of error of the z confidence interval are typical of most confidence intervals.

How do we get a small margin of error? The margin of error for the z confidence interval is

$$\text{margin of error} = z^*\frac{\sigma}{\sqrt{n}}$$

This expression has z^* and σ in the numerator and $\sqrt{n}$ in the denominator. Therefore, the margin of error gets smaller when

- z^* gets smaller. Smaller z^* is the same as lower confidence level C (look again at Figure 16.3). *There is a trade-off between the confidence level and the margin of error. To obtain a smaller margin of error from the same data, you must be willing to accept lower confidence.*
  CAUTION
- σ is smaller. The standard deviation σ measures the variation in the population. You can think of the variation among individuals in the population as noise that obscures the average value μ. It is easier to pin down μ when σ is small.
- n gets larger. Increasing the sample size n reduces the margin of error for any confidence level. Larger samples thus allow more precise estimates. However, *because n appears under a square root sign, we must take four times as many observations to cut the margin of error in half.*
  CAUTION

In practice, we can control the confidence level and sample size, but we can't control σ.

EXAMPLE 16.4 Changing the Margin of Error

In Example 16.3, psychologists recorded the size of the tip of 20 patrons in a restaurant when a message indicating that the next day's weather would be good was written on their bill. The data gave the mean size of the tip, as a percentage of the total bill, as $\bar{x} = 22.21$, and we know that $\sigma = 2$. The 95% confidence interval for the mean percentage tip for all patrons of the restaurant when their bill contains a message that the next day's weather will be good is

$$\bar{x} \pm z^* \frac{\sigma}{\sqrt{n}} = 22.21 \pm 1.960 \frac{2}{\sqrt{20}}$$
$$= 22.21 \pm 0.88$$

The 90% confidence interval based on the same data replaces the 95% critical value $z^* = 1.960$ by the 90% critical value $z^* = 1.645$. This interval is

$$\bar{x} \pm z^* \frac{\sigma}{\sqrt{n}} = 22.21 \pm 1.645 \frac{2}{\sqrt{20}}$$
$$= 22.21 \pm 0.74$$

Lower confidence results in a smaller margin of error, ±0.74 in place of ±0.88. You can calculate that the margin of error for 99% confidence is larger, ±1.15. Figure 16.5 compares these three confidence intervals.

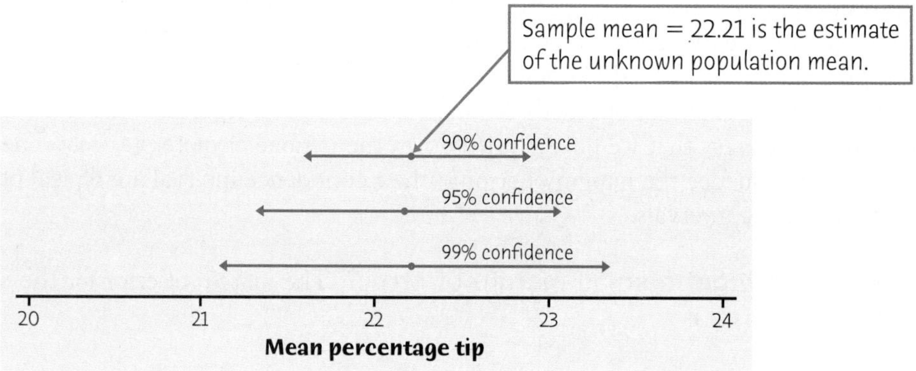

FIGURE 16.5 The lengths of three confidence intervals for Example 16.4. All three are centered at the estimate $\bar{x} = 22.21$. When the data and the sample size remain the same, higher confidence results in a larger margin of error.

If we had a sample of only 10 patrons, you can check that the margin of error for 95% confidence increases from ±0.88 to ±1.24. Cutting the sample size in half does *not* double the margin of error, because the sample size n appears under a square root sign. ■

Apply Your Knowledge

16.8 Confidence Level and Margin of Error. Example 16.1 described NHANES survey data on the body mass index (BMI) of 654 young women. The mean BMI in the sample was $\bar{x} = 26.8$. We treated these data as an SRS from a Normally distributed population with standard deviation $\sigma = 7.5$.

(a) Give three confidence intervals for the mean BMI μ in this population, using 90%, 95%, and 99% confidence.

(b) What are the margins of error for 90%, 95%, and 99% confidence? How does increasing the confidence level change the margin of error of a confidence interval when the sample size and population standard deviation remain the same?

16.9 Sample Size and Margin of Error. Example 16.1 described NHANES survey data on the body mass index (BMI) of 654 young women. The mean BMI in the sample was $\bar{x} = 26.8$. We treated these data as an SRS from a Normally distributed population with standard deviation $\sigma = 7.5$.

(a) Suppose that we had an SRS of just 100 young women. What would be the margin of error for 95% confidence?

(b) Find the margins of error for 95% confidence based on SRSs of 400 young women and 1600 young women.

(c) Compare the three margins of error. How does increasing the sample size change the margin of error of a confidence interval when the confidence level and population standard deviation remain the same?

16.10 Retaking the SAT. In Exercise 16.2, we saw that an SRS of 400 high school seniors gained an average of $\bar{x} = 12$ points in their second attempt at the SAT Mathematics exam. Assuming that the change in score has a Normal distribution with standard deviation $\sigma = 42$, we computed a 95% confidence interval for the mean change in score μ in the population of all high school seniors.

(a) Find a 90% confidence interval for μ based on this sample.

(b) What is the margin of error for 90%? How does decreasing the confidence level change the margin of error of a confidence interval when the sample size and population standard deviation remain the same?

(c) Suppose we had an SRS of just 100 high school seniors. What would be the margin of error for 95% confidence?

(d) How does decreasing the sample size change the margin of error of a confidence interval when the confidence level and population standard deviation remain the same?

LaunchPad Online Resources

- The **StatBoards video**, *Confidence Level Changes the Confidence Interval,* provides examples illustrating the behavior of confidence interval as the confidence level changes.
- The **StatBoards video**, *Sample Size Changes the Confidence Interval,* provides examples illustrating the behavior of confidence interval as the sample size changes.

CHAPTER 16 SUMMARY

Chapter Specifics

- A **confidence interval** uses sample data to estimate an unknown population parameter with an indication of how accurate the estimate is and of how confident we are that the result is correct.
- Any confidence interval has two parts: an interval calculated from the data and a confidence level C. The **confidence interval** often has the form
- The **confidence level** is the success rate of the method that produces the interval. That is, C is the probability that the method will give a correct answer. If you use 95% confidence intervals often, in the long run 95% of your intervals will contain the true parameter value. You do not know whether or not a 95% confidence interval calculated from a particular set of data contains the true parameter value.

- A level C **confidence interval for the mean** μ of a Normal population with known standard deviation σ, based on an SRS of size n, is given by

$$\bar{x} \pm z^* \frac{\sigma}{\sqrt{n}}$$

- The **critical value** z^* is chosen so that the standard Normal curve has area C between $-z^*$ and z^*.
- Other things being equal, the margin of error of a confidence interval gets smaller as
  - the confidence level C decreases,
  - the population standard deviation σ decreases,
  - the sample size n increases.

Link It

The reason we collect data is not to learn about the individuals that we observed, but to infer from the data to some wider population that the individuals represent. Chapters 8 and 9 tell us that the way we produce the data (sampling, experimental design) affects whether we have a good basis for generalizing to some wider population. Chapters 12, 13, 14, and 15 discuss probability, the formal mathematical tool that determines the nature of the inferences we make. In particular, Chapter 15 discusses sampling distributions, which tell us how repeated SRSs behave and hence what a statistic (in particular, a sample mean) computed from our sample is likely to tell us about the corresponding parameter of the population (in particular, a population mean) from which the sample was selected.

In this chapter, we discuss the basic reasoning of statistical estimation, with emphasis on estimating a population mean. To an estimate of the population mean we attach a margin of error and a confidence level. The result is a confidence interval. The sampling distribution of the sample mean, discussed in Chapter 15, provides the mathematical basis for constructing confidence intervals and understanding their properties. Although we apply the reasoning of statistical estimation in a simple and artificial setting (we assume we know the population standard deviation), we will use the same logic in future chapters to construct confidence intervals for population parameters in more realistic settings.

LaunchPad Online Resources

If you are having difficulty with any of the sections of this chapter, these online resources should help prepare you to solve the exercises at the end of this chapter.

- **StatTutor** starts with a video review of each section and asks a series of questions to check your understanding. You may find this helpful if you need additional help understanding any of the material in this chapter.
- **LearningCurve** provides you with a series of questions about the chapter geared to your level of understanding.

CHECK YOUR SKILLS

16.11 To give a 99.9% confidence interval for a population mean μ, you would use the critical value

(a) $z^* = 1.960$.

(b) $z^* = 2.576$.

(c) $z^* = 3.291$.

Use the following information for Exercises 16.12 through 16.14. A laboratory scale is known to have a standard deviation of $\sigma = 0.001$ gram in repeated weighings. Scale readings in repeated weighings are Normally distributed, with mean equal to the true weight of the specimen. Three weighings of a specimen on this scale give 3.412, 3.416, and 3.414 grams. DATA WEIGHTS

16.12 A 95% confidence interval for the true weight of this specimen is

(a) 3.414 ± 0.00113.

(b) 3.414 ± 0.00065.

(c) 3.414 ± 0.00196.

16.13 You want a 99% confidence interval for the true weight of this specimen. The margin of error for this interval will be

(a) smaller than the margin of error for 95% confidence.

(b) greater than the margin of error for 95% confidence.

(c) about the same as the margin of error for 95% confidence.

16.14 Another specimen is weighed eight times on this scale. The average weight is 4.1602 grams. A 99% confidence interval for the true weight of this specimen is

(a) 4.1602 ± 0.00032.

(b) 4.1602 ± 0.00069.

(c) 4.1602 ± 0.00091.

Use the following information for Exercises 16.15 through 16.18. The National Assessment of Educational Progress (NAEP) includes a mathematical test for eighth-grade students. Scores on the test range from 0 to 500. Suppose that you give the NAEP test to an SRS of 900 eighth-graders from a large population in which the scores have mean $\mu = 285$ and standard deviation $\sigma = 125$. The mean $\bar{x}$ will vary if you take repeated samples.

16.15 The sampling distribution of $\bar{x}$ is approximately Normal. It has mean $\mu = 285$. What is its standard deviation?

(a) 125.

(b) 4.167.

(c) 0.139.

16.16 Suppose that an SRS of 900 eighth-graders has $\bar{x} = 288$. Based on this sample, a 95% confidence interval for μ is

(a) 8.17 ± 0.27.

(b) 288 ± 8.17.

(c) 285 ± 8.17.

16.17 In Exercise 16.16, suppose that we computed a 99% confidence interval for μ.

(a) This 99% confidence interval would have a smaller margin of error than the 95% confidence interval.

(b) This 99% confidence interval would have a larger margin of error than the 95% confidence interval.

(c) This 99% confidence interval could have either a smaller or a larger margin of error than the 95% confidence interval. This varies from sample to sample.

16.18 Suppose that we took an SRS of 1600 eighth-graders and found $\bar{x} = 288$. Compared with an SRS of 900 eighth-graders, the margin of error for a 95% confidence interval for μ is

(a) smaller.

(b) larger.

(c) either smaller or larger but we can't say which.

CHAPTER 16 EXERCISES

16.19 Student study times. A class survey in a large class for first-year college students asked, "About how many hours do you study during a typical week?" The mean response of the 463 students was $\bar{x} = 15.3$ hours.[5] Suppose that we know that the study time follows a Normal distribution with standard deviation $\sigma = 8.5$ hours in the population of all first-year students at this university.

(a) Use the survey result to give a 99% confidence interval for the mean study time of all first-year students.

(b) What condition not yet mentioned must be met for your confidence interval to be valid?

16.20 I want more muscle. Young men in North America and Europe (but not in Asia) tend to think they need more muscle to be attractive. One study presented

© Rubberball/age fotostock

200 young American men with 100 images of men with various levels of muscle.[6] Researchers measure level of muscle in kilograms per square meter (kg/m^2) of fat-free body mass. Typical young men have about 20 kg/m^2. Each subject chose two images, one that represented his own level of body muscle and one that he thought represented "what women prefer." The mean gap between self-image and "what women prefer" was 2.35 kg/m^2.

Suppose that the "muscle gap" in the population of all young men has a Normal distribution with standard deviation 2.5 kg/m2. Give a 90% confidence interval for the mean amount of muscle young men think they should add to be attractive to women. (They are wrong: women actually prefer a level close to that of typical men.)

16.21 **An outlier strikes.** There were actually 464 responses to the class survey in Exercise 16.19. One student claimed to study 10,000 hours per week (10,000 is more than the number of hours in a year). We know he's joking, so we left out this value. If we did a calculation without looking at the data, we would get $\bar{x} = 36.8$ hours for all 464 students. Now what is the 99% confidence interval for the population mean? (Continue to use $\sigma = 8.5$.) Compare the new interval with that in Exercise 16.19. The message is clear: always look at your data, because outliers can greatly change your result.

16.22 **Explaining confidence.** A student reads that a recent poll finds that a 95% confidence interval for the mean ideal weight given by adult American women is 139 ± 1.4 pounds. Asked to explain the meaning of this interval, the student says, "95% of all adult American women would say that their ideal weight is between 137.6 and 140.4 pounds." Is the student right? Explain your answer.

16.23 **Explaining confidence.** You ask another student to explain the confidence interval for mean ideal weight described in Exercise 16.22. The student answers, "We can be 95% confident that future samples of adult American women will say that their mean ideal weight is between 137.6 and 140.4 pounds." Is this explanation correct? Explain your answer.

16.24 **Explaining confidence.** Here is an explanation from the Associated Press concerning one of its opinion polls. Explain briefly but clearly in what way this explanation is incorrect.

> *For a poll of 1,600 adults, the variation due to sampling error is no more than three percentage points either way. The error margin is said to be valid at the 95 percent confidence level. This means that, if the same questions were repeated in 20 polls, the results of at least 19 surveys would be within three percentage points of the results of this survey.*

*Exercises 16.25 to 16.27 ask you to answer questions from data. Assume that the "simple conditions" hold in each case. The exercise statements give you the **State** step of the four-step process. In your work, follow the **Plan, Solve,** and **Conclude** steps, illustrated in Example 16.3 for a confidence interval.*

16.25 **Pulling wood apart.** How heavy a load (pounds) is needed to pull apart pieces of Douglas fir 4 inches long and 1.5 inches square? Here are data from students doing a laboratory exercise: DATA WOOD

33,190	31,860	32,590	26,520	33,280
32,320	33,020	32,030	30,460	32,700
23,040	30,930	32,720	33,650	32,340
24,050	30,170	31,300	28,730	31,920

(a) We are willing to regard the wood pieces prepared for the lab session as an SRS of all similar pieces of Douglas fir. Engineers also commonly assume that characteristics of materials vary Normally. Make a graph to show the shape of the distribution for these data. Does it appear safe to assume that the Normality condition is satisfied? Suppose that the strength of pieces of wood like these follows a Normal distribution with standard deviation 3000 pounds.
(b) Give a 95% confidence interval for the mean load required to pull the wood apart.

16.26 **Bone loss by nursing mothers.** Breastfeeding mothers secrete calcium into their milk. Some of the calcium may come from their bones, so mothers may lose bone mineral. Researchers measured the percent change in mineral content of the spines of 47 mothers during three months of breastfeeding.[7] Here are the data: DATA BONELS

−4.7	−2.5	−4.9	−2.7	−0.8	−5.3	−8.3	−2.1	−6.8	−4.3
2.2	−7.8	−3.1	−1.0	−6.5	−1.8	−5.2	−5.7	−7.0	−2.2
−6.5	−1.0	−3.0	−3.6	−5.2	−2.0	−2.1	−5.6	−4.4	−3.3
−4.0	−4.9	−4.7	−3.8	−5.9	−2.5	−0.3	−6.2	−6.8	1.7
0.3	−2.3	0.4	−5.3	0.2	−2.2	−5.1			

Blend Images/Superstock

(a) The researchers are willing to consider these 47 women to be an SRS from the population of all nursing mothers. Suppose that the percent change in this population has standard deviation $\sigma = 2.5\%$. Make a stemplot of the data to verify that the data follow a Normal distribution quite closely. (Don't forget that you need both a 0 and a −0 stem because there are both positive and negative values.)

(b) Use a 99% confidence interval to estimate the mean percent change in the population.

16.27 **This wine stinks.** Sulfur compounds cause "off-odors" in wine, so winemakers want to know the odor threshold, the lowest concentration of a compound that the human nose can detect. The odor threshold for dimethyl sulfide (DMS) in trained wine tasters is about 25 micrograms per liter of wine (μg/L). The untrained noses of consumers may be less sensitive, however. Here are the DMS odor thresholds for 10 untrained students: WINE2

30	30	42	35	22	33	31	29	19	23

(a) Assume that the standard deviation of the odor threshold for untrained noses is known to be $\sigma = 7$ μg/L. Briefly discuss the other two "simple conditions," using a stemplot to verify that the distribution is roughly symmetric with no outliers.

(b) Give a 95% confidence interval for the mean DMS odor threshold among all students.

16.28 **Why are larger samples better?** Statisticians prefer large samples. Describe briefly the effect of increasing the size of a sample on the margin of error of a 95% confidence interval.

Exploring the Web

16.29 **A statistics glossary.** An editorial was published in the *Journal of the National Cancer Institute*, Vol. 101, No. 23 (December 2, 2009) that announced some online resources for journalists, including a statistics glossary. The glossary can be found at `www.oxfordjournals.org/our_journals/jnc/resource/statistics%20glossary.pdf`. Read the definition of a confidence interval. Is this an accurate definition? Explain your answer.

16.30 **Getting around No Child Left Behind.** The PBS website has an interesting article from 2007 discussing how school districts were getting around certain requirements of the No Child Left Behind law. You can find the article at `www.pbs.org/newshour/bb/education-july-dec07-nclb_08-14/`. What does the article have to say about the use of confidence intervals in reporting results about the percent of students passing proficiency tests?

Photographee.eu/Shutterstock

CHAPTER

17

Tests of Significance: The Basics

In this chapter we cover...

17.1 The reasoning of tests of significance
17.2 Stating hypotheses
17.3 *P*-value and statistical significance
17.4 Tests for a population mean
17.5 Significance from a table*
17.6 Resampling: significance from a simulation*

Confidence intervals are one of the two most common types of statistical inference. Use a confidence interval when your goal is to estimate a population parameter. The second common type of inference, called *tests of significance,* has a *different goal:* to assess the evidence provided by data about some claim concerning a population parameter. Here is the reasoning of statistical tests in a nutshell.

EXAMPLE 17.1 I'm a Good Free-Throw Shooter

I claim that I make 75% of my basketball free throws. To test my claim, you ask me to shoot 20 free throws. I make only eight of the 20. "Aha!" you say. "Someone who makes 75% of his free throws would almost never make only eight out of 20. So I don't believe your claim."

Your reasoning is based on asking what would happen if my claim were true and we repeated the sample of 20 free throws many times—I would almost never make as few as eight. This outcome is so unlikely that it gives strong evidence that my claim is not true.

You can say how strong the evidence against my claim is by giving the probability that I would make as few as eight out of 20 free throws if I really make 75% in the long run. This probability is 0.0009. I would make as few as eight of 20 only nine times in 10,000 tries in the long run if my claim to make 75% were true. The small probability convinces you that my claim is false. ■

APPLET

The *Reasoning of a Statistical Test* applet animates Example 17.1. You can ask a player to shoot free throws until the data do (or don't) convince you that he makes fewer than 75%. Significance tests use an elaborate vocabulary, but the basic idea is simple: *an outcome that would rarely happen if a claim were true is good evidence that the claim is not true.*

17.1 The reasoning of tests of significance

The reasoning of statistical tests, like that of confidence intervals, is based on asking what would happen if we repeated the sample or experiment many times. We will act as if the "simple conditions" listed on page 374 are true: we have a perfect SRS from an exactly Normal population with standard deviation σ known to us. Here is an example we will explore.

EXAMPLE 17.2 Sweetening Colas

© Ramin Talaie/Corbis

Diet colas use artificial sweeteners to avoid sugar. These sweeteners gradually lose their sweetness over time. Manufacturers therefore test new colas for loss of sweetness before marketing them. Trained tasters sip the cola along with drinks of standard sweetness and score the cola on a "sweetness score" of 1 to 10, with larger scores corresponding to greater sweetness. The cola is then stored for a month at high temperature to imitate the effect of four months' storage at room temperature. Each taster scores the cola again after storage. This is a matched pairs experiment. Our data are the differences (score before storage minus score after storage) in the tasters' scores. The bigger these differences, the bigger the loss of sweetness.

DATA COLA

Suppose we know that, for any cola, the sweetness loss scores vary from taster to taster according to a Normal distribution with standard deviation $\sigma = 1$. The mean μ for all tasters measures loss of sweetness and is different for different colas.

Here are the sweetness losses for a cola currently on the market, as measured by 10 trained tasters:

1.6 0.4 0.5 −2.0 1.5 −1.1 1.3 −0.1 −0.3 1.2

The average sweetness loss is given by the sample mean $\bar{x} = 0.3$, so that on average the 10 tasters found a small loss of sweetness. Also, more than half (six) of the tasters found a loss of sweetness. Are these data good evidence that the cola lost sweetness in storage? ■

The reasoning is the same as in Example 17.1. We make a claim and ask if the data give evidence *against* it. We seek evidence that there *is* a sweetness loss, so the claim we test is that there *is not* a loss. In that case, the mean loss for the population of all trained testers would be $\mu = 0$.

- If the claim that $\mu = 0$ is true, the sampling distribution of $\bar{x}$ from 10 tasters is Normal with mean $\mu = 0$ and standard deviation

$$\frac{\sigma}{\sqrt{n}} = \frac{1}{\sqrt{10}} = 0.316$$

- This is just like the calculations we did in Chapter 15 (see Example 15.5 on page 354) and Chapter 16 (see Example 16.1 on page 375). Figure 17.1 shows this sampling distribution. We can judge whether any observed $\bar{x}$ is surprising by locating it on this distribution.

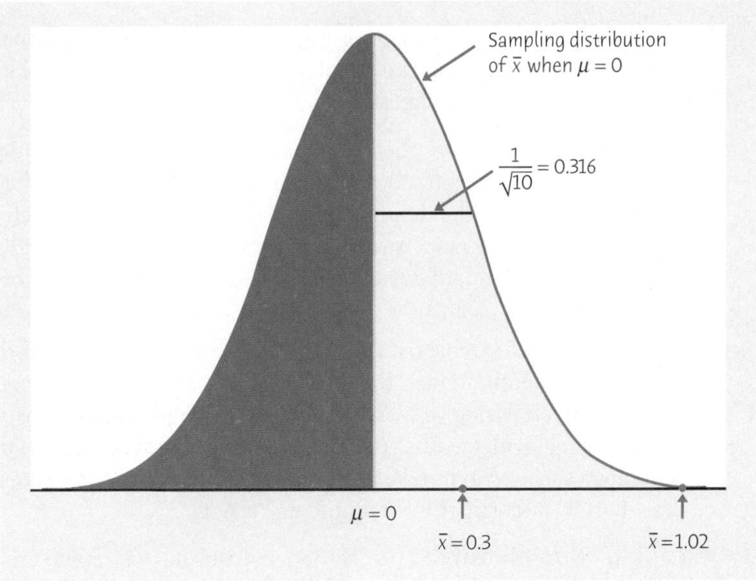

FIGURE 17.1
If the cola does not lose sweetness in storage, the mean score $\bar{x}$ for 10 tasters will have this sampling distribution. The actual result for one cola was $\bar{x} = 0.3$. That could easily happen just by chance. Another cola had $\bar{x} = 1.02$. That's so far out on the Normal curve that it is good evidence that this cola did lose sweetness.

- For this cola, 10 tasters had mean loss $\bar{x} = 0.3$. It is clear from Figure 17.1 that an $\bar{x}$ this large is not particularly surprising. It could easily occur just by chance when the population mean is $\mu = 0$. That 10 tasters found $\bar{x} = 0.3$ is not *strong* evidence that this cola loses sweetness.

EXAMPLE 17.3 Sweetening Colas, Again

Here are the sweetness losses for a new cola, as measured by 10 trained tasters:

2.0 0.4 0.7 2.0 −0.4 2.2 −1.3 1.2 1.1 2.3

DATA COLA2

The average sweetness loss is given by the sample mean $\bar{x} = 1.02$. Most scores are positive. That is, most tasters found a loss of sweetness. But the losses are small, and two tasters (the negative scores) thought the cola gained sweetness. Are these data good evidence that the cola lost sweetness in storage? ■

The taste test for the new cola produced $\bar{x} = 1.02$. That's way out in the tail of the Normal curve in Figure 17.1—so far out that *an observed value this large would rarely occur just by chance if the true μ were 0.* This observed value is good evidence that the true μ is in fact greater than 0, that is, that the cola lost sweetness. The manufacturer must reformulate the new cola and try again.

Apply Your Knowledge

17.1 Student Attitudes. The Survey of Study Habits and Attitudes (SSHA) is a psychological test that measures students' study habits and attitude toward school. The survey yields several scores, one of which measures student attitudes toward studying. The mean student attitude score for college students is about 50, and the standard deviation is about 15. A researcher in the Philippines is concerned about the declining performance of college graduates on professional licensure and board exams. She suspects that poor attitudes of students are partly responsible for the decline and that the mean for college seniors who plan to take professional licensure or board exams is less than 50. She gives the SSHA to an SRS of 225 college seniors in the Philippines who plan to take professional licensure or board exams. Suppose we know that the student attitude scores in the population of such students are Normally distributed with standard deviation $\sigma = 15$.

(a) We seek evidence *against* the claim that $\mu = 50$. What is the sampling distribution of the mean score $\bar{x}$ of a sample of 225 students if the claim is true? Draw the density curve of this distribution. (Sketch a Normal curve, then mark on the axis the values of the mean and one, two, and three standard deviations of the sampling distribution on either side of the mean.)

(b) Suppose that the sample data give $\bar{x} = 48.9$. Mark this point on the axis of your sketch. In fact, the result was $\bar{x} = 47.4$. Mark this point on your sketch. Using your sketch, explain in simple language why one result is good evidence that the mean score of all college seniors in the Philippines who plan to take professional licensure or board exams is less than 50 and why the other outcome is not.

17.2 Measuring Conductivity. The National Institute of Standards and Technology (NIST) supplies a "standard iron rod" whose electrical conductivity is supposed to be exactly 10.1. Is there reason to think that the true conductivity is not 10.1? To find out, NIST measures the conductivity of one rod six times. Repeated measurements of the same thing vary, which is why NIST makes six measurements. These measurements are an SRS from the population of all possible measurements. This population has a Normal distribution with mean μ equal to the true conductivity and standard deviation $\sigma = 0.1$.

(a) We seek evidence *against* the claim that $\mu = 10.1$. What is the sampling distribution of the mean $\bar{x}$ in many samples of six measurements of one rod if the claim is true? Make a sketch of the Normal curve for this distribution. (Draw a Normal curve, then mark on the axis the values of the mean and one, two, and three standard deviations on either side of the mean.)

(b) Suppose that the sample mean is $\bar{x} = 10.09$. Mark this value on the axis of your sketch. Another rod has $\bar{x} = 9.95$ for six measurements. Mark this value on the axis as well. Explain in simple language why one result is good evidence that the true conductivity differs from 10.1 and why the other result gives no reason to doubt that 10.1 is correct.

17.2 Stating hypotheses

A statistical test starts with a careful statement of the claims we want to compare. In Example 17.3, we saw that the taste test data are not plausible if the new cola loses no sweetness. Because the reasoning of tests looks for evidence *against* a claim, we start with the claim we seek evidence against, such as "no loss of sweetness."

Null and Alternative Hypotheses

The claim tested by a statistical test is called the **null hypothesis**. The test is designed to assess the strength of the evidence *against* the null hypothesis. Usually the null hypothesis is a statement of "no effect" or "no difference."

The claim about the population that we are trying to find evidence *for* is the **alternative hypothesis**. The alternative hypothesis is **one-sided** if it states that a parameter is *larger than* or *smaller than* the null hypothesis value. It is **two-sided** if it states that the parameter is *different from* the null value (it could be either smaller or larger).

We abbreviate the null hypothesis as H_0 and the alternative hypothesis as H_a. CAUTION *Hypotheses always refer to a population parameter, not to a particular sample outcome. Be sure to state H_0 and H_a in terms of population parameters.* Because H_a expresses the effect that we hope to find evidence *for,* it is sometimes easier to begin by stating H_a and then set up H_0 as the statement that the hoped-for effect is not present.

In Examples 17.2 and 17.3, we are seeking evidence *for* loss in sweetness. The null hypothesis says "no loss" on the average in a large population of tasters. The alternative hypothesis says "there is a loss." So the hypotheses are

$$H_0: \mu = 0$$
$$H_a: \mu > 0$$

The alternative hypothesis is *one-sided* because we are interested only in whether the cola *lost* sweetness.[1]

EXAMPLE 17.4 Studying Job Satisfaction

Does the job satisfaction of assembly workers differ when their work is machine-paced rather than self-paced? Assign workers either to an assembly line moving at a fixed pace or to a self-paced setting. All subjects work in both settings, in random order. This is a matched pairs design. After two weeks in each work setting, the workers take a test of job satisfaction. The response variable is the difference in satisfaction scores, self-paced minus machine-paced.

The parameter of interest is the mean μ of the differences in scores in the population of all assembly workers. The null hypothesis says that there is no difference between self-paced and machine-paced work, that is,

$$H_0: \mu = 0$$

The authors of the study wanted to know if the two work conditions have different levels of job satisfaction. They did not specify the direction of the difference. The alternative hypothesis is therefore *two-sided*:

$$H_a: \mu \neq 0$$ ■

CAUTION *The hypotheses should express the hopes or suspicions we have* **before** *we see the data. It is cheating to first look at the data and then frame hypotheses to fit what the data show.* For example, the data for the study in Example 17.4 showed that the workers were more satisfied with self-paced work, but this should not influence the choice of H_a. If you do not have a specific direction firmly in mind in advance, use a two-sided alternative.

Statistics In Your World — HONEST HYPOTHESES?

Chinese and Japanese people, for whom the number 4 is unlucky, die more often on the fourth day of the month than on other days. The authors of a study did a statistical test of the claim that the fourth day has more deaths than other days and found good evidence in favor of this claim. Can we trust this? Not if the authors looked at all days, picked the one with the most deaths, then made "this day is different" the claim to be tested. A critic raised that issue, and the authors replied: "No, we had day 4 in mind in advance, so our test was legitimate."

LaunchPad Online Resources

- The **StatBoards video**, *Formulating Hypotheses*, provides examples of formulating null and alternative hypotheses. It will help you understand what type of hypothesis is used for what type of question.

Apply Your Knowledge

17.3 Student Attitudes. State the null and alternative hypotheses for the study of college seniors in the Philippines' attitudes described in Exercise 17.1. (Is the alternative hypothesis one-sided or two-sided?)

17.4 **Measuring Conductivity.** State the null and alternative hypotheses for the study of electrical conductivity described in Exercise 17.2. (Is the alternative hypothesis one-sided or two-sided?)

17.5 **Too Early.** The examinations in a large multi-section statistics class are scaled after grading so that the mean score is 75. The professor thinks that students in the 8:00 A.M. class have trouble paying attention because they are sleepy and suspects that these students have a lower mean score than the class as a whole. The students in the 8:00 A.M. class this semester can be considered a sample from the population of all students in the course, so the professor compares their mean score with 75. State the hypotheses H_0 and H_a.

17.6 **Women's Incomes.** The average income of American women who work full-time and have only a high school degree is $33,230. You wonder whether the mean income of female graduates from your local high school who work full-time but have only a high school degree is different from the national average. You obtain income information from an SRS of 62 female graduates who work full-time and have only a high school degree and find that $\bar{x} = \$32{,}052$. What are your null and alternative hypotheses?

17.7 **Stating Hypotheses.** In planning a study on the number of days in the last 30 that high school students texted while driving sometime during the day, a researcher states the hypotheses as

$$H_0: \bar{x} = 15 \text{ days}$$
$$H_a: \bar{x} > 15 \text{ days}$$

What's wrong with this?

17.3 *P*-value and statistical significance

The idea of stating a null hypothesis that we want to find evidence *against* seems odd at first. It may help to think of a criminal trial. The defendant is "innocent until proven guilty." That is, the null hypothesis is innocence and the prosecution must try to provide convincing evidence against this hypothesis. That's exactly how statistical tests work, though in statistics we deal with evidence provided by data and use a probability to say how strong the evidence is.

The probability that measures the strength of the evidence against a null hypothesis is called a *P-value.* Statistical tests generally work like this:

Test Statistic and *P*-Value

A **test statistic** calculated from the sample data measures how far the data diverge from what we would expect if the null hypothesis H_0 were true. Large values of the statistic show that the data are not consistent with H_0.

The probability, computed assuming that H_0 is true, that the test statistic would take a value as extreme or more extreme than that actually observed is called the ***P*-value** of the test. The smaller the *P*-value, the stronger the evidence against H_0 provided by the data.

Small *P*-values are evidence against H_0 because they say that the observed result would be unlikely to occur if H_0 were true. Large *P*-values fail to give evidence against H_0. Statistical software will give you the *P*-value of a test when you enter your null and alternative hypotheses and your data. So your most important task is to understand what a *P*-value says.

EXAMPLE 17.5 Sweetening Colas: One-Sided *P*-Value

The study of sweetness loss in Examples 17.2 and 17.3 tests the hypotheses

$$H_0: \mu = 0$$
$$H_a: \mu > 0$$

Because the alternative hypothesis says that $\mu > 0$, values of $\bar{x}$ greater than 0 favor H_a over H_0. The test statistic compares the observed $\bar{x}$ with the hypothesized value $\mu = 0$. For now, let's concentrate on the *P*-value. The experiment presented in Examples 17.2 and 17.3 actually compared two colas. For the first cola, the 10 tasters found mean sweetness loss $\bar{x} = 0.3$. For the second, the data gave $\bar{x} = 1.02$. *The P-value for each test is the probability of getting an $\bar{x}$ this large when the mean sweetness loss is really $\mu = 0$.*

The shaded area in Figure 17.2 shows the *P*-value when $\bar{x} = 0.3$. The Normal curve is the sampling distribution of $\bar{x}$ when the null hypothesis $H_0: \mu = 0$ is true. A Normal probability calculation (Exercise 17.8) shows that the *P*-value is $P(\bar{x} \geq 0.3) = 0.1714$.

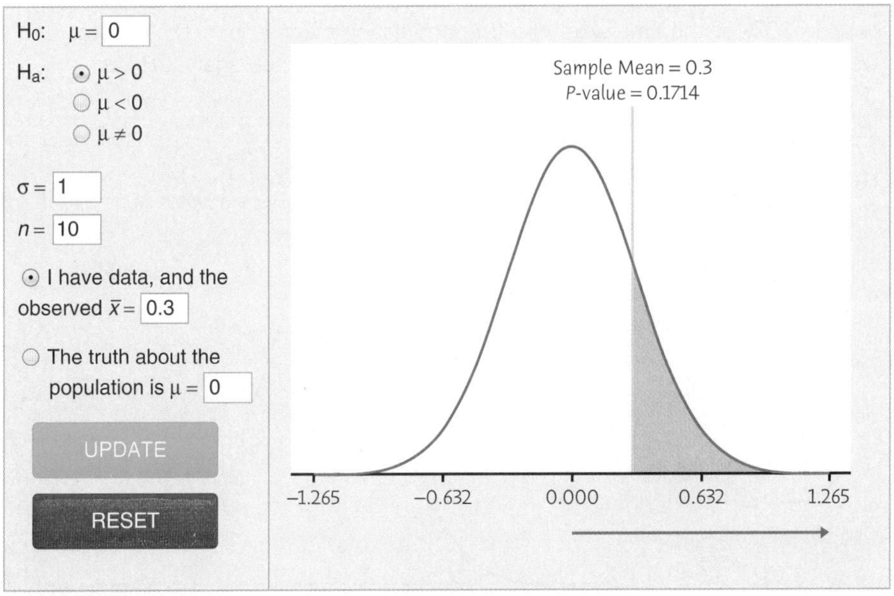

FIGURE 17.2
The one-sided *P*-value for the cola with mean sweetness loss $\bar{x} = 0.3$ in Example 17.4. The figure shows both the input and the output for the *P-Value of a Test of Significance* applet. Note that the *P*-value is the shaded area under the curve, not the unshaded area.

A value as large as $\bar{x} = 0.3$ would occur just by chance in 17% of all samples when $H_0: \mu = 0$ is true. So observing $\bar{x} = 0.3$ is not *strong* evidence against H_0. On the other hand, you can calculate that the probability that $\bar{x}$ is 1.02 or larger when in fact $\mu = 0$ is only 0.0006. We would very rarely observe a mean sweetness loss of 1.02 or larger if H_0 were true. This small *P*-value provides strong evidence against H_0 and in favor of the alternative $H_a: \mu > 0$. ■

Figure 17.2 is actually the output of the *P-Value of a Test of Significance* applet, along with the information we entered into the applet. This applet automates the work of finding *P*-values for samples of size 50 or smaller under the "simple conditions" for inference about a mean.

APPLET

The alternative hypothesis sets the direction that counts as evidence against H_0. In Example 17.5, only large positive values count because the alternative is one-sided on the high side. If the alternative is two-sided, both directions count.

EXAMPLE 17.6 Job Satisfaction: Two-Sided *P*-Value

The study of job satisfaction in Example 17.4 requires that we test

$$H_0: \mu = 0$$
$$H_a: \mu \neq 0$$

Suppose we know that differences in job satisfaction scores (self-paced minus machine-paced) in the population of all workers follow a Normal distribution with standard deviation $\sigma = 60$.

Data from 18 workers give $\bar{x} = 17$. That is, these workers prefer the self-paced environment on the average. Because the alternative is two-sided, the *P*-value is the probability of getting an $\bar{x}$ at least as far from $\mu = 0$ *in either direction* as the observed $\bar{x} = 17$.

APPLET

Enter the information for this example into the *P-Value of a Test of Significance* applet and click "Show P." Figure 17.3 shows the applet output as well as the information we entered. The *P*-value is the sum of the two shaded areas under the Normal curve. It is $P = 0.2293$. Values as far from 0 as $\bar{x} = 17$ (in either direction) would happen 23% of the time when the true population mean is $\mu = 0$. An outcome that would occur so often when H_0 is true is not good evidence against H_0. ■

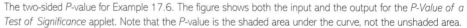

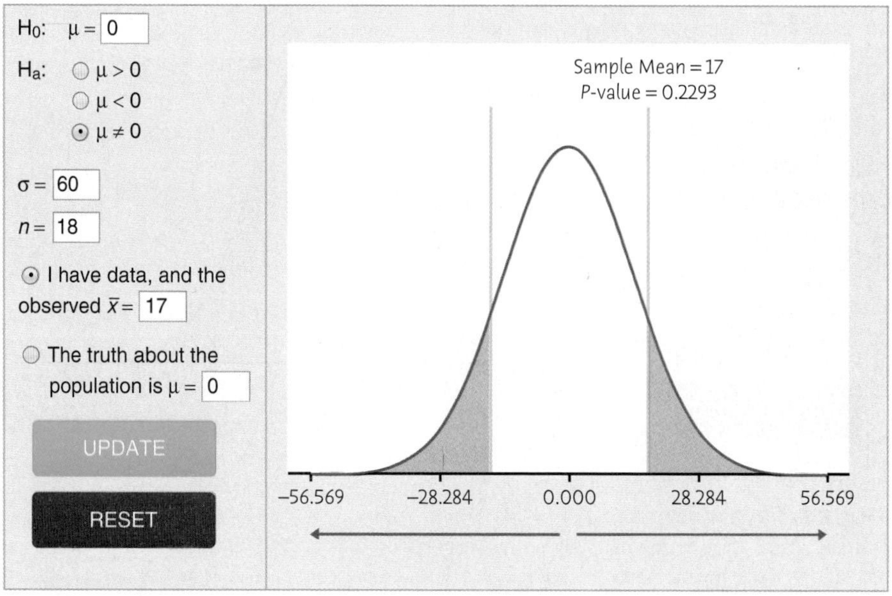

FIGURE 17.3
The two-sided *P*-value for Example 17.6. The figure shows both the input and the output for the *P-Value of a Test of Significance* applet. Note that the *P*-value is the shaded area under the curve, not the unshaded area.

The conclusion of Example 17.6 is *not* that H_0 is true. The study looked for evidence against $H_0: \mu = 0$ and failed to find strong evidence. That is all we can say. No doubt the mean μ for the population of all assembly workers is not exactly equal to 0. A large enough sample would give evidence of the difference, even if it is very small. Tests of significance assess the evidence against H_0. If the evidence is strong, we can confidently reject H_0 in favor of the alternative. *Failing to find evidence against H_0 means only that the data are not inconsistent with H_0, not that we have clear evidence that H_0 is true.* Only data that are inconsistent with H_0 provide evidence against H_0.

CAUTION

In Examples 17.5 and 17.6, we decided that *P*-value $P = 0.0006$ was strong evidence against the null hypothesis and that *P*-values $P = 0.1714$ and $P = 0.2293$

did not give convincing evidence. There is no rule for how small a *P*-value we should require to reject H_0—it's a matter of judgment and depends on the specific circumstances.

Nonetheless, we can compare a *P*-value with some fixed values that are in common use as standards for evidence against H_0. The most common fixed values are 0.05 and 0.01. If $P \leq 0.05$, there is no more than one chance in 20 that a sample would give evidence this strong just by chance when H_0 is actually true. If $P \leq 0.01$, we have a result that in the long run would happen no more than once per 100 samples if H_0 were true. These fixed standards for *P*-values are called **significance levels**. We use α, the Greek letter alpha, to stand for a significance level.

significance levels

Statistical Significance

If the *P*-value is as small as or smaller than α, we say that the data are **statistically significant at level α.** The quantity α is called the **significance level** or the **level of significance.**

CAUTION *"Significant" in the statistical sense does not mean "important." It means simply "not likely to happen just by chance."* The significance level α makes "not likely" more exact. Significance at level 0.01 is often expressed by the statement, "The results were significant ($P < 0.01$)." Here *P* stands for the *P*-value. The actual *P*-value is more informative than a statement of significance because it allows us to assess significance at any level we choose. For example, a result with $P = 0.03$ is significant at the $\alpha = 0.05$ level but is not significant at the $\alpha = 0.01$ level. To avoid confusion, we will use "statistically significant" rather than just "significant" in this chapter. However, in research papers and media publications, you will often see the word "significant" rather than the phrase "statistically significant." In later chapters we will use both.

It is good practice to interpret findings of statistical significance in the context of the problem for which data were collected. For example, in Example 17.5, statistical significance implies something about the loss of sweetness in diet colas. A sample mean of $\bar{x} = 0.3$ is not statistically significant at the $\alpha = 0.05$ level. We would interpret this as meaning that our data did not provide strong evidence that diet cola, on average, loses sweetness after being stored one month at high temperature.

Statistics In Your World

SIGNIFICANCE STRIKES DOWN A NEW DRUG

The pharmaceutical company Pfizer spent \$1 billion developing a new cholesterol-fighting drug. The final test for its effectiveness was a clinical trial with 15,000 subjects. To enforce double-blindness, only an independent group of experts saw the data during the trial. Three years into the trial, the monitors declared that there was a statistically significant excess of deaths and of heart problems in the group assigned to the new drug. Pfizer ended the trial. There went \$1 billion.

LaunchPad Online Resources

- The StatClips video, *P-value Interpretation*, provides examples of interpreting *P*-values.

Apply Your Knowledge

17.8 Sweetening Colas: Find the *P*-Value. The *P*-value for the first cola in Example 17.5 is the probability (taking the null hypothesis $\mu = 0$ to be true) that $\bar{x}$ takes a value at least as large as 0.3.

(a) What is the sampling distribution of $\bar{x}$ when $\mu = 0$? This distribution appears in Figure 17.2.

(b) Do a Normal probability calculation to find the *P*-value. Your result should agree with Example 17.5 up to roundoff error.

17.9 Job Satisfaction: Find the *P*-Value. The *P*-value in Example 17.6 is the probability (taking the null hypothesis $\mu = 0$ to be true) that $\bar{x}$ takes a value at least as far from 0 as 17.

(a) What is the sampling distribution of $\bar{x}$ when $\mu = 0$? This distribution is shown in Figure 17.3.

(b) Do a Normal probability calculation to find the P-value. (Recall that the alternative hypothesis is two-sided.) Your result should agree with Example 17.6 up to roundoff error.

17.10 Lorcaserin and Weight Loss. A double-blind, randomized comparative experiment compared the effect of the drug lorcaserin and a placebo on weight loss in overweight adults. All subjects also underwent diet and exercise counseling. The study reported that, after one year, patients in the lorcaserin group had an average weight loss of 5.8 kilograms (kg), while those on the placebo had an average weight loss of 2.2 kg ($P < 0.001$).[2] Explain to someone who knows no statistics why these results mean that there is good reason to think that lorcaserin works. Include an explanation of what $P < 0.001$ means.

APPLET

17.11 Student Attitudes. Exercise 17.1 describes a study of the attitudes of college seniors in the Philippines. You stated the null and alternative hypotheses in Exercise 17.3.

(a) One sample of 225 students had mean student attitude score $\bar{x} = 48.9$. Enter this $\bar{x}$, along with the other required information, into the *P-Value of a Test of Significance* applet. What is the P-value? Is this outcome statistically significant at the $\alpha = 0.05$ level? At the $\alpha = 0.01$ level?

(b) Another sample of 225 students had $\bar{x} = 47.4$. Use the applet to find the P-value for this outcome. Is it statistically significant at the $\alpha = 0.05$ level? At the $\alpha = 0.01$ level?

(c) Explain briefly why these P-values tell us that one outcome is strong evidence against the null hypothesis and that the other outcome is not.

APPLET

17.12 Measuring Conductivity. Exercise 17.2 describes six measurements of the electrical conductivity of an iron rod. You stated the null and alternative hypotheses in Exercise 17.4.

(a) One set of measurements has mean conductivity $\bar{x} = 10.09$. Enter this $\bar{x}$, along with the other required information, into the *P-Value of a Test of Significance* applet. What is the P-value? Is this outcome statistically significant at the $\alpha = 0.05$ level? At the $\alpha = 0.01$ level?

(b) Another set of measurements has $\bar{x} = 9.95$. Use the applet to find the P-value for this outcome. Is it statistically significant at the $\alpha = 0.05$ level? At the $\alpha = 0.01$ level?

(c) Explain briefly why these P-values tell us that one outcome is strong evidence against the null hypothesis and that the other outcome is not.

17.4 Tests for a population mean

We have used tests for hypotheses about the mean μ of a population, under the "simple conditions," to introduce tests of significance. The big idea is the reasoning of a test: *data that would rarely occur if the null hypothesis H_0 were true provide evidence that H_0 is not true.* The P-value gives us a probability to measure "would rarely occur." In practice, the steps in carrying out a test of significance mirror the overall four-step process for organizing realistic statistical problems.

Tests of Significance: The Four-Step Process

STATE: What is the practical question that requires a statistical test?

PLAN: Identify the parameter, state null and alternative hypotheses, and choose the type of test that fits your situation.

SOLVE: Carry out the test in three phases:

1. **Check the conditions** for the test you plan to use.
2. Calculate the **test statistic.**
3. Find the ***P*-value.**

CONCLUDE: Return to the practical question to describe your results in this setting.

Once you have stated your question, formulated hypotheses, and checked the conditions for your test, you or your software can find the test statistic and *P*-value by following a rule. Here is the rule for the test we have used in our examples.

z Test for a Population Mean

Draw an SRS of size n from a Normal population that has unknown mean μ and known standard deviation σ. To **test the null hypothesis that μ has a specified value,**

$$H_0\colon \mu = \mu_0$$

calculate the **one-sample *z* test statistic**

$$z = \frac{\bar{x} - \mu_0}{\sigma/\sqrt{n}}$$

In terms of a variable Z having the standard Normal distribution, the P-value for a test of H_0 against

$H_a\colon \mu > \mu_0$ is $P(Z \geq z)$

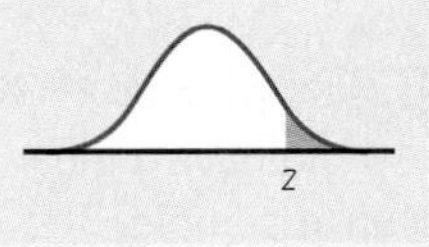

$H_a\colon \mu < \mu_0$ is $P(Z \leq z)$

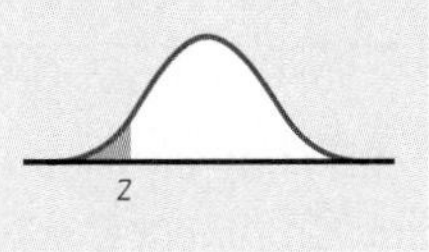

$H_a\colon \mu \neq \mu_0$ is $2P(Z \geq |z|)$

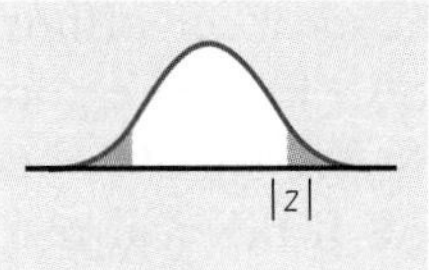

As promised, the test statistic z measures how far the observed sample mean $\bar{x}$ deviates from the hypothesized population value μ_0. The measurement is in the familiar standard scale obtained by dividing by the standard deviation of $\bar{x}$. So we have a common scale for all z tests, and the 68–95–99.7 rule helps us see at once if $\bar{x}$ is far from μ_0. The pictures that illustrate the P-value look just like the curves in Figures 17.2 and 17.3, except that they are in the standard scale.

EXAMPLE 17.7 Executives' Blood Pressures

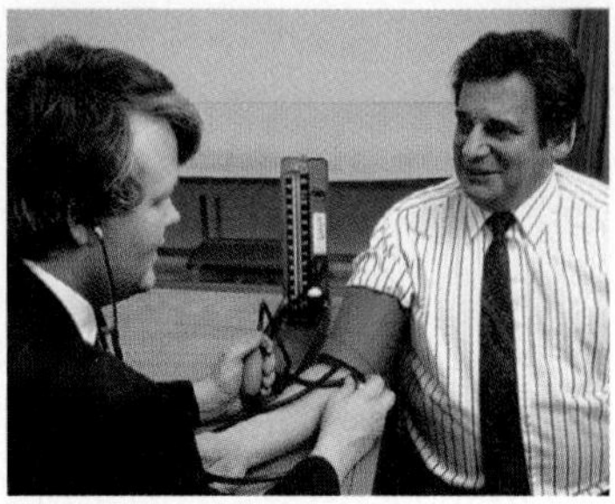

© ImageState/Alamy

STATE: The National Center for Health Statistics reports that the systolic blood pressure for males 35 to 44 years of age has mean 128 and standard deviation 15. The medical director of a large company looks at the medical records of 72 executives in this age group and finds that the mean systolic blood pressure in this sample is $\bar{x} = 126.07$. Is this evidence that the company's executives have a different mean systolic blood pressure from the general population?

PLAN: The null hypothesis is "no difference" from the national mean $\mu_0 = 128$. The alternative is two-sided because the medical director did not have a particular direction in mind before examining the data. So the hypotheses about the unknown mean μ of the executive population are

$$H_0: \mu = 128$$

$$H_a: \mu \neq 128$$

We know that the one-sample z test is appropriate for these hypotheses under the "simple conditions."

SOLVE: As part of the "simple conditions," suppose we are willing to assume that executives' systolic blood pressures follow a Normal distribution with standard deviation $\sigma = 15$. Software can now calculate z and P for you. Going ahead by hand, the **test statistic** is

$$z = \frac{\bar{x} - \mu_0}{\sigma/\sqrt{n}} = \frac{126.07 - 128}{15/\sqrt{72}}$$

$$= -1.09$$

To help find the ***P*-value**, sketch the standard Normal curve and mark on it the observed value of z. Figure 17.4 shows that the P-value is the probability that a standard Normal variable Z takes a value at least 1.09 away from zero. From Table A or software, this probability is

$$P = 2P(Z < -1.09) = (2)(0.1379) = 0.2758$$

CONCLUDE: More than 27% of the time, an SRS of size 72 from the general male population would have a mean systolic blood pressure at least as far from 128 as that of the executive sample. The observed $\bar{x} = 126.07$ is therefore not good evidence that executives differ from other men. ■

In this chapter, we are acting as if the "simple conditions" stated on page 374 are true. In practice, you must verify these conditions.

1. **SRS:** The most important condition is that the 72 executives in the sample are an SRS from the population of all middle-aged male executives in the company. We should check this requirement by asking how the data were produced. If medical records are available only for executives with recent medical problems, for example, the data are of little value for our purpose because of the obvious health bias. It turns out that all executives are given a free annual medical exam, and that the medical director selected 72 exam results at random.
2. **Normal distribution:** We should also examine the distribution of the 72 observations to look for signs that the population distribution is not Normal.
3. **Known σ:** It really is unrealistic to suppose that we know that $\sigma = 15$. We will see in Chapter 20 that it is easy to do away with the need to know σ.

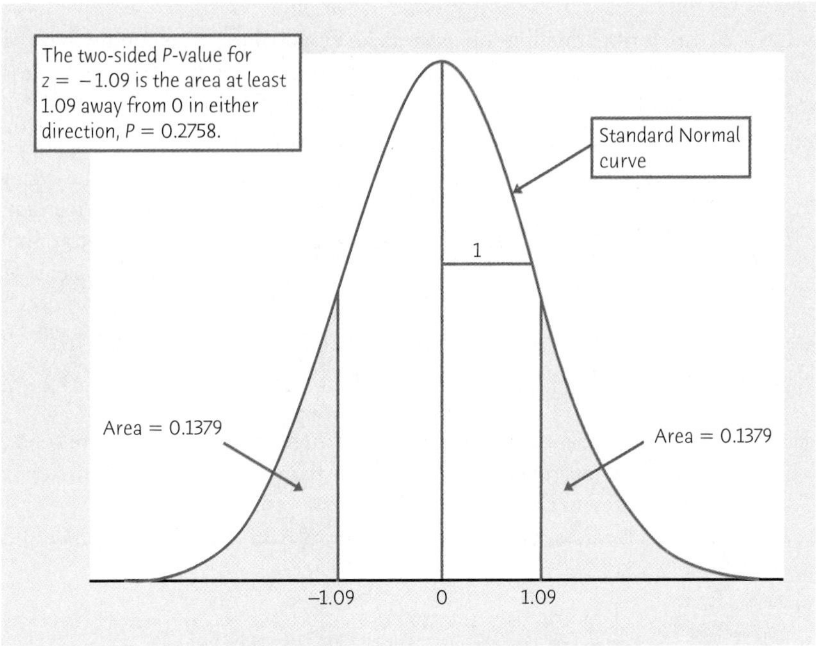
The two-sided P-value for z = −1.09 is the area at least 1.09 away from 0 in either direction, P = 0.2758.
Standard Normal curve
1
Area = 0.1379
Area = 0.1379
−1.09
0
1.09

FIGURE 17.4
The *P*-value for the two-sided test in Example 17.7. The observed value of the test statistic is $z = -1.09$.

LaunchPad Online Resources

- The **StatBoards video**, *Calculating the Test Statistic and Finding the P-value, One Tail,* provides examples of the four-step process for one tail tests for population means.
- The **StatBoards video**, *Calculating the Test Statistic and Finding the P-value, Two Tail,* provides examples of the four-step process for two tail tests for population means.
- The **StatBoards video**, *Interpreting the Result of a Test in Context,* provides examples of the Conclude step of the four-step process in tests for population means.

Apply Your Knowledge

17.13 The *z* Statistic. Published reports of research work are terse. They often report just a test statistic and *P*-value. For example, the conclusion of Example 17.7 might be stated as "($z = -1.09, P = 0.2758$)." Find the values of the one-sample z statistic needed to complete these conclusions:

(a) For the first cola in Example 17.5, $z = ?, P = 0.1714$.

(b) For the second cola in Example 17.5, $z = ?, P = 0.0006$.

(c) For Example 17.6, $z = ?, P = 0.2293$.

17.14 Measuring Conductivity. Here are six measurements of the electrical conductivity of an iron rod: COND

10.08 9.89 10.05 10.16 10.21 10.11

The iron rod is supposed to have conductivity 10.1. Do the measurements give good evidence that the true conductivity is not 10.1?

The six measurements are an SRS from the population of all results we would get if we kept measuring conductivity forever. This population has a Normal distribution with mean equal to the true conductivity of the rod and standard deviation 0.1. Use this information to carry out a test, following the four-step process as illustrated in Example 17.7.

17.15 Bad Weather, Bad Tip? People tend to be more generous after receiving good news. Are they less generous after receiving bad news? The average tip left by adult Americans is 20%. Give 20 patrons of a restaurant a message on their bill warning them that tomorrow's weather will be bad and record the percentage tip they leave. Here are the tips as a percentage of the total bill:[3] TIP3

18.0	19.1	19.2	18.8	18.4	19.0	18.5	16.1	16.8	18.2
14.0	17.0	13.6	17.5	20.0	20.2	18.8	18.0	23.2	19.4

Suppose that percentage tips are Normal with $\sigma = 2$. Is there good evidence that the mean percentage tip is less than 20? Follow the four-step process as illustrated in Example 17.7.

17.5 Significance from a table*

Statistics in practice uses technology (graphing calculator or software) to get *P*-values quickly and accurately. In the absence of suitable technology, you can get approximate *P*-values quickly by comparing the value of your test statistic with critical values from a table. For the z statistic, the table is Table C, the same table we used for confidence intervals.

Look at the bottom row of critical values in Table C, labeled z^*. At the top of the table, you see the confidence level C for each z^*. At the bottom of the table, you see both the one-sided and two-sided *P*-values for each z^*. Values of a test statistic z that are farther out than a z^* (in the direction given by the alternative hypothesis) are statistically significant at the level that matches z^*.

Significance from a Table of Critical Values

To find the approximate *P*-value for any z statistic, compare z (ignoring its sign) with the critical values z^* at the bottom of Table C. If z falls between two values of z^*, the *P*-value falls between the two corresponding values of *P* in the "One-sided *P*" or the "Two-sided *P*" row of Table C.

EXAMPLE 17.8 Is It Statistically Significant?

z^*	2.054	2.326
One-sided *P*	0.02	0.01

The z statistic for a one-sided test is $z = 2.13$. How statistically significant is this result? Compare $z = 2.13$ with the z^* row in Table C.

It lies between $z^* = 2.054$ and $z^* = 2.326$. So the *P*-value lies between the corresponding entries in the "One-sided P" row, which are $P = 0.02$ and $P = 0.01$. This z is statistically significant at the $\alpha = 0.02$ level and *is not* statistically significant at the $\alpha = 0.01$ level.

Figure 17.5 illustrates the situation. The shaded area under the Normal curve is the *P*-value for $z = 2.13$. You can see that *P* falls between the areas to the right of the two critical values, for $P = 0.02$ and $P = 0.01$.

*This material can be skipped if you use software to compute *P*-values.

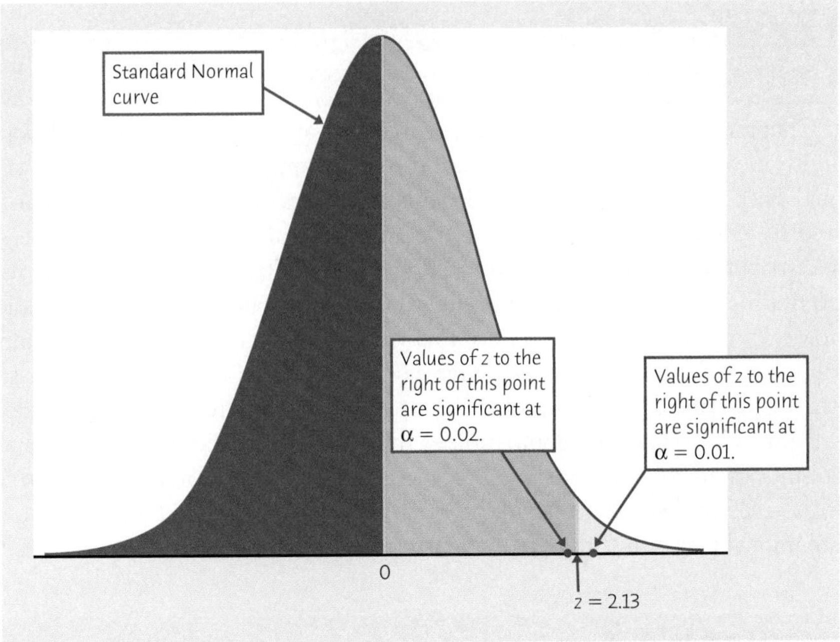

FIGURE 17.5
Is it significant? The test statistic value $z = 2.13$ falls between the critical values required for significance at the $\alpha = 0.02$ and $\alpha = 0.01$ levels. So the test *is* significant at $\alpha = 0.02$ and *is not* significant at $\alpha = 0.01$.

The z statistic in Example 17.7 is $z = -1.09$. The alternative hypothesis is two-sided. Compare $z = -1.09$ (ignoring the minus sign) with the z^* row in Table C. It lies between $z^* = 1.036$ and $z^* = 1.282$. So the P-value lies between the matching entries in the "Two-sided P" row, $P = 0.30$ and $P = 0.20$. This is enough to conclude that the data do not provide good evidence against the null hypothesis. ■

z^*	1.036	1.282
Two-sided P	0.30	0.20

Apply Your Knowledge

17.16 Significance from a Table. A test of H_0: $\mu = 0$ against H_a: $\mu > 0$ has test statistic $z = 1.876$. Is this test statistically significant at the 5% level ($\alpha = 0.05$)? Is it statistically significant at the 1% level ($\alpha = 0.01$)?

17.17 Significance from a Table. A test of H_0: $\mu = 0$ against H_a: $\mu \neq 0$ has test statistic $z = 1.876$. Is this test statistically significant at the 5% level ($\alpha = 0.05$)? Is it statistically significant at the 1% level ($\alpha = 0.01$)?

17.18 Testing a Random Number Generator. A random number generator is supposed to produce random numbers that are uniformly distributed on the interval from 0 to 1. If this is true, the numbers generated come from a population with $\mu = 0.5$ and $\sigma = 0.2887$. A command to generate 100 random numbers gives outcomes with mean $\bar{x} = 0.4365$. Assume that the population σ remains fixed. We want to test

$$H_0: \mu = 0.5$$
$$H_a: \mu \neq 0.5$$

(a) Calculate the value of the z test statistic.

(b) Use Table C: is z statistically significant at the 5% level ($\alpha = 0.05$)?

(c) Use Table C: is z statistically significant at the 1% level ($\alpha = 0.01$)?

(d) Between which two Normal critical values z^* in the bottom row of Table C does z lie? Between what two numbers does the P-value lie? Does the test give good evidence against the null hypothesis?

17.6 Resampling: significance from a simulation*

In Chapter 15 (page 353) we learned that, for an SRS from a Normal population with mean μ and standard deviation σ, the sampling distribution of $\bar{x}$ is also Normal with mean μ and standard deviation $\sigma/\sqrt{n}$. In Section 17.4, this fact allowed us to use software or a table to compute *P*-values. Knowledge of the sampling distribution of $\bar{x}$ computed from SRSs when the null hypothesis is true is the key to finding *P*-values because this allows us to carry out probability calculations.

However, we also saw in Section 15.3 that we can approximate the sampling distribution of $\bar{x}$ by taking a very large number of SRSs of size *n* and constructing the histogram of the values of the sample means, $\bar{x}$. This was done in Example 15.4. Figure 15.2 (page 351) approximates the sampling distribution of $\bar{x}$ by plotting the results of many (1000) samples of size 10 from a Normal population with mean $\mu = 25$ and standard deviation $\sigma = 7$. Can we estimate *P*-values from such a histogram? Here is an example demonstrating how this can be done.

EXAMPLE 17.9 Sweetening Colas

In Examples 17.2, 17.3, and 17.5, we were interested in whether there was evidence of any sweetness loss for a diet cola. In particular, we tested the hypotheses

$$H_0: \mu = 0$$
$$H_a: \mu > 0$$

where sweetness loss score varies from one trained taster to another according to a Normal distribution with mean μ and standard deviation $\sigma = 1$. Based on the mean of the sweetness loss score $\bar{x}$ of 10 trained testers, we computed *P*-values when $\bar{x} = 0.3$ and when $\bar{x} = 1.02$. Now we estimate these *P*-values from the histogram of the sample means of 1000 SRSs of size 10 computed (using software) under the assumption that the null hypothesis is true.

Figure 17.6 is a histogram of the means of 1000 SRSs of size 10 from a Normal population with mean $\mu = 0$ and standard deviation $\sigma = 1$. We find that 163 of the

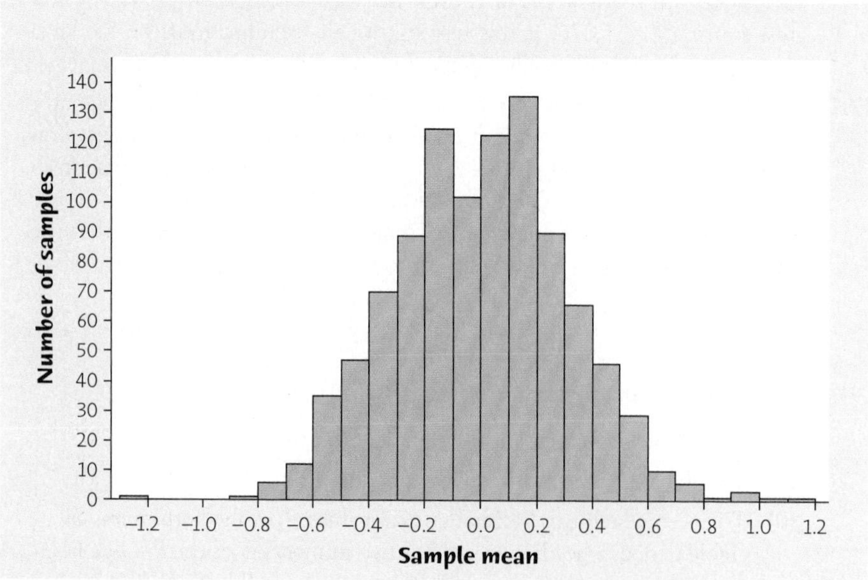

FIGURE 17.6 The histogram of the means of 1000 SRSs of size 10 from a Normal population with mean $\mu = 0$ and standard deviation $\sigma = 1$ for Example 17.9.

*This material can be skipped if you prefer not to cover resampling methods.

samples produced a mean $\bar{x}$ of 0.3 or more. Only two samples produced a mean $\bar{x}$ of 1.02 or more. Thus, based on the histogram, we would estimate the *P*-value of 0.3 to be 163/1000 = 0.163 and the *P*-value of 1.02 to be 2/1000 = 0.002. Compare these estimates to the exact *P*-values of 0.1714 and 0.0006 that we computed in Example 17.5. The simulated values are reasonably good approximations (and would be even better with more samples of size 10). ■

This method of taking a large number of repeated SRSs from the population distribution when the null hypothesis is true and using these to approximate *P*-values is sometimes referred to as resampling.[4] It is a powerful method for estimating *P*-values and can be applied to tests of hypotheses based on test statistics other than sample means. All we need to know is the population distribution under the assumption that the null hypothesis is true. We then resample, using software, many times from this population distribution, compute the value of the sample statistic for each sample, and determine the proportion of times we obtained sample values as or more extreme than our actual data. This proportion is an estimate of the *P*-value. The more times we resample, the better our estimate will be. The next example illustrates the method using the sample median as a test statistic.

EXAMPLE 17.10 Does This Wine Smell Bad?

In Example 15.10 (page 363), we speculated that adults aged 65 and older are less able to detect DMS than other adults. To test this, we selected an SRS of five adults from the population of all adults aged 65 and older. We presented each subject both natural wine and the same wine spiked with DMS at different concentrations to find the lowest concentration (odor threshold) at which they identify the spiked wine. The median of the odor thresholds of the five subjects was 35 micrograms per liter.

Our null hypothesis is that the distribution of odor thresholds for adults aged 65 and older is the same as that for all adults, namely, it is Normal with mean μ = 25 micrograms per liter and standard deviation σ = 7 micrograms per liter. If adults aged 65 and older are less able to detect DMS than other adults, their mean odor threshold should be larger than 25. Thus, our alternative hypothesis is $\mu > 25$ and our hypotheses are

$$H_0: \mu = 25$$
$$H_a: \mu > 25$$

Recall that the mean and median of a normal population are the same. Thus, the sample median of the odor thresholds of the five subjects provides information about the population mean μ and we can use the median of an SRS of size 5 as our test statistic. To determine whether the sample median of 35 is evidence against H_0 we need to know the sampling distribution of a sample median of size 5 computed from an SRS from a Normal population with mean μ = 25 and standard deviation σ = 7. From this sampling distribution, we compute the *P*-value. The *P*-value is the probability, computed under the assumption that H_0 is true, that the test statistic (in this case, the sample median) would take a value as large as or larger than that actually observed (in this case, 35). Unlike the sample mean, we haven't discussed any mathematical results that tell us what this sampling distribution is. However, we simulated this sampling distribution in Example 15.9. Figure 17.7 reproduces this simulation of medians based on 1000 samples of size 5 from a Normal population with mean μ = 25 and

FIGURE 17.7
The histogram of the medians of 1000 SRSs of size 5 from a Normal population with mean $\mu = 25$ and standard deviation $\sigma = 7$ for Example 17.10.

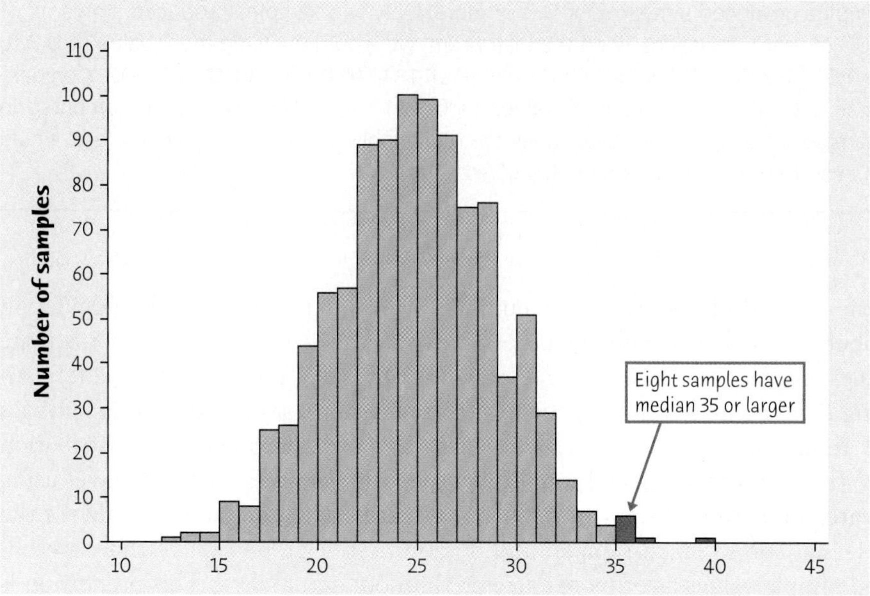

standard deviation $\sigma = 7$. The red bars correspond to samples with medians ≥ 35. There are eight such samples, so we estimate the *P*-value to be $8/1000 = 0.008$. This is strong evidence against the null hypothesis and makes precise our previous statement in Example 15.10 that a median of 35 is statistically significant. In other words, these data provide strong evidence that adults aged 65 and older are less able to detect DMS than other adults. ■

We close with a few remarks. First, we must resample in the same manner as we obtain our data. If our data are obtained by an SRS, we resample by taking repeated SRSs from the population distribution determined by the null hypothesis.

Second, resampling only provided an *estimate* of a *P*-value. Repeat the resampling and you will obtain a different estimate. Accuracy of the estimate is improved by taking a larger number of samples to estimate the sampling distribution.

Finally, resampling requires the use of software. Some software packages are more convenient for resampling than others.

LaunchPad Online Resources

- The **Video Technology Manuals** discuss how to do resampling for several software packages.

Apply Your Knowledge

17.19 Does This Wine Smell Bad? In Example 17.10, suppose the median of the odor thresholds of the sample of five older adults had been 33. Use Figure 17.7 to estimate the *P*-value for testing

$$H_0: \mu = 25$$
$$H_a: \mu > 25$$

What would you estimate the *P*-value to be if the sample median had been 30?

17.20 Sweetening Colas, a Two-Sided *P*-Value. Suppose in Example 17.9 we were interested in whether there was a change in sweetness. In this case, we would test

$$H_0: \mu = 0$$
$$H_a: \mu \neq 0$$

Use Figure 17.6 to estimate the *P*-value if the mean of the scores of 10 tasters is $\bar{x} = 1.02$. Remember, this involves a two-sided *P*-value, so you need to determine the proportion of the 1000 sample means that are ≥ 1.02 or ≤ -1.02.

CHAPTER 17 SUMMARY

Chapter Specifics

- A **test of significance** assesses the evidence provided by data against a **null hypothesis** H_0 in favor of an **alternative hypothesis** H_a.
- Hypotheses are always stated in terms of population parameters. Usually H_0 is a statement that no effect is present, and H_a says that a parameter differs from its null value in a specific direction (**one-sided alternative**) or in either direction (**two-sided alternative**).
- The essential reasoning of a test of significance is as follows. Suppose for the sake of argument that the null hypothesis is true. If we repeated our data production many times, would we often get data as inconsistent with H_0 as the data we actually have? Data that would rarely occur if H_0 were true provide evidence against H_0.
- A test is based on a **test statistic** that measures how far the sample outcome is from the value stated by H_0.
- The **P-value** of a test is the probability, computed supposing H_0 to be true, that the test statistic will take a value at least as extreme as that actually observed. Small *P*-values indicate strong evidence against H_0. To calculate a *P*-value, we must know the sampling distribution of the test statistic when H_0 is true.
- If the *P*-value is as small as or smaller than a specified value α, the data are **statistically significant** at **significance level** α.
- **Tests of significance for the null hypothesis H_0: $\mu = \mu_0$** concerning the unknown mean μ of a population are based on the **one-sample z test statistic**

$$z = \frac{\bar{x} - \mu_0}{\sigma/\sqrt{n}}$$

- The z test assumes an SRS of size n from a Normal population with known population standard deviation σ. *P*-values can be obtained either with computations from the standard Normal distribution or by using technology (applet or software).
- (Optional material) *P*-values can be estimated from a **simulation**. This method can be applied to tests of hypotheses based on test statistics other than the sample mean.

Link It

In this chapter we discuss tests of significance, the second type of statistical inference. The mathematics of probability—in particular, the sampling distributions discussed in Chapter 15—provides the formal basis for a test of significance. The sampling distribution allows us to assess "probabilistically" the strength of evidence against a null hypothesis, either through a level of significance or a *P*-value. The goal of hypothesis testing, which is used to assess the evidence provided by data about some claim concerning a population, is different from the goal of confidence interval estimation, which is used to estimate a population parameter.

Although we apply the reasoning of tests of significance for the mean of a population that has a Normal distribution in a simple and artificial setting (we assume that we know the population standard deviation), we will use the same logic in future chapters to construct tests of significance for population parameters in more realistic settings.

LaunchPad Online Resources

If you are having difficulty with any of the sections of this chapter, these online resources should help prepare you to solve the exercises at the end of this chapter.

- **StatTutor** starts with a video review of each section and asks a series of questions to check your understanding. You may find this helpful if you need additional help understanding any of the material in this chapter.
- **LearningCurve** provides you with a series of questions about the chapter geared to your level of understanding.

CHECK YOUR SKILLS

17.21 You use software to carry out a test of significance. The program tells you that the P-value is $P = 0.008$. You conclude that the probability, computed assuming that H_0 is

(a) true, of the test statistic taking a value as extreme as or more extreme than that actually observed is 0.008.
(b) true, of the test statistic taking a value as extreme as or less extreme than that actually observed is 0.008.
(c) false, of the test statistic taking a value as extreme as or more extreme than that actually observed is 0.008.

17.22 You use software to carry out a test of significance. The program tells you that the P-value is $P = 0.008$. This result is

(a) not statistically significant at either $\alpha = 0.05$ or $\alpha = 0.01$.
(b) statistically significant at $\alpha = 0.05$ but not at $\alpha = 0.01$.
(c) statistically significant at both $\alpha = 0.05$ and $\alpha = 0.01$.

17.23 The z statistic for a one-sided test is $z = 2.41$. This test is

(a) not statistically significant at either $\alpha = 0.05$ or $\alpha = 0.01$.
(b) statistically significant at $\alpha = 0.05$ but not at $\alpha = 0.01$.
(c) statistically significant at both $\alpha = 0.05$ and $\alpha = 0.01$.

17.24 The gas mileage for a particular model car is known to have a standard deviation of $\sigma = 1.0$ mile per gallon in repeated tests in a controlled laboratory environment at a fixed speed. For a fixed speed, gas mileages in repeated tests are Normally distributed. Tests on three cars of this model at 65 miles per hour give gas mileages of 24.3, 24.9, and 24.8 miles per gallon. The z statistic for testing $H_0: \mu = 25$ miles per gallon based on these three measurements is

(a) $z = -0.333$.
(b) $z = -0.577$.
(c) $z = 0.577$.

17.25 Experiments on learning in animals sometimes measure how long it takes mice to find their way through a maze. The mean time is 18 seconds for one particular maze. A researcher thinks that a loud noise will cause the mice to complete the maze faster. She measures how long each of 10 mice takes with a noise as stimulus. The sample mean is $\bar{x} = 16.5$ seconds. The null hypothesis for the test of significance is

Garry Gay/Getty Images

(a) $H_0: \mu = 18$.
(b) $H_0: \mu = 16.5$.
(c) $H_0: \mu < 18$.

17.26 The alternative hypothesis for the test in Exercise 17.25 is

(a) $H_a: \mu \neq 18$.
(b) $H_a: \mu < 18$.
(c) $H_a: \mu = 16.5$.

17.27 Researchers investigated the effectiveness of oral zinc, as compared to a placebo, in reducing the duration of the common cold when taken within 24 hours of the onset of symptoms. The researchers found those taking oral zinc had a statistically significantly shorter duration ($P < 0.05$) than those taking the placebo.[5] This means that

(a) the probability that the null hypothesis is true is less than 0.05.
(b) the value of the test statistic, the mean reduction in duration of the cold, is large.
(c) neither of the above is true.

17.28 You are testing $H_0: \mu = 0$ against $H_0: \mu \neq 0$ based on an SRS of 20 observations from a Normal population. What values of the z statistic are statistically significant at the $\alpha = 0.005$ level?

(a) All values for which $z > 2.576$
(b) All values for which $z > 2.807$
(c) All values for which $|z| > 2.807$

17.29 You are testing $H_0: \mu = 0$ against $H_0: \mu > 0$ based on an SRS of 20 observations from a Normal population. What values of the z statistic are statistically significant at the $\alpha = 0.005$ level?

(a) All values for which $z > 2.576$
(b) All values for which $z > 2.807$
(c) All values for which $|z| > 2.807$

CHAPTER 17 EXERCISES

In all exercises that call for P-values, give the actual value if you use software or the P-Value applet. Otherwise, use Table C to give values between which P must fall.

17.30 Student study times. Exercise 16.19 (page 387) describes a class survey in which students claimed to study an average of $\bar{x} = 15.3$ hours in a typical week. Regard these students as an SRS from the population of all undergraduate students at this university. Does the study give good evidence that students claim to study more than 15 hours per week on the average?

(a) State null and alternative hypotheses in terms of the mean study time in hours for the population.
(b) What is the value of the test statistic z?
(c) What is the P-value of the test? Can you conclude that students do claim to study more than 15 hours per week on the average?

17.31 I want more muscle. If young men thought that their own level of muscle was about what women prefer, the mean "muscle gap" in the study described in Exercise 16.20 (page 387) would be 0. We suspect (before seeing the data) that young men think women prefer more muscle than they themselves have.

(a) State null and alternative hypotheses for testing this suspicion.
(b) What is the value of the test statistic z?
(c) You can tell just from the value of z that the evidence in favor of the alternative is very strong (that is, the P-value is very small). Explain why this is true.

17.32 Hotel managers' personalities. Successful hotel managers must have personality characteristics often thought of as feminine (such as "compassionate") as well as those often thought of as masculine (such as "forceful"). The Bem Sex-Role Inventory (BSRI) is a personality test that gives separate ratings for female and male stereotypes, both on a scale of 1 to 7. A sample of 148 male general managers of three-star and four-star hotels had mean BSRI femininity score $\bar{y} = 5.29$.[6] The mean score for the general male population is $\mu = 5.19$. Do hotel managers, on the average, differ statistically significantly in femininity score from men in general? Assume that the standard deviation of scores in the population of all male hotel managers is the same as the $\sigma = 0.78$ for the adult male population.

(a) State null and alternative hypotheses in terms of the mean femininity score μ for male hotel managers.
(b) Find the z test statistic.
(c) What is the P-value for your z? What do you conclude about male hotel managers?

17.33 Is this what P means? A randomized comparative experiment examined the effect of a technique for improving one's ability to focus one's attention on performance of undergraduate students on the Verbal portion of the Graduate Record Exam (GRE). The researchers found a statistically significant improvement in performance on the Verbal GRE ($P < 0.05$).[7] When asked to explain the meaning of "$P < 0.05$)," a student says, "This means there is only probability less than 0.05 that the null hypothesis is true." Explain what $P < 0.05$ really means in a way that makes it clear that the student's explanation is wrong.

17.34 How to show that you are rich. Every society has its own marks of wealth and prestige. In ancient China, it appears that owning pigs was such a mark. Evidence comes from examining burial sites. The skulls of sacrificed pigs tend to appear along with expensive ornaments, which suggests that the pigs, like the ornaments, signal the wealth and prestige of the person buried. A study of burials from around 3500 B.C. concluded that "there are striking differences in grave goods between burials with pig skulls and burials without them. . . . A test indicates that the two samples of total artifacts are statistically significantly different at the 0.01 level."[8] Explain clearly why "statistically significantly different at the 0.01 level" gives good reason to think that there really is a systematic difference between burials that contain pig skulls and those that lack them.

17.35 Alleviating test anxiety. Research suggests that pressure to perform well can reduce performance on exams. Are there effective strategies to deal with pressure? In an experiment, researchers had students take a test on mathematical skills. The same students were asked to take a second test on the same skills, but now each student was paired with a partner and only if both improved their scores would they receive a monetary reward for participating in the experiment. They were also told that their performance would be videotaped and watched by teachers and students. To help them cope with the pressure, ten minutes before the second exam they were asked to write as candidly as possible about their thoughts and feelings regarding the exam. "Students who expressed their thoughts before the high-pressure test showed a statistically significant 5% math accuracy improvement from the pretest to posttest" ($P < 0.03$).[9] A colleague who knows no statistics says that an increase of 5% isn't a lot—maybe it's just an accident due to natural variation among the students. Explain in simple language how "$P < 0.03$" answers this objection.

17.36 **Healthy body, healthy mind?** In the study described in Exercise 17.33, researchers also examined the effect of a nutrition program on performance of undergraduates on the Verbal GRE. The researchers found no evidence of improvement ($P > 0.5$). The P-value refers to a null hypothesis of "no change" in GRE scores measured before and after the nutrition program. Explain clearly why this value provides no evidence of change.

17.37 **5% versus 1%.** Sketch the standard Normal curve for the z test statistic and mark off areas under the curve to show why a value of z that is statistically significant at the 1% level in a one-sided test is always statistically significant at the 5% level. If z is statistically significant at the 5% level, what can you say about its significance at the 1% level?

17.38 **The wrong alternative.** A graduate student is comparing final-exam test scores of male and female students in an introductory physics class. She starts with no expectations as to which sex will score more highly. After seeing that men did better than women on the first quiz, she tests a one-sided alternative about the mean final-exam scores,

$$H_0: \mu_M = \mu_F$$
$$H_a: \mu_M > \mu_F$$

She finds $z = 1.9$ with one-sided P-value $P = 0.0287$.

(a) Explain why she should have used the two-sided alternative hypothesis.
(b) What is the correct P-value for $z = 1.9$?

17.39 **The wrong P.** The report of a study of seat belt use by drivers says, "Hispanic drivers were not statistically significantly more likely than White/non-Hispanic drivers to overreport safety belt use (27.4 vs. 21.1%, respectively; $z = 1.33$, $P > 1.0$)."[10] How do you know that the P-value given is incorrect? What is the correct one-sided P-value for test statistic $z = 1.33$?

*Exercises 17.40 to 17.43 ask you to answer questions from data. Assume that the "simple conditions" hold in each case. The exercise statements give you the **State** step of the four-step process. In your work, follow the **Plan**, **Solve**, and **Conclude** steps, illustrated in Example 16.3 (page 381) for a confidence interval and in Example 17.7 for a test of significance.*

17.40 4step **Pulling wood apart.** How heavy a load (pounds) is needed to pull apart pieces of Douglas fir 4 inches long and 1.5 inches square? Here are data from students doing a laboratory exercise: DATA WOOD

33,190	31,860	32,590	26,520	33,280
32,320	33,020	32,030	30,460	32,700
23,040	30,930	32,720	33,650	32,340
24,050	30,170	31,300	28,730	31,920

We are willing to regard the wood pieces prepared for the lab session as an SRS of all similar pieces of Douglas fir. Engineers also commonly assume that characteristics of materials vary Normally. Suppose that the strength of pieces of wood like these follows a Normal distribution with standard deviation 3000 pounds.

(a) Is there statistically significant evidence at the $\alpha = 0.10$ level against the hypothesis that the mean is 32,500 pounds for the two-sided alternative?
(b) Is there statistically significant evidence at the $\alpha = 0.10$ level against the hypothesis that the mean is 31,500 pounds for the two-sided alternative?

17.41 4step **Bone loss by nursing mothers.** As discussed in Exercise 16.26 (page 388), breastfeeding mothers secrete calcium into their milk. Some of the calcium may come from their bones, so mothers may lose bone mineral. Researchers measured the percent change in mineral content of the spines of 47 mothers during three months of breastfeeding.[11] Here are the data: DATA BONELS

−4.7	−2.5	−4.9	−2.7	−0.8	−5.3	−8.3	−2.1	−6.8	−4.3
2.2	−7.8	−3.1	−1.0	−6.5	−1.8	−5.2	−5.7	−7.0	−2.2
−6.5	−1.0	−3.0	−3.6	−5.2	−2.0	−2.1	−5.6	−4.4	−3.3
−4.0	−4.9	−4.7	−3.8	−5.9	−2.5	−0.3	−6.2	−6.8	1.7
0.3	−2.3	0.4	−5.3	0.2	−2.2	−5.1			

The researchers are willing to consider these 47 women as an SRS from the population of all nursing mothers. Suppose that the percent change in this population has a Normal distribution with standard deviation $\sigma = 2.5\%$. Do these data give good evidence that, on the average, nursing mothers lose bone mineral?

17.42 4step **This wine stinks.** Sulfur compounds cause "off-odors" in wine, so winemakers want to know the odor threshold, the lowest concentration of a compound that the human nose can detect. The odor threshold for dimethyl sulfide (DMS) in trained wine tasters is about 25 micrograms per liter of wine (μg/L). The untrained noses of consumers may be less sensitive, however. Here are the DMS odor thresholds for 10 untrained students: DATA WINE

30	30	42	35	22	33	31	29	19	23

Assume that the odor threshold for untrained noses is Normally distributed with $\sigma = 7$ μg/L. Is there evidence that the mean threshold for untrained tasters is greater than 25 μg/L?

AP Photo/Kathy Willens

17.43 Eye grease. Athletes performing in bright sunlight often smear black eye grease under their eyes to reduce glare. Does eye grease work? In one study, 16 student subjects took a test of sensitivity to contrast after three hours facing into bright sun, both with and without eye grease. This is a matched pairs design. Here are the differences in sensitivity, with eye grease minus without eye grease:[12] DATA EYEGRS

0.07	0.64	−0.12	−0.05	−0.18	0.14	−0.16	0.03
0.05	0.02	0.43	0.24	−0.11	0.28	0.05	0.29

We want to know whether eye grease increases sensitivity on the average.

(a) What are the null and alternative hypotheses? Say in words what mean μ your hypotheses concern.
(b) Suppose that the subjects are an SRS of all young people with normal vision, that contrast differences follow a Normal distribution in this population, and that the standard deviation of differences is $\sigma = 0.22$. Carry out a test of significance.

17.44 Tests from confidence intervals. A confidence interval for the population mean μ tells us which values of μ are plausible (those inside the interval) and which values are not plausible (those outside the interval) at the chosen level of confidence. You can use this idea to carry out a test of any null hypothesis H_0: $\mu = \mu_0$ starting with a confidence interval: *reject* H_0 *if* μ_0 *is outside the interval and fail to reject if* μ_0 *is inside the interval.*

The alternative hypothesis is always two-sided, H_a: $\mu \neq \mu_0$, because the confidence interval extends in both directions from $\bar{x}$. A 95% confidence interval leads to a test at the 5% significance level because the interval is wrong 5% of the time. In general, confidence level C leads to a test at significance level $\alpha = 1 - C$.

(a) In Example 17.7, a medical director found mean blood pressure $\bar{x} = 126.07$ for an SRS of 72 executives. The standard deviation of the blood pressures of all executives is $\sigma = 15$. Give a 90% confidence interval for the mean blood pressure μ of all executives.
(b) The hypothesized value $\mu_0 = 128$ falls *inside* this confidence interval. Carry out the z test for H_0: $\mu = 128$ against the two-sided alternative. Show that the test is *not statistically significant* at the 10% level.
(c) The hypothesized value $\mu_0 = 129$ falls *outside* this confidence interval. Carry out the z test for H_0: $\mu = 129$ against the two-sided alternative. Show that the test *is statistically significant* at the 10% level.

17.45 Tests from confidence intervals. A 95% confidence interval for a population mean is 30.7 ± 3.2. Use the method described in Exercise 17.44 to answer these questions.

(a) With a two-sided alternative, can you reject the null hypothesis that $\mu = 33$ at the 5% ($\alpha = 0.05$) significance level? Why?
(b) With a two-sided alternative, can you reject the null hypothesis that $\mu = 34$ at the 5% significance level? Why?

Exploring the Web

17.46 Significance in journals. Choose a major journal in your field of study. Use a web search engine to find its website—just search on the journal's name. Find a paper that uses a phrase like "significant ($P = 0.01$)" and summarize the findings in the paper.

17.47 A statistics glossary. An editorial was published in the *Journal of the National Cancer Institute*, Vol. 101, No. 23 (December 2, 2009) that announced some online resources for journalists, including a statistics glossary. The glossary can be found at `www.oxfordjournals.org/our_journals/jnc/resource/statistics%20glossary.pdf`. Read the definition of a P-value. Is this an accurate definition? Explain your answer.

CHAPTER 18

AP Photo/Bebeto Matthews

Inference in Practice

In this chapter we cover...

18.1 Conditions for inference in practice

18.2 Cautions about confidence intervals

18.3 Cautions about significance tests

18.4 Planning studies: sample size for confidence intervals

18.5 Planning studies: the power of a statistical test*

To this point, we have met just two procedures for statistical inference. Both concern inference about the mean μ of a population when the "simple conditions" (page 374) are true: the data are an SRS, the population has a Normal distribution, and we know the standard deviation σ of the population. Under these conditions, a confidence interval for the mean μ is

$$\bar{x} \pm z^* \frac{\sigma}{\sqrt{n}}$$

To test H_0: $\mu = \mu_0$, we use the one-sample z statistic:

$$z = \frac{\bar{x} - \mu_0}{\sigma/\sqrt{n}}$$

z procedures

We call these **z procedures** because they both start with the one-sample z statistic and use the standard Normal distribution.

In later chapters, we will modify these procedures for inference about a population mean to make them useful in practice. We will also introduce procedures for confidence intervals and tests in most of the settings we met in learning to explore data. There are libraries—both of books and of software—full of more elaborate statistical techniques. The reasoning of confidence intervals and tests is the same, no matter how elaborate the details of the procedure are.

There is a saying among statisticians that "mathematical theorems are true; statistical methods are effective when used with judgment." That the one-sample z statistic has the standard Normal distribution when the null hypothesis is true is a mathematical theorem. Effective use of statistical methods requires more than knowing such facts. It requires even more than understanding the underlying reasoning. This chapter begins the process of helping you develop the judgment needed to use statistics in practice. That process will continue in examples and exercises through the rest of this book.

18.1 Conditions for inference in practice

CAUTION *Any confidence interval or significance test can be trusted only under specific conditions.* It's up to you to understand these conditions and judge whether they fit your problem. With that in mind, let's look back at the "simple conditions" for the z procedures.

The final "simple condition," that we know the standard deviation σ of the population, is rarely satisfied in practice. The z procedures are therefore of little practical use. Fortunately, it's easy to remove the "known σ" condition. Chapter 20 shows how. The condition that the size of the population is large compared to the size of the sample is often easy to verify, and when it is not satisfied there are special, advanced methods for inference. The other "simple conditions" (SRS, Normal population) are harder to escape. In fact, they represent the kinds of conditions needed if we are to trust almost any statistical inference. As you plan inference, you should always ask, "Where did the data come from?" and you must often also ask, "What is the shape of the population distribution?" This is the point where knowing mathematical facts gives way to the need for judgment.

Statistics in Your World — DON'T TOUCH THE PLANTS

We know that confounding can distort inference, however, we don't always recognize how easy it is to confound data. Consider the innocent scientist who visits plants in the field once a week to measure their size. To measure the plants, she has to touch them. A study of six plant species found that one touch a week significantly increased leaf damage by insects in two species and significantly decreased damage in another species.

Where did the data come from? *The most important requirement for any inference procedure is that the data come from a process to which the laws of probability apply.* Inference is most reliable when the data come from a random sample or a randomized comparative experiment. Random samples use chance to choose respondents. Randomized comparative experiments use chance to assign subjects to treatments. The deliberate use of chance ensures that the laws of probability apply to the outcomes, and this in turn ensures that statistical inference makes sense.

Where the Data Come from Matters

When you use statistical inference, you are acting as if your data are a random sample or come from a randomized comparative experiment.

CAUTION *If your data don't come from a random sample or a randomized comparative experiment, your conclusions may be challenged.* To answer the challenge, you must usually rely on subject-matter knowledge, not on statistics. It is common to apply statistical inference to data that are not produced by random selection. When you see such a study, ask whether the data can be trusted as a basis for the conclusions of the study.

EXAMPLE 18.1 The Psychologist and the Sociologist

A psychologist is interested in how our visual perception can be fooled by optical illusions. Her subjects are students in Psychology 101 at her university. Most psychologists would agree that it's safe to treat the students as an SRS of all people with normal vision. There is nothing special about being a student that changes visual perception.

A sociologist at the same university uses students in Sociology 101 to examine attitudes toward poor people and antipoverty programs. Students as a group are younger than the adult population as a whole. Even among young people, students as a group come from more prosperous and better-educated homes. Even among students, this university isn't typical of all campuses. Even on this campus, students in a sociology course may have opinions that are quite different from those of engineering students. The sociologist can't reasonably act as if these students are a random sample from any interesting population. ■

Our first examples of inference, using the z procedures, act as if the data are an SRS from the population of interest. Let's look back at the examples in Chapters 16 and 17.

EXAMPLE 18.2 Is It Really an SRS?

The NHANES survey that produced the BMI data for Example 16.1 used a complex multistage sample design, so it's a bit oversimplified to treat the BMI data as coming from an SRS from the population of young women.[1] Although the overall effect of the NHANES sample is close to an SRS, professional statisticians would use more complex inference procedures to match the more complex design of the sample.

The 20 patrons in the tipping study in Example 16.3 were chosen from those eating at a particular restaurant to receive one of several treatments being compared in a randomized comparative experiment. Recall that each treatment group in a completely randomized experiment is an SRS of the available subjects. Researchers sometimes act as if the available subjects are an SRS from some population if there is nothing special about where the subjects came from. In some cases, researchers collect demographic data on subjects to help justify the assumption that the subjects are a representative sample from some population. We are willing to regard the subjects as an SRS from the population of patrons of this particular restaurant, but perhaps this needs to be explored further.

The cola taste test in Examples 17.2 and 17.3 uses scores from 10 tasters. All were examined to be sure that they have no medical condition that interferes with normal taste and then carefully trained to score sweetness using a set of standard drinks. We are willing to take their scores as an SRS from the population of trained tasters.

The medical director who examined executives' blood pressures in Example 17.7 actually chose an SRS from the medical records of all executives in this company. ■

These examples are typical. One is an actual SRS, two are situations in which common practice is to act as if the sample were an SRS, and in the remaining example procedures that assume an SRS are used for a quick analysis of data from a more complex random sample. *There is no simple rule for deciding when you can act as if a sample is an SRS. Pay attention to these cautions:*

CAUTION

- *Practical problems such as nonresponse in samples or dropouts from an experiment can hinder inference even from a well-designed study.* The NHANES survey has about an 80% response rate. This is much higher than opinion polls and most other national surveys, so by realistic standards NHANES data are quite trustworthy. (NHANES uses advanced methods to try to correct for nonresponse, but these methods work a lot better when response is high to start with.)

- *Different methods are needed for different designs.* The z procedures aren't correct for random sampling designs more complex than an SRS. Later chapters give methods for some other designs, but we won't discuss inference for really complex designs like that used by NHANES. Always be sure that you (or your statistical consultant) know how to carry out the inference your design calls for.
- *There is no cure for fundamental flaws like voluntary response surveys or uncontrolled experiments.* Look back at the bad examples in Chapters 8 and 9 and steel yourself to just ignore data from such studies.

What is the shape of the population distribution? Most statistical inference procedures require some conditions on the shape of the population distribution. Many of the most basic methods of inference are designed for Normal populations. That's the case for the z procedures and also for the more practical procedures for inference about means that we will meet in Chapters 20 and 21. Fortunately, this condition is less essential than where the data come from.

This is true because the z procedures and many other procedures designed for Normal distributions are based on Normality of the sample mean $\bar{x}$, not Normality of individual observations. The central limit theorem tells us that $\bar{x}$ is more Normal than the individual observations and that $\bar{x}$ becomes more Normal as the size of the sample increases. In practice, the z procedures are reasonably accurate for any roughly symmetric distribution for samples of even moderate size. If the sample is large, $\bar{x}$ will be close to Normal even if individual measurements are strongly skewed, as Figures 15.4 (pages 356–7) and 15.5 (page 358) illustrate. Later chapters give practical guidelines for specific inference procedures.

There is one important exception to the principle that the shape of the population is less critical than how the data were produced. Outliers can distort the results of inference. *Any inference procedure based on sample statistics like the sample mean $\bar{x}$ that are not resistant to outliers can be strongly influenced by a few extreme observations.*

We rarely know the shape of the population distribution. In practice, we rely on previous studies and on data analysis. Sometimes long experience suggests that our data are likely to come from a roughly Normal distribution, or not. For example, heights of people of the same sex and similar ages are close to Normal, but weights are not. Always explore your data before doing inference. When the data are chosen at random from a population, the shape of the data distribution mirrors the shape of the population distribution. Make a stemplot or histogram of your data and look to see whether the shape is roughly Normal. Remember that small samples have a lot of chance variation, so that Normality is hard to judge from just a few observations. Always look for outliers and try to correct them or justify their removal before performing the z procedures or other inference based on statistics like $\bar{x}$ that are not resistant.

When outliers are present or the data suggest that the population is strongly non-Normal, consider alternative methods that don't require Normality and are not sensitive to outliers. Some of these methods appear in Chapter 28 (available online).

Statistics in Your World — REALLY WRONG NUMBERS

By now you know that "statistics" that don't come from properly designed studies are often dubious and sometimes just made up. It's rare to find wrong numbers that anyone can see are wrong, but it does happen. A German physicist claimed that 2006 was the first year since 1441 with more than one Friday the 13th. Sorry: Friday the 13th occurred in February and August of 2004, which is a bit more recent than 1441.

LaunchPad Online Resources

- The **StatBoards video**, *Effect of Outliers on a Confidence Interval and Test*, provides examples of how outliers can affect inference procedures about a population mean.

Apply Your Knowledge

18.1 Rate the Lecture. A professor is interested in how the 500 students in his class will rate today's lecture. He selects the first 20 students on his class list, reads the names at the beginning of the lecture, and asks them to go online to the course website after class and rate the lecture on a scale of 0 to 5. Which of the following is the most important reason why a confidence interval for the mean rating by all his students based on these data is of little use? Comment briefly on each reason to explain your answer.

(a) The number of students selected is small, so the margin of error will be large.

(b) Many of the students selected may not be in attendance and/or will not respond.

(c) The students selected can't be considered a random sample from the population of all students in the course.

18.2 Running Red Lights. A survey of licensed drivers inquired about running red lights. One question asked, "Of every ten motorists who run a red light, about how many do you think will be caught?" The mean result for 880 respondents was $\bar{x} = 1.92$ and the standard deviation was $s = 1.83$.[2] For this large sample, s will be close to the population standard deviation σ, so suppose we know that $\sigma = 1.83$.

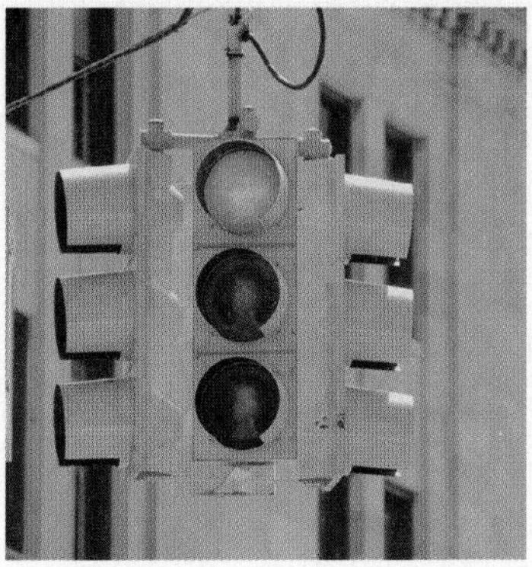

© Ilene MacDonald/Alamy

(a) Give a 95% confidence interval for the mean opinion in the population of all licensed drivers.

(b) The distribution of responses is skewed to the right rather than Normal. This will not strongly affect the z confidence interval for this sample. Why not?

(c) The 880 respondents are an SRS from completed calls among 45,956 calls to randomly chosen residential telephone numbers listed in telephone directories. Only 5029 of the calls were completed. This information gives two reasons to suspect that the sample may not represent all licensed drivers. What are these reasons?

18.3 Sampling Shoppers. A marketing consultant observes 50 consecutive shoppers at a department store the Friday after Thanksgiving, recording how much each shopper spends in the store. Suggest some reasons why it may be risky to act as if 50 consecutive shoppers at this particular time are an SRS of all shoppers at this store.

18.2 Cautions about confidence intervals

The most important caution about confidence intervals in general is a consequence of the use of a sampling distribution. A sampling distribution shows how a statistic such as $\bar{x}$ varies in repeated random sampling. This variation causes *random sampling error* because the statistic misses the true parameter by a random amount.

CAUTION

No other source of variation or bias in the sample data influences the sampling distribution. So *the margin of error in a confidence interval ignores everything except the sample-to-sample variation due to choosing the sample randomly.*

The Margin of Error Doesn't Cover All Errors

The margin of error in a confidence interval covers only random sampling errors. Practical difficulties such as undercoverage and nonresponse are often more serious than random sampling error. The margin of error does not take such difficulties into account.

Recall from Chapter 8 that national opinion polls often have response rates less than 50%, and that even small changes in the wording of questions can strongly influence results. In such cases, the announced margin of error is probably unrealistically small. And of course there is no way to assign a meaningful margin of error to results from voluntary response or convenience samples, because there is no random selection. Look carefully at the details of a study before you trust a confidence interval.

Apply Your Knowledge

18.4 What's Your Weight? A 2013 Gallup Poll asked a national random sample of 477 adult women to state their current weight. The mean weight in the sample was $\bar{x} = 157$. We will treat these data as an SRS from a Normally distributed population with standard deviation $\sigma = 35$.

(a) Give a 95% confidence interval for the mean weight of adult women based on these data.

(b) Do you trust the interval you computed in part (a) as a 95% confidence interval for the mean weight of all U.S. adult women? Why or why not?

18.5 Good Weather, Good Tips? Example 16.3 (page 381) described an experiment exploring the size of the tip in a particular restaurant when a message indicating that the next day's weather would be good was written on the bill. You work part-time as a server in a restaurant. You read a newspaper article about the study that reports that with 95% confidence the mean percentage tip from restaurant patrons will be between 21.33 and 23.09 when the server writes a message on the bill stating that the next day's weather will be good. Can you conclude that if you begin writing a message on patrons' bills that the next day's weather will be good, approximately 95% of the days you work your mean percentage tip will be between 21.33 and 23.09? Why or why not?

18.6 Sample Size and Margin of Error. Example 16.1 (page 375) described NHANES data on the body mass index (BMI) of 654 young women. The mean BMI in the sample was $\bar{x} = 26.8$. We treated these data as an SRS from a Normally distributed population with standard deviation $\sigma = 7.5$.

(a) Suppose that we had an SRS of just 100 young women. What would be the margin of error for 95% confidence?

(b) Find the margins of error for 95% confidence based on SRSs of 400 young women and 1600 young women.

(c) Compare the three margins of error. How does increasing the sample size change the margin of error of a confidence interval when the confidence level and population standard deviation remain the same?

18.7 Do You Eat Fast Food? A July 2013 Gallup poll asked, "How often, if ever, do you eat at fast-food restaurants, including drive-thru, take-out, and sitting down in the restaurant?" Of those sampled, 47% indicated they do so at least weekly. Gallup announced the poll's margin of error for 95% confidence as ±3 percentage points. Which of the following sources of error are included in this margin of error?

(a) Gallup dialed landline telephone numbers at random and so missed all people without landline phones, including people whose only phone is a cell phone.

(b) Some people whose numbers were chosen never answered the phone in several calls or answered but refused to participate in the poll.

(c) There is chance variation in the random selection of telephone numbers.

18.3 Cautions about significance tests

Significance tests are widely used in most areas of statistical work. New pharmaceutical products require significant evidence of effectiveness and safety. Courts inquire about statistical significance in hearing class action discrimination cases. Marketers want to know whether a new package design will significantly increase sales. Medical researchers want to know whether a new therapy performs significantly better. In all these uses, statistical significance is valued because it points to an effect that is unlikely to occur simply by chance. Here are some points to keep in mind when you use or interpret significance tests.

How small a *P* is convincing? The purpose of a test of significance is to describe the degree of evidence provided by the sample against the null hypothesis. The *P*-value does this. But how small a *P*-value is convincing evidence against the null hypothesis? This depends mainly on two circumstances:

- *How plausible is H_0?* If H_0 represents an assumption that the people you must convince have believed for years, strong evidence (small *P*) will be needed to persuade them.
- *What are the consequences of rejecting H_0?* If rejecting H_0 in favor of H_a means making an expensive changeover from one type of product packaging to another, you need strong evidence that the new packaging will boost sales.

These criteria are a bit subjective. Different people will often insist on different levels of significance. Giving the *P*-value allows each of us to decide individually if the evidence is sufficiently strong.

Users of statistics have often emphasized standard levels of significance such as 10%, 5%, and 1%. For example, courts have tended to accept 5% as a standard in discrimination cases.[3] This emphasis reflects the time when tables of critical values rather than software dominated statistical practice. The 5% level ($\alpha = 0.05$) is particularly common. *There is no sharp border between "significant" and "not significant," only increasingly strong evidence as the P-value decreases. There is no practical distinction between the P-values 0.049 and 0.051. It makes no sense to treat $P \leq 0.05$ as a universal rule for what is significant.*

Significance depends on the alternative hypothesis You may have noticed that the *P*-value for a one-sided test is one-half the *P*-value for the two-sided test of the same null hypothesis based on the same data. The two-sided *P*-value combines two equal areas, one in each tail of a Normal curve. The one-sided *P*-value is just one of these areas, in the direction specified by the alternative hypothesis. It makes sense that the evidence against H_0 is stronger when the alternative is one-sided, because it is based on the data *plus* information about the direction of possible deviations from H_0. If you lack this added information, always use a two-sided alternative hypothesis.

Significance depends on sample size A sample survey shows that significantly fewer students are heavy drinkers at colleges that ban alcohol on campus. "Significantly fewer" is not enough information to decide whether there is an *important* difference in drinking behavior at schools that ban alcohol. *How important an effect is depends on the size of the effect as well as on its statistical significance.* If the number of heavy drinkers is only 1% less at colleges that ban alcohol than at other colleges, this is probably not an important effect even if it is statistically significant. (Ask yourself if you would describe 1% as "significantly" fewer when talking with a friend). In fact, the sample survey found that 38% of students at colleges that ban alcohol are "heavy episodic drinkers" compared with 48% at other colleges.[4] That difference is large enough to be important. (Of course, this observational study doesn't prove that an alcohol ban directly reduces drinking; it may be that colleges that ban alcohol attract more students who don't want to drink heavily.)

Such examples remind us to always look at the size of an effect (like 38% versus 48%) as well as its significance. They also raise a question: can a tiny effect really be highly significant? Yes. The behavior of the z test statistic is typical. The statistic is

$$z = \frac{\bar{x} - \mu_0}{\sigma/\sqrt{n}}$$

The numerator measures how far the sample mean deviates from the hypothesized mean μ_0. Larger values of the numerator give stronger evidence against H_0: $\mu = \mu_0$. The denominator is the standard deviation of $\bar{x}$. It measures how much random variation we expect. There is less variation when the number of observations n is large. So z gets larger (more significant) when the estimated effect $\bar{x} - \mu_0$ gets larger *or* when the number of observations n gets larger. Significance depends both on the size of the effect we observe *and* on the size of the sample. Understanding this fact is essential to understanding significance tests.

SHOULD TESTS BE BANNED?

Significance tests don't tell us how large or how important an effect is. Research in psychology has emphasized tests, so much so that some think their weaknesses should ban them from use. The American Psychological Association asked a group of experts. They said: "Use anything that sheds light on your study. Use more data analysis and confidence intervals." But: "The task force does not support any action that could be interpreted as banning the use of null hypothesis significance testing or P-values in psychological research and publication."

Sample Size Affects Statistical Significance

Because large random samples have small chance variation, very small population effects can be highly significant if the sample is large.

Because small random samples have a lot of chance variation, even large population effects can fail to be significant if the sample is small.

Statistical significance does not tell us whether an effect is large enough to be important. That is, **statistical significance is not the same thing as practical significance**.

Keep in mind that "statistical significance" means "the sample showed an effect larger than would often occur just by chance." The extent of chance variation changes with the size of the sample, so the size of the sample does matter. Exercise 18.9 demonstrates in detail how increasing the sample size drives down the P-value. Here is another example.

EXAMPLE 18.3 It's Significant. Or Not. So What?

We are testing the hypothesis of no correlation between two variables. With 1000 observations, an observed correlation of only $r = 0.08$ is significant evidence at the 1% level that the correlation in the population is not zero but positive. *The small P-value does not mean that there is a strong association, only that there is strong evidence of some association.* The true population correlation is probably quite close to the observed sample value, $r = 0.08$. We might well conclude that for practical purposes we can ignore the association between these variables, even though we are confident (at the 1% level) that the correlation is positive.

CAUTION

On the other hand, if we have only 10 observations, a correlation of $r = 0.5$ is not significantly greater than zero even at the 5% level. Small samples vary so much that a large r is needed if we are to be confident that we aren't just seeing chance variation at work. So a small sample will often fall short of significance even if the true population correlation is quite large. ■

Beware of multiple analyses Statistical significance ought to mean that you have found an effect that you were looking for. The reasoning behind statistical significance works well if you decide what effect you are seeking, design a study to search for it, and use a test of significance to weigh the evidence you get. In other settings, significance may have little meaning.

EXAMPLE 18.4 Cell Phones and Brain Cancer

Might the radiation from cell phones be harmful to users? Many studies have found little or no connection between using cell phones and various illnesses. Here is part of a news account of one study:

> *A hospital study that compared brain cancer patients and a similar group without brain cancer found no statistically significant association between cell phone use and a group of brain cancers known as gliomas. But when 20 types of glioma were considered separately an association was found between phone use and one rare form. Puzzlingly, however, this risk appeared to decrease rather than increase with greater mobile phone use.*[5]

Think for a moment. Suppose that the 20 null hypotheses (no association) for these 20 significance tests are all true. Then each test has a 5% chance of being significant at the 5% level. That's what $\alpha = 0.05$ means: results this extreme occur 5% of the time just by chance when the null hypothesis is true. Because 5% is 1/20, we expect about 1 of 20 tests to give a significant result just by chance. That's what the study observed. ■

CAUTION *Running one test and reaching the 5% level of significance is reasonably good evidence that you have found something. Running 20 tests and reaching that level only once is not.* The caution about multiple analyses applies to confidence intervals as well. A single 95% confidence interval has probability 0.95 of capturing the true parameter each time you use it. The probability that all of 20 confidence intervals will capture their parameters is much less than 95%. If you think that multiple tests or intervals may have discovered an important effect, you need to gather new data to make inferences about that specific effect.

Apply Your Knowledge

18.8 Is It Significant? In the absence of special preparation, SAT Mathematics (SATM) scores in 2013 varied Normally with mean $\mu = 514$ and $\sigma = 118$. Fifty students go through a rigorous training program designed to raise their SATM scores by improving their mathematics skills. Either by hand or by using the *P-Value of a Test of Significance* applet, carry out a test of APPLET

$$H_0: \mu = 514$$
$$H_a: \mu > 514$$

(with $\sigma = 118$) in each of the following situations:

(a) The students' average score is $\bar{x} = 541$. Is this result significant at the 5% level?

(b) The average score is $\bar{x} = 542$. Is this result significant at the 5% level?

The difference between the two outcomes in parts (a) and (b) is of no practical importance. Beware attempts to treat $\alpha = 0.05$ as sacred.

18.9 Detecting Acid Rain. Emissions of sulfur dioxide by industry set off chemical changes in the atmosphere that result in "acid rain." The acidity of liquids is measured by pH on a scale of 0 to 14. Distilled water has pH 7.0, and lower pH values indicate acidity. Normal rain is somewhat acidic, so acid rain is sometimes defined as rainfall with a pH below 5.0. Suppose that pH measurements of rainfall on different days in a Canadian forest follow a Normal distribution with standard deviation $\sigma = 0.6$. A sample of n days finds that the mean pH is $\bar{x} = 4.8$. Is this good evidence that the mean pH μ for all rainy days is less than 5.0? The answer depends on the size of the sample.

APPLET

Either by hand or using the *P-Value of a Test of Significance* applet, carry out four tests of

$$H_0: \mu = 5.0$$
$$H_a: \mu < 5.0$$

Use $\sigma = 0.6$ and $\bar{x} = 4.8$ in all four tests. But use four different sample sizes: $n = 9$, $n = 16$, $n = 36$, and $n = 64$.

(a) What are the P-values for the four tests? *The P-value of the same result $\bar{x} = 4.8$ gets smaller (more significant) as the sample size increases.*

(b) For each test, sketch the Normal curve for the sampling distribution of $\bar{x}$ when H_0 is true. This curve has mean 5.0 and standard deviation $0.6/\sqrt{n}$. Mark the observed $\bar{x} = 4.8$ on each curve. (If you use the applet, you can just copy the curves displayed by the applet.) *The same result $\bar{x} = 4.8$ gets more extreme on the sampling distribution as the sample size increases.*

18.10 Confidence Intervals Help. Give a 95% confidence interval for the mean pH μ for each sample size in Exercise 18.9. The intervals, unlike the P-values, give a clear picture of what mean pH values are plausible for each sample.

18.11 Searching for ESP. A researcher looking for evidence of extrasensory perception (ESP) tests 1000 subjects. Nine of these subjects do significantly better ($P < 0.01$) than random guessing.

(a) Nine seems like a lot of people, but you can't conclude that these nine people have ESP. Why not?

(b) What should the researcher now do to test whether any of these nine subjects have ESP?

18.4 Planning studies: sample size for confidence intervals

A wise user of statistics never plans a sample or an experiment without at the same time planning the inference. The number of observations is a critical part of planning a study. Larger samples give smaller margins of error in confidence intervals and make significance tests better able to detect effects in the population. But taking observations costs both time and money. How many observations are enough? We will look at this question first for confidence intervals and then for tests. Planning a confidence interval is much simpler than planning a test. It is also more useful, because estimation is generally more informative than testing. The section on planning tests is therefore optional.

You can arrange to have both high confidence and a small margin of error by taking enough observations. The margin of error of the z confidence interval for the mean of a Normally distributed population is $m = z^*\sigma/\sqrt{n}$. To obtain a desired margin of error m, put in the value of z^* for your desired confidence level, and solve for the sample size n. Here is the result.

Sample Size for Desired Margin of Error

The z confidence interval for the mean of a Normal population will have a specified margin of error m when the sample size is

$$n = \left(\frac{z^*\sigma}{m}\right)^2$$

CAUTION *Notice that it is the size of the sample that determines the margin of error. The size of the population does not influence the sample size we need.* (This is true as long as the population is much larger than the sample.)

EXAMPLE 18.5 How Many Observations?

In Example 16.3 (page 381), psychologists recorded the size of the tip of 20 patrons in a restaurant when a message indicating that the next day's weather would be good was written on their bill. We know that the population standard deviation is $\sigma = 2$. We want to estimate the mean percentage tip μ for patrons of this restaurant who receive this message on their bill within ± 0.5 with 90% confidence. How many patrons must we observe?

The desired margin of error is $m = 0.5$. For 90% confidence, Table C gives $z^* = 1.645$. Therefore,

$$n = \left(\frac{z^*\sigma}{m}\right)^2 = \left(\frac{1.645 \times 2}{0.5}\right)^2 = 43.3$$

Because 43 patrons will give a slightly larger margin of error than desired, and 44 patrons a slightly smaller margin of error, we must observe 44 patrons. *Always round up to the next higher whole number when finding n.* ■

LaunchPad Online Resources

- The **StatBoards video**, *Calculating Sample Size for Estimating a Mean*, provides examples of determining the sample size to achieve a desired confidence level.

Apply Your Knowledge

18.12 Body Mass Index of Young Women. Example 16.1 (page 375) assumed that the body mass index (BMI) of all American young women follows a Normal distribution with standard deviation $\sigma = 7.5$. How large a sample would be needed to estimate the mean BMI μ in this population to within ± 1 with 95% confidence?

18.13 Number Skills of Eighth Graders. Suppose that scores on the mathematics part of the National Assessment of Educational Progress (NAEP) test for eighth-grade students follow a Normal distribution with standard deviation $\sigma = 125$. You want to estimate the mean score within ± 10 with 90% confidence. How large an SRS of scores must you choose?

18.5 Planning studies: the power of a statistical test*

How large a sample should we take when we plan to carry out a test of significance? We know that if our sample is too small, even large effects in the population will often fail to give statistically significant results. Here are the questions we must answer to decide how many observations we need:

- **Significance level.** How much protection do we want against getting a significant result from our sample when there really is no effect in the population?
- **Effect size.** How large an effect in the population is important in practice?
- **Power.** How confident do we want to be that our study will detect an effect of the size we think is important?

The three boldface terms are statistical shorthand for three pieces of information. *Power* is a new idea.

EXAMPLE 18.6 Sweetening Colas: Planning a Study

Let's illustrate typical answers to these questions in the example of testing a new cola for loss of sweetness in storage (Example 17.2, page 392). Ten trained tasters rated the sweetness on a 10-point scale before and after storage, so that we have each taster's judgment of loss of sweetness. From experience, we know that sweetness loss scores vary from taster to taster according to a Normal distribution with standard deviation about $\sigma = 1$. To see if the taste test gives reason to think that the cola does lose sweetness, we will test

$$H_0: \mu = 0$$
$$H_a: \mu > 0$$

Are 10 tasters enough, or should we use more?

Significance level. Requiring significance at the 5% level is enough protection against declaring there is a loss in sweetness when in fact there is no change if we could look at the entire population. This means that when there is no change in sweetness in the population, one out of 20 samples of tasters will wrongly find a significant loss.

Effect size. A mean sweetness loss of 0.8 point on the 10-point scale will be noticed by consumers and so is important in practice.

Power. We want to be 90% confident that our test will detect a mean loss of 0.8 point in the population of all tasters. We agreed to use significance at the 5% level as our standard for detecting an effect. So we want probability at least 0.9 that a test at the $\alpha = 0.05$ level will reject the null hypothesis $H_0: \mu = 0$ when the true population mean is $\mu = 0.8$. ■

The probability that the test successfully detects a sweetness loss of the specified size is the *power* of the test. You can think of tests with high power as being highly sensitive to deviations from the null hypothesis. In Example 18.6, we decided that we want power 90% when the truth about the population is that $\mu = 0.8$.

*Power calculations are important in planning studies, but this more advanced material is not needed to understand the rest of the book.

Power

The **power** of a test against a specific alternative is the probability that the test will reject H_0 at a chosen significance level α when the specified alternative value of the parameter is true.

For most statistical tests, calculating power is a job for comprehensive statistical software. The z test is easier, but we will nonetheless skip the details. The two following examples illustrate two approaches: an applet that shows the meaning of power, and statistical software.

EXAMPLE 18.7 Finding Power: Use an Applet

Finding the power of the z test is less challenging than most other power calculations because it requires only a Normal distribution probability calculation. The *Statistical Power* applet does this and illustrates the calculation with Normal curves. Enter the information from Example 18.6 into the applet: hypotheses, significance level $\alpha = 0.05$, alternative value $\mu = 0.8$, standard deviation $\sigma = 1$, and sample size $n = 10$. Click "Update." The applet output appears in Figure 18.1.

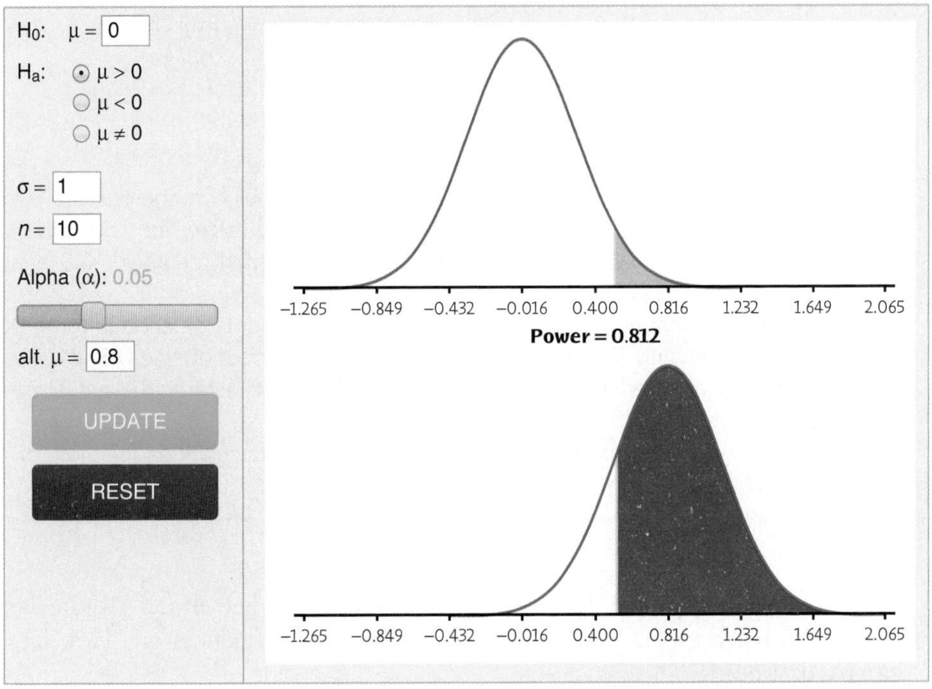

FIGURE 18.1 Output from the *Statistical Power* applet for Example 18.7, along with the information entered into the applet. The top curve shows the behavior of $\bar{x}$ when the null hypothesis is true ($\mu = 0$). The bottom curve shows the distribution of $\bar{x}$ when $\mu = 0.8$.

The power of the test against the specific alternative $\mu = 0.8$ is 0.812. That is, the test will reject H_0 about 81% of the time when this alternative is true. So 10 observations are too few to give power 90%. ■

The two Normal curves in Figure 18.1 show the sampling distribution of $\bar{x}$ under the null hypothesis $\mu = 0$ (top) and also under the specific alternative $\mu = 0.8$ (bottom). The curves have the same shape because σ does not change. The top curve is centered at $\mu = 0$ and the bottom curve at $\mu = 0.8$. The shaded region at the right of the top curve has area 0.05. It marks off values of $\bar{x}$ that are statistically significant at the $\alpha = 0.05$ level. The lower curve shows the probability of these same values when $\mu = 0.8$. This area is the power, 0.812.

The applet will find the power for any given sample size. It's more helpful in practice to turn the process around and learn what sample size we need to achieve a given power. Statistical software will do this but usually doesn't show the helpful Normal curves that are part of the applet's output.

EXAMPLE 18.8 Finding Power: Use Software

Some software packages (for example, SAS, JMP, Minitab, and R) will calculate power. We asked Minitab to find the number of observations needed for the one-sided *z* test to have power 0.9 against several specific alternatives at the 5% significance level when the population standard deviation is $\sigma = 1$. Here is the table that results:

Difference	Sample Size	Target Power	Actual Power
0.1	857	0.9	0.900184
0.2	215	0.9	0.901079
0.3	96	0.9	0.902259
0.4	54	0.9	0.902259
0.5	35	0.9	0.905440
0.6	24	0.9	0.902259
0.7	18	0.9	0.907414
0.8	14	0.9	0.911247
0.9	11	0.9	0.909895
1.0	9	0.9	0.912315

In this output, "Difference" is the difference between the null hypothesis value $\mu = 0$ and the alternative we want to detect. This is the effect size. The "Sample Size" column shows the smallest number of observations needed for power 0.9 against each effect size.

We see again that our earlier sample of 10 tasters is not large enough to be 90% confident of detecting (at the 5% significance level) an effect of size 0.8. If we want power 90% against effect size 0.8, we need at least 14 tasters. The actual power with 14 tasters is 0.911247.

Statistical software, unlike the applet, will do power calculations for most of the tests in this book. ■

Statistics in Your World: FISH, FISHERMEN, AND POWER

Are the stocks of cod in the ocean off eastern Canada declining? Studies over many years failed to find significant evidence of a decline. These studies had low power—that is, they might fail to find a decline even if one were present. When it became clear that the cod were vanishing, quotas on fishing ravaged the economy in parts of Canada. If the earlier studies had had high power, they would likely have detected the decline. Quick action might have reduced the economic and environmental costs.

The table in Example 18.8 makes it clear that smaller effects require larger samples to reach power 90%. Here is an overview of influences on "How large a sample do I need?"

- If you insist on a smaller significance level (such as 1% rather than 5%), you will need a larger sample. A smaller significance level requires stronger evidence to reject the null hypothesis.
- If you insist on higher power (such as 99% rather than 90%), you will need a larger sample. Higher power gives a better chance of detecting an effect when it is really there.
- At any significance level and desired power, a two-sided alternative requires a larger sample than a one-sided alternative.
- At any significance level and desired power, detecting a small effect requires a larger sample than detecting a large effect.

Planning a serious statistical study always requires an answer to the question, "How large a sample do I need?" If you intend to test the hypothesis H_0: $\mu = \mu_0$ about the mean μ of a population, you need at least a rough idea of the size of the population standard deviation σ and of how big a deviation $\mu - \mu_0$ of the population mean from its hypothesized value you want to be able to detect. More elaborate settings, such as comparing the mean effects of several treatments, require more elaborate advance information. You can leave the details to experts, but you should understand the idea of power and the factors that influence how large a sample you need.

To calculate the power of a test, we act as if we are interested in a fixed level of significance such as $\alpha = 0.05$. That's essential to do a power calculation, but remember that in practice we think in terms of *P*-values rather than a fixed level α. To effectively plan a statistical test we must find the power for several significance levels and for a range of sample sizes and effect sizes to get a full picture of how the test will behave.

Type I and Type II errors in significance tests We can assess the performance of a test by giving two probabilities: the significance level α and the power for an alternative that we want to be able to detect. The significance level of a test is the probability of reaching the *wrong* conclusion when the null hypothesis is true. The power for a specific alternative is the probability of reaching the *right* conclusion when that alternative is true. We can just as well describe the test by giving the probabilities of being *wrong* under both conditions.

Type I and Type II Errors

If we reject H_0 when in fact H_0 is true, this is a **Type I error.**

If we fail to reject H_0 when in fact H_a is true, this is a **Type II error.**

The **significance level α** of any fixed level test is the probability of a Type I error.

The **power** of a test against any alternative is the probability of correctly rejecting the null hypothesis for that alternative. It can be calculated as 1 minus the probability of a Type II error for that alternative.

The possibilities are summed up in Figure 18.2. If H_0 is true, our conclusion is correct if we fail to reject H_0 and is a Type I error if we reject H_0. If H_a is true, our conclusion is either correct or a Type II error. Only one error is possible at one time.

		Truth about the population	
		H_0 true	H_a true
Conclusion based on sample	Reject H_0	Type I error	Correct conclusion
	Fail to reject H_0	Correct conclusion	Type II error

FIGURE 18.2
The two types of error in testing hypotheses.

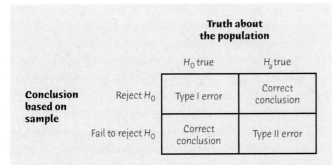

EXAMPLE 18.9 Calculating Error Probabilities

APPLET

Because the probabilities of the two types of error are just a rewording of significance level and power, we can see from Figure 18.1 what the error probabilities are for the test in Example 18.6.

$$P(\text{Type I error}) = P(\text{reject } H_0 \text{ when in fact } \mu = 0)$$
$$= \text{significance level } \alpha = 0.05$$
$$P(\text{Type II error}) = P(\text{fail to reject } H_0 \text{ when in fact } \mu = 0.8)$$
$$= 1 - \text{power} = 1 - 0.812 = 0.188$$

The two Normal curves in Figure 18.1 are used to find the probabilities of a Type I error (top curve, $\mu = 0$) and of a Type II error (bottom curve, $\mu = 0.8$). ■

LaunchPad Online Resources

- The **StatBoards video**, *Power, Type I, and Type II Errors*, provides examples of calculating power, type I, and type II errors.

Apply Your Knowledge

18.14 What Is Power? The Trial Urban District Assessment (TUDA) measures educational progress within participating large urban districts. TUDA gives a reading test scored from 0 to 500. A score of 243 is a "basic" reading level for eighth-graders.[6] Suppose scores on the TUDA reading test for eighth-graders in your district follow a Normal distribution with standard deviation $\sigma = 55$. In 2013 the mean score for eighth-graders in your district was 255. You plan to give the reading test to a random sample of 25 eighth-graders in your district this year to test whether the mean score μ for all eighth-graders in your district is still above the basic level. You will therefore test

$$H_0: \mu = 243$$
$$H_a: \mu > 243$$

If the true mean score is again 255, on average students are performing above the basic level. You learn that the power of your test at the 5% significance level against the alternative $\mu = 255$ is 0.29.

(a) Explain in simple language what "power = 0.29" means.

(b) Explain why the test you plan will not adequately protect you against deciding that average reading scores in your district are not above basic level.

18.15 Thinking about Power. Answer these questions in the setting of Exercise 18.14 about measuring eighth-grade reading scores on TUDA.

(a) You could get higher power against the same alternative with the same α by changing the number of measurements you make. Should you make more measurements or fewer to increase power?

(b) If you decide to use $\alpha = 0.10$ in place of $\alpha = 0.05$, with no other changes in the test, will the power increase or decrease?

(c) If you shift your interest to the alternative $\mu = 260$ with no other changes, will the power increase or decrease?

18.16 How Power Behaves. In the setting of Exercise 18.14, use the *Statistical Power* applet to find the power in each of the following circumstances.

APPLET

(a) Standard deviation $\sigma = 55$, significance level $\alpha = 0.05$, alternative $\mu = 255$, and sample sizes $n = 25$, $n = 50$, and $n = 100$. How does increasing the sample size with no other changes affect the power?

(b) Standard deviation $\sigma = 55$, significance level $\alpha = 0.05$, sample size $n = 25$, and alternatives $\mu = 255$, $\mu = 260$, and $\mu = 265$. How do alternatives more distant from the hypothesis (larger effect sizes) affect the power?

(c) Standard deviation $\sigma = 55$, sample size $n = 25$, alternative $\mu = 255$, and significance levels $\alpha = 0.05$, $\alpha = 0.10$, and $\alpha = 0.20$. How does increasing the desired significance level affect the power?

18.17 How Power Behaves. Use the *Statistical Power* applet to find the power of the test in Exercise 18.14 in each of these circumstances: significance level $\alpha = 0.05$, alternative $\mu = 255$, sample size $n = 25$, and $\sigma = 55$, $\sigma = 40$, and $\sigma = 30$. How does decreasing the variability of the population of measurements affect the power?

APPLET

18.18 Two Types of Error. Your company markets a computerized medical diagnostic program used to evaluate thousands of people. The program scans the results of routine medical tests (pulse rate, blood tests, etc.) and refers the case to a doctor if there is evidence of a medical problem. The program makes a decision about each person.

(a) What are the two hypotheses and the two types of error that the program can make? Describe the two types of error in terms of "false-positive" and "false-negative" test results.

(b) The program can be adjusted to decrease one error probability, at the cost of an increase in the other error probability. Which error probability would you choose to make smaller, and why? (This is a matter of judgment. There is no single correct answer.)

CHAPTER 18 SUMMARY

Chapter Specifics

- A specific confidence interval or test is correct only under specific conditions. The most important conditions concern the method used to produce the data. Other factors such as the shape of the population distribution may also be important.
- Whenever you use statistical inference, you are acting as if your data are a random sample or come from a randomized comparative experiment.
- Always do data analysis before inference to detect outliers or other problems that would make inference untrustworthy.
- The margin of error in a confidence interval accounts for only the chance variation due to random sampling. In practice, errors due to nonresponse or undercoverage are often more serious.
- There is no universal rule for how small a P-value in a test of significance is convincing evidence against the null hypothesis. Beware of placing too much weight on traditional significance levels such as $\alpha = 0.05$.

- Very small effects can be highly significant (small P) when a test is based on a large sample. A statistically significant effect need not be practically important. Plot the data to display the effect you are seeking, and use confidence intervals to estimate the actual values of parameters.
- On the other hand, lack of significance does not imply that H_0 is true. Even a large effect can fail to be significant when a test is based on a small sample.
- Many tests run at once will probably produce some significant results by chance alone, even if all the null hypotheses are true.
- When you plan a statistical study, plan the inference as well. In particular, ask what sample size you need for successful inference.
- The z confidence interval for a Normal mean has specified margin of error m when the sample size is

$$n = \left(\frac{z^*\sigma}{m}\right)^2$$

- Here z^* is the critical value for the desired level of confidence. Always round n up when you use this formula.
- (Optional material) The **power** of a significance test measures its ability to detect the truth of an alternative hypothesis. The power against a specific alternative is the probability that the test will reject H_0 at a particular level α when that alternative is true.
- (Optional material) Increasing the size of the sample increases the power of a significance test. You can use statistical software to find the sample size needed to achieve a desired power.

Link It

In Chapters 16 and 17, we introduced the basic reasoning behind statistical estimation and tests of significance. We applied this reasoning to the problem of making inferences about a mean of a Normally distributed population in a simple and artificial setting (the population standard deviation is known). In this chapter, we begin to move away from our simple and artificial conditions toward the reality of statistical practice. Later chapters deal with inference in fully realistic settings. Here we discuss more carefully conditions under which our procedures for inference do and do not hold. Where the data come from is crucial. Even if the population distribution is not Normal, our procedures are approximately correct if we have a moderate sample size and the shape of the population distribution is roughly symmetric. If the sample size is large, our procedures are approximately correct even if the population distribution is strongly skewed. However, any procedure based on sample statistics like the sample mean that are not resistant to outliers can be influenced by a few extreme observations. This is a caution that we first encountered in Chapter 4.

In this chapter we also provide some cautions about confidence intervals and tests of significance. Some of these echo statements made in Chapters 8 and 9. Some are based on the behavior of confidence intervals and significance tests. These cautions will help you evaluate studies that report confidence intervals or the results of a test of significance.

We conclude this chapter with material about planning a study that will use a confidence interval or a significance test. The fundamental question is, "What sample size do I need for successful inference?"

LaunchPad Online Resources

If you are having difficulty with any of the sections of this chapter, these online resources should help prepare you to solve the exercises at the end of this chapter.

- **StatTutor** starts with a video review of each section and asks a series of questions to check your understanding. You may find this helpful if you need additional help understanding any of the material in this chapter.
- **LearningCurve** provides you with a series of questions about the chapter geared to your level of understanding.

CHECK YOUR SKILLS

18.19 The most important condition for sound conclusions from statistical inference is usually that

(a) the data can be thought of as a random sample from the population of interest.
(b) the population distribution is exactly Normal.
(c) the data contain no outliers.

18.20 The coach of a college men's soccer team records the resting heart rates of the 27 team members. You should not trust a confidence interval for the mean resting heart rate of all male students at this college based on these data because

(a) with only 27 observations, the margin of error will be large.
(b) heart rates may not have a Normal distribution.
(c) the members of the soccer team can't be considered a random sample of all students.

18.21 You turn your web browser to the online Harris Interactive Poll. Based on 2234 responses, the poll reports that 45% of U.S. adults believe global climate change exists and humans are the main cause, 30% believe global climate change exists but that its causes are mainly not related to humans, 13% do not believe global climate change exists, and 12% are undecided.[7] You should refuse to calculate a 95% confidence interval for the proportion of all U.S. adults who do not believe global climate change exists based on this sample because

(a) this percent is too small.
(b) inference from a voluntary response sample can't be trusted.
(c) the sample is too large.

18.22 Many sample surveys use well-designed random samples but half or more of the original sample can't be contacted or refuse to take part. Any errors due to this nonresponse

(a) have no effect on the accuracy of confidence intervals.
(b) are included in the announced margin of error.
(c) are in addition to the random variation accounted for by the announced margin of error.

18.23 A writer in a medical journal says: "An uncontrolled experiment in 37 women found a significantly improved mean clinical symptom score after treatment. Methodologic flaws make it difficult to interpret the results of this study." The writer is skeptical about the significant improvement because

(a) there is no control group, so the improvement might be due to the placebo effect or to the fact that many medical conditions improve over time.
(b) the *P*-value given was $P = 0.048$, which is too large to be convincing.
(c) the response variable might not have an exactly Normal distribution in the population.

18.24 Vigorous exercise is associated with several years of longer life (on the average). Whether mild activities like slow walking are associated with longer life is not clear. Suppose that the added life expectancy associated with slow walking daily for 10 minutes is just one month. A statistical test is more likely to find a significant increase in mean life for those who slow walk daily if

(a) it is based on a very large random sample.
(b) it is based on a very small random sample.
(c) The size of the sample has little effect on significance for such a small increase in life expectancy.

18.25 A medical experiment compared zinc supplements with a placebo for reducing the duration of colds. Let μ denote the mean decrease, in days, in the duration of a cold. A decrease to $\mu = 1$ is a practically important decrease. The significance level of a test of H_0: $\mu = 0$ versus H_a: $\mu > 0$ is defined as

(a) the probability that the test rejects H_0 when $\mu = 0$ is true.
(b) the probability that the test rejects H_0 when $\mu = 1$ is true.
(c) the probability that the test fails to reject H_0 when $\mu = 1$ is true.

18.26 **(Optional topic)** The power of the test in Exercise 18.25 against the specific alternative $\mu = 1$ is defined as

(a) the probability that the test rejects H_0 when $\mu = 0$ is true.
(b) the probability that the test rejects H_0 when $\mu = 1$ is true.
(c) the probability that the test fails to reject H_0 when $\mu = 1$ is true.

18.27 **(Optional topic)** The power of a test is important in practice because power

(a) describes how well the test performs when the null hypothesis is actually true.
(b) describes how sensitive the test is to violations of conditions such as Normal population distribution.
(c) describes how well the test performs when the null hypothesis is actually not true.

CHAPTER 18 EXERCISES

18.28 Hotel managers. In Exercise 17.32 (page 412), you carried out a test of significance based on data from 148 general managers of three-star and four-star hotels. Before you trust your results, you would like more information about the data. What facts would you most like to know?

18.29 Color blindness in Africa. An anthropologist claims that color blindness is less common in societies that live by hunting and gathering than in settled agricultural societies. He tests a number of adults in two populations in Africa, one of each type. The proportion of color-blind people is significantly lower ($P < 0.05$) in the hunter–gatherer population. What additional information would you want to help you decide whether you believe the anthropologist's claim?

18.30 Sampling at the mall. A market researcher chooses at random from women entering a large outlet mall. One outcome of the study is a 95% confidence interval for the mean of "the highest price you would pay for a pair of shoes."

(a) Explain why this confidence interval does not give useful information about the population of all women.

(b) Explain why it may give useful information about the population of women who shop at large outlet malls.

18.31 Sensitive questions. The 2013 Youth Risk Behavior Survey found that 326 individuals in its random sample of 1216 Ohio high school students said that they had had multiple sexual partners during their life. That's 27% of the sample. Why is this estimate likely to be biased? Do you think it is biased high or low? Does the margin of error of a 95% confidence interval for the proportion of all Ohio high school students with multiple partners during their life allow for this bias?

18.32 College degrees. At the Statistics Canada website, www.statcan.gc.ca, you can find the percent of adults in each province or territory who have at least a university certificate, diploma, or degree at bachelor's level or above. It makes no sense to find $\bar{x}$ for these data and use it to get a confidence interval for the mean percent μ in all 13 provinces or territories. Why not?

18.33 An outlier strikes. You have data on an SRS of recent graduates from your college that shows how long each student took to complete a bachelor's degree. The data contain one high outlier. Will this outlier have a greater effect on a confidence interval for mean completion time if your sample is small or if it is large? Why?

18.34 Can we trust this interval? Here are data on the percent change in the total mass (in tons) of wildlife in several West African game preserves in the years 1971 to 1999:[8] WILDMSS

1971	1972	1973	1974	1975	1976	1977	1978	1979	1980
2.9	3.1	−1.2	−1.1	−3.3	−3.7	1.9	−0.3	−5.9	−7.9
1981	**1982**	**1983**	**1984**	**1985**	**1986**	**1987**	**1988**	**1989**	**1990**
−5.5	−7.2	−4.1	−8.6	−5.5	−0.7	−5.1	−7.1	−4.2	0.9
1991	**1992**	**1993**	**1994**	**1995**	**1996**	**1997**	**1998**	**1999**	
−6.1	−4.1	−4.8	−11.3	−9.3	−10.7	−1.8	−7.4	−22.9	

Software gives the 95% confidence interval for the mean annual percent change as −6.66% to −2.55%. There are several reasons why we might not trust this interval.

(a) Examine the distribution of the data. What feature of the distribution throws doubt on the validity of statistical inference?

(b) Plot the percents against year. What trend do you see in this time series? Explain why a trend over time casts doubt on the condition that years 1971 to 1999 can be treated as an SRS from a larger population of years.

18.35 When to use pacemakers. A medical panel prepared guidelines for when cardiac pacemakers should be implanted in patients with heart problems. The panel reviewed a large number of medical studies to judge the strength of the evidence supporting each recommendation. For each recommendation, they ranked the evidence as level A (strongest), B, or C (weakest). Here, in scrambled order, are the panel's descriptions of the three levels of evidence.[9] Which is A, which B, and which C? Explain your ranking.

Evidence was ranked as level — when data were derived from a limited number of trials involving comparatively small numbers of patients or from well-designed data analysis of nonrandomized studies or observational data registries.

Evidence was ranked as level — if the data were derived from multiple randomized clinical trials involving a large number of individuals.

Evidence was ranked as level — when consensus of expert opinion was the primary source of recommendation.

18.36 What is significance good for? Which of the following questions does a test of significance answer? Briefly explain your replies.

(a) Is the sample or experiment properly designed?

(b) Is the observed effect due to chance?

(c) Is the observed effect important?

18.37 **Why are larger samples better?** Statisticians prefer large samples. Describe briefly the effect of increasing the size of a sample (or the number of subjects in an experiment) on each of the following:

(a) The P-value of a test, when H_0 is false and all facts about the population remain unchanged as n increases.

(b) (Optional topic) The power of a fixed level α test, when α, the alternative hypothesis, and all facts about the population remain unchanged.

18.38 **Divorce rates.** Divorce rates vary from city to city in the United States. We have lots of data on many U.S. cities. Statistical software makes it easy to perform dozens of significance tests on dozens of variables to see which ones best predict divorce rate. One interesting finding is that those cities with major league ballparks tend to have significantly lower divorce rates than other cities. To improve your chances of a successful marriage, should you use this "significant" variable to decide where to live? Explain your answer.

18.39 **A test goes wrong.** Software can generate samples from (almost) exactly Normal distributions. Here is a random sample of size 5 from the Normal distribution with mean 10 and standard deviation 2:

DATA RANDOM

6.47   7.51   10.10   13.63   9.91

These data match the conditions for a z test better than real data will: the population is very close to Normal and has known standard deviation $\sigma = 2$, and the population mean is $\mu = 10$. Although we know the true value of μ, suppose we pretend that we do not and we test the hypotheses

$$H_0: \mu = 8$$
$$H_a: \mu \neq 8$$

(a) What are the z statistic and its P-value? Is the test significant at the 5% level?

(b) We know that the null hypothesis does not hold, but the test failed to give strong evidence against H_0. Explain why this is not surprising.

18.40 **Reducing the gender gap.** In many science disciplines, women are outperformed by men on test scores. Will "values affirmation training" improve self-confidence and hence performance of women relative to men in science courses? A study conducted at a large university compares the scores of men and women at the end of a large introductory physics course on a nationally normed standardized test of conceptual physics, the Force and Motion Conceptual Evaluation (FMCE). Half the women in the course were given values affirmation training during the course; the other half received no training. The study reports that there was a significant difference ($P < 0.01$) in the gap between men's and women's scores, although the gap for women who received the values affirmation training was much smaller than that for women who did not receive training. The study also reports that a 95% confidence interval for the mean difference in scores on the FMCE exam between women who received the training and those who didn't is 13 ± 8 points. You are a faculty member in the physics department, and the provost, who is interested in women in science, asks you about the study.

(a) Explain in simple language what "a significant difference ($P < 0.01$)" means.

(b) Explain clearly and briefly what "95% confidence" means.

(c) Is this study good evidence that requiring values affirmation training of all female students would greatly reduce the gender gap in scores on science tests in college courses?

18.41 **How far do rich parents take us?** How much education children get is strongly associated with the wealth and social status of their parents. In social science jargon, this is "socioeconomic status," or SES. But the SES of parents has little influence on whether children who have graduated from college go on to yet more education. One study looked at whether college graduates took the graduate admissions tests for business, law, and other graduate programs. The effects of the parents' SES on taking the LSAT test for law school were "both statistically insignificant and small."

(a) What does "statistically insignificant" mean?

(b) Why is it important that the effects were small in size as well as insignificant?

18.42 **This wine stinks.** How sensitive are the untrained noses of students? Exercise 16.27 (page 389) gives the lowest levels of dimethyl sulfide (DMS) that 10 students could detect. You want to estimate the mean DMS odor threshold among all students and you would be satisfied to estimate the mean to within ± 0.1 with 99% confidence. The standard deviation of the odor threshold for untrained noses is known to be $\sigma = 7$ micrograms per liter of wine. How large an SRS of untrained students do you need?

18.43 **Pulling wood apart.** You want to estimate the mean load needed to pull apart the pieces of wood in Exercise 16.25 (page 388) to within ± 600 pounds with 95% confidence. How large a sample is needed?

The following exercises concern the optional material on the power of a test.

18.44 The first child has higher IQ. Does the birth order of a family's children influence their IQ scores? A careful study of 241,310 Norwegian 18- and 19-year-olds found that firstborn children scored 2.3 points higher on the average than second children in the same family. This difference was highly significant ($P < 0.001$). A commentator said, "One puzzle highlighted by these latest findings is why certain other within-family studies have failed to show equally consistent results. Some of these previous null findings, which have all been obtained in much smaller samples, may be explained by inadequate statistical power."[10] Explain in simple language why tests having low power often fail to give evidence against a null hypothesis even when the hypothesis is really false.

18.45 How valium works. Valium is a common antidepressant and sedative. A study investigated how valium works by comparing its effect on sleep in seven genetically modified mice and eight normal control mice. There was no significant difference between the two groups. The authors say that this lack of significance "is related to the large inter-individual variability that is also reflected in the low power (20%) of the test."[11]

(a) Explain exactly what power 20% against a specific alternative means.
(b) Explain in simple language why tests having low power often fail to give evidence against a null hypothesis even when the null hypothesis is really false.
(c) What fact about this experiment most likely explains the low power?

18.46 Dialysis. An article in the *New England Journal of Medicine* describes a randomized controlled trial that compared early versus late initiation of dialysis on the survival of adults with progressive chronic kidney disease. The experiment found no significant difference between early and late initiation of dialysis. According to the article, the study was designed to have power 80%, with a two-sided Type I error of 0.05, to detect a clinically important difference of approximately 10 percentage points in the absolute risk of death.[12]

(a) What fixed significance level was used in calculating the power?
(b) Explain to someone who knows no statistics why power 80% means that the experiment would probably have been significant if there was a difference between early and late initiation of dialysis.

18.47 APPLET **Power.** In Exercise 18.39, a sample from a Normal population with mean $\mu = 10$ and standard deviation $\sigma = 2$ failed to reject the null hypothesis H_0: $\mu = 8$ at the $\alpha = 0.05$ significance level. Enter the information from this example into the *Statistical Power* applet. (Don't forget that the alternative hypothesis is two-sided.) What is the power of the test against the alternative $\mu = 10$? Because the power is not high, it isn't surprising that the sample in Exercise 18.39 failed to reject H_0.

18.48 Finding power by hand. Even though software is used in practice to calculate power, doing the work by hand builds your understanding. Return to the test in Example 18.6. There are $n = 10$ observations from a population with standard deviation $\sigma = 1$ and unknown mean μ. We will test

$$H_0: \mu = 0$$
$$H_a: \mu > 0$$

with fixed significance level $\alpha = 0.05$. Find the power against the alternative $\mu = 0.8$ by following these steps.

(a) The z test statistic is

$$z = \frac{\bar{x} - \mu_0}{\sigma/\sqrt{n}} = \frac{\bar{x} - 0}{1/\sqrt{10}} = 3.162\bar{x}$$

(Remember that you won't know the numerical value of $\bar{x}$ until you have data.) What values of z lead to rejecting H_0 at the 5% significance level?
(b) Starting from your result in part (a), what values of $\bar{x}$ lead to rejecting H_0? The area above these values is shaded under the top curve in Figure 18.1.
(c) The power is the probability that you observe any of these values of $\bar{x}$ when $\mu = 0.8$. This is the shaded area under the bottom curve in Figure 18.1. What is this probability?

18.49 Finding power by hand: two-sided test. Exercise 18.48 shows how to calculate the power of a one-sided z test. Power calculations for two-sided tests follow the same outline. We will find the power of a test based on a random sample of size 5, discussed in Exercise 18.39. The hypotheses are

$$H_0: \mu = 8$$
$$H_a: \mu \neq 8$$

The population of all measurements is Normal with standard deviation $\sigma = 2$, and the alternative we hope to be able to detect is $\mu = 10$ (the actual population mean). (If you used the *Statistical Power* applet for Exercise 18.47, the two Normal curves for $n = 5$ illustrate parts (a) and (b) below.)

(a) Write the z test statistic in terms of the sample mean $\bar{x}$. For what values of z does this two-sided test reject H_0 at the 5% significance level?

(b) Restate your result from part (a): what values of $\bar{x}$ lead to rejection of H_0?
(c) Now suppose that $\mu = 10$. What is the probability of observing an $\bar{x}$ that leads to rejection of H_0? This is the power of the test.

18.50 **Error probabilities.** You read that a statistical test at significance level $\alpha = 0.01$ has power 0.64. What are the probabilities of Type I and Type II errors for this test?

18.51 **Power.** You read that a statistical test at the $\alpha = 0.05$ level has probability 0.36 of making a Type II error when a specific alternative is true. What is the power of the test against this alternative?

18.52 **Find the error probabilities.** You have an SRS of size $n = 25$ from a Normal distribution with $\sigma = 1$. You wish to test

$$H_0: \mu = 0$$
$$H_a: \mu > 0$$

You decide to reject H_0 if $\bar{x} > 0$ and not reject H_0 otherwise.

(a) Find the probability of a Type I error. That is, find the probability that the test rejects H_0 when in fact $\mu = 0$.
(b) Find the probability of a Type II error when $\mu = 0.25$. This is the probability that the test fails to reject H_0 when in fact $\mu = 0.25$.
(c) Find the probability of a Type II error when $\mu = 0.5$.

18.53 APPLET **Two types of error.** Go to the *Statistical Significance* applet. This applet carries out tests at a fixed significance level. When you arrive, the applet is set for the cola-tasting test of Example 18.6. That is, the hypotheses are

$$H_0: \mu = 0$$
$$H_a: \mu > 0$$

We have an SRS of size 10 from a Normal population with standard deviation $\sigma = 1$, and we will do a test at level $\alpha = 0.05$. At the bottom of the screen, a button allows you to choose a value of the mean μ and then to generate samples from a population with that mean.

(a) Set $\mu = 0$, so that the null hypothesis is true. Each time you click the button, a new sample appears. If the sample $\bar{x}$ lands in the colored region, that sample rejects H_0 at the 5% level. Click 100 times rapidly, keeping track of how many samples reject H_0. Use your results to estimate the probability of a Type I error. If you kept clicking forever, what probability would you get?
(b) Now set $\mu = 0.8$. Example 18.6 shows that the test has power 0.812 against this alternative. Click 100 times rapidly, keeping track of how many samples fail to reject H_0. Use your results to estimate the probability of a Type II error. If you kept clicking forever, what probability would you get?

Exploring the Web

18.54 **Practically important?** Find an example of a study in which a statistically significant result may not be practically important. Summarize the study and its conclusions in your own words. The CHANCE website at `www.causeweb.org/wiki/chance//index.php/Main_Page` is a good place to look for examples.

18.55 **Wise use of inference.** The report of the American Psychological Association's Task Force on Statistical Inference is an excellent brief introduction to wise use of inference. The report appeared in the journal *American Psychologist* in 1999. You can find a copy of the initial report on the web in the list of "TFSI Publications/Links" from this journal at `www.apa.org/science/leadership/bsa/statistical/index.aspx`. Read the initial report. Are the authors of the report opposed to the use of hypothesis testing? Describe one abuse of hypothesis testing that is cited in this report.

18.56 **Spotting bad science.** Professor Pat Humphrey at Georgia Southern University mentioned to us that the website compoundchem.com posts "A Rough Guide to Spotting Bad Science." You can find this at `www.compoundchem.com/2014/04/02/a-rough-guide-to-spotting-bad-science/`. The guide gives 12 warnings. Which of these correspond to warnings given in this chapter? (Give page numbers.)

CHAPTER 19

Chris Clinton/ Getty Images

From Data Production to Inference: Part III Review

In this chapter we cover...

- Part III Summary
- Test Yourself
- Supplementary Exercises

In Part I of this book, you mastered **data analysis**, the use of graphs and numerical summaries to organize and explore any set of data. Part II introduced designs for data production. Part III introduced probability, and the reasoning of statistical inference. Parts IV and V will deal in detail with practical inference.

Statistical inference draws conclusions about a population on the basis of sample data and uses probability to indicate how reliable the conclusions are. A confidence interval estimates an unknown parameter. A significance test shows how strong the evidence is for some claim about a parameter.

The probabilities in both confidence intervals and tests tell us what would happen if we used the method for the interval or test very many times.

- A confidence level is the success rate of the method for a confidence interval. This is the probability that the method actually produces an interval that captures the unknown parameter. A 95% confidence interval gives a correct result (captures the unknown parameter) 95% of the time when we use it repeatedly.

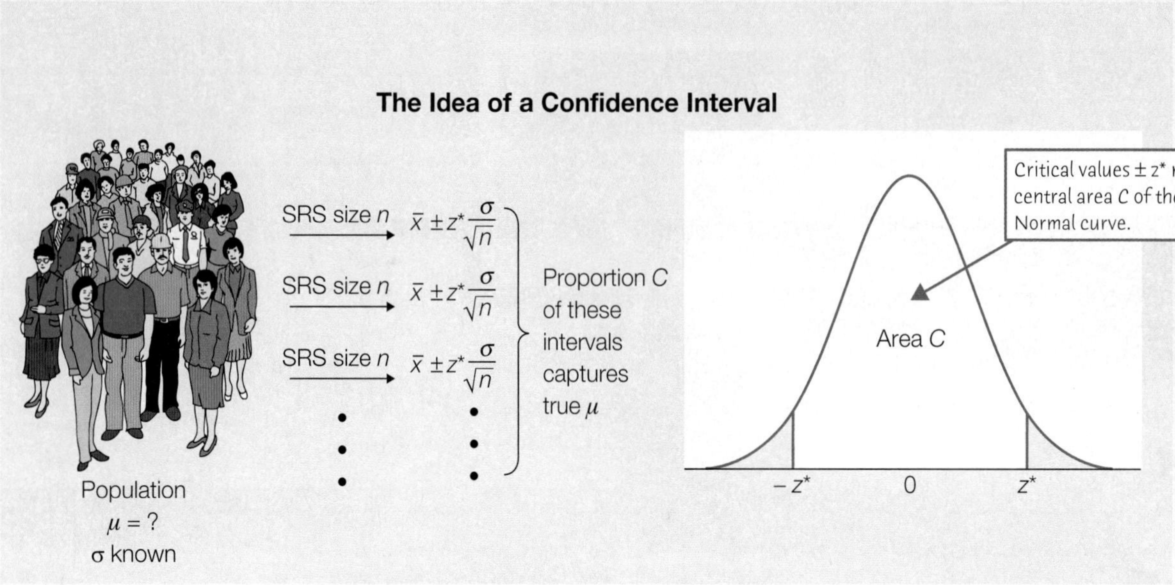

FIGURE 19.1

- A *P*-value tells us how surprising the observed outcome would be if the null hypothesis were true. That is, *P* is the probability that the test would produce a result at least as extreme as the observed result if the null hypothesis really were true. Very surprising outcomes (small *P*-values) are good evidence that the null hypothesis is not true.

Figures 19.1 and 19.2 use the *z* procedures introduced in Chapters 16 and 17 to present in picture form the big ideas of confidence intervals and significance tests. These ideas are the foundation for the rest of this book. We will have much to say about many statistical methods and their use in practice. In every case, the basic reasoning of confidence intervals and significance tests remains the same.

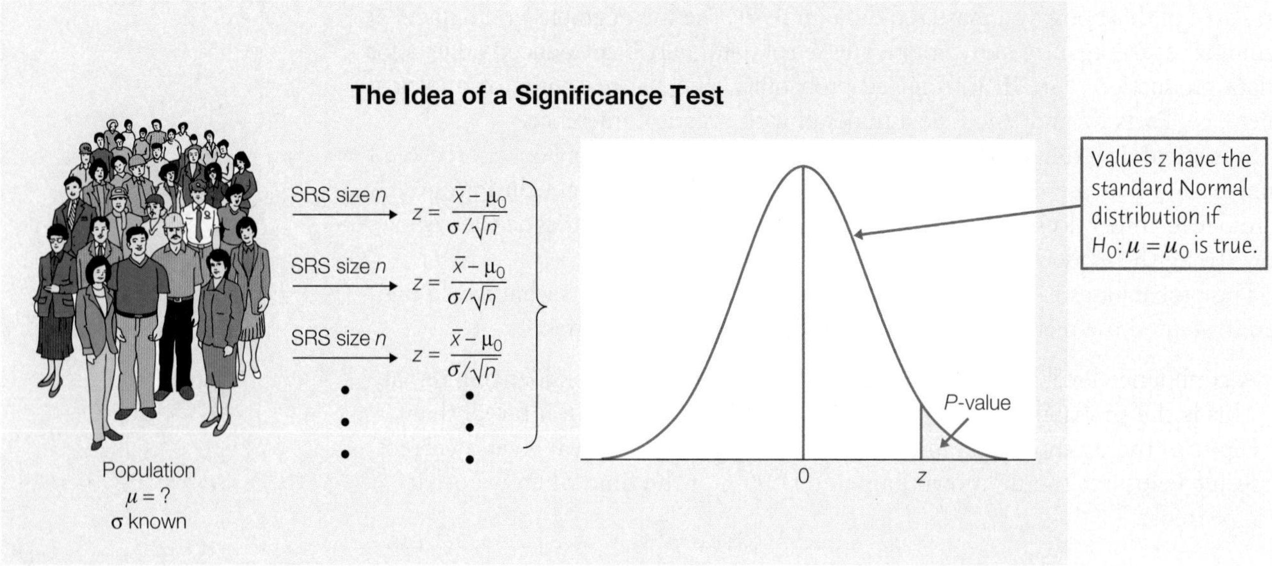

FIGURE 19.2

Part III Summary

Here are the most important skills you should have acquired from reading Chapters 12 to 18.

A. Probability

1. Recognize that the occurrence of some phenomena are random. Probability describes the long-run regularity of random phenomena.
2. Understand that the probability of an event is the proportion of times the event occurs in very many repetitions of a random phenomenon. Use the idea of probability as long-run proportion to think about probability.
3. Use basic probability rules to detect illegitimate assignments of probability: any probability must be a number between 0 and 1, and the total probability assigned to all possible outcomes must be 1.
4. Use basic probability rules to find the probabilities of events that are formed from other events. The probability that an event does not occur is 1 minus the probability it does occur. If two events are disjoint, the probability that one or the other occurs is the sum of their individual probabilities.
5. Find probabilities in a discrete probability model by adding the probabilities of their outcomes. Find probabilities in a continuous probability model as areas under a density curve.
6. Use the notation of random variables to make compact statements about random outcomes, such as $P(\bar{x} \leq 4) = 0.3$. Be able to interpret such statements.

B. Sampling Distributions

1. Identify parameters and statistics in a statistical study.
2. Recognize the fact of sampling variability: a statistic will take different values when you repeat a sample or experiment.
3. Interpret a sampling distribution as describing the values taken by a statistic in all possible repetitions of a sample or experiment under the same conditions.
4. Interpret the sampling distribution of a statistic as describing the probabilities of its possible values.
5. Understand how the sampling distribution of a sample statistic can be used to assess statistical significance.

C. General Rules of Probability (Optional topic)

1. Use Venn diagrams to picture relationships among several events.
2. Use the general addition rule to find probabilities that involve overlapping events.
3. Understand the idea of independence. Judge when it is reasonable to assume independence as part of a probability model.
4. Use the multiplication rule for independent events to find the probability that all of several independent events occur.
5. Use the multiplication rule for independent events in combination with other probability rules to find the probabilities of complex events.
6. Understand the idea of conditional probability. Find conditional probabilities for individuals chosen at random from a table of counts of possible outcomes.

7. Use the general multiplication rule to find $P(A \text{ and } B)$ from $P(A)$ and the conditional probability $P(B \mid A)$.
8. Use tree diagrams to organize several-stage probability models.

D. Binomial Distributions (Optional topic)

1. Recognize the binomial setting: a fixed number n of independent success–failure trials with the same probability p of success on each trial.
2. Recognize and use the binomial distribution of the count of successes in a binomial setting.
3. Use the binomial probability formula to find probabilities of events involving the count X of successes in a binomial setting for small values of n.
4. Find the mean and standard deviation of a binomial count X.
5. Recognize when you can use the Normal approximation to a binomial distribution. Use the Normal approximation to calculate probabilities that concern a binomial count X.

E. The Sampling Distribution of a Sample Mean

1. Recognize when a problem involves the mean $\bar{x}$ of a sample. Understand that $\bar{x}$ estimates the mean μ of the population from which the sample is drawn.
2. Use the law of large numbers to describe the behavior of $\bar{x}$ as the size of the sample increases.
3. Find the mean and standard deviation of a sample mean $\bar{x}$ from an SRS of size n when the mean μ and standard deviation σ of the population are known.
4. Understand that $\bar{x}$ is an unbiased estimator of μ and that the variability of $\bar{x}$ about its mean μ gets smaller as the sample size increases.
5. Understand that $\bar{x}$ has approximately a Normal distribution when the sample is large (central limit theorem). Use this Normal distribution to calculate probabilities that concern $\bar{x}$.

F. Confidence Intervals

1. State in nontechnical language what is meant by "95% confidence" or other statements of confidence in statistical reports.
2. Apply the four-step process (page 381) for any confidence interval. This process will be used more extensively in later chapters. Remember to describe your results in the context of the problem.
3. Calculate a confidence interval for the mean μ of a Normal population with known standard deviation σ, using the formula $\bar{x} \pm z^*\sigma/\sqrt{n}$.
4. Understand how the margin of error of a confidence interval changes with the sample size and the level of confidence C.
5. Find the sample size required to obtain a confidence interval of specified margin of error m when the confidence level and other information are given.
6. Identify sources of error in a study that are *not* included in the margin of error of a confidence interval, such as undercoverage or nonresponse.

G. Significance Tests

1. State the null and alternative hypotheses in a testing situation when the parameter in question is a population mean μ.
2. Explain in nontechnical language the meaning of the P-value when you are given the numerical value of P for a test.

3. Apply the four-step process (page 401) for any significance test. This process will be used more extensively in later chapters. Remember to describe your results in the context of the problem.
4. Calculate the one-sample z test statistic and the P-value for both one-sided and two-sided tests about the mean μ of a Normal population.
5. Assess statistical significance at standard levels α, either by comparing P with α or by comparing z with standard Normal critical values.
6. Recognize that significance testing does not measure the size or importance of an effect. Explain why a small effect can be significant in a large sample and why a large effect can fail to be significant in a small sample.
7. Recognize that any inference procedure acts as if the data were properly produced. The z confidence interval and test require that the data be an SRS from the population.

Test Yourself

The questions below include multiple choice, calculations, and short-answer questions. They will help you review the basic ideas and skills presented in Chapters 12 to 18.

19.1 A randomly chosen subject arrives for a study of exercise and fitness. Describe a sample space for each of the following. (In some cases, you may have some freedom in your choice of S.)

(a) The subject is either female or male.
(b) After 10 minutes on an exercise bicycle, you ask the subject to rate his or her effort on the Rate of Perceived Exertion (RPE) scale. RPE ranges in whole-number steps from 6 (no exertion at all) to 20 (maximal exertion).
(c) You measure VO2 max, the maximum volume of oxygen consumed per minute during exercise. VO2 is generally between 2.5 and 6.1 liters per minute.
(d) You measure the maximum heart rate (beats per minute).

Internet search engines. *Internet search sites compete for users because they sell advertising space on their sites and can charge more if they are heavily used. Choose an Internet search attempt at random. Here is the probability distribution for the site the search uses:*[1]

Site	Google	Microsoft	Yahoo	Ask Network	Others
Probability	0.68	0.18	0.10	0.02	?

Use this information to answer Questions 19.2 to 19.4.

19.2 What is the probability that a search attempt is made at Microsoft or Yahoo?

(a) 0.28
(b) 0.30
(c) 0.96
(d) Cannot be determined from the information given

19.3 What is the probability that a search attempt is made at a site other than the leading four?

(a) 0.02
(b) 0.34
(c) 0.98
(d) Cannot be determined from the information given

19.4 What is the probability that a search attempt is directed to a site other than Google?

(a) 0.02
(b) 0.32
(c) 0.98
(d) Cannot be determined from the information given

How many in the house? *In government data, a household consists of all occupants of a dwelling unit. Here is the distribution of household size in the United States:*

Number of Persons	1	2	3	4	5	6	7
Probability	0.27	0.34	0.16	0.14	0.06	0.02	0.01

Choose an American household at random and let the random variable Y be the number of persons living in the household. Use this information to answer Questions 19.5 to 19.7.

19.5 Express "more than one person lives in this household" in terms of Y. What is the probability of this event?

19.6 What is $P(2 < Y \leq 4)$?

19.7 What is $P(Y \neq 2)$?

(a) 0.27
(b) 0.34
(c) 0.51
(d) 0.66

How many children? *Choose at random an American woman between the ages of 15 and 44. Here is the distribution of the number of children the woman has given birth to:*[2]

X = Number of Children	0	1	2	3	4	5
Probability	0.471	0.169	0.204	0.104	0.034	0.018

(The few women with six or more children are included in the "five children" group.) Use this information to answer Questions 19.8 to 19.11.

19.8 Check that this distribution satisfies the two requirements for a legitimate discrete probability model.

19.9 Describe in words the event $P(X \leq 2)$. What is the probability of this event?

19.10 What is $P(X < 2)$?

(a) 0.169
(b) 0.471
(c) 0.640
(d) 0.844

19.11 Write the event "a woman gives birth to three or more children" in terms of values of X. What is the probability of this event?

Random number generators. *Many random number generators allow users to specify the range of the random numbers to be produced. Suppose you specify that the random number Y can take any value between 0 and 5. The density curve of the outcome has a constant height between 0 and 5, and height 0 elsewhere. Use this information to answer Questions 19.12 to 19.15.*

19.12 The random variable Y is

(a) discrete.
(b) continuous, but not Normal.
(c) continuous and Normal.
(d) none of the above.

19.13 The height of the density curve between 0 and 5 is

(a) 0.2.
(b) 1.
(c) 5.
(d) none of the above.

19.14 Draw a graph of the density curve and find $P(1 \leq Y \leq 3)$.

19.15 Find $P(4 < Y < 7)$.

An IQ test. *The Wechsler Adult Intelligence Scale (WAIS) is a common "IQ test" for adults. The distribution of WAIS scores for persons over 16 years of age is approximately Normal with mean 100 and standard deviation 15. Use this information to answer Questions 19.16 to 19.19.*

19.16 What is the probability that a randomly chosen individual has a WAIS score of 105 or higher?

(a) 0.0005
(b) 0.3707
(c) 0.4400
(d) 0.6293

19.17 What are the mean and standard deviation of the average WAIS score $\bar{x}$ for an SRS of 60 people?

(a) Mean = 13.56, standard deviation = 15
(b) Mean = 100, standard deviation = 15
(c) Mean = 100, standard deviation = 1.94
(d) Mean = 100, standard deviation = 0.25

19.18 What is the probability that the average WAIS score of an SRS of 60 people is 105 or higher?

(a) 0.0049
(b) 0.3707
(c) 0.9738
(d) None of the above

19.19 Would your answers to any of Questions 19.16 to 19.18 be affected if the distribution of WAIS scores in the adult population were distinctly non-Normal? Explain.

Reaction times. *The time that people require to react to a stimulus usually has a right-skewed distribution, as lack of attention or tiredness causes some lengthy reaction times. Reaction times for children with attention deficit hyperactivity disorder (ADHD) are more skewed, as their condition causes more frequent lack of attention. In one study, children with ADHD were asked to press the spacebar on a computer keyboard when any letter other than X appeared on the screen. With 2 seconds between letters, the mean reaction time was 445 milliseconds (ms) and the standard deviation was 82 ms.*[3] *Take these values to be the population μ and σ for ADHD children. Use this information to answer Questions 19.20 to 19.22.*

19.20 What are the mean and standard deviation of the mean reaction time $\bar{x}$ for a randomly chosen group of 15 ADHD children? For a group of 150 such children?

19.21 The distribution of reaction time is strongly skewed. Explain briefly why we hesitate to regard $\bar{x}$ as Normally distributed for 15 children but are willing to use a Normal distribution for the mean reaction time of 150 children.

19.22 What is the approximate probability that the mean reaction time in a group of 150 ADHD children is greater than 450 ms?

Pesticides in whale blubber: estimation. *The level of pesticides found in the blubber of whales is a measure of pollution of the oceans by runoff from land and can also be used to identify different populations of whales. A sample of eight male minke whales in the West Greenland area of the North Atlantic found the mean concentration of the insecticide dieldrin to be $\bar{x}$ = 357 nanograms*

© redbrickstock.com/Alamy

per gram of blubber (ng/g).[4] *Suppose that the concentration in all such whales varies Normally with standard deviation* $\sigma = 50$ *ng/g. Use this information to answer Questions 19.23 to 19.26.*

19.23 A 95% confidence interval to estimate the mean level is

(a) 344.75 to 369.25.
(b) 339.32 to 374.68.
(c) 322.35 to 391.65.
(d) 259.00 to 455.00.

19.24 A 90% confidence interval to estimate the mean level is

(a) 346.72 to 367.28.
(b) 327.92 to 386.08.
(c) 311.36 to 402.54.
(d) 274.75 to 439.25.

19.25 Find an 80% confidence interval for the mean concentration of dieldrin in the minke whale population.

19.26 What general fact about confidence intervals do the margins of error of your three intervals in the previous problems illustrate?

Estimating blood cholesterol. *The distribution of blood cholesterol level in the population of young men aged 20 to 34 years is close to Normal with standard deviation* $\sigma = 41$ *milligrams per deciliter (mg/dL). You measure the blood cholesterol of 14 cross-country runners. The mean level is* $\bar{x} = 172$ *mg/dL. Assume that* σ *is the same as in the general population. Use this information to answer Questions 19.27 to 19.29.*

19.27 A 90% confidence interval for the mean level μ among cross-country runners is

(a) 172 ± 4.82 mg/dL.
(b) 172 ± 18.03 mg/dL.
(c) 172 ± 21.48 mg/dL.
(d) none of the above.

19.28 How large a sample is needed to cut the margin of error in the previous exercise in half?

(a) 2
(b) 4
(c) 28
(d) 56

19.29 How large a sample is needed to cut the margin of error to ± 5 mg/dL?

(a) 14
(b) 68
(c) 182
(d) 259

19.30 The Environmental Protection Agency (EPA) fuel economy ratings say that the 2014 Toyota Prius hybrid car gets 46 miles per gallon (mpg) on the highway. Deborah wonders whether the actual long-term average highway mileage μ of her new Prius is less than 46 mpg. She keeps careful records of gas mileage for 3000 miles of highway driving. Her result is $\bar{x} = 45.2$ mpg. What are her null and alternative hypotheses?

(a) H_0: $\mu = 46$, H_a: $\mu < 46$
(b) H_0: $\mu = 46$, H_a: $\mu > 46$
(c) H_0: $\bar{x} = 46$, H_a: $\bar{x} < 46$
(d) H_0: $\bar{x} = 46$, H_a: $\bar{x} > 46$

19.31 According to the National Survey of Student Engagement (NSSE) the average amount of time that first-year college students spent preparing for class (studying,

Statistics in Your World

HOW MANY MILES PER GALLON?

As gasoline prices rise, more people pay attention to the government's gas mileage ratings of their vehicles. Until recently, these ratings overstated the miles per gallon we can expect in real-world driving. The ratings assumed a top speed of 60 miles per hour, slow acceleration, and no air-conditioning. That doesn't resemble what we see around us on the highway. Maybe it doesn't resemble the way we ourselves drive. Starting with 2008 models, the ratings assume higher speeds (80 miles per hour tops), faster acceleration, and air-conditioning in warm weather. Mileage ratings of the same vehicle dropped by about 12% in the city and 8% on the highway.

reading, writing, doing homework or lab work, analyzing data, rehearsing, and other academic activities) in 2013 was 14.25 hours per week. Your college wonders if the average μ for its first-year students in 2013 differed from the national average. A random sample of 500 students who were first-year students in 2013 claims to have spent an average of $\bar{x} = 13.4$ hours per week on homework in their first year. What are the null and alternative hypotheses for a comparison of first-year students at your college with national first-year students in 2013?

(a) H_0: $\bar{x} = 14.25$, H_a: $\bar{x} \neq 14.25$
(b) H_0: $\bar{x} = 13.4$, H_a: $\bar{x} > 13.4$
(c) H_0: $\mu = 14.25$, H_a: $\mu \neq 14.25$
(d) H_0: $\mu = 13.4$, H_a: $\mu > 13.4$

Testing blood cholesterol. *The distribution of blood cholesterol level in the population of young men aged 20 to 34 years is close to Normal with mean 188 milligrams per deciliter (mg/dL) and standard deviation 41 mg/dL. You measure the blood cholesterol of 14 young men aged 20 to 34 years who have completed at least one marathon. The mean level is $\bar{x} = 172$ mg/dL. Assume that σ is the same as in the general population. Use this information to answer Questions 19.32 to 19.34.*

19.32 We suspect that the mean μ for all young men aged 20 to 34 years who have completed at least one marathon is lower than that of the population of all young men aged 20 to 34 years. Thus, we decide to test the hypotheses H_0: $\mu = 188$, H_a: $\mu < 188$. The z test statistic for testing these hypotheses is

(a) 5.46
(b) −5.46
(c) 1.46
(d) −1.46

19.33 The result is significant at

(a) $\alpha = 0.01$.
(b) $\alpha = 0.05$ but not at $\alpha = 0.01$.
(c) $\alpha = 0.10$ but not at $\alpha = 0.05$.
(d) $\alpha = 0.25$ but not at $\alpha = 0.10$.

19.34 You increase the sample of young men aged 20 to 34 years who have completed at least one marathon from 14 to 56. Suppose that this larger sample gives the same mean level, $\bar{x} = 172$ mg/dL. Redo the test in exercises 19.32 and 19.33. The result is significant at

(a) $\alpha = 0.01$.
(b) $\alpha = 0.05$ but not at $\alpha = 0.01$.
(c) $\alpha = 0.10$ but not at $\alpha = 0.05$.
(d) $\alpha = 0.25$ but not at $\alpha = 0.10$.

19.35 The Food and Drug Administration regulates the amount of dieldrin in raw food. For some foods, no more than 100 ng/g is allowed. Using the information for Exercises 19.23 to 19.26, is there good evidence that the mean concentration μ in whale blubber is above 100 ng/g? Carry out a test of the hypotheses H_0: $\mu = 100$, H_a: $\mu > 100$ assuming that the "simple conditions" (page 374) hold. The P-value of your test is

(a) above 0.10.
(b) less than or equal to 0.10 but greater than 0.05.
(c) less than or equal to 0.05 but greater than 0.01.
(d) no more than 0.01.

19.36 Infants weighing less than 1500 grams at birth are classed as "very low birth weight." Low birth weight carries many risks. One study followed 113 male infants with very low birth weight to adulthood. At age 20, the mean IQ score for these men was $\bar{x} = 87.6$.[5] IQ scores vary Normally with standard deviation $\sigma = 15$. Give a 95% confidence interval for the mean IQ score at age 20 for all very-low-birth-weight males.

19.37 IQ tests are scaled so that the mean score in a large population should be $\mu = 100$. We suspect that the very-low-birth-weight population has mean score less than 100. Does the study described in the previous exercise give good evidence that this is true? State hypotheses, carry out a test assuming that the "simple conditions" (page 374) hold, compute the *P*-value, and give your conclusion in plain language.

19.38 When our brains store information, complicated chemical changes take place. In trying to understand these changes, researchers blocked some processes in brain cells taken from rats and compared these cells with a control group of normal cells. They say that "no differences were seen" between the two groups in four response variables. They give *P*-values 0.45, 0.83, 0.26, and 0.84 for these four comparisons.[6] Which of the following statements is correct?

(a) It is literally true that "no differences were seen." That is, the mean responses were exactly alike in the two groups.
(b) The mean responses were exactly alike in the two groups for at least one of the four response variables measured, but not for all of them.
(c) The statement "no differences were seen" means that the observed differences were not statistically significant at the significance level used by the researchers.
(d) The statement "no differences were seen" means that the observed differences were all less than 1 (and were actually 0.45, 0.83, 0.26, and 0.84 for these four comparisons).

19.39 In a 2013 study, researchers compared various measurements on overweight first-born and second-born middle-aged men.[7] They found that first-borns had a significantly higher weight ($P = 0.013$) than second-borns, but no significant difference in total cholesterol ($P = 0.74$). Explain carefully why $P = 0.013$ means there is evidence that first-born middle-aged men may have higher weights than second-borns and why $P = 0.74$ provides no evidence that first-born middle-aged men may have different total cholesterol levels than second-borns.

19.40 We often see televised reports of brushfires threatening homes in California. Some people argue that the modern practice of quickly putting out small fires allows fuel to accumulate and so increases the damage done by large fires. A detailed study of historical data suggests that this is wrong—the damage has risen simply because there are more houses in risky areas.[8] As usual, the study report gives statistical information tersely. Here is the summary of a regression of number of fires on decade (nine data points, for the 1910s to the 1990s): "Collectively, since 1910, there has been a highly significant increase ($r^2 = 0.61$, $P < 0.01$) in the number of fires per decade." How would you explain this statement to someone who knows no statistics? Include an explanation of both the description given by r^2 and its statistical significance.

19.41 **(Optional topic)** Byron claims that the probability Georgia and Alabama will play in the SEC championship football game this year is 68%. The number 68% is

(a) the proportion of times Georgia and Alabama have played in the championship game in the past.
(b) Byron's personal probability that Georgia and Alabama will play in the SEC championship football game this year.
(c) the area under a Normal density curve.
(d) all of the above.

19.42 **(Optional topic) Causes of death.** *Accidents, suicide, and murder are the leading causes of death for young adults. Here are the counts of violent deaths in a recent year among people 20 to 24 years of age:*

	FEMALE	MALE
Accidents	1818	6457
Homicide	457	2870
Suicide	345	2152

(a) Choose a violent death in this age group at random. What is the probability that the victim was male?
(b) Find the conditional probability that the victim was male, given that the death was accidental.
(c) Use your answers from parts (a) and (b) to explain whether sex and type of death are independent or not.

(Optional topic) Distance learning. *A study of the students taking distance learning courses at a university finds that they are mostly older students not living in the university town. If a distance learning student is chosen at random, the* P(student is over 25 years old) = 0.7, *the* P(student is local) = 0.25 *and the* P(student is over 25 years old and local) = 0.05. *Use this information to answer Questions 19.43 to 19.45.*

19.43 The probability that the randomly selected student is over 25 years old or local is
(a) 0.950.
(b) 0.900.
(c) 0.850.
(d) 0.750.

19.44 The probability that the randomly selected student is over 25 years old and *not* local is
(a) 0.750.
(b) 0.650.
(c) 0.600.
(d) 0.550.

19.45 The conditional probability that the randomly selected student is over 25 years old given that they are not local is
(a) 0.93.
(b) 0.87.
(c) 0.81.
(d) 0.75.

19.46 **(Optional topic)** Alysha makes 40% of her free throws. She takes five free throws in a game. If the shots are independent of each other, the probability that she misses the first two shots but makes the other three is about
(a) 0.230.
(b) 0.115.
(c) 0.023.
(d) 0.600.

19.47 **(Optional topic)** Alysha makes 40% of her free throws. She takes five free throws in a game. If the shots are independent of each other, the probability that she makes *exactly* one of five shots is about
(a) 0.259.
(b) 0.115.
(c) 0.052.
(d) 0.200.

(Optional topic) Have you visited your dentist this year? *Polls find that 62% of American aged 18 to 29 report having seen a dentist in the last 12 months. This percentage varies little among age groups, with the most pronounced differences in dental habits occurring across income groups.*[9] *Use this information to answer Questions 19.48 and 19.49.*

19.48 If you take an SRS of 1000 Americans aged 18 to 29, the approximate distribution of the number of individuals in your sample who would report having seen a dentist in the last 12 months is

(a) $N(0.62, 15.35)$
(b) $N(0.62, 236)$
(c) $N(620, 15.35)$
(d) $N(620, 236)$

19.49 If you take an SRS of 1000 Americans aged 18 to 29, find the approximate probability that fewer than 600 individuals in your sample would report having seen a dentist in the last 12 months.

Supplementary Exercises

Supplementary exercises apply the skills you have learned in ways that require more thought or more elaborate use of technology.

19.50 The addition rule. The addition rule for probabilities, $P(A \text{ or } B) = P(A) + P(B)$, is not always true. Give (in words) an example of real-world events A and B for which this rule is not true.

19.51 Comparing wine tasters. Two wine tasters rate each wine they taste on a scale of 1 to 5. From data on their ratings of a large number of wines, we obtain the following probabilities for both tasters' ratings of a randomly chosen wine:

	TASTER 2				
TASTER 1	**1**	**2**	**3**	**4**	**5**
1	0.05	0.02	0.01	0.00	0.00
2	0.02	0.08	0.04	0.02	0.01
3	0.01	0.04	0.25	0.05	0.01
4	0.00	0.02	0.05	0.18	0.02
5	0.00	0.01	0.01	0.02	0.08

(a) Why is this a legitimate discrete probability model?
(b) What is the probability that the tasters agree when rating a wine?
(c) What is the probability that Taster 1 rates a wine higher than Taster 2? What is the probability that Taster 2 rates a wine higher than Taster 1?

19.52 A 14-sided die. An ancient Korean drinking game involves a 14-sided die. The players roll the die in turn and must submit to whatever humiliation is written on the up-face: something like "Keep still when tickled on face." Six of the 14 faces are squares. Let's call them A, B, C, D, E, and F for short. The other eight faces are triangles, which we will call 1, 2, 3, 4, 5, 6, 7, and 8. Each of the squares is equally likely. Each of the triangles is also equally likely, but the triangle probability differs from the square probability. The probability of getting a triangle is 0.28. Give the probability model for the 14 possible outcomes.

David Moore

19.53 Distributions: means versus individuals. The z confidence interval and test are based on the sampling distribution of the sample mean $\bar{x}$. The National Survey of Student Engagement (NSSE) asks college seniors to rate how much their experience at their institution has contributed to their ability to think critically and analytically. Ratings are on a scale of 1 to 7, with 7 being the highest (best) rating. Suppose scores for all seniors in 2013 are Normal with mean $\mu = 3.3$ and standard deviation $\sigma = 0.8$.

(a) You take an SRS of 100 seniors. According to the 99.7 part of the 68–95–99.7 rule, about what range of ratings do you expect to see in your sample?
(b) You look at many SRSs of size 100. About what range of sample mean ratings $\bar{x}$ do you expect to see?

19.54 Distributions: larger samples. In the setting of the previous exercise, how many seniors must you sample to cut the range of values of $\bar{x}$ in half? This will also cut the margin of error of a confidence interval for μ in half. Do you expect the range of individual scores in the new sample to also be much less than in a sample of size 100? Why?

19.55 Normal body temperature? Here are the daily average body temperatures (degrees Fahrenheit) for 20 healthy adults:[10] BODYTMP

98.74	98.83	96.80	98.12	97.89	98.09
97.87	97.42	97.30	97.84	100.27	97.90
99.64	97.88	98.54	98.33	97.87	97.48
98.92	98.33				

(a) Make a stemplot of the data. The distribution is roughly symmetric and single-peaked. There is one mild outlier. We expect the distribution of the sample mean $\bar{x}$ to be close to Normal.
(b) Do these data give evidence that the mean body temperature for all healthy adults is not equal to the traditional 98.6 degrees? Follow the four-step process for significance tests (page 401). (Suppose that body temperature varies Normally with standard deviation 0.7 degree.)

19.56 Time in a restaurant. The owner of a pizza restaurant in France knows that the time customers spend in the restaurant on Saturday evening has mean 90 minutes and standard deviation 15 minutes. He has read that pleasant odors can influence customers, so he spreads a lavender odor throughout the restaurant. Here are the times (minutes) for customers on the next Saturday evening:[11] RSTRNT

92	126	114	106	89	137
93	76	98	108	124	105
129	103	107	109	94	105
102	108	95	121	109	104
116	88	109	97	101	106

(a) Make a stemplot of the times. The distribution is roughly symmetric and single-peaked, so the distribution of $\bar{x}$ should be close to Normal.

(b) Suppose that the standard deviation $\sigma = 15$ minutes is not changed by the odor. Is there reason to think that the lavender odor has increased the mean time customers spend in the restaurant? Follow the four-step process for significance tests (page 401).

19.57 Normal body temperature. Use the data in Exercise 19.55 to estimate mean body temperature with 90% confidence. Follow the four-step process for confidence intervals (page 381). BODYTMP

19.58 Time in a restaurant. Use the data in Exercise 19.56 to estimate the mean time customers spend in this restaurant on Saturday evenings with 95% confidence. Follow the four-step process for confidence intervals (page 381). RSTRNT

19.59 Tests from confidence intervals. You read in a U.S. Census Bureau report that a 90% confidence interval for the median income in 2012 of American households was $\$51,017 \pm \$343$. Based on this interval, can you reject the null hypothesis that the median income in this group is \$50,000? What is the alternative hypothesis of the test? What is its significance level?

19.60 (Optional topic) Testing for HIV. Enzyme immunoassay tests are used to screen blood specimens for the presence of antibodies to HIV, the virus that causes AIDS. Antibodies indicate the presence of the virus. The test is quite accurate but is not always correct. Here are approximate probabilities of positive and negative test results when the blood tested does and does not actually contain antibodies to HIV:[12]

	TEST RESULT	
	POSITIVE	NEGATIVE
Antibodies present	0.9985	0.0015
Antibodies absent	0.0060	0.9940

Suppose that 1% of a large population carries antibodies to HIV in their blood.

(a) Draw a tree diagram for selecting a person from this population (outcomes: antibodies present or absent) and testing his or her blood (outcomes: test positive or negative).
(b) What is the probability that the test is positive for a randomly chosen person from this population?

19.61 (Optional topic) Type of high school attended. Choose a college freshman at random and ask what type of high school they attended. Here is the distribution of results:[13]

Type	Regular Public	Public Charter	Public Magnet
Probability	0.758	0.029	0.035

Type	Private Religious	Private Independent	Home school
Probability	0.109	0.063	0.006

What is the conditional probability that a college freshman was home schooled, given that he or she did not attend a regular public high school?

19.62 (Optional topic) False HIV positives. Continue your work from Exercise 19.60. What is the probability that a person has the antibody, given that the test is positive? (Your result illustrates a fact that is important when considering proposals for widespread testing for HIV, illegal drugs, or agents of biological warfare: if the condition being tested is uncommon in the population, most positives will be false positives.)

19.63 (Optional topic) On the web. What kinds of websites do males aged 18 to 34 visit? About 50% of male Internet users in this age group visit an auction site such as eBay at least once a month.[14]

(a) If we interview a random sample of 12 male Internet users aged 18 to 34, what is the probability that exactly eight of the 12 have visited an auction site in the past month?

(b) Suppose that we had interviewed a random sample of 500 men aged 18 to 34. What is the probability that at least 235 of the men in the sample visit an online auction site at least once a month? (Check that the Normal approximation is permissible and use it to find this probability.)

19.64 (Optional topic) Low power? It appears that eating oat bran lowers cholesterol slightly. At a time when oat bran was something of a fad, a paper in the *New England Journal of Medicine* found that it had no significant effect on cholesterol.[15] The paper reported a study with just 20 subjects. Letters to the journal denounced publication of a negative finding from a study with very low power. Explain why lack of significance in a study with low power gives no reason to accept the null hypothesis that oat bran has no effect.

19.65 (Optional topic) Type I and Type II errors. Exercise 19.37 asks for a significance test of the null hypothesis that the mean IQ of very-low-birth-weight male babies is 100 against the alternative hypothesis that the mean is less than 100. State in words what it means to make a Type I error and a Type II error in this setting.

© Rosemary Harris/Alamy

PART IV

Inference about Variables

QUANTITATIVE RESPONSE VARIABLE

CHAPTER 20 Inference about a Population Mean

CHAPTER 21 Comparing Two Means

CATEGORICAL RESPONSE VARIABLE

CHAPTER 22 Inference about a Population Proportion

CHAPTER 23 Comparing Two Proportions

CHAPTER 24 Inference about Variables: Part IV Review

With the principles in hand, we proceed to practice, that is, to inference in fully realistic settings. In the remaining chapters of this book, you will meet many of the most commonly used statistical procedures. We have grouped these procedures into two classes, corresponding to our division of data analysis into exploring variables and distributions and exploring relationships. The five chapters of Part IV concern inference about the distribution of a single variable and inference for comparing the distributions of two variables. Part V deals with inference for relationships among variables. In Chapters 20 and 21, we analyze data on quantitative variables. We begin with the familiar Normal distribution for a quantitative variable. Chapters 22 and 23 concern categorical variables, so that inference begins with counts and proportions of outcomes. Chapter 24 reviews this part of the text.

The four-step process for approaching a statistical problem can guide much of your work in these chapters. You should review the outlines of the four-step process for a confidence interval (page 381) and for a test of significance (page 401). The statement of an exercise usually does the State step for you, leaving the Plan, Solve, and Conclude steps for you to complete. It is helpful to first summarize the *State* step in your own words to organize your thinking. Many examples and exercises in these chapters involve both carrying out inference and thinking about inference in practice. Remember that any inference method is useful only under certain conditions, and that you must judge these conditions before rushing to inference.

CHAPTER 20

© Tim Tadder/Corbis

Inference about a Population Mean

In this chapter we cover...

20.1 Conditions for inference about a mean
20.2 The t distributions
20.3 The one-sample t confidence interval
20.4 The one-sample t test
20.5 Using technology
20.6 Matched pairs t procedures
20.7 Robustness of t procedures
20.8 Resampling and standard errors*

This chapter describes confidence intervals and significance tests for the mean μ of a population. We used the z procedures in this same setting to introduce the ideas of confidence intervals and tests. Now we discard the unrealistic condition that we know the population standard deviation σ and present procedures for practical use. We also pay more attention to the real-data setting of our work. The details of confidence intervals and tests change only slightly when you don't know σ. More important, you can interpret your results exactly as before. To illustrate this, Example 20.2 repeats an example from Chapter 16.

20.1 Conditions for inference about a mean

Confidence intervals and tests of significance for the mean μ of a Normal population are based on the sample mean $\bar{x}$. Confidence intervals and P-values involve probabilities calculated from the sampling distribution of $\bar{x}$. Here are the conditions needed for realistic inference about a population mean.

Conditions for Inference about a Mean

We can regard our data as a **simple random sample** (SRS) from the population. This condition is very important.

Observations from the population have a **Normal distribution** with mean μ and standard deviation σ. In practice, it is enough that the distribution be symmetric and single-peaked unless the sample is very small. Both μ and σ are unknown parameters.

There is another condition that applies to all the inference methods in this book: *the population must be much larger than the sample—say, at least 20 times as large.*[1] All our examples and exercises satisfy this condition. Practical settings in which the sample is a large part of the population are rather special, and we will not discuss them.

CAUTION

When the conditions for inference are satisfied, the sample mean $\bar{x}$ has the Normal distribution with mean μ and standard deviation $\sigma/\sqrt{n}$. Because we don't know σ, we estimate it by the sample standard deviation s. We then estimate the standard deviation of $\bar{x}$ by $s/\sqrt{n}$. This quantity is called the *standard error* of the sample mean $\bar{x}$.

Standard Error

When the standard deviation of a statistic is estimated from data, the result is called the **standard error** of the statistic. The standard error of the sample mean $\bar{x}$ is $s/\sqrt{n}$.

Apply Your Knowledge

20.1 Travel Time to Work. A study of commuting times reports the travel times to work of a random sample of 1000 employed adults in Chicago.[2] The mean is $\bar{x} = 40.0$ minutes and the standard deviation is $s = 56.9$ minutes. What is the standard error of the mean?

20.2 Comparing Breathing Frequencies in Swimming. Researchers from the United Kingdom studied the effect of two breathing frequencies on performance times and on several physiological parameters in front crawl swimming. The breathing frequencies were one breath every second stroke (B2) and one breath every fourth stroke (B4). Subjects were 10 male collegiate swimmers. Each subject swam 200 meters using each breathing frequency: once with breathing frequency B2, and once on a different day with breathing frequency B4. A paper states that the results are expressed as mean plus or minus the standard deviation.[3] One result reported in the paper states that the immediate postexercise heart rate for subjects when using breathing frequency B2 was 163 ± 15 beats per minute. What are $\bar{x}$ and the standard error of the mean for these subjects? (This exercise is also a warning to read carefully: that 163 ± 15 is *not* a confidence interval, yet summaries in this form are common in scientific reports.)

20.2 The *t* distributions

If we knew the value of σ, we would base confidence intervals and tests for μ on the one-sample z statistic

$$z = \frac{\bar{x} - \mu}{\sigma/\sqrt{n}}$$

This z statistic has the standard Normal distribution $N(0, 1)$. In practice, we don't know σ, so we substitute the standard error $s/\sqrt{n}$ of $\bar{x}$ for its standard deviation $\sigma/\sqrt{n}$. The statistic that results does not have a Normal distribution. It has a distribution that is new to us, called a *t distribution.*

The One-Sample *t* Statistic and the *t* Distributions

Draw an SRS of size n from a large population that has the Normal distribution with mean μ and standard deviation σ. The **one-sample *t* statistic**

$$t = \frac{\bar{x} - \mu}{s/\sqrt{n}}$$

has the ***t* distribution** with $n - 1$ degrees of freedom.

The t statistic has the same interpretation as any standardized statistic: it says how far $\bar{x}$ is from its mean μ in standard error units. There is a different t distribution for each sample size. We specify a particular t distribution by giving its **degrees of freedom.** The degrees of freedom for the one-sample t statistic come from the sample standard deviation s in the denominator of t. We saw in Chapter 2 (page 58) that s has $n - 1$ degrees of freedom. There are other t statistics with different degrees of freedom, some of which we will meet later. We will write the t distribution with $n - 1$ degrees of freedom as $t(n - 1)$ for short.

degrees of freedom

Figure 20.1 compares the density curves of the standard Normal distribution and the t distributions with two and nine degrees of freedom. The figure illustrates these facts about the t distributions:

- The density curves of the t distributions are similar in shape to the standard Normal curve. They are symmetric about zero, single-peaked, and bell-shaped.
- The variability of the t distributions is a bit greater than that of the standard Normal distribution. The t distributions in Figure 20.1 have more probability in the tails and less in the center than does the standard Normal. This is true because substituting the estimate s for the fixed parameter σ introduces more variation into the statistic.
- As the degrees of freedom increase, the t density curve approaches the $N(0, 1)$ curve ever more closely. This happens because s estimates σ more accurately as the sample size increases. So using s in place of σ causes little extra variation in the statistic when the sample is large.

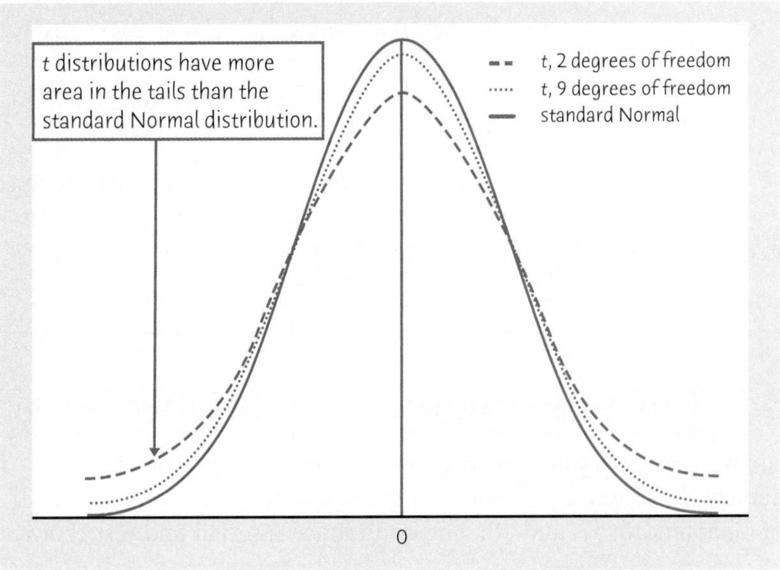

FIGURE 20.1
Density curves for the *t* distributions with two and nine degrees of freedom and for the standard Normal distribution. All are symmetric with center zero. The *t* distributions are somewhat more variable.

Table C in the back of the book gives critical values for the t distributions. Each row in the table contains critical values for the t distribution whose degrees of freedom appear at the left of the row. For convenience, we label the table entries both by the confidence level C (in percent) required for confidence intervals and by the one-sided and two-sided P-values for each critical value. You have already used the standard Normal critical values in the z^* row at the bottom of Table C. By looking down any column, you can check that the t critical values approach the Normal values as the degrees of freedom increase. If you use statistical software, you don't need Table C.

EXAMPLE 20.1 *t* Critical Values

Figure 20.1 shows the density curve for the t distribution with nine degrees of freedom. What point on this distribution has probability 0.05 to its right? In Table C, look in the df = 9 row above one-sided P-value 0.05 and you will find that this critical value is $t^* = 1.833$. To use software, enter the degrees of freedom and the probability you want to the *left*, 0.95 in this case. Here is Minitab's output:

```
Student's t distribution with 9 DF
P( X <= x )          x
    0.95          1.83311
```

For the standard Normal distribution, the point that has probability 0.05 to the right is 1.645 (see the z^* row at the bottom of Table C for the Normal distribution critical values). The value for the standard Normal distribution is smaller than the value for the t distribution. This is an example of what we mean by the statement that t distributions have more probability in the tails than the standard Normal. ■

LaunchPad Online Resources

- The **StatBoards video**, *How to Find t Critical Values*, discusses several examples of finding t critical values.

Apply Your Knowledge

20.3 Critical Values. Use software or Table C to find

(a) the critical value for a one-sided test with level $\alpha = 0.05$ based on the $t(3)$ distribution.

(b) the critical value for a 98% confidence interval based on the $t(25)$ distribution.

20.4 More Critical Values. You have an SRS of size 30 and calculate the one-sample t statistic. What is the critical value t^* such that

(a) t has probability 0.025 to the right of t^*?

(b) t has probability 0.75 to the left of t^*?

20.3 The one-sample *t* confidence interval

To analyze samples from Normal populations with unknown σ, just replace the standard deviation $\sigma/\sqrt{n}$ of $\bar{x}$ by its standard error $s/\sqrt{n}$ in the z procedures of Chapters 16, 17, and 18. The confidence interval and test that result are *one-sample t procedures*. Critical values and *P*-values come from the *t* distribution with $n - 1$ degrees of freedom. The one-sample *t* procedures are similar in both reasoning and computational detail to the *z* procedures.

The One-Sample *t* Confidence Interval

Draw an SRS of size *n* from a large population having unknown mean μ. A level C **confidence interval for μ** is

$$\bar{x} \pm t^* \frac{s}{\sqrt{n}}$$

where t^* is the critical value for the $t(n - 1)$ density curve with area C between $-t^*$ and t^*. This interval is exact when the population distribution is Normal and is approximately correct for large *n* in other cases.

We will provide guidelines on how large a sample size is needed for the *t* confidence interval to be approximately correct in Section 20.7.

BETTER STATISTICS, BETTER BEER

The *t* distribution and the *t* inference procedures were invented by William S. Gosset (1876–1937). Gosset worked for the Guinness brewery, and his goal in life was to make better beer. He used his new *t* procedures to find the best varieties of barley and hops. Gosset's statistical work helped him become head brewer, a more interesting title than professor of statistics. Because Gosset published under the pen name "Student" you will often see the *t* distribution called "Student's *t*" in his honor.

EXAMPLE 20.2 Good Weather, Good Tips?

Let's look again at the study of tipping in a restaurant that we met in Example 16.3. We follow the four-step process for a confidence interval, outlined on page 381.

TIP2

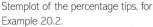

© Ariel Skelley/age fotostock

STATE: Does the expectation of good weather lead to more generous behavior? Psychologists studied the size of the tip in a restaurant when a message indicating that the next day's weather would be good was written on the bill. Here are tips from 20 patrons, measured in percent of the total bill:[4]

20.8	18.7	19.9	20.6	21.9	23.4	22.8	24.9	22.2	20.3
24.9	22.3	27.0	20.4	22.2	24.0	21.1	22.1	22.0	22.7

This is one of three sets of measurements made, the others being tips received when the message on the bill said that the next day's weather would not be good or there was no message on the bill. We want to estimate the mean tip for comparison with tips under the other conditions.

PLAN: We will estimate the mean percentage tip μ for all patrons of this restaurant when they receive a message on their bill indicating that the next day's weather will be good by giving a 95% confidence interval.

SOLVE: We must first check the conditions for inference.

- As in Chapter 16 (page 381), we are willing to regard these patrons as an SRS from all patrons of this restaurant.
- The stemplot in Figure 20.2 does not suggest any strong departures from Normality.

We can proceed to calculation. For these data,

$$\bar{x} = 22.21 \quad \text{and} \quad s = 1.963$$

The degrees of freedom are $n - 1 = 19$. From Table C, we find that for 95% confidence $t^* = 2.093$. The confidence interval is

$$\begin{aligned} \bar{x} \pm t^* \frac{s}{\sqrt{n}} &= 22.21 \pm 2.093 \frac{1.963}{\sqrt{20}} \\ &= 22.21 \pm 0.92 \\ &= 21.29 \text{ to } 23.13 \textit{ percent} \end{aligned}$$

CONCLUDE: We are 95% confident that the mean percentage tip for all patrons of this restaurant when their bill contains a message that the next day's weather will be good is between 21.29 and 23.13. ■

18	7
19	9
20	3468
21	19
22	0122378
23	4
24	099
25	
26	
27	0

FIGURE 20.2
Stemplot of the percentage tips, for Example 20.2.

Our work in Example 20.2 is very similar to what we did in Example 16.3 (page 381). To make the inference realistic, we replaced the assumed $\sigma = 2$ by $s = 1.963$ calculated from the data, and replaced the standard Normal critical value $z^* = 1.960$ by the t critical value $t^* = 2.093$. The resulting confidence interval is slightly wider than the one obtained in Example 16.3 (which was 21.33 to 23.09).

The one-sample t confidence interval has the form

$$\text{estimate} \pm t^*\text{SE}_{\text{estimate}}$$

where "SE" stands for "standard error." We will meet a number of confidence intervals that have this common form. In Example 20.2, the estimate is the sample mean $\bar{x}$, and its standard error is

$$\text{SE}_{\bar{x}} = \frac{s}{\sqrt{n}}$$

$$= \frac{1.963}{\sqrt{20}} = 0.439$$

Software will find $\bar{x}$, s, $\text{SE}_{\bar{x}}$, and the confidence interval from the data. Figure 20.5 (page 465) displays typical software output for Example 20.2.

LaunchPad Online Resources

- The **StatBoards video**, *Calculating a One-Sample t Confidence Interval*, discusses examples of constructing confidence intervals.

Apply Your Knowledge

20.5 Critical Values. What critical value t^* from Table C would you use for a confidence interval for the mean of the population in each of the following situations? (If you have access to software, you can use software to determine the critical values.)

(a) A 95% confidence interval based on $n = 12$ observations.

(b) A 99% confidence interval from an SRS of two observations.

(c) A 90% confidence interval from a sample of size 1001.

20.6 How Much Will I Bet? Our decisions depend on how the options are presented to us. Here's an experiment that illustrates this phenomenon. Tell 20 subjects that they have been given \$50 but can't keep it all. Then present them with a long series of choices among bets they can make with the \$50. Scattered among these choices in random order are 64 choices that ask the subject to choose between betting a fixed amount and an all-or-nothing gamble. The odds for all the bets are the same, but in 32 of the choices, the fixed option reads "Keep \$20," and in the other 32 choices, the fixed option reads "Lose \$30." These two fixed options lead to exactly the same outcome, but people are more likely to choose the fixed option that says they lose money. Here are the percent differences ("Number of times chose 'Lose \$30'" minus "Number of times chose 'Keep \$20'" divided by the number of trials on which the 20 subjects chose the fixed-option gamble rather than the all-or-nothing bet).[5] DATA GAMB1

37.5	30.8	6.2	17.6	14.3	8.3	16.7	20.0	10.5	21.7
30.8	27.3	22.7	38.5	8.3	10.5	8.3	10.5	25.0	7.7

(a) Make a stemplot. Is there any sign of a major deviation from Normality?

(b) All 20 subjects gambled a fixed amount more often when faced with a sure loss than when faced with a sure win. Give a 95% confidence interval for the mean percent increase in gambling a fixed amount when faced with a sure loss.

20.7 She Sounds Tall! Presented with recordings of a pair of people of the same sex speaking the same phrase, can a listener determine which speaker is taller simply from the sound of their voice? Twenty-four young adults at Washington University listened to 100 pairs of speakers, and within each pair were asked to indicate which of the two speakers was the taller. Here are the number correct (out of 100) for each of the 24 participants:[6] DATA TALL

65	61	67	59	58	62	56	67	61	67	63	53
68	49	66	58	69	70	65	56	68	56	58	70

Assume that these young adults can be regarded as an SRS of all young adults in the U.S. Use a 99% confidence interval to estimate the mean number correct in the population of all young adults in the U.S. Follow the four-step process as illustrated in Example 20.2.

20.4 The one-sample *t* test

Like the confidence interval, the t test is very similar to the z test we met earlier.

The One-Sample *t* Test

Draw an SRS of size n from a large population having unknown mean μ. To **test the hypothesis H_0: $\mu = \mu_0$**, compute the **one-sample t statistic**

$$t = \frac{\bar{x} - \mu_0}{s/\sqrt{n}}$$

In terms of a variable T having the $t(n - 1)$ distribution, the P-value for a test of H_0 against

H_a: $\mu > \mu_0$ is $P(T \geq t)$

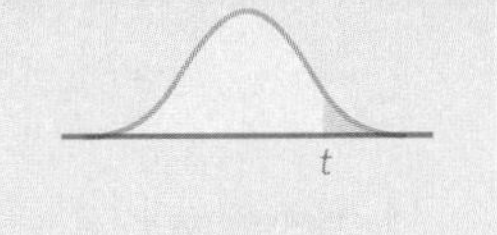

H_a: $\mu < \mu_0$ is $P(T \leq t)$

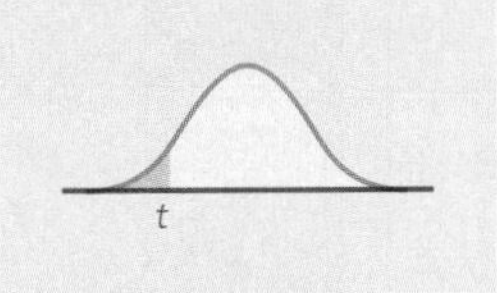

H_a: $\mu \neq \mu_0$ is $2P(T \geq |t|)$

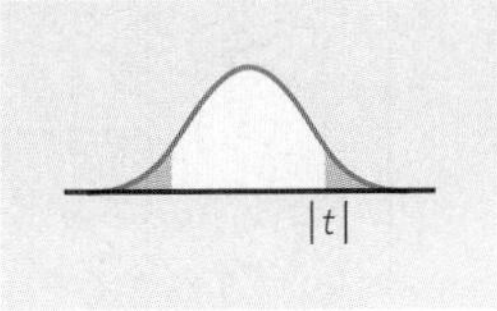

These P-values are exact if the population distribution is Normal and are approximately correct for large n in other cases.

We will provide guidelines on how large a sample size is needed for the *P*-values to be approximately correct in Section 20.7.

EXAMPLE 20.3 Water Quality

4 step

DATA WQUAL

We follow the four-step process for a significance test, outlined on page 401.

STATE: To investigate water quality, on August 8, 2010, the *Columbus Dispatch* took water samples at 20 Ohio State Park swimming areas. Those samples were taken to laboratories and tested for fecal coliform, which are *E. coli* bacteria found in human and animal feces. An unsafe level of fecal coliform means there's a higher chance that disease-causing bacteria are present and more risk that a swimmer will become ill. Ohio considers it unsafe if a 100-milliliter sample (about 3.3 ounces) of water contains more than 400 coliform bacteria.

PLAN: Experts caution that the tests are a snapshot of the quality of the water at the time they were taken. Fecal coliform levels can change as weather and other conditions change. So we ask the question in terms of the mean fecal coliform level μ for all these swimming areas. The null hypothesis is "level is not unsafe," and the alternative hypothesis is "level is unsafe."

$$H_0: \mu = 400$$
$$H_a: \mu > 400$$

These swimming areas had been deemed safe in the past and the reason data were collected was to determine if this continued to be true. So we are looking for evidence that past conditions have changed. If these areas had been found to be unsafe in the past, and we were seeking evidence that they are now safe, our null and alternative hypotheses would be $H_0: \mu = 400$ and $H_a: \mu < 400$.

SOLVE: Here are the fecal coliform levels found by the laboratories:[7]

160	40	2800	80	2000	2000	1500	400	150	500
3000	2200	15	80	2000	2000	2600	600	1000	1500

Are these data good evidence that, on average, the fecal coliform levels in these swimming areas were unsafe?

First, check the conditions for inference. We are willing to regard these particular 20 samples as an SRS from a large population of possible samples. Figure 20.3 is a histogram of the data. We can't accurately judge Normality from 20 observations;

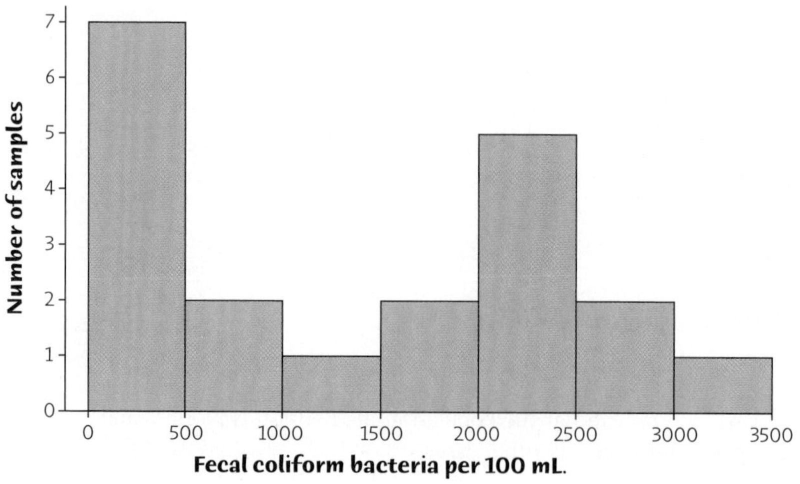

FIGURE 20.3
Histogram of the fecal coliform level in Example 20.3.

there are no outliers but the distribution of fecal coliform levels is somewhat skewed. *P*-values for the *t* test may be only approximately accurate.

The basic statistics are

$$\bar{x} = 1231 \quad \text{and} \quad s = 1038$$

The one-sample *t* statistic is

$$t = \frac{\bar{x} - \mu_0}{s/\sqrt{n}} = \frac{1231 - 400}{1038/\sqrt{20}} = 3.580$$

The *P*-value for $t = 3.580$ is the area to the right of 3.580 under the *t* distribution curve with degrees of freedom $n - 1 = 19$. Figure 20.4 shows this area. Software (see Figure 20.6) tells us that $P = 0.001$.

Without software, we can pin *P* between two values by using Table C. Search the df = 19 row of Table C for entries that bracket $t = 3.580$.

The observed *t* lies between the critical values for one-sided *P*-values 0.001 and 0.0005.

df = 19

t^*	3.579	3.883
One-sided *P*	0.001	0.0005

CONCLUDE: There is quite strong evidence ($P < 0.0025$) that, on average, fecal coliform levels in these Ohio State Park swimming areas are unsafe. ■

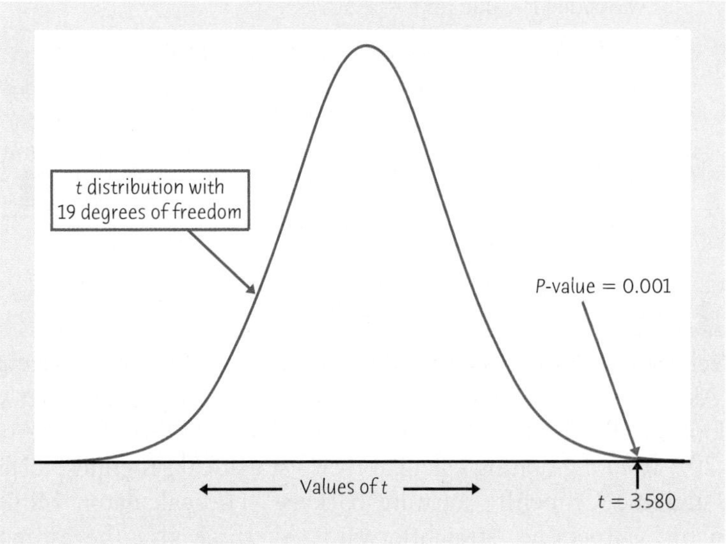

FIGURE 20.4
The *P*-value for the one-sided *t* test in Example 20.3.

LaunchPad Online Resources

- The **StatBoards video**, *Carrying Out a One-Sample t Test,* discusses examples of testing.

Apply Your Knowledge

20.8 Is It Significant? The one-sample *t* statistic for testing

$$H_0: \mu = 0$$
$$H_a: \mu > 0$$

from a sample of $n = 25$ observations has the value $t = 1.75$.

(a) What are the degrees of freedom for this statistic?

(b) Give the two critical values t^* from Table C that bracket t. What are the one-sided P-values for these two entries?

(c) Is the value $t = 1.75$ significant at the 10% level? Is it significant at the 5% level? Is it significant at the 1% level?

(d) (Optional) If you have access to suitable technology, give the exact one-sided P-value for $t = 1.75$.

20.9 Is It Significant? The one-sample t statistic from a sample of $n = 5$ observations for the two-sided test of

$$H_0: \mu = 50$$
$$H_a: \mu \neq 50$$

has the value $t = 2.50$.

(a) What are the degrees of freedom for t?

(b) Locate the two critical values t^* from Table C that bracket t. What are the two-sided P-values for these two entries?

(c) Is the value $t = 2.50$ statistically significant at the 10% level? At the 5% level? At the 1% level?

(d) (Optional) If you have access to suitable technology, give the exact two-sided P-value for $t = 2.50$.

4 step

20.10 She Sounds Tall, Continued. Do the data of Exercise 20.7 give good reason to think that the mean number of correct identifications in the population of all young adults in the U.S. is greater than 50 (the expected number correct if one is just guessing)? Carry out a test of significance, following the four-step process as illustrated in Example 20.3. DATA TALL

20.5 Using technology

Any technology suitable for statistics will implement the one-sample t procedures. As usual, you can read and use almost any output now that you know what to look for. Figure 20.5 displays output for the 95% confidence interval of Example 20.2 from a graphing calculator, two statistical programs, a spreadsheet program, and the CrunchIt! software package. The calculator, Minitab, JMP, and CrunchIt! outputs are straightforward. All three give the estimate $\bar{x}$ and the confidence interval plus a clearly labeled selection of other information. The confidence interval agrees with our hand calculation in Example 20.2. In general, software results are more accurate because of the rounding in hand calculations. Excel gives several descriptive measures but does not give the confidence interval. The entry labeled "Confidence Level (95.0%)" is the margin of error. You can use this together with $\bar{x}$ to get the interval using either a calculator or the spreadsheet's formula capability.

Figure 20.6 displays output for the t test in Example 20.3. The graphing calculator, Minitab, JMP, and CrunchIt! give the sample mean $\bar{x}$, the t statistic, and its P-value. Accurate P-values are the biggest advantage of software for the t procedures. Excel is, as usual, more awkward than software designed for statistics. It lacks a one-sample t test menu selection but does have a function named TDIST for tail areas under t density curves. The Excel output shows functions for the t statistic and its P-value to the right of the main display, along with their values $t = 3.582967678$ and $P = 0.00099192$.

Texas Instruments Graphing Calculator

```
TInterval
 (21.292,23.128)
 x̄=22.2100
 Sx=1.9625
 n=20.0000
```

Minitab

Session

One-Sample T: Percent

Variable	N	Mean	StDev	SE Mean	95% CI
Percent	20	22.210	1.963	0.439	(21.292, 23.128)

JMP

Confidence Intervals

Parameter	Estimate	Lower CI	Upper CI	1-Alpha
Mean	22.21	21.29151	23.12849	0.950

Excel

Microsoft Excel

	B	C	D	E	F
1	Percent				
2					
3	Mean	22.21			
4	Standard Error	0.4388			
5	Median	22.15			
6	Standard Deviation	1.96			
7	Sample Variance	3.85			
8	Range	8.30			
9	Minimum	18.70			
10	Maximum	27			
11	Count	20			
12	Confidence Level(95.0%)	0.9185			
13					

Sheet1 Sheet2 Sheet3

This is an estimate of μ.

This is the margin of error $\pm t^*$ SE.

CrunchIt!

Results - t 1-Sample

Export

n:	20
Sample Mean:	22.21
Standard Error:	0.4388
df:	19
95% ConfInt:	(21.29, 23.13)

FIGURE 20.5
The t confidence interval for Example 20.2: output from a graphing calculator, three statistical programs, and a spreadsheet program.

Texas Instruments Graphing Calculator

```
T-Test
 μ>400.0000
 t=3.5803
 p=.0010
 x̄=1231.2500
 Sx=1037.5374
 n=20.0000
```

Minitab

```
Session

One-Sample T: Fecal coliform per 100 ml

Test of mu = 400 vs > 400

                                                      95%
                                                     Lower
Variable             N   Mean   StDev   SE Mean     Bound      T       P
fecal coliform      20   1231    1038       232       830   3.58   0.001
```

JMP

Test Mean

Hypothesized Value	400
Actual Estimate	1231.25
DF	19
Std Dev	1037.54

t Test

Test Statistic	3.5830
Prob > |t|	0.0020*
Prob > t	0.0010*
Prob < t	0.9990

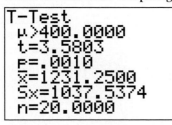

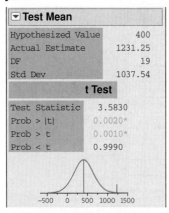

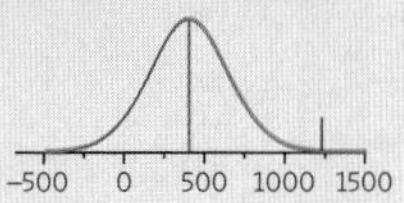

Excel

Microsoft Excel

	B	C	D	E	F
1	fecal coliform per 100 ml				
2			(C3.400)/C4	3.582967678	
3	Mean	1231.25	TDIST (E2,19,1)	0.00099192	
4	Standard Error	232.0004			
5	Median	1250			
6	Standard Deviation	1037.5374			
7	Sample Variance	1076483.88			
8	Range	2985			
9	Minimum	15			
10	Maximum	3000			
11	Count	20			
12	Confidence Level(95%)	485.5825			
13					

Sheet1

This is the *t* statistic.

This is the *P*-value.

FIGURE 20.6
The *t* test for Example 20.3: output from a graphing calculator, three statistical programs, and a spreadsheet program.

CrunchIt!

Results - t 1-Sample

Export ▾

Null hypothesis:	Population mean = 400
Alternative hypothesis:	Population mean > 400

n:	20
Sample Mean:	1231
Standard Error:	232.0
df:	19
t statistic:	3.583
P-value:	0.0009919

FIGURE 20.6
(Continued)

20.6 Matched pairs *t* procedures

Often the goal of an investigation is to demonstrate that a treatment causes an observed effect. In Chapter 9, we learned that randomized comparative studies are more convincing than single-sample investigations for demonstrating causation. For that reason, one-sample inference is less common than comparative inference. One common design to compare two treatments makes use of one-sample procedures. Matched pairs designs were discussed in Chapter 9. In a **matched pairs design,** subjects are matched in pairs and each treatment is given to one subject in each pair. Another situation calling for matched pairs is before-and-after observations on the same subjects.

matched pairs design

Matched Pairs *t* Procedures

To compare the responses to the two treatments in a matched pairs design, find the *difference* between the responses within each pair. Then apply the one-sample *t* procedures to these differences.

The parameter μ in a matched pairs *t* procedure is the mean of the differences in the responses to the two treatments within matched pairs of subjects in the entire population.

EXAMPLE 20.4 Do Chimpanzees Collaborate?

STATE: Humans often collaborate to solve problems. Will chimpanzees recruit another chimp when solving a problem requires collaboration? Researchers presented chimpanzee subjects with food outside their cage that they could bring within reach by pulling two ropes, one attached to each end of the food tray. If a chimp pulled only one rope, the rope came loose and the food was lost. Another chimp was available as a partner, but only if the subject unlocked a door joining two cages. (Chimpanzees learn these things quickly.) The same eight chimpanzee subjects faced this problem in two versions: the two ropes were close enough together that one chimp could pull both (no collaboration needed), or the two ropes were too far apart for one chimp to

CHIMPS

Manoj Shah/Getty Images

pull both (collaboration needed). Table 20.1 shows how often in 24 trials for each version each subject opened the door to recruit another chimp as partner.[8] Is there evidence that chimpanzees recruit partners more often when a problem requires collaboration?

TABLE 20.1 TRIALS (OUT OF 24) ON WHICH CHIMPANZEES RECRUITED A PARTNER

	COLLABORATION NEEDED		
CHIMPANZEE	YES	NO	DIFFERENCE
Namuiska	16	0	16
Kalema	16	1	15
Okech	23	5	18
Baluku	19	3	16
Umugenzi	15	4	11
Indi	20	9	11
Bili	24	16	8
Asega	24	20	4

PLAN: Take μ to be the mean difference (collaboration required minus not) in the number of times a subject recruited a partner. The null hypothesis says that the need for collaboration has no effect, and H_a says that partners are recruited more often when the problem requires collaboration. So we test the hypotheses

$$H_0: \mu = 0$$
$$H_a: \mu > 0$$

SOLVE: The subjects are "semi-free-ranging chimpanzees at Ngamba Island Chimpanzee Sanctuary in Uganda." We are willing to regard them as an SRS from their species. To analyze the data, we examine the difference in the number of times a chimp recruited a partner, so subtract the "no collaboration needed" count from the "collaboration needed" count for each subject. The eight differences form a single sample from a population with unknown mean μ. They appear in the "Difference" column in Table 20.1. All the chimpanzees recruited a partner more often when the ropes were too far apart to be pulled by one chimp.

The stemplot in Figure 20.7 creates the impression of a left-skewed distribution. This is a bit misleading, as the *dotplot* in the bottom part of Figure 20.7 shows. A dotplot simply places the observations on an axis, stacking observations that have the same value. It gives a good picture of distributions with only whole-number values. We know that observations that can take only whole-number values cannot come from a Normal population.[9] In practice, researchers are willing to treat such observations as coming from a Normal population if there are more than just a few possible values and the distribution appears approximately Normal. Of course, we can't assess approximate Normality from just eight observations, but there are no signs of major departures from Normality. The researchers used the matched pairs t test.

The eight differences have

$$\bar{x} = 12.375 \quad \text{and} \quad s = 4.749$$

The one-sample t statistic is therefore

$$t = \frac{\bar{x} - 0}{s/\sqrt{n}} = \frac{12.375 - 0}{4.749/\sqrt{8}} = 7.37$$

```
0 | 4
0 | 8
1 | 11
1 | 5668
```

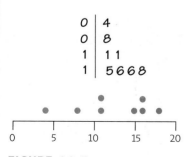

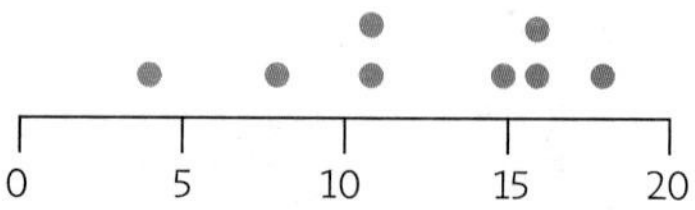

FIGURE 20.7
Stemplot and dotplot of the differences, for Example 20.4.

Find the *P*-value from the $t(7)$ distribution. (Remember that the degrees of freedom are one less than the sample size.) Table C shows that 7.37 is greater than the critical value for one-sided $P = 0.0005$. The *P*-value is therefore less than 0.0005. Software says that $P = 0.000077$.

df = 7

t^*	4.785	5.408
One-sided *P*	0.001	0.0005

CONCLUDE: The data give very strong evidence ($P < 0.0005$) that chimpanzees recruit a collaborator more often when faced with a problem that requires a collaborator to solve. That is, chimpanzees recognize when collaboration is necessary, a skill that they share with humans. ■

Example 20.4 illustrates how to turn matched pairs data into single-sample data by taking differences within each pair. We are making inferences about a single population, the population of all differences within matched pairs. *It is incorrect to ignore the matching and analyze the data as if we had two samples of chimpanzees, one facing ropes close together and the other facing ropes far apart.* Inference procedures for comparing two samples assume that the samples are selected independently of each other. This condition does not hold when the same subjects are measured twice. The proper analysis depends on the design used to produce the data.

CAUTION

LaunchPad Online Resources

- The **StatBoards video**, *Matched Pairs t Procedures,* discusses examples of inference for matched pairs.

Apply Your Knowledge

Many exercises from this point on ask you to give the P-value of a t test. If you have suitable technology, give the exact P-value. Otherwise, use Table C to give two values between which P lies.

20.11 Eye Grease. Athletes performing in bright sunlight often smear black eye grease under their eyes to reduce glare. Does eye grease work? In one study, 16 student subjects took a test of sensitivity to contrast after three hours facing into bright sun, both with and without eye grease. (Greater sensitivity to contrast improves vision and glare reduces sensitivity to contrast.) This is a matched pairs design. Here are the differences in sensitivity, with eye grease minus without eye grease:[10] DATA EYEGRS

4 step

0.07	0.64	−0.12	−0.05	−0.18	0.14	−0.16	0.03
0.05	0.02	0.43	0.24	−0.11	0.28	0.05	0.29

We want to know whether eye grease increases sensitivity to contrast on the average. Do the data support this idea? Complete the Plan, Solve, and Conclude steps of the four-step process, following the model of Example 20.4.

20.12 Eye Grease, Continued. How much more sensitive to contrast are athletes with eye grease than without eye grease? Give a 99% confidence interval to answer this question. DATA EYEGRS

20.7 Robustness of *t* procedures

The *t* confidence interval and test are exactly correct when the distribution of the population is exactly Normal. No real data are exactly Normal. At best, the Normal distribution is an excellent approximation to the actual distribution of data

CATCHING CHEATERS

A certification test for surgeons asks 277 multiple-choice questions. Smith and Jones have 193 common right answers and 53 identical wrong choices. The computer flags their 246 identical answers as evidence of possible cheating. They sue. The court wants to know how unlikely it is that exams this similar would occur just by chance. That is, the court wants a *P*-value. Statisticians offer several *P*-values based on different models for the exam-taking process. They all say that results this similar would almost never happen just by chance. Smith and Jones fail the exam.

from real studies.[11] The usefulness of the *t* procedures in practice therefore depends on how strongly they are affected by lack of Normality.

Robust Procedures

A confidence interval or significance test is called **robust** if the confidence level or *P*-value does not change very much when the conditions for use of the procedure are violated.

The condition that the population is Normal rules out outliers, so the presence of outliers shows that this condition is not fulfilled. The *t* procedures are not robust against outliers unless the sample is large, because $\bar{x}$ and s are not resistant to outliers.

Fortunately, the *t* procedures are quite robust against non-Normality of the population except when outliers or strong skewness are present. (Skewness is more serious than other kinds of non-Normality.) As the size of the sample increases, the central limit theorem ensures that the distribution of the sample mean $\bar{x}$ becomes more nearly Normal, and that the *t* distribution becomes more accurate for critical values and *P*-values of the *t* procedures.

Always make a plot to check for skewness and outliers before you use the *t* procedures for small samples. For most purposes, you can safely use the one-sample *t* procedures when $n \geq 15$ unless an outlier or quite strong skewness is present. Here are practical guidelines for inference on a single mean.[12]

Using the *t* Procedures

Except in the case of small samples, the condition that the data are an SRS from the population of interest is more important than the condition that the population distribution is Normal.

- *Sample size less than 15:* Use *t* procedures if the data appear close to Normal (roughly symmetric, single peak, no outliers). If the data are clearly skewed or if outliers are present, do not use *t*.
- *Sample size at least 15:* The *t* procedures can be used except in the presence of outliers or strong skewness.
- *Large samples:* The *t* procedures can be used even for clearly skewed distributions when the sample is large, roughly $n \geq 40$.

EXAMPLE 20.5 Can We Use *t*?

Figure 20.8 shows plots of several data sets. For which of these can we safely use the *t* procedures?[13]

- Figure 20.8(a) is a histogram of the percent of each state's adult residents who are college graduates. *We have data on the entire population of 50 states, so inference is not needed.* We can calculate the exact mean for the population. There is no uncertainty due to having only a sample from the population, and no need for a confidence interval or test. *If these data were an SRS from a larger population, t inference would be safe despite the mild skewness because* $n = 50$.

CAUTION

- Figure 20.8(b) is a stemplot of the force required to pull apart 20 pieces of Douglas fir. *The data are strongly skewed to the left with possible low outliers, so we cannot trust the t procedures for* $n = 20$.

- Figure 20.8(c) is a stemplot of the lengths of 23 specimens of the red variety of the tropical flower *Heliconia*. *The data are mildly skewed to the right and there are no outliers. We can use the t distributions for such data.*
- Figure 20.8(d) is a histogram of the heights of the female students in a college class. *This distribution is quite symmetric and appears close to Normal. We can use the t procedures for any sample size.* ■

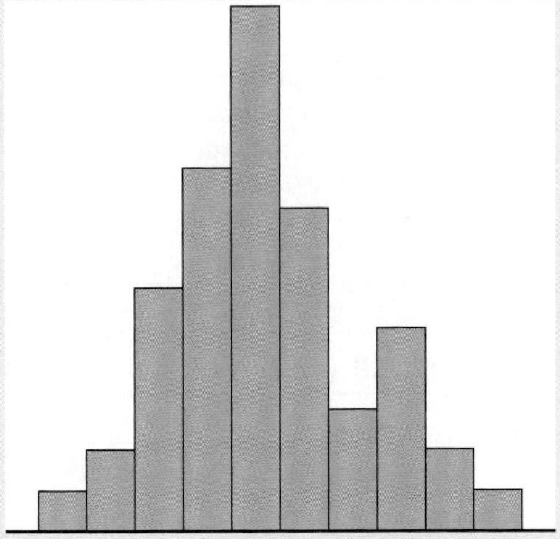

(a)

23	0
24	0
25	
26	5
27	
28	7
29	
30	2 5 9
31	3 9 9
32	0 3 3 6 7 7
33	0 2 3 6

(b)

37	4 8 9
38	0 0 1 1 2 2 8 9
39	2 6 8
40	6 7
41	5 7 9 9
42	0 2
43	1

(c)

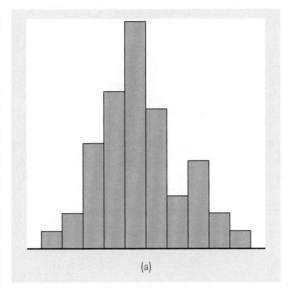

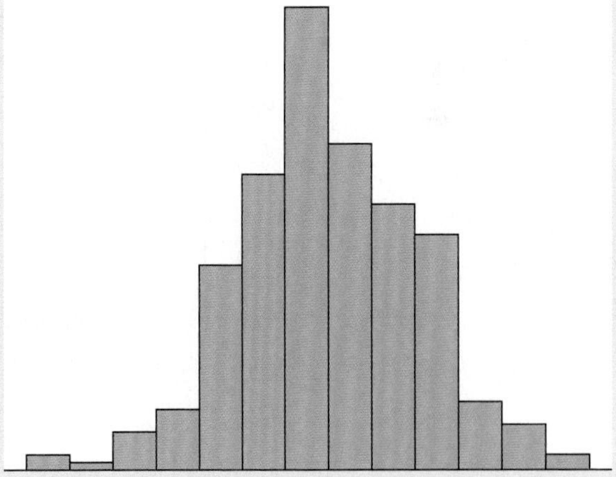

(d)

FIGURE 20.8
Can we use *t* procedures for these data? (a) Percent of adult college graduates in the 50 states. *No,* this is an entire population, not a sample. (b) Force required to pull apart 20 pieces of Douglas fir. *No,* there are just 20 observations and strong skewness. (c) Lengths of 23 tropical flowers of the same variety. *Yes,* the sample is large enough to overcome the mild skewness. (d) Heights of female students. *Yes, for any size sample,* because the distribution is close to Normal.

LaunchPad Online Resources

- The **StatBoards video**, ***When Can We Use t Procedures,*** **discusses examples of checking assumptions and robustness.**

Apply Your Knowledge

© Eric Nathan/Alamy

20.13 Diamonds. A group of earth scientists studied the small diamonds found in a nodule of rock carried up to the earth's surface in surrounding rock. This is an opportunity to examine a sample from a single population of diamonds formed in a single event deep in the earth.[14] Table 20.2 presents data on the nitrogen content (parts per million) and the abundance of carbon 13 in these diamonds. (Carbon has several isotopes, or forms with different numbers of neutrons in the nuclei of their atoms. Carbon 12 makes up almost 99% of natural carbon. The abundance of carbon 13 is measured by the ratio of carbon 13 to carbon 12, in parts per thousand more or less than a standard. The minus signs in the data mean that the ratio is smaller in these diamonds than in standard carbon.) DMNDS

TABLE 20.2 NITROGEN AND CARBON 13 IN A SAMPLE OF DIAMONDS

DIAMOND	NITROGEN (PPM)	CARBON 13 RATIO	DIAMOND	NITROGEN (PPM)	CARBON 13 RATIO
1	487	−2.78	13	273	−2.73
2	1430	−1.39	14	94	−2.33
3	60	−4.26	15	69	−3.83
4	244	−1.19	16	262	−2.04
5	196	−2.12	17	120	−2.82
6	274	−2.87	18	302	−0.84
7	41	−3.68	19	75	−3.57
8	54	−3.29	20	242	−2.42
9	473	−3.79	21	115	−3.89
10	30	−4.06	22	65	−3.87
11	98	−1.83	23	311	−1.58
12	41	−4.03	24	61	−3.97

We would like to estimate the mean abundance of both nitrogen and carbon 13 in the population of diamonds represented by this sample. Examine the data for nitrogen. Can we use a t confidence interval for mean nitrogen? Explain your answer. Give a 90% confidence interval if you think the result can be trusted.

20.14 Diamonds, Continued. Examine the data in Table 20.2 on abundance of carbon 13. Can we use a t confidence interval for mean carbon 13? Explain your answer. Give a 90% confidence interval if you think the result can be trusted. DMNDS

20.8 Resampling and standard errors*

In Section 20.1, we defined the standard error of a statistic to be an estimate of the standard deviation of the statistic. For the sample mean, $\bar{x}$, the standard deviation is $\sigma/\sqrt{n}$. A natural estimate is $s/\sqrt{n}$, which is commonly referred to as the standard error of $\bar{x}$ in textbooks and computer output.

*This material can be skipped if you prefer not to cover resampling methods.

We can also use resampling methods to estimate the standard deviation of many sample statistics, such as the sample mean or sample median. The basic idea is to take many SRSs from a population, compute the sample statistic for each, and then compute the standard deviation of these values. This is an estimate of the standard deviation of the sample statistic, sometimes referred to as the resampling estimate of the standard error. (Remember, the standard deviation of a *sample statistic* is the standard error. There is no need to divide the standard deviation of a *sample statistic* by $\sqrt{n}$.) In practice, we typically do not know the exact distribution of the population. For example, as is the case in this chapter, we may know that the population is (approximately) Normal, but not know the mean μ and standard deviation σ. For purposes of resampling, in such cases we estimate these unknown features of the population distribution from data.[15] The next example demonstrates this process.

EXAMPLE 20.6 Using Resampling to Estimate the Standard Error of $\bar{x}$

In Example 17.3, the sweetness losses for a new cola, as measured by 10 trained tasters, were:

2.0 0.4 0.7 2.0 −0.4 2.2 −1.3 1.2 1.1 2.3

We assumed these values could be regarded as an SRS from a Normal population. The average sweetness loss is given by the sample mean $\bar{x} = 1.02$. The sample standard deviation is $s = 1.20$. For purposes of resampling, we therefore assume the population distribution is Normal with mean $\mu = 1.02$ and standard deviation $\sigma = 1.20$.

Now draw 1000 SRSs of size 10 from a Normal population with mean $\mu = 1.02$ and standard deviation $\sigma = 1.20$. (We used software to do this, and in practice, one uses software to apply resampling methods.) The results are displayed in Figure 20.9. The standard deviation of these 1000 sample means is found to be 0.3786. This is our resampling estimate of the standard error of $\bar{x}$ for an SRS of size 10. The formula in Section 20.1 tells us that the standard error is $s/\sqrt{n} = 1.20/\sqrt{10} = 0.3795$. The values computed by resampling and by the formula are in close agreement. ■

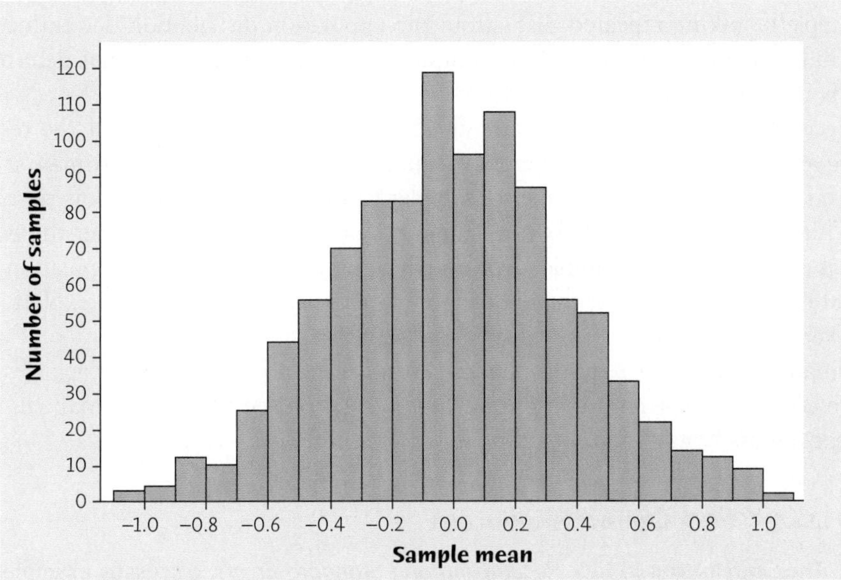

FIGURE 20.9
A histogram, for Example 20.6, of the means of 1000 SRSs of size 10 from a Normal population with mean $\mu = 0$ and standard deviation $\sigma = 1.20$.

For the sample mean $\bar{x}$ we have a simple formula, $s/\sqrt{n}$, for the standard error and there is no need to use resampling to estimate the standard error. However, what can we do when we don't have a simple formula for the standard error of a statistic? For example, suppose we want to estimate the standard error of a sample median. The close agreement between the resampling estimate and the formula in Example 20.6 supports our claim that resampling can be used to estimate standard errors for statistics such as sample medians. The next example illustrates how to do this.

EXAMPLE 20.7 Using Resampling to Estimate the Standard Error of the Sample Median

For the data of Example 20.6, the sample median is 1.15. We have not been given a formula for the standard error of a sample median. Using the 1000 SRSs in Example 20.6, we can calculate the median of each sample and compute the standard deviation of these 1000 medians. We find that this standard deviation is 0.442. This is our resampling estimate of the standard error of the sample median in this cola example. ■

For a Normal population, the mean and median are the same. Thus, the sample median could be used to estimate the center of a Normal population. However, we prefer to use $\bar{x}$ to estimate the center. Example 20.7 helps us see why. The standard error of the sample median in Example 20.7 is larger than the standard error of the sample mean in Example 20.6. In other words, the sample mean produces a "more precise" estimate of the center (in the sense of small standard deviation) than the sample median.

Resampling estimates of standard errors can be used to construct confidence intervals. However, there are several methods for generating resampling-based confidence intervals and these more advanced methods are beyond the scope of this book.

We close with some cautions concerning resampling. First, we must resample in the same manner that we obtain our data. If our data are obtained by an SRS, we resample by taking repeated SRSs from the population distribution determined by the null hypothesis and/or estimates of any population parameters that determine the population distribution.

Second, resampling provides an *estimate* of a standard error. Repeat the resampling and you will obtain a different estimate. The accuracy of the estimate is improved by taking a larger number of samples to estimate the sampling distribution.

Third, we should always check assumptions, such as Normality, about the population distribution. In Examples 20.6 and 20.7, we assumed that the population distribution was Normal. Do the data support this assumption? You will explore this in Exercise 20.44.

Finally, resampling requires the use of software. Some software packages are more convenient for resampling than others. The online technology manuals and videos discuss how to do resampling for several packages.

LaunchPad Online Resources

- The **StatBoards video**, *Resampling and Standard Errors,* discusses examples of using resampling to estimate standard errors.

Apply Your Knowledge

20.15 Estimating a Standard Error. In Example 17.2, we saw that the sweetness losses for a cola currently on the market, as measured by 10 trained tasters, were

1.6 0.4 0.5 −2.0 1.5 −1.1 1.3 −0.1 −0.3 1.2

We assumed these values could be regarded as an SRS from a Normal population and saw that the average sweetness loss was $\bar{x} = 0.3$. The sample standard deviation is $s = 1.2$.

(a) Compute the standard error of $\bar{x} = 0.3$ for these data.

(b) Use the same process as in Example 20.6 to estimate the standard error. Does this agree with your answer in part (a)?

20.16 Estimating a Standard Error. Use your 1000 SRSs in Exercise 20.15 to estimate the standard error of the sample median for the 10 scores. How does this standard error compare with that for $\bar{x}$ in Exercise 20.15?

CHAPTER 20 SUMMARY

Chapter Specifics

- Tests and confidence intervals for the mean μ of a Normal population are based on the sample mean $\bar{x}$ of an SRS. Because of the central limit theorem, the resulting procedures are approximately correct for other population distributions when the sample is large.
- The standardized sample mean is the **one-sample z statistic**

$$z = \frac{\bar{x} - \mu}{\sigma/\sqrt{n}}$$

If we knew σ, we would use the z statistic and the standard Normal distribution.

- In practice, we do not know σ. Replace the standard deviation $\sigma/\sqrt{n}$ of $\bar{x}$ by the **standard error** $s/\sqrt{n}$ to get the **one-sample t statistic**

$$t = \frac{\bar{x} - \mu}{s/\sqrt{n}}$$

The t statistic has the **t distribution** with $n - 1$ degrees of freedom.

- There is a t distribution for every positive **degrees of freedom.** All are symmetric distributions similar in shape to the standard Normal distribution. The t distribution approaches the $N(0, 1)$ distribution as the degrees of freedom increase.
- A level C **confidence interval for the mean μ** of a Normal population is

$$\bar{x} \pm t^* \frac{s}{\sqrt{n}}$$

The **critical value** t^* is chosen so that the t curve with $n - 1$ degrees of freedom has area C between $-t^*$ and t^*.

- **Significance tests** for H_0: $\mu = \mu_0$ are based on the t statistic. Use P-values or fixed significance levels from the $t(n - 1)$ distribution.
- Use these one-sample procedures to analyze **matched pairs** data by first taking the difference within each matched pair to produce a single sample.

- The t procedures are quite **robust** when the population is non-Normal, especially for larger sample sizes. The t procedures are useful for non-Normal data when $n \geq 15$ unless the data show outliers or strong skewness. When $n \geq 40$, the t procedures can be used even for clearly skewed distributions.
- Standard errors can be estimated using **resampling**. This is a useful method when dealing with statistics and parameters other than means.

Link It

In Chapters 16 to 18, we began our study of inference. We focused on inference for a population mean based on a sample from a Normal population. We included the unrealistic assumption that we knew the population standard deviation σ. The reason was to make the underlying mathematics simpler. We were able to use what we learned about the Normal distribution in Chapters 3 and 12 and what we learned about the sampling distribution of the sample mean in Chapter 15 to construct confidence intervals for and conduct hypothesis tests about the population mean. Computing P-values and sample sizes was possible using what we had learned about Normal probability calculations.

In this chapter, we continue our study of inference about a population mean based on a sample from a Normal population in the more realistic setting that we do not know σ. The basic ideas of Chapters 16 to 18 still apply, but now we use the t distribution rather than the Normal distribution. Unfortunately, the mathematics associated with the t distribution is more complicated than that associated with the Normal distribution. We must rely on approximations from tables or statistical software to calculate P-values (and determining sample sizes is also more complicated when we use the t distribution). As we continue our study of statistical inference in realistic settings, statistical software will be invaluable.

As we saw in Chapter 9, statistical studies comparing two or more groups are preferable to one-sample procedures if we want to demonstrate that a treatment causes an observed response. As a first step toward developing methods for comparing means from two populations, we considered matched pairs studies (also discussed in Chapter 9) and saw that by looking at differences, we could use our one-sample t procedures to compare two means. In the next chapter, we consider inference for comparing the means of two populations when we have independent samples from the two populations. This will expand our tools for doing statistical inference in settings that we encounter in practice.

LaunchPad Online Resources

If you are having difficulty with any of the sections of this chapter, these online resources should help prepare you to solve the exercises at the end of this chapter.

- **StatTutor** starts with a video review of each section and asks a series of questions to check your understanding. You may find this helpful if you need additional help understanding any of the material in this chapter.
- **LearningCurve** provides you with a series of questions about the chapter geared to your level of understanding.

CHECK YOUR SKILLS

20.17 We prefer the t procedures to the z procedures for inference about a population mean because

(a) z can be used only for large samples.
(b) z requires that you know the population standard deviation σ.
(c) z requires that you can regard your data as an SRS from the population.

20.18 You are testing H_0: $\mu = 100$ against H_a: $\mu < 100$ based on an SRS of nine observations from a Normal population. The data give $\bar{x} = 98$ and $s = 3$. The value of the t statistic is

(a) −6. (b) −2. (c) −98.

20.19 You are testing H_0: $\mu = 100$ against H_a: $\mu > 100$ based on an SRS of 16 observations from a Normal population. The t statistic is $t = 2.13$. The degrees of freedom for the t statistic are

(a) 17. (b) 16. (c) 15.

20.20 The P-value for the statistic in Exercise 20.19

(a) falls between 0.05 and 0.10.
(b) falls between 0.01 and 0.05.
(c) is less than 0.01.

20.21 You have an SRS of six observations from a Normally distributed population. What critical value would you use to obtain an 80% confidence interval for the mean μ of the population?

(a) 1.440 (b) 1.476 (c) 2.015

20.22 You are testing H_0: $\mu = 0$ against H_a: $\mu \neq 0$ based on an SRS of six observations from a Normal population. What values of the t statistic are statistically significant at the $\alpha = 0.001$ level?

(a) $t > 6.869$
(b) $t < -6.869$ or $t > 6.869$
(c) $t < -5.893$ or $t > 5.893$

20.23 Twenty-five adult citizens of the U.S. were asked to estimate the average income of all U.S. households. The mean estimate was $\bar{x}$ = \$45,000 and s = \$15,000. (*Note:* the actual average household income at the time of the study was about \$68,000.) Assume the 25 adults in the study can be considered an SRS from the population of all adult citizens of the U.S. A 95% confidence interval for the mean estimate of the average income of all U.S. households is

(a) \$38,808 to \$51,192.
(b) \$39,120 to \$50,880.
(c) \$39,867 to \$50,133.

20.24 Which of the following would cause the most worry about the validity of the confidence interval you calculated in Exercise 20.23?

(a) You notice that there is a clear outlier in the data.
(b) A stemplot of the data shows a mild right-skew.
(c) You do not know the population standard deviation σ.

20.25 Which of these settings does *not* allow use of a matched pairs t procedure?

(a) You interview both spouses in 400 married couples and ask each about the average number of minutes each day they spend using social media.
(b) You interview a sample of 225 unmarried male students and another sample of 225 unmarried female students and ask each about the average number of minutes each day they spend using social media.
(c) You interview 100 female students in their freshman year and again in their senior year and ask each about the average number of minutes each day she spends using social media.

20.26 Because the t procedures are robust, the most important condition for their safe use is that

(a) the population standard deviation σ is known.
(b) the population distribution is exactly Normal.
(c) the data can be regarded as an SRS from the population.

CHAPTER 20 EXERCISES

20.27 **Read carefully.** You read in the report of a psychology experiment: "Separate analyses for our two groups of 12 participants revealed no overall placebo effect for our student group (mean = 0.08, SD = 0.37, $t(11) = 0.49$) and a significant effect for our non-student group (mean = 0.35, SD = 0.37, $t(11) = 3.25, p < 0.01$)."[16] The null hypothesis is that the mean effect is zero. What are the correct values of the two t statistics based on the means and standard deviations? Compare each correct t-value with the critical values in Table C. What can you say about the two-sided P-value in each case?

20.28 **Body mass index of young women.** In Example 16.1 (page 375), we developed a 95% z confidence interval for the mean body mass index (BMI) of women aged 20 to 29 years, based on a national random sample of 654 such women. We assumed there that the population standard deviation was known to be $\sigma = 7.5$. In fact, the sample data had mean BMI $\bar{x} = 26.8$ and standard deviation $s = 7.42$. What is the 95% t confidence interval for the mean BMI of all young women?

20.29 Reading scores in Dallas. The Trial Urban District Assessment (TUDA) is a government-sponsored study of student achievement in large urban school districts. TUDA gives a reading test scored from 0 to 500. A score of 243 is a "basic" reading level and a score of 281 is "proficient." Scores for a random sample of 1400 eighth-graders in Dallas had $\bar{x} = 248$ with standard error 1.0.[17]

(a) We don't have the 1400 individual scores, but use of the *t* procedures is surely safe. Why?
(b) Give a 99% confidence interval for the mean score of all Dallas eighth-graders. (Be careful: the report gives the standard error of $\bar{x}$, not the standard deviation *s*.)
(c) Urban children often perform below the basic level. Is there good evidence that the mean for all Dallas eighth-graders is more than the basic level?

20.30 Color and cognition. In a randomized comparative experiment on the effect of color on the performance of a cognitive task, researchers randomly divided 69 subjects (27 males and 42 females ranging in age from 17 to 25 years) into three groups. Participants were asked to solve a series of six anagrams. One group was presented with the anagrams on a blue screen, one group saw them on a red screen, and one group had a neutral screen. The time, in seconds, taken to solve the anagrams was recorded. The paper reporting the study gives $\bar{x} = 11.58$ and $s = 4.37$ for the times of the 23 members of the neutral group.[18]

(a) Give a 95% confidence interval for the mean time in the population from which the subjects were recruited.
(b) What conditions for the population and the study design are required by the procedure you used in part (a)? Which of these conditions are important for the validity of the procedure in this case?

20.31 The placebo effect. The placebo effect is particularly strong in patients with Parkinson's disease. To understand the workings of the placebo effect, scientists measure activity at a key point in the brain when patients receive a placebo that they think is an active drug and also when no treatment is given.[19] The same six patients are measured both with and without the placebo, at different times.

(a) Explain why the proper procedure to compare the mean response to placebo with control (no treatment) is a matched pairs *t* test.
(b) The six differences (treatment minus control) had $\bar{x} = -0.326$ and $s = 0.181$. Is there significant evidence of a difference between treatment and control?

20.32 What do you make of Mitt? Twenty-nine college students, identified as having a positive attitude about Mitt Romney as compared to Barack Obama in the 2012 presidential election, were asked to rate how trustworthy the face of Mitt Romney appeared, as represented in their mental image of Mitt Romney's face. Ratings were on a scale of 0 to 7, with 0 being "not at all trustworthy" and 7 being "extremely trustworthy." Here are the 29 ratings:[20] MITT

2.6	3.2	4.7	3.3	3.4	3.6	3.7	3.8	3.9	4.1
4.2	4.9	5.7	4.2	3.9	3.2	4.1	5.0	5.0	4.6
4.6	3.9	3.9	5.3	2.8	2.6	3.0	3.3	3.7	

(a) Suppose we can consider this an SRS of all U.S. college students. Make a stemplot. Is there any sign of major deviation from Normality?
(b) Give a 95% confidence interval for the mean rating
(c) Is there significant evidence at the 5% level that the mean rating is greater than 3.5 (a neutral rating)?

20.33 Exhaust from school buses. In a study of exhaust emissions from school buses, the pollution intake by passengers was determined for a sample of nine school buses used in the Southern California Air Basin. The pollution intake is the amount of exhaust emissions, in grams per person, that would be inhaled while traveling on the bus during its usual 18-mile trip on congested freeways from South Central LA to a magnet school in West LA. (As a reference, the average intake of motor emissions of carbon monoxide in the LA area is estimated to be about 0.000046 grams per person.) Here are the amounts for the nine buses when driven with the windows open:[21] EMIT

1.15	0.33	0.40	0.33	1.35	0.38	0.25	0.40	0.35

(a) Make a stemplot. Are there outliers or strong skewness that would preclude use of the *t* procedures?
(b) A good way to judge the effect of outliers is to do your analysis twice, once with the outliers and a second time without them. Give two 90% confidence intervals, one with all the data and one with the outliers removed, for the mean pollution intake among all school buses used in the Southern California Air Basin that travel the route investigated in the study.
(c) Compare the two intervals in part (b). What is the most important effect of removing the outliers?

20.34 A big toe problem. Hallux abducto valgus (call it HAV) is a deformation of the big toe that often requires surgery. Doctors used X rays to measure the angle (in degrees) of deformity in 38 consecutive patients under the age of 21 who came to a medical center for surgery to correct HAV. The

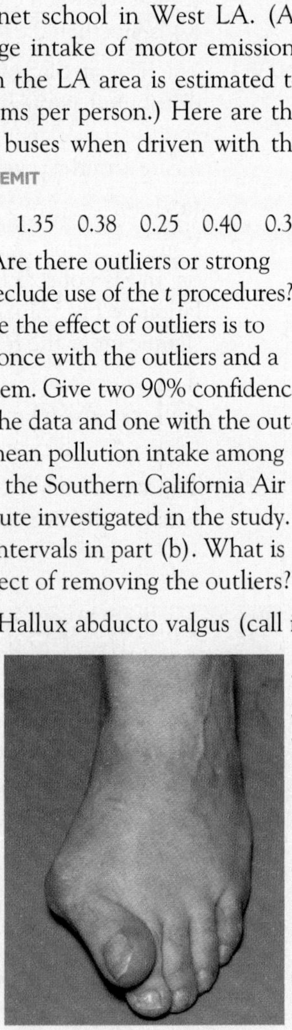

angle is a measure of the seriousness of the deformity. Here are the data:[22] DATA BIGTOE

28	32	25	34	38	26	25	18	30	26	28	13	20
21	17	16	21	23	14	32	25	21	22	20	18	26
16	30	30	20	50	25	26	28	31	38	32	21	

It is reasonable to regard these patients as a random sample of young patients who require HAV surgery. Carry out the Solve and Conclude steps for a 95% confidence interval for the mean HAV angle in the population of all such patients.

20.35 **An outlier's effect.** Our bodies have a natural electrical field that is known to help wounds heal. Does changing the field strength slow healing? A series of experiments with newts investigated this question. In one experiment, the two hind limbs of 12 newts were assigned at random to either experimental or control groups. This is a matched pairs design. The electrical field in the experimental limbs was reduced to zero by applying a voltage. The control limbs were left alone. Here are the rates at which new cells closed a razor cut in each limb, in micrometers per hour:[23] DATA NEWTS

Newt	1	2	3	4	5	6	7	8	9	10	11	12
Control limb	36	41	39	42	44	39	39	56	33	20	49	30
Experimental limb	28	31	27	33	33	38	45	25	28	33	47	23

(a) Why is this a matched pairs design? Explain your answer.
(b) Make a stemplot of the differences between limbs of the same newt (control limb minus experimental limb). There is a high outlier.
(c) A good way to judge the effect of an outlier is to do your analysis twice, once with the outlier and a second time without it. Carry out two *t* tests to see if the mean healing rate is significantly lower in the experimental limbs, with one test including all 12 newts and another omitting the outlier. What are the test statistics and their *P*-values? Does the outlier have a strong influence on your conclusion?

20.36 **An outlier's effect.** A good way to judge the effect of an outlier is to do your analysis twice, once with the outlier and a second time without it. The data in Exercise 20.34 follow a Normal distribution quite closely except for one patient with HAV angle 50 degrees, a high outlier. DATA BIGTOE
(a) Find the 95% confidence interval for the population mean based on the 37 patients who remain after you drop the outlier.
(b) Compare your interval in part (a) with your interval from Exercise 20.34. What is the most important effect of removing the outlier?

20.37 **Men of few words?** Researchers claim that women speak significantly more words per day than men. One estimate is that a woman uses about 20,000 words per day while a man uses about 7,000. To investigate such claims, one study used a special device to record the conversations of male and female university students over a four-day period. From these recordings, the daily word counts of the 20 men in the study were determined. Here are their daily word counts:[24] DATA TALKING

28,408	10,084	15,931	21,688	37,786
10,575	12,880	11,071	17,799	13,182
8,918	6,495	8,153	7,015	4,429
10,054	3,998	12,639	10,974	5,255

(a) Examine the data. Is it reasonable to use the *t* procedures (assume these men are an SRS of all male students at this university)?
(b) If your conclusion in part (a) is "yes," do the data give convincing evidence that the mean number of words per day of men at this university differs from 7,000?

20.38 **Genetic engineering for cancer treatment.** Here's a new idea for treating advanced melanoma, the most serious kind of skin cancer: genetically engineer white blood cells to better recognize and destroy cancer cells, then infuse these cells into patients. The subjects in a small initial study were 11 patients whose melanoma had not responded to existing treatments. One question was how rapidly the new cells would multiply after infusion, as measured by the doubling time in days. Here are the doubling times:[25] DATA CNCRTRT

1.4	1.0	1.3	1.0	1.3	2.0	0.6	0.8	0.7	0.9	1.9

(a) Examine the data. Is it reasonable to use the *t* procedures?
(b) Give a 90% confidence interval for the mean doubling time. Are you willing to use this interval to make an inference about the mean doubling time in a population of similar patients? Explain your reasoning.

20.39 **Genetic engineering for cancer treatment, continued.** Another outcome in the cancer experiment described in Exercise 20.38 is measured by a test for the presence of cells that trigger an immune response in the body and so may help fight cancer. Here are data for the 11 subjects: counts of active cells per 100,000 cells before and after infusion of the modified cells. The difference (after minus before) is the response variable. DATA MORECAN

Before	14	0	1	0	0	0	0	20	1	6	0
After	41	7	1	215	20	700	13	530	35	92	108
Difference	27	7	0	215	20	700	13	510	34	86	108

(a) Explain why this is a matched pairs design.
(b) Examine the data. Is it reasonable to use the *t* procedures?
(c) If your conclusion in part (a) is "yes," do the data give convincing evidence that the count of active cells is higher after treatment?

20.40 Kicking a helium-filled football. Does a football filled with helium travel farther than one filled with ordinary air? To test this, the *Columbus Dispatch* conducted a study. Two identical footballs, one filled with helium and one filled with ordinary air, were used. A casual observer was unable to detect a difference in the two footballs. A novice kicker was used to punt the footballs. A trial consisted of kicking both footballs in a random order. The kicker did not know which football (the helium-filled or the air-filled football) he was kicking. The distance of each punt was recorded. Then another trial was conducted. A total of 39 trials were run. Here are the data for the 39 trials, in yards that the footballs traveled. The difference (helium minus air) is the response variable.[26] DATA FTBALL

Helium	25	16	25	14	23	29	25	26	22	26
Air	25	23	18	16	35	15	26	24	24	28
Difference	0	−7	7	−2	−12	14	−1	2	−2	−2
Helium	12	28	28	31	22	29	23	26	35	24
Air	25	19	27	25	34	26	20	22	33	29
Difference	−13	9	1	6	−12	3	3	4	2	−5
Helium	31	34	39	32	14	28	30	27	33	11
Air	31	27	22	29	28	29	22	31	25	20
Difference	0	7	17	3	−14	−1	8	−4	8	−9
Helium	26	32	30	29	30	29	29	30	26	
Air	27	26	28	32	28	25	31	28	28	
Difference	−1	6	2	−3	2	4	−2	2	−2	

(a) Examine the data. Is it reasonable to use the *t* procedures?
(b) If your conclusion in part (a) is "yes," do the data give convincing evidence that the helium-filled football travels farther than the air-filled football?

20.41 Growing trees faster. The concentration of carbon dioxide (CO_2) in the atmosphere is increasing rapidly due to our use of fossil fuels. Because plants use CO_2 to fuel photosynthesis, more CO_2 may cause trees and other plants to grow faster. An elaborate apparatus allows researchers to pipe extra CO_2 to a 30-meter circle of forest. They selected two nearby circles in each of three parts of a pine forest and randomly chose one of each pair to receive extra CO_2. The response variable is the mean increase in base area for 30 to 40 trees in a circle during a growing season. We measure this in percent increase per year. Here are one year's data:[27] DATA TREES

PAIR	CONTROL PLOT	TREATED PLOT	TREATED − CONTROL
1	9.752	10.587	0.835
2	7.263	9.244	1.981
3	5.742	8.675	2.933

(a) State the null and alternative hypotheses. Explain clearly why the investigators used a one-sided alternative.
(b) Carry out a test and report your conclusion in simple language.
(c) The investigators used the test you just carried out. Any use of the *t* procedures with samples this size is risky. Why?

20.42 Fungus in the air. The air in poultry-processing plants often contains fungus spores. Inadequate ventilation can affect the health of the workers. The problem is most serious during the summer. To measure the presence of spores, air samples are pumped to an agar plate and "colony-forming units (CFUs)" are counted after an incubation period. Here are data from two locations in a plant that processes 37,000 turkeys per day, taken on four days in the summer. The units are CFUs per cubic meter of air.[28] DATA FUNGUS

	DAY 1	DAY 2	DAY 3	DAY 4
Kill room	3175	2526	1763	1090
Processing	529	141	362	224
Kill room − Processing	2646	2385	1401	866

(a) Explain carefully why these are matched pairs data.
(b) The spore count is clearly higher in the kill room. Give sample means and a 90% confidence interval to estimate how much higher. Be sure to state your conclusion in plain language.
(c) You will often see the *t* procedures used for data like these. You should regard the results as only rough approximations. Why?

20.43 Weeds among the corn. Velvetleaf is a particularly annoying weed in corn fields. It produces lots of seeds, and the seeds wait in the soil for years until conditions are right. How many seeds do velvetleaf plants produce? Here are counts from 28 plants that came up in a corn field when no herbicide was used:[29] DATA WEEDS

© CuboImages srl/Alamy

2450	2504	2114	1110	2137	8015	1623	1531	2008	1716
721	863	1136	2819	1911	2101	1051	218	1711	164
2228	363	5973	1050	1961	1809	130	880		

We would like to give a confidence interval for the mean number of seeds produced by velvetleaf plants. Alas, the *t* interval can't be safely used for these data. Why not?

20.44 **Sweetening colas.** Cola makers test new recipes for loss of sweetness during storage. Trained tasters rate the sweetness before and after storage. Here are the sweetness losses (sweetness before storage minus sweetness after storage) found by 10 tasters for one new cola recipe: DATA COLA

2.0	0.4	0.7	2.0	−0.4	2.2	−1.3	1.2	1.1	2.3

Take the data from these 10 carefully trained tasters as an SRS from a large population of all trained tasters.

(a) Use these data to see if there is good evidence that the cola lost sweetness.

(b) It is not uncommon to see the *t* procedures used for data like these. However, you should regard the results as only rough approximations. Why?

20.45 **How much oil?** How much oil will ultimately be produced by wells in a given field is key information in deciding whether to drill more wells. Here are the estimated total amounts of oil recovered from 64 wells in the Devonian Richmond Dolomite area of the Michigan basin, in thousands of barrels:[30] DATA OIL

21.7	53.2	46.4	42.7	50.4	97.7	103.1	51.9
43.4	69.5	156.5	34.6	37.9	12.9	2.5	31.4
79.5	26.9	18.5	14.7	32.9	196	24.9	118.2
82.2	35.1	47.6	54.2	63.1	69.8	57.4	65.6
56.4	49.4	44.9	34.6	92.2	37.0	58.8	21.3
36.6	64.9	14.8	17.6	29.1	61.4	38.6	32.5
12.0	28.3	204.9	44.5	10.3	37.7	33.7	81.1
12.1	20.1	30.5	7.1	10.1	18.0	3.0	2.0

Take these wells to be an SRS of wells in this area.

(a) Give a 95% *t* confidence interval for the mean amount of oil recovered from all wells in this area.

(b) Make a graph of the data. The distribution is very skewed, with several high outliers. A computer-intensive method that gives accurate confidence intervals without assuming any specific shape for the distribution gives a 95% confidence interval of 40.28 to 60.32. How does the *t* interval compare with this? Should the *t* procedures be used with these data?

20.46 ***E. coli* in swimming areas.** To investigate water quality, the *Columbus Dispatch* took water samples at 16 Ohio State Park swimming areas in central Ohio. Those samples were taken to laboratories and tested for *E. coli*, which are bacteria that can cause serious gastrointestinal problems. If a 100-milliliter sample (about 3.3 ounces) of water contains more than 130 *E. coli* bacteria, it is considered unsafe. Here are the *E. coli* levels found by the laboratories:[31] DATA ECOLI

291.0	190.4	47.0	86.0	44.0	18.9	1.0	50.0
10.9	45.7	28.5	8.6	9.6	16.0	34.0	18.9

Take these water samples to be an SRS of the water in all swimming areas in central Ohio.

(a) Are these data good evidence that on average the *E. coli* levels in these swimming areas were unsafe?

(b) Make a graph of the data. The distribution is very skewed. Another method that gives *P*-values without assuming any specific shape for the distribution gives a *P*-value of 0.9997 for the question in part (a). How does the one-sample *t* test compare with this? Should the *t* procedures be used with these data?

The following exercises ask you to answer questions from data without having the details outlined for you. The four-step process is illustrated in Examples 20.2, 20.3, and 20.4. The exercise statements give you the ***State*** *step. Follow the* ***Plan, Solve,*** *and* ***Conclude*** *steps in your work.*

20.47 **Natural weed control?** Fortunately, we aren't really interested in the number of seeds velvetleaf plants produce (see Exercise 20.43). The velvetleaf seed beetle feeds on the seeds and might be a natural weed control. Here are the total seeds, seeds infected by the beetle, and percent of seeds infected for 28 velvetleaf plants: DATA WDCTRL

Seeds	2450	2504	2114	1110	2137	8015	1623	1531	2008	1716
Infected	135	101	76	24	121	189	31	44	73	12
Percent	5.5	4.0	3.6	2.2	5.7	2.4	1.9	2.9	3.6	0.7
Seeds	721	863	1136	2819	1911	2101	1051	218	1711	164
Infected	27	40	41	79	82	85	42	0	64	7
Percent	3.7	4.6	3.6	2.8	4.3	4.0	4.0	0.0	3.7	4.3
Seeds	2228	363	5973	1050	1961	1809	130	880		
Infected	156	31	240	91	137	92	5	23		
Percent	7.0	8.5	4.0	8.7	7.0	5.1	3.8	2.6		

Do a complete analysis of the percent of seeds infected by the beetle. Include a 90% confidence interval for the mean percent infected in the population of all velvetleaf plants. Do you think that the beetle is very helpful in controlling the weed? Why is analyzing percent of seeds infected more useful than analyzing number of seeds infected?

20.48 Recruiting T cells. There is evidence that cytotoxic T lymphocytes (T cells) participate in controlling tumor growth and that they can be harnessed to use the body's immune system to treat cancer. One study investigated the use of a T cell-engaging antibody, blinatumomab, to recruit T cells to control tumor growth. The data below are T cell counts (1000 per microliter) at baseline (beginning of the study) and after 20 days on blinatumomab for six subjects in the study.[32] The difference (after 20 days minus baseline) is the response variable. TCELLS

Baseline	0.04	0.02	0.00	0.02	0.38	0.33
After 20 days	0.28	0.47	1.30	0.25	1.22	0.44
Difference	0.24	0.45	1.30	0.23	0.84	0.11

Do the data give convincing evidence that the mean count of T cells is higher after 20 days on blinatumomab?

20.49 Recruiting T cells, continued. Give a 95% confidence interval for the mean difference in T cell counts (after 20 days minus baseline) in Exercise 20.48 TCELLS

20.50 Mutual funds performance. Mutual funds often compare their performance with a benchmark provided by an "index" that describes the performance of the class of assets in which the fund invests. For example, the Vanguard International Growth Fund benchmarks its performance against the Spliced International Index. Table 20.3 gives annual returns (percent) for the fund and the index. Does the fund's performance differ significantly from that of its benchmark? MFUND

TABLE 20.3 A MUTUAL FUND VERSUS ITS BENCHMARK INDEX

YEAR	FUND RETURN (%)	INDEX RETURN (%)	YEAR	FUND RETURN (%)	INDEX RETURN (%)
1984	−1.02	7.38	1999	26.34	26.96
1985	56.94	56.16	2000	−8.60	−14.17
1986	56.71	69.44	2001	−18.92	−21.44
1987	12.48	24.63	2002	−17.79	−15.94
1988	11.61	28.27	2003	34.45	38.59
1989	24.76	10.54	2004	18.95	20.25
1990	−12.05	−23.45	2005	15.00	13.54
1991	4.74	12.13	2006	25.92	26.34
1992	−5.79	−12.17	2007	15.98	11.17
1993	44.74	32.56	2008	−44.94	−43.38
1994	0.76	7.78	2009	41.63	31.78
1995	14.89	11.21	2010	15.66	8.13
1996	14.65	6.05	2011	−13.68	−13.71
1997	4.12	1.78	2012	20.01	16.83
1998	16.93	20.00	2013	22.95	15.29

(a) Explain clearly why the matched pairs t test is the proper choice to answer this question.
(b) Do a complete analysis that answers the question posed.

20.51 Right versus left. The design of controls and instruments affects how easily people can use them. Timothy Sturm investigated this effect in a course project, asking 25 right-handed students to turn a knob (with their right hands) that moved an indicator by screw action. There were two identical instruments: one with a right-hand thread (the knob turns clockwise), and the other with a left-hand thread (the knob turns counterclockwise). Table 20.4 gives the times in seconds each subject took to move the indicator a fixed distance.[33] RTLFT

(a) Each of the 25 students used both instruments. Explain briefly how you would use randomization in arranging the experiment.
(b) The project hoped to show that right-handed people find right-hand threads easier to use. Do an analysis that leads to a conclusion about this issue.

TABLE 20.4 PERFORMANCE TIMES (SECONDS) USING RIGHT-HAND AND LEFT-HAND THREADS

SUBJECT	RIGHT THREAD	LEFT THREAD	SUBJECT	RIGHT THREAD	LEFT THREAD
1	113	137	14	107	87
2	105	105	15	118	166
3	130	133	16	103	146
4	101	108	17	111	123
5	138	115	18	104	135
6	118	170	19	111	112
7	87	103	20	89	93
8	116	145	21	78	76
9	75	78	22	100	116
10	96	107	23	89	78
11	122	84	24	85	101
12	103	148	25	88	123
13	116	147			

20.52 Comparing two drugs. Makers of generic drugs must show that they do not differ significantly from the "reference" drugs that they imitate. One aspect in which drugs might differ is their extent of absorption in the blood. Table 20.5 gives data taken from 20 healthy nonsmoking male subjects for one pair of drugs.[34] This is a matched pairs design. Numbers 1 to 20 were assigned at random to the subjects. Subjects 1 to 10 received the generic drug first, followed by the reference drug. Subjects 11 to 20 received the reference drug first, followed by the generic drug. In all cases, a washout period separated the two drugs so that the first had disappeared from the blood before the subject took the second. By randomizing the order, we eliminate the order in which the drugs were administered from being confounded with the difference in the absorption in the blood. Do the drugs differ significantly in the amount absorbed in the blood? DATA DRUGS

TABLE 20.5 ABSORPTION EXTENT FOR TWO VERSIONS OF A DRUG

SUBJECT	REFERENCE DRUG	GENERIC DRUG
15	4108	1755
3	2526	1138
9	2779	1613
13	3852	2254
12	1833	1310
8	2463	2120
18	2059	1851
20	1709	1878
17	1829	1682
2	2594	2613
4	2344	2738
16	1864	2302
6	1022	1284
10	2256	3052
5	938	1287
7	1339	1930
14	1262	1964
11	1438	2549
1	1735	3340
19	1020	3050

20.53 Practical significance? Give a 90% confidence interval for the mean time advantage of right-hand over left-hand threads in the setting of Exercise 20.51. Do you think that the time saved would be of practical importance if the task were performed many times—for example, by an assembly-line worker? To help answer this question, find the mean time for right-hand threads as a percent of the mean time for left-hand threads. DATA RTLFT

20.54 Bad weather, bad tips? As part of the study of tipping in a restaurant that we met in Example 16.3 (page 381) the psychologists also studied the size of the tip in a restaurant when a message indicating that the next day's weather would be bad was written on the bill. Here are tips from 20 patrons, measured in percent of the total bill:[35] DATA TIP3

18.0	19.1	19.2	18.8	18.4	19.0	18.5	16.1	16.8	14.0
17.0	13.6	17.5	20.0	20.2	18.8	18.0	23.2	18.2	19.4

Do the data give convincing evidence that the mean percentage tip for all patrons of this restaurant when their bill contains a message that the next day's weather will be bad is less than 20%? (Note that 20% is an often-recommended size for restaurant tips.)

20.55 *t* versus *z*. If you examine Table C, you will notice that critical values of the *t* distribution get closer and closer to the corresponding critical values of the Normal distribution as the number of degrees of freedom increases. You can see this by comparing the *z* critical values at the bottom of Table C with the *t* critical values in the corresponding column. This suggests that, for very large sample sizes, inference based on the Normal probability calculations in Chapters 15 and 16 (pretending σ is known) and inference based on the *t* distribution as discussed in this chapter (σ is not known) may give essentially the same answer if sample sizes are large and we pretend that our estimate of σ is the true value of σ. Many statistical software packages will calculate *t* probabilities. If you have access to such software, answer the following questions.

(a) Use software to determine how large a sample size (or how many degrees of freedom) is needed for the critical value of the *t* distribution to be within 0.01 of the corresponding critical value of the Normal distribution for a 90%, 95%, and 99% confidence interval for a population mean.

(b) Based on your findings, how large a sample size do you think is needed for inference using the Normal distribution and inference using the *t* distribution to give very similar results if σ (both the true value and its estimate) is 1? If σ (both the true value and its estimate) is 100?

Exploring the Web

20.56 Matched pairs. Find an example of a matched pairs study on the web. The *Journal of the American Medical Association* (`jama.amaassn.org`), *Science Magazine* (`www.sciencemag.org`), the *Canadian Medical Association Journal* (`www.cmaj.ca`), the *Journal of Statistics Education* (`www.amstat.org/`

publications/jse), or perhaps the *Journal of Quantitative Analysis in Sports* (www.bepress.com/jqas) are possible sources. To help locate an article, look through the abstracts of articles. Once you find a suitable article, read it, and then briefly describe the study (including why it is a matched pairs study) and its conclusions. If *P*-values, means, standard deviations, and so on are reported, be sure to include them in your summary. Also, be sure to give the reference (either the web link or the journal, issue, year, title of the paper, authors, and page numbers).

20.57 Improper use of a t procedure. Search the web for an example of an improper use of a *t* procedure. You might try using Google to do a search on "improper use of a *t* test." Summarize the study in which the *t* procedure was used and discuss how it was used improperly. Be sure to provide a link to your example or an appropriate reference.

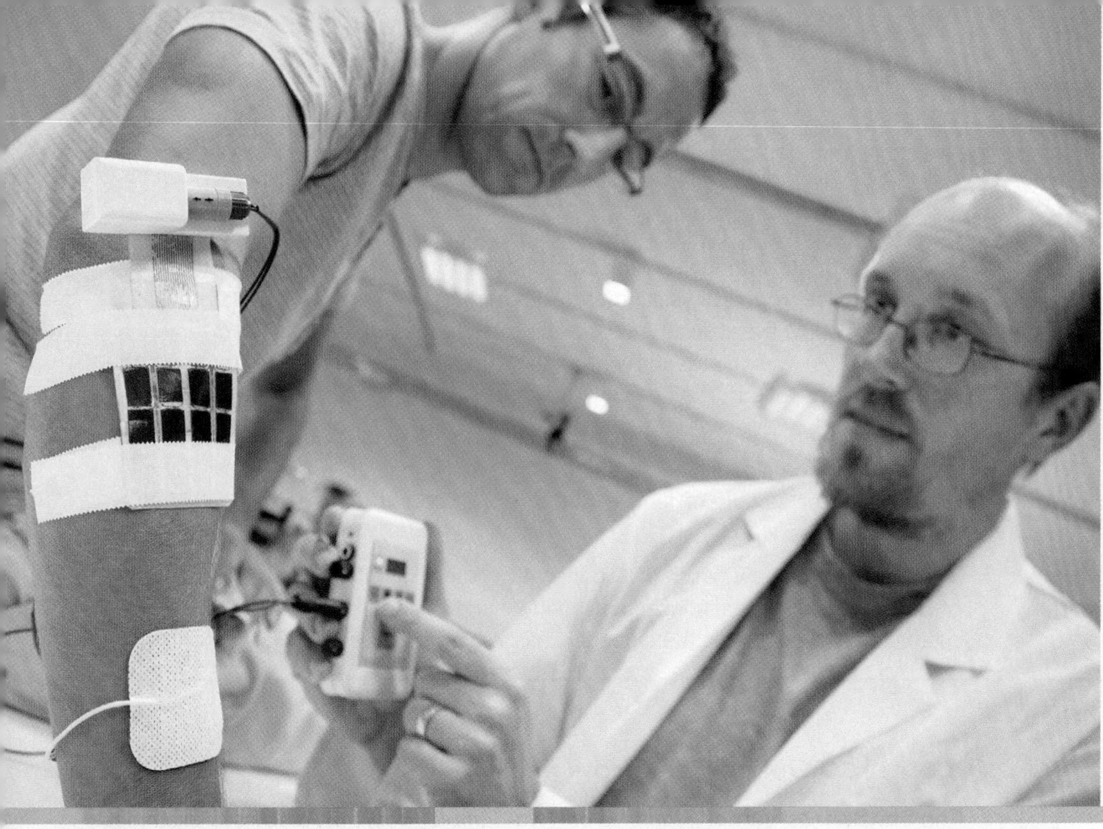

© Javier Larrea/age fotostock

CHAPTER 21

Comparing Two Means

In this chapter we cover...

21.1 Two-sample problems
21.2 Comparing two population means
21.3 Two-sample *t* procedures
21.4 Using technology
21.5 Robustness again
21.6 Details of the *t* approximation*
21.7 Avoid the pooled two-sample *t* procedures*
21.8 Avoid inference about standard deviations*
21.9 Permutation tests*

Comparing two populations or two treatments is one of the most common situations encountered in statistical practice. We call such situations *two-sample problems.*

Two-Sample Problems

- The goal of inference is to compare the responses to two treatments or to compare the characteristics of two populations.
- We have a separate sample from each treatment or each population.

21.1 Two-sample problems

A two-sample problem can arise from a randomized comparative experiment that randomly divides subjects into two groups and exposes each group to a different treatment. Comparing random samples separately selected from two populations is also a two-sample problem. Unlike the matched pairs designs studied earlier, there is no matching of the individuals in the two samples. The two samples are assumed to be independent and can be of different sizes. Inference procedures for two-sample data differ from those for matched pairs. Here are some typical two-sample problems:

EXAMPLE 21.1 Two-Sample Problems

- Does regular physical therapy help lower-back pain? A randomized experiment assigned patients with lower-back pain to two groups: 142 received an examination and advice from a physical therapist; another 144 received regular physical therapy for up to five weeks. After a year, the change in their level of disability (0% to 100%) was assessed by a doctor who did not know which treatment the patients had received.
- A psychologist develops a test that measures social insight. He compares the social insight of female college students with that of male college students by giving the test to a sample of female students and a separate, independent sample of male students.
- A bank wants to know which of two incentive plans will most increase the use of its credit cards. It offers each incentive to independent random samples of credit card customers and compares the amounts charged during the following six months. ■

LaunchPad Online Resources

- The **StatBoards video**, *Examples of Two-Sample Problems*, discusses examples of two-sample problems.

Apply Your Knowledge

Statistics in Your World: DRIVING WHILE FASTING

Muslims fast from sunrise to sunset during the month of Ramadan. Does this affect the rate of traffic accidents? Fasting can improve alertness, reducing accidents. Or it can cause dehydration, increasing accidents. Data from Turkey show a statistically significant increase, starting two weeks into Ramadan. Ah, but because Ramadan follows a lunar calendar, it cycles through the year. Perhaps accidents go down during a winter Ramadan (alertness) but go up during a summer Ramadan (longer fast and dehydration). Ask the statisticians this question and get their favorite answer: we need more data.

Which data design? *Each situation described in Exercises 21.1 to 21.4 requires inference about a mean or means. Identify each as involving (1) a single sample, (2) matched pairs, or (3) two independent samples. The procedures of Chapter 20 apply to designs (1) and (2). We are about to learn procedures for design (3).*

21.1 How Sweet It Is. Choose 10 agricultural plots of land located in different regions of the state. Plant half of each plot with one variety of sweet corn and the other with a different variety. Compare the sugar contents of both varieties of corn.

21.2 Exercise and Memory. To examine the effect of physical exertion on the memory of a witnessed event, researchers randomly divided police officers into two groups. One group completed a high-intensity physical assault exercise and the other did no exercise. Both groups were then placed in a live, occupationally relevant scenario. After the scenario, the mean scores on an exam testing recall of what was witnessed during the scenario were compared.

21.3 Smart Phone Battery Life. To check the lifetime of a new battery in a soon-to-be-released smartphone, an engineer surfed the web over 4G continuously, beginning with a fully charged battery, until the battery was drained. The engineer repeated this for a total of 10 times. He compared the results with the lifetime of 12.8 hours for the current battery used in the company's smartphone.

21.4 Smart Phone Battery Life, Continued. Another engineer is checking the lifetime of the new battery with a prototype that she just developed. She wants to know how the lifetimes of the new battery and her prototype compare. She measures the lifetime of the new battery and her prototype by continuously surfing the web over 4G until the battery is drained. She repeats this a total of 10 times for each type of battery.

21.2 Comparing two population means

Comparing two populations or the responses to two treatments starts with data analysis: make boxplots, stemplots (for small samples), or histograms (for larger samples) and compare the shapes, centers, and variabilities of the two samples. The most common goal of inference is to compare the average or typical responses in the two populations. When data analysis suggests that both population distributions are symmetric, and especially when they are at least approximately Normal, we want to compare the population means. Here are the conditions for inference about means.

Conditions for Inference Comparing Two Means

We have **two SRSs**, from two distinct populations. The samples are **independent**. That is, one sample has no influence on the other. Matching violates independence, for example. We measure the same response variable for both samples.

Both populations are **Normally distributed**. The means and standard deviations of the populations are unknown. In practice, it is enough that the distributions have similar shapes and that the data have no strong outliers.

Call the variable we measure x_1 in the first population and x_2 in the second, because the variable may have different distributions in the two populations. Here is how we describe the two populations:

Population	Variable	Mean	Standard Deviation
1	x_1	μ_1	σ_1
2	x_2	μ_2	σ_2

There are four unknown parameters: the two means and the two standard deviations. The subscripts remind us which population a parameter describes. We want to compare the two population means, either by giving a confidence interval for their difference $\mu_1 - \mu_2$ or by testing the hypothesis of no difference, $H_0: \mu_1 = \mu_2$. The hypothesis of no difference is used when investigating whether a treatment has an effect.

We use the sample means and standard deviations to estimate the unknown parameters. Again, subscripts remind us which sample a statistic comes from. Here is how we describe the samples:

Population	Sample Size	Sample Mean	Sample Standard Deviation
1	n_1	$\bar{x}_1$	s_1
2	n_2	$\bar{x}_2$	s_2

To make inferences about the difference $\mu_1 - \mu_2$ between the means of the two populations, we start from the difference $\bar{x}_1 - \bar{x}_2$ between the means of the two samples.

EXAMPLE 21.2 Daily Activity and Obesity

STATE: People gain weight when they take in more energy from food than they expend. James Levine and his collaborators at the Mayo Clinic investigated the link between obesity and energy spent on daily activity.[1]

Choose 20 healthy volunteers who don't exercise. Deliberately choose 10 who are lean and 10 who are mildly obese but still healthy. Attach sensors that monitor the subjects' every move for 10 days. Table 21.1 presents data on the time (in minutes per day) that the subjects spent standing or walking, sitting, and lying down. Do lean and obese people differ in the average time they spend standing and walking?

TABLE 21.1 TIME (MINUTES PER DAY) SPENT IN THREE DIFFERENT POSTURES BY LEAN AND OBESE SUBJECTS

GROUP	SUBJECT	STAND/WALK	SIT	LIE
Lean	1	511.100	370.300	555.500
Lean	2	607.925	374.512	450.650
Lean	3	319.212	582.138	537.362
Lean	4	584.644	357.144	489.269
Lean	5	578.869	348.994	514.081
Lean	6	543.388	385.312	506.500
Lean	7	677.188	268.188	467.700
Lean	8	555.656	322.219	567.006
Lean	9	374.831	537.031	531.431
Lean	10	504.700	528.838	396.962
Obese	11	260.244	646.281	521.044
Obese	12	464.756	456.644	514.931
Obese	13	367.138	578.662	563.300
Obese	14	413.667	463.333	532.208
Obese	15	347.375	567.556	504.931
Obese	16	416.531	567.556	448.856
Obese	17	358.650	621.262	460.550
Obese	18	267.344	646.181	509.981
Obese	19	410.631	572.769	448.706
Obese	20	426.356	591.369	412.919

PLAN: Examine the data and carry out a test of hypotheses. We suspect in advance that lean subjects (Group 1) are more active than obese subjects (Group 2), so we test the hypotheses

$$H_0: \mu_1 = \mu_2$$
$$H_a: \mu_1 > \mu_2$$

SOLVE (first steps): Are the conditions for inference met? The subjects are volunteers, so they are not SRSs from all lean and mildly obese adults. The study tried to recruit comparable groups: all worked in sedentary jobs, none smoked or were taking medication, and so on. Setting clear standards like these helps make up for the fact that we can't reasonably get SRSs for so invasive a study. The subjects were not told that they were chosen from a larger group of volunteers because they did not exercise and were either lean or mildly obese. Because their willingness to volunteer isn't related to the purpose of the experiment, we will treat them as two independent SRSs.

A back-to-back stemplot (Figure 21.1) displays the data in detail. To make the plot, we rounded the data to the nearest 10 minutes and used 100s as stems and

Lean		Obese
	2	67
72	3	567
	4	11236
886410	5	
81	6	

FIGURE 21.1
Back-to-back stemplot of the times spent walking or standing, for Example 21.2.

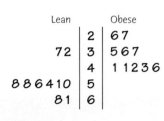

10s as leaves. The distributions are a bit irregular, as we expect with just 10 observations. There are no clear departures from Normality such as extreme outliers or skewness. The lean subjects as a group spend much more time standing and walking than do the obese subjects. Calculating the group means confirms this:

Group	n	Mean $\bar{x}$	Std. Dev. s
Group 1 (lean)	10	525.751	107.121
Group 2 (obese)	10	373.269	67.498

The observed difference in mean time per day spent standing or walking is

$$\bar{x}_1 - \bar{x}_2 = 525.751 - 373.269 = 152.482 \text{ minutes}$$

To complete the Solve step, we must learn the details of inference for comparing two means. ■

21.3 Two-sample *t* procedures

To assess the significance of the observed difference between the means of our two samples, we follow a familiar path. Whether an observed difference is surprising depends on the variability of the observations as well as on the two means. Widely different means can arise just by chance if the individual observations vary a great deal. To take variation into account, we would like to standardize the observed difference $\bar{x}_1 - \bar{x}_2$ by dividing by its standard deviation. This standard deviation of the difference in sample means is

$$\sqrt{\frac{\sigma_1^2}{n_1} + \frac{\sigma_2^2}{n_2}}$$

This standard deviation gets larger as either population gets more variable, that is, as σ_1 or σ_2 increases. It gets smaller as the sample sizes n_1 and n_2 increase.

Because we don't know the population standard deviations, we estimate them by the sample standard deviations from our two samples. The result is the **standard error**, or estimated standard deviation, of the difference in sample means:

standard error

$$\text{SE}_{\bar{x}_1 - \bar{x}_2} = \sqrt{\frac{s_1^2}{n_1} + \frac{s_2^2}{n_2}}$$

When we standardize the estimate by dividing it by its standard error, the result is the **two-sample *t* statistic**:

two-sample t statistic

$$t = \frac{\bar{x}_1 - \bar{x}_2}{\sqrt{\frac{s_1^2}{n_1} + \frac{s_2^2}{n_2}}}$$

The statistic t has the same interpretation as any z or t statistic: it says how far $\bar{x}_1 - \bar{x}_2$ is from zero in standard error units.

The two-sample t statistic has approximately a t distribution. It does not have exactly a t distribution even if the populations are both exactly Normal. In practice, however, the approximation is very accurate. There are two practical options for using the two-sample t procedures:

- **Option 1.** With software, use the statistic t with accurate critical values from the approximating t distribution. The degrees of freedom are calculated from the data by a somewhat messy formula. Moreover, the degrees of freedom may not be a whole number.

- **Option 2.** Without software, use the statistic t with critical values from the t distribution with *degrees of freedom equal to the smaller of* $n_1 - 1$ *and* $n_2 - 1$. These procedures are always conservative for any two Normal populations. The confidence interval has a margin of error *as large as or larger than* is needed for the desired confidence level. The significance test gives a P-value *equal to or greater than* the true P-value.

The two options are exactly the same except for the degrees of freedom used for t critical values and P-values. As the sample sizes increase, confidence levels and P-values from Option 2 become more accurate. The gap between what Option 2 reports and the truth is quite small unless the sample sizes are both small and unequal.[2]

The Two-Sample *t* Procedures

Draw an SRS of size n_1 from a large Normal population with unknown mean μ_1, and draw an independent SRS of size n_2 from another large Normal population with unknown mean μ_2. A level C **confidence interval for $\mu_1 - \mu_2$** is given by

$$(\bar{x}_1 - \bar{x}_2) \pm t^* \sqrt{\frac{s_1^2}{n_1} + \frac{s_2^2}{n_2}}$$

Here t^* is the critical value for confidence level C for the t distribution with degrees of freedom from either Option 1 (software) or Option 2 (the smaller of $n_1 - 1$ and $n_2 - 1$).

To **test the hypothesis H_0: $\mu_1 = \mu_2$**, calculate the **two-sample t statistic**

$$t = \frac{\bar{x}_1 - \bar{x}_2}{\sqrt{\dfrac{s_1^2}{n_1} + \dfrac{s_2^2}{n_2}}}$$

Find P-values from the t distribution with degrees of freedom from either Option 1 (software) or Option 2 (the smaller of $n_1 - 1$ and $n_2 - 1$).

EXAMPLE 21.3 Daily Activity and Obesity

4 step

We can now complete Example 21.2.

SOLVE (inference): The two-sample t statistic comparing the average minutes spent standing and walking in Group 1 (lean) and Group 2 (obese) is

$$t = \frac{\bar{x}_1 - \bar{x}_2}{\sqrt{\dfrac{s_1^2}{n_1} + \dfrac{s_2^2}{n_2}}}$$

$$= \frac{525.751 - 373.269}{\sqrt{\dfrac{107.121^2}{10} + \dfrac{67.498^2}{10}}}$$

$$= \frac{152.482}{40.039} = 3.808$$

Software (Option 1) gives one-sided P-value $P = 0.0008$ based on df $= 15.174$.

Without software, use the conservative Option 2. Because $n_1 - 1 = 9$ and $n_2 - 1 = 9$, there are nine degrees of freedom. Because H_a is one-sided, the P-value is the area to the right of $t = 3.808$ under the $t(9)$ curve. Figure 21.2 illustrates this P-value. Table C shows that $t = 3.808$ lies between the critical values t^* for 0.0025

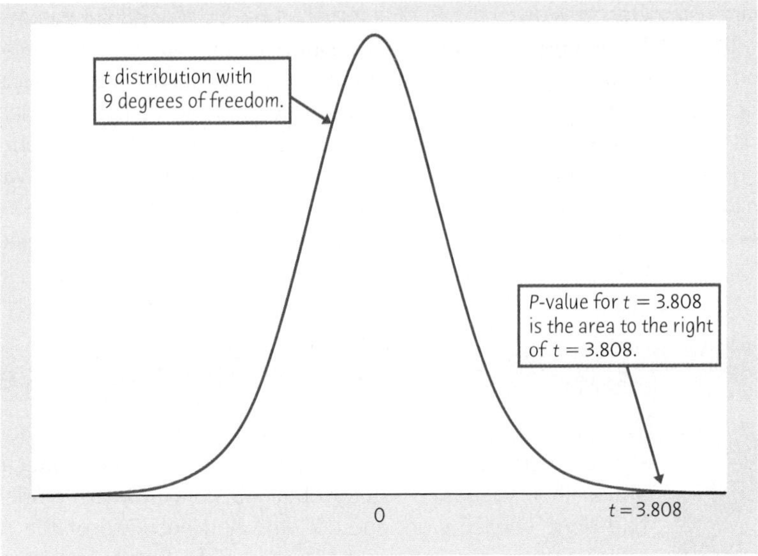

FIGURE 21.2
Using the conservative Option 2, the *P*-value in Example 21.3 comes from the *t* distribution with nine degrees of freedom.

and 0.001. So $0.001 < P < 0.0025$. Option 2 gives a larger (more conservative) *P*-value than Option 1. As usual, the practical conclusion is the same for both versions of the test.

df = 9

t^*	3.690	4.297
One-sided P	0.0025	0.001

CONCLUDE: There is very strong evidence ($P = 0.0008$) that, on average, lean people spend more time walking and standing than do moderately obese people. ■

Does lack of daily activity *cause* obesity? This is an observational study, and that affects our ability to draw cause-and-effect conclusions. It may be that some people are naturally more active and are therefore less likely to gain weight. Or it may be that people who gain weight reduce their activity level. The study went on to enroll most of the obese subjects in a weight-reduction program and most of the lean subjects in a supervised program of overeating. After eight weeks, the obese subjects had lost weight (mean eight kilograms) and the lean subjects had gained weight (mean four kilograms). But both groups kept their original allocation of time to the different postures. This suggests that time allocation is biological and influences weight, rather than the other way around. The authors remark: "It should be emphasized that this was a pilot study and that the results need to be confirmed in larger studies."

EXAMPLE 21.4 How Much More Active Are Lean People?

PLAN: To estimate how much more active lean people are, we give a 90% confidence interval for $\mu_1 - \mu_2$, the difference in average daily minutes spent standing and walking between lean and mildly obese adults.

SOLVE and CONCLUDE: As in Example 21.3, the conservative Option 2 uses nine degrees of freedom. Table C shows that the $t(9)$ critical value is $t^* = 1.833$. We are 90% confident that $\mu_1 - \mu_2$ lies in the interval

$$(\bar{x}_1 - \bar{x}_2) \pm t^*\sqrt{\frac{s_1^2}{n_1} + \frac{s_2^2}{n_2}}$$

$$= (525.751 - 373.269) \pm 1.833\sqrt{\frac{107.121^2}{10} + \frac{67.498^2}{10}}$$

$$= 152.482 \pm 73.390$$

$$= 79.09 \text{ to } 225.87 \text{ minutes}$$

Software using Option 1 gives the 90% interval as 82.35 to 222.62 minutes, based on t with 15.174 degrees of freedom. The Option 2 interval is wider because this method is conservative. Both intervals are quite wide because the samples are small and the variation among individuals, as measured by the two sample standard deviations, is large. Whichever interval we report, we are (at least) 90% confident that the mean difference in average daily minutes spent standing and walking between lean and mildly obese adults lies in this interval. ■

EXAMPLE 21.5 Community Service and Attachment to Friends

Hero Images/Getty Images

STATE: Do college students who have volunteered for community service work and those who have not differ in how attached they are to their friends? A study obtained data from 57 students who had done service work and 17 who had not. One of the response variables was a measure of attachment to friends as measured by the Inventory of Parent and Peer Attachment (larger scores indicate greater attachment). In particular, the response is a score based on the responses to 25 questions. Here are the results:[3]

Group	Condition	n	$\bar{x}$	s
1	Service	57	105.32	14.68
2	No service	17	96.82	14.26

META-ANALYSIS

Small samples have large margins of error. Large samples are expensive. Often we can find several studies of the same issue; if we could combine their results, we would have a large sample with a small margin of error. That is the idea of "meta-analysis." Of course, we can't just lump the studies together, because of differences in design and quality. Statisticians have more sophisticated ways of combining the results. Meta-analysis has been applied to issues ranging from the effect of secondhand smoke to whether coaching improves SAT scores.

PLAN: The investigator had no specific direction for the difference in mind before looking at the data, so the alternative is two-sided. We will test the hypotheses

$$H_0: \mu_1 = \mu_2$$
$$H_a: \mu_1 \neq \mu_2$$

SOLVE: The investigator says that the individual scores, examined separately in the two samples, appear roughly Normal. There is a serious problem with the more important condition that the two samples can be regarded as SRSs from two student populations. We will discuss that after we illustrate the calculations.

The two-sample t statistic is

$$t = \frac{\bar{x}_1 - \bar{x}_2}{\sqrt{\dfrac{s_1^2}{n_1} + \dfrac{s_2^2}{n_2}}} = \frac{105.32 - 96.82}{\sqrt{\dfrac{14.68^2}{57} + \dfrac{14.26^2}{17}}} = \frac{8.5}{3.9677} = 2.142$$

Software (Option 1) says that the two-sided P-value is $P = 0.0414$.

Without software, use Option 2 to find a conservative P-value. There are 16 degrees of freedom, the smaller of

$$n_1 - 1 = 57 - 1 = 56 \quad \text{and} \quad n_2 - 1 = 17 - 1 = 16$$

Figure 21.3 illustrates the P-value. Find it by comparing $t = 2.142$ with the two-sided critical values for the $t(16)$ distribution. Table C shows that the P-value is between 0.05 and 0.04.

df = 16

t^*	2.120	2.235
Two-sided P	0.05	0.04

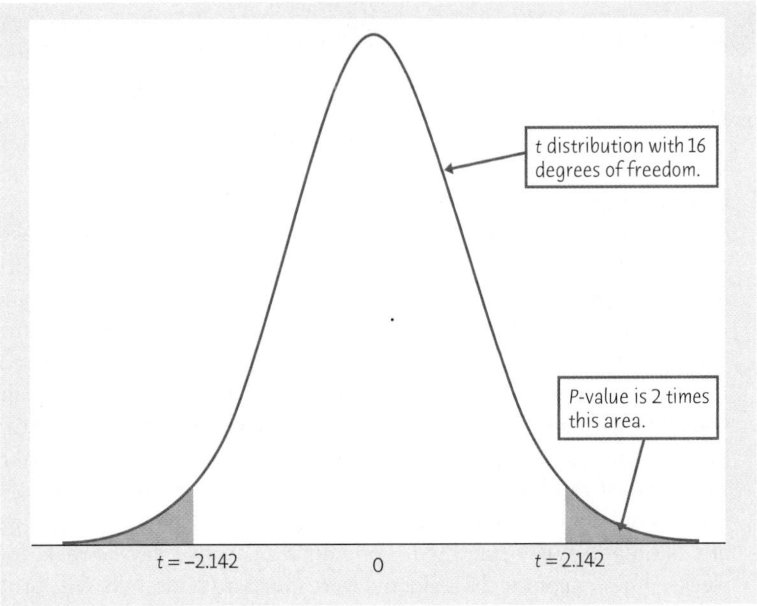

FIGURE 21.3
The *P*-value in Example 21.5. Because the alternative is two-sided, the *P*-value is double the area to the right of $t = 2.142$

CONCLUDE: The data give moderately strong evidence ($P < 0.05$) that students who have engaged in community service do, on the average, differ from those who have not engaged in community service in how attached they are to their friends (and the data suggest they are more attached to their friends). ■

Is the *t* test in Example 21.5 justified? The student subjects were "enrolled in a course on U.S. Diversity at a large mid-western university." Unless this course is required of all students, the subjects cannot be considered a random sample even from this campus. Students were placed in the two groups on the basis of a questionnaire, 39 in the "no service" group and 71 in the "service" group. The data were gathered from a follow-up survey two years later; 17 of the 39 "no service" students responded (44%), compared with 80% response (57 of 71) in the "service" group. Nonresponse is confounded with group: students who had done community service were much more likely to respond. Finally, 75% of the "service" respondents were women, compared with 47% of the "no service" respondents. A person's sex, which can strongly affect attachment, is badly confounded with the presence or absence of community service. The data are so far from meeting the SRS condition for inference that the *t* test is meaningless. Difficulties like these are common in social science research, where confounding variables have stronger effects than is usual when biological or physical variables are measured. This researcher honestly disclosed the weaknesses in data production but left it to readers to decide whether to trust her inferences.

LaunchPad Online Resources

- The **StatBoards video**, *Two-Sample t Confidence Intervals*, discusses examples of constructing confidence intervals using both options.
- The **StatBoards video**, *Two-Sample t Tests,* discusses examples of testing using both options.
- The **Snapshots video**, *Comparing Two Means*, provides further discussion of procedures for comparing two means.
- The **StatClips video**, *Confidence Intervals: Intervals for Differences Example B*, presents an example of a confidence interval for a difference in two means.

Apply Your Knowledge

In exercises that call for two-sample t procedures, use Option 1 if you have technology that implements that method. Otherwise, use Option 2 (degrees of freedom the smaller of $n_1 - 1$ and $n_2 - 1$).

4 step

21.5 Nintendo and Laparoscopic Skills. In laparoscopic surgery, a video camera and several thin instruments are inserted into the patient's abdominal cavity. The surgeon uses the image from the video camera positioned inside the patient's body to perform the procedure by manipulating the instruments that have been inserted. It has been found that the Nintendo Wii™, with its motion-sensing interface, replicates the movements required in laparoscopic surgery more closely than other video games. If training with a Nintendo Wii™ can improve laparoscopic skills, it can complement the more expensive training on a laparoscopic simulator. Forty-two medical residents were chosen and all were tested on a set of basic laparoscopic skills. Twenty-one were selected at random to undergo systematic Nintendo Wii™ training for one hour a day, five days a week, for four weeks. The remaining 21 residents were given no Nintendo Wii™ training and asked to refrain from video games during this period. At the end of four weeks, all 42 residents were tested again on the same set of laparoscopic skills. One of the skills involved a virtual gall bladder removal, with several performance measures including time to complete the task recorded. Here are the improvement (before minus after) times in seconds after four weeks for the two groups:[4] DATA NINT

Treatment						Control					
291	134	186	128	84	243	21	66	54	85	229	92
212	121	134	221	59	244	43	27	77	−29	−14	88
79	333	−13	−16	71	−16	145	110	32	90	45	−81
71	77	144				68	61	44			

Does the Nintendo Wii™ training significantly increase the mean improvement time? Follow the four-step process as illustrated in Examples 21.2 and 21.3.

4 step

21.6 Daily Activity and Obesity. We can conclude from Examples 21.2 and 21.3 that mildly obese people spend less time standing and walking (on the average) than lean people. Is there a significant difference between the mean times the two groups spend lying down? Use the four-step process to answer this question from the data in Table 21.1. Follow the model of Examples 21.2 and 21.3. DATA ACTIVE

21.7 Nintendo and Laparoscopic Skills, Continued. Use the data in Exercise 21.5 to give a 99% confidence interval for the difference in mean improvement times between the treatment and control groups. DATA NINT

21.4 Using technology

Software should use Option 1 for the degrees of freedom to give accurate confidence intervals and *P*-values. Unfortunately, there is variation in how well software implements Option 1. Figure 21.4 displays output from a graphing calculator, three statistical programs, and a spreadsheet program for the test of Example 21.3. All four claim to use Option 1. The two-sample *t* statistic is exactly as in Example 21.3,

Texas Instruments Graphing Calculator

```
2-SampTTest
 μ1>μ2
 t=3.808375604
 p=8.4101904E-4
 df=15.17355038
 x̄1=525.7513
↓x̄2=373.2692
```

Minitab

Session

Two-sample T for stand

```
group          N     Mean    StDev  SE Mean
1             10      526      107       34
2             10    373.3     67.5       21

Difference = mu (1) - mu (2)
Estimate for difference: 152.5
T-Test of difference = 0 (vs >): T-Value = 3.81 P-Value = 0.001 DF = 15
```

Microsoft Excel

Excel

	A	B	C
1	t-Test: Two-Sample Assuming Unequal Variances		
2			
3		*Lean*	*Obese*
4	Mean	525.7513	373.2692
5	Variance	11474.8903	4556.019849
6	Observations	10	10
7	Hypothesized Mean	0	
8	df	15	
9	t Stat	3.808375604	
10	P(T<=t) one-tail	0.000856818	
11	t Critical one-tail	1.753051038	
12	P(T<=t) two-tail	0.001713635	
13	P(T<t) two-tail	2.131450856	
14			

Sheet4

CrunchIt!

Results - t 2-Sample

Export ▾

Null hypothesis:	Difference of means = 0
Alternative hypothesis:	Difference of means > 0

Group	Std Dev	Sample Mean	n
1	107.1	525.8	10
2	67.50	373.3	10

df:	15.17
Difference of means:	152.5
t statistic:	3.808
P-value:	0.0008410

JMP

Assuming unequal variances

Difference	152.482	t Ratio	3.808376
Std Err Dif	40.039	DF	15.17355
Upper CL Dif	237.737	Prob > |t|	0.0017*
Lower CL Dif	67.227	Prob > t	0.0008*
Confidence	0.95	Prob < t	0.9992

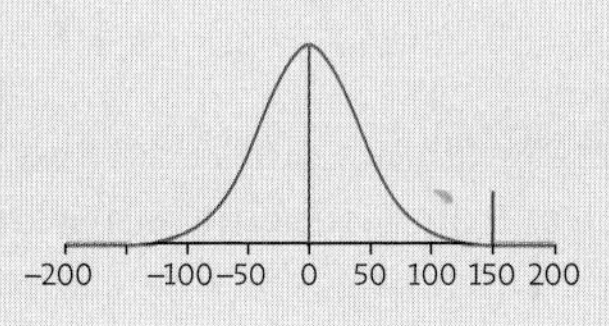

FIGURE 21.4
The two-sample *t* procedures applied to the data on activity and obesity: output from a graphing calculator, three statistical programs, and a spreadsheet program.

$t = 3.808$. You can find this in all four outputs (Minitab rounds to 3.81; Excel and the graphing calculator give additional decimal places). The different technologies use different methods to find the *P*-value for $t = 3.808$.

- CrunchIt!, JMP, and the calculator get Option 1 completely right. The accurate approximation uses the *t* distribution with approximately 15.174 (CrunchIt! rounds this to 15.17) degrees of freedom. The *P*-value is $P = 0.0008$.
- Minitab uses Option 1, but it *truncates* the exact degrees of freedom to the next smaller whole number to get critical values and *P*-values. In this example, the exact df = 15.174 is truncated to df = 15, so that Minitab's results are slightly conservative. That is, Minitab's *P*-value (rounded to $P = 0.001$ in the output) is slightly larger than the full Option 1 *P*-value.
- Excel *rounds* the exact degrees of freedom to the nearest whole number, so that df = 15.174 becomes df = 15. Excel's method agrees with Minitab's in this example. But when rounding moves the degrees of freedom up to the next higher whole number, Excel's *P*-values are slightly smaller than is correct. This is misleading, another illustration of the fact that Excel is substandard as statistical software.

JMP's and Excel's label that the test assumes unequal variances is a bit misleading. *The two-sample t procedures we have described work whether or not the two populations have the same variance.* There is an old-fashioned special procedure that works only when the two variances are equal. We discuss this method in Section 21.7, but there is no need to use it in two-sample problems.

CAUTION

Although different calculators and software give slightly different *P*-values, in practice you can just accept what your technology says. The small differences in *P* don't affect the conclusion. Even "between 0.001 and 0.0025" from Option 2 (Example 21.3) is close enough for practical purposes.

Apply Your Knowledge

21.8 Perception of Life Expectancy. Do women and men differ in how they perceive their life expectancy? A researcher asked a sample of men and women to indicate their life expectancy. This was compared with values from actuarial tables, and the relative percent difference was computed (perceived life expectancy minus life expectancy from actuarial tables was divided by life expectancy from actuarial tables and converted to a percent). Here are the relative percent differences for all men and women over the age of 70 in the sample:[5] DATA LIFEEXP

Men	−28	−23	−20	−19	−14	−13	
Women	−20	−19	−15	−12	−10	−8	−5

Figure 21.5 shows output for the two-sample *t* test using Option 1. (This output is from CrunchIt! software, which does Option 1 without rounding or truncating the degrees of freedom.) Do men and women over 70 years old differ in their perceptions of life expectancy? Using the output in Figure 21.5, write a summary in a sentence or two, including *t*, df, *P*, and a conclusion.

Results - t 2-Sample

Export ▾

Null hypothesis:	Difference of means = 0
Alternative hypothesis:	Difference of means is not 0

Sex	Std Dev	Sample Mean	n
M	5.612	-19.50	6
F	5.589	-12.71	7

df:	10.68
Difference of means:	-6.786
t statistic:	-2.177
P-value:	0.05281

FIGURE 21.5
Two-sample *t* output from CrunchIt! for Exercise 21.8.

21.5 Robustness again

The two-sample *t* procedures are more robust (less sensitive to departures from our conditions for inference for comparing two means) than the one-sample *t* methods, particularly when the distributions are not symmetric. When the sizes of the two samples are equal and the two populations being compared have distributions with similar shapes, probability values from the *t* table are quite accurate for a broad range of distributions when the sample sizes are as small as $n_1 = n_2 = 5$.[6] When the two population distributions have different shapes, larger samples are needed.

As a guide to practice, adapt the guidelines given on page 470 for the use of one-sample *t* procedures to two-sample procedures by replacing "sample size" with the "sum of the sample sizes," $n_1 + n_2$. These guidelines err on the side of safety, especially when the two samples are of equal size. *In planning a two-sample study, choose equal sample sizes whenever possible. The two-sample t procedures are most robust against non-Normality in this case, and the conservative Option 2 probability values are most accurate.*

CAUTION

LaunchPad Online Resources

- The **StatBoards video**, *Robustness and Determining If t Procedures Are Appropriate*, discusses some examples.

Apply Your Knowledge

21.9 **Do Good Smells Bring Good Business?** Businesses know that customers often respond to background music. Do they also respond to odors? One study of this question took place in a small pizza restaurant in France on two Saturday evenings in May. On one of these evenings, a relaxing lavender odor was spread through the restaurant. On the other evening, no scent was used. Table 21.2 gives the time (in minutes) that two samples of 30 customers spent in the restaurant and the amount they spent (in euros).[7] The two evenings were comparable in many ways (weather, customer count, and so on), so we are willing to regard the data as independent SRSs from spring Saturday evenings at this restaurant. The authors say, "Therefore at this stage it would be impossible to generalize the results to other restaurants." DATA ODORS2

TABLE 21.2 TIME (MINUTES) AND SPENDING (EUROS) BY RESTAURANT CUSTOMERS

NO ODOR		LAVENDER	
MINUTES	EUROS SPENT	MINUTES	EUROS SPENT
103	15.9	92	21.9
68	18.5	126	18.5
79	15.9	114	22.3
106	18.5	106	21.9
72	18.5	89	18.5
121	21.9	137	24.9
92	15.9	93	18.5
84	15.9	76	22.5
72	15.9	98	21.5
92	15.9	108	21.9
85	15.9	124	21.5
69	18.5	105	18.5
73	18.5	129	25.5
87	18.5	103	18.5
109	20.5	107	18.5
115	18.5	109	21.9
91	18.5	94	18.5
84	15.9	105	18.5
76	15.9	102	24.9
96	15.9	108	21.9
107	18.5	95	25.9
98	18.5	121	21.9
92	15.9	109	18.5
107	18.5	104	18.5
93	15.9	116	22.8
118	18.5	88	18.5
87	15.9	109	21.9
101	25.5	97	20.7
75	12.9	101	21.9
86	15.9	106	22.5

(a) Does a lavender odor encourage customers to stay longer in the restaurant? Examine the time data and explain why they are suitable for two-sample t procedures. Use the two-sample t test to answer the question posed.

(b) Does a lavender odor encourage customers to spend more while in the restaurant? Examine the spending data. In what ways do these data deviate from Normality? Explain why, with 30 observations, the t procedures are reasonably accurate for these data. Use the two-sample t test to answer the question posed.

21.10 Compressing Soil. Farmers know that driving heavy equipment on wet soil compresses the soil and injures future crops. Here are data on the "penetrability" of the same type of soil at two levels of compression.[8] Penetrability is a measure of how much resistance plant roots will meet when they try to grow through the soil. DATA SOILCM

Compressed Soil									
2.86	2.68	2.92	2.82	2.76	2.81	2.78	3.08	2.94	2.86
3.08	2.82	2.78	2.98	3.00	2.78	2.96	2.90	3.18	3.16

Intermediate Soil									
3.14	3.38	3.10	3.40	3.38	3.14	3.18	3.26	2.96	3.02
3.54	3.36	3.18	3.12	3.86	2.92	3.46	3.44	3.62	4.26

(a) Make stemplots to investigate the shapes of the distributions. The penetrabilities for intermediate soil are skewed to the right and have a high outlier. Returning to the source of the data shows that the outlying sample had unusually low soil density, so that it belongs in the "loose soil" class. We are justified in removing the outlier.

(b) We suspect that the penetrability of compressed soil is less than that of intermediate soil. Do the data (with the outlier removed) support this suspicion?

21.11 Weeds among the Corn. Lamb's-quarter is a common weed that interferes with the growth of corn. An agriculture researcher planted corn at the same rate in 16 small plots of ground, then weeded the plots by hand to allow a fixed number of lamb's-quarter plants to grow in each meter of corn row. No other weeds were allowed to grow. Here are the yields of corn (bushels per acre) for only the experimental plots controlled to have one weed per meter of row and nine weeds per meter of row:[9] DATA WDCORN

One weed/meter	166.2	157.3	166.7	161.1
Nine weeds/meter	162.8	142.4	162.8	162.4

Explain carefully why a two-sample t confidence interval for the difference in mean yields may not be accurate.

21.12 Compressing Soil, Continued. Use the data in Exercise 21.10, omitting the outlier, to give a 90% confidence interval for the decrease in penetrability of compressed soil relative to intermediate soil. DATA SOILCM2

21.6 Details of the *t* approximation*

The exact distribution of the two-sample t statistic is not a t distribution. Moreover, the distribution changes as the unknown population standard deviations σ_1 and σ_2 change. However, an excellent approximation is available. We call this Option 1 for t procedures.

*This section can be omitted unless you are using software and wish to understand what the software does.

Approximate Distribution of the Two-Sample t Statistic

The distribution of the two-sample t statistic is very close to the t distribution with degrees of freedom df given by

$$\text{df} = \frac{\left(\dfrac{s_1^2}{n_1} + \dfrac{s_2^2}{n_2}\right)^2}{\dfrac{1}{n_1 - 1}\left(\dfrac{s_1^2}{n_1}\right)^2 + \dfrac{1}{n_2 - 1}\left(\dfrac{s_2^2}{n_2}\right)^2}$$

This approximation is accurate when both sample sizes n_1 and n_2 are 5 or larger.

EXAMPLE 21.6 Daily Activity and Obesity

DATA ACTIVE

In the experiment of Examples 21.2 and 21.3, the data on minutes per day spent standing and walking give

Group	n	$\bar{x}$	s
Group 1 (lean)	10	525.751	107.121
Group 2 (obese)	10	373.269	67.498

The two-sample t test statistic calculated from these values is $t = 3.808$.

The one-sided P-value is the area to the right of 3.808 under a t density curve, as in Figure 21.2. The conservative Option 2 uses the t distribution with nine degrees of freedom. Option 1 finds a very accurate P-value by using the t distribution with degrees of freedom df given by

$$\text{df} = \frac{\left(\dfrac{107.121^2}{10} + \dfrac{67.498^2}{10}\right)^2}{\dfrac{1}{9}\left(\dfrac{107.121^2}{10}\right)^2 + \dfrac{1}{9}\left(\dfrac{67.498^2}{10}\right)^2}$$

$$= \frac{2{,}569{,}894}{169{,}367.2} = 15.1735$$

These degrees of freedom appear in the graphing calculator output in Figure 21.4. Because the formula is messy and roundoff errors are likely, we don't recommend calculating df by hand. ■

The degrees of freedom df is generally not a whole number. It is always at least as large as the smaller of $n_1 - 1$ and $n_2 - 1$. The larger degrees of freedom that result from Option 1 give slightly shorter confidence intervals and slightly smaller P-values than the conservative Option 2 produces. There is a t distribution for any positive degrees of freedom, even though Table C contains entries only for whole-number degrees of freedom.

The difference between the t procedures using Options 1 and 2 is rarely of practical importance. That is why we recommend the simpler, conservative Option 2 for inference without software. With software, the more accurate Option 1 procedures are painless.

Apply Your Knowledge

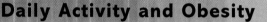

21.13 Friends Shrink Foes. In a study of the perceived formidability of a picture of a possible foe (said to be a terrorist), adult males were presented with the picture and asked to rate the formidability of the foe. Thirty-seven subjects were with friends when asked to provide the rating, while 21 were alone. Is the perceived formidability less in the presence of friends than when alone? Larger scores indicate a greater perceived formidability.[10] Software that uses Option 1 gives these summary results:

Situation	n	Mean	Std dev	Std err	t	df	P
Alone	37	0.29	0.68	0.11	2.255	35.3	<0.02
With friends	21	−0.19	0.83	0.18			

Starting from the sample means and standard deviations, verify each of these entries: the standard errors of the means; the degrees of freedom for the two-sample *t*; the value of *t*.

21.14 Perception of Life Expectancy. Figure 21.5 gives output for the life expectancy data in Exercise 21.8 from software that does Option 1 with the correct degrees of freedom. What are $\bar{x}_i$ and s_i for the two samples? Starting from these values, find the *t* test statistic and its degrees of freedom. Your work should agree with Figure 21.5.

21.15 Friends Shrink Foes, Continued. Write a sentence or two summarizing the comparison of males alone and males with friends in Exercise 21.13, as if you were preparing a report for publication. Use the output in Exercise 21.13.

21.7 Avoid the pooled two-sample *t* procedures*

Most software and graphing calculators, including all four illustrated in Figure 21.4, offer a choice of two-sample *t* statistics. One is often labeled for "unequal" variances, and the other for "equal" variances. The "unequal" variance procedure is our two-sample *t*. *This test is valid whether or not the population variances are equal.* The other choice is a special version of the two-sample *t* statistic that assumes that the two populations have the same variance. This procedure averages (the statistical term is "pools") the two sample variances to estimate the common population variance. The resulting statistic is called the *pooled two-sample t statistic.* It is equal to our *t* statistic if the two sample sizes are the same, but not otherwise. We could choose to use the pooled *t* for tests and confidence intervals.

The pooled *t* statistic has exactly the *t* distribution with $n_1 + n_2 - 2$ degrees of freedom *if* the two population variances really are equal and the population distributions are exactly Normal. The pooled *t* was in common use before software made it easy to use Option 1 for our two-sample *t* statistic. Of course, in the real world distributions are not exactly Normal and population variances are not exactly equal. In practice, the Option 1 two-sample *t* procedures are almost always more accurate than the pooled procedures. Our advice: *Never use the pooled t procedures if you have software that will implement Option 1.*

CAUTION

Statistics in Your World

IS MONEY THE ROOT OF ALL EVIL?

That may go too far, but Kathleen Vohs and her co-workers show that even thinking about money has strong effects. Exercise 21.47 describes a small part of their work. What does Professor Vohs say about the consequences of having money? "Money makes people feel self-sufficient and behave accordingly." With money, you can achieve your goals with less help from others. You feel less dependent on others and more willing to work toward your own goals. Maybe that's good. You also prefer to be less involved with others, so that self-sufficiency is a barrier to close relationships with others. Maybe that's not good. Scientists don't tell us what's good or not good, just that money increases our sense of self-sufficiency.

21.8 Avoid inference about standard deviations*

Two basic features of a distribution are its center and variability. In a Normal population, we measure center by the mean and variability by the standard deviation. We use the *t* procedures for inference about population means for Normal populations,

*This short section offers advice on what not to do. This material is not needed to understand the rest of the book.

and we know that t procedures are widely useful for non-Normal populations as well. It is natural to turn next to inference about the standard deviations of Normal populations. Our advice here is short and clear: don't do it without expert advice.

There are methods for inference about the standard deviations of Normal populations. The most common such method is the "F test" for comparing the standard deviations of two Normal populations. You will find this test in the menus of most statistical software. *Unlike the t procedures for means, the F test for standard deviations is extremely sensitive to non-Normal distributions.* This lack of robustness does not improve in large samples. It is difficult in practice to tell whether a significant test result is evidence of unequal population variability or simply a sign that the populations are not Normal. Because this test is of little use in practice, we don't give its details.

The deeper difficulty underlying the very poor robustness of Normal population procedures for inference about variability already appeared in our work on describing data. The standard deviation is a natural measure of variability for Normal distributions but not for distributions in general. In fact, because skewed distributions have asymmetric tails, no single numerical measure does a good job of describing the variability of a skewed distribution. In summary, the standard deviation is not always a useful parameter, and even when it is (for symmetric distributions), the results of inference about the standard deviation are not trustworthy. Consequently, *we do not recommend trying to make inferences about population standard deviations in basic statistical practice.*[11]

21.9 Permutation tests*

To determine whether the effect of a new treatment differs from a standard treatment or a placebo, we often test the null hypothesis of no treatment effect. When units are assigned to treatments completely at random, this fact allows us to test this hypothesis. This test does not require the assumption that the data come from a Normal population. The reasoning was presented in Example 12.8 (page 287) and Example 15.11 (page 363) and we revisit it here.

EXAMPLE 21.7 Weight Loss

We have six subjects for an experiment: Ari, Luis, Troy, Ana, Deb, and Hui. Three are to be assigned completely at random to a new experimental weight loss treatment and three to a placebo to see if the experimental treatment results in greater weight loss than a placebo. Thus, we test the hypotheses

H_0: no treatment effect
H_a: new treatment results in a greater mean weight loss than a placebo

We will compare mean weight losses for the treatment and placebo groups to test these hypotheses.

Suppose Luis, Deb, and Hui are assigned to the experimental treatment. Here are the weight losses after six weeks.

Treatment Group (Person)	Placebo Group (Person)
15 (Luis)	2 (Ari)
12 (Deb)	8 (Troy)
9 (Hui)	1 (Ana)

*You may prefer to skip this material if you have not discussed resampling methods.

The mean weight loss for the treatment group is $\frac{15 + 12 + 9}{3} = 12.00$, and the mean weight loss for the placebo group is $\frac{2 + 8 + 1}{3} = 3.67$. The difference in mean weight losses (treatment group weight loss minus placebo group weight loss) is $12.00 - 3.67 = 8.33$.

If the null hypothesis of no treatment effect is true, then Bob was going to lose 15 pounds whether he was assigned to the treatment group or the placebo group, and similarly for the other subjects. This fact allows us to compute a *P*-value using only the random assignment of the subjects to the two groups. Here is the reasoning.

If the subjects are assigned to the two treatments completely at random, then all possible assignments of subjects to treatments are equally likely. There are 20 possible ways of assigning three subjects to the treatment group and three to the placebo group, hence each has probability 1/20 of occurring. If the null hypothesis of no treatment effect is true, then the weight loss we observe for any subject will be the same regardless of which treatment the subject receives. Thus, the responses we actually observe for each treatment are due only to the "luck of the draw," namely the particular assignment of subjects to treatments that our randomization produced. For each possible assignment, we can list the weight losses we would observe with each treatment and compute the difference in mean weight losses. We did this in Example 15.11, and we display the results again here. These are all equally likely.

Treatment Group	Placebo Group	Difference in Mean Weight Loss
Ari, Luis, Troy	Ana, Deb, Hui	1.00
Ari, Luis, Ana	Troy, Deb, Hui	−3.67
Ari, Luis, Deb	Troy, Ana, Hui	3.67
Ari, Luis, Hui	Troy, Ana, Deb	1.67
Ari, Troy, Ana	Luis, Deb, Hui	−8.33
Ari, Troy, Deb	Luis, Ana, Hui	−1.00
Ari, Troy, Hui	Luis, Ana, Deb	−3.00
Ari, Ana, Deb	Luis, Troy, Hui	−6.33
Ari, Ana, Hui	Luis, Troy, Deb	−7.67
Ari, Deb, Hui	Luis, Troy, Ana	−0.33
Luis, Troy, Ana	Ari, Deb, Hui	0.33
Luis, Troy, Deb	Ari, Ana, Hui	7.67
Luis, Troy, Hui	Ari, Ana, Deb	6.33
Luis, Ana, Deb	Ari, Troy, Hui	3.00
Luis, Ana, Hui	Ari, Troy, Deb	1.00
Luis, Deb, Hui	Ari, Troy, Ana	8.33
Troy, Ana, Deb	Ari, Luis, Hui	−1.67
Troy, Ana, Hui	Ari, Luis, Deb	−3.67
Troy, Deb, Hui	Ari, Luis, Ana	3.67
Ana, Deb, Hui	Ari, Luis, Troy	−1.00

Ordering the differences in mean weight losses from low to high, recalling each assignment of subjects to the two groups has probability $1/20 = 0.05$, and combining duplicates, we find, as in Example 15.11:

Weight Loss	−8.33	−7.67	−6.33	−3.67	−3.00	−1.67	−1.00	−0.33
Probability	0.05	0.05	0.05	0.10	0.05	0.05	0.10	0.05
Weight Loss	0.33	1.00	1.67	3.00	3.67	6.33	7.67	8.33
Probability	0.05	0.10	0.05	0.05	0.10	0.05	0.05	0.05

The alternative hypothesis is that the new treatment will produce a greater mean weight loss than the placebo, hence the *P*-value (the probability of observing a weight loss as large as or larger than 8.33 if the null hypothesis is true) for the experiment is 0.05.

You can check that if we use the two-sample *t* procedure with Option 1 to test the hypotheses

$$H_0: \mu_1 = \mu_2$$
$$H_a: \mu_1 > \mu_2$$

the *P*-value is 0.0216 for these data. ■

In general, if experimental units are assigned to two treatment groups completely at random, and our null hypothesis is "no treatment effect," we can test hypotheses using sample means as follows. First, list all possible ways units can be assigned to treatment groups.[12] Second, based on the data obtained, for each possible assignment determine what the difference in means (mean for treatment 1 minus mean for treatment 2) would be. Under the null hypothesis, each of these is equally likely. Third, determine the distribution of these possible outcomes by listing all the different possible mean differences and their corresponding probabilities (this can also be represented by a histogram). Using this sampling distribution, determine the *P*-value of the actual mean difference you obtained.

This procedure is sometimes referred to as a **permutation test** and the sampling distribution is referred to as the **permutation distribution**. It assumes units are assigned to treatments completely at random and does not require the assumption of Normality. However, there are some practical issues. First, unless the number of ways units can be assigned to treatments is sufficiently large, *P*-values may not be small. In Example 21.7, the smallest possible *P*-value is 0.05 (and 0.10 for a two-sided alternative). This is because with only 20 possible assignments, the largest and smallest possible outcomes have probability at least 0.05. *If* we can assume Normality and use the two-sample *t* procedure, we can get more accurate *P*-values.

Second, listing all possible outcomes can be quite tedious unless you have special software. There are 70 ways eight units can be divided into two groups of size 4, 252 ways 10 units can be divided into two groups of size 5, 924 ways 12 units can be divided into two groups of size 6, and 184,756 ways 20 units can be divided into two groups of size 10. When the number of units is large, one could use resampling to estimate the sampling distribution. For example, if we have 10 units assigned to each of two treatments, rather than listing all 184,756 possible assignments, we could take 1000 samples of size 10 from the 20 units and determine the difference in means we would observe for each of these 1000 samples.

Third, permutation tests are most likely to be useful with small to moderate size experiments because of the robustness of *t* procedures. In small to moderate size experiments, if there is good reason to assume the sampling distribution of the two-sample *t* statistic has approximately a *t* distribution, the two-sample *t* procedure may be preferable to a permutation test.

LaunchPad Online Resources

- The **StatBoards video**, *Permutation Tests*, discusses additional examples.
- The **Snapshots video**, *Resampling Procedures*, includes some discussion of permutation tests.

Apply Your Knowledge

21.16 A Very Simple Setting. Does a zinc supplement taken at the onset of cold symptoms reduce the time one has a cold? Five volunteers (Dan, Hal, Joe, Kay, and Zoe) agree to take part in an experiment. Three are assigned completely at random to receive the zinc supplement and the other two receive the placebo. The experiment is double-blind. The results are (times are duration of cold in hours):

Zinc Group (Person)	Placebo Group (Person)
38 (Dan)	44 (Hal)
31 (Joe)	47 (Zoe)
35 (Kay)	

(a) There are 10 possible ways the five subjects can be assigned to the two groups, with the zinc group having size 3 and the placebo group size 2. List these.

(b) For each, determine the difference in mean length of cold (mean for the placebo group minus mean for the zinc group). Combine any duplicates and make a table of the possible mean differences and the corresponding probability of each under the null hypothesis of no treatment effect. (Each of the 10 possible assignments of subjects to treatments has probability 1/10 under the null hypothesis.) This is the permutation distribution.

(c) Compute the P-value of the data. Assume the alternative hypothesis is that the mean duration of a cold is less for zinc.

(d) In this example, is it possible to demonstrate significance at the 5% level using the permutation test? Explain.

(e) Assume that cold duration is Normally distributed for both zinc and the placebo. Use the two-sample t procedure to test the hypotheses. Use Option 1 if you have access to software.

21.17 Weeds among the Corn. Exercise 21.11 discusses a completely randomized experiment. Two treatments—one weed per meter and nine weeds per meter—were assigned completely at random to eight plots of land. Each was assigned to four plots. Here are the yields of corn (bushels per acre):[13]

One weed/meter	166.2	157.3	166.7	161.1
Nine weeds/meter	162.8	142.4	162.8	162.4

In Exercise 21.11, you decided that two-sample t procedures might not be accurate. However, the permutation test can be used. Figure 21.6 displays the distribution of all possible mean differences for the 70 different ways treatments could be assigned to plots. Use Figure 21.6 to estimate the P-value of these data.

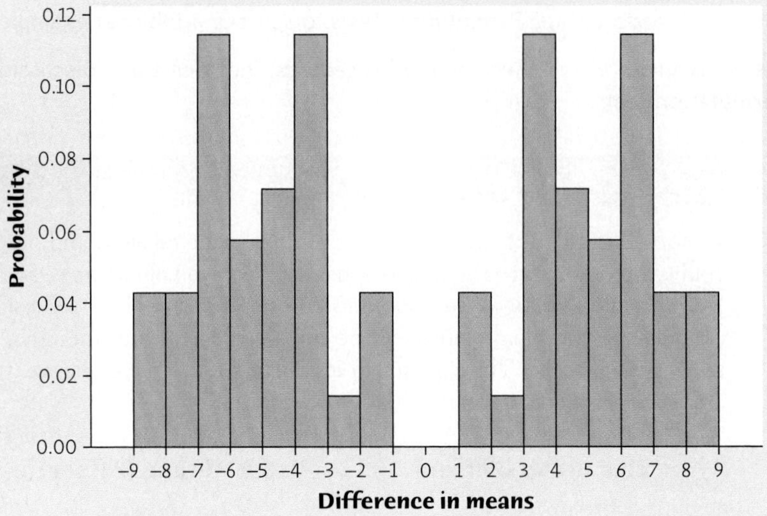

FIGURE 21.6
Permutation distribution for Exercise 21.17.

CHAPTER 21 SUMMARY

Chapter Specifics

- The data in a **two-sample problem** are two independent SRSs, each drawn from a separate population.
- Tests and confidence intervals for the difference between the means μ_1 and μ_2 of two Normal populations start from the difference $\bar{x}_1 - \bar{x}_2$ between the two sample means. Because of the central limit theorem, the resulting procedures are approximately correct for other population distributions when the sample sizes are large.
- Draw independent SRSs of sizes n_1 and n_2 from two Normal populations with parameters μ_1, σ_1, and μ_2, σ_2. The **two-sample *t* statistic** is

$$t = \frac{(\bar{x}_1 - \bar{x}_2) - (\mu_1 - \mu_2)}{\sqrt{\dfrac{s_1^2}{n_1} + \dfrac{s_2^2}{n_2}}}$$

  The statistic t has approximately a t distribution.
- There are two choices for the **degrees of freedom** of the two-sample t statistic. Option 1: software produces accurate probability values using degrees of freedom calculated from the data. Option 2: for conservative inference procedures, use degrees of freedom equal to the smaller of $n_1 - 1$ and $n_2 - 1$.
- The **confidence interval for** $\mu_1 - \mu_2$ is

$$(\bar{x}_1 - \bar{x}_2) \pm t^*\sqrt{\frac{s_1^2}{n_1} + \frac{s_2^2}{n_2}}$$

  The critical value t^* from Option 1 gives a confidence level very close to the desired level C. Option 2 produces a margin of error at least as wide as is needed for the desired level C.
- **Significance tests for** H_0: $\mu_1 = \mu_2$ are based on

$$t = \frac{\bar{x}_1 - \bar{x}_2}{\sqrt{\dfrac{s_1^2}{n_1} + \dfrac{s_2^2}{n_2}}}$$

P-values calculated from Option 1 are very accurate. Option 2 P-values are always at least as large as the true P.

- The two-sample t procedures are quite **robust** against departures from Normality. Guidelines for practical use are similar to those for one-sample t procedures. Equal sample sizes are recommended.
- Procedures for inference about the standard deviations of Normal populations are very sensitive to departures from Normality. Avoid inference about standard deviations unless you have expert advice.
- When experimental units are assigned to treatments completely at random, one can use permutation tests to test the hypothesis of no treatment effect. This test does not require the assumption that the data come from Normal populations.

Link It

In Chapter 20, we studied inference for the mean of a Normal population using procedures based on the t distribution. These procedures are more realistic than those studied in Chapters 15 to 17 because t procedures do not require that we know the population variance. In practice, the most common use of t procedures for a single population mean is with matched pairs data because most research studies make comparisons between two or more populations.

In this chapter, we discuss t procedures for comparing the means of two Normal populations when we have independent samples from these two populations. These t procedures are quite common in practice, and one encounters them in many research papers. Researchers typically report the means and standard deviations for the two samples and, in the case of tests of hypotheses, the value of the t statistic and the corresponding P-value.

In the next two chapters, we extend our procedures for inference to additional settings that occur frequently in practice—in particular, to inference for population proportions. And we will see that the basic ideas about confidence intervals and tests of hypotheses that we learned in Chapters 15 to 17 still apply.

LaunchPad Online Resources

If you are having difficulty with any of the sections of this chapter, these online resources should help prepare you to solve the exercises at the end of this chapter.

- **StatTutor** starts with a video review of each section and asks a series of questions to check your understanding. You may find this helpful if you need additional assistance understanding any of the material in this chapter.
- **LearningCurve** provides you with a series of questions about the chapter geared to your level of understanding.

CHECK YOUR SKILLS

21.18 The 2113 National Assessment of Educational Progress (NAEP) gave a mathematics test to a random sample of twelfth-graders in Michigan. The mean score was 154 out of 300. To give a confidence interval for the mean score of all Michigan twelfth-graders, you would use

(a) the two-sample *t* interval.
(b) the matched pairs *t* interval.
(c) the one-sample *t* interval.

21.19 In the 2013 NAEP sample of Michigan twelfth-graders, the mean mathematics scores were 152 for female students and 156 for male students. To see if this difference is statistically significant, you would use

(a) the two-sample *t* test.
(b) the matched pairs *t* test.
(c) the one-sample *t* test.

21.20 Two new devices for testing blood sugar levels have been developed. How do these devices compare? You test blood sugar levels of 20 diabetics with both devices and use

(a) the two-sample *t* test.
(b) the matched pairs *t* test.
(c) the one-sample *t* test.

21.21 One major reason that the two-sample *t* procedures are widely used is that they are quite *robust.* This means that

(a) *t* procedures do not require that we know the standard deviations of the populations.
(b) confidence levels and *P*-values from the *t* procedures are quite accurate even if the population distribution is not exactly Normal.
(c) *t* procedures compare population means, a comparison that answers many practical questions.

21.22 A study of the effect of exposure to color (red or blue) on the ability to solve puzzles used 42 subjects. Half the subjects (21) were asked to solve a series of puzzles while in a red-colored environment. The other half were asked to solve the same series of puzzles while in a blue-colored environment. The time taken to solve the puzzles was recorded for each subject. To compare the mean times for the two groups of subjects using the two-sample *t* procedures with the conservative Option 2, the correct degrees of freedom is

(a) 41. (b) 40. (c) 20.

21.23 The 21 subjects in the red-colored environment had a mean time for solving the puzzles of 9.64 seconds with standard deviation 3.43; the 21 subjects in the blue-colored environment had a mean time of 15.84 seconds with standard deviation 8.65. The two-sample *t* statistic for comparing the population means has value

(a) 1.50. (b) 3.05. (c) 6.20.

21.24 A study of the use of social media asked a sample of 488 American adults under the age of 40 and a sample of 421 American adults aged 40 or over about their use of social media. Based on their answers, each subject was assigned a social media usage score on a scale of 0 to 25. Higher scores indicate greater usage. The subjects were chosen by random digit dialing of telephone numbers. Are the conditions for two-sample *t* inference satisfied?

(a) Maybe: the SRS condition is OK but we need to look at the data to check Normality.
(b) No: scores in a range between 0 and 25 can't be Normal.
(c) Yes: the SRS condition is OK and large sample sizes make the Normality condition unnecessary.

21.25 We suspect that younger adults use social media more than adults aged 40 or over. To see if this is true, test these hypotheses for the mean social media usage scores of all adults under 40 and all adults 40 and over:

(a) $H_0: \mu_{<40} = \mu_{\geq 40}$ versus $H_a: \mu_{<40} > \mu_{\geq 40}$
(b) $H_0: \mu_{<40} = \mu_{\geq 40}$ versus $H_a: \mu_{<40} \neq \mu_{\geq 40}$
(c) $H_0: \mu_{<40} = \mu_{\geq 40}$ versus $H_a: \mu_{<40} < \mu_{\geq 40}$

21.26 The two-sample *t* statistic for the social media use study ("under 40" mean minus "40 and over" mean) is $t = 3.18$. The *P*-value for testing the hypotheses from the previous exercise satisfies

(a) $0.001 < P < 0.005$.
(b) $0.0005 < P < 0.001$.
(c) $0.001 < P < 0.002$.

CHAPTER 21 EXERCISES

Exercises 21.25 to 21.34 are based on summary statistics rather than raw data. This information is typically all that is presented in published reports. You can perform inference procedures by hand from the summaries. Use the conservative Option 2 (degrees of freedom the smaller of $n_1 - 1$ and $n_2 - 1$) for two-sample t confidence intervals and P-values.

CAUTION

You must trust that the authors understood the conditions for inference and verified that they apply. This isn't always true.

21.27 **Do women talk more than men?** Equip male and female students with a small device that secretly records sound for a random 30 seconds during each

12.5-minute period over two days. Count the words each subject speaks during each recording period, and from this, estimate how many words per day each subject speaks. The published report includes a table summarizing six such studies.[14] Here are two of the six:

	SAMPLE SIZE		ESTIMATED AVERAGE NUMBER (SD) OF WORDS SPOKEN PER DAY	
STUDY	WOMEN	MEN	WOMEN	MEN
1	56	56	16,177 (7520)	16,569 (9108)
2	27	20	16,496 (7914)	12,867 (8343)

Readers are supposed to understand that, for example, the 56 women in the first study had $\bar{x} = 16{,}177$ and $s = 7520$. It is commonly thought that women talk more than men. Does either of the two samples support this idea? For each study:

(a) State hypotheses in terms of the population means for men (μ_M) and women (μ_F).
(b) Find the two-sample t statistic.
(c) What degrees of freedom does Option 2 use to get a conservative P-value?
(d) Compare your value of t with the critical values in Table C. What can you say about the P-value of the test?
(e) What do you conclude from the results of these two studies?

21.28 Alcohol and zoning out. Healthy men aged 21 to 35 were randomly assigned to one of two groups: half received 0.82 grams of alcohol per kilogram of body weight; half received a placebo. Participants were then given 30 minutes to read up to 34 pages of Tolstoy's *War and Peace* (beginning at Chapter 1, with each page containing approximately 22 lines of text). Every two to four minutes participants were prompted to indicate whether they were "zoning out." The proportion of times participants indicated they were zoning out was recorded for each subject. The table below summarizes data on the proportion of episodes of zoning out.[15] (The study report gave the standard error of the mean $s/\sqrt{n}$, abbreviated as SEM, rather than the standard deviation s.)

GROUP	n	$\bar{x}$	SEM
Alcohol	25	0.25	0.05
Placebo	25	0.12	0.03

(a) What are the two sample standard deviations?
(b) What degrees of freedom does the conservative Option 2 use for two-sample t procedures for these samples?
(c) Using Option 2, give a 90% confidence interval for the mean difference between the two groups.

21.29 Stress and weight in rats. In a study of the effects of stress on behavior in rats, 71 rats were randomly assigned to either a stressful environment or a control (nonstressful) environment. After 21 days, the change in weight (in grams) was determined for each rat. The table below summarizes data on weight gain.[16] (The study report gave the standard error of the mean $s/\sqrt{n}$, abbreviated as SEM, rather than the standard deviation s.)

GROUP	n	$\bar{x}$	SEM
Stress	20	26	3
No stress	51	32	2

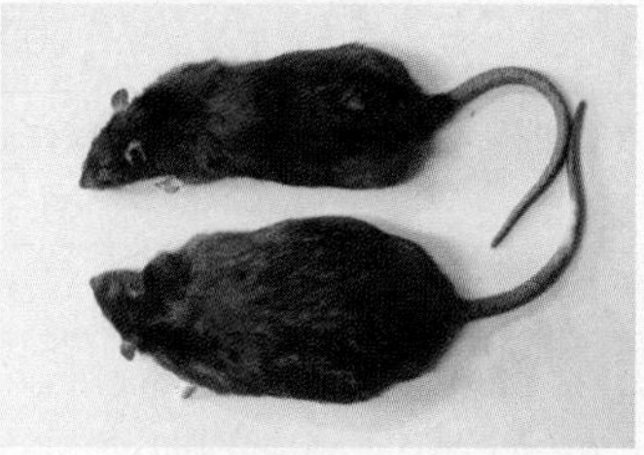

Voisin/Phanie/Science Source

(a) What are the standard deviations for the two groups?
(b) What degrees of freedom does the conservative Option 2 use for two-sample t procedures for these data?
(c) Test the null hypothesis of no difference between the two group means against the two-sided alternative. Use the degrees of freedom from part (b).

21.30 Is Montessori preschool beneficial? Do education programs for preschool children that follow the Montessori method perform better than other programs? A study compared five-year-old children in Milwaukee, Wisconsin, who had been enrolled in preschool programs from the age of three.[17]

(a) Explain why comparing children whose parents chose a Montessori school with children of other parents would not show whether Montessori schools perform better than other programs. (In fact, all the children in the study applied to the Montessori school. The school district assigned students to Montessori or other preschools by a random lottery.)
(b) In all, 54 children were assigned to the Montessori school and 112 to other schools at age three. When the children were five, parents of 30 of the Montessori children and 25 of the others could be located and agreed to and subsequently participated in testing. This information reveals a possible source of bias in the comparison of outcomes. Explain why.
(c) One of the many response variables was score on a test of ability to apply basic mathematics to solve problems. Here are summaries for the children who took this test:

GROUP	n	$\bar{x}$	s
Montessori	30	19	3.11
Control	25	17	4.19

Is there evidence of a difference in the population mean scores? (The researchers used two-sided alternative hypotheses.)

21.31 **Is job satisfaction going down?** In the last 10 years, several authors have stated that people are miserable in their jobs and have become increasingly unhappy over time. Job satisfaction scores from participants in the General Social Survey (GSS) in 1975 and 2006 were studied. Higher scores indicate greater satisfaction, with 2.5 being a neutral score. Here are summaries of scores for 1975 and 2006.[18]

YEAR	SAMPLE SIZE	MEAN	STD. DEV.
1975	1165	3.37	0.81
2006	2177	3.32	0.80

(a) Is there a significant decrease in mean scores from 1975 to 2006? What do these data show about job satisfaction in 1975 compared to 2006?
(b) The paper from which the data came includes several years from 1972 to 2006. 1975 was the year with the highest mean score. What, if any, effect does this information have on your assessment of whether job satisfaction has decreased over time?

21.32 **Illusory pattern perceptions.** When one experiences a lack of control in one's life, does one compensate by seeking structure elsewhere? Assign 36 undergraduate students to one of two conditions. All are asked to identify a concept associated with a series of "grainy" pictures that contain an embedded image and are presented to them sequentially. During the task, they can ask questions to help them determine the associated concept. Half of the subjects (the lack-of-control group) receive feedback that is random and noncontingent on their questions. The other half receive useful feedback (the in-control group). After attempting to complete the task, all subjects are presented with 12 grainy pictures that are similar to those used in the task but lack an embedded image. All are asked whether they perceive an image in the pictures, and the number of pictures identified as containing an image is counted for each subject. Here are the summary statistics:[19]

GROUP	GROUP SIZE	MEAN	STD. DEV.
Lack-of-control	18	5.16	3.5
In-control	18	3.47	2.0

(a) What degrees of freedom would you use in the conservative two-sample t procedures to compare the lack-of-control and in-control groups?
(b) What is the two-sample t test statistic for comparing the mean number of pictures identified as having an image for the two groups?
(c) Test the null hypothesis of no difference between the two population means against the two-sided alternative. Use your statistic from part (b) with degrees of freedom from part (a).

21.33 **Concussions and brain size.** What is the effect of concussions on the brain? Researchers measured the brain sizes (hippocampal volume in microliters) of 25 collegiate football players with a history of clinician-diagnosed concussion and 25 collegiate football players without a history of concussion. Here are the summary statistics:[20]

GROUP	GROUP SIZE	MEAN	STDDEV
Concussion	25	5784	609.3
Non-concussion	25	6489	815.4

(a) Is there evidence of a difference in mean brain size between football players with a history of concussion and those without concussions?
(b) The researchers in this study stated that participants were "consecutive cases of healthy National Collegiate Athletic Association Football Bowl Subdivision Division I football athletes with ($n = 25$) or without ($n = 25$) a history of clinician-diagnosed concussion . . . between June 2011 and August 2013" at a U.S. psychiatric research institute specializing in neuroimaging among collegiate football players. What effect does this information have on your conclusions in part (a)?

21.34 **Coaching and SAT scores.** Coaching companies claim that their courses can raise the SAT scores of high school students. Of course, students who retake the SAT without paying for coaching generally raise their scores, too. A random sample of students who took the SAT twice found 427 who were coached and 2733 who were uncoached.[21] Starting with their Verbal scores on the first and second tries, we have these summary statistics:

		TRY 1		TRY 2		GAIN	
	n	$\bar{x}$	s	$\bar{x}$	s	$\bar{x}$	s
Coached	427	500	92	529	97	29	59
Uncoached	2733	506	101	527	101	21	52

The summary statistics for Gain are based on the changes in the scores of the individual students. Let's first ask if students who are coached increased their scores significantly.

(a) You could use the information on the Coached line to carry out either a two-sample t test comparing Try 1 with Try 2 for coached students, or a matched pairs t test using Gain. Which is the correct test? Why?

(b) Carry out the proper test. What do you conclude?
(c) Give a 99% confidence interval for the mean gain of all students who are coached.

21.35 Coaching and SAT scores, continued. What we really want to know is whether coached students improve more than uncoached students, and whether any advantage is large enough to be worth paying for. Use the information in the previous exercise to answer these questions.

(a) Is there good evidence that coached students gained more on the average than uncoached students?
(b) How much more do coached students gain on the average? Give a 99% confidence interval.
(c) Based on your work, what is your opinion: do you think coaching courses are worth paying for?

21.36 Coaching and SAT scores: critique. The data you used in the previous two problems came from a random sample of students who took the SAT twice. The response rate was 63%, which is pretty good for nongovernment surveys, so let's accept that the respondents do represent all students who took the exam twice. Nonetheless, we can't be sure that coaching actually *caused* the coached students to gain more than the uncoached students. Explain briefly but clearly why this is so.

Exercises 21.37 to 21.44 include the actual data. To apply the two-sample t procedures, use Option 1 if you have technology that implements that method. Otherwise, use Option 2.

21.37 Improving your tips. Researchers gave 40 index cards to a waitress at an Italian restaurant in New Jersey. Before delivering the bill to each customer, the waitress randomly selected a card and wrote on the bill the same message that was printed on the index card. Twenty of the cards had the message, "The weather is supposed to be really good tomorrow. I hope you enjoy the day!" Another 20 cards contained the message, "The weather is supposed to be not so good tomorrow. I hope you enjoy the day anyway!" After the customers left, the waitress recorded the amount of the tip (percent of bill) before taxes. Here are the tips for those receiving the good-weather message:[22] DATA TIP4

20.8 18.7 19.9 20.6 21.9 23.4 22.8 24.9 22.2 20.3
24.9 22.3 27.0 20.5 22.2 24.0 21.2 22.1 22.0 22.7

The tips for the 20 customers who received the bad weather message are

18.0 19.1 19.2 18.8 18.4 19.0 18.5 16.1 16.8 14.0
17.0 13.6 17.5 20.0 20.2 18.8 18.0 23.2 18.2 19.4

(a) Make stemplots or histograms of both sets of data. Because the distributions are reasonably symmetric with no extreme outliers, the *t* procedures will work well.
(b) Is there good evidence that the two different messages produce different percent tips? State hypotheses, carry out a two-sample *t* test, and report your conclusions.

21.38 Do good smells bring good business? In Exercise 21.9, you examined the effects of a lavender odor on customer behavior in a small restaurant. Lavender is a relaxing odor. The researchers also looked at the effects of lemon, a stimulating odor. The design of the study is described in Exercise 21.9. Here are the times in minutes that customers spent in the restaurant when no odor was present: DATA ODORS3

103 68 79 106 72 121 92 84 72 92
85 69 73 87 109 115 91 84 76 96
107 98 92 107 93 118 87 101 75 86

When a lemon odor was present, customers lingered for these times:

78 104 74 75 112 88 105 97 101 89
88 73 94 63 83 108 91 88 83 106
108 60 96 94 56 90 113 97

(a) Examine both samples. Does it appear that use of two-sample *t* procedures is justified? Do the sample means suggest that a lemon odor changes the average length of stay?
(b) Does a lemon odor influence the length of time customers stay in the restaurant? State hypotheses, carry out a *t* test, and report your conclusions.

21.39 Improving your tips, continued. Use the data in Exercise 21.37 to give a 95% confidence interval for the difference between the mean percent tips for the two different messages. DATA TIP4

21.40 The power of positive thinking? Does the way the press depicts the economic future affect the stock market? To investigate this, researchers analyzed the longest article about the crisis from the front page of the Money section of *USA Today* from a randomly chosen weekday of each week between August 2007 and June 2009. Articles were rated as to how positive or negative they were about the economic future. For each week, the change in the Dow Jones Industrial Average (average DJIA value the week after the article appeared minus the average the week the article appeared) was computed. Positive values of the change indicate that the DJIA increased.[23] Here are the changes in DJIA corresponding to very positive articles: DATA DJIA

−325 −200 −225 −75 −25 25 50
225 25 −225 −250 200 250 75

Here are the changes in DJIA values corresponding to very negative articles:

150 300 225 125 −175 −225 −375
−175 0 125 175 475

Is there good evidence that the DJIA performs differently after very positive articles than after very negative articles?

(a) Do the sample means suggest that there is a difference in the change in the DJIA after very positive articles versus after very negative articles?
(b) Make stemplots for both samples. Are there any obvious departures from Normality?
(c) Test the hypothesis H_0: $\mu_1 = \mu_2$ against the two-sided alternative. What do you conclude from part (a) and from the result of your test?
(d) Among a host of different factors that are claimed to have triggered the economic crisis of 2007–09, one was a "culture of irresponsibility" in the way the future was depicted in the press. Do the data provide any evidence that negative articles in the press contributed to poor performance of the DJIA?

21.41 **What do you make of Mitt?** Twenty-nine college students, identified as having a positive attitude about Mitt Romney as compared to Barack Obama in the 2012 presidential election, were asked to rate how trustworthy the face of Mitt Romney appeared, as represented in their mental image of Mitt Romney's face. Ratings were on a scale of 0 to 7, with 0 being "not at all trustworthy" and 7 being "extremely trustworthy." Here are the 29 ratings:[24] DATA MITT2

2.6	3.2	4.7	3.3	3.4	3.6	3.7	3.8	3.9	4.1
4.2	4.9	5.7	4.2	3.9	3.2	4.1	5.0	5.0	4.6
4.6	3.9	3.9	5.3	2.8	2.6	3.0	3.3	3.7	

Twenty-nine college students identified as having a negative attitude about Mitt Romney as compared to Barack Obama in the 2012 presidential election were also asked to rate how trustworthy the face of Mitt Romney appeared. Here are the 29 ratings:

1.8	3.3	4.3	4.4	2.5	2.6	3.5	4.2	4.7	2.5
2.5	3.6	3.9	3.9	4.3	4.3	3.8	3.3	2.9	1.7
3.3	3.3	3.9	4.3	4.1	3.8	3.3	5.3	5.4	

(a) Do the sample means suggest that there is a difference in the mean trustworthy ratings between the two groups?
(b) Make stemplots for both samples. Are there any obvious departures from Normality?
(c) Test the hypothesis H_0: $\mu_1 = \mu_2$ against the one-sided alternative that students with a positive attitude rate Mitt Romney more trustworthy than those with a negative attitude. What do you conclude from part (a) and from the result of your test?

21.42 **How big is the difference?** Continue your work from Exercise 21.40. A researcher wants to know how big a difference there is in the change in the DJIA after reports of a positive economic outlook compared to reports of a negative economic outlook. Give a 90% confidence interval for the difference in mean change in the DJIA. DATA DJIA

21.43 **Do women talk more than men? Another study.** Exercise 21.27 described a series of six studies investigating the number of words women and men speak per day. Exercise 21.27 gives results from two of these studies. Here are the results from another of these studies. The estimated numbers of words spoken per day for 27 women are DATA TALK2

15,357	13,618	9,783	26,451	12,151	8,391	19,763
25,246	8,427	6,998	24,876	6,272	10,047	15,569
39,681	23,079	24,814	19,287	10,351	8,866	10,827
12,584	12,764	19,086	26,852	17,639	16,616	

The estimated numbers of words spoken per day for 20 men are

28,408	10,084	15,931	21,688	37,786	10,575	12,880
11,071	17,799	13,182	8,918	6,495	8,153	7,015
4,429	10,054	3,998	12,639	10,974	5,255	

Does this study provide good evidence that women talk more than men, on average?

(a) Make stemplots for both samples. Are there any obvious deviations from Normality? In spite of these deviations from Normality, it is safe to use the *t* procedures. Explain.
(b) Test the hypothesis H_0: $\mu_1 = \mu_2$ against the one-sided alternative that the mean number of words per day for women (μ_1) is greater than the mean number of words per day for men (μ_2). What do you conclude?

Do birds learn to time their breeding? *Blue titmice eat caterpillars. The birds would like lots of caterpillars around when they have young to feed, but they breed earlier than peak caterpillar season. Do the birds time when they breed based on the previous year's caterpillar supply? Researchers randomly assigned seven pairs of birds to have the natural caterpillar supply supplemented while feeding their young and another six pairs to serve as a control group relying on natural food supply. The next year, they measured how many days after the caterpillar peak the birds produced their nestlings.*[25] *Exercises 21.44 to 21.46 are based on this experiment.*

© *Hugh Clark/Frank Lane Picture Agency/Corbis*

21.44 Did the randomization produce similar groups? The first thing to do is to compare the two groups in the first year. The only difference should be the chance effect of the random assignment. The study report says: "In the experimental year, the degree of synchronization did not differ between food-supplemented and control females." For this comparison, the report gives $t = -1.05$. What type of t statistic (paired or two-sample) is this? What are the degrees of freedom for this statistic? Show that this t leads to the quoted conclusion.

21.45 Did the treatment have an effect? The investigators expected the control group to adjust their breeding date the next year, whereas the well-fed supplemented group had no reason to change. The report continues: "but in the following year food-supplemented females were more out of synchrony with the caterpillar peak than the controls." Here are the data (days behind the caterpillar peak): DATA BREED

Control	4.6	2.3	7.7	6.0	4.6	−1.2	
Supplemented	15.5	11.3	5.4	16.5	11.3	11.4	7.7

Carry out a t test and show that it leads to the quoted conclusion.

21.46 Year-to-year comparison. Rather than comparing the two groups in each year, we could compare the behavior of each group in the first and second years. The study report says: "Our main prediction was that females receiving additional food in the nestling period should not change laying date the next year, whereas controls, which (in our area) breed too late in their first year, were expected to advance their laying date in the second year."

Comparing days behind the caterpillar peak in Years 1 and 2 gave $t = 0.63$ for the control group and $t = -2.63$ for the supplemented group. Are these paired or two-sample t statistics? What are the degrees of freedom for each t? Show that these t-values do *not* agree with the prediction.

The remaining exercises ask you to answer questions from data without having the details outlined for you. The exercise statements give you the ***State*** *step of the four-step process. Follow the* ***Plan, Solve,*** *and* ***Conclude*** *steps as illustrated in Examples 21.2 and 21.3 for significance tests and Example 21.4 for confidence intervals. Remember that examining the data and discussing the conditions for inference are part of the* ***Solve*** *step.*

21.47 Thinking about money changes behavior. 4 step Kathleen Vohs of the University of Minnesota and her coworkers carried out several randomized comparative experiments on the effects of thinking about money. Here's part of one such experiment.[26] Ask student subjects to unscramble 30 sets of five words to make a meaningful phrase from four of the five words. The control group unscrambled phrases like "cold it desk outside is" into "it is cold outside." The treatment group unscrambled phrases that lead to thinking about money, turning "high a salary desk paying" into "a high-paying salary." Then each subject worked a hard puzzle, knowing that he or she could ask for help. Here are the times in seconds until subjects asked for help. For the treatment group: DATA MNYTHNK

609 444 242 199 174 55 251 466 443
531 135 241 476 482 362 69 160

For the control group:

118 272 413 291 140 104 55 189 126
400 92 64 88 142 141 373 156

The researchers suspected that money is connected with self-sufficiency, so that the treatment group will ask for help less quickly on the average. Do the data support this idea?

21.48 Adolescent obesity. 4 step Adolescent obesity is a serious health risk affecting more than 5 million young people in the United States alone. Laparoscopic adjustable gastric banding has the potential to provide a safe and effective treatment. Fifty adolescents between 14 and 18 years old with a body mass index higher than 35 were recruited from the Melbourne, Australia, community for the study.[27] Twenty-five were randomly selected to undergo gastric banding, and the remaining twenty-five were assigned to a supervised lifestyle intervention program involving diet, exercise, and behavior modification. All subjects were followed for two years. Here are the weight losses in kilograms for the subjects who completed the study. In the gastric banding group: DATA ADOBESE

35.6 81.4 57.6 32.8 31.0 37.6 36.5 −5.4 27.9 49.0 64.8 39.0
43.0 33.9 29.7 20.2 15.2 41.7 53.4 13.4 24.8 19.4 32.3 22.0

In the lifestyle intervention group:

6.0 2.0 −3.0 20.6 11.6 15.5 −17.0 1.4 4.0
−4.6 15.8 34.6 6.0 −3.1 −4.3 −16.7 −1.8 −12.8

Is there good evidence that gastric banding is superior to the lifestyle intervention program?

21.49 Active versus traditional learning. 4 step Can active learning improve knowledge retention? Two undergraduate calculus-based engineering statistics courses were taught in different academic quarters, with one employing active-learning methods and the other using traditional learning methods. The traditional class was taught lecture-style with relatively little in-class interaction between peers and with the instructor. The active-learning course integrated four group projects into the curriculum, with in-class time devoted to group work on the projects and fewer homework assignments. To assess knowledge retention, two five-question versions of a test were created. They had similar but not identical questions covering core statistics topics, worth a total of 18 possible points. All students in both sections were randomly given one version of the test as part of their final exam. Then, eight months later, a volunteer subset of the original students were given the version that they had not taken previously. To encourage students to take the second version of the exam, a $10 gift card to the university bookstore was given to each participant. The change in the score from the first version to the second is used to measure a student's long-term ability to retain the course material. DATA ACTVLRN

Here are the changes in exam scores for the 15 students in the active group:[28]

0 5 7 8 0 3 6 2 5 1
3 2 4 3 5

The changes in exam scores for the 23 students in the traditional group are:

7 0 8 2 4 3 1 2 5 8
5 6 3 12 1 6 3 6 7 7
5 6 2

Is there good evidence that active learning is superior to traditional lecturing?

21.50 **Each day I am getting better in math.** A "subliminal" message is below our threshold of awareness but may nonetheless influence us. Can subliminal messages help students learn math? A group of students who had failed the mathematics part of the City University of New York Skills Assessment Test agreed to participate in a study to find out. DATA SUBLIM

All received a daily subliminal message, flashed on a screen too rapidly to be consciously read. The treatment group of 10 students (chosen at random) was exposed to "Each day I am getting better in math." The control group of eight students was exposed to a neutral message, "People are walking on the street." All students participated in a summer program designed to raise their math skills, and all took the assessment test again at the end of the program. Table 21.3 gives data on the subjects' scores before and after the program.[29] Is there good evidence that the treatment brought about a greater improvement in math scores than the neutral message? How large is the mean difference in gains between treatment and control? (Use 95% confidence.)

TABLE 21.3 MATHEMATICS SKILLS SCORES BEFORE AND AFTER A SUBLIMINAL MESSAGE

TREATMENT GROUP		CONTROL GROUP	
BEFORE	AFTER	BEFORE	AFTER
18	24	18	29
18	25	24	29
21	33	20	24
18	29	18	26
18	33	24	38
20	36	22	27
23	34	15	22
23	36	19	31
21	34		
17	27		

21.51 **Active versus traditional learning, continued.**
(a) Use the data in Exercise 21.49 to give a 90% confidence interval for the difference in the mean change in score for students in the active and traditional classes.
(b) Give a 90% confidence interval for the mean change in score of students in the active-learning class. DATA ACTVLRN

21.52 **Tropical flowers.** Different varieties of the tropical flower *Heliconia* are fertilized by different species of hummingbirds. Over time, the lengths of the flowers and the forms of the hummingbirds' beaks have evolved to match each other. Here are data on the lengths in millimeters of two color varieties of the same species of flower on the island of Dominica:[30] DATA FLOWERS

Art Wolfe/Getty Images

H. caribaea RED							
41.90	42.01	41.93	43.09	41.47	41.69	39.78	40.57
39.63	42.18	40.66	37.87	39.16	37.40	38.20	38.07
38.10	37.97	38.79	38.23	38.87	37.78	38.01	

H. caribaea YELLOW							
36.78	37.02	36.52	36.11	36.03	35.45	38.13	37.1
35.17	36.82	36.66	35.68	36.03	34.57	34.63	

Is there good evidence that the mean lengths of the two varieties differ? Estimate the difference between the population means. (Use 95% confidence.)

21.53 **Student drinking.** A professor asked her sophomore students, "How many drinks do you typically have per session? (A drink is defined as one 12 oz beer, one 4 oz glass of wine, or one 1 oz shot of liquor.)" Some of the students didn't drink. Table 21.4 gives the responses of the female and male students who did drink.[31] It is likely that some of the students exaggerated a bit. The sample is all students in one large sophomore-level class. The class is popular, so we are tentatively willing to regard its members as an SRS of sophomore students at this college. Do a complete analysis that reports on DATA DRINKS

(a) the drinking behavior claimed by sophomore women.
(b) the drinking behavior claimed by sophomore men.
(c) a comparison of the behavior of women and men.

TABLE 21.4 DRINKS PER SESSION CLAIMED BY FEMALE AND MALE STUDENTS

FEMALE STUDENTS												
2.5	9	1	3.5	2.5	3	1	3	3	3	3	2.5	2.5
5	3.5	5	1	2	1	7	3	7	4	4	6.5	4
3	6	5	3	8	6	6	3	6	8	3	4	7
4	5	3.5	4	2	1	5	5	3	3	6	4	2
7	7	7	3.5	3	2.5	10	5	4	9	8	1	6
2	5	2.5	3	4.5	9	5	4	4	3	4	6	7
4	5	1	5	3	4	10	7	3	4	4	4	4
2	1	2.5	2.5									
MALE STUDENTS												
7	7.5	8	15	3	4	1	5	11	4.5	6	4	10
16	4	8	5	9	7	7	3	5	6.5	1	12	4
6	8	8	4.5	10.5	8	6	10	1	9	8	7	8
15	3	10	7	4	6	5	2	10	7	9	5	8
7	3	7	6	4	5	2	5	5.5	9	10	10	4
8	4	2	4	12.5	3	15	2	6	3	4	3	10
6	4.5	5										

Exploring the Web

21.54 **A two-sample *t* test example.** Find an example of a two-sample *t* test on the web. The *Journal of the American Medical Association* (`jama.ama-assn.org`), *Science Magazine* (`www.sciencemag.org`), the *Canadian Medical Association Journal* (`www.cmaj.ca`), the *Journal of Statistics Education* (`www.amstat.org/publications/jse`), or perhaps the *Journal of Quantitative Analysis in Sports* (`www.bepress.com/jqas`) are possible sources. To help locate an article, look through the abstracts of articles. Once you find a suitable article, read it and then briefly describe the study (including why it is a two-sample study) and its conclusions. If *P*-values, means, standard deviations, *t* statistics, and the like are reported, be sure to include them in your summary. Also, be sure to give the reference (either the web link or the journal, issue, year, title of the paper, authors, and page numbers).

21.55 Antibiotics after surgery. If your college has online access to the *Archives of Otolaryngology—Head and Neck Surgery,* read the article "Duration-Related Efficacy of Postoperative Antibiotics Following Pediatric Tonsillectomy: A Prospective, Randomized, Placebo-Controlled Trial" (available online at `archotol.ama-assn.org/cgi/content/full/135/10/984`). The authors appear to use two-sample *t* procedures in the paper. After reading the article, read the (brief) discussion on the *Chance* website, `www.causeweb.org/wiki/chance/index.php/Chance_News_57`. What are some criticisms or concerns expressed about the study in the *Chance* article?

Cavan Images/Getty Images

CHAPTER 22

Inference about a Population Proportion

In this chapter we cover...

22.1 The sample proportion $\hat{p}$

22.2 Large-sample confidence intervals for a proportion

22.3 Choosing the sample size

22.4 Significance tests for a proportion

22.5 Plus four confidence intervals for a proportion*

Our discussion of statistical inference to this point has concerned making inferences about population *means*. Interest in the mean response for a population, or a comparison of the means of two populations, is common when the response variable is quantitative and takes a numerical value with some unit of measurement. Now we turn to questions about the *proportion* of some outcome in a population. Even when the original response is a quantitative variable, such as total cholesterol, we might be more interested in whether or not someone has cholesterol greater than 200 mg/dL (outcome of interest). Our proportion of interest is then the population proportion of adults who have cholesterol greater than 200 mg/dL, and the methods of this chapter apply. Here are some further examples that call for inference about population proportions.

EXAMPLE 22.1 Risky Behavior in the Age of AIDS

How common is behavior that puts people at risk of AIDS? In the early 1990s, the landmark National AIDS Behavioral Surveys interviewed a random sample of 2673 adult heterosexuals. Of these, 170 had more than one sexual partner in the past year. That's 6.36% of the sample.[1] Based on these data, what can we say about the percent of all adult heterosexuals who have multiple partners? We want to *estimate a single population proportion*. This chapter concerns inference about one proportion. ■

EXAMPLE 22.2 Young Adults Living at Home

A surprising number of young adults (aged 19 to 25) still live at home with their parents. A random sample of 2253 men and 2629 women in this age group found that 44% of the men but only 35% of the women lived at home. Is this significant evidence that the proportions living at home differ in the populations of all young men and all young women? We want to *compare two population proportions*. This is the topic of Chapter 23. ■

To make inferences about a population mean μ, we use the mean $\bar{x}$ of a random sample from the population. The reasoning of inference starts with the sampling distribution of $\bar{x}$. Now we follow the same pattern, replacing means by proportions.

22.1 The sample proportion $\hat{p}$

We are interested in the unknown proportion p of a population that has some outcome. For convenience, call the outcome we are looking for a "success." In Example 22.1, the population is adult heterosexuals, and the parameter p is the proportion who have had more than one sexual partner in the past year. To estimate p, the National AIDS Behavioral Surveys used random dialing of telephone numbers to contact a sample of 2673 people. Of these, 170 said they had had multiple sexual partners. The statistic that estimates the parameter p is the

sample proportion

sample proportion

$$\hat{p} = \frac{\text{number of successes in the sample}}{\text{total number of individuals in the sample}} = \frac{170}{2673} = 0.0636$$

Read the sample proportion $\hat{p}$ as "p-hat."

How good is the statistic $\hat{p}$ as an estimate of the parameter p? To find out, we ask, "What would happen if we took many samples?" The sampling distribution of $\hat{p}$ answers this question. Here are the facts.[2]

Sampling Distribution of a Sample Proportion

Draw an SRS of size n from a large population that contains proportion p of successes. Let $\hat{p}$ be the **sample proportion** of successes,

$$\hat{p} = \frac{\text{number of successes in the sample}}{n}$$

Then:

- The **mean** of the sampling distribution is p.
- The **standard deviation** of the sampling distribution is

$$\sqrt{\frac{p(1-p)}{n}}$$

- As the sample size increases, the sampling distribution of $\hat{p}$ becomes **approximately Normal**. That is, for large n, $\hat{p}$ has approximately the $N(p, \sqrt{p(1-p)/n})$ distribution.

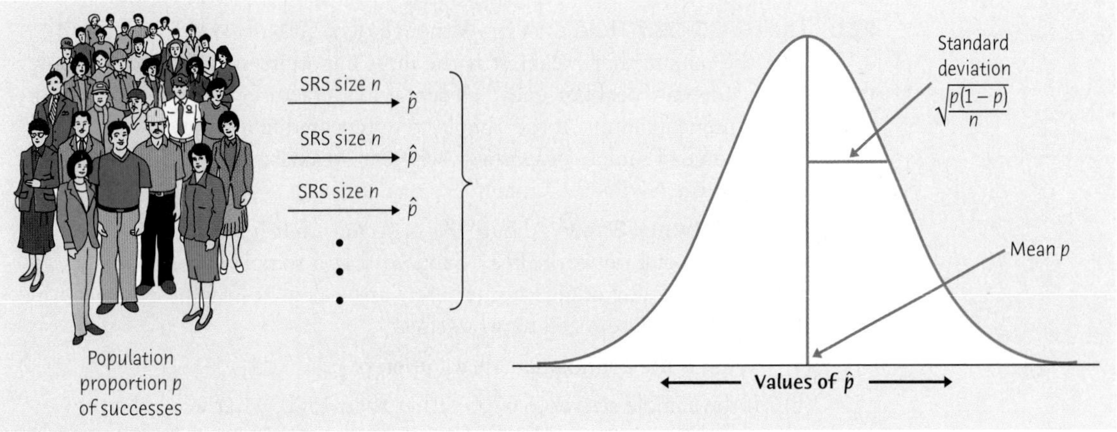

FIGURE 22.1
Select a large SRS from a population in which the proportion p are successes. The sampling distribution of the proportion $\hat{p}$ of successes in the sample is approximately Normal. The mean is p and the standard deviation is $\sqrt{p(1-p)/n}$.

Figure 22.1 summarizes these facts in a form that helps you recall the big idea of a sampling distribution. The behavior of sample proportions $\hat{p}$ is similar to the behavior of sample means $\bar{x}$, except that the distribution of $\hat{p}$ is only approximately Normal. The mean of the sampling distribution of $\hat{p}$ is the true value of the population proportion p. That is, $\hat{p}$ is an unbiased estimator of p. The standard deviation of $\hat{p}$ gets smaller as the sample size n gets larger, so that estimation is likely to be more accurate when the sample is larger. As is the case for $\bar{x}$, the standard deviation gets smaller only at the rate $\sqrt{n}$. We need four times as many observations to cut the standard deviation in half.

EXAMPLE 22.3 Asking about Risky Behavior

Suppose that in fact 6% of all adult heterosexuals had more than one sexual partner in the past year (and would admit it when asked). The National AIDS Behavioral Surveys interviewed a random sample of 2673 people from this population. In many such samples, the proportion $\hat{p}$ of the 2673 people in the sample who had more than one partner would vary according to (approximately) the Normal distribution with mean 0.06 and standard deviation

$$\sqrt{\frac{p(1-p)}{n}} = \sqrt{\frac{(0.06)(0.94)}{2673}}$$
$$= \sqrt{0.0000211} = 0.00459 \blacksquare$$

Apply Your Knowledge

22.1 Staph Infections. A study investigated ways to prevent staph infections in surgery patients. In a first step, the researchers examined the nasal secretions of a random sample of 6771 patients admitted to various hospitals for surgery. They found that 1251 of these patients tested positive for *Staphylococcus aureus*, a bacterium responsible for most staph infections.[3]

(a) Describe the population and explain in words what the parameter p is.

(b) Give the numerical value of the statistic $\hat{p}$ that estimates p.

22.2 The 68–95–99.7 Rule and $\hat{p}$. Although over 50% of American adults believe the maxim that breakfast is the most important meal of the day, only about 30% eat breakfast daily.[4] A cereal manufacturer contacts an SRS of 1000 American adults. If the sample were repeated many times, what would be the range of sample proportions who eat breakfast daily according to the 95 part of the 68–95–99.7 rule?

22.3 Social Network Sites. About 90% of young adult Internet users (aged 18 to 29) use social network sites.[5] Suppose that a sample survey contacts an SRS of 1500 young adult Internet users and calculates the proportion $\hat{p}$ in this sample who use social network sites.

(a) What is the approximate distribution of $\hat{p}$?

(b) If the sample size were 6000 rather than 1500, what would be the approximate distribution of $\hat{p}$?

22.2 Large-sample confidence intervals for a proportion

We can follow the same path from sampling distribution to confidence interval as we did for $\bar{x}$ in Chapter 15. To obtain a level C confidence interval for p, we start by capturing the central probability C in the distribution of $\hat{p}$. To do this, go out z^* standard deviations from the mean p, where z^* is the critical value that captures the central area C under the standard Normal curve. Figure 22.2 shows the result. The confidence interval is

$$\hat{p} \pm z^* \sqrt{\frac{p(1-p)}{n}}$$

standard error of $\hat{p}$

This won't do, because we don't know the value of p. So we replace the standard deviation by the **standard error of $\hat{p}$**

$$SE_{\hat{p}} = \sqrt{\frac{\hat{p}(1-\hat{p})}{n}}$$

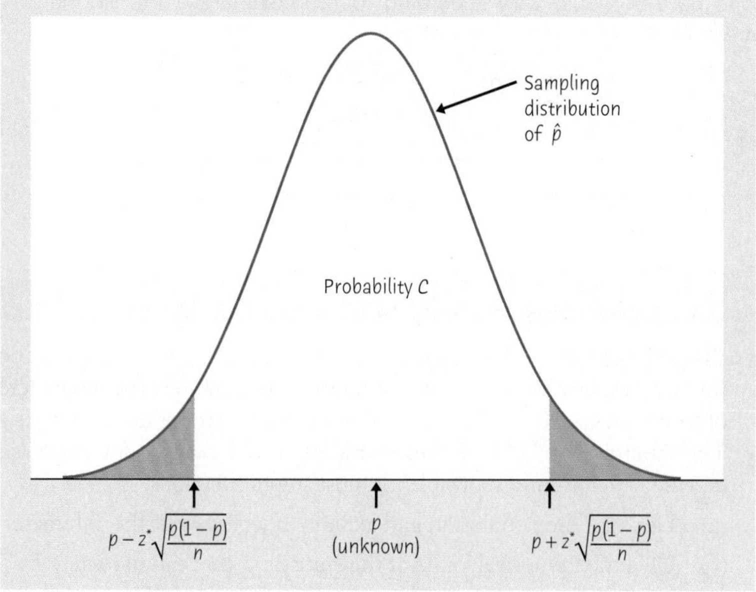

FIGURE 22.2
With probability C, $\hat{p}$ lies within $\pm z^*\sqrt{p(1-p)/n}$ of the unknown population proportion p. That is, in these samples p lies within $\pm z^*\sqrt{p(1-p)/n}$ of $\hat{p}$.

to get the confidence interval

$$\hat{p} \pm z^*\sqrt{\frac{\hat{p}(1-\hat{p})}{n}}$$

As with previous confidence intervals, this interval has the familiar form

$$\text{estimate} \pm z^*\text{SE}_{\text{estimate}}$$

We can trust this confidence interval only for large samples. Because the number of successes must be a whole number, using a continuous Normal distribution to describe the behavior of $\hat{p}$ may not be accurate unless n is large. Because the approximation is least accurate for populations that are almost all successes or almost all failures, we require that the sample have both enough successes and enough failures rather than that the overall sample size be large. *Pay attention to both conditions for inference in the box below that summarizes the confidence interval: we must as usual be willing to regard the sample as an SRS from the population, and the sample must have both enough successes and enough failures. The condition on successes and failures ensures that the sample size is large enough to use the Normal approximation without knowing p.*

CAUTION

Large-Sample Confidence Interval for a Population Proportion

Draw an SRS of size n from a large population that contains an unknown proportion p of successes. An approximate level C **confidence interval for *p*** is

$$\hat{p} \pm z^*\sqrt{\frac{\hat{p}(1-\hat{p})}{n}}$$

where z^* is the critical value for the standard Normal density curve with area C between $-z^*$ and z^*.

Use this interval only when the numbers of successes and failures in the sample are both at least 15.[6]

Why not *t*? Notice that we *don't* change z^* to t^* when we replace the standard deviation by the standard error. When the sample mean $\bar{x}$ estimates the population mean μ, a separate parameter σ describes the variability of the distribution of $\bar{x}$. We separately estimate σ, and this leads to a t distribution. When the sample proportion $\hat{p}$ estimates the population proportion p, the variability depends on p, not on a separate parameter. There is no t distribution—we just make the Normal approximation a bit less accurate when we replace p in the standard deviation by $\hat{p}$.

EXAMPLE 22.4 Estimating Risky Behavior

The four-step process for any confidence interval is outlined on page 381.

STATE: The National AIDS Behavioral Surveys found that 170 of a sample of 2673 adult heterosexuals had had multiple partners. That is,

$$\hat{p} = \frac{170}{2673} = 0.0636$$

What can we say about the population of all adult heterosexuals?

PLAN: We will give a 99% confidence interval to estimate the proportion p of all adult heterosexuals who have multiple partners.

SOLVE: First verify the conditions for inference:

- The sampling design was a complex stratified sample, and the survey used inference procedures for that design. The overall effect is close to an SRS, however.
- The sample is large enough: the numbers of successes (170) and failures (2503) in the sample are both much larger than 15.

The sample size condition is easily satisfied. The condition that the sample be an SRS is only approximately met.

A 99% confidence interval for the proportion p of all adult heterosexuals with multiple partners uses the standard Normal critical value $z^* = 2.576$. The confidence interval is

$$\hat{p} \pm z^* \sqrt{\frac{\hat{p}(1-\hat{p})}{n}} = 0.0636 \pm 2.576\sqrt{\frac{(0.0636)(0.9364)}{2673}}$$
$$= 0.0636 \pm 0.0122$$
$$= 0.0514 \text{ to } 0.0758$$

CONCLUDE: We are 99% confident that the percent of adult heterosexuals who have had more than one sexual partner in the year prior to the survey lies between about 5.1% and 7.6%. ■

CAUTION *As usual, the practical problems of a large sample survey weaken our confidence in the AIDS survey's conclusions.* Only people in households with landline telephones could be reached. Although at the time of the survey about 89% of American households had landline telephones, as the number of cell phone only users increases, using a sample of households with landline phones is becoming less acceptable for surveys of the general population (see page 216). Additionally, some groups at high risk for AIDS, such as people who inject illegal drugs, often don't live in settled households and were therefore underrepresented in the sample. About 30% of the people reached refused to cooperate. A nonresponse rate of 30% is not unusual in large sample surveys, but it may cause some bias if those who refuse differ systematically from those who cooperate. The survey used statistical methods that adjust for unequal response rates in different groups. Finally, some respondents may not have told the truth when asked about their sexual behavior. The survey team tried to make respondents feel comfortable. For example, Hispanic women were interviewed only by Hispanic women, and Spanish speakers were interviewed by Spanish speakers with the same regional accent (Cuban, Mexican, or Puerto Rican). Nonetheless, the survey report says that some bias is probably present:

> *It is more likely that the present figures are underestimates; some respondents may underreport their numbers of sexual partners and intravenous drug use because of embarrassment and fear of reprisal, or they may forget or not know details of their own or of their partner's HIV risk and their antibody testing history.*[7]

Reading the report of a large study like the National AIDS Behavioral Surveys reminds us that statistics in practice involves much more than formulas for inference.

Statistics In Your World

WHO IS A SMOKER?

When estimating a proportion p, be sure you know what counts as a "success." The news says that 20% of adolescents smoke. Shocking. It turns out that this is the percent who smoked at least once in the past month. If we say that a smoker is someone who smoked on at least 20 of the past 30 days and smoked at least half a pack on those days, fewer than 4% of adolescents qualify.

LaunchPad Online Resources

- The **Snapshots video**, *Inference for One Proportion*, provides the details of constructing a large-sample confidence interval for a proportion through an interesting example involving an opinion poll.
- The **StatClips Examples video**, *Confidence Intervals: Intervals for Proportions Example C*, illustrates the computation of a large-sample confidence interval for a proportion.

Apply Your Knowledge

22.4 No Confidence Interval. In the National AIDS Behavioral Surveys sample of 2673 adult heterosexuals, 0.2% (that's 0.002 as a decimal fraction) had both received a blood transfusion and had a sexual partner from a group at high risk of AIDS. Explain why we can't use the large-sample confidence interval to estimate the proportion p in the population who share these two risk factors.

22.5 Canadian Attitudes toward Guns. Canada has much stronger gun control laws than the United States, and Canadians support gun control more strongly than do Americans. A sample survey asked a random sample of 1505 adult Canadians, "Do you agree or disagree that all firearms should be registered?" Of the 1505 people in the sample, 1288 answered either "Agree strongly" or "Agree somewhat."[8]

(a) The survey dialed residential telephone numbers at random in all 10 Canadian provinces (omitting the sparsely populated northern territories). Based on what you know about sample surveys, what is likely to be the biggest weakness in this survey?

(b) Nonetheless, act as if we have an SRS from adults in the Canadian provinces. Give a 95% confidence interval for the proportion who support registration of all firearms.

22.6 Weightlifting Injuries. Resistance training is a popular form of conditioning aimed at enhancing sports performance and is widely used among high school, college, and professional athletes, although its use for younger athletes is controversial. A random sample of 4111 patients aged 8–30 admitted to U.S. emergency rooms with the Consumer Product Safety Commission code "weightlifting" were obtained. These injuries were classified as "accidental" if caused by dropped weight or improper equipment use. Of the 4111 weightlifting injuries, 1552 were classified as accidental.[9] Give a 90% confidence interval for the proportion of weightlifting injuries in this age group that were accidental. Follow the four-step process as illustrated in Example 22.4.

22.3 Choosing the sample size

In planning a study, we may want to choose a sample size that will allow us to estimate the parameter within a given margin of error. We saw earlier (page 425) how to do this for a population mean. The method is similar for estimating a population proportion.

The margin of error in the large-sample confidence interval for p is

$$m = z^* \sqrt{\frac{\hat{p}(1 - \hat{p})}{n}}$$

Statistics in Your World

NEW YORK, NEW YORK

New York City, they say, is bigger, richer, faster, and ruder. Maybe there's something to that. The sample survey firm Zogby International says that as a national average it takes five telephone calls to reach a live person. When calling to New York, it takes 12 calls. Survey firms assign their best interviewers to make calls to New York and often pay them bonuses to cope with the stress.

Here z^* is the standard Normal critical value for the level of confidence we want. Because the margin of error involves the sample proportion of successes $\hat{p}$, we need to guess this value when choosing n. Call our guess p^*. Here are two ways to get p^*:

1. Use a guess p^* based on a pilot study or on past experience with similar studies. You can do several calculations to cover the range of values of $\hat{p}$ you might get.
2. Use $p^* = 0.5$ as the guess. The margin of error m is largest when $p^* = 0.5$, so this guess is conservative in the sense that if we get any other $\hat{p}$ when we do our study, we will get a margin of error smaller than planned.

Once you have a guess p^*, the recipe for the margin of error can be solved to give the sample size n needed. Here is the result for the large-sample confidence interval. For simplicity, use this result even if you plan to use the plus four interval discussed in Section 22.5.

Sample Size for Desired Margin of Error

The level C confidence interval for a population proportion p will have margin of error approximately equal to a specified value m when the sample size is

$$n = \left(\frac{z^*}{m}\right)^2 p^*(1 - p^*)$$

where p^* is a guessed value for the sample proportion. The margin of error will always be less than or equal to m if you take the guess p^* to be 0.5.

Which method for finding the guess p^* should you use? The n you get doesn't change much when you change p^* as long as p^* is not too far from 0.5. You can use the conservative guess $p^* = 0.5$ if you expect the true $\hat{p}$ to be roughly between 0.3 and 0.7. If the true $\hat{p}$ is close to 0 or 1, using $p^* = 0.5$ as your guess will give a sample much larger than you need. Try to use a better guess from a pilot study when you suspect that $\hat{p}$ will be less than 0.3 or greater than 0.7.

EXAMPLE 22.5 Planning a Poll

4 step

Colin Anderson/BrandX/Age fotostock

STATE: Gloria Chavez and Ronald Flynn are the candidates for mayor in a large city. You are planning a sample survey to determine what percent of the voters intend to vote for Chavez. You will contact an SRS of registered voters in the city. You want to estimate the proportion p of Chavez voters with 95% confidence and a margin of error no greater than 3%, or 0.03. How large a sample do you need?

PLAN: Find the sample size n needed for margin of error $m = 0.03$ and 95% confidence. The winner's share in all but the most lopsided elections is between 30% and 70% of the vote. You can use the guess $p^* = 0.5$.

SOLVE: The sample size you need is

$$n = \left(\frac{1.96}{0.03}\right)^2 (0.5)(1 - 0.5) = 1067.1$$

Round the result up to $n = 1068$. (Rounding down would give a margin of error slightly greater than 0.03.)

CONCLUDE: An SRS of 1068 registered voters is adequate for margin of error ±3%. ■

If you want a 2.5% margin of error rather than 3%, then (after rounding up)

$$n = \left(\frac{1.96}{0.025}\right)^2 (0.5)(1 - 0.5) = 1537$$

For a 2% margin of error, the sample size you need is

$$n = \left(\frac{1.96}{0.02}\right)^2 (0.5)(1 - 0.5) = 2401$$

As usual, smaller margins of error call for larger samples.

LaunchPad Online Resources

- The **StatBoards video**, *Computing a Sample Size for One Proportion*, provides an additional example of choosing the sample size to provide a given margin of error.

Apply Your Knowledge

22.7 Did You Use a Mobile Device? The Monterey Bay Aquarium, founded in 1984, is situated on the beautiful coast of Monterey Bay in the historic Cannery Row district. From 2009 to 2013, the aquarium conducted a survey of a random sample of visitors as they exited the museum. The survey includes visitor demographic information, use of social media, and opinions on their aquarium visit. In 2013, the survey included 165 visitors over 65, of which 42 used a mobile device such as an Android phone or iPad during their visit.[10]

(a) What is the margin of error of the large-sample 95% confidence interval for the proportion of visitors over 65 who used a mobile device during their visit?

(b) How large a sample is needed to get the common ± 3 percentage point margin of error? Use the sample from part (a) as a pilot study to get p^*.

22.8 Can You Taste PTC? PTC is a substance that has a strong bitter taste for some people and is tasteless for others. The ability to taste PTC is inherited. About 75% of Italians can taste PTC, for example. You want to estimate the proportion of Americans with at least one Italian grandparent who can taste PTC.

(a) Starting with the 75% estimate for Italians, how large a sample must you collect in order to estimate the proportion of PTC tasters within ± 0.04 with 90% confidence?

(b) Estimate the sample size required if you made no assumptions about the value of the proportion who could taste PTC. How much has the required sample size changed?

22.4 Significance tests for a proportion

The test statistic for the null hypothesis H_0: $p = p_0$ is the sample proportion $\hat{p}$ standardized using the value p_0 specified by H_0,

$$z = \frac{\hat{p} - p_0}{\sqrt{\dfrac{p_0(1 - p_0)}{n}}}$$

This z statistic has approximately the standard Normal distribution when H_0 is true. *P*-values therefore come from the standard Normal distribution. Unlike the confidence interval in which p is unknown and must be estimated by $\hat{p}$ when standardizing the estimate, in the test we can replace p by p_0 when standardizing as p_0 is specified by H_0. Additionally, because H_0 fixes a value of p when standardizing the estimate, the sample size conditions for use of the test are less stringent than for the large-sample confidence interval in which p must be estimated. Here is the procedure for tests.

Significance Tests for a Proportion

Draw an SRS of size n from a large population that contains an unknown proportion p of successes. To **test the hypothesis H_0: $p = p_0$**, compute the z statistic

$$z = \frac{\hat{p} - p_0}{\sqrt{\frac{p_0(1 - p_0)}{n}}}$$

In terms of a variable Z having the standard Normal distribution, the approximate P-value for a test of H_0 against

H_a: $p > p_0$ is $P(Z \geq z)$

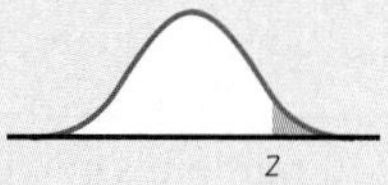

H_a: $p > p_0$ is $P(Z \leq z)$

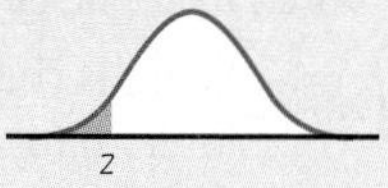

H_a: $p \neq p_0$ is $2P(Z \geq |z|)$

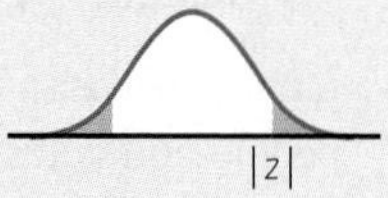

Use this test when the sample size n is so large that both np_0 and $n(1 - p_0)$ are 10 or more.[11]

EXAMPLE 22.6 Are Boys More Likely?

© Blaine Harrington III/Corbis

The four-step process for any significance test is outlined on page 401.

STATE: We hear that newborn babies are more likely to be boys than girls, presumably to compensate for higher mortality among boys in early life. Is this true? A random sample found 13,173 boys among 25,468 firstborn children.[12] The sample proportion of boys was

$$\hat{p} = \frac{13{,}173}{25{,}468} = 0.5172$$

Boys do make up more than half of the sample, but of course we don't expect a perfect 50–50 split in a random sample. Is this sample evidence that boys are more common than girls in the entire population?

PLAN: Take p to be the proportion of boys among all firstborn children of American mothers. (Biology says that this should be the same as the proportion among all children, but the survey data concern first births.)
We want to test the hypotheses

$$H_0: p = 0.5$$
$$H_a: p > 0.5$$

SOLVE: The conditions for inference require that we have a random sample and that $np_0 = (25{,}468)(0.5) = 12{,}734$ and $n(1 - p_0) = (25{,}468)(0.5) = 12{,}734$ are both greater than 10. Since the conditions for inference are met, we can go on to find the z test statistic:

$$z = \frac{\hat{p} - p_0}{\sqrt{\dfrac{p_0(1 - p_0)}{n}}}$$

$$= \frac{0.5172 - 0.5}{\sqrt{\dfrac{(0.5)(0.5)}{25{,}468}}} = 5.49$$

The P-value is the area under the standard Normal curve to the right of $z = 5.49$. We know that this is very small; Table C shows that $P < 0.0005$. Minitab (Figure 22.3) says that P is 0 to three decimal places.

CONCLUDE: There is very strong evidence that more than half of firstborns are boys ($P < 0.001$). ■

```
Session

Test and CI for One Proportion

Test of p = 0.5 p > 0.5

                                   95% lower
Sample          X        N          bound   Z-Value  P-Value
1           13173    25468       0.512087      5.50    0.000

Using the normal approximations.
```

FIGURE 22.3 Minitab output for the significance test of Example 22.6. Roundoff error in Example 22.6 explains the small difference (5.49 versus 5.50) in the values of the z statistic.

EXAMPLE 22.7 Estimating the Chance of a Boy

With 13,173 successes in 25,468 trials, we have at least 15 successes and 15 failures in the sample. The conditions for the large-sample confidence interval are easily met. The 99% confidence interval is

$$\hat{p} \pm z^* \sqrt{\frac{\hat{p}(1 - \hat{p})}{n}} = 0.5172 \pm 2.576 \sqrt{\frac{(0.5172)(0.4828)}{25{,}468}}$$

$$= 0.5172 \pm 0.0081$$

$$= 0.5091 \text{ to } 0.5253$$

We are 99% confident that between about 51% and 52.5% of first children are boys.

The confidence interval is more informative than the test in Example 22.6, which tells us only that more than half are boys. ■

LaunchPad Online Resources

- The **StatBoards video**, *Hypothesis Test for One Proportion*, illustrates the computation of a large-sample test for a proportion through an example.

Apply Your Knowledge

22.9 Spinning Euros. All euros have a national image on the "heads" side and a common design on the "tails" side. Spinning a coin, unlike tossing it, may not give heads and tails equal probabilities. Polish students spun the Belgian euro 250 times, with its portly king, Albert, displayed on the heads side. The result was 140 heads.[13] How significant is this evidence against equal probabilities? Follow the four-step process as illustrated in Example 22.6.

22.10 Vote for the Best Face? We often judge other people by their faces. It appears that some people judge candidates for elected office by their faces. Psychologists showed head-and-shoulders photos of the two main candidates in 32 races for the U.S. Senate to many subjects (dropping subjects who recognized one of the candidates) to see which candidate was rated "more competent" based on nothing but the photos. On election day, the candidates whose faces looked more competent won 22 of the 32 contests.[14] If faces don't influence voting, half of all races in the long run should be won by the candidate with the better face. Is there evidence that the candidate with the better face wins more than half the time? Follow the four-step process as illustrated in Example 22.6.

22.11 No Test. Explain whether we can use the z test for a proportion in these situations.

(a) You toss a coin 10 times in order to test the hypothesis H_0: $p = 0.5$ that the coin is balanced.

(b) A local candidate contacts an SRS of 900 of the registered voters in his district to see if there is evidence that more than half support the bill he is sponsoring.

(c) A college president says, "99% of the alumni support my firing of Coach Boggs." You contact an SRS of 200 of the college's 15,000 living alumni to test the hypothesis H_0: $p = 0.99$.

22.5 Plus four confidence intervals for a proportion*

The large-sample confidence interval $\hat{p} \pm z^*\sqrt{\hat{p}(1 - \hat{p})/n}$ for a sample proportion p is easy to calculate. It is also easy to understand because it rests directly on the approximately Normal distribution of $\hat{p}$. Unfortunately, confidence levels from this interval can be inaccurate, particularly with smaller sample sizes. The actual confidence level is usually *less* than the confidence level you asked for in choosing the critical value z^*. That's bad. What is worse, accuracy does not consistently get better as the sample size n increases. There are "lucky" and "unlucky" combinations of the sample size n and the true population proportion p.

Fortunately, there is a simple modification that is almost magically effective in improving the accuracy of the confidence interval. We call it the "plus four" method

*This section is optional.

because all you need to do is *add four imaginary observations: two successes and two failures.* With the added observations, the **plus four estimate** of p is

plus four estimate

$$\tilde{p} = \frac{\text{number of successes in the sample} + 2}{n + 4}$$

The formula for the confidence interval is exactly as before, with the new sample size and number of successes.[15] You do not need software that offers the plus four interval—just enter the new sample size (actual size + 4) and number of successes (actual number + 2) into the large-sample procedure.

Plus Four Confidence Interval for a Proportion

Draw an SRS of size n from a large population that contains an unknown proportion p of successes. To get the **plus four confidence interval for p**, add four imaginary observations—two successes and two failures. Then use the large-sample confidence interval with the new sample size ($n + 4$) and number of successes (actual number + 2).

Use this interval when the confidence level is at least 90% and the sample size n is at least 10, with any counts of successes and failures.

EXAMPLE 22.8 Cocaine Traces in Spanish Currency

STATE: Cocaine users commonly snort the powder up the nose through a rolled-up paper currency bill. Spain has a high rate of cocaine use, so it's not surprising that euro paper currency in Spain often bears traces of cocaine. Researchers collected 20 euro bills in each of several Spanish cities. In Madrid, 17 out of 20 bore traces of cocaine.[16] The researchers note that we can't tell whether the bills had been used to snort cocaine or had been contaminated in currency-sorting machines. Estimate the proportion of all euro bills in Madrid that have traces of cocaine.

© Andrea Matone/Alamy

PLAN: Take p to be the proportion of bills that show cocaine traces. That is, a "success" is a bill that shows cocaine traces. Give a 95% confidence interval for p.

SOLVE: It is not clear how the bills in the sample were selected, so we don't know if we have an SRS. We will act as though we have an SRS, but we proceed with caution. The conditions for use of the large-sample interval are not met because there are only three failures. To apply the plus four method, add two successes and two failures to the original data. The plus four estimate of p is

$$\tilde{p} = \frac{17 + 2}{20 + 4} = \frac{19}{24} = 0.7917$$

We calculate the plus four confidence interval in the same way as we do the large-sample interval, but we base it on 19 successes in 24 observations. Here it is:

$$\begin{aligned}\tilde{p} \pm z^*\sqrt{\frac{\tilde{p}(1 - \tilde{p})}{n + 4}} &= 0.7917 \pm 1.960\sqrt{\frac{(0.7917)(0.2083)}{24}} \\ &= 0.7917 \pm 0.1625 \\ &= 0.6292 \text{ to } 0.9542\end{aligned}$$

CONCLUDE: Assuming the sample can be regarded as an SRS, we estimate with 95% confidence that between about 63% and 95% of all euro bills in Madrid bear traces of cocaine. ■

For comparison, the ordinary sample proportion is

$$\hat{p} = \frac{17}{20} = 0.85$$

The plus four estimate $\tilde{p} = 0.7917$ in Example 22.8 is farther away from 1 than $\hat{p} = 0.85$. The plus four estimate gains its added accuracy by always moving toward 0.5 and away from 1 or 0, whichever is closer. This is particularly helpful when the sample contains only a few successes or a few failures. The numerical difference between a large-sample interval and the corresponding plus four interval is often small. Remember that the confidence level is the probability that the interval will catch the true population proportion *in very many uses.* Small differences every time can add up to more accurate confidence levels from plus four versus the large-sample interval.

How much more accurate is the plus four interval? Computer studies have asked how large n must be to guarantee that the actual probability that a 95% confidence interval covers the true parameter value is at least 0.94 for all samples of size n or larger. If $p = 0.1$, for example, the answer is $n = 646$ for the large-sample interval and $n = 11$ for the plus four interval.[17] The consensus of computational and theoretical studies is that plus four is better than the large-sample interval for many combinations of n and p.

Apply Your Knowledge

Jennifer Shields/Getty Images

22.12 Black Raspberries and Cancer. Sample surveys usually contact large samples, so we can use the large-sample confidence interval if the sample design is close to an SRS. Scientific studies often use smaller samples that require the plus four method. For example, Familial Adenomatous Polyposis (FAP) is a rare inherited disease characterized by the development of an extreme number of polyps early in life and colon cancer in virtually 100% of patients before the age of 40. A group of 14 people suffering from FAP being treated at the Cleveland Clinic drank black raspberry powder in a slurry of water every day for nine months. The numbers of polyps were reduced in 11 out of 14 of these patients.[18]

(a) Why can't we use the large-sample confidence interval for the proportion p of patients suffering from FAP that will have the number of polyps reduced after nine months of treatment?

(b) The plus four method adds four observations—two successes and two failures. What are the sample size and the number of successes after you do this? What is the plus four estimate $\tilde{p}$ of p?

(c) Give the plus four 90% confidence interval for the proportion of patients suffering from FAP who will have the number of polyps reduced after nine months of treatment.

22.13 Computer/Internet-Based Crime. With over 50% of adults spending more than an hour a day on the Internet, the number experiencing computer or Internet-based crime continues to rise. A survey in 2010 of a random sample of 1025 adults, aged 18 and older, reached by random digit dialing found 113 adults in the sample who said that they or a household member was a victim of a computer or Internet crime on their home computer in the past year.[19]

(a) Give the 95% large-sample confidence interval for the proportion p of all households that have experienced computer or Internet crime during the year before the survey was conducted. Be sure to verify that the sample size is large enough to use the large-sample confidence interval.

(b) Give the plus four 95% confidence interval for p. If you express the two intervals in percents, rounded to the nearest tenth of a percent, how do they differ? (The plus four interval always pulls the results away from 0% or 100%, whichever is closer.)

22.14 **Cocaine Traces in Spanish Currency, Continued.** The plus four method is particularly useful when there are *no* successes or *no* failures in the data. The study of Spanish currency described in Example 22.8 found that in Seville, all 20 of a sample of 20 euro bills had cocaine traces.

(a) What is the sample proportion $\hat{p}$ of contaminated bills? What is the large-sample 95% confidence interval for p? It's not plausible that *every* bill in Seville has cocaine traces, as this interval says.

(b) Find the plus four estimate $\tilde{p}$ and the plus four 95% confidence interval for p. These results are more reasonable in this situation.

CHAPTER 22 SUMMARY

Chapter Specifics

- Tests and confidence intervals for a population proportion p when the data are an SRS of size n are based on the **sample proportion $\hat{p}$**.
- When n is large, $\hat{p}$ has approximately the Normal distribution with mean p and standard deviation $\sqrt{p(1-p)/n}$.
- The level C **large-sample confidence interval for p** is

$$\hat{p} \pm z^* \sqrt{\frac{\hat{p}(1-\hat{p})}{n}}$$

  where z^* is the critical value for the standard Normal curve with area C between $-z^*$ and z^*. Use this interval only when *both* the number of successes and the number of failures in the sample are at least 15.
- The **sample size** needed to obtain a confidence interval with approximate margin of error m for a population proportion is

$$n = \left(\frac{z^*}{m}\right)^2 p^*(1-p^*)$$

  where p^* is a guessed value for the sample proportion $\hat{p}$, and z^* is the standard Normal critical point for the level of confidence you want. If you use $p^* = 0.5$ in this formula, the margin of error of the interval will be less than or equal to m no matter what the value of $\hat{p}$ is.
- **Significance tests for H_0: $p = p_0$** are based on the z statistic

$$z = \frac{\hat{p} - p_0}{\sqrt{\frac{p_0(1-p_0)}{n}}}$$

  with P-values calculated from the standard Normal distribution. Use this test in practice when $np_0 \geq 10$ and $n(1-p_0) \geq 10$.
- **(Optional topic)** To get a more accurate confidence interval for smaller sample sizes, add four imaginary observations, two successes and two failures, to your sample. Then use the same formula for the confidence interval. This is the **plus four confidence interval.** Use this interval in practice for confidence level 90% or higher and sample size n at least 10.

KIDS ON BIKES

In the most recent year for which data are available, 77% of children killed in bicycle accidents were boys. You might take these data as a sample and start from $\hat{p} = 0.77$ to do inference about bicycle deaths in the near future. What you should not do is conclude that boys on bikes are in greater danger than girls. We don't know how many boys and girls ride bikes—it may be that most fatalities are boys because most riders are boys.

Link It

The methods of this chapter can be used to compute confidence intervals and test hypotheses about a population proportion. This may be the proportion in a population with some attribute of interest such as the population proportion of young adults who use social networking sites. The proportion could also correspond to the probability of an outcome in an experiment such as the probability that a spinning coin will land on heads. Since the methods of this chapter are approximations, it is important always to check the conditions required for the approximations to work well, whether you are using the large sample confidence interval, the z test, or the plus four method.

When making inferences about a proportion in a population, an important assumption for the methods of this chapter is that the data are an SRS from the population. This is the most difficult assumption to guarantee, because of the many difficulties associated with obtaining an SRS. These difficulties were described in Chapter 8. With nonresponse rates in many surveys that are over 80%, the final sample may not be representative of the population even when experimenters initially selected an SRS. Because of this, most surveys of large populations need to use more complicated sampling schemes as well as to make modifications in the estimates to adjust for problems such as nonresponse.

In Chapters 20 and 21, we first considered inferences about a single mean, and then turned our attention to situations that required the comparison of two population means. In many instances, we want to compare two population proportions rather than making inferences about a single proportion. Methods to compare two population proportions will be described in the next chapter.

LaunchPad Online Resources

If you are having difficulty with any of the sections of this chapter, these online resources should help prepare you to solve the exercises at the end of this chapter.

- **StatTutor** starts with a video review of each section and asks a series of questions to check your understanding.
- **LearningCurve** provides you with a series of questions about the chapter geared to your level of understanding.

CHECK YOUR SKILLS

22.15 The proportion of drivers who use seat belts depends on characteristics such as age, sex, and ethnicity. As part of a broader study, investigators observed a random sample of 117 female Hispanic drivers in Boston. Suppose that in fact 60% of all female Hispanic drivers in the Boston area wear seat belts. In repeated samples, the sample proportion $\hat{p}$ would follow approximately a Normal distribution with mean

(a) 70.2. (b) 0.6. (c) 0.002.

22.16 The standard deviation of the distribution of $\hat{p}$ in Exercise 22.15 is about

(a) 0.002. (b) 0.045. (c) 0.24.

Use the following information for Exercises 22.17 through 22.19. A 2011 NBC News survey found that 80% of a sample of 4500 American teens said they owned an MP3 player such as an iPod. Assume that the sample was an SRS.

22.17 Based on the sample, the large-sample 90% confidence interval for the proportion of all American teens who own an MP3 player is

(a) 0.80 ± 0.0060. (b) 0.80 ± 0.0098.

(c) 0.80 ± 0.0117.

22.18 In Exercise 22.17, suppose we computed a large-sample 80% confidence interval for the proportion of all American teens who own an MP3 player. This 80% confidence interval

(a) would have a smaller margin of error than the 90% confidence interval.
(b) would have a larger margin of error than the 90% confidence interval.
(c) could have either a smaller or a larger margin of error than the 90% confidence interval. This varies from sample to sample.

22.19 How many American teens must be interviewed to estimate the proportion who own an MP3 player within ± 0.02 with 99% confidence using the large-sample confidence interval? Use 0.5 as the conservative guess for p.

(a) $n = 1692$ (b) $n = 2401$ (c) $n = 4148$

22.20 An opinion poll asks an SRS of 100 college seniors how they view their job prospects. In all, 53 say "Good." The large-sample 95% confidence interval for estimating the proportion of all college seniors who think their job prospects are good is

(a) 0.530 ± 0.082. (b) 0.530 ± 0.098.

(c) 0.530 ± 0.049.

22.21 The sample survey in Exercise 22.20 actually called 130 seniors, but 30 of the seniors refused to answer. This nonresponse could cause the survey result to be in error. The error due to nonresponse

(a) is in addition to the margin of error found in Exercise 22.20.
(b) is included in the margin of error found in Exercise 22.20.
(c) can be ignored because it isn't random.

22.22 Experiments on learning in animals sometimes measure how long it takes mice to find their way through a maze. Only half of all mice complete one particular maze in less than 18 seconds. A researcher thinks that a loud noise will cause the mice to complete the maze faster. She measures the proportion of 40 mice that completed the maze in less than 18 seconds with noise as a stimulus. The proportion of mice that completed the maze in less than 18 seconds is $\hat{p} = 0.7$. The hypotheses for a test to answer the researcher's question are

(a) H_0: $p = 0.5$, H_a: $p > 0.5$.

(b) H_0: $p = 0.5$, H_a: $p < 0.5$.

(c) H_0: $p = 0.5$, H_a: $p \neq 0.5$.

22.23 The value of the z statistic for Exercise 22.22 is 2.53. This test is

(a) not significant at either $\alpha = 0.05$ or $\alpha = 0.01$.
(b) significant at $\alpha = 0.05$ but not at $\alpha = 0.01$.
(c) significant at both $\alpha = 0.05$ and $\alpha = 0.01$.

22.24 A Gallup Poll in November 2012 found that 54% of the people in the sample said they wanted to lose weight. The poll's margin of error for 95% confidence was 4%. This means that

(a) the poll used a method that gets an answer within 4% of the truth about the population 95% of the time.
(b) we can be sure that the percent of all adults who want to lose weight is between 50% and 58%.
(c) if Gallup takes another poll using the same method, the results of the second poll will lie between 50% and 58%.

CHAPTER 22 EXERCISES

22.25 **Do smokers know that smoking is bad for them?** The Harris Poll asked a sample of smokers, "Do you believe that smoking will probably shorten your life, or not?" Of the 1010 people in the sample, 848 said "yes."

(a) Harris called residential telephone numbers at random in an attempt to contact an SRS of smokers. Based on what you know about national sample surveys, what is likely to be the biggest weakness in the survey?

(b) We will nonetheless act as if the people interviewed are an SRS of smokers. Give a 95% confidence interval for the percent of smokers who agree that smoking will probably shorten their lives.

alessandro0770/Getty Images

22.26 Reporting cheating. Students are reluctant to report cheating by other students. A student project put this question to an SRS of 172 undergraduates at a large university: "You witness two students cheating on a quiz. Do you go to the professor?" Only 19 answered "yes."[20] Give a 95% confidence interval for the proportion of all undergraduates at this university who would report cheating.

22.27 Harris announces a margin of error. Exercise 22.25 describes a Harris Poll survey of smokers in which 848 of a sample of 1010 smokers agreed that smoking would probably shorten their lives. Harris announces a margin of error of ±3 percentage points for all samples of about this size. Opinion polls announce the margin of error for 95% confidence.

(a) What is the actual margin of error (in percent) for the large-sample confidence interval from this sample?
(b) The margin of error is largest when $\hat{p} = 0.5$. What would the margin of error (in percent) be if the sample had resulted in $\hat{p} = 0.5$?
(c) Why do you think that Harris announces a ±3% margin of error for all samples of about this size?

22.28 Sampling voters. Your state representative is planning on supporting a bill legalizing same-sex marriage in your state and wants to estimate the proportion of voters in her district who support this bill. She mails a survey to an SRS of 1300 registered voters in her district, and 800 surveys are returned. In the sample of 800 surveys returned to your state representative, 528 are from female voters.

(a) In her district, voter registration records show that 54% of the registered voters are female. If the 800 returned surveys are an SRS of registered voters, do you feel that the difference between the proportion of female voters that responded to the survey and the population proportion of 54% can be easily explained by chance variation? State the hypotheses and give the P-value. Give your conclusion in the context of the problem.

(b) Among the 800 surveys that are returned, 72% support the bill. If women are more likely to support the bill legalizing same-sex marriage than men, what would be the likely direction of the bias in the estimate of the proportion of registered voters who support the bill? Explain your answer.

22.29 Internet searches and cell phones. Pew Research Center's Internet and American Life Project asked a random sample of 2485 cell phone users whether they had used their cell phone to look up health or medical information. Of these, 422 said "yes."[21]

(a) Pew dialed cell phone telephone numbers at random in the continental United States in an attempt to contact a random sample of adults. Based on what you know about national sample surveys, what is likely to be the biggest weakness in the survey?
(b) Act as if the sample is an SRS. Give a large-sample 90% confidence interval for the proportion p of all cell phone users who have used their cell phone to look up health or medical information.
(c) Three out of the five most popular health related searches on cell phones have to do with sex: "pregnancy," "herpes," and "STD" (sexually transmitted diseases). Sex-related queries don't even show up on Google and Yahoo's list of top five health searches on computers. What do you think explains the difference in the topics for health related searches on cell phones versus computers? When drawing conclusions from a sample, you must always be careful about the relevant population.

22.30 Gastric bypass surgery. How effective is gastric bypass surgery in maintaining weight loss in extremely obese people? A Utah-based study conducted between 2000 and 2011 found that 76% of 418 subjects who had received gastric bypass surgery maintained at least a 20% weight loss six years after surgery.[22]

(a) Are the conditions for the use of the large-sample confidence interval met? Explain.
(b) Give a 90% confidence interval for the proportion of those receiving gastric bypass surgery that maintained at least a 20% weight loss six years after surgery.
(c) Interpret your interval in the context of the problem.

22.31 Running red lights. A random digit dialing telephone survey of 880 drivers asked, "Recalling the last ten traffic lights you drove through, how many of them were red when you entered the intersections?" Of the 880 respondents, 171 admitted that at least one light had been red.[23]

(a) Give a 95% confidence interval for the proportion of all drivers who ran one or more of the last 10 red lights they met.

(b) Nonresponse is a practical problem for this survey—only 21.6% of calls that reached a live person were completed. Another practical problem is that people may not give truthful answers. What is the likely direction of the bias: do you think more or fewer than 171 of the 880 respondents really ran a red light? Why?

22.32 **The IRS plans an SRS.** The Internal Revenue Service plans to examine an SRS of individual federal income tax returns from each state. One variable of interest is the proportion of returns claiming itemized deductions. The total number of tax returns in a state varies from more than 15 million in California to fewer than 250,000 in Wyoming.

(a) Will the margin of error for estimating the population proportion change from state to state if an SRS of 2000 tax returns is selected in each state? Explain your answer.

(b) Will the margin of error change from state to state if an SRS of 1% of all tax returns is selected in each state? Explain your answer.

22.33 **Surveying students.** You are planning a survey of students at a large university to determine what proportion favor an increase in student fees to support an expansion of the student newspaper. Using records provided by the registrar, you can select a random sample of students. You will ask each student in the sample whether he or she is in favor of the proposed increase. Your budget will allow a sample of 100 students.

(a) For a sample of size 100, construct a table of the margins of error for 95% confidence intervals when $\hat{p}$ takes the values 0.1, 0.3, 0.5, 0.7, and 0.9.

(b) A former editor of the student newspaper offers to provide funds for a sample of size 500. Repeat the margin of error calculations in part (a) for the larger sample size. Then write a short thank-you note to the former editor describing how the larger sample size will improve the results of the survey.

In responding to Exercises 22.34 to 22.41, follow the ***Plan, Solve,*** *and* ***Conclude*** *steps of the four-step process.*

22.34 **College-educated parents.** The National Assessment of Educational Progress (NAEP) includes a "long-term trend" study that tracks reading and mathematics skills over time, and obtains demographic information. In the 2012 study, a random sample of 9000 17-year-old students was selected.[24] The NAEP sample used a multistage design, but the overall effect is quite similar to an SRS of 17-year-olds who are still in school.

(a) In the sample, 51% of students had at least one parent who was a college graduate. Estimate, with 99% confidence, the proportion of all 17-year-old students in 2012 who had at least one parent graduate from college.

(b) The sample does not include 17-year-olds who dropped out of school, so your estimate is only valid for students. Do you think the proportion of all 17-year-olds with at least one parent who was a college graduate would be higher or lower than 51%? Explain.

22.35 **Downloading music.** A husband and wife, Ed and Rina, share a digital music player that has a feature that randomly selects which song to play. A total of 3476 songs have been loaded into the player, some by Ed and the rest by Rina. They are interested in determining whether they have loaded different proportions of songs into the player. Suppose that when the player was in the random-selection mode, 34 of the first 50 songs selected were songs loaded by Rina. Let p denote the proportion of songs that were loaded by Rina.

(a) State the null and alternative hypotheses to be tested. How strong is the evidence that Ed and Rina have loaded different proportions of songs into the player? Make sure to check the conditions for the use of this test.

(b) Are the conditions for the use of the large-sample confidence interval met? If so, estimate with 95% confidence the proportion of songs that were loaded by Rina.

22.36 **The boomerang generation.** The "boomerang generation" is a term applied to the current generation of young adults. They are so named because of the frequency with which they choose to live with their parents after a brief period of living on their own, thus boomeranging back to their place of origin. A Pew Research Center survey of 808 adults nationwide conducted December 6–19, 2011, found that 39% of 18- to 34-year-olds lived with their parents or had moved back in temporarily because of the economy in the past few years. At the time the poll was conducted, what can you say with 95% confidence about the percent of all 18- to 34-year-olds in the United States who lived with their parents or had moved back in temporarily because of the economy?

22.37 **Do cash incentives improve learning?** A high-school teacher in a low-income urban school in Worcester, Massachusetts, used cash incentives to encourage student learning in his AP statistics class.[25] In 2010, 15 of the 61 students enrolled in his class scored a 5 on the AP statistics exam. Worldwide, the proportion of students who scored a 5 in 2010 was 0.15. Is this evidence that the proportion of students who would score a 5 on the AP statistics exam when taught by the teacher in Worcester using cash incentives is higher than the worldwide proportion of 0.15?

(a) State hypotheses, find the P-value, and give your conclusions in the context of the problem. Do you have any reservations about using the z test for proportions for this data?
(b) Does this study provide evidence that cash incentives cause an increase in the proportion of 5s on the AP statistics exam? Explain your answer.

22.38 Final election poll. Gallup's final presidential election poll was conducted November 1–4, 2012. The poll included a random sample of 2551 adults, aged 18 and over, living in all 50 U.S. states and the District of Columbia, who were identified as likely to vote. Forty-eight percent of these likely voters said they would vote for Barack Obama. Does this provide evidence that, among all likely voters at the time the poll was conducted, less than half supported Barack Obama?

22.39 Order in choice. Does the order in which wine is presented make a difference? Several choices of wine are presented one at a time and in sequence and the subject is then asked to choose the preferred wine at the end of the sequence. In this study, subjects were asked to taste two wine samples in sequence. Both samples given to a subject were the *same* wine, although subjects were expecting to taste two different samples of a particular variety. Of the 32 subjects in the study, 22 selected the wine presented first, when presented with two identical wine samples.[26]

(a) Do the data give good reason to conclude that the subjects are not equally likely to choose either of the two positions when presented with two identical wine samples in sequence?
(b) The subjects were recruited in Ontario, Canada, via advertisements to participate in a study of "attitudes and values towards wine." Can we generalize our conclusions to all wine tasters? Explain.

22.40 Chick-fil-A gets it right. Which fast food chain fills orders most accurately at the drive-thru window? The Quick Service Restaurant (QSR) magazine drive-thru study visits restaurants in the largest fast food chains in all 50 states. All visits occurred during the lunch hours of 11:00 A.M. to 2:30 P.M. or during the dinner hours of 4:00 P.M. to 7:00 P.M. During each visit, the researcher ordered a main item, a side item, and a drink. One item was left off of each order; for example, a field researcher could order a sandwich with no pickles. After receiving the order, all food and drink items were checked for complete accuracy. Any food or drink item received that was not exactly as ordered resulted in the order being classified as inaccurate. Also included in the measurement of accuracy were condiments asked for, napkins, straws, and correct change. Any errors in these resulted in the order being classified as inaccurate. In 2012, Chick-fil-A had the fewest inaccuracies, with only 23 of 274 orders classified as inaccurate.[27] What proportion of orders are filled *accurately* by Chick-fil-A? (Use 95% confidence.)

© Julie Dermansky/Corbis

22.41 Order in choice: planning a study. How large a sample would be needed to obtain margin of error ± 0.05 in the study of choice order for tasting wine? Use the $\hat{p}$ from Exercise 22.39 as your guess for the unknown p.

Exercises 22.42 to 22.44 concern the optional material on the plus four method.

22.42 Order in choice. Does the order in which wine is presented make a difference? Several choices of wine are presented one at a time and in sequence and the subject is then asked to choose the preferred wine at the end of the sequence. In this study, subjects were asked to taste two wine samples in sequence. Both samples given to a subject were the *same* wine, although subjects were expecting to taste two different samples of a particular variety. Of the 32 subjects in the study, 22 selected the wine presented first, when presented with two identical wine samples.[28]

(a) Although the conditions for the large-sample test were met in Exercise 22.39, show that the conditions for the large-sample confidence interval discussed in this chapter are not met.
(b) Are the conditions for the use of the plus four confidence interval met? If so, use the plus four method to give a 90% confidence interval for the proportion of subjects who would select the first choice presented.

22.43 Shrubs that survive fires. Some shrubs have the useful ability to resprout from their roots after their tops are destroyed. Fire is a particular threat to shrubs in dry climates, as it can injure the roots as well as destroy the aboveground material. One study of resprouting took place in a dry area of Mexico.[29] The investigators clipped the tops of samples of several species of shrubs. In some cases, they also applied a propane torch to the stumps to simulate a fire. Of 12 specimens of the shrub *Krameria cytisoides,* five resprouted after fire. Estimate with 90% confidence the proportion of all shrubs of this species that will resprout after fire.

22.44 Prayer among the Millennials. The Millennial generation (so called because they were born after 1980 and began to come of age around the year 2000) are less religiously active than older Americans. One of the questions in the General Social Survey in 2012 was, "How often does the respondent pray?" Among the 457 respondents in the survey between 18 and 32 years of age, 289 prayed at least once a week.[30] Assume that the sample is an SRS.

(a) Verify that the sample size conditions are met for the large-sample confidence interval. What is the large-sample 99% confidence interval for the proportion p of all adults between 18 and 32 years of age who pray at least once a week?

(b) Give the plus four 99% confidence interval for p. If you express the two intervals in percents and round to the nearest tenth of a percent, how do they differ? (As always, the plus four method pulls results away from 0% or 100%, whichever is closer.)

Exploring the Web

22.45 Where are the smokers? The Behavioral Risk Factor Surveillance System (BRFSS) is an ongoing data collection program designed to measure behavioral risk factors for the adult population (18 years of age or older) living in households. Data are collected from a random sample of adults (one per household) through a telephone survey. Go to the website `apps.nccd.cdc.gov/BRFSS/` and under Category go to Tobacco Use. Under the topic Adults Who Are Current Smokers, you will find the percentage of smokers in each state.

(a) Which state has the highest percentage of smokers, and what is the reported value? Which state has the lowest percentage, and what is its value? Are the reported percentages statistics or parameters?

(b) Choose a state of interest to you and click on the link. In the table that opens, there is a line for n and the entries are the numbers who answered "yes" and "no." Find the percentage in the sample who answered "yes." Notice that it is different than the percentage reported in the table. The table estimates are weighted to try to reduce the bias. If it is determined that certain portions of the population are underrepresented in the sample, then that portion of the sample receives more weight when computing the estimate of the percentage. The assumptions for an SRS are rarely met in practice, and more complicated methods are often necessary to estimate proportions.

22.46 More on weighting of estimates. The website for the American Association for Public Opinion Research discusses several issues about polls. Read the discussion of weighting at `http://www.aapor.org/Weighting1.htm`. The necessity of weighting arises in Exercise 22.28 (page 534). Here are the important points.

A state representative wants to estimate the proportion of voters in her district who support a bill legalizing same-sex marriage. In her district, it is known that 54% of the registered voters are female. She mails a survey to an SRS of 1300 registered voters, and 800 surveys are returned. Among the 800 surveys returned, 66% were from women, and the women tended to support the bill more than the men. Because the women are both overrepresented in the sample compared with the population of registered voters and have stronger support for the bill, the proportion in the sample that support the bill will tend to be larger than in the population, producing bias in the estimate.

(a) The online discussion of weighting gives three uses of weighting to adjust poll results. Which of these three uses is relevant in this situation?

Here is a breakdown of the respondents.

	SUPPORT THE BILL		
	YES	NO	TOTAL
Females	422	106	528
Males	154	118	272
Total	576	224	800

(b) What is the proportion of females in the sample, $\hat{p}_F$, that support the bill and the proportion of males in the sample, $\hat{p}_M$, that support the bill?

(c) If the sample proportions $\hat{p}_F$ and $\hat{p}_M$ were equal to the true proportions of females and males that support the bill, what would be the true proportion of all registered voters in the population who support the bill? This is a weighted estimator. (*Hint:* You will need to use the fact that 54% of the population is female and 46% is male in your computation.)

(d) The weighted estimator is smaller than the unweighted estimator $\hat{p} = 576/800 = 0.72$. Why is this the correct direction to adjust the estimator?

Universal Images Group via Getty Images

CHAPTER 23

Comparing Two Proportions

In this chapter we cover...

23.1 Two-sample problems: proportions

23.2 The sampling distribution of a difference between proportions

23.3 Large-sample confidence intervals for comparing proportions

23.4 Using technology

23.5 Significance tests for comparing proportions

23.6 Plus four confidence intervals for comparing proportions*

A two-sample problem can arise from a randomized comparative experiment that randomly divides subjects into two groups and exposes each group to a different treatment. Comparing random samples separately selected from two populations is also a two-sample problem. When the comparison involves the *means* of two populations, we use the two-sample t methods of Chapter 21. In this chapter, we consider two-sample problems in which the measurement on the individual can be categorized as either a success or a failure. Our goal is to compare the *proportions* of successes in the two populations.

23.1 Two-sample problems: proportions

Here are some questions that you will answer in the exercises of this chapter.

EXAMPLE 23.1 Two-Sample Problems for Proportions

- Does the proportion of males who have used the Internet to obtain information about a specific disease or medical problem in the last year differ from the proportion of females who have done so? A sample of males and a sample of females are obtained, and the proportions in each sample who have used the Internet to obtain information about a specific disease or medical problem in the last year are to be compared (see Exercise 23.1).

- Does hand washing with alcohol-based hand sanitizers reduce the risk of infection for the common cold? A randomized experiment assigned 100 subjects to a treatment group that followed a regimen of hand washing with alcohol-based sanitizers and 100 subjects to a control group that used routine hand washing without alcohol-based sanitizers. After 10 weeks, compare the proportions in the two groups who were infected with the common cold virus over the study period (see Exercise 23.35). ■

COMPUTER-ASSISTED INTERVIEWING

The days of the interviewer with a clipboard are past. Interviewers now read questions from a computer screen and use the keyboard to enter responses. The computer skips irrelevant items—once a woman says that she has no children, further questions about her children never appear. The computer can even present questions in random order to avoid bias due to always following the same order. Software keeps records of who has responded and prepares a file of data from the responses. The tedious process of transferring responses from paper to computer, once a source of errors, has disappeared.

We will use notation similar to that used in our study of two-sample t statistics. The groups we want to compare are Population 1 and Population 2. We have a separate SRS from each population or responses form two treatments in a randomized comparative experiment. A subscript shows which group a parameter or statistic describes. Here is our notation:

Population	Population Proportion	Sample Size	Sample Proportion
1	p_1	n_1	$\hat{p}_1$
2	p_2	n_2	$\hat{p}_2$

We compare the populations by doing inference about the difference $p_1 - p_2$ between the population proportions. The statistic that estimates this difference is the difference between the two sample proportions, $\hat{p}_1 - \hat{p}_2$.

EXAMPLE 23.2 Interracial Dating

auremar/Shutterstock

STATE: "Would you date a person of a different race?" Researchers answered this question by collecting data from the Internet dating site Match.com. When people post profiles on the site, they indicate which races they are willing to date. While several races were studied, we focus on the data collected for black people dating white people. A random sample of 100 black males and a random sample of 100 black females were selected from the dating site, with 75 of the black males indicating their willingness to date white females and 56 of the black females indicating their willingness to date white males.[1] Is this good evidence that different proportions of black males and females on this Internet dating site would be willing to date someone who is white? How large is the difference between the proportions of black males and females who would be willing to date someone who is white?

PLAN: Take black males to be Population 1 and black females to be Population 2. The population proportions who are willing to date someone who is white are p_1 for black males and p_2 for black females. We want to test the hypotheses

$$H_0: p_1 = p_2 \quad \text{(the same as } H_0: p_1 - p_2 = 0\text{)}$$
$$H_a: p_1 \neq p_2 \quad \text{(the same as } H_a: p_1 - p_2 \neq 0\text{)}$$

We also want to give a confidence interval for the difference $p_1 - p_2$.

SOLVE: Inference about population proportions is based on the sample proportions

$$\hat{p}_1 = \frac{75}{100} = 0.75 \quad \text{(men)}$$

$$\hat{p}_2 = \frac{56}{100} = 0.56 \quad \text{(women)}$$

We see that 75% of black males but only 56% of black females would be willing to date someone who is white. Because the sample sizes are of moderate size and the sample proportions are quite different, we expect that a test will be highly significant (in fact, $P = 0.0046$). So we concentrate on the confidence interval. To estimate $p_1 - p_2$, start from the difference between sample proportions

$$\hat{p}_1 - \hat{p}_2 = 0.75 - 0.56 = 0.19$$

To complete the Solve step, we must know how this difference behaves. ■

23.2 The sampling distribution of a difference between proportions

To use $\hat{p}_1 - \hat{p}_2$ for inference, we must know its sampling distribution. Here are the facts we need:

- When the samples are large, the distribution of $\hat{p}_1 - \hat{p}_2$ is **approximately Normal.**
- The **mean** of the sampling distribution is $p_1 - p_2$. That is, the difference between sample proportions is an unbiased estimator of the difference between population proportions.
- The **standard deviation** of the distribution is

$$\sqrt{\frac{p_1(1 - p_1)}{n_1} + \frac{p_2(1 - p_2)}{n_2}}$$

Figure 23.1 displays the distribution of $\hat{p}_1 - \hat{p}_2$. The standard deviation of $\hat{p}_1 - \hat{p}_2$ involves the unknown parameters p_1 and p_2. Just as in Chapter 22, we must replace these by estimates to make inferences. And just as in Chapter 22, we do this a bit differently for confidence intervals and for hypothesis tests.

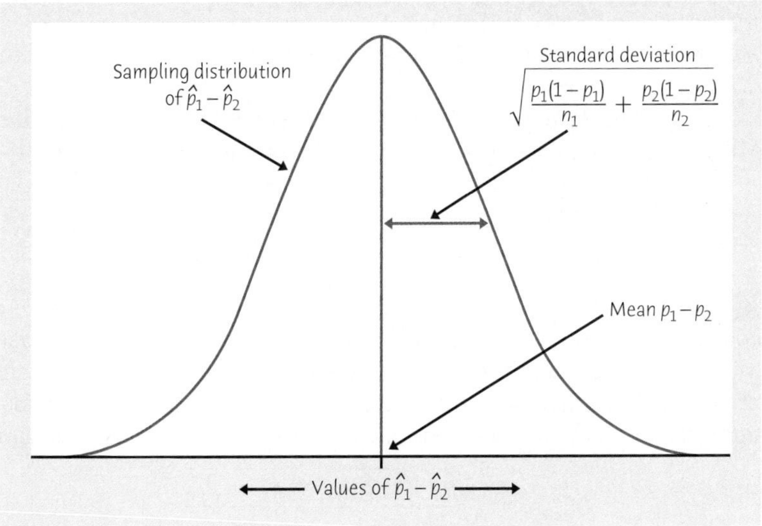

FIGURE 23.1
Select independent SRSs from two populations having proportions of successes p_1 and p_2. The proportions of successes in the two samples are $\hat{p}_1$ and $\hat{p}_2$. When the samples are large, the sampling distribution of the difference $\hat{p}_1 - \hat{p}_2$ is approximately Normal.

23.3 Large-sample confidence intervals for comparing proportions

standard error

To obtain a confidence interval, replace the population proportions p_1 and p_2 in the standard deviation by the sample proportions. The result is the **standard error** of the statistic $\hat{p}_1 - \hat{p}_2$:

$$SE_{\hat{p}_1-\hat{p}_2} = \sqrt{\frac{\hat{p}_1(1-\hat{p}_1)}{n_1} + \frac{\hat{p}_2(1-\hat{p}_2)}{n_2}}$$

The confidence interval has the same form we met in Chapter 22:

$$\text{estimate} \pm z^* SE_{\text{estimate}}$$

Large-Sample Confidence Interval for Comparing Two Proportions

Draw an SRS of size n_1 from a large population having proportion p_1 of successes, and draw an independent SRS of size n_2 from another large population having proportion p_2 of successes. When n_1 and n_2 are large, an approximate level C **confidence interval for $p_1 - p_2$** is

$$(\hat{p}_1 - \hat{p}_2) \pm z^* SE_{\hat{p}_1-\hat{p}_2}$$

In this formula, the standard error $SE_{\hat{p}_1-\hat{p}_2}$ of $\hat{p}_1 - \hat{p}_2$ is

$$SE_{\hat{p}_1-\hat{p}_2} = \sqrt{\frac{\hat{p}_1(1-\hat{p}_1)}{n_1} + \frac{\hat{p}_2(1-\hat{p}_2)}{n_2}}$$

and z^* is the critical value for the standard Normal density curve with area C between $-z^*$ and z^*.

Use this interval only when the numbers of successes and failures are each 10 or more in both samples.

EXAMPLE 23.3 Interracial Dating, Continued

4 step

We can now complete Example 23.2. Here is a summary of the basic information:

Population	Population Description	Sample Size	Number of Successes	Sample Proportion
1	black males	$n_1 = 100$	75	$\hat{p}_1 = 75/100 = 0.75$
2	black females	$n_2 = 100$	56	$\hat{p}_2 = 56/100 = 0.56$

SOLVE: We will give a 95% confidence interval for $p_1 - p_2$, the difference between the proportions of black males and black females who would be willing to date someone who is white. To check that the large-sample confidence interval is safe to use, look at the counts of successes and failures in the two samples. All these four counts are larger than 10, so the large-sample method will be accurate. The standard error is

$$\begin{aligned} SE_{\hat{p}_1-\hat{p}_2} &= \sqrt{\frac{\hat{p}_1(1-\hat{p}_1)}{n_1} + \frac{\hat{p}_2(1-\hat{p}_2)}{n_2}} \\ &= \sqrt{\frac{(0.75)(0.25)}{100} + \frac{(0.56)(0.44)}{100}} \\ &= \sqrt{0.004339} = 0.0659 \end{aligned}$$

The 95% confidence interval is

$$(\hat{p}_1 - \hat{p}_2) \pm z^*SE_{\hat{p}_1 - \hat{p}_2} = (0.75 - 0.56) \pm (1.960)(0.0659)$$
$$= 0.19 \pm 0.13$$
$$= 0.06 \text{ to } 0.32$$

CONCLUDE: We are 95% confident that the percent of black males willing to date white women is between 6 and 32 percentage points higher than the percent of black females who are willing to date white men on comparable Internet dating sites. Even with sample sizes of 100 in each group, the resulting confidence interval 0.06 to 0.32 is quite wide. As with a single proportion, fairly large sample sizes are required to obtain narrow confidence intervals. In a similar study by the authors, it was found that white men were more willing than white women to date someone who is black. Note that our conclusions are restricted to comparable Internet dating sites and don't necessarily reflect dating behavior of the general population, as individuals on dating sites may not be representative of the general population. ■

23.4 Using technology

Figure 23.2 displays software output for Example 23.3 from a graphing calculator and two statistical software programs. As usual, you can understand the output even without knowledge of the program that produced it. Minitab gives the test as

Graphing Calculator

```
2-PropZInt
(.0609,.3191)
p̂1=.75
p̂2=.56
n1=100
n2=100
```

Minitab

Session

Test and CI for Two Proportions

```
Sample   X    N  Sample p
1       75  100  0.750000
2       56  100  0.560000

Difference = p (1) - p (2)
Estimate for difference: 0.19
95% CI for difference: (0.0608950, 0.319105)
Test for difference = 0 (vs not = 0): Z = 2.83  P-Value = 0.005
```

CrunchIt!

Results - Proportion 2-Sample

Export ▾

Null hypothesis:	Difference of proportion = 0
Alternative hypothesis:	Difference of proportion is not 0

	n	Successes	P-hat
Sample 1	100	75	0.7500
Sample 2	100	56	0.5600

Difference:	0.1900
PooledStdErr:	0.06723
z statistic:	2.826
P-value:	0.004710

Results - Proportion 2-Sample

Export ▾

	n	Successes	P-hat
Sample 1	100	75	0.7500
Sample 2	100	56	0.5600

Difference:	0.1900
StdErr:	0.06587
95% ConfInt:	(0.06090, 0.3191)

FIGURE 23.2
Output from a graphing calculator, Minitab, and CrunchIt! for the 95% confidence interval of Example 23.3.

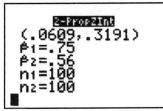

THE COOKIE STRIKES

How many different people clicked on your business website last month? Technology tries to help: when someone visits your site, a little piece of code called a cookie is left on their computer. When the same person clicks again, the cookie says not to count them as a "unique visitor" because this isn't their first visit. But lots of web users delete cookies, either by hand or automatically with software. These people get counted again when they visit your site again. That's bias: your counts of unique visitors are systematically too high. One study found that unique-visitor counts were as much as 50% too high.

well as the confidence interval, confirming that the difference between men and women is highly significant. In CrunchIt!, the test and the confidence interval must be requested using separate commands, resulting in the two outputs in the figure. Excel spreadsheet output is not shown because Excel lacks menu items for inference about proportions. You must use the spreadsheet's formula capability to program the confidence interval or test statistic and then to find the *P*-value of a test. JMP uses a slightly different formula than those given in the text for both the confidence interval and the test and the JMP output is not presented.

LaunchPad Online Resources

- The **Snapshots video**, *Comparing Two Proportions*, provides an example in which the comparison of two proportions is appropriate, and gives some details for computing the large-sample confidence interval.
- The **StatBoards video**, *Confidence Intervals for Comparing Two Proportions*, provides an additional example of computing the large-sample confidence interval which includes the specifics for computing the interval.

Apply Your Knowledge

23.1 Health Online. In 2012, a random sample of 2392 Internet users found approximately 55% had used the Internet to obtain health information about a specific disease or medical problem in the last year. Of the 1084 men in the survey, 520 used the Internet to learn about a specific disease or medical problem, whereas 811 of the 1308 women in the survey had done so.[2] Give a 95% confidence interval for the difference between the proportions of male and female Internet users who used the Internet to obtain health information about a specific disease or medical problem in the last year. Follow the four-step process as illustrated in Examples 23.2 and 23.3.

23.2 Marijuana Use in High School. The Youth Risk Behavior Surveillance System (YRBSS) monitors health risk behaviors among U.S. high school students, which include tobacco use, alcohol and drug use, inadequate physical activity, unhealthy diet, and risky sexual behavior. Approximately 9% of youth have tried marijuana before reaching age 13 years. How does marijuana use change over the high school years? In 2013, the survey randomly selected 3500 ninth-graders and 3497 twelfth-graders and asked them if they had used marijuana one or more times during the previous 30 days.[3] Of these students, 620 ninth-graders and 969 twelfth-graders said "yes." Give a 99% confidence interval for the difference between the proportions of all ninth- and twelfth-grade high school students who smoked marijuana at least once in the 30 days prior to the survey. Follow the four-step process as illustrated in Examples 23.2 and 23.3.

23.3 A Question on the Environment. In 2010, respondents to the General Social Survey were asked how much they agreed or disagreed with the statement: "Many of the claims about the environment are exaggerated."[4] Among the 251 respondents aged 18 to 29, 75 said either "agree" or "strongly agree," whereas among the 376 respondents over age 60, 174 gave one of these responses. Give a 95% confidence interval for the difference between the proportion of people aged 18 to 29 and the proportion of people over 60 who feel that many of the claims about the environment are exaggerated. Follow the four-step process as illustrated in Examples 23.2 and 23.3.

23.5 Significance tests for comparing proportions

An observed difference between two sample proportions can reflect an actual difference between the populations, or it may just be due to chance variation in random sampling. Significance tests help us decide if the effect we see in the samples is really there in the populations. The null hypothesis says that there is no difference between the two populations:

$$H_0\colon p_1 = p_2 \text{ (the same as } H_0\colon p_1 - p_2 = 0)$$

The alternative hypothesis says what kind of difference we expect.

EXAMPLE 23.4 False Memories

STATE: The political event depicted in Figure 23.3 was remembered by about 31% of those surveyed, despite the fact that it never occurred. In 2010, *Slate*, a current affairs and culture magazine, surveyed over 5000 readers regarding their perspective on several past political events and their memories of these events. Unbeknownst to those surveyed, one of the events shown to each participant was fabricated, making this the largest false memory study conducted to date.[5] The hypothesis of interest was whether political preferences guided the formation of false memories, as false memories have been shown to be more easily implanted in memory when they are congruent with a person's preexisting attitudes. Figure 23.3 was viewed by 616 participants who categorized themselves as progressive and 49 participants who categorized themselves as conservative. The event was falsely remembered as having occurred by 212 of the progressives surveyed and by 7 of the conservatives. How strong is the evidence that a larger proportion of progressives have a false memory of this event than conservatives?

Photo illustration by Holly Allen (Slate Magazine)

FIGURE 23.3
September 1, 2005: As parts of New Orleans lie underwater in the wake of Hurricane Katrina, President Bush entertains Houston Astros pitcher Roger Clemens at his ranch in Crawford, Texas, for Example 23.4.

PLAN: Call the population proportions p_1 for progressives and p_2 for conservatives. Because the image of a conservative president relaxing at his ranch during a major crisis is more congruent with the preexisting attitude of progressives, our hypothesis gives a direction for the difference before looking at the data, so we have the one-sided alternative:

$$H_0\colon p_1 = p_2$$
$$H_a\colon p_1 > p_2$$

SOLVE: Consider those who classify themselves as progressives and conservatives as separate SRSs of progressive and conservative *Slate* readers. The sample proportions who falsely remembered the event in Figure 23.3 are

$$\hat{p}_1 = \frac{212}{616} = 0.344 \quad \text{(progressives)}$$

$$\hat{p}_2 = \frac{7}{49} = 0.143 \quad \text{(conservatives)}$$

That is, 34% of those classifying themselves as progressive falsely remembered the event in Figure 23.3, but only 14% of conservatives falsely remembered it. Is this apparent difference statistically significant? To continue the solution, we must learn the proper test. ■

To do a hypothesis test, standardize the difference between the sample proportions $\hat{p}_1 - \hat{p}_2$ to get a z statistic. If H_0 is true, both samples come from populations in which the same unknown proportion p have a false memory of the event depicted in Figure 23.3. We take advantage of this by combining the two samples to estimate

pooled sample proportion

this single p instead of estimating p_1 and p_2 separately. Call this the **pooled sample proportion.** It is

$$\hat{p} = \frac{\text{number of successes in both samples combined}}{\text{number of individuals in both samples combined}}$$

Substituting $\hat{p}$ in place of both $\hat{p}_1$ and $\hat{p}_2$ in the expression for the standard error $SE_{\hat{p}_1 - \hat{p}_2}$ of $\hat{p}_1 - \hat{p}_2$, we get a z statistic that has the standard Normal distribution when H_0 is true. Here is the test.

Significance Test for Comparing Two Proportions

Draw an SRS of size n_1 from a large population having proportion p_1 of successes and draw an independent SRS of size n_2 from another large population having proportion p_2 of successes. To **test the hypothesis H_0: $p_1 = p_2$**, first find the pooled proportion $\hat{p}$ of successes in both samples combined. Then compute the z statistic

$$z = \frac{\hat{p}_1 - \hat{p}_2}{\sqrt{\hat{p}(1 - \hat{p})\left(\frac{1}{n_1} + \frac{1}{n_2}\right)}}$$

In terms of a variable Z having the standard Normal distribution, the P-value for a test of H_0 against

H_a: $p_1 > p_2$ is $P(Z \geq z)$

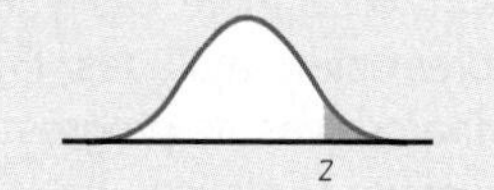

H_a: $p_1 < p_2$ is $P(Z \leq z)$

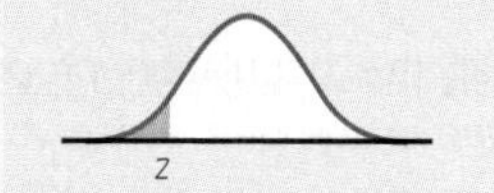

H_a: $p_1 \neq p_2$ is $2P(Z \geq |z|)$

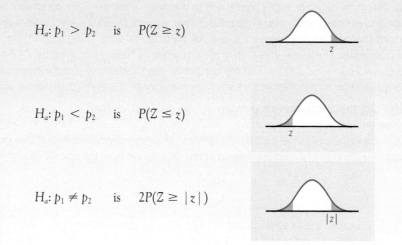

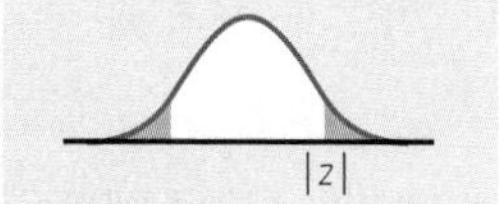

Use this test when the counts of successes and failures are each five or more in both samples.[6]

EXAMPLE 23.5 False Memories, Continued

SOLVE: The data come from an SRS, and the counts of successes and failures are all larger than five. The pooled proportion of progressives and conservatives who falsely remembered this event is

$$\hat{p} = \frac{\text{number "falsely remembered event" among progressives and conservatives combined}}{\text{number of progressives and conservatives combined}}$$

$$= \frac{212 + 7}{616 + 49}$$

$$= \frac{219}{665} = 0.329$$

The z test statistic is

$$z = \frac{\hat{p}_1 - \hat{p}_2}{\sqrt{\hat{p}(1 - \hat{p})\left(\frac{1}{n_1} + \frac{1}{n_2}\right)}}$$

$$= \frac{0.344 - 0.143}{\sqrt{(0.329)(0.671)\left(\frac{1}{616} + \frac{1}{49}\right)}}$$

$$= \frac{0.201}{0.0697} = 2.88$$

The one-sided P-value is the area under the standard Normal curve greater than 2.88. Figure 23.4 shows this area. Software tells us that $P = 0.00199$.

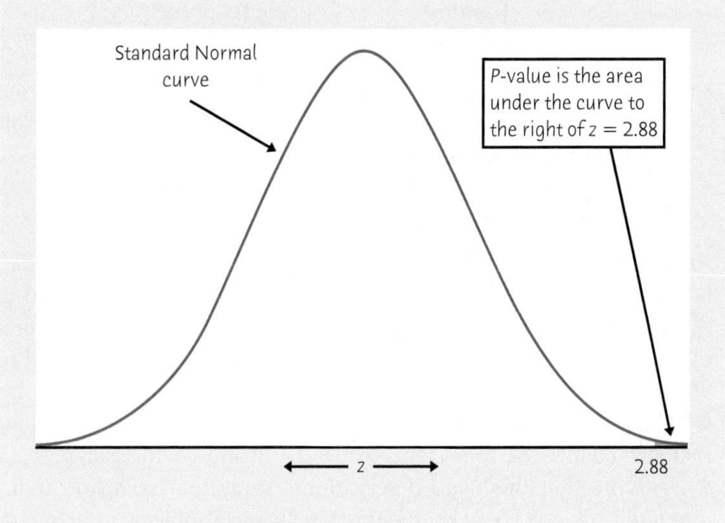

FIGURE 23.4
The P-value for the one-sided test of Example 23.5.

Without software, you can compare $z = 2.88$ with the bottom row of Table C (standard Normal critical values) to approximate P.

It lies between the critical values 2.807 and 3.091 for one-sided P-values 0.0025 and 0.001.

z^*	2.807	3.091
One-sided P	0.0025	0.001

CONCLUDE: There is strong evidence ($P < 0.0025$) that, among *Slate* readers, progressives are more likely than conservatives to have a false memory of Mr. Bush vacationing with Roger Clemens in the wake of the Katrina hurricane. In a second fabricated event by the authors, it was found that conservatives were more likely than progressives to falsely remember President Obama shaking hands with former Iranian President Ahmadinejad at a United Nations conference. ■

The sample survey in this example selected a single sample of *Slate* readers, not two separate samples of progressives and conservatives. To get two samples, we divided the single sample by political orientation. This means that we did not know the two sample sizes n_1 and n_2 until after the data were in hand. The two-sample z procedures for comparing proportions are valid in such situations. This is an important fact about these methods.

LaunchPad Online Resources

- **The StatBoards video, *Hypothesis Testing for a Difference in Proportions*, provides an additional example of the large-sample test for comparing two proportions.**

Apply Your Knowledge

23.4 Seat Belt Use. The proportion of drivers who use seat belts depends on factors such as age (young people are more likely to go unbelted) and sex (women are more likely to use belts). It also depends on local law. In New York City, police can stop a driver who is not belted. In Boston at the time of the survey, police could cite a driver for not wearing a seat belt only if the driver had been stopped for some other violation. Here are data from observing random samples of female Hispanic drivers in these two cities in 2002:[7]

City	Drivers	Belted
New York	220	183
Boston	117	68

© Jupiterimages/AgeFotostock

(a) Is this an experiment or an observational study? Why?

(b) Comparing local laws suggests the hypothesis that a smaller proportion of drivers wear seat belts in Boston than in New York. Do the data give good evidence that this is true for female Hispanic drivers? Follow the four-step process as illustrated in Examples 23.4 and 23.5.

23.5 Protecting Skiers and Snowboarders. Most alpine skiers and snowboarders do not use helmets. Do helmets reduce the risk of head injuries? A study in Norway compared skiers and snowboarders who suffered head injuries with a control group who were not injured. Of 578 injured subjects, 96 had worn a helmet. Of the 2992 in the control group, 656 wore helmets.[8] Is helmet use less common among skiers and snowboarders who have head injuries? Follow the four-step process as illustrated in Examples 23.4 and 23.5. (Note that this is an observational study that compares injured and uninjured subjects. An experiment that assigned subjects to helmet and no-helmet groups would be more convincing.)

23.6 Breast Cancer Treatment. In sentinel lymph node dissection (SLND), surgeons remove two or three lymph nodes close to the breast that are most likely to contain cancer cells. If these "sentinel" lymph nodes are free of tumors, SLND alone is the accepted management for patients. When the sentinel lymph nodes contain metastases, axillary lymph node dissection (ALND, or the removal of further nodes) remains the standard of care, although its contribution to survival is controversial. ALND carries the additional risk of complications such as seroma, infection, and lymphedema. In one study, patients with sentinel metastases identified by SLND were randomized to undergo ALND or no further treatment (SLND alone). Here are the five-year disease-free survival numbers for the two groups:[9]

Group	Sample Size	Disease-Free after Five Years
ALND	420	345
SLND alone	436	366

How strong is the evidence that the proportion of disease-free patients after five years for patients who underwent ALND in addition to SLND differs from that of those who underwent SLND only? Follow the four-step process as illustrated in Examples 23.4 and 23.5.

23.6 Plus four confidence intervals for comparing proportions*

CAUTION *Like the large-sample confidence interval for a single proportion p, the large-sample interval for $p_1 - p_2$ generally has a true confidence level less than the level you asked for.* The inaccuracy is not as serious as in the one-sample case, at least if our guidelines for use are followed. Once again, adding imaginary observations can improve the accuracy.[10]

Plus Four Confidence Interval for Comparing Two Proportions

Draw independent SRSs from two large populations with population proportions of successes p_1 and p_2. To get the **plus four confidence interval for the difference $p_1 - p_2$**, add four imaginary observations: one success, and one failure in each of the two samples. Then use the large-sample confidence interval with the new sample sizes (actual sample sizes + 2) and counts of successes (actual counts + 1).

Use this interval when the sample size is at least 5 in each group, with any counts of successes and failures.

If your software does not offer the plus four method, just enter the new plus four sample sizes and success counts into the large-sample procedure.

EXAMPLE 23.6 Abecedarian Early Childhood Education Program: Adult Outcomes

4 step

STATE: The Abecedarian Project was a randomized controlled study to assess the effects of intensive early childhood education on children who were at high risk based on several sociodemographic indicators. The project randomly assigned children to a treatment group, which was provided with early educational activities before kindergarten, and the remainder to a control group. A recent follow-up study interviewed subjects at age 30 and compared the college graduation rates (earning a four-year degree).[11] Here are the data for two groups:

Population	Population Description	Sample Size	Number of Successes	Sample Proportion
1	treatment	$n_1 = 52$	12	$\hat{p}_1 = 12/52 = 0.2308$
2	control	$n_2 = 49$	3	$\hat{p}_2 = 3/49 = 0.0612$

How much does early childhood education increase the proportion earning a four-year degree by age 30?

PLAN: Give a 90% confidence interval for the difference between population proportions, $p_1 - p_2$.

SOLVE: The conditions for the large-sample interval are not met because there are only three successes in the control group. However, we can use the plus four method because the sample sizes for the treatment and the control group are both at least 5. Add four imaginary observations. The new data summary is

*This section is optional.

Population	Population Description	Sample Size	Number of Successes	Plus Four Sample Proportion
1	treatment	$n_1 + 2 = 54$	$12 + 1 = 13$	$\tilde{p}_1 = 13/54 = 0.2407$
2	control	$n_2 + 2 = 51$	$3 + 1 = 4$	$\tilde{p}_2 = 4/51 = 0.0784$

The standard error based on the new facts is

$$\begin{aligned} SE_{\tilde{p}_1 - \tilde{p}_2} &= \sqrt{\frac{\tilde{p}_1(1-\tilde{p}_1)}{n_1+2} + \frac{\tilde{p}_2(1-\tilde{p}_2)}{n_2+2}} \\ &= \sqrt{\frac{(0.2407)(0.7593)}{54} + \frac{(0.0784)(0.9216)}{51}} \\ &= \sqrt{0.00480} = 0.0693 \end{aligned}$$

The plus four 90% confidence interval is

$$\begin{aligned} (\tilde{p}_1 - \tilde{p}_2) \pm z^* SE_{\tilde{p}_1 - \tilde{p}_2} &= (0.2407 - 0.0784) \pm (1.645)(0.0693) \\ &= 0.1623 \pm 0.1140 \\ &= 0.048 \text{ to } 0.276 \end{aligned}$$

CONCLUDE: We are 90% confident that the early childhood education program increases the proportion of high-risk children who earn a four-year degree by age 30 by between 4.8% and 27.6%. ■

The plus four interval may be conservative (that is, the true confidence level may be *higher* than you asked for) for very small samples and population p values close to 0 or 1. It is generally much more accurate than the large-sample interval when the samples are small. Nevertheless, the plus four interval in Example 23.6 cannot save us from the fact that sample sizes around 50 will produce a wide confidence interval when comparing two proportions.

Apply Your Knowledge

© Jupiterimages/Comstock Images/Alamy

23.7 In-Line Skaters. A study of injuries to in-line skaters used data from the National Electronic Injury Surveillance System, which collects data from a random sample of hospital emergency rooms. The researchers interviewed 161 people who came to emergency rooms with injuries from in-line skating. Wrist injuries (mostly fractures) were the most common.[12]

(a) The interviews found that 53 people were wearing wrist guards, and 6 of these had wrist injuries. Of the 108 who did not wear wrist guards, 45 had wrist injuries. Why should we not use the large-sample confidence interval for these data?

(b) The plus four method adds one success and one failure in each sample. What are the sample sizes and counts of successes after you do this?

(c) Give the plus four 95% confidence interval for the difference between the two population proportions of wrist injuries. State carefully what populations your inference compares.

4 step

23.8 Shrubs that Withstand Fire. Fire is a serious threat to shrubs in dry climates. Some shrubs can resprout from their roots after their tops are destroyed. One study of resprouting took place in a dry area of Mexico.[13] The investigators first clipped the tops of all the shrubs in the study. For the treatment, they then applied a propane torch to the stumps of the shrubs to simulate a fire, and for the control the stumps were left alone. The study included 24 shrubs of which 12 were randomly assigned to the treatment and the remaining 12 to the control. A shrub is a success if it resprouts. For the shrub *Xerospirea hartwegiana*, all 12 shrubs in the control group resprouted, whereas only 8 in the treatment group resprouted. How much does burning reduce the proportion of shrubs of this species that resprout? Give the 95% plus four confidence interval for the amount by which burning reduces the proportion of shrubs of this species that resprout. The plus four method is particularly helpful when, as here, a count of either successes or failures is zero. Follow the four-step process as illustrated in Example 23.6.

CHAPTER 23 SUMMARY

Chapter Specifics

- The data in a **two-sample problem** are two independent SRSs, each drawn from a separate population.
- Tests and confidence intervals to compare the proportions p_1 and p_2 of successes in the two populations are based on the difference $\hat{p}_1 - \hat{p}_2$ between the sample proportions of successes in the two SRSs.
- When the sample sizes n_1 and n_2 are large, the sampling distribution of $\hat{p}_1 - \hat{p}_2$ is close to Normal with mean $p_1 - p_2$.
- The level C **large-sample confidence interval for $p_1 - p_2$** is

$$(\hat{p}_1 - \hat{p}_2) \pm z^* \text{SE}_{\hat{p}_1 - \hat{p}_2}$$

  where the **standard error** of $\hat{p}_1 - \hat{p}_2$ is

$$\text{SE}_{\hat{p}_1 - \hat{p}_2} = \sqrt{\frac{\hat{p}_1(1 - \hat{p}_1)}{n_1} + \frac{\hat{p}_2(1 - \hat{p}_2)}{n_2}}$$

  and z^* is a standard Normal critical value.

  The true confidence level of the large-sample interval can be substantially less than the planned level C. Use this interval only if the counts of successes and failures in both samples are 10 or greater.
- **Significance tests for H_0: $p_1 = p_2$** use the **pooled sample proportion**

$$\hat{p} = \frac{\text{number of successes in both samples combined}}{\text{number of individuals in both samples combined}}$$

  and the z statistic

$$z = \frac{\hat{p}_1 - \hat{p}_2}{\sqrt{\hat{p}(1 - \hat{p})\left(\frac{1}{n_1} + \frac{1}{n_2}\right)}}$$

  P-values come from the standard Normal distribution. Use this test when there are five or more successes and five or more failures in both samples.
- **(Optional topic)** When the conditions for the large-sample confidence interval are not met, to get a more accurate confidence interval, add four imaginary observations, one success and one failure in each sample. Then use the same formula for the confidence interval. This is the **plus four confidence interval**. You can use it whenever both samples have five or more observations.

Link It

Most studies compare two or more treatments rather than investigating a single treatment. These can be observational studies or comparative experiments, and the differences in the types of conclusions that can be reached were described in Chapter 9. This chapter considers the case where the response classifies individuals into two categories, such as black people being willing to date someone who is white or not. The comparison is then a comparison of this proportion for two groups, such as the proportion of black males who are willing to date someone who is white versus the proportion of black females who are willing to date someone who is white. Because the methods of this chapter are approximate, it is important always to check the conditions required for the approximations to work well, whether you are using the large-sample confidence interval, the z test, or the plus four method.

After the appropriate statistical method has been applied, care must be taken when stating the conclusion. The issues described in Chapter 9 are still important. For example, a lack of blinding can result in the expectations of the researcher influencing the results, whereas confounding can mix up the comparison of the two groups with other factors.

When there are two treatments and the response is a quantitative measurement, the comparison between the treatments is often based on a comparison of the treatment means. These methods were described in Chapter 21.

LaunchPad Online Resources

If you are having difficulty with any of the sections of this chapter, these online resources should help prepare you to solve the exercises at the end of this chapter.

- **StatTutor** starts with a video review of each section and asks a series of questions to check your understanding.
- **LearningCurve** provides you with a series of questions about the chapter geared to your level of understanding.

CHECK YOUR SKILLS

In 2012, respondents to the General Social Survey were asked: "There are always some people whose ideas are considered bad or dangerous by other people. For instance, somebody who is against churches and religion. If such a person wanted to make a speech in your (city/town/community) against churches and religion, should he be allowed to speak, or not?"[14]

Among the 573 male respondents, 459 said, "allow," whereas among the 719 female respondents, 552 said, "allow." Is there a difference between the proportion of males and the proportion of females who would allow an anti-religionist to speak? Exercises 23.9 to 23.13 are based on these results.

23.9 Take p_M and p_F to be the proportions of all males and females who would allow an anti-religionist to speak. The hypotheses to be tested are

(a) H_0: $p_M = p_F$ versus H_a: $p_M \neq p_F$.
(b) H_0: $p_M = p_F$ versus H_a: $p_M > p_F$.
(c) H_0: $p_M = p_F$ versus H_a: $p_M < p_F$.

23.10 The sample proportions of males and females who would allow an anti-religionist to speak are

(a) $\hat{p}_M = 0.801$ and $\hat{p}_F = 0.768$.
(b) $\hat{p}_M = 0.832$ and $\hat{p}_F = 0.797$.
(c) $\hat{p}_M = 0.832$ and $\hat{p}_F = 0.801$.

23.11 The pooled sample proportion of respondents who would allow an anti-religionist to speak is

(a) $\hat{p} = 0.768$. (b) $\hat{p} = 0.783$. (c) $\hat{p} = 0.801$.

23.12 The z test for comparing the proportions of males and females who would allow an anti-religionist to speak has

(a) $P < 0.05$.
(b) $0.05 < P < 0.10$.
(c) $P > 0.10$.

23.13 The 90% large-sample confidence interval for the difference $p_M - p_F$ in the proportions of males and females who would allow an anti-religionist to speak is about

(a) 0.033 ± 0.002.
(b) 0.033 ± 0.038.
(c) 0.033 ± 0.045.

23.14 In an experiment to learn if substance M can help restore memory, the brains of 20 rats were treated to damage their memories. The rats were trained to run a maze. After a day, 10 rats were given M, and seven of them succeeded in the maze; only two of the 10 control rats were successful. The z test for "no difference" against "a higher proportion of the M group succeeds" has

(a) $z = 2.25$, $P < 0.02$.
(b) $z = 2.60$, $P < 0.005$.
(c) $z = 2.25$, $0.02 < P < 0.04$.

23.15 The z test in Exercise 23.14

(a) may be inaccurate because the populations are too small.
(b) may be inaccurate because some counts of successes and failures are too small.
(c) is reasonably accurate because the conditions for inference are met.

23.16 **(Optional topic)** The plus four 90% confidence interval for the difference between the proportion of rats that succeed when given M and the proportion that succeed without it is

(a) 0.455 ± 0.312.
(b) 0.417 ± 0.304.
(c) 0.417 ± 0.185.

CHAPTER 23 EXERCISES

When using the large-sample methods of this chapter, be sure to check that the guidelines for their use are met, and to state your conclusions in context.

23.17 **Truthfulness in online profiles.** Many teens have posted profiles on a social-networking website. A sample survey in 2007 asked a random sample of teens with online profiles if they included false information in their profiles. Of 170 younger teens (aged 12 to 14), 117 said "yes." Of 317 older teens (aged 15 to 17), 152 said "yes."[15]

(a) Do these samples satisfy the guidelines for the large-sample confidence interval?
(b) Give a 95% confidence interval for the difference between the proportions of younger and older teens who include false information in their online profiles.

23.18 **Effects of an appetite suppressant.** Subjects with pre-existing cardiovascular symptoms who were receiving sibutramine, an appetite suppressant, were found to be at increased risk of cardiovascular events while taking the drug. The study included 9804 overweight or obese subjects with pre-existing cardiovascular disease and/or type 2 diabetes. The subjects were randomly assigned to sibutramine (4906 subjects) or a placebo (4898 subjects) in a double-blind fashion. The primary outcome measured was the occurrence of any of the following events: nonfatal myocardial infarction or stroke, resuscitation after cardiac arrest, or cardiovascular death. The primary outcome was observed in 561 subjects in the sibutramine group and 490 subjects in the placebo group.[16]

(a) Find the proportion of subjects experiencing the primary outcome for both the sibutramine and placebo groups.
(b) Can we safely use the large-sample confidence interval for comparing the proportions of sibutramine and placebo subjects who experienced the primary outcome? Explain.
(c) Give a 95% confidence interval for the difference between the proportions of sibutramine and placebo subjects who experienced the primary outcome.

23.19 (Optional topic) Genetically altered mice. Genetic influences on cancer can be studied by manipulating the genetic makeup of mice. One of the processes that turn genes on or off (so to speak) in particular locations is called "DNA methylation." Do low levels of this process help cause tumors? Compare mice altered to have low levels with normal mice. Of 33 mice with lowered levels of DNA methylation, 23 developed tumors. None of the control group of 18 normal mice developed tumors in the same time period.[17]

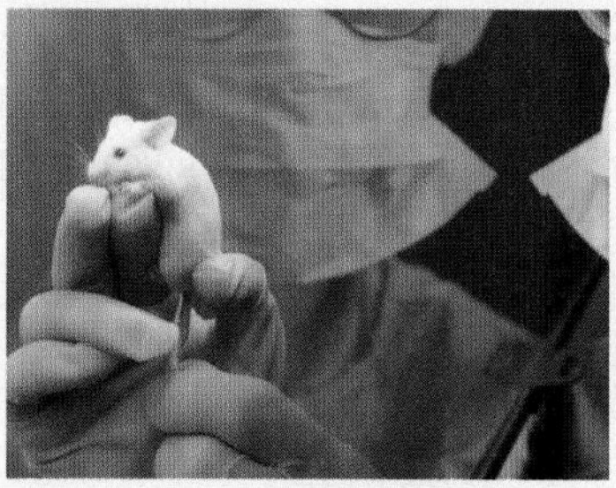

Mark Harmel/Getty Images

(a) Explain why we cannot safely use either the large-sample confidence interval or the test for comparing the proportions of normal and altered mice that develop tumors.
(b) The plus four method adds two observations, a success and a failure, to each sample. What are the sample sizes and the numbers of mice with tumors after you do this?
(c) Give a 99% confidence interval for the difference in the proportions of the two populations that develop tumors.

23.20 Effects of an appetite suppressant, continued. Exercise 23.18 describes a study to determine if subjects with pre-existing cardiovascular symptoms were at an increased risk of cardiovascular events while taking sibutramine. Do the data give good reason to think that there is a difference between the proportions of treatment and placebo subjects who experienced the primary outcome? (Note that sibutramine has not been available in the United States since the end of 2010 due to its manufacturer's concerns over increased risk of heart attack or stroke, although at the present time it can still be purchased in other countries.)

(a) State hypotheses, find the test statistic, and use either software or the bottom row of Table C for the P-value. Be sure to state your conclusion.
(b) Explain simply why it was important to have a placebo group in this study.

Adolescence, music, and algebra. *Research has suggested that musicians process music in the same cortical regions in which adolescents process algebra. When taking introductory algebra, will students who were enrolled in formal instrumental or choral music instruction during middle school outperform those who experienced neither of these modes of musical instruction? The sample consisted of 6026 ninth-grade students in Maryland who had completed introductory algebra. Of these, 3239 students had received formal instrumental or choral instruction during all three years of middle school, whereas the remaining students had not. Of those receiving formal musical instruction, 2818 received a passing grade on the Maryland Algebra/Data Analysis High School Assessment (HSA). In contrast, 2091 of the 2787 students not receiving musical instruction received a passing grade.[18] Exercises 23.21 to 23.23 are based on this study.*

23.21 Does music make a difference?

(a) Is there a significant difference in the proportions of students with and without musical instruction who receive a passing grade on the Maryland HSA? State hypotheses, find the test statistic, and use software or the bottom row of Table C to get a P-value.
(b) Is this an observational study or an experiment? Why?
(c) In view of your answer in part (b), carefully state your conclusions about the relationship between music instruction and success in algebra.

23.22 How many students pass? Give a 95% confidence interval for the proportion of ninth-grade students who receive a passing grade on the HSA.

23.23 How big a difference? Give a 95% confidence interval for the difference between the proportions of students passing the HSA who have received or not received formal musical instruction in middle school.

23.24 The design matters. Due to concerns about the safety of bariatric weight loss surgery, in 2006 the Centers for Medicare and Medicaid Services (CMS) restricted coverage of bariatric surgery to hospitals designated as Centers for Excellence. Did the CMS restriction improve the outcomes of bariatric surgery for Medicare patients? Among the 1847 Medicare patients in the study having surgery in the 18 months preceding the restriction, 270 experienced overall complications, whereas in the 1639 patients having bariatric surgery in the 18 months following the restriction, 170 experienced overall complications.[19]

(a) What are the sample proportions of Medicare patients who experienced overall complications from bariatric surgery before and after the CMS restriction on coverage? How strong is the evidence that the proportions of overall complications are different before and after the CMS restriction? Use an appropriate hypothesis test to answer this question.
(b) Is this an observational study or an experiment? Can we conclude that the CMS restriction has reduced the proportion of overall complications?
(c) Improved outcomes may be due to several factors, including the use of lower-risk bariatric procedures, increased surgeon experience, or healthier patients receiving the surgery. What types of variables are these, and how do they affect the types of conclusions that you can make?
(d) In a second study,[20] a control group consisting of non-Medicare patients obtained before and after the restrictions on coverage was obtained. When compared with this control group, no significant evidence was found that the CMS restriction reduced the proportion of overall complications for the Medicare group. Explain in simple language how comparison with a control group could help reduce the effects of some of the variables described in part (c)?

23.25 **Significant does not mean important.** Never forget that even small effects can be statistically significant if the samples are large. To illustrate this fact, consider a sample of 148 small businesses. During a three-year period, 15 of the 106 headed by men and 7 of the 42 headed by women failed.[21]

(a) Find the proportions of failures for businesses headed by women and businesses headed by men. These sample proportions are quite close to each other. Give the P-value for the z test of the hypothesis that the same proportion of women's and men's businesses fail. (Use the two-sided alternative.) The test is very far from being significant.
(b) Now suppose that the same sample proportions came from a sample 30 times as large. That is, 210 out of 1260 businesses headed by women and 450 out of 3180 businesses headed by men fail. Verify that the proportions of failures are exactly the same as in part (a). Repeat the z test for the new data, and show that it is now significant at the $\alpha = 0.05$ level.
(c) It is wise to use a confidence interval to estimate the size of an effect rather than just giving a P-value. Give the large sample 95% confidence intervals for the difference between the proportions of women's and men's businesses that fail for the settings of both parts (a) and (b). What is the effect of larger samples on the confidence interval?

In responding to Exercises 23.26 to 23.35, follow the ***Plan, Solve,*** *and* ***Conclude*** *steps of the four-step process.*

23.26 **Demographics of social networking sites: sex.** A Pew Internet survey in 2012 examined several demographic variables of users of social networking sites including sex, age, race, education, and income. Of the 1802 Internet users included in the survey, 846 were men and 525 of these used social networking sites. Among the 956 women in the survey, 679 of the women used social networking sites.[22] Is there good evidence that the proportion of Internet users who use social networking sites is different for men and women?

23.27 **Demographics of social networking sites: race.** The survey in Exercise 23.26 also looked at possible differences in the proportions of white and Hispanic Internet users who used social networking sites. They found that 866 of the 1332 white Internet users and 111 of the 154 Hispanics used social networking sites. Is there evidence of a difference between the proportions of white and Hispanic Internet users who use social networking sites?

23.28 **More on social networking and sex.** Continue your work from Exercise 23.26. Estimate the difference between the proportions of male and female Internet users who use social networking sites. (Use 90% confidence.)

23.29 **Smoking cessation.** Chantix is different from most other quit-smoking products in that it targets nicotine receptors in the brain, attaches to them, and blocks nicotine from reaching them. As part of a larger randomized controlled trial, generally healthy smokers who smoked at least 10 cigarettes per day were assigned at random to take Chantix or a placebo. The study was double-blind, with the response measure being continuous absence from smoking for weeks 9 through 12 of the study. Of the 352 subjects taking Chantix, 155 abstained from smoking during weeks 9 through 12, whereas 61 of the 344 subjects taking the placebo abstained during this same time period.[23] Give a 99% confidence interval for the difference (treatment minus placebo) in the proportions of smokers who abstained from smoking during weeks 9 through 12.

23.30 **The Gold Coast.** A historian examining British colonial records for the Gold Coast in Africa suspects that the death rate was higher among African miners than among European miners. In the year 1936, there were 223 deaths among 33,809 African miners and 7 deaths among 1541 European miners on the Gold Coast.[24] (The Gold Coast became the independent nation of Ghana in 1957.) Consider this year as a random sample from the colonial era in West Africa. Is there good evidence that the proportion of African miners who died was higher than the proportion of European miners who died?

23.31 **I refuse!** Do our emotions influence economic decisions? One way to examine the issue is to have subjects play an "ultimatum game" against other people and against a computer. Your partner (person or computer) gets $10, on the condition that it be shared with you. The partner makes you an offer. If you refuse, neither of you gets anything. So it's to your advantage to accept even the unfair offer of $2 out of the $10. Some people get mad and refuse unfair offers. Here are data on the responses of 76 subjects randomly assigned to receive an offer of $2 from either a person they were introduced to or a computer:[25]

	ACCEPT	REJECT
Human offers	20	18
Computer offers	32	6

We suspect that emotion will lead to offers from another person being rejected more often than offers from an impersonal computer. Do a test to assess the evidence for this conjecture.

23.32 **Did the random assignment work?** A large clinical trial of the effect of diet on breast cancer assigned women at random to either a normal diet or a low-fat diet. To check that the random assignment did produce comparable groups, we can compare the two groups at the start of the study. Asked if there is a family history of breast cancer, 3396 of the 19,541 women in the low-fat group and 4929 of the 29,294 women in the control group said "yes."[26] If the random assignment worked well, there should *not* be a significant difference in the proportions with a family history of breast cancer. How significant is the observed difference?

23.33 **(Optional topic) Lyme disease.** Lyme disease is spread in the northeastern United States by infected ticks. The ticks are infected mainly by feeding on mice, so more mice result in more infected ticks. The mouse population in turn rises and falls with the abundance of acorns, their favored food. Experimenters studied two similar forest areas in a year when the acorn crop failed. They added hundreds of thousands of acorns to one area to imitate an abundant acorn crop, while leaving the other area untouched. The next spring, 54 of the 72 mice trapped in the first area were in breeding condition, versus 10 of the 17 mice trapped in the second area.[27] Estimate the difference between the proportions of mice ready to breed in good acorn years and bad acorn years. (Use 90% confidence. Be sure to justify your choice of confidence interval.)

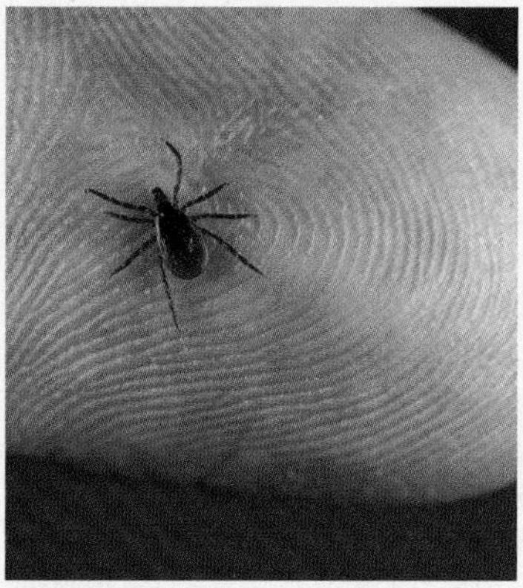

Scott Camazine/Science Source

23.34 **Abecedarian early childhood education program: adult outcomes.** The Abecedarian Project is a randomized controlled study to assess the effects of intensive early childhood education on children who were at high risk based on several sociodemographic indicators.[28] The project randomly assigned some children to a treatment group that was provided with early educational activities before kindergarten, and the remainder to a control group. A recent follow-up study interviewed subjects at age 30 and evaluated educational, economic, and socioemotional outcomes to learn if the positive effects of the program continued into adulthood. The follow-up study included 52 individuals from the treatment group and 49 from the control group. Out of these, 39 from the treatment group and 26 from the control group were considered "consistently" employed (working 30+ hours per week in at least 18 of the 24 months prior to the interview). Does the study provide significant evidence that children who had early childhood education have a higher proportion of consistent employment than those who did not? How large is the difference between the proportions in the two populations that are consistently employed? Do inference to answer both questions. Be sure to explain exactly what inference you choose to do.

23.35 **Hand sanitizers.** Hand disinfection is frequently recommended for prevention of transmission of the rhinovirus that causes the common cold. In particular, hand lotion containing 2% citric acid and 2% malic acid in 70% ethanol (HL+) has been found to have both immediate and persistent ability to inactivate rhinovirus (RV) on the hands in an experimental setting. Is hand disinfection effective in reducing the risk of infection in a natural setting? A total of 212 volunteers were assigned at random to either the HL+ group, which used the hand lotion every three hours or after hand washing, and a control group, which was asked to use routine hand washing but to avoid the use of alcohol-based hand sanitizers. Here are the data on the numbers of subjects with and without RV infection in the two groups over the 10-week study period:[29]

	RV INFECTION	
	YES	NO
HL+	49	67
Control group	49	47

(a) Is this an experiment or an observational study? Why?

(b) Do the data give good evidence that hand sanitizers reduce the chance of an RV infection?

Exploring the Web

23.36 **Hearing loss in adolescents.** Go to the *Journal of the American Medical Association* website, `http://jama.ama-assn.org/content/by/year`, and find the article "Change in Prevalence of Hearing Loss in US Adolescents" by Shargorodsky et al. in the August 18, 2010, issue. If you cannot get the full text of the article, use the information in the abstract plus the information given here to answer the questions. NHANES III is the earlier sample, and NHANES 2005–2006 is the more recent sample.

(a) Is this an observational study or an experiment?

(b) How many people were in the earlier sample, and how many were in the later sample?

(c) If you do not have access to Table 2 of the full article, here are the facts that you will need: in the earlier study, 480 people experienced some hearing loss, whereas in the later study, 333 people experienced some hearing loss. Is there evidence of an increase in hearing loss for children aged 12 to 19 in the later study? State hypotheses, find the test statistic, and use either software or Table A to compute the *P*-value. Although the article used a more sophisticated analysis, your *P*-value should be quite close to the *P*-value of 0.02 reported in the abstract.

23.37 **Compare two surveys.** Go to the website `www.pollingreport.com`, which contains the results of surveys conducted by several survey organizations. Choose a topic of interest to you, and then, to see if attitudes have changed over time, find two surveys that were conducted at two different times but that ask the same question. For example, you might choose the topic of abortion and compare the percents of people who feel abortion should always be legal at points in time separated by several years. State hypotheses to check for a difference over time, find the test statistic, and use either software or Table A to compute the *P*-value. What is your conclusion in context?

© Rosemary Harris/Alamy

CHAPTER 24

Inference about Variables: Part IV Review

In this chapter we cover...

- Part IV Summary
- Test Yourself
- Supplementary Exercises

The procedures of Chapters 20 to 23 are among the most common of all statistical inference methods. Now that you have mastered important ideas and practical methods for inference, it's time to review the big ideas of statistics in outline form. Here is a summary of Parts I, II, and III of this book, leading up to Part IV. The outline contains some important warnings: look for the Caution icon.

1. **Data Production**

- Data basics:
  - Individuals (subjects).
  - Variables: categorical versus quantitative, units of measurement, explanatory versus response.
  - Purpose of study.
- Data production basics:
  - Observation versus experiment.
  - Simple random samples.
  - Completely randomized experiments.
- Beware: bad data production (voluntary response, confounding) can make interpretation impossible.

CAUTION

- Beware: weaknesses in data production (for example, sampling students at only one campus) can make generalizing conclusions difficult.

2. **Data Analysis**

- Plot your data. Look for an overall pattern and striking deviations.
- Add numerical descriptions based on what you see.
- Beware: averages and other simple descriptions can miss the real story.
- One quantitative variable:
  Graphs: stemplot, histogram, boxplot.
  Pattern: distribution shape, center, variability. Outliers?
  Density curves (such as Normal curves) to describe overall pattern.
  Numerical descriptions: five-number summary or $\bar{x}$ and s.
- Relationships between two quantitative variables:
  Graph: scatterplot.
  Pattern: relationship form, direction, strength. Outliers? Influential observations?
  Numerical description for linear relationships: correlation, regression line.
  Beware the lurking variable: correlation does not imply causation.
- Beware the effects of outliers and influential observations.

3. **The Reasoning of Inference**

- Inference uses data to infer conclusions about a wider population.
- When you do inference, you are acting as if your data come from random samples or randomized comparative experiments. Beware: if they don't, you may have "garbage in, garbage out."
- Always examine your data before doing inference. Inference often requires a regular pattern, such as roughly Normal with no strong outliers.
- Key idea: "What would happen if we did this many times?"
- Confidence intervals: estimate a population parameter.
  95% confidence: I used a method that captures the true parameter 95% of the time in repeated use.
  Beware: the margin of error of a confidence interval does not include the effects of practical errors such as undercoverage and nonresponse.
- Significance tests: assess evidence against H_0 in favor of H_a.
  P-value: If H_0 were true, how often would I get an outcome favoring the alternative this strongly? Smaller P = stronger evidence against H_0.
  Statistical significance at the 5% level, $P < 0.05$, means that an outcome this extreme would occur less than 5% of the time if H_0 were true.
  Beware: $P < 0.05$ is not sacred.
  Beware: statistical significance is not the same as practical significance. Large samples can make small effects significant. Small samples can fail to declare large effects significant.
  Always try to estimate the size of an effect (for example, with a confidence interval), not just its significance.

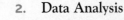

4. **Methods of Inference**

- Choose the right inference procedure.
- Check the conditions for inference.
- Carry out the details.
- State your conclusion in context.

Part IV of this book introduces the fourth and last part of this outline. To actually do inference, you must choose the right procedure, check the conditions for inference, and carry out the details. The Statistics in Summary flowchart on the next page offers a brief guide. It is important to do some of the review exercises because now, for the first time, you must decide which of several inference procedures to use. Learning to recognize problem settings in order to choose the right type of inference is a key step in advancing your mastery of statistics. This is the Plan step in the four-step process, in which you translate the real-world problem from the State step into a specific inference procedure.

The flowchart organizes one way of planning inference problems. Let's go through it from left to right.

1. *Do you want to test a claim or estimate an unknown quantity?* That is, will you need a test of significance or a confidence interval?
2. *Are your data a single sample representing one population or two samples chosen to compare two populations or responses to two treatments in an experiment?* Remember that to work with *matched pairs* data you form one sample from the differences within pairs.
3. *Is the response variable quantitative or categorical?* Quantitative variables take numerical values with some unit of measurement such as inches or grams. The most common inference questions about quantitative variables concern *mean* responses. If the response variable is categorical, inference most often concerns the *proportion* of some category (call it a "success") among the responses.

The flowchart leads you to a specific test or confidence interval, indicated by a formula at the end of each path. The formula is just an aid to guide you toward the Solve and Conclude steps. You (or your technology) will use the formula as part of the Solve step, but don't forget that you must do more.

- *Are the conditions for this procedure met?* Can you act as if the data come from a random sample or randomized comparative experiment? Does data analysis show extreme outliers or strong skewness that forbids use of inference based on Normality? Do you have enough observations for your intended procedure?
- *Do your data come from an experiment or from an observational study?* The details of inference methods are the same for both. But the design of the study determines what conclusions you can reach, because experiments give much better evidence that an effect uncovered by inference can be explained by direct causation.

You may ask, as you study the Statistics in Summary flowchart, "What if I have an experiment comparing four treatments, or samples from three populations?" The flowchart allows only one or two, not three or four or more. Be patient: methods for comparing more than two means or proportions, as well as some other settings for inference, appear in Part V.

Statistics In Your World **HOW MANY WAS THAT?**

Good causes often breed bad statistics. An advocacy group claims, without much evidence, that 150,000 Americans suffer from the eating disorder anorexia nervosa. Soon someone misunderstands and says that 150,000 people *die* from anorexia nervosa each year. This wild number gets repeated in countless books and articles. It really is a wild number: only about 55,000 women aged 15 to 44 (the main group affected) die of *all causes* each year.

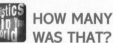

STATISTICS IN SUMMARY

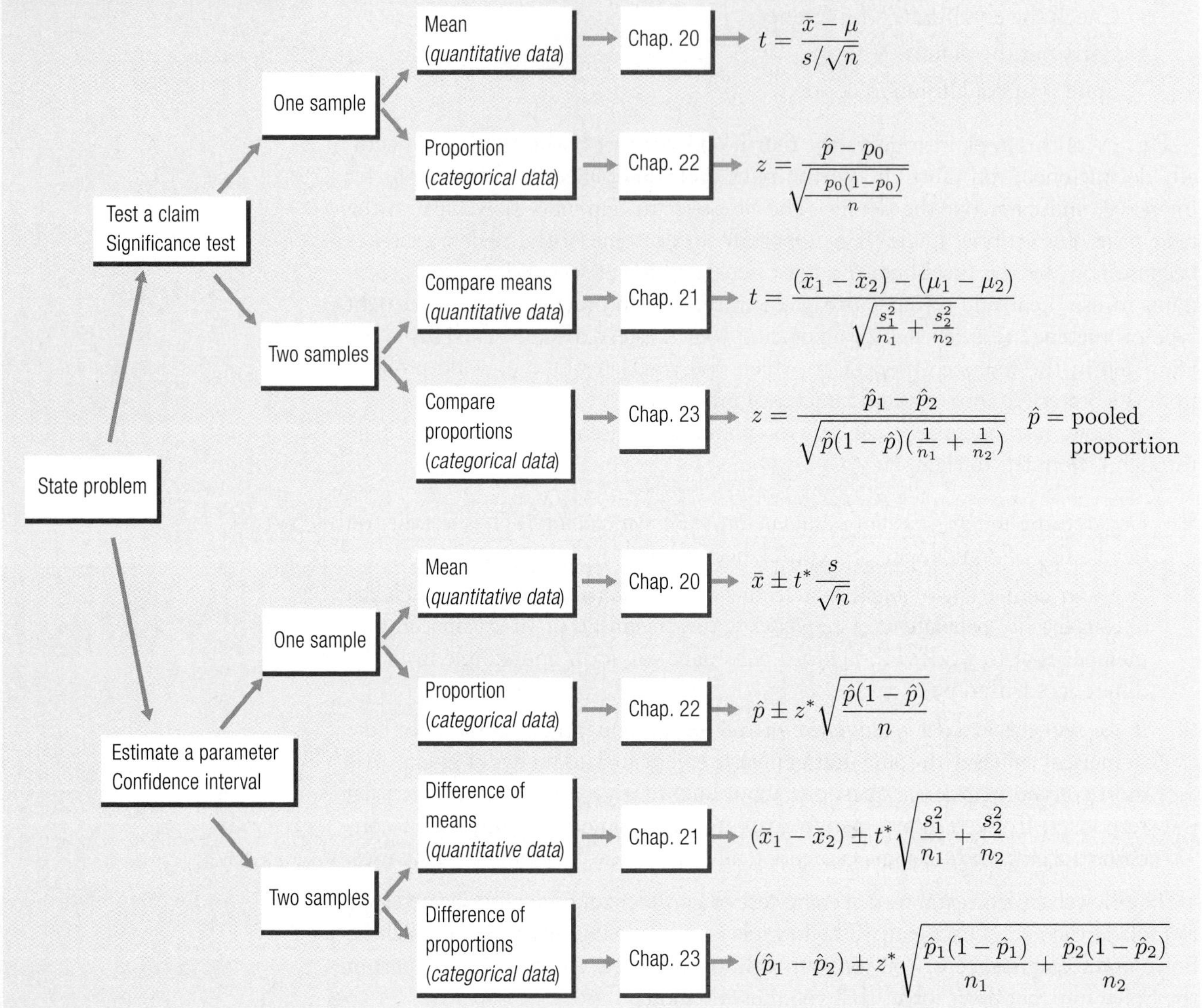

Part IV Summary

Here are the most important skills you should have acquired from reading Chapters 20 to 23.

A. Recognition

1. Recognize when a problem requires inference about population means (quantitative response variable) or population proportions (usually categorical response variable).
2. Recognize from the design of a study whether one-sample, matched pairs, or two-sample procedures are needed.
3. Based on recognizing the problem setting, choose among the one- and two-sample t procedures for means and the one- and two-sample z procedures for proportions.

B. Inference about One Mean

1. Verify that the t procedures are appropriate in a particular setting. Check the study design and the distribution of the data and take advantage of robustness against lack of Normality.
2. Recognize when poor study design, outliers, or a small sample from a skewed distribution make the t procedures risky.
3. Use the one-sample t procedure to obtain a confidence interval at a stated level of confidence for the mean μ of a population.
4. Carry out a one-sample t test for the hypothesis that a population mean μ has a specified value against either a one-sided or a two-sided alternative. Use software to find the P-value or Table C to get an approximate value.
5. Recognize matched pairs data and use the t procedures to obtain confidence intervals and to perform tests of significance for such data.
6. Know that resampling can be used to estimate standard errors.

C. Comparing Two Means

1. Verify that the two-sample t procedures are appropriate in a particular setting. Check the study design and the distribution of the data and take advantage of robustness against lack of Normality.
2. Give a confidence interval for the difference between two means. Use software if you have it. Use the two-sample t statistic with conservative degrees of freedom and Table C if you do not have statistical software.
3. Test the hypothesis that two populations have equal means against either a one-sided or a two-sided alternative. Use software if you have it. Use the two-sample t test with conservative degrees of freedom and Table C if you do not have statistical software.
4. Know that procedures for comparing the standard deviations of two Normal populations are available, but that these procedures are risky because they are not at all robust against non-Normal distributions.
5. Know that permutation tests can be used to test the hypothesis of no treatment effect in experiments in which units have been assigned to treatments completely at random.

D. Inference about One Proportion

1. Verify that you can safely use either the large-sample or the plus four z procedures in a particular setting. Check the study design and the guidelines for sample size.
2. Use the large-sample z procedure to give a confidence interval for a population proportion p when the conditions are met. Otherwise, the plus four modification of the z procedure can give a confidence interval for p that is accurate even for small samples and for any value of p.
3. Use the z statistic to carry out a test of significance for the hypothesis $H_0: p = p_0$ about a population proportion p against either a one-sided or a two-sided alternative. Use software or Table A to find the P-value, or Table C to get an approximate value.

E. Comparing Two Proportions

1. Verify that you can safely use either the large-sample or the plus four z procedures in a particular setting. Check the study design and the guidelines for sample sizes.

2. Use the large-sample z procedure to give a confidence interval for the difference $p_1 - p_2$ between proportions in two populations based on independent samples from the populations when the conditions are met. Otherwise, the plus four modification of the z procedure can give a confidence interval for $p_1 - p_2$ that is accurate even for very small samples and for any values of p_1 and p_2.
3. Use a z statistic to test the hypothesis H_0: $p_1 = p_2$ that proportions in two distinct populations are equal. Use software or Table A to find the P-value, or Table C to get an approximate value.

Test Yourself

The questions below include both multiple-choice, and short-answer questions and calculations. They will help you review the basic ideas and skills presented in Chapters 20 to 23.

Calcium and blood pressure. *In a randomized comparative experiment on the effect of dietary calcium on blood pressure, researchers divided 54 healthy white males at random into two groups. One group received calcium; the other, a placebo. At the beginning of the study, the researchers measured many variables on the subjects. The paper reporting the study gives* $\bar{x} = 114.9$ *and* $s = 9.3$ *for the seated systolic blood pressure of the 27 members of the placebo group. Use this information to answer Questions 24.1 and 24.2.*

24.1 A 95% confidence interval for the mean blood pressure in the population from which the subjects were recruited is

(a) 113.1 to 116.7.
(b) 111.8 to 118.0.
(c) 111.2 to 118.6.
(d) 109.9 to 119.9.

24.2 What conditions for the population and the study design are required by the procedure you used to construct your confidence interval? Which of these conditions are important for the validity of the procedure in this case?

Does nature heal better? *Our bodies have a natural electrical field that is known to help wounds heal. Does changing the field strength slow healing? A series of experiments with newts investigated this question. The data below are the healing rates of cuts (micrometers per hour) in a matched pairs experiment. The pairs are the two hind limbs of the same newt, with the body's natural field in one limb (control) and half the natural value in the other limb (experimental).*[1]

Newt	1	2	3	4	5	6	7	8	9	10	11	12	13	14
Control	25	13	44	45	57	42	50	36	35	38	43	31	26	48
Experimental	24	23	47	42	26	46	38	33	28	28	21	27	25	45
Difference (Control − Experimental)	1	−10	−3	3	31	−4	12	3	7	10	22	4	1	3

The mean and standard deviation of the differences are 5.71 and 10.56 micrometers per hour, respectively. Use this information to answer Questions 24.3 and 24.4.

24.3 Is there good evidence that changing the electrical field from its natural level slows healing? The P-value for your test is

(a) less than 0.01.
(b) between 0.01 and 0.05.
(c) between 0.05 and 0.10.
(d) greater than 0.10.

24.4 Explain why this is a matched pairs experiment. Give a 99% confidence interval for the difference in healing rates (control minus experimental).

Computer use. *A survey asked 3494 randomly selected high school seniors about their computer use.[2] Among the 1,734 females in the sample, 614 used computers three or more hours per day (playing video or computer games or using the computer for something that was not school work). Among the 1,760 men sampled, 676 used computers three or more hours per day. Use this information to answer Questions 24.5 to 24.8.*

24.5 The sample proportion of males who used computers for three or more hours a day is

(a) 0.193.
(b) 0.354.
(c) 0.369.
(d) 0.384.

24.6 The pooled sample proportion of high school seniors who used computers for three or more hours a day is

(a) 0.193.
(b) 0.354.
(c) 0.369.
(d) 0.384.

24.7 The sampling distribution for the difference in the proportions of male and female high school seniors who used computers for three or more hours a day has standard error

(a) 0.0043.
(b) 0.0053.
(c) 0.0081.
(d) 0.0163.

24.8 The z test for comparing the proportions of male and female high school seniors who used computers for three or more hours a day has

(a) $P < 0.01$. (b) $0.01 < P < 0.05$. (c) $0.05 < P < 0.10$.
(d) $P > 0.10$.

24.9 **Wikipedia.** A sample survey of 852 adult Internet users found that 53% look for information on the online collaborative encyclopedia Wikipedia.[3]

(a) Give the standard error SE of $\hat{p}$, the proportion of all adult Internet users who look for information on Wikipedia.
(b) Give a 95% confidence interval for the proportion of all adult Internet users who look for information on Wikipedia.

© Chris Batson/Alamy

Evil genius? *Does cheating enhance creativity? Researchers recruited 178 subjects and asked participants to guess whether the outcome of a virtual coin toss would be heads or tails. After indicating their prediction, participants had to press a button to toss the coin virtually. They were asked to press the button only once, but they were given the opportunity to test the button several times before beginning the experiment. Thus, participants could cheat by pressing the button before making their prediction and appear to have pressed the button one time for each prediction. Participants then reported whether they had guessed correctly and received a \$1 bonus if they had. The program recorded the outcomes of the initial virtual coin tosses so that researchers could tell whether participants cheated. After the coin-toss task, all participants were given a test that measured creativity. Here are the summary statistics for scores on the creativity test:[4]*

Group	n	$\bar{x}$	s
Cheaters	43	3.60	1.26
Noncheaters	135	2.33	1.00

Use this information to answer Questions 24.10 to 24.13.

24.10 A 95% confidence interval for the mean score on the creativity test for those subjects who cheated is

(a) 3.60 ± 0.19.
(b) 3.60 ± 0.39.
(c) 3.60 ± 2.02.
(d) 1.27 ± 0.43.

24.11 A 95% confidence interval for the mean score on the creativity test for those subjects who did not cheat is

(a) 2.33 ± 0.09.
(b) 2.33 ± 0.17.
(c) 2.33 ± 1.98.
(d) 1.27 ± 0.43.

24.12 Is there a significant difference between the mean scores on the creativity test for those who cheated and those who did not? The value of the t statistic for testing the null hypothesis of no difference in the mean scores on the creativity test is

(a) 0.21.
(b) 1.27.
(c) 4.76.
(d) 6.03.

24.13 Is there a significant difference between the mean scores on the creativity test for those who cheated and those who did not? The degrees of freedom using the conservative Option 2 for the t statistic for testing the null hypothesis of no difference in the mean scores on the creativity test is

(a) 42.
(b) 59.8.
(c) 134.
(d) 176.

Michael Lustbader/Science Source

24.14 **Butterflies mating.** Here's how butterflies mate: a male passes to a female a packet of sperm called a spermatophore. Females may mate several times. Will they remate sooner if the first spermatophore they receive is small? Among 20 females who received a large spermatophore (greater than 25 milligrams), the mean time to the next mating was 5.15 days, with standard deviation 0.18 day. For 21 females who received a small spermatophore (about 7 milligrams), the mean was 4.33 days and the standard deviation was 0.31 day.[5] Is the observed difference in means statistically significant? Test using the conservative Option 2 for the degrees of freedom. The P-value is

(a) less than 0.01.
(b) between 0.01 and 0.05.
(c) between 0.05 and 0.10.
(d) greater than 0.10.

Treating migraines. *A randomized clinical trial of 135 youth aged 10 to 17 years diagnosed with chronic migraine were assigned to one of two treatments. One involved 10 cognitive behavioral therapy (CBT) sessions and use of the drug amitriptyline. The other involved 10 headache education sessions plus amitriptyline. Twenty weeks after treatment, the severity of migraines for each subject was assessed using the Pediatric Migraine Disability Assessment Score (PedMIDAS). Here are data on PedMIDAS scores 20 weeks post-treatment for the two groups:*[6]

Group	n	Mean	Standard deviation
CBT	64	15.5	17.4
Education	71	29.6	42.2

Use this information to answer Questions 24.15 to 24.18.

24.15 Both sets of endurance data are skewed to the right. How do we know this (note that PedMIDAS scores are greater than or equal to zero). Why are t procedures nonetheless reasonably accurate for these data?

24.16 A 90% confidence interval for the mean PedMIDAS score for the CBT group is

(a) 15.05 to 15.95. (c) 11.87 to 19.13.
(b) 13.32 to 17.68. (d) 11.15 to 19.85.

24.17 A 90% confidence interval for the mean difference (education minus CBT) in PedMIDAS scores, using the conservative Option 2 for the degrees of freedom, is

(a) 3.18 to 25.02. (c) 8.64 to 19.56.
(b) 4.98 to 23.22. (d) 21.24 to 37.96.

24.18 Do the data show that there is a difference in mean PedMIDAS scores for the two treatments? Carry out an appropriate hypothesis test using Option 1 if you have technology that implements this method. Otherwise, use Option 2.

24.19 **Pre-readers in kindergarten.** A school has two kindergarten classes. There are 21 children in Ms. Hazelcorn's kindergarten class. Of these, 12 are "pre-readers"—children on the verge of reading. There are 19 children in Mr. Shapiro's kindergarten class. Of these, 14 are pre-readers. Is there a difference in the proportions of "pre-readers" in the two classes? The z statistic is computed and the P-value is 0.273.

(a) This test is not reliable because the samples are so small.
(b) This test is of no use because we should be comparing means with a t statistic.
(c) This test is reasonable because the counts of successes and failures are each five or more in both samples.
(d) This test is not appropriate because these samples cannot be viewed as simple random samples taken from a larger population.

24.20 **State of the economy.** If we want to estimate p, the population proportion of likely voters who believe the state of the economy to be the most urgent national concern, with 90% confidence and a margin of error no greater than 2%, how many likely voters need to be surveyed? Assume that you have no idea of the value of p.

(a) 1024 (c) 4096
(b) 1692 (d) 6765

Favoritism for college athletes? Sports Illustrated *surveyed a random sample of 757 Division I college athletes in 36 sports. One question asked was, "Have you ever received preferential treatment from a professor because of your status as an athlete?" Of the athletes polled, 225 said "yes." Use this information to answer Questions 24.21 to 24.23.*

24.21 The sample proportion of athletes who have received preferential treatment from a professor is

(a) 0.160. (c) 0.703.
(b) 0.297. (d) 0.840.

24.22 The standard error SE of $\hat{p}$, the proportion of athletes who have received preferential treatment from a professor, is

(a) 0.003. (c) 0.209.
(b) 0.017. (d) 0.457.

24.23 A 90% confidence interval for the proportion of athletes who have received preferential treatment from a professor is

(a) 0.276 to 0.320. (c) 0.265 to 0.331.
(b) 0.270 to 0.326. (d) 0.255 to 0.342.

Very-low-birth-weight babies. *Starting in the 1970s, medical technology allowed babies with very low birth weight (VLBW, less than 1500 grams, about 3.3 pounds) to survive without major handicaps. It was noticed that these children nonetheless had difficulties in school and as adults. A long-term study has followed 242 VLBW babies to age 20 years, along with a control group of 233 babies from the same population who had normal birth weight.*[7] *At age 20, 179 of the VLBW group and 193 of the control group had graduated from high school. Use this information to answer Questions 24.24 to 24.29.*

24.24 This is an example of

(a) an observational study.
(b) a nonrandomized experiment.
(c) a randomized controlled study.
(d) a matched pairs experiment.

24.25 Take p_{VLBW} and $p_{control}$ to be the proportions of all VLBW and normal-birth-weight (control) babies who would graduate from high school. The hypotheses to be tested are

(a) H_0: $p_{VLBW} = p_{control}$ versus H_a: $p_{VLBW} \neq p_{control}$.
(b) H_0: $p_{VLBW} = p_{control}$ versus H_a: $p_{VLBW} > p_{control}$.
(c) H_0: $p_{VLBW} = p_{control}$ versus H_a: $p_{VLBW} < p_{control}$.
(d) H_0: $p_{VLBW} > p_{control}$ versus H_a: $p_{VLBW} = p_{control}$.

24.26 The pooled sample proportion of babies who would graduate from high school is

(a) $\hat{p} = 0.74$. (c) $\hat{p} = 0.81$.
(b) $\hat{p} = 0.78$. (d) $\hat{p} = 0.83$.

24.27 The numerical value of the z test for comparing the proportions of all VLBW and normal-birth-weight (control) babies who would graduate from high school is

(a) $z = -1.65$. (c) $z = -2.77$.
(b) $z = -2.34$. (d) $z = -3.14$.

24.28 IQ scores were available for 113 men in the VLBW group and for 106 men in the control group. The mean IQ for the 113 men in the VLBW group was 87.6, and the standard deviation was 15.1. The 106 men in the control group had mean IQ 94.7, with standard deviation 14.9. Is there good evidence that mean IQ is lower among VLBW men than among controls from similar backgrounds? To test this with a two-sample t test, the test statistic would be

(a) $t = -1.72$. (c) $t = -5.00$.
(b) $t = -3.50$. (d) $t = -7.10$.

24.29 Of the 126 women in the VLBW group, 38 said they had used illegal drugs; 54 of the 124 control group women had done so. The IQ scores for the VLBW women who had used illegal drugs had mean 86.2 (standard deviation 13.4), and the normal-birth-weight controls who had used illegal drugs had mean IQ 89.8 (standard deviation 14.0). Is there a statistically significant difference between the two groups in mean IQ? The P-value for this test is

(a) less than 0.01.
(b) between 0.01 and 0.05.
(c) between 0.05 and 0.10.
(d) greater than 0.10.

Binge drinking. *According to the National Institute on Alcohol Abuse and Alcoholism (NIAAA) and the National Institutes of Health (NIH), 41% of college students nationwide engage in "binge-drinking" behavior: having five or more drinks on one occasion during the past two weeks. A college president wonders if the proportion of students enrolled at her college who binge drink is actually lower than the national proportion. In a commissioned study, 348 students are selected randomly from a list of all students enrolled at the college. Of these, 132 admit to having engaged in binge drinking. Use this information to answer Questions 24.30 to 24.32.*

24.30 Based on the results of the commissioned study, a 95% confidence interval for the proportion of all students at this college who engage in binge drinking is

(a) 0.328 to 0.430. (c) 0.341 to 0.420.
(b) 0.338 to 0.423. (d) 0.343 to 0.418.

24.31 The college president is more interested in testing her belief that the proportion of students at her college who engage in binge drinking is lower than the national proportion of 0.41. Her staff tests the hypotheses $H_0: p = 0.41$ versus $H_a: p < 0.41$. The *P*-value is

(a) between 0.15 and 0.20.
(b) between 0.10 and 0.15.
(c) between 0.05 and 0.10.
(d) below 0.05.

24.32 Which of the following conclusions is reasonable, based on the *P*-value computed in Exercise 24.31?

(a) There is little evidence to support a conclusion that the proportion of students at this particular college who binge drink is lower than the national proportion of 0.41.
(b) There is moderate but not strong evidence that the proportion of binge-drinking students at this college is lower than the national proportion of 0.41.
(c) There is strong evidence that the proportion of students at this college who binge drink is lower than the national proportion of 0.41.
(d) We can't reach any reasonable conclusion, because the assumptions necessary for a significance test for a proportion are not met in this case.

Who texts? *Younger people use their cell phones to text more often than older people do. A random sample of 625 teens aged 12 to 17 who use their cell phones to text found that 475 sent more than 10 text messages in a typical day. In a random sample of 1917 adults aged 18 and over who use their cell phones to text, 786 sent more than 10 text messages in a typical day.*[8] *Use this information to answer Questions 24.33 and 24.34.*

24.33 Give a 90% confidence interval for the proportion of all teens aged 12 to 17 who use their cell phones to text and send more than 10 text messages in a single day.

24.34 Give a 95% confidence interval for the difference between the proportions of cell phone users who use their cell phones to text and send more than 10 texts a day for these two age groups.

24.35 Smokers in Alaska. In the past decade there have been intensive antismoking campaigns sponsored by both federal and private agencies. The Behavioral Risk Factor Surveillance System (BRFSS) is an ongoing data collection program that monitors behaviors such as smoking on a statewide level using data collected on a random sample of adults through a telephone survey.[9] The first sample, taken in 2002 in Alaska, involved 2045 adults, of which 754 were current smokers. The second sample, taken in 2012 in Alaska, involved 2690 adults, of which 842 were smokers. The samples are to be compared to determine whether the proportion of U.S. adults in Alaska who smoke declined during the 10-year period between the samples. Taking

p_{2002} and p_{2012} to be the proportions of all adults in Alaska over 18 who were current smokers in these years, state hypotheses, find the test statistic, and use either software or the bottom row of Table C for the P-value. Be sure to state your conclusion.

Diet and bone density in cats. *Some dietitians have suggested that highly acidic diets can have an adverse effect on bone density in humans. Alkaline diets have been marketed to avoid or counteract this effect. Is the same thing true for cats, and would an alkaline diet be beneficial? Two groups of four cats were fed diets for 12 months that differed only in acidifying or alkalinizing properties. The bone mineral density (g/cm²) of each cat was measured at the end of 12 months. The summary statistics for bone mineral density appear below.*[10]

Diet	n	$\bar{x}$	s
Acidifying	4	0.63	0.01
Alkalinizing	4	0.64	0.05

Questions 24.36 to 24.39 are based on this study.

24.36 A 95% confidence interval for the mean bone mineral density of cats after 12 months on an acidifying diet is

(a) 0.622 to 0.638. (c) 0.614 to 0.646.
(b) 0.616 to 0.644. (d) 0.620 to 0.650.

24.37 Is there strong evidence that cats on an alkalinizing diet have higher mean bone mineral density after 12 months than cats on an acidifying diet? To test this with a two-sample t test, the values of the t statistic and its degrees of freedom using conservative Option 2 are

(a) $t = 0.39$, df $= 3$. (c) $t = 0.01$, df $= 6$.
(b) $t = 0.39$, df $= 4$. (d) $t = 0.01$, df $= 7$.

24.38 A 95% confidence interval for the difference in mean bone mineral density after 12 months on an alkalinizing diet and an acidifying diet is (use conservative Option 2 for the degrees of freedom)

(a) 0.01 to 0.05. (c) −0.05 to 0.07.
(b) −0.02 to 0.04. (d) −0.07 to 0.09.

24.39 What conditions must be satisfied to justify the procedures you used in Exercise 24.36? In Exercise 24.37? In Exercise 24.38?

Choosing an inference procedure. *In each of Questions 24.40 to 24.45, say which type of inference procedure from the Statistics in Summary flowchart (page 562) you would use, or explain why none of these procedures fits the problem. You do not need to carry out any procedures.*

24.40 Driving too fast. How seriously do people view speeding in comparison with other annoying behaviors? A large random sample of adults was asked to rate a number of behaviors on a scale of 1 (no problem at all) to 5 (very severe problem). Do speeding drivers get a higher average rating than noisy neighbors?

24.41 Preventing drowning. Drowning in bathtubs is a major cause of death in children less than five years old. A random sample of parents was asked many questions related to bathtub safety. Overall, 85% of the sample said they used baby bathtubs for infants. Estimate the percent of all parents of young children who use baby bathtubs.

24.42 Acid rain? You have data on rainwater collected at 16 locations in the Adirondack Mountains of New York State. One measurement is the acidity of the water, measured by pH on a scale of 0 to 14 (the pH of distilled water is 7.0). Estimate the average acidity of rainwater in the Adirondacks.

24.43 Athletes' salaries. Looking online, you find the base salaries of the 25 active players on the roster of the Chicago Cubs as of opening day of the 2014 baseball season. The total salary of these 25 players was $63.8 million, one of the lowest in the major leagues. Estimate the average salary of the 25 active players on the roster.

24.44 Looking back on love. How do young adults look back on adolescent romance? Investigators interviewed 40 couples in their mid-twenties. The female and male partners were interviewed separately. Each was asked about his or her current relationship and also about a romantic relationship that lasted at least two months when they were aged 15 or 16. One response variable was a measure on a numerical scale of how much the attractiveness of the adolescent partner mattered. You want to compare the men and women on this measure.

24.45 Preventing AIDS through education. The Multisite HIV Prevention Trial was a randomized comparative experiment to compare the effects of twice-weekly small-group AIDS discussion sessions (the treatment) with a single one-hour session (the control). Compare the effects of treatment and control on each of the following response variables:

(a) A subject does or does not use condoms six months after the education sessions.
(b) The number of unprotected intercourse acts by a subject between four and eight months after the sessions.
(c) A subject is or is not infected with a sexually transmitted disease six months after the sessions.

Supplementary Exercises

Supplementary exercises apply the skills you have learned in ways that require more thought or more use of technology. Some of these exercises start from actual data rather than from data summaries. Many of these exercises ask you to follow the ***Plan, Solve,*** *and* ***Conclude*** *steps of the four-step process. Remember that the* ***Solve*** *step includes checking the conditions for the inference you plan.*

24.46 Do you have confidence? A report of a survey distributed to randomly selected email addresses at a large university says: "We have collected 427 responses from our sample of 2,100 as of April 30, 2004. This number of responses is large enough to achieve a 95% confidence interval with ±5% margin of sampling error in generalizing the results to our study population."[11] Why would you be reluctant to trust a confidence interval based on these data?

24.47 Treating migraines. Exercises 24.15 to 24.18 involve a randomized clinical trial to investigate the effects of two treatments on migraines. One of the treatments, cognitive behavioral therapy (CBT) along with the drug amitriptyline, was expected to be particularly effective. For each of the 64 subjects on this treatment, severity of migraines was assessed using PedMIDAS scores. Here are summary data for these subjects:

TIME	$\bar{x}$	s
Before treatment	68.2	31.7
After treatment	15.5	17.4

(a) Which t procedures are correct for comparing the mean PedMIDAS scores before and after treatment: one-sample, matched pairs, or two-sample?
(b) The data summary given is not enough information to carry out the correct t procedures. Explain why not.

24.48 Monkeys and music. Humans generally prefer music to silence. What about monkeys? Allow a tamarin monkey to enter a V-shaped cage with food in both arms of the V. After the monkey eats the food, which arm will it prefer? The monkey's location determines what it hears, a lullaby played by a flute in one arm and silence in the other. Each of four monkeys was tested six times, on different days and with the music arm alternating between left and right (in case a monkey prefers one direction). The monkeys chose silence for about 65% of their time in the cage. The researchers reported a one-sample t test for the mean percent of time spent in the music arm, H_0: μ = 50%, against the two-sided alternative, $t = -5.26$, df = 23, $P < 0.0001$.[12]

Although the result is interesting, the statistical analysis is not correct. The degrees of freedom df = 23 show that the researchers assumed that they had 24 independent observations. Explain why the results of the 24 trials are not independent.

24.49 (Optional topic) Drug-detecting rats? Dogs are big and expensive. Rats are small and cheap. Might rats be trained to replace dogs in sniffing out illegal drugs? A first study of this idea trained rats to rear up on their hind legs when they smelled simulated cocaine. To see how well rats performed after training, they were let loose on a surface with many cups sunk in it, one of which contained simulated cocaine. Four out of six trained rats succeeded in 80 out of 80 trials.[13] How should we estimate the long-term success rate p of a rat that succeeds in every one of 80 trials?

(a) What is the rat's sample proportion $\hat{p}$? What is the large-sample 95% confidence interval for p? It's not plausible that the rat will *always* be successful, as this interval says.

(b) Find the plus four estimate $\tilde{p}$ and the plus four 95% confidence interval for p. These results are more reasonable.

24.50 (Optional topic) A new vaccine. In 2006, the pharmaceutical company Merck released a vaccine named Gardasil for human papilloma virus, the most common cause of cervical cancer in young women. The Merck website gives results from "four placebo-controlled, double-blind, randomized clinical studies" with women 16 to 26 years of age, as follows:[14]

	n	CERVICAL CANCER	*n*	GENITAL WARTS
Gardasil	8487	0	7897	1
Placebo	8460	32	7899	91

(a) Give a 99% confidence interval for the difference in the proportions of young women who develop cervical cancer with and without the vaccine.

(b) Do the same for the proportions who develop genital warts.

(c) What do you conclude about the overall effectiveness of the vaccine?

24.51 Starting to talk. At what age do infants speak their first word of English? Here are data on 20 children (ages in months):[15] 1STWORD

15	26	10	9	15	20	18	11	8	20
7	9	10	11	11	10	12	17	11	10

(In fact, the sample contained one more child, who began to speak at 42 months. Child development experts consider this abnormally late, so the investigators dropped the outlier to get a sample of "normal" children. We are willing to treat these data as an SRS.) Is there good evidence that the mean age at first word among all normal children is greater than one year?

24.52 Fertilizing a tropical plant. Bromeliads are tropical flowering plants. Many are epiphytes that attach to trees and obtain moisture and nutrients from air and rain. Their leaf bases form cups that collect water and are home to the larvae of many insects. In an experiment in Costa Rica, Jacqueline Ngai and Diane Srivastava studied whether added nitrogen increases the productivity of bromeliad plants. Bromeliads were randomly assigned to nitrogen or control groups. Here are data on the number of new leaves produced over a seven-month period:[16] FERT

G Newport/Science Source

Control	11	13	16	15	15	11	12	
Nitrogen	15	14	15	16	17	18	17	13

Is there evidence that adding nitrogen increases the mean number of new leaves formed?

24.53 Starting to talk, continued. Use the data in Exercise 24.51 to give a 90% confidence interval for the mean age at which children speak their first word. 1STWORD

24.54 Dyeing fabrics. Different fabrics respond differently when dyed. This matters to clothing manufacturers, who want the color of the fabric to be just right. A researcher dyed fabrics made of cotton and of ramie with the same "procion blue" dye applied in the same way. Then she used a colorimeter to measure the lightness of the color on a scale in which black is 0 and white is 100. Here are the data for eight pieces of each fabric:[17] FBCDYE

Cotton	48.82	48.88	48.98	49.04	48.68	49.34	48.75	49.12
Ramie	41.72	41.83	42.05	41.44	41.27	42.27	41.12	41.49

Is there a significant difference between the fabrics? Which fabric is darker when dyed in this way?

24.55 More on dyeing fabrics. The color of a fabric depends on the dye used and also on how the dye is applied. This matters to clothing manufacturers, who want the color of the fabric to be just right. The study discussed in Exercise 24.54 went on to dye fabric made of ramie with the same procion blue dye applied in two different ways. Here are the lightness scores for eight pieces of identical fabric dyed in each way: FBCDYE2

Method B	40.98	40.88	41.30	41.28	41.66	41.50	41.39	41.27
Method C	42.30	42.20	42.65	42.43	42.50	42.28	43.13	42.45

(a) This is a randomized comparative experiment. Outline the design.

(b) A clothing manufacturer wants to know which method gives the darker color (lower lightness score). Use sample means to answer this question. Is the difference between the two sample means statistically significant? Can you tell from just the *P*-value whether the difference is large enough to be important in practice?

24.56 **Do parents matter?** A professor asked her sophomore students, "Does either of your parents allow you to drink alcohol around him or her?" and "How many drinks do you typically have per session? (A drink is defined as one 12 oz beer, one 4 oz glass of wine, or one 1 oz shot of liquor.)" Table 24.1 contains the responses of the female students who are not abstainers.[18] The sample is all students in one large sophomore-level class. The class is popular, so we are tentatively willing to regard its members as an SRS of sophomore students at this college. Does the behavior of parents make a significant difference in how many drinks students have on the average? FMDRINK

24.57 **Parents' behavior.** We wonder what proportion of female students have at least one parent who allows her to drink in the parent's presence. Table 24.1 contains information about a sample of 94 students. Use this sample to give a 95% confidence interval for this proportion. FMDRINK

24.58 **Diabetic mice.** The body's natural electrical field helps wounds heal. If diabetes changes this field, that might explain why people with diabetes heal slowly. A study of this idea compared normal mice and mice bred to spontaneously develop diabetes. The investigators attached sensors to the right hip and front feet of the mice and measured the difference in electrical potential (millivolts) between these locations. Here are the data:[19] MICE

DIABETIC MICE					
14.70	13.60	7.40	1.05	10.55	16.40
10.00	22.60	15.20	19.60	17.25	18.40
9.80	11.70	14.85	14.45	18.25	10.15
10.85	10.30	10.45	8.55	8.85	19.20
NORMAL MICE					
13.80	9.10	4.95	7.70	9.40	
7.20	10.00	14.55	13.30	6.65	
9.50	10.40	7.75	8.70	8.85	
8.40	8.55	12.60			

(a) Make a stemplot of each sample of potentials. There is a low outlier in the diabetic group. Does it appear that potentials in the two groups differ in a systematic way?

(b) Is there significant evidence of a difference in mean potentials between the two groups?

(c) Repeat your inference without the outlier. Does the outlier affect your conclusion?

24.59 **Keeping crackers from breaking.** We don't like to find broken crackers when we open the package. How can makers reduce breaking? One idea is to microwave the crackers for 30 seconds right after baking them. Analyze the following results from two experiments intended to examine this idea.[20] Does microwaving significantly improve indicators of future breaking? How large is the improvement? What do you conclude about the idea of microwaving crackers?

(a) The experimenter randomly assigned 65 newly baked crackers to be microwaved and another 65 to a control group that is not microwaved. Fourteen days after baking, 3 of the 65 microwaved crackers and 57 of the 65 crackers in the control group showed visible checking, which is the starting point for breaks.

TABLE 24.1 DRINKS PER SESSION BY FEMALE STUDENTS

PARENT ALLOWS STUDENT TO DRINK												
2.5	1	2.5	3	1	3	3	3	2.5	2.5	3.5	5	2
7	7	6.5	4	8	6	6	3	6	3	4	7	5
3.5	2	1	5	3	3	6	4	2	7	5	8	1
6	5	2.5	3	4.5	9	5	4	4	3	4	6	4
5	1	5	3	10	7	4	4	4	4	2	2.5	2.5
PARENT DOES NOT ALLOW STUDENT TO DRINK												
9	3.5	3	5	1	1	3	4	4	3	6	5	3
8	4	4	5	7	7	3.5	3	10	4	9	2	7
4	3	1										

(b) The experimenter randomly assigned 20 crackers to be microwaved and another 20 to a control group. After 14 days, he broke the crackers. Here are summaries of the pressure needed to break them, in pounds per square inch:

	MICROWAVE	CONTROL
Mean	139.6	77.0
Standard deviation	33.6	22.6

24.60 Falling through the ice. Table 7.3 (page 195) gives the dates on which a wooden tripod fell through the ice of the Tanana River in Alaska, thus deciding the winner of the Nenana Ice Classic contest, for the years 1917 to 2013. Give a 95% confidence interval for the mean date on which the tripod falls through the ice. After calculating the interval in the scale used in the table (days from April 20, which is Day 1), translate your result into calendar dates and hours within the dates. (Each hour is 1/24, or 0.042, of a day.) TANANA

24.61 A case for the Supreme Court. In 1986, a Texas jury found a black man guilty of murder. The prosecutors had used "peremptory challenges" to remove 10 of the 11 blacks and 4 of the 31 whites in the pool from which the jury was chosen.[21] The law says that there must be a plausible reason (that is, a reason other than race) for different treatment of blacks and whites in the jury pool. When the case reached the Supreme Court 17 years later, the Court said that "happenstance is unlikely to produce this disparity." Explain why the methods we know can't be safely used to evaluate the inference that lies behind the Court's finding that chance is unlikely to produce so large a black-white difference.

24.62 Mouse genes. A study of genetic influences on diabetes compared normal mice with similar mice genetically altered to remove a gene called *aP2*. Mice of both types were allowed to become obese by eating a high-fat diet. The researchers then measured the levels of insulin and glucose in their blood plasma. Here are some excerpts from their findings.[22] The normal mice are called "wild-type" and the altered mice are called "$aP2^{-/-}$."

*Each value is the mean ± SEM of measurements on at least 10 mice. Mean values of each plasma component are compared between $aP2^{-/-}$ mice and wild-type controls by Student's t test (*P < 0.05 and **P < 0.005).*

PARAMETER	WILD TYPE	$aP2^{-/-}$
Insulin (ng/dL)	5.9 ± 0.9	0.75 ± 0.2**
Glucose (mg/dL)	230 ± 25	150 ± 17*

Despite much greater circulating amounts of insulin, the wild-type mice had higher blood glucose than the $aP2^{-/-}$ animals. These results indicate that the absence of aP2 interferes with the development of dietary obesity-induced insulin resistance.

Other biologists are supposed to understand the statistics reported so tersely.

(a) What does "SEM" mean? What is the expression for SEM based on n, $\bar{x}$, and s from a sample?
(b) Which of the tests we have studied did the researchers apply?
(c) Explain to a biologist who knows no statistics what $P < 0.05$ and $P < 0.005$ mean. Which is stronger evidence of a difference between the two types of mice?

24.63 Mouse genes, continued. The report quoted in Exercise 24.62 says only that the sample sizes were "at least 10." Suppose that the results are based on exactly 10 mice of each type. Use the values in the table to find $\bar{x}$ and s for the insulin concentrations in the two types of mice. Carry out a test to assess the significance of the difference in mean insulin concentration. Does your P-value confirm the claim in the report that $P < 0.005$?

24.64 (Optional topic) Which font? Plain type fonts such as Times New Roman are easier to read than fancy fonts such as Gigi. A group of 25 volunteer subjects read the same text in both fonts. (This is a matched pairs design. One-sample procedures for proportions, like those for means, are used to analyze data from matched pairs designs.) Of the 25 subjects, 17 said that they preferred Times New Roman for web use. But 20 said that Gigi was more attractive.[23]

(a) Because the subjects were volunteers, conclusions from this sample can be challenged. Show that the sample size condition for the large-sample confidence interval is not met, but that the condition for the plus four interval is met.
(b) Give a 95% confidence interval for the proportion of all adults who prefer Times New Roman for web use.
(c) Give a 90% confidence interval for the proportion of all adults who think Gigi is more attractive.

24.65 Opinions about evolution. A sample survey funded by the National Science Foundation asked a random sample of American adults about biological evolution.[24] One question asked subjects to answer "true," "false," or "not sure" to the statement, "Human beings, as we know them today, developed from earlier species of animals." Of the 1484 respondents, 594 said "true." What can you say with 95% confidence about the percent of all American adults who think that humans developed from earlier species of animals?

PART

V

Inference about Relationships

INFERENCE ABOUT RELATIONSHIPS

CHAPTER 25 Two Categorical Variables: The Chi-Square Test

CHAPTER 26 Inference for Regression

CHAPTER 27 One-Way Analysis of Variance: Comparing Several Means

Statistical inference offers more methods than anyone can know well, as a glance at the offerings of any large statistical software package demonstrates. In an introductory text, we must be selective. Parts I to IV have laid a foundation for understanding statistics:

- The nature and purpose of data analysis.
- The central ideas of designs for data production.
- The reasoning behind confidence intervals and significance tests.
- Experience applying these ideas in practice.

Each of the three chapters of Part V offers an introduction to a more advanced topic in statistical inference. You may choose to read any or all of them, in any order.

What makes a statistical method "more advanced"? More complex data, for one thing. In Part IV, we looked only at methods for inference about a single population parameter and for comparing two parameters. All of the chapters in Part V present methods for studying relationships between two variables. In Chapter 25, both variables are categorical, with data given as a two-way table of counts of outcomes. Chapter 26 considers inference in the setting of regressing a response variable on an explanatory variable. This is an important type of relationship between two quantitative variables. In Chapter 27, we meet methods for comparing the mean response in more than two groups. Here, the explanatory variable (group) is categorical and the response variable is quantitative. These chapters together bring our knowledge of inference to the same point that our study of data analysis reached in Chapters 1 to 7.

With greater complexity comes greater reliance on technology. In these final three chapters, you will more often be interpreting the output of statistical software or using software yourself. With effort, you can do the calculations needed in Chapter 25 with a basic calculator. In Chapters 26 and 27, the pain is too great and the contribution to learning too small. Fortunately, you can grasp the ideas without step-by-step arithmetic.

Another aspect of "more advanced" methods is new concepts and ideas. This is where we draw the line in deciding what statistical topics we can master in a first course. Part V builds elaborate methods on the foundation we have laid without introducing fundamentally new concepts. You can see that statistical practice does need additional big ideas by reading the sections on "the problem of multiple comparisons" in Chapters 25 and 27. But the ideas you already know place you among the world's statistical sophisticates.

© Axel Dupeux/Corbis

CHAPTER 25

Two Categorical Variables: The Chi-Square Test

In this chapter we cover...

25.1 Two-way tables
25.2 The problem of multiple comparisons
25.3 Expected counts in two-way tables
25.4 The chi-square test statistic
25.5 Cell counts required for the chi-square test
25.6 Using technology
25.7 Uses of the chi-square test: independence and homogeneity
25.8 The chi-square distributions
25.9 The chi-square test for goodness of fit*

The two-sample z procedures of Chapter 23 allow us to compare the proportions of successes in two groups, either two populations or two treatment groups in an experiment. In the initial example of this chapter, we investigate where young people live. Suppose we want to compare the proportions of men and women who live with their parents. This is a question about the relationship between two categorical variables, sex (female or male) and "Where do you live?" (with parents or not), which can be answered using the two-sample z procedures of Chapter 23. In fact, the data include three more outcomes for "Where do you live?": in another person's home, in your own place, and in group quarters such as a dormitory. When there are more than two outcomes, or when we want to compare more than two groups, we need a new statistical test. The new test addresses a general question: *is there a relationship between two categorical variables?*

25.1 Two-way tables

We saw in Chapter 6 that we can present data on two categorical variables in a **two-way table** of counts. That's our starting point. Let's begin our exploration of where college-age young people live.

two-way table

EXAMPLE 25.1 Where Do Young People Live?

LIVING

A sample survey asked a random sample of young adults, "Where do you live now? That is, where do you stay most often?" Table 25.1 is a two-way table of all 2984 people in the sample (both men and women) classified by their age and by where they lived.[1] Living arrangement is a categorical variable. Even though age is quantitative, the two-way table treats age as dividing young adults into four categories. Table 25.1 gives the counts for all 20 combinations of age and living arrangement. Each of the 20 counts occupies a **cell** of the table. ■

cell

TABLE 25.1 YOUNG ADULTS BY AGE AND LIVING ARRANGEMENT

	AGE (YEARS)				
LIVING ARRANGEMENT	**19**	**20**	**21**	**22**	**TOTAL**
Parents' home	324	378	337	318	1357
Another person's home	37	47	40	38	162
Your own place	116	279	372	487	1254
Group quarters	58	60	49	25	192
Other	5	2	3	9	19
Total	540	766	801	877	2984

As usual, we prepare for inference by first doing data analysis. Because we think that age helps explain where young people live, find the percents of people in each age group who have each living arrangement. The percents appear in Table 25.2. Each column adds to 100% (up to roundoff error) because we are looking at each age group separately. In the language of Chapter 6 (page 166), Table 25.2 shows the four *conditional distributions* of living arrangements given a specific age.

Figure 25.1 is a bar graph comparing the four conditional distributions. The graph shows a strong relationship between age and living arrangement. As young adults age from 19 to 22, the percent living with their parents drops and the percent living in their own place rises. The percent living in group quarters also declines with age as college students move out of dormitories. Are these differences among the four age groups large enough to be statistically significant?

TABLE 25.2 PERCENTS OF EACH AGE GROUP WHO HAVE EACH LIVING ARRANGEMENT (READ DOWN COLUMNS)

	AGE			
	19	**20**	**21**	**22**
Parents' home	60.0%	49.3%	42.1%	36.3%
Another person's home	6.9%	6.1%	5.0%	4.3%
Your own place	21.5%	36.4%	46.4%	55.5%
Group quarters	10.7%	7.8%	6.1%	2.9%
Other	0.9%	0.3%	0.4%	1.0%
Total	100.0%	99.9%	100.0%	100.0%

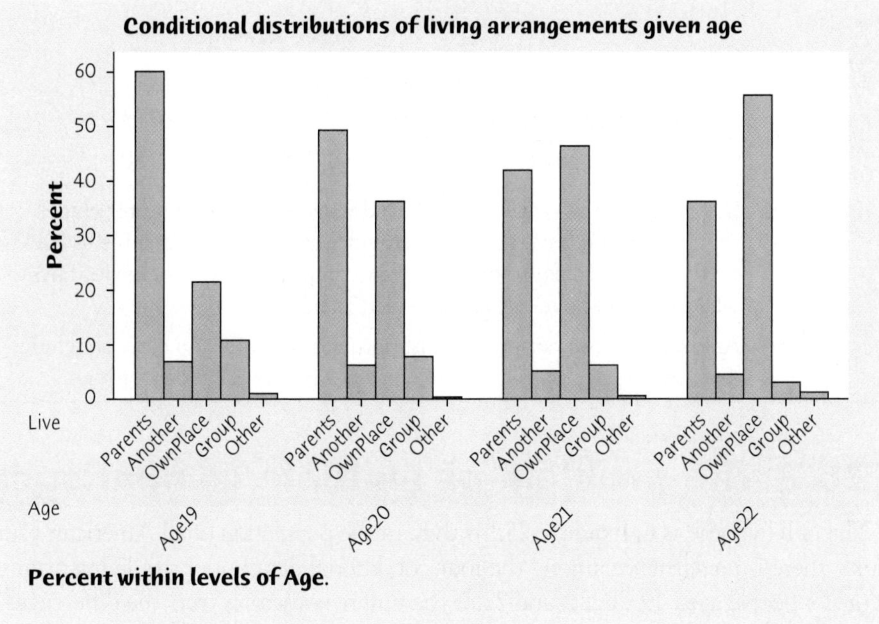

FIGURE 25.1
Bar graph comparing the four conditional distributions of living arrangements given age, for Example 25.1.

Apply Your Knowledge

25.1 Visitors to the Monterey Bay Aquarium. The Monterey Bay Aquarium, founded in 1984, is situated on the beautiful coast of Monterey Bay in the historic Cannery Row district. During 2009–13, the aquarium conducted a survey of a random sample of visitors as they exited the museum. The survey included visitor demographic information, use of social media, and opinions on their aquarium visit. Here is a comparison of the age distribution of Hispanic and Caucasian visitors in 2013:[2] AQUARIUM

Age Category	Caucasian	Hispanic
Under 25	181	79
25 to 34	324	120
35 to 44	232	74
45 and over	571	44

(a) What percent of Caucasian visitors are in each age category? What percent of Hispanic visitors are in each category? Each column should add to 100% (up to roundoff error). These are the conditional distributions of age category given ethnic group.

(b) Make a bar graph that compares the two conditional distributions. What are the most important differences in age between the two ethnic groups?

25.2 Video Gaming and Grades. The popularity of computer, video, online, and virtual reality games has raised concerns about their ability to negatively impact youth, although the existing literature has been inconsistent. The data in this exercise are based on a recent survey of 14- to 18-year-olds in Connecticut high schools. Here are the grade distributions of boys who have and have not played video games.[3] GAMING

	Grade Average		
	As and Bs	Cs	Ds and Fs
Played games	736	450	193
Never played games	205	144	80

(a) It appears that boys who have played video games have better grades than those who have never played video games. Give percents to back up this claim. Make a bar graph that compares your percents for boys who have and have not played video games.

(b) Association does not prove causation. Explain why you can't conclude from this study that playing video games improves grades for boys.

HE STARTED IT! A study of deaths in bar fights showed that, in 90% of the cases, the person who died started the fight. You shouldn't believe this. If you killed someone in a fight, what would you say when the police asked you who started the fight? After all, dead men tell no tales.

25.2 The problem of multiple comparisons

The null hypothesis in Example 25.1 is that, in the population of all American young adults, there is *no difference* among the four conditional distributions of living arrangements for people aged 19, 20, 21, and 22. If the null hypothesis is true, the differences in the sample are just accidents due to random selection of the sample. Put more generally, the null hypothesis is that there is *no relationship* between two categorical variables:

H_0: there is no relationship between age and living arrangement for the population of all American young adults

The alternative hypothesis says that there *is* a relationship but does not specify any particular kind of relationship:

H_a: there is some relationship between age and living arrangement for the population of all American young adults

Any difference among the four distributions of living arrangements in the population of all young adults means that the null hypothesis is false and the alternative hypothesis is true. The alternative hypothesis is not one-sided or two-sided. We might call it "many-sided" because it allows any kind of difference.

With only the methods we already know, we might start by comparing the proportions of people aged 19 and 22 who live with their parents. We could similarly compare other pairs of proportions, ending up with many tests and many *P*-values. This is a bad idea. The *P*-values belong to each test separately, not to the collection of all the tests together. Think of the distinction between the probability that a basketball player makes a free throw and the probability that she makes all of her free throws in a game. *When we do many individual tests or confidence intervals, the individual P-values and confidence levels don't tell us how confident we can be in all of the inferences taken together.*

Because of this, it's cheating to pick out one large difference from Table 25.2 and then test its significance as if it were the only comparison we had in mind. For example, the percents of people aged 19 and 22 who live with their parents are significantly different ($z = 8.72$, $P < 0.001$) if we make just this one comparison. But we could also pick a comparison that is not significant; for example, the proportions of people aged 21 and 22 who live in another person's home do not differ significantly ($z = 0.64$, $P = 0.522$). Individual comparisons can't tell us whether the four distributions, each with five outcomes, are significantly different.

The problem of how to do many comparisons at once with an overall measure of confidence in all our conclusions is common in statistics. This is the problem of **multiple comparisons.** Statistical methods for dealing with multiple comparisons usually have two steps:

multiple comparisons

1. An *overall test* to see if there is good evidence of *any* differences among the parameters that we want to compare.
2. A detailed *follow-up analysis* to decide which of the parameters differ and to estimate how large the differences are.

The overall test, though more complex than the tests we met earlier, is reasonably straightforward. The follow-up analysis can be quite elaborate. We will concentrate on the overall test and use data analysis to describe in detail the nature of the differences.

Apply Your Knowledge

25.3 Video Gaming and Grades. In the setting of Exercise 25.2, we might do several significance tests to compare the grades of boys who have played and never played video games. GAMING

(a) Is there a significant difference between the proportions of Cs gotten by boys who have played and never played video games? Give the P-value.

(b) Is there a significant difference between the proportions of Ds and Fs gotten by boys who have played and never played video games? Give the P-value.

(c) Explain clearly why P-values for individual outcomes like these can't tell us whether the two distributions for all three outcomes for boys who have played and never played differ significantly.

25.4 Is Astrology Scientific? The University of Chicago's General Social Survey (GSS) is the nation's most important social science sample survey. The GSS asked a random sample of adults their opinion about whether astrology is very scientific, sort of scientific, or not at all scientific. Here is a two-way table of counts for people in the sample who had three levels of higher education degrees:[4] ASTRLGY

	Degree Held		
	Junior College	Bachelor	Graduate
Not at all scientific	44	122	71
Very or sort of scientific	31	62	27

(a) Give three 95% confidence intervals, for the percents of people with each degree who think that astrology is not at all scientific.

(b) Explain clearly why we are *not* 95% confident that *all three* of these intervals capture their respective population proportions.

25.3 Expected counts in two-way tables

Our general null hypothesis H_0 is that there is *no relationship* between the two categorical variables that label the rows and columns of a two-way table. To test H_0, we compare the observed counts in the table with the *expected counts,* the counts we would expect—except for random variation—if H_0 were true. If the observed counts are far from the expected counts, that is evidence against H_0. It is easy to find the expected counts.

Expected Counts

The **expected count** in any cell of a two-way table when H_0 is true is

$$\text{expected count} = \frac{\text{row total} \times \text{column total}}{\text{table total}}$$

EXAMPLE 25.2 Where Young People Live: Expected Counts

DATA LIVING

Let's find the expected counts for the study of where young people live. Look back at the two-way table of counts, Table 25.1. That table includes the row and column totals. The expected count of 19-year-olds who live in their parents' home is

$$\frac{\text{row 1 total} \times \text{column 1 total}}{\text{table total}} = \frac{(1357)(540)}{2984} = 245.57$$

The expected count of 22-year-olds who live with their parents is

$$\frac{\text{row 1 total} \times \text{column 4 total}}{\text{table total}} = \frac{(1357)(877)}{2984} = 398.82$$

The actual counts are 324 and 318. More younger people and fewer older people live with their parents than we would expect if there were no relationship between age and living arrangement. Table 25.3 shows all 20 expected counts.

TABLE 25.3 YOUNG ADULTS BY AGE AND LIVING ARRANGEMENT: EXPECTED CELL COUNTS

	AGE				
	19	20	21	22	TOTAL
Parents' home	245.57	348.35	364.26	398.82	1357
Another person's home	29.32	41.59	43.49	47.61	162
Your own place	226.93	321.90	336.61	368.55	1254
Group quarters	34.75	49.29	51.54	56.43	192
Other	3.44	4.88	5.10	5.58	19
Total	540	766	801	877	2984

As this table shows, *the expected counts have exactly the same row and column totals (up to roundoff error) as the observed counts.* That's a good way to check your work. Comparing the actual counts (Table 25.1) and the expected counts (Table 25.3) shows in what ways the data diverge from the null hypothesis. ■

Why the formula works Where does the formula for an expected count come from? Think of a basketball player who makes 70% of her free throws in the long run. If she shoots 10 free throws in a game, we expect her to make 70% of them, or seven of the 10. Of course, she won't make exactly seven every time she shoots 10 free throws in a game. There is chance variation from game to game. But in the long run, seven of 10 is what we expect. In more formal language, if we have n independent tries and the probability of a success on each try is p, we expect np successes.

Now go back to the count of 19-year-olds living in their parents' home. The proportion of all 2984 subjects who live with their parents is

$$\frac{\text{count of successes}}{\text{table total}} = \frac{\text{row 1 total}}{\text{table total}} = \frac{1357}{2984}$$

Think of this as p, the overall proportion of successes. If H_0 is true, we expect (except for random variation) this same proportion of successes in all four age groups. So the expected count of successes among the 540 19-year-olds is

$$np = (540)\left(\frac{1357}{2984}\right) = 245.57$$

That's the formula in the Expected Counts box.

LaunchPad Online Resources

- The **StatBoards video**, *Data Analysis and Expected Counts,* provides an additional example of data analysis in a two-way table and the computation of expected counts.

Apply Your Knowledge

25.5 Visitors to the Monterey Bay Aquarium. The two-way table in Exercise 25.1 displays data on the relationship between age and ethnicity for a random sample of visitors to the Monterey Bay aquarium. The null hypothesis is that there is no relationship between ethnicity (Caucasian vs. Hispanic) and age category. DATA AQUARIUM

(a) If this hypothesis is true, what are the expected counts for the four age categories for Hispanic visitors? This is one column of the two-way table of expected counts. Find the column total and verify that it agrees with the column total for the observed counts.

(b) Hispanic visitors tend to be younger than Caucasian visitors to the aquarium. How is this reflected when comparing the observed and the expected counts for Hispanic visitors?

25.6 Video Gaming and Grades. Exercise 25.2 describes a comparison of the grade distribution of a sample of 14- to 18-year-old boys in Connecticut who do and don't play video games. The null hypothesis "no relationship" says that in the population of all 14- to 18-year-old boys in Connecticut, the proportions who have each grade average are the same for those who play and don't play video games. DATA GAMING

(a) Find the expected cell counts if this hypothesis is true, and display them in a two-way table. Add the row and column totals to your table, and check that they agree with the totals for the observed counts.

(b) Are there any large deviations between the observed counts and the expected counts? What kind of relationship between the two variables do these deviations point to?

25.4 The chi-square test statistic

To test whether the observed differences among the four distributions of living arrangements given age are statistically significant, we compare the observed and expected counts. The test statistic that makes the comparison is the *chi-square statistic.*

Chi-Square Statistic

The **chi-square statistic** is a measure of how far the observed counts in a two-way table are from the expected counts if H_0 were true. The formula for the statistic is

$$\chi^2 = \sum \frac{(\text{observed count} - \text{expected count})^2}{\text{expected count}}$$

The sum is over all cells in the table.

As you might guess, the symbol χ in the box is the Greek letter chi. The chi-square statistic is a sum of terms, one for each cell in the table.

EXAMPLE 25.3 Where Young People Live: The Test Statistic

DATA LIVING

In the study of where young people live, 324 19-year-olds lived with their parents. The expected count for this cell is 245.57. So the term of the chi-square statistic from this cell is

$$\frac{(\text{observed count} - \text{expected count})^2}{\text{expected count}} = \frac{(324 - 245.57)^2}{245.57} = \frac{6151.26}{245.57} = 25.05$$

The chi-square statistic χ^2 is the sum of 20 terms like this one. Here they are, arranged to match the layout of the two-way table:

$$\begin{aligned}\chi^2 &= 25.05 + 2.53 + 2.04 + 16.38\\ &+ 2.01 + 0.71 + 0.28 + 1.94\\ &+ 54.23 + 5.72 + 3.72 + 38.07\\ &+ 15.56 + 2.33 + 0.13 + 17.51\\ &+ 0.71 + 1.70 + 0.87 + 2.09\\ &= 193.58\end{aligned}$$

To find the value $\chi^2 = 193.58$, we had to calculate the 20 expected cell counts in Table 25.3 and then the 20 terms of the sum. Each term in the sum represents the "contribution" to the chi-square statistic for one of the cells in the table. *Moreover, even rounding each term to two decimal places as we have done can still create roundoff error in the sum. Because of this, software is very handy in finding χ^2.* ■

CAUTION

Think of χ^2 as a measure of the distance of the observed counts from the expected counts if H_0 were true. Like any distance, it is always zero or positive, and it is zero only when the observed counts are exactly equal to the expected counts. Large values of χ^2 are evidence against H_0 because they say that the observed counts are far from what we would expect if H_0 were true. *Although the alternative hypothesis H_a is many-sided, the chi-square test is one-sided* because any violation of H_0 tends to produce a large value of χ^2. Small values of χ^2 are not evidence against H_0.

CAUTION

25.5 Cell counts required for the chi-square test

The chi-square test, like the z procedures for comparing two proportions, is an approximate method that becomes more accurate as the counts in the cells of the table get larger. We must therefore check that the counts are large enough to allow us to trust the P-value. Fortunately, the chi-square approximation is accurate for quite modest counts. Here is a practical guideline.[5]

Cell Counts Required for the Chi-Square Test

You can safely use the chi-square test with critical values from the chi-square distribution when no more than 20% of the expected counts are less than 5, and all individual expected counts are 1 or greater. In particular, all four expected counts in a 2 × 2 table should be 5 or greater.

Note that the guideline uses *expected* cell counts. The expected counts for the living arrangements study of Example 25.1 appear in Table 25.3. Only 2 of the 20 expected counts (that's 10%) are less than 5, and all are greater than 1, so the data meet the guideline for safe use of chi-square.

25.6 Using technology

Calculating the expected counts and then the chi-square statistic by hand is time consuming. As usual, software saves time and always gets the arithmetic right. Figure 25.2 shows output for the chi-square test for the living arrangements data from a graphing calculator and three statistical programs.

Texas Instruments Graphing Calculator

```
X²-Test
 X²=193.5482798
 P=6.981157E-35
 df=12
```

```
round([B],2)
[[245.57 348.35...
 [29.32  41.59  ...
 [226.93 321.9  ...
 [34.75  49.29  ...
 [3.44   4.88   ...
```

Minitab

```
Session

Expected counts are printed below observed counts
Chi-Square contributions are printed below expected counts

           Age 19   Age 20   Age 21   Age 22   Total
Parents       324      378      337      318    1357
            245.6    348.3    364.3    398.8
           25.049    2.525    2.040   16.379

Another        37       47       40       38     162
             29.3     41.6     43.5     47.6
            2.014    0.705    0.279    1.940

OwnPlace      116      279      372      487    1254
            226.9    321.9    336.6    368.6
           54.226    5.719    3.720   38.068

Group          58       60       49       25     192
             34.7     49.3     51.5     56.4
           15.564    2.329    0.125   17.505

Other           5        2        3        9      19
              3.4      4.9      5.1      5.6
            0.709    1.697    0.865    2.090

Total         540      766      801      877    2984

Pearson Chi-Square = 193.548,  DF = 12, P-Value = 0.000
2 cells with expected counts less than 5.
```

This key identifies the output for each cell in the table

We have highlighted the six cells with the largest contribution to chi-square in red

FIGURE 25.2
Output from a graphing calculator, Minitab, CrunchIt!, and JMP for the two-way table in the study of where young people live, for Example 25.4.

CrunchIt!

Results - Contingency Table

Export ▾

	Age19	Age20	Age21	Age22	All
Parents	324	378	337	318	1357
	23.88	27.86	24.83	23.43	100
	60	49.35	42.07	36.26	45.48
	10.86	12.67	11.29	10.66	45.48
Another	37	47	40	38	162
	22.84	29.01	24.69	23.46	100
	6.852	6.136	4.994	4.333	5.429
	1.240	1.575	1.340	1.273	5.429
OwnPlace	116	279	372	487	1254
	9.250	22.25	29.67	38.84	100
	21.48	36.42	46.44	55.53	42.02
	3.887	9.350	12.47	16.32	42.02
Group	58	60	49	25	192
	30.21	31.25	25.52	13.02	100
	10.74	7.833	6.117	2.851	6.434
	1.944	2.011	1.642	0.8378	6.434
Other	5	2	3	9	19
	26.32	10.53	15.79	47.37	100
	0.9259	0.2611	0.3745	1.026	0.6367
	0.1676	0.06702	0.1005	0.3016	0.6367
All	540	766	801	877	2984
	18.10	25.67	26.84	29.39	100
	100	100	100	100	100
	18.10	25.67	26.84	29.39	100

Count
% of Row
% of Col
% of Total

This key identifies the output for each cell in the table

Chi-squared statistic:	193.5
df:	12
P-value:	<0.0001

Contingency Table

Age (columns), Live (rows)

Count Col% Expected Cell Chi^2	Age19	Age20	Age21	Age22	
Parents	324	378	337	318	1357
	60.00	49.35	42.07	36.26	
	245.57	348.345	364.262	398.823	
	25.0491	2.5245	2.0403	16.3792	
Another	37	47	40	38	162
	6.85	6.14	4.99	4.33	
	29.3164	41.5858	43.4859	47.6119	
	2.0138	0.7049	0.2794	1.9405	
OwnPlace	116	279	372	487	1254
	21.48	36.42	46.44	55.53	
	226.93	321.905	336.613	368.552	
	54.2260	5.7185	3.7201	38.0680	
Group	58	60	49	25	192
	10.74	7.83	6.12	2.85	
	34.7453	49.2869	51.5389	56.429	
	15.5641	2.3286	0.1251	17.5048	
Other	5	2	3	9	19
	0.93	0.26	0.37	1.03	
	3.43834	4.87735	5.1002	5.58412	
	0.7093	1.6975	0.8648	2.0895	
	540	766	801	877	2984

Tests

N	DF	-LogLike	RSquare (U)
2984	12	101.54293	0.0312

Test	ChiSquare	Prob>ChiSq
Pearson	193.548	<.0001*

FIGURE 25.2
(Continued)

EXAMPLE 25.4 Where Young People Live: Chi-Square Output

DATA LIVING

All four outputs tell us that the chi-square statistic is $\chi^2 = 193.5$, with very small *P*-value. Minitab reports $P = 0.000$, rounded to three decimal places, while CrunchIt! and JMP report $P < 0.0001$. The graphing calculator says that $P = 6.98 \times 10^{-35}$, a very small number indeed. This *P*-value comes from an approximation to the sampling distribution of χ^2. The approximation is less accurate far out in the tail than near the center of the distribution, so you should not take 6.98×10^{-35} literally. Just read it as "*P* is very small." The sample gives very strong evidence that living arrangements differ among the four age groups.

Statistical software generally offers additional information on request. We asked both Minitab and JMP to show the observed counts and expected counts and also the term in the chi-square statistic for each cell, called the "chi-square contribution" or "cell chi^2." We also told JMP to include the column percents associated with each count. The top left cell has expected count 245.57 and contributes 25.049 to the chi-square statistic, as we calculated earlier. (Roundoff errors are smaller with software than in hand calculation.) The graphing calculator also displays the observed and expected cell counts on request. We told the calculator to display the expected counts rounded to two decimal places. To see the remaining columns of expected counts on the calculator's small screen, you must scroll to the right. CrunchIt! displays the counts, the row and column percents associated with each count, and the overall percent of the total associated with the count. Depending on the problem, only one of these percents will generally be of interest.

What about the Excel spreadsheet program? Excel is, in general, a poor choice for statistics. It is particularly awkward for chi-square because its Data Analysis tool pack omits this common test. You will need add-in modules to use Excel effectively for chi-square. ■

The chi-square test is an overall test for detecting relationships between two categorical variables. If the test is significant, it is important to look at the data to learn the nature of the relationship. We have three ways to look at the living arrangements data:

- **Compare selected percents:** which living arrangements occur in quite different percents of the four age groups? This is the method we learned in Chapter 6.
- **Compare observed and expected cell counts:** which cells have more or fewer observations than we would expect if H_0 were true?
- **Look at the terms of the chi-square statistic:** which cells contribute the most to the value of χ^2?

EXAMPLE 25.5 Where Young People Live: Conclusion

DATA LIVING

There is very strong evidence ($\chi^2 = 193.55$, $P < 0.001$) that living arrangements of young people are not the same for ages 19, 20, 21, and 22. Comparing selected percents—specifically, the four conditional distributions of living arrangements for each age in Table 25.2 and Figure 25.1—shows how young people become more independent as they grow older.

The additional information provided by programs like Minitab or JMP shows what differences among the age groups explain the large value of the chi-square statistic. Look at the 20 terms in the chi-square statistic in the Minitab output and compare the observed and expected counts in the cells that contribute most to chi-square. Just six of the 20 cells, those highlighted in red in Figure 25.2, contribute 166.79 of the total chi-square $\chi^2 = 193.55$. These six highlighted cells occur in pairs:

- 54.226 and 38.068: fewer 19-year-olds than expected and more 22-year-olds than expected live in their own place.
- 25.049 and 16.379: more 19-year-olds than expected and fewer 22-year-olds than expected live in their parents' home.
- 15.564 and 17.505: more 19-year-olds than expected and fewer 22-year-olds than expected live in group quarters.

These three trends display the increase in independent living between age 19 and age 22. ■

Apply Your Knowledge

25.7 Visitors to the Monterey Bay Aquarium. Figure 25.3 displays SAS output for the two-way table in Exercise 25.1. The output includes the two-way table of observed counts, the expected counts, and each cell's contribution to the chi-square statistic. DATA AQUARIUM

Table of Age_Group by Race

Frequency / Expected / Cell Chi-Square

Age_Group	Race: Caucasian	Race: Hispanic	Total
Under 25	181	79	260
	209.28	50.72	
	3.8215	15.768	
25 to 34	324	120	444
	357.39	86.614	
	3.1188	12.869	
35 to 44	232	74	306
	246.31	59.694	
	0.831	3.4288	
45 and over	571	44	615
	495.03	119.97	
	11.66	48.109	
Total	1308	317	1625

Statistics for Table of Age_Group by Race

Statistic	DF	Value	Prob
Chi-Square	3	99.6058	<.0001

FIGURE 25.3
SAS output for the two-way table of visitor age by race, for Exercise 25.7.

(a) Verify from the SAS output that the data meet the cell count requirement for use of chi-square.

(b) What hypotheses does chi-square test? What are the test statistic and its P-value?

(c) Which cells contribute the most to χ^2? Compare the observed and expected counts in these cells and comment on the most important differences in the age distribution of Hispanic and Caucasian visitors to the aquarium.

25.8 Video Gaming and Grades. Your data analysis in Exercise 25.2 found that boys who have played video games tend to have higher grades than those who have not. Figure 25.4 gives JMP output for the two-way table in Exercise 25.2. DATA GAMING

Contingency Table

Count / Row% / Expected

Played Games	Grade: A&B	Grade: C	Grade: D&F	
Yes	736	450	193	1379
	53.37	32.63	14.00	
	717.721	453.056	208.223	
No	205	144	80	429
	47.79	33.57	18.65	
	223.279	140.944	64.7771	
	941	594	273	1808

Tests

N	DF	-LogLike	RSquare (U)
1808	2	3.2771979	0.0018

Test	ChiSquare	Prob>ChiSq
Pearson	6.739	0.0344*

FIGURE 25.4
JMP output for the study of video gaming and grades, for Exercise 25.8.

(a) Verify from the output that the data meet the cell count requirement for use of chi-square.

(b) What are the chi-square statistic and its P-value? Explain in simple language what it means to reject H_0 in this setting.

(c) Give an overall conclusion that refers to row percents to describe the nature of the relationship between playing video games and grades.

25.9 Is Astrology Scientific? The General Social Survey asked a random sample of adults about their education and about their view of astrology as scientific or not. Here are the data for people with three levels of higher education degrees: ASTRLGY

	Degree Held		
	Junior College	**Bachelor**	**Graduate**
Not at all scientific	44	122	71
Very or sort of scientific	31	62	27

Figure 25.5 gives Minitab chi-square output for these data. Follow the Plan, Solve, and Conclude steps of the four-step process in using the information in the output to describe how people with these levels of education differ in their opinions about astrology. Be sure that your Solve step includes data analysis and checking conditions for inference as well as a formal test.

```
Session

                 Junior
                college  Bachelor  Graduate       All

Not Science          44       122        71       237
                  58.67     66.30     72.45     66.39
                  49.79    122.15     65.06    237.00
                0.67329   0.00019   0.54255         *

Science              31        62        27       120
                  41.33     33.70     27.55     33.61
                  25.21     61.85     32.94    210.00
                1.32975   0.00037   1.07153         *

All                  75       184        98       357
                 100.00    100.00    100.00    100.00
                  75.00    184.00     98.00    357.00
                      *         *         *         *

Cell Contents:            Count
                          % of Column
                          Expected count
                          Contribution to Chi-square

Pearson Chi-Square = 3.618,  DF = 2, P-Value = 0.16
```

FIGURE 25.5
Minitab output for the two-way table of opinion about astrology by degree held, for Exercise 25.9.

25.7 Uses of the chi-square test: independence and homogeneity

Two-way tables can arise in several ways. Most commonly, the subjects in a single sample are classified by two categorical variables. For example, we classified young adults by their age group and where they lived. The question of whether or not there is a relationship between these two classification variables can be stated in terms of the independence of these variables, using the definition of independence from our study of probability in Chapter 13. Recall that for two events, A and B, to be independent, we must have $P(A) = P(A \mid B)$. In this setting, independence of age and living arrangement would imply that the conditional probability that a randomly selected young adult lives at home given their age would be the same for all ages. That is, knowing a young adult's age gives us no information about the probability of living at home. The test we have been using to this point is generally referred to as the **chi-square test for independence**, as thus far all the examples and exercises have been questions about whether two classification variables are independent or not.

chi-square test for independence

The next example illustrates a different setting for a two-way table, in which we compare separate samples from two or more populations, or from two or more treatments in a randomized controlled experiment. In the setting of a randomized controlled experiment, we think of the different treatment groups as the separate populations. "Which population" is now one of the variables for the two-way table. For each sample, we classify individuals according to one variable, and we are interested in whether or not the probabilities of being classified in each category of this variable are the same for each population. Although our calculations for the chi-square test are unchanged, the method of collecting the data is different. This use of the chi-square test is referred to as the **chi-square test for homogeneity** because we are interested in whether or not the populations from which the samples are selected are homogeneous (the same) with respect to the single classification variable.

chi-square test for homogeneity

EXAMPLE 25.6 Are Cell-Only Telephone Users Different?

4 step

DATA LIVECELL

A.B./Getty Images

STATE: Random digit dialing (RDD) telephone surveys do not call cell phone numbers. We know that cell-only users tend to be younger (see Exercise 25.31), which results in younger adults being underrepresented in an RDD sample. Survey organizations can compensate for this by weighting the younger adults in the RDD sample more heavily, but this assumes that the young adults accessible by landline are similar on the survey issue to their cell-only peers, who are excluded. There is growing evidence that this is not always the case. In young adults between 18 and 25 years of age, Pew interviewed separate random samples of cell-only and landline telephone users and compared them on demographic, lifestyle, and attitudinal issues.[6] Here are the results on the living situations for the two groups:

	Landline Sample	Cell-Only Sample
Live with parents	165	25
Rent	95	74
Own	46	13
Live in a dorm	7	15
Other/refused	16	3
Total	329	130

PLAN: Carry out a chi-square test for

H_0: homogeneity; that is, the distribution of living situation is the same in both populations

H_a: lack of homogeneity; that is, the living situation distribution for the cell-only population differs from that for landline users

Compare column percents or observed versus expected cell counts or terms of chi-square to see the nature of the differences in the distributions of living situations in the two populations.

SOLVE: The Minitab output in Figure 25.6 includes the column percents. These give the distribution of living situation for each type of telephone use. Cell-only users are less likely to live with parents (19.23% versus 50.15% of landline users), more likely to rent (56.92% versus 28.88% of landline users), and more likely to live in a dorm (11.54% versus 2.13% of landline users). There is little difference in the percentages for the other two categories. To see if the differences are significant, first check the guidelines for use of chi-square. The samples can be assumed to be SRSs. The Minitab output shows that the expected counts in all of the 10 cells are greater than 5, so the conditions for use of chi-square (page 584) are satisfied. The chi-square test shows a highly significant difference between the distributions of living situations of the two groups of young adults ($\chi^2 = 61.63$, $P = 0.000$). Comparing observed and expected cell counts again shows that cell-only young adults are less likely to live with parents than would be expected if the null hypothesis of homogeneity were true, and more likely to rent or live in a dorm. The contributions to chi-square for the other two categories are quite small.

CONCLUDE: The association between living situation and type of telephone service is statistically significant. Young adults accessible by landline phone are more likely to live with their parents and less likely to rent or live in a dorm. The Pew study also found young adults with landlines tend to be more likely to attend religious services at least once a week and less likely to report drinking alcohol in the past seven days or to say that it is okay for people to smoke marijuana. They also tend to be

Session

	Landline sample	Cell-only sample	All
Live with parents	165	25	190
	50.15	19.23	41.39
	136.19	53.81	190.00
	6.096	15.427	*
Rent	95	74	169
	28.88	56.92	36.82
	121.14	47.86	169.00
	5.639	14.270	*
Own	46	13	59
	13.98	10.00	12.85
	42.29	16.71	59.00
	0.326	0.824	*
Live in a dorm	7	15	22
	2.13	11.54	4.79
	15.77	6.23	22.00
	4.876	12.341	*
Other/refused	16	3	19
	4.86	2.31	4.14
	13.62	5.38	19.00
	0.416	1.054	*
All	329	130	459
	100.00	100.00	100.00
	329.00	130.00	459.00
	*	*	*

```
Cell Contents:      Count
                    % of Column
                    Expected count
                    Contribution to Chi-square

Pearson Chi-Square = 61.269,  DF = 4,  P-Value = 0.000
```

FIGURE 25.6
Minitab output for the two-way table of living situation and telephone use, for Example 25.6.

less technology savvy using email, texting, and visiting social networking sites less frequently. Because of the rapid increase in cell-only users plus the differences in a number of dimensions between cell-only users and their landline peers, most survey organizations now routinely include a cell-only sample when conducting a survey.[7] ■

One of the most useful properties of chi-square is that it tests the null hypothesis "the row and column variables are not related to each other" whenever this hypothesis makes sense for a two-way table. It makes sense when we are comparing a categorical response in two or more samples, as when we compared people who have only a cell phone with people who have a landline phone. This is the chi-square test for homogeneity. The hypothesis also makes sense when we have data on two categorical variables for the individuals in a single sample, as when we examined age group and living arrangement for a sample of young adults. This is the chi-square test for independence. Statistical significance has the same intuitive meaning in both settings: "A relationship this strong is not likely to happen just by chance."

MORE CHI-SQUARE TESTS

There are other chi-square tests for hypotheses more specific than "no relationship." A sociologist places people in classes by social status, waits 10 years, then classifies the same people again. The row and column variables are the classes at the two times. She might test the hypothesis that there has been no change in the overall distribution of social status in the group. Or she might ask if moves up in status are balanced by matching moves down. These and other null hypotheses can be tested by variations of the chi-square test.

Uses of the Chi-Square Test

Use the chi-square test to test the null hypothesis

H_0: there is no relationship between two categorical variables

when you have a two-way table from one of these situations:

- A single SRS, with each individual classified according to both of two categorical variables. In this case, the null hypothesis of no relationship says that the two categorical variables are independent, and the test is called the chi-square test of independence.
- Independent SRSs from two or more populations, with each individual classified according to one categorical variable. (The other variable says which sample the individual comes from.) In this case, the null hypothesis of no relationship says the populations are homogeneous, and the test is called the chi-square test of homogeneity.

LaunchPad Online Resources

- The **StatBoards videos**, *The Chi-Square Test for Independence* and *The Chi-Square Test for Homogeneity*, provide applications of the chi-square test in the settings of independence and homogeneity.

Apply Your Knowledge

25.10 Do You Use Cocaine? Sample surveys on sensitive issues can give different results depending on how the question is asked. A University of Wisconsin study divided 2400 respondents into three groups at random. All were asked if they had ever used cocaine. One group of 800 was interviewed by phone; 21% said they had used cocaine. Another 800 people were asked the question in a one-on-one personal interview; 25% said "Yes." The remaining 800 were allowed to make an anonymous written response; 28% said "Yes."[8] Are there statistically significant differences among these proportions?

(a) Convert the information given into a two-way table of counts.

(b) Should you use a chi-square test of independence or homogeneity?

(c) Do a complete analysis of these data, following the four-step process.

25.11 Who's More Informed About Politics? In 2012, the General Social Survey asked a random sample of adults, "Compared to most people, how informed are you about politics?" Here are the data classified by their responses to this question and their age group:

	Not at All	A Little	Somewhat	Very	Extremely
Age 20–29	8	29	28	13	0
Age 30–39	15	28	55	23	9
Age 40–49	2	25	49	26	14
Age 50 or older	21	59	120	79	23

(a) Should you use a chi-square test of independence or homogeneity? Explain.

(b) Do a complete analysis of these data, following the four-step process.

25.8 The chi-square distributions

Software usually finds *P*-values for us. The *P*-value for a chi-square test comes from comparing the value of the chi-square statistic with critical values for a *chi-square distribution.*

The Chi-Square Distributions

The **chi-square distributions** are a family of distributions that take only positive values and are skewed to the right. A specific chi-square distribution is specified by giving its **degrees of freedom**.

The chi-square test for a two-way table with r rows and c columns uses critical values from the chi-square distribution with $(r - 1)(c - 1)$ degrees of freedom. The *P*-value is the area under the density curve of this chi-square distribution to the right of the value of the test statistic.

Figure 25.7 shows the density curves for three members of the chi-square family of distributions. As the degrees of freedom increase, the density curves become less

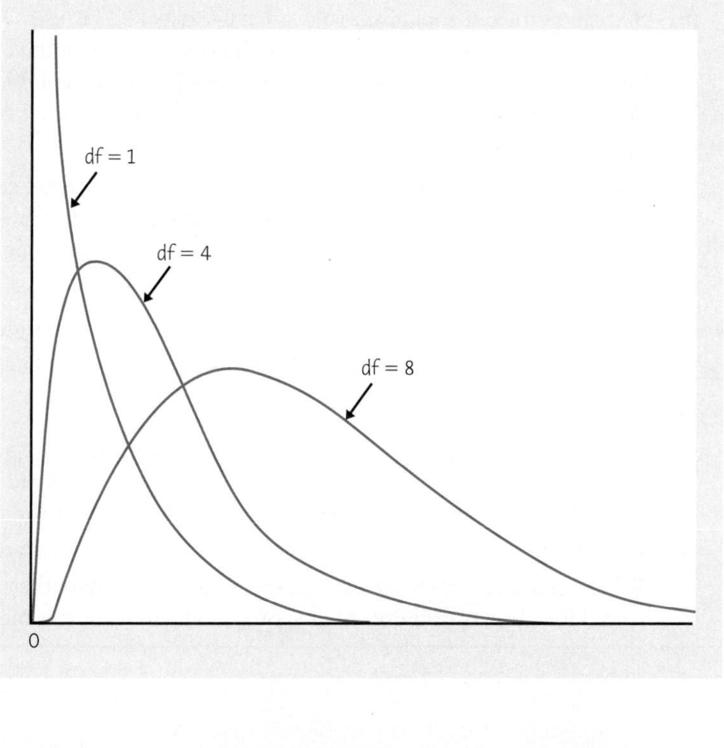

FIGURE 25.7
Density curves for the chi-square distributions with 1, 4, and 8 degrees of freedom. Chi-square distributions take only positive values and are right-skewed.

skewed, and larger values become more probable. Table D in the back of the book gives critical values for chi-square distributions. You can use Table D if you do not have software that gives you *P*-values for a chi-square test.

EXAMPLE 25.7 Using the Chi-Square Table

The two-way table of five outcomes by four age groups for the living arrangements study (Table 25.1) study has five rows and four columns. That is, $r = 5$ and $c = 4$. The chi-square statistic therefore has degrees of freedom

$$(r - 1)(c - 1) = (5 - 1)(4 - 1) = (4)(3) = 12$$

df = 12

p	0.001	0.0005
x^*	32.91	34.82

All four outputs in Figure 25.2 give 12 as the degrees of freedom.

The observed value of the chi-square statistic is $\chi^2 = 193.55$. Look in the df = 12 row of Table D. The value $\chi^2 = 193.55$ falls above the largest critical value in the table, for $P = 0.0005$. Remember that the chi-square test is always one-sided. So the *P*-value of $\chi^2 = 193.55$ is less than 0.0005. ■

We know that all z and t statistics measure the size of an effect in the standard scale centered at zero. We can roughly assess the size of any z or t statistic by the 68–95–99.7 rule, though this is exact only for z. The chi-square statistic does not have any such natural interpretation. But here is a helpful fact: *the mean of any chi-square distribution is equal to its degrees of freedom.* In Example 25.7, χ^2 would have mean 12 if the null hypothesis were true. The observed value $\chi^2 = 193.55$ is so much larger than 12 that we suspect it is significant even before we look at Table D.

Apply Your Knowledge

25.12 Visitors to the Monterey Bay Aquarium. The SAS output in Figure 25.3 gives the degrees of freedom for a table of age category vs. two ethnic groups of visitors (Caucasian and Hispanic) as `DF 3`. DATA AQUARIUM

(a) Show that this is correct for a table with four rows and two columns.

(b) SAS gives the chi-square statistic as `Chi-Sq = 99.6058`. Where does this value fall when compared with critical values of the chi-square distribution with three degrees of freedom in Table D? How does SAS's result `P-Value < 0.0001` compare with the *P*-value from the table?

(c) The original table included four ethnic groups, an additional column for Asians/Pacific Islanders and a second additional column for African Americans, and the same categories for age. What is the proper degrees of freedom for this table?

25.13 Video Gaming and Grades. The JMP output in Figure 25.4 gives two degrees of freedom for the table in Exercise 25.2.

(a) Verify that this is correct.

(b) The computer gives the value of the chi-square statistic as $\chi^2 = 6.739$. Between what two entries in Table D does this value lie? Verify that JMP's *P*-value does fall between the tail probabilities p for these two entries.

(c) What is the mean value of the statistic χ^2 if the null hypothesis is true? How does the observed value of χ^2 compare with this mean?

25.9 The chi-square test for goodness of fit*

The most common and most important use of the chi-square statistic is to test the hypothesis that there is *no relationship between two categorical variables.* A variation of the statistic can be used to test a different kind of null hypothesis: that *a categorical variable has a specified distribution.* Here is an example that illustrates this use of chi-square.

EXAMPLE 25.8 Never on Sunday?

Births are not evenly distributed across the days of the week. Fewer babies are born on Saturday and Sunday than on other days, probably because doctors find weekend births inconvenient.

A random sample of 140 births from local records shows this distribution across the days of the week:

BIRTH140

Day	Sun.	Mon.	Tue.	Wed.	Thu.	Fri.	Sat.
Births	13	23	24	20	27	18	15

Sure enough, the two smallest counts of births are on Saturday and Sunday. Do these data give significant evidence that local births are not equally likely on all days of the week? ■

The chi-square test answers the question of Example 25.8 by comparing observed counts with expected counts under the null hypothesis. The null hypothesis for births says that they *are* evenly distributed. To state the hypotheses carefully, write the discrete probability distribution for days of birth:

Day	Sun.	Mon.	Tue.	Wed.	Thu.	Fri.	Sat.
Probability	p_1	p_2	p_3	p_4	p_5	p_6	p_7

The null hypothesis says that the probabilities are the same on all days. In that case, all seven probabilities must be 1/7. So the null hypothesis is

$$H_0\colon p_1 = p_2 = p_3 = p_4 = p_5 = p_6 = p_7 = \frac{1}{7}$$

The alternative hypothesis says that days are *not* all equally probable:

$$H_a\colon \text{not all } p_i = \frac{1}{7}$$

As usual in chi-square tests, H_a is a "many-sided" hypothesis that simply says that H_0 is not true. The chi-square statistic is also as usual:

$$\chi^2 = \sum \frac{(\text{observed count} - \text{expected count})^2}{\text{expected count}}$$

The expected count for an outcome with probability p is np, as we saw in the discussion following Example 25.2. Under the null hypothesis, all the probabilities p_i are the same, so all seven expected counts are equal to

$$np_i = 140 \times \frac{1}{7} = 20$$

*This special topic is optional.

These expected counts easily satisfy our guidelines for using chi-square. The chi-square statistic is

$$\chi^2 = \sum \frac{(\text{observed count} - 20)^2}{20}$$
$$= \frac{(13-20)^2}{20} + \frac{(23-20)^2}{20} + \cdots + \frac{(15-20)^2}{20}$$
$$= 7.6$$

df = 6

p	0.25	0.20
x^*	7.84	8.56

This new use of χ^2 requires a different degrees of freedom. To find the P-value, compare χ^2 with critical values from the chi-square distribution with degrees of freedom one less than the number of values the birth day can take. That's $7 - 1 = 6$ degrees of freedom. From Table D, we see that $\chi^2 = 7.6$ is smaller than the smallest entry in the df = 6 row, which is the critical value for tail area 0.25.

The P-value is therefore greater than 0.25 (software gives the more exact value $P = 0.269$). These 140 births don't give convincing evidence that births are not equally likely on all days of the week.

The chi-square test applied to the hypothesis that a categorical variable has a specified distribution is called the test for *goodness of fit.* The idea is that the test assesses whether the observed counts "fit" the distribution. The chi-square statistic is the same as for the two-way table test, but the expected counts and degrees of freedom are different. Here are the details.

CHI-SQUARE IN THE CASINO

Gambling devices such as slot machines and roulette wheels are supposed to have a fixed and known distribution of outcomes. Here's a job for the chi-square test of goodness of fit: state gambling regulators use it to verify that casino devices are honest. How much deviation a casino can get away with depends on the state. Nevada cracks down if chi-square is significant at the 5% level. Mississippi gives more leeway, acting only when the 1% level is reached.

The Chi-Square Test for Goodness of Fit

A categorical variable has k possible outcomes, with probabilities $p_1, p_2, p_3, \ldots, p_k$. That is, p_i is the probability of the ith outcome. We have n independent observations from this categorical variable.

To test the null hypothesis that the probabilities have specified values

$$H_0: p_1 = p_{10}, p_2 = p_{20}, \ldots, p_k = p_{k0}$$

find the **expected count** for the ith possible outcome as np_{i0} and use the **chi-square statistic**

$$\chi^2 = \sum \frac{(\text{observed count} - \text{expected count})^2}{\text{expected count}}$$

The sum is over all the possible outcomes.

The P-value is the area to the right of χ^2 under the density curve of the chi-square distribution with $k - 1$ degrees of freedom.

In Example 25.8, the outcomes are days of the week, with $k = 7$. The null hypothesis says that the probability of a birth on the ith day is $p_{i0} = 1/7$ for all days. We observe $n = 140$ births and count how many fall on each day. These are the counts used in the chi-square statistic.

LaunchPad Online Resources

- The **StatBoards video**, *The Chi-Square Test for Goodness of Fit,* provides an application of the chi-square test to the problem of goodness of fit.

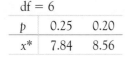

Apply Your Knowledge

25.14 Saving Birds from Windows. Many birds are injured or killed by flying into windows. It appears that birds don't see windows. Can tilting windows down so that they reflect earth rather than sky reduce bird strikes? Place six windows at the edge of a woods: two vertical, two tilted 20 degrees, and two tilted 40 degrees. During the next four months, there were 53 bird strikes: 31 on the vertical windows, 14 on the 20-degree windows, and 8 on the 40-degree windows.[9] If the tilt has no effect, we expect strikes on windows with all three tilts to have equal probability. Test this null hypothesis. What do you conclude?

25.15 More on Birth Days. Births really are not evenly distributed across the days of the week. The data in Example 25.8 failed to reject this null hypothesis because of random variation in a quite small number of births. Here are data on 700 births in the same locale: DATA BIRTH140

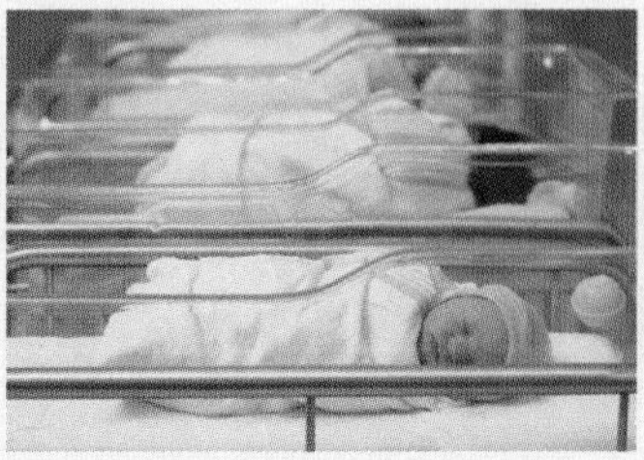
© Blend Images/Alamy

Day	Sun.	Mon.	Tue.	Wed.	Thu.	Fri.	Sat.
Births	84	110	124	104	94	112	72

(a) The null hypothesis is that all days are equally probable. What are the probabilities specified by this null hypothesis? What are the expected counts for each day in 700 births?

(b) Calculate the chi-square statistic for goodness of fit.

(c) What are the degrees of freedom for this statistic? Do these 700 births give significant evidence that births are not equally probable on all days of the week?

25.16 Police Harassment? Police may use minor violations such as not wearing a seat belt to stop motorists for other reasons. A large study in Michigan first studied the population of drivers not wearing seat belts during daylight hours by observation at more than 400 locations around the state. Here is the population distribution of seat belt violators by age group:[10]

Age group	16 to 29	30 to 59	60 or older
Proportion	0.328	0.594	0.078

The researchers then looked at court records and called a random sample of 803 drivers who had actually been cited by police for not wearing a seat belt. Here are the counts:

Age group	16 to 29	30 to 59	60 or older
Count	401	382	20

Does the age distribution of people cited differ significantly from the distribution of ages of all seat belt violators? Which age groups have the largest contributions to chi-square? Are these age groups cited more or less frequently than is justified? (The study found that males, blacks, and younger drivers were all overcited.)

25.17 Order in Choice. Does the order in which wine is presented make a difference? Several choices of wine are presented one at a time and in sequence, and the subject is then asked to choose their preferred wine at the end of the sequence. In this study, subjects were asked to taste four wine samples in sequence. All four samples given to a subject were the *same* wine, although subjects were expecting to taste four different samples of a particular variety.[11]

There were 33 subjects in the study, and the positions in the sequence they selected for their preferred wine were

Position	1	2	3	4
Count	15	5	2	11

(a) What percent of the subjects chose each position?

(b) If the subjects were equally likely to select each position, what are the expected counts for each position?

(c) Does the chi-square test for goodness of fit give good evidence that the subjects were not equally likely to choose each position? (State hypotheses, check the guidelines for using chi-square, give the test statistic and its P-value, and state your conclusion.)

(d) The *primacy* effect is a tendency for subjects to choose the first wine tasted, whereas the *recency* effect is a tendency for subjects to choose the most recent wine tasted. Are either of these effects present in this data?

4 step

25.18 What's Your Sign? For reasons known only to social scientists, the General Social Survey (GSS) regularly asks its subjects their astrological sign. Here are the counts of responses for the 2012 GSS: DATA SIGNS

Sign	Aries	Taurus	Gemini	Cancer	Leo	Virgo
Count	145	162	163	148	186	161
Sign	Libra	Scorpio	Sagittarius	Capricorn	Aquarius	Pisces
Count	178	147	145	151	170	157

If births are spread uniformly across the year, we expect all 12 signs to be equally likely. Are they? Follow the four-step process in your answer.

CHAPTER 25 SUMMARY

Chapter Specifics

- The **chi-square test** for a two-way table tests the null hypothesis H_0 that there is no relationship between the row variable and the column variable. The alternative hypothesis H_a says that there is some relationship but does not say what kind. When the two-way table results from classifying individuals in a single SRS according to two categorical variables, no relationship says the two categorical variables are independent, and the test is called the **chi-square test of independence**. If the two-way table is formed from independent SRSs from two or more populations with each individual classified according to one categorical variable, no relationship says the populations are the same and the test is called the **chi-square test of homogeneity**.
- The test compares the observed counts of observations in the cells of the table with the counts that would be expected if H_0 were true. The **expected count** in any cell is

$$\text{expected count} = \frac{\text{row total} \times \text{column total}}{\text{table total}}$$

- The **chi-square statistic** is

$$\chi^2 = \sum \frac{(\text{observed count} - \text{expected count})^2}{\text{expected count}}$$

- The chi-square test compares the value of the statistic χ^2 with critical values from the **chi-square distribution** with $(r - 1)(c - 1)$ **degrees of freedom**. Large values of χ^2 are evidence against H_0, so the P-value is the area under the chi-square density curve to the right of χ^2.
- The chi-square distribution is an approximation to the distribution of the statistic χ^2. You can safely use this approximation when all expected cell counts are at least 1 and no more than 20% are less than 5.
- If the chi-square test finds a statistically significant relationship between the row and column variables in a two-way table, do data analysis to describe the nature of the relationship. You can do this by comparing well-chosen percents, comparing the observed counts with the expected counts, and looking for the largest **terms of the chi-square statistic**.

STATISTICS IN SUMMARY

Here are the most important skills you should have acquired from reading this chapter.

A. Two-Way Tables

1. Understand that the data for a chi-square test must be presented as a two-way table of counts of outcomes.
2. Recognize that a two-way table can occur from a single SRS with each individual classified according to two categorical variables, or from independent SRSs from two or more populations with each individual classified according to one categorical variable.
3. Use percents to describe the relationship between the row and column variables, starting from the counts in a two-way table.

B. Interpreting Chi-Square Tests

1. Locate the chi-square statistic, its P-value, and other useful facts (row or column percents, expected counts, terms of chi-square) in output from your software or calculator.
2. Use the expected counts to check whether you can safely use the chi-square test.
3. Explain what null hypothesis the chi-square statistic tests in a specific two-way table. Understand the difference between testing for independence or homogeneity in a two-way table.
4. If the test is significant, compare percents, compare observed with expected cell counts, or look for the largest terms of the chi-square statistic to see what deviations from the null hypothesis are most important.

C. Doing Chi-Square Tests by Hand

1. Calculate the expected count for any cell from the observed counts in a two-way table. Check whether you can safely use the chi-square test.
2. Calculate the term of the chi-square statistic for any cell, as well as the overall statistic.
3. Give the degrees of freedom of a chi-square statistic. Make a quick assessment of the significance of the statistic by comparing the observed value with the degrees of freedom.
4. Use the chi-square critical values in Table D to approximate the P-value of a chi-square test.

Link It

The final portion of the text studies relationships between variables. In this chapter, the case of two-way tables is considered and a formal test for answering the question "Is there a relationship between the row and column variables?" is developed. As with procedures described in earlier chapters, we must first consider how the data were produced, as this plays an important role in the conclusions we can reach. In addition, we should begin with data analysis rather than a formal test. In the case of two-way tables, this typically involves looking at percents, both numerically and graphically, to first understand the nature of the relationship. When considering the relationship between the age of young adults and living arrangements in Example 25.1, we can see from our data analysis that as young adults age from 19 to 22, the percent living with their parents drops as the percent living in their own place rises.

Even though there appears to be a clear relationship between age and living arrangements in Example 25.1, we must still answer the question of whether the observed differences are large enough to be statistically significant. The chi-square test can be used for this, but it is an approximate procedure, and the conditions for cell sizes need to be checked before applying the test. Unlike some of the simpler procedures in earlier chapters, if the differences are statistically significant, then we have only determined from the chi-square test that there is evidence of a relationship, not the nature of the relationship. Although there are formal statistical procedures to further investigate the nature of the relationship, at this point we need to be satisfied describing the relationship between the two categorical variables using our data analysis tools.

LaunchPad Online Resources

If you are having difficulty with any of the sections of this chapter, these online resources should help prepare you to solve the exercises at the end of this chapter.

- **StatTutor** starts with a video review of each section and asks a series of questions to check your understanding.
- **LearningCurve** provides you with a series of questions about the chapter geared to your level of understanding.

CHECK YOUR SKILLS

Resistance training is a popular form of conditioning aimed at enhancing sports performance and is widely used among high school, college, and professional athletes, although its use for younger athletes is controversial. A random sample of 4111 patients aged 8–30 admitted to U.S. emergency rooms with the injury code "weightlifting" were obtained. These injuries were classified as "accidental" if caused by dropped weight or improper equipment use. The patients were also classified into the four age categories "8–13," "14–18," "19–22," and "23–30." Here is a two-way table of the results:[12] DATA LIFTING

Corbis/Superstock

	ACCIDENTAL	NOT ACCIDENTAL
8–13	295	102
14–18	655	916
19–22	239	533
23–30	363	1008

25.19 The number of "accidental" injuries in the sample is

(a) 1552. (b) 2559. (c) 4111.

25.20 The percent of the 14- to 18-year-olds in the sample whose injuries were classified as "accidental" is about

(a) 42.2%. (b) 41.7%. (c) 74.3%.

25.21 The percent of the 14- to 18-year-olds in the sample whose injuries were classified as "accidental" is

(a) higher than the percent for the 23- to 30-year-olds.
(b) about the same as the percent for the 23- to 30-year-olds.
(c) lower than the percent for the 23- to 30-year-olds.

25.22 For this two-way table, we should do a chi-square test of

(a) homogeneity. (b) independence.
(c) goodness of fit.

25.23 The expected count of the 14- to 18-year-olds whose injuries were classified as "accidental" is about

(a) 593.09. (b) 655. (c) 977.91.

25.24 The term in the chi-square statistic for the cell of the 14- to 18-year-olds whose injuries were classified as "accidental" is about

(a) 593.09. (b) 3.919. (c) 6.463.

25.25 The degrees of freedom for the chi-square test for this two-way table are

(a) 3. (b) 4. (c) 8.

25.26 The null hypothesis for the chi-square test for this two-way table is

(a) the proportions of "accidental" and "not accidental" injuries are the same.
(b) there is no difference in the conditional probability of an "accidental" injury for each of the four age groups.
(c) "accidental" injuries are more likely for the younger age groups.

25.27 The alternative hypothesis for the chi-square test for this two-way table is

(a) the proportions of "accidental" and "not accidental" injuries are different.
(b) the conditional probabilities of an "accidental" injury for each of the four age groups are not the same.
(c) "accidental" injuries are more likely for the younger age groups.

25.28 Software gives chi-square statistic $\chi^2 = 325.459$ for this table. From the table of critical values, we can say that the P-value is

(a) between 0.0025 and 0.001.
(b) between 0.001 and 0.0005.
(c) less than 0.0005.

25.29 We can use our results to make inference about the relationship between age and type of weightlifting injury in the population of patients aged 18–30 admitted to emergency rooms because

(a) the sample is large, with 4111 weightlifting injuries in all.
(b) the sample is close to an SRS of all weightlifting injuries for this age group.
(c) there are only two types of weightlifting injury.

CHAPTER 25 EXERCISES

If you have access to software or a graphing calculator, use it to speed your analysis of the data in these exercises. Exercises 25.30 to 25.35 are suitable for hand calculation if necessary.

25.30 **Smoking cessation.** A large randomized trial was conducted to assess the efficacy of Chantix® for smoking cessation compared with bupropion (more commonly known as Wellbutrin® or Zyban®) and a placebo. Chantix® is different from most other quit-smoking products, in that it targets nicotine receptors in the brain, attaches to them, and blocks nicotine from reaching them, whereas bupropion is an antidepressant often used to help people stop smoking. Generally healthy smokers who smoked at

4 step

least 10 cigarettes per day were assigned at random to take Chantix® ($n = 352$), bupropion ($n = 329$), or a placebo ($n = 344$). The study was double blind, with the response measure being continuous absence from smoking for weeks 9–12 of the study. Here is a two-way table of the results:[13] DATA SMOKE

	TREATMENT		
	CHANTIX®	BUPROPION	PLACEBO
No smoking in weeks 9–12	155	97	61
Smoked in weeks 9–12	197	232	283

(a) Give a 95% confidence interval for the difference between the proportions of smokers in the bupropion and placebo groups who did not smoke in weeks 9–12 of the study.
(b) What proportion of each of the three groups in the sample did not smoke in weeks 9–12 of the study? Are there statistically significant differences among these proportions? State hypotheses and give a test statistic and its P-value.
(c) In part (b), did you use a chi-square test for independence or homogeneity? Explain.

25.31 4 step **Cell-only versus landline users.** We suspect that people who rely entirely on cell phones will as a group be younger than those who have a landline telephone. Do data confirm this guess? The Pew Research Center interviewed separate random samples of cell-only and landline telephone users and broke down the samples by age group:[14] DATA CELLAGE

	LANDLINE SAMPLE	CELL-ONLY SAMPLE
Age 18–29	335	374
Age 30–49	1242	347
Age 50–64	1625	146
Age 65 or older	1481	36
Total	4683	903

(a) Should you use a chi-square test of independence or homogeneity?
(b) Do a complete analysis of these data, following the four-step process.

25.32 **College students need better sleep!** A random sample of 871 students between the ages of 20 and 24 at a large midwestern university completed a survey including questions about their sleep quality, moods, academic performance, physical health, and psychoactive drug use. Sleep quality was measured using the Pittsburgh Sleep Quality Index (PSQI), with students scoring less than or equal to 5 on the index classified as optimal sleepers, those scoring a 6 or 7 classified as borderline, and those scoring over 7 classified as poor sleepers. The following table looks at the relationship between sleep quality classification and the use of over-the-counter (OTC) or prescription (Rx) stimulant medication more than once a month to help keep awake.[15] DATA SLEEPQ

USE OF OTC/Rx MEDS TO WAKE > 1X/MONTH	SLEEP QUALITY ON PSQI INDEX		
	OPTIMAL	BORDERLINE	POOR
Yes	37	53	84
No	266	186	245

(a) Regarding this survey as approximately an SRS of college students, give a 95% confidence interval for the proportion of college students who use over-the-counter or prescription stimulant medication more than once a month to keep awake.
(b) Find the conditional distributions of sleep quality for students who use over-the-counter or prescription stimulant medication more than once a month to keep awake and those who don't. Make a graph that compares the two conditional distributions. Use your work to describe the overall relationship between students who use and don't use medications to keep awake and their sleep quality.
(c) Do students who use over-the-counter or prescription stimulant medication more than once a month to keep awake differ significantly in their quality of sleep from those who don't? State hypotheses, give the chi-square statistic and its P-value, and state your conclusion.

25.33 **Did the randomization work?** After randomly assigning subjects to treatments in a randomized comparative experiment, we can compare the treatment groups to see how well the randomization worked. We hope to find no significant differences among the groups. A study of how to provide premature infants with a substance essential to their development assigned infants at random to receive one of four types of supplement, called PBM, NLCP, PL-LCP, and TG-LCP.[16]
(a) The subjects were 77 premature infants. Outline the design of the experiment if 20 are assigned to the PBM group and 19 to each of the other treatments.
(b) The random assignment resulted in 9 females in the TG-LCP group and 11 females in each of the other groups. Make a two-way table of group by sex, and do a chi-square test to see if there are significant differences among the groups. What do you find?

25.34 **More on video gaming.** The data for comparing two sample proportions can be presented in a two-way table containing the counts of successes and failures in both samples, with two rows and two columns. In Exercise 25.2, a survey of the consequences of video gaming on 14- to 18-year-olds is described. Here is another question from the survey. The question was about aggressive behavior as evidenced by getting into serious fights, and the comparison was between girls who have and have not played video games. Here are the data:

	SERIOUS FIGHTS	
	YES	NO
Played games	36	55
Never played games	578	1436

(a) Is there evidence that the proportions of all 14- to 18-year-old girls who played or have never played video games and have gotten into serious fights differ? Find the two sample proportions, the z statistic, and its P-value.

(b) Is there evidence that the proportions for 14- to 18-year-old girls who have or have not gotten into serious fights differ between those who have played or have never played video games? Find the chi-square statistic χ^2 and its P-value.

(c) Show that (up to roundoff error) your χ^2 is the same as z^2. The two P-values are also the same. These facts are always true, so you will often see chi-square for 2×2 tables used to compare two proportions.

(d) Suppose we were interested in finding out if the data gave good evidence that video gaming was associated with *increased* aggression in girls as evidenced by getting into serious fights. Can we use the z test for this hypothesis? What about the χ^2 test? What is the important difference between these two procedures?

25.35 **Unhappy rats and tumors.** Some people think that the attitude of cancer patients can influence the progress of their disease. We can't experiment with humans, but here is a rat experiment on this theme. Inject 60 rats with tumor cells, and then divide them at random into two groups of 30. All the rats receive electric shocks, but rats in Group 1 can end the shock by pressing a lever. (Rats learn this sort of thing quickly.) The rats in Group 2 cannot control the shocks, which presumably makes them feel helpless and unhappy. We suspect that the rats in Group 1 will develop fewer tumors. The results: 11 of the Group 1 rats and 22 of the Group 2 rats developed tumors.[17]

(a) Make a two-way table of tumors by group. State the null and alternative hypotheses for this investigation.

(b) Although we have a two-way table, the chi-square test can't test a one-sided alternative. Carry out the z test and report your conclusion.

25.36 **I think I'll be rich by age 30.** A sample survey asked young adults (aged 19–25), "What do you think are the chances you will have much more than a middle-class income at age 30?" The JMP output in Figure 25.8 shows the two-way table and related information, omitting a few subjects who refused to respond or who said they were already rich.[18] RICHBY30

(a) Should you use a chi-square test of independence or homogeneity?

Contingency Table

Sex

Chance: Count / Col% / Expected / Cell Chi^2	Female	Male	
AlmostNo	96 4.06 95.1508 0.0076	98 3.99 98.8492 0.0073	194
Some	426 18.00 349.213 16.8842	286 11.63 362.787 16.2525	712
5050	696 29.40 694.503 0.0032	720 29.28 721.497 0.0031	1416
Good	663 28.01 696.955 1.6543	758 30.83 724.045 1.5924	1421
NearCer	486 20.53 531.177 3.8424	597 24.28 551.823 3.6986	1083
	2367	2459	4826

Tests

N	DF	-LogLike	RSquare (U)
4826	4	22.067196	0.0066

Test	ChiSquare	Prob>ChiSq
Pearson	43.946	<.0001*

FIGURE 25.8
JMP output for the sample survey responses of Exercise 25.36.

(b) Use the output as the basis for a discussion of the relationship between sex and a person's assessment of their chances of being rich by age 30.

25.37 **Sexy magazine ads?** Look at full-page ads in magazines with a young adult readership. Classify ads that show a model as "not sexual" or "sexual" depending on how the model is dressed (or not dressed). Here are data on 1509 ads in magazines aimed at young men, at young women, or at young adults in general:[19]

	READERS		
	MEN	WOMEN	GENERAL
Sexual	105	225	66
Not sexual	514	351	248

Figure 25.9 (page 604) displays Minitab chi-square output. Use the information in the output to describe the relationship between the target audience and the sexual content of ads in magazines for young adults. SEXYADS

Mistakes in using the chi-square test are unusually common. Exercises 25.38 to 25.41 illustrate several kinds of mistakes.

```
Session

                   Men     Women   General       All

Sexual             105       225        66       396
                 16.96     39.06     21.02     26.24
                 162.4     151.2      82.4     396.0
                20.312    36.074     3.265         *

Not sexual         514       351       248      1113
                 83.04     60.94     78.98     73.76
                 456.6     424.8     231.6    1113.0
                 7.227    12.835     1.162         *

All                619       576       314      1509
                100.00    100.00    100.00    100.00
                 619.0     576.0     314.0    1509.0
                     *         *         *         *

Cell Contents:           Count
                         % of Column
                         Expected count
                         Contribution to Chi-square

Pearson Chi-Square = 80.874,  DF = 2, P-Value = 0.00
```

FIGURE 25.9
Minitab output for a study of ads in magazines, Exercise 25.37.

25.38 Sorry, no chi-square. An experimenter hid a toy from a dog behind either Screen A or Screen B. In the first phase the toy was always hidden behind Screen A, whereas in the second phase the toy was always hidden behind Screen B. Will the dog continue to look behind Screen A in the second phase? This was tried under three conditions. In the Social–Communicative condition, the experimenter communicated with the dog by establishing eye contact and addressing the dog while hiding the toy; in the Noncommunicative condition, the toy was hidden without communication; in the Nonsocial condition, the toy was dragged by a string to be hidden without any interaction from the experimenter. There were 12 dogs assigned at random to each condition, and the dog had up to three trials to find the toy hidden behind Screen B in phase 2. An error occurred if the dog continued to search behind Screen A, with the number of errors ranging from zero if the dog found the toy behind Screen B on the initial trial up to three if the dog never correctly chose Screen B. Here are the data:[20] HIDETOY

	NUMBER OF ERRORS			
	0	1	2	3
Social–Communicative	0	3	3	6
Noncommunicative	5	3	1	3
Nonsocial	8	2	2	0

(a) The data do show a difference in the number of errors for the different conditions. Show this by comparing suitable percents.
(b) The researchers used a more complicated but exact procedure rather than chi-square to assess significance for these data. Why can't the chi-square test be trusted in this case?
(c) If you use software, does the chi-square output for these data warn you against using the test?

25.39 Sorry, no chi-square. How do U.S. residents who travel overseas for leisure differ from those who travel for business? Here is the breakdown by occupation:[21]

	LEISURE TRAVELERS	BUSINESS TRAVELERS
Professional/technical	36%	39%
Manager/executive	23%	48%
Retired	14%	3%
Student	7%	3%
Other	20%	7%
Total	100%	100%

Explain why we don't have enough information to use the chi-square test to learn whether these two distributions differ significantly.

25.40 Sorry, no chi-square. Socially isolated individuals have been shown to be at greater risk for the development of a variety of illnesses and increased mortality. It is a more serious problem among the elderly as decreasing economic resources, mobility impairment, and death of contemporaries combine to limit social contact. In a study of 6500 men and women in the English Longitudinal Study of Aging, individuals were classified according to social isolation, with 5269 classified as low/average and the remaining 1231 classified as high social isolation. The following table shows the incidence of several limiting long-standing illnesses for the two groups.[22] Explain why it is not correct to use a chi-square test on this table to compare the "low/average isolation" and "high isolation" groups. Note that to use the chi-square test in a two-way table, each individual must fall into *one* cell of the table.

	LOW/AVERAGE ISOLATION	HIGH ISOLATION
Cancer	168	36
Diabetes	332	89
Coronary heart disease	193	43
Chronic lung disease	65	34
Stroke	134	43
Arthritis	1778	477
Mobility impairment	2940	769
Depression	960	359

25.41 **Sorry, no chi-square.** Does eating chocolate trigger headaches? To find out, women with chronic headaches followed the same diet except for eating chocolate bars and carob bars that looked and tasted the same. Each subject ate both chocolate and carob bars in random order with at least three days between. Each woman then reported whether or not she had a headache within 12 hours of eating the bar. Here is a two-way table of the results for the 64 subjects:[23]

	NO HEADACHE	HEADACHE
Chocolate	53	11
Carob (placebo)	38	26

The researchers carried out a chi-square test on this table to see if the two types of bar differ in triggering headaches. Explain why this test is incorrect. (*Hint:* There are 64 subjects. How many observations appear in the two-way table?)

The remaining exercises concern larger tables that require software for easy analysis. In many cases, you should follow the ***Plan, Solve,*** *and* ***Conclude*** *steps of the four-step process in your answers.*

25.42 **Smokers rate their health.** The University of Michigan Health and Retirement Study (HRS) surveys more than 22,000 Americans over the age of 50 every two years. A subsample of the Health and Retirement Study (HRS) participated in the 2009 Internet-based survey that collected information on a number of topical areas, including health (physical and mental, health behaviors); psychosocial items; economics (income, assets, expectations, and consumption); and retirement.[24] Two of the questions asked on the Internet survey were, "Would you say your health is excellent, very good, good, fair, or poor?" and "Do you smoke cigarettes now?" The two-way table summarizes the answers on these two questions. SMRATE

	CURRENT SMOKER	
	YES	NO
Excellent	25	484
Very good	115	1557
Good	145	1309
Fair	90	545
Poor	29	11

(a) Regarding the HRS Internet survey as approximately an SRS of Americans over the age of 50, give a 99% confidence interval for the proportion of Americans over the age of 50 who are current smokers.
(b) Compare the conditional distributions of self-evaluation of health for current smokers and nonsmokers using both a table and a graph. What are the most important differences?
(c) Carry out the chi-square test for the hypothesis of no difference between the self-evaluation of health for current smokers and nonsmokers. What would be the mean of the test statistic if the null hypothesis were true? The value of the statistic is so far above this mean that you can see at once that it must be highly significant. What is the approximate *P*-value?
(d) Look at the terms of the chi-square statistic and compare observed and expected counts in the cells that contribute the most to chi-square. Based on this and your findings in part (b), write a short comparison of the differences in self-evaluation of health for current smokers and nonsmokers.

25.43 **Who goes to religious services?** The General Social Survey (GSS) asked this question: "Have you attended religious services in the last week?" Here are the responses for those whose highest degree was high school or above: SERVICES

Fotosearch/Superstock

	HIGHEST DEGREE HELD			
	HIGH SCHOOL	JUNIOR COLLEGE	BACHELOR	GRADUATE
Attended services	400	62	146	76
Did not attend services	880	101	232	105

(a) Carry out the chi-square test for the hypothesis of no relationship between the highest degree attained and attendance at religious services in the last week. What do you conclude?

(b) Make a 2 × 3 table by omitting the column corresponding to those whose highest degree was high school. Carry out the chi-square test for the hypothesis of no relationship between the type of advanced degree attained and attendance at religious services in the last week. What do you conclude?

(c) Make a 2 × 2 table by combining the counts in the three columns that have a highest degree beyond high school, so that you are comparing adults whose highest degree was high school against those whose highest degree was beyond high school. Carry out the chi-square test for the hypothesis of no relationship between attaining a degree beyond high school and attendance at religious services for this 2 × 2 table. What do you conclude?

(d) Using the results from these three chi-square tests, write a short report explaining the relationship between attendance at religious services in the last week and the highest degree attained. As part of your report, you should give the percentages who attended religious services for each of the four highest degrees.

25.44 **Condom usage among high school students.** The Centers for Disease Control developed the Youth Risk Behavior Surveillance System (YRBSS) to monitor six categories of priority health-risk behaviors among youth—behaviors that contribute to unintentional injuries and violence; tobacco use; alcohol and other drug use; sexual behaviors that contribute to unintended pregnancy and sexually transmitted diseases; unhealthy dietary behaviors; and physical inactivity. A multistage sample design is used to produce representative samples of students in grades 9–12, who then fill out a questionnaire on these behaviors. The following data are for the question, "Did Not Use A Condom During Last Sexual Intercourse?" The two-way table of grade and condom usage only included students who were currently sexually active. Here are the results:[25] CONDOMS

	CONDOM USED	
	YES	NO
9th grade	248	417
10th grade	341	549
11th grade	467	772
12th grade	815	920

Describe the most important differences between condom usage and grade. Is there a significant overall difference between the proportion who used condoms in the different grades?

25.45 **How are schools doing?** The nonprofit group Public Agenda conducted telephone interviews with a stratified sample of parents of high school children. There were 202 black parents, 202 Hispanic parents, and 201 white parents. One question asked was, "Are the high schools in your state doing an excellent, good, fair, or poor job, or don't you know enough to say?" Here are the survey results:[26] HIGHSCHL

	BLACK PARENTS	HISPANIC PARENTS	WHITE PARENTS
Excellent	12	34	22
Good	69	55	81
Fair	75	61	60
Poor	24	24	24
Don't know	22	28	14
Total	202	202	201

Are the differences in the distributions of responses for the three groups of parents statistically significant? What departures from the null hypothesis "no relationship between group and response" contribute most to the value of the chi-square statistic? Write a brief conclusion based on your analysis.

25.46 **Complications of bariatric surgery.** Bariatric surgery, or weight-loss surgery, includes a variety of procedures performed on people who are obese. Weight loss is achieved by reducing the size of the stomach with an implanted medical device (gastric banding), through removal of a portion of the stomach (sleeve gastrectomy), or by resecting and rerouting the small intestines to a small stomach pouch (gastric bypass surgery). Because there can be complications using any of these methods, the National Institute of Health recommends bariatric surgery for obese people with a body mass index (BMI) of at least 40, and for people with BMI 35 and serious coexisting medical conditions such as diabetes. Serious complications include potentially life-threatening, permanently disabling, and fatal outcomes. Here is a two-way table for data collected in Michigan over several years giving counts of non-life-threatening complications, serious complications, and no complications for these three types of surgeries:[27] BARI

	TYPE OF COMPLICATION			
	NON-LIFE-THREATENING	SERIOUS	NONE	TOTAL
Gastric banding	81	46	5253	5380
Sleeve gastrectomy	31	19	804	854
Gastric bypass	606	325	8110	9041

(a) Is this study an experiment? Explain your answer.
(b) Is there a significant difference in the distribution of type of complication for the three types of surgery? Which surgeries have the greatest chance of complications? Can we conclude that the surgery is more dangerous, or could there be other factors associated with the increased risk?

25.47 **Bring your kids to the Monterey Bay Aquarium!** The Monterey Bay Aquarium, founded in 1984, is situated on the beautiful coast of Monterey Bay in the historic Cannery Row district of Monterey, California. Each year, the aquarium interviews a random sample of visitors as they exit the museum. The survey includes visitor demographic information, use of social media, and opinions on their aquarium visit. For each visitor sampled during the summer months of June, July, and August, here is the distribution of the number of children in their group in each year during 2009–13:[28] ETHNICTY

	YEAR				
NUMBER OF CHILDREN	2009	2010	2011	2012	2013
0	390	330	342	318	268
1	79	94	93	99	131
2	66	96	96	111	137
3 or more	41	37	45	55	70

How has the distribution of the number of children in the group changed over this five year period? Are the differences significant? One of the Aquarium's strategic goals is to attract a younger audience, including more families and intergenerational groups. Does it appear that they are meeting their goal? Explain.

Support for political parties. *Political parties want to know what groups of people support them. The General Social Survey (GSS) asked its 2012 sample, "Generally speaking, do you usually think of yourself as a Republican, Democrat, Independent, or what?" The GSS is essentially an SRS of American adults. Here is a large two-way table breaking down the responses by the highest degree the subject held:*

	NONE	HIGH SCHOOL	JR. COLLEGE	BACHELOR	GRADUATE
Strong Democrat	47	161	35	60	53
Not strong Democrat	43	176	23	63	38
Independent, near Democrat	37	117	10	43	28
Independent	101	186	26	36	24
Independent, near Republican	20	76	16	29	16
Not strong Republican	18	130	16	64	22
Strong Republican	14	101	19	43	15
Other party	5	22	5	14	8

Exercises 25.48 to 25.50 are based on this table.

25.48 **Other parties.** Give a 95% confidence interval for the proportion of adults who are "Independent."

25.49 **Party support in brief.** Make a 2 × 5 table by combining the counts in the three rows that mention Democrat and in the three rows that mention Republican and ignoring strict independents and supporters of other parties. We might think of this table as comparing all adults who lean Democrat and all adults who lean Republican. How does support for the two major parties differ among adults with different levels of education? COMBPART

25.50 **Party support in full.** Use the full table to analyze the differences in political party support among levels of education. The sample is so large that the differences are bound to be highly significant, but give the chi-square statistic and its P-value nonetheless. The main challenge is in seeing what the data say. Does the full table yield any insights not found in the compressed table you analyzed in Exercise 25.49? FULLPART

Exploring the Web

25.51 Make your own table. The Behavioral Risk Factor Surveillance System (BRFSS) is an ongoing data collection program designed to measure behavioral risk factors for the adult population (18 years of age or older) living in households. Data are collected from a random sample of adults (one per household) through a telephone survey. Go to the website `apps.nccd.cdc.gov/BRFSS/` and under BRFSS contents click on *Web Enable Analysis Tool (WEAT)* and then click on *Cross Tabulation Analysis*. After a year is selected, a window will open that will allow you to produce two-way tables.

(a) Choose a state of interest to you and two variables for the two-way table. For example, you could choose Connecticut and look at the relationship between a demographic variable such as education level and a variable such as health care coverage. Once you have chosen your state and two variables, click on *run report* on the bottom of the page. A two-way table will appear in a new window.

(b) Is there a relationship between the two variables you have selected? If the relationship is statistically significant, describe the relationship in a brief report using percentages from the table and an appropriate graph.

25.52 What do the voters think? The American National Election Studies (ANES) is the leading academically run national survey of voters in the United States, conducted before and after every presidential election. SDA (Survey Documentation and Analysis) is a set of programs that allows you to analyze survey data and includes the ANES survey as part of its archive. Go to the website `sda.berkeley.edu/` and click on *Archive*. Go to the 2012 ANES survey.

(a) Open the pre-election questions. Under Liberal/Conservative, choose the variable "liberal/conservative self-placement on a 7 point scale." Use this as your row variable. Under Energy and Climate Change, choose the variable "Anthropogenic climate change," or climate change caused or produced by humans. Use this as your column variable. In the details for the table, set *Weight* to none and for *N of Cases to Display*, make sure the unweighted box is checked. For *Percentaging*, choose row percents. Now click on "run the table."

(b) To analyze the data, make a 3×3 table by combining the rows for extremely liberal and liberal, slightly liberal, middle of the road and slightly conservative, and conservative and extremely conservative. Carry out a formal test to determine if there is a relationship between these two variables, and then describe the relationship in a brief report using percentages from the table or an appropriate graph.

(c) Select two other variables of interest to you, and analyze the relationship between these two variables. If there is a more recent survey than 2012, you should use it.

Barbara Peacock/Getty Images

CHAPTER 26

Inference for Regression

In this chapter we cover...

26.1 Conditions for regression inference

26.2 Estimating the parameters

26.3 Using technology

26.4 Testing the hypothesis of no linear relationship

26.5 Testing lack of correlation

26.6 Confidence intervals for the regression slope

26.7 Inference about prediction

26.8 Checking the conditions for inference

When a scatterplot shows a linear relationship between a quantitative explanatory variable x and a quantitative response variable y, we can use the least-squares line fitted to the data to predict y for a given value of x. When the data are a sample from a larger population, we need statistical inference to answer questions like these about the population:

- Is there really a linear relationship between x and y in the population, or might the pattern we see in the scatterplot plausibly arise just by chance?
- What is the slope (rate of change) that relates y to x in the population, including a margin of error for our estimate of the slope?
- If we use the least-squares line to predict y for a given value of x, how accurate is our prediction (again, with a margin of error)?

This chapter shows you how to answer these questions. Here is an example we will explore.

EXAMPLE 26.1 Crying and IQ

DATA CRYING

4step

STATE: Infants who cry easily may be more easily stimulated than others. This may be a sign of higher IQ. Child development researchers explored the relationship between the crying of infants four to ten days old and their later IQ test scores. A snap of a rubber band on the sole of the foot caused the infants to cry. The researchers recorded the crying and measured its intensity by the number of peaks in the most active 20 seconds. They later measured the children's IQ at age three years using the Stanford-Binet IQ test. Table 26.1 contains data on 38 infants.[1] Do children with higher crying counts tend to have higher IQs?

TABLE 26.1 INFANTS' CRYING (NUMBER OF PEAKS) AND IQ SCORES

CRYING	IQ	CRYING	IQ	CRYING	IQ	CRYING	IQ
10	87	20	90	17	94	12	94
12	97	16	100	19	103	12	103
9	103	23	103	13	104	14	106
16	106	27	108	18	109	10	109
18	109	15	112	18	112	23	113
15	114	21	114	16	118	9	119
12	119	12	120	19	120	16	124
20	132	15	133	22	135	31	135
16	136	17	141	30	155	22	157
33	159	13	162				

PLAN: Make a scatterplot. If the relationship appears linear, use correlation and regression to describe it. Finally, ask whether there is a *statistically significant* linear relationship between crying and IQ.

SOLVE (first steps): Chapters 4 and 5 introduced the data analysis that must come before inference. The first steps we take are a review of this data analysis. Figure 26.1 is a **scatterplot** of the crying data. Plot the explanatory variable (count of crying peaks) horizontally and the response variable (IQ) vertically. Look for the form, direction, and strength of the relationship as well as for outliers or other deviations. There is a moderately strong positive linear relationship, with no extreme outliers or potentially influential observations.

scatterplot

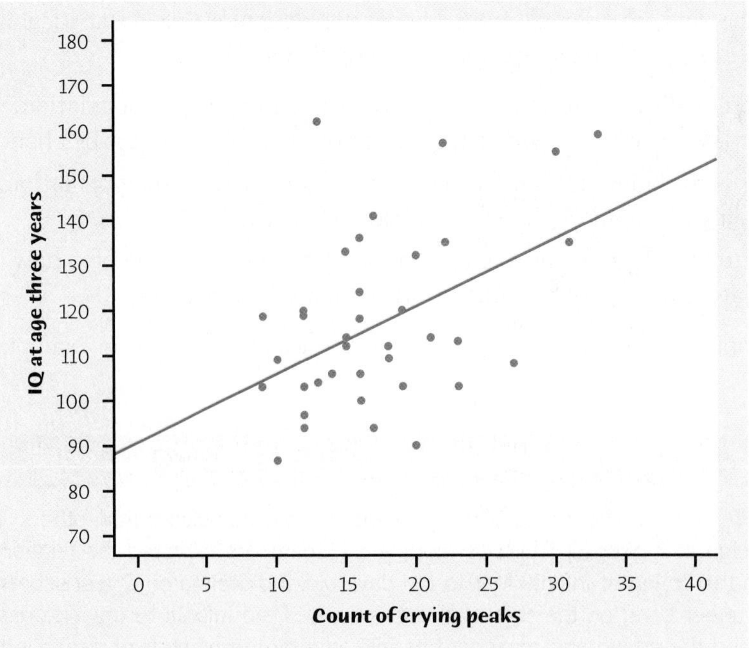

FIGURE 26.1
Scatterplot of the IQ score of infants at age three years against the intensity of their crying soon after birth, with the least-squares regression line, for Example 26.1.

Because the scatterplot shows a roughly linear (straight-line) pattern, the **correlation** describes the direction and strength of the relationship. The correlation between crying and IQ is $r = 0.455$. We are interested in predicting the response from information about the explanatory variable. So we find the **least-squares regression line** for predicting IQ from crying. The equation of the regression line is

correlation

least-squares line

$$\hat{y} = a + bx$$
$$= 91.27 + 1.493x$$

CONCLUDE (first steps): Children who cry more vigorously do tend to have higher IQs. Because $r^2 = 0.207$, only about 21% of the variation in IQ scores is explained by crying intensity. Prediction of IQ will not be very accurate. It is nonetheless impressive that behavior soon after birth can even partly predict IQ three years later. Is this observed relationship statistically significant? We must now develop tools for inference in the regression setting. ■

26.1 Conditions for regression inference

We can fit a regression line to *any* data relating two quantitative variables, though the results are useful only if the scatterplot shows a linear pattern. Statistical inference requires more detailed conditions. Because the conclusions of inference always concern some *population,* the conditions describe the population and how the data are produced from it. The slope b and intercept a of the least-squares line are *statistics.* That is, we calculated them from the sample data. These statistics would take somewhat different values if we repeated the study with different infants. To do inference, think of a and b as estimates of unknown *parameters* that describe the population of all infants.

Conditions for Regression Inference

We have n observations on an explanatory variable x and a response variable y. Our goal is to study or predict the behavior of y for given values of x.

- For any fixed value of x, the response y varies according to a **Normal distribution.** Repeated responses y are **independent** of each other.
- The mean response μ_y has a **straight-line relationship** with x given by a **population regression line**

$$\mu_y = \alpha + \beta x$$

- The slope β and intercept α are unknown parameters.
- The **standard deviation** of y (call it σ) is the same for all values of x. The value of σ is unknown.

There are thus three population parameters that we must estimate from the data: α, β, and σ.

These conditions say that in the population there is an "on the average" straight-line relationship between y and x. The population regression line $\mu_y = \alpha + \beta x$ says that the *mean* response μ_y moves along a straight line as the explanatory variable x changes. We can't observe the population regression line. The values of y that we do observe vary about their means according to a Normal distribution. If we hold x fixed and take many observations on y, the Normal pattern will eventually appear in a stemplot or histogram. In practice, we observe y

for many different values of x, so that we see an overall linear pattern formed by points scattered about the population line. The standard deviation σ determines whether the points fall close to the population regression line (small σ) or are widely scattered (large σ).

Figure 26.2 shows the conditions for regression inference in picture form. The line in the figure is the population regression line. The mean of the response y moves along this line as the explanatory variable x takes different values. The Normal curves show how y will vary when x is held fixed at different values. All the curves have the same σ, so the variability of y is the same for all values of x. You should check the conditions for inference when you do inference about regression. We will see later how to do that.

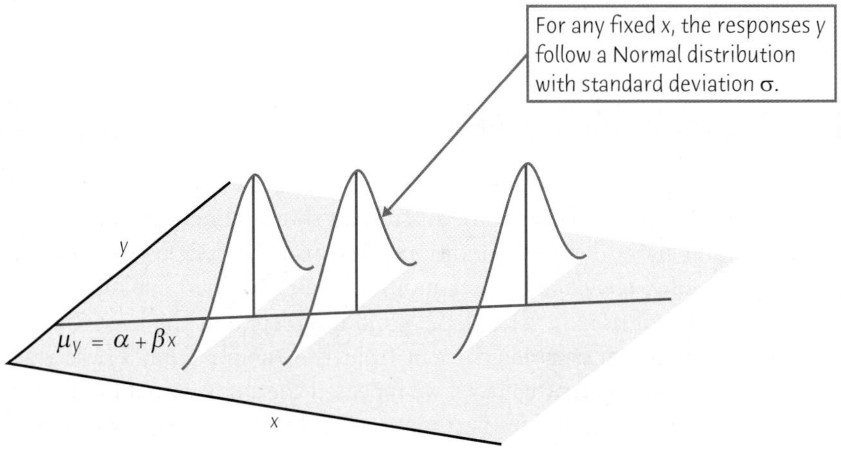
For any fixed x, the responses y follow a Normal distribution with standard deviation σ.
y
μy = α + βx
x

FIGURE 26.2
The nature of regression data when the conditions for inference are met. The line is the population regression line, which shows how the mean response μ_y changes as the explanatory variable x changes. For any fixed value of x, the observed response y varies according to a Normal distribution having mean μ_y and standard deviation σ.

26.2 Estimating the parameters

The first step in inference is to estimate the unknown parameters α, β, and σ.

Estimating the Population Regression Line

When the conditions for regression are met and we calculate the least-squares line $\hat{y} = a + bx$, the slope b of the least-squares line is an unbiased estimator of the population slope β, and the intercept a of the least-squares line is an unbiased estimator of the population intercept α.

EXAMPLE 26.2 Crying and IQ: Slope and Intercept

The data in Figure 26.1 satisfy the condition of scatter about an invisible population regression line reasonably well. The least-squares line is $\hat{y} = 91.27 + 1.493x$. The slope is particularly important. *A slope is a rate of change.* The population slope β says how much higher average IQ is for children with one more peak in their crying measurement. Because $b = 1.493$ estimates the unknown β, we estimate that, on the average, IQ is about 1.5 points higher for each added crying peak.

We need the intercept $a = 91.27$ to draw the line, but it has no statistical meaning in this example. No child had fewer than nine crying peaks, so we have no data near $x = 0$. We suspect that all typical children would cry when snapped with a rubber band, so that we will never observe $x = 0$. ■

The remaining parameter is the standard deviation σ, which describes the variability of the response y about the population regression line. The least-squares line estimates the population regression line. So the **residuals** estimate how much y varies about the population line. Recall that the residuals are the vertical deviations of the data points from the least-squares line:

residuals

$$\text{residual} = \text{observed } y - \text{predicted } y$$
$$= y - \hat{y}$$

There are n residuals, one for each data point. Because σ is the standard deviation of responses about the population regression line, we estimate it by a sample standard deviation of the residuals. We call this sample standard deviation the *regression standard error* to emphasize that it is estimated from data. The residuals from a least-squares line always have mean zero. That simplifies their standard error.

THE JINX!

Athletes are often jinxed. We read of "the rookie of the year jinx," the "cover of *Sports Illustrated* jinx," and many others. That is, athletes who are recognized for an outstanding performance often fail to do as well in the future. No, nature isn't retaliating against them. It's just random variation about their long-term mean performance. They were recognized because they randomly varied above their typical performance, and in the future they return to the mean or randomly vary down from it. If they randomly vary down, they can hope for a "comeback" award the next year.

Regression Standard Error

The **regression standard error** is

$$s = \sqrt{\frac{1}{n-2}\sum \text{residual}^2}$$
$$= \sqrt{\frac{1}{n-2}\sum (y - \hat{y})^2}$$

Use s to estimate the standard deviation σ of responses about the mean given by the population regression line.

Because we use the regression standard error so often, we just call it s. The quantity $\Sigma(y - \hat{y})^2$ is the sum of the squared deviations of the data points from the line. We average the squared deviations by dividing by $n - 2$, the number of data points less 2. It turns out that if we know $n - 2$ of the n residuals, the other two are determined. That is, $n - 2$ are the **degrees of freedom** of s. We first met the idea of degrees of freedom in the case of the ordinary sample standard deviation of n observations, which has $n - 1$ degrees of freedom. Now we observe two variables rather than one, and the proper degrees of freedom are $n - 2$ rather than $n - 1$.

degrees of freedom

Calculating s is unpleasant. You must find the predicted response for each x in your data set, then the residuals, and then s. In practice you will use software that does this arithmetic instantly. Nonetheless, here is an example to help you understand the standard error s.

EXAMPLE 26.3 Crying and IQ: Residuals and Standard Error

CRYINGR

Table 26.1 shows that the first infant studied had 10 crying peaks and a later IQ of 87. The predicted IQ for $x = 10$ is

$$\hat{y} = 91.27 + 1.493x$$
$$= 91.27 + 1.493(10) = 106.2$$

The residual for this observation is

$$\text{residual} = y - \hat{y}$$
$$= 87 - 106.2 = -19.2$$

That is, the observed IQ for this infant lies 19.2 points below the least-squares line on the scatterplot.

Repeat this calculation 37 more times, once for each subject. The 38 residuals are

−19.20	−31.13	−22.65	−15.18	−12.18	−15.15	−16.63	−6.18
−1.70	−22.60	−6.68	−6.17	−9.15	−23.58	−9.14	2.80
−9.14	−1.66	−6.14	−12.60	0.34	−8.62	2.85	14.30
9.82	10.82	0.37	8.85	10.87	19.34	10.89	−2.55
20.85	24.35	18.94	32.89	18.47	51.32		

Check the calculations by verifying that the sum of the residuals is zero. It is 0.04, not quite zero, because of roundoff error. Another reason to use software in regression is that roundoff errors in hand calculation can accumulate to make the results inaccurate.

The variance about the line is

$$s^2 = \frac{1}{n-2}\sum \text{residual}^2$$

$$= \frac{1}{38-2}\left[(-19.20)^2 + (-31.13)^2 + \cdots + (51.32)^2\right]$$

$$= \frac{1}{36}(11{,}023.3) = 306.20$$

Finally, the regression standard error is

$$s = \sqrt{306.20} = 17.50 \ ■$$

We will study several kinds of inference in the regression setting. The regression standard error s is the key measure of the variability of the responses in regression. It is part of the standard error of all the statistics we will use for inference.

LaunchPad Online Resources

- **The StatBoards video, *Fitting the Least-Squares Regression Line*, discusses fitting the least-squares regression line, including residuals and standard errors in an example.**
- **The StatClips Examples video, *Regression: Assumptions Example B*, examines the residuals in an example to assess the regression assumptions.**

Apply Your Knowledge

© Glowimages RM/age fotostock

26.1 Wine and Cancer in Women. Some studies have suggested that a nightly glass of wine may not only take the edge off a day but also improve health. Is wine good for your health? A study of nearly 1.3 million middle-aged British women examined wine consumption and the relative risk of breast cancer. The relative risk is the proportion of those in the study who drank a given amount of wine and who developed breast cancer divided by the proportion of nondrinkers in the study who developed breast cancer. For example, if 10% of the women in the study who drank 10 grams of wine developed breast cancer and 9% of nondrinkers in the study developed breast cancer, the relative risk of breast cancer for women drinking 10 grams of wine per day would be 10%/9% = 1.11. A relative risk greater than 1 indicates a greater proportion of drinkers in the study developed breast cancer than nondrinkers. Wine intake is the mean wine intake, in grams per day, of all women in the study who drank some wine but less than or equal to two drinks per week; who drank between three and six drinks per week; who drank between seven and 14 drinks per week; and who drank 15 or more drinks per week. Here are the data (for drinkers only):[2] CANCER

Wine intake (grams per day) (x)	2.5	8.5	15.5	26.5
Relative risk (y)	1.00	1.08	1.15	1.22

(a) Examine the data. Make a scatterplot with wine intake as the explanatory variable, and find the correlation. There is a strong linear relationship.

(b) Explain in words what the slope β of the population regression line would tell us if we knew it (note that these data represent averages over large numbers of women and are an example of an ecological correlation (see page 146), and one must be careful not to interpret the data as applying to individuals). Based on the data, what are the estimates of β and the intercept α of the population regression line?

(c) Calculate by hand the residuals for the four data points. Check that their sum is 0 (up to roundoff error). Use the residuals to estimate the standard deviation σ that measures variation in the responses (relative risk) about the means given by the population regression line. You have now estimated all three parameters.

26.3 Using technology

Basic "two-variable statistics" calculators will find the slope b and intercept a of the least-squares line from keyed-in data. Inference about regression requires in addition the regression standard error s. At this point, software or a graphing calculator that includes procedures for regression inference becomes almost essential for practical work.

Figure 26.3 shows regression output for the data of Table 26.1 from a graphing calculator, three statistical programs, and a spreadsheet program. When we entered the data into the programs, we called the explanatory variable "Crycount." The software outputs use that label. The graphing calculator just uses "x" and "y" to label the explanatory and response variables. You can locate the basic information in all the outputs. The regression slope is $b = 1.4929$, and the regression intercept is $a = 91.268$. The equation of the least-squares line is therefore (after rounding) just as given in Example 26.1. The regression standard error is $s = 17.4987$, and the squared correlation is $r^2 = 0.207$. Both of these results reflect the rather wide scatter of the points in Figure 26.1 about the least-squares line.

Each output contains other information, some of which we will need shortly and some of which we don't need. In fact, we left out some output to save space. Once you know what to look for, you can find what you want in almost any output and ignore what doesn't interest you.

FIGURE 26.3
Regression of IQ on crying peaks: output from a graphing calculator, three statistical programs, and a spreadsheet program.

Texas Instruments Graphing Calculator

```
LinRegTTest          LinRegTTest
 y=a+bx               y=a+bx
 β≠0 and ρ≠0          β≠0 and ρ≠0
 t=3.0655            ↑b=1.4929
 p=.0041              s=17.4987
 df=36.0000           r²=.2070
↓a=91.2683            r=.4550
```

JMP

Linear Fit

IQ = 91.268299 + 1.4928966*Crycount

Summary of Fit

RSquare	0.207
RSquare Adj	0.184972
Root Mean Square Error	17.49872
Mean of Response	117.2368
Observations (or Sum Wgts)	38

Lack of Fit

Analysis of Variance

Parameter Estimates

Term	Estimate	Std Error	t Ratio	Prob > |t|
Intercept	91.268299	8.934215	10.22	<.0001*
Crycount	1.4928966	0.487001	3.07	0.0041*

Minitab

Session

Regression Analysis: IQ versus Crycount

The regression equation is
IQ = 91.3 + 1.49 Crycount

Predictor	Coef	SE Coef	T	P
Constant	91.268	8.934	10.22	0.000
Crycount	1.4929	0.4870	3.07	0.004

S = 17.50 R-Sq = 20.7% R-Sq(adj) = 18.5%

Excel

Microsoft Excel - Book1

	A	B	C	D	E	F	G
1	SUMMARY OUTPUT						
2							
3	*Regression statistics*						
4	Multiple R	0.4550					
5	R Square	0.2070					
6	Adjusted R Square	0.1850					
7	Standard Error	17.4987					
8	Observations	38					
9							
10		*Coefficients*	*Standard Error*	*t Stat*	*P-value*	*Lower 95%*	*Upper 95%*
11	Intercept	91.2683	8.9342	10.2156	3.5E-12	73.1489	109.3877
12	Crycount	1.4929	0.4870	3.0655	0.004105	0.5052	2.4806
13							

Sheet4 / Sheet1 / Sheet2 / Sheet3

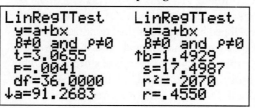

FIGURE 26.3
(Continued)

CrunchIt!

Results - Simple Linear Regression

Export ▾

Fitted Equation: IQ = 91.27 + 1.493 * Crycount

	Estimate	Std. Error	t value	Pr(>ItI)
(Intercept)	91.27	8.934	10.22	<0.0001
Crycount	1.493	0.4870	3.065	0.004105

r-Squared:	0.2070
Adjusted r-Squared:	0.1850
sigma:	17.50

Apply Your Knowledge

26.2 Introspection and Gray Matter. The ability to introspect about self-performance is key to human subjective experience. Accurate introspection requires discriminating correct decisions from incorrect ones, a capacity that varies substantially across individuals. Are individual differences in introspective ability reflected in the anatomy of brain regions responsible for this function? The following data are a measure of introspective ability (labeled Aroc and based on the performance of subjects on a task) and a measure of gray-matter volume (Brodmann area) in the anterior prefrontal cortex of the brain of 29 subjects.[3] DATA INTROSP

Aroc	0.55	0.58	0.59	0.59	0.59	0.61	0.62	0.63	0.63	0.63
Volume	59	62	43	63	83	61	55	57	57	67
Aroc	0.63	0.64	0.65	0.65	0.65	0.65	0.65	0.66	0.66	0.67
Volume	72	62	58	62	65	70	75	60	63	71
Aroc	0.67	0.67	0.68	0.69	0.70	0.70	0.71	0.72	0.75	
Volume	71	80	68	72	66	73	61	80	75	

We want to predict volume from Aroc. Figure 26.4 shows JMP regression output for these data.

(a) Make a scatterplot suitable for predicting volume from Aroc. What is the squared correlation r^2?

(b) For regression inference, we must estimate the three parameters α, β, and σ. From the output, what are the estimates of these parameters?

(c) What is the equation of the least-squares regression line of volume on Aroc? Add this line to your plot. We will continue the analysis of these data in later exercises.

26.3 Great Arctic Rivers. One effect of global warming is to increase the flow of water into the Arctic Ocean from rivers. Such an increase may have major effects on the world's climate. Six rivers (Yenisey, Lena, Ob, Pechora, Kolyma, and Severnaya Dvina) drain two-thirds of the Arctic in Europe and

FIGURE 26.4
JMP output for the introspective ability data, for Exercise 26.2.

Linear Fit

```
Volume = 10.065511 + 86.030829*Aroc
```

Summary of Fit

RSquare	0.200628
RSquare Adj	0.171022
Root Mean Square Error	7.908796
Mean of Response	65.89655
Observations (or Sum Wgts)	29

Lack of Fit

Analysis of Variance

Parameter Estimates

Term	Estimate	Std Error	t Ratio	Prob > |t|
Intercept	10.065511	21.49751	0.47	0.6434
Aroc	86.030829	33.04842	2.60	0.0148*

Asia. Several of these are among the largest rivers on earth. Table 26.2 presents the total discharge (amount of water flowing from these rivers) each year from 1936 to 2010.[4] Discharge is measured in cubic kilometers of water. Use software to analyze these data. DATA ARCTIC

(a) Make a scatterplot of river discharge against time. Is there a clear increasing trend? Calculate r^2 and briefly interpret its value. There is considerable year-to-year variation, so we wonder if the trend is statistically significant.

(b) As a first step, find the least-squares line and draw it on your plot. Then find the regression standard error s, which measures scatter about this line. We will continue the analysis in later exercises.

TABLE 26.2 ARCTIC RIVER DISCHARGE (CUBIC KILOMETERS), 1936 TO 2010

YEAR	DISCHARGE	YEAR	DISCHARGE	YEAR	DISCHARGE	YEAR	DISCHARGE
1936	1721	1955	1656	1974	2000	1993	1845
1937	1713	1956	1721	1975	1928	1994	1902
1938	1860	1957	1762	1976	1653	1995	1842
1939	1739	1958	1936	1977	1698	1996	1849
1940	1615	1959	1906	1978	2008	1997	2007
1941	1838	1960	1736	1979	1970	1998	1903
1942	1762	1961	1970	1980	1758	1999	1970
1943	1709	1962	1849	1981	1774	2000	1905
1944	1921	1963	1774	1982	1728	2001	1890
1945	1581	1964	1606	1983	1920	2002	2085
1946	1834	1965	1735	1984	1823	2003	1780
1947	1890	1966	1883	1985	1822	2004	1900
1948	1898	1967	1642	1986	1860	2005	1930
1949	1958	1968	1713	1987	1732	2006	1910
1950	1830	1969	1742	1988	1906	2007	2270
1951	1864	1970	1751	1989	1932	2008	2078
1952	1829	1971	1879	1990	1861	2009	1900
1953	1652	1972	1736	1991	1801	2010	1813
1954	1589	1973	1861	1992	1793		

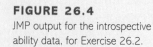

26.4 Testing the hypothesis of no linear relationship

Example 26.1 asked, "Do children with higher crying counts tend to have higher IQs?" Data analysis supports this conjecture. But is the positive association statistically significant? That is, is it too strong to occur often just by chance? To answer this question, test hypotheses about the slope β of the population regression line:

$$H_0\colon \beta = 0$$
$$H_a\colon \beta > 0$$

A regression line with slope 0 is horizontal. That is, the mean of y does not change at all when x changes. So H_0 says that there is *no linear relationship* between x and y in the population. Put another way, H_0 says that *linear regression of y on x is of no value for predicting y*.

The test statistic is just the standardized version of the least-squares slope b, using the hypothesized value $\beta = 0$ for the mean of b. It is another t statistic. Here are the details.

Significance Test for Regression Slope

To **test the hypothesis $H_0\colon \beta = 0$**, compute the t statistic

$$t = \frac{b}{\mathrm{SE}_b}$$

In this formula, the standard error of the least-squares slope b is

$$\mathrm{SE}_b = \frac{s}{\sqrt{\Sigma(x - \bar{x})^2}}$$

The sum runs over all observations on the explanatory variable x. In terms of a random variable T having the $t(n - 2)$ distribution, the P-value for a test of H_0 against

$H_a\colon \beta > 0$ is $P(T \geq t)$

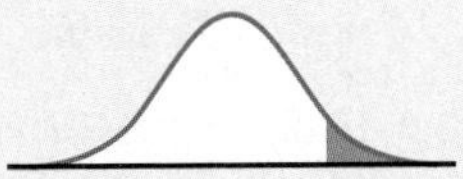

$H_a\colon \beta < 0$ is $P(T \leq t)$

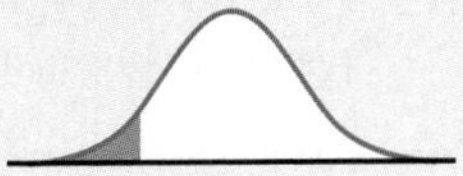

$H_a\colon \beta \neq 0$ is $2P(T \geq |t|)$

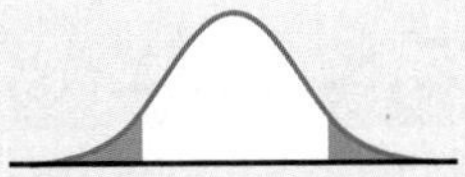

As advertised, the standard error of b is a multiple of the regression standard error s. The degrees of freedom $n - 2$ are the degrees of freedom of s. Although we give the formula for this standard error, you should not try to calculate it by hand. Regression software gives the standard error SE_b along with b itself.

EXAMPLE 26.4 Crying and IQ: Is the Relationship Significant?

The hypothesis H_0: $\beta = 0$ says that crying has no straight-line relationship with IQ. We conjecture that there is a positive relationship, so we use the one-sided alternative H_a: $\beta > 0$.

Figure 26.1 shows that there is a positive relationship, and from Figure 26.3 we see $b = 1.4929$ and $SE_b = 0.4870$. Thus,

$$t = \frac{b}{SE_b} = \frac{1.4929}{0.4870} = 3.07$$

so it is not surprising that all the outputs in Figure 26.3 give $t = 3.07$ with two-sided P-value 0.004. The P-value for the one-sided test is half of this, $P = 0.002$. There is very strong evidence that IQ increases as the intensity of crying increases. ■

Apply Your Knowledge

26.4 Wine and Cancer in Women. Exercise 26.1 gives data on daily wine consumption and the relative risk of breast cancer in women. Software tells us that the least-squares slope is $b = 0.009012$ with standard error $SE_b = 0.001112$.

(a) What is the t statistic for testing H_0: $\beta = 0$?

(b) How many degrees of freedom does t have? Use Table C to approximate the P-value of t against the one-sided alternative H_a: $\beta > 0$. What do you conclude?

26.5 Great Arctic Rivers: Testing. The most important question we ask of the data in Table 26.2 is this: is the increasing trend visible in your plot (Exercise 26.3) statistically significant? If so, changes in the Arctic may already be affecting the earth's climate. Use software to answer this question. Give a test statistic, its P-value, and the conclusion you draw from the test. DATA ARCTIC

26.6 Does Fast Driving Waste Fuel? Exercise 4.8 (page 109) gives data on the fuel consumption of a small car at various speeds from 10 to 150 kilometers per hour. Is there significant evidence of straight-line dependence between speed and fuel use? Make a scatterplot and use it to explain the result of your test. DATA FASTDR

26.5 Testing lack of correlation

The least-squares slope b is closely related to the correlation r between the explanatory and response variables x and y. In the same way, the slope β of the population regression line is closely related to the correlation between x and y in the population. In particular, the slope is 0 exactly when the correlation is 0.

Testing the null hypothesis H_0: $\beta = 0$ is therefore exactly the same as testing that there is *no correlation* between x and y in the population from which we drew our data. You can use the test for zero slope to test the hypothesis of zero correlation between any two quantitative variables. That's a useful trick.

Because correlation also makes sense when there is no explanatory–response distinction, it is handy to be able to test correlation without doing regression. Table E in the back of the book gives critical values of the sample correlation r under the null hypothesis that the correlation is 0 in the population. Use this table when both variables have at least approximately Normal distributions or when the sample size is large.

EXAMPLE 26.5 Testing Lack of Correlation

Figure 26.5 displays two scatterplots that we will use to illustrate testing lack of correlation and also to illustrate once again the need for formal statistical tests. On the left are data from an experiment on the healing of cuts in the limbs of newts. The data are the healing rates (micrometers per hour) for the two front limbs of 18 newts. The right-hand scatterplot shows the first- and second-round scores for the 92 golfers in the 2013 Masters Tournament. (There are fewer than 92 points because of duplicate scores.)

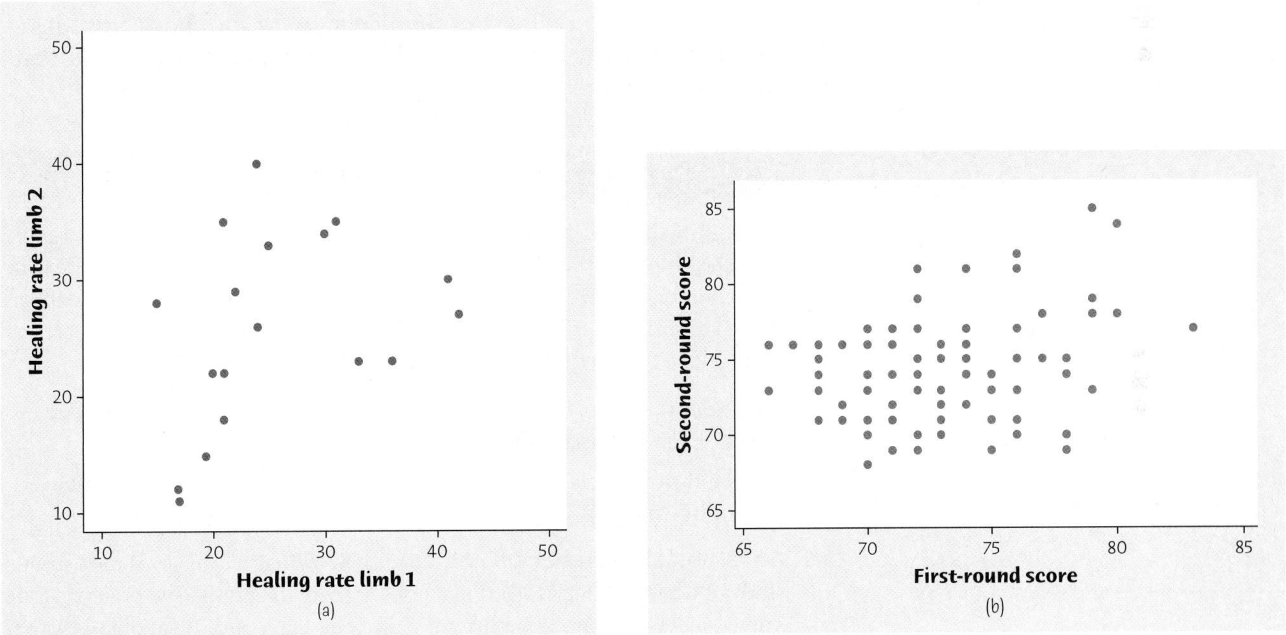

FIGURE 26.5
Two scatterplots for inference about the population correlation, for Example 26.5. (a) Healing rates for the two front limbs of 18 newts. (b) Scores on the first two rounds of the 2013 Masters Tournament.

We will test the hypotheses

$$H_0: \text{population correlation} = 0$$

$$H_a: \text{population correlation} \neq 0$$

for both sets of data. (The Masters scores are all whole numbers, but with $n = 92$ the robustness of t procedures allows their use.) Software gives

```
newts     r = 0.3581   t = 1.5342   P = 0.1445
masters   r = 0.2459   t = 2.41     P = 0.0181
```

The two-sided P-values for the t statistic for testing slope 0 are also the two-sided P-values for testing correlation 0.

Without software, compare the correlation $r = 0.3581$ for newts with the critical values in the $n = 18$ row of Table E. It falls between the table entries for one-tail probabilities 0.05 and 0.10, so the two-sided P-value lies between 0.10 and 0.20. For the Masters data, use the $n = 80$ row of Table E. (There is no table entry for sample size $n = 92$, so we use the next smaller sample size.) The two-sided P-value lies between 0.01 and 0.02. ■

The evidence for nonzero correlation is strong for the Masters scores ($t = 2.41$, $P = 0.0181$) but not for newts ($t = 1.53$, $P = 0.1445$). Yet the correlation for the newts is larger than that for the Masters, and the scatterplots suggest similar linear relationships for both. What happened? The larger sample size for the Masters data is largely responsible. The same r will have a smaller P-value for $n = 92$ than for $n = 18$. Our eyeball impression, even aided by calculating r, can't assess significance. We need the P-value from a formal test to guide us.

LaunchPad Online Resources

- The **StatBoards video**, *Tests of Significance for the Slope*, discusses the test of significance for the slope and its equivalence to the test of correlation in an example.

Apply Your Knowledge

26.7 Wine and Cancer in Women: Testing Correlation. Exercise 26.1 gives data showing that the risk of breast cancer increases linearly with daily wine consumption. There are only four observations, so we worry that the apparent relationship may be just chance. Is the correlation significantly greater than zero? Answer this question in two ways. CANCER

(a) Return to your t statistic from Exercise 26.4. What is the one-sided P-value for this t? Apply your result to test the correlation.

(b) Find the correlation r, and use Table E to approximate the P-value of the one-sided test.

26.8 Does Social Rejection Hurt? Exercise 4.46 (page 125) gives data from a study of whether social rejection causes activity in areas of the brain that are known to be activated by physical pain. The explanatory variable is a subject's score on a test of "social distress" after being excluded from an activity. The response variable is activity in an area of the brain that responds to physical pain. Your scatterplot (Exercise 4.46) shows a positive linear relationship. The research report gives the correlation r and the P-value for a test that r is greater than zero. What are r and the P-value? (You can use Table E or you can get more accurate P-values for the correlation from regression software.) What do you conclude about the relationship? REJECT

26.6 Confidence intervals for the regression slope

The slope β of the population regression line is usually the most important parameter in a regression problem. The slope is the rate of change of the mean response as the explanatory variable increases. We often want to estimate β. The slope b of the least-squares line is an unbiased estimator of β. A confidence interval is more useful because it shows how accurate the estimate b is likely to be. The confidence interval for β has the familiar form

$$\text{estimate} \pm t^* \text{SE}_{\text{estimate}}$$

Because b is our estimate, the confidence interval is $b \pm t^*\text{SE}_b$. Here are the details.

Confidence Interval for Regression Slope

A level C **confidence interval for the slope β** of the population regression line is

$$b \pm t^*\text{SE}_b$$

Here t^* is the critical value for the $t(n-2)$ density curve with area C between $-t^*$ and t^*. The formula for SE_b appears in the box on page 619.

EXAMPLE 26.6 Crying and IQ: Estimating the Slope

All the software outputs in Figure 26.3 give the slope $b = 1.4929$ (or $b = 1.493$), and three also give the standard error $\text{SE}_b = 0.4870$. The outputs giving both use a similar arrangement, a table in which each regression coefficient is followed by its standard error. Excel also gives the lower and upper endpoints of the 95% confidence interval for the population slope β, 0.505 and 2.481.

Once we know b and SE_b, it is easy to find the confidence interval. There are 38 data points, so the degrees of freedom are $n - 2 = 36$. Because Table C does not have a row for df = 36, we must use either software or the next smaller degrees of freedom in the table, df = 30. To use software, enter 36 degrees of freedom. For 95% confidence, enter the cumulative proportion 0.975 that corresponds to upper-tail area 0.025. Minitab gives

```
Student's t distribution with 36 DF
P( X <= x )        x
  0.975          2.02809
```

The 95% confidence interval for the population slope β is

$$\begin{aligned} b \pm t^*\text{SE}_b &= 1.4929 \pm (2.02809)(0.4870) \\ &= 1.4929 \pm 0.9877 \\ &= 0.505 \text{ to } 2.481 \end{aligned}$$

This agrees with Excel's result. We are 95% confident that mean IQ increases by between about 0.5 and 2.5 points for each additional peak in crying. ■

You can find a confidence interval for the intercept α of the population regression line in the same way, using a and SE_a from the "Constant" line of the Minitab output or the "Intercept" line in Excel, JMP, and CrunchIt!. We rarely need to estimate α.

LaunchPad Online Resources

- The **StatBoards video**, *Confidence Intervals for the Slope,* discusses confidence intervals for the slope in an example.

Apply Your Knowledge

26.9 Wine and Cancer in Women: Estimating Slope. Exercise 26.1 gives data on wine consumption and the risk of breast cancer. Software tells us that the least-squares slope is $b = 0.009012$ with standard error $SE_b = 0.001112$. Because there are only four observations, the observed slope b may not be an accurate estimate of the population slope β. Give a 90% confidence interval for β.

26.10 Introspection and Gray Matter: Estimating Slope. Exercise 26.2 gives data on introspective ability and gray-matter volume of the brains of subjects. We want a 95% confidence interval for the slope of the population regression line. Starting from the information in the JMP output in Figure 26.4, find this interval. Say in words what the slope of the population regression line tells us about the relationship between Aroc and gray-matter volume.

26.11 Great Arctic Rivers: Estimating Slope. Use the data in Table 26.2 to give a 90% confidence interval for the slope of the population regression of Arctic river discharge on year. Does this interval convince you that discharge is actually increasing over time? Explain your answer.
ARCTIC

26.7 Inference about prediction

One of the most common reasons to fit a line to data is to predict the response to a particular value of the explanatory variable. This is another setting for regression inference: we want not simply a prediction, but a prediction with a margin of error that describes how accurate the prediction is likely to be.

EXAMPLE 26.7 Beer and Blood Alcohol

STATE: The EESEE story "Blood Alcohol Content" describes a study in which 16 student volunteers at the Ohio State University drank a randomly assigned number of cans of beer. Thirty minutes later, a police officer measured their blood alcohol content (BAC) in grams of alcohol per deciliter of blood. Here are the data:[5]

Student	1	2	3	4	5	6	7	8
Beers	5	2	9	8	3	7	3	5
BAC	0.10	0.03	0.19	0.12	0.04	0.095	0.07	0.06
Student	9	10	11	12	13	14	15	16
Beers	3	5	4	6	5	7	1	4
BAC	0.02	0.05	0.07	0.10	0.085	0.09	0.01	0.05

© James Shaffer/PhotoEdit

The students were equally divided between men and women and differed in weight and usual drinking habits. Because of this variation, many students don't believe that number of drinks predicts blood alcohol well. Steve thinks he can drive legally 30 minutes after he finishes drinking five beers. The legal limit for driving is BAC 0.08 in all states. We want to predict Steve's blood alcohol content, using no information except that he drinks five beers.

PLAN: Regress BAC on number of beers. Use the regression line to predict Steve's BAC. Give a margin of error that allows us to have 95% confidence in our prediction.

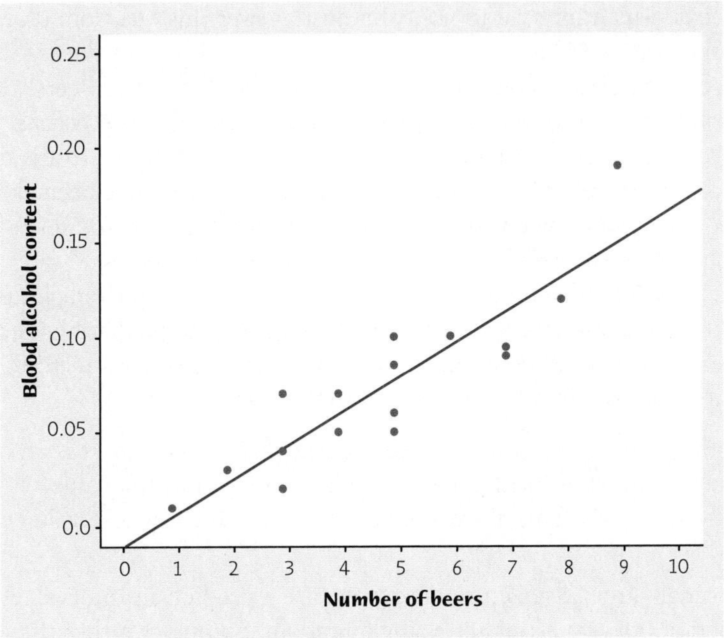

FIGURE 26.6
Scatterplot of students' blood alcohol content against the number of cans of beer consumed, with the least-squares regression line, for Example 26.7.

```
Session

Regression Analysis: BAC versus Beers

The regression equation is
BAC = -0.0127 + 0.0180 Beers

Predictor      Coef    SE Coef      T       P
Constant  -0.01270   0.01264  -1.00   0.332
Beers     0.017964  0.002402   7.48   0.000

S = 0.0204410   R-Sq = 80.0%   R-Sq(adj) = 78.6%

Predicted Values for New Observations

New
Obs      Fit    SE Fit          95% CI                95% PI
  1  0.07712  0.00513  (0.06612, 0.08812)  (0.03192, 0.12232)
```

FIGURE 26.7
Minitab regression output for the blood alcohol content data, for Example 26.7.

SOLVE: The scatterplot in Figure 26.6 and the regression output in Figure 26.7 show that student opinion is wrong: number of beers predicts blood alcohol content quite well. In fact, $r^2 = 0.80$, so that number of beers explains 80% of the observed variation in BAC. To predict Steve's BAC after five beers, use the equation of the regression line:

$$\begin{aligned}\hat{y} &= -0.0127 + 0.0180x \\ &= -0.0127 + 0.0180(5) = 0.077\end{aligned}$$

That's dangerously close to the legal limit of 0.08. What about 95% confidence? The "Predicted Values" part of the output in Figure 26.7 shows *two* 95% intervals. Which should we use? ■

IS REGRESSION GARBAGE?

No—but garbage can be the setting for regression. The U.S. Census Bureau once asked if weighing a neighborhood's garbage would help count its people. So 63 households had their garbage sorted and weighed. It turned out that pounds of plastic in the trash gave the best garbage prediction of the number of people in a neighborhood. The margin of error for a 95% prediction interval in a neighborhood of about 100 households, based on five weeks' worth of garbage, was about ±2.5 people. Alas, that is not accurate enough to help the U.S. Census Bureau.

To decide which interval to use, you must answer this question: do you want to predict the *mean* BAC for *all students* who drink five beers, or do you want to predict the BAC of *one individual student* who drinks five beers? *Both of these predictions may be interesting, but they are two different problems.* The actual prediction is the same, $\hat{y} = 0.077$. But the margin of error is different for the two kinds of prediction. Individual students who drink five beers don't all have the same BAC. So we need a larger margin of error to pin down Steve's result than to estimate the mean BAC for all students who have five beers.

CAUTION

Write the given value of the explanatory variable x as x^*. In Example 26.7, $x^* = 5$. The distinction between predicting a single outcome and predicting the mean of all outcomes when $x = x^*$ determines what margin of error is correct. To emphasize the distinction, we use different terms for the two intervals.

- To estimate the *mean* response, we use a *confidence interval.* It is an ordinary confidence interval for the mean response when x has the value x^*, which is $\mu_y = \alpha + \beta x^*$. This is a parameter, a fixed number whose value we don't know.
- To estimate an *individual* response y, we use a **prediction interval.** A prediction interval estimates a single random response y rather than a parameter like μ_y. The response y is not a fixed number. If we took more observations with $x = x^*$, we would get different responses.

prediction interval

EXAMPLE 26.8 Beer and Blood Alcohol: Conclusion

Steve is one individual, so we must use the prediction interval. The output in Figure 26.7 helpfully labels the confidence interval as "95% CI" and the prediction interval as "95% PI." We are 95% confident that Steve's BAC after five beers will lie between 0.032 and 0.122. The upper part of that range will get him arrested if he drives. The 95% confidence interval for the mean BAC of all students who drink five beers is much narrower, 0.066 to 0.088. ■

MAY THE LONGER NAME WIN!

Regression is far from perfect, but it beats most other ways of predicting. A writer in the early 1960s noted a simple method for predicting U.S. presidential elections: just choose the candidate with the longer name. In the 22 elections from 1876 to 1960, this method failed only once. Let's hope that the writer didn't bet the family silver on this idea. The 13 elections from 1964 to 2012 presented 11 tests of the "long name wins" method (the 1980 candidates and the 2000 candidates had names of the same length). The longer name lost eight of the 11.

The meaning of a prediction interval is very much like the meaning of a confidence interval. A 95% prediction interval, like a 95% confidence interval, is right 95% of the time in repeated use. "Repeated use" now means that we take an observation on y for each of the n values of x in the original data and then take one more observation y with $x = x^*$. Form the prediction interval from the n observations, then see if it covers the one more y. It will in 95% of all repetitions.

The interpretation of prediction intervals is a minor point. The main point is that it is harder to predict one response than to predict a mean response. Both intervals have the usual form

$$\hat{y} \pm t^*\text{SE}$$

but the prediction interval is wider than the confidence interval because individuals are more variable than averages. As a further illustration, consider a professional athlete such as the basketball player Kobe Bryant. You can find his career statistics online. During his career as a starter, his season scoring *average* ranges from 7.6 to 35.4 points per game, but individual game performances range from 0 points to a career high of 81 points. Individual game performances of professional athletes are more variable than season average performances. So the margin of error for predictions of individual game performances will be larger than for predictions of season averages. You will rarely need to know the details because software automates the calculation, but here they are.

Confidence and Prediction Intervals for Regression Response

A level C **confidence interval for the mean response μ_y** when x takes the value x^* is

$$\hat{y} \pm t^*\mathrm{SE}_{\hat{\mu}}$$

The standard error $\mathrm{SE}_{\hat{\mu}}$ is

$$\mathrm{SE}_{\hat{\mu}} = s\sqrt{\frac{1}{n} + \frac{(x^* - \bar{x})^2}{\Sigma(x - \bar{x})^2}}$$

A level C **prediction interval for a single observation y** when x takes the value x^* is

$$\hat{y} \pm t^*\mathrm{SE}_{\hat{y}}$$

The standard error for prediction $\mathrm{SE}_{\hat{y}}$ is

$$\mathrm{SE}_{\hat{y}} = s\sqrt{1 + \frac{1}{n} + \frac{(x^* - \bar{x})^2}{\Sigma(x - \bar{x})^2}}$$

In both intervals, t^* is the critical value for the $t(n - 2)$ density curve with area C between $-t^*$ and t^*.

There are two standard errors: $\mathrm{SE}_{\hat{\mu}}$ for estimating the mean response μ_y, and $\mathrm{SE}_{\hat{y}}$ for predicting an individual response y. The only difference between the two standard errors is the extra 1 under the square root sign in the standard error for prediction. The extra 1 makes the prediction interval wider. Both standard errors are multiples of the regression standard error s. The degrees of freedom are again $n - 2$, the degrees of freedom of s.

LaunchPad Online Resources

- The **StatBoards video**, *Confidence and Prediction Intervals,* uses an example to make clear the difference in these two intervals.
- The **Snapshots video**, *Regression Inference,* discusses inference for regression.

Apply Your Knowledge

26.12 Wine and Cancer in Women: Prediction. Exercise 26.1 gives data on wine consumption and the risk of breast cancer. For a new group of women who drink an average of 10 grams of wine per day, predict their relative risk of breast cancer.

(a) Figure 26.8 is part of the output from Minitab for prediction when $x^* = 10.0$. Which interval in the output is the proper 95% interval for predicting the relative risk?

(b) Minitab gives only one of the two standard errors used in prediction. It is $\mathrm{SE}_{\hat{\mu}}$, the standard error for estimating the mean response. Use this fact along with the output to give a 90% confidence interval for the mean relative risk of breast cancer in all women who drink an average of 10 grams of wine per day.

FIGURE 26.8
Partial Minitab output for regressing relative risk of breast cancer on mean daily intake of wine, for Exercise 26.12.

```
Session

Predictor        Coef      SE Coef        T        P
Constant      0.99309      0.01777    55.88    0.000
Intake       0.009012     0.001112     8.10    0.015


Predicted Values for New Observations

New
Obs      Fit   SE Fit          95% CI                  95% PI
  1 1.08321  0.01057  (1.03775, 1.12868)(0.98643, 1.18000)
```

26.13 Introspection and Gray Matter: Prediction. Analysis of the data in Exercise 26.2 shows that the relationship of introspective ability, as measured by Aroc, and gray-matter volume is roughly linear. We might want to predict the mean gray-matter volume of a person with an Aroc of 0.65. Here is the JMP output for prediction when $x^* = 0.65$:

	Aroc	Volume	Predicted Volume	StdErr Pred Volume	Lower 95% Mean Volume	Upper 95% Mean Volume	StdErr Indiv Volume	Lower 95% Indiv Volume	Upper 95% Indiv Volume
1	0.65	•	65.985549133	1.4690245109	62.971359812	68.999738454	8.044072016	49.480476694	82.490621572

(a) Use the regression line from Figure 26.4 to verify that "Predicted volume" is the predicted value for $x^* = 0.65$. (Start with the results in the "Estimate" column of Figure 26.4 to reduce roundoff error.)

(b) What is the 95% interval we want?

26.8 Checking the conditions for inference

You can fit a least-squares line to any set of explanatory-response data when both variables are quantitative. If the scatterplot doesn't show a roughly linear pattern, the fitted line may be almost useless. But it is still the line that fits the data best in the least-squares sense. To use regression inference, however, the data must satisfy additional conditions. *Before you can trust the results of inference, you must check the conditions for inference one by one.* There are ways to deal with violations of any of the conditions. If you see a clear violation, get expert advice.

CAUTION

Although the conditions for regression inference are a bit elaborate, it is not hard to check for major violations. The conditions involve the population regression line and the deviations of responses from this line. We can't observe the population line, but the least-squares line estimates it and the residuals estimate the deviations from the population line. *You can check all the conditions for regression inference by looking at graphs of the residuals.* This is what we recommend in practice (and a failure to do so can sometimes lead to unjustified conclusions), and most regression software will calculate and save the residuals for you. Start by making a stemplot or histogram of the residuals and also a **residual plot**, a plot of the residuals against the explanatory variable x, with a horizontal line at the "residual = 0" position. The "residual = 0"

residual plot

line represents the position of the least-squares line in the scatterplot of y against x. Let's look at each condition in turn.

- **The relationship is linear in the population.** Look for curved patterns or other departures from a straight-line overall pattern in the residual plot. You can also use the original scatterplot, but the residual plot magnifies any effects.
- **The response varies Normally about the population regression line.** Because different y-values usually come from different x-values, the responses themselves need not be Normal. It is the deviations from the population line—estimated by the residuals—that must be Normal. Check for clear skewness or other major departures from Normality in your stemplot or histogram of the residuals.
- **Observations are independent.** In particular, repeated observations on the same individual are not allowed. You should not use ordinary regression to make inferences about the growth of a single child over time, for example. Signs of dependence in the residual plot are a bit subtle, so we usually rely on common sense.
- **The standard deviation of the responses is the same for all values of x.** Look at the scatter of the residuals above and below the "residual = 0" line in the residual plot. The scatter should be roughly the same from one end to the other. You will sometimes find that, as the response y gets larger, so does the scatter of the residuals. Rather than remaining fixed, the standard deviation σ about the line changes with x as the mean response changes with x. There is no fixed σ for s to estimate. You cannot trust the results of inference when this happens.

You will always see some irregularity when you look for Normality and fixed standard deviation in the residuals, especially when you have few observations. Don't overreact to minor violations of the conditions. Like other t procedures, inference for regression is (with one exception) not very sensitive to lack of Normality, especially when we have many observations. Do beware of influential observations, which can greatly affect the results of inference.

The exception is the prediction interval for a single response y. This interval relies on Normality of individual observations, not just on the approximate Normality of statistics like the slope a and intercept b of the least-squares line. The statistics a and b become more Normal as we take more observations. This contributes to the robustness of regression inference, but it isn't enough for the prediction interval. We will not study methods that carefully check Normality of the residuals, so *you should regard prediction intervals as rough approximations.*

CAUTION

EXAMPLE 26.9 Climate Change Chases Fish North

STATE: As the climate grows warmer, we expect many animal species to move toward the poles in an attempt to maintain their preferred temperature range. Do data on fish in the North Sea confirm this expectation? Table 26.3 gives data for 25 years on mean winter temperatures at the bottom of the North Sea (degrees Celsius) and the center of the distribution of anglerfish in degrees of north latitude.[6]

DATA ANGFISH

PLAN: Regress latitude on temperature. Look for a positive linear relationship and assess its significance. Be sure to check the conditions for regression inference.

SOLVE: The scatterplot in Figure 26.9 shows a clear positive linear relationship. The solid line in the plot is the least-squares regression line of the center of the fish distribution (north latitude) on winter ocean temperature. Software shows that the slope is $b = 0.818$. That is, each degree of ocean warming moves the fish about

© Jurgen Freund/naturepl.com

TABLE 26.3 WINTER TEMPERATURE (°C) AND ANGLERFISH LATITUDE, 1977 TO 2001

YEAR	TEMP.	LATITUDE	YEAR	TEMP.	LATITUDE	YEAR	TEMP.	LATITUDE
1977	6.26	57.20	1986	6.52	57.72	1994	7.02	58.71
1978	6.26	57.96	1987	6.68	57.83	1995	7.09	58.07
1979	6.27	57.65	1988	6.76	57.87	1996	7.13	58.49
1980	6.31	57.59	1989	6.78	57.48	1997	7.15	58.28
1981	6.34	58.01	1990	6.89	58.13	1998	7.29	58.49
1982	6.32	59.06	1991	6.90	58.52	1999	7.34	58.01
1983	6.37	56.85	1992	6.93	58.48	2000	7.57	58.57
1984	6.39	56.87	1993	6.98	57.89	2001	7.65	58.90
1985	6.42	57.43						

0.8 degree of latitude farther north. The t statistic for testing H_0: $\beta = 0$ is $t = 3.6287$ with one-sided P-value $P = 0.0007$. There is very strong evidence that the population slope is positive, $\beta > 0$.

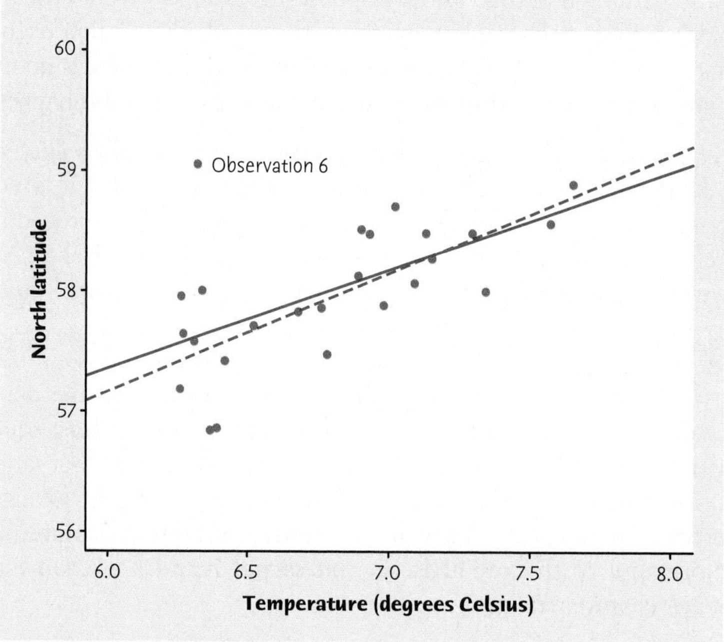

FIGURE 26.9
Scatterplot of the latitude of the center of the distribution of anglerfish in the North Sea against mean winter temperature at the bottom of the sea, for Example 26.9. The two regression lines are for the data with (solid) and without (dashed) Observation 6.

CONCLUDE: The data give highly significant evidence that anglerfish have moved north as the ocean has grown warmer. Before relying on this conclusion, we must check the conditions for inference. ■

The software that did the regression calculations also finds the 25 residuals. In the same order as the observations in Example 26.9, they are

```
-0.3731   0.3869   0.0687  -0.0240   0.3714   1.4378  -0.8131
-0.8095  -0.2740  -0.0658  -0.0867  -0.1121  -0.5185   0.0415
 0.4234   0.3588  -0.2721   0.5152  -0.1821   0.2052  -0.0211
 0.0743  -0.4466  -0.0747   0.1899
```

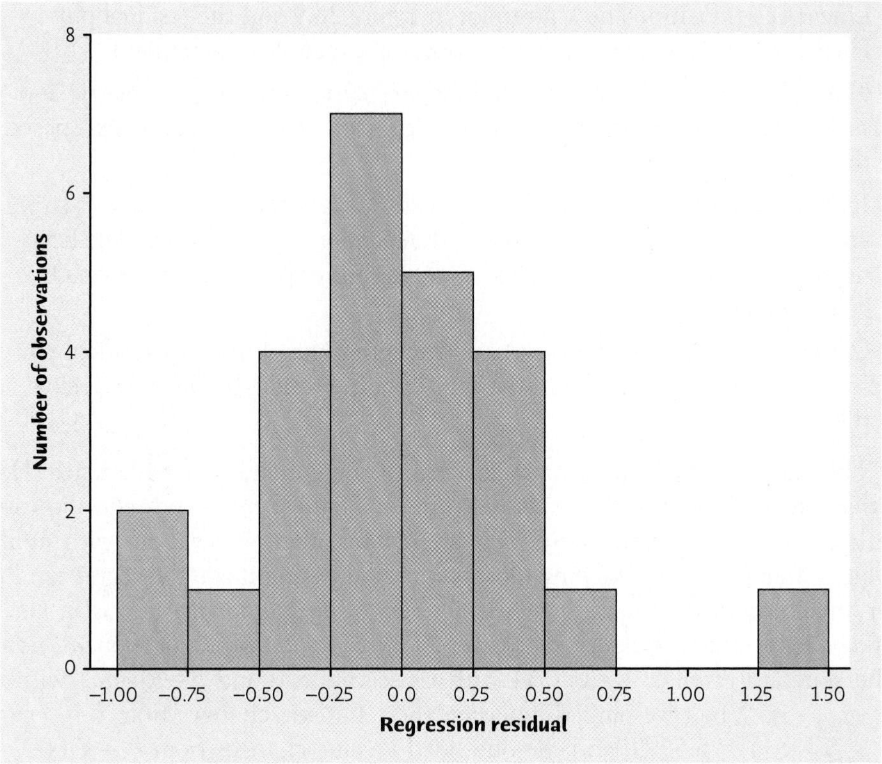

FIGURE 26.10
Histogram of the residuals from the regression of latitude on temperature in Example 26.9.

Begin by making two graphs of the residuals. Figure 26.10 is a histogram of the residuals. Figure 26.11 is the residual plot, a plot of the residuals against the explanatory variable, sea-bottom temperature. The "residual = 0" line marks the position of the regression line. Patterns in residual plots are often easier to see if you use a wider vertical scale than your software's default plot, and we suggest you do so if possible. Both graphs show that Observation 6 is a high outlier. Let's check the conditions for regression inference.

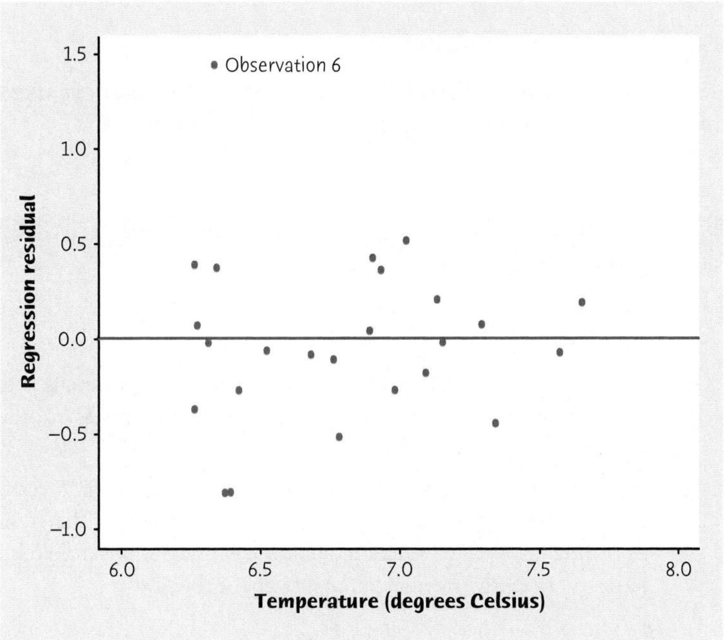

FIGURE 26.11
Residual plot for the regression of latitude on temperature in Example 26.9.

- **Linear relationship.** The scatterplot in Figure 26.9 and the residual plot in Figure 26.11 both show a linear relationship except for the outlier.
- **Normal residuals.** The histogram in Figure 26.10 is roughly symmetric and single peaked. There are no important departures from Normality except for the outlier.
- **Independent observations.** The observations were taken a year apart, so we are willing to regard them as close to independent. The residual plot shows no obvious pattern of dependence, such as runs of points all above or all below the line.
- **Constant standard deviation.** Again excepting the outlier, the residual plot shows no unusual variation in the scatter of the residuals above and below the line as x varies.

The outlier is the only serious violation of the conditions for inference. How influential is the outlier? The dashed line in Figure 26.9 is the regression line without Observation 6. Because there are several other observations with similar values of temperature, dropping Observation 6 does not move the regression line very much. *Even though the outlier is not very influential for the regression line, it influences regression inference because of its effect on the regression standard error.* The standard error is $s = 0.4734$ with Observation 6 and $s = 0.3622$ without it. When we omit the outlier, the t statistic changes from $t = 3.6287$ to $t = 5.5599$, and the one-sided P-value changes from $P = 0.0007$ to $P < 0.00001$. Fortunately, the outlier does not affect the conclusion we drew from the data. Dropping Observation 6 makes the test for the population slope *more* significant and *increases* the percent of variation in fish location explained by ocean temperature.

CAUTION

One more caution about inference in this example: as usual in an observational study, the possibility of lurking variables makes us hesitant to conclude that rising temperature is *causing* anglerfish to move north. Ocean temperature was steadily rising during these years. The effect on fish latitude of any lurking variable that increased over time—perhaps increased commercial fishing—is confounded with the effect of temperature.

LaunchPad Online Resources

- The **StatBoards video**, *Checking Conditions for Inference,* uses an example to illustrate checking conditions for inference.

Apply Your Knowledge

26.14 Crying and IQ: Residuals. The residuals for the study of crying and IQ appear in Example 26.3. DATA CRYINGR

(a) Make a stemplot to display the distribution of the residuals (round to the nearest whole number). Are there strong outliers or other signs of departures from Normality?

(b) Make a residual plot of residuals against crying peaks. Try a vertical scale of −60 to 60 to show patterns more clearly. Draw the "residual = 0" line. Does the residual plot show clear deviations from a linear pattern or clearly unequal variability about the line?

(c) Using the information given in Example 26.1, explain why the 38 observations are independent.

26.15 Introspection and Gray Matter: Residuals. Figure 26.4 gives part of the JMP output for the data on introspective ability and gray-matter volume in Exercise 26.2. Figure 26.12 comes from another part of the output. It gives x, y, the predicted response $\hat{y}$, the residual $y - \hat{y}$, and related quantities for each of the 29 observations. Most statistical software provides similar output. Examine the conditions for regression inference one by one. This example illustrates mild violations of the conditions that did not prevent the researchers from doing inference. DATA INTRSPR

(a) **Linear relationship.** Your scatterplot and r^2 from Exercise 26.2 show that the relationship is roughly linear. Plot the residuals against Aroc. Are any deviations from a straight line apparent?

(b) **Normal variation about the line.** Make a stemplot of the residuals (round to the third decimal place and don't forget that −0.00 and 0.00 are separate stems). With only 29 observations, a small amount of skew is not disturbing. Does your plot indicate that any of the observations are outliers? If so, which?

(c) **Independent observations.** The data come from 29 different subjects who were each measured separately. Can you conclude that the observations are independent?

(d) **Variability about the line stays the same.** Is there any evidence that the variability may be larger at one end?

	Aroc	Volume	Predicted Volume	StdErr Pred Volume	Residual Volume
1	0.55	59	57.382466281	3.5852535563	1.6175337187
2	0.58	62	59.963391137	2.7113878012	2.0366088632
3	0.59	43	60.823699422	2.4401564173	-17.82369942
4	0.59	63	60.823699422	2.4401564173	2.176300578
5	0.59	83	60.823699422	2.4401564173	22.176300578
6	0.61	61	62.544315992	1.9532437971	-1.544315992
7	0.62	55	63.404624277	1.7530601213	-8.404624277
8	0.63	57	64.264932563	1.5967836073	-7.264932563
9	0.63	57	64.264932563	1.5967836073	-7.264932563
10	0.63	67	64.264932563	1.5967836073	2.7350674374
11	0.63	72	64.264932563	1.5967836073	7.7350674374
12	0.64	62	65.125240848	1.4982174843	-3.125240848
13	0.65	58	65.985549133	1.4690245109	-7.985549133
14	0.65	62	65.985549133	1.4690245109	-3.985549133
15	0.65	65	65.985549133	1.4690245109	-0.985549133
16	0.65	70	65.985549133	1.4690245109	4.0144508671
17	0.65	75	65.985549133	1.4690245109	9.0144508671
18	0.66	60	66.845857418	1.5132250456	-6.845857418
19	0.66	63	66.845857418	1.5132250456	-3.845857418
20	0.67	71	67.706165703	1.6248405171	3.2938342967
21	0.67	71	67.706165703	1.6248405171	3.2938342967
22	0.67	80	67.706165703	1.6248405171	12.293834297
23	0.68	68	68.566473988	1.7913132096	-0.566473988
24	0.69	72	69.426782274	1.9989844836	2.5732177264
25	0.7	66	70.287090559	2.2364066174	-4.287090559
26	0.7	73	70.287090559	2.2364066174	2.7129094412
27	0.71	61	71.147398844	2.4951011589	-10.14739884
28	0.72	80	72.007707129	2.7691126141	7.9922928709
29	0.75	75	74.588631985	3.6477372643	0.4113680154

FIGURE 26.12 Residuals from JMP for Exercise 26.15. The table gives the predicted volume and the residual for each observation.

CHAPTER 26 SUMMARY

Chapter Specifics

- **Least-squares regression** fits a straight line to data to predict a response variable y from an explanatory variable x. Inference about regression requires more conditions.
- The **conditions for regression inference** say that there is a **population regression line** $\mu_y = \alpha + \beta x$ that describes how the mean response varies as x changes. The observed response y for any x has a Normal distribution with mean given by the population regression line and with the same standard deviation σ for any value of x. Observations on y are independent.
- The **parameters to be estimated** are the intercept α and the slope β of the population regression line and also the standard deviation σ. The slope a and intercept b of the least-squares line estimate α and β. Use the **regression standard error s** to estimate σ.
- The regression standard error s has $n - 2$ **degrees of freedom.** All t procedures for regression inference have $n - 2$ degrees of freedom.
- To test **the hypothesis that the slope is zero in the population,** use the t statistic $t = b/\text{SE}_b$. This null hypothesis says that straight-line dependence on x has no value for predicting y. In practice, use software to find the slope b of the least-squares line, its standard error SE_b, and the t statistic.
- The t test for regression slope is also a test for **the hypothesis that the population correlation between x and y is zero.** To do this test without software, use the sample correlation r and Table E.
- **Confidence intervals for the slope** of the population regression line have the form $b \pm t^*\text{SE}_b$.
- **Confidence intervals for the mean response** when x has value x^* have the form $\hat{y} \pm t^*\text{SE}_{\hat{\mu}}$. **Prediction intervals** for an individual future response y have a similar form with a larger standard error, $\hat{y} \pm t^*\text{SE}_{\hat{y}}$. Software often gives these intervals.

STATISTICS IN SUMMARY

Here are the most important skills you should have acquired from reading this chapter.

A. Preliminaries

1. Make a scatterplot to show the relationship between an explanatory and a response variable.
2. Use software or a calculator to find the correlation and the equation of the least-squares regression line.
3. Recognize which type of inference you need in a particular regression setting.

B. Inference Using Software Output

1. Explain in any specific regression setting the meaning of the slope β of the population regression line.
2. Understand software output for regression. Find in the output the slope and intercept of the least-squares line, their standard errors, and the regression standard error.
3. Use that information to carry out tests of H_0: $\beta = 0$ and calculate confidence intervals for β.
4. Explain the distinction between a confidence interval for the mean response and a prediction interval for an individual response.
5. If software gives output for prediction, use that output to give either confidence or prediction intervals.

C. **Checking the Conditions for Regression Inference**

1. Make a stemplot or histogram of the residuals and look for strong departures from Normality.
2. Make a residual plot and look for departures from a linear pattern or unequal variability about the "residual = 0" line.
3. Ask whether the study design suggests that observations are independent.

Link It

In Chapters 4 and 5, we studied scatterplots, correlation, and the least-squares regression line as methods for exploring data. In Chapter 5, we mentioned that we must exercise caution in how we interpret any relationships we observe through such exploratory analyses. We also mentioned that such interpretations rest on the assumption that the relationship is valid in some broader sense. And we promised to explore this more carefully later in this book. In this chapter, we did so by considering inference for regression. Inference for regression allows us to determine whether the relationship we observe in a scatterplot is valid for some larger population. It also allows us to attach margins of error to our estimates of the slope and intercept, as well as to make predictions based on the least-squares regression line.

In Chapters 4 and 5, we discussed several cautions about correlation and the least-squares regression line. These same cautions apply to inference for regression. We discussed a systematic approach using the residuals and what we know of the study design to determine if the assumptions behind inference for regression are reasonable.

This chapter considers inference for relationships between two quantitative variables. Chapter 25 considered inference for relationships between two categorical variables. In the next chapter, we consider inference for relationships between a quantitative and a categorical variable—in particular, for deciding whether the mean of a response is the same for more than two categories.

LaunchPad Online Resources

If you are having difficulty with any of the sections of this chapter, these online resources should help prepare you to solve the exercises at the end of this chapter.

- **StatTutor** starts with a video review of each section and asks a series of questions to check your understanding. You may find this helpful if you need additional help understanding any of the material in this chapter.
- **LearningCurve** provides you with a series of questions about the chapter geared to your level of understanding.

CHECK YOUR SKILLS

Florida reappraises real estate every year, so the county appraiser's website lists the current "fair market value" of each piece of property. Property usually sells for somewhat more than the appraised market value. Here are the appraised market values and actual selling prices (in thousands of dollars) of condominium units sold in a beachfront building in a 137-month period between 2003 and 2014:[7]

© Franz Marc Frei/Corbis

SELLING PRICE	APPRAISED VALUE	MONTH	SELLING PRICE	APPRAISED VALUE	MONTH	SELLING PRICE	APPRAISED VALUE	MONTH
825	626	0	1164	953	59	1083	601	101
590	492	1	1425	922	64	1200	904	109
1075	930	3	1865	1190	64	650	505	110
890	790	9	1450	610	64	1000	877	113
845	648	13	875	806	64	1100	895	118
1100	942	14	1510	1241	70	549	481	119
715	345	15	1375	813	70	1405	822	120
1325	1032	19	560	496	73	1000	870	121
700	556	21	1050	774	79	1200	752	124
1322	879	26	605	470	86	1100	678	125
1900	1016	26	675	545	88	955	591	129
1600	1040	28	693	690	93	1200	888	129
980	442	34	360	690	94	900	850	129
940	771	37	1455	1132	96	825	516	134
850	715	47	1068	867	99	886	626	137
1100	997	54	1235	1009	100			

Here is part of the Minitab output for regressing selling price on appraised value, along with prediction for a unit with appraised value $800,000:

```
Predictor    Coef  SE Coef     T      P
Constant    111.9    123.2  0.91  0.368
appraised  1.2103   0.1540  7.86  0.000

S = 221.341  R-Sq = 57.9%  R-Sq(adj) = 56.9%

Predicted Values for New Observations

New
Obs     Fit  SE Fit       95% CI             95% PI
  1  1080.1    32.6  (1014.5, 1145.7)  (629.5, 1530)
```

Exercises 26.16 to 26.24 are based on this information.

26.16 The equation of the least-squares regression line for predicting selling price from appraised value is

(a) price = 1.2103 + 111.9 × appraised value.
(b) price = 111.9 + 1.2103 × appraised value.
(c) price = 123.2 + 0.1540 × appraised value.

26.17 What is the correlation between selling price and appraised value?

(a) 0.761 (b) 0.579 (c) 0.569

26.18 The slope β of the population regression line describes

(a) the average selling price in a population of units when a unit's appraised value is 0.
(b) the average increase in selling price in a population of units when appraised value increases by $1000.
(c) the exact increase in the selling price of an individual unit when its appraised value increases by $1000.

26.19 Is there significant evidence that selling price increases as appraised value increases? To answer this question, test the hypotheses

(a) H_0: $\beta = 0$ versus H_a: $\beta > 0$.
(b) H_0: $\beta = 0$ versus H_a: $\beta \neq 0$.
(c) H_0: $\alpha = 0$ versus H_a: $\alpha > 0$.

26.20 Minitab shows that the P-value for this test is

(a) 0.368.
(b) 0.1540
(c) less than 0.001.

26.21 The regression standard error for these data is

(a) 0.1540.
(b) 123.2.
(c) 221.341.

26.22 Confidence intervals and tests for these data use the t distribution with degrees of freedom

(a) 47. (b) 46. (c) 45.

26.23 A 95% confidence interval for the population slope β is

(a) 1.2103 ± 447.3302.
(b) 1.2103 ± 0.3112.
(c) 1.2103 ± 0.1540.

26.24 Louisa owns a unit in this building appraised at \$800,000. The Minitab output includes prediction for this appraised value. She can be 95% confident that her unit would sell for between

(a) \$629,500 and \$1,530,000.
(b) \$1,014,500 and \$1,145,700.
(c) \$1,080,100 and \$1,112,700.

CHAPTER 26 EXERCISES

26.25 Genetically engineered cotton. A strain of genetically engineered cotton, known as Bt cotton, is resistant to certain insects, which results in larger yields of cotton. Farmers in northern China have increased the number of acres planted in Bt cotton. Because Bt cotton in resistant to certain pests, farmers have also reduced their use of insecticide. Scientists in China were interested in the long-term effects of Bt cotton cultivation and decreased insecticide use on insect populations that are not affected by Bt cotton. One such insect is the mirid bug. Scientists measured the number of mirid bugs per 100 plants and the proportion of Bt cotton planted at 38 locations in northern China for the 12-year period from 1997 to 2008. The scientists reported a regression analysis as follows:[8]

© Softdreams/Dreamstime.com

number of mirid bugs per 100 plants

$$= 0.54 + 6.81 \times \text{Bt cotton planting proportion}$$

$r^2 = 0.90, \quad P < 0.0001$

(a) What does the slope $b = 6.81$ say about the relation between Bt cotton planting proportion and number of mirid bugs per 100 plants?
(b) What does $r^2 = 0.90$ add to the information given by the equation of the least-squares line?
(c) What null and alternative hypotheses do you think the P-value refers to? What does this P-value tell you?
(d) Does the large value of r^2 and the small P-value indicate that increasing the proportion of acres planted in Bt cotton causes an increase in mirid bugs?

Exercise 7.51 (page 197) gives data from a study of the "gate velocity" of molten metal that experienced foundry workers choose based on the thickness of the aluminum piston being cast. Gate velocity is measured in feet per second, and the piston wall thickness is in inches. A scatterplot (you need not make one) shows no outliers and a moderately strong positive linear relationship. Figure 26.13 displays part of the Minitab regression output. Exercises 26.26 to 26.28 analyze these data.

26.26 Casting aluminum: Is there a relationship? Figure 26.13 leaves out the t statistics and their P-values. Based on the information in the output, test the hypothesis that there is no straight-line relationship between thickness and gate velocity. State hypotheses, give a test statistic and its approximate P-value, and state your conclusion in the context of the problem.

26.27 Casting aluminum: intervals. The output in Figure 26.13 includes prediction for piston wall thickness $x^* = 0.5$ inch. Use the output to give 90% intervals for

(a) the slope of the population regression line of gate velocity on piston thickness.
(b) the average gate velocity for a type of piston with thickness 0.5 inch.

26.28 Casting aluminum: residuals. The output in Figure 26.13 includes a table of the x and y variables, the fitted values $\hat{y}$ for each x, the residuals, and some related quantities. DATA ALUMRES

(a) Plot the residuals against thickness (the explanatory variable). Use vertical scale -200 to 200 so that the pattern is clearer. Add the "residual = 0" line. Does your plot show a systematically nonlinear relationship? Does it show systematic change in the variability about the regression line?
(b) Make a histogram of the residuals. Minitab identifies the residual for Observation 9 as a suspected outlier. Does your histogram agree?

FIGURE 26.13
Minitab output for the regression of gate velocity on piston thickness in casting aluminum parts, for Exercises 26.26 to 26.28.

Session

Regression Analysis: Veloc versus Thick

```
The regression equation is
Veloc = 70.4 + 275 Thick


Predictor      Coef   SE Coef        T         P
Constant      70.44     52.90
Thick        274.78     88.18

S = 56.3641    R-Sq = 49.3%    R-Sq(adj) = 44.2%

Obs   Thick   Veloc     Fit  SE Fit  Residual  St Resid
  1   0.248   123.8   138.6    32.8     -14.8     -0.32
  2   0.359   223.9   169.1    24.8      54.8      1.08
  3   0.366   180.9   171.0    24.3       9.9      0.19
  4   0.400   104.8   180.3    22.2     -75.5     -1.46
  5   0.524   228.6   214.4    16.8      14.2      0.26
  6   0.552   223.8   222.1    16.4       1.7      0.03
  7   0.628   326.2   243.0    17.0      83.2      1.55
  8   0.697   302.4   262.0    19.7      40.4      0.77
  9   0.697   145.2   262.0    19.7    -116.8     -2.21R
 10   0.752   263.1   277.1    22.8     -14.0     -0.27
 11   0.806   302.4   291.9    26.4      10.5      0.21
 12   0.821   302.4   296.0    27.4       6.4      0.13

R denotes an observation with a large standardized residual.

Predicted Values for New Observations

New
Obs     Fit    SE Fit         90% CI              90% PI
  1   207.8      17.4   (176.2, 239.4)    (100.9, 314.8)
```

(c) Redoing the regression without Observation 9 gives regression standard error $s = 42.4725$ and predicted mean velocity 216 feet per second (90% confidence interval 191.4 to 240.6) for piston walls 0.5 inch thick. Compare these values with those in Figure 26.13. Is Observation 9 influential for inference?

Table 4.1 (page 107) gives 37 years' data on boats registered in Florida and manatees killed by boats. Figure 4.2 (page 108) shows a strong linear relationship. The correlation is $r = 0.954$. Figure 26.14 shows part of the Minitab regression output. Exercises 26.29 to 26.31 analyze the manatee data.

26.29 Manatees: Conditions for inference. We know that there is a strong linear relationship. Let's check the other conditions for inference. Figure 26.14 includes a table of the two variables, the predicted values $\hat{y}$ for each x in the data, the residuals, and related quantities. MANATRS

(a) Round the residuals to the nearest whole number and make a stemplot. The distribution is single peaked and symmetric and appears close to Normal.

(b) Make a residual plot of residuals against boats registered. Use a vertical scale from −20 to 20 to show the pattern more clearly. Add the "residual = 0" line. There is no clearly nonlinear pattern. The variability about the line may be a bit greater for larger values of the explanatory variable, but the effect is not large.

(c) It is reasonable to regard the number of manatees killed by boats in successive years as independent. The number of boats grew over time. Someone says that pollution also grew over time and may explain more manatee deaths. How would you respond to this idea?

26.30 Manatees: Do more boats bring more kills? The output in Figure 26.14 omits the t statistics and their P-values. Based on the information in the output, is there good evidence that the number of manatees killed increases as the number of boats registered increases? State hypotheses and give a test statistic and its approximate P-value. What do you conclude?

Session

Regression Analysis: Kills versus Boats

```
The regression equation is
Kills = -44.8 + 0.132 Boats


Predictor      Coef    SE Coef      T         P
Constant   -44.831      5.502
Boats     0.132259   0.007067

S = 7.93911    R-Sq = 90.9%    R-Sq(adj) = 90.7%

Obs   Boats   Kills      Fit  SE Fit  Residual  St Resid
  1     447   13.00    14.29    2.55     -1.29     -0.17
  2     460   21.00    16.01    2.47      4.99      0.66
  3     481   24.00    18.79    2.34      5.21      0.69
  4     498   16.00    21.03    2.24     -5.03     -0.66
  5     513   24.00    23.02    2.16      0.98      0.13
  6     512   20.00    22.89    2.16     -2.89     -0.38
  7     526   15.00    24.74    2.09     -9.74     -1.27
  8     559   34.00    29.10    1.91      4.90      0.64
  9     585   33.00    32.54    1.78      0.46      0.06
 10     614   33.00    36.38    1.65     -3.38     -0.43
 11     645   39.00    40.48    1.52     -1.48     -0.19
 12     675   43.00    44.44    1.43     -1.44     -0.18
 13     711   50.00    49.21    1.34      0.79      0.10
 14     719   47.00    50.26    1.33     -3.26     -0.42
 15     681   53.00    45.24    1.41      7.76      0.99
 16     679   38.00    44.97    1.41     -6.97     -0.89
 17     678   35.00    44.84    1.42     -9.84     -1.26
 18     696   49.00    47.22    1.37      1.78      0.23
 19     713   42.00    49.47    1.34     -7.47     -0.95
 20     732   60.00    51.98    1.32      8.02      1.02
 21     755   54.00    55.02    1.31     -1.02     -0.13
 22     809   66.00    62.17    1.36      3.83      0.49
 23     830   82.00    64.94    1.41     17.06      2.18R
 24     880   78.00    71.56    1.57      6.44      0.83
 25     944   81.00    80.02    1.86      0.98      0.13
 26     962   95.00    82.40    1.95     12.60      1.64
 27     978   73.00    84.52    2.04    -11.52     -1.50
 28     983   69.00    85.18    2.07    -16.18     -2.11R
 29    1010   79.00    88.75    2.22     -9.75     -1.28
 30    1024   92.00    90.60    2.30      1.40      0.18
 31    1027   73.00    91.00    2.32    -18.00     -2.37R
 32    1010   90.00    88.75    2.22      1.25      0.16
 33     982   97.00    85.05    2.06     11.95      1.56
 34     942   83.00    79.76    1.85      3.24      0.42
 35     922   88.00    77.11    1.75     10.89      1.41
 36     902   81.00    74.47    1.66      6.53      0.84
 37     897   72.00    73.81    1.64     -1.81     -0.23

R denotes an observation with a large standardized residual.
Predicted Values for New Observations
New
Obs     Fit    SE Fit        95% CI              95% PI
  1   74.20     1.65    (70.84, 77.56)    (57.74, 90.67)
```

FIGURE 26.14
Minitab output for the regression of number of manatees killed by boats on the number of boats (in thousands) registered in Florida, for Exercises 26.29 to 26.31.

26.31 Manatees: estimation. The output in Figure 26.14 includes prediction of the number of manatees killed when there are 900,000 boats registered in Florida. Give 95% intervals for

(a) the increase in the number of manatees killed for each additional 1000 boats registered.
(b) the number of manatees that will be killed next year if there are 900,000 boats registered next year.

26.32 Fidgeting keeps you slim: inference. Our first example of regression (Example 5.1, page 128) presented data showing that people who increased their nonexercise activity (NEA) when they were deliberately overfed gained less fat than other people. Use software to add formal inference to the data analysis for these data. DATA FATGAIN

(a) Based on 16 subjects, the correlation between NEA increase and fat gain was $r = -0.7786$. Is this significant evidence that people with higher NEA increase gain less fat? (Report a t statistic from regression output and give the one-sided P-value.)
(b) The slope of the least-squares regression line was $b = -0.00344$, so that fat gain decreased by 0.00344 kilogram for each added calorie of NEA. Give a 90% confidence interval for the slope of the population regression line. This rate of change is the most important parameter to be estimated.
(c) Sam's NEA increases by 400 calories. His predicted fat gain is 2.13 kilograms. Give a 95% interval for predicting Sam's fat gain.

26.33 Predicting tropical storms. Exercise 5.61 (page 160) gives data on William Gray's predictions of the number of named tropical storms in Atlantic hurricane seasons from 1984 to 2013. Use these data for regression inference as follows. DATA STORMS2

(a) Does Professor Gray do better than random guessing? That is, is there a significantly positive correlation between his forecasts and the actual number of storms? (Report a t statistic from regression output, and give the one-sided P-value.)
(b) Give a 95% confidence interval for the mean number of storms in years when Professor Gray forecasts 16 storms.

26.34 Coral growth. Sea surface temperatures across much of the tropics have been increasing since the mid-1970s. At the same time, the growth of coral has been decreasing. Scientists examined data on mean sea surface temperatures (SST) in degrees Celsius and mean coral growth in millimeters (mm) per year over a several-year period at locations in the Red Sea. Here are the data:[9] DATA CORAL

SST	29.68	29.87	30.16	30.22	30.48	30.65	30.90
Growth	2.63	2.58	2.68	2.60	2.48	2.38	2.26

(a) Do the data indicate that coral growth decreases linearly as SST increases? Is this change statistically significant?
(b) Use the data to predict with 95% confidence the mean coral growth (millimeters per year) when SST is 30.0 degrees Celsius.

26.35 Predicting tropical storms: residuals. Make a stemplot of the residuals (round to the nearest tenth) from your regression in Exercise 26.33. Explain why your plot suggests that we should not use these data to get a prediction interval for the number of storms in a single year. DATA STORMS2

26.36 Coral growth: residuals. Do the data in Exercise 26.34 on mean sea surface temperatures and coral growth in the Red Sea satisfy the conditions for regression inference? To examine this, here are the residuals: DATA CORALR

SST	29.68	29.87	30.16	30.22	30.48	30.65	30.90
Residual	−0.067	−0.060	0.128	0.066	0.025	−0.024	−0.068

(a) **Linear relationship.** A plot of the residuals against the explanatory variable x magnifies the deviations from the least-squares line. Does the plot show any systematic deviation from a roughly linear pattern?
(b) **Normal variation about the line.** Make a histogram of the residuals. With only seven observations, no clear shape emerges. Do strong skewness or outliers suggest lack of Normality?
(c) **Independent observations.** Why are the seven observations independent?
(d) **Variability about the line stays the same.** Does your plot in part (a) show any systematic change in variability as x changes?

26.37 Our brains don't like losses. Exercise 4.29 (page 121) describes an experiment that showed a linear relationship between how sensitive people are to monetary losses ("behavioral loss aversion") and activity in one part of their brains ("neural loss aversion"). DATA LOSSES

(a) Make a scatterplot with neural loss aversion as x and behavioral loss aversion as y. One point is a high outlier in both the x and y directions. In Exercise 5.41 (page 156) you found that this outlier is not influential for the least-squares line.
(b) The research report says that $r = 0.85$ and that the test for regression slope has $P < 0.001$. Verify these results, using all the observations.
(c) The report recognizes the outlier and says, "However, this regression also remained highly significant ($P = 0.004$) when the extreme data point (top right corner) was removed from the analysis." Repeat your analysis omitting the outlier. Show

that the outlier influences regression inference by comparing the t statistic for testing slope with and without the outlier. Then verify the report's claim about the P-value of this test.

26.38 **Time at the table.** Does how long young children remain at the lunch table help predict how much they eat? Here are data on 20 toddlers observed over several months at a nursery school.[10] "Time" is the average number of minutes a child spent at the table when lunch was served. "Calories" is the average number of calories the child consumed during lunch, calculated from careful observation of what the child ate each day. DATA TABTIME

Time	21.4	30.8	37.7	33.5	32.8	39.5	22.8	34.1	33.9	43.8
Calories	472	498	465	456	423	437	508	431	479	454
Time	42.4	43.1	29.2	31.3	28.6	32.9	30.6	35.1	33.0	43.7
Calories	450	410	504	437	489	436	480	439	444	408

(a) Make a scatterplot. Find the correlation and the least-squares regression line. (Be sure to save the regression residuals.) Based on your work, describe the direction, form, and strength of the relationship.
(b) Check the conditions for regression inference. Parts (a) to (d) of Exercise 26.36 provide a handy outline. Use vertical limits -100 to 100 in your plot of the residuals against time to help you see the pattern. What do you conclude?
(c) Is there significant evidence that more time at the table is associated with more calories consumed? Give a 95% confidence interval to estimate how rapidly calories consumed changes as time at the table increases.

26.39 **DNA on the ocean floor.** We think of DNA as the stuff that stores the genetic code. It turns out that DNA occurs, mainly outside living cells, on the ocean floor. It is important in nourishing seafloor life. Scientists think that this DNA comes from organic matter that settles to the bottom from the top layers of the ocean. "Phytopigments," which come mainly from algae, are a measure of the amount of organic matter that has settled to the bottom. The data contains concentrations of DNA and phytopigments (both in grams per square meter) in 116 ocean locations around the world.[11] Look first at DNA alone. Describe the distribution of DNA concentration and give a confidence interval for the mean concentration. Be sure to explain why your confidence interval is trustworthy in the light of the shape of the distribution. The data show surprisingly high DNA concentration, and this by itself was an important finding. DATA DNA

26.40 **Time at the table: prediction.** Rachel attends the nursery school of Exercise 26.38. Over several months, Rachel averages 40 minutes at the lunch table. Give a 95% interval to predict Rachel's average calorie consumption at lunch. DATA TABTIME

Exercises 26.41 to 26.45 ask practical questions involving regression inference without step-by-step instructions. Do complete regression analyses, using the Plan, Solve, and Conclude steps of the four-step process to organize your answers. Follow the model of Example 26.9 and the subsequent discussion, and check the conditions as part of the Solve step.

26.41 4 step **Yukon squirrels.** The introduction to Exercise 4.47 (page 125) gives data on the abundance of the pinecones that red squirrels feed on and the population density of red squirrels over 23 years. How significant is the evidence that more cones lead to higher population density? (Use a vertical scale from -1.5 to 1.5 in your residual plot to show the pattern more clearly.) DATA SQRLCO

26.42 4 step **(Optional topic) A big-toe problem.** Table 7.4 (page 197) and Exercises 7.47 and 7.49 describe the relationship between two deformities of the feet in young patients. Metatarsus adductus (MA) may help predict the severity of hallux abducto valgus (HAV). The paper that reports this study says, "Linear regression analysis, using the hallux abducto angle as the response variable, demonstrated a significant correlation between the metatarsus adductus and hallux abducto angles."[12] Do a suitable analysis to verify this finding. The study authors note that the scatterplot suggests that the variation in y may change as x changes, so they offer a more elaborate analysis as well. DATA BIGTOE

26.43 4 step **Beavers and beetles.** Exercise 5.59 (page 159) describes a study that found that the number of stumps from trees felled by beavers predicts the abundance of beetle larvae. Is there good evidence that more beetle larvae clusters are present when beavers have left more tree stumps? Estimate how many more clusters accompany each additional stump, with 95% confidence. DATA BEAVERS

26.44 4 step **Sulfur, the ocean, and the sun.** Sulfur in the atmosphere affects climate by influencing formation of clouds. The main natural source of sulfur is dimethylsulfide (DMS) produced by small organisms in the upper layers of the oceans. DMS production is in turn influenced by the amount of energy the upper ocean receives from sunlight. Exercise 4.30 (page 122) gives monthly data on solar radiation dose (SRD, in watts per square meter) and surface DMS concentration (in nanomolars) for a region in the Mediterranean. Do the data provide convincing evidence that DMS increases as SRD increases? We also want to estimate the rate of increase, with 90% confidence. DATA SULFUR

26.45 DNA on the ocean floor. Another conclusion of the study introduced in Exercise 26.39 was that organic matter settling down from the top layers of the ocean is the main source of DNA on the seafloor. An important piece of evidence is the relationship between DNA and phytopigments. Do the data give good reason to think that phytopigment concentration helps explain DNA concentration? (Try vertical limits −1 to 1 to make the pattern of your residual plot clearer.) DATA DNA

26.46 (Optional topic) A lurking variable. Return to the data on selling price versus appraised value for beachfront condominiums that are the basis for the Check Your Skills Exercises 26.16 to 26.24. The data are in order by date of the sale, and the data table includes the number of months from the start of the data period. Here are the residuals from the regression of selling price on appraised value (rounded) ordered from left to right: DATA CONDRES

−44.54	−117.36	−162.48	−178.04	−51.17	−152.00	185.56	−35.93
−84.82	146.25	558.43	229.38	333.15	−105.04	−127.26	−218.57
−101.32	197.20	312.84	599.82	−212.40	−103.89	279.13	−152.20
1.33	−75.73	−96.51	−254.00	−587.00	−26.97	−93.23	−98.10
243.71	−6.01	−73.10	−173.33	−95.12	−145.05	298.23	−164.86
177.96	167.52	127.82	13.35	−240.66	88.59	16.46	

(a) Plot the residuals against the explanatory variable (appraised value). To make the pattern clearer, use vertical limits −600 to 600. Does the pattern you see agree with the conditions of linear relationship and constant standard deviation needed for regression inference?
(b) Make a stemplot of the residuals. Are there strong deviations from Normality that would prevent regression inference?
(c) Next, plot the residuals against month. Are the positive and negative residuals randomly scattered, as would be the case if the conditions for regression inference are satisfied? (*Comment:* Suppose prices for beachfront property rise rapidly during any period. Because property is reassessed just once a year, selling prices might pull away from appraised values over time in the period, creating a pattern of many negative residuals followed by several positive residuals. As this example illustrates, it is often wise to plot residuals against important lurking variables as well as against the explanatory variable.)

26.47 (Optional topic) Standardized residuals. Software often calculates **standardized residuals** as well as the actual residuals from regression. Because the standardized residuals have the standard z-score scale, it is easier to judge whether any are extreme. Figure 26.13 and the associated data include the standardized residuals for the regression of gate velocity on piston wall thickness.

standardized residuals

(a) Find the mean and standard deviation of the standardized residuals. Why do you expect values close to those you obtain?
(b) Make a stemplot of the standardized residuals. Are there any striking deviations from Normality?
(c) The most extreme standardized residual is $z = -2.21$. Minitab flags this as "large." What is the probability that a standard Normal variable takes a value this extreme (that is, less than −2.21 or greater than 2.21)? Your result suggests that a residual this extreme would be a bit unusual when there are only 12 observations. That's why we examined Observation 9 in Exercise 26.28.

26.48 (Optional topic) Tests for the intercept. Figure 26.7 gives Minitab output for the regression of blood alcohol content (BAC) on number of beers consumed. The t test for the hypothesis that the population regression line has *slope* $\beta = 0$ has $P < 0.001$. The data show a positive linear relationship between BAC and beers. We might expect the *intercept* α of the population regression line to be 0, because no beers ($x = 0$) should produce no alcohol in the blood ($y = 0$). To test

$$H_0: \alpha = 0$$
$$H_a: \alpha \neq 0$$

we use a t statistic formed by dividing the least-squares intercept a by its standard error SE_a. Locate this statistic in the output of Figure 26.7 and verify that it is in fact a divided by its standard error. What is the P-value? Do the data suggest that the intercept is not zero?

26.49 (Optional topic) Confidence intervals for the intercept. The output in Figure 26.7 allows you to calculate confidence intervals for both the slope β and the intercept α of the population regression line of BAC on beers in the population of all students. Confidence intervals for the intercept α have the familiar form $a \pm t^*SE_a$ with degrees of freedom $n - 2$. What is the 95% confidence interval for the intercept? Does it contain zero, the value we might guess for α?

Exploring the Web

26.50 Predicting batting averages. As you did in Exercise 5.66 (page 161), go to www.mlb.com/ and find the batting averages for a diverse set of 30 players for both the 2012 and 2013 seasons. You can click on the "Stats" tab to find the results for the current season as well as historical data. You should select only players who played in at least 50 games both seasons. Find the least-squares regression line for predicting batting average in 2013 from that in 2012 based on your sample of 30 players. In 2012, the major league leader in batting was Buster Posey, who had a batting average of 0.336. Find a 95% prediction interval for the 2013 batting average of someone who hit 0.336 in 2012. How does this prediction compare with Buster Posey's 2013 batting average?

26.51 Olympic medal counts. In Exercise 4.50 (page 126) you made a scatterplot of the Winter Olympics medal counts for 2010 and 2014. We investigate these medal counts further. Go to the *Chance News* website at www.causeweb.org/wiki/chance/index.php/Chance_News_61#Predicting_medal_counts and read the article "Predicting Medal Counts." Next, search the web (as you did in Chapter 4) and locate the Winter Olympics medal counts for 2010 and 2014 (I found Winter Olympics medal counts on Wikipedia). Find the equation of the least-squares regression line for predicting the 2014 medal counts from the 2010 counts. Compute 95% confidence intervals for the slope and intercept of your regression line. Are your results consistent with the comment in the *Chance News* article that states "we would have done well simply predicting that the Vancouver totals would match the Torino totals"?

Anthony Mercieca/Science Source

CHAPTER 27

One-Way Analysis of Variance: Comparing Several Means

In this chapter we cover...

27.1 Comparing several means

27.2 The analysis of variance *F* test

27.3 Using technology

27.4 The idea of analysis of variance

27.5 Conditions for ANOVA

27.6 *F* distributions and degrees of freedom

27.7 Some details of ANOVA*

The two-sample t procedures of Chapter 21 compare the means of two populations or the mean responses to two treatments in an experiment. Of course, studies don't always compare just two groups. We need a method for comparing any number of means.

EXAMPLE 27.1 Comparing Tropical Flowers

STATE: Ethan Temeles and W. John Kress of Amherst College studied the relationship between varieties of the tropical flower *Heliconia* on the island of Dominica and the different species of hummingbirds that fertilize the flowers.[1] Over time, the researchers believe, the lengths of the flowers and the forms of the hummingbirds' beaks have evolved to match each other. If that is true, flower varieties fertilized by different hummingbird species should have distinct distributions of length.

FLOWER

Table 27.1 gives length measurements (in millimeters) for samples of three varieties of *Heliconia*, each fertilized by a different species of hummingbird. Do the three varieties display distinct distributions of length? In particular, are the mean lengths of their flowers different?

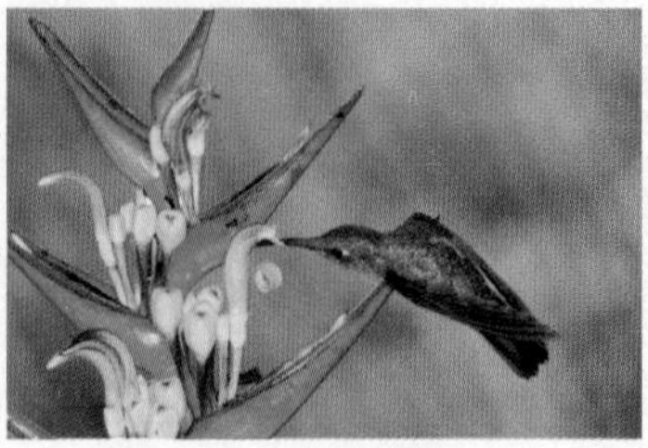

TABLE 27.1 FLOWER LENGTHS (MILLIMETERS) FOR THREE *HELICONIA* VARIETIES

H. BIHAI							
47.12	46.75	46.81	47.12	46.67	47.43	46.44	46.64
48.07	48.34	48.15	50.26	50.12	46.34	46.94	48.36
H. CARIBAEA RED							
41.90	42.01	41.93	43.09	41.47	41.69	39.78	40.57
39.63	42.18	40.66	37.87	39.16	37.40	38.20	38.07
38.10	37.97	38.79	38.23	38.87	37.78	38.01	
H. CARIBAEA YELLOW							
36.78	37.02	36.52	36.11	36.03	35.45	38.13	37.10
35.17	36.82	36.66	35.68	36.03	34.57	34.63	

PLAN: Use graphs and numerical descriptions to describe and compare the three distributions of flower length. Finally, ask whether the differences among the mean lengths of the three varieties are *statistically significant*.

SOLVE (first steps): Figure 27.1 displays side-by-side stemplots with the stems lined up for easy comparison. The lengths have been rounded to the nearest tenth of a millimeter. Here are the summary measures we will use in further analysis:

Sample	Variety	Sample Size	Mean Length	Standard Deviation
1	*H. bihai*	16	47.60	1.213
2	*H. caribaea* red	23	39.71	1.799
3	*H. caribaea* yellow	15	36.18	0.975

CONCLUDE (first steps): The three varieties differ so much in flower length that there is little overlap among them. In particular, the flowers of *H. bihai* are longer than either *H. caribaea* red or *H. caribaea* yellow. The mean lengths are 47.6 mm for *H. bihai*, 39.7 mm for *H. caribaea* red, and 36.2 mm for *H. caribaea* yellow. Are these observed differences in sample means statistically significant? We must develop a test for comparing more than two population means. ■

H. bihai		*H. caribaea red*		*H. caribaea* yellow	
34		34		34	6 6
35		35		35	2 5 7
36		36		36	0 0 1 5 7 8 8
37		37	4 8 9	37	0 1
38		38	0 0 1 1 2 2 8 9	38	1
39		39	2 6 8	39	
40		40	6 7	40	
41		41	5 7 9 9	41	
42		42	0 2	42	
43		43	1	43	
44		44		44	
45		45		45	
46	3 4 6 7 8 8 9	46		46	
47	1 1 4	47		47	
48	1 2 3 4	48		48	
49		49		49	
50	1 3	50		50	

FIGURE 27.1
Side-by-side stemplots comparing the lengths in millimeters of samples of flowers from three varieties of *Heliconia*, from Table 27.1.

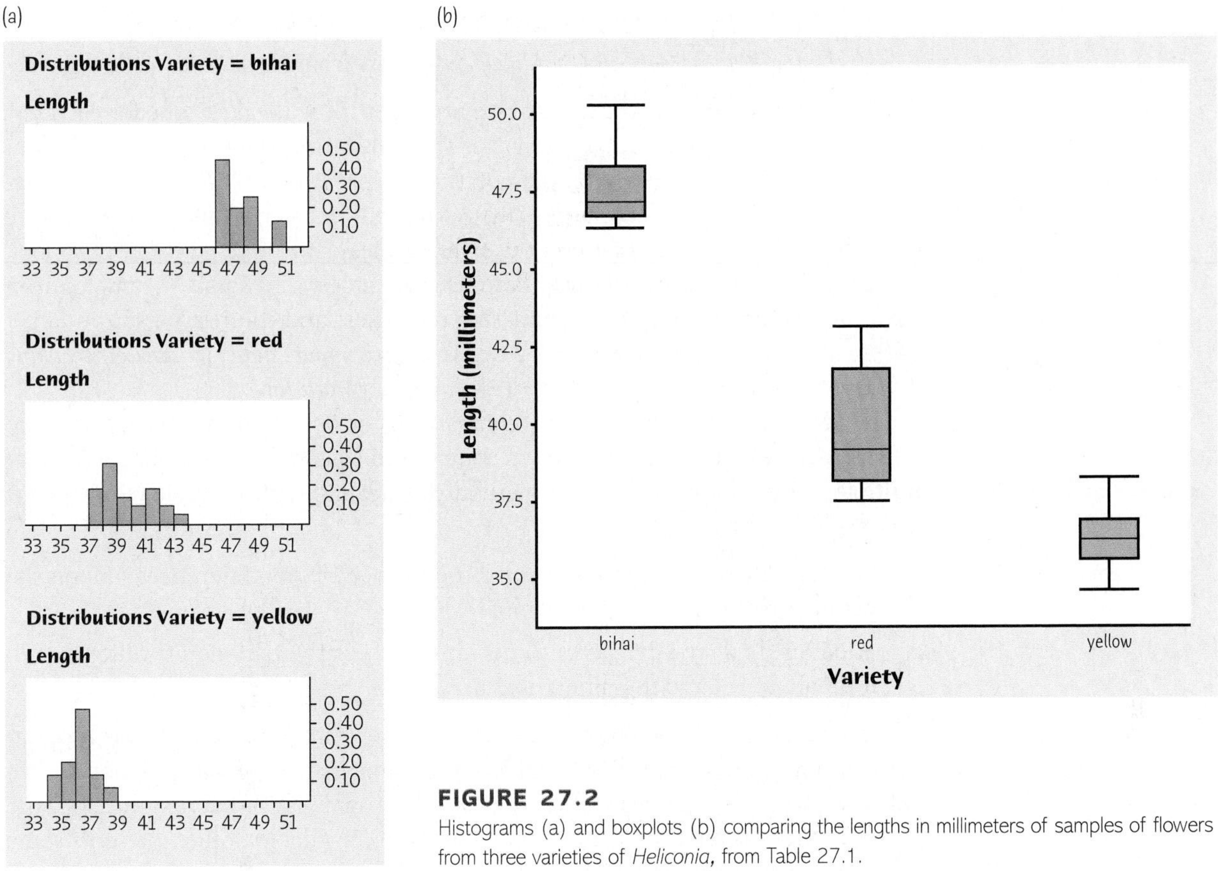

FIGURE 27.2
Histograms (a) and boxplots (b) comparing the lengths in millimeters of samples of flowers from three varieties of *Heliconia*, from Table 27.1.

In the Solve step, we have chosen to draw side-by-side stemplots to compare the three distributions. The three distributions can also be compared using histograms or boxplots. Figure 27.2(a) provides comparative histograms and Figure 27.2(b) comparative boxplots. Because we want to compare the size of the responses for the three distributions as well as examine the general shape, it is best to align the response axis for the three graphs for ease of comparison. If the histograms in Figure 27.2(a) were side by side rather than above each other as in the figure, comparisons of the lengths for the three varieties would be more difficult. For comparison purposes, it is also best to use the percent scale for the histograms when the sample sizes differ. Figure 27.2(b) shows the comparative boxplots. Our impression of the distributions is similar for each of the three graphical displays. The flowers of *H. bihai* are longer than either red or yellow, and the variability for the lengths of the red flowers is somewhat larger than for the other two varieties. The choice of an appropriate graph is based on your software capabilities and, to some extent, on personal preference. Although boxplots show less detail, they are useful for comparison as long as the sample sizes are not very small.

27.1 Comparing several means

Call the mean lengths for the three populations of flowers μ_1 for *H. bihai*, μ_2 for *H. caribaea* red, and μ_3 for *H. caribaea* yellow. The subscript reminds us which group a parameter or statistic describes. To compare these three population means, we might use the two-sample t test several times:

- Test H_0: $\mu_1 = \mu_2$ to see if the mean length for *H. bihai* differs from the mean for *H. caribaea* red.

- Test H_0: $\mu_1 = \mu_3$ to see if *H. bihai* differs from *H. caribaea* yellow.
- Test H_0: $\mu_2 = \mu_3$ to see if *H. caribaea* red differs from *H. caribaea* yellow.

The weakness of doing three tests is that we get three *P*-values, one for each test alone. That doesn't tell us how likely it is that *three* sample means are as different from each other as these are. It may be that $\bar{x}_1 = 47.60$ and $\bar{x}_3 = 36.18$ are significantly different if we look at just two groups, but not significantly different if we know that they are the largest and the smallest means in three groups. As we look at more groups, we expect the gap between the largest and smallest sample mean to get larger. (Think of comparing the tallest and shortest person in larger and larger groups of people.) *We can't safely compare many parameters by doing tests or confidence intervals for two parameters at a time.*

CAUTION

The problem of how to do many comparisons at once with an overall measure of confidence in all our conclusions is common in statistics. This is the problem of **multiple comparisons**. Statistical methods for dealing with multiple comparisons usually have two steps:

multiple comparisons

1. An *overall test* to see if there is good evidence of *any* differences among the parameters that we want to compare, and
2. A detailed *follow-up analysis* to decide which of the parameters differ and to estimate how large the differences are.

The overall test, though more complex than the tests we have met to this point, is reasonably straightforward. Formal follow-up analysis can be quite elaborate. We will concentrate on the overall test and use data analysis to describe in detail the nature of the differences. Companion Chapter 30, available online, presents some details of follow-up inference.

LaunchPad Online Resources

- The **StatBoards video**, *Inference for More than Two Sample Means,* provides further discussion of comparing more than two means and multiple comparisons.

27.2 The analysis of variance *F* test

We want to test the null hypothesis that there are *no differences* among the mean lengths for the three populations of flowers:

$$H_0: \mu_1 = \mu_2 = \mu_3$$

The basic *conditions for inference* (more detail later) are that we have random samples from the three populations and that flower lengths are Normally distributed in each population.

The alternative hypothesis is that there is *some difference.* That is, not all three population means are equal:

$$H_a: \text{not all of } \mu_1, \mu_2, \text{ and } \mu_3 \text{ are equal}$$

The alternative hypothesis is no longer one-sided or two-sided. It is "many-sided" because it allows any relationship other than "all three equal." For example, H_a includes the case in which $\mu_2 = \mu_3$ but μ_1 has a different value. The test of H_0 against H_a is called the **analysis of variance F test**. Analysis of variance is usually abbreviated as ANOVA. The ANOVA *F* test is almost always carried out by software that reports the test statistic and its *P*-value.

analysis of variance F test

EXAMPLE 27.2 Comparing Tropical Flowers: ANOVA

SOLVE (inference): Software tells us that, for the flower length data in Table 27.1, the test statistic is $F = 259.12$ with *P*-value $P < 0.0001$. There is very strong evidence that the three varieties of flowers do not all have the same mean length.

The *F* test does not say *which* of the three means are significantly different. It appears from our preliminary data analysis that *H. bihai* flowers are distinctly longer than either *H. caribaea* red or *H. caribaea* yellow. *H. caribaea* red and *H. caribaea* yellow are closer together, but the *H. caribaea* red flowers tend to be longer.

CONCLUDE: There is strong evidence ($P < 0.0001$) that the population means are not all equal. The most important difference among the means is that the *H. bihai* variety has longer flowers than the *H. caribaea* red and yellow varieties. ■

Example 27.2 illustrates our approach to comparing means. The ANOVA *F* test (done by software) assesses the evidence for *some* difference among the population means. Formal follow-up analysis would allow us to say which means differ and by how much, with (say) 95% confidence that *all* our conclusions are correct. We rely instead on examination of the data to show what differences are present and whether they are large enough to be interesting.

Apply Your Knowledge

27.1 Road Rage. "The phenomenon of road rage has been frequently discussed but infrequently examined." So begins a report based on interviews with 1382 randomly selected drivers.[2] The respondents' answers to interview questions produced scores on an "angry/threatening driving scale" with values between 0 and 19. What driver characteristics go with road rage? There were no significant differences among races or levels of education. What about the effect of the driver's age? Here are the mean responses for three age groups:

<30 yr	30–55 yr	>55 yr
2.22	1.33	0.66

The report says that $F = 34.96$, with $P < 0.01$.

(a) What are the null and alternative hypotheses for the ANOVA *F* test? Be sure to explain what means the test compares.

(b) Based on the sample means and the *F* test, what do you conclude?

27.2 Angry Women, Sad Men. What are the relationships among the portrayal of anger or sadness, sex, and status conferred? Sixty-eight subjects were randomly assigned to view a videotaped job interview in which either a male or a female professional described feeling either anger or sadness. The targets being interviewed wore professional attire and were ostensibly being interviewed for a job while sitting at a table. The targets described an incident in which they and a colleague lost an account, and when asked by the interviewer how it made them feel, responded that the incident made them feel either angry or sad. The subjects were divided into four groups and evaluated one of the following four types of interviews:[3]

	Male Target	Female Target
Expressed anger	Group A	Group C
Expressed sadness	Group B	Group D

After watching the interview, subjects evaluated the target on a composite measure of status conferral which included items assessing how much status, power, and independence the target deserved in his or her future job. The measure of status ranged from 1 = none to 11 = a great deal.

(a) What are the null and alternative hypotheses for the ANOVA F test? Be sure to explain what means the test compares.

(b) Figure 27.3 is a bar graph displaying the means for the four groups. What is the approximate size of the mean difference in status conferred on angry men versus angry women? Which of these two groups has the higher mean status?

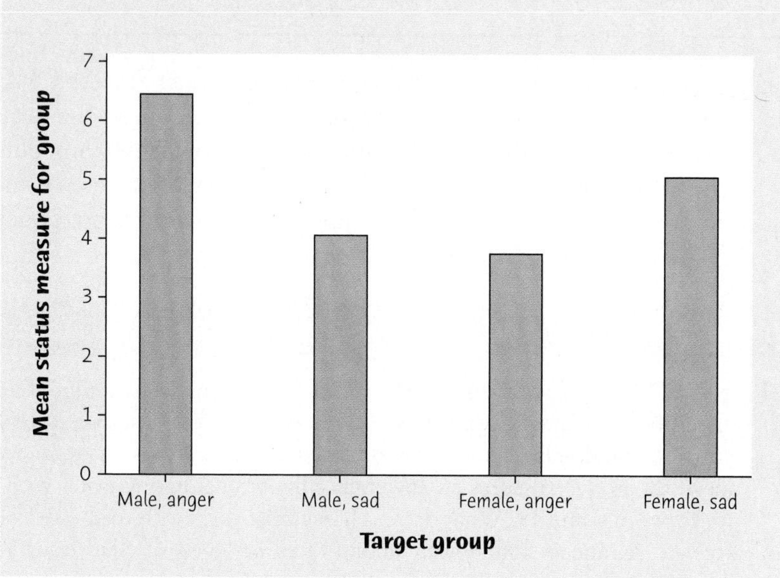

FIGURE 27.3
Bar graph comparing the mean status conferred for the four types of targets, for Exercise 27.2.

27.3 Using technology

Any technology used for statistics should perform analysis of variance. Figure 27.4 displays ANOVA output for the data of Table 27.1 from a graphing calculator, three statistical programs, and a spreadsheet program.

Minitab, JMP, and Excel give the sizes of the three samples and their means. These agree with those in Example 27.1. Minitab also gives the standard deviations, Excel gives the variances, and JMP gives the standard errors. All five outputs report the F test statistic, $F = 259.12$, and its P-value. Minitab reports the P-value as zero to three decimal places, while CrunchIt! and JMP report that $P < 0.0001$. This is all we need to know about the P-value in practice. Excel and the graphing calculator offer the specific value 1.92×10^{-27}. (This would be correct if the population distributions were exactly Normal. In practice, read such values simply as "P is very small.") There is very strong evidence that the three varieties of flowers do not all have the same mean length.

Texas Instruments Graphing Calculator

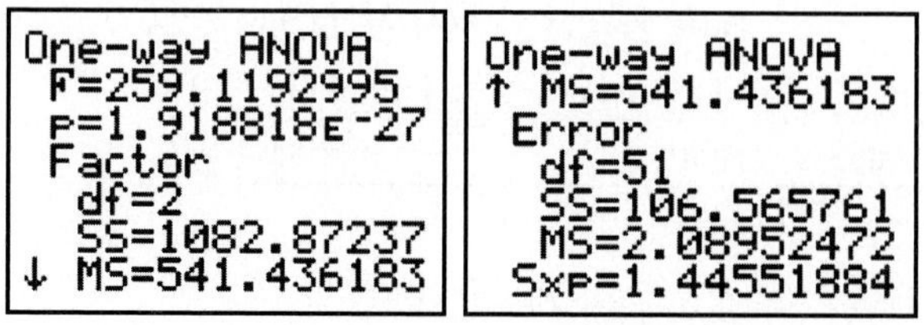

FIGURE 27.4
ANOVA for the flower length data: output from a graphing calculator, three statistical programs, and a spreadsheet program.

FIGURE 27.4
(Continued).

Minitab

Session

One-way ANOVA: length versus variety

```
Source    DF       SS      MS       F      P
Variety    2  1082.87  541.44  259.12  0.000
Error     51   106.57    2.09
Total     53  1189.44

S = 1.446   R-Sq = 91.04%   R-Sq(adj) = 90.69%

                          Individual 95% CIs For Mean Based on
                          Pooled StDev
Level    N    Mean  StDev  --------+---------+---------+---------+
Bihai   16  47.598  1.213                                    (-*-)
Red     23  39.711  1.799                (*-)
Yellow  15  36.180  0.975  (-*--)
                           --------+---------+---------+---------+
                                38.5      42.0      45.5      49.0

Pooled StDev = 1.446
```

CrunchIt!

Results - One-Way ANOVA

Export ▾

Source	Sum of Squares	df	Mean Square	F-value	P-value
Variety	1083	2	541.4	259.1	<0.0001
Error	106.6	51	2.090		
Total	1189	53			

JMP

Analysis of Variance

Source	DF	Sum of Squares	Mean Square	F Ratio	Prob > F
Variety	2	1082.8724	541.436	259.1193	<.0001*
Error	51	106.5658	2.090		
C. Total	53	1189.4381			

Means for Oneway Anova

Level	Number	Mean	Std Error	Lower 95%	Upper 95%
bihai	16	47.5975	0.36138	46.872	48.323
red	23	39.7113	0.30141	39.106	40.316
yellow	15	36.1800	0.37323	35.431	36.929

Std Error uses a pooled estimate of error variance

Excel

Microsoft Excel - ta25-01.dat

	A	B	C	D	E	F	G
1	Anova: Single Factor						
2							
3	SUMMARY						
4	*Groups*	*Count*	*Sum*	*Average*	*Variance*		
5	bihai	16	761.56	47.5975	1.471073		
6	red	23	913.36	39.7113	3.235548		
7	yellow	15	542.7	36.18	0.951257		
8							
9							
10	ANOVA						
11	*Source of variation*	*SS*	*df*	*MS*	*F*	*P-value*	*F crit*
12	Between Groups	1082.872	2	541.4362	259.1193	1.92E-27	3.178799
13	Within Groups	106.5658	51	2.089525			
14							
15	Total	1189.438	53				

Sheet4 / ta25-01

All five outputs report degrees of freedom (df), sums of squares (SS), and mean squares (MS). We don't need this information now. Minitab and JMP also give confidence intervals for all three means that help us see which means differ and by how much. None of the intervals overlap, and *H. bihai* is much above the other two. These are 95% confidence intervals for each mean separately. We are *not* 95% confident that *all three* intervals cover the three means. This is another example of the peril of multiple comparisons.

Apply Your Knowledge

27.3 Logging in the Rain Forest. How does logging in a tropical rain forest affect the forest in later years? Researchers compared forest plots in Borneo that had never been logged (Group 1) with similar plots nearby that had been logged one year earlier (Group 2) and eight years earlier (Group 3). Although the study was not an experiment, the authors explained why we can consider the plots to be randomly selected. The data appear in Table 27.2. The variable Trees is the count of trees in a plot; Species is the count of tree species in a plot. The variable Richness is Species/Trees, the number of species divided by the number of individual trees.[4] DATA LOGFULL

TABLE 27.2 DATA FROM A STUDY OF LOGGING IN BORNEO

GROUP	TREES	SPECIES	RICHNESS	GROUP	TREES	SPECIES	RICHNESS
1	27	22	0.81481	2	18	15	0.83333
1	22	18	0.81818	2	17	15	0.88235
1	29	22	0.75862	2	14	12	0.85714
1	21	20	0.95238	2	14	13	0.92857
1	19	15	0.78947	2	2	2	1.00000
1	33	21	0.63636	2	17	15	0.88235
1	16	13	0.81250	2	19	8	0.42105
1	20	13	0.65000	3	18	17	0.94444
1	24	19	0.79167	3	4	4	1.00000
1	27	13	0.48148	3	22	18	0.81818
1	28	19	0.67857	3	15	14	0.93333
1	19	15	0.78947	3	18	18	1.00000
2	12	11	0.91667	3	19	15	0.78947
2	12	11	0.91667	3	22	15	0.68182
2	15	14	0.93333	3	12	10	0.83333
2	9	7	0.77778	3	12	12	1.00000
2	20	18	0.90000				

(a) Make an appropriate comparative graph of the variable Trees for the three groups. What effects of logging are visible?

(b) Figure 27.5 shows Excel ANOVA output for Trees. What do the group means show about the effects of logging?

(c) What are the values of the ANOVA F statistic and its P-value? What hypotheses does F test? What conclusions about the effects of logging on number of trees do the data lead to?

Microsoft Excel Book 1

	A	B	C	D	E	F	G
1	Anova: Single Factor						
2							
3	SUMMARY						
4	*Groups*	*Count*	*Sum*	*Average*	*Variance*		
5	Group 1	12	285	23.75	25.6591		
6	Group 2	12	169	14.0833	24.8106		
7	Group 3	9	142	15.7778	33.1944		
8							
9							
10	ANOVA						
11	*Source of variation*	*SS*	*df*	*MS*	*F*	*P-value*	*F crit*
12	Between Groups	625.1566	2	312.57828	11.4257	0.000205	3.31583
13	Within Groups	820.7222	30	27.3574			
14							
15	Total	1445.879	32				

Sheet4 / Sheet1 / Sheet2 / Sheet3

FIGURE 27.5
Excel output for analysis of variance on the number of trees in forest plots, for Exercise 27.3.

27.4 Political Views and Age. The University of Chicago's General Social Survey (GSS) is the nation's most important social science sample survey. The GSS asked a random sample of adults in 2012 their age and where they placed themselves on the political spectrum from extremely liberal to extremely conservative. The categories in the original survey included slightly liberal, liberal, and extremely liberal, but these have been combined into the single category liberal, and similarly with conservative.[5]

(a) Figure 27.6 gives the Minitab ANOVA output for these data. What do the mean ages say about the relationship between age and political views?

(b) What are the values of the ANOVA F statistic and its P-value? What hypotheses does F test? Briefly describe the conclusions you draw from these data.

Session

One-way ANOVA: Politics vs Age

```
Source    DF      SS    MS      F      P
Factor     2    7319  3659  11.69  0.000
Error   1871  585591   313
Total   1873  592910

S = 17.69   R-Sq = 1.23%   R-Sq(adj) = 1.13%

                                     Individual 95% CIs For Mean Based on
                                     Pooled StDev
Level           N   Mean  StDev   ---+---------+---------+---------+-----
Conservative  628  51.09  17.42                           (-----*------)
Moderate      713  46.98  18.13    (------*-----)
Liberal       533  46.80  17.41   (------*-------)
                                  ---+---------+---------+---------+-----
                                   46.0      48.0      50.0      52.0

Pooled StDev = 17.69
```

FIGURE 27.6
Minitab output for the data on respondents' ages for three different political viewpoints, for Exercise 27.4.

27.4 The idea of analysis of variance

The details of ANOVA are a bit daunting (they appear in the optional Section 27.7 at the end of this chapter). The main idea of ANOVA is more accessible and much more important. Here it is: when we ask if a set of sample means gives evidence for differences among the population means, what matters is not how far apart the sample means are but how far apart they are *relative to the variability of individual observations.*

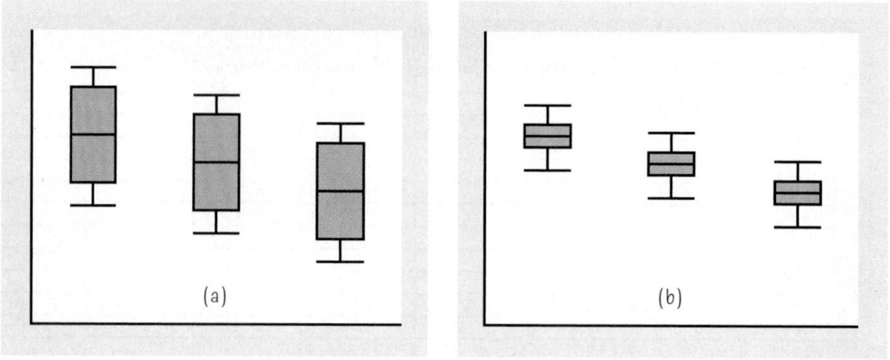

FIGURE 27.7
Boxplots for two sets of three samples each. The sample means are the same in parts (a) and (b). Analysis of variance will find a more significant difference among the means in part (b) because there is less variation among the individuals within those samples.

Look at the two sets of boxplots in Figure 27.7. For simplicity, these distributions are all symmetric, so that the mean and median are the same. The center line in each boxplot is therefore the sample mean. Both sets of boxplots compare three samples with the same three means. Could differences this large easily arise just due to chance, or are they statistically significant?

- The boxplots in Figure 27.7(a) have tall boxes, which show lots of variation among the individuals in each group. With this much variation among individuals, we would not be surprised if another set of samples gave quite different sample means. The observed differences among the sample means could easily happen just by chance.
- The boxplots in Figure 27.7(b) have the same centers as those in Figure 27.7(a), but the boxes are much shorter. That is, there is much less variation among the individuals in each group. It is unlikely that any sample from the first group would have a mean as small as the mean of the second group. Because means as far apart as those observed would rarely arise just by chance in repeated sampling, they are good evidence of real differences among the means of the three populations we are sampling from.

You can use the *One-Way ANOVA* applet to demonstrate the analysis of variance idea for yourself. The applet allows you to change both the group means and the variability within groups. You can watch the ANOVA F statistic and its P-value change as you work.

This comparison of the two parts of Figure 27.7 is too simple in one way. It ignores the effect of the sample sizes, an effect that boxplots do not show. *Small differences among sample means can be significant if the samples are large. Large differences among sample means can fail to be significant if the samples are small.* All we can be sure of is that for the same sample size, Figure 27.7(b) will give a much smaller P-value than Figure 27.7(a). Despite this qualification, the big idea remains: if sample means are far apart relative to the variation among individuals in the same groups, that's evidence that something other than chance is at work.

The Analysis of Variance Idea

Analysis of variance compares the variation due to specific sources with the variation among individuals who should be similar. In particular, ANOVA tests whether several populations have the same mean by comparing how far apart the sample means are with how much variation there is within the samples.

It is one of the oddities of statistical language that methods for comparing means are named after the variance. The reason is that the test works by comparing two kinds of variation. Analysis of variance is a general method for studying sources of variation in responses. Comparing several means is the simplest form of ANOVA, called **one-way ANOVA.**

one-way ANOVA

The ANOVA *F* Statistic

The **analysis of variance *F* statistic** for testing the equality of several means has this form:

$$F = \frac{\text{variation among the sample means}}{\text{variation among individuals in the same sample}}$$

If you want more detail, read the optional material in Section 27.7 at the end of this chapter. The *F* statistic can take only values that are zero or positive. It is zero only when all the sample means are identical and gets larger as they move farther apart. Large values of *F* are evidence against the null hypothesis H_0 that all population means are the same. Although the alternative hypothesis H_a is many-sided, the ANOVA *F* test is one-sided because any violation of H_0 tends to produce a large value of *F*.

LaunchPad Online Resources

- The **StatClips video**, *ANOVA–Background and Introduction*, and the **Snapshots video**, *Introduction to ANOVA*, both discuss the rationale for ANOVA and the hypotheses being tested with the ANOVA *F* statistic.
- The **StatClips Examples video**, *ANOVA–Background and Introduction Example A*, provides an additional example of the idea of ANOVA.

Apply Your Knowledge

27.5 ANOVA Compares Several Means. The *One-Way ANOVA* applet displays the observations in three groups, with the group means highlighted by black dots.

APPLET

(a) What are the *F* and *P* value for these three samples? (The *P*-value is marked by a red dot that will move along the scale as you modify the samples.)

(b) Which group has the largest mean? If you grab its mean point with the mouse and begin to move it in either direction, the *F* value will change. How small can you make *F*? What did you do to this mean to make *F* small? Roughly how significant is your small *F*?

(c) Starting with your configuration at the end of part (b), drag any one of the group means either up or down as far as they will go. What happens to *F*? What happens to the *P*-value? How large can you make *F*? What did you do to the three means to make *F* large?

27.6 ANOVA Uses Within-Group Variation. If the *One-Way ANOVA* applet is open from Exercise 27.5, click on New Samples. Otherwise, open the *One-Way ANOVA* applet which displays the observations in three groups, with the group means highlighted by black dots. You are going to use the Applet to investigate the effects described in Figure 27.7 (page 654).

APPLET

(a) Use the mouse to slide the Standard Deviation at the top of the display to the right. You see that the group means do not change, but the variability of the observations in each group increases. What happens to *F* and *P* as the variability among the observations in each group increases? What are the values of *F* and *P* when the slider is all the way to the right? This is similar to Figure 27.7(a): variation within groups hides the differences among the group means.

(b) Leave the Standard Deviation slider at the extreme right of its scale, so that variability within groups stays fixed. Use the mouse to move the group means apart. What happens to F and P as you do this?

27.5 Conditions for ANOVA

Like all inference procedures, ANOVA is valid only in some circumstances. Here are the conditions under which we can use ANOVA to compare population means.

Conditions for ANOVA Inference

- We have ***I* independent SRSs**, one from each of I populations. We measure the same quantitative response variable for each sample.
- The ith population has a **Normal distribution** with unknown mean μ_i. One-way ANOVA tests the null hypothesis that all the population means are the same.
- All the populations have the **same standard deviation** σ, whose value is unknown.

WE WEREN'T WORKING ANYWAY

A "consultant" estimated that the annual NCAA men's basketball tournament costs employers \$3.8 billion in time wasted by workers participating in office pools, checking game scores, and so on. That's unlikely. Most of the games are played outside work hours, at night and on weekends. More to the point, economists note that workers waste lots of time every workday by talking to other employees, chatting on the phone, shopping online, and so on. The dollar value of time spent on the basketball tournament probably comes in large part from time we were wasting anyway.

The first two conditions are familiar from our study of the two-sample t procedures for comparing two means. As usual, the design of the data production is the most important condition for inference. Biased sampling or confounding can make any inference meaningless. *If we do not actually draw separate SRSs from each population or carry out a randomized comparative experiment, it may be unclear to what population the conclusions of inference apply.* ANOVA, like other inference procedures, is often used when random samples are not available. You must judge each use on its merits, a judgment that usually requires some knowledge of the subject of the study in addition to some knowledge of statistics.

CAUTION

Because no real population has exactly a Normal distribution, the usefulness of inference procedures that assume Normality depends on how sensitive they are to departures from Normality. Fortunately, procedures for comparing means are not very sensitive to lack of Normality. The ANOVA F test, like the t procedures, shows **robustness**. What matters is Normality of the sample means, so ANOVA becomes safer as the sample sizes get larger because of the central limit theorem effect. Remember to check for outliers that change the value of sample means and for extreme skewness. When there are no outliers and the distributions are roughly symmetric, you can safely use ANOVA for sample sizes as small as 4 or 5.

robustness

The third condition is annoying: ANOVA assumes that the variability of observations, measured by the standard deviation, is the same in all populations. The t test for comparing two means (Chapter 21) does not require equal standard deviations. Unfortunately, the ANOVA F for comparing more than two means is less broadly valid. It is not easy to check the condition that the populations have equal standard deviations. Statistical tests for equality of standard deviations are very sensitive to lack of Normality, so much so that they are of little practical value. You must either seek expert advice or rely on the robustness of ANOVA.

How serious are unequal standard deviations? ANOVA is not too sensitive to violations of the condition, especially when all samples have the same or similar sizes and no sample is very small. When designing a study, try to take samples of about the same size from all the groups you want to compare. The sample standard deviations estimate the population standard deviations, so check before doing ANOVA that the sample standard deviations are similar to each other. We expect some variation among them due to chance. Here is a rule of thumb that is safe in almost all situations.

Checking Standard Deviations in ANOVA

The results of the ANOVA F test are approximately correct when the largest sample standard deviation is no more than twice as large as the smallest sample standard deviation.

EXAMPLE 27.3 Comparing Tropical Flowers: Conditions for ANOVA

The study of *Heliconia* blossoms is based on three independent samples that the researchers consider to be random samples from all flowers of these varieties in Dominica. The stemplots in Figure 27.1 show that the *H. bihai* and *H. caribaea* red varieties have slightly skewed distributions, but the sample means of samples of sizes 16 and 23 will have distributions that are close to Normal. The sample standard deviations for the three varieties are

$$s_1 = 1.213 \quad s_2 = 1.799 \quad s_3 = 0.975$$

These standard deviations satisfy our rule of thumb:

$$\frac{\text{largest } s}{\text{smallest } s} = \frac{1.799}{0.975} = 1.85 \quad \text{(less than 2)}$$

We can safely use ANOVA to compare the mean lengths for the three populations. ■

EXAMPLE 27.4 Thinking about Money Changes Behavior

MONEY

STATE: Kathleen Vohs of the University of Minnesota and her co-workers carried out several randomized comparative experiments on the effects of thinking about money. Here's an outline of one of the experiments. Ask student subjects to unscramble 30 sets of five words to make a meaningful phrase from four of the five. The control group unscrambled phrases like "cold it desk outside is" into "it is cold outside." The "play money" group unscrambled similar sets of words, but a stack of Monopoly money was placed nearby. The "money prime" group unscrambled phrases that led to thinking about money, turning "high a salary desk paying" into "a high-paying salary." Then each subject worked a hard puzzle, knowing that they could ask for help. Table 27.3 shows the time in seconds that each subject worked on the puzzle before asking for help.[6] Psychologists think that money tends to make people self-sufficient. If so, the two groups that were encouraged in different ways to think about money should take longer on the average to ask for help. Do the data support this idea?

PLAN: Examine the data to compare the effect of the treatments and check that we can safely use ANOVA. If the data allow ANOVA, assess the significance of observed differences in mean times to ask for help.

SOLVE: Figure 27.8 compares the histograms of the data in the three groups. We expect some irregularity in small samples, but there are no outliers or strong skewness that would hinder use of ANOVA. The Minitab ANOVA output in Figure 27.9 shows that the group standard deviations easily satisfy our rule of thumb. The control group subjects asked for help much sooner (mean 186.1 seconds) than did subjects in the two money groups (means 305.2 seconds and 314.1 seconds). The three means are significantly different ($F = 3.73$, $P = 0.031$).

CONCLUDE: The experiment gives good evidence that reminding people of money in either of two ways does make them less willing to ask others for help. This is consistent with the idea that money makes people feel more self-sufficient. ■

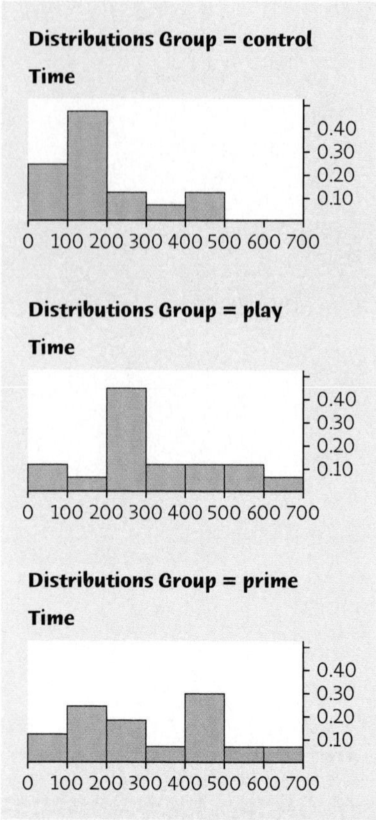

FIGURE 27.8
Histograms comparing the time until subjects asked for help with a puzzle, for Example 27.4.

TABLE 27.3 TIME (SECONDS) UNTIL SUBJECTS ASK FOR HELP WITH A PUZZLE

GROUP	TIME	GROUP	TIME	GROUP	TIME
Prime	609	Play	455	Control	118
Prime	444	Play	100	Control	272
Prime	242	Play	238	Control	413
Prime	199	Play	243	Control	291
Prime	174	Play	500	Control	140
Prime	55	Play	570	Control	104
Prime	251	Play	231	Control	55
Prime	466	Play	380	Control	189
Prime	443	Play	222	Control	126
Prime	531	Play	71	Control	400
Prime	135	Play	232	Control	92
Prime	241	Play	219	Control	64
Prime	476	Play	320	Control	88
Prime	482	Play	261	Control	142
Prime	362	Play	290	Control	141
Prime	69	Play	495	Control	373
Prime	160	Play	600	Control	156
		Play	67		

One-way ANOVA: time versus group

```
Source  DF       SS     MS     F      P
Group    2   174912  87456  3.73  0.031
Error   49  1149568  23461
Total   51  1234479

S = 153.2   R-Sq = 13.21%    R-Sq(adj) = 9.66%

                                Individual 95% CIs For Mean Based on
                                Pooled StDev
Level     N   Mean  StDev  ----+---------+---------+---------+----
control  17  186.1  118.1  (----------*----------)
play     18  305.2  162.5                  (----------*----------)
prime    17  314.1  172.8                    (----------*----------)
                           ----+---------+---------+---------+----
                             140       210       280       350
Pooled StDev = 153.2
```

FIGURE 27.9
Minitab ANOVA output for comparing the three treatments in Example 27.4.

LaunchPad Online Resources

- The **Snapshots video**, *ANOVA Inference*, discusses the details of using the ANOVA *F* statistic for inference through an example and the **StatBoards video**, *Checking Conditions for ANOVA*, provides an additional example on checking the conditions for inference in ANOVA.

Apply Your Knowledge

27.7 Checking Standard Deviations. Verify that the sample standard deviations for these sets of data do allow use of ANOVA to compare the population means.

(a) The counts of trees in Exercise 27.3 and Figure 27.5

(b) The ages of Exercise 27.4 and Figure 27.6

27.8 Species Richness after Logging. Table 27.2 gives data on the species richness in rain forest plots, defined as the number of tree species in a plot divided by the number of trees in the plot. ANOVA may not be trustworthy for the richness data. Do data analysis: make an appropriate graph to examine the distributions of the response variable in the three groups, and also compare the standard deviations. What characteristic of the data makes ANOVA risky? DATA LOGFULL

27.9 Fertilizing Bromeliads. Bromeliads are tropical flowering plants. Many are epiphytes that attach to trees and obtain moisture and nutrients from air and rain. Their leaf bases form cups that collect water and are home to the larvae of many insects. Preliminary to a study of changes in the nutrient cycle, Jacqueline Ngai and Diane Srivastava examined the effects of adding nitrogen, phosphorus, or both to the cups. They randomly assigned eight bromeliads growing in Costa Rica to each of four treatment groups, including an unfertilized control group. A monkey destroyed one of the plants in the control group, leaving seven bromeliads in that group. Here are the numbers of new leaves on each plant over the seven months following fertilization:[7] DATA BROMLIAD

4 step

Nitrogen	Phosphorus	Both	Neither
15	14	14	11
14	14	16	13
15	14	15	16
16	11	14	15
17	13	14	15
18	12	13	11
17	15	17	12
13	15	14	

Analyze these data and discuss the results. Does nitrogen or phosphorus have a greater effect on the growth of bromeliads? Follow the four-step process as illustrated in Example 27.4.

27.6 *F* distributions and degrees of freedom

The ANOVA *F* statistic is

$$F = \frac{\text{variation among the sample means}}{\text{variation among individuals in the same sample}}$$

To find the *P*-value for this statistic, we must know the sampling distribution of *F* when the null hypothesis (all population means equal) is true. This sampling distribution is an ***F* distribution.**

F distribution

The *F* distributions are a family of right-skewed distributions that take only values greater than zero. The density curves in Figure 27.10 illustrate their shapes. A specific *F* distribution is determined by the *degrees of freedom* of the numerator and denominator of the *F* statistic. You may have noticed that all our software outputs include degrees of freedom, labeled either "df" or "DF." Optional Section 27.7 "Some details of ANOVA" shows where the degrees of freedom come from. When describing an *F* distribution, always give the numerator degrees of freedom first. Our brief notation will be $F(\text{df1}, \text{df2})$ for the *F* distribution with df1 degrees of freedom in the numerator and df2 in the denominator. *Interchanging the degrees of freedom changes the distribution, so the order is important.*

CAUTION

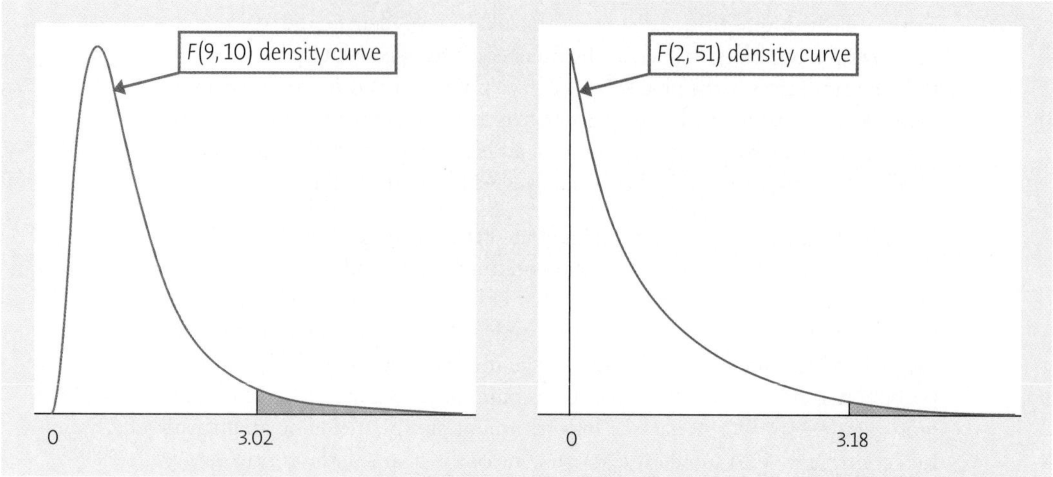

FIGURE 27.10
Density curves for two F distributions. Both are right-skewed and take only positive values. The upper 5% critical values are marked under the curves.

Tables of F critical points are awkward because we need a separate table for every pair of degrees of freedom df1 and df2. Fortunately, software gives you P-values for the ANOVA F test without the need for a table.

EXAMPLE 27.5 Comparing Flowers: The *F* Distribution

Look again at the software output in Figure 27.4 for the flower length data. All three outputs give the degrees of freedom for the F test, labeled "df" or "DF." There are two degrees of freedom in the numerator and 51 in the denominator. P-values for the F test, therefore, come from the F distribution $F(2, 51)$ with two and 51 degrees of freedom. The right-hand curve in Figure 25.10 is the density curve of this distribution. The 5% critical value marked on that curve is 3.18, and the 1% critical value is 5.05. The observed value $F = 259.12$ of the ANOVA F statistic lies far to the right of these values, so the P-value is extremely small. ■

The degrees of freedom of the ANOVA F statistic depend on the number of means we are comparing and the number of observations in each sample. That is, the F test takes into account the number of observations. Here are the details.

Degrees of Freedom for the *F* Test

We want to compare the means of I populations. We have an SRS of size n_i from the ith population, so that the total number of observations in all samples combined is

$$N = n_1 + n_2 + \cdots + n_I$$

If the null hypothesis that all population means are equal is true, the ANOVA F statistic has the F distribution with $I - 1$ degrees of freedom in the numerator and $N - I$ degrees of freedom in the denominator.

EXAMPLE 27.6 Degrees of Freedom for *F*

In Examples 27.1 and 27.2, we compared the mean lengths for three varieties of flowers, so $I = 3$. The three sample sizes are

$$n_1 = 16 \quad n_2 = 23 \quad n_3 = 15$$

The total number of observations is therefore

$$N = 16 + 23 + 15 = 54$$

The ANOVA **F** test has numerator degrees of freedom

$$I - 1 = 3 - 1 = 2$$

and denominator degrees of freedom

$$N - I = 54 - 3 = 51$$

These are the degrees of freedom given in the outputs in Figure 27.3. ■

Apply Your Knowledge

27.10 Logging in the Rain Forest, Continued. Exercise 27.3 compares the number of tree species in rain forest plots that had never been logged (Group 1) with similar plots nearby that had been logged one year earlier (Group 2) and eight years earlier (Group 3).

(a) What are I, n_i, and N for these data? Identify these quantities in words and give their numerical values.

(b) Find the degrees of freedom for the ANOVA F statistic. Check your work against the Excel output in Figure 27.5.

27.11 What Music Will You Play? People often match their behavior to their social environment. One study of this idea first established that the type of music most preferred by black college students is R&B and that whites' most preferred music is rock. Will students hosting a small group of other students choose music that matches the makeup of the people attending? Assign 90 black business students at random to three equal-sized groups. Do the same for 96 white students. Each student sees a picture of the people he or she will host. Group 1 sees six blacks, Group 2 sees three whites and three blacks, and Group 3 sees six whites. Ask how likely the host is to play the type of music preferred by the other race. Use ANOVA to compare the three groups to see whether the racial mix of the gathering affects the choice of music.[8]

(a) For the white subjects, $F = 16.48$. What are the degrees of freedom?

(b) For the black subjects, $F = 2.47$. What are the degrees of freedom?

© PhotoAlto/Alamy

27.7 Some details of ANOVA*

Now we will give the actual formula for the ANOVA F statistic. We have SRSs from each of I populations. Subscripts from 1 to I tell us to which sample a statistic refers to:

Population	Sample Size	Sample Mean	Sample Std. Dev.
1	n_1	$\bar{x}_1$	s_1
2	n_2	$\bar{x}_2$	s_2
⋮	⋮	⋮	⋮
I	n_I	$\bar{x}_I$	s_I

You can find the F statistic from just the sample sizes n_i, the sample means $\bar{x}_i$, and the sample standard deviations s_i. You don't need to go back to the individual observations.

*This more advanced section is optional if you are using software to find the F statistic.

The ANOVA F statistic has the form

$$F = \frac{\text{variation among the sample means}}{\text{variation among individuals in the same sample}}$$

mean squares

The measures of variation in the numerator and denominator of F are called **mean squares**. A mean square is a more general form of a sample variance. An ordinary sample variance s^2 is an average (or mean) of the squared deviations of observations from their mean, so it qualifies as a "mean square."

Call the overall mean response $\bar{x}$. That is, $\bar{x}$ is the mean of all N observations together. You can find $\bar{x}$ from the I sample means by

$$\bar{x} = \frac{\text{sum of all observations}}{N} = \frac{n_1\bar{x}_1 + n_2\bar{x}_2 + \cdots + n_I\bar{x}_I}{N}$$

(This expression works because multiplying a group mean $\bar{x}_i$ by the number of observations n_i it represents gives the sum of the observations in that group.)

The numerator of F is a mean square that measures variation among the I sample means $\bar{x}_1, \bar{x}_2, \ldots, \bar{x}_I$. To measure this variation, look at the I deviations of the means of the samples from $\bar{x}$,

$$\bar{x}_1 - \bar{x}, \bar{x}_2 - \bar{x}, \ldots, \bar{x}_I - \bar{x}$$

mean square for groups (MSG)

The mean square in the numerator of F is an average of the squares of these deviations. We call it the **mean square for groups (MSG)**:

$$\text{MSG} = \frac{n_1(\bar{x}_1 - \bar{x})^2 + n_2(\bar{x}_2 - \bar{x})^2 + \cdots + n_I(\bar{x}_I - \bar{x})^2}{I - 1}$$

Each squared deviation is weighted by n_i, the number of observations it represents.

mean square for error (MSE)

The mean square in the denominator of F measures variation among individual observations in the same sample. For any one sample, the sample variance s_i^2 does this job. For all I samples together, we use an average of the individual sample variances. It is another weighted average, in which each s_i^2 is weighted by its degrees of freedom $n_i - 1$. The resulting mean square is called the **mean square for error (MSE)**:

$$\text{MSE} = \frac{(n_1 - 1)s_1^2 + (n_2 - 1)s_2^2 + \cdots + (n_I - 1)s_I^2}{N - I}$$

"Error" doesn't mean a mistake has been made. It's a traditional term for chance variation. Here is a summary of the ANOVA test.

The ANOVA *F* Test

Draw an independent SRS from each of I Normal populations that have a common standard deviation but may have different means. The sample from the ith population has size n_i, sample mean $\bar{x}_i$, and sample standard deviation s_i.

To test the null hypothesis that all I populations have the same mean against the alternative hypothesis that not all the means are equal, calculate the **ANOVA *F* statistic**

$$F = \frac{\text{MSG}}{\text{MSE}}$$

The numerator of F is the **mean square for groups**

$$\text{MSG} = \frac{n_1(\bar{x}_1 - \bar{x})^2 + n_2(\bar{x}_2 - \bar{x})^2 + \cdots + n_I(\bar{x}_I - \bar{x})^2}{I - 1}$$

The denominator of F is the **mean square for error**

$$\text{MSE} = \frac{(n_1 - 1)s_1^2 + (n_2 - 1)s_2^2 + \cdots + (n_I - 1)s_I^2}{N - I}$$

When H_0 is true, F has the ***F* distribution** with $I - 1$ and $N - I$ degrees of freedom.

The denominators in the formulas for MSG and MSE are the two degrees of freedom $I - 1$ and $N - I$ of the F test. The numerators are called **sums of squares**, from their algebraic form. It is usual to present the results of ANOVA in an **ANOVA table**. Output from software usually includes an ANOVA table.

sums of squares

ANOVA table

EXAMPLE 27.7 ANOVA Calculations: Software

Look again at the three outputs in Figure 27.4. The three software outputs give the ANOVA table. The calculator, with its small screen, gives the degrees of freedom, sums of squares, and mean squares separately. Each output uses slightly different language to identify the two sources of variation. The basic ANOVA table is

Source of Variation	df	SS	MS	F Statistic
Variation among samples	2	1082.87	MSG = 541.44	259.12
Variation within samples	51	106.57	MSE = 2.09	

You can check that each mean square MS is the corresponding sum of squares SS divided by its degrees of freedom df. The F statistic is MSG divided by MSE. ■

Because MSE is an average of the individual sample variances, it is also called the *pooled sample variance,* written as s_p^2. When all I populations have the same population variance σ^2, as ANOVA assumes that they do, s_p^2 estimates the common variance σ^2. The square root of MSE is the **pooled standard deviation,** s_p. It estimates the common standard deviation σ of observations in each group. The Minitab and calculator outputs in Figure 27.4 give the value $s_p = 1.446$.

pooled standard deviation

The pooled standard deviation s_p is a better estimator of the common σ than any individual sample standard deviation s_i because it combines (pools) the information in all I samples. We can get a confidence interval for any one of the means μ_i from the usual form

$$\text{estimate} \pm t^*\text{SE}_{\text{estimate}}$$

using s_p to estimate σ. The confidence interval for μ_i is

$$\bar{x}_i \pm t^*\frac{s_p}{\sqrt{n_i}}$$

Use the critical value t^* from the t distribution with $N - I$ degrees of freedom because s_p has $N - I$ degrees of freedom. These are the confidence intervals that appear in Minitab ANOVA output.

EXAMPLE 27.8 ANOVA Calculations: Without Software

We can do the ANOVA test comparing the mean lengths of *H. bihai, H. caribaea* red, and *H. caribaea* yellow flower varieties using only the sample sizes, sample means, and sample standard deviations. These appear in Example 27.1, but it is easy to find them with a calculator. There are $I = 3$ groups with a total of $N = 54$ flowers.

The overall mean of the 54 lengths in Table 27.1 is

$$\begin{aligned}\bar{x} &= \frac{n_1\bar{x}_1 + n_2\bar{x}_2 + n_3\bar{x}_3}{N} \\ &= \frac{(16)(47.598) + (23)(39.711) + (15)(36.180)}{54} \\ &= \frac{2217.621}{54} = 41.067\end{aligned}$$

The mean square for groups is

$$\begin{aligned}\text{MSG} &= \frac{n_1(\bar{x}_1 - \bar{x})^2 + n_2(\bar{x}_2 - \bar{x})^2 + n_3(\bar{x}_3 - \bar{x})^2}{I - 1} \\ &= \frac{1}{3-1}[(16)(47.598 - 41.067)^2 + (23)(39.711 - 41.067)^2 \\ &\quad + (15)(36.180 - 41.067)^2] \\ &= \frac{1082.996}{2} = 541.50\end{aligned}$$

The mean square for error is

$$\begin{aligned}\text{MSE} &= \frac{(n_1 - 1)s_1^2 + (n_2 - 1)s_2^2 + (n_3 - 1)s_3^2}{N - I} \\ &= \frac{(15)(1.213^2) + (22)(1.799^2) + (14)(0.975^2)}{51} \\ &= \frac{106.580}{51} = 2.09\end{aligned}$$

Finally, the ANOVA test statistic is

$$F = \frac{\text{MSG}}{\text{MSE}} = \frac{541.50}{2.09} = 259.09$$

Our work differs slightly from the output in Figure 27.4 because of roundoff error. We don't recommend doing these calculations, because tedium and roundoff errors cause frequent mistakes. ■

LaunchPad Online Resources

- The **StatBoards video**, *Understanding the ANOVA Table*, reviews the details of the ANOVA table.

Apply Your Knowledge

The calculations of ANOVA use only the sample sizes n_i, the sample means $\bar{x}_i$, and the sample standard deviations s_i. You can therefore re-create the ANOVA calculations when a report gives these summaries but does not give the actual data. These optional exercises ask you to do the ANOVA calculations starting with the summary statistics. P-values require either a table or software for the F distributions.

27.12 Road Rage. Exercise 27.1 describes a study of road rage. Here are the means and standard deviations for a measure of "angry/threatening driving" for random samples of drivers in three age groups:

Age Group	n	$\bar{x}$	s
Less than 30 years	244	2.22	3.11
30 to 55 years	734	1.33	2.21
Over 55 years	364	0.66	1.60

(a) The distributions of responses are somewhat right-skewed. ANOVA is nonetheless safe for these data. Why?

(b) Check that the standard deviations satisfy the guideline for ANOVA inference.

(c) Calculate the overall mean response $\bar{x}$, the mean squares MSG and MSE, and the ANOVA F statistic.

(d) Which F distribution would you use to find the P-value of the ANOVA F test? Software gives $P < 0.001$. Write a brief conclusion based on the sample means and the ANOVA F test.

27.13 Angry Women, Sad Men. Exercise 27.2 describes a study in which subjects conferred status on men and women who were displaying either anger or sadness. Subjects were randomly assigned to the four treatments, with 17 subjects assigned to each treatment. The study report contains the following information about status measure conferred for each of the four groups:

Treatment	n	$\bar{x}$	s
Males expressing anger	17	6.47	2.25
Females expressing anger	17	3.75	1.77
Males expressing sadness	17	4.05	1.61
Females expressing sadness	17	5.02	1.80

(a) Do the standard deviations satisfy the rule of thumb for safe use of ANOVA?

(b) Calculate the overall mean response $\bar{x}$, the mean squares MSG and MSE, and the F statistic.

(c) Which F distribution would you use to find the P-value of the ANOVA F test? Write a brief conclusion based on the sample means and the ANOVA F test.

27.14 Attitudes toward Math. Do high school students from different racial/ethnic groups have different attitudes toward mathematics? Measure the level of interest in mathematics on a five-point scale for a national random sample of students. Here are summaries for students who were taking math at the time of the survey:[9]

Racial/Ethnic Group	n	$\bar{x}$	s
African American	809	2.57	1.40
White	1860	2.32	1.36
Asian/Pacific Islander	654	2.63	1.32
Hispanic	883	2.51	1.31
Native American	207	2.51	1.28

(a) The conditions for ANOVA are clearly satisfied. Explain why.

(b) Calculate the ANOVA table and the F statistic.

(c) Software gives $P < 0.001$. What explains the small P-value? Do you think the differences are large enough to be important?

CHAPTER 27 SUMMARY

Chapter Specifics

- **One-way analysis of variance (ANOVA)** compares the means of several populations. The **ANOVA F test** tests the null hypothesis that all the populations have the same mean. If the F test shows significant differences, examine the data to see where the differences lie and whether they are large enough to be important.

- The **conditions for ANOVA** state that we have an **independent SRS** from each population; that each population has a **Normal distribution**; and that all populations have the **same standard deviation**.
- In practice, ANOVA inference is relatively **robust** when the populations are non-Normal, especially when the samples are large. Before doing the F test, check the observations in each sample for outliers or strong skewness. Also verify that the largest sample standard deviation is no more than twice as large as the smallest standard deviation.
- When the null hypothesis is true, the **ANOVA F statistic** for comparing I means from a total of N observations in all samples combined has the **F distribution** with $I - 1$ and $N - I$ degrees of freedom.
- ANOVA calculations are reported in an **ANOVA table** that gives sums of squares, mean squares, and degrees of freedom for variation among groups and for variation within groups. In practice, we use software to do the calculations.

STATISTICS IN SUMMARY

Here are the most important skills you should have acquired from reading this chapter.

A. Recognition

1. Recognize when testing the equality of several means is helpful in understanding data.
2. Recognize that the statistical significance of differences among sample means depends on the sizes of the samples and on how much variation there is within the samples.
3. Recognize when you can safely use ANOVA to compare means. Check the data production, the presence of outliers, and the sample standard deviations for the groups you want to compare.

B. Interpreting ANOVA

1. Explain what null hypothesis F tests in a specific setting.
2. Locate the F statistic and its P-value on the output of analysis of variance software.
3. Find the degrees of freedom for the F statistic from the number and sizes of the samples.
4. If the test is significant, use graphs and descriptive statistics to see what differences among the means are most important.

Link It

Analysis of variance is a general statistical method for studying sources of variation in a response. In this chapter, we have studied one-way ANOVA, which is a specific statistical technique designed to test the null hypothesis of the equality of the means of several populations. As such, it is an extension of the two sample t test of Chapter 21 which tested the null hypothesis of the equality of *two* population means.

When we reject the null hypothesis of equality of two means with the two-sample t test, then we have evidence that the two means are different. Rejection of the null hypothesis with a one-way ANOVA is a more ambiguous conclusion—it is evidence of a difference in the means of the populations, but the result does not tell us *which* differences between the means are statistically significant. This is similar to the chi-square test in Chapter 25, where rejection of the null hypothesis indicates that there is a relationship between the two categorical variables, but says nothing about the nature of that relationship. Typically, after rejection of the null hypothesis in a one-way ANOVA, we perform a more detailed *follow-up analysis* to decide which of the means are different and to estimate how large these differences are. The companion Chapter 30, available online, presents some details of this follow-up inference.

LaunchPad Online Resources

If you are having difficulty with any of the sections of this chapter, these online resources should help prepare you to solve the exercises at the end of this chapter.

- **StatTutor** starts with a video review of each section and asks a series of questions to check your understanding.
- **LearningCurve** provides you with a series of questions about the chapter geared to your level of understanding.

CHECK YOUR SKILLS

27.15 The purpose of analysis of variance is to compare

(a) the variances of several populations.
(b) the proportions of successes in several populations.
(c) the means of several populations.

27.16 A study examined the effects of Vitamin E and Memantine on the functional decline of patients in the early stages of Alzheimer's disease. The investigators took 450 patients and randomly divided them into three groups, each containing 150 patients. One group was assigned to Vitamin E, one group to Memantine, and the last group to a placebo. The primary response was their score on the Alzheimer's Disease Cooperative Study/Activities of Daily Living (ADCS-ADL) Inventory, which is designed to assess a patient's functional ability to perform a range of daily living activities, with lower scores indicating worse function. The degrees of freedom for the ANOVA F statistic comparing the mean ADCS-ADL inventory scores are

(a) 2 and 147. (b) 2 and 447. (c) 3 and 147.

27.17 The alternative hypothesis for the ANOVA F test in Exercise 27.16 is

(a) either the Vitamin E or the Memantine treatment has higher mean ADCS-ADL inventory scores than the placebo treatment.
(b) both the Vitamin E and Memantine treatments have higher mean ADCS-ADL inventory scores than the placebo treatment.
(c) the mean ADCS-AD inventory scores for the three groups are not all the same.

27.18 The F distributions are

(a) a family of distributions with bell-shaped density curves centered at 0.
(b) a family of distributions that are right-skewed and take only values greater than 0.
(c) a family of distributions that are left-skewed and take values between 0 and 1.

"Durable press" cotton fabrics are treated to improve their recovery from wrinkles after washing. Unfortunately, the treatment also reduces the strength of the fabric. This study compared the breaking strength of fabrics treated by three commercial durable press processes. Five specimens of the same fabric were assigned at random to each group. Here are the data, in pounds of pull needed to tear the fabric:[10] DATA WEAKFAB

Permafresh 55	29.9	30.7	30.0	29.5	27.6
Permafresh 48	24.8	24.6	27.3	28.1	30.3
Hylite LF	28.8	23.9	27.0	22.1	24.2

Here is a partial Minitab output including the ANOVA table (several numbers have been omitted), along with the means and standard deviations of the breaking strengths for the three durable press processes:

```
Source   DF      SS        MS   F       P
Process   2    47.50                0.026
Error    12    56.74
Total    14   104.24

Level     N     Mean    StDev
Perm55    5    25.20     2.67
Perm48    5    27.02     2.39
Hylite    5    29.54     1.17
```

Exercises 27.19 to 27.22 are based on this study.

27.19 The conclusion of the ANOVA test is that

(a) there is strong evidence ($P = 0.026$) that the mean breaking strength is not the same for all three treatments.
(b) there is strong evidence ($P = 0.026$) that the mean breaking strength is higher for the Hylite process than for the other two processes.
(c) the data give no evidence ($P = 0.026$) to suggest that mean breaking strength differs among the three processes.

27.20 The P-value in the output may not be accurate because

(a) there are several extreme outliers in the data.
(b) three treatments are too few for ANOVA.
(c) the data show some evidence of a violation of the assumption that the three populations have the same standard deviation.

27.21 To compare the treatments, we might use three 90% two-sample t confidence intervals to compare each pair of treatments: Perm55 versus Perm48, Perm55 versus Hylite, and Perm48 versus Hylite. The weakness of doing this is that

(a) we don't know how confident we can be that all three intervals cover the true differences in means.
(b) 90% confidence is okay for one comparison, but it isn't high enough for six comparisons done at once.
(c) we can't compare two treatments that use different processes.

27.22 **(Optional topic)** The value of the ANOVA F-statistic for testing equality of the population means of the three processes is

(a) 0.84. (b) 5.02. (c) 10.04.

27.23 A company runs a three-day workshop on strategies for working effectively in teams. On each day, a different strategy is presented. Forty-eight employees of the company attend the workshop. At the outset, all 48 are divided into 12 teams of four. The teams remain the same for the entire workshop. Strategies are presented in the morning. In the afternoon, the teams are presented with a series of small tasks, and the number of these completed successfully using the strategy taught that morning is recorded for each team. The mean number of tasks completed successfully by all teams each day and the standard deviation follow:

DAY	n	$\bar{x}$	s
1	12	17.25	7.10
2	12	17.64	14.14
3	12	17.21	14.03

In this example, we notice

(a) the data show very strong evidence of a violation of the assumption that the three populations have the same standard deviation.
(b) ANOVA cannot be used on these data because the sample sizes are less than 20.
(c) the assumption that the data are independent for the three days is unreasonable because the same teams were observed each day.

CHAPTER 27 EXERCISES

Exercises 27.24 to 27.27 describe situations in which we want to compare the mean responses in several populations. For each setting, identify the populations and the response variable. Then give I, n_i, and N. Finally, give the degrees of freedom of the ANOVA F statistic.

27.24 Morning or evening? Are you a morning person, an evening person, or neither? Does this personality trait affect how well you perform? A sample of 100 students took a psychological test that found 16 morning people, 30 evening people, and 54 who were neither. All the students then took a test of their ability to memorize at 8 A.M. and again at 9 P.M. The response variable is the score at 8 A.M. minus the score at 9 P.M.

27.25 Does art sell products? How does visual art affect the perception and evaluation of consumer products? Subjects were asked to evaluate an advertisement for bathroom fittings that contained an art image, a nonart image, or no image. The art image was Vermeer's painting *Girl with a Pearl Earring*, and the nonart image was a photograph of the actress Scarlett Johansson in the same pose wearing the same garments as the girl in the painting and was taken from the motion picture *Girl with a Pearl Earring*. Thus the art and nonart image were a match on content. College students were divided at random into three groups of 39 each, with each group assigned to one of the three types of advertisements. Students evaluated the product in the advertisement on a scale of 1 to 7, with 1 being the most unfavorable rating and 7 being the most favorable. The paper reported a one way ANOVA on the product evaluation index had $F = 6.29$ with $P < 0.05$.[11]

27.26 Test accommodations. Many states require schoolchildren to take regular statewide tests to assess their progress. Children with learning disabilities who read poorly may not do well on mathematics tests because they can't read the problems. Most states allow "accommodations" for learning-disabled children. Randomly assign 100 learning-disabled children in equal numbers to three types of accommodation and a control group: math problems are read by a teacher, math problems are read by a computer, math problems are read by a computer that also shows a video, and standard test conditions. The researcher would like to compare the mean scores on the state mathematics assessment.

27.27 Exercise and type 2 diabetes. It is generally accepted that regular exercise provides health benefits to individuals with type 2 diabetes, although the exact exercise regimen (aerobic vs. resistance vs. both) is unclear. The subjects in this study were sedentary 30- to 75-year-old adults with type 2 diabetes and elevated hemoglobin A1c levels above 6.5%. The level of hemoglobin A1c correlates very well with a person's recent overall blood sugar levels. If the blood sugars have generally been running high during the previous few months, the level of hemoglobin A1c will be high. In a randomized controlled study, 41 subjects were assigned to a nonexercise control group, 73 to resistance training only, 72 to aerobic exercise only, and 76 to combined aerobic and resistance training. The weekly duration of exercise was similar for all three exercise groups, and subjects remained on the exercise regimens for nine months. At the end of nine months, the hemoglobin A1c levels of subjects were measured.[12]

27.28 Don't handle the merchandise? Although consumers often want to touch products before purchasing them, they generally prefer that others have not touched products they would like to buy. Can another person touching a product create a positive reaction? Subjects were given instructions to contact a sales associate at a university bookstore who would provide them with a shirt to try on. When meeting the sales associate, subjects were told there was only one shirt left and it was being tried on by another "customer." The other customer trying on the shirt was a confederate of the experimenter and was either an attractive, well-dressed professional female model or an average-looking female college student wearing jeans and a tee shirt. Subjects, who were either males or females, saw the confederate leaving the dressing room where the shirt was left for them to try on. There was also a control group of subjects who were handed the shirt directly off the rack by the sales associate. Thus there were five treatments: male subjects seeing a model, female subjects seeing a model, male subjects seeing a college student, female subjects seeing a college student, and the control group. Subjects evaluated the product on five dimensions, each dimension on a seven-point scale, with the five scores then averaged to give the subject's evaluation measure, with higher numbers indicating a more positive evaluation. Here are the sample sizes, means, and standard deviations for the five groups:[13]

TREATMENT GROUP	n	$\bar{x}$	s
Males seeing a model	22	5.34	0.87
Males seeing a student	23	3.32	1.21
Females seeing a model	24	4.10	1.32
Females seeing a student	23	3.50	1.43
Controls	27	4.17	1.50

(a) Verify that the sample standard deviations allow the use of ANOVA to compare the population means. What do the means suggest about the effect of the subject's sex and the attractiveness of the confederate on the evaluation of the product?
(b) The paper reports the ANOVA $F = 8.30$. What are the degrees of freedom for the ANOVA F statistic and the P-value? State your conclusions.

27.29 Plants defend themselves. When some plants are attacked by leaf-eating insects, they release chemical compounds that attract other insects that prey on the leaf-eaters. A study carried out on plants growing naturally in the Utah desert demonstrated both the release of the compounds and that they not only repel the leaf-eaters but also attract predators that act as the plants' bodyguards.[14] The investigators chose eight plants attacked by each of three leaf-eaters and eight more that were undamaged, 32 plants of the same species in all. They then measured emissions of several compounds during seven hours. Here are data (mean ± standard error of the mean for eight plants) for one compound. The emission rate is measured in nanograms (ng) per hour.

GROUP	EMISSION RATE (ng/h)
Control	9.22 ± 5.93
Hornworm	31.03 ± 8.75
Leaf bug	18.97 ± 6.64
Flea beetle	27.12 ± 8.62

(a) Make a graph that compares the mean emission rates for the four groups. Does it appear that emissions increase when the plant is attacked?
(b) What hypotheses does ANOVA test in this setting?
(c) We do not have the full data. What would you look for in deciding whether you can safely use ANOVA?
(d) What is the relationship between the standard error of the mean (SEM) and the standard deviation for a sample? What are the four sample standard deviations? Do they satisfy our rule of thumb for safe use of ANOVA?

27.30 Can you hear these words? To test whether a hearing aid is right for a patient, audiologists play a tape on which words are pronounced at low volume. The patient tries to repeat the words. There are several different lists of words that are supposed to be equally difficult. Are the lists equally difficult when there is background noise? To find out, an experimenter had subjects with normal hearing listen to four lists with a noisy background. The response variable was the percent of the 50 words in a list that the subject repeated correctly. The data set contains 96 responses.[15] Here are two study designs that could produce these data:

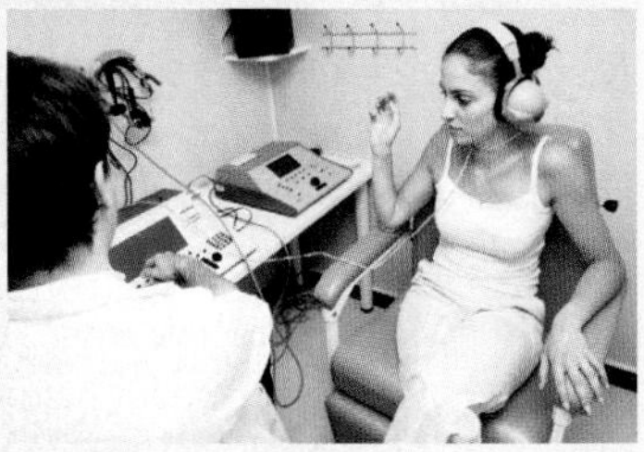

Phanie/Science Source

Design A. The experimenter assigns 96 subjects to four groups at random. Each group of 24 subjects listens to one of the lists. All individuals listen and respond separately.

Design B. The experimenter has 24 subjects. Each subject listens to all four lists in random order. All individuals listen and respond separately.

Does Design A allow use of one-way ANOVA to compare the lists? Does Design B allow use of one-way ANOVA to compare the lists? Briefly explain your answers.

27.31 More rain for California? The changing climate will probably bring more rain to California, but we don't know whether the additional rain will come during the winter wet season or extend into the long dry season in spring and summer. Kenwyn Suttle of the University of California at Berkeley and his co-workers randomly assigned plots of open grassland to three treatments: added water equal to 20% of annual rainfall either during January to March (winter) or during April to June (spring), and no added water

FIGURE 27.11
Minitab ANOVA output for comparing the total plant biomass of grassland plots under different water conditions, for Exercise 27.31.

```
Session
One-way ANOVA: biomass versus treatment

Source     DF      SS     MS      F      P
Treatment   2   97583  48792  27.52  0.000
Error      15   26593   1773
Total      17  124176

S = 42.11   R-Sq = 78.58%   R-Sq(adj) = 75.73%

                                 Individual 95% CIs For Mean Based on
                                 Pooled StDev
Level    N    Mean   StDev   -----+---------+---------+---------+---
Control  6  136.65   21.69   (-----*----)
Spring   6  315.39   37.34                                (----*----)
Winter   6  205.17   58.77              (----*----)
                             ----+---------+---------+---------+----
                               140       210       280       350

Pooled StDev = 42.11
```

(control). Here are some of the data, for plant biomass (in grams per square meter) produced by each plot in a single year.[16] MASS2003

WINTER	SPRING	CONTROL
264.1514	318.4182	129.0538
187.7312	281.6830	144.6578
291.1431	288.8433	172.7772
176.2879	382.6673	113.2813
141.7525	326.8877	142.1562
169.9737	293.8502	117.9808

Figure 27.11 shows Minitab ANOVA output for these data.

(a) Make side-by-side stemplots of plant biomass for the three treatments, as well as a table of the sample means and standard deviations. What do the data appear to show about the effect of extra water in winter and in spring on biomass? Do these data satisfy the conditions for ANOVA?

(b) State H_0 and H_a for the ANOVA F test, and explain in words what ANOVA tests in this setting.

(c) Report your overall conclusions about the effect of added water on plant growth in California.

27.32 **Can you hear these words?** Figure 27.12 displays the Minitab output for one-way ANOVA applied to the hearing data described in Design A in Exercise 27.30. The response variable is "Percent," and "List" identifies the four lists of words. Based on this analysis, is there good reason to think that the four lists are not all equally difficult? Write a brief summary of the study findings.

27.33 **College students need better sleep!** A random sample of 898 students between the ages of 20 and 24 at a large midwestern university completed a survey including questions about their sleep quality, moods, academic performance, physical health, and psychoactive drug use. Sleep quality was measured using the Pittsburgh Sleep Quality Index (PSQI), with students scoring less than or equal to 5 on the index classified as optimal sleepers, those scoring a 6 or 7

```
Session
Analysis of Variance for Percent
Source   DF       SS      MS      F      P
List      3    920.5   306.8   4.92  0.003
Error    92   5738.2    62.4
Total    95   6658.6

                                  Individual 95% CIs For Mean
                                  Based on Pooled StDev
Level     N     Mean   StDev   ---+---------+---------+---------+---
1        24   32.750   7.409                       (------*------)
2        24   29.667   8.058                (------*------)
3        24   25.250   8.316    (------*------)
4        24   25.583   7.779     (------*------)
                               ---+---------+---------+---------+---
Pooled StDev =   7.898           24.0      28.0      32.0      36.0
```

FIGURE 27.12
Minitab ANOVA output for comparing the percents heard correctly in four lists of words, for Exercise 27.32.

classified as borderline, and those scoring over 7 classified as poor sleepers. The depression subscale of the Profile of Moods State (POMS) was used to assess how severely students experienced depression on a typical day, with high scores indicating greater levels of depression. We want to know if there is a significant difference in depression scores among the three classifications of sleep.[17] The full data set is too large to print here, but here are the first seven individuals: DEPRESED

Quality of sleep:	poor	poor	border	poor	poor	optimal	border
Depression score:	5	8	5	11	7	7	10

(a) Follow the four-step process in data analysis and ANOVA. Be sure to check the conditions for ANOVA and to include an appropriate graph that compares the depression scores for the three qualities of sleep.
(b) Explain why we can trust the ANOVA *F* test to give valid results for these data.
(c) This is an observational study. Explain why. In this study, we can see why association does not prove causation. Explain how poor-quality sleep might lead to higher levels of depression. Then explain how higher depression scores might affect quality of sleep. You see that a cause-and-effect relationship might go in either direction.

27.34 **Do good smells bring good business?** Businesses know that customers often respond to background music. Do they also respond to odors? Nicolas Guéguen and his colleagues studied this question in a small pizza restaurant in France on Saturday evenings in May. On one evening, a relaxing lavender odor was spread through the restaurant; on another evening, a stimulating lemon odor; a third evening served as a control, with no odor. The three evenings were comparable in many ways (weather, customer count, and so on), so we are willing to regard the data as independent SRSs from spring Saturday evenings at this restaurant. Table 27.4 contains data on how long (in minutes) customers stayed in the restaurant on each of the three evenings.[18] ODORS

(a) Make an appropriate graph comparing the customer times for each evening. Do any of the distributions show outliers, strong skewness, or other clear deviations from Normality?
(b) Do a complete analysis to see whether the groups differ in the average amount of time spent in the restaurant. Follow the four-step process in your work. Did you find anything surprising?

TABLE 27.4 TIME (MINUTES) THAT CUSTOMERS REMAIN IN A RESTAURANT WHEN EXPOSED TO ODORS

LAVENDER ODOR									
92	126	114	106	89	137	93	76	98	108
124	105	129	103	107	109	94	105	102	108
95	121	109	104	116	88	109	97	101	106
LEMON ODOR									
78	104	74	75	112	88	105	97	101	89
88	73	94	63	83	108	91	88	83	106
108	60	96	94	56	90	113	97		
NO ODOR									
103	68	79	106	72	121	92	84	72	92
85	69	73	87	109	115	91	84	76	96
107	98	92	107	93	118	87	101	75	86

27.35 **Good weather and tipping.** Favorable weather has been shown to be associated with increased tipping. Will just the belief that future weather will be favorable lead to higher tips? The researchers gave 60 index cards to a waitress at an Italian restaurant in New Jersey. Before delivering the bill to each customer, the waitress randomly selected a card and wrote on the bill the same message that was printed on the index card. Twenty of the cards had the message "The weather is supposed to be really good tomorrow. I hope you enjoy the day!" Another 20 cards contained the message, "The weather is supposed to be not so good tomorrow. I hope you enjoy the day anyway!" The remaining 20 cards were blank, indicating that the waitress was not supposed to write any message. Choosing a card at random ensured that there was a random assignment of the diners to the three experimental conditions. Here are the percentage tips for the three messages:[19] TIPPING

Good weather report	20.8	18.7	19.9	20.6	22.0	23.4	22.8
	24.9	22.2	20.3	24.9	22.3	27.0	20.4
	22.2	24.0	21.2	22.1	22.0	22.7	
Bad weather report	18.0	19.0	19.2	18.8	18.4	19.0	18.5
	16.1	16.8	14.0	17.0	13.6	17.5	19.9
	20.2	18.8	18.0	23.2	18.2	19.4	
No weather report	19.9	16.0	15.0	20.1	19.3	19.2	18.0
	19.2	21.2	18.8	18.5	19.3	19.3	19.4
	10.8	19.1	19.7	19.8	21.3	20.6	

Do the data support the hypothesis that there are differences among the tipping percentages for the three experimental conditions? Does a prediction of good weather seem to increase the tip percentage? Follow the four-step process in data analysis and ANOVA. Be sure to check the conditions for ANOVA and to include an appropriate graph which compares the tipping percentages for the three conditions.

27.36 Bring your pen to class! The use of laptops for taking notes is becoming more common despite the fact that there is some evidence that it is less effective than longhand notes for learning. In this study, 151 subjects were randomly assigned to three groups and each group was asked to take notes in a different manner during a lecture lasting approximately 15 minutes. Forty-eight subjects were randomly assigned to take notes using longhand in the usual way they would take notes in a class when they expected to be tested on the material, and fifty-one subjects received the same instructions but were told to take their notes on a laptop. The third "laptop-intervention" group contained fifty-two students. This group used a laptop but were told that those using a laptop often transcribed what they were hearing rather than thinking about it, and they were asked to try and write the notes in their own words. The study found little difference in factual learning, but those using longhand had greater conceptual learning than either of the other groups. Here we focus on the content of what they wrote, specifically the number of words.[20] DATA WRDCOUNT

(a) Begin by comparing the distribution of word count for the three treatment groups using both graphical and numerical measures.

(b) Are the differences in the mean word counts for the three groups statistically significant? Carry out an appropriate analysis, making sure to check conditions, find the P-value, and state your conclusions in context.

27.37 Logging in the rain forest: Species counts. Table 27.2 gives data on the number of trees per forest plot, the number of species per plot, and species richness. Exercise 27.3 analyzed the effect of logging on number of trees. Exercise 27.8 concludes that it would be risky to use ANOVA to analyze richness. Use software to analyze the effect of logging on the number of species. DATA LOGFULL

(a) Make a table of the group means and standard deviations. Do the standard deviations satisfy our rule of thumb for safe use of ANOVA? What do the means suggest about the effect of logging on the number of species?

(b) Carry out the ANOVA. Report the F statistic and its P-value and state your conclusion.

27.38 In your own words, please. (4 step) The data in Exercise 27.36 show that those using longhand write fewer words than those in either of the laptop groups. Do those using laptops have greater verbatim overlap with the lecture? For each set of notes, they compared three-word chunks of text in the notes with three-word chunks of text in the lecture transcription and reported a percentage of matches, with a higher percentage suggesting that the note taker had a greater tendency to transcribe the lecture verbatim rather than writing the notes in their own words. Give a complete analysis to compare the means of the percentage of matches for the three groups. DATA VERBATIM

More rain for California? *Exercise 27.31 describes a randomized experiment carried out by Kenwyn Suttle and his co-workers to examine the effects of additional water on California grassland. The experimental units are 18 plots of grassland, assigned at random among three treatments: added water in the winter wet season, added water in the spring dry season, and no added water (control group). Field experiments, unlike laboratory experiments, are exposed to variations in the natural environment. The experiment therefore continued over five years, 2001 to 2005. Table 27.5 (page 674) gives data on the total plant biomass (grams per square meter) that grew on each plot during each year.[21] The "Plot" column shows how the random assignment of 18 of the 36 available plots worked. Exercises 27.39 to 27.41 are based on this information.*

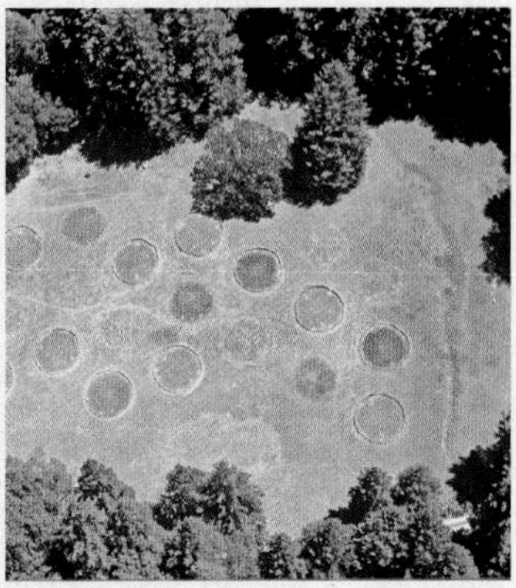
Suttle, K. B., M. A. Thomsen, and M. E. Power. "Species Interactions Reverse Grassland Responses to Changing Climate." Science 315.5812 (2007): 640–42.

27.39 Plot the means. Starting from the data in Table 27.5, you can calculate the mean plant biomass for each treatment in each year as follows: DATA MASSMEAN

	YEAR				
	2001	**2002**	**2003**	**2004**	**2005**
Winter	132.58	203.33	205.17	223.58	332.84
Spring	257.69	388.85	315.39	299.54	289.66
Control	81.67	180.31	136.65	201.07	257.37

(a) Plot the means for each of the three treatments against year, connecting the yearly means for each treatment by lines to show the pattern over time. Use the same plot for all three treatments, with a different color for each treatment. From this plot, you can get an overall picture of the experiment's results.

(b) Across all five years, does more water in the wet season increase plant growth? What about more water in the dry season? Which seasonal addition of water has the larger effect?

(c) One-way ANOVAs comparing the mean plant biomass separately in each year find significant differences in three years and no significant difference in two years. Based on your plot, in which three years do you think the treatment means differ significantly?

TABLE 27.5 PLANT BIOMASS FOR THREE WATER CONDITIONS OVER FIVE YEARS

		YEAR				
TREATMENT	PLOT	2001	2002	2003	2004	2005
Winter	3	136.8358	228.0717	264.1514	254.6453	344.3933
Winter	8	151.4154	189.9505	187.7312	233.8155	203.3908
Winter	14	136.1536	209.0485	291.1431	253.4506	331.9724
Winter	20	121.6323	189.6755	176.2879	228.5882	388.1056
Winter	27	124.1459	188.0090	141.7525	158.6675	382.8617
Winter	32	125.2986	215.2174	169.9737	212.3232	346.3042
Spring	4	338.1301	422.7411	318.4182	517.6650	344.0489
Spring	11	291.8597	339.8243	281.6830	342.2825	261.8016
Spring	18	244.8727	398.7296	288.8433	270.5785	262.7238
Spring	22	234.6599	400.6878	382.6673	212.5324	316.9683
Spring	25	197.5830	326.9497	326.8877	213.9879	224.1109
Spring	35	239.0122	444.1556	293.8502	240.1927	328.2783
Control	6	73.4288	148.8907	129.0538	178.9988	237.6596
Control	7	110.6306	182.6762	144.6578	205.5165	281.1442
Control	17	95.3405	196.8303	172.7772	242.6795	313.7242
Control	24	83.0584	186.1953	113.2813	231.7639	258.3631
Control	28	30.5886	154.0401	142.1562	134.9847	235.8320
Control	33	96.9709	213.2537	117.9808	212.4862	217.5060

(d) In 2005, there were unusual late rains during the spring. How does the effect of this natural rainfall show up in your plot? (You see that it would not be wise to do an experiment like this in just one year.)

27.40 4 step **The results for 2001.** Your work in Exercise 27.31 shows that there were significant differences in mean plant biomass among the three treatments in 2003. Do a complete analysis of the data for 2001 and report your conclusions. DATA MASSALL

27.41 Conditions for ANOVA. Examine the data for the year 2004. The conditions for ANOVA inference are not met. In what way do these data fail to meet the conditions? (It is not very surprising that, in five ANOVAs, one will fail to satisfy our quite conservative conditions.) DATA MASSALL

27.42 4 step **Does playing video games make better surgeons?** In laparoscopic surgery, a video camera and several thin instruments are inserted into the patient's abdominal cavity. The surgeon uses the image from the video camera positioned inside the patient's body to perform the procedure by manipulating the instruments that have been inserted. The Top Gun Laparoscopic Skills and Suturing Program was developed to help surgeons develop the skill set necessary for laparoscopic surgery. Because of the similarity in many of the skills involved in video games and laparoscopic surgery, it was hypothesized that surgeons with greater prior video game experience might acquire the skills required in laparoscopic surgery more easily. Thirty-three surgeons participated in the study and were classified into the three categories—never used, under three hours, and more than three hours—depending on the number of hours they played video games at the height of their video game use. They also performed Top Gun drills, and received a score based on the time to complete the drill and the number of errors made, with lower scores indicating better performance. Here are the Top Gun scores and video game categories for the 33 participants.[22] DATA TOPGUN

Never played	9379 8302 5489 5334 4605 4789 9185 7216 9930 4828 5655 4623 7778 8837 5947
Under three hours	5540 6259 5163 6149 4398 3968 7367 4217 5716
Three or more hours	7288 4010 4859 4432 4845 5394 2703 5797 3758

Give a complete analysis to compare the means of the Top Gun performance scores for the three groups. This is an observational study. Explain why and indicate how this affects the conclusions that can be drawn.

27.43 Which test? Example 27.4 describes one of the experiments done by Kathleen Vohs and her coworkers to demonstrate that even being reminded of money makes people more self-sufficient and less involved with other people. Here are three more of these experiments. For each experiment, which statistical test from Chapters 20 to 27 would you use, and why?

(a) Randomly assign student subjects to money and control groups. The control group unscrambles neutral phrases, and the money group unscrambles money-oriented phrases, as described in Example 27.4. Then ask the subjects to volunteer to help the experimenter by coding data sheets, about five minutes per sheet. Subjects said how many sheets they would volunteer to code. Participants in the money condition volunteered to help code fewer data sheets than did participants in the control condition.

(b) Randomly assign student subjects to high-money, low-money, and control groups. After playing Monopoly for a short time, the high-money group is left with \$4000 in Monopoly money, the low-money group with \$200, and the control group with no money. Each subject is asked to imagine a future with lots of money (high-money group), a little money (low-money group), or just their future plans (control group). Another student walks in and spills a box of 27 pencils. How many pencils does the subject pick up? Participants in the high-money condition gathered fewer pencils than subjects in the other two groups.

(c) Randomly assign student subjects to three groups. All do paperwork while a computer on the desk shows a screensaver of currency floating underwater (Group 1), a screensaver of fish swimming underwater (Group 2), or a blank screen (Group 3). Each subject must now develop an advertisement and can choose whether to work alone or with a partner. Count how many in each group make each choice. Choosing to perform the task with a co-worker was reduced among money condition participants.

Exploring the Web

27.44 Confidence in the banking system. The General Social Survey (GSS) is a sociological survey used to collect data on demographic characteristics and attitudes of residents of the United States. The survey is conducted face-to-face with an in-person interview by the National Opinion Research Center at the University of Chicago, of a randomly selected sample of adults (18+). SDA (Survey Documentation and Analysis) is a set of programs that allows you to analyze survey data and includes the GSS survey as part of its archive. Go to the website sda.berkeley.edu/ and click on Archive. Unless there is a more recent file, open the 1972–2012 cumulative data file (without the quick tables option).

(a) In the Analysis tab on the top of the page, click on Comparison of Means. You are to do an ANOVA that examines how the mean age of the respondent varies with the person's confidence in the banking and financial systems. To do this, type in the dependent variable as "Age" and the row (treatment) variable as "Confinan." For the selection filter, type in "Year (2012)," or the most recent year available. For weight, change it to "noweight." Finally, in table options, the *only* boxes that should be checked are "Std dev," "N," and "ANOVA stats." Make sure checks are removed from the other boxes. Now click on Run the Table.

(b) How many respondents are included in the analysis? What are the three means and standard deviations? In the ANOVA table, explain how the degrees of freedom were obtained. What are the F- and P-values? Write a brief report explaining the relationship between the average respondent's age and their confidence in the banking system.

27.45 Confidence in the banking system, continued. This exercise is a continuation of Exercise 27.44. You are going to download the data set and reproduce the analysis, as well as provide some additional plots. First open the 1972–2012 cumulative data file following the instructions in Exercise 27.44.

(a) In the Download tab on the top of the page, click on Customized Subset. For the data file, if you highlight the CSV bubble, an Excel spreadsheet will be downloaded. For the selection filter, again type in "year (2012)" or the year used in Exercise 27.44. In the box for entering the names of individual variables, enter "age" and "confinan." Click Continue on the bottom of the page. In the new window, click on Create the Files and in the next window click on Data Files. You can now either open or save the data file to your computer.

(b) Import the data into your statistical software package. You first need to "clean" the data a little because there are observations for which either the "age" or "confinan" variable are missing. For the "confinan" variable, any value other than a 1, 2, or 3 is a missing value code. Delete these observations. For the "age" variable, the missing value codes are 0, 98, and 99. Eliminate any observations with these values for "age." You should now have the same number of observations as in Exercise 27.44.

(c) Draw comparative boxplots of the age distribution for the three values of "confinan." Describe the shapes of the three distributions. What information can you obtain from the boxplots that was not included in the output for Exercise 27.44?

(d) Reproduce the one-way ANOVA table using your software. Your results should agree with Exercise 27.44.

NOTES AND DATA SOURCES

Chapter 0 Notes

1. Parts of this essay are shared with David S. Moore, "Introduction: learning from data," in Roxy Peck et al. (eds.), *Statistics: A Guide to the Unknown*, 4th ed., Thomson, 2006.
2. See, for example, Martin Enserink, "The vanishing promises of hormone replacement," *Science*, 297 (2002), pp. 325–326; and Brian Vastag, "Hormone replacement therapy falls out of favor with expert committee," *Journal of the American Medical Association*, 287 (2002), pp. 1923–1924. A National Institutes of Health panel's comprehensive report is *International Position Paper on Women's Health and Menopause*, NIH Publication 02-3284, 2002.
3. A. C. Nielsen, Jr., "Statistics in marketing," in *Making Statistics More Effective in Schools of Business*, Graduate School of Business, University of Chicago, 1986.
4. The data in Figure 0.2 were found online at `www.eia.gov/dnav/pet/hist/LeafHandler.ashx?n=PET&s=EMM_EPMR_PTE_NUS_DPG&f=W`.
5. FUTURE II Study Group, "Quadrivalent vaccine against human papillomavirus to prevent high-grade cervical lesions," *New England Journal of Medicine*, 356 (2007), pp. 1915–1927. We have simplified the conclusions so that students with as yet no statistics background can better follow the essay.
6. Rita F. Redburg, "Vitamin E and cardiovascular health," *Journal of the American Medical Association*, 294 (2005), pp. 107–109.
7. The data in Figure 0.3 were found online at `www1.eere.energy.gov/vehiclesandfuels/facts/2012_fotw741.html`.
8. Gerd Gigerenzer, "Dread risk, September 11, and fatal traffic accidents," *Psychological Science*, 15 (2004), pp. 286–287. The graph in Figure 0.3 was adapted from a graph in the article.

Chapter 1 Notes

1. Higher Education Research Institute 2013 Freshman Survey, at `www.heri.ucla.edu`.
2. *The Infinite Dial 2013: Navigating Digital Platforms,* at `www.arbitron.com`.
3. *Radio Today 2013: How America Listens to Radio,* at `www.arbitron.com`.
4. See Note 1.
5. Centers for Disease Control and Prevention, National Center for Health Statistics, *Births: Final data for 2012,* National Vital Statistics Reports, 62, No. 9, December 2013 at `www.cdc.gov/nchs/products/nvsr.htm`. The data for day of the week are given in the Supplemental Tables.
6. Marilyn M. Seastrom et al., "User's guide to computing high school graduation rates, Volume 1," from the National Center for Education Statistics, 2006, and the Department of Education website for the 2010 graduation rates at `ed.gov`.
7. Our eyes do respond to area, but not quite linearly. It appears that we perceive the ratio of two bars to be about the 0.7 power of the ratio of their actual areas. See W. S. Cleveland, *The Elements of Graphing Data,* Wadsworth, 1985, pp. 278–284.
8. From the 2011 American Community Survey at `www.pewhispanic.org`.
9. From the Gary Community School Corporation, courtesy of Celeste Foster, Purdue University.
10. The 2013 state information was compiled from the College Board website at `http://www.commonwealthfoundation.org/policyblog/detail/sat-scores-by-state-2013`.
11. From the Centers for Disease Control website at `cdc.gov/lyme/stats/index.html`. This is the chart of cases by age and sex, 2001–2010. The number of confirmed cases is from the chart of cases by symptom, 2001–2010.
12. The per capita total health expenditures in 2011 were obtained from the Country Data at the Global Health Observatory Data Repository of the World Health Organization at `http://apps.who.int/gho/data/`. All amounts are in international dollars at purchasing power parity. That is, the exchange rate between a currency and the dollar is set not at the fluctuating market rate but at the rate that gives a dollar the same buying power in each country.
13. The U.S. Geological Survey maintains data for various water parameters at monitoring sites throughout the United States at `waterdata.usgs.gov/nwis`. The data can be graphed or downloaded. The data in Figure 1.12 are for USGS 254754080344300 SHARK RIVER SLOUGH NO.1.
14. College Entrance Examination Board, *Trends in College Pricing, 2013,* at `www.trends.collegeboard.org`. The averages are "enrollment weighted," so that they give average tuition over *students* rather than over *colleges.* The reported averages have been adjusted to constant 2013 dollars.
15. The percentages were obtained from the figure for out-of-state tuition and fees at public four-year institutions at `www.trends.collegeboard.org`.
16. PPG Industries annual automotive color trend data `www.ppg.com/en/newsroom/news/Pages/20131024A.aspx`.
17. Data on the distribution of age for social networking sites was taken from a graph whose original source was `www.comscore.com`.
18. Centers for Disease Control and Prevention, National Center for Health Statistics, *Deaths: Preliminary Data for 2011,* 61, No. 6, October 2012, at `www.cdc.gov/nchs`.

19. Sharon R. Ennis et al., "The Hispanic Population: 2010," 2010 U.S. Census Briefs at `www.census.gov`.
20. "2008 Student surveys: Complete results," *Macleans.ca*, February 19, 2008, at `oncampus.macleans.ca`.
21. Tom Lloyd et al., "Fruit consumption, fitness, and cardiovascular health in female adolescents: the Penn State Young Women's Health Study," *American Journal of Clinical Nutrition*, 67 (1998), pp. 624–630.
22. Data provided by Darlene Gordon from her PhD thesis, "Relationships among academic self-concept, academic achievement, and persistence with self-attribution, study habits, and perceived school environment," Purdue University, 1997.
23. Monthly stock returns from the website of Professor Kenneth French of Dartmouth, `mba.tuck.dartmouth.edu/pages/faculty/ken.french`. A fine point: the data are actually the "excess returns" on stocks, the actual returns less the small monthly returns on Treasury bills. The data is in the file Fama/French Benchmark Factors.
24. National Institutes of Health, Essential Fatty Acids Education site, `efaeducation.nih.gov`.
25. 2010 *Statistical Abstract of the United States*, Table 159, at `www.census.gov`.
26. As of the beginning of 2014, yearly data were available through 2010 at the World Bank website `data.worldbank.org/indicator/EN.ATM.CO2E.PC`.
27. J. Ward Testa (editor), "Fur Seal Investigations, 2008–2009," *NOAA Technical Memorandum NMFS-AFSC-226*, (2011), p. 74. We would like to thank Rod Towell of NOAA Federal for supplying recent data not contained in the report.
28. Domenico Giannotti et al., "Play to become a surgeon: Impact of Nintendo Wii training on laparoscopic skills," *PLOS ONE*, V8, e5272, February 2013 at `www.plosone.org`.
29. David M. Fergusson and L. John Horwood, "Cannabis use and traffic accidents in a birth cohort of young adults," *Accident Analysis and Prevention*, 33 (2001), pp. 703–711.
30. John Morton, et al., "Acoustic features mediating height estimation from human speech," Abstract from the 166th meeting of the Acoustical Society of America. We would like to thank the authors for supplying the data and additional details of the study.
31. See Note 14.
32. U.S. Census Bureau, New Residential Construction page, at `http://www.census.gov/construction/nrc/`. Go to the *Search Database* link to download the data. These are monthly data that are not seasonally adjusted.
33. Ozone Hole Watch, at `ozonewatch.gsfc.nasa.gov/index.html`.

Chapter 2 Notes

1. From the 2003 American Community Survey, at the U.S. Census Bureau website, `www.census.gov`. The data are a subsample of the 13,194 individuals in the ACS North Carolina sample who had travel times greater than zero.
2. This study is available online at `www.dispatch.com/live/content/databases/index.html`.
3. This isn't a mathematical theorem. The mean can be less than the median in right-skewed distributions that take only a few values, many of which lie exactly at the median. The rule almost never fails for distributions taking many values, and most counterexamples don't appear clearly skewed in graphs even though they may be slightly skewed according to technical measures of skewness. See Paul T. von Hippel, "Mean, median, and skew: Correcting a textbook rule," *Journal of Statistics Education*, 13, No. 2 (2005), online journal,
4. National Association of College and University Business Officers, Commonfund study of endowments for 2013, at `www.nacubo.org`.
5. From the U.S. Census Bureau, `www.census.gov/construction/nrs/pdf/uspricemon.pdf`.
6. U.S. Census Bureau, *Historical Income Tables: Households*, at `www.census.gov`.
7. The U.S. Department of Energy, `www.fueleconomy.gov/feg/download.shtml`.
8. We would like to thank Patricia Humphrey for supplying the test scores for students at Georgia Southern University.
9. From the Environmental Protection Agency, `www.epa.gov/radon/pubs/consguid.html`.
10. C. H. Cannon, D. R. Peart, and M. Leighton, "Tree species diversity in commercially logged Bornean rainforest," *Science*, 281 (1998), pp. 1366–1367. We thank Charles Cannon for providing the data.
11. Raymond Fisman and Edward Miguel, "Cultures of corruption: Evidence from diplomatic parking tickets," National Bureau of Economic Research Working Paper 12312, June 2006, at `www.nber.org`.
12. D. G. Jakovljevic and A. K. McConnell, "Influence of different breathing frequencies on the severity of inspiratory muscle fatigue induced by high-intensity front crawl swimming," *Journal of Strength and Conditioning Research*, 23, No. 4 (2009), pp. 1169–1174.
13. Current Population Survey 2012 Annual Social and Economic Supplement, at `www.census.gov`.
14. John J. Topoleski, *Retirement Savings and Household Wealth in 2010*, Congressional Research Service, April 2013.
15. T. Bjerkedal, "Acquisition of resistance in guinea pigs infected with different doses of virulent tubercle bacilli," *American Journal of Hygiene*, 72 (1960), pp. 130–148.
16. See Note 5 for Chapter 1.
17. Data for 1986 from David Brillinger, University of California, Berkeley. See David R. Brillinger, "Mapping aggregate birth data," in A. C. Singh and P. Whitridge (eds.), *Analysis of Data in Time*, Statistics Canada, 1990, pp. 77–83. A boxplot similar to Figure 2.6 appears in David R. Brillinger, "Some examples of random process environmental data

analysis," in P. K. Sen and C. R. Rao (eds.), *Handbook of Statistics*, Vol. 18, *Bioenvironmental and Public Health Statistics*, North Holland, 2000.

18. Paul E. O'Brien et al., "Laparascopic adjustable gastric banding in severely obese adolescents," *Journal of the American Medical Association*, 303 (2010), pp. 519–526. We thank the authors for providing the data.
19. The current roster is from `canadiens.nhl.com`, and their salaries were obtained from `usatoday.com/sports/nhl`.
20. Nicolas Guéguen and Christine Petr, "Odors and consumer behavior in a restaurant," *Journal of Hospitality Management*, 25 (2006), pp. 335–339. We thank Nicolas Guéguen for providing the data.
21. Bruce Rind and David Strohmetz, "Effects of beliefs about future weather conditions on restaurant tipping," *Journal of Applied Social Psychology*, 31 (2001), pp. 2160–2164. We thank the authors for supplying the original data.
22. James C. Rosser et al., "The impact of video games on training surgeons in the 21st century," *Archives of Surgery*, 142 (2007), pp. 181–186. We thank Douglas Gentile for providing the data.
23. Information and data from the NHANES survey can be found at `cdc.gov/nchs/nhanes.htm`.
24. Revenues for the Global 500 companies can be found at `http://money.cnn.com/magazines/fortune/global500/2013/full_list/`.

Chapter 3 Notes

1. See Note 9 for Chapter 1.
2. Cheryl D. Fryar et al., "Anthropometric reference data for children and adults: United States, 2007–2010," *Vital and Health Statistics*, Series 11, Number 252, (October, 2012), at `www.cdc.gov/nchs`. This report provides the means of various anthropometric measurements. Standard deviations were computed from the first and third quartiles assuming Normality. Instructions on measuring upper arm length are in the *National Health and Nutrition Examination Survey: Anthropometry Procedures Manual*, January 2007.
3. Monsoon rainfall from B. Parthasarathy, Indian Institute of Tropical Meterology, at `www.iges.org`. The data cover the years 1871 to 2000.
4. See Note 2.
5. All SAT facts are from the College Board website, `www.collegeboard.com`, and all ACT facts are from the ACT website, `www.act.org`.
6. See Note 2.
7. From the 2009–10 Guide for the College Bound Student Athlete at `www.ncaastudent.org/NCAA_Guide.pdf`.
8. All MCAT facts are from the Medical College Admissions Test website, `www.aamc.org/applying/students/mcat/admissionsadvisors/mcat_stats/`.
9. Detailed data appear in P. S. Levy et al., "Total Serum Cholesterol Values for Youths 12–17 Years" *Vital and Health Statistics*, Series 11, No. 155, National Center for Health Statistics, 1976.
10. James A. Levine et al., "Inter-individual variation in posture allocation: Possible role in human obesity," *Science*, 307 (2005), pp. 584–586. We thank James Levine for providing the data.
11. See Note 2.
12. See Note 22 for Chapter 1.
13. The data were provided by Nicolas Fisher.
14. See Note 8 for Chapter 2.
15. See Note 3.
16. See Note 23 for Chapter 2.

Chapter 4 Notes

1. Neal E. Cantin et al., "Ocean warming slows coral growth in the Central Red Sea," *Science*, 329 (2010), pp. 322–325.
2. Data for 2013 college bound seniors, SAT scores are from the College Board website, `www.collegeboard.org`. Data for the percent of college bound seniors taking the SAT in each state are from the website `http://www.commonwealthfoundation.org/policyblog/detail/sat-scores-by-state-2013`.
3. The homicide and suicide rates were found online at the website `http://www.healthy.ohio.gov/vipp/data/county.aspx`.
4. Initial concerns were based on government data for 2005, presented in "An accident waiting to happen?" *Consumer Reports*, March 2007, pp. 16–19. Data for 2012 was found online at `www.transtats.bts.gov` (for delay percent by airline) and `web.mit.edu/airlinedata/www/default.html` (outsource).
5. The Florida Department of Highway Safety and Motor Vehicles (at `www.flhsmv.gov/dmv/vslfacts.html`) gives the number of registered vessels. The Florida Wildlife Commission maintains a manatee death database at `research.myfwc.com/manatees`.
6. Based on T. N. Lam, "Estimating fuel consumption from engine size," *Journal of Transportation Engineering*, 111 (1985), pp. 339–357. The data for 10 to 50 km/h are measured; those for 60 and higher are calculated from a model given in the paper and are therefore smoothed.
7. A careful study of this phenomenon is W. S. Cleveland, P. Diaconis, and R. McGill, "Variables on scatterplots look more highly correlated when the scales are increased," *Science*, 216 (1982), pp. 1138–1141.
8. See Note 1.
9. Data for Figure 4.6(b) come from William Gray's website, at `hurricane.atmos.colostate.edu`. Data for Figure 4.6(c) were provided by Drina Iglesia, Purdue University, from a study reported in D. D. S. Iglesia, E. J. Cragoe, Jr., and J. W. Vanable, "Electric field strength and epithelization in the newt (*Notophthalmus viridescens*)," *Journal of Experimental Zoology*, 274 (1996), pp. 56–62. Data for Figure 4.6(d) are for

the Wilshire 5000 stock index. As a fine point, plots (b), (c), and (d) are square with the same scales on both axes because both variables measure similar quantities in the same units.

10. This exercise is based on a study by Gareth Hagger-Johnson et al., "Reaction times and mortality from the major causes of death: The NHANES-III study," *PLOS ONE*, 9(1) (2014), pp. 1–6.
11. This exercise is based on a study by Shakira F. Suglia et al., "Soft drinks consumption is associated with behavior problems in 5-year-olds," *The Journal of Pediatrics*, 163, No. 5 (2013), pp. 1323–1328. The actual study was more involved than indicated in the exercise. Researchers did find an association between soft drinks consumed and behavior problems, but did not compute a correlation.
12. The data are a sample of seven values in a graph in Julie A. Mennella et al. "Preferences for salty and sweet tastes are elevated and related to each other during child-hood," *PLOS ONE*, 2014. The paper is available online at `www.plosone.org/article/info:doi/10.1371/journal.pone.0092201#pone-0092201-g003`. The seven values selected produce a correlation similar to that found for all the data presented in the paper.
13. This exercise is motivated by Scott Berry, "Statistical fallacies in sports," *Chance*, 19, No. 4 (2006), pp. 50–56, where scores from the 2006 Masters are analyzed. Masters scores for 2013 were found online at `www.masters.com/en_US/discover/past_winners.html`.
14. Andrew J. Oswald, et al., "Objective confirmation of subjective measures of human well-being: evidence from the U.S.A.," *Science*, 327 (2010), pp. 576–579.
15. From a graph in Naomi E. Allen et al., "Moderate alcohol intake and cancer incidence in women," *Journal of the National Cancer Institute*, 101 (2009), pp. 296–305.
16. From a graph in Magdalena Bermejo et al., "Ebola outbreak killed 5000 gorillas," *Science*, 314 (2006), p. 1564.
17. From a graph in Bernt-Erik Saether, Steiner Engen, and Erik Mattysen, "Demographic characteristics and population dynamical patterns of solitary birds," *Science*, 295 (2002), pp. 2070–2073.
18. From a graph in Sabrina M. Tom et al., "The neural basis of loss aversion in decision-making under risk," *Science*, 315 (2007), pp. 515–518.
19. From a graph in Martin Wild et al., "From dimming to brightening: Decadal changes in solar radiation at Earth's surface," *Science*, 308 (2005), pp. 847–850.
20. From a graph in Camilla A. Hinde et al., "Parent-offspring conflict and coadaptation," *Science*, 327 (2010), pp. 1373–1376.
21. See Note 21 for Chapter 2.
22. The data are available online at `ncdc.noaa.gov/sotc/global/2013/13`. The website explains that the 20th century average global temperature (combined land and ocean surface temperature) was 13.9 degrees Celsius.
23. Brian J. Whipp and Susan A. Ward, "Will women soon outrun men?" *Nature*, 355 (1992), p. 25. An article in *Scientific American* in 2004 made a similar prediction that women would outrun men in the 100-meter dash in 2156. This article can be found online at `www.scientificamerican.com/article.cfm?id=data-trends-suggest-women`.
24. From a graph in Glenn J. Tattersall et al., "Heat exchange from the toucan bill reveals a controllable vascular thermal radiator" *Science*, 325 (2009), pp. 468–470.
25. From a graph in Naomi I. Eisenberger, Matthew D. Lieberman, and Kipling D. Williams, "Does rejection hurt? An fMRI study of social exclusion," *Science*, 302 (2003), pp. 290–292.
26. Ben Dantzer et al., "Density triggers maternal hormones that increase adaptive off-spring growth in a wild mammal," *Science*, 340 (2013), pp. 1215–1217. The data used here are estimated from a figure in the paper.
27. Data for 2013 college bound seniors SAT scores are from the College Board website, `www.collegeboard.org`. Average teacher salaries are available at `www.washingtonpost.com/blogs/answer-sheet/wp/2013/12/15/how-much-teachers-get-paid-state-by-state/`.

Chapter 5 Notes

1. From a graph in James A. Levine, Norman L. Eberhardt, and Michael D. Jensen, "Role of nonexercise activity thermogenesis in resistance to fat gain in humans," *Science*, 283 (1999), pp. 212–214.
2. The online article was found at `www.livestrong.com/blog/everyone-u-s-exercise-wed-lose-2-billion-pounds/?utm source=newsletter&utm_medium=email&utm_campaign=0312`.
3. M. C. Hansen et al., "High-resolution global maps of 21st-century forest cover change," *Science*, 342 (2013), pp. 850–853.
4. See Note 1 for Chapter 4.
5. The homicide and suicide rates were found online at the website `http://www.healthy.ohio.gov/vipp/data/county.aspx`.
6. From a graph in Tania Singer et al., "Empathy for pain involves the affective but not sensory components of pain," *Science*, 303 (2004), pp. 1157–1162. Data for other brain regions showed a stronger correlation and no outliers.
7. Contributed by Marigene Arnold, Kalamazoo College.
8. Gannett News Service article appearing in the *Lafayette (Ind.) Journal and Courier*, April 23, 1994.
9. P. Goldblatt (ed.), *Longitudinal Study: Mortality and Social Organization*, Her Majesty's Stationery Office, 1990. At least, so claims Richard Conniff, *The Natural History of the Rich*, Norton, 2002, p. 45. The Goldblatt report is not available to us.
10. Laura L. Calderon et al., "Risk factors for obesity in Mexican-American girls: Dietary factors, anthropometric factors, physical activity, and hours of television viewing," *Journal of the American Dietetic Association*, 96 (1996), pp. 1177–1179.

11. *The Health Consequences of Smoking: 1983,* Public Health Service, Washington, D.C., 1983.
12. Data provided by Robert Dale, Purdue University.
13. The data are a sample of seven values in a graph in Julie A. Mennella et al. "Preferences for salty and sweet tastes are elevated and related to each other during childhood," *PLOS ONE,* 2014. The paper is available online at `www.plosone.org/article/info:doi/10.1371/journal.pone.0092201#pone-0092201-g003`. The seven values selected produce a correlation similar to that found for all the data presented in the paper.
14. G. L. Kooyman et al., "Diving behavior and energetics during foraging cycles in king penguins," *Ecological Monographs,* 62 (1992), pp. 143–163.
15. Chu, S., "Diamond ring pricing using simple linear regression," *Journal of Statistics Education,* 4 (1996), available online at `www.amstat.org/publications/jse/v4n3/datasets.chu.html`.
16. The last data pair is the height of one of the authors and his sister. The first 11 data pairs are from Karl Pearson and A. Lee, "On the laws of inheritance in man," *Biometrika,* 2 (1902), p. 357. These first 11 data also appear in D. J. Hand et al., *A Handbook of Small Data Sets,* Chapman & Hall, 1994. This book offers more than 500 data sets that can be used in statistical exercises.
17. From a presentation by Charles Knauf, Monroe County (N.Y.) Environmental Health Laboratory.
18. See Note 17 for Chapter 4.
19. Frank J. Anscombe, "Graphs in statistical analysis," *The American Statistician,* 27 (1973), pp. 17–21.
20. Debora L. Arsenau, "Comparison of diet management instruction for patients with non-insulin dependent diabetes mellitus: Learning activity package vs. group instruction," MS thesis, Purdue University, 1993.
21. See Note 2 for Chapter 4.
22. Gary Smith, "Do statistics test scores regress toward the mean?" *Chance,* 10, No. 4 (1997), pp. 42–45.
23. From a graph in G. D. Martinsen, E. M. Driebe, and T. G. Whitham, "Indirect interactions mediated by changing plant chemistry: Beaver browsing benefits beetles," *Ecology,* 79 (1998), pp. 192–200.
24. P. Velleman, *ActivStats 2.0,* Addison Wesley Interactive, 1997.
25. From William Gray's website, `hurricane.atmos.colostate.edu`. Forecasts are those made each June.
26. Data for 1936–1999 are from a graph in Bruce J. Peterson et al., "Increasing river discharge to the Arctic Ocean," *Science,* 298 (2002), pp. 2171–2173. Data for 2000–2008 are from a graph in I. Ashik et al., "Arctic report card: Update for 2010," available online at `http://www.arctic.noaa.gov/reportcard_previous.html`. The graph is on page 41 of the report. Data for 2009–2010 are from a graph in K. R. Arrigo et al., "Arctic report card: update for 2011," available online at `http://www.arctic.noaa.gov/reportcard_previous.html`. The graph is on page 157 of the report.
27. See Note 23 for Chapter 4.

Chapter 6 Notes

1. Lydia Said, "In U.S., half of women prefer a job outside the home," Gallup, September 7, 2012. Found online, `www.gallup.com`.
2. Rani A. Desai et al., "Video-gaming among high school students: Health correlates, gender differences, and problematic gaming," *Pediatrics,* 126 (2010), pp. 1416–1424.
3. From the October 2009 Current Population Survey, at `www.census.gov`.
4. Siem Oppe and Frank De Charro, "The effect of medical care by a helicopter trauma team on the probability of survival and the quality of life of hospitalized victims," *Accident Analysis and Prevention,* 33 (2001), pp. 129–138. The authors give the data in Example 6.4 as a "theoretical example" to illustrate the need for their more elaborate analysis of actual data using severity scores for each victim.
5. Found online at `www.benedicttigers.com/cumestats.aspx?path=mbball&year=2012`.
6. I. Westbrooke, "Simpson's paradox: An example in a New Zealand survey of jury composition," *Chance,* 11 (1998), pp. 40–42.
7. These data are from an April 20, 2010, and report, "Teens and mobile phones," by Amanda Lenhart, Rich Ling, Scott Campbell, and Kristen Purcell of the Pew Internet and American Life Project. It can be found online at `pewinternet.org/Reports/2010/Teens-and-Mobile-Phones.aspx`.
8. This General Social Survey exercise presents a table constructed using the search function at the GSS archive, `sda.berkeley.edu/archive.htm`. These data are from the 2012 GSS.
9. Gregory D. Myer et al., "Youth versus adult weightlifting injuries presenting to United States emergency rooms: Accidental versus nonaccidental injury mechanisms," *Journal of Strength and Conditioning Research,* 23 (2009), pp. 2054–2060.
10. These data are available at the U.S. Census Bureau website, online at `www.census.gov/hhes/www/cpstables/032012/perinc/pinc02_000.htm`.
11. M. Radelet, "Racial characteristics and imposition of the death penalty," *American Sociological Review,* 46 (1981), pp. 918–927.
12. D. Gonzales et al., "Journal of the American Medical Association Varenicline, an $\alpha 4\beta 2$ Nicotinic Acetylcholine Receptor Partial Agonist, vs Sustained-Release Bupropion and Placebo for Smoking Cessation," *New England Journal of Medicine,* 340 (1999), pp. 685–691.
13. Michael Gurian, "Where have the men gone? No place good," *Washington Post,* December 4, 2005, at `www.washingtonpost.com`. These data are from the website of the National Center for Education Statistics, `http://nces.ed.gov/pubs2013/2013008.pdf`.
14. Nancy J. O. Birkmeyer, "Hospital complication rates with bariatric surgery in Michigan," *Journal of the American Medical Association,* 304 (2010), pp. 435–442.

15. The data for the University of Michigan Health and Retirement Study (HRS) can be downloaded from the website `ssl.isr.umich.edu/hrs/start.php`.
16. Data on Phil's predictions can be found online at `www.groundhog.org`. Historical data on March temperatures can be found online at `www.ncdc.noaa.gov/cag/`.
17. Hannah Lund et al., "Sleep patterns and predictors of disturbed sleep in a large population of college students," *Journal of Adolescent Health,* 46 (2010), 124–132. We would like to thank the authors for supplying the data.

Chapter 7 Notes

1. Harry B. Meyers, "Investigations of the life history of the velvetleaf seed beetle, *Althaeus folkertsi Kingsolver*," MS thesis, Purdue University, 1996.
2. From a plot in K. Krishna Kumar et al., "Unraveling the mystery of Indian monsoon failure during El Niño," *Science,* 314 (2006), pp. 115–119.
3. From the 2006 American Community Survey, at `factfinder.census.gov`.
4. Kristen Purcell et al., "Search engine use 2012," March 2012, Pew Internet and American Life Project at `www.pewinternet.org`.
5. See Note 8 for Chapter 3.
6. Data were provided by Brigitte Baldi, University of California at Irvine.
7. Data on spending were found online at `www.capgeek.com/archive/`. Points earned by each team were found online at `www.hockey-reference.com/leagues/NHL_2012.html`.
8. J. T. Dwyer et al., "Memory of food intake in the distant past," *American Journal of Epidemiology,* 130 (1989), pp. 1033–1046.
9. Data from a plot in Josef P. Rauschecker, Biao Tian, and Marc Hauser, "Processing of complex sounds in the macaque nonprimary auditory cortex," *Science,* 268 (1995), pp. 111–114. The paper states that there are $n = 41$ observations, but only $n = 37$ can be read accurately from the plot.
10. Mei-Hui Chen, "An exploratory comparison of American and Asian consumers' catalog patronage behavior," MS thesis, Purdue University, 1994.
11. "Dancing in step," *Economist,* March 22, 2001.
12. From a plot in Jon J. Ramsey et al., "Energy expenditure, body composition, and glucose metabolism in lean and obese rhesus monkeys treated with ephedrine and caffeine," *American Journal of Clinical Nutrition,* 68 (1998), pp. 42–51.
13. These data were found online at `www.cdc.gov/tobacco/data_statistics/tables/trends/cig_smoking/index.htm`.
14. Janice E. Williams et al., "Anger proneness predicts coronary heart disease risk," *Circulation,* 101 (2000), pp. 2034–2039.
15. From a graph in Peter A. Raymond and Jonathan J. Cole, "Increase in the export of alkalinity from North America's largest river," *Science,* 301 (2003), pp. 88–91.
16. From the Nenana Ice Classic website, `www.nenanaakiceclassic.com`. See Raphael Sagarin and Fiorenza Micheli, "Climate change in nontraditional data sets," *Science,* 294 (2001), p. 811, for a careful discussion.
17. Data for 2012 from the Organization for Economic Cooperation and Development website at `oecd-library.org/statistics`.
18. Louie H. Yang, "Periodical cicadas as resource pulses in North American forests," *Science,* 306 (2004), pp. 1565–1567. The data are simulated Normal values that match the means and standard deviations reported in this article.
19. Alan S. Banks et al., "Juvenile hallux abducto valgus association with metatarsus adductus," *Journal of the American Podiatric Medical Association,* 84 (1994), pp. 219–224.
20. Todd W. Anderson, "Predator responses, prey refuges, and density-dependent mortality of a marine fish," *Ecology,* 81 (2001), pp. 245–257.
21. From a graph in Craig Packer et al., "Ecological change, group territoriality, and population dynamics in Serengeti lions," *Science,* 307 (2005), pp. 390–393.
22. Peter H. Chen, Neftali Herrera, and Darren Christiansen, "Relationships between gate velocity and casting features among aluminum round castings," no date. Provided by Darren Christiansen.
23. These and other data are available at the Ohio Department of Health website at `http://www.odh.ohio.gov/odhprograms/chss/ad_hlth/youthrsk/youthrsk1.aspx`.
24. R. Shine, T. R. L. Madsen, M. J. Elphick, and P. S. Harlow, "The influence of nest temperatures and maternal brooding on hatchling phenotypes in water pythons," *Ecology,* 78 (1997), pp. 1713–1721.

Chapter 8 Notes

1. From the *New York Times*/CBS News poll at `www.nytimes.com`. The methodological statement is similar for most polls listed.
2. Gary S. Foster and Craig M. Eckert, "Up from the grave: a sociohistorical reconstruction of an African American community from cemetary data in the rural Midwest," *Journal of Black Studies,* 33 (2003), pp. 468–489.
3. See Note 1.
4. The regulations that govern seat belt survey design can be found at `http://www.nhtsa.gov/Research/`. Details on the Hawaii survey are in Karl Kim et al., *Results of the 2002 Highway Seat Belt Use Survey,* at `www.state.hi.us/dot`.
5. Donald L. McCabe, Linda Klebe Trevino, and Kenneth D. Buttereld, "Dishonesty in academic environments," *Journal of Higher Education,* 72 (2001), pp. 29–45.
6. For information about the response rates for the American Community Survey of households(there is a separate sample of group quarters), go to `www.census.gov/acs/www/methodology/response_rates_data/`.
7. Maeve Duggan and Aaron Smith, "Social media update 2013," (2013). The Pew press release and the full report

are at pewinternet.org/files/2013/12/PIP_Social-Networking-2013.pdf.

8. For more detail on the limits of memory in surveys, see N. M. Bradburn, L. J. Rips, and S. K. Shevell, "Answering autobiographical questions: The impact of memory and inference on surveys," *Science,* 236 (1987), pp. 157–161.
9. The immigration questions are from the New York Times/CBS News Poll taken May 18 to 23, 2007, found at www.pollingreport.com. The responses on welfare are from a *New York Times*/CBS News Poll reported in the *New York Times,* July 5, 1992. Many other examples appear in T. W. Smith, "That which we call welfare by any other name would smell sweeter," *Public Opinion Quarterly,* 51 (1987), pp. 75–83. The example on the effect of question order is cited in Daniel Kahnemann et al., "Would you be happier if you were richer? A focusing illusion," *Science,* 312 (2006), pp. 1908–1910.
10. The research study "Health reform and the decline of physician private practice" (2010) can be found at www.physiciansfoundation.org.
11. You can go to www.pollingreport.com to see the results of many polling agencies compiled on a variety of issues.
12. Information from various articles in the special issue on cell phone surveys, *Public Opinion Quarterly,* 71, No. 5 (2007). The 2012 cell phone use numbers were obtained from the CDC website, www.cdc.gov.
13. See Mick P. Couper, "Web surveys: A review of issues and approaches," *Public Opinion Quarterly,* 64 (2000), pp. 464–494.
14. The Renfew Center Foundation, "New survey indicates there's more to makeup use than meets the eye," at www.renfrewcenter.com.
15. The American Association for Public Opinion Research, "AAPOR report on online panels," 2010, at www.aapor.org.
16. Rachel Sherman and John Hickner, "Academic physicians use placebos in clinical practice and believe in the mind-body connection," *Journal of General Internal Medicine,* 23 (2008), pp. 7–10.
17. The National Public Radio Fans on Facebook 2010 survey can be found at www.slideshare.net/nprresearch/npr-facebook-fans-survey-findings-overview.
18. Stephen J. Blumberg and Julian V. Luke, "Wireless substitution: Early release of estimates from the national health interview survey, January–June 2013," at www.cdc.gov.
19. From the website of the Gallup Organization, www.gallup.com. Individual poll reports remain on this site for only a limited time.
20. Information about area codes can be found on the North American Numbering Plan Administration website at, www.nanpa.com.
21. The questions are available from a Gallup poll, April 4–7, 2013, at www.gallup.com/poll/1714/taxes.aspx.
22. Josephine Mazzuca, "The Right to Bear Arms: U.S. and Canada," http://www.gallup.com/poll/7381/Right-Bear-Arms-US-Canada.aspx.
23. Bryan E. Porter and Thomas D. Berry, "A nationwide survey of self-reported red light running: Measuring prevalence, predictors, and perceived consequences," *Accident Analysis and Prevention,* 33 (2001), pp. 735–741.
24. Mario A. Parada et al., "The validity of self-reported seatbelt use: Hispanic and non-Hispanic drivers in El Paso," *Accident Analysis and Prevention,* 33 (2001), pp. 139–143.
25. Giuliana Coccia, "An overview of non-response in Italian telephone surveys," *Proceedings of the 99th Session of the International Statistical Institute,* 1993, Book 3, pp. 271–272.
26. "Amplifiers study: The Twitter users who are most likely to retweet and how to engage them," January, 2013, posted by Taylor Schreiner at blog.twitter.com/2013/.
27. Information about the Ontario College of Pharmacists obtained from its website at www.ocpinfo.com. The numbers reported are from the 2013 Annual Report.
28. The Health Care in Canada Survey can be found by going to survey reports and presentations at the website www.hcic-sssc.ca.
29. Clyde O. McDaniel, Jr., "Dating roles and reasons for dating," *Journal of Marriage and the Family,* 31 (1969), pp. 97–107.
30. Robert C. Parker and Patrick A. Glass, "Preliminary results of double-sample forest inventory of pine and mixed stands with high- and low-density LiDAR," in Kristina F. Connoe (ed.), *Proceedings of the 12th Biennial Southern Silvicultural Research Conference,* U.S. Department of Agriculture, Forest Service, Southern Research Station, 2004.
31. The question is available from a Gallup poll, March 6–9, 2014, at www.gallup.com/poll/168176/americans-favor-energy-conservation-production.aspx.
32. The article can be found at www2.macleans.ca/2010/07/16/sometimes-a-gaffe-is-more-than-a-gaffe/.

Chapter 9 Notes

1. James C. Rosser et al., "The impact of video games on training surgeons in the 21st century," *Archives of Surgery,* 142 (2007), pp. 181–186. We thank Douglas Gentile for providing the data.
2. I. J. Goldberg et al., "Wine and your heart: A science advisory for healthcare professionals from the Nutrition Committee, Council on Epidemiology and Prevention, and Council on Cardiovascular Nursing of the American Heart Association," *Circulation,* 103 (2001), pp. 472–475.
3. Hyunjin Song and Norbert Schwarz, "If it's hard to read, it's hard to do: Processing fluency affects effort prediction and motivation," *Psychological Science,* 19 (2008), pp. 986–988.
4. Hsin-Chieh Yeh et al., "Smoking, smoking cessation, and risk for type 2 diabetes mellitus: A cohort study," *Annals of Internal Medicine,* 152 (2010), pp. 10–17.

5. Charles A. Nelson III et al., "Cognitive recovery in socially deprived young children: The Bucharest Early Intervention Project," *Science*, 318 (2007), pp. 1937–1940.
6. See Note 18 for Chapter 2.
7. The description of the factors and the response is based on a portion of the study by Alice Healy et al., "Terrorism after 9/11: Reactions to simulated news reports," *American Journal of Psychology*, 122 (2009), pp. 153–165.
8. The description of the factors and the response is based on a study by M. K. Baloch and F. Bibi, "Effect of harvesting and storage conditions on the post harvest fruit quality and shelf life of mango (*Mangifera indica* L.) fruit," *South African Journal of Botany*, 83 (2012), pp. 109–116.
9. See Note 18 for Chapter 2.
10. K. B. Suttle, Meredith A. Thomsen, and Mary E. Power, "Species interactions reverse grassland responses to changing climate," *Science*, 315 (2007), pp. 640–642. See Chapter 25 for an analysis of some data from this experiment.
11. Julie Mares et al., "Healthy diets and the subsequent prevalence of nuclear cataract in women," *Archives of Opthalmology*, 128 (2010), pp. 738–749.
12. Raine Sihvonen et al., "Arthroscopic partial meniscectomy versus sham surgery for a degenerative meniscal tear," *New England Journal of Medicine*, 369 (2013), pp. 2515–2524.
13. Libby Moulton, "The active placebo effect: Patent eligible subject matter," in the *The Columbia Science and Technology Law Review*, December, 2010.
14. Marielle H. Emmelot-Vonk et al., "Effect of testosterone supplementation on functional mobility, cognition, and other parameters in older men," *Journal of the American Medical Association*, 299 (2008), pp. 39–52.
15. David L. Strayer, Frank A. Drews, and William A. Johnston, "Cell phone-induced failures of visual attention during simulated driving," *Journal of Experimental Psychology: Applied*, 9 (2003), pp. 23–32.
16. See Note 12 for Chapter 2.
17. Alysha Fligner and Xiaoyan Deng, "The effect of font naturalness on perceived healthiness of food products," Senior thesis, The Ohio State University.
18. Sterling C. Hilton et al., "A randomized controlled experiment to assess technological innovations in the classroom on student outcomes: An overview of a clinical trial in education," manuscript, no date. A brief report is Sterling C. Hilton and Howard B. Christensen, "Evaluating the impact of multimedia lectures on student learning and attitudes," *Proceedings of the 6th International Conference on the Teaching of Statistics*, at `www.stat.aukland.ac.nz`.
19. S. Katherine Nelson et al., "In defense of parenthood: Children are associated with more joy than misery," *Psychological Science*, 24 (2013), pp. 3–10.
20. Brad J. Bushman, "Violence and sex in television programs do not sell products in advertisements," *Psychological Science*, 16 (2005), pp. 702–707.
21. An Pan et al., "Red meat consumption and mortality," *Archives of Internal Medicine*, 172 (2012), pp. 555–563.
22. N. Kalak et al., "Daily morning running for 3 weeks improved sleep and psychological functioning in healthy adolescents compared with controls," *Journal of Adolescent Health*, 51 (2012), pp. 615–622.
23. Margot Roosevelt, "Study finds traffic pollution can speed hardening of arteries," *LA Times*, February 14, 2010.
24. Jo Phelan et al., "The stigma of homelessness: The impact of the label 'homeless' on attitudes towards poor persons," *Social Psychology Quarterly*, 60 (1997), pp. 323–337.
25. Esther Duflo, Rema Hanna, and Stephan Ryan, "Monitoring works: Getting teachers to come to school," report dated November 21, 2007, at `econ-mit.edu/files/2066`.
26. B. Sekalska et al., "Effect of ibuprofen on the development of fat-induced atherosclerosis in New Zealand rabbits," *Journal of Experimental Animal Science*, 43 (2007), pp. 283–299. In the article, two of the treatment groups had eight rabbits and two had seven rabbits, but for simplicity we assumed eight rabbits had been assigned to all four treatments.
27. John H. Kagel, Raymond C. Battalio, and C. G. Miles, "Marijuana and work performance: Results from an experiment," *Journal of Human Resources*, 15 (1980), pp. 373–395.
28. N. Kemp et al., "Children's text messaging: abbreviations, input methods and links with literacy," *Journal of Adolescent Computer Assisted Learning*, 27 (2011), pp. 18–27.
29. The description of the factors and the response is based on a portion of the study by Brian Wnasik and Perre Chandon, "Can 'low-fat' nutrition labels lead to obesity," *Journal of Marketing Research*, 43 (2006), pp. 605–617.
30. Ian G. Williamson et al., "Antibiotics and topical nasal steroid for treatment of acute maxillary sinusitis," *Journal of the American Medical Association*, 298 (2007), pp. 2487–2496.
31. Found at `www.forbes.com`, January 14. 2014.
32. Based on Evan H. DeLucia et al., "Net primary production of a forest ecosystem with experimental CO_2 enhancement," *Science*, 284 (1999), pp. 1177–1179. The investigators used the block design.
33. E. M. Peters et al., "Vitamin C supplementation reduces the incidence of postrace symptoms of upper-respiratory tract infection in ultramarathon runners," *American Journal of Clinical Nutrition*, 57 (1993), pp. 170–174.
34. The study is described in Gina Kolata, "New study finds vitamins are not cancer preventers," *New York Times*, July 21, 1994. Look in the *Journal of the American Medical Association* of the same date for the details.
35. George I. Papakostas et al., "S-Adenosyl methionine (SAMe) augmentation of serotonin reuptake inhibitors for antidepressant nonresponders with major depressive disorder: A double-blind, randomized clinical trial," *American Journal of Psychiatry*, 167 (2010), pp. 942–948.

Chapter 10 Notes

1. The U.S. Department of Health and Human Services maintains a website that reports cases of misconduct. The

web address is ori.hhs.gov. If you visit the site, you will see reports of actual cases of misconduct. Many of these involve handling of data.

2. John C. Bailar III, "The real threats to the integrity of science," *Chronicle of Higher Education*, April 21, 1995, pp. B1–B2.
3. For more on the problems with open-access publishing, see Gina Kolata, "Scientific articles accepted (personal checks, too)," *The New York Times*, April 7, 2013, and the March 28, 2013, issue of *Nature*.
4. This statement can be found at the *JAMA* website: jama.jamanetwork.com/public/instructionsForAuthors.aspx.
5. See the details on the website of the Office for Human Research Protections of the Department of Health and Human Services, www.hhs.gov/ohrp.
6. The difficulties of interpreting guidelines for informed consent and for the work of institutional review boards in medical research are a main theme of Beverly Woodward, "Challenges to human subject protections in U.S. medical research," *JAMA*, 282 (1999), pp. 1947–1952. The references in this paper point to other discussions. Updated regulations and guidelines appear on the OHRP website (see Note 5).
7. See Note 5.
8. See, for example, www.lc.org/hotissues/2001/aba_1-18/public_records_laws_by_state.htm.
9. The Heartbleed Bug, which came to public attention in early 2014, is a security bug that affects open-source OpenSSL cryptographic software. This bug renders many websites that were thought to be secure vulnerable to attacks by hackers. You can read more about this online at heartbleed.com.
10. For more discussion of the dangers of observational studies in clinical trials, see David Madigan et al., "A Systematic Statistical Approach to Evaluating Evidence from Observational Studies," *Annual Review of Statistics and Its Application*, 1 (2014), pp. 11–39.
11. Quotation from the *Report of the Tuskegee Syphilis Study Legacy Committee*, May 20, 1996. A detailed history is James H. Jones, *Bad Blood: The Tuskegee Syphilis Experiment*, New York: Free Press, 1993.
12. Dr. Hennekens's words are from an interview in the Annenberg/Corporation for Public Broadcasting video series *Against All Odds: Inside Statistics*. The lack of certainty that Dr. Hennekens refers to is now called "clinical equipoise" in discussions of ethics.
13. For more discussion of the ethics of the use of placebos in clinical trials, see http://www.ama-assn.org/ama/pub/physician-resources/medical-ethics/code-medical-ethics/opinion8083.page on the American Medical Association website.
14. Ezekial J. Emanuel, David Wendler, and Christine Grady, "What makes clinical research ethical?" *JAMA*, 283 (2000) pp. 2701–2711.
15. R. D. Middlemist, E. S. Knowles, and C. F. Matter, "Personal space invasions in the lavatory: Suggestive evidence for arousal," *Journal of Personality and Social Psychology*, 33 (1976), pp. 541–546.
16. For a review of domestic violence experiments, see C. D. Maxwell et al., *The Effects of Arrest on Intimate Partner Violence: New Evidence from the Spouse Assault Replication Program*, U.S. Department of Justice, NCH188199, 2001. Available online at www.ojp.usdoj.gov/nij/pubs-sum/188199.htm.
17. Joseph Millum and Ezekial J. Emanuel, "The ethics of international research with abandoned children," *Science*, 318 (2007), pp. 1874–1875. This paper has some useful comments on international research in general.

Chapter 11 Notes

1. Based on a news item "Bee off with you," *Economist*, November 2, 2002, p. 78.
2. Votes as of June 27, 2007, at www.pbs.org/wgbh/nova/sciencenow.
3. Seang-Mei Saw et al., "IQ and the association with myopia in children," *Investigative Opthalmology & Visual Science*, 45 (2004), pp. 2943–2948.
4. Simplified from Sanjay K. Dhar, Claudia González-Vallejo, and Dilip Soman, "Modeling the effects of advertised price claims: tensile versus precise pricing," *Marketing Science*, 18 (1999), pp. 154–177.
5. K. J. Mukamal et al., "Prior alcohol consumption and mortality following acute myocardial infarction," *Journal of the American Medical Association*, 285 (2001), pp. 1965–1970.

Chapter 12 Notes

1. The Gallup Poll is based on telephone interviews conducted October 3–6, 2013. Each adult interviewed by Gallup had a known chance of being among those selected, but this chance depended on characteristics such as gender, age, race, Hispanic ethnicity, education, region, adults in the household, and phone status (cell phone only/landline only/both, cell phone mostly, and having an unlisted landline number). Gallup used special weights to adjust for differences in the probability of being selected to obtain an estimate of the proportion of all adults who owned a gun at the time of the poll in the population of all U.S. adults. The actual estimate used by Gallup was close to 37% and uses the result from the sample to estimate what is true for the population.
2. Note that pennies have rims that make spinning more stable. The probability of a head in spinning a coin depends on the type of coin and also on the surface. See Exercise 21.3 for an account of 56% of heads in spinning a Belgian 1-euro coin. *Chance News 11.02* at www.dartmouth.edu/~chance reports about 45% heads in more than 20,000 spins of American pennies by Robin Lock's students at Saint Lawrence University.
3. The percentages were found at the GMAT website, www.gmac.com/market-intelligence-and-

research/gmat-statistics/profile-of-gmat-candidates.aspx.

4. Data for 2011 from the website of Statistics Canada, www.statcan.gc.ca.
5. You can find a mathematical explanation of Benford's law in Ted Hill, "The first-digit phenomenon," *American Scientist*, 86 (1996), pp. 358–363; and Ted Hill, "The difficulty of faking data," *Chance*, 12, No. 3 (1999), pp. 27–31. Applications in fraud detection are discussed in the second paper by Hill and in Mark A. Nigrini, "I've got your number," *Journal of Accountancy*, May 1999, available online at www.aicpa.org/pubs/jofa/joaiss.htm.
6. Based on a December 2013 Gallup poll, found at gallup.com/poll/1588/Children-Violence.aspx.
7. These are the mean and standard deviation of all test takers in 2013. These statistics are available online at www.aamc.org/students/applying/mcat/admissionsadvisors/mcat_stats/.
8. Information from http://gradedistribution.registrar.indiana.edu/reports.php.
9. Thomas K. Cureton et al., *Endurance of Young Men*, Monographs of the Society for Research in Child Development, Vol. 10, No. 1, 1945.
10. Based on a March 2014 Gallup poll, found at http://www.gallup.com/poll/15370/Party-Affiliation.aspx.
11. Data from 2009 found online at http://nhts.ornl.gov/tables09/fatcat/2009/household_HHVEHCNT.html.
12. See Note 16 for Chapter 1.
13. National population estimates for July 1, 2012, at the Census Bureau website www.census.gov. The table omits people who consider themselves to belong to more than one race.
14. Based on data from the *2012 Statistical Abstract of the United States*, Table 58, at www.census.gov.
15. This exercise is based on a famous example discussed in Part II of *The Design of Experiments* by Sir R. A. Fisher. The example involves a lady tasting tea. Fisher used this example to illustrate the logic of hypothesis testing. Presumably, if milk is added to hot tea, the milk can curdle and produce a slightly sour taste. This is avoided by adding the milk first.

Chapter 13 Notes

1. This is one of several tests discussed in Bernard M. Branson, "Rapid HIV testing: 2005 update," a presentation by the Centers for Disease Control and Prevention, at www.cdc.gov. The Malawi clinic result is reported by Bernard M. Branson, "Point-of-care rapid tests for HIV antibody," *Journal of Laboratory Medicine*, 27 (2003), pp. 288–295.
2. Robert P. Dellavalle et al., "Going, going, gone: Lost Internet references," *Science*, 302 (2003), pp. 787–788.
3. From the U.S. Department of Commerce Bureau of Economic Analysis at www.bea.gov. Motor vehicle sales information is included with the National Economic Accounts, and the numbers reported are based on sales during April 2012–March 2013.
4. Maeve Duggan, "Photo and video sharing grow online," October 28, 2013, from the Pew Internet and American Life Project, at www.pewinternet.org.
5. Sales in 2006 from the website of the Entertainment Software Association, at www.theesa.com.
6. From the Teens Fact Sheet which highlights the Pew Internet Project's research on teens, at www.pewinternet.org. The information is updated whenever new data is available. The data in the example are from September 2012.
7. See Note 4.
8. Hal R. Arkes and Wolfgang Gaissmaier, "Psychological research and the prostate-cancer screening controversy," *Journal of Psychological Science*, 23(6) (2012), pp. 547–553.
9. W. Uter et al., "Inter-relation between variables determining constitutional UV sensitivity in Caucasian children," *Photodermatology, Photoimmunology and Photmeicine*, 20, (2004), pp. 9–13.
10. F. H. Schroder et al., "Screening and prostate-cancer mortality in a randomized European study," *New England Journal of Medicine*, 360 (2009) pp. 1320–1328.
11. From the statistics page of the National Science Foundation website at, www.nsf.gov/statistics.
12. M. McCulloch et al., "Diagnostic accuracy of canine scent detection in early- and late-stage lung and breast cancers," *Integrative Cancer Therapies*, 5 (2006), pp. 30–39.
13. From the Internal Revenue Service website, at www.irs.gov/taxstats.
14. Projections from U.S. Department of Education, *Projections of Education Statistics to 2016*, December 2007, at nces.ed.gov.
15. Data provided by Patricia Heithaus and the Department of Biology at Kenyon College.
16. F. J. G. M. Klaassen and J. R. Magnus, "How to reduce the service dominance in tennis? Empirical results from four years at Wimbledon," in S. J. Haake and A. O. Coe (eds.), *Tennis Science and Technology*, Blackwell, 2000, pp. 277–284.
17. R. S. Gupta et al., "The Prevalence, Severity, and Distribution of Childhood Food Allergy in the United States," *Journal of Pediatrics*, 128 (2011), pp. e9–e17.
18. From the National Institutes of Health's National Digestive Diseases Information Clearinghouse, found at wrongdiagnosis.com.
19. The probabilities given are realistic, according to the fundraising firm SCM Associates, at scmassoc.com.
20. B. Budowle et al., "Population Data on the Thirteen CODIS Core Short Tandem Repeat Loci in African Americans, U.S. Caucasians, Hispanics, Bahamians, Jamaicans, and Trinidadians," *Journal of Forensic Sciences*, (1999), pp. 1277–1286.
21. Marilyn vos Savant, "Ask Marilyn column," *Parade Magazine*, p. 16, September 9, 1990.

Chapter 14 Notes

1. Matthew A. Carlton and William D. Stansfield, "Making babies by the flip of a coin?" *American Statistician*, 59 (2005), pp. 180–182.
2. Richard F. MacLeod et al., "Damage to fresh tomatoes can be reduced" at ucce.ucdavis.edu/files/repositoryfiles/ca3012p10-72079.pdf.
3. Pew Internet at www.pewinternet.org provides a full disposition of all sampled numbers in many of its reports. The response rates vary somewhat, but the percent of working numbers is fairly consistent at around 30%.
4. From Statistics Canada, Smoking 2011, at www.statcan.gc.ca.
5. The survey question is reported in Susanah Fox and Maeve Duncan, "Tracking for Health" January, 2013 at pewinternet.org. In fact, the estimate in the survey was 60% saying "Yes."
6. Information obtained from the Centers for Disease Control website, at www.cdc.gov.
7. Information obtained from the Planned Parenthood website, at www.plannedparenthood.org.
8. Results for the full 2013 season, at www.pgatour.com.
9. Sales data from www.hyundainews.com/us/en-us/Corporate/SalesReleases/PressReleases.aspx.
10. Corby K. Martin et al., "Weight loss and retention rates and weight loss in a commercial weight loss program and the effect of corporate partnership," *International Journal of Obesity*, 34 (2010), pp. 742–750. The retention rate after four weeks was read from a graph in the paper.
11. Associated Press news item dated December 9, 2007, found at www.msnbc.msn.com.
12. See demonstrations.wolfram.com/MonteCarloEstimateForPi/ for an online demonstration of this idea.

Chapter 15 Notes

1. U.S. Census Bureau, *Income, Poverty, and Health Insurance in the United States: 2013*, Current Population Reports P60-243. Available online at www.census.gov/prod/2013pubs/p60-245.pdf.
2. The data can be found at the Bureau of Labor Statistics website and are available for download at http://www.bls.gov/tus/home.htm.
3. Strictly speaking, the formula $\sigma/\sqrt{n}$ for the standard deviation of $\bar{x}$ assumes that we draw an SRS of size n from an *infinite* population. If the population has finite size N, this standard deviation is multiplied by $\sqrt{1 - (n - 1)/(N - 1)}$. This "finite population correction" approaches 1 as N increases. When the population is at least 20 times as large as the sample, the correction factor is between about 0.97 and 1. It is reasonable to use the simpler form $\sigma/\sqrt{n}$ in these settings.
4. Earnings for all 98,095 households were downloaded using the Census Bureau's Data Ferret software. The histograms in Figure 15.4 were produced from the downloaded data.
5. See Note 10 for Chapter 3.
6. Found online at pages.stern.nyu.edu/adamodar/New_Home_Page/datafile/histret.html. Sophisticates will note that for compounding over several years we want the geometric mean return, which was 9.38%.

Chapter 16 Notes

1. Margaret A. McDowell et al., "Anthropometric reference data for children and adults: U.S. population, 1999–2002," National Center for Health Statistics, Advance Data from Vital and Health Statistics, No. 361, 2005, at www.cdc.gov/nchs.
2. Information about the NAEP test can be found online at nationsreportcard.gov/math_2013/.
3. B. Rind and D. Strohmetz, "Effect of beliefs about future weather conditions on restaurant tipping," *Journal of Applied Social Psychology*, 31 (2001), pp. 2160–2164.
4. See Note 22 for Chapter 1.
5. The National Survey of Student Engagement conducts an annual survey that includes hours spent preparing for class each week. The numbers in this exercise are based on the results of the 2013 survey. You can find survey results at nsse.iub.edu/html/findings.cfm.
6. Chi-Fu Jeffrey Yang, Peter Gray, Harrison G. Pope, Jr., "Male body image in Taiwan versus the West," *American Journal of Psychiatry*, 162 (2005), pp. 263–269.
7. M. Ann Laskey et al., "Bone changes after 3 mo. of lactation: influence of calcium intake, breast-milk output, and vitamin D–receptor genotype," *American Journal of Clinical Nutrition*, 67 (1998), pp. 685–692.

Chapter 17 Notes

1. For one-sided alternatives, some statisticians prefer to write the null hypothesis so that it includes all possible values of μ that are not specified by the alternative. They would write either

$$H_0: \mu \leq 0$$
$$H_a: \mu > 0$$

or

$$H_0: \mu \geq 0$$
$$H_a: \mu < 0$$

and interpret a null hypothesis such as $H_0: \mu = 0$ as implying either $H_0: \mu \leq 0$ or $H_0: \mu \geq 0$ for one-sided alternatives.

Sir R. A. Fisher, one of the greatest statisticians and scientists of the 20th century, argued that "the null hypothesis must be exact, that is free from vagueness or ambiguity . . ." (see his book *The Design of Experiments*, Section 8 of Part II). His reason was that the null hypothesis must determine an exact distribution from which probability calculations can be made. He would advocate the use of $H_0: \mu = 0$ and we have followed Fisher's formulation in this textbook.

2. Steven R. Smith et al., "Multicenter, placebo-controlled trial of lorcaserin for weight management," *The New England Journal of Medicine*, 363, No. 3 (2010), pp. 245–256.
3. See Note 3 for Chapter 16.
4. This procedure is also referred to as the parametric bootstrap.
5. Rashmi Ranjan Das and Meenu Singh, "Oral zinc for the common cold," *JAMA*, 311 No. 14 (2014), pp. 1440–1441.
6. Ajay Ghei, "An empirical analysis of psychological androgeny in the personality profile of the successful hotel manager," MS thesis, Purdue University, 1992.
7. Michael D. Mrazek et al., "Mindfulness training improves working memory capacity and GRE performance while deducting mind wandering," *Psychological Science*, 24, No. 5 (2013), pp. 776–781.
8. Seung-Ok Kim, "Burials, pigs, and political prestige in Neolithic China," *Current Anthropology*, 35 (1994), pp. 119–141.
9. Gerardo Ramirez and Sian L. Bellock, "Writing about testing worries boosts exam performance in the classroom," *Science*, 331 (2011), pp. 211–213.
10. Mario A. Parada et al., "The validity of self-reported seatbelt use: Hispanic and non-Hispanic drivers in El Paso," *Accident Analysis and Prevention*, 33 (2001), pp. 139–143.
11. See Note 7 for Chapter 16.
12. Data simulated from a Normal distribution based on information in Brian M. DeBroff and Patricia J. Pahk, "The ability of periorbitally applied antiglare products to improve contrast sensitivity in conditions of sunlight exposure," *Archives of Ophthalmology*, 121 (2003), pp. 997–1001.

Chapter 18 Notes

1. See www.cdc.gov/nchs/tutorials/currentnhanes/SurveyDesign/SampleDesign/intro.htm.
2. Bryan E. Porter and Thomas D. Berry, "A nationwide survey of self-reported red light running: measuring prevalence, predictors, and perceived consequences," *Accident Analysis and Prevention*, 33 (2001), pp. 735–741.
3. For a discussion of statistical significance in the legal setting, see D. H. Kaye, "Is proof of statistical significance relevant?" *Washington Law Review*, 61 (1986), pp. 1333–1365. Kaye argues: "Presenting the *P*-value without characterizing the evidence by a significance test is a step in the right direction. Interval estimation, in turn, is an improvement over *P*-values."
4. From a press release from the Harvard School of Public Health College Alcohol Study, April 12, 2001, at www.hsph.harvard.edu/cas/.
5. Warren E. Leary, "Cell phones: questions but no answers," *New York Times*, October 26, 1999.
6. Information about TUDA can be found online at nationsreportcard.gov/reading_math_tuda_2013/#/.
7. Poll published April 10, 2014, at www.harrisinteractive.com/NewsRoom/HarrisPolls/tabid/447/ctl/ReadCustom%20Default/mid/1508/ArticleId/1412/Default.aspx. A note at the bottom of the page says: "Because the sample is based on those who agreed to participate in the Harris Interactive panel, no estimates of theoretical sampling error can be calculated."
8. Justin S. Brashares et al., "Bushmeat hunting, wildlife declines, and fish supply in West Africa," *Science*, 306 (2004), pp. 1180–1183. The data used here (and in Figure 1B of the article) are found in the online supplementary material.
9. Gabriel Gregoratos et al., "ACC/AHA guidelines for implantation of cardiac pacemakers and antiarrhythmia devices: executive summary," *Circulation*, 97 (1998), pp. 1325–1335.
10. From the commentary by Frank J. Sulloway, "Birth order and intelligence," *Science*, 316 (2007), pp. 1711–1712. The study report appears in the same issue, Petter Kristensen and Tor Bjerkedal, "Explaining the relation between birth order and intelligence," *Science*, 316 (2007), p. 1717.
11. C. Kopp et al., "Modulation of rhythmic brain activity by diazepam: GABAA receptor subtype and state specificity," *Proceedings of the National Academy of Sciences*, 101 (2004), pp. 3674–3679.
12. Bruce A. Cooper et al., "A randomized, controlled trial of early versus late initiation of dialysis," *New England Journal of Medicine*, 363, No. 7 (2010), pp. 609–619.

Chapter 19 Notes

1. Data for U.S. searches in January 2014 from www.comscore.com/Insights/Press_Releases/2014/2/comScore_Releases_January_2014_US_Search_Engine_Rankings.
2. U.S. Census Bureau, *Fertility of American Women: June 2010*, at www.census.gov.
3. Aaron S. Hervey et al., "Reaction time distribution analysis of neuropsychological performance in an ADHD sample," *Child Neuropsychology*, 12 (2006), pp. 125–140.
4. K. E. Hobbs et al., "Levels and patterns of persistent organochlorines in minke whale (*Balaenoptera acutorostrata*) stocks from the North Atlantic and European Arctic," *Environmental Pollution*, 121 (2003), pp. 239–252.
5. Maureen Hack et al., "Outcomes in young adulthood for very-low-birth-weight infants," *New England Journal of Medicine*, 346 (2002), pp. 149–157.
6. Mikyoung Park et al., "Recycling endosomes supply AMPA receptors for LTP," *Science*, 305 (2004), pp. 1972–1975.
7. Benjamin B. Albert et al., "Among overweight middle-aged men, first-borns have lower insulin sensitivity than second-borns," *Scientific Reports*, 4 (2014), article number 3906. Available online at www.nature.com/srep/index.html.
8. Jon E. Keeley, C. J. Fotheringham, and Marco Morais, "Reexamining fire suppression impacts on brushland fire regimes," *Science*, 284 (1999), pp. 1829–1831.
9. From the Gallup-Healthways Well-Being index survey in 2013, www.gallup.com.

10. Data simulated from a Normal distribution with $\mu = 98.2$ and $\sigma = 0.7$. These values are based on P. A. Mackowiak, S. S. Wasserman, and M. M. Levine, "A critical appraisal of 98.6 degrees F, the upper limit of the normal body temperature, and other legacies of Carl Reinhold August Wunderlich," *Journal of the American Medical Association*, 268 (1992), pp. 1578–1580.
11. Nicolas Guéguen and Christine Petr, "Odors and consumer behavior in a restaurant," *International Journal of Hospitality Management*, 25 (2006), pp. 335–339. We thank Nicolas Guéguen for providing the data.
12. Probabilities from trials with 2897 people known to be free of HIV antibodies and 673 people known to be infected, reported in J. Richard George, "Alternative specimen sources: methods for confirming positives," 1998 Conference on the Laboratory Science of HIV, found online at the Centers for Disease Control and Prevention, `www.cdc.gov`.
13. Higher Education Research Institute 2013 Freshman Survey, at `www.heri.ucla.edu`.
14. John Schwartz, "Leisure pursuits of today's young men," *New York Times*, March 29, 2004. The source cited is comScore Media Matrix.
15. J. F. Swain et al., "Comparison of the effects of oat bran and low-fiber wheat on serum lipoprotein levels and blood pressure," *New England Journal of Medicine*, 322 (1990), pp. 147–152.

Chapter 20 Notes

1. Note 3 for Chapter 15 explains the reason for this condition in the case of inference about a population mean.
2. The website `project.wnyc.org/commute-times-us/embed.html#5.00/42.000/-89.500` has an interactive map that gives the average commute time for any region in the U.S.
3. D. G. Jakovljevic and A. K. McConnell, "Influence of different breathing frequencies on the severity of inspiratory muscle fatigue induced by high-intensity front crawl swimming," *Journal of Strength and Conditioning Research*, 23(4) (2009), pp. 1169–1174.
4. See Note 3 for Chapter 16.
5. From a graph in Benedetto De Martino et al., "Frames, biases, and rational decision-making in the human brain," *Science*, 313 (2006), pp. 684–687. We simplified the design a bit for easier comprehension: the starting amounts and gambles offered differed from trial to trial, though still matched in pairs; 32 very unbalanced "catch trials" were mixed with the 64 experimental trials to be sure subjects were paying attention; and all money amounts were in British pounds, not dollars.
6. John Morton, et al., "Acoustic features mediating height estimation from human speech," Abstract from the 166th meeting of the Acoustical Society of America. We would like to thank the authors for supplying the data and additional details of the study.
7. This study was found online at `www.dispatch.com/live/content/databases/index.html`.
8. Alice P. Melis, Brian Hare, and Michael Tomasello, "Chimpanzees recruit the best collaborators," *Science*, 311 (2006), pp. 1297–1300. A Normal quantile plot does not show major lack of Normality and a saddlepoint approximation that allows for skew gives $P = 0.0039$. So the t test is reasonably accurate despite the skew and small sample size.
9. Data that take only whole-number values are discrete. Thus, technically, they cannot come from a population with a continuous distribution, such as the Normal.
10. Data simulated from a Normal distribution based on information in Brian M. DeBroff and Patricia J. Pahk, "The ability of periorbitally applied antiglare products to improve contrast sensitivity in conditions of sunlight exposure," *Archives of Ophthalmology*, 121 (2003), pp. 997–1001.
11. The fact that all data must be truncated to a finite number of decimal places implies that real data can never exactly follow any continuous distribution. For example, integer-valued data are discrete, so that at best, such data can only be roughly approximated by a continuous distribution.
12. For a qualitative discussion explaining why skewness is the most serious violation of the Normal shape condition, see Dennis D. Boos and Jacqueline M. Hughes-Oliver, "How large does n have to be for the Z and t intervals?" *American Statistician*, 54 (2000), pp. 121–128. Our recommendations are based on extensive computer work. See, for example, Harry O. Posten, "The robustness of the one-sample t-test over the Pearson system," *Journal of Statistical Computation and Simulation*, 9 (1979), pp. 133–149; and E. S. Pearson and N. W. Please, "Relation between the shape of population distribution and the robustness of four simple test statistics," *Biometrika*, 62 (1975), pp. 223–241.
13. For more advanced users, a good way to ascertain if the t procedures are safe is to compare the 95% confidence interval produced by t with the BCa interval from a bootstrap with at least 1000 resamples. For part (b), the t interval is 29,428 to 32,254 and a BCa interval is 29,106 to 31,894. For part (c), on the other hand, t gives 38.93 to 40.49 and BCa gives 38.97 to 40.44. These results confirm the judgment that t is safe for part (c) but not for part (b).
14. Table 1 of E. Thomassot et al., "Methane-related diamond crystallization in the earth's mantle: stable isotopes evidence from a single diamond-bearing xenolith," *Earth and Planetary Science Letters*, 257 (2007), pp. 362–371.
15. Assuming that we know the population distribution, or estimate features of it from data, and then resampling from this distribution is sometimes referred to as the parametric bootstrap. There are resampling methods that do not require us to know the population distribution. These methods are referred to as the (nonparametric) bootstrap. They are very useful, but beyond the scope of this book.
16. From the online supplement to Tor D.Wager et al., "Placebo-induced changes in fMRI in the anticipation and experience of pain," *Science*, 303 (2004), pp. 1162–1167.

17. TUDA results for 2013 from the National Center for Education Statistics, at nationsreportcard.gov/tuda.asp.
18. Ravi Mehta and Rui Zhu, "Blue or red? Exploring the effect of color on cognitive task performances," *Science*, 323 (2009), pp. 1226–1229.
19. Raul de la Fuente-Fernandez et al., "Expectation and dopamine release: mechanism of the placebo effect in Parkinson's disease," *Science*, 293 (2001), pp. 1164–1166.
20. Allison I. Young, Kyle G. Ratner, and Russell H. Fazio, "Political attitudes bias the mental representation of a presidential candidate's face," *Psychological Science*, 25 (2) (2014). Available online at pss.sagepub.com. The data are estimated from a graph in the paper and only students with a support score of 1 or greater were included.
21. J. D. Marshall et al., "Vehicle self-pollution intake fraction: children's exposure to school bus emissions," *Environmental Science and Technologhy*, 39 (2005), pp. 2559–2563.
22. See Note 19 for Chapter 7.
23. Data provided by Drina Iglesia, Purdue University. The data are part of a larger study reported in D. D. S. Iglesia, E. J. Cragoe, Jr., and J. W. Vanable, "Electric field strength and epithelization in the newt (*Notophthalmus viridescens*)," *Journal of Experimental Zoology*, 274 (1996), pp. 56–62.
24. Matthias R. Mehl et al., "Are women really more talkative than men?" *Science*, 317 (2007), p. 82.
25. Data for Cohort 2 in Richard A. Morgan et al., "Cancer regression in patients after transfer of genetically engineered lymphocytes," *Science*, 314 (2006), pp. 126–129. The doubling time data are given in the paper and the immune response data appear in the supplementary online material.
26. M. B. Laferty "OSU scientist gets a kick out of sports controversy," *The Columbus Dispatch*, November 21, 1993.
27. I thank Jason Hamilton, University of Illinois, for providing the data. The study is reported in Evan H. DeLucia et al., "Net primary production of a forest ecosystem with experimental CO_2 enhancement," *Science*, 284 (1999), pp. 1177–1179. No method for inference can be trusted with $n = 3$. In this study, each observation is very costly, so the small n is inevitable.
28. Michael W. Peugh, "Field investigation of ventilation and air quality in duck and turkey slaughter plants," MS thesis, Purdue University, 1996.
29. Harry B. Meyers, "Investigations of the life history of the velvetleaf seed beetle, *Althaeus folkertsi* Kingsolver," MS thesis, Purdue University, 1996. The 95% t interval is 1227.9 to 2507.6. A 95% bootstrap BCa interval is 1444 to 2718, confirming that t inference is inaccurate for these data.
30. J. Marcus Jobe and Hutch Jobe, "A statistical approach for additional infill development," *Energy Exploration and Exploitation*, 18 (2000), pp. 89–103. The comparison interval is the BCa interval based on 1000 bootstrap resamples.
31. This study is available online at www.dispatch.com/live/content/databases/index.html.
32. Ralf Bargou et al., "Tumor regression in cancer patients by very low doses of a T cell engaging antibody," *Science*, 321 (2008), pp. 974–977.
33. Data provided by Timothy Sturm.
34. Lianng Yuh, "A biopharmaceutical example for undergraduate students," manuscript, no date.
35. See Note 3 for Chapter 16.

Chapter 21 Notes

1. James A. Levine et al., "Inter-individual variation in posture allocation: possible role in human obesity," *Science*, 307 (2005), pp. 584–586. We thank James Levine for providing the data.
2. Detailed information about the conservative t procedures can be found in Paul Leaverton and John J. Birch, "Small sample power curves for the two sample location problem," *Technometrics*, 11 (1969), pp. 299–307; Henry Scheffé, "Practical solutions of the Behrens-Fisher problem," *Journal of the American Statistical Association*, 65 (1970), pp. 1501–1508; and D. J. Best and J. C. W. Rayner, "Welch's approximate solution for the Behrens-Fisher problem," *Technometrics*, 29 (1987), pp. 205–210.
3. Kathleen G. McKinney, "Engagement in community service among college students: is it affected by significant attachment relationships?" *Journal of Adolescence*, 25 (2002), pp. 139–154. To see the questions in the Inventory of Parent and Peer Attachments, go to chipts.cch.ucla.edu/assessment/IB/List_Scales/inventory%20parent%20and%20peer%20attachment.htm.
4. Domenico Giannotti et al., "Play to become a surgeon: Impact of Nintendo Wii training on laparoscopic skills," *PLOS ONE*, V8, e5272, February 2013 at www.plosone.org.
5. P. A. Handcock, "The effect of age and sex on the perception of time in life," *The American Journal of Psychology*, 123 (2010), pp. 1–13.
6. See the extensive simulation studies in Harry O. Posten, "The robustness of the two-sample t-test over the Pearson system," *Journal of Statistical Computation and Simulation*, 6 (1978), pp. 295–311; and Harry O. Posten, H. Yeh, and Donald B. Owen, "Robustness of the two-sample t-test under violations of the homogeneity assumption," *Communications in Statistics*, 11 (1982), pp. 109–126.
7. Nicolas Guéguen and Christine Petr, "Odors and consumer behavior in a restaurant," *Journal of Hospitality Management*, 25 (2006), pp. 335–339. I thank Nicolas Guéguen for providing the data. Although the spending data are quite discrete, a bootstrap BCa 95% confidence interval for the difference in means based on 1000 resamples is 2.394 to 4.826, close to the Option 1 95% interval 2.209 to 4.736. So the sample means are sufficiently Normal to allow use of t procedures.
8. Parmeshwar S. Gupta, "Reaction of plants to the density of soil," *Journal of Ecology*, 21 (1933), pp. 452–474.
9. Data provided by Samuel Phillips, Purdue University.

10. Daniel M. T. Foster and Colin Holbrook, "Friends shrink foes: The presence of comrades decreases envisioned physical formidability of an opponent," *Psychological Science,* 24 (2013). Available online at pss.sagepub.com.
11. The problem of comparing variabilities is difficult even with advanced methods. Common distribution-free procedures do not offer a satisfactory alternative to the F test, because they are sensitive to unequal shapes when comparing two distributions. A recent survey of possible approaches is Dennis D. Boos and Cavell Brownie, "Comparing variances and other measures of dispersion," *Statistical Science,* 19 (2005), pp. 571–578.
12. If there are n_1 units assigned to treatment 1 and n_2 assigned to treatment 2, from Chapter 14 we know that the total number of possible assignments is

$$\binom{n_1 + n_2}{n_2}$$

13. See Note 9.
14. Matthias R. Mehl et al., "Are women really more talkative than men?" *Science,* 317 (2007), p. 82.
15. Michael A. Sayette et al., "Lost in the sauce, the effects of alcohol on mind wandering," *Psychological Science,* 20 (2009), pp. 747–752.
16. Eduardo Dias-Ferreira et al., "Chronic stress causes frontostriatal reorganization and affects decision-making," *Science,* 325 (2009), pp. 621–625. Many of the details appear in the supporting online material.
17. Angeline Lillard and Nicole Else-Quest, "Evaluating Montessori education," *Science,* 313 (2006), pp. 1893–1894. Many of the details appear in the supporting online material.
18. Nathan A. Bowling et al., "Mean job satisfaction levels over time: are things bad and getting worse?" *TIP* (online journal at www.siop.org/tip), April 2013.
19. Jennifer A. Whitson and Adam D. Galinsky, "Lacking control increases illusory pattern perception," *Science,* 322 (2008), pp. 115–117.
20. Rashmi Singh et al., "Relationship of collegiate football experience and concussion with hippocampal volume and cognitive outcomes," *JAMA,* 311 (2014), pp. 1883–1888.
21. Wayne J. Camera and Donald Powers, "Coaching and the SAT I," *TIP* (online journal at www.siop.org/tip), July 1999.
22. See Note 3 for Chapter 16.
23. A. Timur Sevincer et al., "Positive thinking about the future in newspaper reports and presidential addresses predicts economic downturn," *Psychological Science,* 25 (2014). Available online at pss.sagepub.com. Data are estimates from a plot in the article, with positive articles corresponding to articles with a positive thinking z score above 2 and negative articles corresponding to a positive thinking z score below -2.
24. Allison I. Young, Kyle G. Ratner, and Russell H. Fazio, "Political attitudes bias the mental representation of a presidential candidate's face," *Psychological Science,* 25 (2) (2014). Available online at pss.sagepub.com. The data are estimated from a graph in the paper and only students with support scores greater than 1 or less than -1 were included.
25. Fabrizio Grieco, Arie J. van Noordwijk, and Marcel E. Visser, "Evidence for the Effect of learning on timing of reproduction in blue tits," *Science,* 296 (2002), pp. 136–138. The data in Exercise 21.45 are from a graph in this paper.
26. Kathleen D. Vohs, Nicole L. Mead, and Miranda R. Goode, "The psychological consequences of money," *Science,* 314 (2006), pp. 1154–1156. I thank Kathleen Vohs for supplying the data.
27. Paul E. O'Brien et al., "Laparascopic adjustable gastric banding in severely obese adolescents," *Journal of the American Medical Association,* 303 (2010), pp. 519–526. I thank the authors for providing the data.
28. Paul Kvam, "The Effect of active learning methods on student retention in engineering statistics," *The American Statistician,* 54 (2000), pp. 136–140.
29. Data provided by Warren Page, New York City Technical College, from a study done by John Hudesman.
30. Ethan J. Temeles and W. John Kress, "Adaptation in a plant-hummingbird association," *Science,* 300 (2003), pp. 630–633. We thank Ethan J. Temeles for providing the data.
31. Data provided by Marigene Arnold, Kalamazoo College.

Chapter 22 Notes

1. Joseph H. Catania et al., "Prevalence of AIDS-related risk factors and condom use in the United States," *Science*, 258 (1992), pp. 1101–1106.
2. Strictly speaking, the formula $\sqrt{p(1-p)/n}$ for the standard deviation of $\hat{p}$ assumes that we draw an SRS of size n from an *infinite* population. If the population has finite size N, this standard deviation is multiplied by $\sqrt{1-(n-1)/(N-1)}$. This "finite population correction" approaches 1 as N increases. When the population is at least 20 times as large as the sample, the correction factor is between about 0.97 and 1. It is reasonable to use the simpler form $\sqrt{p(1-p)/n}$ in these settings. See also Note 3 for Chapter 15.
3. L. G. M. Bode, "Preventing surgical-site infections in nasal carriers of *Staphylococcus aureus*," *New England Journal of Medicine*, 362 (2010), pp. 9–17.
4. From the article "Kellogg Survey: Most Americans Don't Eat Breakfast Daily," June 2011 at detroitcbslocal.com.
5. Joanne Brenner, "Pew Internet: Social Networking (full detail)," Pew Internet and American Life Project, 2012 at pewinternet.org.
6. This rule of thumb is based on study of computational results in the papers cited in Note 15 and discussion with Alan Agresti.
7. The quotation is from page 1104 of the article cited in Note 1.
8. G. A. Mauser and H. Taylor Buckner, "Canadian attitudes toward gun control: the real story," The Mackenzie Institute, 1997, at teapot.usask.ca/cdn-firearms/Mauser/gunstory.html.

9. Myer, Gregory D. et al., "Youth versus Adult Weightlifting Injuries Presenting to United States Emergency Rooms: Accidental versus Nonaccidental Injury Mechanisms," *Journal of Strength and Conditioning Research*, 23 (2009), pp. 2054–2060.
10. We would like to thank the Monterey Bay Aquarium for supplying the data from its visitor survey.
11. In fact, *P*-values for two-sided tests are more accurate than those for one-sided tests. Our rule of thumb is a compromise to avoid the confusion of too many rules.
12. See Note 1 for Chapter 14.
13. Data found on the New Scientist website at `www.newscientist.com/article/dn1748-ero-coin-accused`.
14. Alexander Todorov et al., "Inferences of competence from faces predict election outcomes," *Science*, 308 (2005), pp. 1623–1626.
15. This interval is proposed by Alan Agresti and Brent A. Coull, "Approximate is better than 'exact' for interval estimation of binomial proportions," *The American Statistician*, 52 (1998), pp. 119–126. Note in particular that the plus four interval is often more accurate than the Clopper-Pearson "exact interval" based on the binomial distribution of the sample count and implemented by, for example, Minitab.

    There are several even more accurate but considerably more complex intervals for p that might be used in professional practice. See Lawrence D. Brown, Tony Cai, and Anirban DasGupta, "Interval estimation for a binomial proportion," *Statistical Science*, 16 (2001), pp. 101–133. A detailed theoretical study that uncovers the reason the large-sample interval is inaccurate is Lawrence D. Brown, Tony Cai, and Anirban DasGupta, "Confidence intervals for a binomial proportion and asymptotic expansions," *Annals of Statistics*, 30 (2002), pp. 160–201.
16. BBC News, December 25, 2006, at `news.bbc.co.uk`.
17. From Alan Agresti and Brian Caffo, "Simple and effective confidence intervals for proportions and differences of proportions result from adding two successes and two failures," *The American Statistician*, 45 (2000), pp. 280–288. When can the plus four interval be safely used? The answer depends on just how much accuracy you insist on. Brown and coauthors (see Note 15) recommend $n \geq 40$. Agresti and Coull (see Note 15) demonstrate that performance is almost always satisfactory in their eyes when $n \geq 5$. Our rule of thumb $n \geq 10$ allows for confidence levels C other than 95% and fits our philosophy of not insisting on more exact results than practice requires. The big point is that plus four is very much more accurate than the standard interval for most values of p and all but very large n.
18. Gary Stoner et al., "Regression of rectal polyps in familial adenomatous polyposis patients with freeze dried black raspberries," *Cancer Prev Res* (2008), 1(7 Suppl): PR-14.
19. Lydia Saad, "In U.S., 11% of households report computer crimes, a new high," December 2010, at `www.gallup.com`. The sampling scheme was more complex than an SRS, so the computation of the number in the sample reporting crimes and acting as if it were an SRS is oversimplified.
20. Michele L. Head, "Examining college students' ethical values," Consumer Science and Retailing honors project, Purdue University, 2003.
21. Elizabeth Cohen, "Your top health searches, asked and answered," Pew Internet and American Life Project, 2010, at `pewinternet.org`. The cell phone sample used random digit dialing drawn through a systematic sampling from dedicated wireless 100-blocks and shared service 100-blocks with no directory-listed landline numbers, so acting as if we have an SRS is oversimplified.
22. Ted D. Adams et al., "Health benefits of gastric bypass surgery after 6 years," *JAMA*, 308 (2012), pp. 1122–1131.
23. See Note 23 for Chapter 8.
24. *The Nation's Report Card: Trends in Academic Progress 2012 (NCES 2013 456)* can be found on the website `nces.ed.gov`.
25. Sam Dillon, "Incentives for advanced work let pupils and teachers cash in," *New York Times*, October 2, 2011.
26. A. Mantonakis et al., "Order in Choice: Effects of Serial Position on Preferences," *Psychological Science*, 20 (2009), pp. 1309–1312.
27. Sam Oches "The drive-through performance study," October 2013 on the website `www.qsrmagazine.com/reports`.
28. See Note 26.
29. Francisco Lloret et al., "Fire and resprouting in Mediterranean ecosystems: insights from an external biogeographical region, the Mexican shrubland," *American Journal of Botany*, 88 (1999), pp. 1655–1661.
30. The data were obtained from the GSS Cumulative Datafile 1972–2012 at `sda.berkeley.edu/archive.htm`. The data were restricted to the year 2012 and respondents between 18- and 32 years of age.

Chapter 23 Notes

1. Shauna B. Wilson et al., "Dating across race: An examination of African American Internet personal advertisements," *Journal of Black Studies*, 37 (2007), pp. 964–982.
2. Based on data in Susannah Fox et al., "Health online 2012," January 2013, at `pewinternet.org`.
3. From the 2013 Youth Risk Behavior Surveillance System at `apps.nccd.cdc.gov/youthonline/App/Default.aspx`. The data are from a complex multistage sample, so that acting as if we have SRSs is oversimplified.
4. The data was obtained from the GSS Cumulative Datafile 1972–2010 at `sda.berkeley.edu/archive.htm`.
5. Steven J. Frenda et al., "False memories of fabricated political events," *Journal of Experimental Social Psychology*, 49 (2013), pp. 280–286. We would like to thank the authors for providing the data and the photo in Figure 23.3.

6. This rule of thumb is quite conservative. It is in fact safe to arrange the data as a 2×2 table and apply the rule of thumb from Chapter 23 that all four *expected* counts must be five or greater. We give the conservative rule here because expected counts are messy to explain in the present context.
7. JoAnn K. Wells, Allan F. Williams, and Charles M. Farmer, "Seat belt use among African Americans, Hispanics, and whites," *Accident Analysis and Prevention,* 34 (2002), pp. 523–529.
8. Steiner Sulheim et al., "Helmet use and risk of head injuries in alpine skiers and snowboarders," *Journal of the American Medical Association,* 295 (2006), pp. 919–924.
9. Armando E. Giuliano MD et al., "Axillary dissection vs no axillary dissection in women with invasive breast cancer and sentinel node metastasis," *Journal of the American Medical Association,* 305 (2011), pp. 569–575. The sample sizes for the two groups and the proportions of patients in each group that are disease-free after five years have been chosen to match those in the paper.
10. The plus four method is due to Alan Agresti and Brian Caffo. See Note 15 for Chapter 22.
11. Frances A. Campbell et al., "Adult outcomes as a function of an early childhood educational program: An Abecedarian Project follow up," *Developmental Psychology,* 48 (2012), pp. 1033–1043.
12. Modified from Richard A. Schieber et al., "Risk factors for injuries from in-line skating and the effectiveness of safety gear," *New England Journal of Medicine,* 335 (1996), Internet summary at `content.nejm.org`.
13. See Note 29 for Chapter 22.
14. See Note 4.
15. See Amanda Lenhart and Mary Madden, "Teens, Privacy and Online Social Networks," Pew Internet and American Life Project, 2007, at `www.pewinternet.org`.
16. W. P. T. James et al., "Effect of Sibutramine on cardiovascular outcomes in overweight and obese subjects," *New England Journal of Medicine,* 363 (2010), pp. 905–917.
17. François Gaudet et al., "Induction of tumors in mice by genomic hypomethylation," *Science,* 300 (2003), pp. 489–492.
18. Barbara Helmrich, "Window of Opportunity? Adolescence, Music and Algebra," *Journal of Adolescent Research,* 25 (2010), pp. 557–577.
19. Ninh T. Nguyen et al., "Improved bariatric surgery outcomes for medicare beneficiaries after implementation of the medicare national coverage determination," *Archives of Surgery,* 145 (2010), pp. 72–78.
20. Justin Dimick et al., "Bariatric surgery complications before vs after implementation of a national policy restricting coverage to centers of excellence," *Journal of the American Medical Association,* 309 (2013), pp. 792–799.
21. Arne L. Kalleberg and Kevin T. Leicht, "Gender and organizational performance: Determinants of small business survival and success," *The Academy of Management Journal,* 34 (1991), pp. 136–161.
22. See Maeve Duggan and Joanna Brenner, "The demographics of social media users–2012," Pew Internet and American Life Project, 2013, at `www.pewinternet.org`.
23. D. Gonzales et al., "Varenicline, an $\alpha 4\beta 2$ nicotinic acetylcholine receptor partial agonist, vs sustained-release bupropion and placebo for smoking cessation," *Journal of the American Medical Association,* 340 (2006), pp. 47–55.
24. Data courtesy of Raymond Dumett, Purdue University.
25. Based on Alan G. Sanfey et al., "The neural basis of economic decision-making in the ultimatum game," *Science,* 300 (2003), pp. 1755–1758. The paper reports a chi-square test (equivalent to a two-sided z test). This analysis is incorrect for the paper's data, as there were in fact only 19 participants, each appearing twice in each row of the table given in the exercise. Exercise 23.31 therefore amends the data, assuming 76 participants, so that the elementary analysis is correct.
26. Ross L. Prentice et al., "Low-fat dietary pattern and risk of invasive breast cancer," *Journal of the American Medical Association,* 295 (2006), pp. 629–642.
27. Clive G. Jones et al., "Chain reactions linking acorns to gypsy moth outbreaks and Lyme disease risk," *Science,* 279 (1998), pp. 1023–1026.
28. See Note 11.
29. R.B. Turner et al., "A randomized trial of the efficacy of hand disinfection for prevention of rhinovirus infection," *Clinical Infectious Diseases,* 54 (2012), pp. 1422–1426.

Chapter 24 Notes

1. Data provided by Drina Iglesia, Purdue University. The data are part of a larger study reported in D. D. S. Iglesia, E. J. Cragoe, Jr., and J. W. Vanable, "Electric field strength and epithelization in the newt (*Notophthalmus viridescens*)," *Journal of Experimental Zoology,* 274 (1996), pp. 56–62.
2. See Note 3 for Chapter 23.
3. Kathryn Zickhur and Lee Rainie, "Wikipedia, past and present," at `www.pewinternet.org`.
4. Francesca Gino and Scott S. Wilermuth, "Evil genius? How dishonesty can lead to greater creativity," *Psychological Science,* 25 (2014), pp. 973–981.
5. K. S. Oberhauser, "Fecundity, lifespan and egg mass in butterflies: effects of malederived nutrients and female size," *Functional Ecology,* 11 (1997), pp. 166–175.
6. Scott W. Powers et al., "Cognitive behavioral therapy plus amitriptyline for chronic migraine in children and adolescents," *JAMA,* 310 (2013), pp. 2622–2630.
7. Maureen Hack et al., "Outcomes in young adulthood for very-low-birth-weight infants," *New England Journal of Medicine,* 346 (2002), pp. 149–157. The exercises are simplified, in that the measures reported in this paper have been statistically adjusted for "sociodemographic status."
8. Based on data in Amanda Lenhart, "Cell phones and American adults," Pew Internet and American Life Project, September 2010, at `pewinternet.org`.

9. From the Prevalence and Trends Data of the Behavioral Risk Factor Surveillance System (BRFSS), at `www.cdc.gov/BRFSS/`.
10. Joseph W. Bartges et al., "Influence of acidifying or alkalizing diets on bone mineral density and urine relative supersaturation with calcium oxalate and struvite in healthy cats," *American Journal of Veterinary Research*, 74 (2013) pp. 1347–1352.
11. Jin Ha Lee and J. Stephen Downie, "Survey of music information needs, uses, and seeking behaviors: preliminary findings," online Proceedings of the 5th International Conference on Music Information Retrieval, 2004, at `ismir2004.ismir.net`.
12. Josh McDermott and Marc D. Hauser, "Nonhuman primates prefer slow tempos but dislike music overall," *Cognition*, 104 (2007), pp. 654–668. Failure to take account of repeated measures on the same subjects is one of the most common errors observed in statistical analysis.
13. James Otto, Michael F. Brown, and William Long III, "Training rats to search and alert on contraband odors," *Applied Animal Behaviour Science*, 77 (2002), pp. 217–232.
14. From the Merck website, `www.merckvaccines.com/gardasilProductPage_frmst.html`.
15. These data were originally collected by L. M. Linde of UCLA but were first published by M. R. Mickey, O. J. Dunn, and V. Clark, "Note on the use of stepwise regression in detecting outliers," *Computers and Biomedical Research*, 1 (1967), pp. 105–111. The data have been used by several authors. We found them in N. R. Draper and J. A. John, "Influential observations and outliers in regression," *Technometrics*, 23 (1981), pp. 21–26.
16. Jacqueline T. Ngai and Diane S. Srivastava, "Predators accelerate nutrient cycling in a bromeliad ecosystem," *Science*, 314 (2006), p. 963. I thank Jacqueline Ngai for providing the data.
17. Yvan R. Germain, "The dyeing of ramie with fiber reactive dyes using the cold pad-batch method," MS thesis, Purdue University, 1988.
18. Data provided by Marigene Arnold, Kalamazoo College.
19. Data provided by Corinne Lim, Purdue University, from a student project supervised by Professor Joseph Vanable.
20. Saiyad S. Ahmed, "Effect of microwave drying on checking and mechanical strength of low-moisture baked products," MS thesis, Purdue University, 1994.
21. Michael O. Finkelstein and Bruce Levin, "Statistical proof of discrimination in peremptory challenges," *Chance*, 17, No. 1 (2004), pp. 35–38.
22. G. S. Hotamisligil et al., "Uncoupling of obesity from insulin resistance through a targeted mutation in *aP2*, the adipocyte fatty acid binding protein," *Science*, 274 (1996), pp. 1377–1379.
23. Data simulated from a Normal distribution with the mean and standard deviation reported by Sarah Morrison and Jan Noyes, "A comparison of two computer fonts: serif versus ornate sans serif," *Usability News*, 5.2 (2003) at `psychology.wichita.edu/surl/usability_news.html`.
24. John D. Miller et al., "Public acceptance of evolution," *Science*, 313 (2006), pp. 765–766. The information in the exercise appears in the supplementary online material.

Chapter 25 Notes

1. See Note 1 for Chapter 6.
2. Pennsylvania State University Division of Student Affairs, "Net behaviors November 2006," *Penn State Pulse*, at `www.sa.psu.edu`.
3. See Note 2 for Chapter 6.
4. All General Social Survey exercises in this chapter present tables constructed using the search function at the GSS archive, `sda.berkeley.edu/archive.htm`. Most concern data from the 2012 GSS.
5. There are many computer studies of the accuracy of chi-square critical values for X^2.Our guideline goes back to W. G. Cochran (1954). Later work has shown that it is often conservative in the sense that, if the expected cell counts are all similar and the degrees of freedom exceed 1, the chi-square approximation works well for an average expected count as small as 1 or 2. Our guideline protects against dissimilar expected counts. It has the added advantage that it is safe in the 2×2 case, where the chi-square approximation is least good. So our guideline is helpful for beginners—there is no single condition that is not conservative and applies to 2×2 and larger tables with similar and dissimilar expected cell counts. There are exact procedures that (with software) should be used for tables that do not satisfy our guideline. For a survey, see Alan Agresti, "A survey of exact inference for contingency tables," *Statistical Science*, 7 (1992), pp. 131–177.
6. Scott Keeter, et al., "What's missing from national RDD surveys? The impact of the growing cell-only population," (2007) at `www.pewresearch.org`.
7. Leah Christian, et al., "Assessing the cell phone challenge to survey research in 2010," at `www.pewresearch.org`.
8. Modified from Felicity Barringer, "Measuring sexuality through polls can be shaky," *New York Times*, April 25, 1993.
9. Based on a news item in *Science*, 305 (2004), p. 1560. The study, by Daniel Klem, appeared in the *Wilson Journal*.
10. David W. Eby et al., "The effect of changing from secondary to primary safety belt enforcement on police harassment," *Accident Analysis and Prevention*, 36 (2000), pp. 819–828.
11. See Note 2 from Chapter 17.
12. See Note 9 for Chapter 6.
13. See Note 12 for Chapter 6.
14. See Note 7.
15. See Note 17 for Chapter 6.

16. Virgilio P. Carnielli et al., "Intestinal absorption of long-chain polyunsaturated fatty acids in preterm infants fed breast milk or formula," *American Journal of Clinical Nutrition*, 67 (1998), pp. 97–103.
17. Adapted from M. A. Visintainer, J. R. Volpicelli, and M. E. P. Seligman, "Tumor rejection in rats after inescapable or escapable shock," *Science*, 216 (1982), pp. 437–439.
18. The National Longitudinal Study of Adolescent Health interviewed a stratified random sample of 27,000 adolescents, then reinterviewed many of the subjects six years later, when most were aged 19–25. These data are from the Wave III reinterviews in 2000 and 2001, found at the Web site of the Carolina Population Center, `www.cpc.unc.edu`.
19. Tom Reichert, "The prevalence of sexual imagery in ads targeted to young adults," *Journal of Consumer Affairs*, 37 (2003), pp. 403–412.
20. József Topál et al., "Differential sensitivity to human communication in dogs, wolves and human infants," *Science*, 325 (2009), pp. 1269–1272. Many statistical software packages offer "exact tests" that are valid even when there are small expected counts.
21. U.S. Department of Commerce, Office of Travel and Tourism Industries, in-flight survey, 2007, at `tinet.ita.doc.gov`.
22. Andrew Steptoe, et al., "Social isolation, loneliness, and all-cause mortality in older men and women," *Proceedings of the National Academy of Sciences*, 110 (2013), pp. 5797–5801. Two categories of depression were combined in the table.
23. See Note 2 for Chapter 16. We have simplified slightly: the table in the paper is exactly as in the exercise but contains data for 63 subjects plus data from one type of bar for three subjects who dropped out. Although the authors say that their chi-square refers to this table, they give a nonsignificant value that contradicts what the table shows.
24. See Note 15 for Chapter 6.
25. Two way tables from the Youth Risk Behavior Surveillance System can be constructed from the website `apps.nccd.cdc.gov/youthonline/App/Default.aspx`.
26. Data compiled from a table of percents in "Americans view higher education as key to the American dream," press release by the National Center for Public Policy and Higher Education, May 3, 2000, at `www.highereducation.org`.
27. See Note 14 for Chapter 6.
28. Data produced by Ries and Smith, found in William D. Johnson and Gary G. Koch, "A note on the weighted least squares analysis of the Ries-Smith contingency table data," *Technometrics*, 13 (1971), pp. 438–447.

Chapter 26 Notes

1. Samuel Karelitz et al., "Relation of crying activity in early infancy to speech and intellectual development at age three years," *Child Development*, 35 (1964), pp. 769–777.
2. From a graph in Naomi E. Allen et al., "Moderate alcohol intake and cancer incidence in women," *Journal of the National Cancer Institute*, 101 (2009), pp. 296–305. These data represent averages over large numbers of women and are an example of an ecological correlation (see page 146 in Chapter 5); one must be careful not to interpret the data as applying to individuals.
3. From a graph in Stephen M. Fleming et al., "Relating introspective accuracy to individual differences in brain structure," *Science*, 329 (2010), pp. 1541–1543.
4. See Note 26 for Chapter 5.
5. Electronic Encyclopedia of Statistical Examples and Exercises (EESEE) at the text website, `www.whfreeman.com/bps`.
6. From a graph in Allison L. Perry et al., "Climate change and distribution shifts in marine fishes," *Science*, 308 (2005), pp. 1912–1915. The explanatory variable is the five-year running mean of winter (December to March) sea-bottom temperature.
7. Data for the building at 1800 Ben Franklin Drive, Sarasota, Florida, starting in February 2003. From the website of the Sarasota County Property Appraiser, `www.sc-pa.com`.
8. Yanhui Lu et al., "Mirid bug outbreaks in multiple crops correlated with wide-scale adoption of Bt cotton in China," *Science*, 328 (2010), pp. 1151–1154.
9. See Note 1 for Chapter 4.
10. Based on Marion E. Dunshee, "A study of factors affecting the amount and kind of food eaten by nursery school children," *Child Development*, 2 (1931), pp. 163–183. This article gives the means, standard deviations, and correlation for 37 children, from which the data in the exercise are simulated.
11. From Table S2 in the online supplement to Antonio Dell'Anno and Roberto Danovaro, "Extracellular DNA plays a key role in deep-sea ecosystem functioning," *Science*, 309 (2005), p. 2179.
12. See Note 19 for Chapter 7.

Chapter 27 Notes

1. Ethan J. Temeles and W. John Kress, "Adaptation in a plant-hummingbird association," *Science*, 300 (2003), pp. 630–633. We thank Ethan J. Temeles for providing the data.
2. Elisabeth Wells-Parker et al., "An exploratory study of the relationship between road rage and crash experience in a representative sample of US drivers," *Accident Analysis and Prevention*, 34 (2002), pp. 271–278.
3. Victoria L. Brescoll and Eric L. Uhlmann, "Can an angry woman get ahead? Status conferral, gender and expression of emotion in the workplace," *Psychological Science*, 19 (2008), pp. 268–273. The description and data are based on study 1 in this article.
4. See Note 10 for Chapter 2.
5. The data from the General Social Survey for this exercise was constructed using the search function and download

capabilities at the GSS archive, sda.berkeley.edu/archive.htm.

6. A. Timur Sevincer et al., "Positive thinking about the future in newspaper reports and presidential addresses predicts economic downturn," *Psychological Science,* 25 (2014). Available online at pss.sagepub.com. Data are estimates from a plot in the article, with positive articles corresponding to articles with a positive thinking z score above 2 and negative articles corresponding to a positive thinking z score below −2.
7. See Note 16 for Chapter 24.
8. David B. Wooten, "One-of-a-kind in a full house: some consequences of ethnic and gender distinctiveness," *Journal of Consumer Psychology,* 4 (1995), 205–224.
9. John P. Thomas, "Influences on mathematics learning and attitudes among African American high school students," *Journal of Negro Education,* 69 (2000), pp. 165–183.
10. Wayne J. Camera and Donald Powers, "Coaching and the SAT I," *TIP* (online journal at www.siop.org/tip), July 1999.
11. Henrik Hagvedt and Vanessa M. Patrick, "Art infusion: The influence of visual art on the perception and evaluation of consumer products," *Journal of Marketing Research,* XLV (2008), pp. 379–389.
12. Timothy Church et al., "Effects of aerobic and resistance training on hemoglobin A1c levels in patients with type 2 diabetes: A randomized controlled trial," *Journal of the American Medical Association,* 304 (2010), pp. 2253–2262.
13. Jennifer J. Argo et al., "Positive consumer contagion: Responses to attractive others in a retail context," *Journal of Marketing Research,* XLV (2008), pp. 690–701.
14. Data from the online supplement to André Kessler and Ian T. Baldwin, "Defensive function of herbivore-induced plant volatile emissions in nature," *Science,* 291 (2001), pp. 2141–2144.
15. The data and the full story can be found in the Data and Story Library at lib.stat.cmu.edu. The original study is by Faith Loven, "A study of interlist equivalency of the CID W-22 word list presented in quiet and in noise," MS thesis, University of Iowa, 1981.
16. See Note 10 for Chapter 9. We thank Kenwyn Suttle for providing these data, for the year 2003.
17. See Note 17 for Chapter 6.
18. See Note 20 for Chapter 2.
19. See Note 21 for Chapter 2.
20. Pam A. Mueller and Daniel M. Oppenheimer, "The pen is mightier than the keyboard: Advantages of longhand over laptop note taking," *Psychological Science,* 25(6) (2014), pp. 1159–1168.
21. See Note 10 for Chapter 9.

TABLES

Table A Standard Normal Cumulative Proportions

Table B Random Digits

Table C *t* Distribution Critical Values

Table D Chi-square Distribution Critical Values

Table E Critical Values of the Correlation *r*

Table entry for z is the area under the standard Normal curve to the left of z.

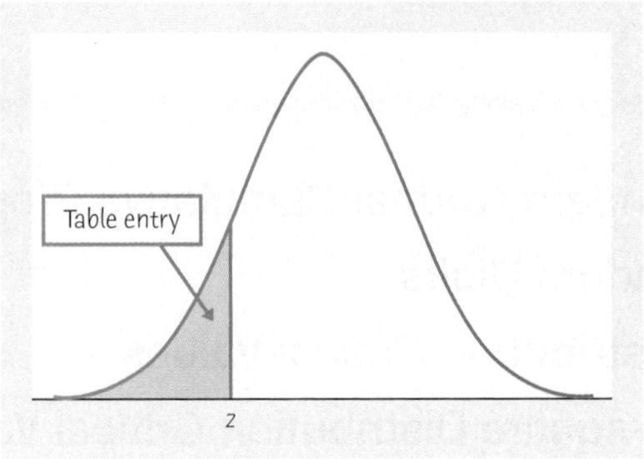

TABLE A STANDARD NORMAL CUMULATIVE PROPORTIONS

z	.00	.01	.02	.03	.04	.05	.06	.07	.08	.09
−3.4	.0003	.0003	.0003	.0003	.0003	.0003	.0003	.0003	.0003	.0002
−3.3	.0005	.0005	.0005	.0004	.0004	.0004	.0004	.0004	.0004	.0003
−3.2	.0007	.0007	.0006	.0006	.0006	.0006	.0006	.0005	.0005	.0005
−3.1	.0010	.0009	.0009	.0009	.0008	.0008	.0008	.0008	.0007	.0007
−3.0	.0013	.0013	.0013	.0012	.0012	.0011	.0011	.0011	.0010	.0010
−2.9	.0019	.0018	.0018	.0017	.0016	.0016	.0015	.0015	.0014	.0014
−2.8	.0026	.0025	.0024	.0023	.0023	.0022	.0021	.0021	.0020	.0019
−2.7	.0035	.0034	.0033	.0032	.0031	.0030	.0029	.0028	.0027	.0026
−2.6	.0047	.0045	.0044	.0043	.0041	.0040	.0039	.0038	.0037	.0036
−2.5	.0062	.0060	.0059	.0057	.0055	.0054	.0052	.0051	.0049	.0048
−2.4	.0082	.0080	.0078	.0075	.0073	.0071	.0069	.0068	.0066	.0064
−2.3	.0107	.0104	.0102	.0099	.0096	.0094	.0091	.0089	.0087	.0084
−2.2	.0139	.0136	.0132	.0129	.0125	.0122	.0119	.0116	.0113	.0110
−2.1	.0179	.0174	.0170	.0166	.0162	.0158	.0154	.0150	.0146	.0143
−2.0	.0228	.0222	.0217	.0212	.0207	.0202	.0197	.0192	.0188	.0183
−1.9	.0287	.0281	.0274	.0268	.0262	.0256	.0250	.0244	.0239	.0233
−1.8	.0359	.0351	.0344	.0336	.0329	.0322	.0314	.0307	.0301	.0294
−1.7	.0446	.0436	.0427	.0418	.0409	.0401	.0392	.0384	.0375	.0367
−1.6	.0548	.0537	.0526	.0516	.0505	.0495	.0485	.0475	.0465	.0455
−1.5	.0668	.0655	.0643	.0630	.0618	.0606	.0594	.0582	.0571	.0559
−1.4	.0808	.0793	.0778	.0764	.0749	.0735	.0721	.0708	.0694	.0681
−1.3	.0968	.0951	.0934	.0918	.0901	.0885	.0869	.0853	.0838	.0823
−1.2	.1151	.1131	.1112	.1093	.1075	.1056	.1038	.1020	.1003	.0985
−1.1	.1357	.1335	.1314	.1292	.1271	.1251	.1230	.1210	.1190	.1170
−1.0	.1587	.1562	.1539	.1515	.1492	.1469	.1446	.1423	.1401	.1379
−0.9	.1841	.1814	.1788	.1762	.1736	.1711	.1685	.1660	.1635	.1611
−0.8	.2119	.2090	.2061	.2033	.2005	.1977	.1949	.1922	.1894	.1867
−0.7	.2420	.2389	.2358	.2327	.2296	.2266	.2236	.2206	.2177	.2148
−0.6	.2743	.2709	.2676	.2643	.2611	.2578	.2546	.2514	.2483	.2451
−0.5	.3085	.3050	.3015	.2981	.2946	.2912	.2877	.2843	.2810	.2776
−0.4	.3446	.3409	.3372	.3336	.3300	.3264	.3228	.3192	.3156	.3121
−0.3	.3821	.3783	.3745	.3707	.3669	.3632	.3594	.3557	.3520	.3483
−0.2	.4207	.4168	.4129	.4090	.4052	.4013	.3974	.3936	.3897	.3859
−0.1	.4602	.4562	.4522	.4483	.4443	.4404	.4364	.4325	.4286	.4247
−0.0	.5000	.4960	.4920	.4880	.4840	.4801	.4761	.4721	.4681	.4641

Table entry for z is the area under the standard Normal curve to the left of z.

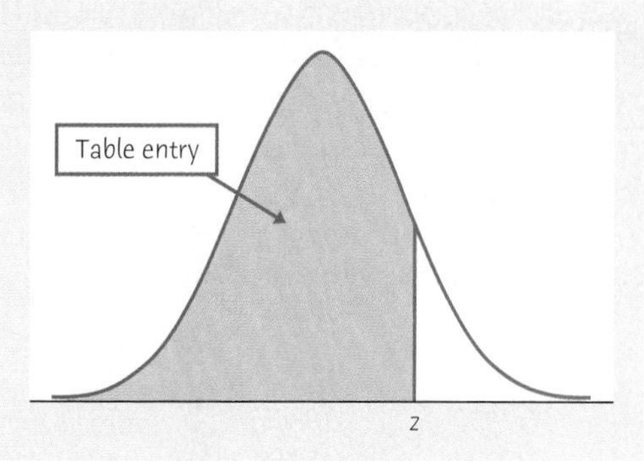

TABLE A STANDARD NORMAL CUMULATIVE PROPORTIONS (*CONTINUED*)

z	.00	.01	.02	.03	.04	.05	.06	.07	.08	.09
0.0	.5000	.5040	.5080	.5120	.5160	.5199	.5239	.5279	.5319	.5359
0.1	.5398	.5438	.5478	.5517	.5557	.5596	.5636	.5675	.5714	.5753
0.2	.5793	.5832	.5871	.5910	.5948	.5987	.6026	.6064	.6103	.6141
0.3	.6179	.6217	.6255	.6293	.6331	.6368	.6406	.6443	.6480	.6517
0.4	.6554	.6591	.6628	.6664	.6700	.6736	.6772	.6808	.6844	.6879
0.5	.6915	.6950	.6985	.7019	.7054	.7088	.7123	.7157	.7190	.7224
0.6	.7257	.7291	.7324	.7357	.7389	.7422	.7454	.7486	.7517	.7549
0.7	.7580	.7611	.7642	.7673	.7704	.7734	.7764	.7794	.7823	.7852
0.8	.7881	.7910	.7939	.7967	.7995	.8023	.8051	.8078	.8106	.8133
0.9	.8159	.8186	.8212	.8238	.8264	.8289	.8315	.8340	.8365	.8389
1.0	.8413	.8438	.8461	.8485	.8508	.8531	.8554	.8577	.8599	.8621
1.1	.8643	.8665	.8686	.8708	.8729	.8749	.8770	.8790	.8810	.8830
1.2	.8849	.8869	.8888	.8907	.8925	.8944	.8962	.8980	.8997	.9015
1.3	.9032	.9049	.9066	.9082	.9099	.9115	.9131	.9147	.9162	.9177
1.4	.9192	.9207	.9222	.9236	.9251	.9265	.9279	.9292	.9306	.9319
1.5	.9332	.9345	.9357	.9370	.9382	.9394	.9406	.9418	.9429	.9441
1.6	.9452	.9463	.9474	.9484	.9495	.9505	.9515	.9525	.9535	.9545
1.7	.9554	.9564	.9573	.9582	.9591	.9599	.9608	.9616	.9625	.9633
1.8	.9641	.9649	.9656	.9664	.9671	.9678	.9686	.9693	.9699	.9706
1.9	.9713	.9719	.9726	.9732	.9738	.9744	.9750	.9756	.9761	.9767
2.0	.9772	.9778	.9783	.9788	.9793	.9798	.9803	.9808	.9812	.9817
2.1	.9821	.9826	.9830	.9834	.9838	.9842	.9846	.9850	.9854	.9857
2.2	.9861	.9864	.9868	.9871	.9875	.9878	.9881	.9884	.9887	.9890
2.3	.9893	.9896	.9898	.9901	.9904	.9906	.9909	.9911	.9913	.9916
2.4	.9918	.9920	.9922	.9925	.9927	.9929	.9931	.9932	.9934	.9936
2.5	.9938	.9940	.9941	.9943	.9945	.9946	.9948	.9949	.9951	.9952
2.6	.9953	.9955	.9956	.9957	.9959	.9960	.9961	.9962	.9963	.9964
2.7	.9965	.9966	.9967	.9968	.9969	.9970	.9971	.9972	.9973	.9974
2.8	.9974	.9975	.9976	.9977	.9977	.9978	.9979	.9979	.9980	.9981
2.9	.9981	.9982	.9982	.9983	.9984	.9984	.9985	.9985	.9986	.9986
3.0	.9987	.9987	.9987	.9988	.9988	.9989	.9989	.9989	.9990	.9990
3.1	.9990	.9991	.9991	.9991	.9992	.9992	.9992	.9992	.9993	.9993
3.2	.9993	.9993	.9994	.9994	.9994	.9994	.9994	.9995	.9995	.9995
3.3	.9995	.9995	.9995	.9996	.9996	.9996	.9996	.9996	.9996	.9997
3.4	.9997	.9997	.9997	.9997	.9997	.9997	.9997	.9997	.9997	.9998

TABLE B RANDOM DIGITS

LINE								
101	19223	95034	05756	28713	96409	12531	42544	82853
102	73676	47150	99400	01927	27754	42648	82425	36290
103	45467	71709	77558	00095	32863	29485	82226	90056
104	52711	38889	93074	60227	40011	85848	48767	52573
105	95592	94007	69971	91481	60779	53791	17297	59335
106	68417	35013	15529	72765	85089	57067	50211	47487
107	82739	57890	20807	47511	81676	55300	94383	14893
108	60940	72024	17868	24943	61790	90656	87964	18883
109	36009	19365	15412	39638	85453	46816	83485	41979
110	38448	48789	18338	24697	39364	42006	76688	08708
111	81486	69487	60513	09297	00412	71238	27649	39950
112	59636	88804	04634	71197	19352	73089	84898	45785
113	62568	70206	40325	03699	71080	22553	11486	11776
114	45149	32992	75730	66280	03819	56202	02938	70915
115	61041	77684	94322	24709	73698	14526	31893	32592
116	14459	26056	31424	80371	65103	62253	50490	61181
117	38167	98532	62183	70632	23417	26185	41448	75532
118	73190	32533	04470	29669	84407	90785	65956	86382
119	95857	07118	87664	92099	58806	66979	98624	84826
120	35476	55972	39421	65850	04266	35435	43742	11937
121	71487	09984	29077	14863	61683	47052	62224	51025
122	13873	81598	95052	90908	73592	75186	87136	95761
123	54580	81507	27102	56027	55892	33063	41842	81868
124	71035	09001	43367	49497	72719	96758	27611	91596
125	96746	12149	37823	71868	18442	35119	62103	39244
126	96927	19931	36809	74192	77567	88741	48409	41903
127	43909	99477	25330	64359	40085	16925	85117	36071
128	15689	14227	06565	14374	13352	49367	81982	87209
129	36759	58984	68288	22913	18638	54303	00795	08727
130	69051	64817	87174	09517	84534	06489	87201	97245
131	05007	16632	81194	14873	04197	85576	45195	96565
132	68732	55259	84292	08796	43165	93739	31685	97150
133	45740	41807	65561	33302	07051	93623	18132	09547
134	27816	78416	18329	21337	35213	37741	04312	68508
135	66925	55658	39100	78458	11206	19876	87151	31260
136	08421	44753	77377	28744	75592	08563	79140	92454
137	53645	66812	61421	47836	12609	15373	98481	14592
138	66831	68908	40772	21558	47781	33586	79177	06928
139	55588	99404	70708	41098	43563	56934	48394	51719
140	12975	13258	13048	45144	72321	81940	00360	02428
141	96767	35964	23822	96012	94591	65194	50842	53372
142	72829	50232	97892	63408	77919	44575	24870	04178
143	88565	42628	17797	49376	61762	16953	88604	12724
144	62964	88145	83083	69453	46109	59505	69680	00900
145	19687	12633	57857	95806	09931	02150	43163	58636
146	37609	59057	66967	83401	60705	02384	90597	93600
147	54973	86278	88737	74351	47500	84552	19909	67181
148	00694	05977	19664	65441	20903	62371	22725	53340
149	71546	05233	53946	68743	72460	27601	45403	88692
150	07511	88915	41267	16853	84569	79367	32337	03316

Table entry for C is the critical value t^* required for confidence level C. To approximate one- and two-sided P-values, compare the value of the t statistic with the critical values of t^* that match the P-values given at the bottom of the table.

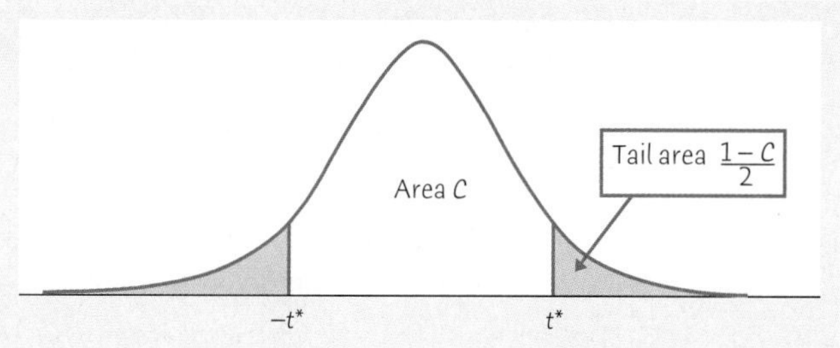

TABLE C *t* DISTRIBUTION CRITICAL VALUES

DEGREES OF FREEDOM	CONFIDENCE LEVEL *C*											
	50%	60%	70%	80%	90%	95%	96%	98%	99%	99.5%	99.8%	99.9%
1	1.000	1.376	1.963	3.078	6.314	12.71	15.89	31.82	63.66	127.3	318.3	636.6
2	0.816	1.061	1.386	1.886	2.920	4.303	4.849	6.965	9.925	14.09	22.33	31.60
3	0.765	0.978	1.250	1.638	2.353	3.182	3.482	4.541	5.841	7.453	10.21	12.92
4	0.741	0.941	1.190	1.533	2.132	2.776	2.999	3.747	4.604	5.598	7.173	8.610
5	0.727	0.920	1.156	1.476	2.015	2.571	2.757	3.365	4.032	4.773	5.893	6.869
6	0.718	0.906	1.134	1.440	1.943	2.447	2.612	3.143	3.707	4.317	5.208	5.959
7	0.711	0.896	1.119	1.415	1.895	2.365	2.517	2.998	3.499	4.029	4.785	5.408
8	0.706	0.889	1.108	1.397	1.860	2.306	2.449	2.896	3.355	3.833	4.501	5.041
9	0.703	0.883	1.100	1.383	1.833	2.262	2.398	2.821	3.250	3.690	4.297	4.781
10	0.700	0.879	1.093	1.372	1.812	2.228	2.359	2.764	3.169	3.581	4.144	4.587
11	0.697	0.876	1.088	1.363	1.796	2.201	2.328	2.718	3.106	3.497	4.025	4.437
12	0.695	0.873	1.083	1.356	1.782	2.179	2.303	2.681	3.055	3.428	3.930	4.318
13	0.694	0.870	1.079	1.350	1.771	2.160	2.282	2.650	3.012	3.372	3.852	4.221
14	0.692	0.868	1.076	1.345	1.761	2.145	2.264	2.624	2.977	3.326	3.787	4.140
15	0.691	0.866	1.074	1.341	1.753	2.131	2.249	2.602	2.947	3.286	3.733	4.073
16	0.690	0.865	1.071	1.337	1.746	2.120	2.235	2.583	2.921	3.252	3.686	4.015
17	0.689	0.863	1.069	1.333	1.740	2.110	2.224	2.567	2.898	3.222	3.646	3.965
18	0.688	0.862	1.067	1.330	1.734	2.101	2.214	2.552	2.878	3.197	3.611	3.922
19	0.688	0.861	1.066	1.328	1.729	2.093	2.205	2.539	2.861	3.174	3.579	3.883
20	0.687	0.860	1.064	1.325	1.725	2.086	2.197	2.528	2.845	3.153	3.552	3.850
21	0.686	0.859	1.063	1.323	1.721	2.080	2.189	2.518	2.831	3.135	3.527	3.819
22	0.686	0.858	1.061	1.321	1.717	2.074	2.183	2.508	2.819	3.119	3.505	3.792
23	0.685	0.858	1.060	1.319	1.714	2.069	2.177	2.500	2.807	3.104	3.485	3.768
24	0.685	0.857	1.059	1.318	1.711	2.064	2.172	2.492	2.797	3.091	3.467	3.745
25	0.684	0.856	1.058	1.316	1.708	2.060	2.167	2.485	2.787	3.078	3.450	3.725
26	0.684	0.856	1.058	1.315	1.706	2.056	2.162	2.479	2.779	3.067	3.435	3.707
27	0.684	0.855	1.057	1.314	1.703	2.052	2.158	2.473	2.771	3.057	3.421	3.690
28	0.683	0.855	1.056	1.313	1.701	2.048	2.154	2.467	2.763	3.047	3.408	3.674
29	0.683	0.854	1.055	1.311	1.699	2.045	2.150	2.462	2.756	3.038	3.396	3.659
30	0.683	0.854	1.055	1.310	1.697	2.042	2.147	2.457	2.750	3.030	3.385	3.646
40	0.681	0.851	1.050	1.303	1.684	2.021	2.123	2.423	2.704	2.971	3.307	3.551
50	0.679	0.849	1.047	1.299	1.676	2.009	2.109	2.403	2.678	2.937	3.261	3.496
60	0.679	0.848	1.045	1.296	1.671	2.000	2.099	2.390	2.660	2.915	3.232	3.460
80	0.678	0.846	1.043	1.292	1.664	1.990	2.088	2.374	2.639	2.887	3.195	3.416
100	0.677	0.845	1.042	1.290	1.660	1.984	2.081	2.364	2.626	2.871	3.174	3.390
1000	0.675	0.842	1.037	1.282	1.646	1.962	2.056	2.330	2.581	2.813	3.098	3.300
z^*	0.674	0.841	1.036	1.282	1.645	1.960	2.054	2.326	2.576	2.807	3.091	3.291
One-sided P	.25	.20	.15	.10	.05	.025	.02	.01	.005	.0025	.001	.0005
Two-sided P	.50	.40	.30	.20	.10	.05	.04	.02	.01	.005	.002	.001

Table entry for p is the critical value χ^* with probability p lying to its right.

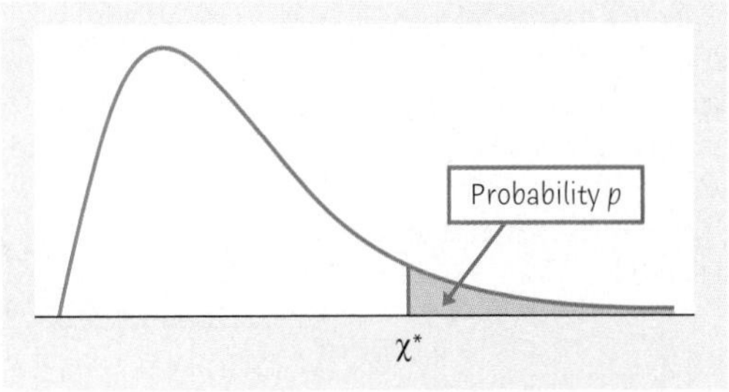

TABLE D CHI-SQUARE DISTRIBUTION CRITICAL VALUES

	p											
df	**.25**	**.20**	**.15**	**.10**	**.05**	**.025**	**.02**	**.01**	**.005**	**.0025**	**.001**	**.0005**
1	1.32	1.64	2.07	2.71	3.84	5.02	5.41	6.63	7.88	9.14	10.83	12.12
2	2.77	3.22	3.79	4.61	5.99	7.38	7.82	9.21	10.60	11.98	13.82	15.20
3	4.11	4.64	5.32	6.25	7.81	9.35	9.84	11.34	12.84	14.32	16.27	17.73
4	5.39	5.99	6.74	7.78	9.49	11.14	11.67	13.28	14.86	16.42	18.47	20.00
5	6.63	7.29	8.12	9.24	11.07	12.83	13.39	15.09	16.75	18.39	20.51	22.11
6	7.84	8.56	9.45	10.64	12.59	14.45	15.03	16.81	18.55	20.25	22.46	24.10
7	9.04	9.80	10.75	12.02	14.07	16.01	16.62	18.48	20.28	22.04	24.32	26.02
8	10.22	11.03	12.03	13.36	15.51	17.53	18.17	20.09	21.95	23.77	26.12	27.87
9	11.39	12.24	13.29	14.68	16.92	19.02	19.68	21.67	23.59	25.46	27.88	29.67
10	12.55	13.44	14.53	15.99	18.31	20.48	21.16	23.21	25.19	27.11	29.59	31.42
11	13.70	14.63	15.77	17.28	19.68	21.92	22.62	24.72	26.76	28.73	31.26	33.14
12	14.85	15.81	16.99	18.55	21.03	23.34	24.05	26.22	28.30	30.32	32.91	34.82
13	15.98	16.98	18.20	19.81	22.36	24.74	25.47	27.69	29.82	31.88	34.53	36.48
14	17.12	18.15	19.41	21.06	23.68	26.12	26.87	29.14	31.32	33.43	36.12	38.11
15	18.25	19.31	20.60	22.31	25.00	27.49	28.26	30.58	32.80	34.95	37.70	39.72
16	19.37	20.47	21.79	23.54	26.30	28.85	29.63	32.00	34.27	36.46	39.25	41.31
17	20.49	21.61	22.98	24.77	27.59	30.19	31.00	33.41	35.72	37.95	40.79	42.88
18	21.60	22.76	24.16	25.99	28.87	31.53	32.35	34.81	37.16	39.42	42.31	44.43
19	22.72	23.90	25.33	27.20	30.14	32.85	33.69	36.19	38.58	40.88	43.82	45.97
20	23.83	25.04	26.50	28.41	31.41	34.17	35.02	37.57	40.00	42.34	45.31	47.50
21	24.93	26.17	27.66	29.62	32.67	35.48	36.34	38.93	41.40	43.78	46.80	49.01
22	26.04	27.30	28.82	30.81	33.92	36.78	37.66	40.29	42.80	45.20	48.27	50.51
23	27.14	28.43	29.98	32.01	35.17	38.08	38.97	41.64	44.18	46.62	49.73	52.00
24	28.24	29.55	31.13	33.20	36.42	39.36	40.27	42.98	45.56	48.03	51.18	53.48
25	29.34	30.68	32.28	34.38	37.65	40.65	41.57	44.31	46.93	49.44	52.62	54.95
26	30.43	31.79	33.43	35.56	38.89	41.92	42.86	45.64	48.29	50.83	54.05	56.41
27	31.53	32.91	34.57	36.74	40.11	43.19	44.14	46.96	49.64	52.22	55.48	57.86
28	32.62	34.03	35.71	37.92	41.34	44.46	45.42	48.28	50.99	53.59	56.89	59.30
29	33.71	35.14	36.85	39.09	42.56	45.72	46.69	49.59	52.34	54.97	58.30	60.73
30	34.80	36.25	37.99	40.26	43.77	46.98	47.96	50.89	53.67	56.33	59.70	62.16
40	45.62	47.27	49.24	51.81	55.76	59.34	60.44	63.69	66.77	69.70	73.40	76.09
50	56.33	58.16	60.35	63.17	67.50	71.42	72.61	76.15	79.49	82.66	86.66	89.56
60	66.98	68.97	71.34	74.40	79.08	83.30	84.58	88.38	91.95	95.34	99.61	102.7
80	88.13	90.41	93.11	96.58	101.9	106.6	108.1	112.3	116.3	120.1	124.8	128.3
100	109.1	111.7	114.7	118.5	124.3	129.6	131.1	135.8	140.2	144.3	149.4	153.2

Table entry for p is the critical value r^* of the correlation coefficient r with probability p lying to its right.

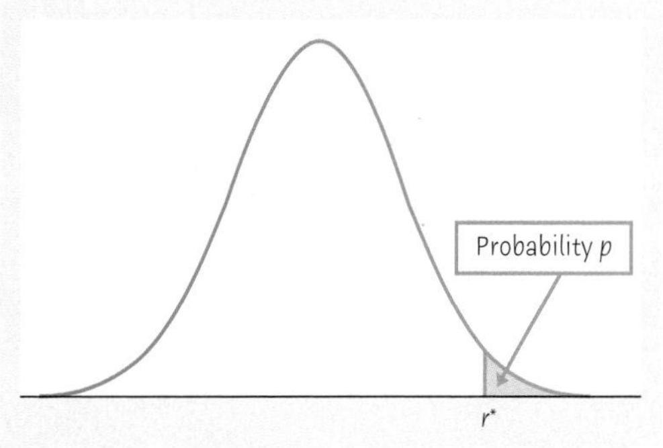

TABLE E CRITICAL VALUES OF THE CORRELATION r

	UPPER TAIL PROBABILITY p									
n	.20	.10	.05	.025	.02	.01	.005	.0025	.001	.0005
3	0.8090	0.9511	0.9877	0.9969	0.9980	0.9995	0.9999	1.0000	1.0000	1.0000
4	0.6000	0.8000	0.9000	0.9500	0.9600	0.9800	0.9900	0.9950	0.9980	0.9990
5	0.4919	0.6870	0.8054	0.8783	0.8953	0.9343	0.9587	0.9740	0.9859	0.9911
6	0.4257	0.6084	0.7293	0.8114	0.8319	0.8822	0.9172	0.9417	0.9633	0.9741
7	0.3803	0.5509	0.6694	0.7545	0.7766	0.8329	0.8745	0.9056	0.9350	0.9509
8	0.3468	0.5067	0.6215	0.7067	0.7295	0.7887	0.8343	0.8697	0.9049	0.9249
9	0.3208	0.4716	0.5822	0.6664	0.6892	0.7498	0.7977	0.8359	0.8751	0.8983
10	0.2998	0.4428	0.5494	0.6319	0.6546	0.7155	0.7646	0.8046	0.8467	0.8721
11	0.2825	0.4187	0.5214	0.6021	0.6244	0.6851	0.7348	0.7759	0.8199	0.8470
12	0.2678	0.3981	0.4973	0.5760	0.5980	0.6581	0.7079	0.7496	0.7950	0.8233
13	0.2552	0.3802	0.4762	0.5529	0.5745	0.6339	0.6835	0.7255	0.7717	0.8010
14	0.2443	0.3646	0.4575	0.5324	0.5536	0.6120	0.6614	0.7034	0.7501	0.7800
15	0.2346	0.3507	0.4409	0.5140	0.5347	0.5923	0.6411	0.6831	0.7301	0.7604
16	0.2260	0.3383	0.4259	0.4973	0.5177	0.5742	0.6226	0.6643	0.7114	0.7419
17	0.2183	0.3271	0.4124	0.4821	0.5021	0.5577	0.6055	0.6470	0.6940	0.7247
18	0.2113	0.3170	0.4000	0.4683	0.4878	0.5425	0.5897	0.6308	0.6777	0.7084
19	0.2049	0.3077	0.3887	0.4555	0.4747	0.5285	0.5751	0.6158	0.6624	0.6932
20	0.1991	0.2992	0.3783	0.4438	0.4626	0.5155	0.5614	0.6018	0.6481	0.6788
21	0.1938	0.2914	0.3687	0.4329	0.4513	0.5034	0.5487	0.5886	0.6346	0.6652
22	0.1888	0.2841	0.3598	0.4227	0.4409	0.4921	0.5368	0.5763	0.6219	0.6524
23	0.1843	0.2774	0.3515	0.4132	0.4311	0.4815	0.5256	0.5647	0.6099	0.6402
24	0.1800	0.2711	0.3438	0.4044	0.4219	0.4716	0.5151	0.5537	0.5986	0.6287
25	0.1760	0.2653	0.3365	0.3961	0.4133	0.4622	0.5052	0.5434	0.5879	0.6178
26	0.1723	0.2598	0.3297	0.3882	0.4052	0.4534	0.4958	0.5336	0.5776	0.6074
27	0.1688	0.2546	0.3233	0.3809	0.3976	0.4451	0.4869	0.5243	0.5679	0.5974
28	0.1655	0.2497	0.3172	0.3739	0.3904	0.4372	0.4785	0.5154	0.5587	0.5880
29	0.1624	0.2451	0.3115	0.3673	0.3835	0.4297	0.4705	0.5070	0.5499	0.5790
30	0.1594	0.2407	0.3061	0.3610	0.3770	0.4226	0.4629	0.4990	0.5415	0.5703
40	0.1368	0.2070	0.2638	0.3120	0.3261	0.3665	0.4026	0.4353	0.4741	0.5007
50	0.1217	0.1843	0.2353	0.2787	0.2915	0.3281	0.3610	0.3909	0.4267	0.4514
60	0.1106	0.1678	0.2144	0.2542	0.2659	0.2997	0.3301	0.3578	0.3912	0.4143
80	0.0954	0.1448	0.1852	0.2199	0.2301	0.2597	0.2864	0.3109	0.3405	0.3611
100	0.0851	0.1292	0.1654	0.1966	0.2058	0.2324	0.2565	0.2786	0.3054	0.3242
1000	0.0266	0.0406	0.0520	0.0620	0.0650	0.0736	0.0814	0.0887	0.0976	0.1039

ANSWERS TO ODD-NUMBERED EXERCISES

Chapter 0 Getting Started

0.1: (a) More than likely the individuals who chose to take vitamin E were more health-conscious in general than those who didn't. They may also have been more affluent (i.e., had money available to purchase the vitamins). (b) In a randomized experiment, people of all types are randomly assigned to the treatments.

0.3: (a) The proportion of Americans who feel this way is likely much lower. The "survey" used voluntary response (people made their own decision to participate or not). Also, those who knew about the "poll" had just watched Mr. Schultz's long monologue. (b) As long as the poll was voluntary response, the sample size (868, 2500, or 100,000) doesn't matter.

Chapter 1 Picturing Distributions with Graphs

1.1: (a) The individuals are the car makes and models. (b) The variables are vehicle class (categorical), transmission type (categorical), number of cylinders (usually treated as quantitative), city mpg (quantitative), highway mpg (quantitative), and annual fuel cost (dollars, quantitative).

1.3: (a) The given shares sum to 76.4%. 100% – 76.4% = 23.6% listen to stations with other formats. (b) The bar graph is shown.

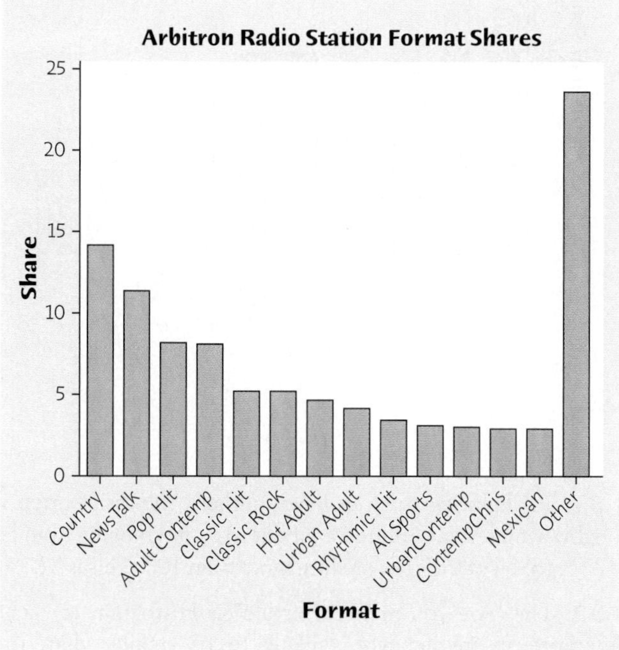

(c) If you include a wedge for "Other format" that accounts for 23.6% of the total, a pie chart is reasonable.

1.5: A pie chart would make it more difficult to distinguish between the weekend days and the weekdays. Some births are scheduled (induced labor, for example), and probably are scheduled for weekdays.

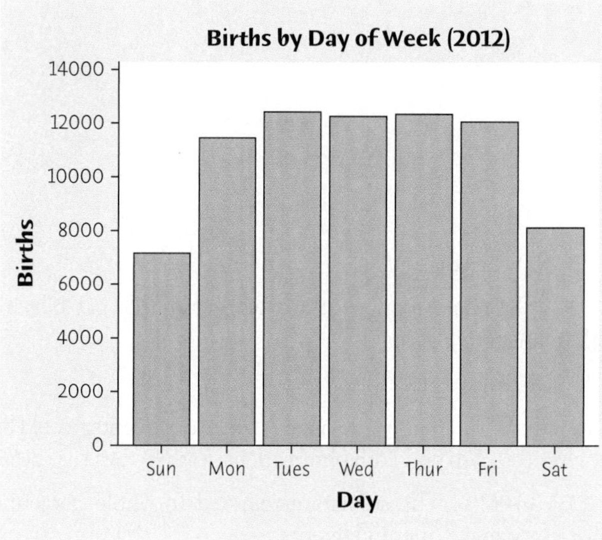

1.7: Use the applet to answer these questions.

1.9: (a) There are two clear peaks in the distribution. If we gave only one center, it would most likely not be truly representative. (b) Young boys might spend a lot of time outdoors playing and engaged in sports; their time in places where they would encounter ticks might well be less as they get older. With families and yard work, their time outside might increase. (c) Hiking in the woods at any age will make a person more likely to encounter ticks. (d) The histograms have the same shapes, but females have a slightly lower incidence rate at every age. Females possibly spend less time outdoors.

1.11: A stemplot for health expenditure per capita (in PPP) is given. Stems are thousands, and are split. This distribution is right-skewed, with a single high outlier (United States). There seem to be two clusters of countries. The center of this distribution is around 20 ($2000 spent per capita), ignoring the outlier. The distribution varies from 0|1 ($100 spent per capita) to 8|6 (about $8600 spent per capita).

0	1 1 4 4
0	6 6 7 9 9 9 9
1	0 2 3 4 4
1	7
2	2
2	9
3	0 1 2 3
3	7 9
4	1 1 4
4	5 5 6
5	1
5	6 7
6	
6	
7	
7	
8	
8	6

1.13: (a) the students

1.15: (b) Square footage and average monthly gas bill are both quantitative variables.

1.17: (b) 58% to 61%

1.19: (b) 80%. There are 50 observations, so the center would be between the 25th and 26th observations; both of these are 80%.

1.21: (b) 92%. The stems are rounded to whole percents; you cannot make finer judgments.

1.23: (a) Individuals are students who have finished medical school. (b) Five, in addition to "Name." "Age" (in years) and "USMLE" (score points) are quantitative. The others are categorical.

1.25: "Other colors" account for 3%. A bar graph would be an appropriate display. If you included the "other" category, a pie chart could be made.

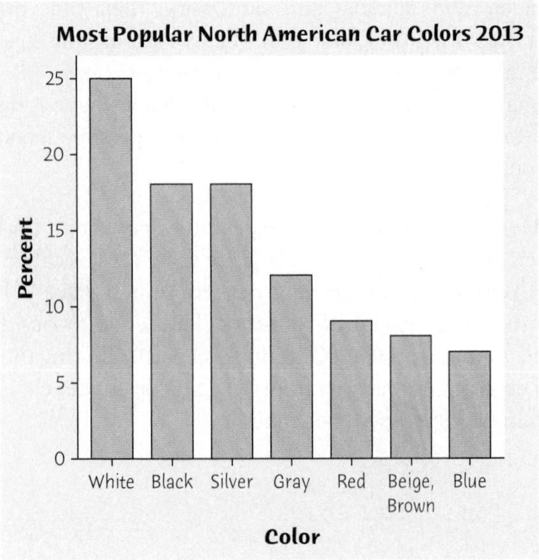

1.27: (a) A bar graph is given. (b) To make a pie chart, you would need to know the total number of deaths in this age group, or the number of deaths due to "other" causes.

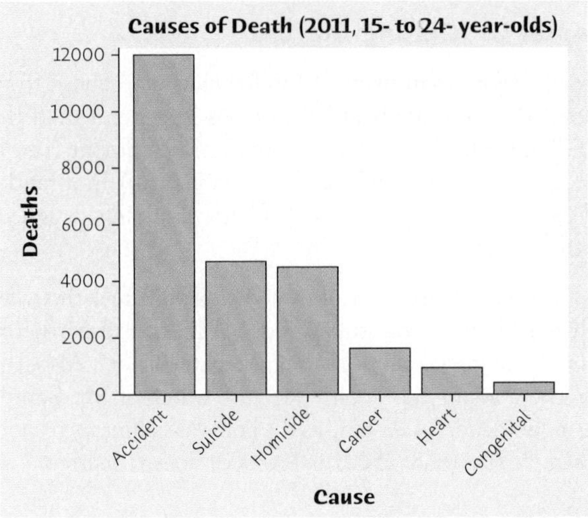

1.29: (a) A bar graph is provided. (b) A pie chart would be inappropriate; these percentages aren't parts of a single "whole."

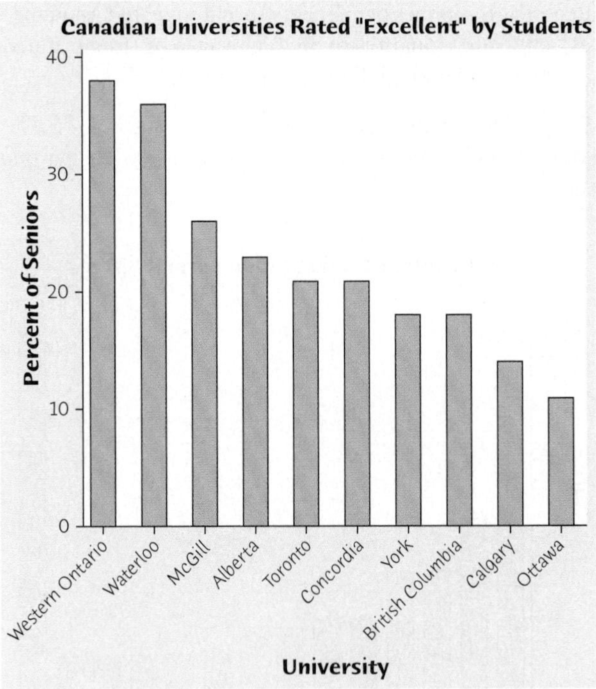

1.31: (a) Ignoring the four lower outliers, the distribution is roughly symmetric, centered at about 110, and ranging from 86 to 136. (b) 64 of the 78 scores are more than 100. This is 82.1%.

1.33: (1.) "Are you male or female" is Histogram (c). The difference in frequencies is likely to be smaller than the right-handed/left-handed. (2.) "Are you right-handed or left-handed?" is Histogram (b) because there are many more right-handed people than left handed people. (3.) "What is your height in inches?" is Histogram (d). Height distribution is likely to be symmetric. (4.) "How many minutes do you study on a typical weeknight?" is Histogram (a). Time spent studying may well be right-skewed, with most students spending less time studying, but some students studying a lot.

1.35: (a) States vary in population. Nurses per 100,000 provides a better measure of how many nurses are available to serve a state. (b) A histogram is provided. The District of Columbia, South Dakota, and Massachusetts are the three states different from the others. Washington, D.C., and Massachusetts are populous with large medical communities. It's difficult to know why South Dakota would also have an unusually high number of nurses.

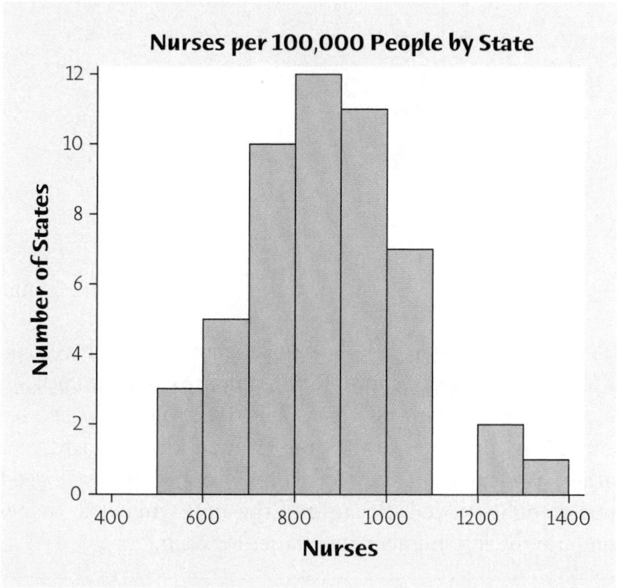

1.37: The shape of the distribution is roughly symmetric (maybe left skewed if we ignore the high outlier); 245 seems to be a high outlier. The center is about 171. The data range from about 94 to about 245.

9	4 6
10	2 9
11	
12	2
13	
14	5
15	8
16	5 7
17	0 1 1 3 9 9
18	2 2
19	2
20	1 2 3 3
21	
22	
23	
24	5

1.39: The decline in population is not seen in the stemplot made in Exercise 1.37.

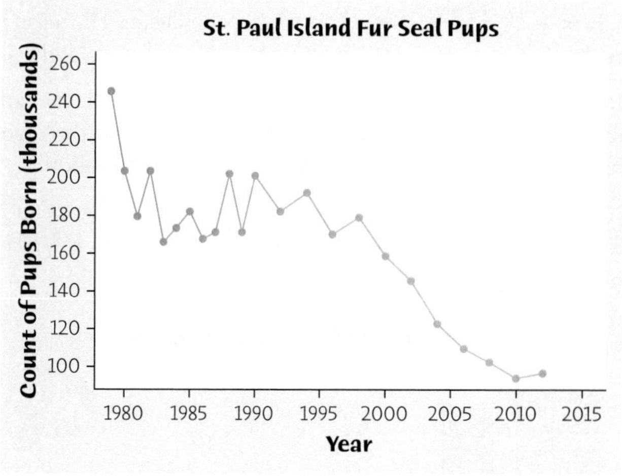

1.41: Coins with earlier (lower) dates are rarer. There are more coins with larger dates (newer coins) than with smaller dates.

1.43: (a) Graph (a) appears to show the greatest increase. Vertical scaling can impact one's perception of the data. (b) In 2000, tuition was a bit more than \$4000 and rises to almost \$9000; this is an increase of almost \$5000. Both plots describe the same data.

1.45: (a) A time plot of ozone hole size (area) is provided. There was a trend until about 1995; after that, we see only year-to-year variability. The hole may have leveled out in recent years; the overall trend since about 2006 might be a decrease in size.

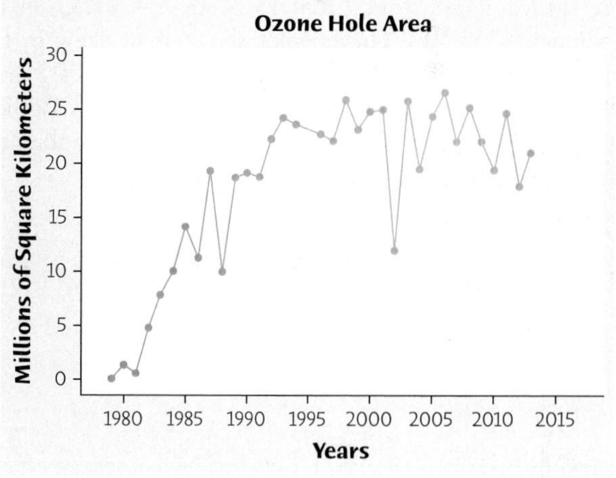

(b) A stemplot of ozone hole size (area) is provided. The midpoint is about 20 million square kilometers. A stemplot fails to capture the relationship between size of hole and year.

0	0 0 1 4
0	7
1	0 0 1 2 4
1	7 8 8 9 9 9 9
2	1 2 2 2 2 2 3 3 4 4 4 4
2	5 5 5 5 6

Chapter 2 Describing Distributions with Numbers

2.1: $\bar{x} = \frac{291.0 + 10.9 + \ldots + 9.6}{16} = 56.28$ per milliliter. The mean is greater than most of the observations because of the two outliers.

2.3: The mean travel time is $\bar{x} = 31.25$ minutes. The median travel time is 22.5 minutes. The mean is significantly larger than the median due to the right skew in the distribution.

2.5: A histogram is given. Note the right skew. The mean is larger than the median. $\bar{x} = 4.60566$ and the median is 3.7034 tons per person.

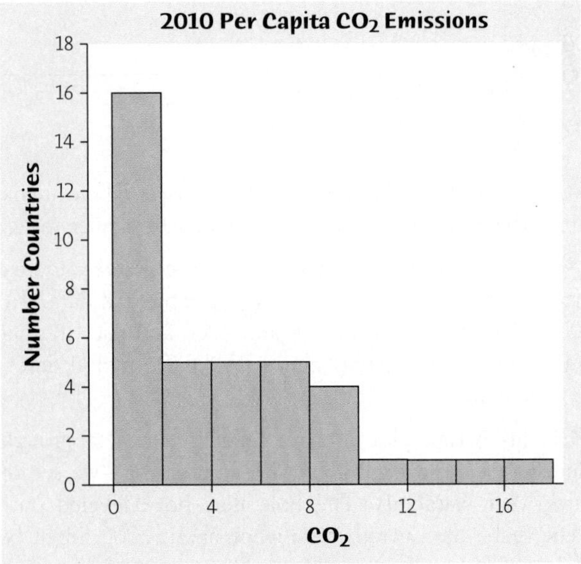

2.7: (a) Minimum = 11, $Q_1 = 18$, Median = 21, $Q_3 = 26$, Maximum = 51. (b) The boxplot shows right skew in the distribution. There are several high outliers (which are most likely hybrid cars). Most software will identify the outliers automatically (if you do not want them identified, they are omitted from the graph).

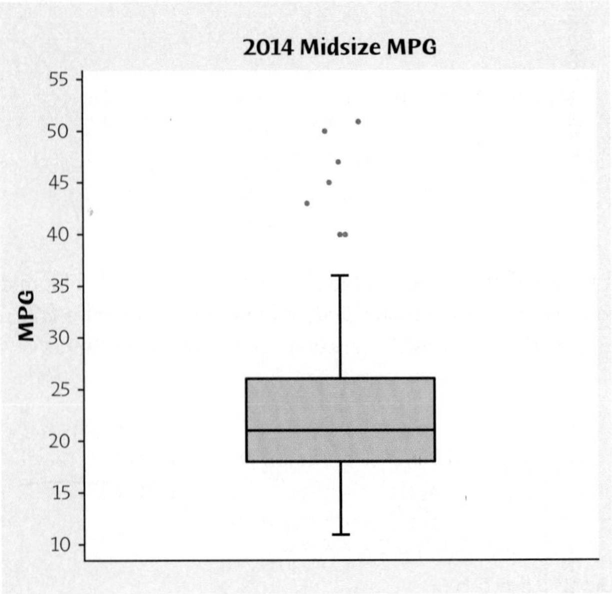

2.9: $IQR = 26 - 18 = 8$, so $Q_3 + 1.5 \times IQR = 26 + 1.5 \times 8 = 38$. There are eight values greater than 38 that would be identified as potential outliers (40, 40, 40, 43, 45, 47, 50, and 51). Because $Q_1 - 1.5 \times IQR = 18 - 1.5 \times 8 = 6$, there are no potential outliers on the low end.

2.11: Both data sets have the same mean and standard deviation (about 7.5 and 2.0, respectively). Stemplots reveal that Data A has a very left-skewed distribution, while Data B has a slightly right-skewed distribution with a high outlier.

A		B
1	3	
7	4	
	5	257
1	6	58
2	7	079
7711	8	48
211	9	
	10	
	11	
	12	5

2.13: PLAN: Create side-by-side boxplots for the three types of plots and compute appropriate summary statistics. SOLVE: None of the distributions are symmetric; Group 2 (logged one year earlier) has a low outlier and Group 3 (logged eight years earlier) is clearly left-skewed, while Group 1 (never logged) appears right-skewed. Because of the non-symmetric shapes, compute the five-number summaries for each.

	Min	Q_1	M	Q_3	Max
Group 1 (never logged)	16	19.25	23	27.75	33
Group 2 (logged 1 year earlier)	2	12	14.5	17.75	20
Group 3 (logged 8 years earlier)	4	12	18	20.5	22

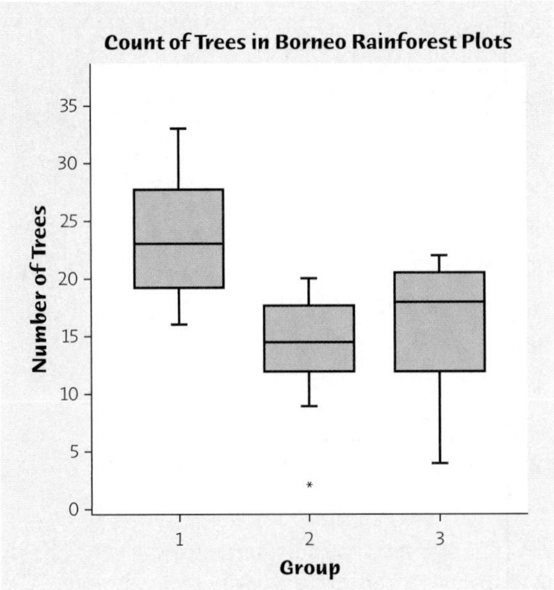

(If you compute the means and standard deviations, they are: Group 1: $\bar{x} = 23.75$, $s = 5.07$; Group 2: $\bar{x} = 14.08$, $s = 4.98$; and Group 3: $\bar{x} = 15.78$, $s = 5.76$.) CONCLUDE: It is clear from the boxplots and summary statistics that plots that have never been logged have more trees than either type of logged plot. Further, if we compare the distributions and summary statistics for the two different types of logged plots, it takes a long time for the rain forest to recover from having been logged.

2.15: (b) 167.48

2.17: (b) 151.6, 163.5, 168.25, 174.3, 177.6

2.19: (a) 25%. Q_3 has 75% of observations equal to or less than its value.

2.21: (c) 8.2

2.23: (b) seconds

2.25: The distribution of incomes in this group is almost certainly right-skewed. The mean is \$62,597 and the median is \$50,281.

2.27: With 849 colleges, the median location is $(849 + 1)/2 = 425$, so the median is the 425th (ordered) endowment. The first quartile, Q_1, is found by taking the median of the first 424 endowments. This would be the $(424 + 1)/2 = 212.5$th endowment. Similarly, Q_3 is found as the 637.5th endowment (212.5 above the median).

2.29: The boxplots do not reveal the gap in the South between the rates for Georgia and the District of Columbia.

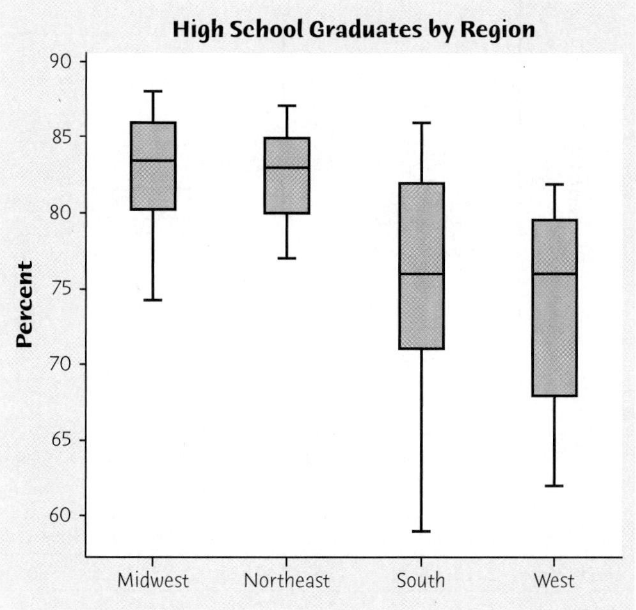

2.31: A histogram of the survival times is given. The distribution is strongly right-skewed, with center around 100 days, and range from about 0 to about 600 days. (b) Because of the extreme right skew, use the five-number summary: 43, 82.5, 102.5, 151.5, 598 days. Notice that the median is closer to Q_1 than to Q_3.

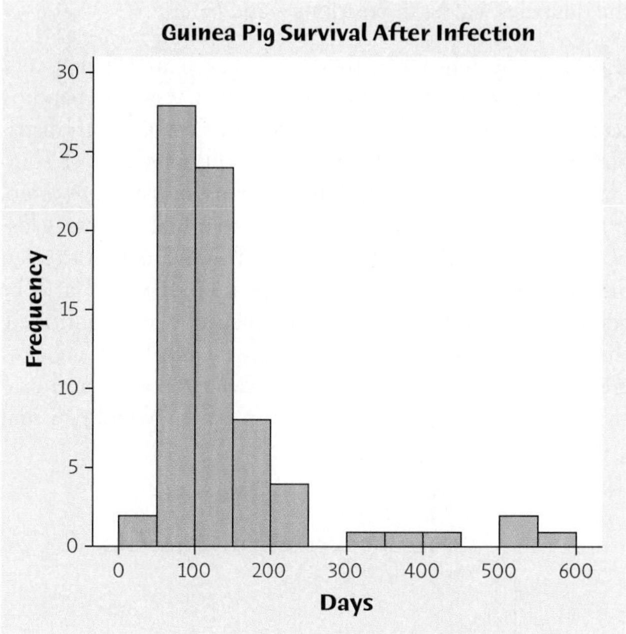

2.33: (a) Symmetric distributions are best summarized using $\bar{x}$ and s. The distribution for the treatment group was right-skewed. The control distribution could be called symmetric, but it has a high outlier. (b) Removing the outliers reduces all three statistics. The mean decreased by 8.45 seconds, about double the decrease in the median (3.5 seconds).

	With outlier	Without outlier
Mean	59.7	51.25
Standard deviation	63.0	50.97
Median	61	57.5

2.35: (a) The sixth observation must be placed at the median for the original five observations. (b) No matter where you put the seventh observation, the median is one of the two repeated values above, because it will be the fourth (ordered) observation.

2.37: $\bar{x} = 8.83\%$, far from the national percentage of 12.5%. You can't average averages. Some states, such as California and Florida, are larger and should carry more weight in the national percentage.

2.39: Answers will vary, but a raise in the minimum wage will probably have a greater impact on the median income. Most Americans earn "middle income" or less; a few people earn huge amounts each year. The few large amounts will still pull the mean toward that end of the distribution.

2.41: Many answers are possible. Start by ensuring that the median is 12, by "locking" 12 as the fourth smallest value. We also have 4 specified as the minimum and 19 as the maximum, so the seven numbers must be 4, ___, ___, 12, ___, ___, 19. With three numbers either side of the median, the quartiles will be in positions 2 and 6.

2.43: (a) Weight losses that are negative are weight *gains*. (b) A side-by-side boxplot is provided. Gastric banding seems to produce higher weight losses. Because both distributions are somewhat right-skewed, the five-number summary would be appropriate. The summary statistics are given. (c) It's better to measure loss relative to initial weight. A loss of 5 kg would not mean the same if individuals started at different weights. *Percent* reduction in BMI would also be good. (d) If the subjects who dropped out had continued, the difference between these groups would be as great or greater because many of the "lifestyle" dropouts had negative weight losses (i.e., weight gains), which would pull that group down.

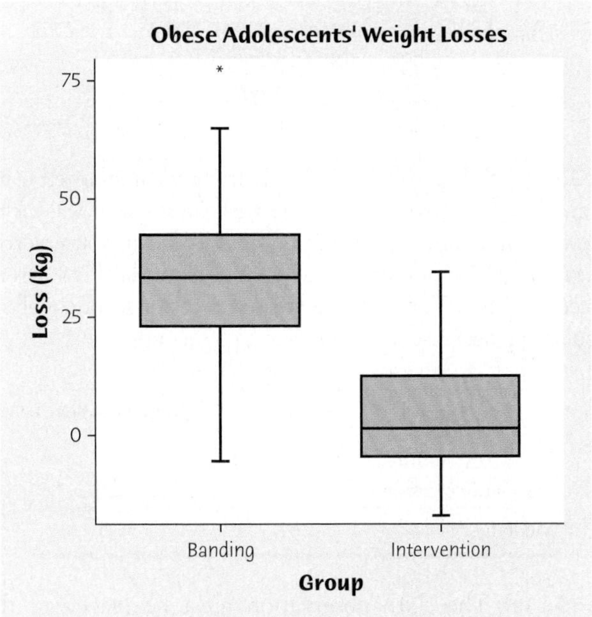

	Min	Q_1	M	Q_3	Max
Banding	−5.40	22.70	33.35	42.67	81.40
Intervention	−17.00	−4.38	1.70	12.57	34.60

2.45: PLAN: Graph the returns with a histogram and a time plot. Based on those, compute and report appropriate summary statistics. SOLVE: The histogram and summary statistics are given. CONCLUDE: The distribution of average returns is left-skewed. Most years, the average return is positive. Returns range from about −40% to 40%, with median return about 16%.

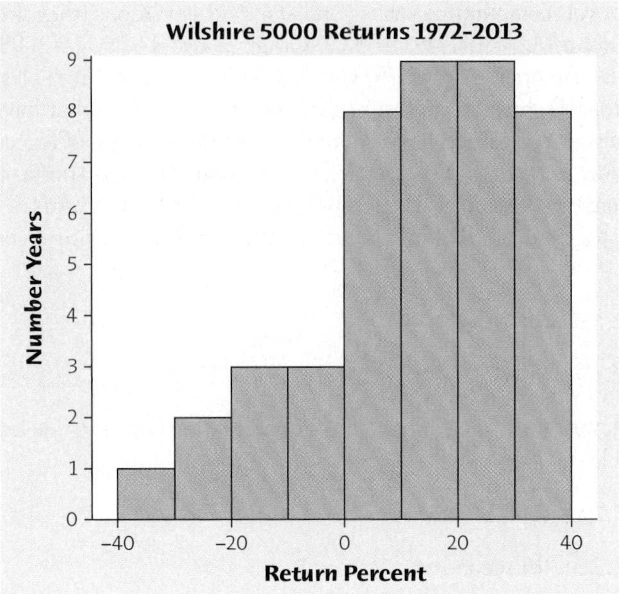

Mean	St. Dev.	Minimum	Q_1	Median	Q_3	Maximum
12.19	18.17	−37.23	0.98	16.06	26.77	37.38

2.47: PLAN: Create side-by-side boxplots of the three distributions, and compute appropriate summary statistics. SOLVE: Side-by-side boxplots of tip results are given. Good weather forecasts generally yielded better tips than the other two, and a relatively symmetric distribution. The bad weather forecast had an outlier on each end, while the no-weather message had three low outliers.

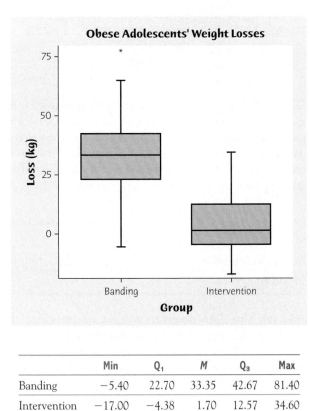

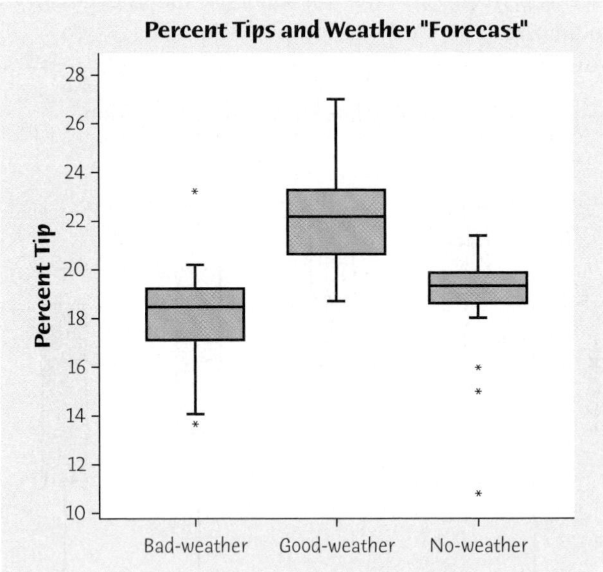

	Mean	St. Dev.	Minimum	Q_1	Median	Q_3	Maximum
Good weather	22.22	1.96	13.6	17.125	22.15	23.25	27.0
Bad weather	18.18	2.10	13.6	17.125	18.45	19.15	23.2
No weather	18.725	2.39	10.8	18.57	19.3	19.87	21.3

2.49: (a) Do side-by-side boxplots to compare. All three distributions are right-skewed with high outliers. The median increases slightly with increasing age (from 173 to 190 to 204). We also see an increase in variability as people age. (b) 25% or more of the individuals in each age group had total cholesterol levels above 200. Unless their original cholesterol levels were *extremely* high, the 4 or 24 people on medication in their 20s and 30s probably wouldn't affect these distributions a great deal because there were roughly 950 people in each group. However, there were 1139 people in their 40s; 117 of those on medication is more than 10% of this group. If those 117 had not been on medication, that distribution would likely show more variability and higher cholesterol readings.

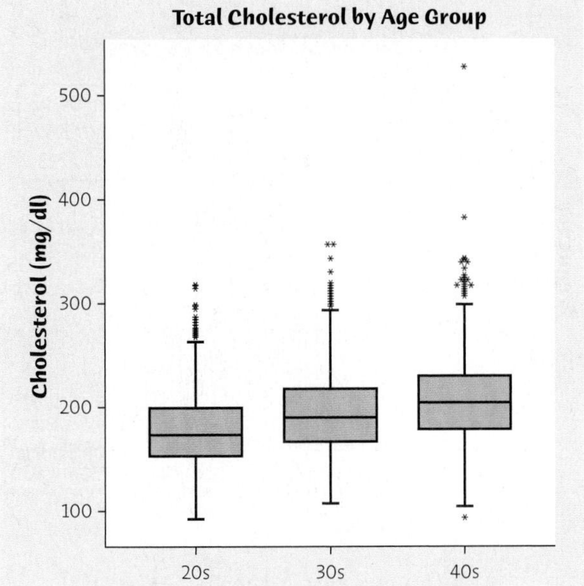

2.51: (a) Min = 1.3, Q_1 = 4.1, Median = 6.2, Q_3 = 13.4, Max = 27.1. (b) High end outliers are larger than 13.4 + 1.5 × (13.4 − 4.1) = 27.35. California (27.1% foreign-born) is not an outlier.

2.53: Using Minitab, the five-number summary of cholesterol levels for people in their 20s is 92, 154, 173, 199, 318. We have *IQR* = 199 − 154 = 45. Outliers would be values smaller than 154 − 1.5 × 45 = 86.5 or larger than 199 + 1.5 × 45 = 266.5. Using this criterion, there are no low outliers, but there are high end outliers.

Chapter 3 The Normal Distributions

3.1: Sketches will vary. (a) Symmetric distributions are mirror images on either side of the center. This distribution should have two humps (bimodal) (b) A distribution that is skewed to the left has a long *left* tail.

3.3: $\mu = 2.5$, the balance point. The median is also 2.5 because the distribution is symmetric.

3.5: A sketch of the Normal curve is given. The tick marks are placed at the mean, and at one, two, and three standard deviations above and below the mean.

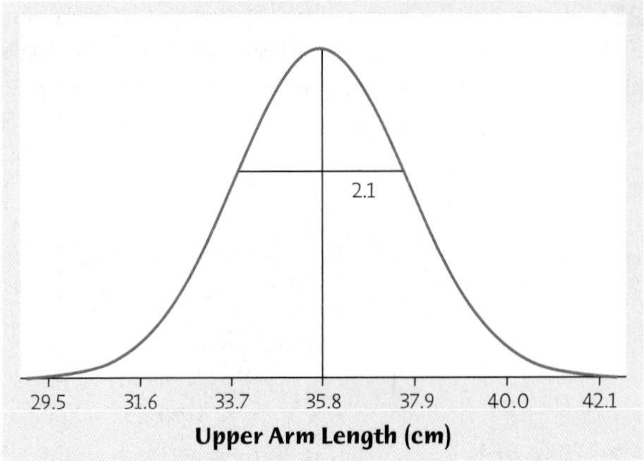

3.7: (a) In 95% of all years, monsoon rain levels are between 852 ± 2(82) = 688 and 1016 mm. (b) The driest 2.5% of monsoon rainfalls are less than 688 mm (more than 2σ below μ).

3.9: A woman 6 feet tall has $z = \frac{72 - 64.2}{2.8} = 2.79$. A man 6 feet tall has $z = \frac{72 - 69.4}{3.0} = 0.87$. The woman is much taller relative to other women.

3.11: Let x be the monsoon rainfall in a given year. (a) $x \le 697$ mm corresponds to $z \le \frac{697 - 852}{82} = -1.89$. Table A gives 0.0294 = 2.94%. (b) $682 < x < 1022$ corresponds to $\frac{682 - 852}{82} < z < \frac{1022 - 852}{82}$, or $-2.07 < z < 2.07$. From Table A, 0.9808 − 0.0192 = 0.9616 = 96.16%.

3.13: (a) Using Table A, we find this value has $z = 0.39$ (software gives $z = 0.3853$). (b) We want a proportion of 0.80 below. Using Table A, $z = 0.84$ (software gives $z = 0.8416$).

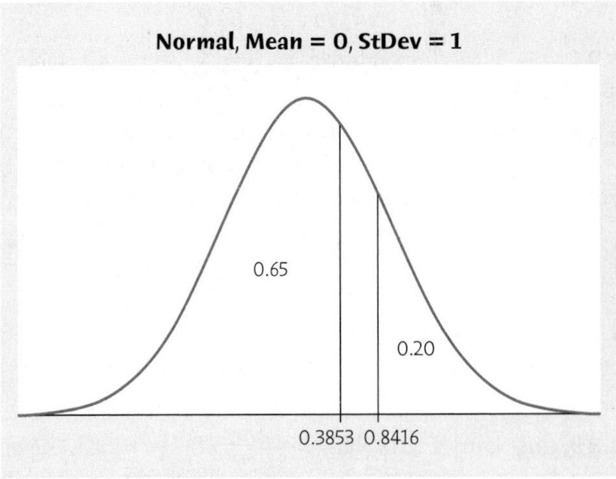

3.15: (c) Economic variables such as income and house prices are usually right-skewed.

3.17: (b) The curve is centered at 2.

3.19: (b) 266 ± 2(16) = 234 to 298 days.

3.21: (a) $z = \frac{132 - 100}{15} = 2.13$

3.23: (b) 0.1056

3.25: Sketches will vary, but should be some variation on the one shown here: the peak at 0 should be "tall and skinny," while near 1, the curve should be "short and fat."

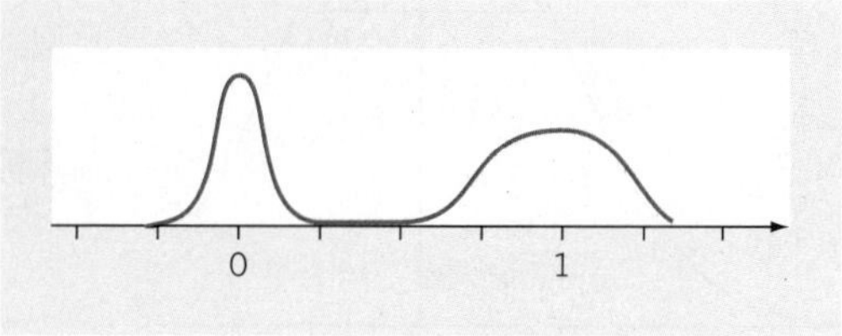

3.27: 70 is 2σ below μ (that is, has $z = -2$), so about 2.5% of adults have WAIS scores below 70.

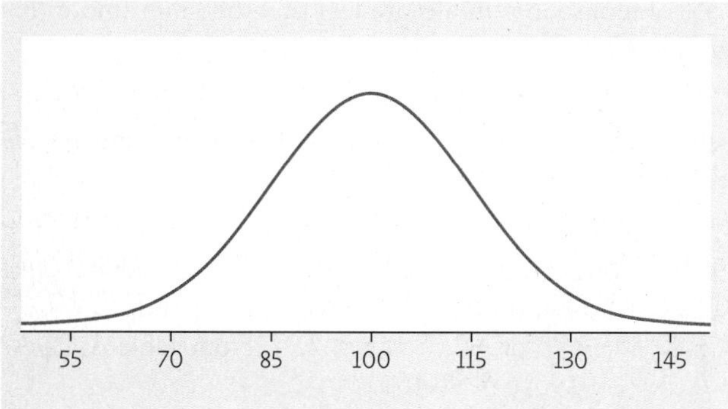

3.29: (a) Looking up 0.3000 in Table A, $z = -0.52$. (Software gives $z = -0.5244$.) (b) If 35% are more than z, 65% are less than or equal to z; $z = 0.39$. (Software gives $z = 0.3853$.)

3.31: $x < 5.0$ corresponds to $z < \frac{5.0 - 5.43}{0.54} = -0.80$; Table A gives 0.2119.

3.33: $0.8725 < x < 0.8775$ corresponds to $\frac{0.8725 - 0.8750}{0.0012} < z < \frac{0.8775 - 0.8750}{0.0012}$, or $-2.08 < z < 2.08$. From Table A, $0.9812 - 0.0188 = 0.9624$.

3.35: Cars with better mileage correspond to $x > 28$, $z > \frac{28 - 22.2}{5.2} = 1.12$. $1 - 0.8686 = 0.1314$, or 13.14%.

3.37: Q_1 and Q_3 have $z = -0.67$ and $z = 0.67$, respectively. $Q_1 = 22.2 - (0.67)(5.2) = 18.72$ mpg and $Q_3 = 22.2 + (0.67)(5.2) = 25.68$ mpg.

3.39: (a) Larry's arm has $z = \frac{37.2 - 39.1}{2.3} = -0.83$. Using Table A, his percentile is 0.2033 (about the 20th). (b) Answers will vary due to variation in students' arm lengths.

3.41: We want the proportion corresponding to $x > 69.4$ inches. This corresponds to $z > \frac{69.4 - 64.2}{2.8} = 1.86$. From Table A, $1 - 0.9686 = 0.0314$, or 3.14%.

3.43: (a) For men, $X > 750$ corresponds to $z > \frac{750 - 531}{121} = 1.81$. $1 - 0.9649 = 0.0351$. (b) For women, $X > 750$ corresponds to $z > \frac{750 - 499}{112} = 2.24$. $1 - 0.9875 = 0.0125$.

3.45: (a) About 0.6% of healthy young adults have osteoporosis (the cumulative probability below $z = -2.5$). (b) The BMD level 2.5σ below the young adult mean would be $z = -0.5$ for these older women. Table A gives 0.3085.

3.47: (a) 170,777/1,799,243 = 0.0949, or 9.49%. (b) There are 56,351 + 170,777 = 227,128 students with ACT score 28 or higher. 227,128/1,799,243 = 0.1262, or 12.62%. (c) If x is the ACT score, $x > 28$ corresponds to $z > \frac{28 - 20.9}{5.4} = 1.31$. The proportion is $1 - 0.9049 = 0.0951$, or 9.51%.

3.49: (a) A histogram is provided and is roughly symmetric.

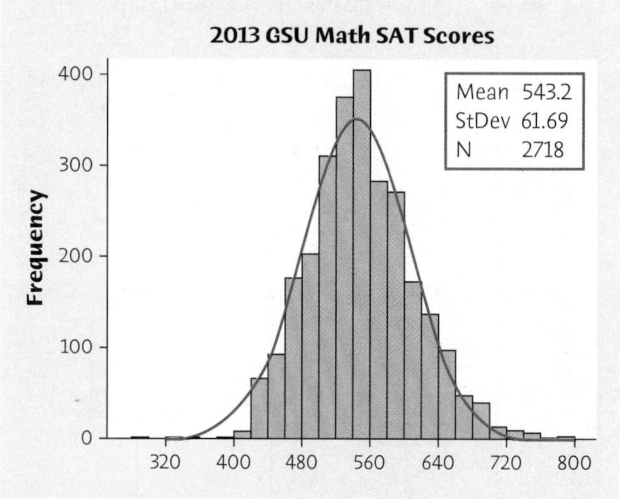

(b) Mean = 543.2, Median =540, Standard deviation = 61.69, $Q_1 = 500$, $Q_3 = 580$. The mean and median are close, and the distances from each quartile to the median are equal. These results are consistent with a Normal distribution. (c) If x is the score of a randomly selected GSU entering student, then we are assuming x has the $N(543.2, 61.69)$ distribution. The proportion of GSU students scoring higher than 514 corresponds to $x > 514$, or $z > \frac{514 - 543.2}{61.69} = -0.47$, or $1 - 0.3192 = 0.6808$, or 68.08%. (d) 859 scored 510 or less, so $2718 - 859 = 1859$ entering GSU students scored higher than 514, which is 1859/2718 = 0.6840, or 68.4%. The nominal Normal probability in part (c) fits the actual data well.

3.51: (a) 14/548 = 0.0255 (2.55%) weighed less than 100 pounds. $x < 100$ corresponds to $z < \frac{100 - 161.58}{48.96} = -1.26$. Using Table A, the area is 0.1038 (10.38%). (b) 33/548 = 0.0602 (6.02%) weighed more than 250 pounds. $x > 250$ corresponds to $z > \frac{250 - 161.58}{48.96} = 1.81$. Using Table A, about $1 - 0.9649 = 0.0351$ (3.51%) would weigh more than 250 pounds. (c) The Normal distribution model predicts 10.38% of women weigh less than 100 pounds, while actually about 2.55% do. This is a substantial error since the Normal model also predicts 3.51% of values more than 250, where we actually observed 6.02% more than 250. These data seem to be far from Normal.

3.53: Because the quartiles of any distribution have 50% of observations between them, place the flags so that the reported area is 0.5. The closest the applet gets is an area of 0.4978, between −0.671 and 0.671. The quartiles of any Normal distribution are about 0.67σ above and below μ.

Chapter 4 Scatterplots and Correlation

4.1: (a) Explanatory: number of lectures attended; response: grade on final exam. (b) Explanatory: time exercising; response is calories burned. (c) Explanatory: Time spent online using Facebook; GPA is the response. (d) Explore the relationship.

4.3: For example: weight, sex, other food eaten by the students, type of beer (light, imported, . . .).

4.5: Outsource percent is the explanatory variable. These data do not support concerns of the critics.

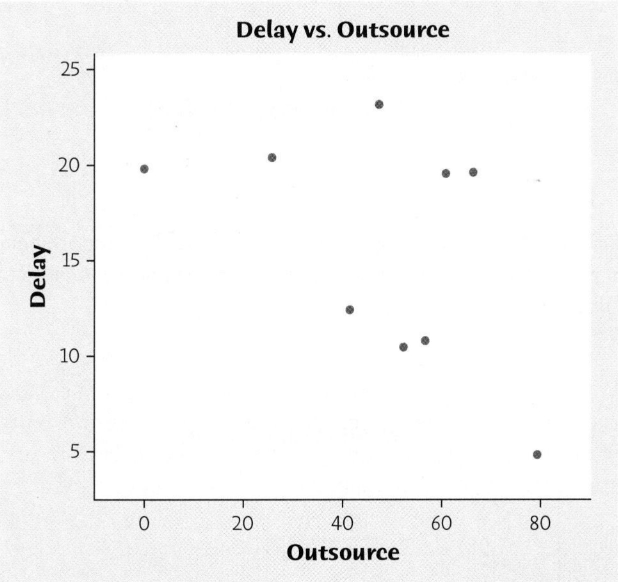

4.7: One could consider there to be two outliers; Frontier has an unusually low outsourcing percent and a high delay percent; Hawaiian has a very high outsourcing percent and a very low delay percent. Without Frontier, there would be a decreasing relationship that is approximately linear and moderately strong. Without Hawaiian, there is really no relationship, but two sets of points: five airlines with high delays, which don't seem to depend on outsourcing, and three airlines with low delay percentages that again don't seem to depend on outsourcing.

4.9: (a) Caution counties are marked with squares, the others with circles. (b) For both types of counties, there appears to be no relationship. The caution counties form a band in the lower portion of the graph, while the non-caution counties form a band in the upper portion.

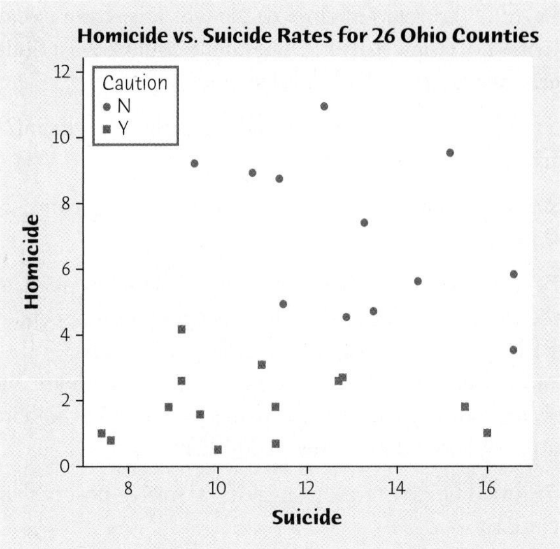

4.11: r would not change; units do not affect correlation.

4.13: $\bar{x} = 50$ mph, $s_x = 15.8114$ mph, $\bar{y} = 26.8$ mpg, and $s_y = 2.6833$ mpg. Refer to the table: note that $r = 0/4 = 0$. $r = 0$ because these variables do not have a straight-line relationship. Correlation only measures the strength and direction of a *linear* relationship between two variables.

z_x	z_y	$z_x z_y$
−1.2649	−1.0435	1.3199
−0.6325	0.4472	−0.2828
0	1.19269	0
0.6325	0.4472	0.2828
1.2649	−1.0435	−1.3199
		0

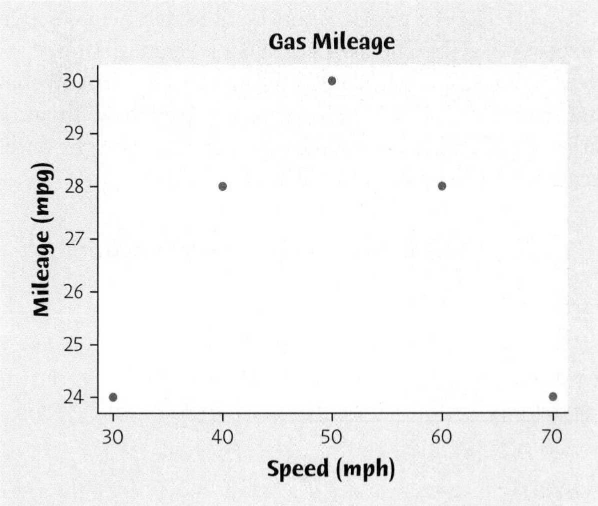

4.15: (c) The association should be negative (if slower reaction times mean less time to death).

4.17: (a) 0.9. Without the outlier, there is a strong positive linear relationship.

4.19: (c) A correlation close to 0 might arise from a scatterplot with no visible pattern, but there could be a nonlinear pattern. See Exercise 4.13, for example.

4.21: (a) 1. The line would exactly be Exam2 = Exam1 − 10.

4.23: (b) Computation with calculator or software gives $r = 0.298$.

4.25: (a) Overall, there is a slightly negative association between these variables. (b) There is general disagreement—low BRFSS scores correspond to greater happiness, and these are associated with higher-ranked states (the least happy states, according to the objective measure). (c) It is hard to declare any of the data values as "outliers."

4.27: (a) The scatterplot suggests a strong positive linear association.

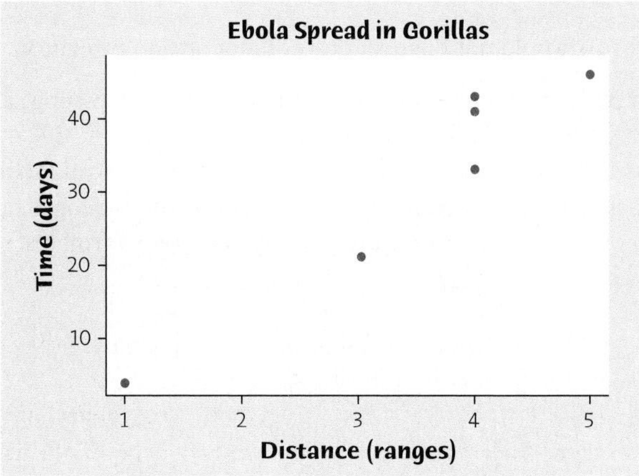

(b) $r = 0.9623$. This is consistent with the pattern seen in part (a). (c) Correlation would not change, since it does not depend on units.

4.29: (a) The scatterplot is shown; note that neural activity is explanatory. (b) The association is moderately strong, positive, and linear. The outlier is in the upper right corner (behavioral score 155.2). (c) For all points, $r = 0.8486$. Without the outlier, $r = 0.7015$. The correlation is greater with the outlier because it fits the pattern of the other points.

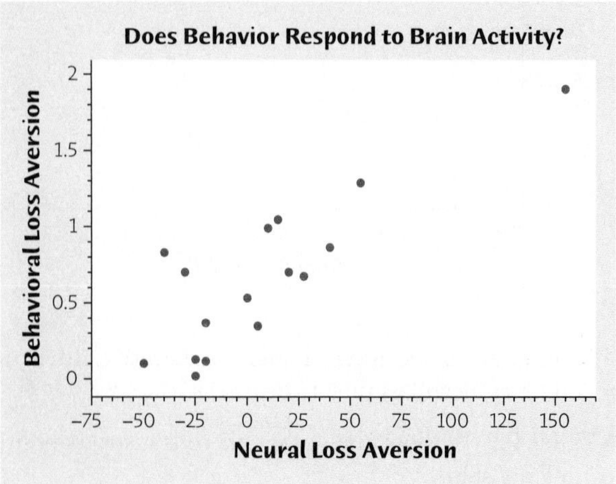

4.31: (a) The scatterplot is shown. (b) The plot suggests that there is a strong relationship between alcohol intake and relative risk of breast cancer. It seems that type of alcohol has nothing to do with the increase since the same pattern and rate of increase is seen for both groups.

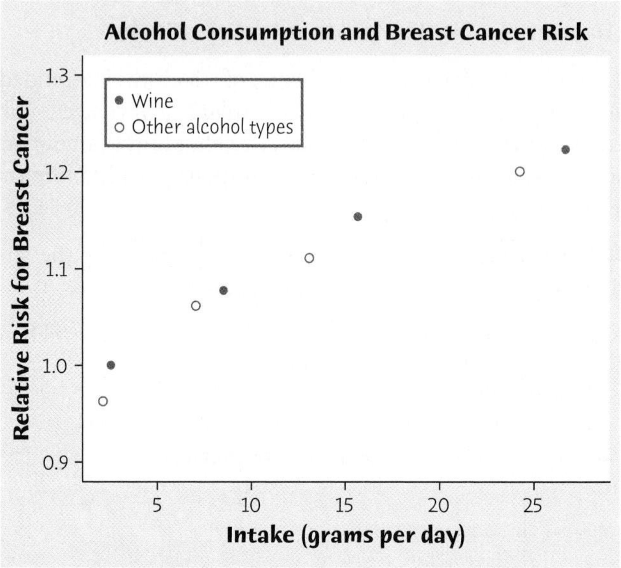

4.33: (a) The plot suggests that "Good" weather reports tend to yield higher tips. (b) The explanatory variable is categorical, so r cannot be used.

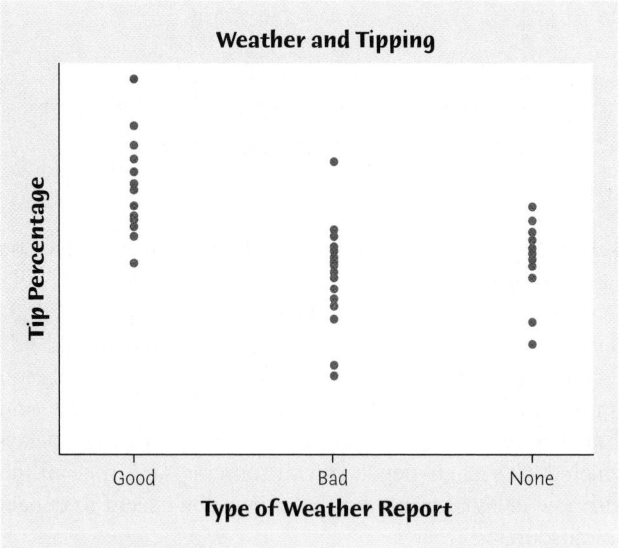

4.35: (a) The scatterplot is provided. Set B has stretched out the x-values, but the pattern is still the same. (b) Units do not impact correlation. For both data sets, $r = 0.298$.

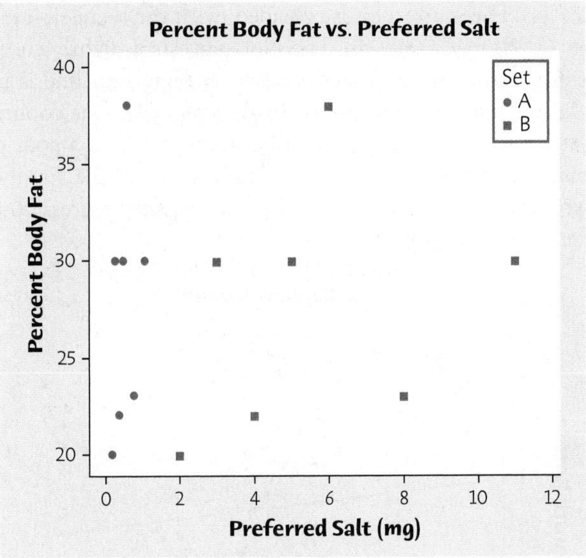

4.37: (a) Small-cap stocks have a lower correlation with municipal bonds, so the relationship is weaker. (b) She should look for a negative correlation (although this would also mean that this investment tends to *decrease* when bond prices rise).

4.39: (a) Because sex has a nominal scale, we cannot compute r. There is a strong *association* between sex and income. (b) $r = 1.09$ is impossible, because r is restricted to be between -1 and 1. (c) Correlation has no units, so "$r = 0.63$ centimeter" is incorrect.

4.41: (a) The correlation will be closer to 1. (b) Answers will vary, but r will decrease, and can be made negative by dragging the point down far enough.

4.43: PLAN: To investigate global warming, create a scatterplot and look for an increasing (positive) pattern. SOLVE: The plot suggests that temperatures have been increasing overall, but there seems to have been a slowing in the past few years; this graph looks curved. Correlation may not be a useful measure here; $r = 0.6334$. CONCLUDE: Over time, average global temperatures have increased, but the increase may not be linear.

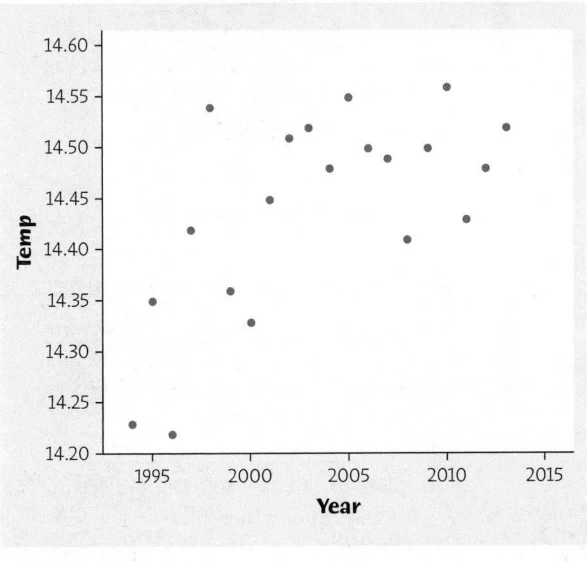

4.45: PLAN: Plot heat loss from the beak against temperature. Compute the correlation, if the relationship is reasonably linear. SOLVE: The plot is shown. There is a reasonably strong linear relationship, so we can use correlation to describe this relationship. $r = 0.9143$. CONCLUDE: When the temperature increases, a greater percentage of total heat loss is due to beak heat loss. That is, the beak plays a more important role in cooling down the toco toucan as the weather becomes hotter.

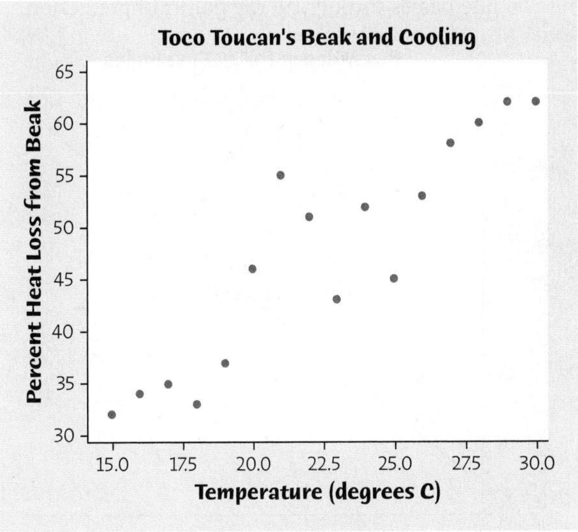

4.47: PLAN: Begin with a scatterplot, and compute the correlation if appropriate. SOLVE: A scatterplot shows a moderately strong, positive, linear association. The point at the upper right (5.3, 3.4) may be an outlier. This point seems to make the linear relationship appear more positive. Including the possible outlier, $r = 0.564$. If the outlier is omitted, $r = 0.4406$. CONCLUDE: The positive association supports the idea that squirrel populations increase when the pine cone supply is higher in the previous autumn. However, the relationship is somewhat weak; squirrels in the Yukon may have other good food sources.

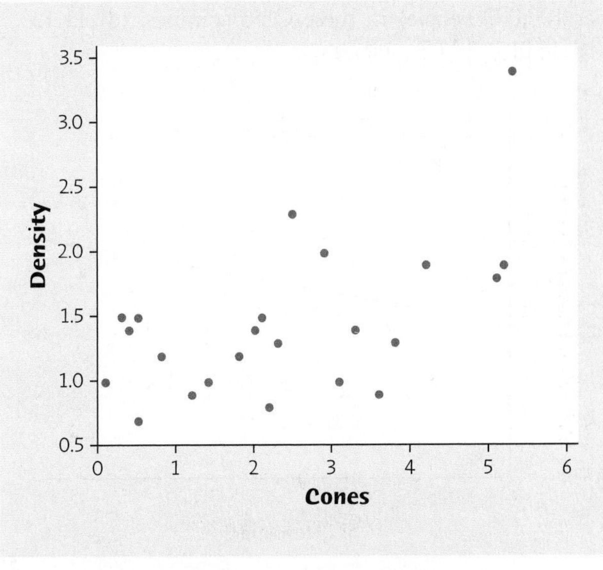

Chapter 5 Regression

5.1: (a) The slope is 1.033. On average, highway mileage increases by 1.033 mpg for each additional 1 mpg change in city mileage. (b) The intercept is 6.785 mpg. This is the highway mileage for a nonexistent car that gets 0 mpg in the city. (c) For a car that gets 16 mpg in the city, we predict 6.785 + (1.033)(16) = 23.31 mpg. For a car that gets 28 mpg in the city, we predict 6.785 + (1.033)(28) = 35.71 mpg. (d) The regression line passes through all the points of prediction.

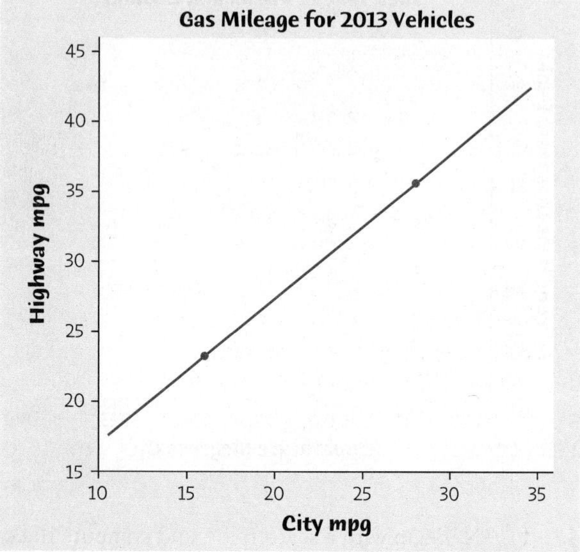

5.3: (a) The slope is 1021. For each year since 2000, forest loss averages about 1021 km^2. (b) In square meters, the slope would be 1021×10^6 = 1,021,000,000, a loss of 1 billion square meters per year (on average). In thousands of square kilometers, the slope would be 1.021; a loss of a bit more than 1000 km^2 per year (on average). Units matter in regression.

5.5: (a) The scatterplot (with the regression line) is shown. This relationship is certainly weak. (b) $\widehat{\text{Suicide}}$ = 11.125 + 0.195* Homicide (from software). (c) The slope means that for every suicide (per 100,000 people), there are about 0.195 homicides (per 100,000 people) in these Ohio counties. (d) 11.125 + 0.195(8.0) = 12.685 suicides.

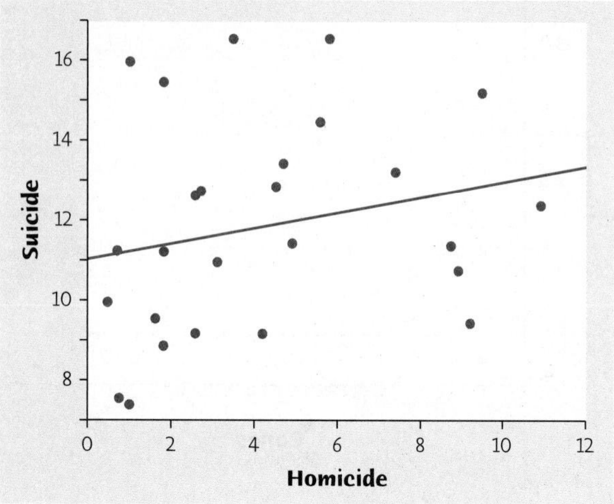

5.7: (a) The scatterplot is provided, with the regression line. $\hat{y} = 1.0284 - 0.004498x$. The plot suggests a slightly curved pattern, not a strong linear pattern. A regression line is not useful for making predictions. (b) $r^2 = 0.031$. This confirms what we see in the graph: the regression line does a poor job summarizing the relationship. Only about 3% of the variation in growth rate is explained by the least-squares regression on difference in begging intensity.

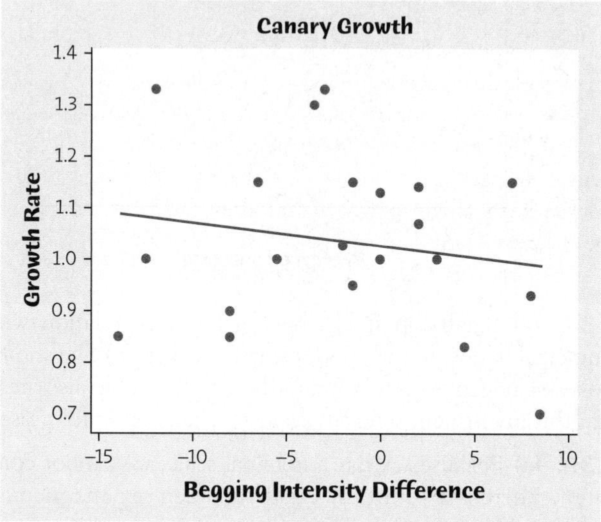

5.9: (a) Plot is provided following, top. (b) The pattern is curved, so linear regression is not appropriate. (c) For $x = 10$, $\hat{y} = 11.058 - 0.01466(10) = 10.91$, so the residual is $21.00 - 10.91 = 10.09$. The sum of the residuals is -0.01. (d) The first two and last four residuals are positive, and those in the middle are negative. Plot following, bottom.

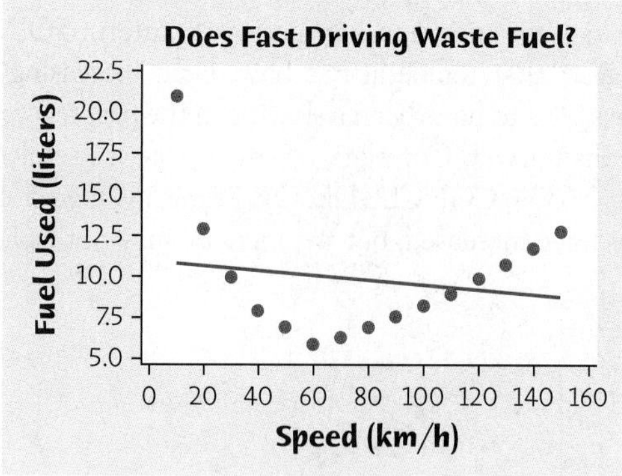

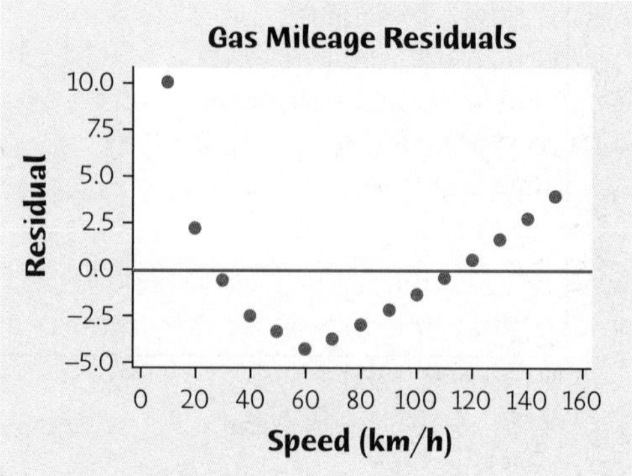

5.11: (a) Any point that falls exactly on the regression line will not increase the sum of squared vertical distances, so the regression line does not change. Any other line (even if it passes through this new point) will necessarily have a higher total sum of squared prediction errors. r increases because the new point reduces the relative scatter about the regression line. (b) Influential points are those whose x-coordinates are outliers. The regression line will "follow" an influential point if it is moved up or down in the y direction.

5.13: (a) Hawaiian Airlines is the point identified with "H." Since this point is an outlier and falls outside the x range of the other data points, it is influential. (b) With the outlier, $r = -0.488$. If the outlier is deleted, $r = -0.241$. (c) The two regression lines are plotted. The line based on the full data set has been pulled down toward the outlier, indicating that the outlier is influential. Based on the complete data set, $\hat{y} = 21.815 - 0.12851x$. Using this, when $x = 79.1$, we predict 11.65% delays. Omitting the outlier, $\hat{y} = 19.460 - 0.05528x$, so we predict 15.09% delays. The outlier impacts predictions because it impacts the regression line.

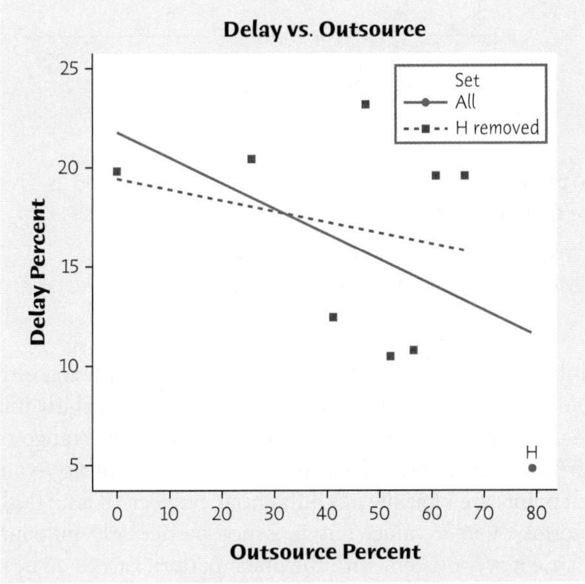

5.15: (a) $\hat{y} = -44.831 + 0.1323x$ (or, $\widehat{\text{Kills}} = -44.831 + 0.1323\text{ Boats}$). (b) If 890,000 boats are registered, $x = 890$, and $\hat{y} = -44.831 + (0.1323)(890) = 72.92$ manatees killed. The prediction seems reasonable, as long as conditions remain the same, because "890" is within the space of observed values of x. (c) If $x = 0$, then we would "predict" -44.831 manatees to be killed by boats. This is absurd, because it is clearly impossible. This illustrates the folly of extrapolation . . . $x = 0$ is well outside the range of observed values of x on which the regression line was based.

5.17: Possible lurking variables include the IQ and socioeconomic status of the mother. These variables are associated with smoking in various ways, and are also predictive of a child's IQ.

5.19: One example might be that men who are married, widowed, or divorced may be more "invested" in their careers than men who are single. There is still a feeling of societal pressure for a man to "provide" for his family.

5.21: (b) 0.2. Consider two points on the regression line—say, (90, 4) and (130, 11). The slope of the line segment connecting these points is $\frac{11-4}{130-90} = 7/40 = 0.175$.

5.23: (a) $y = 1000 + 100x$

5.25: (c) 16 cubic feet

5.27: (a) The slope of the line is positive.

5.29: (a) $\hat{y} = 24.2 + 6.0x$

5.31: (a) The least-squares regression line says that increasing the size of a diamond by 1 carat increases its price by 3721.02 Singapore dollars, on average. (b) A diamond of size 0 carats would have a predicted price of 259.63 Singapore dollars. This is probably an extrapolation, since the data set on which the line was constructed almost certainly had no rings with diamonds of size 0 carats.

5.33: (a) $\hat{y} = 0.919 + 2.0647x$. For every degree Celsius, the toucan will lose about 2.06% more heat through its beak. (b) $\hat{y} = 0.919 + 2.0647(25) = 52.5$. At a temperature of 25 degrees Celsius, we predict a toucan to lose 52.5% more heat through its beak, on average. (c) $r = \sqrt{r^2} = \sqrt{0.836} = 0.914$. Correlation is positive because the regression line has a positive slope.

5.35: (a) $b = rs_y/s_x = (0.5)(\frac{8}{40}) = 0.1$, and $a = \bar{y} - b\bar{x} = 75 - (0.1)(280) = 47$. The regression equation is $\hat{y} = 47 + 0.1x$. Each point of pre-exam total score means an additional 0.1 points on the final exam, on average. (b) $\hat{y} = 47 + (0.1)(300) = 77$. (c) With $r = 0.5$, $r^2 = (0.5)^2 = 0.25$, so the regression line accounts for only 25% of the variability in student final exam scores; the regression line doesn't predict final exam scores very well.

5.37: (a) $\hat{y} = 28.037 + 0.521x$. $r = 0.555$. The plot is provided. (b) $\hat{y} = 28.037 + (0.521)(70) = 64.5$ inches (rounded). This prediction isn't expected to be very accurate because the correlation isn't very large; $r^2 = (0.555)^2 = 0.308$.

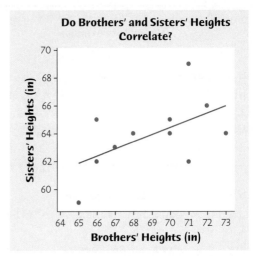

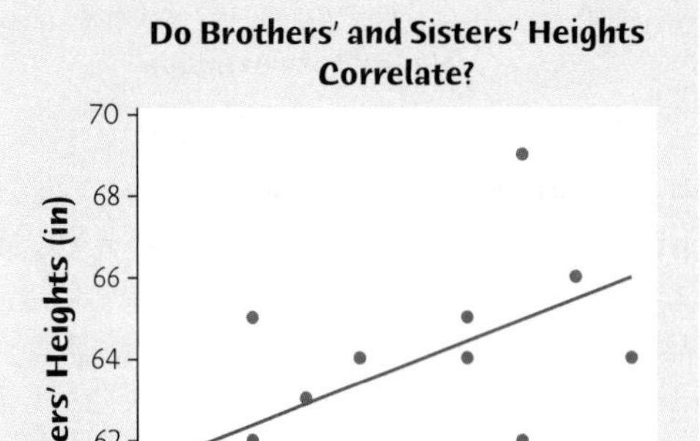

5.39: (a) $\hat{y} = 31.934 - 0.304x$. (b) On the average, for each additional 1% increase in returning birds, the number of new birds joining the colony decreases by 0.304. (c) When $x = 60$, we predict $\hat{y} = 13.69$ new birds.

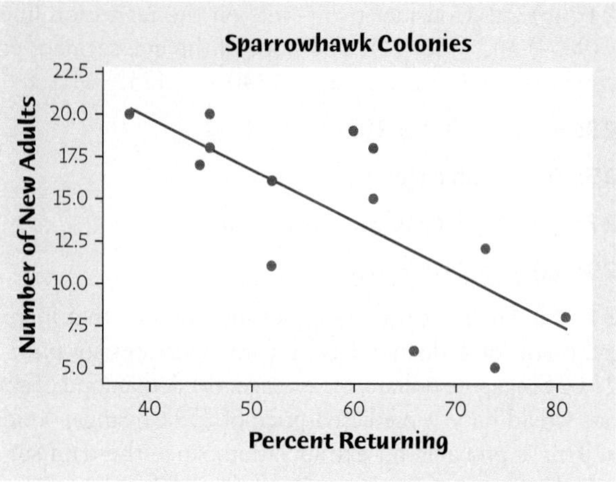

5.41: (a) The outlier is in the upper-right corner. (b) With the outlier omitted, $\hat{y} = 0.586 + 0.00891x$. (This is the solid line in the plot.) (c) The line does not change much because the outlier fits the pattern of the other points; r changes because the scatter (relative to the line) is greater with the outlier removed. (d) The correlation changes from 0.8486 (with all points) to 0.7015 (without the outlier). With all points included, the regression line is $\hat{y} = 0.585 + 0.0879x$ (nearly indistinguishable from the other regression line).

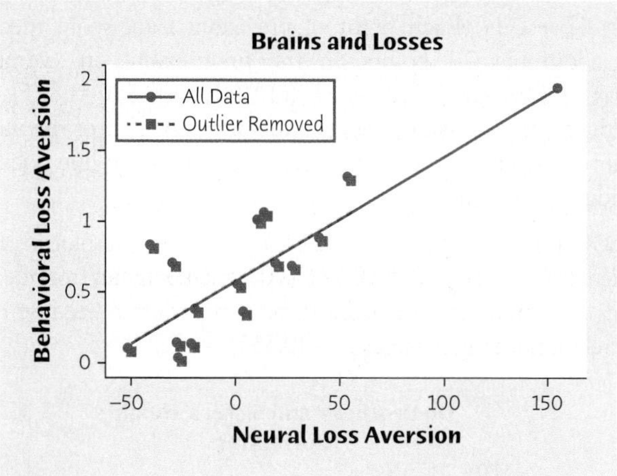

Minitab output - all points

The regression equation is Behave = 0.585 + 0.00879 Neural

Predictor	Coef	SE Coef	T	P
Constant	0.58496	0.07093	8.25	0.000
Neural	0.008794	0.001465	6.00	0.000

Minitab output - outlier removed

The regression equation is Behave = 0.586 + 0.00891 Neural

Predictor	Coef	SE Coef	T	P
Constant	0.58581	0.07506	7.80	0.000
Neural	0.008909	0.002510	3.55	0.004

5.43: (a) The two unusual observations are indicated on the scatterplot. (b) The correlations are

$r_1 = 0.4819$ (all observations)
$r_2 = 0.5684$ (without Subject 15)
$r_3 = 0.3837$ (without Subject 18)

Both outliers change the correlation. Removing Subject 15 increases r because its presence makes the scatterplot less linear. Removing Subject 18 decreases r because its presence decreases the relative scatter about the linear pattern.

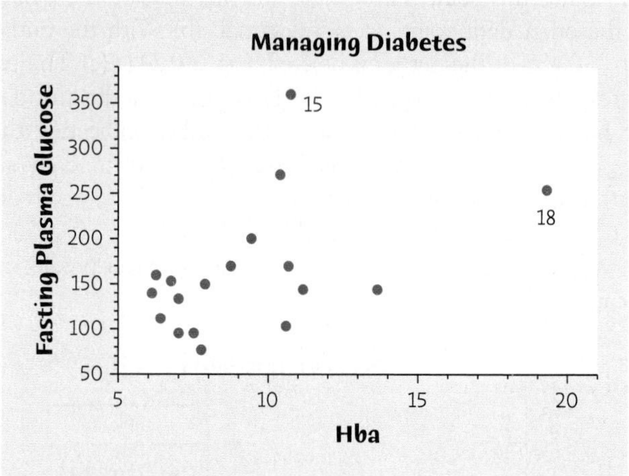

5.45: The scatterplot with regression lines added is given. The equations are

$\hat{y} = 66.4 + 10.4x$ (all observations)
$\hat{y} = 69.5 + 8.92x$ (without #15)
$\hat{y} = 52.3 + 12.1x$ (without #18)

While the equation changes in response to removing either subject, one could argue that neither one is particularly influential, because the line moves very little over the range of x (HbA) values. Subject 15 is an outlier in terms of its y-value; such points are typically not influential. Subject 18 is an outlier in terms of its x-value, but it is not particularly influential because it is consistent with the linear pattern suggested by the other points.

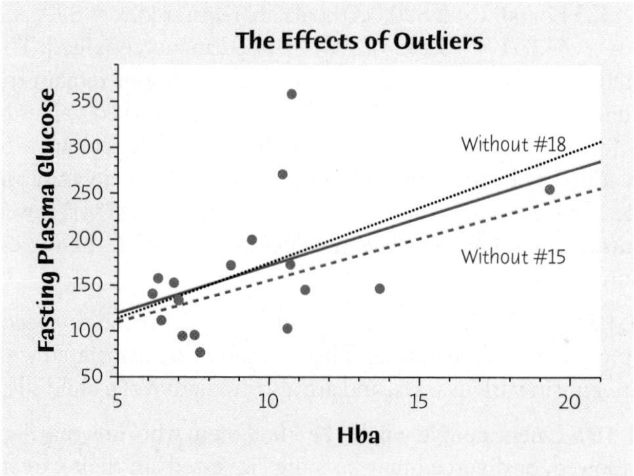

5.47: The correlation would be smaller. Individual weight will vary much more than the average weight for a given height.

5.49: Responses will vary. For example, students who choose the online course might have more self-motivation or have better computer skills.

5.51: (a) $\widehat{\text{MathSAT}} = 471.82 + 0.00048(\text{TeachSal})$. For these states, increasing the average teacher salary increases the mean Math SAT score by 0.00048 points, on average. (b) $\widehat{\text{MathSAT}} = 472.8 + 0.0020(\text{TeachSal})$. For these states, increasing the average teacher salary increases the mean Math SAT score by 0.0020 points, on average. (c) The slopes here have opposite signs from that found in Exercise 5.50. Consideration of a third (lurking) variable changed the relationship.

5.53: Here is a simple example to show how this can happen: suppose that most workers are currently 30 to 50 years old. Suppose further that each worker's current salary is his/her age (in thousands of dollars). Over the next 10 years, all workers age, and their salaries increase. Suppose every worker's salary increases by between \$4000 and \$8000. Then every worker will be making *more* money than he/she did 10 years before, but *less* money than a worker of that same age 10 years before.

5.55: For a player who shot 80, we predict $\hat{y} = 56.47 + (0.243)(80) = 75.91$. For a player who shot 70, $\hat{y} = 56.47 + (0.243)(70) = 73.48$. The player who shot 80 the first round (worse than average) is predicted to have a worse-than-average score the second round, but better than the first round. The player who shot 70 the first round (better than average) is predicted to do better than average in the second round, but not as well (relatively) as in the first round. Both players are predicted to "regress" to the mean.

5.57: A regression line is appropriate only for data set (b). For data set (c), the point not in the vertical stack is very influential—the stacked points alone give no indication of slope for the line. The curved relationship exhibited by (d) clearly indicates that predictions based on a straight line are not appropriate.

5.59: PLAN: We construct a scatterplot (with beaver stumps as the explanatory variable), and, if appropriate, find the regression line and correlation. SOLVE: The scatterplot shows a positive linear association. Regression seems to be appropriate; $\hat{y} = -1.286 + 11.89x$. The straight-line relationship explains $r^2 = 83.9\%$ of the variation in beetle larvae. CONCLUDE: The strong positive association supports the idea that beavers benefit beetles.

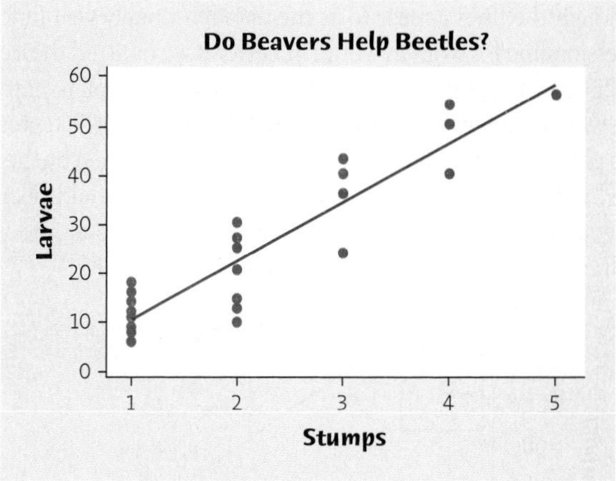

5.61: PLAN: We construct a scatterplot, with Forecast as the explanatory variable, and Actual as the response variable. If appropriate, we find the least-squares regression line. We consider the impact of the potential outlier (2005 season). SOLVE: The scatterplot shows a reasonable but not very strong linear relationship. In recent years, it seems that there is no relationship between forecast and actual (a flat line). The 2005 season is influential, pulling the regression line somewhat. We might consider deleting this point and fitting both lines. Deleting the point, we obtain the solid regression line, $\hat{y} = 2.753 + 0.7964$ Forecast when the original equation is $\hat{y} = 1.668 + 0.920$ Forecast. If the forecasts were perfect, the intercept of this line would be 0, and the slope would be 1. Deleting the 2005 season, $r = 0.628$, and $r^2 = 39.4\%$. Even after deleting the outlier, the regression line explains only 39.4% of variation in number of hurricanes. CONCLUDE: Predictions using the regression line are not very accurate. However, there is a positive association . . . so a forecast of many hurricanes may reasonably be expected to forebode a heavy season for hurricanes.

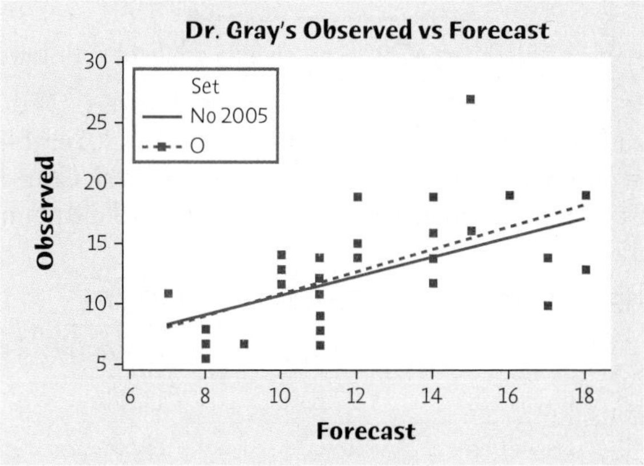

5.63: PLAN: We plot marathon times by year for each sex, using different symbols. If appropriate, we fit least-squares regression lines for predicting time from year for each gender. We then use these lines to guess when the times will agree. SOLVE: The scatterplot is provided with regression lines plotted. The regression lines are:

For men: $\hat{y} = 66{,}072 - 29.535x$
For women: $\hat{y} = 182{,}976.15 - 87.73x$

Although the lines appear to fit the data reasonably well (and the regression line for women would fit better if we omitted the outlier associated with year 1926), this analysis is inviting you to extrapolate, which is never advisable. CONCLUDE: Using the regression lines plotted, we might expect women to "outrun" men by the year 2009. Omitting the outlier, the line for women would decrease more steeply, and the intersection would occur sooner, by 1995. We'll note that as of July 2014, this has not happened.

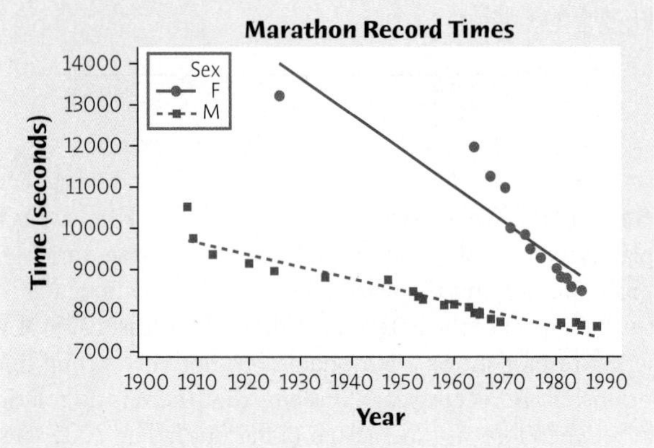

Chapter 6 Two-Way Tables

6.1: (a) 736 + 450 + 193 + 205 + 144 + 80 = 1808 people. 736 + 450 + 193 = 1379 played video games. (b) (736 + 205)/1808 = 0.5205 = 52.05% earned As and Bs. Do this for all three grade levels. The complete marginal distribution for grades is

GRADE	PERCENT
As and Bs	52.05%
Cs	32.85%
Ds and Fs	15.10%

32.85% + 15.10% = 47.95% received a grade of C or lower.

6.3: There are 736 + 450 + 193 = 1379 players. Of these, 736/1379 = 53.37% earned As or Bs. There are 205 + 144 + 80 = 429 nonplayers. Of these, 205/429 = 47.79% earned As or Bs. Continuing in like manner, the conditional distributions of grades follows:

GRADES	PLAYERS	NONPLAYERS
As and Bs	53.37%	47.79%
Cs	32.63%	33.57%
Ds and Fs	14.00%	18.65%

If anything, players have slightly higher grades than nonplayers, but this could be due to chance.

6.5: Two examples are shown. In general, choose a to be any number from 10 to 50, and then all the other entries can be determined.

30	20
30	20

50	0
10	40

6.7: (a) For Rotorua district, 79/8889 = 0.0089, or 0.9%, of Maori are in the jury pool, while 258/24,009 = 0.0107, or 1.07%, of the non-Maori are in the jury pool. For Nelson district, the corresponding percents are 0.08% for Maori and 0.17% for non-Maori. In each district, the percent of non-Maori in the jury pool exceeds the percent of Maori in the jury pool.

	MAORI	NON-MAORI
In jury pool	80	314
Not in jury pool	10,138	56,353
Total	10,218	56,667

(b) Overall, 80/10,218 = 0.0078, or 0.78%, of Maori are in the pool, while 314/56,667 = 0.0055, or 0.55%, of non-Maori are in the pool. Overall the Maori have a larger percent in the jury pool, but in each region they have a lower percent in the jury pool. (c) The reason is that the Maori constitute a large proportion of Rotorua's population, while in Nelson they are a small minority community.

6.9: (b) 150 teens in schools that forbid cell phones

6.11: (a) the marginal distribution of school permissiveness

6.13: (c) the conditional distribution of the frequency that a teen brings a cell phone to school among the schools that forbid cell phones

6.15: (b) the conditional distribution of school permissiveness among those who brought their cell phone to school every day

6.17: (b) an example of Simpson's paradox

6.19: For each type of injury (accidental, not accidental), the distribution of ages follows.

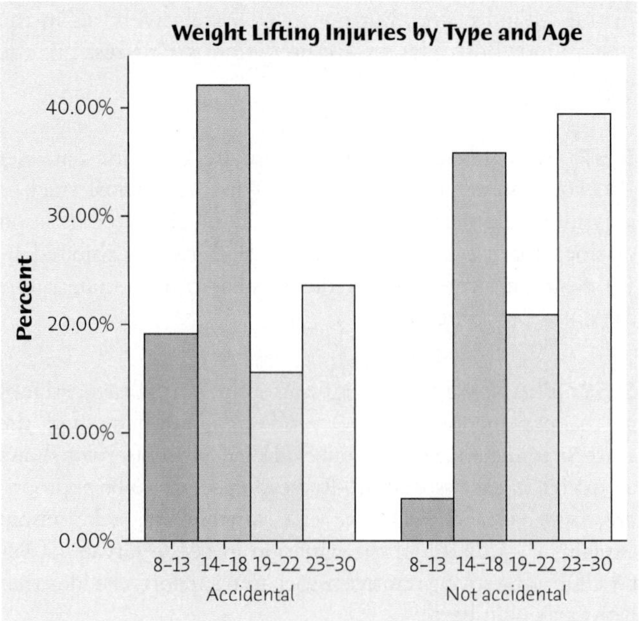

	ACCIDENTAL	NOT ACCIDENTAL
8–13	19.0%	4.0%
14–18	42.2%	35.8%
19–22	15.4%	20.8%
23–30	23.4%	39.4%

Among accidental weight-lifting injuries, the percentage of younger lifters is larger, while among the injuries that are not accidental, the percentage of older lifters is larger.

6.21: The percent of single men with no income is 513/4938 = 0.1039, or 10.39%. The percent of men with no income that are single is 513/1918 = 0.2675, or 26.75%.

6.23: (a) We need to compute percents to account for more married men than single men. (b) A table is provided; descriptions of the relationship may vary. For example, married men made up less of the no income distribution compared to the $100,000 and up distribution.

	SINGLE	MARRIED	DIVORCED	WIDOWED
No income	26.75%	51.87%	20.07%	1.30%
$100,000 and up	4.76%	85.43%	8.99%	0.81%

6.25: (a) The two-way table of race (white, black) versus death penalty (death penalty, no death penalty) follows:

	WHITE DEFENDANT	BLACK DEFENDANT
Death penalty	19	17
No death penalty	141	149

(b) For black victims, white defendants were given the death penalty in 0/9 = 0, or 0%, of cases. Black defendants were given the death penalty in 6/103 = 0.058, or 5.8%. For white victims, white defendants were given the death penalty in 19/151 = 0.126, or 12.6%, of cases, and black defendants were given the death penalty 11/63 = 0.175, or 17.5%, of the time. For both victim races, black defendants are given the death penalty relatively more often than white defendants. However, overall, 19/160 = 0.119, or 11.9%, of white defendants got the death penalty, while 17/166 = 0.102, or 10.2%, of black defendants got the death penalty. (c) For white defendants, (19 + 132)/(19 + 132 + 0 + 9) = 0.9438 = 94.4% of victims were white. For black defendants, (11 + 52)/(11 + 52 + 6 + 97) = 0.3795, or 37.95%, of victims were white. The death penalty was predominantly assigned to cases involving white victims: 14.0% of all cases with a white victim, and only 5.5% of all cases with a black victim, had a death penalty assigned. Because most white defendants' victims are white, and cases with white victims carry additional risk of a death penalty, white defendants are being assigned the death penalty more often overall.

6.27: PLAN: Find and compare the conditional distributions of outcome for each treatment. SOLVE: The percentages for each column are provided. For example, for Chantix, the percentage of successes is 155/(155 + 197) = 0.4403, or 44.0%.

	CHANTIX	BUPROPION	PLACEBO
Percent not smoking in weeks 9–12	44.0%	29.5%	17.7%

CONCLUDE: Clearly, a larger percent of subjects using Chantix were not smoking during weeks 9–12, compared with results for either of the other treatments.

6.29: PLAN: Calculate and compare the conditional distributions of sex for each degree level. SOLVE: For example, 646/(646 + 383) = 0.6278, or 62.78% of associate's degrees went to women. The table shows the percent of women at each degree level, which is all we need. CONCLUDE: Women constitute a substantial majority of associate's, bachelor's, and master's degrees, and a small majority of doctor's and professional degrees.

DEGREE	% FEMALE
Associate's	62.78%
Bachelor's	57.88%
Master's	61.94%
Doctor's and Professional	53.53%

6.31: PLAN: Find and compare the conditional distributions for health for each group. SOLVE: The table provides the percent of subjects with various health outlooks for each group. CONCLUDE: Clearly, the outlooks of current smokers are generally bleaker than those of current nonsmokers. Much larger percentages of nonsmokers reported being in "excellent" or "very good" health, while much larger percentages of smokers reported being in "fair" or "poor" health.

	HEALTH OUTLOOK				
	EXCELLENT	VERY GOOD	GOOD	FAIR	POOR
Current smoker	6.2%	28.5%	35.9%	22.3%	7.2%
Current nonsmoker	12.4%	39.9%	33.5%	14.0%	0.3%

6.33: Because the numbers of students who use (or do not use) medications are different, we find the conditional distributions of those who do and do not use medications. Those who use medications are less likely to have optimal sleep. This is a case where one would not want to ascribe causation: Do those who use medications to stay awake have poor sleep quality because they use the medication, or do they use the medications to stay awake because they had poor sleep quality before using them?

	SLEEP QUALITY		
	OPTIMAL	BORDERLINE	POOR
Use medications	21.3%	30.5%	48.3%
Do not use medications	38.2%	26.7%	35.2%

Chapter 7 Exploring Data: Part I Review

7.1: (c)

7.3: (d)

7.5: (c)

7.7: (c)

7.9: (c)

7.11: (a) centimeters; (b) centimeters; (c) centimeters; (d) grams^2

7.13: (d) 43 of 51 observations at least 20

7.15: (c) About 54% use a search engine at least once a day; an additional 15–16% use one at least three times per week.

7.17: (a) $0.800 - 3(0.078) = 0.566$ to $0.800 + 3(0.078) = 1.034$ mm (b) 0.878 is 1σ above the mean; so about 16% (15.87%)

7.19: (a) Minimum = 7.2, $Q_1 = 8.5$, $M = 9.3$, $Q_3 = 10.9$, Maximum = 12.8 (b) $M = 27$ (c) 25% of values exceed $Q_3 = 30$ (d) Yes. Virtually all Torrey pine needles are longer than virtually all Aleppo pine needles.

7.21: (a)

7.23: (c)

7.25: (d)

7.27: (d)

7.29: (d)

7.31: (c)

7.33: (a)

7.35: (a) No (b) $r^2 = 0.64$, or 64%

7.37: (a) 8.683 kg (b) 10.517 kg (c) Such a comparison is unreasonable because the lean group is less massive, and therefore would be expected to burn less energy on average. (d) See scatterplot. (e) It appears that the rate of increase in energy burned per kilogram of mass is about the same for both groups. The obese monkeys burn less energy than the lean monkeys, because their points tend to be below the others.

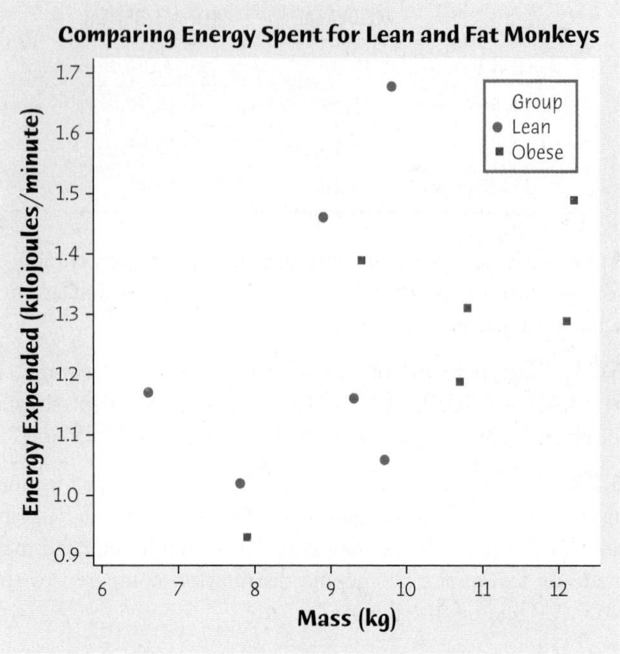

7.39: (a) 190/8474 = 0.0224, or 2.24% (b) 633/8474 = 0.0747, or 7.47% (c) 27/633 = 0.0427, or 4.27% (d) 4621/8284 = 0.5578, or 55.78% (e) The conditional distribution of CHD for each level of anger is tabulated below. The result for the high anger group was computed in part (c). Clearly, angrier people are at greater risk of CHD.

LOW ANGER	MODERATE ANGER	HIGH ANGER
1.70%	2.33%	4.27%

7.41: The time plot shows a lot of fluctuation from year to year, but also shows a recent increase: prior to 1972, the discharge rarely rose above 600 km^3, but since then, it has exceeded that level more than half the time. A histogram or stemplot cannot show this change over time.

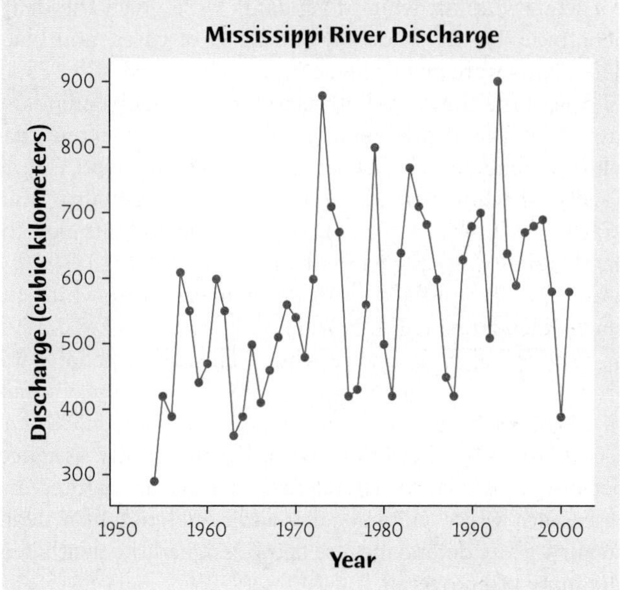

7.43: (a) The plot is provided. (b) $\hat{y} = 144.79 - 0.0659x$. The slope is negative, suggesting that the ice breakup day is decreasing (by about 0.0659 days per year). (c) The regression line is not very useful for prediction, as it accounts for only about 9% ($r^2 = 0.0903$) of the variation in ice breakup time.

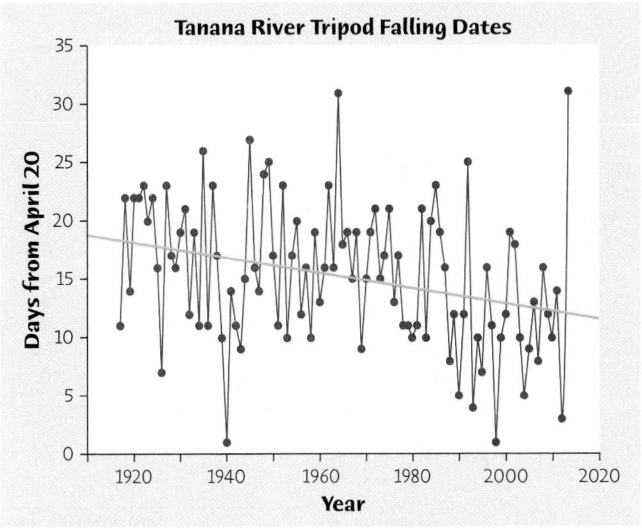

7.45: (a) A histogram is provided. (b) Luxembourg is the outlier with average starting salary \$65,171. Otherwise, the distribution is fairly symmetric. (c) $\bar{x} = \$27,381.4$ and $s = \$9651.78$. The five-number summary is Min = \$9,526, Q_1 = \$23,130, M = \$28,328, Q_3 = \$32,629, Max = \$46,456. (d) The U.S. average starting salary (\$36,858) is above the third quartile. Only six countries pay their teachers a higher starting salary.

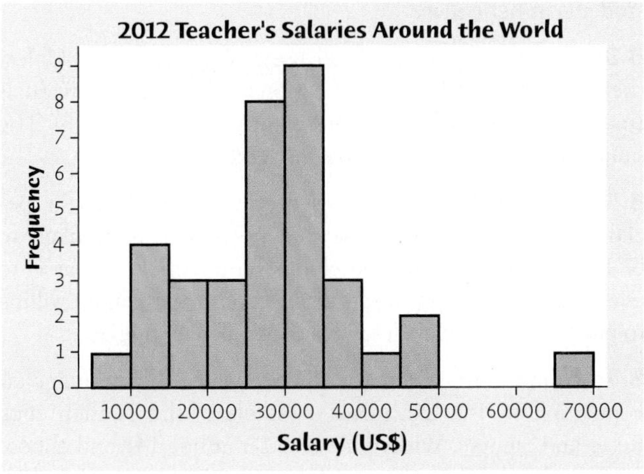

7.47: PLAN: Graph the data and compute appropriate numerical summaries. SOLVE: A stemplot is shown. The distribution seems to be fairly Normal apart from a high outlier of 50°. The five-number summary is preferred because of the outlier: Min = 13°, Q_1 = 20°, M = 25°, Q_3 = 30°, Max = 50°. CONCLUDE: Student descriptions of the distribution will vary. Most patients have a deformity angle in the range of 15° to 35°.

1	3 4
1	6 6 7 8 8
2	0 0 0 1 1 1 1 2 3
2	5 5 5 5 6 6 6 6 8 8 8
3	0 0 0 1 2 2 2 4
3	8 8
4	
4	
5	0

7.49: PLAN: We examine the relationship with a scatterplot and (if appropriate) correlation and regression line. SOLVE: MA angle is the explanatory variable. The scatterplot shows a moderate to weak positive linear association, with one clear outlier (HAV angle 50°). $r = 0.302$, and $\hat{y} = 19.723 + 0.3388x$. CONCLUDE: MA angle can be used to give estimates of HAV angle, but the variability is so large that estimates would not be very reliable. The linear relationship explains only $r^2 = 9.1\%$ of the variation in HAV angle.

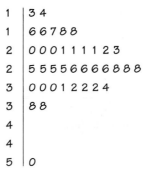

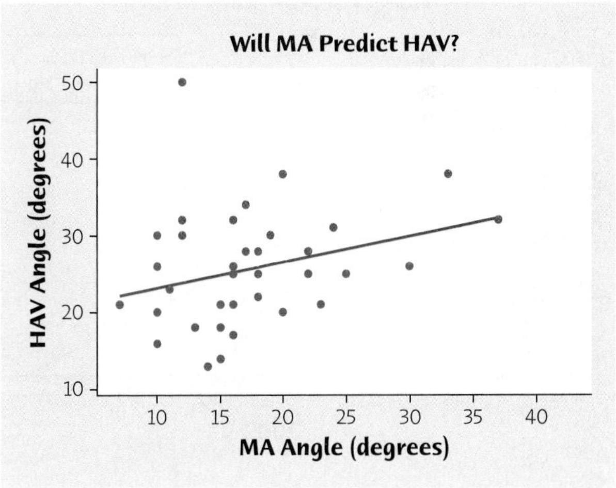

7.51: PLAN: We will examine the relationship with a scatterplot and (if appropriate) correlation and regression line. SOLVE: The scatterplot, shown with the line $\hat{y} = 70.44 + 274.78x$, shows a moderate, positive linear relationship. The linear relationship explains about $r^2 = 49.3\%$ of the variation in gate velocity. CONCLUDE: The regression formula might be used as a rule of thumb for new workers to follow, but the large variability in the scatterplot suggests that there may be other factors that should be taken into account.

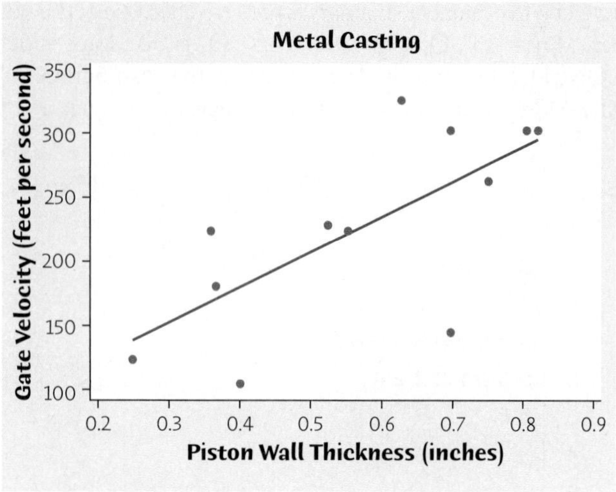

7.53: (a) The scatterplot of 2003 returns against 2002 returns shows (ignoring the outlier) a strong negative association. (b) For all 23 points, $r = -0.616$; with the outlier removed, $r = -0.838$. The outlier deviates from the linear pattern of the other points; removing it makes the negative association stronger, and so r moves closer to -1. (c) Regression formulas are given. The first line is solid; the second is dashed. The line must pivot up toward Fidelity Gold in order to minimize the sum of squares for all 23 deviations. Fidelity Gold is very influential.

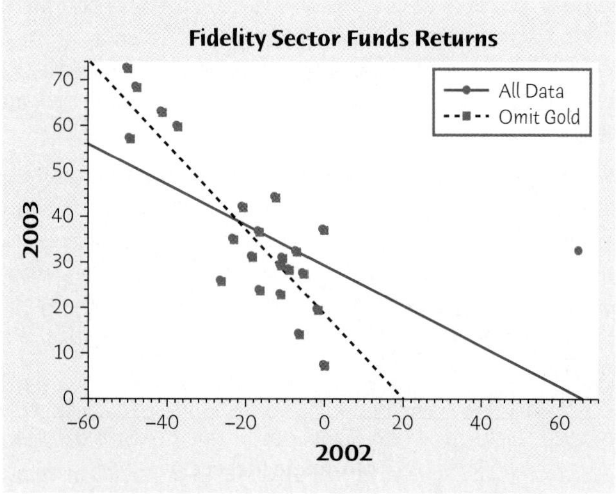

	r	EQUATION
All 23 funds	−0.616	$\hat{y} = 31.117 - 0.4132x$
Without Gold	−0.838	$\hat{y} = 21.462 - 0.8403x$

7.55: (a) Fish catch is the explanatory variable. The point for 1999 is at the bottom of the plot. (b) Correlations are given in the table provided. The outlier decreases r because it weakens the strength of the association. (c) The two regression lines are given; the solid line uses all points, while the dashed line omits the outlier. The effect of the outlier on the line is small: there are several other years with similar changes in bushmeat biomass. Also, this year was not particularly extreme in the amount of fish caught.

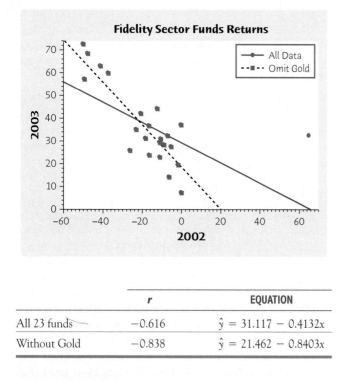

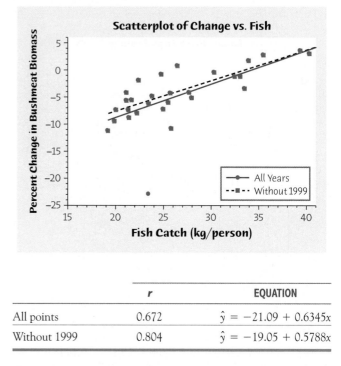

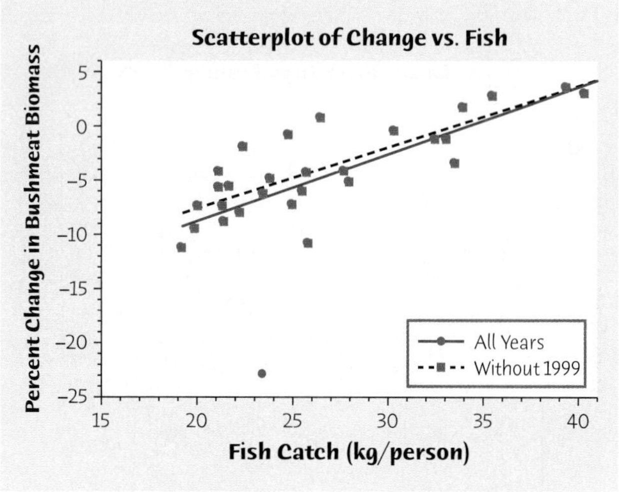

	r	EQUATION
All points	0.672	$\hat{y} = -21.09 + 0.6345x$
Without 1999	0.804	$\hat{y} = -19.05 + 0.5788x$

Chapter 8 Producing Data: Sampling

8.1: (a) The population is (all) college students. (She wanted to know about *all* college students.) (b) The sample is the 104 students at the researcher's college who returned the questionnaire. Because she only has her own college to sample from, the population she can make conclusions about is students at her college.

8.3: (a) The population is all users of the software. Unless their market is primarily educational, the 1100 individuals (mostly faculty) will not represent the population. (b) The sample is the 186 people who completed the survey.

8.5: Because all the students surveyed are enrolled in a special senior honors class, these students may be more likely to be interested in joining the club. The direction of bias is to overestimate the proportion of all psychology majors willing to pay to join this club. This is a convenience sample.

8.7: Number from 01 to 26 alphabetically. With the applet: Population = 1 to 26, select a sample of size 5, then click Reset and Sample. With Table B, enter at line 141 and choose 23 = Rodriguez, 12 = Gemayel, 16 = Ippolito, 25 = Sgambellone, and 02 = Ahmadiani.

8.9: With the election soon, the polling organization wants to increase the accuracy of its results. Larger samples provide better information.

8.11: Label the suburban townships from 01 to 30 alphabetically. With Table B line 116, choose 14 = New Trier, 03 = Bloom, 10 = Lemont, 22 = Proviso, and 06 = Cicero. Next, label the Chicago townships from 1 to 8, down the columns. With Table B, enter at line 126 and choose 6 = Rogers Park, 2 = Jefferson, and 7 = South Chicago.

8.13: (a) The population is all physicians practicing in the United States. The sample size is $n = 2379$. If the 2379 were randomly selected, we could draw conclusions, but there was too much nonresponse. (b) The nonresponse rate is $\frac{100{,}000 - 2379}{100{,}000} = 97.62\%$. We don't know the attitudes of the nonrespondents, so the results may not be credible. (c) They only received 2379 responses.

8.15: (a) The sample was not randomly selected. (b) The true percentage is most likely lower. People who don't listen often probably don't visit the NPR Facebook page.

8.17: (a)

8.19: (b)

8.21: (b) Plots are stratified by terrain.

8.23: (c) In part (b) "07" appears in the sample twice.

8.25: (b) People over 65 are a subset of the original sample.

8.27: The population is the 1000 envelopes stuffed during a given hour. The sample is the 40 envelopes selected.

8.29: With the applet: Population = 1 to <u>287</u>, select a sample of size <u>20</u>, then click <u>Reset</u> and <u>Sample</u>. Using Table B, number the area codes 001 to 287. Enter at line 122, and select area codes labeled 138, 159, 052, 087, and 275.

8.31: The questions were worded very differently. The U.S. has the second Amendment that allows guns. Canada has very different gun laws; gun ownership is generally forbidden. Given these facts, opinions would differ in any case (even if the questions were worded the same).

8.33: Online polls, call-in polls, and voluntary response polls in general tend to attract responses from those who have strong opinions. On the other hand, there is no reason to believe that randomly chosen adults would overrepresent any particular group; those give a more reliable picture of public opinion.

8.35: (a) Assign labels 0001 through 5024, enter the table at line 114, and select: 4514, 0381, 0202, 0915, and 1776. (b) More than 171 respondents have run red lights. We would not expect very many people to claim they *have* run red lights when they have not, but some people will deny running red lights when they have.

8.37: (a) Each person has a 10% chance: 4 of 40 men, and 3 of 30 women. (b) This is not an SRS because not every group of 7 people can be chosen; the only possible samples are those with 4 men and 3 women.

8.39: (a) How the sample was obtained can contribute to bias in the results, if not done randomly and fairly. We need to try to avoid undercoverage, for example. (b) Answers will vary. For example, exactly how the 655 Internet users were selected is not given.

8.41: Sample separately in each stratum; that is, assign separate labels, choose the first sample, then continue on in the table to choose the next sample, etc. Beginning with line 112 in Table B, we choose:

FOREST TYPE	LABELS	PARCELS SELECTED
Climax 1	01 to 36	04, 11, 19, 35
Climax 2	01 to 72	27, 30, 57, 62, 56, 02, 06
Climax 3	01 to 31	08, 02, 25
Secondary	01 to 42	11, 17, 14, 29

8.43: (a) Because 200/5 = 40, choose one of the first 40 names at random. Beginning on line 120, the addresses selected are 35, 75, 115, 155, and 195. (Only the first number is chosen from the table.) (b) All addresses are equally likely; each has chance 1/40 of being selected. This is not an SRS because the only possible samples have exactly one address from the first 40, one address from the second 40, and so on. An SRS could contain any five of the 200 addresses in the population.

8.45: (a) This design omits households without telephones, those with only cell phones, and those with unlisted numbers. Such households would likely be made up of poor individuals (who cannot afford a phone), those who choose not to have landline phones, and those who do not wish to have their phone numbers published. (b) Those with unlisted landline numbers would be included in the sampling frame.

8.47: (a) The wording is clear, but will almost certainly be slanted toward a high positive response. (b) The wording is clear, and it makes the case for a national health care system, and so will slant responses toward "yes." (c) This survey question is most likely to produce a response similar to: "Uhh . . . yes? I mean, no? I'm sorry, could you repeat the question?"

8.49: In Canada, as in many places, elected officials aren't necessarily qualified to make decisions about statistics. Critics of the proposal are worried that the sample will not be representative of the population—presumably because people that fill out the optional long-form questions will be systematically different from those that don't. Larger samples do not address such problems of bias.

Chapter 9 Producing Data: Experiments

9.1: (a) The explanatory variable is the amount of alcohol drunk. The response is heart disease. (b) Weight, sleep, exercise, etc., would be considered to be lurking (or confounding) variables. (c) Without an experiment, no causal relationship can be identified.

9.3: This is an observational study; it is not reasonable to conclude any cause-and-effect relationship.

9.5: Subjects: the students. Factors: type of attack, and prime used. Treatments: four combinations of *love thy neighbor* prime, or *eye-for-an-eye* prime and attack on military or cultural/educational target. Response variable: rating of U.S. reaction to attack.

		PRIME USED	
		LOVE-THY-NEIGHBOR	EYE-FOR-AN-EYE
Target	Military	1	2
	Cultural	3	4

9.7: Making a comparison between the treatment group and the percent finding work *last year* is not helpful. Over a year, many things can change. In order to draw conclusions, we need to make the $500 bonus offer to some people and not to others during the same time period, and compare the two groups.

9.9: (a) Diagram follows. (b) If using Table B, label 01 to 36 and take two digits at a time.

9.11: In a controlled scientific study the effects of factors other than the nonphysical treatment (e.g., the placebo effect, differences in the prior health of the subjects) can be eliminated or accounted for, so that the differences in improvement observed between the subjects can be attributed to the differences in treatments.

9.13: (a) The researchers did not alter the diets. (b) Such language is reasonable because with observational studies; no "cause and effect" conclusion would be reasonable.

9.15: "Lack of blindness" means that the experimenter knows which subjects were taught to meditate. He or she may have some expectation about whether meditation will lower anxiety; this could (unconsciously) influence the end-of-month assessment.

9.17: (a) The explanatory variable is the font; the response is the degree of agreement with the statement about healthiness. (b) *Completely randomized design:* 50 students are randomly assigned to each of the two fonts, and then rate the healthiness of the product. (c) *Matched pairs design:* each student sees both fonts in random order, then rates the product's perceived healthiness for each font.

9.19: (a) Life satisfaction is observed; no treatments were imposed.

9.21: (b) All participating students had the same treatment.

9.23: (b)

9.25: (a) The researchers did not randomly assign where the people lived, so no treatments were actively imposed.

9.27: (a) The choice should be made randomly.

9.29: (a) This is an observational study; the subjects chose their own "treatments." The explanatory variable is red meat consumption, and the response variable is whether or not a subject dies. (b) Many answers are possible. For example, smoking is known to increase the risk of cancer. (c) Many answers are possible. For example, how many servings of fruits and vegetables were consumed along with the red meat?

9.31: This is an experiment, because the treatment is selected (randomly, we assume) by the interviewer. The explanatory variable (factor) is the level of identification, and the response variable is whether or not the interview is completed.

9.33: (a) In an observational study, we simply observe subjects who live near highways and compare them with others who do not live near highways. In an experiment, we would *assign* where the subjects live. (b) Answers will vary. For example, those who live near highways might have less money and be attracted to the (possibly) lower housing costs near a highway. (c) Not feasibly. Who would want their housing situation randomly assigned?

9.35: (a) Diagram follows. (b) Assign labels 001 to 120. If using Table B, line 108 gives 090, 009, 067, 092, 041, 059, 040, 080, 029, 091.

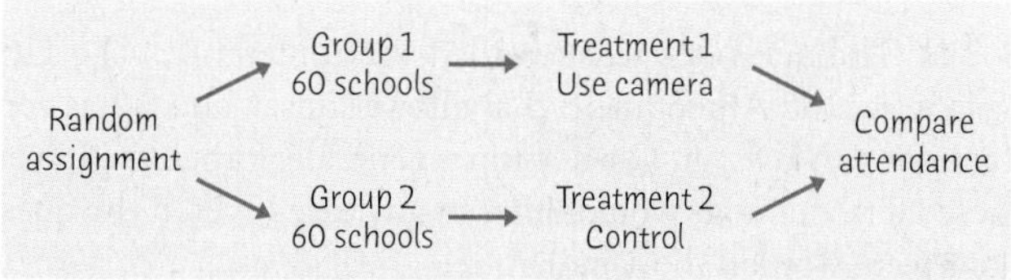

9.37: (a) The outline is shown. (b) From line 120, we choose subjects corresponding to the numbers 16, 04, 26, 21, 19, 07, 22, 10, 25, 13, 15, 05, 29, 09, 08 for the first group, and the rest for group 2. Thus, the marijuana group consists of Mattos, Bower, Williams, Sawant, Reichert, DeVore, Scannell, Giriunas, Stout, Kennedy, Mani, Burke, Zaccai, Fritz, and Fleming. All other subjects are assigned to the non-marijuana group. (c) This could be a double-blind experiment, assuming that subjects can't distinguish between the types of marijuana smoked and the persons measuring output and earnings of subjects don't know what kind of marijuana a subject smoked. Ensuring these conditions is very unlikely.

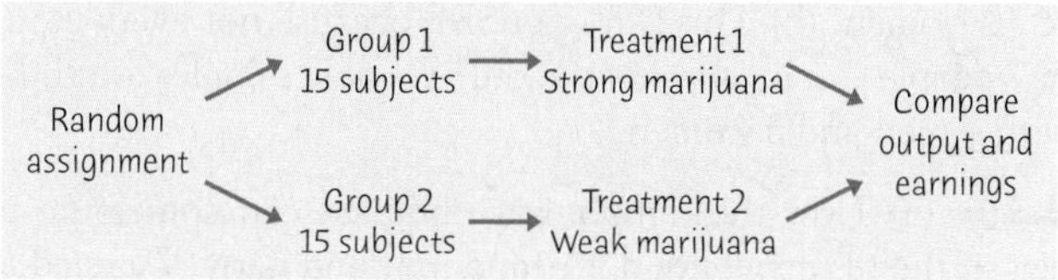

9.39: (a) There are two factors. The first factor is type of granola, and has two levels (regular and low-fat). The second factor is serving size label, and has three levels (2 servings, 1 serving, and no label). There are six treatment combinations (regular granola at 2 servings, regular granola at 1 serving, regular granola with no serving label, low-fat granola with 2 servings, low-fat granola with 1 serving, and low-fat granola with no serving label). At 20 subjects per treatment, there were 120 subjects in the experiment. (b)

	GRANOLA TYPE	
SERVING SIZE LABEL	REGULAR	LOW-FAT
2 servings	20 subjects	20 subjects
1 serving	20 subjects	20 subjects
No label	20 subjects	20 subjects

9.41: The factors are pill type and spray type. "Double-blind" means that the treatment assigned to a patient was unknown to both the patient and those responsible for assessing the effectiveness of that treatment. "Placebo-controlled" means that some of the subjects were given placebos. Even though placebos possess no medical properties, some subjects may show improvement or benefits just as a result of participating in the experiment.

9.43: (a) The subjects are randomly chosen (preferably people who like flavored water). Each subject tastes two cups of flavored water, in identical unlabeled cups. One contains MiO, the other the ready to drink product, in random order. Preference is the response variable. (b) We must assign 10 customers to get regular coffee first. Label the subjects 01 to 20. Starting at line 138, the "MiO first" group is: 16, 08, 15, 13, 17, 04, 10, 19, 12, and 18.

9.45: Each player will be put through the sequence (100 yards, four times) twice—once with oxygen and once without. For each player, randomly determine whether to use oxygen on the first or second trial. Allow ample time (perhaps a day or two) between trials for full recovery.

9.47: The diagram is shown. The last stage ("Observe heart health") might be described in more detail.

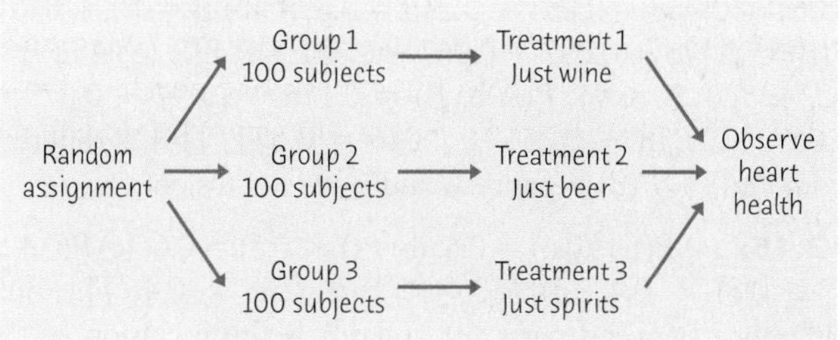

9.49: Any experiment randomized in this way assigns all the women to one treatment and all the men to the other. Sex is completely confounded with treatment. If women and men respond differently, the experiment will be strongly biased. The direction of the bias is random, depending on the coin toss.

9.51: (a) "Randomized" means that patients were randomly assigned to receive either SAMe or a placebo. "Double-blind" means that the treatment assigned to a patient was unknown to both the patient and those responsible for assessing the effectiveness of that treatment. Even though placebos possess no medical properties, some subjects may show improvement or benefits just as a result of participating in the experiment; the placebos allow those doing the study to observe this effect. (b) Statistical significance means that the SAMe group had a difference in response (more had a positive response) that could *not* be attributed to chance. This means that it appears SAMe helps reduce depression when used with standard treatment. (c) Diagram follows.

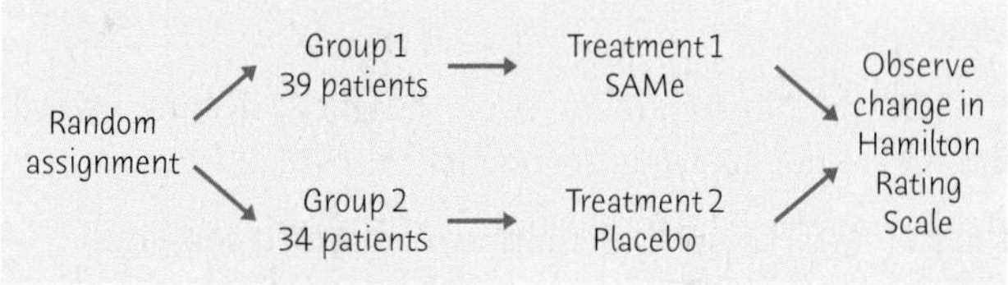

Chapter 10 Data Ethics

As the text states, "Most of these exercises pose issues for discussion. There are no right or wrong answers, but there are more and less thoughtful answers." We have not tried to supply answers for exercises that are largely matters of opinion. For that reason, only a few are provided.

10.1: These six proposals are clearly in increasing order of risk. Most students will consider that option (a) qualifies as minimal risk, and most will agree that option (e) goes beyond minimal risk.

10.3: Certainly the prospect of losing needed health services would induce most subjects to agree to participate. Likewise, an employer's "voluntary" participation might be rewarded (or penalized) in job performance evaluations. Both of these could be termed coercive.

10.11: Most students will see option (a) as allowable, option (b) as questionable even though the meetings are public, and option (c) as not allowable due to the psychologist "pretending to be converted."

Chapter 11 Producing Data: Part II Review

11.1: (c) hives with bees; hives with no bees; no hives

11.3: (a) Label the students 01 through 35 alphabetically. (b) Starting on line 115, the selected students are 04 = Bower, 17 = Huling, 32 = Vore, 22 = Newburg, 09 = Ding, and 26 = Pulak. (c) The response variable is "How much I trust the Internet for health information."

11.5: (d)

11.7: (a) No treatments were assigned by the researchers.

11.9: Many answers are possible. One possible lurking variable is "student attitude about the purpose of college" (students with a view that college is about partying, rather than studying may be more likely to binge drink and more likely to have lower grades). A correct example of a lurking variable *must* be a variable that simultaneously drives both "GPA" and "binge drinking" together.

11.11: (d) 30-second once, 30-second twice, 45-second once, 45-second twice, 60-second once, and 60-second twice

11.13: People who visit the *NOVA Science Now* website don't represent American adults broadly. Those people taking the survey went out of their way to participate in this online poll, and they read pro and con arguments after watching a program about the issue. It seems reasonable to believe that these people understand the issues better than most American adults.

11.15: (b) No treatments were assigned. Testing eyesight and IQ are not applying treatments.

11.17: (a) This is the definition of informed consent.

11.19: Placebos do work with real pain, so the placebo response tells nothing about physical basis of the pain. In fact, placebos work poorly in hypochondriacs. The survey is described in the April 3, 1979, edition of the *New York Times*.

11.21: (a) increase (b) decrease (c) increase (d) decrease

Note: *The first and third statements make an argument in favor of a national health insurance system, while the second and fourth suggest reasons to oppose it.*

11.23: (a) The factors are storage method (three levels: fresh, room temperature for one month, refrigerated for one month) and preparation method (two levels: cooked immediately or after one hour). There are therefore six treatments. The response variables are the tasters' color and flavor ratings. (b) Randomly allocate n potatoes to each of the six groups, then compare ratings. (c) For each taster, randomly choose the order in which the fries are tasted. We'll note that tasters may become confused with six different batches to taste, however.

	Cooked immediately	Wait one hour
Fresh	1	2
Stored	3	4
Refrigerated	5	6

11.25: (a) This is an observational study; the subjects chose their own "treatments" (how much to drink). The explanatory variable is alcohol consumption, and the response variable is whether or not a subject dies. (b) Many answers are possible. For example, some nondrinkers might avoid drinking because of other health concerns. We do not know what kind of alcohol (beer? wine? whiskey?) the subjects were drinking.

Chapter 12 Introducing Probability

12.1: In a *large* number of five-card poker hands, the fraction in which you will be dealt a straight flush is about 1/64,974. It *does not* mean that exactly 1 out of 64,974 such hands would yield a straight flush.

12.3: (a) There are 21 zeroes among the first 200 digits of the table (rows 101–105), for a proportion of 0.105. (b) Answers will vary, but more than 99% of all students should get between 7 and 33 heads out of 200 flips when $p = 0.1$.

12.5: (a) S = {lives on campus, lives off campus}. (b) S = {All numbers between ________ and ________ years}. (Choices of upper and lower limits will vary, due to characteristics of your institution.) (c) S = {0000, 0001, 0002, . . . , 9999}. (d) S = {A, B, C, D, F} (students might also include W, "+" and "−", depending on your institution).

12.7: Add 1 to each pair-total found in the solution to Exercise 12.6: S = {3, 4, 5, 6, 7, 8, 9}. Each of the 16 possible pairings is equally likely, so (for example) the probability of a total of 5 is 3/16, because 3 pairings add to 4 (and then we add 1). The complete set of probabilities is shown.

TOTAL	PROBABILITY
3	1/16 = 0.0625
4	2/16 = 0.125
5	3/16 = 0.1875
6	4/16 = 0.25
7	3/16 = 0.1875
8	2/16 = 0.125
9	1/16 = 0.0625

12.9: (a) Event B specifically rules out obese subjects, so there is no overlap with event A. (b) A or B is the event "The person chosen is overweight or obese." $P(A \text{ or } B) = P(A) + P(B) = 0.36 + 0.33 = 0.69$. (c) $P(C) = 1 - P(A \text{ or } B) = 1 - 0.69 = 0.31$.

12.11: (a) Disjoint. (b) Not disjoint. $300,000 is more than $100,000 and more than $250,000. (c) Disjoint. $3 + x$ cannot equal 3.

12.13: (a) $A = \{4, 5, 6, 7, 8, 9\}$, $P(A) = 0.097 + 0.079 + 0.067 + 0.058 + 0.051 + 0.046 = 0.398$. (b) $B = \{2, 4, 6, 8\}$, $P(B) = 0.176 + 0.097 + 0.067 + 0.051 = 0.391$. (c) A or $B = \{2, 4, 5, 6, 7, 8, 9\}$, $P(A \text{ or } B) = 0.176 + 0.097 + 0.079 + 0.067 + 0.058 + 0.051 + 0.046 = 0.574$. This is different from $P(A) + P(B)$ because A and B are not disjoint.

12.15: (a) $P(Y \leq 0.6) = 0.6$. (b) $P(Y < 0.6) = 0.6$. (c) $P(0.4 \leq Y \leq 0.8) = 0.4$. (d) $P(0.4 < Y \leq 0.8) = 0.4$. The only difference between parts (c) and (d) is the inclusion of the point $Y = 0.4$. This has 0 probability for a continuous variable.

12.17: (a) $P(X \geq 35)$ (b) $P(X \geq 35) = P(Z > \frac{35 - 25.3}{6.5}) = P(Z \geq 1.49) = 1 - 0.9319 = 0.0681$ (using Table A)

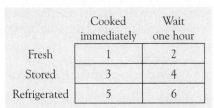

12.19: (a) $Y \geq 8$ means the student runs the mile in 8 minutes or more. $P(Y \geq 8) = P(Z \geq \frac{8-7.11}{0.74}) = 1 - 0.8849 = 0.1151$ (using Table A). (b) "The student could run mile in less than 6 minutes" is the event $Y < 6$. $P(Y < 6) = P(Z < \frac{6-7.11}{0.74}) = 0.0668$ (using Table A).

12.21: (a) If Joe says $P(\text{Syracuse wins}) = 0.1$, then he believes $P(\text{Duke wins}) = 0.3$ and $P(\text{North Carolina wins}) = 0.2$. (b) Joe's probabilities for Duke, Syracuse, and North Carolina add up to 0.6, so that leaves probability 0.4 for all other teams.

12.23: (b) The set $\{0, 1, 2, 3, 4, 5\}$ lists all possible counts.

12.25: (b) The other probabilities add to 0.97.

12.27: (b) $P(\text{not Republican}) = 1 - P(\text{Republican}) = 1 - 0.25 = 0.75$.

12.29: (c) "7 or greater" means 7, 8, or 9—three of the ten possibilities.

12.31: (c) $Y > 1$ standardizes to $Z > 2.56$. Table A gives 0.0052.

12.33: (a) legitimate (even though it is not a "fair" die) (b) legitimate (even if the deck of cards is not!) (c) not legitimate

12.35: We have dropped the trailing zeroes from the land area figures. (a) $P(\text{area is forested}) = 4176/9094 = 0.4592$ (b) $P(\text{area is not forested}) = 1 - 0.4592 = 0.5408$

12.37: (a) The given probabilities add to 0.97, so other colors must account for the remaining 0.03. (b) $P(\text{white or silver}) = 0.25 + 0.18 = 0.43$, $P(\text{neither white nor silver}) = 1 - 0.43 = 0.57$

12.39: The probabilities of 2, 3, 4, and 5 are unchanged (1/6), so $P(1 \text{ or } 6)$ must still be 1/3. If $P(6) = 0.2$, then $P(1) = 1/3 - 0.2 = 2/15$.

Face	⚀	⚁	⚂	⚃	⚄	⚅
Probability	0.13	1/6	1/6	1/6	1/6	0.2

12.41: (a) It is legitimate because every person must fall into exactly one category, the probabilities are all between 0 and 1, and they add up to 1. (b) $0.169 = 0.002 + 0.008 + 0.153 + 0.006$ (c) $0.355 = 1 - 0.645$

12.43: (a) A corresponds to the outcomes in the first column or the third row. (b) $P(A) = 0.550$. This is different from the sum of the probabilities in parts (c) and (d) of Exercise 12.42 because that sum counts the overlap (0.168) twice.

12.45: (a) X is discrete, because it has a finite sample space. (b) "At least one nonword error" is the event $\{X \geq 1\}$ or $\{X > 0\}$. $P(X \geq 1) = 1 - P(X = 0) = 0.9$. (c) $\{X \leq 2\}$ is "no more than two nonword errors," or "fewer than three nonword errors." $P(X \leq 2) = P(X = 0) + P(X = 1) + P(X = 2) = 0.1 + 0.2 + 0.3 = 0.6$. $P(X < 2) = P(X = 0) + P(X = 1) = 0.1 + 0.2 = 0.3$.

12.47: (a) Just using initials: {(A, D), (A, M), (A, S), (A, R), (D, M), (D, S), (D, R), (M, S), (M, R), (S, R)}. (b) Each has probability 1/10 = 10%. (c) Mei-Ling is chosen in 4 of the 10 possible outcomes: 4/10 = 40%. (d) There are 3 pairs with neither Sam nor Roberto, $P = 3/10$.

12.49: The possible values of Y are 1, 2, 3, . . . , 12, each with probability 1/12. Aside from drawing a diagram showing all the possible combinations, one can reason that the first (regular) die is equally likely to show any number from 1 through 6. Half of the time, the second roll shows 0, and the other half it shows 6. Each possible outcome therefore has probability $(1/6)(1/2) = 1/12$.

12.51: (a) This is a continuous random variable because the set of possible values is an interval. (b) The height should be 1/2 because the area under the curve must be 1. (For a rectangle, area = L × W.) The density curve is illustrated. (c) $P(Y \leq 1) = 1/2$

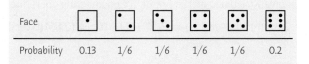

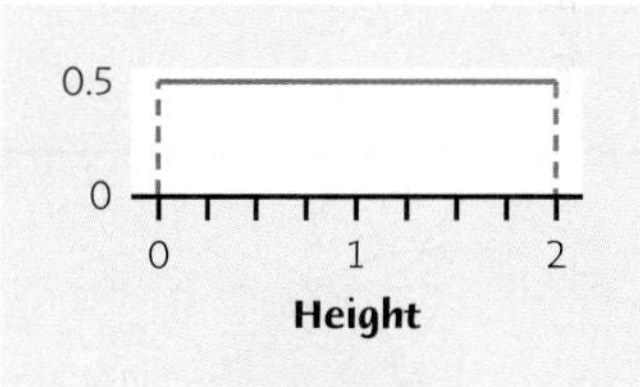

12.53: (a) $P(0.49 \leq V \leq 0.53) = P(\frac{0.49-0.51}{0.009} \leq Z \leq \frac{0.53-0.51}{0.009}) = P(-2.22 \leq Z \leq 2.22) = 0.9868 - 0.0132 = 0.9736$ (b) $P(V \leq 0.49) = P(Z \leq \frac{0.49-0.51}{0.009}) = P(Z \leq 2.22) = 0.0132$.

12.55: (a) Because there are 10,000 equally likely four-digit numbers (0000 through 9999), the probability of an exact match is 1/10,000 = 0.0001. (b) There are a total of $24 = 4 \times 3 \times 2 \times 1$ arrangements of the four digits 5, 9, 7, and 4 (there are four choices for the first digit, three for the second, two for the third), so the probability of a match in any order is 24/10,000 = 0.0024.

12.57: (a)–(c) Results will vary, but after n tosses, the distribution of the proportion (call it $\hat{p}$) is approximately Normal with mean 0.5 and standard deviation $1/(2\sqrt{n})$, while the distribution of the count of heads is approximately Normal with mean $0.5n$ and standard deviation $\sqrt{n}/2$, so using the 68–95–99.7 rule, we have the results shown in the table. Note that the range for the proportion $\hat{p}$ gets narrower, while the range for the count gets wider.

n	99.7% RANGE FOR $\hat{p}$	99.7% RANGE FOR COUNT
40	0.5 ± 0.237	20 ± 9.5
120	0.5 ± 0.137	60 ± 16.4
240	0.5 ± 0.097	120 ± 23.2
480	0.5 ± 0.068	240 ± 32.9

12.59: (a) With $n = 50$, the variability in the proportion (call it $\hat{p}$) is larger. With $n = 100$, nearly all answers will be between 0.19 and 0.47. With $n = 400$, nearly all answers will be between 0.26 and 0.40. (b) Results will vary.

Chapter 13 General Rules of Probability

13.1: It is unlikely that these events are independent. In particular, it is reasonable to expect that younger adults are more likely than older adults to be college students.

13.3: If we assume that each site is independent of the others, P(all seven are still good) $= (1 - 0.13)^7 = 0.3773$.

13.5: A Venn diagram is provided. (a) P(degree was earned by a man) $= P(\text{not } W) = 1 - 0.59 = 0.41$, or 41%. (b) $P(B \text{ and not } W) = 0.50 - 0.29 = 0.21$, or 21%. (c) Because $P(B \text{ and } M) = 0.21$, but $P(B) \times P(M) = (0.50)(0.41) = 0.205$, $P(B \text{ and } M) \neq P(B) \times P(M)$. So, B and M are not independent.

Note: *While the multiplication here results in a "close" result to P(B and M), in the probabilistic sense, these must be* ***exactly*** *equal to have independence.*

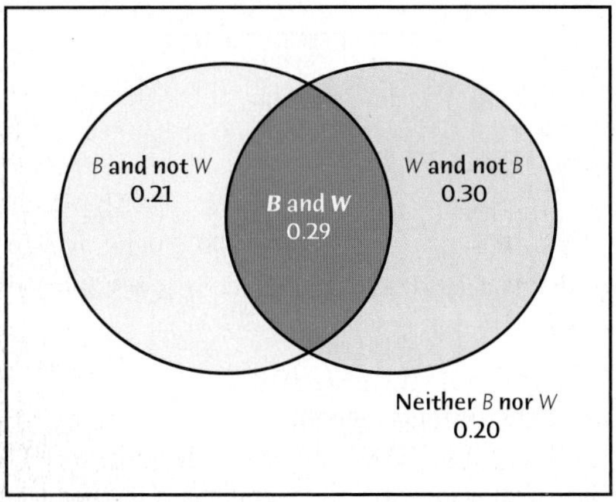

13.7: $P(W \mid B) = \frac{P(B \text{ and } W)}{P(B)} = \frac{0.29}{0.50} = 0.58$

13.9: Let H be the event that an adult belongs to a club, and T be that he/she goes at least twice a week. Note that $P(T \text{ and } H) = P(T)$, because one has to be a member to attend. $P(T) = P(H)P(T \mid H) = (0.10)(0.40) = 0.04$.

13.11: (a) and (b) These probabilities are provided following. (c) The product of these conditional probabilities gives the probability of a flush in spades. The product is about 0.0004952. (d) There are four possible suits in which to have a flush, so P(flush) is four times that found in part (c), or about 0.001981.

P(1st card is ♠)	$\frac{13}{52} = 0.25$
P(2nd card is ♠ | first card is ♠)	$\frac{12}{51} = 0.2353$
P(3rd card is ♠ | first two are ♠)	$\frac{11}{50} = 0.22$
P(4th card is ♠ | first three are ♠)	$\frac{10}{49} = 0.2041$
P(5th card is ♠ | first four are ♠)	$\frac{9}{48} = 0.1875$

13.13: (a) The tree diagram is shown. (b) Two branches result in a positive test result. P(Positive test) $= (0.063)(0.21) + (0.937)(0.06) = 0.01323 + 0.05622 = 0.06945$.

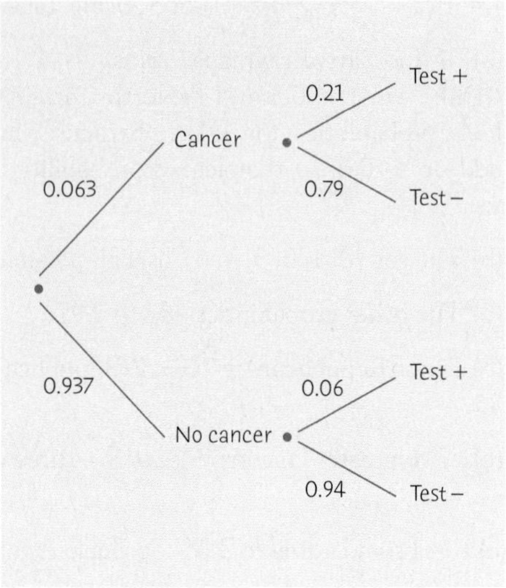

13.15: (a) $P(\text{no cancer} \mid \text{positive test}) = \frac{P(\text{no cancer and positive test})}{P(\text{positive test})} = \frac{(0.937)(0.06)}{0.06945} = \frac{0.05622}{0.06945} = 0.8095$. The 0.06945 in the denominator was found in Exercise 13.13. (b) About 81% of positive test results come from people without prostate cancer. Most of those being treated do not have the disease, and might suffer those serious side effects.

13.17: (b) $P = (1 - 0.02)^3 = (0.98)^3 = 0.9412$

13.19: (a) Because the tests are independent, P(at least one positive) $= 1 - P$(both negative) $= 1 - (0.1)(0.2) = 0.98$.

13.21: (c) Of 4798 education doctorates, 1501 were awarded to males. $1501/4798 = 0.3128$.

13.23: (c) Outcome A (engineering doctorate) is what has been given (signaled by the phrase "of engineering doctorates").

13.25: (c) $P(W \text{ and } D) = P(W)P(D \mid W) = (0.86)(0.028) = 0.024$

13.27: P(8 losses) $= (1 - 0.25)^8 = 0.1001$

13.29: (a) P(win the jackpot) $= (\frac{1}{20})(\frac{9}{20})(\frac{1}{20}) = 0.001125$. (b) The other (non-cherry) symbol can show up on the middle wheel, with probability $(\frac{1}{20})(\frac{11}{20})(\frac{1}{20}) = 0.001375$, or on either of the outside wheels, with probability $= (\frac{19}{20})(\frac{9}{20})(\frac{1}{20})$ (each). (c) Combining all three cases from part (b), we have P(exactly two cherries) $= 0.001375 + 2(0.021375) = 0.044125$.

13.31: PLAN: Let I be the event "infection occurs" and let F be "the repair fails." We have been given $P(I) = 0.03$, $P(F) = 0.14$, and $P(I \text{ and } F) = 0.01$. We want $P(\text{not } I \text{ and not } F)$. SOLVE: First, $P(I \text{ or } F) = P(I) + P(F) - P(I \text{ and } F) = 0.03 + 0.14 - 0.01 = 0.16$. Observe that the desired probability is the complement of "I or F." $P(\text{not } I \text{ and not } F) = 1 - P(I \text{ or } F) = 0.84$. CONCLUDE: 84% of operations succeed and are free from infection.

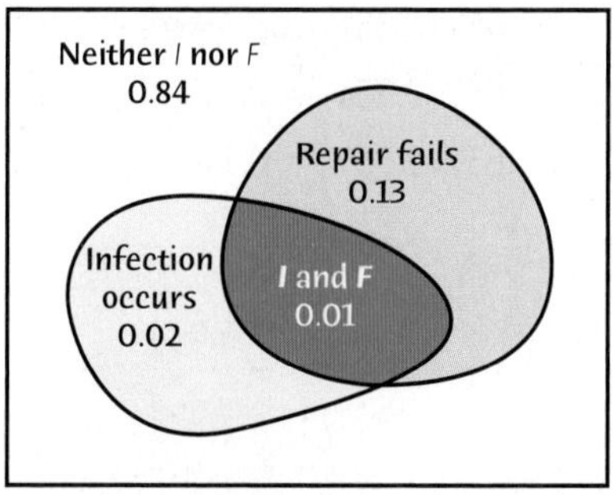

13.33: Let I be the event "infection occurs" and let F be "the repair fails." Refer to the Venn diagram in Exercise 13.31. $P(I \mid \text{not } F) = \frac{P(I \text{ and not} F)}{P(\text{not} F)} = \frac{0.02}{0.86} = 0.0233$

13.35: (a) $P(\text{positive} \mid \text{disease}) = 564/574 = 0.9826$ (b) $P(\text{no disease} \mid \text{negative}) = 708/718 = 0.9861$

13.37: (a) These events are not independent; P(pizza with mushrooms) = 4/7, but P (mushrooms | thick crust) = 2/3. (b) With the eighth pizza, P (mushrooms) = 4/8 = 1/2, and P (mushrooms | thick crust) = 2/4 = 1/2, so these events are independent.

13.39: Let W be the event "the person is a woman" and M be "the person earned a Master's degree." (a) $P(\text{not } W) = 1569/3825 = 0.4102$ (b) $P(\text{not } W \mid M) = 322/816 = 0.3946$ (c) The events "choose a man" and "choose a Master's degree recipient" are not independent. If they were, the probabilities in parts (a) and (b) would be equal.

13.41: Let D be the event "a seedling was damaged by a deer." (a) $P(D) = 209/871 = 0.2400$ (b) The conditional probabilities are

$P(D \mid \text{no cover}) = 60/211 = 0.2844$
$P(D \mid \text{cover} < 1/3) = 76/234 = 0.3248$
$P(D \mid 1/3 \text{ to } 2/3 \text{ cover}) = 44/221 = 0.1991$
$P(D \mid \text{cover} > 2/3) = 29/205 = 0.1415$

(c) Cover and damage are not independent; $P(D)$ decreases noticeably when thorny cover is 1/3 or more.

13.43: Having no thorny cover means there is less than 1/3 thorny cover. This conditional probability is $P(\text{cover} < 1/3 \mid D) = (60 + 76)/(60 + 76 + 44 + 29) = 136/209 = 0.6507$, or 65.07%.

13.45: This is $P(A \text{ and } (\text{not } B) \text{ and } (\text{not } C)) = 0.35$.

13.47: (a) P(doubles on first toss) = 1/6 (b) We need no doubles on the first roll, then doubles on the second toss. P(first doubles appears on toss 2) = (5/6)(1/6) = 5/36 (c) Similarly, P(first doubles appears on toss 3) = $(5/6)^2(1/6) = 25/216$ (d) P(first doubles appears on toss 4) = $(5/6)^3(1/6)$, etc. In general, P(first doubles appears on toss k) = $(5/6)k^{-1}(1/6)$. (e) P(go again within 3 turns) = P(roll doubles in 3 or fewer rolls) = $(1/6) + (5/6)(1/6) + (5/6)^2(1/6) = 0.4213$.

13.49: PLAN: We construct a tree diagram showing the results (allergic or not) for each of the three individuals. SOLVE: In the tree diagram, each "up-step" represents an allergic individual. At the end of each of the eight complete branches is the value of X. Any branch with two up-steps and one down-step has probability $0.02^2 \times 0.98^1 = 0.000392$, and yields $X = 2$. Any branch with one up-step and two down-steps has probability $0.02^1 \times 0.98^2 = 0.019208$, and yields $X = 1$.

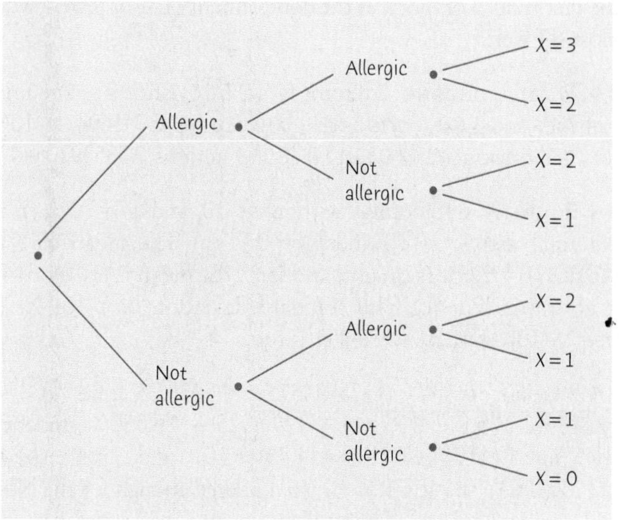

There are three branches each corresponding to $X = 2$ and $X = 1$, and only one branch each for $X = 3$ and $X = 0$. $P(X = 0) = 0.98^3 = 0.941192$ and $P(X = 3) = 0.02^3 = 0.000008$. Meanwhile, $P(X = 1) = 3(0.02)^1(0.98)^2 = 0.057624$, and $P(X = 2) = 3(0.02)^2(0.98)^1 = 0.001176$. CONCLUDE: $P(X = 0) = 0.941192$, $P(X = 1) = 0.057624$, $P(X = 2) = 0.001176$, and $P(X = 3) = 0.000008$.

13.51: $P(X = 2 \mid X \geq 1) = \frac{P(X = 2 \text{ and } X \geq 1)}{P(X \geq 1)} = \frac{P(X = 2)}{P(X \geq 1)} = \frac{0.001176}{1 - 0.941192} = 0.020$.

13.53: Let R = {recent donor}, P = {pledged}, and C = {contributed}. (a) $P(C) = (0.5)(0.4)(0.8) + (0.3)(0.3)(0.6) + (0.2)(0.1)(0.5) = 0.224$, or 22.4% (b) $P(R \mid C) = \frac{P(R \text{ and } C)}{P(C)} = \frac{(0.5)(0.4)(0.8)}{0.224} = 0.7143$, or 71.4%

13.55: The proportion having combination (16, 17) is $2(0.232)(0.213) = 0.098832$.

13.57: If the DNA profile found on the hair is possessed by 1 in 1.6 million individuals, then we would expect about 3 individuals in the database of 4.5 million convicted felons to demonstrate a match. (4.5 million)/(1.6 million) = 2.8125, which was rounded up to 3.

Chapter 14 Binomial Distributions

14.1: Binomial. (1) We have a fixed number of observations ($n = 20$). (2) It is reasonable to believe that each call is independent of the others. (3) "Success" means reaching a working residential number; "failure" is any other outcome. (4) Each randomly dialed number has chance $p = 0.3$ of reaching a live person.

14.3: Not binomial. The trials aren't independent. If one tile in a box is cracked, there are likely more tiles cracked.

14.5: (a) C, the number caught, is binomial with $n = 10$ and $p = 0.7$. M, the number missed, is binomial with $n = 10$ and $p = 0.3$. (b) $P(M = 3) = \binom{10}{3}(0.3)^3(0.7)^7 = (120)(0.027)(0.08235) = 0.2668$. With software, $P(M \geq 3) = 0.6172$. Use the fact that "3 or more" is the complement of "2 or fewer" with most software.

14.7: (a) `5 choose 2` returns 10. (b) `500 choose 2` returns 124,750, and `500 choose 100` returns 2.041694×10^{107} (c) `(10 choose 1)*0.11*0.89^9` returns 0.38539204407.

14.9: (a) X is binomial with $n = 10$ and $p = 0.3$; Y is binomial with $n = 10$ and $p = 0.7$. (b) The mean of Y is $(10)(0.7) = 7$ errors caught, and for X the mean is $(10)(0.3) = 3$ errors missed. (c) The standard deviation of Y (or X) is $\sigma = \sqrt{10(0.7)(0.3)} = 1.4491$ errors.

14.11: (a) $\mu = (1175)(0.37) = 434.75$ and $\sigma = \sqrt{1175(0.37)(1 - 0.37)} = \sqrt{273.8925} = 16.550$ students. (b) $np = (1175)(0.37) = 434.75 \geq 10$ and $n(1 - p) = (1175)(0.63) = 740.25 \geq 10$, so n is large enough for the Normal approximation. The college wants 450 students, so $P(X \geq 451) = P(Z \geq \frac{451 - 434.75}{16.550}) = P(Z \geq 0.98) = 0.1635$. (c) The exact binomial probability is 0.1706; the Normal approximation is 0.0071 too low. (d) To decrease the chance of more students than they want, they need to decrease the number admitted. If $n = 1145$, we have $P(\text{more than } 450) = 0.047$; with $n = 1146$, we have $P = 0.0494$.

14.13: (b) He has three independent eggs, each with probability 1/4 of containing salmonella.

14.15: (c) The selections are not independent; once we choose one student, it changes the probability that the next student is a business major.

14.17: (b) $(0.25)(0.75)^3(0.25) = 0.0264$

14.19: (b) 8 or 9 are two of the ten possible digits, so 0.20.

14.21: (a) $\mu = np = (80)(0.20) = 16$

14.23: (a) A binomial distribution is *not* appropriate, because the kicker faces different situations each time (wind, distance, etc.); his probability of success is likely to change from one attempt to another. (b) It would be reasonable to use a binomial distribution for free throws made because each is from the same position with respect to the basket with no interference allowed for the shot, and presumably each is independent of any others.

14.25: (a) $n = 5$ and $p = 0.65$. (b) $S = \{0, 1, 2, 3, 4, 5\}$ (c) All cases computed.

$P(X = 0) = \binom{5}{0}(0.65)^0(0.35)^5 = 0.00525$

$P(X = 1) = \binom{5}{1}(0.65)^1(0.35)^4 = 0.04877$

$P(X = 2) = \binom{5}{2}(0.65)^2(0.35)^3 = 0.18115$

$P(X = 3) = \binom{5}{3}(0.65)^3(0.35)^2 = 0.33642$

$P(X = 4) = \binom{5}{4}(0.65)^4(0.35)^1 = 0.31239$

$P(X = 5) = \binom{5}{5}(0.65)^5(0.35)^0 = 0.11603$

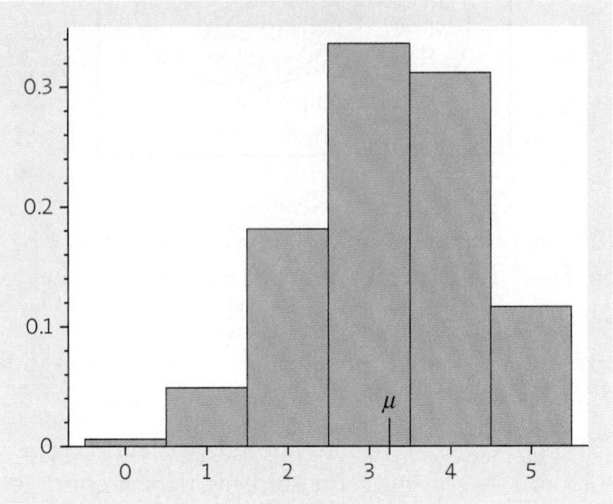

(d) $\mu = np = (5)(0.65) = 3.25$ and $\sigma = \sqrt{5(0.65)(1 - 0.65)} = 1.0665$ years. The mean μ is indicated on the probability histogram.

14.27: (a) One woman getting pregnant will not affect whether another gets pregnant (independent trials), and (we assume) each has the same probability of getting pregnant. (b) Under ideal conditions, the number who get pregnant is binomial with $n = 20$ and $p = 0.01$; $P(N \geq 1) = 1 - P(N = 0) = 1 - 0.8179 = 0.1821$. In typical use, $p = 0.03$, and $P(N \geq 1) = 1 - 0.5438 = 0.4562$.

14.29: (a) X is binomial with $n = 600$ and $p = 0.03$. The Normal approximation is safe: $np = 18$ and $n(1 - p) = 582$ are both at least 10. The mean is 18, and $\sigma = 4.1785$, so $P(X \geq 20) = P(Z \geq \frac{20 - 18}{4.1785}) = P(Z \geq 0.48) = 1 - 0.6844 = 0.3156$. The exact binomial probability is 0.3477. (b) Under ideal conditions, $p = 0.01$, so $np = 6$ is too small.

14.31: (a) If R is the number of red-blossomed plants, then $P(R = 3) = \binom{4}{3}(0.75)^3(0.25)^1 = 0.4219$. (b) With $n = 60$, $\mu = np = (60)(0.75) = 45$. (c) $P(R \geq 45) = P(Z \geq 0) = 0.5000$ (software gives 0.5688 as the exact binomial probability).

14.33: (a) Of 720,783 total vehicles, Elantras accounted for a proportion of 247,912/720,783 = 0.3439. (b) If E is the number of Elantra buyers, E has the binomial distribution with $n = 1000$, and $p = 0.3439$. $\mu = np = (1000)(0.3439) = 343.9$ and $\sigma = \sqrt{np(1 - p)} = \sqrt{1000(0.3439)(1 - 0.3439)} = 15.021$ Elantra buyers. (c) $P(E < 400) = P(E \leq 399) = P(Z \leq -3.67) = 0.0001$

14.35: (a) With $n = 100$, $\mu = 75$ and $\sigma = 4.3301$ questions, so $P(70 \leq X \leq 80) = P(-1.15 \leq Z \leq 1.15) = 0.7498$ (software gives 0.7518). (b) With $n = 250$, $\mu = 187.5$ and $\sigma = 6.8465$ questions, and a score between 70% and 80%

means 175 to 200 correct answers, so $P(175 \le X \le 200) = P(-1.83 \le Z \le 1.83) = 0.9328$ (software gives 0.9428). If one used the more mathematical idea that $P(70 < X < 80) = P(71 \le X \le 79) = 0.6424$ for the 100 question test and $P(176 < X < 199) = 0.9328$ for the 250 question test.

14.37: (a) Answers will vary, but over 99.8% of samples should have 0 to 4 bad tomatoes. (b) Each time we choose a sample of size 10, the probability that we have exactly 1 bad tomato is 0.3854; out of 20 samples, the number of times that we have exactly 1 bad tomato has a binomial distribution with parameters $n = 20$ and $p = 0.3854$. Most students will find that between 2 and 14 of their 20 samples have exactly 1 bad tomato.

14.39: The number N of infections among untreated BJU students is binomial with $n = 1400$ and $p = 0.80$, so $\mu = 1120$ and $\sigma = 14.9666$ students. 75% of that group is 1050; the Normal approximation is safe because $(1400)(0.80) = 1120$ and $(1400)(0.20) = 280$ are both at least 10. $P(N \ge 1050) = P(Z \ge \frac{1050 - 1120}{14.9666}) = P(Z \ge -4.68)$, which is very near to 1. (Exact binomial computation gives 0.999998.)

14.41: Let V and U be (respectively) the number of new infections among the vaccinated and unvaccinated children. (a) $P(V = 1) = 0.3741$ and $P(U = 1) = 0.0960$. Because these events are (assumed) independent, $P(V = 1 \text{ and } U = 1) = (0.3741)(0.0960) = 0.0359$. (b) $P(2 \text{ infections}) = P(V = 0 \text{ and } U = 2) + P(V = 1 \text{ and } U = 1) + P(V = 2 \text{ and } U = 0) = P(V = 0)P(U = 2) + P(V = 1)P(U = 1) + P(V = 2)P(U = 0) = (0.4181)(0.3840) + (0.3741)(0.0960) + (0.1575)(0.0080) = 0.1977$

14.43: The number X of fairways hit is binomial with $n = 30$ and $p = 0.63$. (a) $np = 18.9$ and $n(1 - p) = 11.1$, so the Normal approximation is safe. (b) $\mu = np = 18.9$, and $\sigma = \sqrt{30(0.63)(0.37)} = 2.6444$. Using the Normal approximation, $P(X \ge 23) = P(Z \ge 1.55) = 0.0606$. (c) With the continuity correction, $P(X \ge 23) = P(X \ge 22.5) = P(Z \ge 1.36) = 0.0869$ (using Table A). The answer using the continuity correction is closer to the exact answer (0.0838).

Chapter 15 Sampling Distributions

15.1: Both are statistics; they came from the 11 subjects in the experiment.

15.3: 98.8% and 22 are statistics; they are based on the survey of 500 American anabolic androgenic steroid users. 50.0% is a parameter.

15.5: Although the probability of having to pay for a total loss for is very small, if this were to happen, it would be financially disastrous. For thousands of policies, the law of large numbers says that the average claim on many policies will be close to the mean, so the insurance company can be assured that the premiums collected will (almost certainly) cover the claims.

15.7: (a) Histogram is shown. (b) The mean is $\mu = 69.4$ (c) and (d) Results will vary. The results of one sample selected students 5, 2, 9, and 4 with scores of 72, 63, 75, and 55. Their mean is $\bar{x} = (72 + 63 + 75 + 55)/4 = 66.25$. Students should repeat this process until they have 10 sample means.

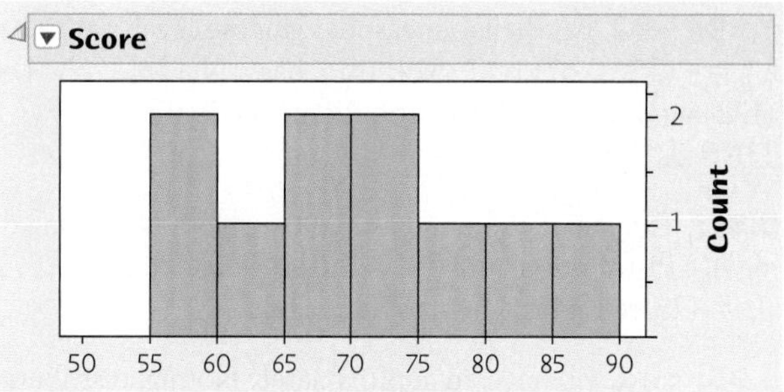

15.9: (a) The sampling distribution of $\bar{x}$ is $N(115, 25/\sqrt{100}) = N(115 \text{ mg/dL}, 2.5 \text{ mg/dL})$. $P(112 < \bar{x} < 118) = P(-1.2 < Z < 1.2) = 0.8849 - 0.1151 = 0.7698$, using Table A. (b) With $n = 1000$, the sample mean has the $N(115 \text{ mg/dL}, 0.7906 \text{ mg/dL})$ distribution, so $P(112 < x < 118) = P(-3.79 < Z < 3.79) = 0.9998$.

15.11: No: the histogram of the sample values will look like the *population* distribution, whatever it might happen to be. The central limit theorem says that the histogram of *sample means* (from many large samples) will look more and more Normal.

15.13: PLAN: Use the central limit theorem to approximate the probability that the mean loss for 10,000 insurance policies is no more than \$135. SOLVE: In spite of the skewness of the population distribution, the average loss among 10,000 policies will be approximately $N(\$125, \$300/\sqrt{10{,}000}) = N(\$125, \$3)$. $P(\bar{x} \le \$135) = P(Z \le \frac{135 - 125}{3}) = P(Z \le 3.33) = 0.9996$. CONCLUDE: We can be about 99.96% certain that average losses will not exceed \$135 per policy.

15.15: Answers will vary due to randomness. Students should find that a sample mean of 76 will almost never occur; this indicates that the mean for the four honors students is unusually high.

15.17: (b) statistic; this is a proportion of the people interviewed in the sample of 60,000 households

15.19: (b) The law of large numbers says that the mean from a large sample is close to the population mean. Statement (c) is also true, but is based on the central limit theorem, not on the law of large numbers.

15.21: (c) The standard deviation of the distribution of $\bar{x}$ is $\sigma/\sqrt{n}$.

15.23: (c) The central limit theorem says that the mean from a large sample has (approximately) a Normal distribution. Statement (a) is also true, but is based on the law of large numbers, not on the central limit theorem.

15.25: 1 is a parameter; 1.07 is a statistic (the mean of the 10 measurements in the sample).

15.27: In the long run, the gambler earns an average of 94.7 cents per bet. In other words, the gambler loses (and the house gains) an average of 5.3 cents for each $1 bet.

15.29: Let X be Shelia's measured glucose level. (a) $P(X > 130) = P(Z > 0.67) = 0.2514$. (b) $\bar{x}$ has a $N(122, 12/\sqrt{4}) = N(122 \text{ mg/dL}, 6 \text{ mg/dL})$ distribution; $P(\bar{x} > 130) = P(Z > 1.33) = 0.0918$.

15.31: The mean of four measurements has a $N(122 \text{ mg/dL}, 6 \text{ mg/dL})$ distribution, and $P(Z > 1.645) = 0.05$ if Z is $N(0, 1)$, so $L = 122 + 1.645 \times 6 = 131.87$ mg/dL.

15.33: (a) $\bar{x}$ will have an approximately Normal distribution with mean 8.8 beats per five seconds, and standard deviation $1/\sqrt{24} = 0.288675$ beats per five seconds. (b) $P(\bar{x} < 8) = P(Z < -3.92) =$ essentially 0. (c) The average over 12 five-second intervals needs to be less than $100/12 = 8.333$ beats per five seconds. $P(\bar{x} < 8.333) = P(Z < -1.62) = 0.0526$.

15.35: PLAN: Use the central limit theorem to approximate this probability. SOLVE: Over 40 years, $\bar{x}$ (the mean return) is approximately Normal with $\mu = 13.3\%$ and $\sigma = 17.0\%/\sqrt{40} = 2.688\%$. $P(\bar{x} > 10\%) = P(Z > -1.23) = 0.8907$, and $P(\bar{x} < 5\%) = P(Z < -3.09) = 0.0010$. CONCLUDE: There is about an 89% chance of getting average returns over 10%, and a 0.1% chance of getting average returns less than 5%.

15.37: Choose n so that $6.5/\sqrt{n} = 1$. That means $\sqrt{n} = 6.5$, so $n = 42.25$. Because n must be a whole number, take $n = 43$.

15.39: On the average, Joe loses 40 cents each time he plays (that is, he spends $1 and gets back 60 cents).

15.41: (a) With $n = 150{,}000$, $\mu_{\bar{x}} = \$0.40$ and $\sigma_{\bar{x}} = \$18.96/\sqrt{150{,}000} = \$0.0490$. (b) $P(\$0.30 < \bar{x} < \$0.50) = P(-2.04 < Z < 2.04) = 0.9586$

15.43: The mean is $10.5 = (3)(3.5)$ because a single die has a mean of 3.5. Sketches will vary, as will the number of rolls.

Chapter 16 Confidence Intervals: The Basics

16.1: (a) The sampling distribution of $\bar{x}$ has mean μ and standard deviation $\frac{\sigma}{\sqrt{n}} = \frac{125}{\sqrt{170{,}100}} = 0.3031$. (b) 95% of all values of $\bar{x}$ fall within two standard deviations of the sampling distribution of μ—that is, within $2(0.3031) = 0.6062$. (c) 285 ± 0.6062, or between 284.3938 and 285.6062.

16.3: Answers will vary due to randomness. In 99.7% of all repetitions of part (a), students should see between 5 and 10 hits. Out of 1000 80% confidence intervals, nearly all students will observe between 76% and 84% capturing the mean.

16.5: Search Table A for 0.1250. This area corresponds to $-z^* = -1.15$, or $z^* = 1.15$.

16.7: (a) A stemplot shows two low scores (72 and 74) that are both possible outliers, but there are no other apparent deviations from Normality. (b) PLAN: We estimate μ by giving a 99% confidence interval. SOLVE: The problem states that these girls are an SRS of the population, which is very large. We saw that the scores are consistent with having come from a Normal population, so conditions are met. With $\bar{x} = 105.84$, and $z^* = 2.576$, our 99% confidence interval for μ is given by $105.84 \pm 2.576\frac{15}{\sqrt{31}} = 105.84 \pm 6.94$. CONCLUDE: We are 99% confident that the mean IQ of seventh-grade girls in this district is between 98.90 and 112.78 points.

16.9: With $z^* = 1.96$ and $\sigma = 7.5$, m.e. $= z^*\frac{\sigma}{\sqrt{n}} = \frac{14.7}{\sqrt{n}}$. (a) and (b) The margins of error are given in the table. (c) Margin of error decreases as n increases.

n	m.e.
100	1.47
400	0.735
1600	0.3675

16.11: (c) $z = 3.291$. Using Table A, search for 0.9995.

16.13: (b) As the confidence level increases, z^* increases.

16.15: (b) $\sigma(\bar{x})$ is $\frac{\sigma}{\sqrt{n}} = \frac{125}{\sqrt{900}} = 4.167$.

16.17: (b) As the confidence level increases, z^* increases. This makes the margin of error larger.

16.19: (a) We use $\bar{x} \pm z^*\frac{\sigma}{\sqrt{n}} = 15.3 \pm 2.576\frac{8.5}{\sqrt{463}} = 15.3 \pm 1.018 = 14.282$ to 16.318 hours. (b) The 463 students in this class must be a random sample of all of the first-year students at this university to satisfy conditions for inference.

16.21: The margin of error is $2.576\frac{8.5}{\sqrt{464}} = 1.02$, so the extra observation has minimal impact on the margin of error. If $\bar{x} = 36.8$, then the 99% confidence interval for average amount of time spent studying becomes $36.8 \pm 1.02 = 35.78$ to 37.82 hours.

16.23: This student is confused. If we repeated the sample over and over, 95% of all future sample means would be within 1.96 standard deviations (that is, within $1.96\frac{\sigma}{\sqrt{n}}$) of the true, unknown value of μ. Future samples will have no memory of this sample.

16.25: (a) A stemplot of the data is provided. Notice that the distribution is noticeably skewed to the left. The data do not appear to follow a Normal distribution. (b) PLAN: We will estimate μ by giving a 95% confidence interval. SOLVE: The problem states that we are willing to take this sample to be an SRS of the population. In spite of the shape of the stemplot, we are told to assume that this distribution is Normal with standard deviation $\sigma = 3000$ lb. $\bar{x} = 30{,}841$ lb, so the 95% confidence interval for μ is given by 30,841 ± $1.96\frac{3000}{\sqrt{20}}$. CONCLUDE: With 95% confidence, the mean load μ required to break apart pieces of Douglas fir is between 29,526.2 and 32,155.8 pounds; however, given the shape of the distribution of the data, we cannot rely much on this interval.

Stem and Leaf

Stem	Leaf	Count
33	0 2 3 7	4
32	0 3 3 6 7 7	6
31	3 9 9	3
30	2 5 9	3
29		
28	7	1
27		
26	5	1
25		
24	1	1
23	0	1

23 | 0 represents 23000

16.27: (a) A stemplot is given. There is little evidence that the sample does not come from a Normal distribution. For inference, we must assume that the 10 untrained students were selected randomly from the population of all untrained people. (b) PLAN: We will estimate μ with a 95% confidence interval. SOLVE: We have assumed that we have a random sample, and that the population we're sampling from is Normal. We obtain $\bar{x} = 29.4\ \mu g/L$. Our 95% confidence interval for μ is given by $29.4 \pm 1.96\frac{7}{\sqrt{10}} = 25.06$ to $33.74\ \mu g/L$.

Stem-and-Leaf DMS N = 10
Leaf Unit = 1

1	9
2	2 3
2	9
3	0 0 1 3
3	5
4	2

CONCLUDE: With 95% confidence, the mean sensitivity for all untrained people is between 25.06 and 33.74 μg/L.

Chapter 17 Tests of Significance: The Basics

17.1: (a) If $\mu = 50$, the sampling distribution is approximately Normal with $\mu = 50$ and standard deviation $\frac{\sigma}{\sqrt{n}} = \frac{15}{\sqrt{225}} = 1$. (b) The actual result lies toward the low tail of the curve, while 48.9 is fairly close to the middle. If $\mu = 50$, observing a value similar to 48.9 would not be too surprising, but 47.4 is much less likely, and therefore provides some evidence that $\mu < 50$.

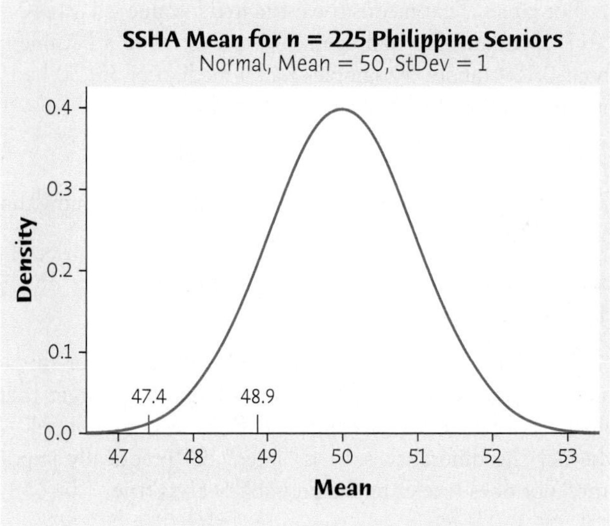

17.3: H_0: $\mu = 50$ vs. H_a: $\mu < 50$. The teacher suspects that poor attitudes are in part responsible for the decline in scores.

17.5: H_0: $\mu = 75$ vs. H_a: $\mu < 75$. The professor suspects this section's students perform worse than the population of all students in the class on average.

17.7: Hypotheses are statements about parameters, not statistics. The research question is not about the sample mean ($\bar{x}$), but should be about the population mean, μ.

17.9: (a) With $\sigma = 60$ and $n = 18$, the standard deviation is $\frac{\sigma}{\sqrt{n}} = \frac{60}{\sqrt{18}} = 14.1421$, so the distribution of $\bar{x}$ is $N(0, 14.1421)$. (b) $P = 2P(\bar{x} \geq 17) = 2P(Z \geq \frac{17-0}{14.1421}) = 0.2302$

17.11: (a) The P-value for $\bar{x} = 48.7$ is 0.1357. This is not significant at either $\alpha = 0.05$ or $\alpha = 0.01$. (b) The P-value for $\bar{x} = 47.4$ is 0.0047. This is significant at both $\alpha = 0.05$ and $\alpha = 0.01$. (c) If $\mu = 50$ (if H_0 were true), observing a value similar to 48.7 would not be too surprising, but 47.4 is much less likely, and it therefore provides strong evidence that $\mu < 50$.

17.13: (a) $z = \frac{0.3-0}{1/\sqrt{10}} = \frac{0.3-0}{0.3162} = 0.9488$ (b) $z = \frac{1.02-0}{1/\sqrt{10}} = \frac{1.02-0}{0.3162} = 3.226$ (c) $z = \frac{17-0}{60/\sqrt{18}} = \frac{17-0}{14.1421} = 1.2021$. Note that in part (c) the test is two-sided, while in parts (a) and (b), it is one-sided.

17.15: PLAN: Let μ be the average percentage tip for all customers receiving bad news. We test H_0: $\mu = 20$ against H_a: $\mu < 20$. SOLVE: We have a random sample of $n = 20$ customers and were told to assume tips have a Normal distribution. $\bar{x} = 18.19\%$. $\sigma(\bar{x}) = \frac{2}{\sqrt{20}} = 0.4472$, so $z = \frac{18.19-20}{0.4472} = -4.05$. The P-value is $P(Z \leq -4.05) \approx 0$. CONCLUDE: There is overwhelming evidence that the average tip percentage when bad news is delivered is lower than the average tip percentage overall (20%).

17.17: $z = 1.876$ is not significant at the $\alpha = 0.05$ level because z is not larger than 1.96 or less than −1.96. It is also not significant at $\alpha = 0.01$ because $|z| < 2.576$.

17.19: It appears that six samples had a median of 33, three had a median of 34, and we were told that eight had a median of 35 or larger. That means the estimated P-value is $17/1000 = 0.017$. If the median had been 30, we'd estimate a P-value of about 0.109 (about 50 samples had a median of 30, 30 had a median of 31, and 12 had a median of 32).

17.21: (a) This is the definition of a P-value.

17.23: (c) The P-value for $z = 2.41$ is 0.0080 (assuming that the difference is in the correct direction).

17.25: (a) The null hypothesis states that μ takes on the "default" value, 18 seconds.

17.27: (c) A small P-value means we should not (or should rarely) find an observed difference as large as or larger than what was seen in H_0 is true. The P-value does not tell us whether the difference seen is "large" or "practically important," nor does it refer to the probability H_0 is true.

17.29: (a) This is a one-sided alternative, so we have 0.005 in the right tail of the Normal distribution, leading to $z > 2.576$.

17.31: (a) H_0: $\mu = 0$ vs. H_a: $\mu > 0$. (b) $z = \frac{2.35 - 0}{2.5/\sqrt{200}} = 13.29$ (c) The P-value is essentially 0. Under H_0, it would be virtually impossible to observe a sample mean as large as 2.35 based on a sample of 200 men. This sample mean cannot explained by random chance, and we would easily reject H_0.

17.33: "$P < 0.05$" *does* mean that H_0 is not likely to be correct . . . but only in the sense that it provides a poor explanation of the data observed. It means that if H_0 is true, a sample as contrary to H_0 as ours would occur by chance alone less than 5% of the time if the experiment was repeated over and over. It does *not* mean that there is a less than 5% chance that H_0 is true.

17.35: The person is confusing practical significance with statistical significance. In fact, a 5% increase isn't a lot in a pragmatic sense. However, $P = 0.03$ means that random chance does not easily explain the difference observed.

17.37: In the sketch, the "significant at 1%" region includes only the dark shading ($z > 2.326$). The "significant at 5%" region of the sketch includes both the light and dark shading ($z > 1.645$). If $P < 0.01$, we must have $P < 0.05$. The converse is false: something that occurs "less than 5 times in 100 repetitions" is not necessarily as rare as something that happens "less than once in 100 repetitions," so a test that is significant at the 5% level is not necessarily significant at the 1% level.

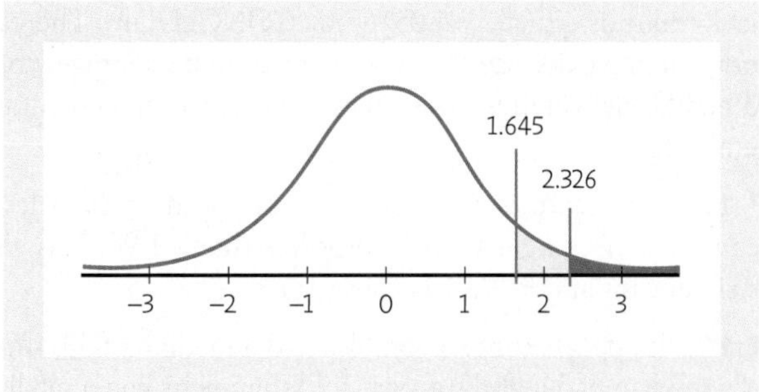

17.39: Because a P-value is a probability, it can never be greater than 1. The correct P-value is $P(Z \geq 1.33) = 0.0918$.

17.41: PLAN: We will test the hypotheses H_0: $\mu = 0\%$ against H_a: $\mu < 0\%$. SOLVE: $\bar{x} = -3.587\%$, $z = \frac{-3.587 - 0}{2.5/\sqrt{47}} = -9.84$, and $P = P(Z \leq -9.84) \approx 0$. CONCLUDE: There is overwhelming evidence that, on average, nursing mothers lose bone mineral.

17.43: (a) We test H_0: $\mu = 0$ vs. H_a: $\mu > 0$, where μ is the mean sensitivity difference in the population. (b) PLAN: We test the hypotheses stated in part (a). SOLVE: $\bar{x} = 0.10125$, $z = \frac{0.10125 - 0}{0.22/\sqrt{16}} = 1.84$ and $P = 0.0329$. CONCLUDE: The sample gives significant evidence (at $\alpha = 0.05$) that eye grease increases sensitivity.

17.45: (a) No, because 33 falls in the 95% confidence interval, which is (27.5, 33.9). (b) Yes, because 34 does not fall in the 95% confidence interval.

Chapter 18 Inference in Practice

18.1: The most important reason is (c); this is a convenience sample consisting of the first 20 students on a list. This is not an SRS. The other two reasons are valid, but less important issues. Reason (a)—the size of the sample and large margin of error—would make the interval less informative, even if the sample were representative of the population. Reason (b)—nonresponse—is a potential problem with every survey, but there is no particular reason to believe it is more likely in this situation.

18.3: The day after Thanksgiving is widely regarded as a day on which retailers offer great deals—the kinds of shoppers found that day probably don't represent shoppers generally. Also, the sample isn't random.

18.5: You cannot conclude this. The restaurant you work at is most likely different in many ways from the one where the experiment took place. We cannot talk about 95% of individual days; the confidence interval is for the average tip in a long sequence of tips.

18.7: The only source of error included in the margin of error is that due to random sampling variability, so (c).

18.9: (a) and (b) The results and the curve for $n = 9$ are shown. We see that as the sample size increases, the same difference between μ_0 and $\bar{x}$ goes from being not at all significant to highly significant.

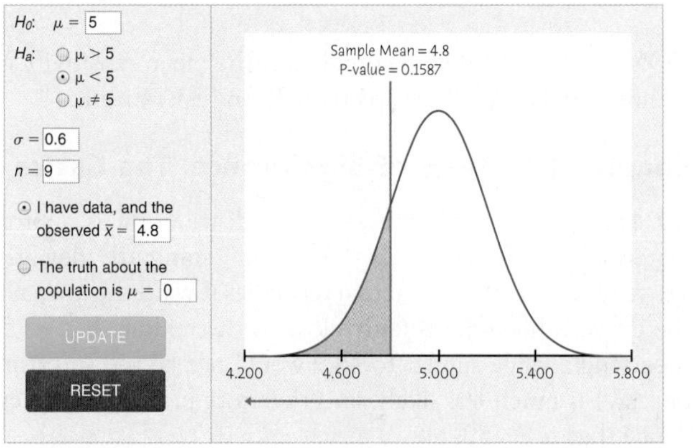

n	P-value
9	0.1587
16	0.0912
36	0.0228
64	0.0038

18.11: (a) Each test (subject) has a 5% chance of being deemed "significant" at the 5% level when the null hypothesis (no ESP) is true. With 1000 tests, we'd expect 50 such occurrences. (b) Retest the nine promising subjects with a different version of the test.

18.13: For a margin of error ± 10, we need at least $n = \left(\frac{(1.645)(125)}{10}\right)^2 = 422.82$, so $n = 423$.

18.15: (a) Increase power by taking more measurements. (b) If you increase α, you make it easier to reject H_0, and so increase power. (c) A value of $\mu = 260$ is even further from the stated value of $\mu = 243$ under H_0, so power increases.

18.17: As σ decreases, power increases. More precise measurements increase the researcher's ability to recognize a false null hypothesis.

σ	55	40	30
Power	0.290	0.442	0.639

18.19: (a) All statistical methods are based on probability samples. We must have a random sample in order to apply them.

18.21: (b) Inference from a voluntary response sample is never reasonable. Online web surveys are voluntary response surveys.

18.23: (a) There is no control group. Any observed improvement may be due to the treatment, or may be due to some other cause.

18.25: (a) The significance level (α) is the probability of rejecting H_0 when H_0 is true.

18.27: (c) Power describes the test's ability to reject a false H_0.

18.29: We need to know that the samples taken from both populations are random. Are the samples large?

18.31: Many people might be reluctant to relate details of their sex lives, or perhaps some will be inclined to exaggerate. It would not be surprising that such an estimate would be biased, but this author cannot guess the direction.

18.33: The effect is greater if the sample is small. With a larger sample, the impact of any one value is small.

18.35: Opinion—even expert opinion—unsupported by data is the weakest type of evidence, so the third description is level C. The second description refers to experiments (clinical trials) and large samples; that is the strongest evidence (level A).

18.37: (a) The P-value decreases. (b) Power increases.

18.39: (a) $\bar{x} = 9.524$, $z = \frac{9.524 - 8}{2/\sqrt{5}} = 1.704$, $P = 2P(Z \geq 1.704) = 0.0884$ (using software). This is not significant at the 5% level of significance. We would not reject 8 as a plausible value of μ, even though $\mu = 10$. (b) The small sample size makes it difficult to detect a difference that is really there.

18.41: (a) "Statistically insignificant" means that the differences observed were no more than might have been expected to occur by chance even if SES had no effect on LSAT results. (b) If the results are based on a small sample, then even if the null hypothesis were not true, the test might not be sensitive enough to detect the effect (i.e., have low power). Knowing the effects were small tells us that the test was not insignificant merely because of a small sample size.

18.43: $n = \left(\frac{(1.96)(3000)}{600}\right)^2 = 96.04$. Take $n = 97$.

18.45: (a) This test has a 20% chance of rejecting H_0 when the alternative is true. (b) If the test has 20% power, then when the alternative is true, it will fail to reject H_0 80% of the time. (c) The sample sizes are very small, which typically leads to low-power.

18.47: From the applet, against the alternative $\mu = 10$, power = 0.609.

18.49: (a) Because the alternative is $\mu \neq 8$, we reject H_0 at the 5% level when $z \geq 1.96$ or $z \leq -1.96$. (b) $z = \frac{\bar{x} - 8}{2/\sqrt{5}} = 1.118(\bar{x} - 8)$. We reject H_0 if $1.118(\bar{x} - 8) \leq -1.96$ or if $1.118(\bar{x} - 8) \geq 1.96$. Solving for $\bar{x}$, we reject H_0 if $\bar{x} \leq 6.247$ or $\bar{x} \geq 9.753$. (c) When $\mu = 10$, the power is $P(\bar{x} \leq 6.247) + P(\bar{x} \geq 9.753) = P(Z \leq \frac{6.247 - 10}{2/\sqrt{5}}) + P(Z \geq \frac{9.753 - 10}{2/\sqrt{5}}) = P(Z \leq -4.20) + P(Z \geq -0.28) = 0.0000 + 0.6103 = 0.6103$.

18.51: Power $= 1 - P(\text{Type II error}) = 1 - 0.36 = 0.64$.

18.53: (a) In the long run, this probability should be 0.05. Out of 100 simulated tests, the number of false rejections will have a binomial distribution with $n = 100$ and $p = 0.05$. Most students will see between 0 and 10 rejections. (b) If the power is 0.812, the probability of a Type II error is 0.188. Out of 100 simulated tests, the number of false non-rejections will have a binomial distribution with $n = 100$ and $p = 0.188$. Most students will see between 10 and 29 non-rejections.

Chapter 19 From Data Production to Inference: Part III Review

19.1: (a) $S = \{\text{Male, Female}\}$. (b) $S = \{6, 7, 8, \ldots, 19, 20\}$. (c) $S = \{\text{All values } 2.5 \leq \text{VOP} \leq 6.1 \text{ liters per minute}\}$. (d) $S = \{\text{All heart rates such that heart rate} > \_\_\_ \text{ bpm}\}$; (students may select a reasonable minimum).

19.3: (a) $1 - (0.68 + 0.18 + 0.10 + 0.02) = 0.02$

19.5: $Y > 1$, or $Y \geq 2$; $P(Y \geq 2) = 1 - 0.27 = 0.73$

19.7: (d) $1 - 0.34 = 0.66$

19.9: $P(X \leq 2)$ is the probability of randomly choosing a woman between the ages of 15 and 44 who has given birth to two or fewer children. $P(X \leq 2) = 0.471 + 0.169 + 0.204 = 0.844$.

19.11: $P(X \geq 3) = 0.104 + 0.034 + 0.018 = 0.156$

19.13: (a) The height of the density curve is $1/5 = 0.2$, because the area under the density function must be 1.

19.15: There is no area above $X = 5$ so $P(4 < X < 7) = P(4 < X \leq 5) = 1/5 = 0.2$.

19.17: (c) mean = 100, standard deviation = $15/\sqrt{60} = 1.94$.

19.19: The answer in Exercise 19.16 would change, since this refers to the population distribution, which is now non-Normal. The answer in Exercise 19.17 would not change—the mean of $\bar{x}$ is 100, and the standard deviation of $\bar{x}$ is 1.94, regardless of the population distribution. The answer in Exercise 19.18 would, essentially, not change. The central limit theorem tells us that the sampling distribution of $\bar{x}$ is approximately Normal when n is large enough, no matter what the population distribution.

19.21: If the population we're sampling from is heavily skewed, then a larger sample is required for the central limit theorem to apply. If $n = 15$, the sampling distribution of $\bar{x}$ may not be approximately Normal, but if $n = 150$, it will surely be approximately Normal.

19.23: (c) $357 \pm 1.96\frac{50}{\sqrt{8}} = 322.35$ to 391.65

19.25: $357 \pm 1.282\frac{50}{\sqrt{8}} = 334.34$ to 379.66

19.27: (b) $172 \pm 1.645\frac{41}{\sqrt{14}} = 172 \pm 18.03$ mg/dL.

19.29: (c) $n \geq (\frac{(1.645)(41)}{5})^2 = 181.95$, which rounds up to 182.

19.31: (c) We want to know if your college differed. Hypotheses are in terms of population parameters, not sample statistics.

19.33: (c) $\alpha = 0.10$ but not at $\alpha = 0.05$. The P-value is 0.0721.

19.35: (d) $z = \frac{357 - 100}{50/\sqrt{8}} = 14.538$. The actual P-value is essentially 0.

19.37: H_0: $\mu = 100$ vs. H_a: $\mu < 100$; $z = \frac{87.6 - 100}{15/\sqrt{113}} = -8.79$; P-value ≈ 0. This is overwhelming evidence that the mean IQ for the very-low-birth-weight population is less than 100.

19.39: $P = 0.74$ means that the observed difference is easily explained by random chance. $P = 0.013$ means that the observed difference was unlikely to have occurred by chance alone; such a result (or something more extreme) would be expected only 13 times in 1000 repetitions of this study.

19.41: (b) This is a personal probability (Byron's opinion), and not something based on many repetitions of a football season (which would be impossible).

19.43: (b) $P(\text{over 25 or local}) = P(\text{over 25}) + P(\text{local}) - P(\text{over 25 and local}) = 0.7 + 0.25 - 0.05 = 0.90$

19.45: (b) $P(\text{over 25} \mid \text{not local}) = P(\text{over 25 and not local})/P(\text{not local}) = 0.65/0.75 = 0.867$

19.47: (a) This is binomial with $n = 5$ and $p = 0.4$.

19.49: $P(X < 600) = P(X \leq 599) = P(Z < -1.37) = 0.0853$.

19.51: (a) All probabilities are between 0 and 1, and their sum is 1. (b) Let R_1 be Taster 1's rating and R_2 be Taster 2's rating. Add the probabilities on the diagonal (upper left to lower right): $P(R_1 = R_2) = 0.05 + 0.08 + 0.25 + 0.18 + 0.08 = 0.64$. (c) $P(R_1 > R_2) = 0.18$. This is the sum of the ten numbers in the "lower left" part of the table. $P(R_2 > R_1) = 0.18$; this is the sum of the ten numbers in the "upper right" part of the table.

19.53: (a) Nearly all the *individual students* should be in the range $\mu \pm 3\sigma = 3.3 \pm 3(0.8) = 0.9$ to 5.7. (b) The sample mean $\bar{x}$ has a $N(\mu, \sigma/\sqrt{100}) = N(3.3, 0.08)$ distribution, so nearly all such *means* should be in the range $3.3 \pm 3(0.08) = 3.3 \pm 0.24$, or 3.06 to 3.54.

19.55: (a) The stemplot confirms the description given in the text. (Arguably, there are two high "mild outliers," although the $1.5 \times IQR$ criterion only flags the highest as an outlier.) (b) PLAN: Let μ be the mean body temperature. H_0: $\mu = 98.6°$ vs. H_a: $\mu \neq 98.6°$. SOLVE: Assume we have a Normal distribution and an SRS. $\bar{x} = 98.203°$, so $z = \frac{98.203 - 98.6}{0.7/\sqrt{20}} = -2.54$. $P = 2P(Z < -2.54) = 0.0110$. CONCLUDE: We have fairly strong evidence—significant at $\alpha = 0.05$, but not at $\alpha = 0.01$—that mean body temperature is not equal to 98.6°. (Specifically, the data suggest that mean body temperature is lower.)

19.57: PLAN: We will estimate μ by giving a 90% confidence interval. SOLVE: Assume we have a Normal distribution and an SRS (conditions were checked in Exercise 19.55). With $\bar{x} = 98.203$, our 90% confidence interval for μ is $98.203 \pm 1.645(\frac{0.7}{\sqrt{20}}) = 98.203 \pm 0.257$. CONCLUDE: We are 90% confident that the mean body temperature for healthy adults is between 97.95° and 98.46°.

19.59: For the two-sided test H_0: M = \$50,000 vs. H_a: M $\neq$ \$50,000 with significance level $\alpha = 0.10$, we can reject H_0 because \$50,000 falls outside the 90% confidence interval.

19.61: Let H be the event "student was home schooled." Let R be the event "student attended a regular public school." We want $P(H \mid \text{not } R)$. Note that the event "H and not R" = "H." Then $P(H \mid \text{not } R) = \frac{P(H)}{P(\text{not } R)} = \frac{0.006}{1 - 0.758} = 0.025$.

19.63: (a) For $n = 12$ such users, the distribution of those who have visited an auction site in the past month is binomial with $n = 12$ and $p = 0.50$. $P(X = 8) = \binom{12}{8}(0.5)^8 (1 - 0.5)^{12-8} = 0.1208$. (b) With $n = 500$, the distribution of those who have visited online auction sites will be approximately Normal because $np = n(1 - p) = 500(0.5) = 250$, which is at least 10. The standard deviation is $\sigma = \sqrt{500(0.5)(1 - 0.5)} = 11.18$. $P(X \geq 235) = P(Z \geq -1.34) = 0.9099$.

19.65: A Type I error means that we conclude the mean IQ is less than 100 when it really is 100 (or more). A Type II error means that we conclude the mean IQ is 100 (or more) when it really is less than 100.

Chapter 20 Inference about a Population Mean

20.1: $SE = s/\sqrt{n} = 56.9/\sqrt{1000} = 1.7993$ minutes

20.3: (a) $t^* = 2.353$ (b) $t^* = 2.485$

20.5: (a) $df = 12 - 1 = 11$, so $t^* = 2.201$ (b) $df = 2 - 1 = 1$, so $t^* = 63.657$ (c) $df = 1001 - 1 = 1000$, so $t^* = 1.646$

20.7: PLAN: We will estimate μ with a 99% confidence interval. SOLVE: We are told to view the observations as an SRS. A stemplot shows some possible bimodality, but no outliers. With $\bar{x} = 62.1667\%$ and $s = 5.806\%$ correct identifications, and $t^* = 2.807$ ($df = 23$), the 99% confidence interval for μ is $62.1667 \pm 2.807\frac{5.806}{\sqrt{24}} = 62.1667 \pm 3.3267$. CONCLUDE: We are 99% confident that the mean percent of correct answers to identifying the taller of two people by voice is between 58.84% and 65.49%.

20.9: (a) $df = 5 - 1 = 4$ (b) $t = 2.50$ is bracketed by $t^* = 2.132$ (with two-tail probability 0.10) and $t^* = 2.776$ (with two-tail probability 0.05). $0.05 < P < 0.10$. (c) This test is significant at the 10% level since $P < 0.10$. It is not significant at either the 5% or 1% levels. (d) From software, $P = 0.0668$.

20.11: PLAN: Take μ to be the mean difference (with eye grease minus without) in sensitivity. We test H_0: $\mu = 0$ versus H_a: $\mu > 0$, using a one-sided alternative because if the eye grease works, it should increase sensitivity. SOLVE: We must assume that the athletes in the experiment can be regarded as an SRS of all such athletes and that the treatments were randomized. We were provided the differences for each athlete; a stemplot of these differences (provided) seems to show two outliers; checking with the $1.5 \times IQR$ rule, these are not outliers ($Q_1 = -8$, $Q_3 = 26$, and the upper fence is 77). However, P-values will only be approximate due to the skew and relatively small sample size. $\bar{x} = 0.1012$ and $s = 0.2263$, so $t = \frac{0.1012 - 0}{0.2263/\sqrt{16}} = 1.79$ with $df = 15$. Using Table C, $P < 0.05$ (software gives 0.0469). CONCLUDE: We have evidence that eye grease does increase sensitivity to contrast. Due to the skew in the data, we may not want to place a lot of emphasis on this result.

-1	8 6 2 1
-0	5
0	2 3 5 5 7
1	4
2	4 8 9
3	
4	3
5	
6	4

20.13: The stemplot suggests that the distribution of nitrogen contents is heavily skewed with a strong outlier. Although t procedures are robust, they should not be used if the population being sampled is this heavily skewed.

20.15: (a) $SE = 1.2/\sqrt{10} = 0.379$. (b) If you use Minitab, create a command file with a .MTB extension using a text editor such as Notepad. This author's commands are shown.

```
random 10 c2;
Normal 0.3 1.2.
let c3(k1)=mean(c2)
let k1=k1+1
```

From descriptive statistics on the randomization distribution, the standard deviation of the 1000 random sample means is 0.3861; this is our estimate of the standard error, which is close to the value from the typical formula.

```
Descriptive Statistics: C3

Variable       N    N*     Mean   SE Mean   StDev
C3          1000     0   0.3149    0.0122  0.3861
```

20.17: (b) We virtually never know the value of σ.

20.19: (c) $df = 16 - 1 = 15$

20.21: (b) 1.476. Here, $df = 5$.

20.23: (a) \$38,808 to \$51,192. The interval is computed as $45{,}000 \pm 2.064\frac{15{,}000}{\sqrt{25}}$.

20.25: (b) If you sample 225 unmarried male students, and then sample 225 unmarried female students, no matching is present.

20.27: For the student group: $t = \frac{0.08 - 0}{0.37/\sqrt{12}} = 0.749$. For the non-student group: $t = \frac{0.35 - 0}{0.37/\sqrt{12}} = 3.277$. From Table C, the first P-value (assuming a two-sided alternate hypothesis) is between 0.4 and 0.5 (software gives 0.47), and the second P-value is between 0.005 and 0.01 (software gives 0.007).

20.29: (a) With $n = 1400$, the CLT says the sample mean is approximately Normal, no matter the original distribution. (b) From Table C, take $t^* = 2.581$ ($df = 1000$), or using software take $t^* = 2.579$. For either value, the 99% confidence interval is $248 \pm t^*(1) = 245.4$ to 250.6, rounded to one decimal place. (c) Because the 99% confidence interval for μ is entirely above 243, we can believe the mean for all Dallas children is above basic.

20.31: (a) A subject's responses to the two treatments would not be independent. (b) This is a two-sided test because the placebo could stimulate activity or suppress activity at this point in the brain. We have $t = \frac{-0.326 - 0}{0.181/\sqrt{6}} = -4.41$. With $df = 5$, $P = 0.0069$, there is significant evidence of a difference.

20.33: (a) A stemplot suggests the presence of outliers. The sample is small and the stemplot is skewed, so use of t procedures is not appropriate. (b) In the first interval, using all nine observations, we have $df = 8$ and $t^* = 1.860$. For the second interval, removing the two outliers (1.15 and 1.35), df = 6 and $t^* = 1.943$. The two 90% confidence intervals are

$$0.549 \pm 1.860\left(\frac{0.403}{\sqrt{9}}\right) = 0.299 \text{ to } 0.799 \text{ grams,}$$

and

$$0.349 \pm 1.943\left(\frac{0.053}{\sqrt{7}}\right) = 0.310 \text{ to } 0.388 \text{ grams.}$$

(c) The confidence interval computed without the two outliers is much narrower and has a lower center. Using fewer data values reduces degrees of freedom (and gives a larger value of t^*). By removing two values far from the others, s reduces from 0.403 to 0.053.

20.35: (a) The stemplot clearly shows the high outlier mentioned in the text. (b) Let μ be the mean difference (control minus experimental) in healing rates. We test H_0: $\mu = 0$ versus H_a: $\mu > 0$. The alternative hypothesis says that the control limb has a faster healing rate. With all 12 differences: $\bar{x} = 6.417$ and $s = 10.7065$, so $t = \frac{6.417 - 0}{10.7065/\sqrt{12}} = 2.08$. With df = 11, $P = 0.0311$ (using software). Omitting the outlier: $\bar{x} = 4.182$ and $s = 7.7565$, so $t = \frac{4.182 - 0}{7.7565/\sqrt{11}} = 1.79$. With df = 10, $P = 0.052$. With all 12 differences there is greater evidence that the mean healing time is greater for the control limb. When we omit the outlier, the evidence is weaker.

20.37: (a) A histogram of the sample shows a significant outlier, and indicates skew. We might consider applying t procedures to the sample after removing the most extreme observation (37, 786). (b) If we remove the largest observation, the remaining sample is not heavily skewed and has no outliers. Now, we test H_0: $\mu = 7000$ versus H_a: $\mu \neq 7000$. With the outlier removed, $\bar{x} = 11{,}555.16$ and $s = 6{,}095.015$. $t = \frac{11{,}555.16 - 7000}{6095.015/\sqrt{19}} = 3.258$. With df = 18 with software, $P = 0.0044$. There is overwhelming evidence that the mean number of words per day for men at this university differs from 7000.

20.39: (a) Each patient was measured before and after treatment. (b) The stemplot of differences shows an extreme right-skew, and one or two high outliers. The t procedures should not be used. (c) Some students might perform the test (H_0: $\mu = 0$ versus H_a: $\mu > 0$) using t procedures, despite the presence of strong skew and outliers in the sample. If so, they should find $\bar{x} = 156.36$, $s = 234.2952$, and $t = 2.213$, yielding $P = 0.0256$.

20.41: (a) We test H_0: $\mu = 0$ versus H_a: $\mu > 0$, where μ is the mean difference (treated minus control). This is one-sided because the researchers have reason to believe that CO_2 will increase growth rate. (b) $\bar{x} = 1.916$ and $s = 1.050$, so $t = \frac{1.916 - 0}{1.050/\sqrt{3}} = 3.16$ with df = 2, $P = 0.044$. This is significant at the 5% significance level. (c) For very small samples, t procedures should only be used when we can assume that the population is Normal. We have no way to assess the Normality of the population based on these three observations.

20.43: A stemplot reveals that these data contain two extreme high outliers (5973 and 8015). t procedures are not appropriate.

20.45: (a) $\bar{x} = 48.25$ and $s = 40.24$ thousand barrels. From Table C, $t^* = 2.000$ (df = 60). Using software, with df = 63, $t^* = 1.998$. The 95% confidence interval for μ is $48.25 \pm 2.000\left(\frac{40.24}{\sqrt{64}}\right) = 48.25 \pm 10.06 = 38.19$ to 58.31 thousand barrels. (Using software, the confidence interval is 38.20 to 58.30 thousand barrels.) (b) A stemplot confirms the skewness and outliers described. The two intervals have similar widths, but the new interval is shifted higher by about 2000 barrels. Although t procedures are fairly robust, we should be cautious about trusting the result in part (a) because of the strong skew and outliers. The computer-intensive method may produce a more reliable interval.

20.47: PLAN: We will construct a 90% confidence interval for μ, the mean percent of beetle-infected seeds. SOLVE: A stemplot shows a single-peaked and roughly symmetric distribution. We assume that the 28 plants can be viewed as an SRS of the population, so t procedures are appropriate. $\bar{x} = 4.0786$ and $s = 2.0135$ percent. Using df = 27, the 90% confidence interval for μ is $4.0786 \pm 1.703\left(\frac{2.0135}{\sqrt{28}}\right) = 4.0786 \pm 0.648 =$ 3.43% to 4.73%. CONCLUDE: The beetle infects less than 5% of seeds, so it is unlikely to be effective in controlling velvetleaf.

20.49: $\bar{x} = 0.5283$, $s = 0.4574$, and df = 5. A 95% confidence interval for the mean difference in T cell counts after 20 days on blinatumomab is $0.5283 \pm 2.571\left(\frac{0.4574}{\sqrt{6}}\right) = 0.5283 \pm 0.4801 = 0.0482$ to 1.0084 thousand cells.

20.51: (a) For each subject, randomly select which knob (right or left) that subject should use first. (b) PLAN: We test H_0: $\mu = 0$ versus H_a: $\mu < 0$, where μ denotes the mean difference in time (right-thread time − left-thread time). SOLVE: A stemplot of the differences gives no reason that t procedures are not appropriate. We assume our sample can be viewed as an SRS. $\bar{x} = -13.32$ seconds and $s = 22.936$ seconds, so $t = \frac{-13.32 - 0}{22.936/\sqrt{25}} = -2.90$. With df = 24 we find $P = 0.0039$. CONCLUDE: We have good evidence (significant at the 1% level) that the mean difference really is negative—that is, the mean time for right-hand-thread knobs is less than the mean time for left-hand-thread knobs.

20.53: Refer to Exercise 20.51. With df = 24, $t^* = 1.711$, so the confidence interval for μ is given by $-13.32 \pm 1.711\left(\frac{22.936}{\sqrt{25}}\right) = -13.32 \pm 7.85 = -21.2$ to -5.5 seconds. Now, $\bar{x}_{RH}/\bar{x}_{LH} = 104.12/117.44 = 0.887$. Right-handers working with right-handed knobs can accomplish the task in about 89% of the time needed by those working with left-handed knobs.

20.55: (a) Starting with Table C values, for 90% confidence, $t(100) = 1.660$ and $z^* = 1.645$. For 95% confidence, $t(100) = 1.984$ and $z^* = 1.96$. Similarly, for 99% confidence, $t(100) = 2.626$ and $z^* = 2.576$. The differences are 0.015, 0.024, and 0.05. Larger confidence levels will need more observations. We note that $t(1000)$ is within 0.01 of z^* for all these confidence levels. Using software, we find that $t(150) = 1.655$ for 90% confidence, $t(240) = 1.9699$ for 95% confidence, and $t(485) = 2.586$ for 99.5% confidence. (b) Answers will vary. We'll note that the effect of the standard deviation difference multiplies the margin of error in the calculation by 100, which implies that "similar" takes more observations with $\sigma = 100$

than for $\sigma = 1$. Using $n = 485$ with $\sigma = 100$, the 99% t margin of error is 11.74, compared to a 99% z margin of error of 11.69. Using $\sigma = 1$, the margins of error are both 0.117, rounding to three decimal places.

Chapter 21 Comparing Two Means

21.1: This is a matched-pairs design. Each plot is a matched pair.

21.3: This involves a single sample.

21.5: PLAN: Test H_0: $\mu_{Wii} = \mu_{NoWii}$ versus H_a: $\mu_{Wii} > \mu_{NoWii}$. We use a one-sided alternative that practice with the Wii™ should result in more improvement than just performing the same operation again. SOLVE: These data came from participants in a randomized experiment, so the two groups are independent. Stemplots suggest some deviation from Normality, and a possible high outlier for the No Wii™ group. Boxplots (not shown) indicate no outliers and a relatively symmetric distribution for the Wii™ group, but both the −88 and 229 are outliers for the No Wii™ group. We proceed with the t test for two samples appealing to robustness (especially good with equal sample sizes). With $\bar{x}_{Wii} = 132.71$, $\bar{x}_{NoWii} = 59.67$, $s_{Wii} = 98.44$, $s_2 = 63.04$, $n_{Wii} = 21$, and $n_{NoWii} = 21$, $SE = \sqrt{\frac{s_1^2}{n_1} + \frac{s_2^2}{n_2}} = 25.509$ and $t = \frac{\bar{x}_1 - \bar{x}_2}{SE} = 2.86$. Using df as the smaller of $21 - 1$ and $21 - 1$, we have df = 20, and $0.0025 < P < 0.005$. Using software, df = 34.04 and $P = 0.0036$. CONCLUDE: There is very strong evidence that playing with a Nintendo Wii™ does help improve the skills of student doctors, at least in terms of the mean time to complete a virtual gall bladder operation.

21.7: From Exercise 21.5, $\bar{x}_{Wii} = 132.71$, $\bar{x}_{NoWii} = 59.67$, $n_{Wii} = n_{NoWii} = 21$, SE = 25.509. A 99% confidence interval for the mean difference in improvement in time to complete the virtual gall bladder operation is $\bar{x}_{Wii} - \bar{x}_{NoWii} \pm t^*SE = 73.04 \pm 2.845(25.509) = 0.467$ to 145.613 seconds. Software uses df = 34.04 and gives an interval of 3.45 to 142.65 seconds.

21.9: (a) Back-to-back stemplots appear to be reasonably Normal, and the discussion in the exercise justifies our treating the data as independent SRSs, so we can use t procedures. We test H_0: $\mu_1 = \mu_2$ versus H_a: $\mu_1 < \mu_2$, where μ_1 is the population mean time in the restaurant with no scent, and μ_2 is the mean time in the restaurant with a lavender odor. Here, with $\bar{x}_1 = 91.27$, $\bar{x}_2 = 105.700$, $s_1 = 14.930$, $s_2 = 13.105$, $n_1 = 30$, and $n_2 = 30$: $SE = \sqrt{\frac{s_1^2}{n_1} + \frac{s_2^2}{n_2}} = 3.627$ and $t = \frac{\bar{x}_1 - \bar{x}_2}{SE} = -3.98$. Using software, df = 57.041, and $P = 0.0001$. Using the more conservative df = 29 (lesser of $30 - 1$ and $30 - 1$) and Table C, $P < 0.0005$. There is very strong evidence that customers spend more time on average in the restaurant when the lavender scent is present. (b) Back-to-back stemplots of the spending data are skewed and have many gaps (perhaps due to pricing). We test H_0: $\mu_1 = \mu_2$ versus H_a: $\mu_1 < \mu_2$, where μ_1 is the population mean amount spent in the restaurant with no scent, and μ_2 is the mean amount spent in the restaurant with lavender odor. Here, with $\bar{x}_1 = €17.5133$, $\bar{x}_2 = €21.1233$, $s_1 = €2.3588$, $s_2 = €2.3450$, $n_1 = 30$, and $n_2 = 30$: $SE = \sqrt{\frac{s_1^2}{n_1} + \frac{s_2^2}{n_2}} = €0.6073$ and $t = \frac{\bar{x}_1 - \bar{x}_2}{SE} = -5.94$. Using software, df = 57.998 and $P < 0.0001$. Using the more conservative df = 29 and Table C, $P < 0.0005$. There is very strong evidence that customers spend more money on average when the lavender scent is present.

21.11: We have two small samples ($n_1 = n_2 = 4$), so the t procedures are not reliable unless both distributions are Normal.

21.13: Here are the details of the calculations:

$$SE_{Alone} = \frac{0.68}{\sqrt{37}} = 0.1118$$

$$SE_{Friends} = \frac{0.83}{\sqrt{21}} = 0.1811$$

$$SE = \sqrt{SE_{Alone}^2 + SE_{Friends}^2} = 0.21283$$

$$df = \frac{SE^4}{\frac{1}{36}\left(\frac{0.68^2}{37}\right)^2 + \frac{1}{20}\left(\frac{0.83^2}{21}\right)^2} = \frac{0.00205}{0.00005815} = 35.284$$

$$t = \frac{0.29 - (-0.19)}{0.21283} = 2.255$$

21.15: Reading from the software output shown in the statement of Exercise 21.13, we find that there is a significant difference in mean perceived formidability for men alone and with friends ($t = 2.255$, df = 35.3, $P < 0.02$). Because larger scores indicate greater perceived formidability, it appears that foes appear more formidable when alone as opposed to when with friends.

21.17: The observed means are $\bar{x}_1 = 162.825$ and $\bar{x}_9 = 157.6$. The observed difference in means is 5.225. From Figure 21.6, the one-sided P-value is about $0.06 + 0.11 + 0.04 + 0.04 = 0.25$. We have no evidence that an increase in weeds from one per meter to nine per meter decreases the crop yield.

21.19: (a) a two-sample t test. We have two independent populations: females and males.

21.21: (b) Confidence levels and P-values from the t procedures are quite accurate even if the population distributions are not exactly Normal.

21.23: (b) $\frac{15.84 - 9.64}{\sqrt{\frac{8.65^2}{21} + \frac{3.43^2}{21}}} = 3.05$

21.25: (a) We suspect that younger people use social networks more than older people, so this is a one-sided alternative.

21.27: (a) To test the belief that women talk more than men, test H_0: $\mu_F = \mu_M$ versus H_a: $\mu_F > \mu_M$. (b)–(d) The table provides a summary of t statistics, degrees of freedom, and P-values for both studies. We take the conservative approach for computing df as the smaller sample size, minus 1.

STUDY	t	df	TABLE C VALUES	P-VALUE
1	−0.248	55	$\lvert t \rvert < 0.679$	$P > 0.25$
2	1.507	19	$1.328 < t < 1.729$	$0.05 < P < 0.10$

Note that for Study 1 we referenced df = 50 in Table C. (e) The first study gives no support to the belief that women talk more than men; the second study gives weak support, significant only at a relatively high significance level ($\alpha = 0.10$).

21.29: (a) Call group 1 the Stress group, and group 2 the No stress group. $s = \text{SEM}\sqrt{n}$, so $s_1 = 3\sqrt{20} = 13.416$ and $s_2 = 2\sqrt{51} = 14.283$. (b) Using conservative Option 2, df = 19 (the lesser of 20 − 1 and 51 − 1). (c) We test H_0: $\mu_1 = \mu_2$ versus H_a: $\mu_1 \neq \mu_2$. With $n_1 = 20$ and $n_2 = 51$, $\text{SE} = \sqrt{\frac{s_1^2}{n_1} + \frac{s_2^2}{n_2}} = 3.605$, and $t = \frac{\bar{x}_1 - \bar{x}_2}{\text{SE}} = \frac{26 - 32}{3.605} = -1.664$. With df = 19, using Table C, $0.10 < P < 0.20$. There is little evidence in support of a conclusion that mean weights of rats in stressful environments differ from those of rats without stress.

21.31: (a) We test H_0: $\mu_{1975} = \mu_{2006}$ versus H_a: $\mu_{1975} > \mu_{2006}$. $\text{SE} = \sqrt{\frac{0.81^2}{1165} + \frac{0.80^2}{2177}} = 0.02928$, $t = \frac{3.37 - 3.32}{0.0293} = 1.708$. This is significant at the 5% level: $P = 0.0439$ (df = 2353.38) or $0.025 < P < 0.05$ (df = 1000). There is good evidence that mean job satisfaction decreased from 1975 to 2006. (b) The difference is barely significant at the 0.05 level (most likely due to the large sample sizes). Knowing that 1975 had the highest mean job satisfaction score in this time period casts doubt about whether this is actually decreasing. Also, a difference of 0.05 in the means may not be of practical importance.

21.33: Let μ_C be the mean brain size for players who have had concussions, and let μ_{NC} be the mean for those who have not had concussions. We test H_0: $\mu_C = \mu_{NC}$ versus H_a: $\mu_C \neq \mu_{NC}$. This is a two-sided test, as we simply want to know if there is a difference in mean brain size. $\text{SE} = \sqrt{\frac{609.3^2}{25} + \frac{815.4^2}{25}} = 203.5803$, and $t = \frac{5784 - 6489}{203.5803} = -3.463$. Using the conservative version for df (Option 2), df = 24, and $0.002 < P < 0.005$. Using software, df = 44.43, and $P = 0.0012$. There is strong evidence that the mean brain size is different for football players who have had concussions as opposed to those who have not had concussions. (b) The fact that these were consecutive cases indicates they are not a random sample of all football players who have or have not had concussions. That could weaken or negate the results of the test. We'd need more information about how and why these players were referred to the institute.

21.35: (a) The hypotheses are H_0: $\mu_1 = \mu_2$ versus H_a: $\mu_1 > \mu_2$, where μ_1 is the mean gain among all coached students, and μ_2 is the mean gain among uncoached students. We find $\text{SE} = \sqrt{\frac{59^2}{427} + \frac{52^2}{2733}} = 3.0235$, and $t = \frac{29 - 21}{3.0235} = 2.646$. Using the conservative approach, df = 100 in Table C, and we obtain $0.0025 < P < 0.005$. Using software, df = 534.45, and $P = 0.0042$. There is evidence that coached students had a greater average increase than uncoached students. (b) The 99% confidence interval is $8 \pm t^*(3.0235)$ where t^* equals 2.626 (using df = 100 with Table C) or 2.585 (df = 534.45 with software). This gives either 0.06 to 15.94 points, or 0.184 to 15.816 points, respectively. (c) Increasing one's score by 0 to 16 points is not likely to make a difference in being granted admission or scholarships from any colleges.

21.37: (a) Neither sample histogram suggests strong skew or presence of strong outliers; *t* procedures are reasonable. (b) Let μ_1 be the mean tip percentage when the forecast is good, and μ_2 be the mean tip percentage when the forecast is bad. We have $\bar{x}_1 = 22.22$, $\bar{x}_2 = 18.19$, $s_1 = 1.955$, $s_2 = 2.105$, $n_1 = 20$, and $n_2 = 20$. We test H_0: $\mu_1 = \mu_2$ versus H_a: $\mu_1 \neq \mu_2$. Here, $\text{SE} = \sqrt{\frac{s_1^2}{n_1} + \frac{s_2^2}{n_2}} = 0.642$ and $t = \frac{\bar{x}_1 - \bar{x}_2}{\text{SE}} = 6.274$. Using df = 19 (the conservative Option 2) and Table C, we have $P < 0.001$. Using software, df = 37.8, and $P < 0.00001$. There is overwhelming evidence that the mean tip percentage differs between the two types of forecasts presented to patrons.

21.39: Refer to results in Exercise 21.37. Using df = 19, $t^* = 2.093$ and the 95% confidence interval for the difference in mean tip percentages between these two populations is $22.22 - 18.19 \pm 2.093(0.642) = 4.03 \pm 1.34 = 2.69$ to 5.37 percent. Using df = 37.8 with software, $t^* = 2.025$ and the corresponding 95% confidence interval is 2.73% to 5.33%.

21.41: (a) The mean rating for those with a positive attitude toward Mitt is larger than the mean for those with a negative attitude; the standard deviations are relatively large, however. (b) The distribution of ratings for those with positive attitudes toward Mitt is somewhat right skewed (but not extremely so). The distribution of ratings for those with negative attitudes toward Mitt is fairly symmetric. A check with a boxplot indicates the two lowest values are not outliers. (c) We find $\text{SE} = \sqrt{\frac{0.8015^2}{29} + \frac{0.9127^2}{29}} = 0.22556$ and $t = \frac{3.9379 - 3.6103}{0.22556} = 1.452$, for which the *P*-value is $0.10 < P < 0.20$ (using df = 28) or 0.1521 (using software, with df = 55.08). There is no evidence of a difference in the mean trustworthiness rating of Mitt Romney's face according to whether or not college students had a positive or negative attitude about Mitt (as compared to Barack Obama).

	n	$\bar{x}$	s
Positive	29	3.9379	0.8015
Negative	29	3.6103	0.9127

21.43: (a) Stemplots suggest that there is some skew in both populations, but the sample sizes should be large enough to overcome this problem. (b) With subscripts as assigned in the statement of the problem (Group 1 = Women), we test H_0: $\mu_1 = \mu_2$ versus H_a: $\mu_1 > \mu_2$. We have $\bar{x}_1 = 16{,}496.1$, $\bar{x}_2 = 12{,}866.7$, $s_1 = 7914.35$, $s_2 = 8342.47$, $n_1 = 27$, and $n_2 = 20$; we find $\text{SE} = \sqrt{\frac{7914.35^2}{27} + \frac{8342.47^2}{20}} = 2408.26$, and $t = \frac{16{,}496.1 - 12{,}866.7}{2408.26} = 1.51$. With df = 39.8 (software), $P = 0.070$. Using Table C with df = 19, $0.05 < P < 0.10$. There is some evidence that, on average, women say more words than men, but the evidence is not particularly strong.

21.45: We test H_0: $\mu_1 = \mu_2$ versus H_a: $\mu_1 \neq \mu_2$, where μ_1 is the mean days behind caterpillar peak for the control group, and μ_2 is the mean days for the supplemented group. Now, with $\bar{x}_1 = 4.0$, $\bar{x}_2 = 11.3$, $s_1 = 3.10934$, $s_2 = 3.92556$, $n_1 = 6$, and $n_2 = 7$, we find $\text{SE} = \sqrt{\frac{s_1^2}{n_1} + \frac{s_2^2}{n_2}} = 1.95263$, $t = \frac{4.0 - 11.3}{\text{SE}} = -3.74$. The two-sided *P*-value is either $0.01 < P < 0.02$ (using df = 5) or 0.0033 (df = 10.96 with software), agreeing with the stated conclusion (a significant difference).

21.47: PLAN: We test $H_0: \mu_1 = \mu_2$ versus $H_a: \mu_1 > \mu_2$, where μ_1 is the mean time for the treatment group, and μ_2 is the mean time for the control group. The alternative hypothesis is one-sided because the researcher suspects that the treatment group will wait longer before asking for help. SOLVE: We must assume that the data comes from an SRS of the intended populations. A back-to-back stemplot shows some irregularity in the treatment times and skewness in the control times. We hope that our equal and moderately large sample sizes will overcome any deviation from Normality. With $\bar{x}_1 = 314.0588$, $\bar{x}_2 = 186.1176$, $s_1 = 172.7898$, $s_2 = 118.0926$, $n_1 = 17$, and $n_2 = 17$, we find $SE = \sqrt{\frac{s_1^2}{n_1} + \frac{s_2^2}{n_2}} = 50.7602$, and $t = \frac{314.0588 - 186.1176}{SE} = 2.521$, for which $0.01 < P < 0.02$ (df = 16) or $P = 0.0088$ (df = 28.27). CONCLUDE: There is strong evidence that the treatment group waited longer to ask for help on average.

21.49: PLAN: We test $H_0: \mu_1 = \mu_2$ versus $H_a: \mu_1 > \mu_2$, where μ_1 is the mean score for the Active group, and μ_2 is the mean score for the Traditional group. SOLVE: We must assume that the data comes from an SRS of the intended populations. The stemplots for each sample show no heavy skew and no outliers. $\bar{x}_1 = 3.6$, $\bar{x}_2 = 4.74$, $s_1 = 2.41$, $s_2 = 2.85$, $n_1 = 15$, and $n_2 = 23$. Of course, because $\bar{x}_1 < \bar{x}_2$, we will not conclude that $\mu_1 > \mu_2$. We find $SE = \sqrt{\frac{s_1^2}{n_1} + \frac{s_2^2}{n_2}} = 0.86$ and $t = \frac{3.6 - 4.74}{SE} = -1.32$, for which $P > 0.50$, regardless of df. If you use software, df = 33.42, and $P = 0.9026$. CONCLUDE: There is no support for a conclusion that active learning yields a higher average score than traditional learning.

21.51: (a) Refer to Exercise 21.49 for details. For 90% confidence, $t^* = 1.761$ (using df = 14) or $t^* = 1.692$ (using df = 33.42). A 90% confidence interval for $\mu_1 - \mu_2$ is $3.6 - 4.74 \pm t^*(0.86)$, or -2.65 to 0.37 (using df = 14) or -2.60 to 0.32 (using df = 33.42). (b) Now we want a 90% confidence interval for the mean change in score for the active class. That is, we construct a 90% confidence interval for μ_1. With 15 observations, we have df = 14, and $t^* = 1.761$. The confidence interval is given by $3.6 \pm 1.761\frac{2.41}{\sqrt{15}} = 2.50$ to 4.70.

21.53: Because this exercise asks for a "complete analysis" without suggesting hypotheses or confidence levels, student responses may vary. This solution gives 95% confidence intervals for the means in parts (a) and (b), and performs a hypothesis test and gives a 95% confidence interval for part (c). Note that the first two problems call for single-sample t procedures (Chapter 20), while the last uses the Chapter 21 procedures. Student answers should be formatted according to the "four-step process" of the text; these answers are not formatted as such, but can be used to check student results. We begin with summary statistics.

	n	$\bar{x}$	s
Women	95	4.2737	2.1472
Men	81	6.5185	3.3471

A back-to-back stemplot of responses for men and women reveals that the distribution of claimed drinks per day for women is slightly skewed, but has no outliers. For men, the distribution is only slightly skewed but contains four outliers. However, these outliers are not too extreme. In all problems, it seems use of t procedures is reasonable.

(a) We construct a 95% confidence interval for μ_w, the mean number of claimed drinks for women. Here, $t^* = 1.990$ (df = 80 in Table C) or $t^* = 1.9855$ (df = 94, software), and $SE = 2.1472/\sqrt{95} = 0.2203$. A 95% confidence interval for μ_w is $4.2737 \pm 1.990(0.223) = 3.84$ to 4.71 drinks. The interval using software is virtually the same. With 95% confidence, the mean number of claimed drinks for women is between 3.84 and 4.71 drinks. (b) We construct a 95% confidence interval for μ_m, the mean number of claimed drinks for men. Here, $t^* = 1.990$ (df = 80 in Table C or software), and $SE = 3.3471/\sqrt{81} = 0.3719$. A 95% confidence interval for μ_m is $6.5185 \pm 1.990(0.3719) = 5.78$ to 7.26 drinks. With 95% confidence, the mean number of claimed drinks for men is between 5.78 and 7.26 drinks. (c) We test $H_0: \mu_m = \mu_w$ versus $H_a: \mu_m \neq \mu_w$. We have $SE = \sqrt{\frac{2.1472^2}{95} + \frac{3.3471^2}{81}} = 0.4322$, and $t = \frac{4.2737 - 6.5185}{SE} = -5.193$. Regardless of the choice of df (80 or 132.15), this is highly significant ($P < 0.001$). We have very strong evidence that the claimed number of drinks is different for men and women. To construct a 95% confidence interval for $\mu_m - \mu_w$, we use $t^* = 1.990$ (df = 80) or $t^* = 1.9781$ (df = 132.15). Using $\bar{x}_1 - \bar{x}_2 \pm t^*\sqrt{\frac{s_1^2}{n_1} + \frac{s_2^2}{n_2}}$, we obtain either 2.2448 ± 0.8601 or 2.2448 ± 0.8549. After rounding either interval, we report that with 95% confidence, on average, sophomore men who drink claim an additional 1.4 to 3.1 drinks per day compared with sophomore women who drink.

Chapter 22 Inference for a Population Proportion

22.1: (a) The population is surgical patients. p is the proportion of all surgical patients who will test positive for *Staphylococcus aureus*. (b) $\hat{p} = \frac{1251}{6771} = 0.185$, or 18.5%.

22.3: (a) Because $np = 0.90(1500) = 1350$ and $n(1 - p) = 0.10(1500) = 150$ and both are at least 10, the sampling distribution of $\hat{p}$ is approximately Normal with mean $p = 0.90$ and standard deviation $\sqrt{\frac{p(1-p)}{n}} = \sqrt{\frac{0.90(1-.90)}{1500}} = 0.0077$. (b) If $n = 6000$, the sampling distribution of $\hat{p}$ is approximately Normal with mean $p = 0.90$ and standard deviation $\sqrt{\frac{p(1-p)}{n}} = \sqrt{\frac{0.90(1-0.90)}{6000}} = 0.0039$.

22.5: (a) The survey excludes residents of the northern territories, as well as those who have no phones or have only cell phone service. (b) $\hat{p} = \frac{1288}{1505} = 0.8558$ so $SE = \sqrt{\frac{\hat{p}(1-\hat{p})}{n}} = 0.009050$; the 95% confidence interval is $0.8558 \pm (1.96)(0.009055) = 0.8381$ to 0.8735, or 83.8% to 87.4%.

22.7: (a) $\hat{p} = \frac{42}{165} = 0.255$, so the margin of error is $1.96\sqrt{\frac{0.255(1-0.255)}{165}} = 0.0665$. (b) For a $\pm 3\%$ margin of error, we'll need at least $n = \left(\frac{1.96}{0.03}\right)^2(0.255)(1 - 0.255) = 810.898$, so 811 visitors over age 65.

22.9: SOLVE: Let p be the proportion of times a spun Belgian euro coin lands heads. We test $H_0: p = 0.50$ vs. $H_a: p \neq 0.50$. Because the sample consists of 250 trials, we expect 125 "successes" (heads) and 125 "failures" (tails); both of these are at least 10 and we assume the sample represents an SRS of all possible coin spins, so conditions are met. $\hat{p} = \frac{140}{250} = 0.56$ $SE = \sqrt{\frac{p_0(1-p_0)}{n}} = \sqrt{\frac{0.50(1-0.50)}{250}} = 0.0316$. $z = \frac{\hat{p}-p_0}{SE} = \frac{0.56-0.50}{0.0316} = 1.90$, and $P = 0.0574$. CONCLUDE: There is some evidence that the proportion of times a Belgian euro coin spins heads is not 0.50; the P-value is close to 0.05, but not less than 0.05. Perhaps more spins would be conclusive.

22.11: (a) The number of trials is not large enough. The expected number of successes (heads) and the expected number of failures (tails) are both 5. These should be 10 or more. (b) As long as the sample can be viewed as an SRS, a z test for a proportion can be used. (c) Under the null hypothesis, we expect only $200(0.01) = 2$ failures. We should have at least 10 expected failures and at least 10 expected successes.

22.13: The large-sample conditions are met because we had 113 people who have experienced computer crime and $1025 - 113 = 912$ people who have not; both are at least 15. (a) $\hat{p} = \frac{113}{1025} = 0.1102$, and $SE_{\hat{p}} = \sqrt{\frac{\hat{p}(1-\hat{p})}{1025}} = 0.0098$. A 95% confidence interval for p is given by $0.1102 \pm 1.96(0.0098) = 0.0910$ to 0.1294. (b) Using the plus four method, $\tilde{p} = \frac{113+2}{1025+4} = 0.1118$, and $SE_{\tilde{p}} = \sqrt{\frac{\tilde{p}(1-\tilde{p})}{1029}} = 0.0098$. A 95% confidence interval for p is given by $0.1118 \pm 1.96(0.0098) = 0.0926$ to 0.1310, or 9.3% to 13.1%. These intervals are virtually identical, but the plus four confidence interval is very slightly shifted to the right.

22.15: (b) The sampling distribution of $\hat{p}$ has mean $p = 0.60$.

22.17: (b) The 90% confidence interval is $0.80 \pm 1.645\sqrt{\frac{0.80(1-0.80)}{4500}}$.

22.19: (c) $n = \left(\frac{z^*}{m}\right)^2 p^*(1-p^*) = \left(\frac{2.576}{0.02}\right)^2(0.5)(0.5) = 4147.36$, so we would need at least 4148.

22.21: (a) Sources of bias are not accounted for in a margin of error.

22.23: (c) The P-value is 0.0057.

22.25: (a) The survey excludes those who have no phones or have only cell-phone service. (b) We have 848 "Yes" answers and 162 "No" answers; both of these are at least 15. With the sample proportion $\hat{p} = \frac{848}{1010} = 0.8396$, the large sample 95% confidence interval is $0.8396 \pm 1.96\sqrt{\frac{0.8396(1-0.8396)}{1010}} = 0.8170$ to 0.8622.

22.27: (a) $\hat{p} = \frac{848}{1010} = 0.8396$, $SE = 0.01155$, so the margin of error is $1.96SE = 0.02263 = 2.26\%$. (b) If instead $\hat{p} = 0.50$, then $SE = 0.01573$ and the margin of error for 95% confidence would be $1.96\ SE = 0.03084 = 3.08\%$. (c) For samples of about this size, the margin of error is no more than about ±3% no matter what $\hat{p}$ is.

22.29: (a) The survey excludes residents of Alaska and Hawaii, and those who do not have cell-phone service. (b) We have 422 successes and 2063 failures (both at least 15), so the sample is large enough to use the large-sample confidence interval. We have $\hat{p} = \frac{422}{2485} = 0.1698$, and $SE = \sqrt{\frac{0.1698(1-0.1698)}{2485}} = 0.0075$. For 90% confidence, the margin of error is $1.645SE = 0.0124$ and the confidence interval is 0.1574 to 0.1822, or 15.7% to 18.2%. (c) Perhaps people who use the cell phone to search for information online are younger, and more interested in sexually related topics.

22.31: (a) Large-sample methods are safe because we have 171 successes and $880 - 171 = 709$ failures. We have $\hat{p} = \frac{171}{880} = 0.1943$, $SE = \sqrt{\frac{0.1943(1-0.1943)}{880}} = 0.01334$, the margin of error is $1.96SE = 0.02614$, and the 95% confidence interval is 0.1682 to 0.2204. (b) It is likely that more than 171 respondents have run red lights. We would not expect people to claim that they have run red lights when they have not, but some people will deny running red lights when they have.

22.33: (a) The margins of error are $1.96\sqrt{\frac{\hat{p}(1-\hat{p})}{100}} = 0.196\sqrt{\hat{p}(1-\hat{p})}$ (below). (b) With $n = 500$, the margins of error are $1.96\sqrt{\frac{\hat{p}(1-\hat{p})}{500}} = 0.088\sqrt{\hat{p}(1-\hat{p})}$. The new margins of error are less than half their former size.

	$\hat{p}$	0.1	0.3	0.5	0.7	0.9
(a)	m.e.	0.0588	0.0898	0.0980	0.0898	0.0588
(b)	m.e.	0.0263	0.0402	0.0438	0.0402	0.0263

22.35: PLAN: With p representing the proportion of songs downloaded by Rina, we test H_0: $p = 0.50$ versus H_a: $p \neq 0.50$. The test is two-sided because we wonder if the proportion loaded by Rina *differs* from that loaded by Ed. SOLVE: We assume that the 50 songs sampled are an SRS. With 50 songs, we expect $50(0.50) = 25$ each successes and failures (both at least 10), so conditions for use of the test are satisfied. $\hat{p} = \frac{34}{50} = 0.68$, so $z = \frac{0.68-0.50}{\sqrt{\frac{0.50(1-0.50)}{50}}} = 2.55$ and $P = 0.0108$. CONCLUDE: There is strong evidence that the proportion of songs downloaded by Rina differs from 0.50. In fact, it seems that Rina downloaded more than Ed. (b) The conditions for a large sample confidence interval are met because 34 of the sample's 50 songs were downloaded by Rina and 16 by Ed; both of these are larger than 15. The 95% confidence interval for the proportion downloaded by Rina is $0.68 \pm 1.96\sqrt{\frac{0.68(1-0.68)}{50}} = 0.5507$ to 0.8093. At 95% confidence, Rina has downloaded between about 55% and 81% of the songs on their player.

22.37: (a) H_0: $p = 0.15$ and H_a: $p > 0.15$. Here, we expect $0.15(61) = 9.15 < 10$ "successes"; there is a possible problem with the use of the z test in this case. Whether we can view this particular class as a simple random sample (of all this teacher's students? of all AP statistics students?) is questionable. The chance of obtaining a sample proportion at least as large as $\hat{p} = \frac{15}{61} = 0.2459$ if the true proportion is $p = 0.15$ is 0.0178 (1.78%) because $z = \frac{0.2459-0.15}{\sqrt{0.15(1-0.15)/61}} = 2.10$. (c) Answers will vary. This was not a designed, randomized experiment, so we cannot say the cash incentive "caused" the increase in 5s.

22.39: (a) PLAN: H_0: $p = 0.50$ and H_a: $p \neq 0.50$. The alternate is two-sided because we want to know if subjects are *not* equally likely to choose either of the two positions. SOLVE: We assume we have an SRS from the population. With 32 subjects, we expect 16 successes (people that picked the first wine) and 16 failures (people that picked the second wine). $\hat{p} = \frac{22}{32} = 0.6875$ and $z = \frac{0.6875 - 0.50}{\sqrt{0.50(0.50)/32}} = 2.12$ with P-value $2(0.0170) = 0.0340$. CONCLUDE: We have strong evidence that people are not equally likely to choose either of two options. It appears they are more likely to select the first presented wine as their preference. (b) People that would go out of their way to participate in such a study are presumed to represent the population of all wine drinkers (or adults). The assumption that we have a simple random sample may not be reasonable.

22.41: PLAN: We obtain the sample size required to estimate the proportion of wine tasters that select the first choice to within ±0.05 with 95% confidence. SOLVE: We guess that the unknown value of p is 0.6875, as computed in Exercise 22.39. $n = (\frac{z^*}{m})^2 p^*(1 - p^*) = (\frac{1.96}{0.05})^2(0.6875)(1 - 0.6875) = 330.14$, so take $n = 331$. CONCLUDE: We require a sample of at least 331 wine tasters in order to estimate the proportion that would choose the first option to within 0.05 with 95% confidence.

22.43: PLAN: We will give a 90% confidence interval for the proportion of all *Krameria cytisoides* shrubs that will resprout after fire. SOLVE: We assume that the 12 shrubs in the sample can be treated as an SRS. Because the number of resprouting shrubs is just 5, the conditions for a large sample interval are not met. Using the plus four method: $\tilde{p} = \frac{5 + 2}{12 + 4} = 0.4375$, SE $= 0.1240$, the margin of error is $1.645\text{SE} = 0.2040$, and the 90% confidence interval is 0.2335 to 0.6415. CONCLUDE: We are 90% confident that the proportion of *Krameria cytisoides* shrubs that will resprout after fire is between about 0.23 and 0.64.

Chapter 23 Comparing Two Proportions

23.1: PLAN: Let p_M be the proportion of all males who have used the Internet to search for health information and p_F be the proportion of females who have done so. We want a 95% confidence interval for the difference in these proportions. SOLVE: The samples were large with clearly more than 10 "successes" and "failures" in each sample. $\hat{p}_F = \frac{811}{1308} = 0.6200$ and $\hat{p}_M = \frac{520}{1084} = 0.4797$. SE $= \sqrt{\frac{0.62(1 - 0.62)}{1308} + \frac{0.4797(1 - 0.4797)}{1084}} = 0.0203$. $(0.62 - 0.4797) \pm 1.96(0.0203) = 0.1403 \pm 0.0398$. CONCLUDE: We are 95% confident that between 10% and 18% more women than men have looked for health information on the Internet.

23.3: PLAN: Let p_1 be the proportion of 18- to 29-year-olds who think claims about the environment are exaggerated and p_2 be the proportion of those 60 and older who think this. We want a 95% confidence interval for the differences in these proportions. SOLVE: The samples were large with clearly more than 10 "successes" and "failures" in each sample. $\hat{p}_2 = \frac{174}{376} = 0.4628$ and $\hat{p}_1 = \frac{75}{251} = 0.2988$. SE $= \sqrt{\frac{0.4628(1 - 0.4628)}{376} + \frac{0.2988(1 - 0.2988)}{251}} = 0.0387$. $(0.4628 - 0.2988) \pm 1.96(0.0387) = 0.164 \pm 0.0759$. CONCLUDE: Based on these samples, we are 95% confident that between about 8.8% and 24.0% more people 60 and older believe that claims about the environment are exaggerated than people 18 to 29 years old.

23.5: PLAN: Let p_1 and p_2 be (respectively) the proportions of injured skiers and injured snowboarders who wear helmets. H_0: $p_1 = p_2$ versus H_a: $p_1 < p_2$. SOLVE: The smallest count is 96, so conditions are met. $\hat{p}_1 = \frac{96}{578} = 0.1661$ and $\hat{p}_2 = \frac{656}{2992} = 0.2193$. $\hat{p} = \frac{96 + 656}{578 + 2992} = 0.2106$. SE $= \sqrt{\hat{p}(1 - \hat{p})(\frac{1}{578} + \frac{1}{2992})} = 0.01853$. $z = \frac{0.1661 - 0.2193}{0.01853} = -2.87$, and $P = 0.0021$. CONCLUDE: We have strong evidence (significant at $\alpha = 0.01$) that skiers and snowboarders with head injuries are less likely to use helmets than skiers and snowboarders without head injuries.

23.7: (a) One count is only 6, and guidelines for using the large-sample confidence interval call for all counts to be at least 10. (b) The new sample sizes are 55 and 110, and success counts are 7 and 46. (c) $\tilde{p}_1 = \frac{6 + 1}{53 + 2} = 0.1273$ and $\tilde{p}_2 = \frac{45 + 1}{108 + 2} = 0.4182$. A plus four 95% confidence interval for $p_1 - p_2$ is $(0.1273 - 0.4182) \pm 1.96\sqrt{\frac{\tilde{p}_1(1 - \tilde{p}_1)}{55} + \frac{\tilde{p}_2(1 - \tilde{p}_2)}{110}} = -0.2909 \pm 0.1275 = -0.4184$ to -0.1634. With 95% confidence, among injured skaters the difference between proportion with wrist guards and those without is between −41.8% and −16.3%. It appears that more injured skaters fail to wear wrist guards.

23.9: (a) The question is, "Is there a difference?"

23.11: (b) $\hat{p} = \frac{459 + 552}{573 + 719}$

23.13: (b) SE $= \sqrt{\frac{0.801(1 - 0.801)}{573} + \frac{0.768(1 - 0.768)}{719}} = 0.0229$. The margin of error is $1.645(0.0229) = 0.0377$.

23.15: (b) We have only three failures in the treatment group and only two successes in the control group.

23.17: (a) The four counts are 117, 53, 152, and 165, so all counts are large enough. (b) $\hat{p}_1 = \frac{117}{170} = 0.6882$, and $\hat{p}_2 = \frac{152}{317} = 0.4795$; the 95% confidence interval is $\hat{p}_1 - \hat{p}_2 \pm 1.96\sqrt{\frac{\hat{p}_1(1 - \hat{p}_1)}{170} + \frac{\hat{p}_2(1 - \hat{p}_2)}{317}} = 0.2087 \pm 0.0887$. Based on these samples, between 12% and 29.7% more younger teens than older teens have posted false information in their online profiles, at 95% confidence.

23.19: (a) One of the counts is 0. (b) The new sample size for the treatment group is 35, 24 of which have tumors; the new sample size for the control group is 20, 1 of which has a tumor. (c) $\tilde{p}_1 = \frac{23 + 1}{33 + 2} = 0.6857$ and $\tilde{p}_2 = \frac{0 + 1}{18 + 2} = 0.05$. The plus four 99% confidence interval is $\tilde{p}_1 - \tilde{p}_2 \pm 2.576\sqrt{\frac{\tilde{p}_1(1 - \tilde{p}_1)}{35} + \frac{\tilde{p}_2(1 - \tilde{p}_2)}{20}} = 0.6357 \pm 0.2380$. We are 99% confident that lowering DNA methylation increases the incidence of tumors by between about 40% and 87%.

23.21: (a) Let p_1 and p_2 be (respectively) the proportions of subjects in the music and no music groups that receive a passing grade on the Maryland HSA. H_0: $p_1 = p_2$ versus H_a: $p_1 \neq p_2$. $\hat{p}_1 = \frac{2818}{3239} = 0.870$, $\hat{p}_2 = \frac{2091}{2787} = 0.750$, $\hat{p} = \frac{2818 + 2091}{3239 + 2787} = 0.815$. $z = \frac{\hat{p}_1 - \hat{p}_2}{\sqrt{\hat{p}(1-\hat{p})\left(\frac{1}{3239} + \frac{1}{2787}\right)}} = 11.94$. $P < 0.0001$. We have overwhelming evidence that the proportion of music students passing the Maryland HSA is greater than that for the no music group. (b) and (c) This is an observational study—people who choose to (or can afford to) take music lessons differ in many ways from those who do not. We cannot conclude that music causes an improvement.

23.23: We have at least 10 successes and 10 failures in both samples. For the music group, $\hat{p}_1 = \frac{2818}{3239} = 0.870$. For the no music group, $\hat{p}_2 = \frac{2091}{2787} = 0.750$. $\hat{p}_1 - \hat{p}_2 \pm 1.96\sqrt{\frac{\hat{p}_1(1-\hat{p}_1)}{3239} + \frac{\hat{p}_2(1-\hat{p}_2)}{2787}} = 0.100$ to 0.140, or 10.0% to 14.0%.

23.25: (a) H_0: $p_M = p_F$ versus H_a: $p_M \neq p_F$. $\hat{p}_M = \frac{15}{106} = 0.1415$, $\hat{p}_F = \frac{7}{42} = 0.1667$, and $\hat{p} = 0.1486$. Then, $SE = \sqrt{\hat{p}(1-\hat{p})(\frac{1}{106} + \frac{1}{42})} = 0.06485$, so $z = \frac{\hat{p}_M - \hat{p}_F}{0.06485} = -0.39$. $P = 0.6966$, which provides no evidence of a difference in failure rates. (b) $\hat{p}_M = \frac{450}{3180} = 0.1415$, $\hat{p}_F = \frac{210}{1260} = 0.1667$, and $\hat{p} = 0.1486$, but now $SE = \sqrt{\hat{p}(1-\hat{p})(\frac{1}{3180} + \frac{1}{1260})} = 0.01184$, so $z = \frac{\hat{p}_M - \hat{p}_F}{0.01184} = -2.13$ and $P = 0.0332$. (c) We are asked to construct two confidence intervals—one based on the smaller samples of part (a) and one based on the larger samples of part (b). For case (a), $\hat{p}_M = 0.1415$ and $\hat{p}_F = 0.1667$, so a 95% confidence interval $\hat{p}_M - \hat{p}_F \pm 1.96\sqrt{\frac{\hat{p}_M(1-\hat{p}_M)}{106} + \frac{\hat{p}_F(1-\hat{p}_F)}{42}} = -0.156$ to 0.1056. We note that because there were only seven business failures in those headed by women, this interval is not really appropriate. For case (b), $\hat{p}_M = 0.1415$ and $\hat{p}_F = 0.1667$. The resulting confidence interval is $\hat{p}_M - \hat{p}_F \pm 1.96\sqrt{\frac{\hat{p}_M(1-\hat{p}_M)}{3180} + \frac{\hat{p}_F(1-\hat{p}_F)}{1260}} = -0.0491$ to -0.0013.

23.27: PLAN: Let p_W be the proportion of whites who use social networking sites and p_H be the proportion of Hispanics who use them. We want to know if the proportions are different, so we will test H_0: $p_W = p_H$ versus H_a: $p_W \neq p_H$. SOLVE: All sample counts are larger than 10 (the smallest is 43). $\hat{p}_W = \frac{866}{1332} = 0.6502$ and $\hat{p}_H = \frac{111}{154} = 0.7208$. $\hat{p} = \frac{866 + 111}{1332 + 154} = 0.6575$.

$$z = \frac{0.6502 - 0.7208}{\sqrt{0.6575(1 - 0.6575)\left(\frac{1}{1332} + \frac{1}{154}\right)}} = -1.75. \quad P\text{-value} = 0.0801.$$

CONCLUDE: At the 0.05 level, we fail to reject H_0. This survey has failed to find a difference in the proportions of whites and Hispanics who use social networking sites.

23.29: PLAN: We construct a 99% confidence interval for $p_1 - p_2$, where p_1 denotes the proportion of people on Chantix who abstained from smoking, and p_2 is the corresponding proportion for the placebo population. SOLVE: the smallest count was 61, so the large-sample procedures are safe. $\hat{p}_1 = \frac{155}{352} = 0.4403$, and $\hat{p}_2 = \frac{61}{344} = 0.1773$. The 99% confidence interval is $\hat{p}_1 - \hat{p}_2 \pm 2.576\sqrt{\frac{\hat{p}_1(1-\hat{p}_1)}{352} + \frac{\hat{p}_2(1-\hat{p}_2)}{344}} = 0.2630 \pm 0.0864$. CONCLUDE: We are 99% confident that the success rate for abstaining from smoking is between 17.7 and 34.9 percentage points higher for smokers using Chantix than for smokers on a placebo.

23.31: PLAN: Let p_1 be the proportion of people that will reject an unfair offer from another person, and p_2 be the proportion for offers from a computer. We test H_0: $p_1 = p_2$ versus H_a: $p_1 > p_2$. SOLVE: All counts are greater than 5, so the conditions are met. $\hat{p}_1 = \frac{18}{38} = 0.4737$ and $\hat{p}_2 = \frac{6}{38} = 0.1579$. $\hat{p} = \frac{18 + 6}{38 + 38} = 0.3158$, and $SE = \sqrt{\hat{p}(1-\hat{p})(\frac{1}{38} + \frac{1}{38})} = 0.1066$. $z = \frac{0.4737 - 0.1579}{0.1066} = 2.96$, $P = 0.0015$. CONCLUDE: There is very strong evidence that people are more likely to reject an unfair offer from another person than from a computer.

23.33: PLAN: Let p_1 and p_2 be (respectively) the proportions of mice ready to breed in good acorn years and bad acorn years. We give a 90% confidence interval for $p_1 - p_2$. SOLVE: One count is only 7, so we use the plus four method. $\tilde{p}_1 = \frac{54 + 1}{72 + 2} = 0.7432$, and $\tilde{p}_2 = \frac{10 + 1}{17 + 2} = 0.5789$, $\tilde{p}_1 - \tilde{p}_2 \pm 1.645\sqrt{\frac{\tilde{p}_1(1-\tilde{p}_1)}{74} + \frac{\tilde{p}_2(1-\tilde{p}_2)}{19}} = 0.1643 \pm 0.2042 = -0.0399$ to 0.3685. CONCLUDE: We are 90% confident that the proportion of mice ready to breed in good acorn years is between 0.04 lower than and 0.37 higher than the proportion in bad acorn years.

23.35: (a) This is an experiment because the researchers assigned subjects to the groups. (b) PLAN: Let p_1 and p_2 be (respectively) the proportions that have an RV infection for the HL+ group and control group. H_0: $p_1 = p_2$ versus H_a: $p_1 < p_2$. SOLVE: We have large enough counts (49, 67, 49, 47). $\hat{p}_1 = \frac{49}{49 + 67} = 0.4224$, $\hat{p}_2 = \frac{49}{49 + 47} = 0.5104$, and $\hat{p} = \frac{49 + 49}{116 + 96} = 0.4623$. $SE = \sqrt{\hat{p}(1-\hat{p})(\frac{1}{116} + \frac{1}{96})} = 0.0688$. $z = \frac{0.4224 - 0.5104}{0.0688} = -1.28$, $P = 0.1003$. CONCLUDE: We do not have enough evidence to reject the null hypothesis; there is little evidence to conclude that the proportion of HL+ users with a rhinovirus infection is less than that for non-HL+ users.

Chapter 24 Inference about Variables: Part IV Review

24.1: (c) $ME = 2.056(9.3)/\sqrt{27} = 3.7$

24.3: (b) $t = 2.023$, $df = 13$

24.5: (d) Our estimate is $\hat{p} = 676/1760 = 0.384$.

24.7: (d) The standard error is

$$SE = \sqrt{\frac{0.354(1 - 0.354)}{1734} + \frac{0.384(1 - 0.384)}{1760}} = 0.0163.$$

24.9: (a) The standard error is 0.0171. (b) A 95% confidence interval is 0.497 to 0.563.

24.11: (b) $2.33 \pm 1.984\frac{1.00}{\sqrt{135}}$ using 100 df from Table C.

24.13: (b) This value is from software.

24.15: The standard deviations are larger than the means. Because PedMIDAS scores must be greater than or equal to 0, the distributions must be right skewed. The sample sizes are fairly large ($n = 64$ and 71), so the sample means should be approximately Normal by the central limit theorem.

24.17: (b) Use technology.

24.19: (c) We would have to view these children as random samples from the larger population of children who could be in her class.

24.21: (b) $\hat{p} = 225/757 = 0.297$

24.23: (b) $0.297 \pm 1.645(0.017)$

24.25: (c) It seems reasonable that the researchers suspect that VLBW babies are less likely to graduate from high school.

24.27: (b) $z = \dfrac{0.7397 - 0.8283}{\sqrt{\hat{p}(1-\hat{p})\left(\frac{1}{242}+\frac{1}{233}\right)}} = -2.34$

24.29: (d) $t = \dfrac{86.2 - 89.8}{\sqrt{\frac{13.4^2}{38}+\frac{14^2}{54}}} = -1.25$, and the test is two-sided.

24.31: (b) $z = \dfrac{0.379 - 0.41}{\sqrt{0.41(1-0.41)/348}} = -1.18$, so $P = 0.1190$.

24.33: $\hat{p} = \frac{475}{625} = 0.76$. The interval is $0.76 \pm 1.645 \sqrt{\frac{0.76(1-0.76)}{625}} = 0.732$ to 0.788.

24.35: $H_0: p_{2002} = p_{2012}$ versus $H_a: p_{2002} > p_{2012}$. $\hat{p}_{2002} = \frac{754}{2045} = 0.3687$ and $\hat{p}_{2012} = \frac{842}{2690} = 0.3130$, $\hat{p} = \frac{754+842}{2045+2690} = 0.3371$.

$$z = \frac{0.3687 - 0.3130}{\sqrt{0.3371(1-0.3371)\left(\frac{1}{2045}+\frac{1}{2690}\right)}} = 4.02 \text{ and } P = P(Z > 4.02),$$

which is close to 0. There is extremely strong evidence that the proportion of smokers in Alaska has decreased. How much of the decrease can be attributed to the campaign is uncertain.

24.37: (a) Using technology, $t = 0.392$, df = 3.24.

24.39: In all three cases, the observations must be able to be seen as random and representative samples of both types of diets. Also, the populations must be Normally distributed. Because the sample sizes are very small, this is almost impossible to check with typical graphical methods.

24.41: A large-sample (or plus four) confidence interval for a population proportion

24.43: This is the entire population of Chicago Cubs players. Statistical inference is not appropriate.

24.45: (a) two-sample test or confidence interval for difference in proportions (b) two-sample test or confidence interval for difference in means (c) two-sample test or confidence interval for difference in proportions

24.47: (a) This is a matched pairs situation; the responses of each subject before and after treatment are not independent. (b) We need to know the standard deviation of the differences, not the two, individual, sample standard deviations. (Note that the mean difference is equal to the difference in the means, which is why we only need to know the standard deviation of the differences.)

24.49: (a) $\hat{p} = 80/80 = 1$; the margin of error for 95% confidence (or any level of confidence) is 0 because $z^*\sqrt{\frac{(1)(1-1)}{n}} = 0$. Almost certainly, if more trials were performed, a rat would eventually make a mistake, so the actual success rate is less than 1. (b) The plus four estimate is $\tilde{p} = 82/84 = 0.9762$, and the plus four 95% confidence interval is $\tilde{p} \pm z^*\sqrt{\frac{\tilde{p}(1-\tilde{p})}{n+4}} = 0.9762 \pm 0.0326 = 0.9436$ to 1.0088. Ignoring the upper limit, we are 95% confident that the actual success rate is 0.9436 or greater.

24.51: PLAN: We test $H_0: \mu = 12$ versus $H_a: \mu > 12$, where μ denotes the mean age at first word, in months. SOLVE: We regard the sample as an SRS; a stemplot (not shown) shows that the data are right-skewed with a high outlier (26 months). If we proceed with the t procedures despite this, we find $\bar{x} = 13$ and $s = 4.9311$ months. $t = \frac{13-12}{4.9311/\sqrt{20}} = 0.907$, with df = 19, and $P = 0.1879$. If you delete the outlier mentioned above, $\bar{x} = 12.3158$, $s = 3.9729$, and $t = 0.346$, yielding $P = 0.3665$; the conclusion will not change. CONCLUDE: We cannot conclude that the mean age at first word is greater than one year.

24.53: PLAN: We give a 90% confidence interval for μ, the mean age at first word, measured in months. SOLVE: See results from Exercise 24.51. For df = 19, $t^* = 1.729$, so the 90% confidence interval is $13 \pm 1.729 \frac{4.9311}{\sqrt{20}} = 11.09$ to 14.91 months. CONCLUDE: We are 90% confident that the mean age at first word for normal children is between about 11 and 15 months.

24.55: (a) The design is shown. (b) PLAN: We test $H_0: \mu_B = \mu_C$ versus $H_a: \mu_B \neq \mu_C$. SOLVE: We have $\bar{x}_B = 41.2825$ and $s_B = 0.2550$, and $\bar{x}_C = 42.4925$ and $s_C = 0.2939$; $n_B = n_C = 8$. SE = 0.1376, and $t = \frac{\bar{x}_B - \bar{x}_C}{SE} = -8.79$. With df = 7 (or 13.73 from software), $P < 0.001$. Using software, with df = 13.73, $P = 0.0000$ to four places. There is overwhelming evidence that method B gives darker color on average. However, the magnitude of this difference may be too small to be important in practice.

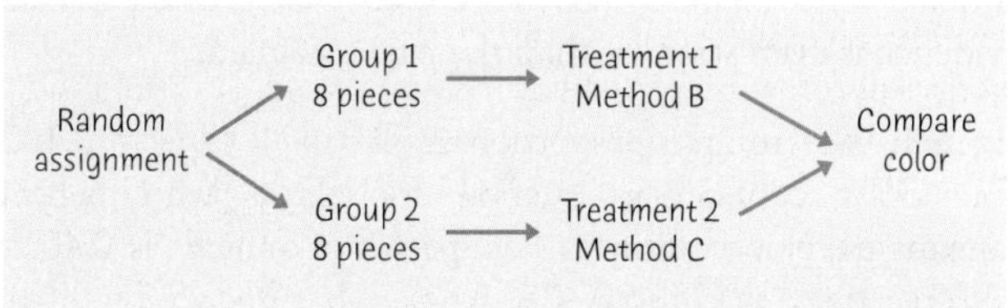

24.57: PLAN: We give a 95% confidence interval for p, the proportion of female students with at least one parent who allows drinking. SOLVE: We are told that the sample represents an SRS. Large sample methods may be used because the number of successes and failures are both greater than 15. With $\hat{p} = 65/94 = 0.6915$, we have SE $= 0.04764$, so the margin of error is 1.96 SE $= 0.09337$, and the interval is 0.5981 to 0.7849. CONCLUDE: With 95% confidence, the proportion of female students who have at least one parent who allows drinking is 0.598 to 0.785.

24.59: (a) PLAN: We want to compare the proportions p_1 (microwaved crackers that show checking) and p_2 (control crackers that show checking). We can do this either by testing hypotheses or with a confidence interval, but because the "microwave checked" count is only 3, significance tests are not appropriate. We will use the plus four procedure and construct a confidence interval for $p_1 - p_2$. SOLVE: We find that $\tilde{p}_1 = \frac{3+1}{65+2} = 0.0597$ and $\tilde{p}_2 = \frac{57+1}{65+2} = 0.8657$. SE $= \sqrt{\frac{\tilde{p}_1(1-\tilde{p}_1)}{67} + \frac{\tilde{p}_2(1-\tilde{p}_2)}{67}} = 0.05073$, and a 95% confidence interval is given by $\tilde{p}_1 - \tilde{p}_2 \pm 1.96(0.0507) = -0.9054$ to -0.7066. CONCLUDE: We are 95% confident that microwaving reduces the percentage of checked crackers by between 70.7% and 90.5%. (b) PLAN: We want to compare μ_1 and μ_2, the mean breaking pressures of microwaved and control crackers. We test H_0: $\mu_1 = \mu_2$ versus H_a: $\mu_1 \neq \mu_2$ and construct a 95% confidence interval for $\mu_1 - \mu_2$. SOLVE: We assume the data can be considered SRSs from the two populations, and that the population distributions are not far from Normal. Now, SE $= 9.0546$ and $t = 6.914$, so the P-value is very small, regardless of whether we use df $= 19$ or df $= 33.27$. A 95% confidence interval for the difference in mean breaking pressures between these cracker types is 43.65 to 81.55 psi (using df $= 19$ and $t^* = 2.093$), or 44.18 to 81.02 psi (using df $= 33.27$ and $t^* = 2.0339$). CONCLUDE: There is very strong evidence that microwaving crackers changes their mean breaking strength. We are 95% confident that microwaving crackers increases their mean breaking strength by between 43.65 and 81.55 psi.

24.61: Two of the counts are too small to perform a significance test safely.

24.63: The group means are $\bar{x}_1 = 5.9$ (insulin), $\bar{x}_2 = 0.75$ (glucose) ng/mL, and the standard deviations are $s_1 = 0.9\sqrt{10} = 2.85$ and $s_2 = 0.2\sqrt{10} = 0.632$ ng/mL. PLAN: We test H_0: $\mu_1 = \mu_2$ versus H_a: $\mu_1 \neq \mu_2$. SOLVE: The estimated standard error of the difference in sample means is SE $= \sqrt{0.9^2 + 0.2^2} = 0.9220$, so $t = \frac{5.9 - 0.75}{\text{SE}} = 5.59$. With either df $= 9$ or df $= 9.89$, $P < 0.001$. CONCLUDE: The evidence is even stronger than the paper claimed.

24.65: The sample proportion is $\hat{p} = 594/1484 = 0.400$. The 95% confidence interval for those who believe humans developed from earlier species of animals is $0.40 \pm 1.96\sqrt{\frac{0.40(0.60)}{1484}}$, 0.375 to 0.425.

Chapter 25 Two Categorical Variables: The Chi-Square Test

25.1: (a) The table provided gives percentages in each category. As an example, there were a total of 1308 surveyed Caucasians. Of these, 181 were under 25; the proportion of Caucasians surveyed who were under 25 is $181/1308 = 0.1384$, which is represented as 13.8% in the table.

	CAUCASIAN	HISPANIC
Under 25	13.8%	24.9%
25 to 34	24.8%	37.9%
35 to 44	17.7%	23.3%
45 and over	43.7%	13.9%

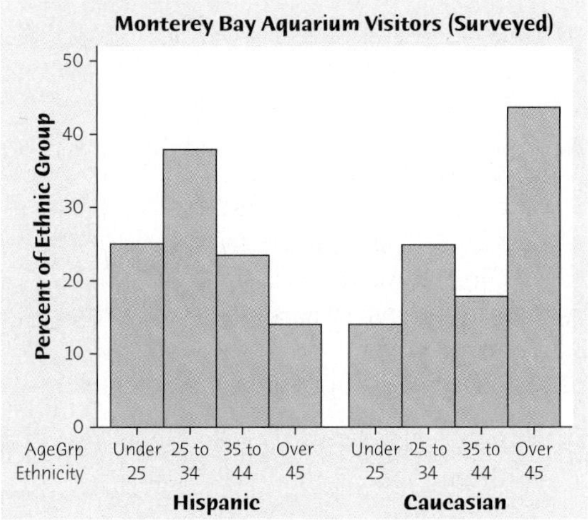

(b) The bar graph clearly reveals that Hispanic visitors tend to be younger; over 40% of Caucasian visitors are older than 45.

25.3: (a) To test H_0: $p_1 = p_2$ versus H_a: $p_1 \neq p_2$ for the proportions of boys who did and did not play video games, we have $\hat{p}_1 = 0.326$ and $\hat{p}_2 = 0.336$. The pooled proportion is $\hat{p} = \frac{450 + 144}{1379 + 429} = 0.3285$, and SE $= 0.02596$, so $z = 0.39$, for which $P = 0.6966$. We conclude that we have no evidence to say the proportion of C students is different for boys who have (and have not) played video games. (b) To test H_0: $p_1 = p_2$ versus H_a: $p_1 \neq p_2$ for the proportions who earn Ds and Fs, we have $\hat{p}_1 = 0.1400$ and $\hat{p}_2 = 0.1865$. The pooled proportion is $\hat{p} = 0.1510$, and SE $= 0.01979$, so $z = -2.35$, for which $P = 0.0188$. There is strong evidence that playing video games changes the proportion of boys who would earn D or F grade averages. (c) If we did three individual tests, we would not know how confident we could be in all three results when taken together.

25.5: (a) Expected counts are below observed counts in the table provided. For example, for Hispanics under 25, the expected count is $\frac{(260)(317)}{1625} = 50.72$. (b) We find that the observed counts are much greater than the expected counts for all three younger age groups of Hispanics. The actual count of Hispanics over 45 (44) is much smaller than the expected count (120).

	Caucasian	Hispanic	All
Under25	181	79	260
	209.3	50.7	
25to34	324	120	444
	357.4	86.6	
35to44	232	74	306
	246.3	59.7	
Over45	571	44	615
	495.0	120.0	
All	1308	317	1625

25.7: (a) All expected counts are well above 5 (the smallest is 50.72). (b) We test H_0: there is no relationship between age and ethnic group for Monterey Bay Aquarium visitors versus H_a: there is a relationship between age and ethnic group. From the SAS output, we have $\chi^2 = 99.6058$ and $P < 0.0001$. (c) The largest contributions generally come from the Hispanics, reflecting that group tends to visit the aquarium at younger ages.

25.9: PLAN: We will carry out a chi-square test for association between education level and opinion about astrology. We test H_0: there is no relationship between education level and astrology opinion versus H_a: there is some relationship between education level and astrology opinion. SOLVE: Examining the output provided in Figure 25.5, we see that all expected cell counts are greater than 5 (the smallest is 25.21), so conditions for use of the chi-square test are satisfied. We see that $\chi^2 = 3.618$ and $P = 0.16$. CONCLUDE: There no evidence of an association between education level and opinion of astrology. Examining the table, we note that for people with graduate degrees, more than expected felt that astrology is not scientific. For people with a junior college degree, more than expected believed that astrology is scientific; these differences were not large enough to believe that the differences were due to something other than sampling variability.

25.11: (a) Because there was one sample that was later categorized by two variables, this is a test of independence. (b) PLAN: We will test H_0: there is no relationship between age and how politically informed the person is versus H_a: there is a relationship between age and how politically informed the person is. SOLVE: Minitab output is provided. We see that all expected cell counts are greater than 5 (the smallest is 5.73), so conditions are satisfied. $\chi^2 = 32.057$ on df = 12, and $P = 0.001$ to three decimal places. CONCLUDE: Age and being informed about politics are related. In particular, we can see that people 20 to 29 years old are less likely to be informed (observed counts are larger than expected for the cells with little awareness and smaller than expected for cells with much awareness). The other large contribution to the chi-square statistic came from people 40 to 49 years old: fewer than expected said they were not at all informed.

	Not at all	A little	Somewhat	Very	Extremely	All
20-29	8	29	28	13	0	78
	10.26	37.18	35.90	16.67	0.00	100.00
	5.73	17.57	31.40	17.57	5.73	78.00
	0.8977	7.4379	0.3680	1.1881	5.7316	*
30-39	15	28	55	23	9	130
	11.54	21.54	42.31	17.69	6.92	100.00
	9.55	29.28	52.33	29.28	9.55	130.00
	3.1062	0.0561	0.1360	1.3474	0.0320	*
40-49	2	25	49	26	14	116
	1.72	21.55	42.24	22.41	12.07	100.00
	8.52	26.13	46.70	26.13	8.52	116.00
	4.9932	0.0487	0.1136	0.0006	3.5180	*
50 and older	21	59	120	79	23	302
	6.95	19.54	39.74	26.16	7.62	100.00
	22.19	68.02	121.57	68.02	22.19	302.00
	0.0640	1.1967	0.0203	1.7716	0.0294	*
All	46	141	252	141	46	626
	7.35	22.52	40.26	22.52	7.35	100.00
	46.00	141.00	252.00	141.00	46.00	626.00
	*	*	*	*	*	*

Cell Contents: Count
% of Row
Expected count
Contribution to Chi-square

Pearson Chi-Square = 32.057, DF = 12, P-Value = 0.001

25.13: (a) df $= (r - 1)(c - 1) = (2 - 1)(3 - 1) = 2$ (b) The computed value (6.739) is between the table values 5.99 and 7.38; we conclude that $0.025 < P < 0.05$, which is consistent with output's reported $P = 0.034$. (c) Under the null hypothesis of no association, the mean value of χ^2 is df = 2. Our computed value is larger than this. The small P-value suggests that random chance does not easily explain the larger than expected value of χ^2.

25.15: (a) If all days were equally likely, we would have $p_1 = p_2 = \ldots = p_7 = \frac{1}{7}$, and we would expect 100 births on each day. (b) The chi-square statistic is then $X^2 = \frac{(84 - 100)^2}{100} + \frac{(110 - 100)^2}{100} + \cdots + \frac{(72 - 100)^2}{100} = 19.12$. (c) df $= 7 - 1 = 6$. From Table D, $\chi^2 = 19.12$ yields $0.0025 < P < 0.005$. Software gives $P = 0.004$. We have strong evidence that births are not spread evenly across the week.

25.17: (a) $15/33 = 0.455$, or 45.5% of subjects chose position 1. Similarly, the percentages for each position are 15.2% for position 2, 6.1% for position 3, and 33.3% for position 4. (b) If subjects are equally likely to select any position, then we would expect $33/4 = 8.25$ subjects in each position. (c) PLAN: We test H_0: $p_1 = p_2 = p_3 = p_4 = \frac{1}{4}$ versus H_a: the four order selection probabilities are not equally likely. SOLVE: As computed above, we expect 8.25 subjects per cell under the null hypothesis, so all expected cell counts exceed 5. Also, we have at least one observation per cell. Conditions for the chi-square test are satisfied. Now, $\chi^2 = \frac{(15 - 8.25)^2}{8.25} + \frac{(5 - 8.25)^2}{8.25} + \frac{(2 - 8.25)^2}{8.25} + \frac{(11 - 8.25)^2}{8.25} = 12.45$. df $= 4 - 1 = 3$. From Table D, $\chi^2 = 12.45$ yields $0.005 < P < 0.01$ ($P = 0.006$ from technology). CONCLUDE: There is strong evidence that positions are not selected with equal probability—some positions are more likely to be selected than others. (d) We see that the largest contributions to χ^2 are from the first and third positions. In the first and fourth cases, we have more observations than expected, and in the second and third positions we have fewer observations than expected. There is evidence of *both* primacy and recency effects.

25.19: (a) 295 + 655 + 239 + 363 = 1552

25.21: (a) For those 23 to 30 years old, the percentage is 26.5%.

25.23: (a) The expected cell count is (1571)(1552)/4111 = 593.09.

25.25: (a) df = $(r - 1)(c - 1) = (4 - 1)(2 - 1) = 3$

25.27: (b) This is the hypothesis of association between "age" and "type of injury."

25.29: (b) We assume that the sample is an SRS, or essentially an SRS from all weightlifting injuries.

25.31: (a) These were separate random samples, so this is a test of homogeneity. (b) PLAN: We test H_0: the distribution of age groups is the same for landline and cell-only individuals versus H_a: the distributions are different. SOLVE: All expected cell counts are more than 5, so the guidelines for the chi-square test are satisfied. We have $\chi^2 = 1032.892$, df = 3, and $P < 0.0005$. CONCLUDE: There is strong evidence of an association between age group and the type of telephone. In fact, the younger age groups were much more likely than expected to have only cellphones.

	Cell Only	Landline	All
Age18-29	374	335	709
	52.75	47.25	100.00
	115	594	709
	587.04	113.20	*
Age30-49	347	1242	1589
	21.84	78.16	100.00
	257	1332	1589
	31.63	6.10	*
Age50-64	146	1625	1771
	8.24	91.76	100.00
	286	1485	1771
	68.75	13.26	*
Age65up	36	1481	1517
	2.37	97.63	100.00
	245	1272	1517
	178.51	34.42	*
All	903	4683	5586
	16.17	83.83	100.00
	903	4683	5586
	*	*	*

Cell Contents: Count
% of Row
Expected count
Contribution to Chi-square

Pearson Chi-Square = 1032.892, DF = 3, P-Value = 0.000

25.33: (a) The diagram is shown. To perform the randomization, label the infants 01 to 77, and choose pairs of random digits. (b) See the Minitab output for the two-way table. We find $\chi^2 = 0.568$, df = 3, and $P = 0.904$. There is no reason to doubt that the randomization "worked."

	Female	Male
NLCP	11	8
	10.36	8.64
	0.03907	0.04689
PBM	11	9
	10.91	9.09
	0.00076	0.00091
PL-LCP	11	8
	10.36	8.64
	0.03907	0.04689
TG-LCP	9	10
	10.36	8.64
	0.17943	0.21531

Cell Contents: Count
Expected count
Contribution to Chi-square

Pearson Chi-Square = 0.568, DF = 3, P-Value = 0.904
Likelihood Ratio Chi-Square = 0.567, DF = 3, P-Value = 0.904

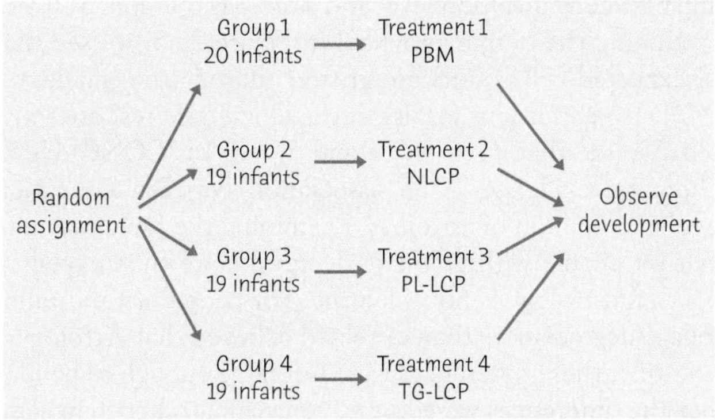

25.35: (a) The two-way table follows. We test $H_0: p_1 = p_2$ versus $H_a: p_1 < p_2$. (b) The z test must be used because the chi-square procedure measures evidence in support of any association and is implicitly two-sided. We have $\hat{p}_1 = 0.3667$ and $\hat{p}_2 = 0.7333$. The pooled proportion is $\hat{p} = \frac{11 + 22}{30 + 30} = 0.55$, and the standard error is SE = 0.12845, so $z = -2.85$ and $P = 0.0022$. We have strong evidence that rats that can stop the shock (and therefore presumably have better attitudes) develop tumors less often than rats that cannot (and therefore are presumably depressed).

	TUMOR	NO TUMOR
Group 1	11	19
Group 2	22	8

25.37: PLAN: We test H_0: there is no relationship between sexual content of ads and magazine audience versus H_a:

there is some relationship between sexual content of ads and magazine audience. SOLVE: Examining the Minitab output in Figure 25.9, we see that conditions for use of the chi-square test are satisfied because all expected cell counts exceed 5 (the smallest is 82.4). We have $\chi^2 = 80.874$ with df = 2, leading to $P < 0.0005$. CONCLUDE: Magazines aimed at women are much more likely to have sexual depictions of models than the other two types of magazines. Specifically, about 39% of ads in women's magazines show sexual depictions of models, compared with 21% and 17% of ads in general-audience and men's magazines, respectively. The two women's chi-squared terms account for over half of the total chi-square value.

25.39: We need cell counts, not just percentages. If we had been given the number of travelers in each group—leisure and business—we could have estimated this.

25.41: In order to do a chi-square test, each subject can only be counted once. In this experiment, each individual is represented for both treatments (carob and chocolate).

25.43: (a) We test H_0: there is no relationship between degree held and service attendance versus H_a: there is some relationship between degree held and service attendance. Examining the Minitab output, $\chi^2 = 14.19$ with df = 3, *P*-value = 0.003. There is strong evidence of an association between degree held and service attendance.

	HS	JColl	Bachelor	Graduate	All
No	880 842.7	101 107.3	232 248.9	105 119.2	1318 1318.0
Yes	400 437.3	62 55.7	146 129.1	76 61.8	684 684.0

Cell Contents: Count
Expected count

Pearson Chi-Square = 14.190, DF = 3, P-Value = 0.003

(b) The new table is shown. We find $\chi^2 = 0.73$ on df = 2. *P* = 0.694. In this table, we find no evidence of association between religious service attendance and degree held.

	JColl	Bachelor	Graduate	All
No	101 98.9	232 229.3	105 109.8	438 438.0
Yes	62 64.1	146 148.7	76 71.2	284 284.0

Cell Contents: Count
Expected count

Pearson Chi-Square = 0.729, DF = 2, P-Value = 0.694

(c) The new table is shown. We have $\chi^2 = 13.40$ with df = 1. *P* = 0.000 to three decimal places (it's actually 0.0002). There is overwhelming evidence of association between level of education (high school versus beyond high school) and religious service attendance.

	BeyondHS	HSchool	All
Attend	284 246.7	400 437.3	684 684.0
NoAttend	438 475.3	880 842.7	1318 1318.0

Cell Contents: Count
Expected count

Pearson Chi-Square = 13.416, DF = 1, P-Value = 0.000

(d) In general, we find that people with degrees beyond high school attend services more often than expected; people with high school degrees attend services less often than expected. Of those with high school degrees, 31.3% attended services, and the percentages are 38.0%, 38.6%, and 42.0%, respectively, for people with junior college, bachelor's, and graduate degrees.

25.45: PLAN: We use a chi-square test to test H_0: there is no relationship between race and opinion about schools versus H_a: there is some relationship between race and opinion about schools. SOLVE: All expected cell counts exceed 5 (the smallest is 21.26), so use of a chi-square test is appropriate. We find that $\chi^2 = 22.426$ with df = 8, and *P* = 0.004. Nearly half of the total chi-square comes from the first two terms; most of the rest comes from the second and fifth rows. CONCLUDE: We have strong evidence of a relationship between race and opinion of schools. Specifically, according to the sample (as illustrated in the table), blacks are less likely and Hispanics are more likely to consider schools to be excellent, while Hispanics and whites differ in the percent considering schools good. Also, a higher percentage of blacks rated schools as "fair."

	Black	Other	White	All
Exclnt	12 22.70	22 22.59	34 22.70	68 68.00
Good	69 68.45	81 68.11	55 68.45	205 205.00
Fair	75 65.44	60 65.12	61 65.44	196 196.00
Poor	24 24.04	24 23.92	24 24.04	72 72.00
Don't Know	22 21.37	14 21.26	28 21.37	64 64.00

Cell Contents: Count
Expected count

Pearson Chi-Square = 22.426, DF = 8, P-Value = 0.004

25.47: PLAN: We compare how the number of children per group has changed from 2009 through 2013 at the Monterey Bay Aquarium. We will create a bar graph and do a chi-square test of homogeneity (each year is a separate sample). SOLVE: To examine any possible change in the number of children per group, we first look at a segmented bar graph of the data. The graph indicates that the number of groups with no children has been steadily decreasing and the number of groups with three or more children has been increasing slightly. There is some fluctuation in the number of groups with one or two children, but 2013 definitely has more of these groups than 2009.

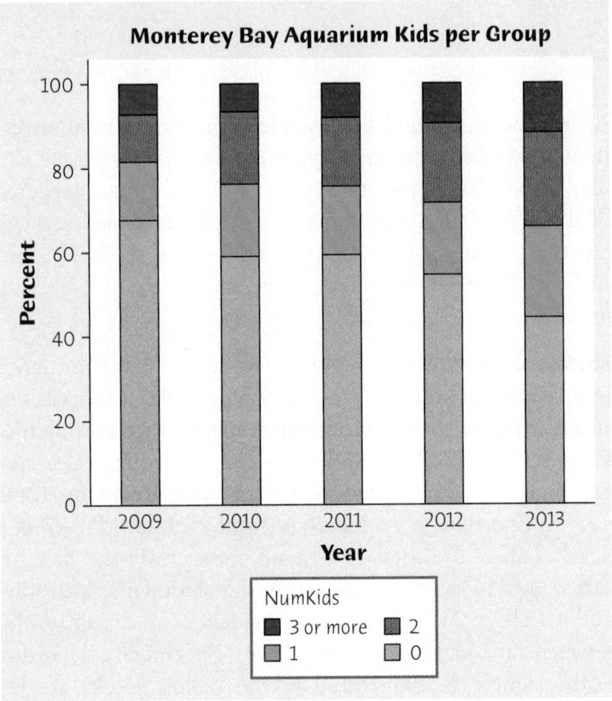

First, we note that all expected counts are above 5 (the smallest is 49.3). The test statistic is $\chi^2 = 75.248$ with df = 12. With P-value < 0.0005, we reject the null hypothesis that there has been no change in the distribution of the number of children per group. It seems clear (especially when comparing 2009 to 2013) that they are meeting their goal of attracting a younger audience.

Rows: NumKids Columns: Year

	2009	2010	2011	2012	2013	All
0	390	330	342	318	268	1648
	327.6	316.7	327.6	331.5	344.6	1648.0
1	79	94	93	99	131	496
	98.6	95.3	98.6	99.8	103.7	496.0
2	66	96	96	111	137	506
	100.6	97.3	100.6	101.8	105.8	506.0
3 or more	41	37	45	55	70	248
	49.3	47.7	49.3	49.9	51.9	248.0
All	576	557	576	583	606	2898
	576.0	557.0	576.0	583.0	606.0	2898.0

Cell Contents: Count
Expected count

Pearson Chi-Square = 75.248, DF = 12, P-Value = 0.000

25.49: PLAN: We compare the percentages leaning toward each party within each education group. SOLVE: The requested table is provided. At each education level, we compute the percentage leaning toward each party. For example, among bachelor's degree holders, 166/(166 + 136) = 54.97% lean Democrat, while the other 45.03% lean Republican.

	Bachelor	Graduate	Highschool	JrCollege	None	All
Dem	166	119	454	68	127	934
	54.97	69.19	59.66	57.14	70.95	60.93
Rep	136	53	307	51	52	599
	45.03	30.81	40.34	42.86	29.05	39.07

Cell Contents: Count
% of Column

CONCLUDE: At every education level, people leaning Democrat outweigh people leaning Republican. The difference is greatest at the "None" level of education, then decreases until the party support is nearly equal for bachelor's degree holders. Among graduate degree holders, Democrats strongly outnumber Republicans.

Chapter 26 Inference for Regression

26.1: (a) A scatterplot of the data is provided, along with the least-squares regression line. From software, $r = 0.985$.

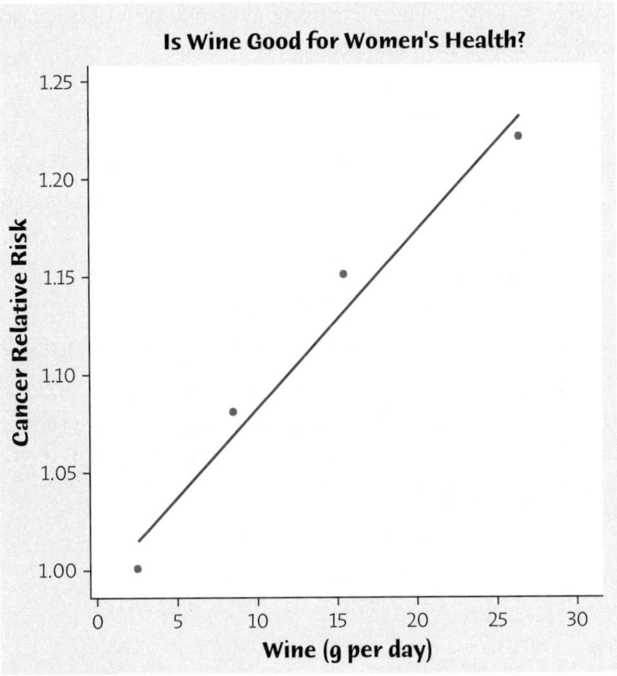

(b) If we knew it, the slope β would tell us how much the relative risk of breast cancer changes in women for each increase of 1 gram of wine per day (on average). We estimate that an increase in intake of 1 gram per day increases relative risk of breast cancer by about 0.009. According to our estimate, wine intake of 0 grams per day is associated with a relative risk of breast cancer of 0.9931 (about 1). (c) $\hat{y} = 0.9931 + 0.0090x$. $s = \sqrt{\frac{0.00079}{4-2}} = 0.01987$.

x	y	$\hat{y}$	RESIDUAL $y - \hat{y}$	$(y - \hat{y})^2$
2.5	1.00	1.0156	−0.0156	0.00024
8.5	1.08	1.0697	0.0103	0.00011
15.5	1.15	1.1328	0.0172	0.00030
26.5	1.22	1.2319	−0.0119	0.00014
			0	0.00079

26.3: (a) A scatterplot of discharge by year is provided, along with the fitted regression line. Discharge seems to be increasing over time, but there is also a lot of variation in this trend, and our impression is easily influenced by the most recent years' data. $r^2 = 0.215$, so the least-squares regression line explains 21.5% of the total observed variability in Arctic discharge. (b) $\hat{y} = -3362 + 2.6327x$, $s = 110.477$.

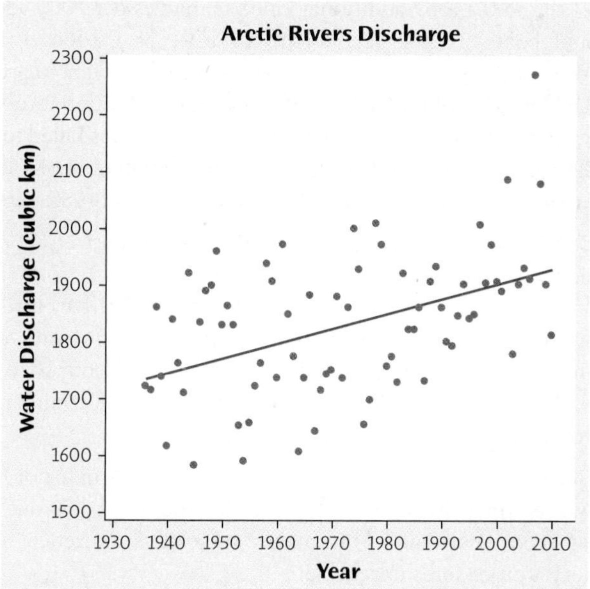

26.5: $H_0: \beta = 0$ versus $H_a: \beta > 0$. We compute $t = \frac{b}{SE_b} = \frac{2.6327}{0.5893} = 4.47$. df $= n - 2 = 75 - 2 = 73$. In referring to Table C, we round df down to df $= 60$; $P < 0.0005$. Using software, $P = 0.000$ There is strong evidence of an increase in Arctic discharge over time.

26.7: (a) $H_0: \beta = 0$ versus $H_a: \beta > 0$, $t = 8.104$ with df $= 2$ (all from software). $0.005 < P < 0.01$. This test is equivalent to testing H_0: population correlation $= 0$ versus H_a: population correlation > 0. (b) $r = 0.985$. From Table E with $n = 4$, $0.005 < P < 0.01$.

26.9: $t^* = 2.920$ (df $= 4 - 2 = 2$, with 90% confidence). $0.009012 \pm 2.920(0.001112) = 0.009012 \pm 0.003247 = 0.00577$ to 0.01226. With 90% confidence, the expected increase in relative risk of breast cancer associated with an increase in alcohol consumption by 1 gram per day is between 0.00577 and 0.01226.

26.11: $b = 2.6327$ and $SE_b = 0.5893$. With $n = 75$, df $= 73$. Using Table C, use df $= 60$. $t^* = 1.671$ ($t^* = 1.666$ from technology). $2.6327 \pm 1.671(0.5893) = 2.6327 \pm 0.9847 = 1.6480$ to 3.6174 cubic kilometers per year (technology: 1.6509 to 3.6145). With 90% confidence, the yearly increase in Arctic discharge is between 1.6480 and 3.6174 cubic kilometers. This confidence interval excludes "0," so there is evidence that Arctic discharge is increasing over time.

26.13: (a) If $x^* = 0.65$, $\hat{\mu} = 10.0655 + 86.0308\ (0.65) = 65.98552$. (b) $SE_{\hat{\mu}} = 1.47$. For df $= 29 - 2 = 27$ and 95% confidence, $t^* = 2.052$. A 95% confidence interval for mean brain gray matter volume in people with 0.65 AROC is given by $65.98552 \pm 2.052(1.469) = 62.971$ to 69.000.

Fit	SE Fit	95% CI	95% PI
65.9855	1.46902	(62.9714, 68.9997)	(49.4805, 82.4906)

26.15: (a) The residual plot provided does not suggest any deviation from a straight-line relationship between brain volume and Aroc score, although there are two large (in absolute value) residuals near the left end of the plot. Both had AROC scores of about 0.63, but one had his/her volume underpredicted (the positive residual) and one was overpredicted (the negative residual).

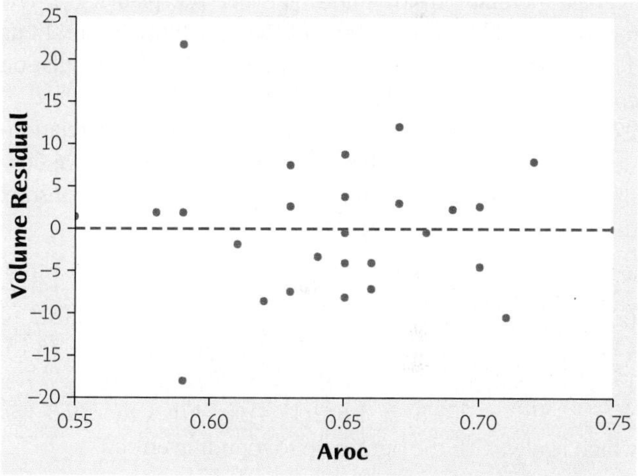

(b) A stemplot of residuals, provided, does not suggest that the distribution of residuals departs strongly from Normality. The value 22 from observation 5 may be an outlier; other than that, the residuals are symmetric and mound-shaped.

Stem and Leaf

Stem	Leaf	Count
2	2	1
1		
1	2	1
0	889	3
0	222333334	9
-0	44432110	8
-0	88777	5
-1	0	1
-1	8	1
-2		

-1|8 represents -18

(c) It is reasonable to assume that observations are independent because we have 29 different subjects, measured separately. (d) Other than the large residuals noted in part (a), there is no indication that variability changes; there are fewer individuals with low Aroc scores, so there is naturally less variability on the left end of the plot.

26.17: (a) With a positive association, $r = +\sqrt{r^2} = +\sqrt{0.579} = 0.761$.

26.19: (a) This is a one-sided alternative because we wonder if larger appraisal values are associated with larger selling prices.

26.21: (c) $s = 221.341$

26.23: (b) With 45 degrees of freedom, $t^* = 2.014$, so the margin of error is $2.014(0.1540) = 0.3102$. Using Table C and 40 df, $t^* = 2.021$ and ME = 0.3112.

26.25: (a) Scientists estimate that each additional 1% increase in the percentage of Bt cotton plants results in an average increase of 6.81 mirid bugs per 100 plants. (b) The regression model explains 90% of the variability in mirid bug density. (c) $H_0: \beta = 0$ versus $H_a: \beta > 0$ (H_0: population correlation = 0 versus H_a: population correlation > 0). $P < 0.0001$; there is strong evidence of a positive linear relationship between the proportion of Bt cotton plants and the density of mirid bugs. (d) We cannot conclude a causal relationship; this was not a designed experiment.

26.27: For 90% intervals with df = 10, $t^* = 1.812$. (a) $274.78 \pm 1.812(88.18) = 274.78 \pm 159.78 = 115.0$ to 434.6 fps/inch. (b) This is the "90% CI" given in Figure 26.13: 176.2 to 239.4 fps. To confirm this, $\hat{y} = 207.8$ and $SE_{\hat{\mu}} = 17.4$. $\hat{y} \pm t^* SE_{\hat{\mu}} = 207.8 \pm 1.812(17.4) = 176.3$ to 239.3 fps, which agrees with the output up to rounding error.

26.29: (a) There is little evidence of non-Normality in the residuals. (b) The scatterplot confirms the comments in the text: There is no clear pattern, but the variability about the "residual = 0" line may be slightly greater when x is larger. (c) Presumably, close inspection of a manatee's corpse will reveal nonsubtle clues when cause of death is from collision with a boat propeller. It seems reasonable that the kills listed in the table are mostly not caused by pollution.

−1	8 6
−1	2 0 0 0
−0	7 7 5
−0	3 3 3 2 1 1 1 1
0	0 1 1 1 1 1 2 3 4
0	5 5 5 6 7 8 8
1	1 2 3
1	7

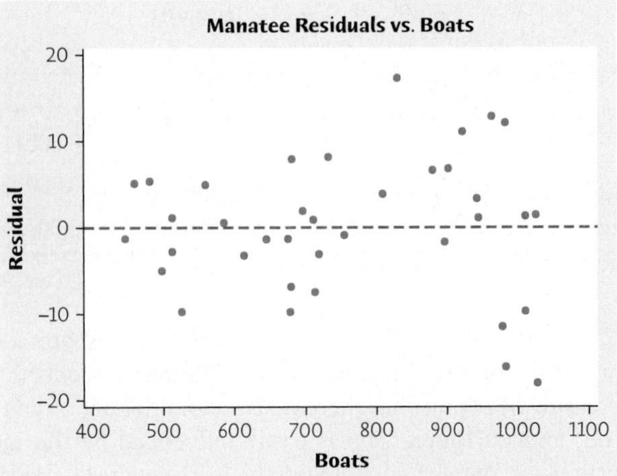

26.31: (a) With df = 35 (rounding df down to 30), $t^* = 2.042$. $0.132259 \pm 2.042(0.007067) = 0.132259 \pm 0.014431 = 0.11783$ to 0.14669 additional killed manatees per 1000 additional boats. (Using technology, 0.11791 to 0.14661.) (b) $\hat{y} = -44.831 + 0.132259(900) = 74.2021$, which agrees with Figure 26.14 under "Fit." We need the prediction interval because we are forecasting the number of manatees killed for a single year. A 95% prediction interval for the number of killed manatees is 57.74 to 90.67 kills if 900,000 boats are registered.

26.33: (a) H_0: Population correlation = 0 against H_a: Population correlation > 0; $t = 4.06$ with df = 30 − 2 = 28, $P = 0.0004$. There is very strong evidence of a positive correlation between Gray's forecasted number of storms and the number of storms that actually occur. (b) $\hat{\mu} = 1.6681676 + 0.9199861(16) = 16.388$, and JMP gives the 95% confidence interval for the mean as 14.077 to 18.699.

26.35: The stemplot suggests that the residuals may not follow a Normal distribution. Specifically, there is a low outlier that seems extreme. This makes regression inference and interval procedures unreliable.

26.37: (a) Shown is the scatterplot with two (nearly identical) regression lines. (b) For all points, $r = 0.8486$. $t = 6.00$, $P < 0.0005$. (c) Without the outlier, $r = 0.7014$, $t = 3.55$, and $P = 0.004$. In both cases there is strong evidence of a linear relationship between neural loss aversion and behavioral loss aversion. Omitting the outlier weakens this evidence somewhat.

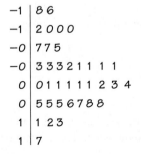

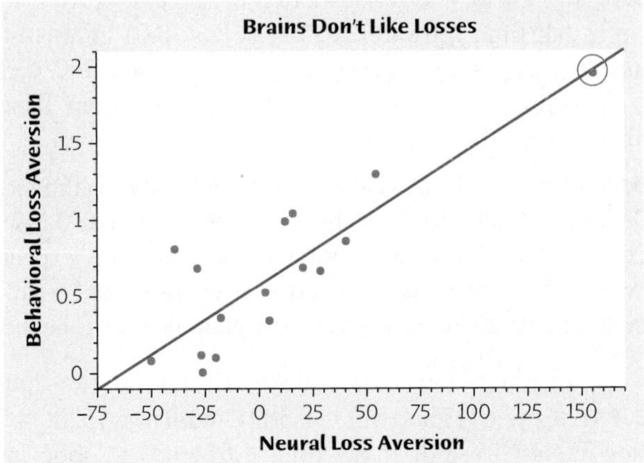

26.39: The distribution is skewed right, but the sample is large, so t procedures should be safe. $\bar{x} = 0.2781$ g/m^2 and $s = 0.1803$ g/m^2. Table C gives $t^* = 1.984$ for df $= 100$ (rounded down from 115). $0.2871 \pm (1.981)(0.183/\sqrt{116}) = 0.2449$ to 0.3113 g/m^2.

26.41: PLAN: We examine the relationship between pine cone abundance and squirrel density using a scatterplot and regression. SOLVE: A scatterplot indicates a positive relationship that is roughly linear with what appears to be an outlier at the upper right of the graph. Regression gives predicted squirrel density as $\hat{y} = 0.961 + 0.205x$. The slope is significantly different from zero ($t = 3.13$, $P = 0.005$). For one-sided alternative H_a: $\beta > 0$, $P = 0.0025$. Conditions for inference seem to be violated. The residual plot shows what appears to be increasing variability with increasing cone values. The stemplot of the residuals indicates two large positive outliers; the distribution may be right-skewed. CONCLUDE: We seem to have strong evidence of a positive linear relationship between cone abundance and squirrel density; however, conditions for inference may not be satisfied.

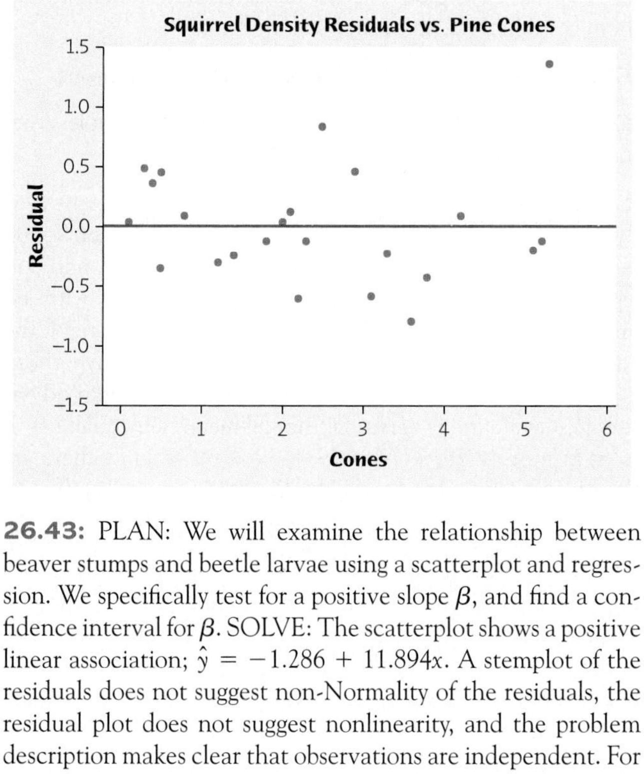

26.43: PLAN: We will examine the relationship between beaver stumps and beetle larvae using a scatterplot and regression. We specifically test for a positive slope β, and find a confidence interval for β. SOLVE: The scatterplot shows a positive linear association; $\hat{y} = -1.286 + 11.894x$. A stemplot of the residuals does not suggest non-Normality of the residuals, the residual plot does not suggest nonlinearity, and the problem description makes clear that observations are independent. For H_0: $\beta = 0$ versus H_a: $\beta > 0$, $t = 10.47$ (df $= 21$), $P < 0.0005$. $t^* = 2.080$ for 95% confidence. We are 95% confident that β is between $11.894 \pm (2.080)(1.136) = 9.531$ and 14.257. CONCLUDE: We have strong evidence that beetle larvae counts increase with beaver stump counts. Specifically, we are 95% confident that each additional stump is (on average) accompanied by between 9.5 and 14.3 additional larvae clusters.

26.45: PLAN: Using a scatterplot and regression, we examine how well phytopigment concentration explains DNA concentration. SOLVE: The scatterplot shows a fairly strong, linear, positive association; $\hat{y} = 0.1523 + 8.1676x$. A stemplot of the residuals looks reasonably Normal, but the scatterplot suggests that the variability about the line is greater when phytopigment concentration is greater. This may make regression inference unreliable. Finally, observations are independent, from the context of the problem. The slope is significantly different from 0 ($t = 13.25$, df $= 114$, $P < 0.001$). A 95% confidence interval for β: $8.1676 \pm 1.984(0.6163) = 6.95$ to 9.39. CONCLUDE: The significant linear relationship between phytopigment and DNA concentrations is consistent with the belief that organic matter settling is a primary source of DNA.

26.47: (a) $\bar{x} = -0.00333$, and $s = 1.0233$. For a standardized set of values, we expect the mean and standard deviation to be 0 and 1, respectively. (b) The stemplot does not look particularly symmetric, but it is not strikingly non-Normal for such a small sample. (c) The probability is about 0.0272.

26.49: For df $= 14$ and a 95% confidence interval, $t^* = 2.145$, the interval is $-0.01270 \pm 2.145(0.01264) = -0.0398$ to 0.0144. This interval does contain 0.

Chapter 27 One-Way Analysis of Variance: Comparing Several Means

27.1: (a) The null hypothesis is "all age groups have the same population mean road-rage measurement," and the alternative is "at least one group has a different mean." (b) The F test is quite significant, giving strong evidence that the means are different. The sample means suggest that the degree of road rage decreases with age. (We assume that higher numbers indicate *more* road rage.)

27.3: (a) The stemplots suggest that logging reduces the number of trees per plot and that recovery is slow (the one-year-after and eight-years-after logging stemplots are similar). (b) The means lead one to the same conclusion as in part (a): The first mean (23.75) is much larger than the other two (14.08 and 15.78). (c) In testing H_0: $\mu_1 = \mu_2 = \mu_3$ versus H_a: not all means are the same, we find that $F = 11.43$ with df $=$ 2 and 30, which has $P = 0.0002$, so we conclude that these differences are significant: The mean number of trees per plot is significantly lower in logged areas.

Never logged		1 year ago		8 years ago	
0		0	2	0	4
0		0	9	0	
1		1	2244	1	22
1	699	1	57789	1	5889
2	0124	2	0	2	22
2	7789	2		2	
3	3	3		3	

27.5: (a) Answers will vary due to randomness. (b) By moving the middle mean to the same level as the other two, it is possible to reduce F to about 0.02, which has a P-value very close to the left end of the scale (near 1). (c) By moving any mean up or down (or any two means in opposite directions), the value of F increases (and P decreases) until it moves to the right end of the scale.

27.7: (a) We have $s_1^2 = 25.6591$, $s_2^2 = 24.8106$, and $s_3^2 = 33.1944$, so $s_1 = 5.065$, $s_2 = 4.981$ and $s_3 = 5.761$. The ratio of largest to smallest standard deviation is 5.761/4.981 = 1.16, which is less than 2. Conditions are satisfied. (b) The three standard deviations are $s_L = 17.41$, $s_M = 18.13$ and $s_C = 17.42$. The ratio of largest to smallest standard deviation is 18.13/17.41 = 1.04, which is less than 2. Conditions are satisfied.

27.9: PLAN: Examine the data to compare the effect of the treatments and check that we can safely use ANOVA. If the data allow ANOVA, assess the significance of observed differences in mean numbers of new leaves. SOLVE: Side-by-side stemplots shows some irregularity but no outliers or strong skewness. The Minitab ANOVA output shows that the group standard deviations easily satisfy our rule of thumb (2.059/1.302 = 1.58 < 2). The differences among the groups were significant at $\alpha = 0.05$: $F = 3.44$, df = 3 and 27, $P = 0.031$. CONCLUDE: Nitrogen had a positive effect, the phosphorus and control groups were similar, and the plants that got both nutrients fell between the others.

Control		Nitrogen		Phosphorus		Both	
11	00	11		11	0	11	
12	0	12		12	0	12	
13	0	13	0	13	0	13	0
14		14	0	14	000	14	0000
15	00	15	00	15	00	15	0
16	0	16	0	16		16	0
17		17	00	17		17	0
18		18	0	18		18	

Minitab output

```
Source     DF        SS        MS       F       P
Sleep       3     27.21      9.07    3.44   0.031
Error      27     71.18      2.64
Total      30     98.39
                     Individual 95% CIs For Mean
                     Based on Pooled StDev
Level  N    Mean   StDev  --------+---------+---------+--------
    C  7  13.286   2.059  (-----*-----)
    N  8  15.625   1.685             (-----*-----)
    P  8  13.500   1.414  (-----*-----)
   NP  8  14.625   1.302        (-----*-----)
                          --------+---------+---------+--------
Pooled StDev = 1.624          13.5      15.0      16.5
```

27.11: (a) $I = 3$ and $N = 96$, so df = 2 and 93. (b) $I = 3$ and $N = 90$, so df = 2 and 87.

27.13: (a) No sample standard deviation is larger than twice any other. Specifically, the ratio of largest to smallest standard deviation is 2.25/1.61 = 1.40, which is less than 2. Conditions are safe for use of ANOVA. (b) Calculations are provided:

$$\bar{x} = \frac{17 \times 6.47 + 17 \times 3.75 + 17 \times 4.05 + 17 \times 5.02}{68} = 4.8225$$

$$\text{MSG} = \frac{17(6.47 - 4.8225)^2 + 17(3.75 - 4.8225)^2 + 17(4.05 - 4.8225)^2 + 17(5.02 - 4.8225)^2}{4 - 1}$$

$$= 25.502$$

$$\text{MSE} = \frac{(17 - 1)2.25^2 + (17 - 1)1.77^2(17 - 1)1.61^2 + (17 - 1)1.80^2}{68 - 4} = 3.507$$

$$F = \frac{\text{MSG}}{\text{MSE}} = 7.272$$

(c) We have df = 4 − 1 = 3 and 68 − 4 = 64. $P = 0.0003$ (obtained using software). There is strong evidence that the mean status scores between the four groups studied are not equal—a conclusion consistent with the solution to Exercise 27.2.

27.15: (c) the means of several populations

27.17: (c) The alternate hypothesis for ANOVA is always that there is some difference in the means (but it does not specify the type of difference).

27.19: (a) $P = 0.026$, so we reject H_0 and conclude that there is a difference in mean breaking strength.

27.21: (a) This is the problem of multiple comparisons.

27.23: (c) We do not have three independent samples from three populations.

27.25: The populations are college students that might view the advertisement with an art image, college students that might view the advertisement with a non-art image, and college students that might view the advertisement with no image. The response variable is student evaluation of the advertisement on the 1–7 scale. We test the hypothesis H_0: $\mu_1 = \mu_2 = \mu_3$ (all three groups have equal mean advertisement evaluation) versus H_a: not all means are equal. $I = 3$, $n_1 = n_2 = n_3 = 39$, so $N = 39 + 39 + 39 = 117$. There are then $I - 1 = 3 - 1 = 2$ and $N - I = 117 - 3 = 114$ df.

27.27: The response variable is hemoglobin A1c level. $I = 4$: a control (sedentary) population, an aerobic exercise population, a resistance training population, and a combined aerobic and resistance training population. We test H_0: $\mu_1 = \mu_2 = \mu_3 = \mu_4$ (all four groups have equal mean hemoglobin A1c levels) versus H_a: not all means are equal. $n_1 = 41$, $n_2 = 73$, $n_3 = 72$, and $n_4 = 76$. $N = 41 + 73 + 72 + 76 = 262$. We have $I - 1 = 4 - 1 = 3$ and $N - I = 262 - 4 = 258$ df.

27.29: (a) The graph suggests that emissions rise when a plant is attacked because the mean control emission rate is half the smallest of the other rates. (b) The null hypothesis is "all groups have the same mean emission rate." The alternative is "at least one group has a different mean emission rate." (c) The most important piece of additional information would be whether the data are sufficiently close to Normally distributed. (From the description, it seems reasonably safe to assume that these are more or less random samples.) (d) SEM = $s/\sqrt{8}$, so

$s = \text{SEM}\sqrt{8}$; they are 16.77, 24.75, 18.78, and 24.38. This factor cancels out in the process of finding the ratio, so we can find this ratio directly from the SEMs: $\frac{8.75}{5.93} = \frac{24.75}{16.77} = 1.48$, which satisfies our rule of thumb.

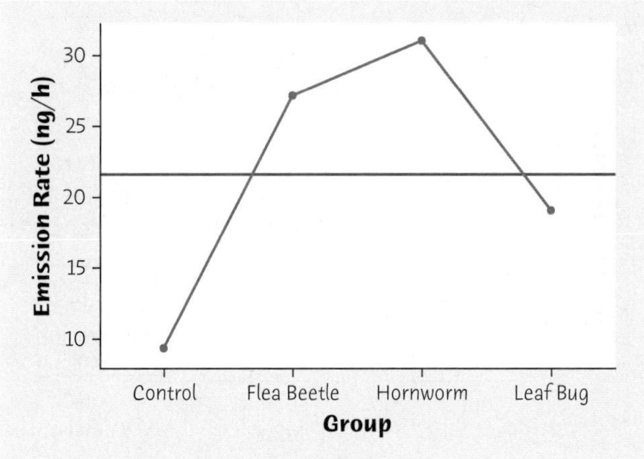

27.31: (a) The treatment means suggest that extra water in the spring has the greatest effect on biomass, with a lesser effect from added water in the winter. ANOVA is risky with these data; the standard deviation ratio is nearly 3 (58.77/21.69 = 2.71), and the winter and spring distributions may have skewness or outliers (although it is difficult to judge with such small samples).

Level	N	Mean	StDev
Control	6	136.65	21.69
Spring	6	315.39	37.34
Winter	6	205.17	58.77

(b) We test H_0: $\mu_w = \mu_s = \mu_c$ (all treatments have the same mean) versus H_a: at least one mean is different. (c) ANOVA gives a statistically significant result ($F = 27.52$, df = 2 and 15, $P < 0.0005$), but as noted in part (a), the conditions for ANOVA are not satisfied. Based on stemplots and the means, however, we should still be safe in concluding that added water increases biomass.

One-Way ANOVA: Biomass versus Treatment

Source	DF	SS	MS	F	P
Treatment	2	97583	48792	27.52	0.000
Error	15	26593	1773		
Total	17	124176			

S = 42.11 R-Sq = 78.58% R-Sq(adj) = 75.73%

27.33: (a) PLAN: We have data on sleep quality and depression scores for 898 students at a large midwestern university. We'll have to assume these students are close to a random sample of college students and that the students are independent of one another. SOLVE: Side-by-side boxplots are used to examine the distributions. All three groups show outliers at the high end of the depression score range, but with such large samples (the smallest is 246), it is reasonable to believe the sample means have Normal distributions (using the CLT). The condition on standard deviations is satisfied because 4.719/2.560 = 1.84 < 2. We have $F = 75.52$ with df = 2 and 895, giving $P = 0.000$ (to three decimal places). CONCLUDE: The mean depression scores for the three levels of sleep quality are not the same. From the output and graphs, it appears the mean depression score for poor sleepers is highest; the mean depression score for optimal sleepers is lowest. (b) Assuming the students were randomly selected, the large sample size would lead us to believe these students are most likely representative of other college students. (c) Students were not randomly assigned to sleep conditions. Explanations about causation may vary, but this might well be a case of one condition (poor sleep) feeding the other (depression) in a "vicious cycle."

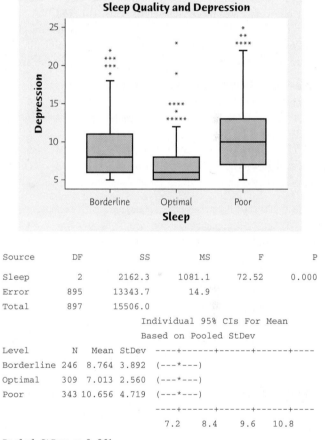

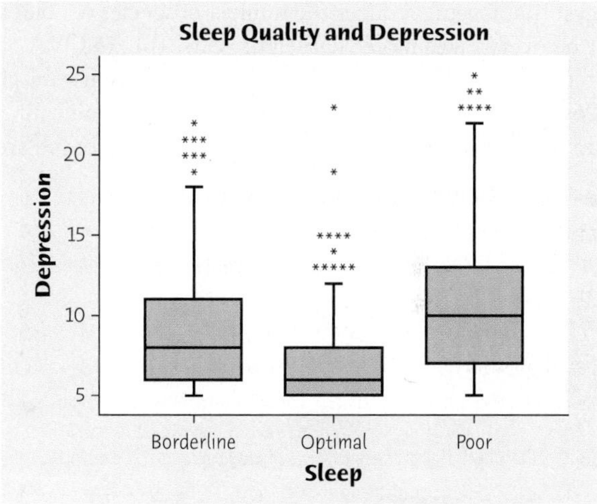

Source	DF	SS	MS	F	P
Sleep	2	2162.3	1081.1	72.52	0.000
Error	895	13343.7	14.9		
Total	897	15506.0			

```
                             Individual 95% CIs For Mean
                             Based on Pooled StDev
Level        N    Mean  StDev  ----+------+------+------+----
Borderline 246   8.764  3.892  (---*---)
Optimal    309   7.013  2.560  (---*---)
Poor       343  10.656  4.719  (---*---)
                               ----+------+------+------+----
                                  7.2    8.4    9.6   10.8

Pooled StDev = 3.861
```

27.35: PLAN: We will carry out an ANOVA test for the equality of means. SOLVE: The ratio of largest standard deviation to smallest standard deviation is 2.388/1.959 = 1.22, which is less than 2. Histograms show some evidence of non-Normality, and perhaps one outlier in the "No Weather Report" group. We proceed, as the samples are reasonably large. From the output, we have $F = 20.679$ with $3 - 1 = 2$ and $60 - 3 = 57$ df with $P = 0.000$. CONCLUDE: There is overwhelming evidence that the mean tip percentages are not the same for all three groups. Examination of the summary statistics and the histograms suggests that while mean tip for the bad-report group is similar to that of the no-report group, the mean tip for the good-weather report is higher.

One-way ANOVA: Percent versus Report

Source	DF	SS	MS	F	P
Report	2	192.22	96.11	20.68	0.000
Error	57	264.92	4.65		
Total	59	457.15			

S = 2.156 R-Sq = 42.05% R-Sq(adj) = 40.02%

Level	N	Mean	StDev
Bad	20	18.180	2.098
Good	20	22.220	1.959
None	20	18.725	2.388

27.37: (a) The table is given in the Minitab output; because $4.500/3.529 = 1.28 < 2$, ANOVA should be safe. The means suggest that logging reduces the number of species per plot and that recovery takes more than eight years. (b) ANOVA gives $F = 6.02$ with df = 2 and 30, so $P < 0.01$ (software gives 0.006). We conclude that these differences are significant; the mean number of species per plot really is lower in logged areas.

One-Way ANOVA: Species versus Group

Source	DF	SS	MS	F	P
Group	2	204.4	102.2	6.02	0.006
Error	30	509.2	17.0		
Total	32	713.6			

S = 4.120 R-Sq = 28.64% R-Sq(adj) = 23.88%

```
                              Individual 95% CIs For Mean
                              Based on Pooled StDev
Level   N   Mean  StDev  ------+------+------+-------+
1      12 17.500  3.529                (------*------)
2      12 11.750  4.372  (------*------)
3       9 13.667  4.500      (------*------)
                         ------+------+------+-------+
                            12.0   15.0   18.0    21.0

Pooled StDev = 4.120
```

27.39: (a) See plot. (b) There is a slight increase in growth when water is added in the wet season, but there is a much greater increase when it is added during the dry season. (c) The means differ significantly during the first three years. (d) The year 2005 is the only one for which the winter biomass was higher than the spring biomass.

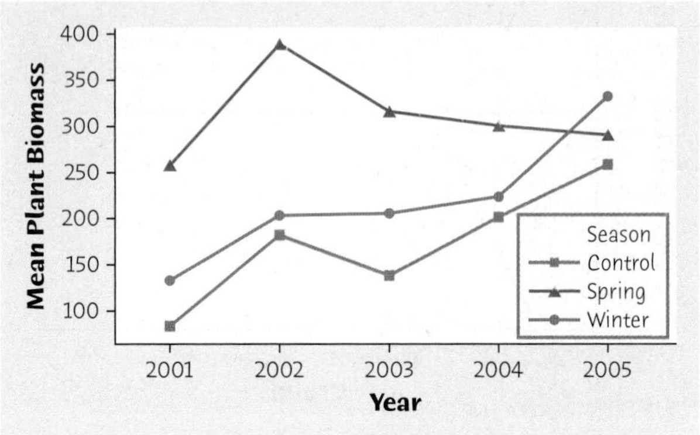

27.41: In addition to a high standard deviation ratio ($117.18/35.57 = 3.29$), the spring biomass distribution has a high outlier.

27.43: (a) This is a comparison of two means, so it requires a two-sample t test. (b) This is a comparison of three means, so it requires ANOVA. (c) This is a comparison of three proportions, so it requires a chi-square test of homogeneity.

INDEX

Bolded page numbers indicate material available in companion Chapters 28–31; available online.

Absolute values, **28-16**
Addition rule
 for any two events, 307–309
 general, 307–309
 Venn diagram, 308
Addition rule for disjoint events
 defined, 284, 296, 318
 description, 304
 occurrence of, 307
Alternative hypothesis. *See also* Hypotheses
 abbreviation, 395
 chi-square test for goodness of fit, 595
 comparing proportions, 545
 defined, 394, 409
 Kruskal-Wallis test, **28-24, 28-29**
 one-sided, 394, 395, 409
 power and, 428
 statistical significance and, 421
 two categorical variables and, 580
 two-sided, 394, 409
Analysis of variance. *See* ANOVA; One-way ANOVA; Two-way ANOVA
Analysis of variance *F* test. *See* ANOVA *F* test
Anonymity, 258
ANOVA
 balanced designs, **30-15**
 comparisons, **30-1**
 conditions for, **30-1, 30-32**
 defined, 654, **30-1**
 idea of, 653–656
 multiple comparisons, **30-2**
 one-way. *See* one-way ANOVA
 robustness against violations, **30-1**
 two-way. *See* two-way ANOVA
ANOVA F statistic
 calculation of, 662
 defined, 655, 666, **30-2**
 degrees of freedom, 660
 form of, 659, 662, **30-2**
 mean squares, 662
 P-values, 654
 regression coefficients test, **29-56**
 for regression model, **29-15**
 for two-way ANOVA, **30-31**
 values, 655
ANOVA F test
 defined, 648, 665
 degrees of freedom for, 660
 example, 649
 follow-up analysis, **30-4**
 means of several populations, **28-24**
 multiple regression models, **29-49**
 one-way ANOVA, **30-29**
 for parallel lines, **29-15**
 process, 662
 P-values, 660
 results, 649, 657
 robustness, 656, **30-15**
 for standard deviations, 502
 two-way ANOVA, **30-28, 30-31, 30-32**
ANOVA tables
 defined, 663
 general form, **29-48**
 in output, **30-29**
 software output, 663
 squared multiple correlation coefficient (R^2) interpretation with, **29-12**
 sum of squares row, **29-55**
Associations
 causation and, 148–151
 consistent, 150
 defined, 106
 linear, 107, 111–113
 negative, 106, 107
 positive, 107
 strong, 150
Automated algorithms, **29-41**

Back-to-back stemplots, 43
Balanced designs
 advantages of, **30-15**
 defined, **30-14, 30-32**
 two-way ANOVA, **30-14–30-15, 30-32**
Bar graphs
 defined, 18, 35
 flexibility of, 18
 function of, 35
 as having limited use, 19
 illustrated, 18
 in presenting categorical data, 173
 segmented, 167, 169
Behavior
 chance, 278
 chaotic, 278
 of confidence intervals, 383–385
 of sample proportions, 519
 z test statistic, 422
Behavioral experiments
 challenges of, 262
 conducting, 261–263
 ethical principles, 262
 informed consent, 262
Benford's law, 287
Bias
 convenience samples, 206
 defined, 206, 219
 nonresponse, 214–215, 219
 online polls and mall interviews and, 209
 random sampling and, 211
 randomization and, 245
 response, 215, 219
 simple random samples (SRS) and, 209
 undercoverage, 214, 219
Binomial coefficients
 in counting successes, 331
 defined, 330, 339
 in finding binomial probabilities, 331
Binomial distributions
 defined, 328, 338
 inferences, 339
 mean, 334, 339
 Normal approximation to, 335–338, 339
 observations, 339
 overview of, 327–328
 probability histogram for, 336
 simple random samples (SRS) and, 329
 standard deviation, 334, 339
 in statistical sampling, 328–329
 summary, 442
 use of, 339
Binomial mean, 334, 339
Binomial probabilities
 binomial coefficients in, 330, 331
 defined, 339
 factorial notation, 331
 finding, 330–332
 formula, 331
 Normal approximation to, 337
 technology in finding, 332–333
Binomial setting, 327, 328, 338
Binomial standard deviation, 334, 339
Block design
 conclusions, 243
 defined, 242, 245
 examples of, 242–243
 randomization, 243
 treatment groups, 242
Bootstrap methods, **28-1**
Boxplots
 defined, 53, 66
 five-number summary, 53–55
 illustrated, 54
 modified, 56
 for side-by-side comparisons, 53
 two sets of three samples, 654

Calculators. *See also* Texas Instruments graphing calculator
 graphing, 61
 requirements, 8
 with "two-variable statistics" functions, 61
Capability
 control versus, **31-30–31-31**
 defined, **31-30, 31-37**
 example, **31-30–31-31**
 summary, **31-38**
Categorical data
 summary, 183
 two-way tables for, 163–173
Categorical variables
 defined, 14, 35
 distribution of, 17
 inference about relationship, 577–600
 lurking variables and, 171
 relationships between, 163, 170, 173
 in scatterplots, 109–110
 in two-way tables, 577–580
Causation
 association and, 148–151
 criteria for establishing, 150

Causation (*continued*)
direct, 149
high correlation and, 152
Cause-and-effect diagrams
defined, **31-2**
illustrated, **31-4**
outline for, **31-3**
use of, **31-37**
Cautions
confidence intervals, 419–420
correlation and regression, 183
experiments, 239–241
least-squares regression line, 152
sample surveys, 214–216
tests of significance, 421–424
Center
defined, 25, 35
distributions, 501
measures, choosing, 59–60
measuring, 48–50
in numerical summary, 66
resistant measure of, 48
sampling distribution, 352
Center line
control charts, **31-8**
defined, **31-37**
p charts, **31-33**
s charts, **31-14, 31-24**
three-sigma control charts, **31-13**
$\bar{x}$ charts, **31-14, 31-24**
Central limit theorem
defined, 355–356, 367
examples, 356–357, 358
general versions, 356
Normal approximation, 360
Normal distributions and, 356
Normal probability calculations and, 356
watching in action, 359
Central Limit Theorem applet, 359
Chance
behavior, 278
in choosing samples, 278
only operating, 363
in random sample results, 345
randomization, 245
in randomized comparative experiments results, 345
statistical inference and, 416
Chart setup
defined, **31-7, 31-37**
estimating process mean, **31-22–31-23**
estimating process standard deviation, **31-23–31-24**
procedure, **31-24–31-26**
s charts, **31-24–31-25**
$\bar{x}$ charts, **31-25–31-26**
Chi-square distributions
as approximation, 599
critical values, 584, 594
defined, 593, 599
degrees of freedom, 593, 599
density curves, 593
Kruskal-Wallis test statistic, **28-30**
mean, 594
table, 594
Chi-square statistic
chi-square test for goodness of fit, 596
defined, 583, 598
example, 584
formula, 583, 598
technology in calculation, 585–586
terms of, 587, 599
Chi-square table, 594
Chi-square test for goodness of fit
alternative hypothesis, 595
chi-square statistic, 596
example, 595
expected counts, 595–596
null hypothesis, 595, 596
Chi-square test for homogeneity
defined, 590, 598
null hypothesis, 592
use of, 592
Chi-square test for independence
defined, 590, 598
null hypothesis, 592
use of, 592
Chi-square tests
calculation by hand, 599
cell counts required for, 584–585
defined, 587, 598
hypotheses, 591–592
interpreting, 599
null hypothesis, 592
as one-sided, 584
P-values, 593
relationship between rows and columns, 599
as significant, 587
for two-way tables, 598
uses of, 589–593, 600
Classes
histograms, 25, 26
large data sets and, 30
stemplots, 30
Clinical trials. *See also* Data ethics
benefits of, 260
defined, 260
issues, 260
randomized, 261
subjects, 260, 261
treatments nested in, 261
Clusters, 106
Coin tossing, 281
Coins, 278
Column variables, 164, 173
Common cause variation, **31-6**
Comparative experiments, 234
Comparing mean and median, 50–51
Comparing several means, 648
Comparing several samples, **28-23–28-24**
Comparing two means
conditions for, 487
confidence interval, 506
example, 487–489
Normally distributed populations, 487
summary, 563
tests of significance, 506–507
two SRSs, 487
Comparing two proportions
alternative hypothesis, 545
difference, 540
large-sample confidence intervals for, 542–543, 551
notation, 540
null hypothesis, 545
overview of, 539
plus four confidence intervals, 549–551
sample survey, 547
sampling distribution of difference, 541
standard error, 542, 551
summary, 563–564
technology and, 543–545
tests of significance, 545–548, 551
as two-sample problem, 539–541, 551
two-sample z procedure, 547
Comparing two samples, **28-2–28-6**
Completely randomized design, 235
Conditional distributions
comparing, 578–579
defined, 166, 173
example of, 578
marginal distributions versus, 166
response variables, comparing, 170
roundoff error, 167
sets of, 167
Conditional probability
concept, 309
defined, 310, 318
equation, 310
examples, 309–310
multiplication rule for any two events and, 311
not confusing, 310
tree diagrams and, 316
Confidence Interval applet, 378
Confidence intervals
basics of, 373–386
behavior of, 383–385
cautions about, 419–420
comparing two means, 506
construction of, 386
defined, 376, 385
estimation goal, 410
idea of, 440
individual t tests for coefficients and, **29-50**
intercept of population regression line, 623
large-sample, for comparing proportions, 542–543
large-sample, for proportions, 520–523
margin of error, 383, 419–420, 431
for mean of Normal population, 380
for mean response, 627, 634
multiple comparisons and, 580
for multiple regression models, **29-50, 29-56**
95%, 423
one-sample t, 458–461
parts of, 376, 385
planning, 424
plus four, 531
plus four, for comparing two proportions, 549–551
plus four, for proportions, 528–531
for population mean, 375, 379–383, 386, 415, 455, 475
for population proportions, 524, 528–531
for population regression line slope, 622–624, 634
for regression response, 627
for regression slope, 622–624
robust, 470
sample size of, 424–425
summary, 442

t, 458–461, 464, 465
trust of, 416
Tukey simultaneous, **30-7**
use of, 373, 391
watching, 378
z, 425, 432
Confidence levels
critical values, 379, 432
defined, 377, 385, 439
interpreting, 378
margin of error and, 376
95%, 376, 377
overall, **30-7, 30-32**
user selection of, 378
for *z* statistics, 404
Confidentiality. *See also* Data ethics
anonymity versus, 258
breach of, 259
defined, 258, 263
privacy policy and, 259
Confounding
defined, 229, 244
experiments and, 231
illustrated, 229
Continuity correction, **28-6**
Continuous distributions, **28-2**
Continuous probability models. *See also* Probability models
defined, 289, 296
Normal distributions as, 291
probability assignment, 291
Continuous random variables, 293, 296
Contrasts
defined, **30-10**
estimating, **30-11**
example, **30-11**
inference for, **30-12–30-13**
pairwise differences, **30-11**
population, **30-10**
sample, **30-11**
tests for, **30-13**
Control
advantage of, **31-31**
capability versus, **31-30–31-31**
control charts in, **31-26**
defined, **31-37**
as experimental principle, 238, 245
Control charts. *See also* *s* charts; *R* charts; $\bar{x}$ charts
application of, **31-7**
constants, **31-15**
creating, **31-8**
defined, **31-37**
estimating process mean, **31-22–31-23**
estimating process standard deviation, **31-23–31-24**
false alarms, **31-20, 31-21**
following process forward, **31-22**
functioning of, **31-7**
graphical and numerical descriptions, **31-7**
out-of-control signal, **31-20**
for process monitoring, **31-7–31-19**
runs signal, **31-20–31-21**
for sample proportions, **31-32–31-33**
setup, **31-7, 31-22**
setup procedure, **31-24–31-26**
summary, **31-38**
three-sigma, **31-13**
using, **31-19–31-21**
using past data, **31-24, 31-37**
Control groups
defined, 234
randomized comparative experiments, 239
Control limits
defined, **31-8, 31-37**
lower, **31-11**
p charts, **31-33–31-36**
for process going forward, **31-26**
s charts, **31-14, 31-24**
sample mean outside of, **31-9**
three-sigma control charts, **31-13**
upper, **31-11**
$\bar{x}$ charts, **31-14, 31-24**
Convenience samples
bias, 206
defined, 206
margin of error and, 420
Correlation
based on averages, 146
calculating, 117
causation and, 152
cautions, 183
defined, 111, 117
direction of linear relationship, 118
ecological, 146, 151
explanatory and response variables and, 113, 118
facts about, 113–117
formula, 112
lack of, testing, 620–622
least-squares regression line slope and, 135
least-squares regression lines and, 146
least-squares slope and, 620
lurking variables and, 146–147, 152
mean and, 116
measuring linear association with, 111–113
nonsense, 148
nonzero, 622
as not complete summary, 116
as not resistant, 116
quantitative variable requirement, 114, 115
regression and, 151
regression inference, 611
regression lines and, 146
square of, 151
standardizing observations and, 112
strength measurement, 114, 115, 116
summary, 182
use of, 111
values of, 113
Correlation and Regression applet, 145
Correlation coefficients, **29-38–29-39**
Critical values
central probability under Normal curve, 380
chi-square distributions, 584, 594
for confidence levels, 379, 432
defined, 379
selection of, 386
t, 458, 475, 506
table, significance from, 404–405
Crossed designs, **30-14, 30-32**
Cross-sectional data, 34
CrunchIt!
ANOVA, 651
binomial probabilities, 332
chi-square statistic, 586
comparing two proportions, 543, 544
descriptive measures, 62
expected counts, 586
Kruskal-Wallis test, **28-27**
least-squares regression, 133
Mann-Whitney test, **28-8**
multiple regression models, **29-40, 29-41, 29-42, 29-45**
parallel regression lines, **29-10, 29-11**
regression inference, 617
t confidence interval, 465
t test, 467
two-sample *t* procedures, 495, 496
two-way table output, 168
Wilcoxon signed rank test, **28-19**
Cumulative distribution function, 88
Cumulative proportions
defined, 86–87, 94
illustrated, 87
Cycles, 33, 35

Data
always looking at, 3–4
categorical, 183
cross-sectional, 34
defined, 2
exploring, 16, 179–183
as numbers with a context, 2
plotting, 35, 60
science of learning from, 2
source of, 2
statistics as science of, 13
summary, 181
time-series, 34
Data analysis
defined, 7, 179
exploratory, 16, 35
before inference, 431
organizing principles, 11
from sampling designs, 213
summary, 560
where to begin, 4
Data ethics
behavioral and social science experiments and, 261–263
clinical trials and, 260–261
complex issues of, 254
confidentiality and, 258–260
importance of, 253
informed consent and, 256–258
institutional review boards and, 254–256
summary, 269
Data production
data ethics and, 253–263
defined, 7
designs for, 267
experiments, 227–245
introduction to, 201
observational studies in, 245
review, 267–269
sampling, 203–220
summary, 268–269, 559–560
Data sets, 35
Degrees of freedom
for ANOVA *F* test, 660
chi-square distributions, 593, 599
defined, 58

Degrees of freedom (*continued*)
  F distributions, 659
  individual *t* procedures, **29-56**
  multiple linear regression model, **29-30**
  population regression lines, 613
  regression standard error, 634, **29-6**
  sums of squares, **30-30, 30-31**
  t approximation, 500
  t distributions, 457, 475
  two-sample *t* statistic, 506
Density curves. *See also* Normal curves
  areas under, 78
  in assigning probabilities, 291, 296
  chi-square distributions, 593
  defined, 77, 93
  describing, 78–80
  F distributions, 660
  finding with exponential distribution, 360
  histograms and, 76–77
  illustrated, 76, 77
  mean of, 79, 80
  median of, 79, 80
  outliers and, 78
  for overall pattern description, 78
  probability as area under, 290
  shapes, 77
  standard deviation of, 80
  summary, 181–182
  symmetric, 79
  t distributions, 457
  uniform, 290
Dependent variables. *See* Response variables
Deviations. *See also* Standard deviations
  from pattern, 25, 35
  squared, 59
Dice, 278
Direction
  defined, 117
  negative association and, 107, 117
  positive association and, 107, 117
  relationships, 106, 117
  scatterplots, 105–106
Discrete random variables, 293, 296
Disjoint events
  addition rule for, 284, 296, 304, 318
  defined, 284, 296, 318
  independent events and, 307
Distributions. *See also* Normal distributions
  binomial, 327–339, 442
  of categorical variable, 17
  center, 25, 501
  chi-square, 584, 593–594
  conditional, 578
  continuous, **28-2**
  defined, 17, 35
  describing (quantitative variable), 181
  describing with numbers, 47–66
  displaying, 181
  exponential, 358
  F, 659–661
  marginal, 164–166
  mean, 48–49
  median, 49–50
  numerical summaries of, 181
  outliers, 25–26
  permutation, 504
  picturing with graphs, 13–36
  population, 351, 366
  probability, 296
  sampling, 345–367, 441, 442
  shape, 25, 28
  skewed, 25, 35
  standard, **28-1**
  symmetric, 25, 35
  t, 456–458
  technology and, 61–63
  uniform, 290
  variability, 25, 501
Double-blind experiments, 240, 245
Dropouts, 417
Dummy variables. *See* Indicator variables

Ecological correlation
  beware of, 146
  defined, 146, 151
Effect size
  in number of observations, 426
  power and, 428
  small, 432
Errors. *See also* Margin of error
  probabilities, calculating, 430
  roundoff, 17
  standard, 456
  Type I, 429
  Type II, 429
Estimated regression model, **29-55**
Events
  defined, 282, 296
  disjoint, 284, 318
  independent, 305–307, 318
  nonoverlapping intervals, 291
  outcomes, 284
  probability of, 282, 284
  in Venn diagram, 304
Excel. *See* Microsoft Excel
Expected counts
  chi-square test, 584–585
  chi-square test for goodness of fit, 595–596
  defined, 581, 598
  example, 582
  formula, 581, 582
  technology in calculation, 585–586
  in two-way tables, 581–583
Experimental design
  block design, 242–243, 245
  completely randomized, 235
  control, 238, 245
  defined, 244
  matched pairs design, 241–242, 245
  principles of, 238, 245
  randomization, 238, 245
  replication, 238, 245
Experiments
  advantages of, 231–232
  attention to detail and, 245
  behavioral, 261–263
  cautions about, 239–241
  clinical trial, 260–261
  comparative, 234
  confounding and, 231
  control group, 234
  defined, 149, 152, 244
  double-blind, 240, 245
  dropouts from, 417
  factors, 230, 231–232, 244
  field, 233
  laboratory, 233
  lack of realism, 240–241, 245
  with living subjects, 233
  lurking variables and, 149
  observational studies versus, 227–230
  permutation tests, 504
  placebos, 239, 245
  poor, 233–234
  producing data with, 227–245
  randomized comparative, 234–239
  social science, 261–263
  statistical analysis of, 241
  subjects, 230, 231, 244
  summary, 268–269
  treatments, 230, 231, 244
  two-factor, 232
  uncontrolled, 233
  in understanding cause and effect, 228
  use of, 228, 245
  vocabulary of, 230
Explanatory variables
  confounded, 229
  correlation and, 113, 118
  creating new, **29-43**
  defined, 102, 117
  identifying, 102
  interaction between, **29-22–29-28**
  least-squares regression line and, 152
  multiple regression model and, **29-16–29-17**
  plotting, 104
  regression and, 135
  two interacting, **29-29–29-30**
Exploratory data analysis
  conclusions, 201
  defined, 16, 35
  purpose of, 201
Exponential distributions
  in finding density curves, 360
  use of, 358
Extrapolation
  beware of, 146
  defined, 146, 151

F distributions
  defined, 659, 666
  degrees of freedom, 659, 662
  density curves, 660
  values, 659
F tests. *See* ANOVA *F* test
Factorial notation, 331, 339
Factors. *See also* Interactions; Main effects
  combining effects of, 231–232
  constant, 231
  defined, 230, 244, **30-13**
  two-way ANOVA, **30-13, 30-28, 30-33**
Field experiments, 233
Finite probability models, 286, 296
First quartile. *See also* Quartiles
  defined, 52, 66
  finding, 92
  in five-number summary, 53
Five-number summary
  boxplots and, 53–55
  defined, 53, 66
  use of, 59

Flowcharts
defined, **31-2**
illustrated, **31-3**
outline, 234, 235
use of, **31-37**
Follow-up analysis
contrasts, **30-10–30-13**
F test, **30-4**
for multiple comparisons, 581, 648
one-way ANOVA, **30-4, 30-32**
two-way ANOVA, **30-32, 30-33**
Form
defined, 106, 117
of relationships, 106, 117
scatterplots, 106
Four-step process
definition, 63
for confidence intervals, 381
for tests of significance, 401

Galton, Sir Francis, 129
General addition rule
for any two events, 307–309
defined, 307–308, 319
Venn diagram, 308
General multiplication rule, 311–313, 319
Graphing calculators, 61. *See also* Texas Instruments graphing calculator
Graphs. *See also specific types of graphs*
as best overall picture of distributions, 60
marginal distribution, 164–165
purpose of, 24
as visual tools, 35

Histograms
center, 25
classes, choosing, 21–22
creation example, 21–23
cross-sectional data, 34
defined, 21, 35
density curves and, 76–77
drawing, 22–23
examining, 25
function of, 35
illustrated, 26, 27, 31
interpreting, 24–29
outliers, 25
overall pattern, 25
probability, 336
shape, 25
spread, 25
variability, 25
Hypotheses. *See also* Alternative hypothesis; Null hypothesis
chi-square test, 591–592
before data, 395
Kruskal-Wallis test, **28-24**
population parameter reference, 395
simultaneous tests of, **30-7**
stating, 394–396
Wilcoxon rank sum test, **28-10–28-11**
Hypothesis testing
error types, 429
goal of, 409–410

Independent events
defined, 305, 318
disjoint events and, 307
multiplication rule for, 305, 313, 318
positive probability, 313
Independent observations
defined, 327
failure of, 327
regression inference, 632
residual plots, 629
success of, 327
Independent variables. *See* Explanatory variables
Indicator variables
defined, **29-3, 29-4, 29-55**
as dummy variables, **29-4**
use of, **29-4**
Individual *t* procedures
comparing groups with, **30-6**
degrees of freedom, **29-56**
Individual *t* tests for coefficients. *See also t* tests
confidence intervals and, **29-50**
form, **29-50**
inference for multiple regression, **29-16**
model in use and, **29-49**
multiple regression models, **29-49–29-50**
results interpretation caution, **29-49**
Individuals
defined, 14, 34
as subjects, 230
variables and, 13–16
Inference. *See* Statistical inference
Inference about a population contrast, **30-11**
Inference about a population mean
conditions for, 455–456
matched pairs *t* procedures, 467–469
Normal distribution, 455
one-sample *t* confidence interval and, 458–461
one-sample *t* test and, 461–464
population as much larger than sample, 456
resampling and standard errors and, 472–475
simple random samples (SRS), 456
summary, 563
t distributions and, 456–458
t procedures robustness and, 469–472
technology and, 464–467
Inference about a population proportion
assumption, 532
large confidence intervals and, 520–523
overview of, 517–518
plus four confidence intervals, 528–531
sample proportion, 518–520
sample size selection, 523–525
summary, 563
tests of significance, 525–528
Inference about relationships
overview, 575–576
regression, 609–635
two categorical variables, 577–600
Inference about variables
comparing two means, 485–516
comparing two proportions, 539–557
overview of, 453
population mean, 455–484
population proportion, 517–538
review, 559–562
summary, 562–564
Inference for multiple regression. *See also* Multiple regression models
conditions for, **29-15**
constant variance, **29-14**
examples, **29-17–29-20**
F statistic, **29-15**
independence, **29-14–29-15**
individual *t* tests for coefficients, **29-16**
linear trend, **29-14**
Normality, **29-14**
Inference for two-way ANOVA
examples, **30-20–30-27**
outline, **30-20**
Influential observations
defined, 143, 151
example, 143
multiple regression model, **29-53**
outliers in *x* direction and, 145
subjective nature of change, 143
Informed consent. *See also* Data ethics
behavioral experiments, 262
defined, 256, 263
difficulties of, 257
example, 256–257
forms, 257
Institutional review boards. *See also* Data ethics
defined, 263
functions of, 254–256
purpose of, 254
web page example, 255
work load size and, 256
Interactions. *See also* Multiple regression models
defined, **29-22**
between factors, **30-17**
main effects and, **30-16–30-17**
model development and, **29-25**
plots, **30-24, 30-25**
product term, **29-22**
two-way ANOVA, **30-24, 30-25**
two-way ANOVA *F* tests, **30-31**
Intercept
defined, 129, 151
least-squares regression line, 132
population regression lines, 612
regression inference, 611
regression line, 130, 151
Interquartile range (*IQR*)
defined, 55
formula, 55
in 1.5 × *IQR* rule, 56

JMP
ANOVA, 651
descriptive measures, 62
least-squares regression, 133
parallel regression lines, **29-10, 29-11**
regression inference, 616, 618
t confidence interval, 465
t test, 466
two-sample *t* procedures, 495, 496

Kruskal-Wallis test
alternative hypothesis, **28-24, 28-29**
comparing several samples, **28-23–28-24**
conditions for, **28-24**
defined, **28-24, 28-29**
hypotheses, **28-24**
idea of, **28-24**
large sample sizes and, **28-25**
null hypothesis, **28-24, 28-25, 28-29**

Kruskal-Wallis test (*continued*)
technology and, **28-26–28-27**
ties in, **28-25**
use of, **28-30**
Kruskal-Wallis test statistic
chi-square distribution, **28-30**
defined, **28-25, 28-29**
exact distribution of, **28-25**
formula, **28-25**
sums of ranks, **28-25**

Laboratory experiments, 233
Lack of realism, 240–241, 245
Large-sample confidence intervals
for comparing two proportions, 542–543, 551
for population proportions, 520–523, 531
for sample proportions, 528
Law of large numbers
defined, 348, 366
example, 348–349
gambling and, 349
mathematical proof of, 348
statistical estimation and, 347–350
true probability and, 337
use of, 349
Law of Large Numbers applet, 349
Least-squares method, 131
Least-squares regression
defined, 634
facts about, 135–138
technology and, 132–135
Least-squares regression lines
cause and effect and, 148
caution, 152
correlation and, 146
defined, 131
equation, 132
explanatory variables and, 152
illustrated, 131, 136
intercept, 132
outliers in x direction and, 145
passing through point, 136
popularity of, 131
predicted response, 132
regression inference, 611
slope, 132, 135
technology and, 132–135
Least-squares residuals, 140
Least-squares slope
correlation and, 620
standard error, 619
standardized version, 619
Leaves, 29
Left-skewed distribution, 25
Level of significance. *See* Significance level
Linear regression, 619
Linear relationships
defined, 107, 117
direction of, 118
measuring with correlation, 111–113
regression inference, 632
residual plots, 629
Lines
plotting, 129
straight, 129, 136, 611
Lurking variables
beware of, 146–147
categorical variable relationships and, 171
confounded, 229
defined, 147, 152
experiments and, 149
explanatory and response variables and, 152
in Simpson's paradox, 172

Main effects
as "on the average" effect, **30-17**
defined, **30-15, 30-17, 30-32**
interactions and, **30-16–30-17**
with no interaction, **30-15–30-16**
of repetitions, **30-16**
two-way ANOVA F tests, **30-31**
Mann-Whitney test, **28-8, 28-29**
Margin of error
caution, 419
confidence intervals, 383, 419–420, 431
convenience samples and, 420
defined, 376
population proportion, 523–524, 531
prediction with, 624
sample size and, 383, 425
small, 424–425
smaller, 383
voluntary response samples and, 420
Marginal distributions
calculating, 164–165
conditional distributions versus, 166
defined, 164, 173
graph of, 165
relationships among variables and, 166
Matched pairs design
as block design, 242
defined, 241, 245, 467
mean difference in, **28-16**
order of treatments, 241–242
subjects, 241
Wilcoxon signed rank test, **28-16–28-18**
Matched pairs t procedures
defined, 467, 475
example, 467–469
population mean, 467
Matched pairs t test, **28-29**
Mean and Median applet, 50
Mean of a density curve
as balance point, 79, 80, 93
defined, 79, 80, 93
in Normal distributions, 81
notation, 80
Mean response
confidence intervals for, 627, 634
estimation of, 626
standard errors for, 627
Mean square for error (MSE)
in ANOVA F statistic calculation, 662
defined, 662
denominator, 663
formula, 662
Mean square for groups (MSG)
in ANOVA F statistic calculation, 662
defined, 662
denominator, 663
formula, 662
Mean squares, 662
Means
binomial, 334–335, 339
chi-square distributions, 594
comparing several, 647–648
comparing two, 485
contrasts among, **30-10**
correlation and, 116
defined, 48, 66
example of, 48
extreme observations and, 59
finding, 48
inference about a, 374
of least-squares residuals, 140
median comparison, 50–51
as not a resistant measure, 48
outliers and, 59
population. *See* population mean
process, **31-8**
of random samples, 358
regression to the, 158
regression toward, 129
sample. *See* sample means
of sample mean, 353
of sampling distribution, 366, 518
for symmetric density curves, 93
use of, 59
variability about the, 59
Wilcoxon signed rank statistic, **28-17**
Median
defined, 49, 66
of density curves, 79, 80
finding, 49–50
in five-number summary, 53
location in ordered list, 49
mean comparison, 50–51
for symmetric density curves, 93
Microsoft Excel
ANOVA, 651, 653
binomial probabilities, 333
descriptive measures, 62
least-squares regression, 133
regression inference, 616
t confidence interval, 465
t test, 466
two-sample t procedures, 495, 496
Minitab
ANOVA, 651, 653, 658, **30-3, 30-8**
binomial probabilities, 332
chi-square statistic, 585
comparing two proportions, 543–544
descriptive measures, 62
expected counts, 585
interaction plots, **30-24, 30-25**
Kruskal-Wallis test, **28-26**
least-squares regression, 132
multiple regression models, **29-45**
one-way ANOVA, **28-26, 30-39**
parallel regression lines, **29-10, 29-11**
power calculation, 428
probability histogram, 336
regression inference, 616
regression model with two regression lines, **29-24**
simple linear regression, **29-35**
t confidence interval, 465
t test, 466
Tukey pairwise multiple comparisons, **30-39**
two-sample t procedures, 495, 496
two-way ANOVA, **30-22, 30-26**

two-way table output, 168, 591
Wilcoxon signed rank test, **28-19**
Modified boxplots, 56
Mosaic plots
defined, 169
illustrated, 169
information in, 169–170
Multiple analyses, 423
Multiple comparisons
alternative hypothesis and, 580
conclusions, **30-10**
confidence intervals and, 580
defined, **30-2**
follow-up analysis, 581, 648
null hypothesis and, 580
overall test, 581, 648
problem definition, 580
problem of, 580–581, 648
P-values and, 580
statistical methods for dealing with, 580–581, 648
steps for dealing with, 648
Tukey pairwise, **30-7, 30-32, 30-33**
Multiple linear regression model. *See also* Multiple regression models
β parameters estimation, **29-30**
defined, **29-28**
degrees of freedom, **29-30**
examples, **29-28–29-32**
interaction of explanatory variables, **29-29–29-30**
Normal distribution, **29-28**
population standard deviation estimation, **29-30**
standard deviation, **29-28**
Multiple regression case study
automated algorithms, **29-41**
creating new explanatory variable, **29-43**
CrunchIt! output, **29-40, 29-41, 29-45**
CrunchIt! parameter estimates, **29-42**
descriptive statistics and correlation coefficients, **29-39**
final model, **29-44**
goal, **29-36**
highest correlation, **29-41**
including all explanatory variables, **29-40–29-41**
including other explanatory variables, **29-42–29-43**
marketing data, **29-36–29-38**
Minitab output, **29-45**
overview of, **29-36**
regression model construction, **29-36**
relationships among variables, **29-38–29-39**
Multiple regression models
ANOVA *F* test, **29-49**
automated algorithms, **29-41**
building, **29-36**
case study, **29-36–29-48**
choosing, **29-25–29-26**
conditions for inference, **29-53–29-55**
confidence intervals for, **29-50, 29-56**
defined, **29-2**
estimated, **29-55**
explanatory variables and, **29-16–29-17**
F statistic for, **29-15**
flexibility of, **29-28–29-32**
general multiple linear regression model, **29-28–29-34**
illustrated, **29-5**
individual *t* tests for coefficients, **29-49–29-50**
inference for, **29-13–29-22**
inference for regression parameters, **29-48–29-52**
influential observations, **29-53**
interaction, **29-22–29-28**
interpreting, **29-4**
outliers, **29-53**
overview of, **29-1–29-2**
parallel regression lines, **29-2–29-5**
parameter estimation, **29-5–29-10**
predication intervals for, **29-50, 29-56**
prediction and, **29-50**
quadratic regression model, **29-30–29-32**
regression coefficients and, **29-34–29-36**
residual plots, **29-53, 29-54**
residuals, **29-5–29-6**
simple linear regression model and, **29-57**
technology and, **29-10**
with two regression lines, **29-23, 29-55**
Multiplication rule
for any two events, 311
extended, 312
general, 311–313, 319
for independent events, 305, 313, 318
occurrence of, 307
Multistage samples, 213

Natural tolerances, **31-29–31-30, 31-37**
Negative association
defined, 106, 107
direction and, 117
indication, 113
NHANES survey, 417–418
95% confidence interval, 423
No linear relationship, testing hypothesis of, 619–620
Nonnumerical outcomes, 294
Nonparametric tests
comparison of, **28-2**
defined, **28-2, 28-29**
Kruskal-Wallis test, **28-23–28-29**
Mann-Whitney test, **28-8**
overview of, **28-1**
replacing data by ranks, **28-30**
Wilcoxon rank sum test, **28-2–28-15**
Wilcoxon signed rank test, **28-16–28-23**
Nonresponse
defined, 214, 219
inference and, 417
rates of, 214–215
web surveys and, 217
Nonsense correlation, 148
Normal approximation
accuracy of, 336
to binomial distribution, 335–338, 339
of binomial probability, 337
central limit theorem, 360
rank sum statistic, **28-6–28-8**
remembering, 335
rule of thumb, 337
Wilcoxon rank sum test, **28-11**
Wilcoxon signed rank statistic, **28-18–28-20**
Normal Approximation to Binomial applet, 337
Normal curves. *See also* Density curves
central area under, 379
characteristics of, 80–81
critical value, 380
defined, 80, 93
illustrated, 81
locating points on, 91
Normal Density Curve applet, 88
Normal distributions
ANOVA, 656, 666
calculating percentage in interval for, 94
central limit theorem and, 356
in comparing two means, 487
as condition for inference about a mean, 456
as continuous probability models, 291
defined, 80, 81, 93
importance of, 82
multiple linear regression model, **29-28**
process monitoring, **31-8**
sample mean and, 355, 367
sample standard deviation and, **31-13**
as simple condition, 374
68–95–99.7 rule, 82–85, 93–94
special properties of, 81
specification of, 81
standard, 85–86, 94
standard deviation, 502
summary, 181–182
t confidence interval and test and, 469–470
tests for population mean, 402
two-way ANOVA, **30-32**
use of, 94
Normal population procedures, 502
Normal proportions
finding, 86–88
using a table to find, 89
Null hypothesis. *See also* Hypotheses
abbreviation, 395
chi-square test, 592
chi-square test for goodness of fit, 595, 596
chi-square test for homogeneity, 592
chi-square test for independence, 592
comparing proportions, 545
defined, 394, 409
evidence against, 398–399
Kruskal-Wallis test, **28-24, 28-25, 28-29**
of no treatment effect, 502
one-way ANOVA, 666
plausibility of, 421
P-values and, 398–399
rejection consequences, 421
testing, 401
tests of significance for, 409
two categorical variables and, 580
Null model, **29-15**
Numbers
with context, data as, 7
describing distributions with, 47–66
faked, 287
random, 278
written differences of, 21
Numerical outcomes, 293
Numerical summaries
center in, 66
choosing, 59–60
of distributions, 181
as misleading, 60, 66
plotting data and, 60, 66
variability in, 66

Observational studies
cause and effect and, 228
confounding and, 229
defined, 228, 244
experiments versus, 227–230
learning from, 3
planning, 424–431
in producing data, 245
sample surveys, 204–205, 214–216, 219, 227
Observations
in binomial distributions, 328
binomial distributions, 339
cost and time of taking, 424
in data sets, 35
extreme, 59
in five-number summary, 53
independent, 327
influential, 143–145
number determination factors, 426
ranks of, **28-3, 28-29**
standardizing, 85–86
1.5 × *IQR* rule, 56
One-sample *t* confidence interval, 458–461
One-sample *t* statistic, 457, 475
One-sample *t* test, 461–464
One-sample *z* test statistic, 401, 409, 475
One-sided alternative hypothesis, 394, 395, 409
One-sided *P*-value, 397, 458
One-Variable Statistical Calculator applet, 31
One-way ANOVA. *See also* ANOVA
ANOVA tables, **30-29**
beyond, **30-1–30-5**
calculations, 663–664, 666
comparisons, **30-1**
conditions for, 656–659, 666
defined, 655, 666, **30-1**
degrees of freedom for error, **30-11**
details of, 653, 661–665
example, **30-2–30-4**
F test, **30-29**
follow-up analysis, **30-4, 30-32**
independent SRSs, 656, 666
interpreting, 666
mean response for several treatments, **28-23**
Normal distribution, 656, 666
rejection of null hypothesis, 666
robustness, 666
standard deviation, 656, 666
technology and, 650–653
total sum of squares, **30-29**
two-way ANOVA output comparison, **30-28, 30-30–30-31**
variability of observations, 656
One-Way ANOVA applet, 654
Ordered lists, 49
Outcomes
event, 284
nonnumerical, 294
numerical, 293
personal probability of, 294–296
probability, 284
proportion, 517
random number generator, 289
sample space, 282
tree diagrams and, 316
Outliers
checking for, 470
defined, 25, 35
density curves and, 78
explanations for, 59
identifying, 55–57
mean and, 59
multiple regression model, **29-53**
not hiding, 48
1.5 × *IQR* rule for, 56
regression inference and, 632
removing, **28-1**
sample mean values and, 59–60
scatterplots, 106, 117
standard deviation and, 59
statistical inference and, 418
in *x* direction, 144, 145
Out-of-control signals
defined, **31-20, 31-37**
false alarms, **31-20, 31-21**
p charts, **31-33**
Overall confidence level, **30-7, 30-32**
Overall patterns
defined, 25, 35
density curves for, 78
examples, 25–28
histograms, 25
residual plots, 140–141
in scatterplots, 106
time plots, 33, 34
Overall significance level, **30-7**
Overall test, for multiple comparisons, 581, 648

p charts
center line, **31-33**
control limits, **31-33–31-36**
defined, **31-32, 31-37**
illustrated, **31-35**
interpretation of, **31-37**
out-of-control signals, **31-33**
settings, **31-32**
usefulness examples, **31-32**
using past data, **31-33**
Pairwise differences
contrast, **30-11**
defined, **30-6**
Parallel regression lines. *See also* Multiple regression models
defined, **29-2, 29-55**
example, **29-2–29-3**
illustrated, **29-5**
indicator variable, **29-3–29-4**
overall *F* test for, **29-15**
residual plot, **29-19**
technology and, **29-10–29-13**
Parameters
defined, 346, 366
estimating with sample statistic, 367
population, hypotheses and, 395
regression inference, 612–615
sample statistics and, 346–347
Pareto charts, **31-5, 31-34, 31-37**
Past data
control charts, **31-24–31-26, 31-33, 31-37**
p charts, **31-33**
s charts, **31-24, 31-25**
$\bar{x}$ charts, **31-24, 31-26**
Patterns
deviations, 25, 35
faked numbers, 287
overall, 25, 33, 34, 35
probability description of, 280
in residual plots, 631
underlying, process of identifying, 118
Permutation distribution, 504
Permutation tests
defined, 504
small samples and, **28-2**
use of, 504, 507
Personal probabilities
defined, 295
outcomes, 294–296
set of, 295
Pie charts
defined, 17, 35
function of, 35
as having limited use, 19
illustrated, 18
requirements of, 17
Placebo effect, 239–240
Placebos, 239, 245
Planning studies
planning inference and, 432
power of statistical test and, 426–431
sample size for confidence intervals and, 424–425
two-sample problems, 497
Plotting a line, 129
Plotting your data, 35, 60
Plus four confidence intervals
for comparing two proportions, 549–551
for population proportions, 528–531
Plus four estimate
accuracy, 530
for comparing proportions, 549–551
confidence intervals for proportions, 528–531
defined, 528–529, 531
example, 529
formula, 529
Pooled sample deviation, 663
Pooled sample proportion
defined, 546, 551
significance tests for comparing two proportions, 545–546, 551
Pooled sample variance, 663
Pooled two-sample *t* statistic, 501
Population contrasts
defined, **30-10**
estimating, **30-11**
inference about, **30-11**
Population distributions
defined, 351, 366
description, 351
Normality assumption, checking, 474
sampling distribution and, 354
shape of, 418
Population mean
confidence intervals for, 375, 379–383, 386, 415, 455, 475
defined, 366
estimation of, 350
inference about, 455–476
matched pairs *t* procedures, 467
notation, 346
pairwise differences, **30-6**
sample mean and, 350

simple random samples (SRS), 406
statistical inference, 415
tests for, 400–404
tests of significance for, 455
two, comparing, 487–489
unbiased estimator of, 353
z test for, 401
Population proportions
comparing, 539–552
confidence interval, 524
defined, 517
difference between sample proportions and, 545
estimating, 523
inference about, 517–532
large-sample confidence intervals for, 520–523, 531
margin of error, 523–524, 531
plus four confidence intervals for, 528–531
P-values, 526
sample size and, 523–525, 531
sampling distribution of difference between, 541
significance tests for, 525–528, 531
of successes, 526
Population regression lines
defined, 634, **29-4**
degrees of freedom, 613
estimating, 612
intercept, 612, 623
residuals, 613
slope, 612
slope, confidence intervals for, 622–624, 634
straight-line relationship with, 611
Population regression model, **29-28**
Population standard deviation
confidence interval for the mean and, 386
notation, 346
as simple condition, 374, 416
simple random samples (SRS), 406
tests for population mean, 402
Populations
defined, 203, 204, 219
inference about, 211–212
sample surveys and, 204
samples versus, 204–206
size, sample size and, 425
Positive association
defined, 107
direction and, 117
indication, 113
Power
alternative hypothesis and, 428
calculating, 427
defined, 427
finding with applet, 427
finding with software, 428
in number of observations, 426
in planning studies, 426–431
sample size and, 428, 432
significance tests, 426–431, 432
Predication intervals
defined, 626, 634
interpretation of, 626
meaning of, 626
for multiple regression models, **29-50, 29-56**
for regression response, 627
as rough approximations, 629
for single observation, 627
for single response, 629
Predicator variables. *See* Explanatory variables
Prediction
inference about, 624–628
with margin of error, 624
regression models and, **29-50**
Privacy policies, 258, 259
Probability
as area under density curve, 290
assigning in finite model, 286
assigning in infinite discrete sample space, 286
binomial, 330–332
conditional, 309–311
defined, 267, 277, 296
idea of, 278–279
introduction to, 277–297
law of large numbers and, 337
long-run regular pattern, 280
of observing sample statistic, 362
outcomes, 284
personal, 294–296
proportions and, 279
random samples and, 297
randomized comparative experiments and, 297
randomness and, 279
success, 339
summary, 441
of Type I and Type II errors, 429
Probability applet, 279, 280
Probability distribution, 296
Probability models
continuous, 289–293, 296
defined, 282, 296
in describing long-run behavior, 296
description, 281
discrete, 286
finite, 286, 296
list of, 303
tree diagrams, 314–318
uses for, 303
Probability rules
addition rule for disjoint events, 307–309
conditional probability, 309–311
in describing long-run behavior, 296
general, 303–319
general addition rule, 307–309
general multiplication rule, 311
list of, 284
multiplication rule, 307
summary, 441–442
using, 284–285
Procedures
based on sample statistics, 418
Normality basis, 418
t, 459, 467–472
z, 415–418, 440
Process mean
defined, **31-8**
estimating, **31-22–31-23**
process in control and, **31-9**
Process monitoring
conditions for, **31-8**
defined, **31-8, 31-37**
Normal distribution, **31-8**
s charts for, **31-13–31-19**
$\bar{x}$ charts for, **31-7–31-13**
Process proportions, **31-33**
Process standard deviation
defined, **31-8**
estimating, **31-23–31-24**
process in control and, **31-9**
target value and, **31-23**
Processes. *See also* Statistical process control
capability, **31-30–31-31, 31-37, 31-38**
cause-and-effect diagrams, **31-2, 31-3, 31-4**
common cause variation, **31-6**
in control, **31-6, 31-31, 31-37**
defined, **31-2**
describing, **31-2–31-6**
examples of, **31-2**
flowcharts, **31-2, 31-3**
as focus, **31-28**
measurements made on, **31-4**
operations, **31-2**
out of control, **31-6, 31-37**
in sample-versus-population framework, **31-2**
special cause variation, **31-6**
summary, **31-37–31-38**
Proportions
comparing, 539–552
confidence intervals for, 521
cumulative, 87, 94
finding values given, 91–93
large-sample confidence intervals for, 520–523
Normal, 86–88
outcome, 517
plus four confidence intervals for, 528–531
population, 517–532
probability and, 279
sample, 518–520, 531
sampling distribution of difference between, 541
significance tests for, 525–528
of successes, 339
Pseudorandom numbers, 280
P-Value of a Test of Significance applet, 397
P-values
ANOVA F test, 660
approximation of, 407
chi-square test, 593
defined, 396, 409, 440
finding, 406
finding, automating, 397
how small as convincing, 421, 431
large, 396
multiple comparisons and, 580
null hypothesis and, 398–399
one-sample t test, 461–462
one-sided, 397, 458
population proportion, 526
resampling, 406, 407
statistical significance and, 396–400
t distributions, 476
technology for getting, 404
two-sided, 398, 458
Wilcoxon rank sum test, **28-8, 28-29**
Wilcoxon signed rank test, **28-29**

Quadratic regression model, **29-30–29-32**
Quantitative variables
in correlation, 114, 115
defined, 14, 35
histograms and, 21–24
as measure of quality, **31-8**

Quantitative variables (*continued*)
  stemplots and, 29–32
  units of measurement, 14
  values, 21
Quartiles
  area under the curve, 93
  calculating, 52
  defined, 52, 66
  finding, 52–53
  first, 52, 66
  third, 52, 66

R charts
  defined, **31-20, 31-37**
  out-of-control signal, **31-20**
  using, **31-19–31-20**
Random digit dialing
  call screening and, 216
  cell phones and, 216
  defined, 210, 219
  landline future and, 217
Random digits
  in sample selection, 219
  selecting, 289
  table of, 208, 245, 289
Random number generator, 289
Random numbers, from computer program, 280
Random samples
  chance and results, 345
  large versus small, 211–212
  means of, 358
  multistage, 213
  probability and, 297
  simple (SRS). *See* simple random samples (SRS)
  stratified, 212, 219
  variability and, 278
Random sampling
  bias and, 211
  defined, 212, 219
  designs for, 212–213
  rational subgroups versus, **31-28–31-29**
  reasons to use, 211
Random sampling error, 419
Random variables
  continuous, 293, 296
  defined, 293, 296
  discrete, 293, 296
  notation, 293
Randomization
  chance, 245
  as experimental principle, 238, 245
  in prevention of bias, 245
Randomized comparative experiments
  bias and, 245
  chance and results, 345
  completely randomized design, 235
  control groups, 239
  defined, 234
  double-blind, 240
  illustrated, 268
  lack of realism, 240–241, 245
  logic of, 237–239
  for medical advances, 237
  placebos, 239, 245
  probability and, 297
  replication, 238
  statistically significant, 238
  subject assignment, 237
  subjects, 239
  treatments, 237
  use of, 245
Randomized trials, 261
Randomness
  defined, 279
  probability and, 279
  search for, 280–281
  understanding, 279
Rank tests
  center of populations, **28-2**
  comparison of, **28-2**
  continuous distributions requirement, **28-2**
  defined, **28-2, 28-29**
  idea of, **28-3**
  Kruskal-Wallis test, **28-23–28-29**
  ranks of observations, **28-3, 28-29**
  statistics summary, **28-30**
  summary, **28-30**
  ties in, **28-11–28-15, 28-20–28-23**
  Wilcoxon rank sum test, **28-2–28-15**
  Wilcoxon signed rank test, **28-16–28-23**
Rational subgroups
  choosing samples for, **31-29**
  defined, **31-28, 31-37**
  random sampling versus, **31-28–31-29**
Realism, lack of, 240–241, 245
Reasoning of a Statistical Test applet, 392
Regression. *See also* Multiple regression models
  cautions, 183
  correlation and, 151
  defined, 127
  explanatory and response variables
    and, 135
  least-squares, 132
  linear, 619
  lurking variables and, 146–147, 152
  to the mean, 158
  outliers in *x* direction and, 144
  residuals, 138–143, 151
  simple linear, **29-2**
  toward the mean, 129
Regression coefficients
  hypothesis test, **29-56**
  woes of, **29-34–29-36**
Regression inference
  about prediction, 624–628
  checking conditions for, 628–633, 635
  condition violations, 628
  conditions for, 611–612, 633, 634
  confidence intervals, 622
  correlation, 611
  example, 609–611
  illustration of conditions, 612
  least-squares regression line, 611
  least-squares slope, 619, 620
  nature of data, 612
  outliers and, 632
  parameter estimation, 612–615
  questions answered by, 609
  residual plots, 628–629
  scatterplots, 610
  slope and intercept, 611
  standard deviation, 611
  straight-line relationship, 611
  technology, 615–618
  testing hypothesis of no linear relationship and,
    619–620
  testing lack of correlation and, 620–622
  using software output, 634
Regression lines
  correlation and, 146
  defined, 127, 151
  extrapolation, 146, 151
  finding for relationships, 137
  good, 131
  intercept, 130, 151
  least-squares, 131–132
  parallel, **29-2–29-5**
  population, 611, 612, 634
  for prediction, 137
  residual plots and, 140
  residuals and, 138–143
  slope of, 130, 151
  slope size, 130
  summary, 182–183
  using, 129
Regression parameters, inference for, 612–615,
    29-48–29-52
Regression slope
  confidence intervals for, 622–624
  defined, 622
  estimation of, 622–623
  example, 623
  significance test for, 619
  t test, 634
Regression standard error
  calculation of, 613
  defined, 613, 634, **29-6, 29-55**
  degrees of freedom, 634, **29-6**
  as key measure, 614
Relationships. *See also* Associations
  between categorical variables, 163
  defined, 107
  direction, 106, 117
  displaying with scatterplots, 103–105
  explanatory-response, 101–102
  finding regression line for, 137
  form of, 106, 117
  inference about, 575–576
  linear, 107, 117
  row and column variables, 599
  strength of, 106, 117
Replication, 238, 245
Resampling
  accuracy, 408
  defined, 407
  in estimating standard deviation, 473
  in estimating standard error of the sample
    mean, 473
  in estimating standard error of the sample
    median, 474
  example, 407–408
  repeated SRSs and, 408
  standard errors and, 472–475, 476
Residual plots
  conditions for, 629
  defined, 140, 628
  illustrated, 141, 631
  independent observations and, 629, 632
  linear relationship and, 629, 632
  multiple regression model, **29-53, 29-54**
  Normal residuals and, 629, 632

overall patterns, 140–141
parallel regression lines, **29-19**
patterns in, 631
quadratic regression model, **29-31**
for regression inference, 628–629
regression line and, 140
standard deviation and, 629, 632
Residuals
defined, 138, 151, 613
finding, 140
multiple regression model, **29-5–29-6**
Normal, 629, 632
population regression lines, 613
standardized, 642
Resistant measure, 48, 66
Response bias, 215, 219
Response variables
correlation and, 113, 118
defined, 102, 117
identifying, 102
plotting, 104
regression and, 135
Right-skewed distribution
defined, 25
density curve, 77
histogram, 77
Robustness
ANOVA *F* test, 656, **30-15**
defined, **28-1**
Normal population procedures, 502
t procedures, 469–472
two-sample *t* procedures, 497–499, 507
Rounding, 30
Roundoff errors, 17, 167
Row totals, 173
Row variables, 163, 173
Runs signal, **31-20–31-21**

s charts. *See also* Control charts
center line, **31-14, 31-24**
chart setup, **31-24–31-25**
constants, **31-14, 31-15**
in control, **31-17**
control limits, **31-14, 31-24**
defined, **31-37**
example, **31-15–31-16**
illustrated, **31-15, 31-16**
lack of control on, **31-17**
Normal distribution and, **31-13**
out-of-control signal, **31-20**
for process monitoring, **31-13–31-19**
special causes, **31-17**
using past data, **31-24, 31-25**
Sample contrasts, **30-11**
Sample means
behavior of, 348
computed from SRS, 367
defined, 366
mean of, 353, **31-23**
Normal distribution and, 355, 367
notation, 346
outside control limits, **31-9**
population mean and, 350
resampling for estimating standard error of, 473
sample number versus, **31-10**
sample size and, 353, 355
sampling distribution of, 352–355, 375, 406, 442
standard deviation of, 353
standard error of, 456
trust of, 353
two, comparing, 487
watching change, 349
Sample median, 474
Sample proportions
behavior of, 519
control charts for, **31-32–31-36**
defined, 518, 531
difference between, 540, 545
equation, 518
large-sample confidence intervals for, 528
notation, 518
out-of-control signals, **31-33**
pooled, 546, 551
sampling distribution of, 518
standard error of, 520, 551
of successes, 518, 519
Sample size
of confidence intervals, 424–425
importance of, 353
large, 432
margin of error and, 383, 425
population proportion and, 523–525, 531
population size and, 425
power and, 428, 432
sample mean and, 353, 355
significance and, 421–422
t procedures and, 470
Sample space
defined, 282, 296
infinite discrete, 286
interval of numbers, 289
outcomes of, 282
simplicity or complexity, 282
Sample standard deviation. *See also* *s* charts
facts, **31-14**
Normal distribution and, **31-13**
notation, 346
Sample statistics
defined, 346, 366
in estimating unknown parameter, 367
inference procedures based on, 418
parameters and, 346–347
probability of observing, 362
sampling distribution of, 362
standard deviation, 473
Sample surveys
cautions, 214–216
comparing two proportions, 547
defined, 219
example, 204–205
nonresponse, 214–215, 219
as observational study, 227
planning, 204
populations and, 204
response, 219
response bias, 215, 219
undercoverage, 214
wording effects, 215, 219
Samples
chance in choosing, 278
choosing, 203
comparing several, **28-23–28-24**
comparing two, **28-2–28-6**
convenience, 206
defined, 204, 219
imitating, 350
multistage, 213
nonresponse in, 417
populations versus, 204–206
simple random, 207–211, 219
voluntary response, 207, 219
Sampling
biased, 220
binomial distributions in, 328–329
impact of technology, 216–219
meaningful conclusions and, 220
poor, 206–207
in producing data, 203–220
random. *See* random sampling
summary, 268
Sampling designs
data analysis from, 213
defined, 204, 219
random samples, 212–213
Sampling distributions
center, 352
central limit theorem and, 355–361
of a count, 329
defined, 351, 366
difference between proportions, 541
as idea pattern, 351
interpretation of, 351
mean of, 366
mean of a density curve, 518
obtaining without simulation, 351
population distribution and, 354
probability of observing values, 363
repeated SRSs and, 386
of sample mean, 352–355, 375, 406, 442
of sample proportions, 518
of sample statistic, 362
shape, 351, 353
standard deviation of, 353, 366, 518
statistical significance and, 361–366
summary, 441
variability, 352
Scatterplots
background variables and, 108
categorical variables in, 109–110
defined, 103, 104, 117
deviations in, 106
direction, 105–106, 117
displaying relationships with, 103–105
examining, 106
form, 106, 117
illustrated, 104
interpreting, 105–109
outliers, 106, 117
overall patterns, 105–106
plotting values in, 104
regression inference, 610
residual plot, 140
strength, 106, 117
summary, 182
Segmented bar graphs
defined, 167
illustrated, 169
mosaic plot and, 169–170
Shape. *See also* Distributions
defined, 25, 35
importance of, 28

Shape. *See also* Distributions (*continued*)
population distribution, 418
sampling distribution, 351, 353
Significance level
defined, 399
in number of observations, 426
overall, **30-7**
P-value and, 399
Significance tests. *See* Tests of significance
Simple conditions
defined, 374
Normal distribution, 374
population standard deviation, 374, 416
SRS, 374
tests for population mean, 402
verification of, 402
Simple linear regression
conditions for inference, **29-4**
CrunchIt! output, **29-41**
defined, **29-2**
multiple regression and, **29-57**
partial regression output, **29-35**
Simple Random Sample applet, 208, 209, 235
Simple random samples (SRS)
benefits of, 220
bias and, 209
binomial distributions and, 329
as building block, 213
choosing, 209
concept, 207–208
conclusions, 220
as condition for inference about a mean, 456
defined, 208, 219
illustrated, 268
labels, 208, 209
population mean, 406
population standard deviation, 406
results, trusting, 209
sampling distributions and, 386
as simple condition, 374
tests for population mean, 402
using tables to choose, 208
Simpson's paradox
defined, 171, 172, 173
lurking variables and, 172
Simulations
defined, 350, 409
obtaining sampling distributions without, 351
significance from, 406–409
Simultaneous tests, **30-7**
68–95–99.7 rule
defined, 82, 93–94
examples, 83–84
illustrated, 82
Skewed distributions
defined, 25, 35
help gained from, 30
types of, 25
Skewed to the left, 25
Skewed to the right, 25
Skewness, checking for, 470
Slope
defined, 129, 151
least-squares, 619, 620
least-squares regression line, 132, 135
population regression lines, 612, 634
regression, 622–624
regression inference, 611
regression line, 130, 151
size of, 130
straight lines, 129
as zero hypothesis, 634
Social science experiments, 262–263
Special causes
defined, **31-6**
s charts, **31-17**
$\bar{x}$ charts, **31-17**
Splitting stems, 31
Spreadsheets
defined, 15
illustrated example, 15
use of, 14
Square of the correlation, 151
Squared deviations, 59
Squared multiple correlation coefficient (R^2)
defined, **29-12, 29-55**
form, **29-12**
interpretation with ANOVA table, **29-12**
using, **29-12**
Standard deviation of a density curve
defined, 93
locating, 93
in Normal distributions, 81
notation, 80
as variability measure for Normal distribution, 81
Standard deviations
ANOVA, 656, 657, 666
binomial, 334–335, 339
calculating, 57–58
defined, 57, 66
degrees of freedom, 58
extreme observations and, 59
F test for, 502
formula, 58
inference about, avoiding, 501–502
multiple linear regression model, **29-28**
Normal distributions, 502
as not a resistant measure, 59
outliers and, 59
pooled, 663
population, 346
process, **31-8**
regression inference, 611, 632
resampling and, 473
residual plots, 629
sample, 346
of sample mean, 353
sample statistic, 473
of sampling distribution, 353, 366, 518
two-way ANOVA, **30-32**
units of measurement, 59
use of, 59
usefulness of, 59
variability about the mean, 59
Wilcoxon signed rank statistic, **28-17**
Standard distributions, **28-1**
Standard errors
comparing two proportions, 520, 542, 551
defined, 456, 472, 475
least-squares slope, 619
for mean response, 627
in one-sample *t* confidence interval, 460
regression, 613, 634, **29-55**
resampling and, 472–475, 476
sample contrasts, **30-11**
of sample mean, 473–475
of sample proportions, 520, 551
two-sample *t* procedures, 489
Standard Normal distribution, 85, 86, 94
Standard Normal table
defined, 88
example, 88, 89–90
to find Normal proportions, 89
using, 88–91
Standardized residuals, 642
Standardized value, 85
Standardized variable, 94
Standardizing
defined, 85, 93
example, 85, 89, 90
Statistical estimation
law of large numbers and, 347–350
reasoning of, 374–376
Statistical inference. *See also* Confidence intervals; Tests of significance
binomial distributions, 339
chance and, 416
checking conditions for, 381–382
conclusions, 201, 439
conditions for, 416–419
data analysis before, 431
data source and, 416
defined, 7, 11, 211, 373, 439
dropouts and, 417
goals of, 275
methods of, 561
multiple regression, **29-14–29-20**
nonresponse and, 417
outliers and, 418
planning, 432
population mean, 374, 415, 455–476
population proportion, 517–532
populations, 211–212
in practice, 415–433
probability ideas, 347–348
purpose of, 201
reasoning of, 560
regression, 619–635
reliability, 416
shape of population distribution and, 418
simple conditions, 374
standard deviations, avoiding, 501–502
two-way ANOVA, **30-20–30-27**
use of, 431
variables, 453–564
Statistical Power applet, 427
Statistical problems
conclusion, 7, 63, 66
four-step process, 63
organizing, 63–65
planning, 7, 63, 66
solving, 7, 63, 66
stating, 7, 63, 66
two-sample, 485–486
Statistical process control
comments on, **31-28–31-30**
common cause, **31-6**
control charts, **31-6**
defined, **31-6**
idea of, **31-6–31-7**

introduction to, **31-1–31-2**
natural tolerances, **31-29–31-30**
process focus, **31-28**
processes and, **31-2–31-6**
rational subgroups, **31-28–31-29**
special cause, **31-6**
Statistical significance
alternative hypothesis and, 421
defined, 238, 363, 399, 422
example, 363
findings, interpretation of, 399
lack of, 432
multiple analyses and, 423
practical significance versus, 422
P-value and, 396–400
reasoning behind, 423
sample size and, 421–422
sampling distributions and, 361–366
from a simulation, 406–409
standard levels of, 421
from a table, 404–405
Statistical tests, reasoning of, 391–392
Statistics
in dealing with variation, 6
defined, 2, 13
elements of, 7
learning, 8
main ideas of, 8
as science of data, 13
Statistics in Summary flowchart
defined, 561
illustrated, 562
in planning inference problems, 561
treatment comparison and, 561
Stemplots
back-to-back, 43, **28-3**
classes, 30
creating, 29
creation example, 29–30
defined, 35
function of, 35
illustrated, 30
leaves, 29
rounding, 30
side-by-side, 646, **30-3**
splitting stems, 31
stems, 29
use of, 29
Stems
defined, 29
reducing with rounding, 30
splitting, 31
Straight lines
correlation and strength, 136
equation, 129
intercept, 129
slope, 129
Strata, 212, 219
Stratified random samples, 212, 219
Strength
correlation measurement of, 114, 115, 116
defined, 106, 117
of relationships, 106, 117
scatterplots, 106
Subjects
clinical trials, 260, 261
controlling environment of, 231
defined, 230, 244
informed consent, 256
in matched pairs design, 241
randomized comparative experiments, 237, 239
Sum of squares for groups (SSG), **28-24**
Sums of squares
defined, 663
degree of freedom, **30-30, 30-31**
total, **30-29–30-31**
Symmetric density curves
defined, 79
mean for, 93
median for, 93
Symmetric distributions, 25, 35

t approximation
defined, 500
degrees of freedom, 500
details of, 499–501
example, 500
t confidence intervals
Normal distribution and, 469–470
output, 465
technology and, 464–467
t critical values, 458, 475, 506
t distributions
defined, 457, 475
degrees of freedom, 457, 475
density curves, 457
mathematics associated with, 476
P-values, 476
variability, 457
t procedures
matched pairs, 467–469, 475
one-sample, 459
robustness of, 469–472, 476
sample size and, 470
scale of measurement and, **28-13**
two-sample, 489–494
using, 470
t statistic
interpretation of, 457
one-sample, 457, 475
pooled, 501
significance tests and, 475
in testing slope is zero hypothesis, 634
two-sample, 489, 506
t tests
individual, for coefficients, **29-16**
matched pairs, **28-29**
Normal distribution and, 469–470
one-sample, 461–464
output, 466
regression slope, 634
technology and, 464–467
two-sample, **28-29**
Tables
critical values, significance from, 404–405
of random digits, 208, 245, 289
two-way, 163–173, 577–580, 581–583
Technology
ANOVA, 650
comparing two proportions, 543–545
in describing distributions with numbers, 61–63
in expected counts and chi-square statistic calculation, 585–586
Kruskal-Wallis test, **28-26–28-27**
in least-squares regression, 132–135
parallel regression lines, **29-10–29-13**
regression inference, 615–618
sampling and, 216–219
t test and, 464–467
Wilcoxon rank sum test, **28-8–28-10**
Wilcoxon signed rank test, **28-19**
Test statistic. *See also* *t* statistic; *z* statistic
calculation of, 401
chi-square, 583–584
defined, 409
Kruskal-Wallis, **28-24–28-29**
one-sample, 401, 409
for SSG, **28-24**
Tests of significance
basics of, 391–410
cautions about, 421–424
comparing two means, 506–507
comparing two proportions, 545–548, 551
concept of, 392
defined, 391, 409
four-step process, 401
idea of, 440
interpreting, 421–422
for null hypothesis, 409
for population mean, 400–404, 455
for population proportion, 525–528, 531
power of, 426–431, 432
P-value, 396–400
reasoning of, 392–394, 409
for regression slope, 619
robust, 470
summary, 442–443
t statistic and, 475
trust of, 416
Type I error, 429
Type II error, 429
use of, 373
Texas Instruments graphing calculator
ANOVA, 650
binomial probabilities, 332
chi-square statistic, 585
comparing two proportions, 543
descriptive measures, 62
expected counts, 585
least-squares regression, 132
regression inference, 616
t confidence interval, 465
t test, 466
two-sample *t* procedures, 495
Third quartile. *See also* Quartiles
defined, 52, 66
in five-number summary, 53
Ties
in Kruskal-Wallis test, **28-25**
in Wilcoxon signed rank test, **28-20–28-23**
in Wilcoxon sum rank test, **28-11–28-13**
Time plots
cycles, 33, 35
defined, 32, 33, 35
example, 33
illustrated, 33
overall pattern, 33, 34
time series data, 34
trends, 34, 35
use of, 32, 36

Time-series data, 34
Total sum of squares
  defined, **30-29**
  one-way ANOVA, **30-29**
  two-way ANOVA, **30-29**
Treatments
  block design, 242
  in clinical trials, 261
  defined, 230, 244
  matched pairs design, 241–242
  no effect, null hypothesis of, 502
  randomized comparative experiments, 237
  studying effects of, 231
Tree diagrams
  conditional probability and, 316
  defined, 314
  example illustration, 315, 317
  outcomes and, 316
Trends, 34, 35
Tukey pairwise multiple comparisons, **30-7, 30-32, 30-33, 30-39**
Tukey simultaneous confidence intervals, **30-7**
Two-sample problems
  comparing two population means, 487–489
  comparing two proportions, 539–541, 551
  defined, 485, 506
  example, 486
  inference about standard deviations and, 501–502
  occurrence of, 485
  pooled two-sample *t* procedures and, 501
  sample independence, 485
  study planning, 497
  technology and, 494–497
  Wilcoxon rank sum test, **28-2–28-6**
Two-sample *t* procedures
  defined, 490
  examples, 490–493
  output, 495
  pooled, avoiding, 501
  population variance and, 496
  practical options, 489–490
  robustness, 497–499, 507
  standard error, 489
  technology and, 494–497
  use of, 507
Two-sample *t* statistic
  approximate distribution of, 500
  calculation of, 490
  defined, 489, 506
  degrees of freedom, 506
  exact distribution of, 499
Two-sample *t* test, **28-29**
Two-sample *z* procedure, 547
Two-sided alternative hypothesis, 394, 409
Two-sided *P*-values, 398, 458
Two-way ANOVA. *See also* ANOVA; One-way ANOVA
  ANOVA *F* statistic for, **30-31**
  balanced designs, **30-14, 30-32**
  characteristics of, **30-4**
  conditions for, **30-14, 30-32**
  crossed designs, **30-14, 30-32**
  defined, **30-4, 30-32**
  details, **30-28–30-32**
  F tests, **30-28, 30-31, 30-32**
  factors, **30-13, 30-28, 30-33**
  follow-up analysis, **30-32, 30-33**
  independent SRSs, **30-32**
  inference, **30-20–30-27**
  interpreting, **30-18, 30-33**
  Minitab output, **30-22, 30-26**
  Normal distribution, **30-32**
  one-way ANOVA output comparison, **30-28, 30-30–30-31**
  standard deviation, **30-32**
  strong interactions, **30-24**
  total sum of squares, **30-29**
  variation breakdown, **30-30**
Two-way tables
  chi-square test, 598
  column totals, 173
  column variables, 164, 173
  conditional distribution, 166–170
  defined, 167–168, 173
  expected counts in, 581–583
  illustrated, 164
  marginal distributions from, 164–165
  output, 168
  percent calculation, 165
  row totals, 173
  row variables, 163, 173
  Simpson's paradox and, 171–172
  working with, 599
Type I errors, 429
Type II errors, 429

Unbiased estimator
  defined, 353, 366
  of population mean, 353
Uncontrolled experiments, 233
Undercoverage
  defined, 214, 219
  web surveys and, 217–218
Uniform distribution, 290
Units of measurement
  defined, 14, 35
  standard deviation, 59

Values
  absolute, **28-16**
  critical, 379, 386
  finding, given a proportion, 91–93
  in scatterplots, 104
Variability
  about the mean, 59
  defined, 25, 35
  distributions, 501
  measures, choosing, 59–60
  measuring with quartiles, 51–53
  measuring with standard deviation, 57–59
  in numerical summary, 66
  random samples and, 278
  sampling distribution, 352
  t distributions, 457
Variables. *See also* Distributions
  average value versus typical value, 51
  background, in scatterplots, 108
  categorical, 14, 16–21, 35, 109–110
  column, 164, 173
  confounded, 229, 244
  defined, 3, 14, 34
  distribution of, 17, 35
  effects on another, 233
  explanatory, 101–103, 117
  indicator, **29-3–29-4, 29-55**
  individuals and, 13–16
  lurking, 146–147, 152, 171–172
  negative association, 106, 107
  positive association, 107
  quantitative, 14, 21–32, 35
  random, 293–294
  relationships between, 3, 101
  response, 101–103, 117
  row, 163, 173
  standardized, 94
Variance
  defined, 57, 66
  degrees of freedom, 58
  formula, 58
Variation
  as everywhere, 5–6
  patterns behind, 5
  statistical tools for understanding, 6
Venn diagrams
  defined, 304
  general addition rule, 308
  illustrated, 304
Voluntary response samples
  defined, 207, 219
  as fundamental flaw, 418
  margin of error and, 420

Web surveys
  defined, 217, 219
  problems with, 217
  undercoverage and, 217–218
  use of, 218
Wilcoxon rank sum statistic
  continuity correction, **28-6**
  defined, **28-6, 28-29**
  Normal approximation, **28-6–28-8**
  standardized, **28-6**
Wilcoxon rank sum test. *See also* Nonparametric tests
  defined, **28-4, 28-29**
  hypotheses, **28-10–28-11**
  Normal approximation, **28-11**
  P-values, **28-8, 28-29**
  rank sum statistic, **28-6**
  ranks, **28-3–28-4**
  same-shape condition, **28-10**
  sum of ranks, **28-4**
  technology and, **28-8–28-10**
  ties in, **28-11–28-15**
  two-sample problems, **28-2–28-6**
  use of, **28-30**
Wilcoxon signed rank statistic. *See also* Nonparametric tests
  defined, **28-17, 28-29**
  exact distribution of, **28-20**
  mean, **28-17**
  Normal approximation, **28-18–28-20**
  standard deviation, **28-17**
Wilcoxon signed rank test
  defined, **28-17, 28-29**
  P-values, **28-29**
  ranks for positive differences, **28-17**
  technology and, **28-19**

ties in, **28-20–28-23**
use of, **28-30**
Wording effects, 215

$\bar{x}$ charts. *See also* Control charts
center line, **31-8, 31-14, 31-24**
chart setup, **31-25–31-26**
control limits, **31-8, 31-14, 31-24**
creating, **31-8**
defined, **31-8, 31-37**
example, **31-9–31-10, 31-15–31-16**
illustrated, **31-16**
interpreting, **31-11–31-12**
out-of-control signal, **31-20**
for process monitoring, **31-7–31-13**
runs signal, **31-20–31-21**
special causes, **31-17**
using, **31-19–31-20**
using past data, **31-24, 31-26**

z confidence interval, 425, 432
z procedures
defined, 415
example, 416–417
Normality basis, 418
use of, 418, 440
z statistic
assumption, 409
behavior of, 422
confidence levels for, 404
defined, 401
one-sample, 401, 409
for population mean, 401
scale, 401
z-score, 85, 93

Table entry for z is the area under the standard Normal curve to the left of z.

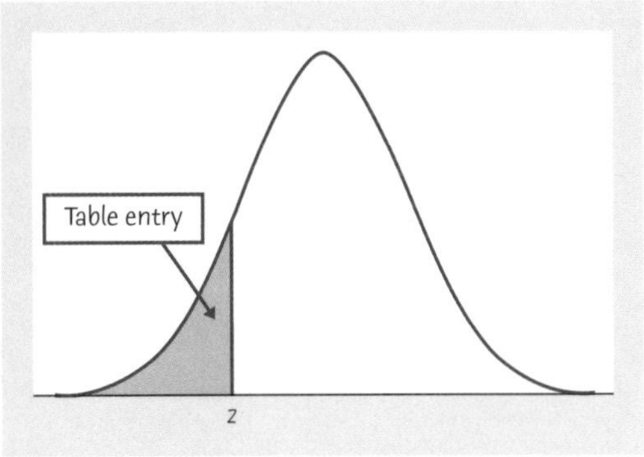

TABLE A STANDARD NORMAL CUMULATIVE PROPORTIONS

z	.00	.01	.02	.03	.04	.05	.06	.07	.08	.09
−3.4	.0003	.0003	.0003	.0003	.0003	.0003	.0003	.0003	.0003	.0002
−3.3	.0005	.0005	.0005	.0004	.0004	.0004	.0004	.0004	.0004	.0003
−3.2	.0007	.0007	.0006	.0006	.0006	.0006	.0006	.0005	.0005	.0005
−3.1	.0010	.0009	.0009	.0009	.0008	.0008	.0008	.0008	.0007	.0007
−3.0	.0013	.0013	.0013	.0012	.0012	.0011	.0011	.0011	.0010	.0010
−2.9	.0019	.0018	.0018	.0017	.0016	.0016	.0015	.0015	.0014	.0014
−2.8	.0026	.0025	.0024	.0023	.0023	.0022	.0021	.0021	.0020	.0019
−2.7	.0035	.0034	.0033	.0032	.0031	.0030	.0029	.0028	.0027	.0026
−2.6	.0047	.0045	.0044	.0043	.0041	.0040	.0039	.0038	.0037	.0036
−2.5	.0062	.0060	.0059	.0057	.0055	.0054	.0052	.0051	.0049	.0048
−2.4	.0082	.0080	.0078	.0075	.0073	.0071	.0069	.0068	.0066	.0064
−2.3	.0107	.0104	.0102	.0099	.0096	.0094	.0091	.0089	.0087	.0084
−2.2	.0139	.0136	.0132	.0129	.0125	.0122	.0119	.0116	.0113	.0110
−2.1	.0179	.0174	.0170	.0166	.0162	.0158	.0154	.0150	.0146	.0143
−2.0	.0228	.0222	.0217	.0212	.0207	.0202	.0197	.0192	.0188	.0183
−1.9	.0287	.0281	.0274	.0268	.0262	.0256	.0250	.0244	.0239	.0233
−1.8	.0359	.0351	.0344	.0336	.0329	.0322	.0314	.0307	.0301	.0294
−1.7	.0446	.0436	.0427	.0418	.0409	.0401	.0392	.0384	.0375	.0367
−1.6	.0548	.0537	.0526	.0516	.0505	.0495	.0485	.0475	.0465	.0455
−1.5	.0668	.0655	.0643	.0630	.0618	.0606	.0594	.0582	.0571	.0559
−1.4	.0808	.0793	.0778	.0764	.0749	.0735	.0721	.0708	.0694	.0681
−1.3	.0968	.0951	.0934	.0918	.0901	.0885	.0869	.0853	.0838	.0823
−1.2	.1151	.1131	.1112	.1093	.1075	.1056	.1038	.1020	.1003	.0985
−1.1	.1357	.1335	.1314	.1292	.1271	.1251	.1230	.1210	.1190	.1170
−1.0	.1587	.1562	.1539	.1515	.1492	.1469	.1446	.1423	.1401	.1379
−0.9	.1841	.1814	.1788	.1762	.1736	.1711	.1685	.1660	.1635	.1611
−0.8	.2119	.2090	.2061	.2033	.2005	.1977	.1949	.1922	.1894	.1867
−0.7	.2420	.2389	.2358	.2327	.2296	.2266	.2236	.2206	.2177	.2148
−0.6	.2743	.2709	.2676	.2643	.2611	.2578	.2546	.2514	.2483	.2451
−0.5	.3085	.3050	.3015	.2981	.2946	.2912	.2877	.2843	.2810	.2776
−0.4	.3446	.3409	.3372	.3336	.3300	.3264	.3228	.3192	.3156	.3121
−0.3	.3821	.3783	.3745	.3707	.3669	.3632	.3594	.3557	.3520	.3483
−0.2	.4207	.4168	.4129	.4090	.4052	.4013	.3974	.3936	.3897	.3859
−0.1	.4602	.4562	.4522	.4483	.4443	.4404	.4364	.4325	.4286	.4247
−0.0	.5000	.4960	.4920	.4880	.4840	.4801	.4761	.4721	.4681	.4641